Proceedings of the 2015 International Conference on

Electronics, Electrical Engineering and Information Science

(EEEIS2015)

Proceedings of the 2015 International Conference on

Electronics, Electrical Engineering and Information Science

(EEEIS2015)

Guangzhou, China   7 – 9 August 2015

Editors

Jian Wang

Huazhong University of Science and Technology, China

Xiaolong Li

Indiana State University, USA

World Scientific

NEW JERSEY • LONDON • SINGAPORE • BEIJING • SHANGHAI • HONG KONG • TAIPEI • CHENNAI • TOKYO

Published by

World Scientific Publishing Co. Pte. Ltd.
5 Toh Tuck Link, Singapore 596224
USA office: 27 Warren Street, Suite 401-402, Hackensack, NJ 07601
UK office: 57 Shelton Street, Covent Garden, London WC2H 9HE

British Library Cataloguing-in-Publication Data
A catalogue record for this book is available from the British Library.

ELECTRONICS, ELECTRICAL ENGINEERING AND INFORMATION SCIENCE (EEEIS2015)
Proceedings of the 2015 International Conference on Electronics, Electrical Engineering and Information Science

ISBN 978-981-4740-12-8

Printed in Singapore

Editorial

The 2015 International Conference on Electronics. Electrical Engineering and Information Science (EEEIS2015) has been held successfully in Guangzhou, China during August 7~9, 2015. The EEEIS2015 was sponsored by Xi'an University of Science and Technology, Anhui University of Science and Technology, Guangdong University of Technology. It aims to bring together researchers, developers, and users from around the world in fields of Electronics Engineering, Electrical Engineering and Information Science for sharing the latest research results, and for exploring new areas of research and development.

The organizing committee received a total of 252 submissions from various parts of the world. The Technical Program Committee worked very hard to have all papers reviewed before the review deadline. The final technical program consists of 117 papers. There are two keynote speeches and seven invited sessions. All the keynote speakers are internationally recognized leading experts in their research fields, who have demonstrated outstanding proficiency and have achieved distinction in their profession. The proceedings were published by World Scientific Publishing Company and submitted to CPCI and Ei Compendex databases for indexing. We would like to mention that, due to the limitation of the conference venue capacity, we are not able to include many fine papers in the technical program. Our apology goes to those authors.

We would like to express our sincere gratitude to all the members of Technical Program committee and organizers for their enthusiasm, time, and expertise. Our deep thanks also go to many volunteers and staff for the long hours and hard work they have generously given to EEEIS2015. Finally, we would like to thank all authors, speakers, and participants of this conference for their contributions to EEEIS2015, and also look forward to welcoming you to the EEEIS2016.

2015 International Conference on Electronics, Electrical Engineering and Information Science [EEEIS2015]

Table of Contents

Chapter 1

Electronics Engineering

Research on the characteristics of inductance of interior permanent magnet synchronous motor

X. Wu[†], S.H. Wang and Z.M. Gao
College of Electrical and Power Engineering, Taiyuan University of Technology, Taiyuan, 030024 China
[†]*Email: wuxiao01234@126.com*

Y.T. Wang
State Grid of Shandong, Binzhou, 256600 China

Corresponding related expressions of current between d-q axis and stator three phase were derived, then to analyze the impact of characteristic of inductance when there were armature excitation, permanent magnet and incorporating of them. On this basis, influence to inductance considering iron saturation and cross effect was investigated. This paper takes an Interior Permanent Magnet Synchronous Motor (IPMSM) with 11kW, 380V as example, firstly Ansoft/Maxwell was used to establish the finite element simulation model, then Fixed Permeability Method (FPM) was applied to analyze laws between d-q axis inductance and current which include quadrate axis current, direct axis current and both of them, by the way, laws of d-q axis inductance and rotor position was obtained. This paper is significant for analyzing and designing of PMSM.

Keywords: D-q axis inductance; Fixed permeability method; Finite element; Iron saturation.

1. Introduction

Permanent magnet synchronous motor has advantages such as small power density, high efficiency and widely speed range [1-2], which makes it widely used in daily life. One of the most significant aspects is that inductance parameters of the motor were calculated accurately before prototype production.

Magnetic conductivity analysis [3-4] and magnetic fields analysis methods are mainly used to calculate inductance parameters. For magnetic conductivity analysis, multi-loop analysis way [5] was employed mostly because of the completed theory. Literature [6] derived the unified expressions for calculating inductance by simplifying the multi-loop method and only considering fundamental air-gap MMF, but this expression is only fit for the situation when the rotor is at a particular position or the stator current is a certain number. Magnetic fields analysis method is a new method with the development of finite element software. Literature [7] obtained rules of stator self inductance and mutual inductance with rotor position by adopting AC standstill method, but the influence of saturation and cross effect were not considered in the passage. KANG G H analyzed the effect of saturation in finite element software using mathematical way in [8], later on, NAJAFI S did the simulation with the model [9]. Literature [8] certificated correctness of FPM by comparing the result of multi-loop method and FPM, and operating the no-load and full-load experiments. Literature [10-11] got the steady parameters of inductance considering saturation and cross effect. *d-q* axis inductance was solved through Park transformation in literature [7-11], which can't reflect the influence of *d* axis and *q* axis current on inductance intuitively.

In this paper, the coordinate transformation formula was fatherly deduced to obtain corresponding related expressions of current between *d-q* axis and stator three phase. *d* and *q* axis current and rotor position were parameterized to obtain laws between the characteristic of inductance and them, which considering saturation and cross effect by using FPM.

2. Current and Inductance of D Axis and Q Axis

Formula of transformation from static ABC axis to *d-q* axis rotating at synchronous speed is as following [12]:

$$\begin{pmatrix} i_d \\ i_q \end{pmatrix} = \sqrt{\frac{2}{3}} \begin{pmatrix} \cos\theta & \cos(\theta - \frac{2\pi}{3}) & \cos(\theta - \frac{4\pi}{3}) \\ -\sin\theta & -\sin(\theta - \frac{2\pi}{3}) & -\sin(\theta - \frac{4\pi}{3}) \end{pmatrix} \begin{pmatrix} i_A \\ i_B \\ i_C \end{pmatrix} \tag{1}$$

Where θ is expressed in electric angle between *d* axis and Phase A axis. The inverse matrix expression of formula(1) is:

$$\begin{pmatrix} i_A \\ i_B \\ i_C \end{pmatrix} = \sqrt{\frac{2}{3}} C \begin{pmatrix} i_d \\ i_q \end{pmatrix} \tag{2}$$

Where

$$C = \begin{pmatrix} \cos\theta & -\sin\theta \\ \cos(\theta - \frac{2\pi}{3}) & -\sin(\theta - \frac{2\pi}{3}) \\ \cos(\theta - \frac{4\pi}{3}) & -\sin(\theta - \frac{4\pi}{3}) \end{pmatrix}$$

The following relation between *d-q* axis current and stator current is existed:

$$\begin{cases} i_d = \sqrt{3} i_s \sin\beta \\ i_q = \sqrt{3} i_s \cos\beta \end{cases} \tag{3}$$

Where i_s is stator current; β is the angle between i_q and i_s. Assuming that $NI = \sqrt{2} i_s$, then formula(2) and (3) can be rewritten as formula(4).

$$\begin{cases} i_A = NI\sin(\beta - \theta) \\ i_B = NI\sin(\beta - \theta + \frac{2\pi}{3}) \\ i_C = NI\sin(\beta - \theta + \frac{4\pi}{3}) \end{cases} \tag{4}$$

Where N is turns of each floor of per slot; I is an intermediate variable.

Formula (4) shows that stator current and d-q axis current can be adjusted by changing β when θ=0. Such as following:

i. When β=0, stator current can be calculated according to formula (4):

$$\begin{cases} i_A = 0 \\ i_B = \frac{\sqrt{3}}{2} NI \\ i_C = -\frac{\sqrt{3}}{2} NI \end{cases} \tag{5}$$

Thus d axis current and q axis current can be calculated:

$$\begin{cases} i_d = 0 \\ i_q = \sqrt{\frac{3}{2}} NI \end{cases} \tag{6}$$

Formula (6) shows that only a certain q axis current value is existed when stator current is just as formula (5).

ii. When β=π/2, stator current and d-q axis current can be calculated:

$$\begin{cases} i_A = NI \\ i_B = -\frac{1}{2} NI \\ i_C = -\frac{1}{2} NI \end{cases} \tag{7}$$

$$\begin{cases} i_d = \sqrt{\dfrac{3}{2}} NI \\ i_q = 0 \end{cases} \tag{8}$$

At this time, it is equal to that only a certain *d* axis current value is existed.

iii. $\beta=\pi/3$

$$\begin{cases} i_A = \dfrac{\sqrt{3}}{2} NI \\ i_B = 0 \\ i_C = -\dfrac{\sqrt{3}}{2} NI \end{cases} \tag{9}$$

$$\begin{cases} i_d = \dfrac{3\sqrt{2}}{4} NI \\ i_q = \dfrac{\sqrt{6}}{4} NI \end{cases} \tag{10}$$

iv. $\beta=\pi/6$

$$\begin{cases} i_d = \dfrac{\sqrt{6}}{4} NI \\ i_q = \dfrac{3\sqrt{2}}{4} NI \end{cases} \tag{11}$$

At the situation iii & iv, it is equal to that both *d* axis and *q* axis current were existed. In the formula (6) (8) (10) (11), the following equation is existed that $\sqrt{i_d^2 + i_q^2} = \sqrt{\frac{3}{2}} NI = \sqrt{3} i_s$, from which the previous statement can be certified rightly.

Above all, by changing θ and β can not only realize *q* axis current and *d* axis current alone and together, but also can be intuitive shows the percentage of them. Three phase inductance matrix can be obtained immediately in Ansoft, in order to get the *d-q* axis inductance parameters, formula (12) is used.

$$L_{dq} = C^{\mathrm{T}} \cdot L_{ABC} \cdot C \cdot L_{\mathrm{ef}} \cdot \left(\frac{N}{a} \right)^2 \cdot M \tag{12}$$

Where C^{T} is the transposed matrix of C; L_{ef} is the effective length of iron core; M is the symmetric number of the finite element model.

3. Fixed Permeability Method

Fixed permeability method (FPM) is a new type of solving flux linkage with the development of finite software. It can disintegrate all of the flux linkage in motor to linkage caused by permanent magnet and stator current [13], thus inductance parameters will be obtained. There is no saturated for linear material since that the permeability is constant. For nonlinear ferromagnetic materials, the value of inductance will be different at different current because of different permeability. FPM changes a nonlinear problem into a linear problem by saving the value of permeability of per element after each calculation. The precision of meshing is very high when implementing FPM, procedures of FPM is as following:

Step 1: Calculating the rated conditions with nonlinear equations;

Step 2: Calculating the permeability of each element in step1;

Step 3: Frozen permeability of all elements in step2, then the value of permeability is constant;

Step 4: Calculating the condition without excitation and changing material to what in step3;

Step 5: Calculating the condition after applying excitation to the motor with linear equations;

Step 6: Calculating the value of inductance on the basis of energy method.

FPM was applied to calculate inductance parameters in the following content of this passage.

4. Rules of Inductance with D-Q Axis Current

4.1 *Finite element model and excitation*

This passage takes an IPMSM with 4 poles, 11kW, lap coil, Y connection, N=56, a=1, Lef=140mm, M=1,based on which the finite element model was established as Fig.2. In the subsequent analysis, in order to get inductance curves along with current, "*I*" was parameterized to make $\sqrt{\frac{3}{2}}NI$ changed from -80A to 80A and θ=0.

4.2 *Inductance curves with q axis current*

Current value shown in formula (5) was applied to stator to obtain curves along with *q* axis current. Fig. 1 shows waveforms of L_q and L_d.

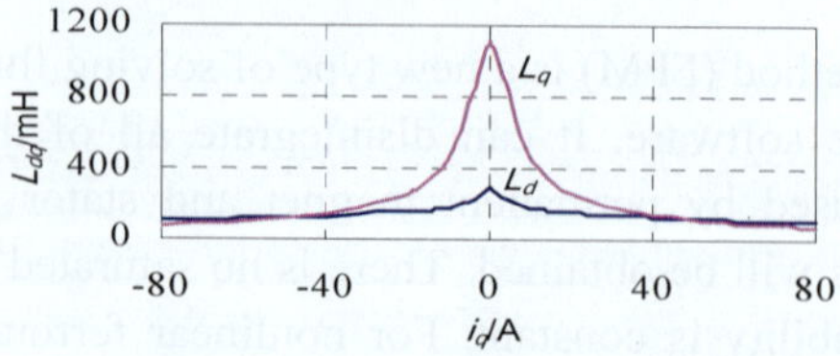

Fig.1 Curves of L_q and L_d along with i_q.

Fig. 1 shows that curves of L_q and L_d are symmetrical about i_q=0. With the increase of i_q, the decrease range of L_q is greater than L_d, when the current is bigger, the two curves are comparable. That means L_q is more sensitive to the change of i_q than L_d. Fig. 2 shows the distribution of flux lines when i_q=1A, ignoring the application of permanent magnet.

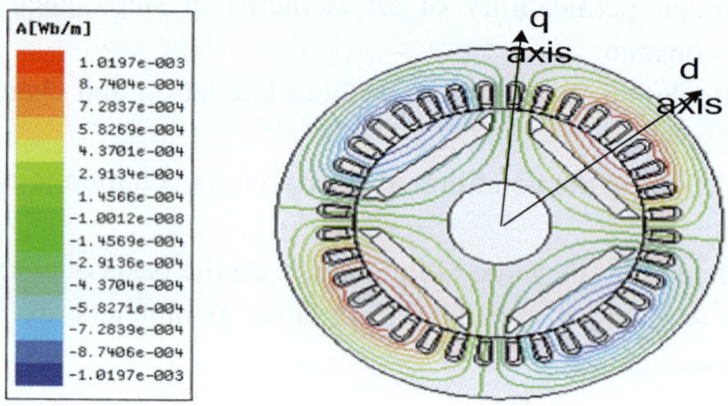

Fig. 2 Distribution of flux lines when iq=1A,id=0.

4.3 *Inductance curves with d axis current*

Curves of L_q and L_d along with i_d shows as Fig. 3.

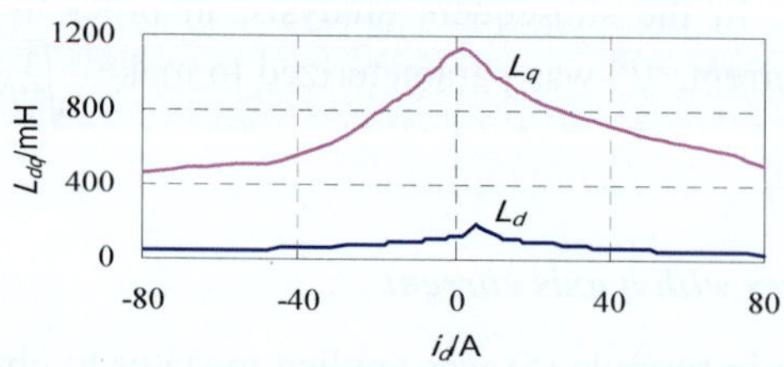

Fig. 3 Curves of Lq and Ld along with id.

Trends of the curves are similar to Fig. 1, but without cross when the current is bigger. The symmetrical axis of L_d is not i_d=0, but has a tiny positive deviation.

That means L_q is more sensitive to the change of i_d than L_d. Combined with Fig. 1 and Fig. 3, the influence of i_q to L_q is greater than the influence of i_d, the influence of i_q to L_d is about the same with i_d. Fig. 4 shows the distribution of flux lines when i_d=1A, ignoring the application of permanent magnet.

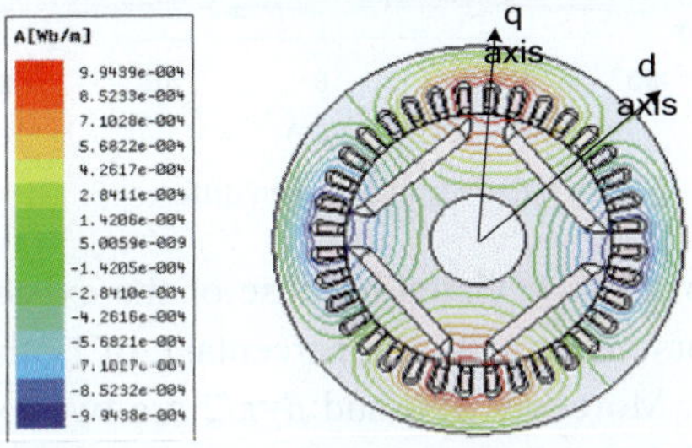

Fig. 4 Distribution of flux lines when id=1A,iq=0.

Fig. 2 and Fig. 4 shows flux lines under i_d and i_q. Flux lines through q axis all when there is only i_q, which is same to d axis. When both the i_q and i_d are existed, there will occurs deviation.

4.4 *Inductance curves with q axis and d axis current*

The shape of curves of L_q and L_d is determined by the percentage of i_q and i_d when both the i_q and i_d are existed. When β=π/3 or π/6, the percentage of i_d and i_q was given by formula (10) and (11).Fig.5 and Fig.6 show curves of L_q and L_d at different β.

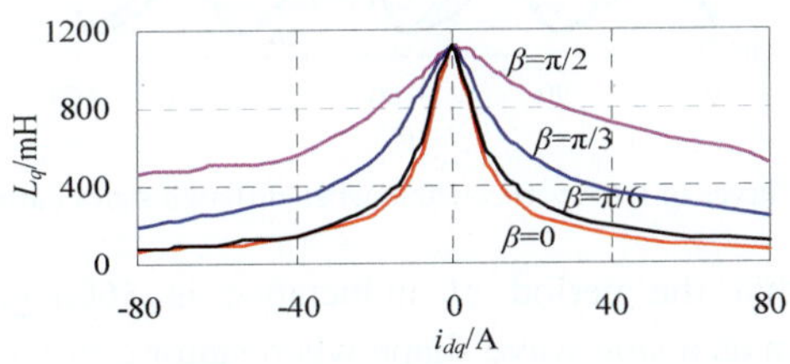

Fig. 5 Curves of Lq when different β.

No matter what the content of i_d and i_q, curves of L_q are symmetrical about i_{dq}=0 and the maximum value keeps same. Through the above analysis, L_q is strongly influenced by i_q, so value of L_q will reduce more since the more percentage of i_q. All of those curves are limited, ranging from β=0 to β=π/2.

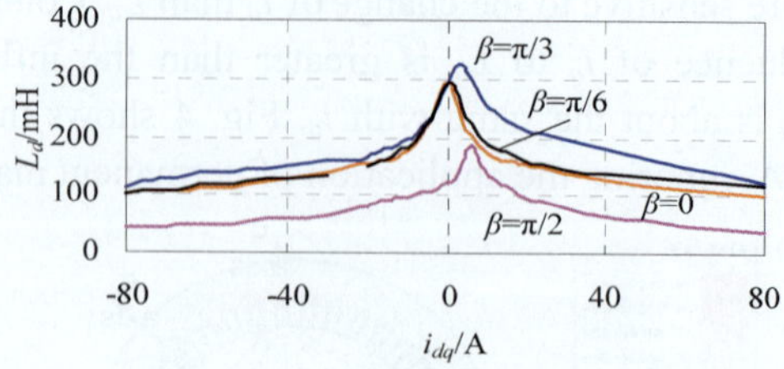

Fig. 6 Curves of Ld when different β.

Symmetric axis of L_d is deviated because of the existence of i_d. Where the maximum of L_d appears varies with the percentage of i_d, the more content of i_d, the greater value of s. Moreover β=0 and β=π/2 are two extremes for all of the symmetric axises. Fig. 6 also shows that the maximum value of L_d varies with β. The value of L_d caused by i_d and i_q is bigger than that of i_q.

5. Rules of Inductance with Rotor Position

Magnetic circuit in the motor will be different with the rotation of rotor, which leads to the change of characteristic of inductance. In order to obtain rules of inductance varies with rotor position, θ was parameterized from 0 to 720deg. Curves of self inductance of phase A at different stator current are given in Fig. 7.

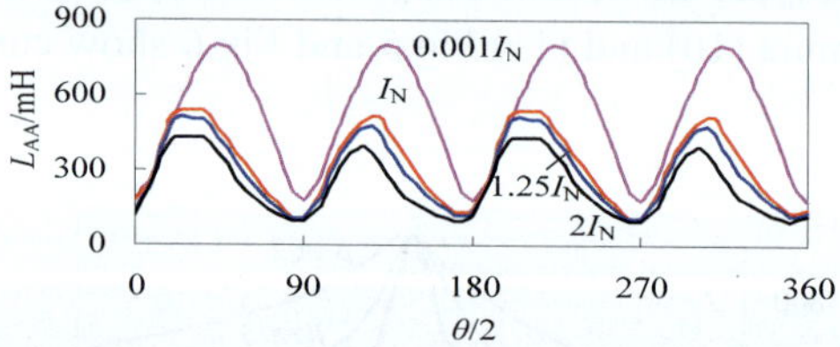

Fig. 7 Curves of LAA varies with θ when different stator current.

Fig. 7 shows that the period of inductance is 360deg. Curves of self inductance are shown as a sine wave shape when stator current is little, and they will show distortion with the increase of stator current, which is because the saturation of the magnetic circuit. Characteristics of the mutual inductance are similar with that of self inductance. Fig. 8 shows curves of L_q and L_d when stator current is $0.25I_N$.

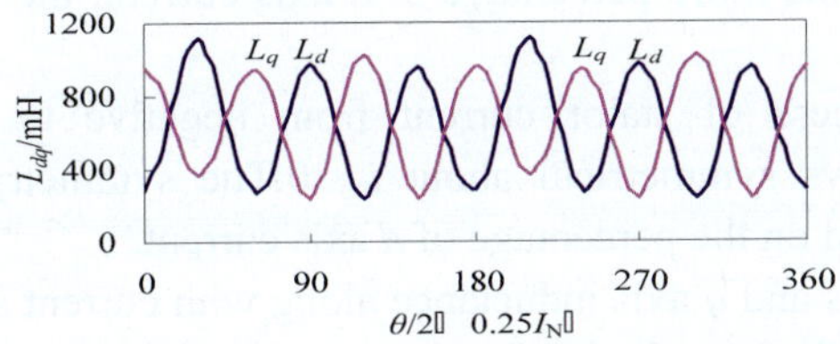

Fig. 8 Curves of Lq and Ld varies with θ when stator current is 0.25IN.

Fig.8 shows that the period of d axis and q axis inductance is 360deg too. It is similar to self inductance that L_q and L_d is also affected by stator current.

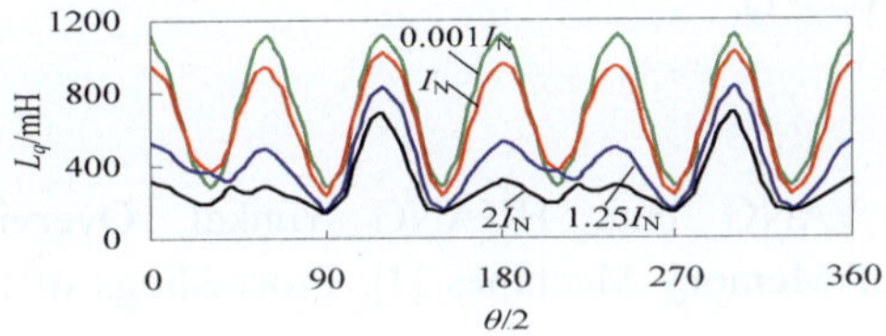

(a) Curves of Lq

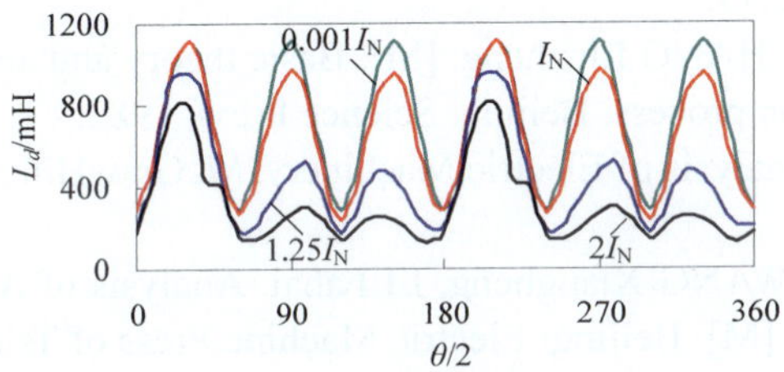

(b) Curves of Ld

Fig. 9 Curves of Lq and Ld varies with θ when different stator current.

Fig. 9 shows that curves of L_q and L_d is distorted since the bigger stator current.

6. Conclusion

Characteristic of inductance was analyzed, following conclusions was proposed:

1. q axis inductance is sensitive to the change of d axis and q axis current, and is subject to q axis current ; influence of current to d axis inductance is less then q axis, and d axis inductance is smaller than q axis inductance in numerical.

2. When the stator current is a certain number, q axis inductance is subject to the percentage of q axis current, and is no more than inductance value which caused only by q axis current; d axis inductance is more than that caused only by

q axis current, and the more percentage of *d* axis current, the larger inductance of *d* axis.

3. In the process of stator current from negative to positive, *q* axis inductance is always symmetrical about i_{dq}=0.The symmetry axis of *d* axis inductance is depend on the percentage of *d* axis current.

4. Law of *d* axis and *q* axis inductance along with current and rotor position is very complex, analysis can't simply adopt average instead.

Acknowledgement

This work was supported by the Natural Science Foundation of Shanxi Province (Grant No. 2013011035-1).

References

1. LIN Heyun, YANG Hui, HUANG Yunkai. Overview and recent development of Memory Machines [J]. Proceedings of the CSEE, 2013, 33(33): 57-65.
2. WANG Xiuhe. Permanent Magnet Machines [M]. Beijing, China Electric Power Press, 2011.
3. GAO Jingde, ZHANG Linzheng. [M]. Basic theory and analysis method of motor transition process. Beijing, Science Press, 1982.
4. Krause P C. Analysis of Electric Machinery. McGrawHill Book Comapany, 1986.
5. GAO Jingde, WANG Xiangheng, LI Fahai. Analysis of AC Machines and Their Systems [M]. Beijing: Electric Machine Press of Tsinghua University, 1993.
6. WU Xinzhen, WANG Xiangheng. Unified Expression for Calculating Inductances of AC Machines [J]. 2004, 19(1): 65-68.
7. CHENG Shu-kang, YU Yan-jun, CHAI Feng. Analysis of the Inductance of Interior Permanent Magnet Synchronous Motor [J]. Proceedings of CSEE, 2009.29(18): 94-99.
8. KANG G H, HONG J P, KIM G T. Improved Parameters Modeling of Interior Permanent Magnet Synchronous Motor Based on Finite Element Analysis [J]. IEEE Transactions on Magnetics, 2000, 36(4): 1867-1870.
9. NAJAFI S, KAR N C. Impact of Saturation on Transient Behavior of Interior Permanent Magnet Synchronous Motor [C] IEEE CCECE/CC GEI, May 7-10, 2006, Ottawa, Canada. [S.1.]:IEEE, 2006:807-810.

10. Li Heming, ZHANG Jian, LUO Yingli. Finite Element Analysis of PMSM Steady State Parameters Considering Cross-Saturation Effect [J]. Proceedings of CSEE, 2012, 32(12): 104-110.
11. LI Jingcan, LIAO Yun, YAO Jun. Impact of saturation on the characteristic of inductance ot permanent magnet synchronous motors [J]. Jounal of Chongqing University, 2009, 32(12): 1285-1289.
12. CHEN Boshi. Electric drive automatic control system [M]. Beijing: Mechanical Press, 2010.
13. Sang-Yeop KWAK, Jae-Kwang KIM. Characteristic Analysis of Multilayer-Buried Magnet Synchronous Motor Using Fixed Permeability Method [J]. IEEE Transactions on Energy Conversion, 2005, 20(3): 549-555.

The research of power cable sheath circulation online monitoring system

C. Zhong[1]

*Graduate Department, Shenyang Institute of Engineering,
Shenyang 110136, Liaoning Province, China
Email: 285210072@qq.com*

X.H. Yuan

*School of Electrical Engineering,
Shenyang Institute of Engineering,
Shenyang 110136, Liaoning Province, 436100,
Liaoning Province, China*

X. Lu

*Power Company Control Center,
Hubei Provence School of Electrical Engineering,
Shenyang Institute of Engineering, Shenyang 110136,
Liaoning Province, China*

N. Zhao and H.Y. Zhong

*Graduate Department, Shenyang Institute of Engineering,
Shenyang 110136, Liaoning Province, China*

Power cable has good electrical properties and has been widely applied in the power grid, and the condition of the cable insulation directly affects the security and stability of power system. To meet the needs of the power system, effectively monitor the cable insulation condition, according to the study of power cable metal sheath layer circulation online monitoring system to monitor the insulation condition of the cable. Analyzing and verifying the feasibility of the system through specific case.

Keywords: Power cable ; Protecting layer circulation; Online monitoring.

1. Introduction

With the rapid development of power grid construction, power cable is used more and more widely, performance monitoring and historical data on the new

---

[1] Chao Zhong

cable collection demand will be more and more. At the same time, cable laying now early succession to normal life, the problem will be increasing insulation cable, so the power system requires a set of perfect cable performance monitoring system [6], the protective layer circulation online monitoring device is one of them. Power cable online monitoring and fault diagnosis task is to understand the cable insulation state, and the state of the display and recording, alarm for abnormal state, so that operation personnel to timely treatment, and provides information and data for the equipment insulation evaluation, rational use and safe operation [8,9].

2. The Principle of Monitoring System

Many factors in the operation of cable in cable sheath insulation may cause damage or ground tank inlet, resulting in metal protecting layer multi point grounding. And if the earth to form a loop, circulating current increased significantly, seriously can reach above the main current of 90%, therefore, needs to carry on the online monitoring of cable. The monitoring principle is: the collection and measurement of grounding current and main cable current effective value, and through the GPRS or optical fiber mode, the remote transmission of data to the computer background service program at a certain time interval, a background service program collects data after the establishment of historical data files, and these data plotted curves, cable operation and maintenance personnel to understand the long-term running state of whole cable according to these curves provide information [1-10].

3. Modules and Functions of the System

3.1 *System modules*

Modular design to adopt the system, the whole system is composed of a sheath circulation signal acquisition module, wireless transmission module and circulation signal post processing platform module three part. Among them, the sheath circulation signal acquisition module mainly consists of a sensor signal transmission module composition, circulation mainly from centralized data collector (RTU) and GSM/GPRS communication networks; post processing platform module mainly consists of the background server, client software, debugging software and engineering equipment debugging etc. This monitoring system consists of frame as shown in Figure 1.

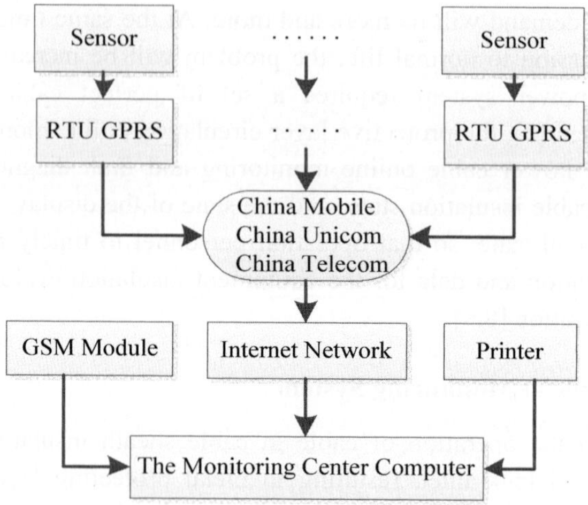

Fig. 1. Monitoring system framework.

3.2 *The functions of compositional modules*

1) The acquisition module of protective layer circulating current signal: cable sheath connected online monitoring system of circulation through the current sensor installed on the cable channel, cable connector and cable terminal tower cable terminal joint of the temperature sensor, temperature sensor, pressure sensor fault vibration of cable metal sheath layer circulation collection, cable joint fault vibration and cable surface temperature, middle and terminal joint pressure information data, information and data collected through the flame signal cable transmission to RTU.

2) The wireless transmission module of circulation signal: centralized data collector which is RTU, is a kind of remote control intelligent device. RTU real-time processing, storage of the signal from the sensor. The GPRS module which is general packet radio service, is a kind of new mobile data service development based on GSM. The main use of wireless operators (China Mobile, Unicom, and Telecom) to provide paid through the network, the network operators to establish the background and the communication channel between the centralized data collector. Needing to provide the operator or the power sector SIM card supporting the implementation of online monitoring wireless network, making the system more practical.

3) Post processing platform module: the back-end server is a background service software and hardware platform, client software is the establishment of historical data files after collecting data, and these data plotted curves, cable operation and maintenance personnel according to these curves provide information to the long running state solution of whole cable. At the

same time, background service program to provide cable current abnormal alarm and alarm equipment state. Debugging software installation and testing of software used for the production of debugging and on-site, capable of measuring parameters and network parameter setting and reading unit, at the same time to observe the running state of the equipment. Engineering debugging equipment is generally the mobile phone or laptop, to receive the field measurement data.

4. Testing and Analysis of System Performance

In order to test the comprehensive performance of the monitoring system, needing to make the actual evaluation system mainly according to the monitoring accuracy and stability, cable metal sheath grounding ways mainly have the single end grounding and cross connection grounding, integrity to test results, verify the accuracy and stability of the system were sampling. Short circuit metal layer of single core cable protection is usually adopts the end is directly connected to the other end of the gap or protection and resistance grounding way (as shown in Figure 2a), long line cable adopts three-phase piecewise cross connection at both ends of the grounding way (as shown in Figure 2b). In order to protect layer circulation theory of short circuit monitoring as compared to verify its accuracy and practical value, with multi protective layer circulation long line monitoring results to verify its stability.

1) Verify the accuracy accuracy: online monitoring of short line verification, need to get the actual value and theoretical value respectively, therefore, need to establish mathematical model is shown in Figure 2a, shown in Figure 2b, the equivalent circuit [4,5,10] shown in Figure 3.

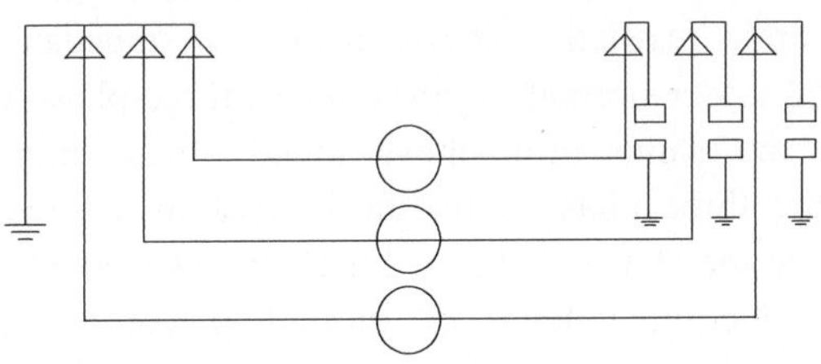

Fig. 2a. Metal protecting layer single earthing system.

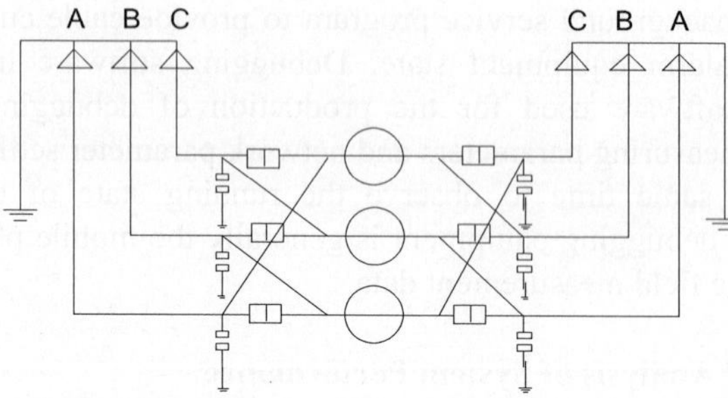

Fig. 2b. Metal protecting layer cross connection grounding system.

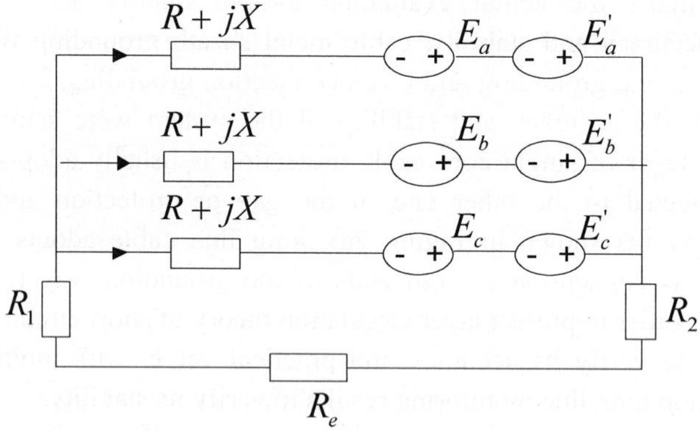

Fig. 3. The cable protecting layer circulation equivalent circuit.

Two kind of grounding method of the cable's equivalent circuit shown in Figure 3, of which R_1,R_2 are sheathed at both ends of the grounding resistance; R_e is large floor drain resistance; R+jX is the self impedance of three-phase cable metal sheath, under normal circumstances, three-phase cable structure is the same, since the same three-phase metal shield impedance; E_a,E_b,E_c respectively for the three phase core line current in a three-phase induction voltage on the cable metal sheath layer; Ea',E_b',E_c' respectively are other sheath circulation and induction voltage of grounding current generated in the three-phase metal protecting layer; I_{sa}, I_{sb}, I_{sc} respectively standing for A,B,C three-phase cable metal sheath layer circulation, are unknown; the I_s is a metal protective layer general circulation, $I_s=I_{sa}+I_{sb}+I_{sc}$. According to Figure 3 equivalent circuit can get the loop equation of the protective layer circulation [10,11] as shown in Eq. (1):

$$\begin{aligned} I_{sa} &= (R+jX)+I_s(R_1+R_2+R_e)+E_a' = E_a \\ I_{sb} &= (R+jX)+I_s(R_1+R_2+R_e)+E_b' = E_b \\ I_{sc} &= (R+jX)+I_s(R_1+R_2+R_e)+E_c' = E_c \end{aligned} \quad (1)$$

After the derivation of deformation matrix equations are obtained protective layer circulation as shown in Eq. (2):

$$\begin{bmatrix} Z_{11} & Z_{12} & Z_{13} \\ Z_{21} & Z_{22} & Z_{23} \\ Z_{31} & Z_{32} & Z_{33} \end{bmatrix} \begin{bmatrix} I_{sa} \\ I_{sb} \\ I_{sc} \end{bmatrix} = \begin{bmatrix} E_a \\ E_b \\ E_c \end{bmatrix} \quad (2)$$

Among them, the leftmost coefficient matrix can be calculated according to the structural parameters of cable, two kinds of various elements of the coefficient matrix of the grounding mode of Z_{ii} have different calculation methods with Z_{ij}. The right side is the induced voltage in each phase of metal shield in the three-phase I_{load}, if I_{load} is known, can find the right side matrix. The calculated fault phase shield circulating current as shown in Eq. (3):

$$I_{S1} = 2\omega I\left(\frac{\sqrt{3}}{2}\ln_2 + j\ln\frac{\sqrt{2s}}{GMR_s}\right) \Big/ (R+jX+R_e) \quad (3)$$

Among them, X=ΨL, R_s is a DC resistance of the metal protecting layer; R_e is the leakage resistance of the earth and the two grounding resistance's addition; ω is AC electric angle; G is conductance; S is the cable cross-sectional area, M is the magnetic moment. In order to validate the analysis conclusion of the above theory, selected three loop cable as an actual measurement, in single end grounded and grounded at both ends (fault) are compared with the case, get the shield circulating current theoretical and measured values, the results as indicated in Table 1.

Table 1. Single earthing system support layer circulation theoretical calculations and experimental comparison.

Line number		I	II	III	IV
Operating voltage(KV)		110	110	110	110
Line length(m)		520	530	720	510
Power section(mm^2)		500	400	500	500
Line's operational current(A)		72.6	99.4	46.7	50.9
Current of direct earthing extremiy(A)	Theoretical values	3.2	3.0	4.5	3.2
	Actual measured values	3.3	3.2	4.1	3.1
Sheath side protectical current(A)	Theoretical values	0	0	0	0
	Actual measured values	A:0.3	A:0.3	A:0.1	A:0.2
		B:0.3	B:0.2	B:0.1	B:0.2
		C:0.3	C:0.3	C:0.1	C:0.3
Both ends after grounding shield circulating current(A)	Actual measured values	A:38	A:19.8	A:25	A:32.8
		B:28.5	B:27.4	B:15.3	B:27.6
		C:23.1	C:23.5	C:9.1	C:24.2
The maximum phase shield circulating current/core current	52.3%	27.6	53.5%	64.4%	

As can be seen from the table, under normal circumstances, for the single end earthing system, if the sheath cable insulation is good, the direct grounding end of current is very small, generally no more than 10% core current. When the cable protective layer is damaged, shield circulating current will increase, and the maximum term of protection layer circulation reached 64.4% core line current, through the comparison, the feasibility and accuracy of the system is very well.

2) Stability validation: in order to test the stability of the monitoring system, experiments were carried out to select a running cable line of 110kV. The whole line is 2.5km long, every 500m after grounding box cross connection, each ground box set a monitoring point for online monitoring, signal analysis and processing of the circulation of the collected metal shield. Considering the factors that influence the operation of the system, make the system run continuously for a period of time, at the same time for data transmission, storage and analysis. Monitoring center in a certain time monitoring data such as shown in Table 2.

Table 2 .The monitoring center monitoring data.

Monitoring points		1	2	3	4	5
Protective layer circulation(A)	PhaseA	9.73	9.55	9.59	9.74	9.65
	PhaseB	9.27	9.33	9.15	9.23	9.26
	PhaseC	11.4	11.6	10.7	11.4	12.0

The tested results show that: the quickly data transmission speed of this monitoring system, timely data processing, table running, and monitoring data error meet the accuracy requirements, in accordance with the relevant industry standards.

5. Conclusion

The research in this paper support layer circulation online monitoring system for power cable, and the test result has fully demonstrated its feasibility. The system without changing the wiring ways of the original to collect field data, and then through the system software to get the data analysis and processing, to realize online monitoring cable grounding circulation. With the far monitoring distance, the signal transmission is not limited by geography and climate, data processing timely, and has the advantages of automatic alarm when appears abnormal situation etc.

References

1. P. Sheng, J. Li. Some measures of 110KV cable sheath grounding and protective layer protection [J]. Sichuan Electric Power Technology, 2008, 31:91-94.

2. S.M. Chen, L.J. Yang, Q.G. Zhang, H.B. Su, Y.F. Li. Effect of 110KV cable sheath connection on the sheath voltage [J]. High Voltage Technology, 2006, 32(3):37-39.
3. L.L. Zhang. Study of 110 ~ 220kV single core high voltage XLPE cable sheath insulation on-line monitoring system[C] In 2011 the national power cable installation and operation of the exchange of experience will be excellent papers set: Guangdong Weiheng Power Transmission Engineering Co. Ltd., 2011:154-158.
4. X. Jia, X.L. Cao, M. Yu. Load flow circulation and sheath of single core cable [J]. High Voltage Technology. 2001, 27(1):25-26.
5. Z.Y. Liu, H.M. Wang. The design principle of power cable structure. Xi'an. Xi'an Jiao Tong University Press. 1995.
6. Z. Zhang, Z.Y. Zhao. Research on a new method of on-line monitoring and experiment of high voltage insulation power cable [J]. Power Supply, 2012, 29(2):63-66.
7. Z.C. Zhou, J.J. He, J.W. Chen, Y.P. Zhou. Research of on-line insulation monitoring of single core XLPE cable metal sheath [J]. High Voltage Apparatus. 2006, 42(2):90-92.
8. Y.S. Liu, L. Sun, Z.H. Li. Power cable on-line monitoring platform to current development[J]. Technology Guide. 2012, 30:7-8.
9. Y.H. Zeng, G. Liu, X.Y. Luo, Daolong Wang. High voltage cable metal sheath grounding current on-line monitoring system [J]. Application of Computer System. 2013, 22(10):51-53.
10. B. Wang, J.S. Luo, H.X. Huang, J. Lv, H. Zhou. Analysis of 220 metal sheath KV high voltage single core power cable circulation [J]. High Voltage Apparatus. 2009, 45(5):141-145.
11. C.T. Chen, G.S. Zhang, Z.C. Zhou, D.D. Zhang, J.W. Chen, J.J. He, Y. Wang, Z.Y. Li. Analysis of multi-point grounding metal sheath circulation method in monitoring the XLPE cable. High Voltage Technology. 2001, 28(7):28-30.

Study of on-line partial discharge monitoring for XLPE power cable

N. Zhao
Graduate Department, Shenyang Institute of Engineering,
Shenyang 110136, Liaoning Province, China
Email: 1012224262@qq.com

X. H. Yuan
School of Electrical Engineering, Shenyang Institute of Engineering,
Shenyang 110136, Liaoning Province, China

X. Lu
Grid Corporation of Hubei, China

C. Zhong
Graduate Department, Shenyang Institute of Engineering,
Shenyang 110136, Liaoning Province, China

Based on analyzing the main causes and several basic detecting methods for the on-line partial discharge (PD) monitoring of the power cable, the importance of the on-line partial discharge monitoring was strictly confirmed. The difference of the electromagnetic wave transmission characteristic between the cable body and the intermediate joint was discussed in this work, and the transmittal characteristic was studied by analogue simulation. Results indicate that the propagation laws of the electromagnetic wave generated by partial discharge between the cable body and the intermediate joint was different. Therefore, when it comes to the location and detecting of the partial discharge, the influence of the intermediate joint should be considered. Effective measures should be taken to reduce the operation risk of power supply system, and ensure the safety and stability of power system.

Keywords: XLPE power cable; Cable joint; Partial discharge (PD); On-line monitoring.

1. Introduction

XLPE power cable was very suitable to be used in different voltage levels of the power system for its various advantages such as good electrical and thermal performance, simple structure, high temperature resistance and convenient installation. Thus, XLPE is extensively used in urban power grid transmission line and power distribution network recently. However, practical applications proved that partial discharge is the main cause for the damage of power cable insulation. Hence, on-line partial discharge monitoring is essential for monitoring the working condition of XLPE cable insulation. Also, it is of great importance for predicting the operating life of cable and ensuring the security of the working cables [1].

The cable joint are the weakest part in the whole cable, and the insulation breakdown is most likely to be happened in this part. Due to the difference between the cable body and the intermediate joint, the transmittal characteristic is distinctive. So, the on-line partial discharge monitoring will be different in these two parts. Therefore, the research on the transmission characteristic of the cable body and the joints is essential.

2. Generation of Cable Fault and Partial Discharge

It is generally believed that the working life of the power cable in common environment is 20-30 years. But manufacturing process, product quality and installation and running environment will greatly affect the running safety and lifetime of the cable [2].

Partial discharge is a localized dielectric breakdown of a small portion of the insulation system under high voltage stress, without perfoliate breakdown of the whole insulating parts. As time accumulation and the lasting destruction, partial discharge will cause material carbonization. Thus, partial discharge will gradually developed into electrical tree and fill in the cable inside, eventually lead to thoroughly breakdown.

Operating experience at home and abroad show that the majority of the cable accident (especially at the cable joint and termination) is caused by the problems of cable insulation. For the small discharge energy of the partial discharge, it won't affect the insulation of the cable in a short time. However, too many weak partial discharges can damage the insulation slowly and lead to the insulation breakdown over a long period. Namely, although the cables have passed the voltage test from the factory, the small defects existing in the cables can result in thoroughly breakdown in the long time running.

3. Methods for PD Detecting in the Cable

At present, some methods for PD detecting in the XLPE cable are: difference method, direction coupling method, electromagnetic coupling method, ultra high frequency capacitance method, ultra high frequency inductance method, ultrasound detection method and so on [3-5].

Difference method for high voltage cable on-line monitoring was raised in Japan in 1989. In this method, two platinum electrodes are respectively fixed on the left and right side of the outer sheath of the insulation joint connection box and the metal sheath of the insulation joint section. Appropriate impedance was chosen to be connected externally, and the equivalent capacitance of the cable insulation layer was used as the coupling capacitance. In this way, the PD signal can be detected by the impedance coupling.

In the direction coupling method, direction coupling sensor was installed on both sides of the intermediate joints of the cable. The sensor was installed between the outer semi-conducting layer and the metal sheath. The insulation performance of the cable won't be influenced by this installation method. Four detecting ports were extracted from the sensors, and signals will be detected from these ports. Whether the discharge signal is from the middle joint or from outer space will be judged in this way.

Electromagnetic coupling method is based on the pulse current method. Rogowski coil was installed on the grounding line of the cable shield, and the PD is detected through inducing the partial discharge pulse in the cable shielding layer.

Capacitance coupling method is put forward by xi 'an jiaotong university and the university of Southampton. This approach to detect the partial charge is described as follows: the sheath of the XLPE cable was removed, and foil strips were sticked on the outer semi-conducting layer as detecting electrodes. The metal shielding layer which was cut off was connected by wire.

In the inductance coupling method, the partial discharge information was obtained through the magnetic field generated by the discharge signals. This method was firstly put forward in Netherlands. It is always applicable as an on-line monitoring method for the partial discharge of the armored cable. The monitoring principle is: the pulse caused by partial discharge in the cable can be decomposed as radial and tangential component while it is propagated along the cable shielding, and an axial magnetic field will be caused by the tangential component of the pulse current. Thus, the sensor in the magnetic field will give a signal of bipolar pulse voltage caused by the variation of the magnetic flux. The variation of the partial discharge will be reflected by the variation of the pulse voltage.

Ultrasonic detection technology is one of the common used methods to detect partial discharge. The ultrasonic signals generated during the partial discharge of the power cable, and this signal can be detected by the ultrasonic sensor. Therefore, whether the partial discharge happened in the cable body or its attachment can be judged.

Ultra high frequency method is a new method for detecting partial discharge recently, and it is firstly used in the partial discharge detecting in GIS. Rapid development has happened on this method for its kinds of advantages such as high sensitivity, strong anti-jamming capability, ability to locate the partial discharge power and identify different types of defects etc.

4. Influences on PD Monitoring by Intermediate Joints of the Cable

As is listed in Figure 1 and Table 1, the structure and some parameters needed in simulations are pointed out. The cable used is YJY-30 with a maximum voltage of 30kv which is made of XLPE.

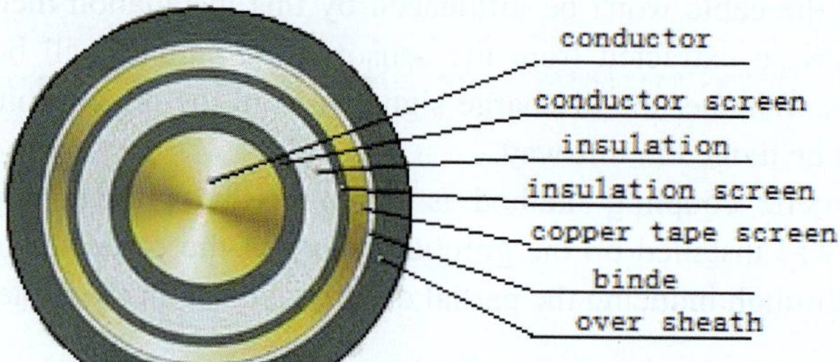

Fig. 1. Structure of cable.

Table 1. Parameters of the electrical performance for dielectric.

Material	Relative dielectric constant	Conductivity
XPLE	2.3	0
Semiconductor	30.0	2
Silicone Rubber	3.2	0
Conductive silicon rubber	90.0	0.5
Copper	—	5.8×10^{7}
PVC	3.0	1×10^{-13}

The model above is used to calculate the transmitting characteristic of the partial discharge in the cable and intermediate joints. The difference of the characteristics was analyzed for the transmission of the electromagnetic wave. Partial discharge source is simulated by a 1 mm dipole antenna. The partial discharge current pulse is set as gaussian pulse with the amplitude being 10 mA, and pulse width being 1 ns.

The same partial discharge signal propagate along the cable joint and cable body.

As shown in Figure 2, there are two cables with the length of 200 cm, and a cable has the intermediate joint, and another have no joint. the distance between the partial discharge source and the cable joint or the cable body are 100cm. The distribution of the field strength of observation point is as listed in Figure 3. Seen form Fig. 3, The same partial discharge signal propagate along the cable joint and cable body, and the electric field detected is of great difference.

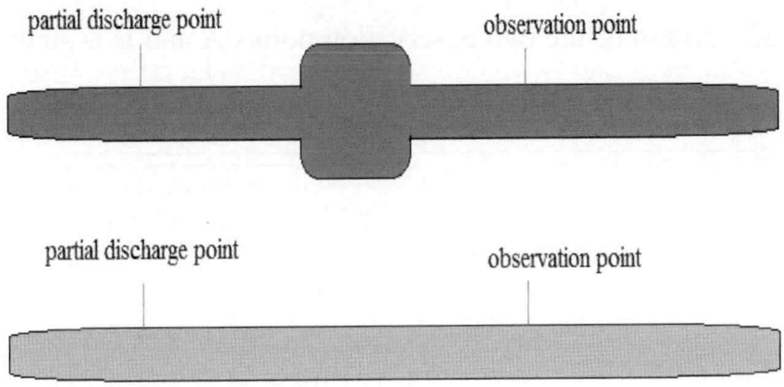

Fig. 2. The location of the partial discharge source and the observation point.

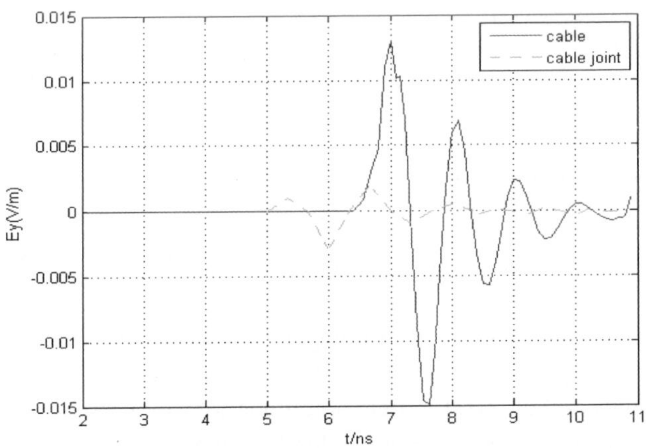

Fig. 3. The field strength of observation point.

Also, it is obviously that the attenuation of the electromagnetic wave transmitting through the cable joints is relatively more serious, while the transmission speed is greater.

As shown in Figure 4, the partial discharge source was placed on one side of the cable joint, and two observation points A and B, with a gap of 30cm, were placed respectively on the other side of the joint being 100cm apart.

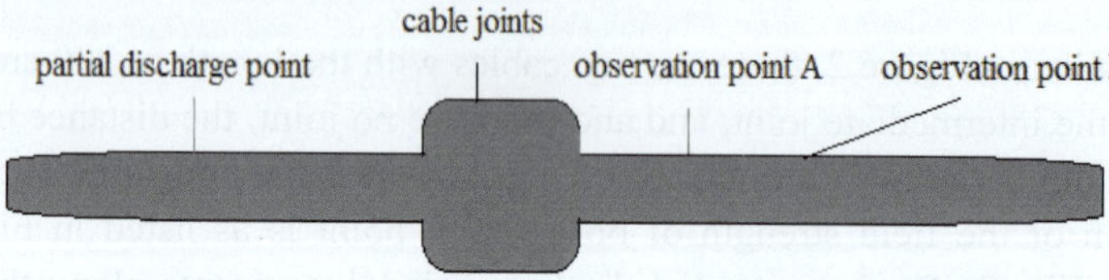

Fig. 4. The location of the partial discharge source and the observation point.

The field strength of the two observation points A and B is shown in Figure 5.

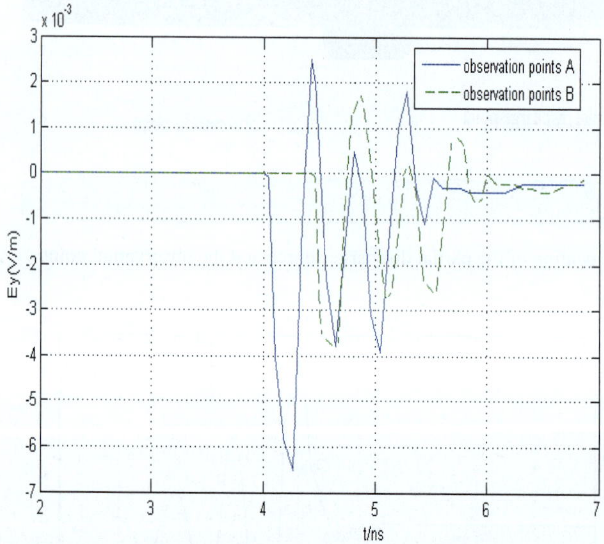

Fig. 5. The field strength of each observation point.

Seen from Fig. 5, while the electromagnetic waves transmitted through the intermediate joints, great difference is obviously noted down in the different observation points, and the electric field amplitude is also different.

Signal attenuation and propagation velocity is the key point in PD detection and location. In currently detecting method of partial discharge, the influence taken by the attachment of the cable is always ignored, thus lead to the inaccuracy of the detection. Therefore, the influence of the cable joint should not be ignored in the simulation. Due to the reason that the propagation rules of pulse signal in the intermediate joint and the propagation rules of cable body is quite different, the effect of cable joint must be fully considered while determining the cable partial discharge detection and location in actual.

5. Conclusions

This paper briefly describes the causes of the partial discharge in cable and several basic methods for detecting the discharge. The effect taken by the cable joint for the partial discharge detecting is mainly discussed.

While the electromagnetic wave caused by the partial discharge in cable transmits in different medium, the attenuation characteristics will not be identical. For example, it will show a much more severe attenuation while transmitted through the cable joint than in the cable body, and the transmission speed will be much faster in the cable joint. Therefore, when it comes to the partial discharge detection and location, the effects on the electromagnetic pulse waves taken by the intermediate joint should be taken fully considered.

The parameters of cable joint in actual installation should be chosen according to the actual status in order to avoid too large error, and avoid the influences on the accuracy of the partial detecting and location. Thus, the safety and stability can be ensured in this way.

References

1. CIGRE Working Group. After laying tests on high voltage extruded insulation cable system [J] Electra,1997(173) :33-41. Insulation Magazine, 2000, 16(5) : 13-18.
2. Tang Ju, Li Wei, Yang Hao, et al. Ultra-high frequency detection and analysis for partial discharge in high voltage cable accessories [J]. High Voltage Engineering, 2009, 35(7):1571-1577.
3. Guo Canxin, Zhang Li, Qian Yong. Currentstatus of partial discharge detection and location techniquesin XLPE power cable [J]. High Voltage Apparatus, 2009, 45(3) : 56-60.
4. Katsuta G, Toya A, Muraoka K, et al. Development of a method of partial discharge detection in extra high voltage cross-linked polyethylene insulated cable lines [J]. IEEE Transactions on Power Delivery, 1992, 7(3):1068-1079.
5. Wang Kai, Yang Juanjuan, Xu Yang. Studies on the VHF detection of PD in HV cable and joint [J]. Electric Wires & Cable, 2002(3):35-37, 46.

BP neural network for real-time load forecasting analysis and application of power system

Z.T. Guo[1†] and H. Song
Artificial Intelligence of Key Laboratory of Sichuan Province, Sichuan University of Science & Engineering, 643000, Zigong, Sichuan, China.
[†]Email:gzhiting@163.com.

P. G. Wen and J. Wang
Sichuan University of Science & Engineering, Zigong, China

With respect to the traditional power load prediction method, characterize data uncertainty as randomness, use the linear expression and mathematical tools such as the probability theory and mathematical statistics to handle and analyze historical data. The model has such problems as incompatibility with complex power systems, low prediction accuracy and slow speed. Analyze the data with the BP neural network method, and predict the power system load by using the feature of approaching nonlinear system [2]. Meanwhile, modularize the network and collect data to realize the effect of real-time prediction so as to make the power load prediction more accurate and quick.

Keywords: BP neural network; Nonlinear system; Load of forecasting; Real-time prediction.

1. Introduction

Power system and national economy are closely related to national security and people's daily life. Because of the easier conversion between the energy of electricity and other energies, it has been widely used. Therefore, the quality of power is an essential index of the reliability of power system. and reasonable distribution of load capacity is an important guarantee for the security of power system [7]. The scientific and reasonable mechanism of load forecasting which load distribution can be implemented effectively, ensures the safety and reliability of power system. The load of Power system is due to various complicated factors, it is difficult to use mathematical inductionary law [8], the traditional

[1] Zhi-Ting Guo

way of the established model is difficult to achieve the requirements of load forecast, especially for short-term load forecasting. Neural network load forecasting model is based on no limit, the nonlinear relationship of complex power system load data can be better adapted to, in the process of learning to achieve a good prediction of the load.

When used for load forecasting, artificial neutral networks using neural networks can be any desperate characteristics of nonlinear systems, during the training process, by learning to extract and approximate implicit relationship between input and output.In this paper,dealt with the historical data of the power system and related information, exclude information of infrequent events, provide reliable information for power system load forecasting. According to the forecast target and data to establish the appropriate load forecasting method. Establish appropriate mathematical model. BP network can achieve input - output nonlinear relationship.And has the ability to approximate any nonlinear function. So it is good for power load forecasting. After setting the mathematical model, by rolling the predictive mode, continuous extraction of the load on a moment of history, and then put into the mathematical models,to achieve a real-time load forecasting.

2. Load Forecasting Based on BP Neural Network Algorithm

2.1. *Principle analysis of power load forecast based on BP neural network*

BP network is a multilayer feed forward neural network, if the transfer function of the neuron is the S function; the output is continuous between the 0-1, from any nonlinear mapping from input to output. Weights are adjusted using the back learning algorithm, it is usually referred to as BP (Back Propagation Network)[1]. At present, in the practice application of the artificial neural network, BP neural network is widely used. The prediction of BP neural network is to use the approximation function of BP to the system load data analysis, analysis of the output, to achieve the purpose of load forecasting.

BP network is composed of input layer, hidden layer and output layer is a multilayer feed forward neural network, between layer and layer through transfer functions, there is no relationship between each layer of neurons. Establishing the neural network structure, first of all it is to the network training, network training refers to the continuous input vector, comparing the output vector and the expectation vector, constantly adjusting the weights and threshold, which makes the structure of neural network, adaptive predictive data requirements, performance function to achieve the minimum [5]. Correction of a simple BP neural network weights and threshold can be represented by the following formula:

$$x_{k+1} = x_k - ag_k \tag{1}$$

In the formula, xk+1 is the new modified threshold value and weight matrix; xk is the history of the threshold value and weight matrix; a is the learning rate as a function of the gradient; gk.

$$g_k = \frac{\partial E(k)}{\partial x(k)} \tag{2}$$

The E (k) for the total error performance function in the iteration of the network output k. For example, when a BP neural network with two layers, and only one input samples, error performance function using the following formula to express it [3]:

$$\begin{aligned} a_i^2(k) &= \mathrm{f}^2\left\{\sum_{j=1}^{s^2}\left[w_{i,j}^2(k)a_i(k) - b_i^2(k)\right]\right\} \\ &= f^2\left\{\sum_{j=1}^{s^2}\left[w_{i,j}^2(k)f\left(\sum_{j=1}^{s^2}\left(iw_{i,j}(k)p_i + ib_i(\mathrm{k})\right)\right) + b_i^2(k)\right]\right\} \end{aligned} \tag{3}$$

Similarly, if the input is the n sample, then:

$$\Delta x(k+1) = \eta \Delta x(k) + a(1-\eta)\frac{\partial E(k)}{\partial x(k)} \tag{4}$$

$$\eta\left(0 < \eta < 1\right)$$

$$x\left(k+1\right) = x\left(k\right) + \Delta x\left(k+1\right) \tag{5}$$

This algorithm is to determine the correction next by correcting the results first, if the correction on a large, (4) in the symbol in contrast with the correction of a symbol, in order to make correction of a reduced, to reduce the role of shocks such as correction; fruit a small (4) type, the symbol and a correction in the same sign, to accelerate the convergence effect. In this paper, the power system load forecasting process, can be introduced into the momentum factor mc to achieve MOBP algorithm.

2.2. *Standardization of BP neural network input and output data*

In the data samples, the input of the network is a three-dimensional vector, the data needs to be normalized treatment before entering the neural network [4].

Process the data between the interval [0 1]. Set the maximum and minimum date of input and out be pmax and pmin in the bp network, raw data of input and output be pi, Standardized input and output data be pi', then:

$$p'_i = \frac{p_i - p_{\min}}{p_{\max} - BXp_{\min}} \tag{6}$$

$$p_i = p_{\min} + p'_i(p_{\max} - p_{\min}) \tag{7}$$

3. Examples of Simulation

Through the neural network module into the Matlab tool, to realize the power system load forecasting, the module in the Simulink mode set aside, and through the real-time sampling data, which can simulate the real time forecast.

3.1. *The design of the network*

According to the algorithm determines the characteristics of the network structure, the network structure of 3 layers, is composed of input layer, hidden layer and output layer, according to the historical load data, using the rolling forecasts way, namely use the first three historical load to predict the next load. determine the number of neurons in the input is 3, the number of neurons in the output is 1, the number of neurons in the hidden layer is determined in a variety of empirical formula, this paper the formula is as follows:

$$n = \sqrt{p+t+\alpha} \quad \alpha \in [1,10] \tag{8}$$

P is the number of input neurons, t is the number of output neurons, alpha is a constant value, calculated by 1 to 10, the number of neurons in the hidden layer, between 3 to 13, after the experiment, when the number of neurons in the hidden layer is 8, the error is smaller, better prediction effect.

3.2. *The forecast*

In order to make a more accurate forecast.Make the historical date normalized at first. Input vector becomes p=[0.5775 0.6386 0.5770];[0.6386 0.5770 0.5380];[0.5770 0.5380 0.5456] ....[0.8071 0.7129 0.8734].

The use of MATLAB simulation, the establishment of the network using the latest version of the BP network net=feedforwardnet code (8); the number of the figures in brackets represent 8 of neurons in the hidden layer. As figure 1 shows:

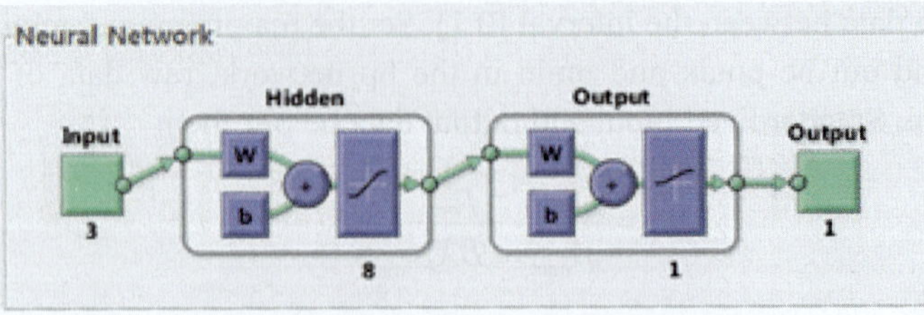

Fig. 1. Structure of neural BP network.

The use of MOBP algorithm, mc=0.09 to achieve load prediction by introducing the momentum factor, the results are as follows:

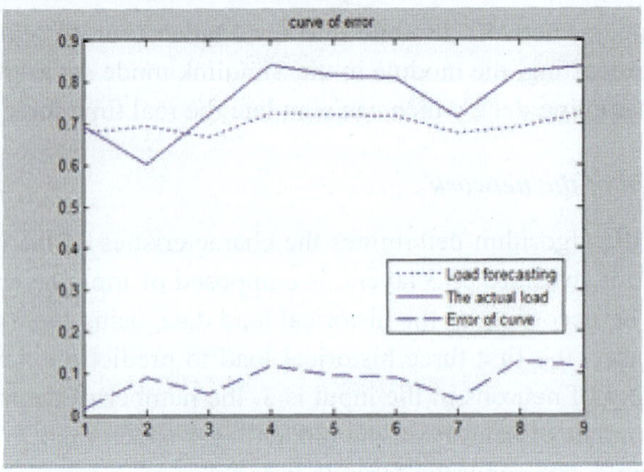

Fig. 2. Load forecasting simulation image.

Due to historical causes fewer samples, training is not enough, there will be a big error, increase training load data, at the same time, pretreatment of incomplete, inaccurate data, can improve the prediction accuracy, to achieve better results. The trained neural network module out into the Simulink interface, real-time acquisition of the data, the data acquisition module, into the input, so as to achieve the effect of real-time prediction [6]. With the using of command of 'gensim (net) [3], ('gensim' is a function in matlab which can make the network modular), can make the load forcasting model modular, put the historical data into the inputs, to achieve real-time load forecasting.

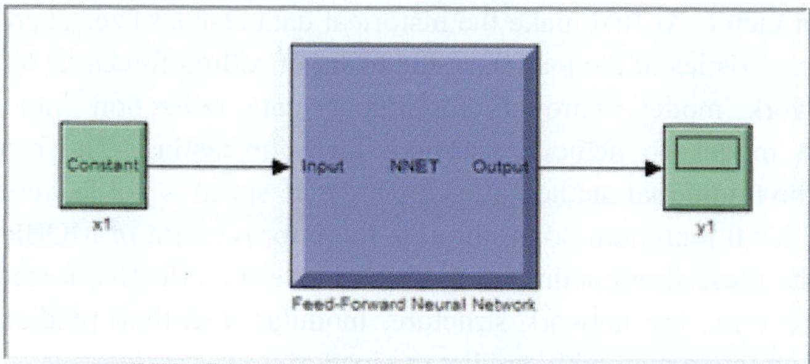

Fig. 3. Modularization of load forecasting.

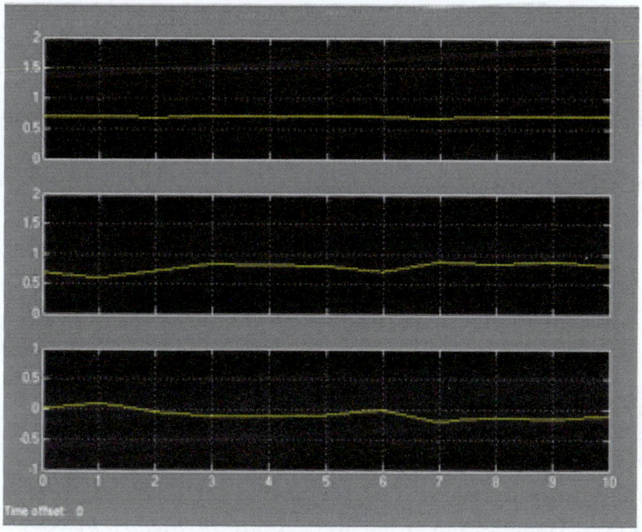

Fig. 4. Real time image load of forecasting.

Figure 4 is in the Simulink environment, the simulation of the load forecasting function of the trained neural network load forecasting based on representing the actual load curve, and the actual error. Simulation results show that BP neural network load forecasting has a good applicability based on modular processing is convenient, can achieve real-time forecasting purposes, than the traditional mathematical model prediction method is more accurate and efficient.

4. Conclusion

In this paper, through the analysis of the BP neural network in power system load forecasting applicability analysis, the neural network structure, the real-time prediction of power system load, and the improved BP algorithm, adding momentum factor. At first make the historical data standardize. Then according to the characteristics of the load data, the using of rolling forecasts, to build up a good network model. Through continuous data collection into the well-established model, to achieve real-time load forecasting. The mathematical model of the traditional method, the convergence speed is not fast enough, easy to fall into local minimum point, through the improvement of MOBP algorithm to overcome these shortcomings, so as to achieve the effective forecast of load at the same time, the network structure, modular, real-time prediction of the load, is an effective method to predict the load of power system.

References

1. Shuang Cong. *With the application of neural network based on MATLAB toolbox* [M]. Hefei, University of Science & Technology China Press,1998.
2. Huanglin Zeng. *Intelligent Computing* [M]. Chongqing, Chongqing University press, 1998.
3. Defeng Zhang. *Simulation and application of MATLAB neural network* [M]. Beijing, Publishing House of Electronics Industry, 2009.
4. Hui Hu, Hua Yang. *Application of Artificial ANN to Short-Term Load Forecasting in Power System* [J]. Hunan University, 2004, 31(5):51-53.
5. Qiang Gao, Shenghui Wang. *Study of mid-term electric load forecasting based on ANN* [J]. Journal of Shenyang University of Technology, 2004, 26(1):41-43.
6. Xiuyan Zhang, Guobin Tao, Qingqiang Liu. *The Explores of the BP Neural Network Implement Based on Simulink* [J]. Journal of Jiamusi University (Natural Science Edition), 2007, 25(5): 587~589.
7. Ying Zhou, Bangde Yin, Ling Ren. *Study of Electricity Short-term Load Forecast Based on BP Neural Network* [J]. Electrical Measurement & Instrumentation, 2011, 02:68-71.
8. Wei Cui, Mei Zhang. *Prediction of load based on grey theory and BP networks* [J]. Journal of Liaoning Technical University (Natural Science Edition). 2011, 04:631-633.

STM32-based digital charging power supply

K.Y. Wu[†], T. Zhang and Z.W. He
South China University of Technology, Guangzhou, China
[†]*Email: wuky@scut.edu.cn*

X.M. Jiang and X.G. Liu
Guangdong Institute of Automation, Guangzhou, China

To improve the flexibility of control of charging power supply, a STM32-based digital charging power supply is proposed. Utilizing the PWM generator integrated in STM32 to conduct phase-shift control of the PWM waveform by software, it realizes direct digital control of the PWM signal, zero-voltage-switching of the switch and high-frequency inverter modulation of charging power supply. The constant-current control of charging output is realized through anti-windup PI control algorithm. The experiment result shows that the STM32-based digital charging power supply has the advantage of soft-switching realization, good charging performance, stability and robustness, and fully shows the benefit of STM32 digital control.

Keywords: Charging power supply; Digital control; Soft-switching; STM32.

1. Introduction

Pure electric vehicle has become the trend of new energy automobile development. Currently, the development of two critical parts, battery and charger, cannot meet the requirement of electric vehicles, and has already become the bottleneck of electric vehicle development. The charging technology for electric vehicles is one of the requisite technologies. It has direct impact on the efficiency and lifetime of battery, and also as a key factor, it limits the electric vehicles' fast development and popularization [1]. With the development of inverter and digital control technology, the main circuit of switching mode power supply is developing from hard-switching to soft-switching and its control circuit is changing from analog control to digital control. The analog control system is excessively complex; whilst processors of the Single Chip Microcomputer(SCM) family lack the performance to achieve real-time control [2] and the cost of utilizing a DSP is too high [3]. Recently, with the development of STM32 designed for embedded applications that require high performance, low cost and low power consumption, STM32 has come very easy to use and can process analog and digital signal in real time [4]. Therefore, STM32 can be used for digital control of charging power supply. The latest technology of charging power supply is the combination of soft-switching and digital control, that is, the digital control technology of charging power supply, which makes the traditional charging power supply more stable and have better performance and extensive functions. Due to the existing problems of charging power supply and its control, combining digital control technology, STM32-based digital charging power supply is proposed for the high frequency inverter modulation of charging power supply.

2. Soft-switching Main Circuit and Its Working Principle

The main circuit of traditional charging power supply utilizes hard-switching and power components endure high stress of switch. During switching, the superposition of the voltage and current of the switch brings high switching loss, serious electromagnetic pollution accompanying high peak voltage and current [5]. To overcome the hard-switching problem stated above, this paper proposes the topology of phase-shift full-bridge soft-switching main circuit to fulfill the requirement of charging power supply. Considering real status of charging power supply, in the topology of phase-shift full-bridge soft-switching main circuit, saturated inductor and resonance capacitor in parallel with MOSFET are added to the traditional hard-switching circuit topology. It makes good use of magnetizing inductor energy plus leakage inductor energy and the resonance of saturated inductor plus capacitor to achieve the purpose of soft-switching [6]. It has overcome the problems of low efficiency, low switching frequency in the traditional hard-switching circuit topology, and fulfills the requirement of charging power supply. The topology of phase-shift full-bridge switch circuit is shown in Figure 1.

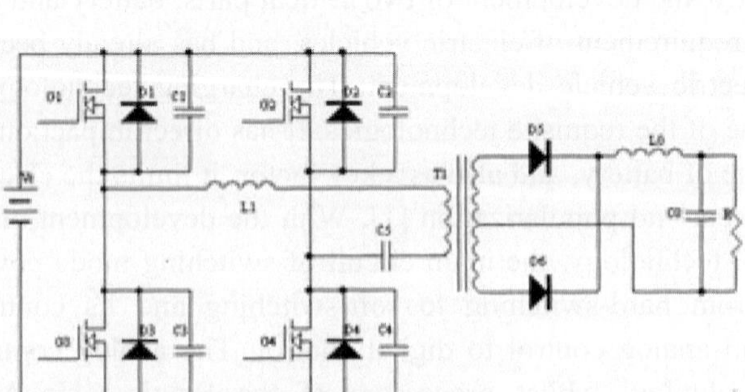

Fig. 1. The schematic of phase-shift full-bridge switch main circuit.

The working principle of phase-shift full-bridge switch main circuit is stated as following: G1, G3 as well as G2, G4 switches on in turn for 180°. G1 switches on first, after phase-shift of θ,G4 switches on, at this time, inductor L1 saturates, G1 switches off, the transformer primary current flows from G1 to shunt capacitor C1 and C3 of the leading-leg, charges C1, discharges C3 and G1 is zero-voltage switch-off. When C3 voltage drops to near zero, the G3's anti-parallel diode D3 conducts naturally, and clamps the voltage across G3 at zero volts, after that, G3 switches on, so zero-voltage switch-on is realized for G3. The switching mechanism is similar for other switches. The essence of the soft-switching is to charge or discharge the shunt capacitor in the process of resonance to make the voltage of one leg quickly go up to the voltage of power supply or down to zero, so that the shunt diode of the same leg to conduct soon conducts, and the voltage over the switch is clamped to zero to provide conditions for zero-voltage switch-on; when the switch switches off, the shut capacitor voltage can't jump and remains at zero, and zero voltage switch-off is realized.

3. STM32 Digital Control System

3.1. *Hardware of control system*

The structure of the designed STM32-based digital control system is shown in Figure 2. The control system comprises of STM32 digital control minimum system, voltage and current detection circuit, fault protection circuit, high frequency driving circuit as well as Human-Machine interface.

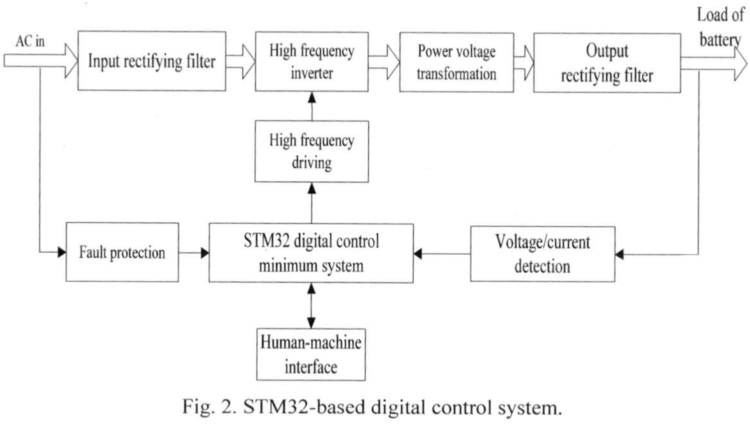

Fig. 2. STM32-based digital control system.

The STM32F103ZET6 is the core of the STM32-based digital control system. STM32 chipset series are suitable for industry control. It has high speed calculation, high efficiency instruction and is good for real-time control. The STM32 chipset has functions of 2 advanced-control timer, TIM1 and TIM8. Each advanced-control timer can be seen as three-phase PWM generators and has complementary PWM output with programmable inserted dead times. It can generate phase-shift PWM by setting the mode of advanced-control timer TIM1 or TIM8[7].

3.2. *Software of control system*

The control system software includes main modules of main program, sub-program of current anti-windup PI and sub-program of timing control.

3.2.1. *Main program*

According to STM32 hardware circuit and charging process feature of charging power supply, C language is used for programming. Before charging, the charging parameters like charging current, voltage are set. According to what is set, the program controls the phase-shift of the PWM waveform and controls the switch components in the main switch circuit of phase-shift full-bridge. During charging, STM32 samples the parameters of charging voltage and current. In order to make charging power supply have better constant-current external characteristic and dynamic characteristic to ensure that the control system has suitable dynamic response time during charging, the program has to process the samples fast. The main program mainly has the functions of system initialization, related registers configuration and main loop entrance. The process of main program is shown in Figure 3.

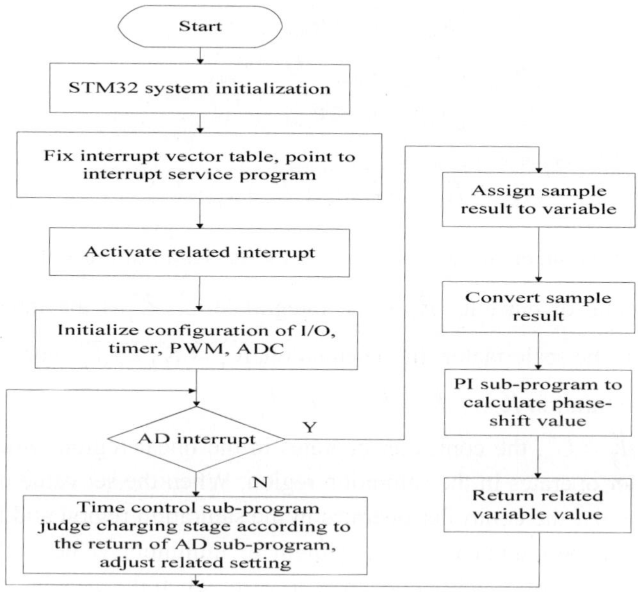

Fig. 3. The flowchart of control system main program.

3.2.2. *Sub-program of current anti-windup PI*

The constant-current of traditional charging power supply is mainly achieved by means of PI controller. The integral term aims to eliminate the static errors and improve the control precision. When the set values drastically increase or decrease, great errors will arise in the system in a short time. The errors accumulated by PI calculation integral term will cause the increased control output which can easily exceed the limit control quantity of controlled objects resulting in strong integral saturation effects such as system oscillation and prolonged adjusting time. PI control algorithm of anti-windup can eliminate the adverse effects caused by the integral saturation of traditional PI control algorithm[8]. PI control algorithm of anti-windup is:

$$e_k = I_g(k) - I_f(k) \tag{1}$$

$$U = R_{K-1} + K_P e_K \tag{2}$$

$$u_k = \begin{cases} u_{\max} & U > u_{\max} \\ U & u_{\min} \le U \le u_{\max} \\ u_{\min} & U < u_{\min} \end{cases} \tag{3}$$

$$R_k = R_{k-1} + K_I e_k + K_C (u_k - U) \tag{4}$$

Where e_k is the error value, $I_g(k)$ is the set value of current, $I_f(k)$ is the feedback value of current, R_k is the integral term, K_I is the integral coefficient, K_P is the scale factor. In the equation, $K_I = K_P T / T_I$, correction factor of integral saturation $K_C = K_I / K_P = T / T_I$.

When $u_k = U$, the controller operates in the linear region; when $u_k \neq U$, the controller operates in the saturation region. When the set value of current is changed greatly, the controller operates in the saturation region and the input of controller will be clamped by a maximum or minimum value predetermined, then the integral term quickly converges to zero; when the current error is within the range of a fixed error band, the controller operates in the linear region, with the integral term in PI controller activated. Therefore, the system has asymptotic stability in the linear region in control of the controller and can converge to the linear region in the saturation region, thus guaranteeing the stability of current control for charging power supply.

3.2.3. *Sub-program of timing control*

When charging power supply is connected to the battery, once the system powers up, STM32 program initializes, when the charging button is pressed, after system confirm the battery is connected correctly, the power supply start to charge the battery, simultaneously the charging current is sampled, the anti-windup PI control algorithm is applied on current to realize constant-current output. When the system samples the current of very low, it shows that the battery is charged to full, the system stops output, stops running, and the charging process ends. The process of timing control is shown in Figure 4.

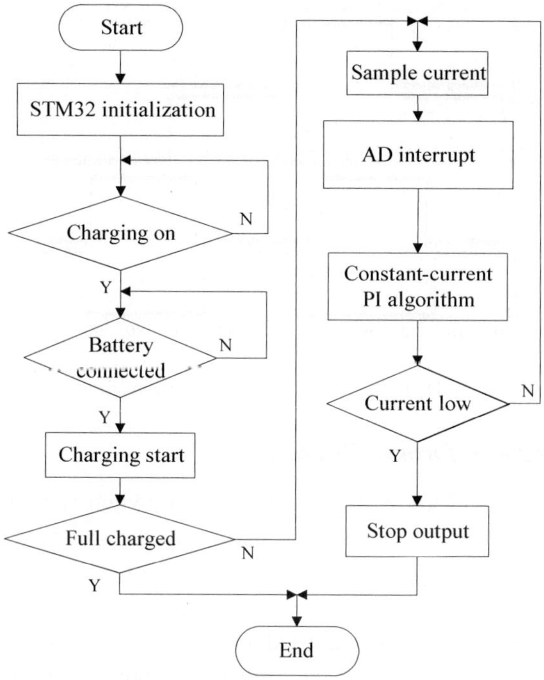

Fig. 4. The flowchart of charging timing control sub-program.

4. The Study of Soft-switching Power Supply Experiment

All key nets of main circuit and PWM output circuit are tested to verify the design intention of the control system hardware and software.

4.1. *Software-based phase-shift PWM waveform generator*

The phase-shift modulation uses cycle interrupt and underflow interrupt of timer. When the cycle interrupt of timer is triggered, the original matching value of count-ups is changed into the matching value required by count-downs in the cycle interrupt service program; When the underflow interrupt of timer is triggered, the original matching value of count-downs is changed into the matching value required by count-ups of next cycle in the underflow interrupt service program thus to realize the control of phase-shift full-bridge PWM soft-switching.

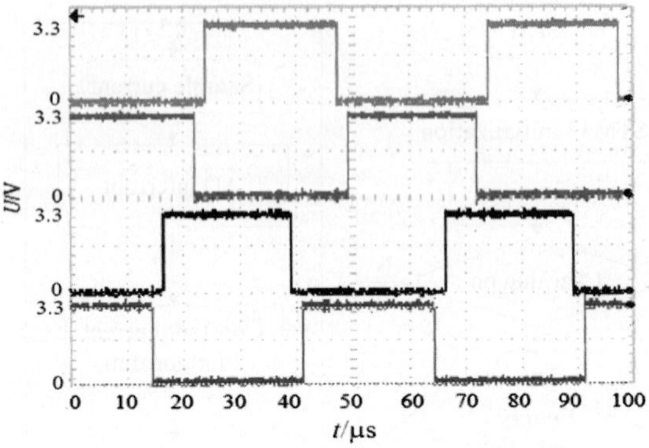

Fig. 5. STM32-based phase-shift PWM waveform.

4.2. *Soft-switching waveform verification*

The soft-switching waveform of lagging leg power switch is shown in Figure 6 and Figure 7 respectively. It is relatively narrow of the soft-switching scope of the lagging leg of full-bridge inverter power supply of phase-shift control soft-switching, it is more difficult to realize soft-switching than that on leading-leg, in lagging-leg resonance only leakage inductor energy takes part in the process of current exchanges with lagging-leg, and the resonance status is worst. As shown in Figure 6, before the MOSFET driving voltage rise to the MOSFET threshold, the MOSFET collector-to-emitter voltage already falls to zero, it shows the realization of zero-voltage-switching, as shown in Figure 7, after MOSFET driving voltage falls to zero, the MOSFET collector-to-emitter voltage rises from zero gradually, and it shows the realization of zero-voltage switch-off. Thus we can see that the lagging-leg switch-on is full scope soft-switching-on and the output duty ratio is the overlap of up/down phase-shift PWM.

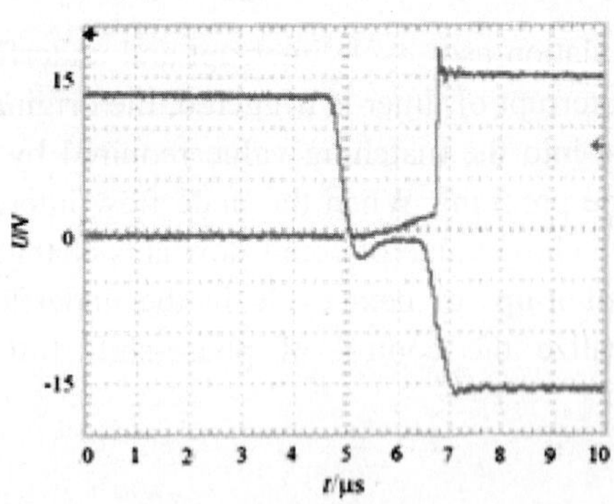

Fig. 6. The switch-on transient waveform of switch.

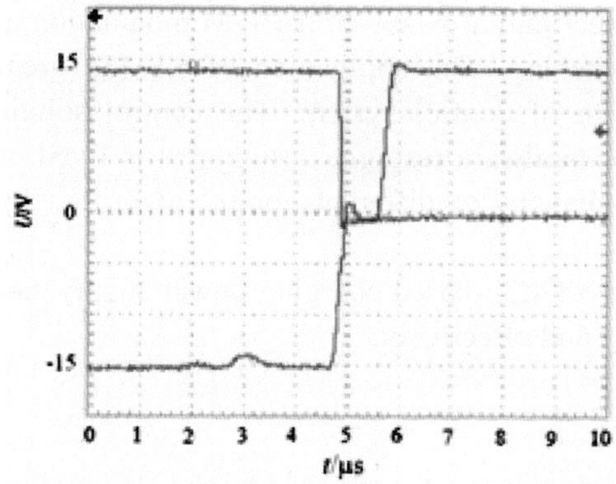

Fig. 7. The switch-off transient waveform of switch.

4.3. *Charging waveform experiment*

The capacity of the lead-acid battery used in the experiment is 100AH. Set charging current to 0.15C, 15A, and the charging current waveform is as shown in Figure 8.

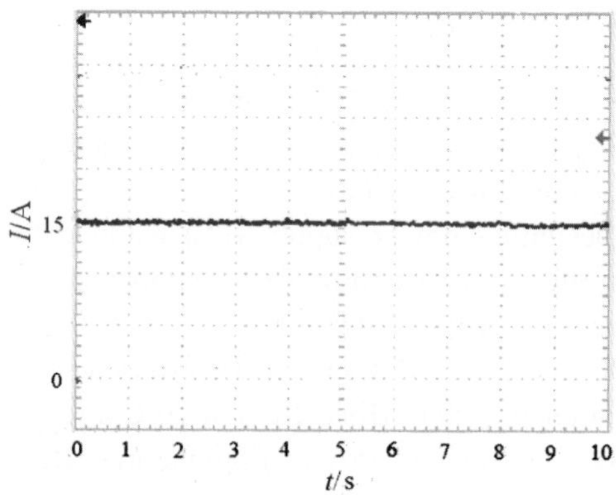

Fig. 8. The charging current waveform of the lead-acid battery.

As Figure 8 shows, the charging current is constant stably, as the battery impedance is not pure resistive, there is some glitch, and the charging requirements is fulfilled.

5. Conclusion

(1) Phase-shift full-bridge main circuit topology can be used to well achieve the switch zero-voltage-switching, which improves the charging power supply efficiency and reduces electromagnetic pollution.

(2) The PWM module integrated in STM32 can be used to easily and reliably achieve the direct digital phase-shift PWM modulation of phase-shift full-bridge soft-switching. The application of STM32 greatly simplifies the hardware design of control system. The control solution is realized by software. After hardware resource assignment is fixed, it can realize real-time, on-line changes of different control algorithm by only modifying software.

(3) The designed STM32-based charging power supply can charge lead-acid battery steadily and effectively.

Acknowledgements

This material is based upon work supported by the National Science Foundation of China (Grant No. 51205136), the Science and Technology Planning Project of Guangdong Province, China (Grant No. 2013B010402002) and the Competitive Allocation Project Special Fund of Guangdong Province-Chinese Academy of Sciences Comprehensive Strategic Cooperation (Grant No. 2013B09150 0082).

References

1. Yilmaz M, Krein P T. Review of battery charger topologies, charging power levels, and infrastructure for plug-in electric and hybrid vehicles [J]. IEEE Transactions on Power Electronics, 2013, 28(5): 2151-2169.
2. Mohamadi T. Designing power supply (PS) using digital PID based on AVR microcontrollers [C] Design & Test Symposium (EWDTS), 2011 9th East-West. Sevastopol: IEEE, 2011: 1-5.
3. Chan Ee Lyn, Rahim N A, Mekhilef S. DSP-based fuzzy logic controller for a battery charger [C] 2002 IEEE Region 10 Conference on Computers, Communications, Control and Power Engineering, Beijing: IEEE, 2002, (3): 1512-1515.
4. Limian Ren. Research on power harmonic detection system based on STM32 [C] IEEE Workshop on Advanced Research and Technology in Industry Applications (WARTIA), Ottawa: 2014: 1142-1144.
5. Liang Y C, Ng T K. Design of battery charging system with fuzzy logic controller [J]. International Journal of Electronics, 2003, 75(1): 75-86.
6. Pahlevaninezhad M, Pan S Z, Jain P. A ZVS phase-shift full-bridge DC/DC converter with optimized reactive current used for electric vehicles [C] Industrial Electronics Society, IECON 2013-39th Annual Conference of the IEEE, Vienna: IEEE, 2013: 4546-4551.

7. STM32F103xC, STM32F103xD, STM32F103xE datasheet (Rev9) [R]. ST Mircoelectronics Inc., 2014.
8. Choi J W, Lee S C. Antiwindup strategy for PI-type speed controller [J]. IEEE Transactions on Industrial Electronics, 2009, 56(6): 2039-2046.

Calculation of TTC with the power transmission method

Y. Liu[1†] and H. Bao[2]

School of Electrical & Electronic Engineering,
North China Electric Power University, China
[†]*Email: liuyangbyq1@163.com*

Total Transfer Capability (TTC) is an important indicator to characterize system security. It's in an important technique value to ensure the safe and reliable operation of the system, and to guide the operation of the system dispatcher. This paper introduces the basic concepts of TTC, first by using power tracing method, the electricity network is divided into several single power source. Then use the power transmission method to verify the relationship between equivalent node impedance and the power, By the power transmission method to establish the transmission relation between nodes and branches, This paper proposes a new approach to calculate TTC.

Keywords: Total Transfer Capability (TTC); Continuation Power Flow (CPF); Power Transmission; Power Tracing.

1. General Instructions

1.1. *Technical value of TTC*

For large interconnected power system, TTC of its area has important technical value. Because of the competitive pressures of modern power system in the economic, environmental and power market, the power system is approaching the limit. So how to safe the operational constraints of the power system, to definite the exact factors that affecting the ability of the power exchange between the regional power system, to meet the requirements of the regional electricity has became an important issue needs to be resolved in power systems [1].

1.2. *Calculation methods for TTC*

And now there are three main methods for solving this: continuation power flow (CPF), optimal power flow (OPF) and sensitivity analysis. CPF includes a

---

[1] Yang Liu
[2] Hai Bao

variety of complex constraints, but slows in the calculation, and is conservative in the result [2]; Results of OPF are only an ideal case, the robustness of the system is not strong, with slow calculation speed and limited constraints include [3-5]. Calculation of sensitivity analysis method is faster, but considering less constraints with a lower accuracy [6] .When the system transmission reaching its limit, by using power tracing formula [7-8], the multiple power supply network is divided into several single power system, and each equivalent impedance of branches can be calculated in it. Then, use impedance analysis to verify the relationship between the impedance and trends and establish the relationship between impedance and power transfer through the power transmission method.

2. Calculation of TTC with CPF

2.1. *Concept of TTC*

Total Transfer Capability is regional exchange capacity, it's defined by the North American Electricity Reliability Council system (NERC).Power exchange capacity between two regions of a system refers to the maximum power that can transfer from zone 1 to zone 2 though all transmission circuits between the two regions [9-12].

2.2. *Calculation steps of CPF*

When using CPF methods and exchange sensitive method to calculate TTC, both need to use the CPF to calculate the base power flow. Just consider the following three static physical constraints: the constrains of branch load; 2 the active and reactive power of generator node; 3 node voltage constraints [12].

To calculate the TTC between power generating area A and the reception area B , increase injection power of generation nodes in area A and the power of load nodes in area B node simultaneously at the initial base point until reaching the physical constraint. In this state, the exchange of power between the two regions is the desired TTC.

At the base power flow, assuming the system net injection power of each node is given as the following mode:

$$\begin{cases} P_i = \Delta P_i^0 + tK_{Pi} & i \in N \\ Q_i = \Delta Q_i^0 + tK_{Qi} & i \in n_1 \end{cases} \tag{1}$$

In the formula: P_i is net active node injection; Qi is net reactive node injection. Superscript 0 indicates that the value of the base flow; Kpi is collection for the growth direction of power flow between the two regions, and kpi are zeros for the non-power node area, n_1 is the total number of PQ nodes in the system; N is the total number of nodes in the system. With t growing, note the t for t^*;

The polar form power flow equation is:

$$\begin{cases} \Delta P_i = P_{Gi} - P_{Li} - V_i \sum_{j \in i} V_j (G_{ij} \cos\theta_{ij} + B_{ij} \sin\theta_{ij}) = 0 & i \in N \\ \Delta Q_i = Q_{Gi} - Q_{Li} - V_i \sum_{j \in i} V_j (G_{ij} \sin\theta_{ij} - B_{ij} \cos\theta_{ij}) = 0 & i \in n_1 \end{cases} \tag{2}$$

Simple equation expressed as

$$(x,t,p) = 0 \tag{3}$$

The calculate process is as following:

1. Calculate the base power flow in current system under study; 2.Set up the growth direction of power flow; 3. Gradually increasing t from 0, calculate the corresponding power flow with CPF, until reaching a certain constrain, At this time t is noted as t_*.; 4.The limiting case that obtained from Step 3 is the TTC of the base power flow .

3. Divide the Network with Power Tracing

3.1. *The method of power tracing*

To calculate the equivalent impedance in single supply system with power point tracing method, first of all, get the power corresponding to each supply component, providing the basis for power tracing.

Take a linear network of n nodes and q current source. The load of network is changed into the equivalent impedance.

Establish the nodal admittance matrix Y and the nodal voltage equation as following:

$$Y * \dot{U} = \dot{I}_s \tag{4}$$

Thus available:

$$\dot{U} = \dot{I}_s * Y^{-1} = \dot{I}_s * Z_{\mathrm{m}} \tag{5}$$

Any nodal voltage U_i can be expressed as:

$$\dot{U}_{\mathrm{i}} = \sum_{m=1}^{q} z_{mi} \dot{I}_{ms}, (i = 1,2\ldots,\mathrm{n}) \tag{6}$$

In the formul : z_{mi}, i_{ms} are the element of Z_m and I_s

With Circuit basic formula, the calculated Injector power S_{ks} is shown in the following formula:

$$\dot{S}_{ks} = \dot{U}_k * \dot{I}_{ks} = \sum_{m=1}^{q} z_{mk} \dot{I}_{ms} \overset{*}{\dot{I}}_{ks} \ (k = 1,2,\ldots,q) \tag{7}$$

While the power of branch to the ground S_j and tie line S_{ij} is different, they will inevitably mix with components of other supply. Respectively as the following:

$$\dot{S}_i = \sum_{m=1}^{q} z_{mi} y_i \dot{U}_i \overset{*}{\dot{I}}_{ms} \tag{8}$$

$$\dot{S}_{ij} = -\sum_{m=1}^{q} (\dot{U}_i - \dot{U}_j)(z_{mi} - z_{mj}) y_{ij} \overset{*}{\dot{I}}_{ms}, (i, j = 1,2,3\ldots,n) \tag{9}$$

Get equation (6) into the equation (8), (9):

$$\dot{S}_i = y \sum_{m=1}^{q} \sum_{i=1}^{q} z_{mi} z_{ii}^{*} \dot{I}_{ms} \overset{*}{\dot{I}}_{is} \tag{10}$$

$$\dot{S}_{ij} = -y_{ij} \sum_{i=1}^{q} \sum_{m=1}^{q} (z_{mi} - z_{mj})(z_{ii} - z_{ij})^{*} \overset{*}{\dot{I}}_{ms} I_{is}^{*}, (i, j = 1,2,3\ldots,n) \tag{11}$$

It can obtained after the comparison of formula (7),(8), (9), (10) and (11), the S_{ks} that included in power of the branch to the ground and tie line is:

$$\dot{S}_i = z_{ki} y_i \dot{U}_i \overset{*}{\dot{I}}_{ks} \tag{12}$$

$$\dot{S}_{ij} = -(\dot{U}_i - \dot{U}_j)(z_{ki} - z_{kj}) y i_j \overset{*}{\dot{I}}_{ks} \tag{13}$$

$$\dot{S}_i = y \sum_{i=1}^{q} z_{ki} z_{ii}^{*} \dot{I}_{ks} \overset{*}{\dot{I}}_{is} \tag{14}$$

$$\dot{S}_{ij} = -y_{ij} \sum_{i=1}^{q} (z_{ki} - z_{kj})(z_{ii} - z_{ij})^{*} \overset{*}{\dot{I}}_{ks} I_{is}^{*}, \tag{15}$$

Among them $i,j=1,2,3\ldots,n$.

By equation (12), (13), (14) and (15), the power components that belongs to each power supply can be separated, then the equivalent impedance that corresponding to each power supply can be got.

4. The Method of Power Transmission

4.1. *Series case*

When two nodes in series, the power through the node *m* and *I* is of the same value, therefore, the power ratio can be expressed directly by the form of impedance.

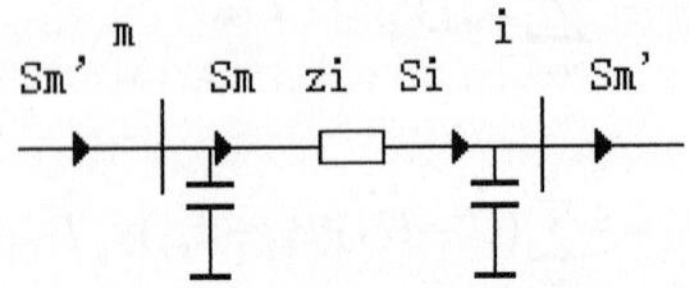

Fig. 1. Schematic of node m and i.

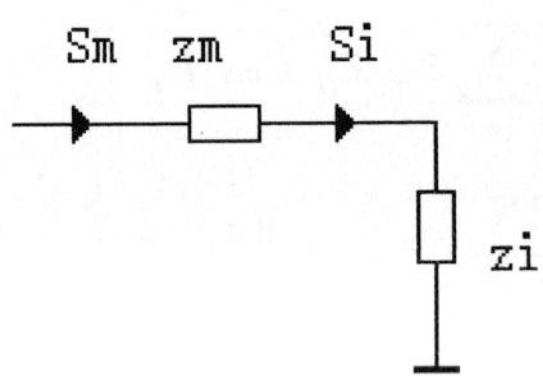

Fig. 2. Equivalent impedance diagram of node m and i.

It can be obtained by equivalent impedance diagram

$$\frac{S_i}{S_m} = \frac{I * z_i * I^*}{I * (z_m + z_i) * I^*} = \frac{z_i}{z_m + z_i} = \frac{1}{1 + \frac{z_m}{z_i}} \tag{16}$$

When two nodes in series, equation (16) describes the relationship between the impedance and power.

4.2. *Parallel case*

In parallel branches, the branch S_l and $S_l^{'}$ is of the same node voltages. Figure 3 is the of parallel case when branches to the ground presence in the network.

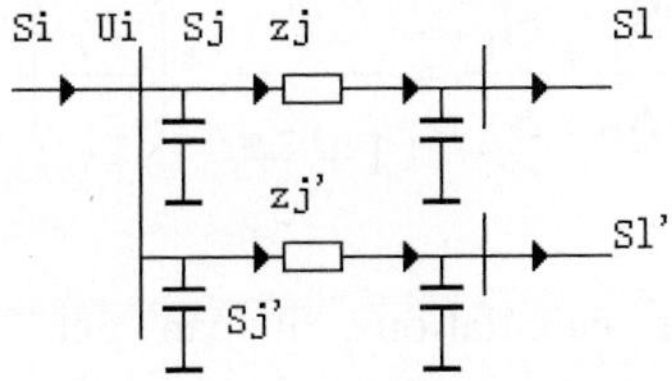

Fig. 3. Schematic of branch Sl and Sl'.

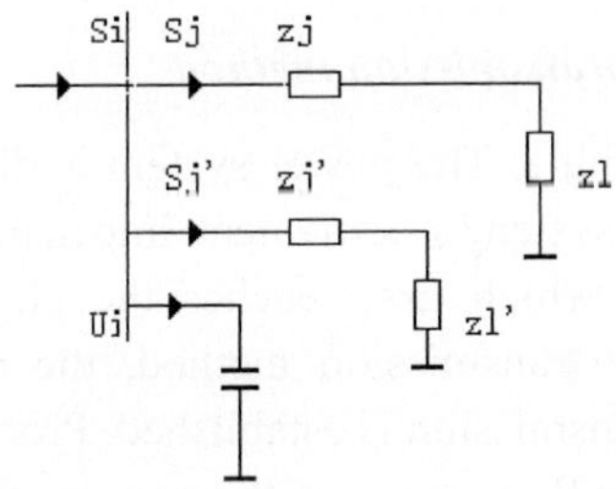

Fig. 4. Equivalent impedance diagram of branch Sl and Sl'.

With the condition of the same node voltages, it can be obtained:

$$\begin{cases} S_i = U_i * (\frac{U_i}{z_\Sigma})^* = U_i * U_i^* / z_\Sigma^* \\ \mathrm{S_j} = U_i * (\frac{U_i}{z_j})^* = U_i * U_i^* / z_j^* \\ \mathrm{S_{j'}} = U_i * (\frac{U_i}{z_j})^* = U_i * U_i^* / z_{j'}^* \end{cases} \tag{17}$$

Resulting:

$$\begin{cases} \frac{S_j}{S_i} = (\frac{z_j}{z_\Sigma})^* \\ \frac{S_{j'}}{S_i} = (\frac{z_{j'}}{z_\Sigma})^* \end{cases} \tag{18}$$

(18) describes the relationship between the power transmission and the impedance in parallel circuit.

4.3. *Complex series and parallel case*

When node *m* and *i* in series and then separate into S_l and S_l' branches, by the formula (16), (18) the relationship between S_m and S_j can be quickly got.

$$\frac{S_j}{S_m} = \frac{S_j}{S_i} * \frac{S_i}{S_m} = \frac{1}{1 + \frac{z_m}{z_Z}} * \left(\frac{z_j}{z_Z}\right)^* \tag{19}$$

Through the above calculations, it can get, for any power system, relationship between the power transmission and the equivalent impedance can be established by using power transmission method.

4.4. *Analysis with power transmission method*

By the power tracing formula, The power system is divided into various single supply system and each system's equivalent impedance can be got. and then analyze Branch or node which first reaches the physical limits. Finally, by application of the power transmission method, the relationship between the impedance and power Transmission is established. Proceed as follows:

(1) Calculate the power flow status of the systems that reaches its limit, with the application of CPF method.

(2) Calculate each equivalent impedance when all supplies work alone and the system meets its limit by the power transmission method.

(3) Transfer the single supply systems into circuit and form a new nodal admittance matrix.

(4) Establish the relationship between each nodes.

(5) analyze the relationship between the equivalent impedance and the power Transmission.

5. Analysis of Example

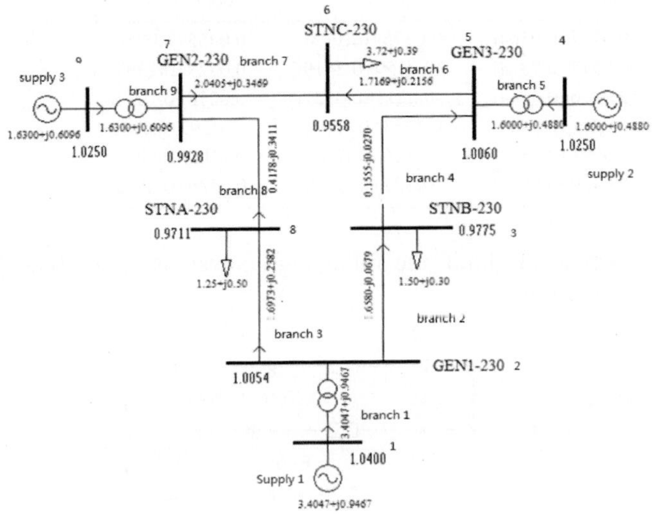

Fig. 5. The power flow when system meets its limits.

Use IEEE 9 as an example. by CPF method, the flow interface can be obtained at the system limits, Figure 4 shows each state variable and branch power.

Each branch will be split into three separate part which corresponds to a supply by power tracing method. Table 1 is the node equivalent impedance to ground corresponding to the tie line branch. Table 2 shows the equivalent impedance of each node directly to the ground.

Table 1. The node equivalent impedance to ground corresponding to the tie line branch.

Node	Supply1(z1)	Supply2(z2)	Supply3(z3)
1	0.276-0.077i	/	/
2	0.561-0.033i	3.779-2.226i	-4.354+2.101i
3	0.527-0.085i	-3.994+1.453i	3.738-2.825i
4	-1.128+0.03i	4.728-3.487i	1.614-0.386i
5	0.579-0.177i	/	/
6	1.030-0.194i	-5.735+1.067i	0.797-0.238i
7	0.887-0.103i	0.750-0.214i	-8.668+1.681i
8	-0.936+0.009i	1.669-0.558i	6.361-5.221i
9	/	/	0.530-0.198i

Table 2. The equivalent impedance of each node directly to the ground.

Node	Supply1	Supply2	Supply3
2	-0.0177+0.0364i	0.0395+0.1008i	0.0465+0.095i
3	-0.0574+0.0874i	-0.0159+0.2419i	0.0427+0.2397i
4	-0.0076+0.0544i	-0.0211+0.0927i	0.0278+0.1034i
6	-0.0291+0.0780i	0.0353+0.1558i	-0.0299+0.1510i
7	-0.0091+0.0572i	0.0186+0.1141i	-0.0151+0.0993i
8	-0.0469+0.0839i	0.0387+0.2269i	0.0226+0.2198i

When reaching the limit, the following analysis is of the branch 7 which first meets the constraint.

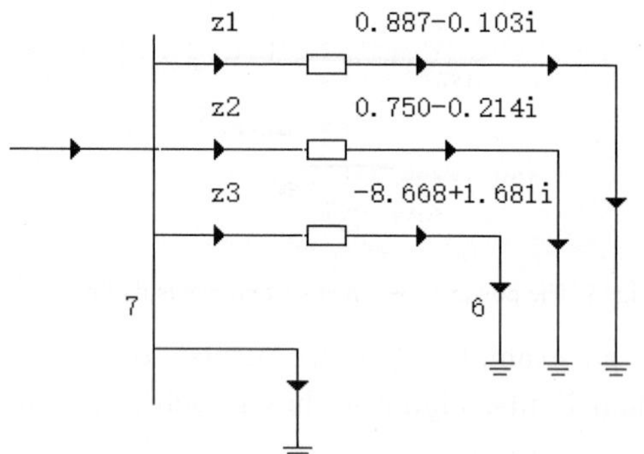

Fig. 6. The separated equivalent impedance of node 7.

It can be obtained from the equivalent impedance diagram, when the generator 1 and generation 2 each works alone, node 7 normally provides power to node 6,with z1, z2 of smaller value which makes transmission hinder smaller. When the generator 3 works alone, the power should be conveyed from node 6 to node 7, but for the high resistance of z3 which hinders the transmission from node 6 to node 7, it changes the power direction. When combined effect of the three generators, the transmission direction is from node 7 to node 6, although generator 3 provides hinder.

Figure 7 shows the analysis of branch 8: generator 1 and generation 2 each works alone, there is a high positive impedance transmission between node 7 to node 8, which increases transmission hinder from node 7 to 8; when generating 1 works alone, the power should be conveyed from node 7 to node 8, while the equivalent impedance of z1 is very small, which will enhance the power transmission from node 8 to 7. Thus, when three generator work together, although there are two generators supply power to node 8, the generator 1 gets a really small equivalent impedance, therefore, the finial direction is reversed, from node 8 to node 7.

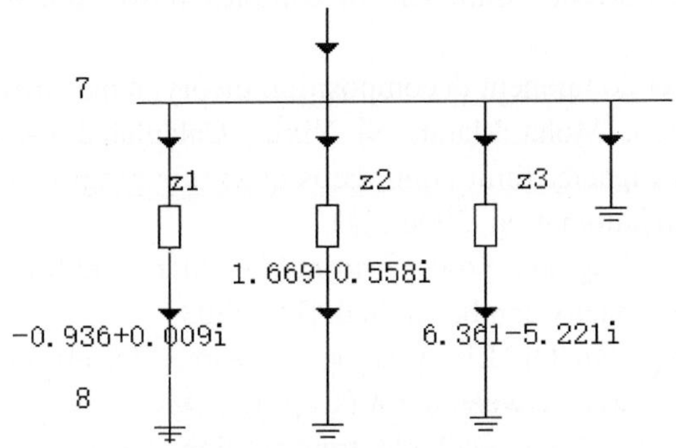

Fig. 7. The separated equivalent impedance of node 8.

After the equivalent impedance to ground obtained, the power transmission can be described by the equivalent impedance though the method of power transmission.

6. Conclusion

Through the above analysis, it can be concluded that power transmission is affected by the equivalent impedance in the electric network, power will give priority to the direction whose the equivalent impedance to ground is smaller. If the relationship between the equivalent impedance and the power flow, which reaches the system limits, can be described by the power transmission method, then, a new approach can be applied to sovle the problems of TTC.

References

1. Li Guoqing, Wang Shan, Yu Yixin. Large interconnected power system studies reviewed power exchange capacity between regions [J]. Chinese CSEE, 2001, 21 (4).
2. Jiangwei Yong, Zhang Boming, WU Wen-chuan, etc. Transfer Capability Calculation application and online scheduling decisions of [J]. Automation of Electric Power Systems Journal, 2008, 32 (10).
3. Wang Shan, Wei Wei. Transmission System Available Transfer Capability [J]. Automation of Electric Power Systems Journal, 2005, 17 (5).
4. Xu Qi research. Power transmission capacity available to calculate the electricity market environment. [D], 2005.
5. Li Guoqing Research based on large interconnected power system area continuous method of transmission capacity. [D]. 1998.
6. Cui Yali, Transmission system available transmission capacity. [J] Electric Power Automation Equipment, 2003 (4).

7. Wang Xiuli, Current component of complex power tracing [Proceedings], 2003.
8. Li Wei, Power component decomposition theory: a new method [D], 2007.
9. Qin Hua Diao, Moha Maud, Ni Yixin, Calculated using inter-regional transmission capacity limit continuous quadratic programming method. [J], power system automation, 2000 (24).
10. Liu Huan Zhi. Regional power grid market share and Sun Grid computing transmission capacity can be studied. [D], 2003.
11. Wang Yurong, Xu Qi, Liu Wenjuan research. Deterministic method for calculating the area between the ATC. [C], 2008.
12. Yongsheng Qiu. Area available transmission capacity between the grid computing. [D]. MA, 2004.

The analysis of static frequency converter (SFC) in pumped storage power station (G/M)

S. X. Jin[1†]

State Grid Liaoning Electric Power Research Institute, Shenyang, China
[†]Email: jsxmumu@126.com

H. Li[2]

State Grid Liaoning Economic Research Institute, Shenyang, China

X. F. Jin[3]

State Grid Dalian Electric Power Supply Company, Dalian, China,
Emails: dwghlih@163.com, 174840173@qq.com

Static Frequency Converter (SFC) is one of the important devices in reversible pumped storage power station (G/M), it can drive the speed of the dragged unit from zero to rating smoothly in a short time (usually about 240s), this paper introduce the form, working principle and sequence of the SFC in pumped storage power station, record and analysis the electrical parameters curves of the dragged unit under speed in pump condition.

Keywords: Static Frequency Converter; Thyristor; Initial Rotor Position; Firing Angle.

1. The Composition and Structure of SFC

1.1. *The composition*

Pushier Pumped Storage Power Station consists of four 300 MW units, with a total installed capacity of 1200 MW, configures a SFC supplied power from the low voltage side of the 2nd and the 4th main transformers in the manner of a preparation, with higher reliability. As Figure 1, SFC system is composed of input transformer (TR1), output transformer (TR2), input switchgear, measurement and control holder, the rectifier bridge of grid side, the inverter bridge of unit side, smoothing reactors, output switchgear, controllers, current-limiting reactor and cooling system.

---

[1] Shi-Xin Jin
[2] Hua Li
[3] Xiao-Fei Jin

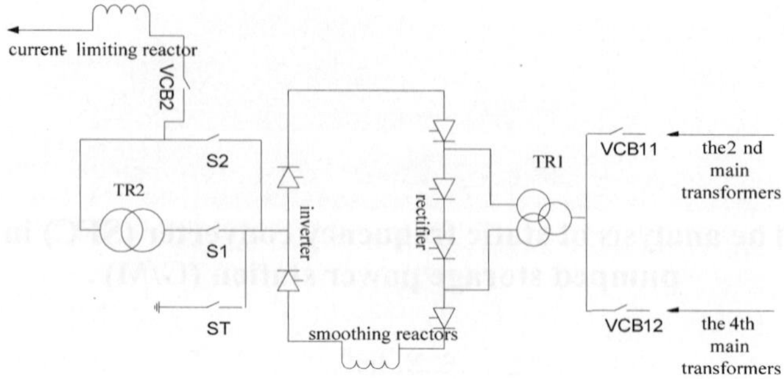

Fig.1 The composition of SFC.

1.2. *The structure*

The input transformer is three- winding, transformation ratio is 18000/2400/2400, wiring group is Dd0, Dy1. Two bridge arms of low-voltage side string together, take advantage of this method is that it can effectively remove the system third harmonics, the fifth and seventh harmonics produced by recitfication.In addition, the DC reactor of DC side can effectively improve the smoothness of the DC side waveform, and suppress the harmonic produced by thyristor rectification. Setting Rectifier Bridge and Inverter Bridge redundantly according to N+ 1, each bridge arm of network bridge side is composed of three thyristors, each bridge arm of machine bridge side is composed of four thyristors, thyristors total of 60 to enhance the system reliability and withstand voltage. When SFC drags, the controller sends optical signal through fiber to the pulse control panel of thyristor, the pulse control panel will change optical signals into electrical signals and the signal is applied to the gate of the thyristor, thyristor control achieved. SFC cooling water system is a combination of inside and outside cooling water, the cooling water inside the thyristor is deionized water, external cooling water (crude water) through a heat exchanger to cool the deionized water to ensure that the thyristor insulation is not damaged.

2. The Working Principle and Dragging State of SFC

2.1. *The working principle*

The working principle of SFC is to change the low voltage side voltage of the main transformer through inverter into alternating current with frequency 0 ~ 50 Hz, the rotating magnetic field produced by ac current drag the rotor from zero to rated speed, due to the voltage with low frequency is close to DC, so need to change the output into the bypass through isolation switch S2, the output will be changed into input when the unit frequency is greater than 5Hz, SFC into the linear growth stage, when the output Driving Force Moment of SFC is greater than the resistance moment, the output current of SFC is the rated current of TR2, when the unit speed reaches 333 r / min, the speed will not continue to increase, the output Driving Force Moment of SFC is equal to resistance moment that rotate the rotor uniformly required, the output current of SFC reduce to 1/3 of the rated current.

Because need to conduct two thyristors each time to form a loop, the voltage applied to network bridge side is line voltage, SFC controller transmits pulses every electrical angle of 60° to the other group of thyristors, 6 groups of trigger pulse per cycle (360° electrical angle), the thyristor conduction sequence is 1-2,2-3,3-4,4-5,5-6,6-1, one group of thyristor conduct a 120°each time. Setting the trigger angle of network bridge side is α, the trigger angle of machine bridge side is β, the average voltage of DC side is U_d, line voltage's effective value of the main transformer's low-voltage side is *UT*, effective value of generator terminal's line voltage is *US*, Eq.1 shows the relationship between DC side voltage and AC side voltage.

$$U_d = \frac{3\sqrt{2}}{\pi} * \mathrm{UT} * \cos\alpha = \frac{3\sqrt{2}}{\pi} * \mathrm{US} * \cos(\pi - \beta) \tag{1}$$

When $\alpha = 0°$, U_d is $U_{dmax} = 1.35*4800 = 6480\text{V}$, when $\alpha = 90°$, $U_d = 0$, when $\alpha = 180°$, Ud is $U_{dmin} =$ -6480 V.

When $0 \leq \alpha \leq 90°$, $90° < \beta < 180°$, $U_d > 0$, the unit works in the pump mode, the network bridge NB1, NB2 work in the rectifier state, the machine bridge MB works in the inverter state. When $90 \leq \alpha \leq 180°$, $0° < \beta < 90°$, $U_d < 0$, the unit works in the Power Generation state, network bridges NB1, NB2 work in the forced inverter state, the machine bridge MB works in the rectifier state, and by adjusting the angle of α and β, the uniform frequency conversion can be realized.

2.2. *The dragging state*

Due to the automatic reversing of thyristor rectifier bridge requires a suitable AC voltage, but unit's speed is increased from zero to the rated speed, when the unit speed is less than 95%, excitation works in constant current mode, so the rise of terminal voltage is proportional to the rise of terminal frequency, resulting SFC's drag is bound to two states. When f <5 Hz, the voltage applied on thyristor bridge arm is too small to turn-off it, this moment SFC works in the pulse reversing stage, change the output into by-pass through S2, use a lower torque to drag. This process is realized as follows: SFC controller sets the bridge NB1, NB2 to work in the whole inverter mode to cut off the DC circuit current Id, when Id is equal to 0, the SCR pulse of bridge NB1, and NB2 is blocked.

When the controller detects a loop current is 0, the pulse will be delivered to the next group of triggered SCR in accordance with the position calculated by rotor signals, at the same time cancel the full inverter state of bridge NB1, NB2 to recovery loop current, so that a new set of SCR guide circuit, at low speed stage, the minimum time to turn off the DC current forcibly needs 4ms, with the increase in motor frequency, the time used to cancel or recovery the loop current will be shorter and shorter, eventually forced reversing cannot continue to run [1-2]. Therefore, there must be an upper limit frequency in pulse coupling mode, generally no more than 10% of the rated frequency. When the unit frequency is greater than 5 Hz, machine terminal has established adequate reversing voltage, then the SFC disconnect S2, closing S1, the voltage of machine bridge side applied to the stator through output transformer, then the SFC reaches the rated acceleration torque, when thyristor reversing, no need force inverters, and no need to know the initial position of the rotor, then SFC operates in synchronous reversing stage, the controller only needs to send pulse to next group of thyristor once every 60° electrical angle, when the next group of thyristor conduction is realized, can naturally turn off the thyristor by reverse voltage[3]. The whole dragging state of SFC is composed of these two phases, in order to complete the drag , the upper limit frequency of the pulse should be slightly higher than the lower limit frequency of synchronous conversing process [4].

3. The Calculation of The Rotor Initial Position

3.1. *The relationship between the rotor's initial position angles and the stator current's direction and conducting thyristor number*

The initial position of the rotor is the position that compared with the stator coordinate, at the moment that pumping conditions of unit start, use the sine and cosine values of the angle λ between MMF generated by the rotor and direct-axis [5] to measure, calculating position initial of rotor is in order to determine two group of thyristors which should be conduct when SFC works in pulse reverse stage. The relationship between the initial position of the rotor and the conducted thyristors within a cycle is shown in Table 1.

Table 1. The relationship.

mode	thyristor		current direction		Rotor position
1		6	U->X	Y->V	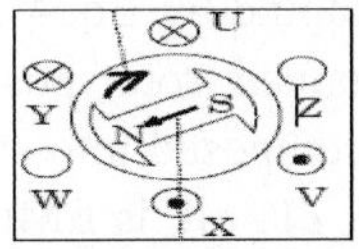
	1				
2			U->X	Z->W	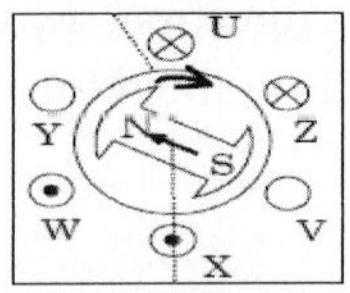
		2			
3			V->Y	Z->W	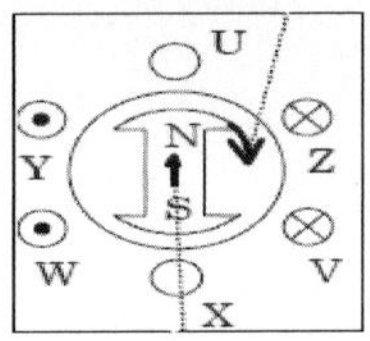
	3				
4			V->Y	X->U	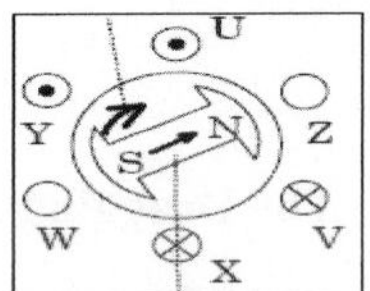
		4			
5			W->Z	X->U	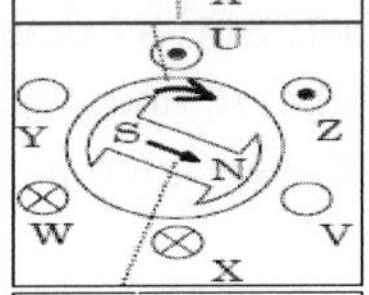
	5				
6		6	W->Z	Y->V	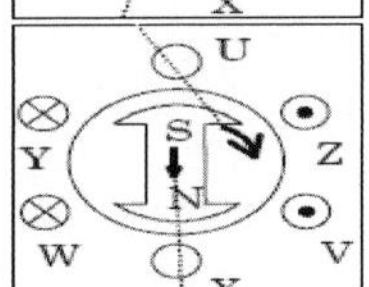

3.2. *The calculation method of the rotor's initial position angle*

Pushier station SFC uses electric angle measurement method namely uses induction electromotive force of the armature winding to detect the rotor position. start excitation before SFC put into the use, during the process that magnetic field in the rotor is quickly established, the stator windings induce three-phase electromotive force *eA*, *eB*, *eC*, due to the different spatial position of the rotor, the size and orientation of three-phase induction electromotive force is also different, the precision digital voltage transformer (PIB) located on the machine bridge side detects the stator EMF by starting bus, and transfers it to the controller to calculate. To decompose the induction electromotive force of stator to the rotor coordinate system dq, such as Eq. 2and Eq.3, so that it can be transformed into a input voltage model of vector ed, eq in dq coordinate system, compute by Eq. 4 using the voltage model, the size of the integral flux can be obtained, rotor position angle can be calculated using the Eq. 5, *ed* is the EMF decomposed into straight axis, *eq* is the EMF decomposed into quadrature axis, *eU*, *eV*, *eW* is EMF induced on the stator three-phase, ψd is the flux in straight axis direction, ψq is the flux in quadrature axis direction, λ is the initial position angle of the rotor.

$$ed = eU - eV\cos\pi/3 - eW\cos\pi/3 \tag{2}$$

$$eq = eV\cos\pi/6 - eV\cos\pi/6 \tag{3}$$

$$\varphi d = \int eddt \quad \varphi q = \int eqdt \tag{4}$$

$$\psi = \sqrt{\psi d^2 + \Psi q^2} \quad \cos\lambda = \psi q/\psi \tag{5}$$

$$\sin\lambda = \psi d/\psi \tag{6}$$

4. The Dragging Procedure of SFC

When SFC is dragging, monitoring system (CSCS) to complete the dragging procedure under unit pumps conditions by calling the control procedure of SFC controller internal, dragging procedure is shown in Figure 2.

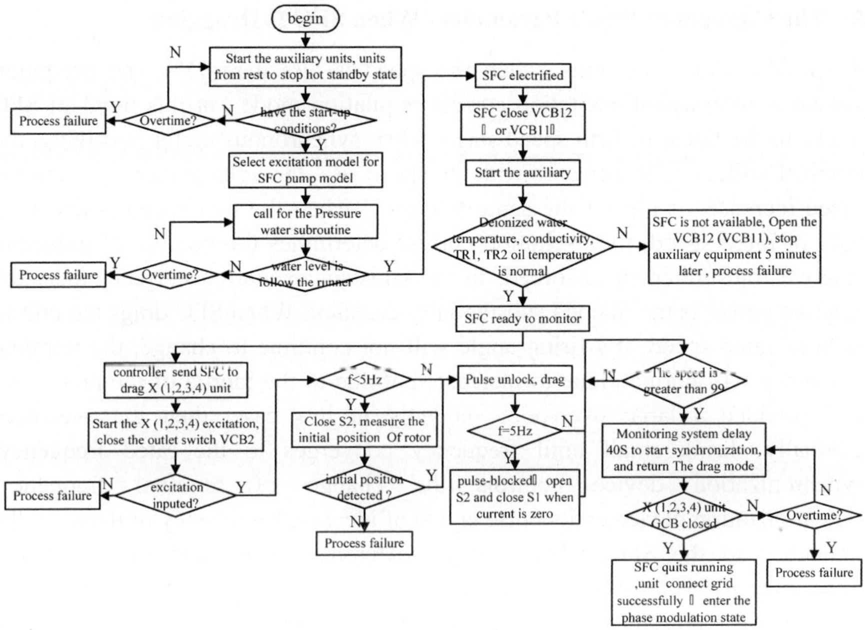

Fig. 2. The dragging procedure of SFC.

While SFC is dragging the unit, first select by monitoring the unit to operate in SFC pumping conditions, select the excitation regulator to operate in SFC dragging mode, and then call the inflation pressure water program, after the success of inflation pressure water, monitoring starts SFC, SFC controller will start a variety of associated auxiliaries (deionized water pump, input and output variable oil pump and thyristor cooler fan) Orderly upon receipt the start order from the monitoring, if the operating parameters of SFC are normal, issue to monitor "SFC ready" signal and close the switch of input transformer's high voltage side at the same time, monitoring send a control signal to SFC of dragging a units after received the feedback from SFC, SFC enters sequence dragging procedure after signal reception, drags up the unit to rated speed with a constant acceleration, the drag torque of SFC is set by the internal controller, the setting principle is that increase torque as far as possible so as to reduce the dragging time under the condition that the maximum output current of machine bridge side is not greater than the rated current of output transformer in the units accelerated stage. After unit grid connected, SFC quit running into hot standby or quiescent state, waiting for the next drag commands.

5. The Changes of Unit's Parameters When SFC Is Dragging

When SFC drags the units, when the speed is less than 95%, the excitation system is in constant excitation current regulation mode , at this moment SFC works in the linear growth speed stage, when synchronous motor is driving, the terminal voltage will increases with the speed rise, the voltage increases and the speed increases is almost the same tendency, when the unit speed is less than 95% of rated speed, the excitation current determines the number of inductive reactive transmitted or absorbed, in the constant excitation current mode, the reactive power is maintained substantially constant. When SFC drags the unit to 99% of rated speed, it's firing angle will not continue to change, the terminal frequency will not continue to increase, but due to the inertia of the unit, there will produce a large overshoot suddenly in this case, then the overshoot gradually shrink small until frequency converges to the rated frequency, synchronization device controls the change of terminal frequency, synchronization device emits pulse signal of increased frequency or decrease the frequency to the SFC, when voltage difference, angle difference and slip difference of the terminal voltage and the main transformer low voltage side satisfies the requirements of the synchronization device, close synchronization device and relay starts, the unit synchronous grid connected, after the generator outlet switch (GCB) is closed, in order to avoid the closing loop operation, SFC exit immediately, variable frequency device seal the pulse, open the VCB2, S2 switch, SFC is in hot standby state , is able to drag the next unit, if there is no unit to be drag or the SFC continuous operation for over 24 min, VCB11, VCB12 open normally, 5 min later , SFC stops auxiliaries, enters the stopped state.

6. Conclusions

SFC with step-less speed regulation, fast reaction, stable start, high starting success rate and easy maintenance, etc. has been widely used in Pumped Storage Power Station. The article introduced the composition, working principle, workflow of SFC system in Pushier Pumped Storage Power Station. Analyzed changes of various electrical parameters vary with the frequency during the unit dragging by SFC, hope to have some guidance and help for the work on SFC debugging.

References

1. Hu Xueqin, Summary and Exploration of Static Frequency Converter, Yunnan Water Power, vol. 23, no. 1, pp.109-113, 2007.

2. Zhou Jiayuan, Zheng Huijuan, Several Problems of the Start-up of Pumped-storage Power Unit, Water Power, vol. 59, no.10, pp.59-62, 2007.
3. Wang Zhaoan, Huang Jun, Power Electronics, Beijing, China Machine Press, 2000.
4. Ge Baojun, Li Yuling, Study of Process of Starting Pumped Storage Machine by Static Frequency Converter with Field Current Controlled, Electric Machines and Control, vol. 7, no. 3, pp.187-190, 2002.
5. Li Yuling, Wang Zitao, Ge Baojun, Study of Static Variable Frequency Starting Process of Synchronous Motors, Journal of Harbin University of Science and Technology, vol. 5, no. 1, pp.79-82, 2000.

Optimized design and finite element verification for the new type permanent magnet moving-coil linear motor

P. Wu[†]
Shanxi Electric Power Research Institute
Taiyuan, 030024 China
[†]Email: kien_viv@163.com

S.H. Wang and Y. Song
Taiyuan University of technology

An optimized design for the permanent magnet moving-coil linear motor which as the electro-hydraulic proportional valve drive, developing this special motor optimal design program using Visual Basie 6.0. Comparative analysis the force power ration,the flux density, the electromagnetic force and velocity with time before and after the optimization process of this motor, in the Maxwell 2D environment. The Structural parameters are obtained of this motor, FEM results show that the combination of the genetic algorithm and finite element numerical calculation has a very good effect on the optimized design of the permanent magnet moving-coil linear motor.

Keywords: Permanent magnet moving-coil linear motor; Genetic algorithm; Optimized design; Force power ratio; FEM.

1. Introduction

Electro-hydraulic proportional valve is an important element of precision control in a hydraulic drive system. Permanent magnet moving-coil linear motor, with a small in size, simple structure, ease of assembly and to avoid the use of special processing and custom radially magnetized ring poles, Thrust / volume ratio performance, its thrust is proportional to the size of the current,easy to achieve precise control of thrust and bidirectional drive,and its electromagnetic transient time constant is small, can be directly converted into electrical signals for linear mechanical motion, suitable as a proportional valve actuators [1].

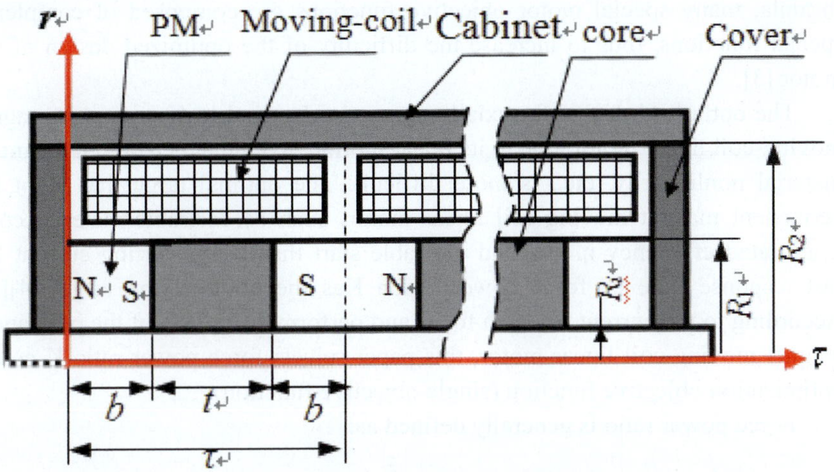

Fig. 1. Schematic diagram of the axially magnetized permanent magnet moving-coil linear motor.

Cylindrical permanent magnet moving-coil linear motor of axially magnetized structure is shown in figure 1. This motor is characterized by the use of ring-type (or cylindrical) axially magnetized permanent magnet, forming a plurality of magnetic poles combined with the inner core, generating a magnetic field in the air gap of the cylindrical space and acting on the current carrying movable-coil, thus producing an axially magnetic force [2].

In this paper, using the genetic algorithms to improve the force power ration of the motor, an axially magnetized permanent magnet moving-coil linear motor for the electro-hydraulic proportional value drive is optimized design, meanwhile two-dimensional magnetized model is established for two size permanent magnet moving-coil linear motor before and after optimized using genetic algorithm at Ansoft Mawell 2D environment, and get the static and dynamic simulated characteristics of the two prototype.

2. Establishing the Mathematical Model of The Optimal Design

Building the mathematical model of the motor optimal design mainly includes three aspects: selection of the optimal target, selection of the optimal variables and determine the constraints.

2.1. *Selection of the optimal target*

Optimized design of the motor is carried around the target function and the objective function directly affects the optimization of the motor. Because the objective function of the motor is generally unable to express a simple analytical formula, many special motor objective functions are composed of complexly special functions, thus to increase the difficulty of the optimized design of the motor [3].

The optimized design for axially magnetized cylindrical permanent magnet moving-coil linear motor, due to its linear motion, axial magnetization, magnetic material nonlinearity, etc., is more difficult. The optimal target has a lot for permanent magnet moving-coil linear motor, such as desirable material costs Cost, rated efficiency η, etc. and desirable start thrust Fst, starting current Ist, fast response time ts, force power ration Kas the optimization target [4][5]. According to the current research focus and performance index of the permanent magnet moving-coil linear motor, this paper selects force power ratio K as the optimization objective function (single objective function).

Force power ratio is generally defined as:

$$K = \frac{F_e}{P_1} \tag{1}$$

Where: K is force power ratio; P1 is input power; Fe is electromagnetic force.

Using genetic algorithms to optimize the design, the fitness value is generally used to reflect the merits of the individual fitness, and fitness value mainly depends on the fitness function to strike. In the optimal design of the new permanent magnet moving-coil linear motor, classified method can be used for transformation the objective function into the fitness function. When the issue is the maximum value of the objective function, the objective function can be translated into a fitness function by the formula:

$$\mathrm{Fit}(f(x)) = \begin{cases} f(x) - c_{\min} & f(x) > c_{\min} \\ 0 & \text{other} \end{cases} \tag{2}$$

Where: Cmin is a minimum estimate for.

2.2. *Selection of the optimal variables*

When the motor is optimized designed, each design variables selected must be independent, that is to say, the optimization variables can be used to express other variables. Meanwhile it requires that the optimal design variables have a greater impact on the objective function and the constraints. Range of the optimization variables should be easy to determine.

Combined with the previous experience in the design, the optimal variables of the axially magnetized cylindrical permanent magnet moving-coil linear motor can be selected as follows: permanent magnet thickness b, number of pole pairs p, inner core length t, armature aspect ratio λ, the inner diameter of the outer core R2, the outer diameter of the outer core R1, gap height g, coil conductor Z1, pole arc αp ,etc.. In this paper, the optimal design variable X is:

$$X = \begin{bmatrix} x_1 \\ x_2 \\ x_3 \\ x_4 \end{bmatrix} = \begin{bmatrix} b \\ g \\ a_p \\ t \end{bmatrix} \tag{3}$$

In the use of genetic algorithms, the optimal design variables should be coded first, coding with the binary one-dimensional multi-parameter. When checking performance, it can decode the binary coded into actual decimal value.

2.3. *Determine the constraints*

Similar to the optimal design of the rotary motor, the constraints of the permanent magnet moving-coil linear motor can be divided into four categories: structural parameter constraints, which includes the permanent magnet thickness b, the wire diameter d, the gap height g, the inner diameter of the outer core R2, the shaft diameter Ri, and the tank full rate Sf, etc.; electromagnetic parameter constraints, which includes the air-gap magnetic field density Bg, the outer core magnetic field density Bj, the line load A, and the winding current density J, etc.; performance constraints, which includes the efficiency η, the starting thrust Fst and the starting current Ist, etc.; cost constraints, which includes the motor manufacturing costs.

$$\begin{cases} g_1(x) = \dfrac{B_g - B_{g0}}{B_{g0}} \le 0 \\ g_2(x) = \dfrac{B_j - B_{j0}}{B_{j0}} \le 0 \\ g_3(x) = \dfrac{\eta_0 - \eta}{\eta_0} \le 0 \\ g_4(x) = \dfrac{F_{st0} - F_{st}}{F_{st0}} \le 0 \\ g_5(x) = \dfrac{S_f - S_{f0}}{S_{f0}} \le 0 \\ g_6(x) = \dfrac{J - J_{max}}{J_{max}} \le 0 \end{cases} \tag{4}$$

Because the magnitudes of each constraint function values have a large difference, in the actual optimization, the various constraints must be normalized. The constraints of the permanent magnet moving-coil linear motor can be expressed as Formula (4):

Where: x is the constraint value; the subscript 0 represents the standard value of the corresponding constraint value; the *max* represents the upper limit of the constraint value.

2.4. *Operator choice of the selection, crossover and variation*

Genetic algorithms are optimal algorithms that simulate genetic selection and the natural elimination process of biological evolution, through computer simulations to achieve competitive groups of biological evolution, natural selection, and other genetic and variation, etc..

According to the evolutionary principle of survival of Darwinian, "natural selection, survival of the fittest", selection probability pi can be expressed as:

$$p_i = f_i \Big/ \sum_{j=1}^{N} f_i \qquad (5)$$

Where: fi is the i-th individual fitness value; pi is the probability of the i-th individual selected; $\sum_{j=1}^{N} f_i$ is total fitness for the population, where N is the population size. The choice probability size represents the proportion of individual fitness. Individual fitness value is higher, the higher the probability of being selected.

Crossover can improve the average and maximum population of individual fitness significantly. Mutation can increase the diversity of the population of individuals, in order to achieve the purposes of the global optimization.

If the crossover probability pc and mutation probability pm are made too large, the search of the algorithm is capable and easy to develop a new search area; but in the latter part of the optimal search, individual fitness in the population is too large, thus outstanding individual can easily be destroyed. If the crossover probability and mutation probability are made too small, it is not easy to produce a new individual in the optimal process, and the search process is slow, greatly reducing the efficiency of genetic algorithm [6].

Adaptive genetic algorithm maintains the diversity of the population and also ensures a better convergence property of genetic algorithms. In the adaptive genetic algorithm:

$$p_c = \begin{cases} \dfrac{k_1(f_{max} - f')}{f_{max} - f_{avg}} & f \geq f_{avg} \\ k_2 & f < f_{avg} \end{cases} \tag{6}$$

$$p_m = \begin{cases} \dfrac{k_3(f_{max} - f)}{f_{max} - f_{avg}} & f \geq f_{avg} \\ k_4 & f < f_{avg} \end{cases} \tag{7}$$

Where: fmax is the largest group fitness value; favg is the average fitness value of each generation group; f' is the larger fitness value of the two individual to crossover; f is the fitness value of individual variation; k1,k2,k3,k4 takes a value of (0,1) interval.

In order that the largest individual crossover probability of the group fitness value and the mutation probability of the group fitness value is not zero, operators of formula (6) and formula (7) need to be improved. Improved, crossover probability pc and mutation probability pm are expressed as follows:

$$p_c = \begin{cases} p_{c1} - \dfrac{(p_{c1} - p_{c2})(f' - f_{avg})}{f_{max} - f_{avg}} & f' \geq f_{avg} \\ p_{c1} & f' < f_{avg} \end{cases} \tag{8}$$

$$p_m = \begin{cases} p_{m1} - \dfrac{(p_{m1} - p_{m2})(f_{max} - f)}{f_{max} - f_{avg}} & f \geq f_{avg} \\ p_{m1} & f < f_{avg} \end{cases} \tag{9}$$

2.5. *Termination condition*

When using genetic algorithms to optimize the permanent magnet moving-coil linear motor, it would need to be set a termination condition and approximately think that the optimal solution meet the conditions set by the termination is the global optimal solution in the search space. In this paper, the maximum termination algebra is given in advance. When genetic algorithms iterate to the maximum termination algebra, the iterations is stopped, and the default solution in this case is the global optimal solution in the entire search space.

3. Optimized Design Results

Optimized design results of the motor are shown in Table 1.

Table 1. Optimized design results.

	mm	before optimization	after optimization
structure	thickness of PM	12	12.2
	gap length(mm)	6.5	6.3
	inner core length(mm	8	7.8
	polar distance(mm)	20	20
	coil outer diameter(mm)	16.2	16
	coil inner diameter(mm)	10.8	10.6
	the outer diameter of the outer core(mm)	17.5	17.5
	the inner diameter of the outer core(mm)	16.5	16.3
	pole arc	0.4	0.36
	tank full rate	0.6646	0.6857
performance	electromagnetic force (N)	80.9	87.9
	input power(W)	55.812	55.812
	force power ratio	1.45	1.57

Seen from Table 1, the force power ratio from 1.45 changes to 1.57 after optimization, and force power ration is improved obviously. This means that using the genetic algorithm has a very positive effect on improving the force power ratio of the axially magnetized permanent magnet moving-coil linear motor used for the electro-hydraulic proportional value drive. The entire calculation process getting the satisfactory optimal solution just uses two minutes, far less than the time required for traditional optimization method.

4. Finite Element Simulations

Establishing the model of the permanent magnet moving-coil linear motor before and after optimization using genetic algorithm; analysis and calculation of the magnetic field distribution, parameters change and the electromagnetic force under the same circumstances around the motor before and after optimization; simulating and analysis of the dynamic characteristics of the motor before and after optimization using the time stepping element analysis software, velocity and electromagnetic force, etc. versus time curves are obtained. Finite element analysis model used for simulation prototype is shown in Figure 2. Axially magnetized permanent magnet moving-coil linear motor is quadrupole and axisymmetric structure, whose geometric structure is shown in Figure 1 and whose frame size is shown in Table 1.

Flux density of the permanent magnet moving-coil linear motor has two components: radial magnetic flux density and axial magnetic flux density. Moving-coil is not cutting axial flux density; the interaction between axial flux density and coil current generates the electromagnetic force in the radial direction and not be considered here. This article focuses on the radial flux density of the permanent magnet moving-coil linear motor.

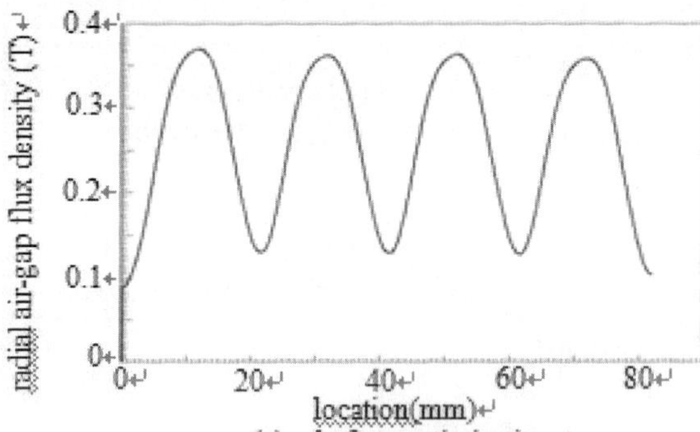

(a) before optimization

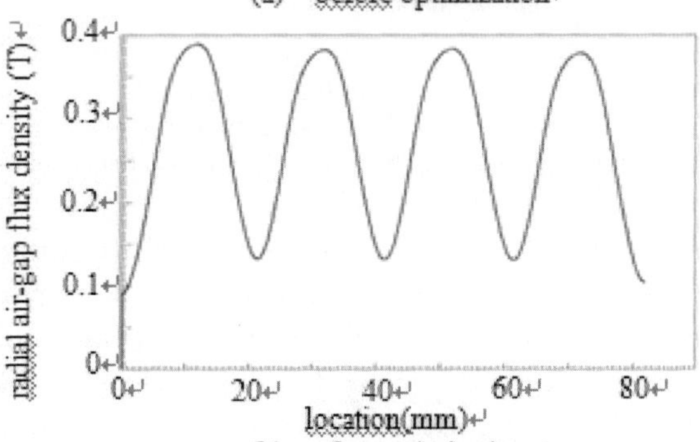

(b) after optimization

Fig. 2. Radial air-gap flux density before and after optimization when J is 4A/mm^2.

Figure 2 is a radial flux density distribution of the air gap magnetic when the coil current density J is 4A / mm2. From Figure 2 we can see that the peak of radial flux density is 0.37 ~ 0.4T before optimization and after optimization is 0.38 ~ 0.42T, overall flux.

4.1. *The static characteristic simulation results*

Density has improved, thus electromagnetic force of the permanent magnet moving-coil linear motor is increased and the force power ratio of the motor is also increased. The iron core of the motor shell has a signal of magnetic saturation before optimization and saturation level decreased after optimization [7].

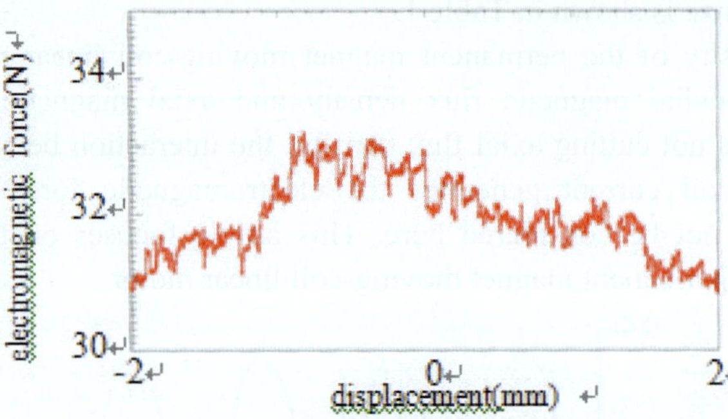

(a) before optimization

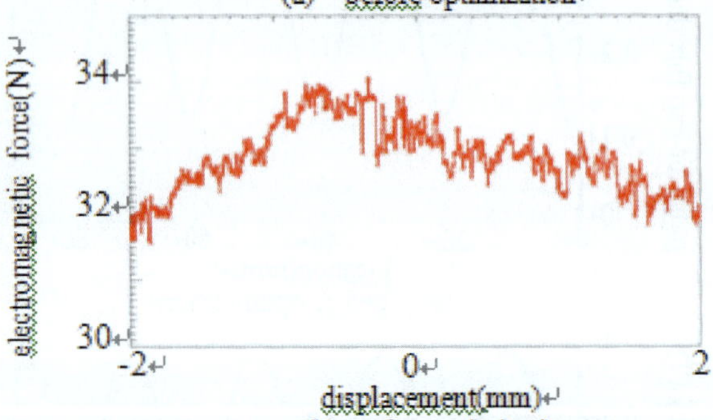

(b) after optimization

Fig. 3. Eelectromagnetic force of the motor before and after optimization.

Figure 3 is the electromagnetic force with the mover displacement curve when the coil current density is 4A / mm2 before and after optimization. From the figure we can see, the electromagnetic force has increased after optimization, and the force power ratio of the motor has also increased. These are due to increase thickness of the permanent magnet and reduce the gap height of the motor.

Figure 4 is a self-inductance coefficient (self-inductance ratio) of the first single coil with unit length with displacement curve before and after optimization. The figure shows, the self-inductance ratio of the first coil has a little decline, but little change, this is due to the well-distribution of the air gap magnetic field of the motor. After optimization, the self-inductance ratio of the motor is a whole down, which can improve the dynamic response of the motor.

Figure 5 is the self-inductance ratio of the third coil with the mover displacement curve before and after optimization.

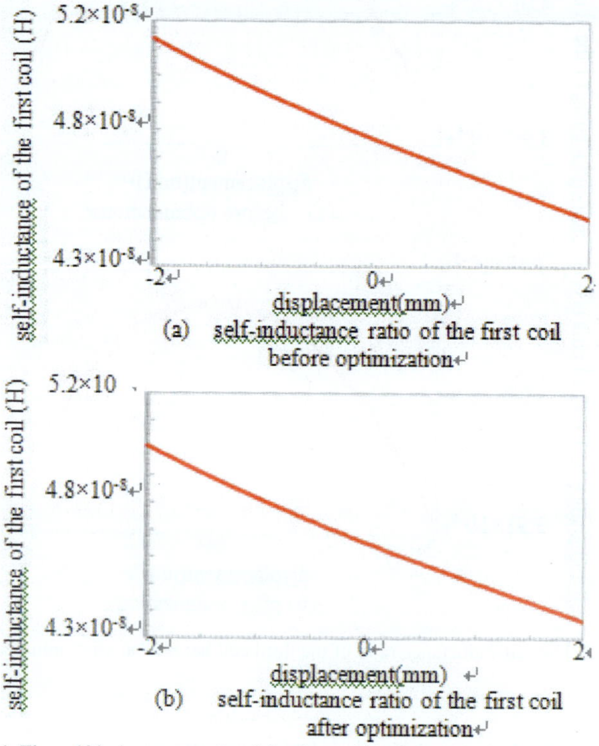

Fig. 4. The self-inductance ratio of the first coil before and after optimization.

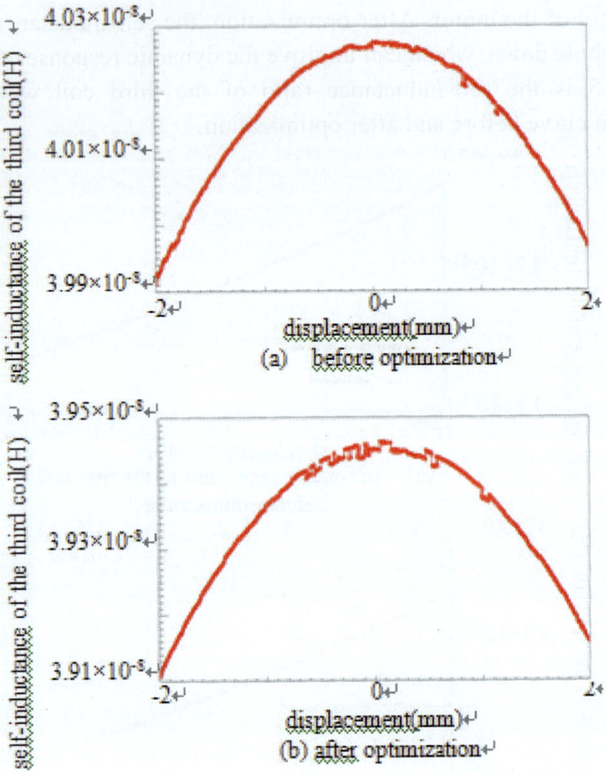

Fig. 5. The self-inductance ratio of the third coil before and after optimization.

4.2. *The dynamic characteristic simulation results*

The motor is always in the dynamic process in normal operation, so only analysis of the static is not enough. Before and after optimization, the input voltage of the running permanent magnet moving-coil linear motor is 24V and motor is no-load running. Before optimization, the speed of the motor is up to 0.489m/s, but speed of the optimized motor reaches 0.54m/s. The overall speed of the motor has improved, and the dynamic performance has also improved.

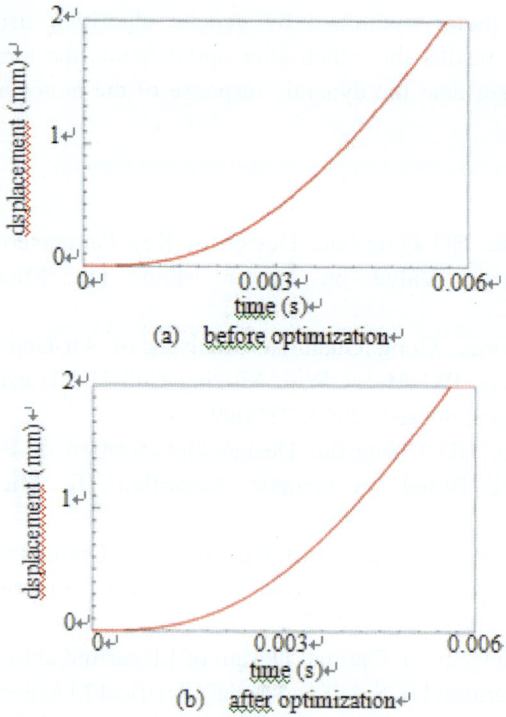

Fig. 6. Displacement curve before and after optimization.

Figure 6 is a displacement curve of the operating motor before and after optimization. Maximum displacement of the running mover is 2mm. Motor cover has stalled action to mover coil when the mover displacement more than 2mm and it forces the speed of the mover to decrease to zero. From the figure we can see, the permanent magnet moving-coil linear motor before optimization needs 0.0098s to complete a trip, but the optimized permanent magnet moving-coil linear motor needs 0.0089s to complete a trip. After optimization, the time which the motor needs to complete a trip is shortened.

5. Conclusion

At present, the permanent magnet moving-coil linear motor used for the electro-hydraulic proportional valve control system can improve the control performance of the electro-hydraulic drive system. Based on establishing the mathematical model of optimal design of the permanent magnet moving-coil linear motor, using self-adaptive genetic algorithm, developing the motor optimal design program in the VB language environment, to optimize the design of the body of the axial magnetized permanent magnet moving-coil linear motor used for electro-hydraulic proportional valve drive system, simulation and analysis of the motor optimized by genetic algorithm using the software Maxwell 2D, the results show that, after optimization, the force power ratio of the motor increases, and the dynamic response of the motor have been greatly improved

References

1. WANG Ai-le, BU Qing-hua. Design of Key Parameters of Proportional Electromagnetic Valve on Linear Motor [J]. Micromotors, 2006, 39(1):24-27.
2. Wang Shuhong, Xiong Guangyu. Analysis of Air-Gap Field in a New Tubular Linear PM Motor With Moving Coil [J]. Transactions of China Electrotechnical Society, 2007, 22(5):40-44.
3. Chen Jin-tao, GU Cheng-lin. Design Optimization of Permanent-magnet DC Machine Based on Genetic Algorithm [J]. Micromotors, 2006, 39(8):17-20.
4. Kang Yanqin, Shangguan Xuanfeng, Xiao Jiale. Optimal Design of Tubular Linear Induction Motor Based on Genetic Algorithm [J]. Motor Technology, 2008, 25(8):72-75.
5. Wang Le-qiang, Ji Qi. Optimal Design of Linear Induction Motor Based on Genetic Algorithm [J]. Small & Special Electrical Machines, 2008, 3:14-17.
6. Qian Qiang, Wang Shu-hong. Optimal Design of Permanent Magnet Brushless DC Motor Based on Genetic Algorithm [J]. Small & Special Electrical Machines, 2010, 2:18-21.
7. Song Yuan. Design Optimization and Analysis of Permanent Magnet Linear Motor with Moving Coil [D]. Taiyuan: Taiyuan University of Technology, 2011.

A summary of the research on the single terminal electrical data protection of HVDC transmission

S. P. Gao, J. Xie[†] and Z. X. Fu
Xi'an University of Science and Technology,
Xi'an, Shaanxi, 710054, China
[†] *Email: 527149583@qq.com*

Z. Wang
State Grid Jiangxi Ji'an Electric Power Supply Company,
Ji'an, Jiangxi, 343009, China

R. Y. Zhang
State Grid Jiangxi Baoji Electric Power Supply Company, Baoji,
Shaanxi, 721004, China

Based on the analysis of the HVDC line, at a given frequency, a great difference of impedance characteristics of DC line area can be observed between external and internal fault. Occurred in the region and external fault, because of the function of the wave reactor and the DC filter, the difference of the current characteristic is significant. In this paper, we did some analysis and comparison of some traditional protection methods and the wavelet protection method after the improvement. This wavelet transform method is highly efficient and reliable, which can provide reference value for subsequent single terminal data protection.

Keywords: High voltage direct current; Single terminal electrical data; Characteristic frequency signal; Protection.

1. Introduction

In China, the development of energy and economy is not balanced, HVDC has several advantages such as long-distance, large capacity transmission and so forth. In the "power transmission from west to East, the national network" strategy plays an important role and is a major regional non synchronous network and the national unified network lays a good foundation. MTDC system refers to the DC transmission system containing three or more than three of the stations. Compared with the two terminal HVDC system, the operation more flexible, more economical, while achieving multi power and multi HVDC power [1].Compared with the two terminal HVDC system, the operation is more flexible, more economical, while achieving multi power and multi HVDC power [1] and widely used in the remote power transmission, asynchronous networking, island power and other fields [1-3].Relay protection of power system is an important part of power system, its role is the power system occurs fault of electrical components from the system removed as soon as possible and make the fault from suffered greater damage, and to ensure the power system as soon as possible to restore the normal operation of [3]. DC transmission line protection is the safe and stable operation of the DC line protection, at present DC line main protection are used travelling wave protection as the main protection. At the same time with the low voltage protection, line of transverse differential protection, longitudinal differential protection as a back-up protection [6].

2. Traditional Protection Method

2.1. *Traveling wave protection*

The function of the relay protection of the DC line is to remove the danger as soon as possible when the transmission line is faulty, so as to ensure the stable operation of the system, and avoid the impact on the whole system. The disadvantages of power frequency protection is time-consuming, it is difficult to meet the basic requirements of medium line protection of rapidity. Therefore, the traveling wave protection has generated. In the paper, the transient traveling wave component generated during the DC system is constructed to protect the traveling wave, which can satisfy the speed of the relay protection. In the paper, the transient traveling wave component generated during the DC system is constructed to protect the traveling wave, which can satisfy the speed of the relay protection. At present, the protection of the ultra fast wave traveling wave, which is based on the fault current and voltage traveling wave, is the principle of high speed traveling wave protection, and the protection of the high speed traveling wave is constructed by the transient structure of the traveling wave [13].The protection of traveling wave is mainly of the following types [14], traveling wave polarity direction protection, traveling wave direction protection, traveling wave differential protection and traveling wave distance protection. The principle of the protection of the traveling wave level is based on the comparison of the original polarity of the fault transient components after the fault transient components. Line on the other end of the fault component in transmission line occur internal faults with the same polarity, when external fault occurs in the transmission line, with the same polarity of fault voltage and current and there is in near fault end. In AC system exists polarity judgment error problem, mainly due to the zero voltage and phase coupling factors, in DC system does not have this situation, so traveling wave directional protection in DC protection has great advantages. Traveling wave directional comparison protection can work in two ways, respectively is the single ended independence and both ends of the longitudinal mode, under traveling wave feature invariance theory support can also get forward and reverse fault criterion, the basic principle of this method is: the direction of fault is judged by the direction of the traveling wave fault, then by judging the action at both ends of the line direction element to determine whether the relay protection action. The protection method is theoretically analyzed, and its direction is not related to the initial phase, the position and the type of the fault. But is to protect the existing problems that the fault detection time is not long, the reason is when traveling wave in the propagation process happens repeatedly exhumation, which will lead to the traveling wave does not meet the established direction criterion. Fault results detection and judgment to consider when dorsal fault, but this method can use only the first traveling wave detection and judgment, so when the back side of the fault before wave used, emission rules it does not necessarily apply to reverse traveling wave, which may lead to malfunction of miscarriage of justice and relay protection of the results and the method is very sensitive to noise and interference, protecting it is very easy to misled. Traveling wave differential is based on traveling wave feature invariance, the principle is line from one end of the moving traveling wave at the other end reaches the line after the communication, then detected the size and shape of the traveling wave and the beginning of a traveling wave is the same. The protection method mainly through digital carrier channel will be at both ends of the transmission line traveling wave signal sampled value transmission, the utility model has the advantages of is a complete fault information is used, the disadvantage is very high requirements on the digital carrier channel, the network topology can impact protection performance, parameter selection inappropriate will affect the calculation accuracy. The Distance wave protection based on travelling wave principle according to the principle of reflected waves, its principle is through the detection of the original wave and reflected wave arrive at the protected lines at a time difference and through the operation to complete the fault location. The advantage of this method is that the data requests are very low, and the single ended fault is used, and the relay protection is fast, and the principle is simple and easy to implement. The main problems are: bus structure of relay measurement results have great influence, delay measurement will because waveform distortion and attenuation is not accurate, at the same time in the fault location error will increase. The method of linear programming

The maximum power supply capability of distribution network linear programming algorithm is mainly the maximum power supply capability of distribution network computing constraints in the process of main transformer, feeder capacity constraints as constraint conditions, the main transformer in distribution network load maximum as optimization objective, using linear programming software (such as lingo) method for solving the development process of this kind of method. Mainly as follows:

The original model based calculation of distribution network [16-17] given in maximum power supply capacity [13], proposed a new calculation model and the model considers the following constraints, including substation load transfer for equality constraint, the substation load transfer for equality constraints, the main transformer load rate of inequality constraints, contact feeder capacity inequality constraints and regional load constraints, to the main variable for the maximum load and as optimization objectives, and establish the linear programming the model of maximum power supply capability of distribution network, the model is used to solve the linear programming software lingo. This model includes two steps, the goal for the first step in solving the largest distribution network power supply ability, but the results of each main transformer load rate will appear the phenomenon of imbalance; the second step is the first step to optimize the maximum power supply capacity as an equality constraint condition, the main transformer with load balancing as optimization the goal again is solved, the purpose is to find in ensuring a load rate the most balanced solution vector invariant maximum power supply capability of distribution network. The model is compared with the literature methods, has also increased communications between each main transformer capacity constraints[13], also considered constraints regional load, running overload region during the process of appear to consider the actual distribution network. But this model did not consider occurred in the N-1 fault, load transfer of main transformer capacity of short-time overload occurs in the process . The actual operation of distribution network feeder in the contact relationship between often perplexing, moreover feeder N-1 constraints on the distribution network planning are need to consider, in order to solve the two problems, the reference literature presented calculation model of distribution network feeder interconnection based on the maximum power supply capability, solve the correlation problem feeder and the constraint conditions of feeder N-1, makes the results more close to the actual situation, the same document like the models, the model also includes two sub models with a maximum, to solve the distribution network power supply ability value, one for the solution satisfies the maximum power supply capability of distribution network value equilibrium under the load rate.

According to the paper reference model proposed by the nature of law model of maximum power supply capability of distribution network calculated model solution is discussed and analyzed. It is pointed out that the maximum power supply capacity of distribution network model solution can be achieved in a variety of operating mode of distribution network, calculate namely model method of maximum power supply capability of value can be achieved, thus proves that it has certain application value [18]. In addition, this paper also has been given in order to achieve maximum power supply capability of distribution net value of main transformer capacity and the matching relationship between the feed lines, to give the distribution network planning and optimization method has some guidance.

2.2. *Differential protection for HVDC line under differential pressure*

The differential under voltage protection equation for DC line is:

$$\begin{Bmatrix} du/dt > k_1 \\ |U_{dl}| < k_2 \end{Bmatrix}$$

From the type, we can see that the voltage change rate differential criterion, the judgement of the fault is in the transmission line, the line low voltage levels constitute low voltage criterion, has the effect of electrode selection, can be the fault and the fault distinguish, as after the DC line protection, the differential voltage set value and traveling wave protection are the same [15].Differential owe voltage protection as a back-up protection, with a certain delay, need to consider fault control system of protection feature transformation and electrical effects, in terms of reliability is higher than the travelling wave protection [16].

In summary, the differential under voltage protection action speed slightly slower than the traveling wave protection, sensitivity and reliability is better than the traveling wave protection, but there are still high fault resistance ability is poor, low sensitivity, setting on the basis of the lack of need through the simulation experiment of setting, etc.

2.3. *Low voltage protection*

The line through the grounding impedance, DC voltage decreased more slowly, traveling wave protection often cannot detect this kind of fault, line of low voltage protection as the high resistance grounding of DC line main protection, through reduction of the DC voltage detection is to judge the fault. Low voltage protection circuit is mainly used for high resistance grounding fault and fault detection of DC metal on the line, the tuning principle is: DC voltage to maintain the DC line fault, if the exclusion of other fault and DC voltage is lower than a certain duration setting value, then the DC line fault restart. However, the low voltage protection in HVDC projects is high load, and may cause errors when low load.. There is a need to adjust protection criteria of different system load, to ensure fast, accurate operation of fault [17]. It pointed out that the literature has a close relationship with low pressure and low voltage protection control system The higher the transition resistance, the smaller the detection line, the lower the voltage condition, and the error and the rejection [18].

In summary, low voltage protection is simple in theory, but can't tell the difference between area fault and fault and does not meet the protection speed, high resistance easily and refuse to move. Although the principle of low voltage protection is simple, it is lack of setting basis, and it cannot distinguish the area of fault and the high resistance fault in the area, and the selectivity is bad and the speed of action is slow.

2.4. *Differential protection*

Differential protection action criterion mainly according to the difference between DC current on the side converter station, on line DC current as criterion makes differential protection must consider two stations between communication of protection effect, and also by PLC transmission constraints, a certain delay [19].

Through paper [20] analysis differential current of DC line current under different conditions, pointed out that the DC line longitudinal differential protection is completely independent of the channel, can be avoided only a DC system data acquisition problems led to error, also can avoid the DC protection and conversion device between the pole control problems. The article [21] put forward to improve its performance in differential protection by traveling wave differential principle, respectively with the modulus and modulus of fault structure and fault identification criterion.

In the AC current system, the differential protection is the main protection, and its sensitivity is high and the action is rapid. However, the DC system differential protection does not consider the problem of the capacitor current, in order to prevent the error need to have a time delay, so the DC system differential protection performance and a lot of room for improvement

3. Method and Algorithm

3.1. *Transverse differential protection principle*

In the DC filter specific frequency, the impedance characteristics of DC line fault zone, the fault current characteristics of different, more easily in the region, identifying external fault. In order to improve the reliability [23], the concept of single terminal electric quantity and the concept of the criterion is constructed.. The calculation of the specific frequency characteristics of current (voltage), select the integral absolute value within a period of time, the time the transmission time is usually 5ms or two times the length of line The amount of electrical extraction of positive and negative ends, and addition operations, if greater than a threshold value, for the DC line area fault The internal fault by literature [24], is on both sides of the DC line protection (current) will have a mutation in the fault, under different fault conditions corresponding to the different mutation direction The mutation direction is the same in the region when the fault is the same as the positive mutation, while the mutation direction is opposite to the outer fault of the DC line.. According to this feature, the fault direction can be constructed. As indicated by the literature [25], line monopole fault, the non fault very due to the effect of electromagnetic coupling effect will also be subjected to a fault, but very sound not coupled over a certain threshold value, the characteristics of this fault pole is selected. In the selection of data window, taking into account the lightning disturbance and fault regulation process, we select the time for 5ms.

3.2. *Improved recursive wavelet transform*

Practical application in power system has a variety of fault signals, these signals of amplitude frequency and phase frequency characteristics in power system has very important role, can reflect the different characteristics of the fault signal, wavelet transform corresponding to the various waveforms and fault signal amplitude and phase relationship is very close, which on phase transformation and contains important information of fault signal analysis.

Currently used in the protection algorithm in wavelet based and the disadvantage is cannot effectively detect the fault information of phase transformation relationship by using the improved recursive wavelet algorithm can make full use of the signal phase relationship [22].

The improved recursive wavelet algorithm in order to meet the power system medium speed dynamic requirements, the algorithm only by a single direction wavelet transform. The choice of wavelet base has a great influence on the performance of the improved recursive wavelet transform.

4. Conclusions

Grid speed protection fault will greatly enhance. In this context, combined with the current distribution network, of high voltage direct current (HVDC) single ended electrical protection are summarized the development of future HVDC single end technology, mainly including wavelet algorithm and its application. This work can provide reference information for the research on the future smart distribution network in HVDC protection.

References

1. Zhao Wan Jun. HVDC transmission engineering technology [M]. Beijing: China Electric Power Press, 2004.
2. Song Guobing, Gao Shuping, Cai Xinlei, et al. Survey of relay protection technology for HVDC transmission lines [J]. Automation of Electric Power Systems, 201236(22):123-129 (in Chinese).
3. Pan Weiming. The study of traveling wave protection of HVDC transmission line [D]. Nanjing: Nanjing University of Science & Technology, Mar, 2013.
4. Li Xuepeng, Study on travelling wave protection and fault location for HVDC transmission line [D]. Baoding: North China Electric Power University, Dec, 2005.
5. Dong Xin. Research on theory of traveling wave protecion of HVDC lines [D]. Jilin: Northeast Dianli University, Mar, 2008.
6. Zhang Baohui, Zhang Cong, You Min, et al. Research on transient-based protection for HVDC lines [J]. Power System Protection and Control, 2010, 38(15):18-23 (in Chinese).
7. Shu Hongchun, Tian Xincui, Dong Jun, et al. ±800 kV DC transmission line boundary protection using model matching algorithms in time-domain [J]. Transactions of China Electrotechnical Society, 2013, 28(10):320-327 (in Chinese).
8. Chen Shilong, Zhang Jie, Bi Guihong. Wavelet analysis based on two-terminal transient voltage protection for UHVDC transmission lines [J]. Power System Technology, 2013, 37(10):2019-2725 (in Chinese).
9. Yan Xu, Boundary Protection Strategy for the VSC-MTDC under DC Faults [J]. Applied Mechanics and Materials Vols. 448-453 (2014) pp. 2030-2035.

10. Zhang Baohui, Kong Fei, Zhang Song, et al. Technical development of non-unit protection devices based on transient signals for HVDC transmission lines [J]. Proceedings of the CESS, 2013, 33(4):179-185 (in Chinese).
11. Shu Hongchun, Liu Kezhen, Zhu Shengqiang, et al. ±800kV UHVDC transmission line protection based on single end electrical transient signal [J]. Proceeding of the CESS, 2010, 30(13):108-117 (in Chinese).
12. Gao Shuping, Suonan Jiale, Song Guobing, et al. A new whole-line quick-action protection principle for HVDC transmission lines using one-end current [J]. Proceedings of the CSEE,2012,32(7):107-113 (in Chinese).
13. Li Aimin, Cai Zexiang, et al. Study on the Dynamic Performance Characteristics of HVDC Control and Protections for the HVDC Line Fault [C]. 2009 IEEE Power and Energy Society General Meeting, 2009, 1-5.
14. Cai Zexiang, Han Kunlun, Xu Min, "High voltage direct current transmission line differential under voltage protection characteristics of the dynamic analysis and tuning" Power Automation Equipment 2014, 34 (2): 114-119.
15. Cai Xinlei "VSC-HVDC DC transmission line longitudinal protection and single ended fault locating new principle research" Xi'an Jiao Tong University (D). 2012.
16. Zhu Taoxi, Chen Xilei, Zhou Hao, "Low voltage protection criteria for HVDC," Power System Protection and Control, 2011, 39 (9): 125-130.
17. Li Aimin, Wang Zhibin, Based on EMTDC DC line fault characteristics simulation analysis of Southern Power Grid technology, 2009, 12 (3): 125-130.
18. He Jie, Zhang Peidong, "The research on the longitudinal differential protection of HVDC transmission line", 2008.
19. Zhou Hongyang, Yujiang, Huang Jiayin, Zhao Manyong, "The relevant problem of the DC link longitudinal differential protection" Southern Power Network Technology 2008, 2 (3): 17-21.
20. Ha Hengxu, Yu Yang, Yi Ruipeng, et al. Novel scheme of travelling wave based differential protection for bipolar HVDC transmission lines [C]. 2010 International Conference on Power System Technology (POWERCON), 2010: 1-6.
21. J. Serra. Image Analysis and Mathematical Morphology. London: Academic Press, 1982.
22. He Zhengyou "Wavelet analysis for power system transient signal processing applications" China Electric Power Press, 2010.
23. Zhang Baohui, Lv Zhilai, "The theory of boundary protection based on third part of the fault transient spectrum difference EMTP simulation" relay 2002, 30 (11): 1-4.

24. Wang Chao, Shu Hongchun, Zhang Jie "Based on mathematical morphology of transmission line traveling wave protection research of Yunnan electric power technology, 2005 (33) 5:11-15.
25. Yang Xiaobo. Based on the continuous wavelet transform unstructured distortion of fabric defects automatic detection "Journal of Southern Yangtze University (Natural Science Edition), 2005, 4 (6): 630-633.

New energy lithium battery charge and discharge characteristics analysis of power drive systems

P.J. Mao[†], Y. H. Ding, S. H. Zhang, Y. L. Zhang and J. Wang
Henan University of Science and Technology, Luoyang, 471003
[†]*Email:* 790137007@qq.com

K. Jiang
Luoyang, Henan Province Tobacco Company, Luoyang, 471003
Email: zyl217907@126.com

In this paper, through a series of experiments of power drive systems 18650 lithium battery charge and discharge characteristics, charging experiment by constant current module, discharge experiments by constant digital electronic load, through 18650 lithium battery charge and discharge curves and analysis of relevant data , we can get the rules and factors of the 18650 lithium battery charge and discharge characteristics . Finally, analysis problems and solutions of the 18650 lithium battery charge and discharge test .

Keywords: 18650 lithium battery; charging characteristics; discharge characteristics.

1. Introduction

In recent years, with the promotion and development of electric vehicles, the performance studying of lithium-ion battery become one of the focus [1]. Lithium-ion battery has drawn great attention due to its high energy density, long cycle life, no memory effect, low self-discharge and environmental friendliness and so on [2-3]. With the extensive application of lithium-ion battery, the performance requirements are also increasing [4]. Fully grasp the performance of the lithium battery, developed a lithium battery for electric vehicles, thereby increasing the mileage and power of electric vehicles. The main performance of the battery is its charge and discharge characteristics [5]. In this paper, main from the constant current charging module, constant digital electronic load to proceed, the completion of 18650 lithium battery charge and discharge experiments, analysis of experimental results.

2. The Design of The lithium battery charge and discharge experiment

2.1. *The 18650 lithium battery*

18650 battery included lithium-ion batteries and nickel-metal hydride batteries, in this article only studied lithium-ion batteries .This experiment is that the experiment of windfire 4000mAh / 3.7v 18650 single lithium battery charge and discharge characteristics . Part technical parameters of the battery are as follows: rated voltage 3.7v, rated capacity4000mah,diameter of 18650 lithium battery 18.0mm, height 65.0mm. Experiments carried out under room temperature (25 ℃), charging cut-off voltage 4.25v , discharging cut-off voltage 3.0v.

2.2. *The design of the lithium battery charge and discharge experiment*

2.2.1. *The design of the lithium battery charge experiment*

According to18650 lithium battery the characteristics and the Maas optimal charging current curve, in this study, using "constant current - voltage" charging method,this can avoid the overcharge phenomenon of the post-constant current charging method, in addition to, avoid the current is too big phenomenon of the pre-charget constant voltage charging method .

In this study, constant current module for single 18650 lithium battery charging experiment, with voltage, current, power display and USB interfaces, you can set the float voltage and charge current ,what's more can judge charging or fully charged by the indicator shows. In the initial stage of charging, first low current pre-charge for the battery, and then use a larger constant current fast charge for the battery, when the battery voltage is close to its rated voltage, charging changed into a constant voltage. Due to constant power charger, it would gradually reduce the charging current, at last, trickle charge for the battery until it reaches the rated voltage [6-7]. The charging curve of the single 18650 lithium battery "constant current - voltage," as shown in Figure 1. Before the "constant current - voltage" charging, first using a small current to the precharge treatment for 18650 lithium battery, this play a repair protection role if lithium battery occurred over-discharge phenomenon. Then "constant current - voltage" charging , using a large constant current charging ,shorten the charging time . Last constant voltage charging, when the battery reaches the termination voltage, constant voltage charging method, make sure the battery is fully charged.

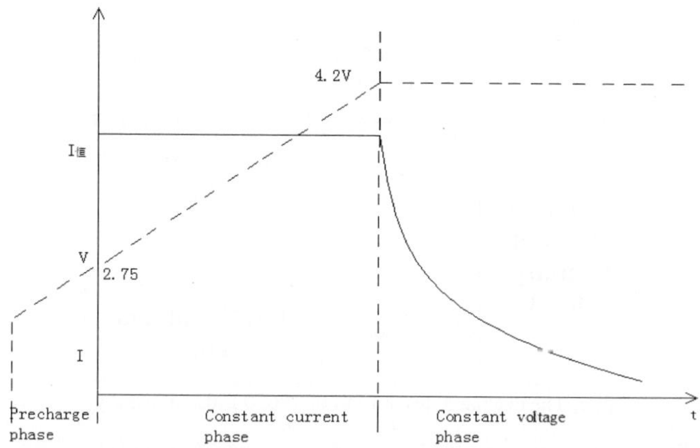

Fig. 1. 18650 lithium battery the charging curve of constant - constant .

2.2.2. *The design of the lithium battery discharge experiment*

When the 18650 lithium battery discharge, the discharge current is not too large, if too large, the battery current might cause internal heat and permanent damage to the battery. When the discharge cut-off voltage is too low, if the battery voltage is lower than the discharge cut-off voltage, continued discharge, can cause over-discharge phenomenon and permanent damage [8]. Therefore , in this study, low current and high current constant current discharge mode for the single 18650 lithium battery, discharge rate control in the range of the nominal capacity (1C),When the terminal voltage drops below 3.0V, stop discharge [9-10].

In this study, a constant current digital electronic load, 12V DC power supply and heat sink to achieve 18650 lithium battery constant current discharge experiments.12V DC constant current load to provide power, heat dissipation to cool the MOS radiator fins, as shown below. It also achieve the lithium battery capacity test, and can set the cut-off voltage, resistance can be displayed. Fully charged lithium battery is automatically stopped when reached a preset constant discharge cut-off voltage , accumulated out the real capacity of the battery, and can communicate with the computer via the COM interface, recorded discharge curve and other information by the computer. By constant current discharge can not only test the real capacity of the battery, and the battery easy pairing, can be applied to the battery repair, if appropriate current repeatedly charging and discharging ,the battery will have a repair.

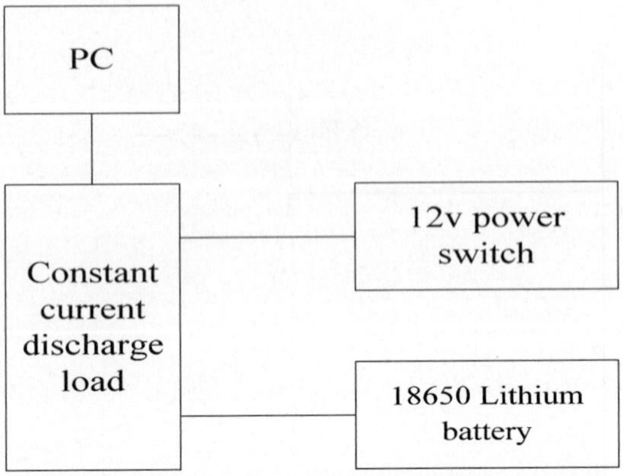

Fig. 2. The discharge test circuit of 18650 Lithium battery.

3. The Results and Analysis of the Lithium Battery Charge and Discharge Experiment

3.1. *The results and analysis of the lithium battery charge experiment*

In accordance with the technical requirements, single 18650 lithium battery terminal voltage up to 4.25V, according to the actual requirements, power drive systems single 18650 lithium battery do a "constant current - voltage" charge in the range of 1C charge rate. The experiment results after pretreatment, the experimental results after data processing, the results are shown below.

As can be seen from Figure 3, 4, 5, charging time and capacity of the 18650 lithium battery is different at a constant current stage. Time of the constant voltage stage is relatively long, in this stage the charging current drops rapidly, so the time of the constant voltage stage is limited. In order to ensure a sufficient amount of capacity, trickle charge at the final stage, improve the charging efficiency.

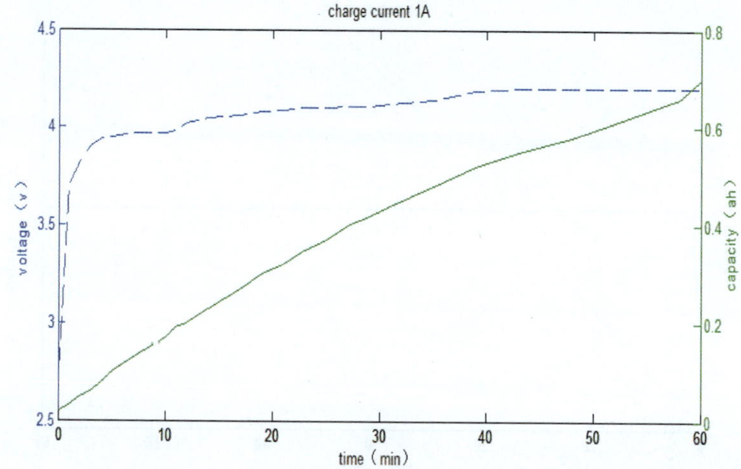

Fig. 3. 18650 lithium battery the charging curve of constant – constant.

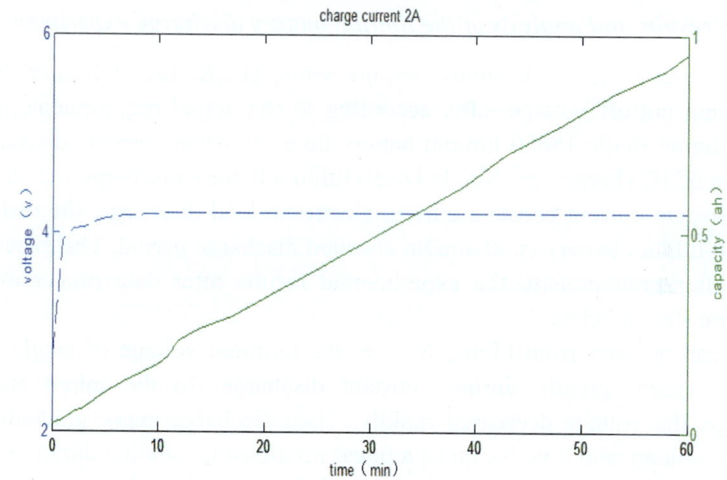

Fig. 4. 18650 lithium battery the charging curve of constant – constant.

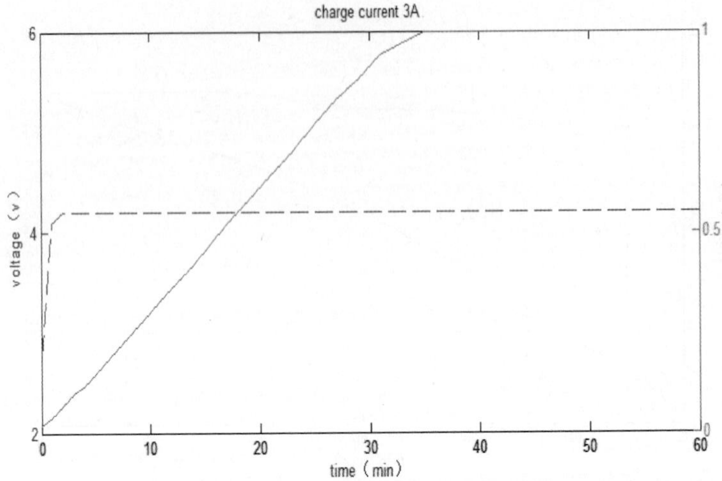

Fig. 5. 18650 lithium battery the charging curve of constant – constant.

3.2. *The results and analysis of the lithium battery discharge experiment*

In accordance with the technical requirements, single 18650 lithium battery discharging cut-off voltage 3.0v, according to the actual requirements, power drive systems single 18650 lithium battery do a "constant current" discharge in the range of 1C charge rate. Single 18650 lithium battery discharge experiments, using constant current constant current electronic load discharge, the resistance of 18650 lithium battery is 30.4mΩin constant discharge period. The experiment results after pretreatment, the experimental results after data processing, the results are shown below.

As can be seen from Figure 6, 7, 8, the terminal voltage of single 18650 lithium changes greatly during constant discharge. In the initial stage of discharge, the voltage decreased rapidly. Then discharge curve gradually flat, capacity, voltage and time is almost a linear relationship, and the duration of this period related to quality of the battery, discharge rate, voltage and ambient temperature and so on. In the final stage of the discharge, when the battery terminal voltage drops to about 3.0V, if you continue to discharge, the curve will become vertical plummet.

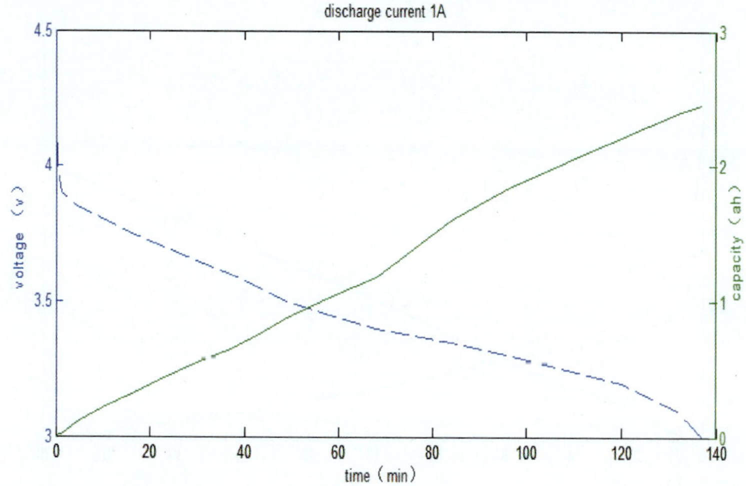

Fig. 6. 18650 lithium battery the discharging curve of 1A constant current.

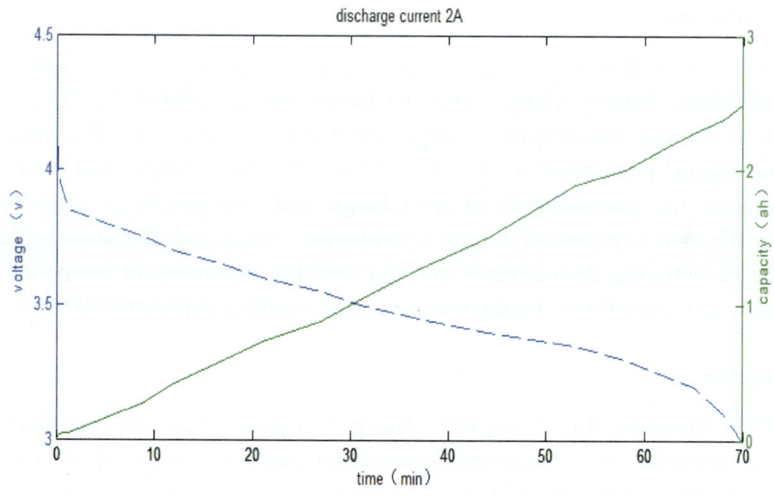

Fig. 7. 18650 lithium battery the discharging curve of 2A constant current .

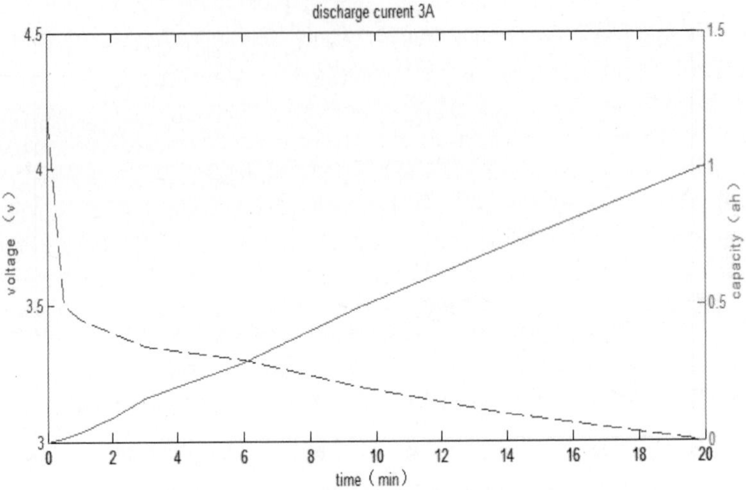

Fig. 8. 18650 lithium battery the discharging curve of 3A constant current.

4. Conclusion

Through 18650 lithium battery charge and discharge curves and analysis of relevant data, battery charge and discharge curves related to charge and discharge current, temperature, charge and discharge times, self-discharge and battery aging phenomenon. In the lithium battery charge and discharge experiment, the phenomenon of overcharge and over-discharge affected the battery life. The experiment did not consider the charge and discharge cycle life of lithium batteries, also assume that the working environment temperature is constant, had some flaws, laying a more comprehensive experimental basis.

References

1. Chen Quanshi, Lin Yongmin, Zhang Dongmin. Research on discharge characteristics of the lead-acid batteries for electric vehicles [J]. Automotive Technology,1996, (8):7-11.
2. Zhang Qing, Li Gechen. Lithium-ion battery charge and discharge characteristics [J]. Automation technology and application, 2008, 27(12):107-108.
3. Bo Lin, Wang Quandi. Electrical characterization and modeling studies of electric vehicle batteries [D]. 2013:1-2.
4. Jia Yaoqin. Microprocessor-based lithium battery charge and discharge control with high precision research [J]. Power Electronics Technology, 2004, 38(5):6-8.

5. Wang Hongliang, Cui Shengmin. Establish test lead-acid battery charge and discharge characteristics model [J]. Batteries, 2005, (1):38-39.
6. Fan Yuchao. Electric Vehicle Battery Management System [D]. Henan: Zhengzhou University, 2013:6-15.
7. Liu Zaifei. Electric Vehicle Lithium Battery Management System [D]. Shandong: Qingdao University, 2013:2-5.
8. Xu Jin. The analysis and testing of Lithium battery charge and discharge characteristics [J]. Western China Science and Technology, 2011, 10(33): 3-4.
9. Wang Zhifu, Peng Lianyun, Sun Fengchun, Zhang Chenning. Electric vehicles using lithium-ion battery charge and discharge characteristics [J]. Batteries, 2003, 33(3):167-168.
10. He Hongwen, Sun Fengchun, Zhang Chenguang, Bai Wenjie. Experimental study on lithium-ion battery charge and discharge characteristics [J]. Beijing University of Technology, 2002, 22(5):578-581.

Reliability centered maintenance for condition based maintenance application on transformation equipment

D.J. Ren[1†]
Zhangjiakou City, Power Supply Bureau
Zhangjiakou, Hebei 075000, China
†Email:14592318@163.com

Now electricity distribution systems operate in a liberalized market. These systems should therefore be able to provide electricity to customers with a high degree of reliability and be cost-effective for suppliers. RCM (Reliability Centered Maintenance) was invented by the aircraft industry in the 1960s, to organize the increasing need for maintenance for reducing costs without reducing b safety. Today RCM methods invented by transformation equipment are seen as very complex and are not fully accepted by the Algerian power industry. The extensive need of human and capital resources in the introduction phase is also a negative factor that could be one of the reasons of why RCM methods are not used in our branch. This article provides a discussion of the two primary objectives of RCM: to ensure safety through preventive maintenance actions, and, when safety is not a concern, preserve functionality in the most economical manner. For the power distribution systems facilities, the mission should be considered at the same level as safety.

Keywords: RCM; FMECA; Transformation equipment.

1. Introduction

Many electric power utilities install Reliability-centred maintenance (RCM) programs for establishing maintenance requirements [1,2]. Much data and long (vast) experience is needed to success-fully conduct the RCM [3]. Despite being around for a long time, its application in the power sector is still at an infancy stage [4]. Significant changes in business environment make electric power utility decision making and risk management, especially the need for improved risk characterisation, increasingly challenging [5].

Energy development and utilisation are facing two problems in China: the energy crisis and the effects of global warming.

An important means to solve the above problems effectively is to increase the intensity of the development and utilisation of photovoltaic system, which directly transforms the inexhaustible solar energy into the powerful green electrical energy.

However, some problems plague the development of the photovoltaic industry. Due to these problems, the photovoltaic industry is seriously hindered in

---

[1] Da-Jiang Ren

its efforts to develop sustainable. Therefore, accurately understanding the developing situation of the photovoltaic industry and analysing and solving the problems that occur during the development process of the photovoltaic industry in China are important topics of current research. This paper presents research results regarding these topics.

First, this paper considers the present situation of the photovoltaic industry in China from three aspects: the power market, power generation industry policy and power generation technology level.

Second, a discussion of the existing problems of the above three aspects is presented; in addition, based on the developing trends of renewable energy in China, development orientation of the photovoltaic industry are identified for years to come. Finally, strategies are proposed to promote the rapid development of the distributed photovoltaic industry in China

2. Reliability Centered Maintenance (RCM)

Reliability centered maintenance (RCM) is a systematic process used to determine what has to be accomplished to ensure that any physical facility is able to meet continuously its designed functions in its current operating context [2]. RCM leads to a maintenance program that focuses preventive maintenance (PM) on specific failure modes likely to occur. Any organization can benefit from RCM if its breakdowns account for more than 20-25% of the total maintenance Workload [3].

This paper proposes a practical procedure to develop a cost effective maintenance program for electric power distribution systems. The procedure is mainly based on the reliability centered maintenance (RCM) method that prioritizes maintenance requirement of failure modes, and selects the effective maintenance activity for the critical failure modes. Reliability centered maintenance (RCM) is a decision making process in the selection of a cost-effective maintenance program to improve the reliability, based on determined criticality of failure modes. It prioritizes the maintenance requirement of all failure modes, and selects the effective maintenance activity for the critical failure modes [4,5]. In this paper, the RCM based maintenance program is developed using the proposed procedure for an electric utility in ALGERIA

2.1. *Methods of RCM*

RCM is a systematic method to keep a balance between preventive and corrective maintenance. This method chooses the right preventive maintenance activities for the right component at the right time to reach the most cost-efficient solution [6].

The first description came in 1978 by Nowlan. It was introduced in nuclear power in 1980 and in hydro power in 1990 [7]. RCM is characterized by maintaining system function, identifying failure modes, prioritizing functions, and choosing efficient maintenance.

RCM is a technique that is used to develop cost effective maintenance plans and criteria so the operational capability of equipment is achieved, restored, or maintained. The main objective of RCM is to reduce the maintenance cost by focusing on the most important functions of the system. There are several different formulations of RCM processes in the literature. According to [8] an RCM analysis basically provides answers to the following questions.

Shows a detailed logic diagram of the RCM method:

1. What are the functions and associated performance standards of the equipment in its present operating context?

2. In what ways can it fail to full fill its functions?

3. What is the cause of each functional failure?

4. What happens when each failure occurs?

5. In what way does each failure matter?

6. What can be done to prevent each failure?

7. What should be done if a suitable preventive task cannot be found?

The RCM analysis may be carried out as a sequence of activities or steps. In IEC standards for RCM analysis [9] following basic steps in an RCM analysis are listed, Fig. 1 shows a detailed logic diagram of the RCM method.

1.What are the functions and associated performance standards of the equipment in its present operating context?

2. In what ways can it fail to full fill its functions?

3. What is the cause of each functional failure?

4. What happens when each failure occurs?

5. In what way does each failure matter?

6. What can be done to prevent each failure?

7. What should be done if a suitable preventive task cannot be found?

The RCM analysis may be carried out as a sequence of activities or steps. In IEC standards for RCM analysis [9] following basic steps in an RCM analysis are listed, Fig. 1 shows a detailed logic diagram of the RCM method.

1.Defining the system and/or subsystems and boundaries.

2.Defining the functions of each system or subsystem identifying functionally significant item (FSI).

3.I dentifying the pertinent FSI functional failure causes.

4.Predicting the effects and probability of these failures.

5.Using a decision logic tree to categorize the effects of the FSI failures.

6. Identifying applicable and effective maintenance tasks which comprise the initial maintenance program.

7. Redesign of the equipment or process, if no applicable tasks can be identified.

8. Establishing a dynamic maintenance program, which results from a routine and systematic update of the initial maintenance program and its revision assisted by the monitoring, collection and analysis of in-service data?

To avoid waiting again for several years prior to application to obtain this information, it seems necessary to predict the evolution of equipment reliability and so the consequences of implementing the new program. This step involves the modeling and simulation program, before its application [10].

2.2. *Practical method of FMECA*

FMECA is a useful tool when performing an RCM analysis. FMECA is a way to evaluate potential failure modes and their effects and causes in a systematic and structured manner. Failure modes means the ways in which something could fail. Effects analysis refers to studying the consequences of those failures. The purpose of the FMECA is to take actions to eliminate or reduce failures, starting with the highest-priority ones. By itself, an FMECA is not a problem solver; it should be used in combination with other problem solving tools. The analysis can be done either in a qualitatively or quantitatively way. Basic steps in performing a FMECA could be [11-12]:

FMECA is a useful tool when performing an RCM analysis. FMECA is a way to evaluate potential failure modes and their effects and causes in a systematic and structured manner. Failure modes means the ways in which something could fail. Effects analysis refers to studying the consequences of those failures.

The purpose of the FMECA is to take actions to eliminate or reduce failures, starting with the highest-priority ones. By itself, an FMECA is not a problem solver; it should be used in combination with other problem solving tools. The analysis can be done either in a qualitatively or quantitatively way. Basic steps in performing a FMECA could be [13]:

1. Define the system to be analyzed. Complete system definition includes defining of system boundaries, identification of internal and interface functions, expected performance, and failure definitions.

2. Identify failure modes associated with system failures. For each function, identify all the ways failure could happen. These are potential failure modes.

3. Identify potential effects of failure modes. For each failure mode, identify all the consequences on the system. "What happens when the failure occurs?"

4. Determine and rank how serious each effect is. The most critical pieces of equipment which affected the overall function of the system need to identified and determined.

5. For each failure mode, determine all the potential root causes.

6. For each cause, identify available detection methods.

7. Identify recommended actions for each cause that can reduce the severity of each failure.

Then, a block diagram of the system needs to be created. This diagram gives an overview of the major components or process steps and how they are related. These are called logical relations around which the FMECA can be developed. It is useful to create a coding system to identify the different system elements. The block diagram should always be included with the FMECA.

3. Summary

In the study the literature review highlighted major challenges in the current approaches for conducting the RCM and risk assess-ment. Since average annual costs, and repair and failure rates are used to produce the KPI (Key Performance Indicator), the approach is a convenient and cost effective way of providing a broad view of the effectiveness of the strategies on a group of network assets.

Second, the method was applied to statistical data to generate a failure probability distribution benchmarking model in addition to the current practices for conducting the FMECA where failure probabilities are based only on condition monitoring a notion that describes as inadequate.

This paper presents the application of RCM model to optimize the maintenance management of equipment EFS. The main results that we had are recommendations for the re-integration of this maintenance strategy, including an optimal implementation of this approach. This work showed the feasibility of conducting an optimization method of maintenance RCM. This approach is based on the analysis FMECA. The implementation of this approach shows its contribution in reducing maintenance costs. Indeed it can.

Define the requirements of dependability in a precise.

Identify critical functions for the system.

Define the maintenance policy for the system and its.

At the level of system reliability, we have identified the components on which special attention should be paid. The example discussed in the context of this work was developed following a logical and structured analysis. It allowed better control the system under study while identifying weak links and knows the types of maintenance applied to each subsystem and component. In the end, it is a real process of optimizing maintenance costs.

References

1. Svensk Energi – Swedenergy AB. RCM for electrical distribution systems – a simplified decision model for maintenance planning Part 1, 2001 [in Swedish].
2. McKenna T, Oliverson R. Glossary of reliability and maintenance terms. Houston, Texas: Gulf Publishing Co.; 1997.
3. Smith Anthony M, Hinchcliffe Glenn R. RCM-Gateway World Class Maintenance, 2004).
4. Bertling L, Eriksson R, Allan RN. Relation between preventive maintenance and reliability for a cost-effective distribution system. IEEE Porto Power Tech Conference 2001.
5. Picknel ll J, Steel KA. Using a CMMS to support RCM. Maintenance Technology 1997:110–7. October.
6. Bertling L. Reliability assessment of electrical power systems course material for Vattenfall El distribution. KTH Elec. Eng., Stockholm, Sweden, Paper A-ETS/ EEK-0504, August 2005.
7. Bertling L. Reliability Centred Maintenance for electric power distribution.
8. Moubray J. Reliablity-centred Maintenance. Butterworth-Heinemann, 1991.
9. IEC Standards 60300-3-11. Application guide – Reliability centered maintenance, _rst edition, Mars 1999.
10. Despujols, A. Optimisation de la maintenance par la fiabilite. Techniques l'ingenieur, dossier MT9310, 2004.
11. Life management for generator circuit-breakers D. BRAUN ABB High Voltage, Technologies Ltd., A. GUERIG Nordostschweizerische Kraftwerke AG (Switzerland) 1994.
12. Mohamed HAMADACHE. determination des besoins pour l'elaboration d'une strategie de maintenance des équipements automatisés du réseau aérien de distfubution d'électricité à moyenne tension montreal, le 30 avril 2010.
13. Tutorials on Life Cycle Costing and Reliability Engineering Course Material Professor U Dinesh Kumar Indian Institute of Management Bangalore. 2005.

Study on feature fusion based electric motor faults diagnosis

L.J. Xu and Z.L. Kang†
College of Mechanical and Electronic Engineering,
Sichuan Agricultural University, Ya'an, Sichuan Province, 625014, China
†Email: *kangzhiliang96@163.com*

T. Long
School of Control Engineering,
Chengdu University of Information Technology, Chengdu, China

Z.Y. Zou
College of Mechanical and Electronic Engineering,
Sichuan Agricultural University, Ya'an, Sichuan Province, 625014, China

For the characteristic that a single class of signal contains insufficient fault information, a method for diagnosing motor faults with multi-class signal fusion that combines generalized discriminant analysis (GDA) and support vector machine (SVM) is designed. First, corresponding spectrum features are respectively extracted from the collected rotor vibration signals and the current signals; the spectrum features of the 2 kinds of signals are combined and the combined feature is performed nonlinear fusion with GDA to obtain the fusion feature; input the GDA fusion feature into the SVM to carry out motor fault diagnosis. Test results show that the fusion feature obtained from this method have the most fault type information as well as strong robustness to both the training sample number and the kernel parameters of classifier such that it has high value of practical application.

Keywords: Motor; Generalized discriminant analysis; Features fusion; Fault diagnosis.

1. Introduction

Electric motors are the most widely applied power supply equipment and driving units. At present, the most of equipment is driven by all kinds of motor and any industrial sector can hardly do without motors. The complexity of motors and their operation environments decides that the diagnosis of motor faults is very complicated, while the losses incurred by motor faults are immeasurable. In order to ensure reliable operation of motors and carry out in-depth analysis to the fault mechanisms, it has very important significant to study the diagnostic techniques of motor faults. Traditionally, the faults are diagnosed by subjective experience judgment and analytical approach of contact sensor detection that detects physical amounts such as noise, vibration and venting and carries out experience judgment. However, this kind of method greatly depends on experience, so its effects are very limited [1-2].

The motor fault diagnosis technology has made great progress since 1970s to now. In recent years, with the development of artificial intelligent, neural network, rough set, wavelet analysis, fuzzy theory, SVM and so on are more and more applied in motor fault diagnosis, thus initiating the development of motor maintenance from breakdown maintenance to status based predictive maintenance so as to ensure the stability and reliability of the system containing motor and the economic efficiency of maintenance [3-6]. The most of motor fault diagnosis technologies are now based on a single parameter which could be any of motor related parameters that may carry fault features and can be detected such as current, vibration, temperature, noise and component of lubricating oil etc. However, is has uncertainty or even result in wrong diagnostic result to carry out motor fault diagnosis with a single parameter. This paper plans to establish corresponding fault mode library with three kinds of fault, i.e. rotor imbalance, rotor end ring breaking and rotor bar breaking, collect rotor's current and vibration signals and extract their spectrum features to design effective feature fusion method with GDA (generalized discriminant analysis), and then input the fusion feature into the classifier that is SVM so as to obtain higher motor fault identification rate.one space after each other.

2. Feature Fusion and Diagnosis Design

The fault identification rate would be low if the spectrum feature of a single class signal is inputted into the classifier SVM since it contains insufficient fault information, this paper will combine the spectrum features of the rotor current signals and rotor vibration signals of the motor and carry out fusion extraction by GDA to the combined feature to make the extracted low-dimensional fusion feature contain more information of fault types. And it will surely achieve higher fault identification rate that such fusion feature is inputted into the classifier SVM. A single class signal contains very limited information, while spectrum features of different signals contain fault information of motor in different status and spaces, so fusion design is planned to carry out to the spectrum feature of above two classes of signal, see figure 1 for feature fusion and diagnosis flow.

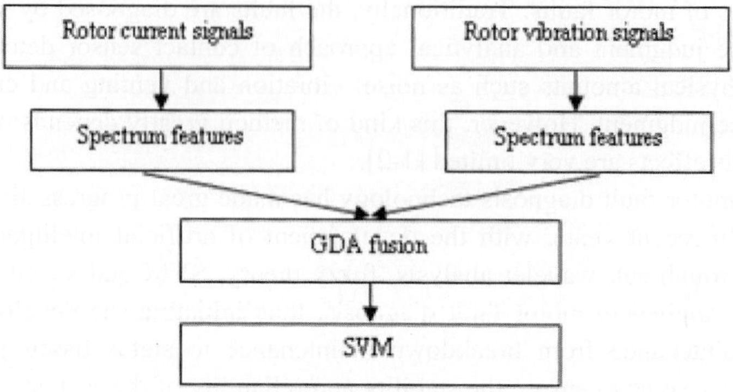

Fig. 1. Feature fusion and diagnosis flow chart.

2.1. *Generalized discriminant analysis (GDA)*

GDA[7] is essentially a kind of kernel mapping, which maps the input vectors of linear inseparability in low-dimensional input space to a high-dimensional feature space through a nonlinear mapping function, and then linear separability can be carried out in such high-dimensional space and the inner product of the samples in the feature space can be calculated with kernel function. Set the original feature parameter $X = \{x_1, x_2, \cdots x_m\}$, $x \in R^n$, the sample class is c. The subset of the ith class of sample is X_i, the sample number is m_i, and the mean value of the ith class of sample is $u_i = \frac{1}{n}\sum_{j=1}^{n_i} x_j^i$. The sample vector after non-linear mapping Φ is $\Phi(x) \in H$ (i.e. high-dimensional space), the intra-class discrete matrix of samples in the space H is S_W^Φ, the inter-class discrete matrix is S_B^Φ, and then the discriminant criterion of Fisher in the high-dimensional space H is:

$$J(\omega_{opt}) = \arg\max \frac{\omega^T S_B^\Phi \omega}{\omega^T S_W^\Phi \omega} \tag{1}$$

Where, $\omega \in H$ is the projection direction of features after kernel mapping, ω_{opt} is the first r main feature vectors obtained by solving the feature values of $(S_W^\Phi)^{-1} S_B^\Phi$ and sequencing from large to small, thus realizing dimensionality reduction, and ω meets $\omega_j^T \omega = 0, \quad j = 1, \cdots, r$, and it is surely in the space expanded by all training samples $\Phi(x)$ in space H. According to the theory of reproducing kernel, ω is solved as

$$\omega = \sum_{x_j \in X_i} \alpha_i \Phi(x_j) = \Phi\alpha \tag{2}$$

Where, α is the optimal kernel discriminant direction of the optimal projection direction ω in space H. Therefore, it derives

$$< \omega, \Phi(x) >= \sum_{i=1}^{n} \alpha_i \Phi(x_i) \Phi(x) = \sum_{i=1}^{n} \alpha_i K(x_i, x) \quad (3)$$

Where, $K(x_i, x)$ is the core function and selected as basis function in this paper

$$K(x_i, x) = \exp(-\frac{\|x_i - x\|^2}{\sigma^2}) \quad (4)$$

2.2. *Support vector machine (SVM)*

SVM, proposed by Vapnik in middle 1990s, is a kind of machine learning algorithm established on the basis of statistical learning VC theory and structural risk minimization principle. This algorithm adopts kernel function method to map the inputted space to a high dimensional space and obtains the global optimum solution through a linear constraint secondary planning problem. Set a training set S={(x_k, y_k)|k=1,2,...,N}, of which $x_k \in R^n, y_k \in R$ are input and output data respectively, the optimization problem in the original space and its constraint conditions are:

$$\min J(w, e) = C \sum_{i=1}^{N} e_i^2 + \frac{1}{2} \|w\|^2 \quad (5)$$

$$s.t: y(x) = w^T \varphi(x_i) + b + e_i \quad i = 1, 2, \cdots, N \quad (6)$$

Where, e_i is error, w is weight, $\phi(.)$ refers to nonlinear function, b is deviation. Importing Lagrangian α, it obtains:

$$L(w, e, b; \alpha) = J(w, e) - \sum_{i=1}^{N} \alpha_i \{ w^T \varphi(x_i) + b + e_i - y_i \} \quad (7)$$

According to KKT conditions, it determines parameters a_i, b and substitutes into the regression model of the vector machine:

$$y(x) = \sum_{i=1}^{N} \alpha_i K(x_i \bullet x) + b \quad (8)$$

$K(x_i \bullet x)$ is the inner product kernel function meeting Mercer condition, which normally values the radial basis function (RBF).

3. Testing and Analysis

This paper adopts Y160M2-2 model three-phase asynchronous motor as the research subject to establish the motor testing platform. The load is DC electric generator, and it can conveniently change the load of Y160M2-2 by regulating the exciting voltage of DC generator. The rotor vibration signal and the rotor current signals are respectively converted into 1~5V electrical signals by corresponding signal regulating circuit, the electrical signals are converted into digital signal by signal capture card and sent into PC, and 50 spectrum feature samples of each of 3 classes of faults that are rotor imbalance, rotor end ring breaking and rotor bar breaking as test data.

3.1. *Fault discrimination under different features*

The classifier SVM adopts the one-to-many classification strategy, the kernel parameter $\sigma = 2$, the regular parameter $C = 1$, and the training precision $e = 0.001$. 30 samples of each class of fault are taken as training samples and the rest samples are taken as testing samples. See table 1 for the fault identification rate on the basis of all kinds of spectrum feature. It can see from the data in table 1 that the identification rate when spectrum feature of a single class of signal is inputted into SVM is low, the identification rate when combined feature (i.e. direct combination of the spectrum features of the 2nd class signals) is inputted into SVM is slightly higher, and the identification rate when GDA fusion feature is inputted into SVM is the highest. This shows GDA fusion feature contains the most information of fault type and it is more beneficial for improving the accuracy of fault diagnosis.

Table 1. Fault identification rate under different features (%).

Feature of fault	Rotor imbalance	Rotor end ring breaking	Rotor bar breaking	Ave. identification rate
Spectrum feature of vibration	100.0	91.35	87.59	92.98
Spectrum feature of current	95.24	87.20	85.40	89.28
Combined feature	100.0	91.40	87.90	93.10
GDA fusion feature	100.0	95.24	92.30	95.85

3.2. *The relationship between training sample number, kernel parameters and fault identification rate*

The training time of GDA depends on the number of training sample, and the calculation time increases dramatically with the increase of sample number. See table 2 for motor fault identification rate by selecting GDA fusion features and under different training sample number. It can see from the data in table 2 that the sample number has less influence on fault identification rate, which shows the method of this paper also has high fault identification rate when the number of training sample is small.

Table 2. Fault identification rate at different training sample number (%).

Training sample quantity	Rotor imbalance	Rotor end ring breaking	Rotor bar breaking	Ave. identification rate
20	99.25	95.10	91.90	95.42
25	100.0	95.16	92.05	95.74
30	100.0	95.24	92.30	95.85

The kernel parameter σ of the classifier SVM, as an important parameter, often has influence on the identification performance of SVM. See table 3 for the average fault identification rate obtained by inputting all kinds of feature into SVM when σ takes different values and the training sample number is 30. It can see from the data in table 3 that with σ taking different values, the average identification rate of all kinds of fault obtained when GDA fusion feature is inputted into SVM is 95.85%, while the identification rates when other features are inputted into SVM are influenced by the value of σ. This shows that the GDA fusion feature does not only contain the most fault type information, but also free of the influence of σ, which is beneficial for popularization of this method.

Table 3. The average fault identification rate at different σ value (%).

σ	Vibration spectrum feature	Current spectrum feature	Combined feature	GDA fusion feature
0.25	92.98	86.43	93.50	95.85
5	92.98	89.28	92.13	95.85
10	93.40	88.20	90.10	95.85

4. Conclusion

This paper designed the GDA and SVM integrated motor fusion fault diagnosis method. First, the spectrum features are respectively extracted from the rotor vibration signals and the current signals and combined, and then GDA is utilized to carry out nonlinear fusion to the combined feature and SVM is taken as the classifier and successfully applied in motor fault diagnosis. It is known from the study that the GDA fusion feature contains the most fault type information and is beneficial for improving fault identification rate. Under different training sample number, the method in this paper still has favorable fault identification ability even there are less training sample number and is almost unaffected by the kernel parameter σ of SVM, which shows the method in this paper has extensive value of practical application.

Acknowledgement

The paper was supported by the project (No. 12ZA277) of Natural Science Foundation of Sichuan Education Department in China.

References

1. Han Tian, Yin Zhongjun &Yang Shaowei (2009). Discussion of diagnostic methods for broken rotor bar of induction motor. Proceedings of the CSU-EPSA, 21(1): 93-117.
2. Cao Haiyang, Xu Xinghua & Zhu Qiwei (2013). The design and implementation of experiment platform for motor fault diagnosis based on data comprehensive acquisition system. Engine Technical, (5): 38-40, 54.
3. Xu Yunzhi, Fang Yong-li & Zhang Xiao (2013). The motor fault diagnosis of turn-to-turn short circuit based on LS-SVM and energy spectrum analysis. Journal of North China Electric Power University, 40(02): 30-33.
4. Gan Chun, Wu Jianhua & Yang Shiyou (2014). Fault diagnosis of power converters for switched reluctance motors based on wavelet packet energy analysis. Proceedings of the CSEE, 34(9): 1415-1422.
5. Shi Liping, Wang Panpan & Hu Yongjun (2014). Broken rotor bar fault diagnosis of induction motors based on bare-bone particle swarm optimization and support vector machine. Transactions of China Electro Technical Society, 29(01): 147-155.
6. Zhang Niaona, Wang Yongqing & Li Jingshuai (2012). Fault diagnosis of permanent magnet synchronous motor based on SOM neural networks. Journal of Jilin University (information science edition), 30(06): 555-560.
7. Baudat G, Fatiha Anouar (2000). Generalized discriminant analysis using a kernel approach. Neural Computation, 1(12): 2385-2404.

The optimal utilization model of power resource based on peak-valley time-of-use tariff

L. Chen[1]
Beijing University of Agriculture, Changping District, Beijing, 102206, China
Email: 408511915@qq.com

H. Xin[2†], Q. K. Tan[3], G. Pan[4], S. K. Yu[5] and Z. F. Tan[6]
School of Economics and Management,
North China Electric Power University, Changping District, Beijing, 102206, China
Emails: [2]1943096760@qq.com, [3]314219858@qq.com,
Email: xinhe624@foxmail.com, [†]xinhe624@163.com

In order to optimize system power efficiency, reduce the average cost of power generation and both the supply and users be able to benefit from it, we implement the peak-valley time-of-use tariff. In this paper, based on the response of demand side, we apply semi trapezoid membership function to divide the peak, valley and flat periods of power load curve. Considering the problem of distribution of on-grid electricity in peak, flat, valley period, we should make sure that the supply and the users can both benefit from it. Aiming at cutting a peak to fill valley and minimizing the power load in peak period and the difference between peak and valley, we build the optimal utilization model. The calculation results show the feasibility and effectiveness of the optimization model.

Keywords: Peak-valley; Time-of-use tariff; Semi trapezoid membership function; Cutting a peak to fill valley; Cost savings.

1. Introdcution

The primary goal of power industry system reform of our country is " separation of plant and network, bidding ingeneration side", surf the net at a competitive price, electricity price, fluctuate in line with market conditions, operating cost free fluctuation based on the user side, and implement the system of peak-valley

---

[1] Li Chen
[2] He Xin
[3] Qing-Kun Tan
[4] Ge Pan
[5] Shun-Kun Yu
[6] Zhong-Fu Tan

time-of-use tariff. How to construct a reasonable time of peak-valley time-of-use tariff system, the use of demand side management means to guide users to adjust the power consumption patterns of behavior, improving the utilization rate of operation benefit and comprehensive system of resources, cut a peak to fill valley and reasonable pricing has become a key.

At present, the domestic research on the pricing system is mainly used the method of economics to obtain different electricity price model to reflect the supply or production cost. The literature on [1] is hourly bidding of power market are analyzed, discussed the short comings the bidding of power market based on time, put forward the concept of block bidding power market and the corresponding mechanism, put forward the method of block bidding power market and to block bidding based, and discussed the electric power market mechanism resource the role of literature configuration; [2-5] to resolve the difficulties in power industry marketization reform of our country, the electric power market is analyzed dynamically, based on generation side, through the analysis of the running rule of power plant, and the electric power market, bidding mode is suitable for the construction Chinese present situation of generation market, established a dynamic model the market as a whole; the literature of [6-7] according to the marginal cost pricing principle to commodity production, the basic rule of market economy, to consider the spatial and temporal characteristics and calculation method of electricity from the power company business operation angle, time of use electricity price mathematical model is established from the viewpoint of engineering; The literature of [8] based on the theory of marginal cost, considering the annual balance of payment in the grid, to retain the marginal cost pricing of economic information at the same time. Through the grid current and trace method, to determine the distribution of the active and reactive power price and fixed cost each between users, a new method to calculate the transmission price; [9-10] estimation of Internet Literature peak-valley time-of-use tariff with marginal cost method and cost allocation method, to determine the mathematical model of peak-valley time-of-use tariff and sales side peak valley electricity generation side time-sharing; literature [11-18] using price theory and economics theory describes the structure, pricing strategy of the electricity price of peak and valley time, based on the user side, put forward the concept of user reaction degree description the impact of price on load curve of the peak and valley, and proposes a method to determine the basic peak-valley time-of-use tariff and electricity price to cut a peak to fill valley, as the target of optimization is established based on the peak-valley time-of-use tariff model of DSM.

2. The Peak-Valley Time-of-Use Price Design Model

To be taken the electricity as a special commodity, the characteristic of balance between supply and demand always determines that electricity generation costs in the peak period are generally higher than the other periods, the less load demand in low valley period would lead to the low efficiency in units which also will increase the costs of electricity production per quantity. In order to optimize the system of power generation efficiency and to reduce the average cost of power generation, we should introduce some measures for load curve to shift peak, according to the load curve of the peak and valley, formulating the corresponding price in different period, guiding users to make a reasonable use of electricity, transferring the distribution of electricity load. The rationality of the design of the price in peak and valley time depends on the partition of valley and peak and formulation of electricity price. So we need to carry out a further research on it.

To divide peak and valley period in the load curve, we need to make an analysis by combining the peak point and the lowest valley in the load curve, which provide a certain basis for the division of peak and valley, specific classification principle is as follows:

A. The highest peak points in the load curve must belong to the peak period, possibility to 1;
B. The lowest valley points in the load curve must belong to the valley period, possibility to 1;
C. The other points in the load curve may belong to the peak period or the valley period. In order to analyze the possibility of other points in the load curve belonging to peak or valley period, we introduce half trapezoid membership function in this paper, and apply the partial small and large semi trapezoidal distribution function to simulate the possibility of each time point in valley period and peak time, the partial small and large semi trapezoidal distribution function is shown below:

Partial small semi trapezoidal distribution function

$$A(x)=\begin{cases}1, & x\le a\\ \dfrac{b-x}{b-a}, & a<x\le b\\ 0, & x>b\end{cases} \tag{1}$$

Partial large semi trapezoidal distribution function

$$A(x)=\begin{cases}0, & x \le a \\ \dfrac{x-a}{b-a}, & a < x \le b \\ 1, & x > b\end{cases} \tag{2}$$

Partial small and large semi trapezoidal distribution function are used to determine the comparability of each points in the load curve related with the highest and lowest load. We do not need to consider the specific load data, therefore, for convenience, we this article only select the linear part of the membership functions to analyze, the specific distribution function is shown in figure 1 and 2.

If dividing the peak and valley time of load curve by semi trapezoidal distribution function, we could sum up n kinds of schemes, each scheme marked as f_i, then the total set of divides of peak and valley is $F = \{ f_i, i = 1,2,\ldots,n\}$, we separately choose f_i as the optimization model to simulate, then comparing all the results, and choose the best scheme f_i^*, so peak-valley time-of-use tariff in this scheme is the relative actual price.

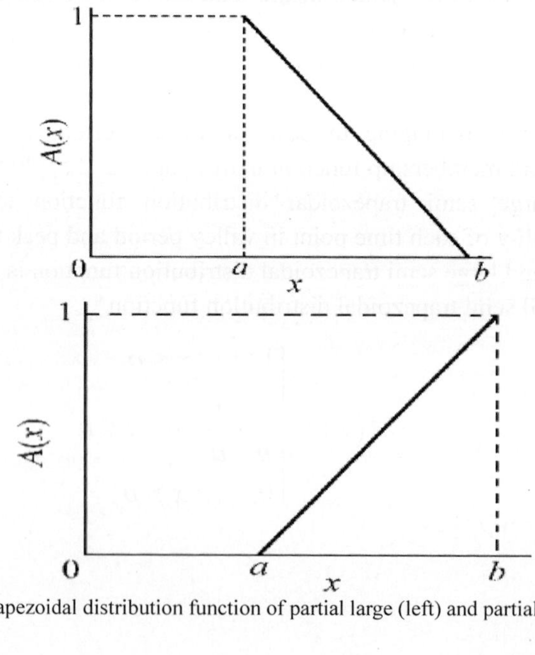

Fig.1. Semi trapezoidal distribution function of partial large (left) and partial small (right).

We should ensure that the supply side and the consumer side can make a benefit from the implementation of peak-valley time-of-use tariff. The supply side mainly is to reduce the cost of electricity production, the consumer side is to reduce electricity costs. Assuming that before implementing the scheme, the electricity price for the user is P_0, the amount for the user is Q_0, then we can get the cost before the implementation of peak-valley time-of-use tariff, is also the income of the supply side, i.e.:

$$R_0 = P_0 Q_0 \tag{3}$$

We assume that after the implementation of peak-valley time-of-use tariff, the users' electricity consumption is unchanged, a part of load in peak time transfer to valley period. The amount of electricity load in the same part changes in the same proportion. Assuming that after the implementation, the price in peak ,valley, usual time is P_p、P_v、P_f, the amount respectively is Q_p、Q_v、Q_f, then the income of the supply side is:

$$R_1 \quad P_p Q_p \quad P_v Q_v \quad P_f Q_f \tag{4}$$

The implementation of peak-valley time-of-use tariff can achieve cutting a peak to fill up the valley in load curve, reducing power investment. For the supply side, in order to obtain the benefits in peak-valley time-of-use tariff, the post income should be higher than the past income. The saving investment cost is C, i.e.:

$$R_1 \quad R_0 \quad C \tag{5}$$

The implementation of peak-valley time-of-use tariff can stimulate the user to transfer electricity usage periods: a decresing amount in peak time, a increasing amount in valley time. In general, there is an inverse relationship between the price and the consumption. So we assume that the variation of load curve in scheme f_1 is:

$$L_{f_1} \quad L_{f_1} \; P_p, P_f, P_v, t \tag{6}$$

Implementation of peak-valley time-of-use tariff is mainly to reduce the load in peak period, increase the load in valley time. So the goal of system optimization should consist of objective functions: one is the minimized load in peak time, the other is minimized load difference between peak and valley, as follows:

$$\min_{f_1 \in F} Q_1 f_1 = \min_{P_p, P_f, P_v} \left[\max_{0 \le t \le 24} L_{f_1} \left(P_p, P_f, P_v, t \right) \right] \tag{7}$$

$$\min_{f_1 \in F} Q_1 f_1 = \min_{P_p, P_f, P_v} \left[\max_{0 \le t \le 24} L_{f_1}\left(P_p, P_f, P_v, t\right) - \min_{0 \le t \le 24} L_{f_1}\left(P_p, P_f, P_v, t\right) \right] \tag{8}$$

We need to ensure that the supply side obtain excess profit by implementing peak-valley time-of-use tariff. The users' cost is lower than the past. The specific constraints are as follows:

The profits of the supply side:

$$R_1 \ge R_0 - C \tag{9}$$

The cost of the users:

$$R_1 \ge R_0 - C \tag{10}$$

The cost contrains:

The implementation of peak-valley time-of-use tariff should ensure that the price in valley period is higher than the marginal cost of power generation, but not higher than the past average electricity price, the price in peak period should be higher than the average one, but not infinitely high, need control in a reasonable ratio range, as follows:

$$P_e \le P_v \le P_0 \tag{11}$$

$$P_0 \le P_p \le P_0 \ 1 \tag{12}$$

Where, P_e represents the past marginal cost; represents the post price rise in peak period.

3. The Optimized Result of Peak-Valley Time-of-Use Tariff

In order to simulate the optimization model of peak-valley time-of-use tariff, we take typical daily electric load data in a certain area as the basis, assuming that the average electricity price is 425 yuan / MW·h before implementation, the marginal cost of power generation in the valley periods is 120 yuan / MW·h, the most high and the minimum power load are respectively 1.4938GW and 0.9427GW. After implementing peak-valley time-of-use tariff, we assume that the price in usual time is in line with the past average price. According to the model above, we can get the most optimized division of peak and valley and scheme of price. The specific period division is shown in Table 1:

Table 1. TOU optimization program.

	peak	flat	valley
period	9:00-11:0017:00-21:00	7:00-90;11:00-17:00	21:00-24:00;0:00-7:00
price(yuan/MW·h)	664.4	425	140

According to the optimization scheme, we can get the load curve changes from the original to the final and dividing the peak and valley part in load curve, then we can obtain load period distribution in different schemes, the specific is shown in Table 2.

It is seen from Table 2, we are able to get a smooth load curve by adopting peak-valley time-of-use tariff, descending the highest load in peak period, ascending load in valley period, reducing the proportion. It is helpful to provide good spare space for random electric power grid.

Table 2.Load optimization results of TOU system.

		Division result	
	Original load	Original division	Optimal division
Maximum load/GW	1.4938	1.3673	1.3414
Minimum load/GW	0.9427	1.0876	1.0418
Difference between peak and valley/GW	0.5251	0.2789	0.2996
Ratio in peak and valley/%	46.13	12	34.74

4. Conclusion

In this paper, we apply semi trapezoid membership function to divide the peak and valley periods of power load curve, aiming at cutting a peak to fill valley and minimizing the power load in peak period and the difference between peak and valley, we build the optimal utilization model. We get the more feasible price and ensure that both the supply and demand benefit from it. The conclusion is as follows:

1) This paper constructs optimization model of electricity price of peak valley time-sharing user response curve based on the research on the application of, not only peak, flat, valley period; also applies to the study on each time the use response curve, can obtain the more accurate load, it can lay a foundation for multi period time of use electricity price.

2) The use of fuzzy optimization to solve the model of peak-valley time-of-use tariff, it can not only avoid the structural evaluation function; and can reflect all optimal solution of single objective and multi-objective satisfactory solution of the relationship between, considering all the objective conditions, seeking a suitable optimization scheme, so that all targets are possible in the better condition, in order to obtain more objective peak-valley time-of-use tariff.

3) The implementation of peak valley time of use electricity price is not only advantageous measures of demand side management, but also the social energy conservation and optimization of the important measures. Therefore,

the government through tax cuts incentive power supply, in order to promote the smooth implementation of the measures.

References

1. Wang Xifan. Block bidding model based power market [J]. Proceedings of the CSEE, 2001, 21(12):1-6 (in Chinese).
2. Yao Jiangang, Fu Weisheng, Chen Qingqi, et al. Research about pricing, energy sources and environmental protection for electricity market [J]. Proceedings of the CSEE, 2000, 20(5):71-75 (in Chinese).
3. Tang Zhenfei, Tang Guoqing, Yu Erkeng, et al. Power market dynamic analysis [J]. Proceedings of the CSEE, 2001, 21(2):88-92 (in Chinese).
4. Ding Junwei, Xia Qing, Hu Yang, et al. A new model for gradually establishing competitive generation market [J]. Proceedings of the CSEE, 2003, 23(3):10-15 (in Chinese).
5. Yao Jiangang, Gao Jie, Li Xiquan, et al. Research on bidding model in a generation-side power market [J]. Proceedings of the CSEE, 2004, 24(5):79-83 (in Chinese).
6. Ma Jinming, Qin Leiming, Zhu Zhenqing. Studies on spot sale pricing for regional energy utility [J]. Automation of Electric Power System, 2002, (3):9-12 (in Chinese).
7. Huo Yisong, Jin Yimin. Research of the Time-of-use Pricing Model in Electric Power Market [J]. Jiangsu Electrical Engineering, 2003, 22(2):17-19 (in Chinese).
8. Ren Zhen, Wu Guoyue, Huang Wenying. A new method to transmission pricing in electricity market [J]. Proceedings of the CSEE, 2003, 23(1):37-40 (in Chinese).
9. Zeng Ming, Liu Ming, Zhao Qingbo, et al. Theory and application of change of mutual relation between selling side and generation side use time pricing [J]. Electric Power, 2003, 16(9):70-74 (in Chinese).
10. Zhang Wei. The Pricing Method of Power Plant Based on Load and Season [J]. Central China Electric Power, 2004, 17(2):42-44 (in Chinese).
11. Li Yang, Wang Zhihua, Lu Yi, et al. The implementation of peak and valley time price for electricity and the response of large industries [J]. Automation of Electric Power System, 2001, 25(8):45-58 (in Chinese).
12. Li Hui, Kang Chongqing, Xia Qing. Price based decision making for demand side management considering customer satisfaction index [J]. Power System Technology, 2004, 28(23):1-6 (in Chinese).

Research on boost converter with linear outer loop controller and nonlinear MIT MRAC inner loop controller

P. Mao[†], W.J. Wang, Y.H. Zhang, L. Zhou, M. Zhang and J. Cai
C-MEIC & CICAEET,
Nanjing University of Information Science & Technology, China
[†]*Email: maomao9724@163.com*

H.F. Xiao
Southeast University, China

To improve overall performance, average model of the boost converter was built and the principle of the MIT model reference adaptive control (MRAC) was analyzed firstly, based on this, a new control structure of linear outer loop with PI controller and nonlinear inner loop with MIT MRAC controller was proposed. Design method of 2nd-order reference model was proposed, and other key problems in design of the control system were discussed also. All conclusions and the proposed control scheme were verified by simulation results finally.

Keywords: MIT MRAC; Two loop control; Boost converter; Reference model.

1. Introduction

Boost converter is widely used in industry field [1], and control scheme greatly effects overall performance of the converter. Up to now, linear proportional-integral (PI) controller is popular for this kind of converter [2]. Parameter design of PI controller is relatively easy, however, this traditional controller also has its own limitations, e.g. the system performance will change with the fluctuation of circuit parameters because controller parameters are fixed usually.

Adaptive control is advanced nonlinear control method [3][4], which can make sure the system has expected performance by adjusting controller parameters dynamically with adaptive law even though the system parameters change significantly. In view of the good control performance, adaptive control has been used successfully in space control, motor control and so on [5].

MIT MRAC belongs to adaptive control method, and it is also one of relatively early MRAC methods. With adaptive law, the system can achieve same performance with reference model [6]. Although MIT MRAC cannot make sure stability of the system, it is still a practical adaptive control method for its simplicity and convenience of algorithm implementation, specially, not like other MRAC methods, MIT MRAC is also suitable for non minimum phase system boost converter belongs to.

Based on above consideration, a new two-loop control scheme with linear PI voltage controller and nonlinear MIT MRAC current controller is proposed. Design method of reference model is analyzed, all conclusions and design method are verified through simulation results.

2. Modeling of Boost Converter

2.1. *Time domain model*

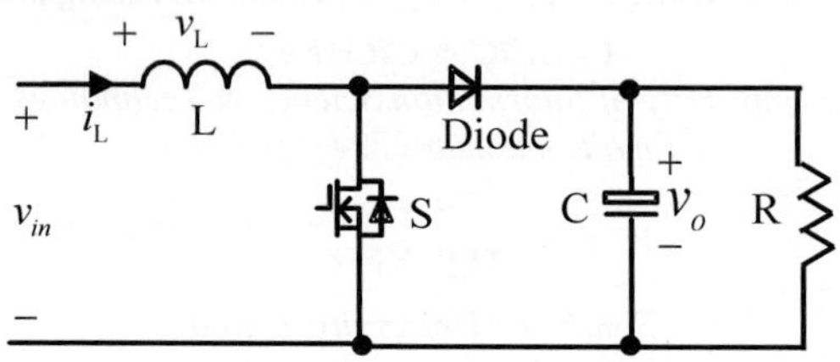

Fig. 1. Topology of boost converter.

Topology of boost converter is shown in Figure 1, v_{in}, v_o are input and output voltage, *L*, *C* are input and output side filter separately. Suppose switching frequency f_s is high enough, the large signal average model [2] of the converter can be yield as following when the inductor current is continuous

$$\begin{cases} L\dfrac{d(t)\langle i_L(t)\rangle_{T_s}}{dt} = \langle v_{in}(t)\rangle_{T_s} - d'(t)\langle v_o(t)\rangle_{T_s} \\ C\dfrac{d(t)\langle v_o(t)\rangle_{T_s}}{dt} = d'(t)\langle i_L(t)\rangle_{T_s} - \dfrac{\langle v_o(t)\rangle_{T_s}}{R} \end{cases} \tag{1}$$

Switching period $T_s=1/f_s$, $d(t)$, $d'(t)$ are conduction and close-off duty cycle separately, $\langle i_L(t)\rangle_{Ts}$ is the average inductor current of a switching period, other expressions are defined similarly. Equation (2) can be got when small-signal perturbation is given at the moment of *t*.

$$\begin{cases} \langle v_{in}(t)\rangle_{T_s} = V_{in} + \hat{v}_{in}(t) \\ \langle v_o(t)\rangle_{T_s} = V_o + \hat{v}_o(t) \\ \langle i_L(t)\rangle_{T_s} = I_L + \hat{i}_L(t) \\ d(t) = D + \hat{d}(t) \end{cases} \tag{2}$$

Combined with steady state functions between input and output voltage, time domain model of the system can be deduced as (3) when the product term of small-signal is neglected.

$$\begin{cases} L\dfrac{d\hat{i}_L(t)}{dt} = \left[\hat{v}_{in}(t) - D'\hat{v}_o(t) + V_o\hat{d}(t)\right] \\ C\dfrac{d\hat{v}_o(t)}{dt} = \left[D'\hat{i}_L(t) - \dfrac{\hat{v}_o(t)}{R} - I_L\hat{d}(t)\right] \end{cases} \tag{3}$$

V_{in}, V_o, I_L and D are steady state value at the moment of t。

2.2. *Frequency domain model*

Frequency domain model can be obtained as shown in (4) and (5) with laplace transformation of (3).

$$sL\hat{i}_L(s) = -D'\hat{v}_o(s) + \hat{v}_{in}(s) + \hat{v}_{in} \bullet \hat{d}(s) / D' \tag{4}$$

$$sC\hat{v}_o(s) = D'\hat{i}_L(s) - \hat{v}_o(s)/R - V_g \bullet \hat{d}(s) / D'^2 \bullet R \tag{5}$$

From (4) and (5)

$$sL\hat{i}_L(s) = \frac{-D'^2 R}{sCR+1}\hat{i}_L(s) + \frac{V_{in}RD'}{D'^2R(sCR+1)}\hat{d}(s) + \hat{v}_{in}(s) + \frac{V_{in}}{D'}\hat{d}(s) \tag{6}$$

Form (6), transfer function of control to inductor current can be obtained as following when v_{in}(t) is set to zero.

$$\left.\frac{\hat{i}_L(s)}{\hat{d}(s)}\right|_{\hat{v}_{in}(s)=0} = \frac{V_{in}CRs + 2V_{in}}{D'LCRs^2 + D'Ls + D'^3R} \tag{7}$$

(7) is the key transfer function which will be used to control the inductor current with the two-loop control scheme.

3. Design of MRAC Controller

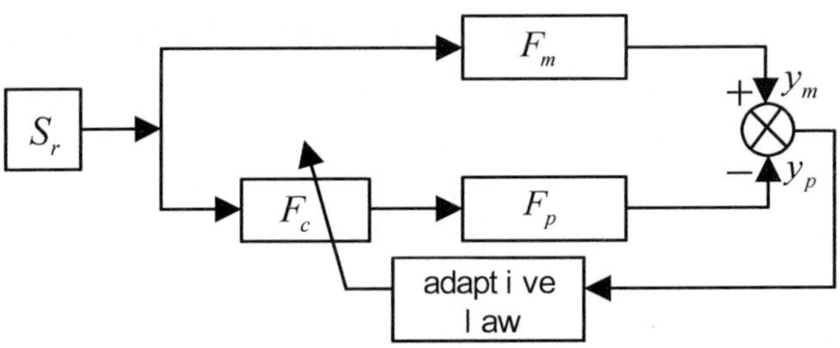

Fig. 2. Control diagram of MRAC.

Parallel type MRAC system diagram is shown in Figure 2, where S_r is reference signal, F_p is transfer function of the plant, that is transfer function from the output of controller to the output of the system, F_m is reference model, y_m and y_p are ideal output and real output separately. The object of MRAC is to let $F_c \bullet F_p$ approach F_m by adjusting controller parameters dynamically according to the adaptive law. From figure 2, it can be seen that MRAC scheme is still based on output error, which is same with traditional linear control method.

3.1. *Principle of MIT MRAC and its parameter design method*

As shown in Figure 2, reference model and plant model are F_m and F_p separately, generalized error $\varepsilon(t)$ is defined as (8)

$$\varepsilon(t) = y_m(t) - y_p(t) \tag{8}$$

Performance function J is defined as (9)

$$J = \frac{1}{2}\int_{t_0}^{t} \varepsilon^2(\tau)d\tau \tag{9}$$

Suppose controller transfer function is k_c

$$F_c = k_c \tag{10}$$

Adaptive law of k_c can be derived with gradient method, which can make sure that J become minimum and

$$\lim_{t\to\infty} \varepsilon(t) = 0 \tag{11}$$

The above design principle proves that MRAC controller can make sure J as small as possible, instead of realizing $F_C \bullet F_p = F_m$. According to gradient method, expression that can make sure that J decrease gradually is

$$k_c(t) = k_{c0}(0) - \lambda \int_{t0}^{t} \varepsilon(t)\frac{\partial \varepsilon(t)}{\partial k_c}dt \tag{12}$$

k_{c0} and λ are initial value and parameter adjustment step separately, and λ is usually a small value. Take derivation of (12), can we get

$$\dot{k}_c(t) = -\lambda \varepsilon(t)\frac{\partial \varepsilon(t)}{\partial k_c} \tag{13}$$

(13) shows adjustment law of k_c. Because y_m and k_c are independent, the following expression holds

$$\frac{\partial \varepsilon(t)}{\partial k_c} = \frac{\partial y_m}{\partial k_c} - \frac{\partial y_p}{\partial k_c} = -\frac{\partial y_p}{\partial k_c} \tag{14}$$

Substituting (14) into (13), can we get

$$\dot{k}_c(t) = -\lambda \varepsilon(t)\frac{\partial y_p}{\partial k_c} \tag{15}$$

Combined with the above analysis, (16) can be derived

$$\dot{k}_c(t) = \lambda k_p / k_m \, \varepsilon(t) y_m(t) = \mu \varepsilon(t) y_m(t) \tag{16}$$

k_p, k_m are gain parameter of plant and reference model separately, μ is the final parameter adjustment step. k_p is usually unknown, and μ can be set as a small constant in real control. From (16), the final adjustment law of k_c can be drawn as in (17).

$$k_c(t) = k_c(0) + \mu\int_0^t \varepsilon(t) y_m(t) dt \tag{17}$$

Diagram of MIT MRAC is shown in Figure 3, $k_c(0)$ is the initial value of k_c. It can be seen from Figure 3 that reference signal, output of reference model and the real output are all used to adjust controller parameter, which proves that MIT MRAC also belongs to error based control scheme.

3.2. *Design of reference model*

Reference model has important effect on system performance. From Figure 3, it can be seen that the system dynamic and static performance can be changed by the adjustment of controller parameter, to get small $\varepsilon(t)$ during the transient process, F_m and F_p should keep the same order, which means F_m had better been set to two-order system, same with F_p shown in (7). Parameter design method of two-order system will be discussed in the following paragraph.

The standard two-order system is shown in (18), damping coefficient ξ and natural frequency ω_n are to determined. For this model, the DC steady state gain is 1, and steady state error e_{ss} is 0.

$$G(s) = \frac{\omega_n{}^2}{s^2 + 2\xi\omega_n s + \omega_n{}^2} \tag{18}$$

For the above classical two-order system shown in (18), overshot σ, rising time t_r and regulate time t_s with step up input are the key property indexes, which can also be used as design basis. The parameter design step is summarized as following.

1) Substituting given ξ into (19), we obtain σ.

$$\sigma = e^{-\pi\xi/\sqrt{1-\xi^2}} \times 100\% \tag{19}$$

2) By substituting the given rising time t_r and ξ calculated by (1) into (20), we can obtain the natural frequency ω_n

$$t_r = \frac{\pi - \arccos\xi}{\omega_n\sqrt{1-\xi^2}} \tag{20}$$

3) Substitute ξ and ω_n into (21), and confirm whether regulate time t_s satisfy demand, where Δ is width of permitted error band.

$$t_s = \frac{1}{\xi\omega_n}\left(\ln\frac{1}{\Delta} + \ln\frac{1}{\sqrt{1-\xi^2}}\right) \tag{21}$$

According to design method proposed above, if ξ is selected as 0.6877, σ will be 5.1% calculated by (19),which can meet for most of demand in real applications. Let t_r be 0.5s, and substitute the calculated ξ into (20), can we obtain the natural frequency ω_n 6.42rad/s. Substitute ξ and ω_n into (21), and get ts 0.959s when Δ is 0.02. The final reference model is shown as (22).

$$G_m(s) = \frac{41.2164}{s^2 + 8.83s + 41.2164} \tag{22}$$

3.3. *Novel two loop controlled boost converter*

Two-loop control scheme with inner current loop and outer voltage loop are widely used in converter control field, and linear PI controller is popular. However, linear controller also has its own limitation. Based on the above consideration, a new control scheme with outer linear PI voltage controller and inner nonlinear MRAC controller are proposed. Diagram of the new control scheme used on boost converter are shown in Figure 3.

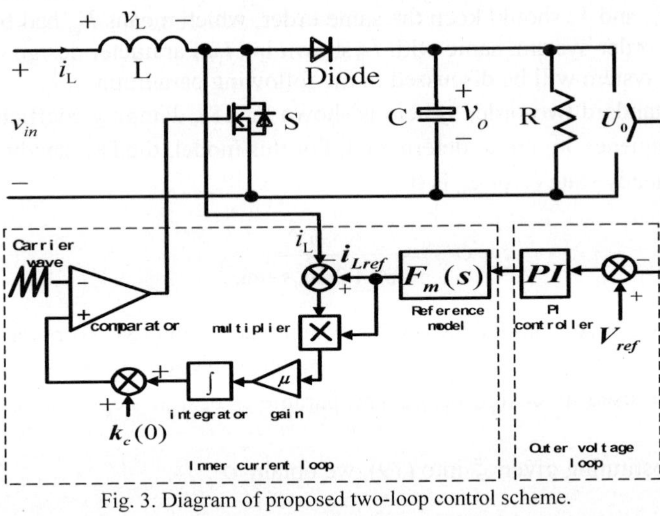

Fig. 3. Diagram of proposed two-loop control scheme.

System configuration of Figure 3 is listed as follows: input voltage v_{in} is 200V, reference output voltage v_{ref} is 400V, input side filter inductance L is 1mH, output side filter capacitor C is 2000uF, load resistor R is 10Ω, switching frequency f_s is 20kHz.

Substitute the above system parameters into (7), and the plant transfer function can be yield as (23)

$$F_p(s) = \frac{4s + 400}{0.00001s^2 + 0.0005s + 1.25} \tag{23}$$

4. Simulation Results

To prove the feasibility of parameter design method for two-order system, and verify the proposed two-loop control scheme for boost converter, simulation was made with the software of matlab / simulink.

4.1. *Verify the two-order system parameter design method*

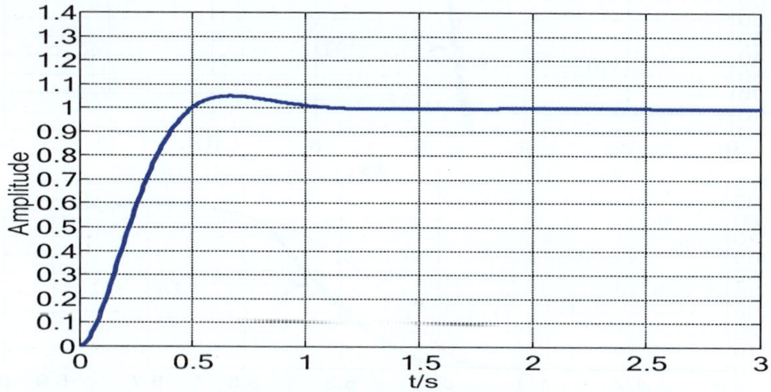

Fig. 4. Step response of the two-order system shown in (22).

To verify the parameter design method proposed in 3.2, step response of the two-order system shown in (22) is given in Figure 4. Simulation result shows that σ is about 5% t_r is 0.5s, and t_s is about 0.7s. The above results are basically same with data given in the last paragraph of 3.2, which verify the proposed parameter design method.

4.2. *Verify the proposed two-loop control scheme*

Simulation results will be given in the following sections to verify the MRAC based voltage controller and the two-loop control scheme. Simulations include two parts, (1) frequency domain simulation to verify performance of MRAC based voltage controller; (2) time domain simulation to verify closed-loop performance of the system shown in Figure 4. System configuration is given in 3.3, and the reference model of inner current loop is shown in (22).

4.2.1. *Frequency domain simulation*

With the control scheme shown in Figure 3, where the plant model of F_s is shown in (7) and the reference model F_m is shown in (22), frequency domain stimulation is made with step input signal of 80A, and the output of reference model and plant is shown in figure 5, the corresponding partial enlarged detail is also given. What needs to be explained is that, the input current at steady state is 80A, which can be calculated with the system configuration in 3.3.

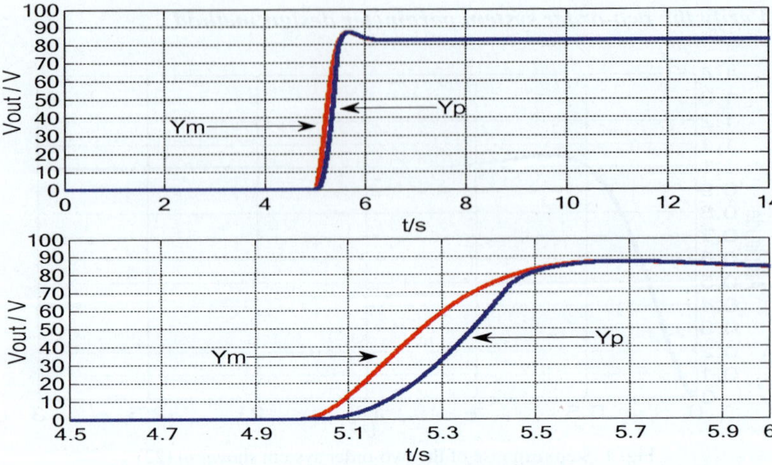

Fig. 5. Step response of inner current loop.

Simulation waveforms show the output current can well follow the reference current with MRAC controller.

4.2.2. *Time domain simulation*

Time domain simulation is made with the control scheme shown in Figure 3 to verify the stability of two-loop system, and the simulation conditions include steady load, sudden increase and decrease of load.

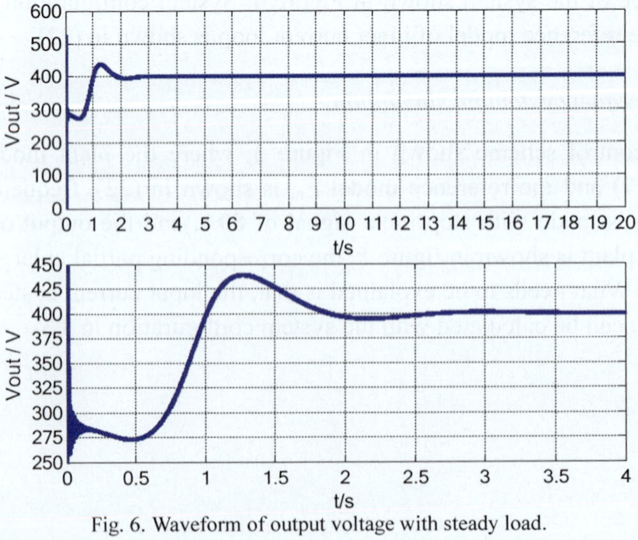

Fig. 6. Waveform of output voltage with steady load.

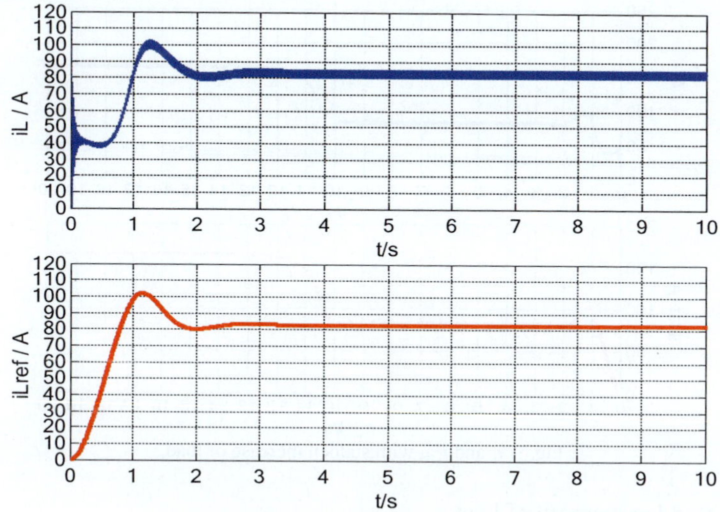

Fig. 7. Waveform of i_L and i_{Lref} with steady load.

(1) Steady load

The second waveform in Figure 6 is partial enlarged detail of v_{out}. In Figure 7, i_{Lref} is output of reference model. Simulation results show that the outer loop has good performance and i_L can well follow i_{Lref}, which verify the two-loop control scheme and the current controller based on MRAC.

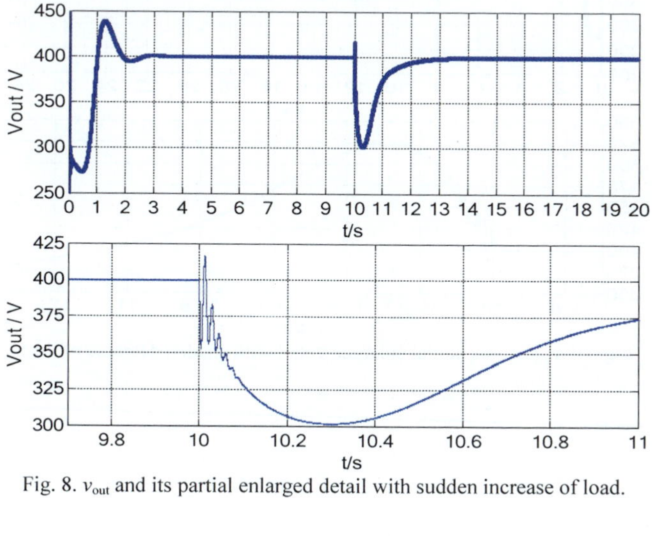

Fig. 8. v_{out} and its partial enlarged detail with sudden increase of load.

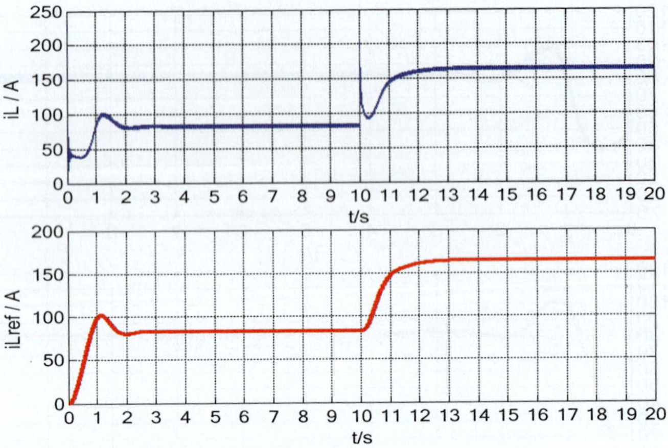

Fig. 9. i_L and i_{Lref} with sudden increase of load.

(2) Sudden increase of load

Figure 8 and 9 show v_{out} , i_L and i_{Lref} separately when the load resistance R decreases from 10Ω to 5Ω when t=10s. Simulation results show v_{out} keeps 400V again after transient process of 3s, and i_L can well follow i_{Lref} also.

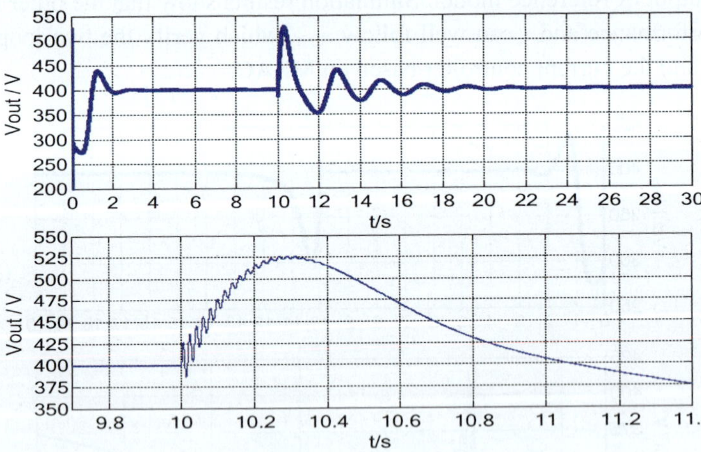

Fig. 10. v_{out} and its partial enlarged detail with sudden decrease of load.

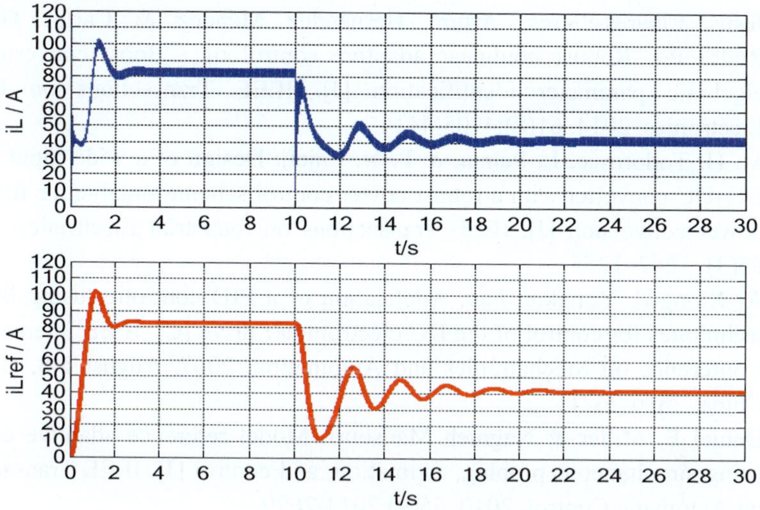

Fig. 11. i_L and i_{Lref} with sudden decrease of load.

(3) Sudden decrease of load

Figure 10 and 11 give simulation results when load resistance R increases from 10Ω to 20Ω when t=10s. Simulation results show v_{out} comes to 400V again after transient process of 9s.

5. Conclusion

A novel two-loop control scheme with linear and nonlinear controller is proposed. All conclusions and parameter design method are verified through simulation results. The control scheme proposed in this paper can also be used for other kind of converters.

Acknowledge

This research is sponsored by "National Natural Science Foundation of China (No. 51407203)" and "Laboratory Opening Project of C_MIEC (No. KCMEIC01)".

Reference

1. B. Singh, B.N. Singh. A review of single-phase improved power quality AC-DC converters [J]. IEEE Transactions on Industrial Electronics, 2004, 51(3): 641-660.
2. Robert W. Erickson & Dragan Maksimovic. Fundamentals of power electronics- 2nd edition. Springer, 2001.

3. Jesús Linares-Flores, Arturo Hernández Méndez & Carlos García-Rodríguez. Robust nonlinear adaptive control of a Boost converter via algebraic parameter identification [J]. IEEE Transactions on Power Electronics, 2014, 61(8):4105-4114.
4. M. H. Todorovic, L. Palma, & P. N. Enjeti, Design of a wide input range DC–DC converter with a robust power control scheme suitable for fuel cell power conversion [J], IEEE Transactions on Industrial Electronics, 2009, 55(3): 1247–1255.
5. Ai Xiong & Yongkun Fan. Application of a PID controller using MRAC techniques for control of the DC electromotor drive [C], IEEE International Conference on Mechatronics and Automation, 2007, August 5-8, Harbin, China: 2616-2621.
6. Daniel E. Miller & Nagmeh Mansouri. Model reference adaptive control using simultaneous probing, estimation, and control [J], IEEE Transactions on Automatic Control, 2010, 55(9):2014-2029.

Design and realization of a new power quality monitoring device

Z. Jiang[1†], Y. Zhou[2] and X. M. Chen[3]
School of Electrical Engineering, Zhengzhou University, Zhengzhou, Henan, 450001, China
†Email: jzqzcg@163.com

To improve accuracy and speed of a power quality monitoring device, the monitoring device with field-programmable gate array (FPGA) as the core of processing and controlling is designed. The hardware of the device includes signal acquisition unit, communication unit, power supply unit, etc; the software includes an NIOS II embedded core, optimizing the flow of data processed parallel, and Phase-locked frequency multiplication module and FFT harmonic analysis module implementing the synchronous sampling and analysis of power quality data are built. Test results show that the device has advantages of fast response, high accuracy and satisfactory real-time performance, and conforms to the national standards of power quality monitoring.

Keywords: Power quality; Field-programmable gate array (FPGA); Fast fourier transform (FFT).

1. Introduction

In recent years, increasing asymmetric, impact and nonlinear loads in the power grid have affected power quality to varying degrees, and aroused concern from both power supply enterprises and sensitive users [1-3]. Real-time continuous power quality monitoring is the prerequisite for both the identification and improvement of power quality problems, with high requirements for accuracy and speed of the device in order to guarantee the reliability of test results. Early designs based on acquisition card + industrial personal computer (IPC) have advantages of comprehensive functions and friendly user interface, but are large and expensive and thus not suitable for large-scale on-site application. New designs of DSP + ARM combine the advantages of both acquisition card and IPC and are capable of complex calculation, but complex structure limits their extendability and upgradability [4-5].

---

[1] Zheng Jiang
[2] Yong Zhou
[3] Xue-Mei Chen

As a semi-custom type of ASIC, field-programmable gate array (FPGA) not only improves the disadvantages of custom circuits but also enables more gates compated to existing programmable devices [6]. Application of FPGA in signal processing can effectively solve the problem of parallelism and speed, meeting the high requirements of operational capacity and real-time performance with relatively low cost, good modifiability, testability and upgradability. In this paper, such a type of FPGA-based power quality monitoring device will be introduced, and test results will be given.

2. Hardware Frame

The device adopts a modular structure composed of a signal acquisition unit, a core processing unit, a storage unit, a communication unit and a human-computer interaction unit, as shown in Figure 1.

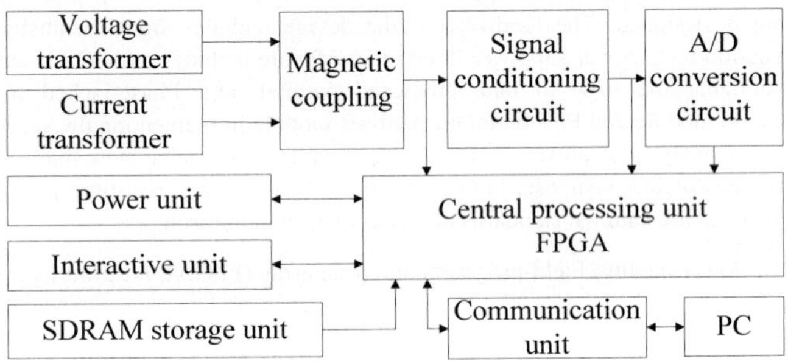

Fig. 1. Diagram of the device.

The workflow of the device is:

Voltage and current signals are acquired by transformers and transmitted through the magnetic coupling isolator to the signal conditioning circuit, where they are conditioned to meet the input requirements and then inputted into the A/D conversion circuit. The FPGA unit analyses the data acquired, calculating various power quality indexes. Results are real-time displayed on the man-machine interface and written into the storage unit to be packaged and transmitted to the host computer. If the results show the existence of a fault, related information will be transmitted to the host computer as well, and alarm will be given.

2.1. *Signal acquisition unit*

In order to guarantee the reliability of the device, phase voltage and phase current are sampled and converted respectively by voltage transformer (PT) and current transformer (CT), which are required to have uniform frequency response characteristics between 50Hz and 2500Hz to ensure accuracy. High-precision modular electromagnetic isolation sensors developed by Sichuan Mianyang Weibo Electronics Co. Ltd are adopted, part of the circuit is shown in Figure 2.

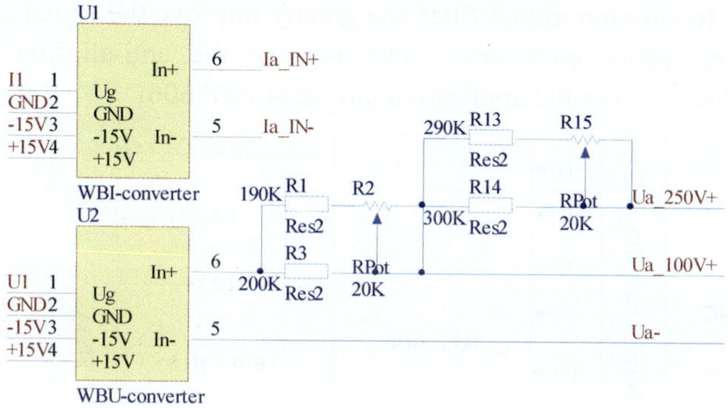

Fig. 2. Voltage and current transformer application circuit diagram.

In order to adapt to the measurement of both PT secondary side voltage (such as substations) and mains voltage, two different input voltage ranges are set: 100V and 250V. External resistors are connected in series on a proportion of 1kΩ / V, and measurement is calibrated by adjusting the potentiometers. Take the link of R1, R2 and R3 as an example:

When R2 is adjusted to the left end, in-connection resistance of the Ua_100V + measurement port is:

$$\frac{R_1 R_3}{R_1 + R_3} = \frac{190 \times 200}{190 + 200} = 97.44(\Omega) \tag{1}$$

When R2 is adjusted to the right end, in-connection resistance of the Ua_100V + port is:

$$\frac{(R_1 + R_2) R_3}{R_1 + R_2 + R_3} = \frac{(190 + 20) \times 200}{190 + 20 + 200} = 102.44(\Omega) \tag{2}$$

It can be seen that adjustment range of the resistors in series connection with the Ua_100V+ port is -2.56%~2.44%. With 1% resistors selected, voltage sensors can be calibrated by adjusting potentiometers, according to the circuit design and parameters given above. The tracking voltage output range of voltage and current sensors is AC 0V~5V.

The 16-bit, 8-channel, simultaneous sampling, bipolar input, single power supply A/D converter AD7606 developed by ADI is adopted. The chip integrates analog input clamp protection circuit, second-order anti-aliasing filter, track-and-hold amplifier and low-noise high-impedance input signal conditioning amplifier, and the maximum frequency of the input signal is 22KHz [7]. Its on-chip digital filter can greatly improve the signal to noise ratio performance, reduce code error, and improve the anti-aliasing suppression ability. Figure 3 shows the application circuit of AD7606.

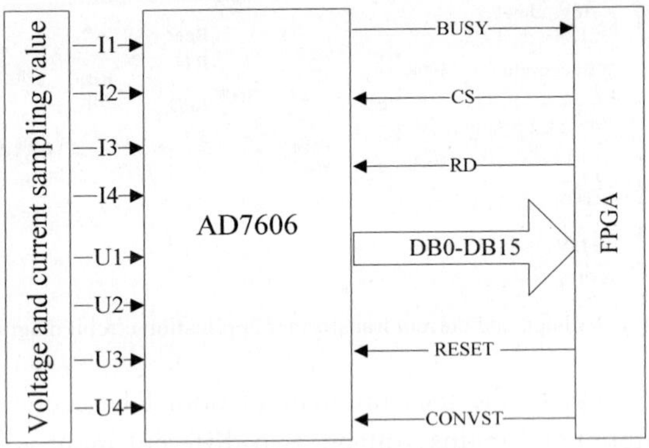

Fig. 3. AD7606 wiring diagram.

CONVST_A and CONVST_B are sampling trigger pins of AD7606, which can control four channels respectively. Both pins are connected together so that sampling of 8 channels can be triggered simultaneously to ensure accuracy.

All data ports of AD7606 (such as the data reading control pin $\overline{RD}$ and the chip-selecting pin $\overline{CS}$) are connected to the I/O ports of FPGA chip, which facilitates data reading operation.

2.2. *Core processing unit*

The core processing unit consists of the frequency synchronization circuit (including a zero-cross detection circuit, a phase-selection circuit and a phase-lock frequency multiplier circuit), NIOS II core module, data calculation module and bidirectional memory integrated in the FPGA.

The device adopts Altera's EP4CE6E22C8N, which has the following characteristics: (1) low consumption dual power system with an I/O voltage of 3.3V and a core operating voltage of 1.2V; (2) ample system resources including 6272 logic units (LES), 270 000 bytes of memory, 2 PLLs and 15 18*18 embedded multipliers; (3) high level of integration allowing one or more 32-bit NIOS II soft core processors or IP core controllers such as SRAM and FLASH to be built in; (4) easy migration of application programs (merely changing the allocation of the pins) without modification of underlying drivers.

Phase locked frequency multiplier circuit is used to reduce the impact of period measurement errors and the picket fence effect with FFT algorithm caused by asynchronous sampling, and improve sampling accuracy by tracking the frequency changes of the signal and adjusting sampling frequency in real time. The sampling rate of the device is set to $f_s = 5kHz$, the calculation uses data from 10 cycles (time interval T=200ms) at a time. The number of sampling channels is $N = 8$, including 3 channels for voltage signals, 3 for current signals and 2 as alternate channels, and the total number of data calculated at a time is:

$$M = N * f * T = 8000 \tag{3}$$

NIOS II core module is used to store the results of calculation and coordinate the interaction control of various units of the device. Date calculation module adopts radix-2 decimation-in-time FFT algorithm to calculate various power quality indexes such as the RMS of voltage and current, active and reactive power, voltage deviation, three-phase imbalance and harmonic distortion, the results are displayed in real time and stored for transmission.

2.3. *Communication unit*

The device has a USB port, through which data can be stored in flash disks or transmitted to the host computer for further processing and analysis. The device adopts CY7C68013A developed by CYPRESS, which integrates USB 2.0 transceiver, enhanced 8051 microcontroller, serial interface engine (SIE) and programmable peripheral interface and has 4 USB transmission modes (interrupt transmission, controlled transmission, synchronous transmission and mass transmission) at 2 transmission rates (12Mbps and 480Mbps). Figure 4 shows the USB communication module.

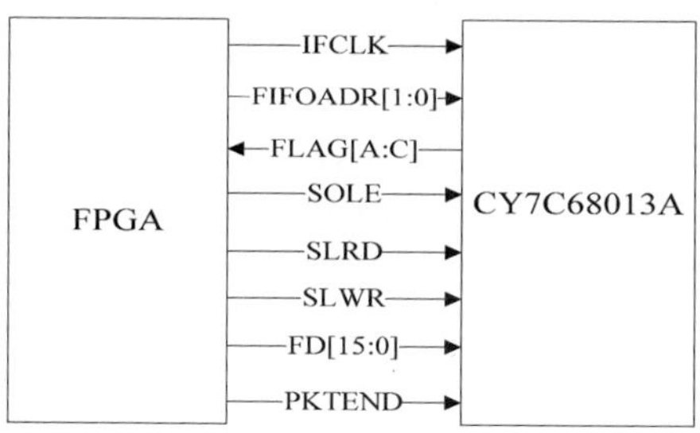

Fig. 4. USB communication module.

The USB interface chip adopts synchronous Slave FIFO mode, and CY7C68013A chip is connected to FPGA through the 16-bit bidirectional data bus (FD[15: 0]). FPGA provides a 25MHz external clock signal (IFCLK), and the maximum transmission rate reaches 200Mbps. FIFOADR[1: 0] is used to select the FIFO endpoint buffer, FLAGA, FLAGB and FLAGC are used to show the state of FIFO, SLOE can enable FD output, SLRD and SLWR are respectively used as reading and writing strobes of FIFO.

MAX232 is adopted as the RS232 module of the device, which is used for TTL level conversion at 115.2 kbps, as shown in Figure 5.

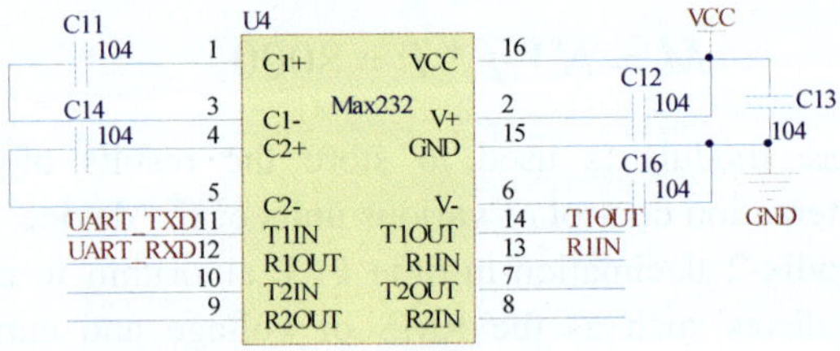

Fig. 5. RS232 module.

SIM300 developed by SIMcom is adopted as the GPRS module of the device, which works in E-GSM900 mode at 3 bands. The DTR pin of SIM300 is set to be at high level, the module is in sleep mode when without transmission requests or hardware interrupts and aroused when a message is received. In this way, power consumption of the device is reduced and the service life is prolonged.

3. Software Design

3.1. *Terminal software design*

The comprehensive hardware and software solution based on NIOS II soft core developed by Altera has brought new concepts of embedded system design [8]. As a IP core of the processor, NIOS II can be embedded into FPGA chip with very few logical units used, thus the development cost is quite low. Like other processor systems, NIOS II has its own command system, register group, cache and interrupt handling system; besides, it also has a custom instruction system and a hardware accelerator. The development of NIOS II core is based on HAL (hardware abstraction layer), an interface of communication program library for underlying hardware that provides API functions closely integrated with C Language.

Program of FPGA includes a phase-lock frequency multiplier, a bidirectional storage, a zero-cross detection module and a power quality index program. Workflow of the program is shown in Figure 6.

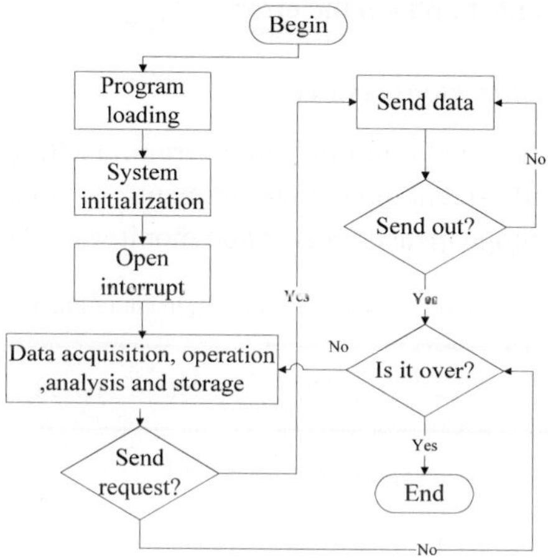

Fig. 6. The main program block diagram.

When the device is powered on, the FPGA program is loaded and the operating environment is initialized, then under the control of NIOS II the A/D conversion circuit collects data. Based on the data acquired, FFT calculation and calculation of various power quality indexes are conducted, with results stored and sent upon request.

3.2. *FFT algorithm design*

Fast Fourier Transform (FFT) is the key to harmonic analysis [9]. With MCU and DSP, FFT algorithm are both coded through software programming; though DSP accelerates the calculation with its MAC unit, such type of algorithm coding is essentially serial operation and the algorithm complexity is $O(n\log n)$. With FPGA, FFT calculation is conducted by parallel multiply-add units and merely requires $\log n$ levels of multiplication and addition, and thus the operation is accelerated by n times. Take 8-point FFT calculation as an example: DSP or MCU requires 12 multiply/add operations, while FPGA parallel FFT circuit only needs 3 levels of multiply-add operations, and significantly accelerates the calculation of power quality data.

FFT algorithm coded in Verilog HDL in FPGA is capable of fast harmonic analysis, providing results of amplitude and phase. Parallel hardware structure and parallel calculation process fully utilize the logic resources of FPGA, overcoming the disadvantages of serial operation algorithm.

4. Test Results and Analysis

Three-phase AC voltage signals are generated by JCD4060 power supply, results given by the device are compared with those of the theoretical calculation or standard instruments to obtain the errors.

4.1. *Frequency measurement results*

Table 1 shows the test data and absolute errors. Tektronix TDS2012C digital oscilloscope is used to compare measurements of frequency, f_1 and f_2 are respectively oscilloscope measurements and monitoring device measurements.

Table 1. Frequency measurement test table.

f_1 /Hz	49.997	49.996	49.998	49.996
f_2 /Hz	49.990	49.991	49.994	49.997
Absolute error /Hz	0.007	0.005	0.004	0.001

As shown in Table 1, the deivce's maximum absolute error of frequency measurement is 0.005Hz, the accuracy meets the national standard [10] requirements.

4.2. *Voltage effective value measurement results*

As the voltage display of JCD4046 is 3.5- digit, VC86E voltmeter with 4.5-digit voltage display is adopted as test reference. Results are shown in Table 2.

Table 2. Three-phase voltage measurement test table.

Supply voltage /V	Monitoring device measurement value /V			Relative error /%		
	A	B	C	A	B	C
120.00	120.15	119.98	120.16	0.13	-0.02	0.13
110.00	110.14	109.94	110.13	0.13	-0.05	0.12
100.00	100.11	99.98	100.08	0.11	-0.02	0.08
90.00	90.05	89.95	90.07	0.06	-0.06	0.08
80.00	80.03	79.88	80.08	0.04	-0.03	0.10
60.00	60.05	60.01	60.03	0.08	0.02	0.05
40.00	40.05	39.96	40.04	0.13	-0.10	0.10
20.00	19.97	19.98	20.03	-0.15	-0.10	0.15

As shown in Table 2, the device's relative measurement error of three-phase voltage ranges between -0.15%~0.15%, the accuracy meets the national standard [11] requirements for A class instruments.

4.3. *Voltage harmonic measurement results*

The order of harmonic components of JCD4060's voltage output ranges between 2-21, thus the test does not involve higher orders. The fundamental voltage is set to the nominal voltage 100V, and Fluke345 is adopted as test reference. Results are shown in Table 3.

Table 3. Three phase voltage 2~21 harmonics measuring meter.

Harmonics order	Monitoring device measurement value /%			Fluke measurements %		
	A	B	C	A	B	C
2	9.58	9.58	9.63	9.4	9.4	9.4
3	9.43	9.41	9.31	9.3	9.3	9.2
4	9.08	9.07	8.97	8.9	8.9	8.8
5	8.42	8.39	8.31	8.3	8.3	8.2
6	6.79	6.84	6.86	6.6	6.7	6.7
7	4.84	4.90	4.99	4.7	4.8	4.9
8	3.12	3.17	3.12	3.1	3.1	3.1
9	1.78	1.73	1.64	1.6	1.6	1.8
10	0.75	0.73	0.79	0.6	0.6	0.7
11	0.06	0.05	0.03	0	0	0
12	5.21	5.33	5.59	5.1	5.2	5.5
13	4.15	4.25	4.19	4	4	4.1
14	3.29	3.35	3.38	3.2	3.3	3.4
15	2.62	2.68	2.84	2.4	2.5	2.9
16	2.00	2.05	2.18	2	1.9	2.1
17	1.31	1.36	1.35	1.3	1.3	1.4
18	1.05	1.06	1.14	0.8	0.8	0.9
19	0.33	0.09	0.32	0.2	0.2	0.3
20	0.11	0.03	0.14	0	0	0
21	0.01	0.00	0.01	0	0	0

Measurement errors of the ratios of 2nd-21st order harmonics are shown in Table 4, where U_N is the nominal voltage 100V, U_b is the harmonic voltage and δ is the three-phase maximum error.

Table 4. Three phase voltage 2~21 harmonics error.

Harmonics	U_b/V	δ/%
2	9.4	2.4
3	9.3	1.4
4	8.9	2.1
5	8.3	1.4
6	6.7	2.8
7	4.8	2.9
8	3.1	2.3
9	1.7	4.7
10	0.7	0.09
11	0	0.06
12	5.2	2.5
13	4.1	3.7
14	3.3	2.4
15	2.6	3.1
16	2	4.0
17	1.3	4.6
18	0.8	0.14
19	0.2	0.11
20	0	0.14
21	0	0.01

As shown in Table 3 and Table 4, three-phase voltage values of 2nd-21st order harmonics are all given. When harmonic voltage $U_b \geq 1\%U_N$, the maximum error is $4.7\%U_b$, less than $5\%U_b$; when the harmonic voltage $U_b < 1\%U_N$, the maximum error is $0.14\%U_N$, less than $0.15\%U_N$. The accuracy meets the national standard [12] requirements for B class instruments.

4.4. *Voltage sequence component measurement results*

JCD4060 outputs unbalanced three-phase voltage, and the device calculates the amplitude and phase angle of the fundamental wave of each phase through FFT and then calculates corresponding voltage sequence components. Results are shown in Table 5.

Table 5. Three phase voltage and sequence component testing table

Times	Three-phase voltage values /V		
	A	B	C
1	120.08 ∠0 °	119.93 ∠-119.5 °	120.02 ∠121 °
2	109.14 ∠0 °	109.91 ∠-125.3 °	113.04 ∠129.4 °
3	99.88 ∠0 °	105.98 ∠-127.2 °	99.32 ∠121.1 °
Times	Sequence component test results /V		
	Positive sequence	Negative sequence	Zero sequence
1	121.14 ∠0.20 °	0.69 ∠-58.43 °	0.54 ∠-119.71 °
2	112.45 ∠0.70 °	8.37 ∠-15.13 °	8.81 ∠-178.28 °
3	103.39 ∠-1.50 °	5.37 ∠63.57 °	5.09 ∠166.94 °
Times	Theory calculation results /V		
	Positive sequence	Negative sequence	Zero sequence
1	120.01 ∠0.50 °	0.64 ∠-57.97 °	0.57 ∠-117.74 °
2	110.07 ∠1.44 °	8.06 ∠-14.26 °	8.74 ∠-174.86 °
3	101.51 ∠-2.14 °	5.08 ∠44.83 °	5.17 ∠177.68 °

As shown in Table 5, the device's measurement error of voltage unbalance ranges between -0.2%~0.2%, the accuracy meets the national standard [13] requirements.

5. Conclusion

This paper introduces the hardware design and software programming of a power quality monitoring device based on FPGA. Test results show that the device is reliable, accurate and easy to operate, well meeting the requirements of power quality monitoring with high practical value. Highly integrated with few peripheral extensions and little hardware limitations, the device can be easily upgraded with new algorithms and modules.

References

1. J. Chu & J.W. Ma. Detection of Power Quality Disturbed Signals Based on Wavelet Transform [J]. Power System Protection and Control, 2009, 37(5): 34-36, 64.
2. H.X. Lin. Main Problems of Modern Power Quality [J]. Power System Technology, 2001.10, 25(10):5-12.
3. X.N. Xiao. Power Quality Analysis and Control [M]. Bei Jing: China Electric Power Press, 2004.
4. X.Y. Chen, T. Pang & L.J. Xu. Design of Laboratory Power Quality Monitoring System Based on Embedded Platform [J]. Computer Measurement & Control,2013, 21 (6): 1501-1504.
5. Z.H. Tang, Y.C. Long & J. Xie. Power Quality Monitoring System Based on Dynamics Measure [J]. Low Voltage Apparatus, 2013 (9): 50-54.
6. F. Zhao & D.M. Ma. Examples of Embedded Systems Design on FPGA [M]. Xi An: Xidian University Publisher, 2008: 26-28.
7. Y. Zhou & M.L. Zhu. Design of Online Power Quality Monitoring Equitment Based on DSP and GPRS [J]. Journal of Zhengzhou University (Engineering Science), 2014, 35(5): 23-26.
8. Altera Corporation. Nios II The World's Most Versatile Embedded Processors [DB/OL]. 2008.
9. L. Xie, D. Xie & Y.C. Zhang. The Algorithm and Realization of The New Type Meter Measuring Power Quality, 2008, 45 (513): 1-4, 10.
10. General Administration of Quality Supervision, Inspection and Quarantine of the People's Republic of China (AQSIQ), Standardization Administration of the People's Republic of China. GB/T 15945-2008 Power quality-frequency deviation for power system [S]. Bei Jing: China Standard Press, 2008.
11. General Administration of Quality Supervision, Inspection and Quarantine of the People's Republic of China (AQSIQ), Standardization Administration of the People's Republic of China. GB/T 12325-2008

Power quality-Deviation of supply voltage [S]. Bei Jing: China Standard Press, 2008.

12. General Administration of Quality Supervision, Inspection and Quarantine of the People's Republic of China (AQSIQ), Standardization Administration of the People's Republic of China. GB/T 14549-1993 Quality of electric energy supply harmonics in public supply network [S]. Bei Jing: China Standard Press, 1993.
13. General Administration of Quality Supervision, Inspection and Quarantine of the People's Republic of China (AQSIQ), Standardization Administration of the People's Republic of China. GB/T 15543-2008 Power quality—Three-phase voltage unbalance [S]. Bei Jing: China Standard Press, 2008.

The simulation study for islanding detection based on the active power change of improved active frequency shift method

Y.Y. Ge
State Grid Liaoning Province Power co., Ltd
Electric Power Research Institute,
Shenyang, China

J.J Sun
School of Electrical Engineering, Wuhan University,
Wuhan, Hubei province, China

G. Wang
State Grid Liaoning Province Power Co., Ltd
Electric Power Research Institute,
Shenyang, China

Z.Q. Gao [†]and L. Yu
School of Electric Power, Shenyang Institute of Engineering,
No. 18, Puchang Road, Shenbei New District, Shenyang,
Liaoning Province, 110136, China
Emails: [†]gaozhiqianghit@163.com.cn, yulixxx@126.com

Besides the function of keeping the current injected into the grid having the same frequency and phase with the voltage of grid, photovoltaic grid-connected system must have the ability of perfect islanding detection. Once the island phenomenon occurs and wasn't cut off from the photovoltaic grid-connected system in time, it may result in damage of the equipment and threaten personal safety. Considering the severity of island phenomenon, the islanding detection should be rapid and accurate, as well as taking into account of the power quality. This paper proposes a novel feedback active frequency drift method (AFD), which is based on positive feedback of the change of active power. This method adds the change of active power into the frequency deviation, thus, the frequency deviation of point of common coupling is accelerated, and the time of islanding detection is shortened.

Keywords: PV grid-connected; Islanding detection; Active frequency shift; Gird connected inverter.

1. Introduction

With the development of economy and society, electricity demand is increasing and fossil and other non-renewable energy consumption is a large number [1-2]. Solar energy as a renewable energy power generation is gaining more and more attention and application [3].

However, before a large number of solar powers into the grid system needs to ensure safe transmission power system and reliable operation. Among them, the most prominent influence on stable operation is the occurrence of islanding.

Island phenomenon arises when the grid system failure or line maintenance, solar power plants continue to remain powered state line, at this time, photovoltaic power plants and electric power supply system load formed islands that cannot control.

IEEE 929-2000 standard specifies a distributed power generation system after islanding phenomenon must have anti-islanding detection functions [4]. Islanding detection is focused on within the specified time to maintain voltage and frequency within the normal range.

In order to formulate appropriate method to check the island phenomenon, two aspects must be considered. On the one hand detection methods can be applied to various types of island phenomenon. On the other hand the detection methods can be recycled with the testing procedures, so it can not only reduce costs, but also effectively reduce the detection time.

In this paper, the active variable is introduced into active the positive feedback active frequency shift islanding detection method. The active power changes are introduced frequency increment positive feedback function in this method, and common point voltage frequency shift speed is accelerated, the detection of islanding time is shortened. Matlab simulation results show the effectiveness of the proposed islanding detection method.

2. Active Frequency Offset Method

In order to detective the phenomena of island, a small frequency offset is introduced periodically to inverter output current in the method of active frequency offset method (AFD) [5].

Frequency offset rule is implemented in the locked program. Achieve specific process is as follows: by the PLL to determine the frequency and phase of the voltage at the PCC, adjusted and net inverter output current of a given frequency, so that the current frequency is closer than the PCC voltage frequency. When the grid is operating normally, the voltage at the PCC is limited by grid voltage and frequency to stabilize at rated operating point, calculate PLL lock angle will also be stable. When the grid and PV power plant is disconnected, then frequency will be disturbed to offset rated operating point, so that the frequency out of control. When the grid frequency deviation is greater than 0.5Hz, it is considered to produce island phenomenon, protection is triggered.

When the inverter is controlled by the microprocessor controller, active frequency offset islanding detection method is very easy to implement [6]. When the inverter is connected with the distribution network, the system frequency cannot be changed. When the inverter and the power distribution network disconnect, the frequency of the inverter output voltage will be forced to rise or fall, increasing the natural frequency offset of the island system to seek and load frequency resonance.

Figure 1 shows that the frequency of the inverter output current waveform with AFD, and the comparison with the non distortion of the sinusoidal waveform.

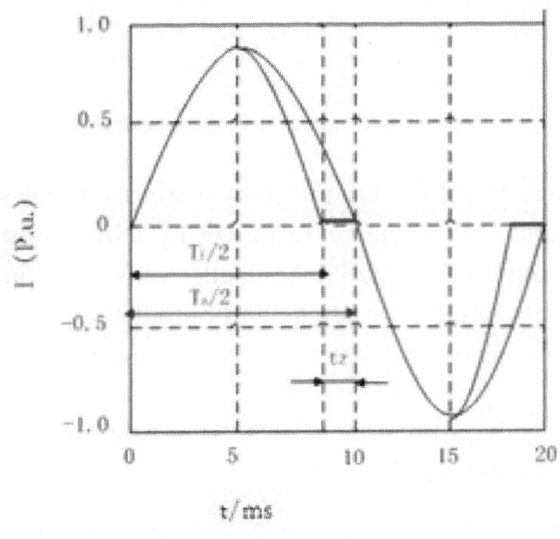

Fig. 1. AFD islanding detection method the output current waveform.

3. Photovoltaic Maximum Power Point Tracking

Photovoltaic battery is a very unstable nonlinear DC power supply, the output voltage and current is affected by the light intensity and environmental temperature. Figure 2 shows the characteristic curve of the solar cell.

As can be seen from the Fig. 2, the different light intensities produce different power, but each one will have a maximum power output, the maximum power value corresponding to the amplitude of the voltage is almost the same. Therefore, how to improve the efficiency of the photovoltaic cells, and then to achieve maximum output power photovoltaic cells, which will play an important role and significance to improve the overall efficiency of photovoltaic power generation systems and photovoltaic (PV) grid.

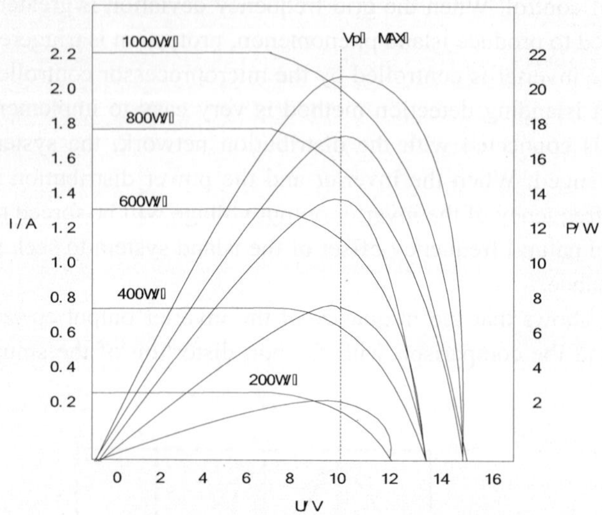

Fig. 2. Generation characteristic curve of solar photovoltaic cells.

Maximum power point tracking system is an electrical system, which can make the photovoltaic cell output more power by adjusting the working condition of electrical modules [7]. MPPT controller main functions are as follows: detection of the main circuit DC voltage and output current, calculation the output power of the solar array, and maximum power point tracking.

4. New Type of Active Frequency Shift Island Detection Method Based on Active Power Change

In order to verify the validity of the proposed islanding detection method, the simulation is studied using Matlab/simulink software.

Figure 3 is a photovoltaic grid islanding detection simulation map AFD algorithm based on active power change. The parameters of model are as follows: the module of PV+boost is 380V DC, filter inductor L=8mH and filter capacitor C=10PF, R=R1=5Ω, RMS of ac voltage source is 220Vand frequency is 50Hz.

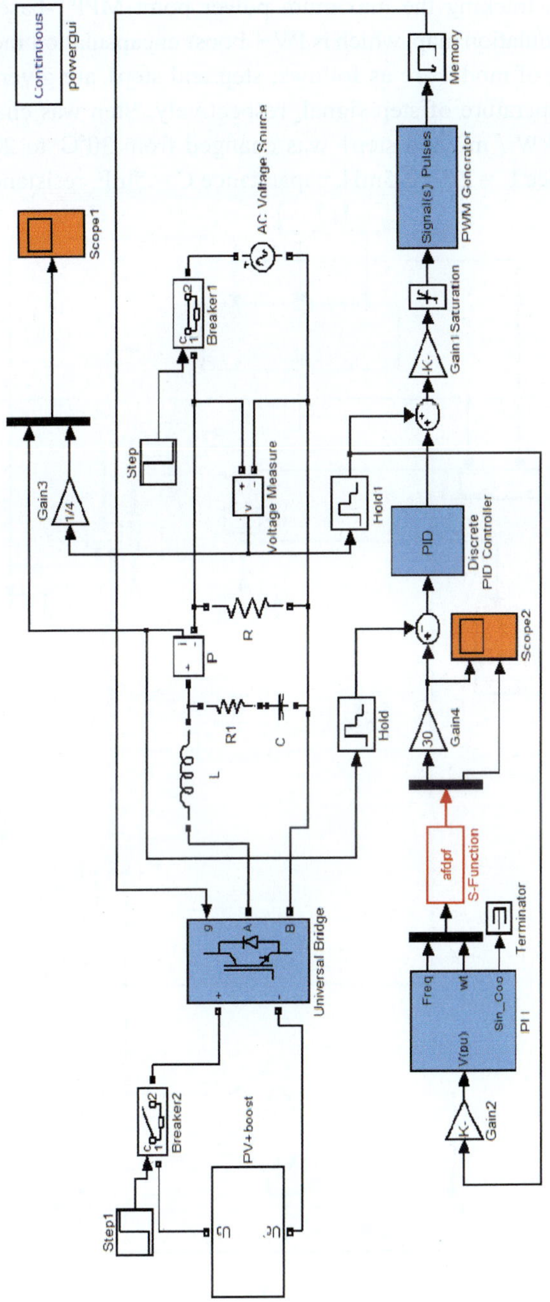

Fig. 3. Photovoltaic grid islanding detection simulation map based on AFD.

Figure 4 is tracking the maximum power point MPPT-based single-phase photovoltaic simulation map, which is PV + boost encapsulation module of Fig. 3. The parameters of model are as follows: step and step1 are given solar light intensity and temperature of step signal, respectively. Step was changed from 0.5 kW / m2 to 1 kW / m2 and step1 was changed from 30℃ to 28°C. The boost circuit inductance L = 2 * 10-5mH, capacitance C = 5mF, resistance R = 1.8kΩ.

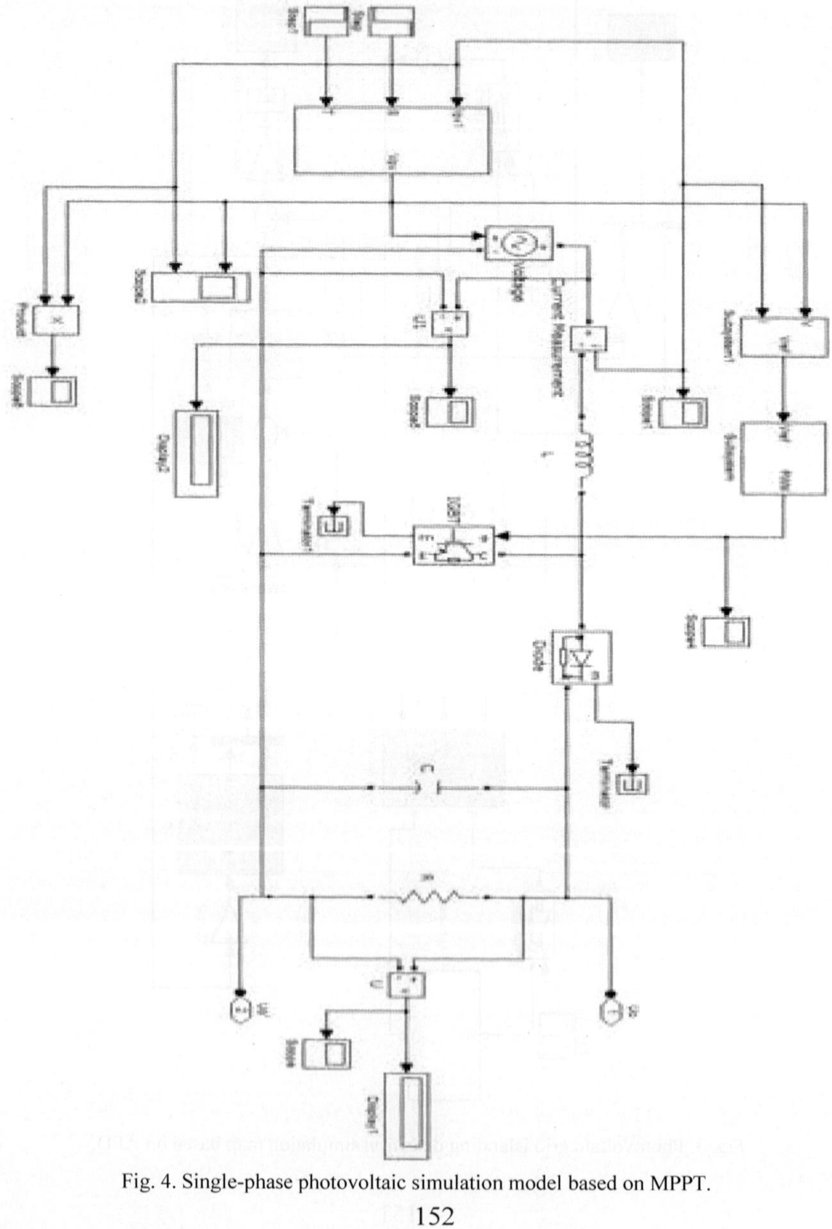

Fig. 4. Single-phase photovoltaic simulation model based on MPPT.

Figure 5 is photovoltaic (PV) grid system output voltage and current waveforms based on the positive feedback AFD islanding detection with pure impedance load.

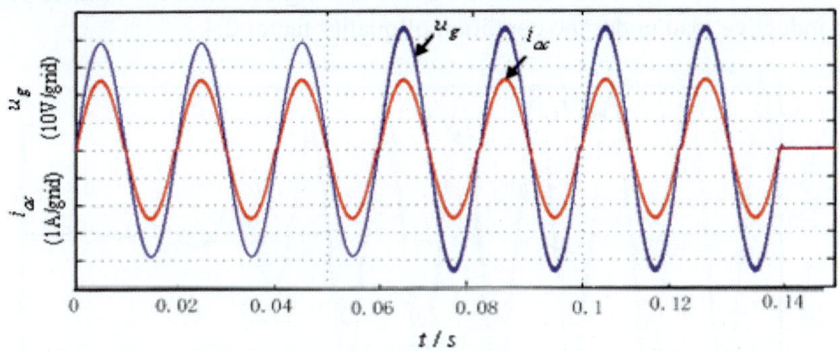

Fig. 5. Output waveforms of AFD islanding detection with resistive load.

From figure 5, it can be seen that when 0.06s the grid is off, frequency deviation is more and more big by the positive feedback effect. Around 140 ms, output frequency reaches the protection threshold, the islanding is detected, and photovoltaic system will stop working.

Figure 6 is photovoltaic (PV) grid system output voltage and current waveforms based on the positive feedback AFD islanding detection on variable power with pure impedance load.

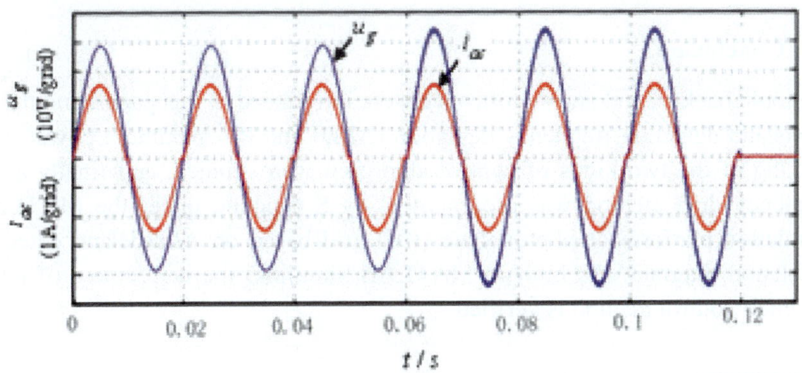

Fig. 6. Output waveforms of AFD islanding detection on variable power with resistive load.

From figure 6, it can be seen that when 0.06s the grid is off, frequency deviation is increased due to the change of active power and frequency deviation is accelerated. Around 120 ms, output frequency reaches the protection threshold, the islanding is detected, and photovoltaic system will stop working. Compared with the positive feedback active frequency shift island detection method, island detection time is shortened about 20ms, and the output current waveform distortion is a little changed during normal operation.

Figure 7 is photovoltaic (PV) grid system output voltage and current waveforms based on the positive feedback AFD islanding detection on variable power with inductive load under the condition of quality factor 2.4.

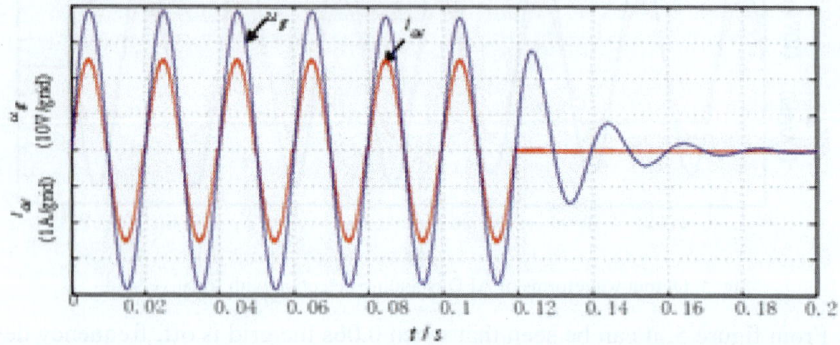

Fig. 7. Output waveforms of AFD islanding detection on variable power with inductive load.

From Figure 7, it can be seen that when the grid is disconnected, the frequency of output voltage change is bigger and the increased of deviation accelerates frequency protection threshold speed which make detection island time shorter based on the positive feedback AFD islanding detection on variable power.

5. Conclusion

The topic background and significance were introduced in this paper, and the active island detection frequency deviation (AFD) and PV maximum power point tracking of the two kinds of control algorithm were studied, and finally active frequency drift method was programmed by S function under the MATLAB. Simulation platform of photovoltaic (PV) grid based on AFD island detection simulation diagram was built in MATLAB/Simulink, the correctness of island detection control method is verified.

Preferences

1. Y. Jung et al; A novel active frequency drift method of islanding prevention for the grid-connected photovoltaic inverter [C]; Proc. of 36th Annual Power Electronics Specialists Conference (PESC). 2005:1915-1921.
2. B.Yu et al; A robust anti-islanding method for grid-connected photovoltaic inverter [C]; The 4th World Conference on Photovoltaic Energy Conversion. 2006:2242-2245.

3. Picault et al; Guidelines for evaluating grid connected PV system topologies [C]. IEEE industrial electronics society. 2009 IEEE International Conference on Industrial Technology. 2009:1-5.
4. IEEE recommended practice for utility interface of photovoltaic (PV) systems. IEEE std. 929-2000, Institute of Electrical and Electronics Engineers, 2000.
5. Y Jin et al; Anti-islanding protection for distributed generation systems based on reactive power drift [C]. 35th Annual Conf. IEEE Ind. Electron, 2009.
6. Photong C et al; Evaluation of single-stage power converter topologies for grid-connected photovoltaic [C]. The Institute of Electrical and Electronics. Proceedings of the IEEE International Conference on Industrial Technology. 2010:1161-1168.
7. Sera deaso et al; Optimized maximum power point tracker for fast-changing environmental conditions [J]. IEEE Transactions on Industrial Electronics, 2008, 55(7):2629-2637.

An overview on smart distribution grid

X.S. Zhou[1], P.D. Xiang[2]† and Y.J. Ma[3]

Key Research Laboratory for Control Theory & Applications in Complicated Systems, Tianjin University of Technology, Tianjin, China
† Email: xpd880725@126.com

The concept of smart distribution grid (SDG) is put forward to cope with environmental problems and provide higher quality electrical energy to users. To begin with, three main characteristics of SDG including the granted access to a large number of distribution generations, self healing, and active interaction with various power users are introduced in this paper which are superior to those of the traditional distribution grid. Then, some key technologies applied to achieve those characteristics are analyzed. Finally, the key challenges and development tendencies of SDG are discussed which refer the universal energy network and the active distribution network as the two main development tendencies. SDG is the result of the intellectualization of the traditional distribution grid, and it will be the inexorable trend in the future development of distribution grid.

Keywords: Smart distribution grid; Micro grid; Universal energy network; Active distribution network.

1. Introduction

Western countries firstly put forward the concept of smart distribution grid (SDG) in order to cope with the global environmental problems and meet the needs of customers for safe, reliable, high quality and economic power supply. Meanwhile, China came up with the concept "strong transmission grid, intelligent distribution grid". As the representative of intelligence, the granted access to a large number of distribution generations, self healing, and active interaction with various power users serve as the three major characteristics to realize the intellectualization of SDG [1-3].

---

[1] Xue-Song Zhou
[2] Peng-Da Xiang
[3] You-Jie Ma

Generally, distribution generation refers to small, modular, dispersive, and reliable generation unit arranged in the vicinity of the user with the generation output ranging from a few kilowatts to hundreds of megawatts, including wind power, solar power, biomass power, geothermal power, etc. Meanwhile, micro grid (MG) refers to the system assembled by distribution generation, energy storage device, energy conversion device, related loads and monitoring and protection device with the function of self control, protection and administration. Renewable energy is applied in MG, which is conducive to environmental protection. Distribution generation can be operated under the mode of joint-grid operation and isolated operation, and the isolated operation system can form a micro grid. As for the grid, the joint-grid operation of distribution generation can realize the purpose of balance power, which can support the grid and vice versa. Moreover, distribution generation can be disconnected from the grid and thus enter into an isolated operation when grid gets a large scale blackout, which is flexible and efficient so as to improve the security level of power supply [4].

Self-healing control of distribution grid can predict all kinds of potential safety hazard and impending disturbance in real time by sharing and calling all available grid resources, and then take optimization control strategy under the normal operation of distribution grid as well as the control strategy of prevention and correction, emergency recovery, and maintenance under the abnormal operation of distribution grid, thus to make the power grid transfer from the abnormal running state into normal operation as soon as possible so as to reduce human intervention during the runtime of distribution grid and the influence resulting from the disturbance or fault of grid on users. It can achieve optimization control selectively and purposely, improve the grid performance, boost the system stability margin and enhance the ability to resist disturbance while working normally. Meanwhile, it regards prevention as the major means of control to timely detect, diagnose and eliminate the hidden problems. Moreover, it can maintain the system to run continuously under the faulty condition without causing loss to the system operation, and can recover from faults through the autonomous repair function. The control principle of Self-healing in the distribution grid is to ensure uninterrupted power supply. The goal is that the faults of distribution grid should be avoided through the control and prevention and if the faults occur, as less as the loss of load should be guaranteed through the emergency recovery control and maintenance control [5].

Since there isn't enough interaction between distribution grid and the user, power shortage frequently occurs. However, SDG can be applied to form interaction and immediate connection between users and grid corporation as well as users and users, which can realize the real-time, high speed and bidirectional read of data. It can not only improve the efficiency of distribution grid but also provide a better approach with less energy waste for the entire society.

Traditional distribution grid has no the three characteristics of SDG above, while SDG has more advantages than traditional distribution grid. In summary, SDG is the upgraded version of traditional distribution grid.

2. Key Technologies of SDG

As for SDG, application of advanced technology is indispensable to exhibit its three characteristics of the granted access to a large number of distribution generations, self healing, and active interaction with various power users.

2.1. *Distribution generation joint-grid control and micro grid technology*

The access of distribution generation changes the single power supply structure, and the current in all branches of distribution grid may no longer flow in one direction. Under such circumstance, the dispersed access of lots of distribution generations will definitely influence the operation of distribution grid. From the angle of SDG, these two technologies need be researched: (1) The technology of joint-grid between distribution grid and SDG; (2) The advanced distribution automation technology after joint-grid; (3) The technology of micro grid.

At present, most of distribution generations are connected to distribution grid through power electron converter. As typical joint-grid of distribution generation structure diagram is shown in Fig. 1, electric energy generated from powers such as wind power, solar power and so on is converted to the electric energy that meets the requirement of joint-grid operation or load. The energy storage device can enhance the quality of the electric energy further. When the grid runs regularly, the joint-grid switch is closed. Then distribution generation run in the joint-grid operation state, transporting extra or all the electric energy into the grid. However, when the grid runs irregularly, the joint-grid switch is opened. Then distribution generation can be quickly disconnected from grid forming a MG with the load, providing electric energy for load continuously. In terms of the control methods of power electron converter, advanced control methods applied involves PI control, repeat control and other methods, the introduction of which can significantly enhance the conversion efficiency and the quality of output electric energy.

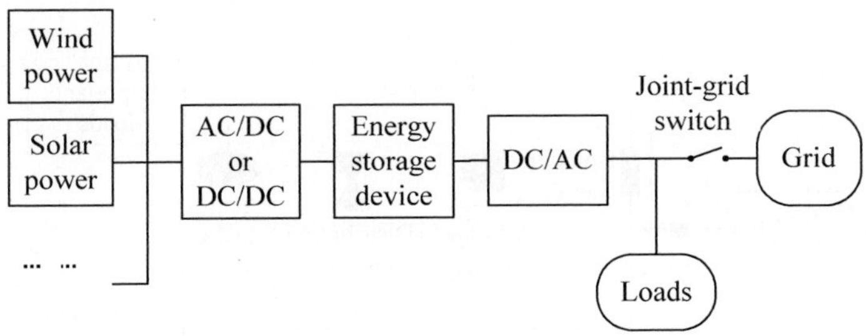

Fig. 1. Typical joint-grid of distribution generation structure diagram.

The access of distribution generation will exert influence on node voltage, short-circuit current, reliability of power supply, etc. Therefore, effective control of system operation is required after joint-grid. As an important technology used in the management and control of the distribution grid, advanced distribution automation (ADA) technology achieves the comprehensive automation control of the distributed generation and the power distribution system. With the use of computer, automation control technology, data communication and information management technology, ADA can integrate information like real time running state of grid, grid structure, device and users, and then carries out control through analysis. The ADA of smart distribution grid contains all data flow related to power distribution system in the electric power enterprises. Compared with the traditional distribution automation technology, ADA supports the access of distributed generation, provides the function of real-time simulation analysis and auxiliary decision-making for the smart distribution network, and supports advanced application software as well as the technology of smart distribution control, all of which make the ADA good openness and expansibility in the application of smart distribution grid [6].

Micro grid technology is the integration of new technology of power electronics and distributed generation, technology of renewable energy generation, and technology of energy storage. As is shown in Fig. 2, MG is connected to grid via Point of Common Coupling (PCC), and energy manager can make adjustment for the system based on the pre-set value of voltage and power. When MG is disconnected from grid and runs independently, the structure of MG should be specially designed to prevent the strike to the inner load of MG. Unimportant loads are put on the same feeder line such as Feeder Line B, while important or sensitive loads are put on another feeder line such as Feeder Line A, on which there are also distribution generation, energy storage and control device and other devices installed to guarantee that unimportant loads can be removed by isolating device while important loads for regular operation can be ensured when distribution generation is disconnected from grid.

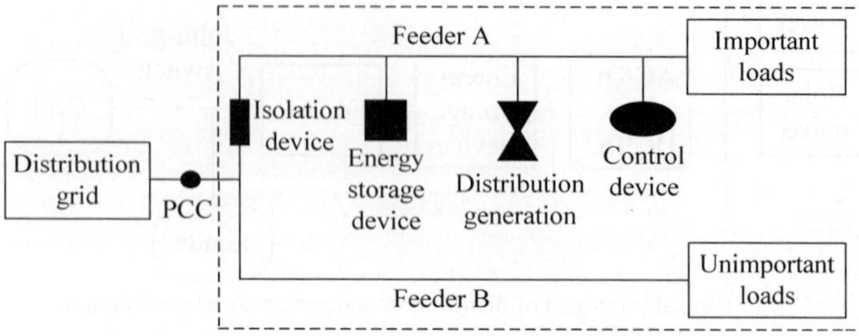

Fig. 2. Typical micro grid structure diagram.

2.2. *Distribution fast simulation and modeling technology*

Distribution fast simulation and modeling technology is an important tool to realize self-healing of the distribution grid, whose main function involves adaptive automatic protection, fault automatic locating and resolution, grid reconfiguration, automatic voltage and reactive power control. Distribution fast simulation and modeling technology makes full use of mathematical analysis tools and advanced forecasting techniques based on real-time software platform and analysis of the grid operation state and its physical structure to accurately estimate the state of the distribution grid, optimize the real time operation and predict potential events of the distribution grid so as to provide auxiliary decision-making advice and optimal decision services for system operators. The simulation tools mainly include evaluation of the state of the distribution grid, optimization of power flow, assessment of the grid dynamic security, and load forecasting. Meanwhile, the modeling tools include network topology analysis, equipment model, load model and generation model. When fault or abnormal operating state occurs in grid, the application of distribution fast simulation and modeling technology provides reference to network reconstruction, which is the key technology in achieving self-healing of distribution grid.

2.3. *Advanced metering infrastructure (AMI)*

Measurement and communication technology is the basis of the informatization and digitization of the distribution grid, and is also the premise to enhance the overall operation and management of distribution grid automation. AMI based on advanced measurement and communication technology is the basis of active interaction between distribution grid and users. AMI mainly consists of the smart meter, the two-way communication network, the metering data management system (MDMS) and home area network (HAN) [7].

Smart meter can obtain the information of users during certain period such as power consumption, voltage and current, serving as the basic data measuring sensor in the distribution grid. It can be used as the communication gateway between power companies and users indoor network, enabling users to check the electricity information in real time and receive electricity price signal from the power companies.

Two-way communication network can transfer the information recorded by smart meters including faults warning and device interference warning to the data center, laying a solid foundation for the realization of the advanced application functions of distribution information system. At the same time, by means of two-way communication network, related information such as electricity price information and demand side response can be timely passed to the users, which can realize two-way information interaction transfer between users and the distribution grid, and thus realizes the interaction between distribution grid and users.

MDMS is a database management system with an analysis tool, which can be combined with the AMI automatic data collection system to handle and storage the measure data of smart meters effectively. The basic function of MDMS is to validate, edit and estimate the AMI data. Even under the condition that communication network is interrupted and faults occur at the user side, MDMS can also guarantee the integrity and the accuracy of the data flow that flows to the mentioned information system or software.

HAN of users is the network that can connect the controllable electric appliance or device indoor, which can be connected to the smart meter through the gateway so as to accept the measuring data of meters and the price information of the electricity company. It can also adjust the power consumption according to the requirement of the market or the system.

Operational efficiency can be substantially elevated and source deployment can be significantly optimized with the coordination within all components in AMI and the cooperation between AMI and external measurement and information system.

There will be a series of advanced technologies applied in SDG. However, there is still no unified view about whether some technology must be applied or not. Besides, those technologies are constantly updated. In a word, above mentioned three characteristics are recognized worldwide, which means that the technologies which helps better realize the three characteristics are the key ones we should focus on.

3. Key Challenges of SDG

Communication technology is the basis for SDG to realize intelligentization. However, there is still lack of communication standardization in domestic distribution grid. Specifically, the lack of relevant standardization will limit the data exchange between different devices, which will reduce the operational efficiency of distribution grid and thus hamper the intelligentization process of distribution grid [8].

The construction of SDG requires a great number of intelligent devices, interoperability among different devices which is very limited under current circumstances such as the interoperability between distribution generation and storage device. With the absence of interoperability, distribution generation cannot carry out joint-grid operation. Therefore, the complete interoperability among different devices is the major issue to be solved in the construction of SDG [9].

Lots of intelligent devices are equipped on SDG to realize intelligentization, which will generate huge amount of data every day. Thus, the storage and administration of data is the new challenge occurring in the construction of SDG [10].

4. Tendency

4.1. *Universal energy network*

Universal energy network refers to energy network, Internet of things and Internet which is synthesized with Internet of Energy, and adjusts measures to local conditions. It adopts various kinds of energy such as wind power and solar power and makes matching and dispatching according to customer's demands. In addition, with the utilization of universal energy network, not only the distributive and centralized supply can be coordinated, but the energy allocation between multiple universal energy station can be formed. Thus, universal energy network can optimize energy efficiency essentially, reflecting in not only realizing the comprehensive use of resource, but also achieving the intelligent allocation among state grid, municipal gas network, municipal heating network, etc, which can significantly support the future development of SDG [11].

4.2. *Active distribution network (ADN)*

ADN is the essential component of SDG as well as the advance stages of the technology of SDG. In such case, it can be a power distribution system able to control and manage the distribution generation actively, which can optimize the operation of distribution generations. Operators are allowed to change the network topology so as to control the inner-net power flow while distribution generation can also provide support for the distribution grid under proper surveillance and joint-grid protocol. The application of ADN can significantly enhance the consumptive ability of renewable resource and elevate the level of load management, which is the key to solve the problem of granted access to a large number of distribution generations [12-13].

5. Conclusions

As an essential component of smart grid, the study and application of SDG possesses great influence on global economic development, which serves as the necessary approach to cope with the challenge of future energy, environment, economy as well as the effective method to elevate the automation level. Meanwhile, it's a systematic project with high complexity, which requires close cooperation within the field of electricity, electronic, control, information and so on and thus to accelerate the innovation of SDG.

References

1. Lijun, Qin & Qiyan, Ma 2010. Distribution Smart Grid and Its Key Technologies. Beijing: Beijing Power Press.
2. Richard E. Brown 2008. Impact of Smart Grid on Distribution System design. IEEE Power and Energy Society 2008 General Meeting: Conversion and Delivery of Electrical Energy in the 21st Century, PES.
3. R.W. Uluski 2010. The role of advanced distribution automation in the smart grid. IEEE PES General Meeting, PES 2010.
4. Chengshan, Wang & Peng, Li 2010. Development and challenge of distribution generation, micro grid and smart distribution grid. Automation of Electric Power Systems.
5. Zhizhong, Guo 2005. The self-healing control scheme of the grid. Automation of Electric Power Systems.
6. Jiaquan, Yang & Shi, Su 2011. Research and application design of advanced distribution automation technology. 2011 Yunnan Power Technology Forum Meeting.
7. Wenpeng, Luan 2009. Advanced measurement system. Southern Power System Technology.
8. Gao, Niu 2013. Overview of communication technology development in smart distribution grid. Computer CD Software and Applications.
9. Yixin, Yu 2015. Challenges and opportunities of smart distribution grid. Distribution & Utilization.

10. Keyan, Liu et al. 2015. The research of smart distribution grid big data applications and scene analysis. Proceedings of the CSEE.
11. Junqing, Liu 2010. Universal energy network appeared at the World Expo: system energy efficiency lead future technological innovation. China Electric Power Education.
12. Dong, Liu 2014. Progress of domestic technology for active distribution network. Distribution & Utilization.
13. Yi, You et al. 2012. Active distribution network technology and its development. Automation of Electric Power Systems.

Design and test of the multi-direction vibration energy harvester

K. Liu[1], S.Y. Wang[2], X.H. Yin[3], Q.Q. Fu[4], Z. H. Zhang[5], and J. W. Kan[6†]

Institute of Precision Machinery, Zhejiang Normal University, Jinhua, China

Emails: [1]lk679759@163.com,[2] jutwsy@163.com,

Emails: [3]yxh4609@zjnu.cn, [4]zsdzjz@126.com,

Emails: [5]zhangzhh@zjnu.cn,[†6]jutkjw@163.com

In order to achieve low-frequency, multi-direction vibration energy reliably recovery and utilization, a multi-direction vibration energy harvester (MDVEH) based on the piezoelectric effect is presented. This paper introduces mainly the structural design, working principle, and test of the prototype. Theoretical analysis shows that the MDVEH's electric performance is commonly decided by the environmental vibration frequency, amplitude, proof mass, structure parameters of piezoelectric vibrator, spring stiffness and so on. It can achieve its intended function only when all factors are configured appropriately. Test prototype is fitted to the shaker table and tested in different excitation frequencies, amplitudes and proof masses. The test shows that the MDVEH can work in different directions of vibration, there is an optimum operating frequency about 13Hz, in which MDVEH can generate the maximum voltage about 15V, and within a certain range, the output voltage increases with the rise of the proof mass and the amplitude.

Keyword: Piezoelectric; Multi-direction vibration; Energy harvester; Proof mass.

1. Introduction

To provide real-time and endless energy directly for low-power electric devices, the researches devoted to energy harvesting are increasing rapidly [1-4]. The successful development of piezoelectric vibration energy harvester and its application can not only solve the problem of power supply for portable micro power electronic products and remote sensing monitoring system, but also help to reduce the battery pollution. As we all know, for piezoelectric vibration energy

---

[1] Kun Liu

[2] Shu-Yun Wang

[3] Xiao-Hong Yin

[4] Qing-Qing Fu

[5] Zhong-Hua Zhang

[6] Jun-Wu Kan

harvester, only when the natural frequency and vibration frequency is equal, its power and the efficiency can become the highest. But the resonant frequency of piezoelectric vibrator is usually several hundreds, and even thousands of Hertz, while the environmental vibration frequency generally is more than 10Hz, even lower, thus it isn't feasible to directly use the piezoelectric vibrator to collect environmental vibration energy. According to the ordinary method, mass block installed at the end of cantilever beam or at the central of the piezoelectric disk can reduce the natural frequency of the piezoelectric vibrator [5,6], but the piezoelectric vibration will deform greatly in non-work state, and damages in work due to the heavy mass. On the other hand, the nature vibrating source is a multi-direction [7], and the vibrating direction and amplitude changes with time in the environment, such as all kinds of vehicles driving in, the vibration source includes longitudinal vibration caused by bumpy road, front and rear side of vibration caused by the car, lateral vibration caused by drifting in the direction of travel. Vibratory coupling and superposition exist in different vibration sources. Thus, from a single direction or in multiple fixed directions energy recovery can't meet the effective recovery of the multi-direction vibration energy.

Aiming at improving the recyclable efficiency of environmental vibration energy, a multi-direction vibration energy harvester based on piezoelectric plate materials is propose, When it starts to work, environmental vibration source cause vibrating of the inertia mass and then power is transmitted to the piezoelectric vibrator by springs, finally vibration energy is converted into electrical energy. Thus, the design can be achieved with a low-frequency, multi-direction and reliable vibration energy harvester.

2. Structure and Work Principle

The presented MDVEH (as showed in Fig. 1) consists of six single piezoelectric plates, a proof mass fixed on the center of the piezoelectric plate using a spring. The piezoelectric plate consists of substrate plate and a PZT plate bonded on it. When MDVEH vibrates, mass changes its position, which causes the spring extruded or compressed. Then vibration moved to the piezoelectric plate center, and the piezoelectric plate center is compressed. Basing on the theory of elasticity and piezoelectricity, this paper aims at the analysis of piezoelectric vibration energy harvester multi-direction.

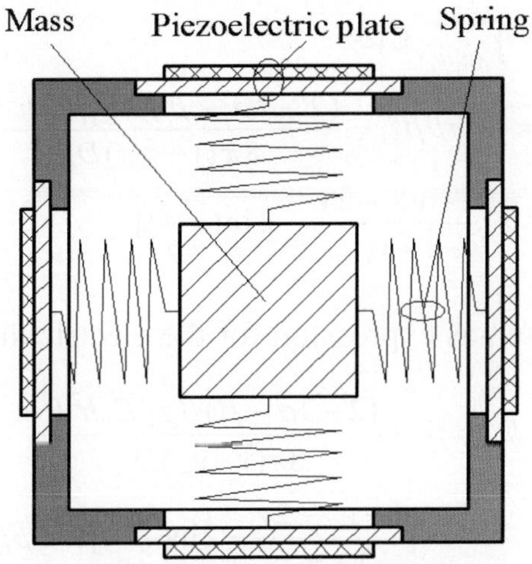

Fig. 1. The structure of MDVEH.

When circular piezoelectric plate is bent, the relationship among stress, strain and the electric field intensity is [8]:

$$T_1 = \frac{E_P}{1-v_p^2}(S_1 + v_p S_2) - \frac{g_{31}E_P}{1-v_p}D_3 \tag{1}$$

$$T_2 = \frac{E_P}{1-v_p^2}(S_2 + v_p S_1) - \frac{g_{31}E_P}{1-v_p}D_3 \tag{2}$$

$$E_3 = -g_{31}(T_1 + T_2) + \beta_{33}^T D_3 \tag{3}$$

In the formula, T_1 and T_2, radial and tangential stresses respectively; $S_1 = zd^2w/dr^2$ and $S_2 = zdw/rdr$, the radial and tangential strains respectively; g_{31} for the piezoelectric voltage constant; D_3 is the electric displacement in the Z direction, E_3 is the electric field strength in the Z direction; v_p is the Poisson's ratio of the piezoelectric materials, E_n is the elasticity modulus of piezoelectric materials; $\beta_{33}^T = 1/\varepsilon_{33}^T$ is the dielectric isolation rate; $\varepsilon_{33}^T = 1300\varepsilon_0$ is the permittivity in the Z direction.

For single-chip piezoelectric plates used in this paper, the center bears the concentrated load. The Eq. 3 for Z integrals:

$$V = \int_{-(h-h_z)}^{h_z - h_m} E_3 dz$$

$$= D_3 h\beta\beta_{33}^S - \frac{(2-2\alpha-\beta)\beta g_{31} E_P h^2 P}{8\pi(1-\nu_P)D_C}$$

$$\times[\frac{1-\nu_C-2\gamma}{1+\nu_C} - 2\ln(\frac{r}{\lambda a})] \tag{4}$$

Eq.4 can be written as expression for the electric displacement：

$$D_3 = \{V + \frac{(2-2\alpha-\beta)\beta g_{31} E_P h^2 P}{8\pi(1-\nu_P)D_C}$$

$$\times[\frac{1-\nu_C-2\gamma}{1+\nu_C} - 2\ln(\frac{r}{\lambda a})]\} / h\beta\beta_{33}^S \tag{5}$$

The electric displacement for area integration, charge generated from piezoelectric plate electrode can be calculated. Because the electrode is in the equipotential surface, voltage does not change with the piezoelectric vibrator radius, the amount of charge is generated for the：

$$Q_0 = 2\pi\int_0^b D_3 rd_r = C_f V$$

$$+\frac{(2-2\alpha-\beta)(1-\gamma)\lambda^2 a^2 g_{31} E_P hP}{4(1-\nu_P)(1+\nu_C)D_C\beta_{33}^S} \tag{6}$$

$C_f = \lambda^2 a^2 \pi / (\beta_{33}^S h)$ is the free capacitance of piezoelectric vibrator. Eq.6 is to calculate the charge when the external force and external electric field exists at the same time. For the piezoelectric power generator, external electric field of V = 0, so the charge produced by force alone is：

$$Q_g = \frac{(2-2\alpha-\beta)(1-\gamma)\lambda^2 a^2 g_{31} E_P hP}{4(1-\nu_P)(1+\nu_C)D_C\beta_{33}^S} \tag{7}$$

According to the relationship between the charge and voltage, it gets an open-circuit voltage that piezoelectric vibrator produces under external force:

$$V_g = \frac{Q_g}{C_f} = \frac{(2-2\alpha-\beta)(1-\gamma)\beta g_{31} E_P h^2 P}{4(1-\nu_P)(1+\nu_C)D_C} \tag{8}$$

Finally, the energy is generated from piezoelectric vibrator under concentrated force:

$$U_g = \frac{1}{2}C_f V_g^2 = \frac{\beta h^3}{32\pi\beta_{33}^S} \times [\frac{(2-2\alpha-\beta)(1-\gamma)\lambda a g_{31} E_P P}{D_C(1-\nu_P)(1+\nu_C)}]^2 \tag{9}$$

Eq. 7, Eq. 8 and Eq. 9 gives the relationship between the charge, voltage, energy and related device elements. The output voltage is proportional to the square of the pressure *P*. And the pressure *P* is connected with spring stiffness of *K* and the quality M, so the performance of MDVEH is common decided by the piezo-plate, environmental vibration frequency / amplitude, mass, spring stiffness and other factors. Therefore, it can achieve its intended function only when all factors are configured appropriately. By the following experiments, it can be seen in the related factors how to influence the MDVEH's electric performance.

3. Experiment and Analysis

In order to verify the feasibility of the presented MDVEH (as showed in Fig. 2), and obtain the effects of related factors on its power performance, an experimental prototype is fabricated and tested (as showed in Fig. 3). The test equipment includes mainly a shaker table, a power amplifier and an oscillograph. The experimental prototype consists of six ø54 × $1.6mm^3$ circular PZT plates, acrylic boards with 6mm thickness, springs, plastic gaskets, mass and other components. PZT plate 1/2/3/4 is bent on the vertical plane and PZT plate 5/6 is on the horizontal plane. The prototype is set on the shaker table, and tested under different excitation frequencies, amplitudes, directions and proof masses.

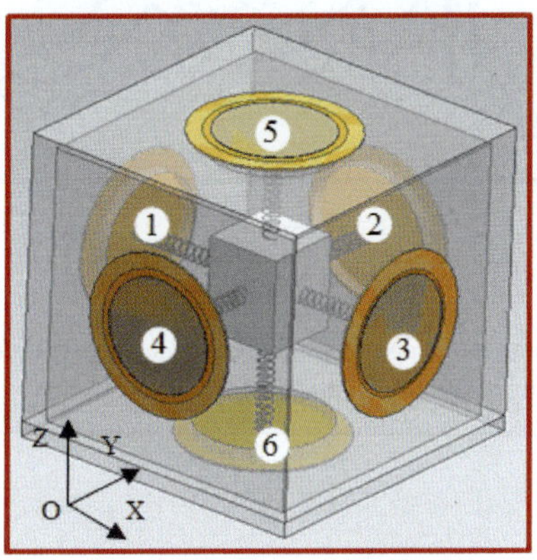

Fig. 2. The model of MDVEH.

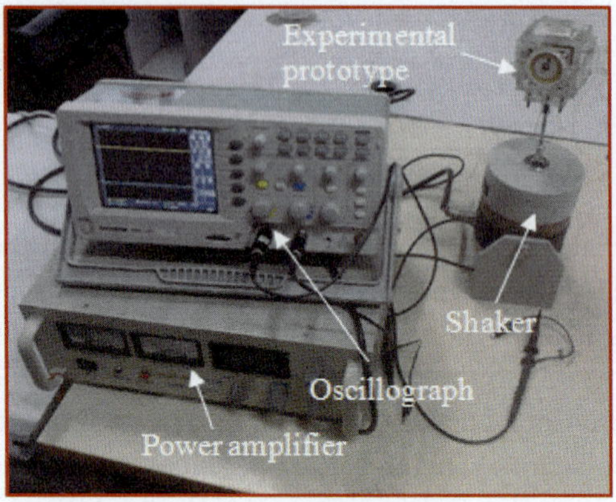

Fig. 3. Test system of experimental prototype .

Figure 4 shows the relationship between output voltage and excited frequency under the vertical vibration. The curves in Figure 4 show that output voltage of piezoelectric plate 1/2/3/4 has the similar variation trend along with the change of vibration frequency. When excited frequency reaches around 13Hz, output voltage of piezoelectric plates 1/2/3/4 is maximum, and voltages are 6.2/4.4/8.6/9.8V respectively.

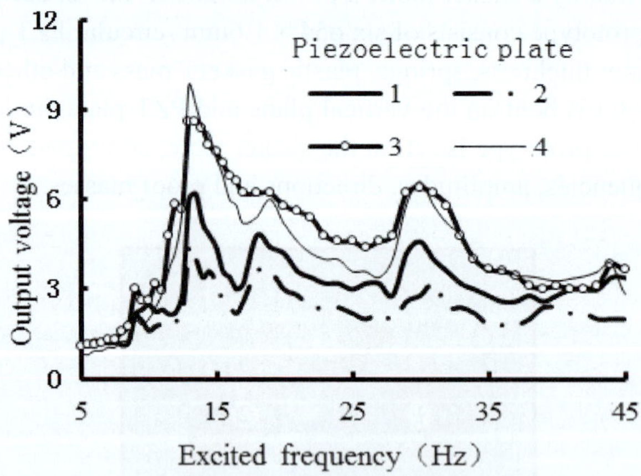

Fig. 4. Output voltage vs frequency under vertical vibration. (Amplitude=1.0mm / Mass=120g).

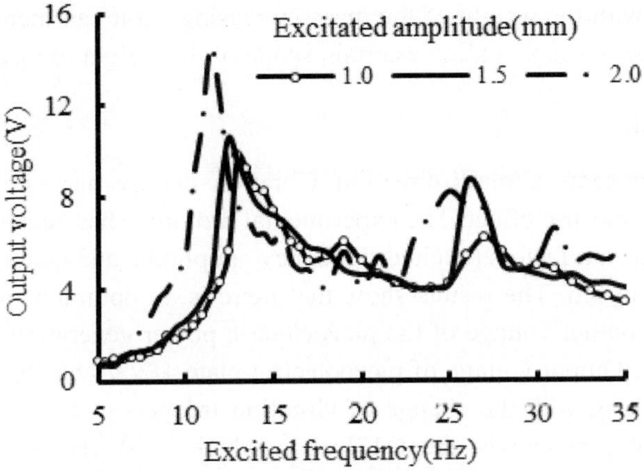

Fig. 5. Output voltage vs frequency of PZT plate 4 (Mass=100g).

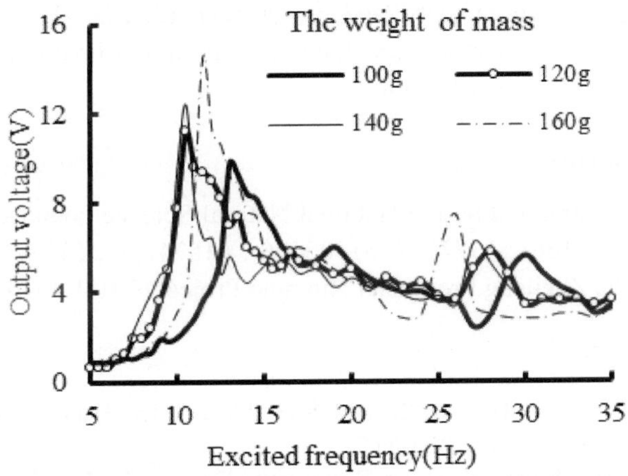

Fig. 6. Output voltage vs frequency of PZT plate 4 (Amplitude=2.0mm).

Figure 5 shows the relationship between output voltage and excited frequency under vertical vibration. From curves can be seen that when the experimental prototype vibrates vertically, several resonance frequencies decrease with the rise of excited amplitude, but peak voltages increase. The optimum frequency is at range of 10-15Hz in the test, which is much lower than the natural frequency of piezoelectric plate (above 1000Hz). MDVEH can generate the maximum voltage about 15V. To sum up, the presented experimental prototype can complete the recycle of low-frequency vibration energy.

Figure 6 shows the relationship between output voltage and excited frequency under different mass. It is clearly that within a certain range, the peak voltages rise with the weight of the mass increasing, so it can increase the generating capacity through adding a certain scope of the weight of mass.

4. Conclusions

This paper presents a multi-direction vibration energy harvester (MDVEH) based on piezoelectric effect. The experimental prototype has been designed and made. The relevant factors (excited frequency, amplitude and quality of mass) is tested and analyzed. The results show that there is an optimum frequency that can make the output voltage of the piezoelectric power generation device reach the maximum. Output voltage of piezoelectric plate 1/2/3/4 has the similar variation trend along with the change of vibration frequency. The best frequency measured in experiment is about 13Hz, in which MDVEH can generate the maximum voltage about 15V. So MDVEH can be used in low-frequency vibration energy recycling. Increasing in a certain scope of the weight of mass, can increase the generating capacity, and power generation is relatively stable, so MDVEH can realize low-frequency, high strength, multi-direction vibration energy reliably recovery and utilization.

Acknowledgements

This work was supported by the National Natural Science Foundation of China (Grant No. 51277166, 51377147, 51305406 , 61574128 , 51577173) and was supported by the Zhejiang Provincial Xinmiao Project (2014R404035).

References

1. Szarka G.D., Stark B.H. & Burrow S.G. 2012. IEEE Transactions on Power Electronics, Vol.27, pp. 803-815.
2. Harb A. 2011. State-of-the-art. Renewable Energy, 36(10): 2641-2654.
3. Khaligh A., Zeng P., Zheng C. 2010. state of the art. Industrial Electronics, IEEE Transactions on, 57(3): 850-860.
4. Saadon S. & Sidek O. 2011. Energy Conversion and Management, Vol.52, pp. 500-504.
5. Roundy S., Leland E.S., Baker J, et al. 2005. Pervasive Computing, IEEE, 4(1): 28-36.
6. Fang H.B., Liu J.Q., Xu Z.Y., et al. 2006. Microelectronics Journal, 37(11): 1280-1284.
7. Hou Z.W., Chen R.W., Liu X.J. 2012. Zhendong yu Chongji (Journal of Vibration and Shock), 31(16): 33-37.
8. Kan J.W., Xu H.L., Wang S.Y., et al. 2013. Journal of vibration. The test and diagnosis, 33 (5) : 814-818.

Implementation of current-mode low bandgap reference voltage circuit for power management applications

M.C. Lee[1†] and M. C. Hsieh[2]
Department of Electronic Engineering, Oriental Institute of Technology, New Taipei City, Taiwan, R.O.C.
†Email. lmchin@mail.oit.edu.tw

B. L. Chen[3]
Department of Communication Engineering, Oriental Institute of Technology, New Taipei City, Taiwan, R.O.C.

This paper proposes a current-mode low bandgap reference voltage circuit with low temperature coefficient and independent of supply voltage for applications to low voltage power management integrated circuits. This proposed circuit is design and implemented using the TSMC 0.18μm 1P6M CMOS process technology. Based on simulated and measured results , the chip size is $0.58303 \times 0.50206mm^2$ with power dissipation about $0.43mW$, and the operation temperature range form $-20°C$ to $100°C$ with temperature coefficient about $15ppm/°C$. The chip supply voltage can from $1.0V$ to $1.8V$ with PSRR about 85dB, and its output reference voltage can stable at $0.59mV$.

Keywords: Bandgap reference voltage circuit; Temperature coefficient; Power supply rejection ratio; Start-up circuit; Bias current source circuit.

1. Introduction

All kinds of portable electronic products in the current market have an orientation toward light, slim and small. They have both a variety of applications and long operation time. Thus a high-efficiency power management module is necessary such as low-power power converter or low dropout regulator to integrate with mixed-mode signal circuit into a single chip to achieve the different kinds of low-power applications [1]. Therefore it requires a low bandgap reference voltage circuit to offer different reference bias voltage for power management circuits and mixed-mode IC. The conventional bandgap reference voltage circuit such as the use of NMOS OPA architecture requires a high input of the power supply voltage and a higher consumption of whole circuit, and the output reference voltage of more than 1V [2]. In this paper we design a current-mode low bandgap reference voltage circuit consisting of low bandgap reference voltage core circuit, PMOS two-stage operational amplifier, reference current source

---

[1] Min-Chin Lee
[2] Ming-Chia Hsieh
[3] Ban-Li Chen

circuit and start up circuit. This circuit architecture using PMOS input of OPA, with the larger of the input common-mode voltage range, and can reduce the power consumption of the whole circuit. This proposed circuit is design and implemented using the TSMC 0.18μm CMOS 1P6M process. The chip can output $0.59mV$ reference voltage source stably, and this circuit can be used in various application of low-power power management module.

2. Principle and Structure of Current-Mode Low Bandgap Reference Voltage Circuit

The architecture of the proposed current-mode low bandgap reference voltage circuit (BGRV) is shown in Figure 1. The input supply voltage range of this circuit is from 1.0V to 1.8V and the output reference voltage can stable at 590mV. This BGRV consists of the start-up circuit, two-stage OPA, bias current source circuit, current- mode low BGRV core circuit (bandgap core).

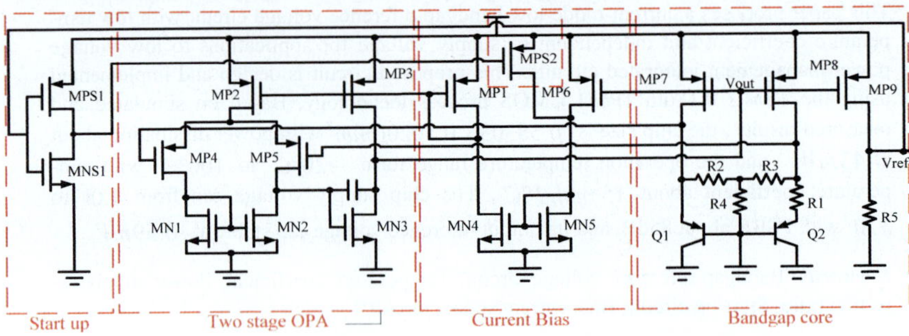

Fig. 1. Proposed current-mode low bandgap reference voltage (BGRV) circuit.

The current-mode low BGRV circuit architecture using PMOS input OPA, with the larger of the input common-mode voltage range, and can reduce the power consumption of the whole circuit.

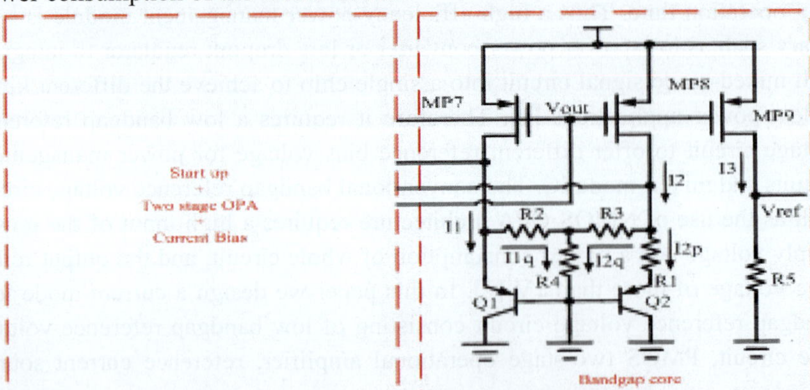

Fig. 2. Core circuit of current-mode low-BGRV circuit.

Figure 2 shows the core circuit of current-mode low BGRV circuit and its two-stage PMOS input OPA that is composed of the PMOS differential pair stage (MP31, MP32 MP35,MN33,MN34) and CS amplifier stage (MN36,MP37). Besides the lower power consumption compared with traditional bandgap reference voltage circuit, the output value of V_{ref} can be changed by adjust the value of R1, R3, R5 and R6, and its output reference voltage can be as low as 1V [3, 4]. The following is the current- mode low BGRV circuit output reference voltage source of derivation, as follows:

$$I_3 = I_2 = I_1 \tag{1}$$

$$\therefore I_3 = I_D + I_C = I_B + I_A \tag{2}$$

$$\because V_{EB2} = V_{EB1} + I_C R_1 \tag{3}$$

$$\because V_{EB} = V_{EB2} - V_{EB1} = I_C R_1$$

$$\therefore I_C = \Delta V_{EB} / R_1 = V_T \ln(n) / R_1 \text{, and } I_2 = I_C + I_D \tag{4}$$

$$I_2 = \frac{V_T \ln(n)}{R_1} + \frac{V_{EB1} + I_C R_1}{R_2 + R_4} \text{, and } I_1 = I_B + I_A \tag{5}$$

$$I_1 = \frac{V_{EB2}}{R_5 + R_6} + \frac{V_T \ln(n)}{R_1} \text{, and } V_{EB2} = V_{EB1} + I_C R_1 \tag{6}$$

Therefore, the reference Output Voltage V_{ref} as :

$$V_{ref} = R_3 I_3 = R_3 \left(\frac{V_{EB2}}{R_5 + R_6} + \frac{V_T \ln(n)}{R_1} \right) \tag{7}$$

In this paper we focus on the analysis and design of component units of the current-mode low BGRV circuit.

3. Analysis and Design of Current-Mode Low BGRV Circuit

3.1. *Start- up circuit*

The start-up circuit of the proposed current-mode low BGRV circuit is shown in Figure 3, it is composed of MPS1, MNS1 and MPS2. We used a inverter (MPS1, MNS1) that is temporarily closed at power on, thus forcing the gate of the MPS2 to a low voltage causing a current to flowing the bias current circuit, the current-mode low BGRV core circuit and the two-stage OPA, and the core circuit starts operating.

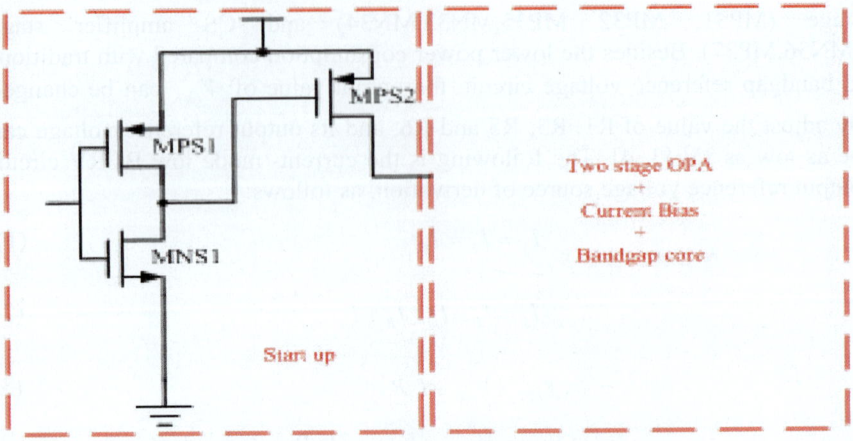

Fig. 3. Start-up circuit of current-mode low-BGRV circuit.

Later, when the current-mode low BGRV core circuit starts operating, it will slowly reach the desired low reference output voltage back to the inverter will be forced to make MPS1 of gate voltage drop, so MPS1 turn ON, and MNS1 turn OFF, at the gate of MPS2 generate a high voltage, MPS2 turn OFF, then this start-up circuit is turned off. [4].

3.2. *PMOS two stage operational amplifier*

The two stage PMOS input OPA plays an important role in current-mode low BGRV circuit, shown in Figure 4 [5, 6]. The greater the operational amplifier output swing the greater the ability to adapt the change of the input power supply. The high gain range of amplifier is also greater. In order to make the overall OPA have sufficient phase margin, Miller compensation capacitor must be added in the path of second stage CS output feedback to the first PMOS input differential stage output.

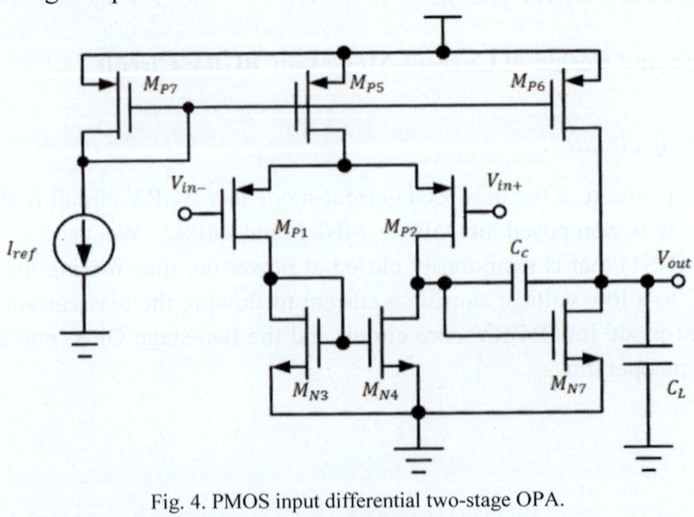

Fig. 4. PMOS input differential two-stage OPA.

Although by capacitance Miller compensation allows phase margin increase, the circuit becomes more stable, but relatively unity gain bandwidth will narrow, which is to be noted that, in Table 1 is the proposed current-mode low BGRV circuit two stage OPA's specifications.

Table 1. PMOS input differential two-stage OPA specification.

Low Frequency Gain, AV(dB)	83.9
Phase Margin PM, (degree)	55.1
Unit Gain Bandwidth, UGB(MHz)	69.1
Common Mode Rejection Ratio, CMRR (dB)	89.2
Power Supply Rejection Ratio, PSRR (dB)	88.3

3.3. *Bias current source circuit*

The PMOS two-stage OPA requires a stable tail bias current source I_{REF}. It is realized through a bias current source circuit which is independent of supply power and environment temperature [5, 6]. As in Figure 5 its output reference bias current I_{out} is shown below, and the start-up circuit is composed of MPS1, MNS1 and MPS2 as shown Figure 3.

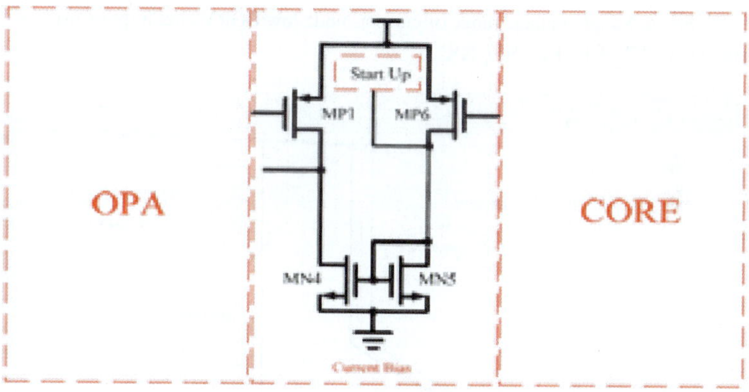

Fig. 5. Bias current source circuit of PMOS two-stage OPA.

When the start-up circuit (MPS2) starts to drive the current mirror circuit, so that the bias reference current source circuit starts to operate, then driven two stage OPA and the current-mode low BGRV core circuit, respectively.

4. Simulation and Measured Result

By using TSMC 0.18μm CMOS 1P6M process to simulate the designed current-mode low BGRV circuit, the result of the pre-sim and post-sim under five different corners is shown below

4.1. *Supply voltage range*

Figure 6 and 7 show the results of pre-sim and post-sim that the reference output voltage V_{ref} stable at $0.596V$ steadily under five different corners with the range of input voltage from $1.2V$ to $1.8V$.

Pre-sim: TT, FF, FS, SF, SS

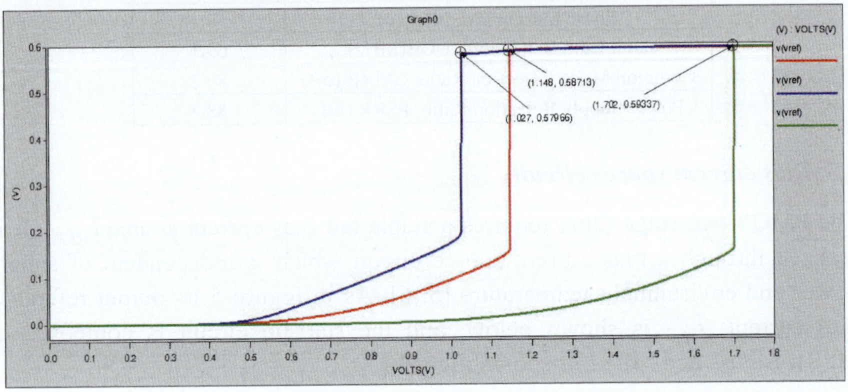

Fig. 6. Supply voltage range of current-mode low BGRV circuit (pre-sim).

Post-sim: TT, FF, FS, SF, SS

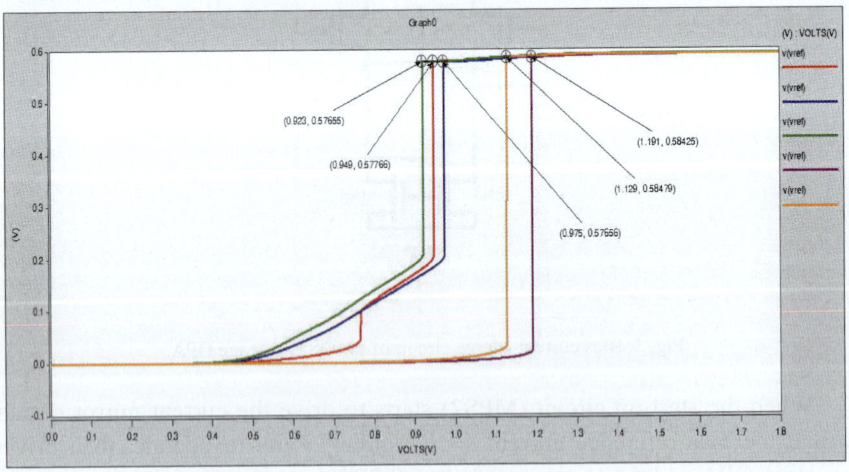

Fig. 7. Supply voltage range of current-mode low BGRV circuit (post-sim).

4.2. *Temperature coefficient*

Figure 8 and 9 show the results of pre-sim and post-sim that temperature coefficient of circuit about $10 \sim 20\,ppm/°C$ under the operation temperature from $-20°C$ to $100°C$. The variation of reference output voltage under five different corners at $25°C$ is about $1.84mV$.

Pre-sim: TT(19.43 ppm), FF(36.45 ppm), FS(13 ppm), SF(33.76 ppm), SS(12.26 ppm)

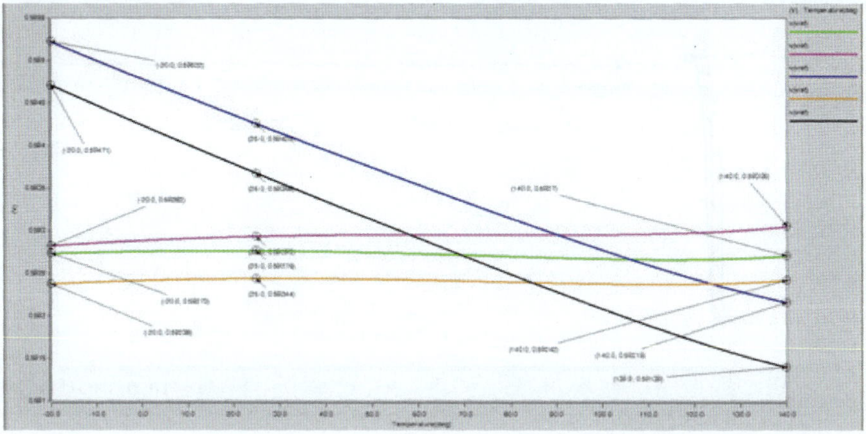

Fig. 8. Temperature coefficient of current-mode low BGRV circuit (Pre-sim), vref=0.59V.

Post-si: TT(11.1 ppm), FF(25.5 ppm), FS(11.8 ppm), SF(22.8 ppm), SS(13.8 ppm)

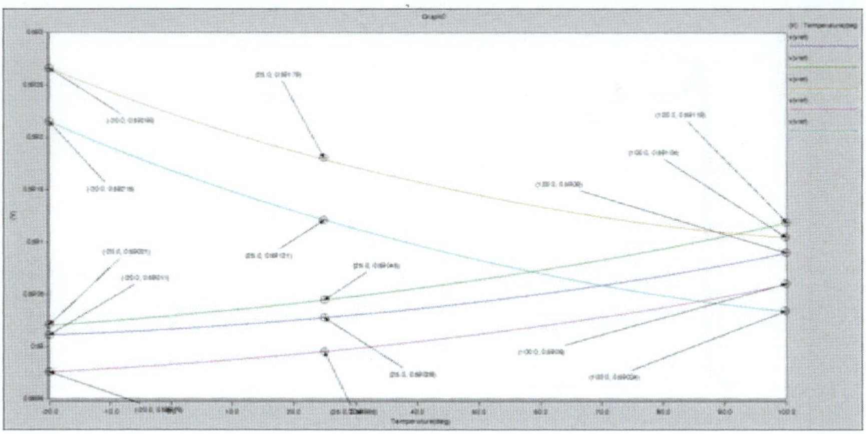

Fig. 9. Temperature coefficient of current-mode low BGRV circuit (post-sim), vref=0.59V.

4.3. *Transient response*

Figure 10 and 11 show the output voltage of the designed current-mode low BGRV circuit can stable at $0.596V$ steadily under five different corners with input source voltage at $1.8V$. And the both steady-state time are about $50us$ such that the output voltages are all at $0.596V$ steadily.

Pre-sim：

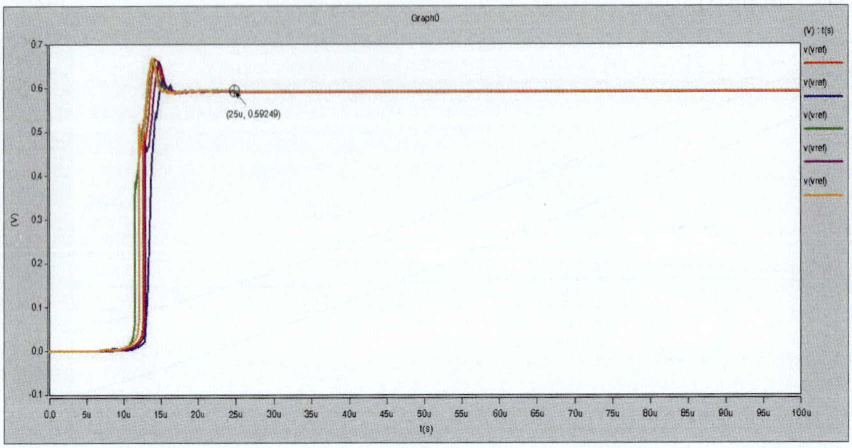

Fig. 10. Transient response of current-mode low BGRV circuit (pre-sim).

Post-sim：

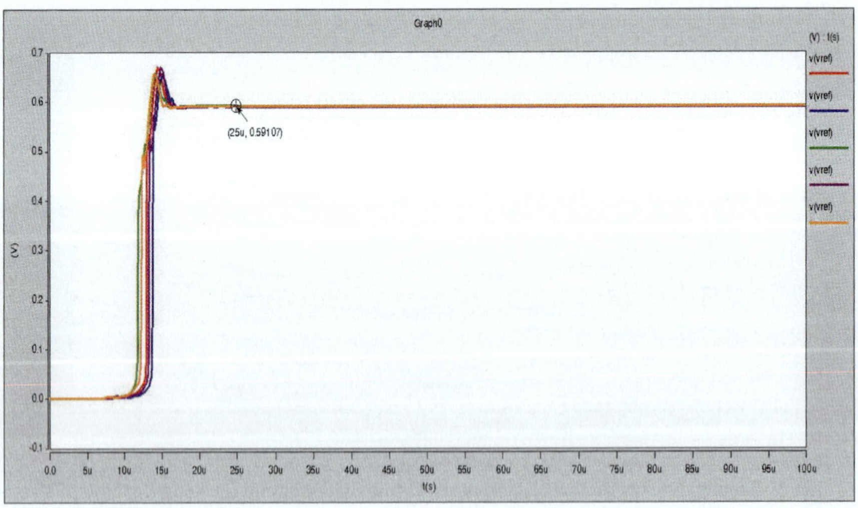

Fig. 11. Transient response of current-mode low BGRV circuit (post-sim).

4.4. *Measurement results*

After summarizing the analysis and design of all component circuits, the physical layout of the proposed current-mode low bandgap reference voltage circuit is shown in Figure 12 and the chip size is about $0.58303 \times 0.50206mm^2$ [7, 8]. The PCB board for chip characteristics measurement is shown in Figure 13.

Fig. 12. Layout of current-mode low BGRV circuit.

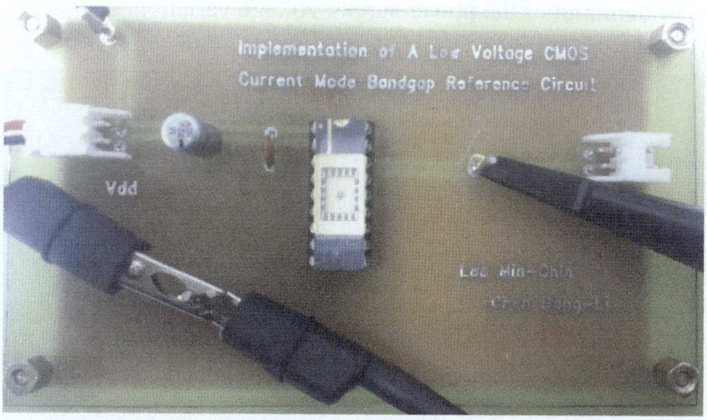

Fig. 13. Measurement PCB of current-mode low BGRV circuit.

From the measurement result in Figure 14, it shows that the range of input power supply voltage (yellow) is from $1.0V$ to $1.8V$. The output reference voltage V_{ref} (blue) of this current-mode low BGRV circuit stable at $0.597V$.

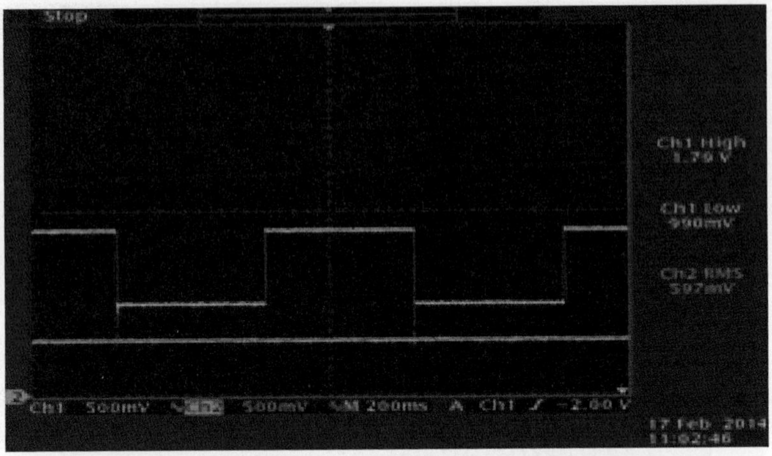

Fig. 14. Output reference voltage of chip measurement.

From the measurement result in Figure 15, it shows that the input power supply voltage (yellow) is changed to ± 10%, measured supply voltage variation, the results show the chip output reference voltage (blue) stable at 0.597V, regardless of the input of the power supply voltage variation.

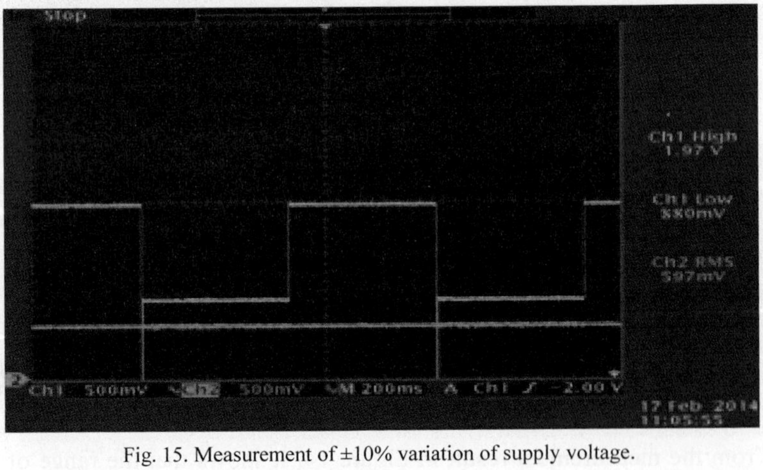

Fig. 15. Measurement of ±10% variation of supply voltage.

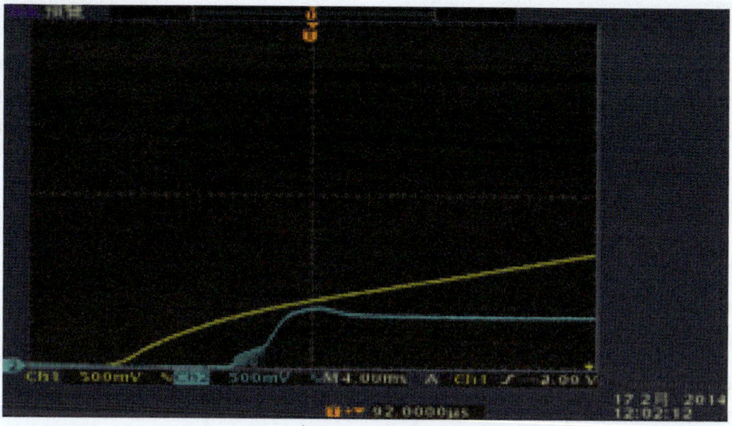

Fig. 16. Transient response of chip measurement.

Figure 16 is the transient response measurements of the chip, from the left that when input power supply voltage (yellow) is from $0V$ to $1.8V$ time to be $45ms$, and from the right is amplified can be seen more clearly in this current-mode low BGRV circuit with the output reference voltage V_{ref} (blue) stable at $0.597V$, and steady-state time is about $7ms$.

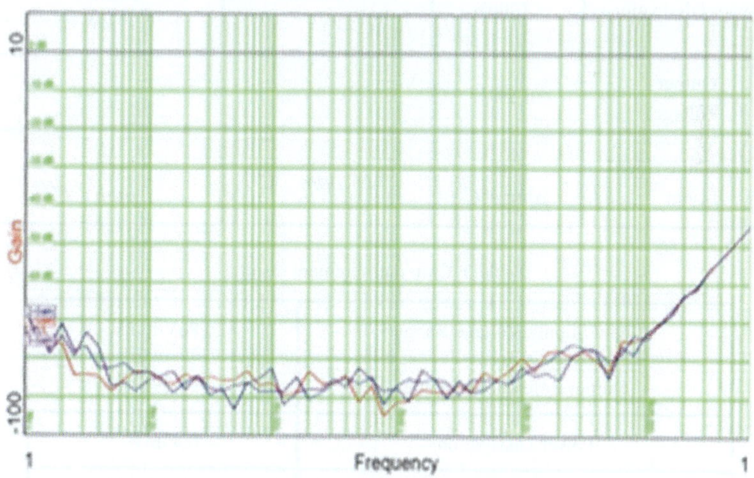

Fig. 17. Power supply reject ratio (PSRR) of chip measurement.

Figure 17 shows the measured result of power supply reject ratio (PSRR) of the current-mode low BGRV circuit for frequency domain. When the input supply voltage $1.8V$ is series connection with an sinusoidal signal having $50mV_{p\ p}$, the output reference voltage of this chip is also at $0.597V$ stably. Figure 10 is the result that we use frequency response analyzer to measure the PSRR, the measured result shows the chip suppress noise frequency range of about within 1Hz ~ 100KHz, PSRR are about -85 dB.

5. Conclusion

Based on the aforementioned discussions, we can conclude that the implemented current-mode low bandgap reference voltage circuit has a range of input supply source voltage $1.0V$ to $1.8V$, 85dB PSRR, and temperature coefficient about $15 ppm/°C$ under the operation temperature range form $-20°C$ to $100°C$. The chip size is about $0.58303 \times 0.50206 mm^2$ with power dissipation about 0.43mW and the output reference voltage maintain a stable value of $0.59V$. Finally, simulation and measured results are summarized in Table 2 and 3 that shows the current-mode low bandgap reference voltage circuit suitable for the low power management module applications

Table 2. Comparison between expected specification and pre-sim, post-sim result.

Parameter	Specification.	Pre-sim	Post-sim
Technology	TSMC 0.18μm 1P6M CMOS		
Input Supply Voltage Range	1.2V to 1.8V	1.2V to 1.8V	1.2V to 1.8V
Power Dissipation	< 100uW	575.5949 uW	572.8377 uW
Output Voltage	0.59V	0.59V	0.59V
Temperature Coefficient	< 15 $ppm/°C$	12.44 $ppm/°C$	14.1 $ppm/°C$

Table 3. Chip measurement result.

Input Supply Voltage Range	0.86V to 2V
Output Reference Voltage	0.59V
Power Dissipation	0.43mW
PSRR	-85dB

References

1. K. N. Leung, P. K. T. Mok, and S. K. Lau, "A Low-Voltage CMOS Low-Dropout Regulator with Enhanced Loop Response" IEEE International Symposium on Circuits and System, vol.1, pp.385-388, May 2004.

2. Ka Nang Leung, and Philip K. T. Mok, "A Sub-1-V 15-ppm C CMOS Bandgap Voltage Reference Without Requiring Low Threshold Voltage Device, " IEEE J. Solid-State Circuits, vol. 37, pp. 526-530, April 2002.
3. A. Boni, "OP-amps and Startup Circuits for CMOS Bandgap References with Near 1-V Supply", IEEE J. Solid-State Circuits, vol. 37, no. 10, pp. 1339-1343, Oct. 2002.
4. Mok, P.K.T and Ka Nang Leung, "Design Considerations of Recent Advanced Low-Voltage Low-Temperature-Coefficient CMOS Bandgap Voltage Reference", IEEE Custom Integrated Circuits Conference, pp. 635-643, Oct. 2004.
5. Phillip E. Allen and Douglas R., Holberg, "Analog Circuit Design", Saunders College Publishing HBJ, 2004.
6. Behazad Razavi, "Design of Analog COMS Integrated Circuits", International edition, McGraw-Hill, 2001.
7. Hong-Yi Huang, "Lab for Analog Integrated Circuits Design", 2003.
8. Hong-Yi Huang, "Analysis of Mixed-Signal IC Layout", 2006.

The analysis and calculation of the safety of the human body in the DC earth electrode field

T. T. Guo[1†] and X. S. Wen[2]

School of Electrical Engineering, Wuhan University, Wuhan, China
†Email: 1316714851@qq.com

Due to the rigor of location condition of DC earth electrode field and the difficulty of the site conditions and the difficulty of the land requisition, there may be rivers and lakes near the DC earth electrode field. In this paper, a DC pole which has a river near the DC earth electrode field is established in CDEGS, the electric field intensity and the current flowing through human body in the river are calculated, the safety of the human in the river is analyzed, and the influence of the electrical resistivity of water in the river on the electric field strength of the water is analyzed. According to the results of calculation, when HVDC system under the operation of mono-polar grounding circuit, the electric field intensity in the rivers and lakes can't meet the requirements of regulation, but it can guarantee the safety of human body, because of the difficulty of relocation, it is recommended that the land of DC earth electrode field can be enclosed in a railing and the underwater part can be enclosed in the net.

Keywords: DC earth electrode field; River; Human body; Safety.

1. Introduction

When HVDC system under the operation of monopolar grounding circuit, there will be great DC current flowing into the earth, and a large potential gradient is formed near the DC pole. Due to the rigor of location condition of DC earth electrode field and the difficulty of the site conditions and the difficulty of the land requisition, not much of the DC earth electrode field meet the requirement, and some DC earth electrode field have rivers flowing through the field. According to regulations of DL/T 5224-2005, the electric field intensity in the river near the DC pole can't be more than 1.25V/m [1-4]. With the increase of DC transmission capacity, when HVDC system under the operation of monopolar grounding circuit, it is difficult to meet the requirements of the electric field intensity in the river, and because of the difficulty of relocation, therefore, under the premise of ensure the safety of human body in the river, take isolation measures to protect creatures in the river is also a kind of way [5-8].

---

[1] Ting-Ting Guo
[2] Xi-Shan Wen

There is a DC earth electrode field with a river flowing around the DC pole three surfaces, because of the abrupt change of the soil resistivity between the water and soil interface, the potential and electric field intensity in river will be distorted. Considering that there may be human in the river, such as swimming in the river, at this time, the human resistance will greatly reduce, and there is no contact resistance between the feet and ground when walking which can greatly reduce the current flowing through the human body, so, the current flowing through the body may exceed the safety value. It is necessary to consider the influence of river on the soil model when establishing the DC pole model, and the electric potential and electric field intensity in river need to be analyzed, and the safety of human in the river is also need to be checked.

2. The Establishment of a DC Pole Model with River

The parameter of DC pole: the DC pole is vertical grounding pole electrode, and the buried depth is 5m, the length of each vertical grounding electrode is 30m, the spacing between each vertical grounding electrode is 20m, each vertical grounding electrodes are connected by feeder cable. The grounding electrodes are double loop, and the outer and inner ring respectively has 38 and 23 grounding electrodes. The feed rod is high silicon iron, the outer diameter of the coke is 1.2m. The river wide is 20m, the river deep is 5m, the resistivity of the water is 50Ω•m, the distance between river and DC pole is 10m. The final CDEGS model is shown in Figure 1:

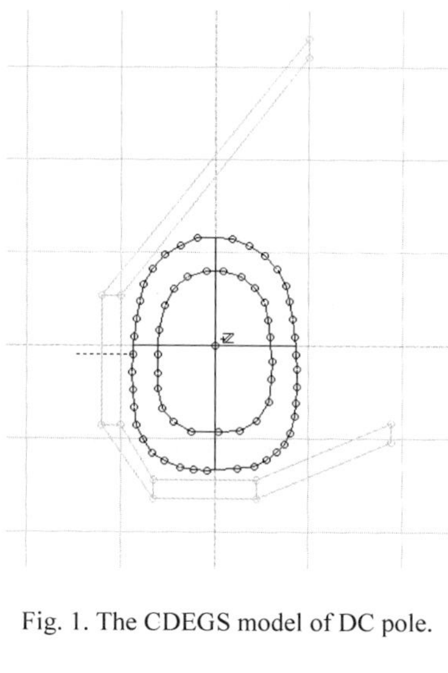

Fig. 1. The CDEGS model of DC pole.

The soil model is layered, and the parameters of the soil model are shown in Table 1:

Table 1. The test results of soil resistivity.

The number of layers	Thickness(m)	soil resistivity(Ω•m)
1	3	600.00
2	7	120.00
3	140	75.00
4	150	105.00
5	7000	98.00
6	30000	18.00
7	∞	83.00

In the CDEGS, the human body is simulated by a coated conductor, the resistance of human body is 500Ω, the height of human body is 1.8m, the coating radius is 0.25m, so, the resistivity of the coating is:

$$\rho=\frac{R\times S}{l}=\frac{500\times\pi\times 0.25^2}{1.8}=54.542\Omega\cdot m$$

To ensure the safety of the human, the calculation current of the DC pole is the maximum transient current, the size of which is 4539A and the function time is 3s.

If the human body height is h, the body resistance is R, the electric field intensity in the river is E, then the current I of the human body is:

$$I=\frac{E\times h}{R}$$

According to the formula, the higher person is, the greater the human body withstand the potential difference, and the body resistance is smaller, the larger the current is flowing through the body, for safety reasons, to take the body height of 1.8m, the human resistance is 500Ω.

3. The Analysis of Simulation Result

3.1. *The simulation result*

The potential distribution of different depth in horizontal direction is shown in Fig. 2:

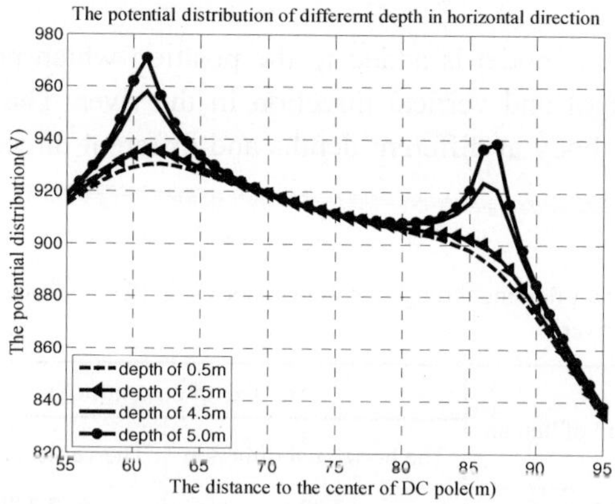

Fig. 2. The potential distribution of different depth in horizontal direction.

As shown in Fig. 2, the DC pole center is located in the origin of X axis, and the inner and outer rings of DC pole is located in the two spikes of the potential distribution, and the river is located in the region of the X axis direction of 100 meters to 120 meters. We can see in Fig. 2 that the potential of the interface between the river and the soil has a small margin of the mutation, and the potential of the river near the DC pole is less than that without the river and the potential of the river far from the DC pole is increased compared to that without the river, moreover, the deeper the depth, this kind of mutation is more obvious.

The potential distribution of different depth in vertical direction is shown in Fig. 3, compared to Fig. 2, the potential of the DC pole in the vertical direction is smaller than that in horizontal direction, this is mainly due to the DC pole is not the standard ring, what's more, the length of the vertical direction is longer and there are rivers flowing through in both directions which is supportive of bulk flow.

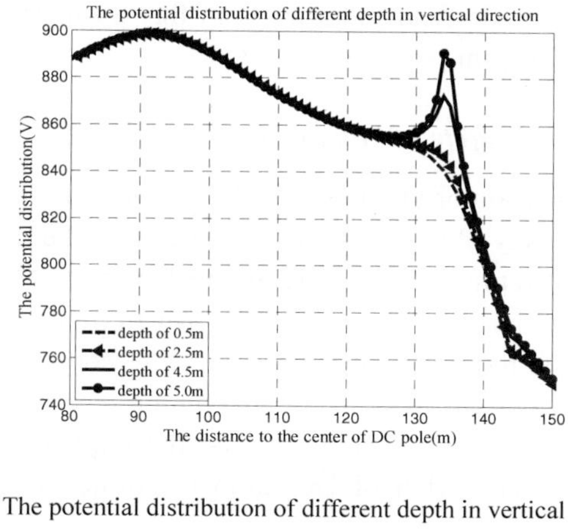

Fig. 3. The potential distribution of different depth in vertical direction.

The human body model is added to the position which nearest to the DC pole in the horizontal and vertical direction in the river. The current flowing through the human body at different depths and different directions in the river is shown in Table 2:

Table 2. The current flowing through the human body at different depths and different directions in the river.

The position of human	The size of current(mA)	
	The horizontal direction	The vertical direction
0.5	7.870	7.234
2.5	8.905	8.752
4.5	15.914	17.237

The largest current flowing through the human body is in the 4.5 meters deep at vertical direction, reached 17.237mA.

When the calculated current is rated current, namely 3125A, the electric field intensity at different depths and different directions in the river is shown in Table 3:

Table 3. The electric field intensity at different depths and different directions in the river.

The position of human	The electric field intensity(v/m)	
	The horizontal direction	The vertical direction
0.5	1.525	2.231
2.5	1.626	2.404
4.5	1.921	2.890

The maximum electric field intensity is in the 4.5 meters deep at vertical direction, reached 2.890V/m.

3.2. *The analysis of the safety of human body*

According to the provision of DL/T 5224-2005 that for DC pole near the pond, the electric field intensity of any point in the water can't be more than 1.25V/m under normal current rating. This rule is in consideration of the safety of fish, according to the experiment, the voltage of a fish from start to finish is up to 1V~5V (depending on the size of the fish), will lose consciousness. According to the results of previous calculations, the maximum electric field intensity has reached 2.890V/m, which significantly exceed the regulations. Therefore, it's necessary to enclose the region of DC earth electrode field in water by the network to reduce damage to fish. Compared to fish, the ability to tolerance DC current of human is stronger; next, the safety of human will be analyzed.

Usually, the minimum current which the human body can sense is 2mA, the relationship of the allowable DC current and the time is shown in Fig. 4.

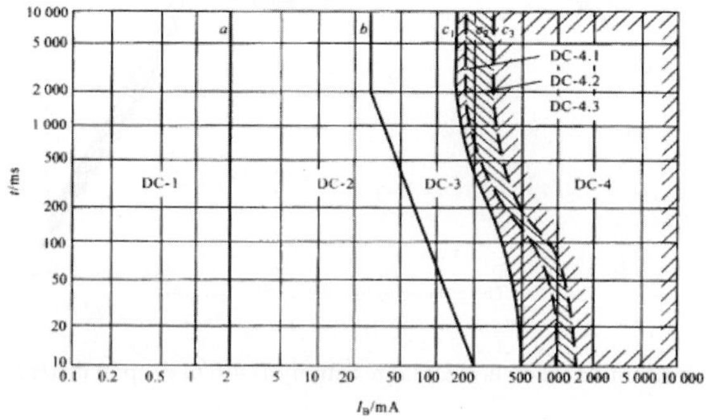

Fig. 4. The relationship of the allowable DC current and the time.

The region shown in the DC-1 is generally not any response area for the general case, DC-2 is generally without harmful physiological function region, DC-3 is not causing death, but it can cause muscle contraction, dyspnea and blood pressure and other symptoms, the region above the curve C1, namely the DC-4, will cause ventricular fibrillation, cardiac arrest and breathing stop and severe burns etc. pathological effects.

As we can see, the max current flowing through the human body is 17.237mA, and the function time is 3s, which belong to the DC-3. Obviously, although there is no danger of life, human will have muscle contraction, dyspnea and blood pressure and other symptoms. To ensure the safety of human, it's also necessary to enclose the region of DC earth electrode field on land by the fence.

3.3. *The influence of water resistance rate on the electric field intensity in the water*

The resistivity of the river water is changed with the change of the factors such as the weather and the temperature, in this section, the resistivity of the river is changed, the influence of the electric potential distribution and the current flowing through the human body is observed. Fig. 4 respects when the water resistance is 20Ω.m, 50Ω.m and 80Ω.m, the potential distribution when horizontal and vertical is 4.5 m deep.

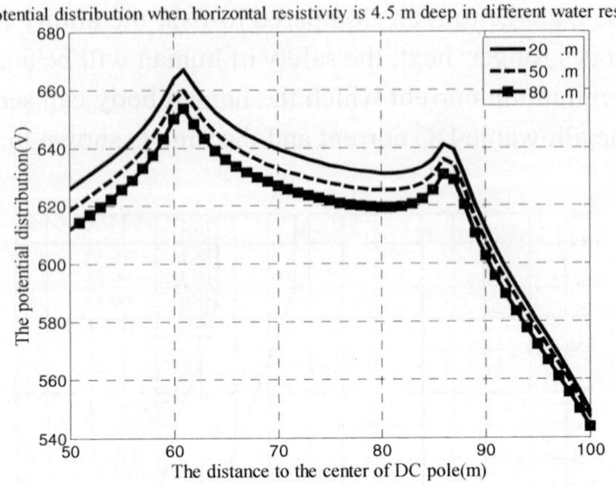

Fig. 5. Potential distribution when horizontal resistivity is 4.5 m deep in different water resistance.

As seen in Fig. 5, the smaller the resistivity of the water is, the potential is higher of the DC pole river side at 4.5 meters deep. This is mainly due to the lower resistivity of the water, will attract a deep current flowing from the surface, more current will flow from the river resulting in potential near the river has increased. Change the resistivity, the potential trend same.

In different water resistivity, potential difference and electric field strength between the two sides of the river as shown in Table 4:

Table 4. Potential difference and electric field strength between the two sides of the river at a depth of 4.5 m at different water resistivity.

Water resistivity (Ω·m)	Horizontal direction		Vertical direction	
	Potential difference (V)	Electric field strength (V/m)	Potential difference (V)	Electric field strength (V/m)
20	35.575	1.779	36.894	1.845
50	38.410	1.921	39.808	1.990
80	46.265	2.313	42.575	2.129

As can be seen from Table 4, although the smaller water resistivity is, the potential is higher, but the potential difference between the two sides of the river is smaller, the electric field intensity is also smaller, so that the current flowing through the human body is also smaller, people will be safer.

4. Conclusion

In this paper, a DC pole which has a river near the DC earth electrode field is established in CDEGS, the calculation results show that the maximum electric field intensity has exceed the current which the fish can tolerance, and the max current flowing through the human body although is no danger to life, human will have muscle contraction, dyspnea and blood pressure and other symptoms. It's necessary to take isolation measures to protect the safety of human and fish.

Due to the low resistivity of the river, will attract a deep current flowing from the surface, more current will flow from the river before, and resulting in potential near the river has increased. Although the smaller water resistivity is, the potential is higher, but the potential difference between the two sides of the river is smaller, the electric field intensity is also smaller, so that the current flowing through the human body is also smaller, people are safer.

References

1. Jinxi Ma, Farid P. Dawalibi. Analysis of Grounding Systems in Soils with Finite Volumes of Different Resistivities [J]. IEEE Transaction on Power Delivery, 2002, 17(4): 596-602.
2. Robert J. Heppe. Computation of potential at surface above an energized grid or other electrode, allowing for non-uniform current [J]. IEEE Transactions on Power Apparatus and System, 1979, 98(6): 1978-1989.
3. Roger. F. Harrington. Field Computation by Moment Method [M]. New York: Macmillan, 1968.
4. IEEE Transmission and Distribution Committee Working Group. Geomagnetic Disturbance Effects on Power Systems. IEEE Trans. on Power Delivery, 1993, 8(3): 1206-1216.
5. R.Pijola. Effects of space weather on high-latitude ground systems. Advances in Space Research. 2005, 36:2231-2240.
6. Boteler D H, Bui-Van Q.; Lemay J. Directional sensitivity to geomagnetically induced currents of the Hydro-Quebec 735 kV power system [J]. IEEE Tran. On Power Delivery, 1994, 9(4):1963-1971.
7. Risto Pirjola. Study of effects of changes of earthing resistances on geomagnetically induced currents in an electric power transmission system [J]. Radio Science, 2008, 43(1):RS1004.
8. Thomas J. Overbye, Komal S. Shetye, et, al. Power Grid Sensitivity Analysis of Geomagnetically Induced Currents [J]. IEEE Tran. On Power Delivery, 2013, 28(4):4821-4828.

Some methods for error estimation for large current transformer

T. Yao[1†], Q.C. Zeng[2], B. Jiang[3] and G.C Lei[4]

University of Electronic Science and Technology of China, Chengdu, China
[†]*Email: tengshaofly@163.com*

Several methods for measuring or estimating error for large current transformer are analyzed in this paper. The methods investigated include the equal ampere-turns method, the low-current deduction method, the low-voltage deduction method and the computer simulation method based on artificial neural network (ANN). Subsequently the advantages and disadvantages, the mathematical modelling and the scope of application of these methods are analyzed and compared. Meanwhile, the prospects of development and the problems to be solved at present of different measurement methods are identified. The analysis and results will provide a reference for the development of the relevant standards in the future.

Keywords: Large current transformer; Error test; Mechanical analysis.

1. Introduction

Large current transformer, short for the LCT, refers to the current transformer with tens of thousands of ampere. As an extensive use of the current transformation equipment in the power system, current transformer transforms the large current of a bus into secondary currents with low amperes, which can supply measuring instruments and relay protection devices. At the same time, the high voltage line current is measured with insulation high-voltage lines, which ensure the safety of staff members.

As the system current becomes much larger in power production domain, particularly the terminal current in generator bus which can reach heavy amperes. The 600MW units have becomes the major unit of new power plants, and 1000MW generators have also begun to be put into operation in our country. However, subject to technical reasons, the rated voltage of the generator bus bar is constant, which makes a larger rated current, brings new problems and great challenges to current transformer.

---

[1] Teng Yao
[2] Qing-Chuan Zeng
[3] Bo Jiang
[4] Guo-Ci Lei

Nowadays, because of the commercial operation mode of separation of power production and transmission, it becomes very important to calculate electric power accurately. For example, in a 600MW generating units, assuming the metering current transformer with 0.1% error under the condition of units full one hour, the economic loss caused by 600kWh, and this situation in the actual operation of the power grid is easily to happen and not that easily to find. Therefore, it is necessary to ensure the accuracy of power measurement with the use of large current transformer accurately.

The secondary metering differential method has been widely used to test large current transformer at an early stage. However, with the development of power grid, the common current transformer cannot meet the demand for electricity. The large current transformer for measurement and protection systems in large generating units has been widely used.

High current generator and high capacity power device that are required for testing current transformer with primary current reaches tens of thousands of ampere in this approach. The total weight of the test equipment is not less than 200kg for a transformer with a rated current of 2000A. The field tests are unlikely to be conducted if the rated primary current is too large. Therefore, it is necessary to study the detection technique of large current transformer.

According to the definition of verification by the international metrology organization, it is required "to find out and to confirm whether the measurement apparatus meets the statutory requirements, including inspection, adding of tags and (or) issuance of inspection certificate." As a compulsory verification of measuring equipment, transformer must be certified, to be traced to the higher accuracy of the electric energy measurement standards before it entered service. Considering that there are no relevant standards to stipulate the verification of large current transformer, the verification shall be replaced with testing or measurement in this article.

Due to the advancement of scientific research, the equal ampere-turns method, the low-current deduction method, the low-voltage deduction method and the computer simulation verification method based on artificial neural network (ANN) have been developed for large current transformer with tens of thousands of amperes. Nowadays, these verification methods based on different principles have been widely used by transformer manufacturers, power sector and research institutions.

2. Error measurement methods for current transformer

2.1. *Equal ampere-turns method*

As the most widely used method, the equal ampere-turns method is based on the law of equal ampere-turns. This traditional "equal ampere-turns method" does not need a large primary current by increasing the number of turns of the primary winding. A magnetic field generated by the primary current that is uniform distributed in the core of large current transformer by twining the primary turns on the current transformer core uniformly. The winding turns equal to the ratio of rated transformation ratio of current transformer and rated transformation ratio of standard current transformer. The primary current value can be sensed by the secondary winding proportionally before the magnetic field of the primary winding of the transformer causes core saturation. It is used to measure the error of current transformer in a real working environment. However, it is unable to simulate unevenly distributed core magnetic field that is caused by a sum of stray magnetic field generated by the external current in a working environment and main magnetic field. Hence, the traditional "equal ampere-turns method" cannot accurately measure the performance parameters of the large current transformer in a real working environment.

For this reason, some researchers presented the non-uniformly-distributed equal ampere-turns verification method. As a result, the traditional "equal ampere-turns method" is called the "uniformity-distributed equal ampere-turns method".

Unlike the uniformity-distributed equal ampere-turns method, the non-uniformly-distributed equal ampere-turns method simulates a transformer of a core with non-uniform magnetic field in an actual working environment. It is done by having concentrated primary turns on one part of the current transformer core, to generate the leakage magnetic field, to also obtain more reliable verification result. Hence, this method is able to reduce the possibility of accident and to ensure the safety and stability of power system operation. Using this method, however, a quantitative link could not be established between the concentrated angle of primary turns on the current transformer core and magnetic field distribution in the actual work environment. Comparing to the uniformity-distributed equal ampere-turns method, the errors tested by the non-uniformly-distributed method is usually very unsatisfactory. And it does not need the large capacity of current source by using equal ampere-turns method to test large current transformer, makes the verification equipment volume decreases, and bring strong operability. Hence, most manufacturers still prefer the traditional ampere-turns method.

2.2. *Low-current deduction method*

The low-current deduction method is also called the load error extrapolation method. It is used for testing error of large current transformer at low current by increasing the secondary load. By estimating the error of the current transformer, we can calculate the actual error by measuring the no-load error, the secondary impedance Z_2 and the excitation admittance in the case of enthetic impedance when the detected current transformer has no nonlinear error compensation [1].

The excitation for core and power loss are provided by the inrush current I_0 in an actual working environment of current transformer. Magnetizing ampere-turns causes the error of the transformer. The composite error $\widetilde{\varepsilon}$ of the current transformer can be expressed by

$$\widetilde{\varepsilon} = \frac{-K_n \dot{I}_2 - \dot{I}_1}{\dot{I}_1} \times 100\% = -\frac{\dot{I}_0}{\dot{I}_1} \times 100\% \tag{1}$$

Where $K_n=N_2/N_1$ is the current transformer turn ratio and I_0 is the excitation current of current transformer. As shown in Eq. (1), the current transformer error is determined by the primary current and the excitation current which are generated by the inrush current.

When the primary current and the secondary current reach certain specified values, the induced electromotive force of the secondary coil E_{2n} is

$$\dot{E}_{2n} = \dot{I}_{2n}(Z + Z_2) \tag{2}$$

Where I2n is the rated secondary current of current transformer, Z, Z2 are respectively the rated load impedance and the equivalent impedance of the secondary coil winding of current transformer.

When the actual primary current Ik1n, the actual secondary current and the actual inrush current are, respectively, k times the rated primary current, the rated secondary current and the rated inrush current, then the actual induced electromotive force Ekn is

$$\dot{E}_{\mathrm{kn}} = \dot{I}_{\mathrm{k}2n}(Z_2 + Z_L) \tag{3}$$

$$\dot{I}_{\mathrm{k}2n} = k\,\dot{I}_{2n} \tag{4}$$

$$Z_L' = Z_L = \frac{1-k}{k} Z_2 + \frac{1}{k} Z \tag{5}$$

Where Ik2n is the actual secondary current of current transformer and ZL is the actual load impedance of secondary winding of current transformer.

As shown in Eq. (2) through Eq. (4), the actual induced electromotive force equals to the rated induced electromotive force when Eq. (5) is established. Hence, we can obtain the same induced electromotive force at the rated primary current for any multiples of the rated primary current by changing the actual load of secondary winding. Here, the magnetizing impedance is substantially unchanged when the core is not saturated. Therefore, for an actual load Z_L and a primary current I_{k1n}, the error $\tilde{\varepsilon}'$ is

$$\tilde{\varepsilon}' = -\frac{\dot{I}_0}{\dot{I}_1} \times 100\% = -\frac{1}{k}\tilde{\varepsilon}$$

Through proper derivation and by choosing an appropriate secondary load, we can calculate the error for different primary current measurement point without a high-current switching device when testing a large current transformer. This method is able to result in considerable savings in man power and financial resources.

2.3. *Other error measurement methods*

In addition to the common ampere turns method and the low-current extrapolation method, some other error test methods are analyzed in the study, including the low-voltage extrapolation method and the computer simulation method

Based on the third kind of the reciprocity theorem, the low-voltage deduction method requires the current transformer as a voltage transformer with the same ratio and accuracy level in the case of magnetic field distribution of the core remains constant [2]. This method is developed by an error test in secondary side method [3]. The problem of special current transformer with no compensation and no internal impedance is solved. High current power supply and large standard current transformer are not required in this method. However, the test signal is vulnerable to an external magnetic field. There is no literature at present for the detection device based on this method that can detect the current transformer with tens of thousands of ampere.

Using measurement and experimental data, the computer simulation method is able to analyze the error of current transformer. It is also able to establish an error mathematical model, to deduce the magnetic field distribution of the running state and to control the measurement process. The artificial neural network (ANN) based on radial basis function neural network (RBF) is used to establish a mathematical model accurately, to substitute measurement error of different secondary loads of the low current in the error mathematical model, and other results from the high current measuring point is obtained.

Considerable number of experimental samples and special calibration devices are required in the computer simulation measurement method. This article will explore whether it can be applied for error test of large current transformer.

3. Traceability

In this paper, a large current transformer refers to a transformer with the primary current at tens of thousands amperes. There is no relevant verification procedure available at present. It is a common practice to refer to the national measurement procedures JJG 313 Verification Procedures of Measuring Current Transformers, JJG 169 Verification Procedures of Transformer Calibrator and JJG 1264 Calibration Rules of Transformer Load Box for testing standard transformer, load box and calibrator. The test device is considered traceability as long as it satisfies the above requirements.

4. Conclusion

1) Comparing to the uniformity-distributed equal ampere-turns method, the transformer error tested by the non-uniformly-distributed method simulate unevenly distributed core magnetic field that is caused by a sum of stray magnetic field generated by the external current in a working environment and main magnetic field , more closes to the working environment and better reflects the actual working condition of current transformer. However, this method does not have a quantitative interpretation and complete theoretical analysis and is more difficult to meet the national standard. As a result, most of manufacturers still prefer the traditional ampere-turns method.

However, should the concentrated angle of primary turns on the current transformer core and the magnetic field distribution in the actual work environment are able to establish a quantitative link and to give a reasonable explanation, it will then be valuable and significant for verification, design and improvement of large current transformer.

2) Similar to the equal ampere-turns method, the detecting device is lighter in the low-current deduction method by eliminating high current power supply. However, this method is only suitable for a transformer with turn compensation. In actual measurement, we can combine the equal ampere-turns method and the low-current deduction method [4], or the low-current extrapolation and the computer simulation method [5].

3) At present, it is yet to be verified whether the low-voltage deduction method is able to provide test error for transformer whose primary current reaches tens of thousands ampere. Currently computer simulation based on ANN is only at a theory stage. Therefore, the establishment of an error model of current transformer is still a subject that worth studying.

Because there is no verification procedure for large current transformer, these methods are widely used by most manufacturers, power sector and research institutes. In this paper, the advantages, disadvantages and prospects of these different methods are discussed as a reference for developing the procedure in the future.

References

1. J.L. Zhang. 2011. Study on calibration methods of current transformer field tester [D]. *Xiamen University.*
2. DL/T668-1999. Instrument transformer testing equipment for measuring service [S].
3. X.M. Zhao. 1988. Large Current Transformer with Error Test in Secondary Side [J]. *Electrical Measurement & Instrumentation* (12).
4. Y.J. Zhou. & L.J. Xu. 2011. On-site error verification method for high current transformer of generator outlet [J]. *Zhejiang electric power* (11).
5. J.X. Wang & L.R. Wang. 2003. Error of huge current transformer verified with computer simulation technology [J]. *Jilin Electric Power* (17).

Research on coordinated charging of electric vehicle based on GA-BP power quality prediction

Z.M. Li
State Grid Electric Power Research Institute, Nanjing, China

T. Lin
Songyuan Power Supply Company, Jilin Electric Power Co., Ltd, Songyuan, China

Y.H. Ma and C. Dong
State Grid Electric Power Research Institute, Nanjing, China

J. Zhou, Y.L. Li†
School of Automation, Nanjing University of Science and Technology, Nanjing, China
†*Email: liyalu2100@163.com*

As the research, the factors affecting power quality includes the parameters of distribution network and the charging mode. In the paper, we make the establishment of a charging station and analyze power quality via Matlab/Simulink. Considering the problem of large workload of Matlab/Simulink and different distribution network, we use GA-BPalgorithm to predict power quality. In the end, we use the analytic hierarchy process to make evaluation of power quality prediction and give coordinated charging suggestion. The results of example show the rationality of this model and method.

Keywords: Electric vehicle; Neural network; Coordinated charging.

1. Introduction

Electric vehicle (EV) is increasingly watched around the world for its significant advantages like high energy efficiency, low noise and zero emissions [1]. As a nonlinear load, EV charger makes a lot of pollution like harmonic current to power grid [2-3]. The researches of coordinated charging mainly focus on grid economy at present. An optimization model was built with the purpose of minimizing the peak-valley ratio based on genetic algorithm [4]. Based on comprehensive analysis of the EV users' requirements, the stochastic dispatching optimal power flow model of power grid by considering of the uncertainty of EVs charging was built [5]. Before the construction of EV charging station, it is necessary to do some calculation and evaluation.

In the paper, we used GA-BP algorithm to predict the changes of power quality under different charging modes. Then, we used the AHP to make evaluation of power quality prediction results. The expert estimation method was proposed to determine index weights. By comparing and analyzing, we give some coordinated charging suggestions in the end.

2. The Model of EV Charging Station

The EV charging station under juxtaposition mode should be inspected of the changes in harmonic of PPC when access the power grid. The power grid with EV charging station is diagrammed in Figure 1.

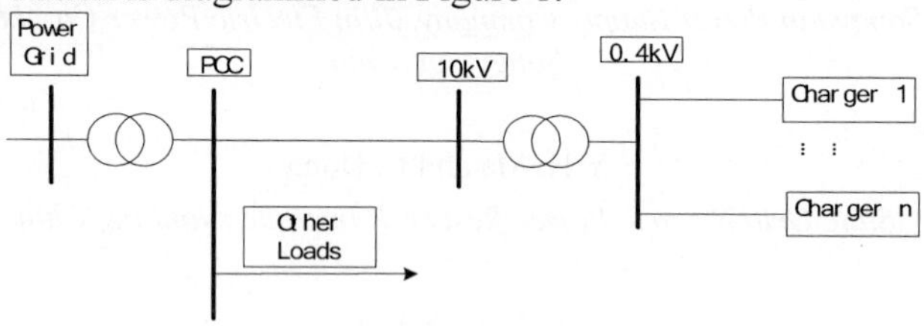

Fig. 1. Power grid with electric vehicle charging station.

Set up the model by Matlab/Simulink. The charging station chooses type I charger and sets up a 6-pulse diode uncontrolled rectifier charger. Adopt 10kv power supply. The line of transformer is LGJ-120/25, whose per kilometer parameters include: r=0.223, x=0.348. Transformers are S9 series, with the voltage of 10kV/0.4kV and Dyn11 connection. Finally, the model of charging station with 6 chargers is established which is shown in Figure 2.

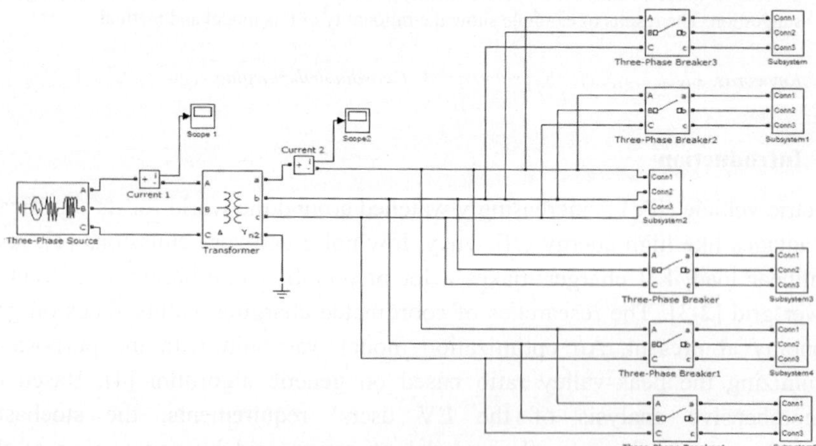

Fig. 2. The model of electric vehicle charging station with 6 chargers.

Through the simulation results, we can obtain the influence of EV chargers to power quality such as harmonic current, voltage deviation and voltage imbalance. The change of harmonic current (5th and 7th) under different size of EV charging station is shown in Table 1.

Table 1. Harmonic current values.

Harmonic Number	Harmonic Limit /A	Harmonic current value/A		
		1 Charger	3Chargers	6Chargers
5	0.76	0.81	2.5	4.8
7	1.06	0.49	1.3	2.35

3. The Prediction of the Change of Harmonic Current by Using GA-BP Network

Considering the large workload and long computing time of Matlab/Simulink, we determine the power grid parameters and predict the change of harmonic current by using GA-BP neural network according to the known quality data.

3.1. *BP neural network and GA*

BP network is also called Back-Propagation Network. With the ongoing of the training of error back propagation, the accuracy of the network response to the input mode is also rising [6]. BP is widely used in large data prediction, but has the drawbacks of easy to fall into local minimum and slow convergence speed.

Genetic Algorithm (GA) can simulate the principle of biological evolution, and can also obtain the optimal solution of global random search by selection, crossover and mutation. Thus there will be widely practical application in the engineering field.

BP algorithm is suitable for local search, while GA is better for global search. So in this paper, we train the initial data of weights and thresholds by GA, and then put them into BP network to form the GA-BP prediction model.

3.2. *Prediction model*

(1) *Coding scheme*

In order to guarantee the stability of the group, we use real-coded schema. That is, each individual is a real number string containing the weight and threshold of each layer.

(2) *Generation of initial population*

Generate an initial population at random:

$$X = \{X_1, X_2, ..., X_n\}^T \tag{1}$$

Where $X_i=\{x_1, x_2, ..., x_n\}$ represents the distribution of initial weights and thresholds. Hence, the length of individuals can be expressed as:

$$n = r \times s_1 + s_1 \times s_2 + s_1 + s_2 \tag{2}$$

Where r is node number of input layer, s_1 is node number of hidden layer, s_2 is node number of output layer.

(3) *Fitness function*

Fitnessindicates the assistance of individual to find the optimal solution. Individuals with higher fitness are more prone to genetic to the next generation. Fitness function is introduced to evaluate the fitness of the individual, according to which we can determine the criteria for individuals. Using the simulation data trains BP network, and the fitness value F can be calculated by the sum of error absolute value between desired output and predicted output:

$$F = k(\sum_{i=1}^{n} abs(y_i - o_i)) \tag{3}$$

Where n is the number of output nodes, y_i is the desired output, o_i is the predicted output, k is a certain coefficient.

(4) *Selection*

Through selecting strong individuals from the initial population, we can generate a new population. Selection can avoid the loss of useful genetic information. The genetic operation of the survival of the fittest can make the fitness of individuals closer to the optimal solution.

This study selected the roulette method. The fitness-proportionate is the probability of it into the next generation.

(5) *Crossover*

Evolution cannot leave the chromosome crossover and recombination. To simulate this phenomenon, crossover operator is proposed to generate new individual. Crossover probability in this study is P_C=0.4. Choose any two individuals and randomly set an intersection in the string, after which exchange part of the structure, so as to produce a new individual. For instance, the crossover operation at intersection j between the kth and the lth chromosome can be expressed as:

$$a_{kj} = a_{kj}(1-b) + a_{lj}b\,, a_{lj} = a_{lj}(1-b) + a_{kj}b \tag{4}$$

Where b is a random value between [0,1].

(6) *Mutation*

GA can also simulate the evolution process of mutation. Mutation probability in this study is P_m=0.2. Select a number of individuals, from which we choose aberrant point at random and do mutation operation. Select the jth gene of the ith individual to do mutation operation:

$$a_{ij} = \begin{cases} a_{ij} + (a_{ij} - a_{\max}) f(g) & r \geq 0.5 \\ a_{ij} + (a_{\min} - a_{ij}) f(g) & r < 0.5 \end{cases} \tag{5}$$

Where $a_{\max}$is the upper bound of the gene, $a_{\min}$is the lower bound of the gene, $f(g) = r_2 (1- g / G_{\max})$, g is the current number of iterations, $G_{\max}$is the maximum number, r is a random value between [0,1].

(7) *Repeat steps from (3) to (6)*

The graph of the process of learning and prediction of GA-BP neural network is shown in Figure 3.

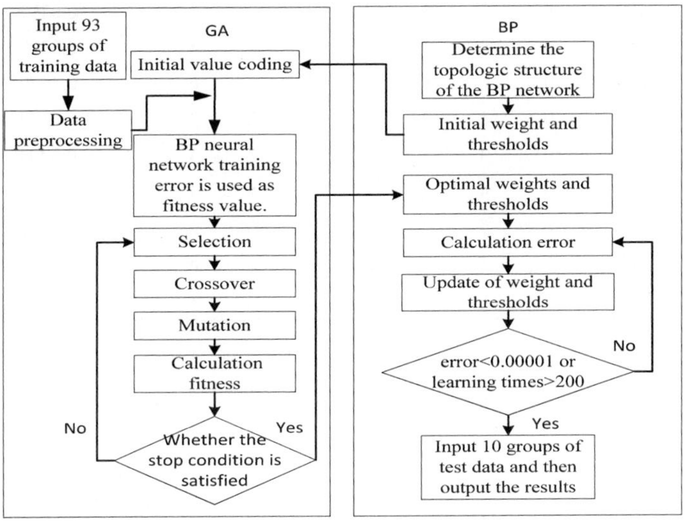

Fig. 3. The graph of the process of learning and prediction of GA-BP network.

3.3. *Prediction results analysis*

In order to prove the prediction effect of GA-BP, we simulate the charging condition of five hours of the EV charging station. Change the number of chargers every 50 minutes and the whole process has a total of 6 time periods (T1~T6).

Choose the capacity of transformer and system, the length of lines and the number of chargers accessed during the six periods as the inputs. Choose the maximum value of the fifth harmonic current in the process of charging as the output. Select 103 groups of data obtained by simulation, among which 93 groups are treated as training data to establish the learning mode of GA-BP and the other 10 groups are treated as test data. After we get the inputs and the outputs, we normalize the data between [0, 1].

The prediction results are shown in Table 2. The comparison between the predicted output and the desired output is shown in Figure 4, so we can see the prediction errors in Figure 5. It can be concluded that GA-BP neural network shows its better predictive ability, the prediction errors of which are all below 0.7% (Table 2). And the improving neural network prediction model has the advantage of better stability, fast convergences and high precision.

Table 2. The prediction result of the fifth harmonic current.

S_T/ MVA	S_N /MVA	L /kM	T1	T2	T3	T4	T5	T6	GA-BP prediction error /%
0.8	289	0	1	1	1	1	1	1	0.6412
1.25	289	0	3	3	3	3	3	3	0.0212
1.6	289	0	4	4	4	4	4	4	0.3777
2.5	289	0	6	6	6	6	6	6	0.3979
1.6	150	0	2	2	2	2	2	2	0.0122
1.6	200	0	3	3	3	3	3	3	0.0903
1.6	350	0	4	4	4	4	4	4	0.4211
1.6	500	0	5	5	5	5	5	5	0.1913
1.6	289	5	1	2	3	4	5	6	0.0031
1.6	289	0	6	5	4	3	2	1	0.0078

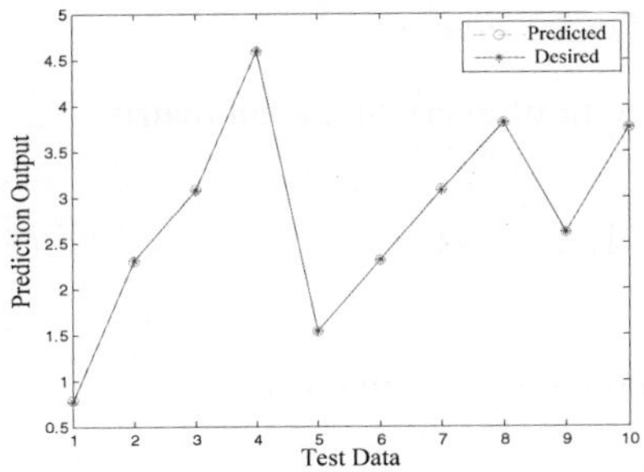

Fig. 4. The prediction output of GA-BP.

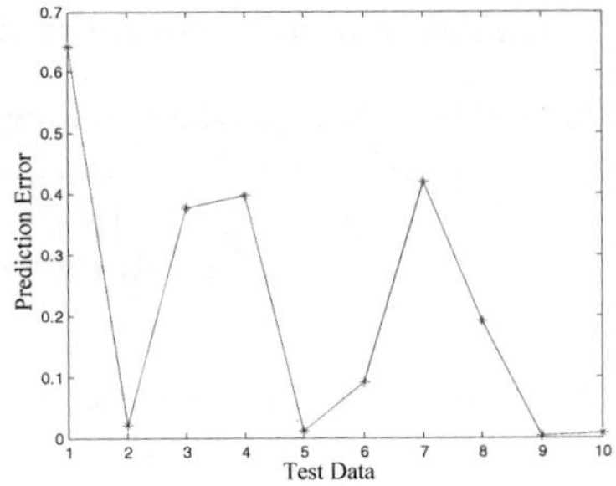

Fig. 5. The predictionerrors of GA-BP.

4. Scheme for Coordinated Charging Based on Analytic Hierarchy Process

As we can see in Section 3, the power quality will change under different charging order. We can find the coordinated charging of EV by unitary quantificationand evaluation.

Analytic Hierarchy Process (AHP) can split the complex problem into a number of layers and indexes. We can build the evaluation system and confirm weights by AHP according to the function and hierarchical structure of charging station [7-8].

4.1. *Evaluation model*

4.1.1. *AHP structure*

The AHP structure is shown in Figure 6.

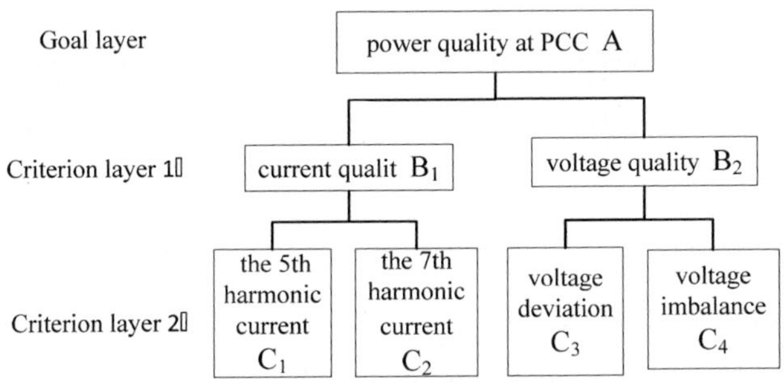

Fig. 6. AHP model of power quality.

4.1.2. *Judgment matrix*

Establish judgment matrix $C = (C_{ij})_{n \times n}$ and find the weight vector in each level through experts' grading by criteria of 1-9 scale method.

4.1.3. *Calculation and consistency test of index weight*

Calculate the largest eigenvalue and eigenvector of every judgment matrix:

$$\overline{W_i} = \sqrt[n]{\prod_{j=1}^{n} a_{ij}} \; (i = 1, 2, ..., n) \tag{6}$$

Hence, we can get the eigenvector W and the largest eigenvalue λ_{max}:

$$W_i = \frac{\overline{W_i}}{\sum_{j=1}^{n} \overline{W_j}} \tag{7}$$

$$\lambda_{max} = \sum_{i=1}^{n} \frac{(AW)_i}{nW_i} \tag{8}$$

When the judgment matrix is not completely consistent, the corresponding value will also change. So we set the consistency index as:

$$CI = \frac{\lambda_{max} - n}{n - 1} \tag{9}$$

The smaller the CI value, the better the consistency. Hence, the index weights of power quality at PCC are shown as follows:

Table 3. Weights of power quality at PCC.

Index	A		Weight of C
	B1 (0.9)	B2 (0.1)	
C_1	0.75		0.675
C_2	0.25		0.225
C_3		0.667	0.067
C_4		0.333	0.033

4.2. *Coordinated charging*

The GA-BP prediction model can predict the power quality under different charging order. In the paper, we transform the quality data into corresponding scores (the total score is 100), such as the 5th and 7th harmonic current and voltage deviation and voltage imbalance. The smaller the data value, the higher the score. At the same time, the weight of the power quality can be obtained by multiplying calculation.

4.3. *Computational examples and analysis*

We simulate the charging condition of five hours of the EV charging station. During the charging period, we cannot remove the chargers. Choose the capacity of transformer as ST=1.6MVA, the capacity of system as SN=289.3MVA, the length of lines as L=0km and the number of chargers accessed during the six time periods as the inputs. The prediction result of 252 access modes is shown as follows:

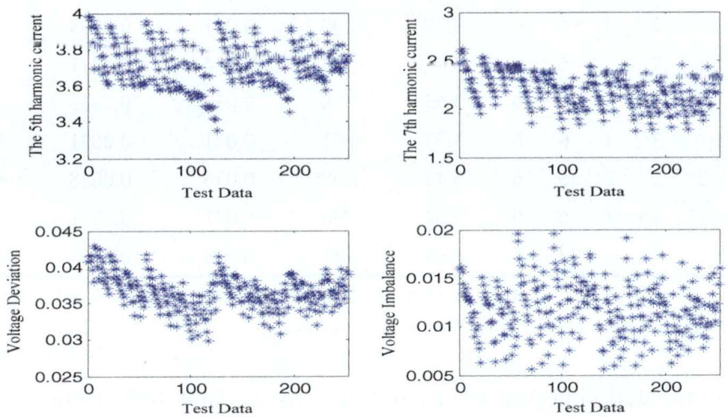

Fig. 7. Harmonic prediction of PPC under 255 access modes.

Calculate the weights of power quality at PCC by using AHP, and then multiply the points of each group, so we obtain the power quality scores in 252 different charge order after relative stacking. The higher score represents the better performance. The result is shown in Table 4, choosing the first 12 groups. By comparison and analysis it can be known that the coordinated charging is:

(1) During the interval T1~T2, we should access charger as little as possible.

(2) During the interval T3~T6, we should access charger as much as possible. Another thing to note is during T5~T6 all the chargers must be accessed.

In the first 150 minutes charging algorithm for EVs is constant current and limited voltage. The equivalent resistance value of the charger is very small and the relative changing is slow. At this time less accessing of EV chargers can restrain the harmonic current, reduce voltage deviation and unbalanced degree.

In the next 150 minutes charging algorithm for EVs is constant voltage and limited current. The equivalent resistance value of the charger is large and the relative changing is fast. At this time more accessing of EV chargers will not result in increase of harmonic current, on the contrary it can improve the power quality of PCC.

Table 4. Power quality under coordinated charging modes.

No.	T 1	T 2	T 3	T 4	T 5	T 6	5th current /A	7th current /A	Voltage deviation	Voltage imbalance	Score
101	1	1	4	4	6	6	3.53	1.80	0.030	0.6412	95.98
117	1	1	3	3	6	6	3.48	1.84	0.030	0.0212	95.63
187	1	1	3	3	5	6	3.57	1.78	0.031	0.3777	94.82
171	1	1	4	4	5	6	3.59	1.78	0.032	0.3979	94.82
71	1	1	5	5	6	6	3.57	1.86	0.034	0.0122	94.08
116	1	2	3	3	6	6	3.49	1.90	0.031	0.0903	93.9
124	1	1	2	2	6	6	3.41	1.98	0.034	0.4211	93.82
222	1	1	3	3	4	6	3.63	1.78	0.032	0.1913	93.41
107	1	1	3	4	6	6	3.53	1.97	0.031	0.0031	93.28
123	1	2	2	2	6	6	3.42	1.98	0.036	0.0078	93.28
100	1	2	4	4	6	6	3.54	1.90	0.031	0.0031	93.25
206	1	1	4	4	4	6	3.64	1.82	0.034	0.0078	93.07

Note: T1 is 0 to 50 minutes, T2 is 50 to 100 minutes, T3 is 100 to 150 minutes, T4 is 150 to 200 minutes. T5 is 200 to 250 minutes. T6 is 250 to 300 minutes.

The coordinated charging are as follows: In the first 100 minutes chargers should be accessed as little as possible, in the following 200 minutes chargers should be accessed as much as possible, in the last 100 minutes, all the chargers must be accessed.

5. Conclusion

As the research, the factors affecting power quality includes two aspects, the parameters of distribution network and the charging mode. The proposed coordinated charging of EV based on GA-BP prediction conforms to practical law, which can be used as theory basis and construction reference of EV charging stations.

Acknowledgment

This project was supported by National High Technology Research and Development Program of China (863 Program) (2011AA05A110), National Energy Application Project For Research And Demonstration (NY20110702-1), as well as State Grid Corporation of China (2013-2014 Research on Experiment and Detection Technology of Electric Vehicle Chargers.

References

1. Cui, Y.F. & Yang Q. 2010. *Research on the Domestic and Foreign Development Status of EV and the Charging Technology*. Yunnan Electric Power 38(2): 9-12.
2. Gomez J.C. & Morcos M.M. 2003. *Impact of EV Battery Chargers on the Power Quality of Distribution Systems*. Power Delivery, IEEE Transactions on 18(3):975-981.
3. Zhen P. et al. 2013. *Harmonic Analysis of High-Frequency Electric Vehicle Charger*. Journal of Electric Power 28(2):100-113.
4. Ge S.Y. et al. 2012. *Optimization of Peak-valley TOU Power Price Time-period in Ordered Charging Mode of Electric Vehicle*. Power System Protection and Control 4(10):1-5.
5. Huang R. & Zhou X. 2012. *The Controlled Charging Dispatch Strategy by Considering of the Uncertainty of Electric Vehicles*. Modern Electric Power 29(3):57-63.
6. Zhou Y. 2011. *Study of Electricity Short-term Load Forecast Based on BP Neural Network*. Electrical Measurement & Instrumentation 48(542):68-71.
7. Zhao X. et al. 2005. *Fuzzy Synthetic Evaluation of Power Quality Based on Changeable Weight*. Power System Technology 29(6):11-16.
8. Li M.S. & Yang F. 2013. *Energy Efficiency Evaluation Method of Electric Vehicle Charging Station Based on Analytic Hierarchy Process and Fuzzy Synthetic Evaluation*. Electrical Measurement & Instrumentation 50(9): 69-73.

Study on battery charging method for electric vehicle

H.B. Yu[1] and X.J. Yang[2]
China Electric Power Research Institute, Beijing, China

Y.P. Wang[3]
China North Vehicle Research Institute, Beijing, China

Z.F. Wang[4†]
Collaborative Innovation Center of Electric Vehicles in Beijing, China
Emails: †ahu_bit@163.com, zhifu-wang@sohu.com

The charging method of batteries used in electric vehicles (EVs) significantly affects its commercial application. This paper aims to make some contributions to the existing literature. This paper analyzed the rapid charging method about the Lead-acid battery. Firstly the battery model was established with the experiment result. The rapid charging method with fuzzy control was applied in the paper and the charging experiment result shows that the batteries can be changed more 14Ah than the normal charging method.

Keywords: Electric vehicles (EVs); Battery; Charging strategies.

1. Introduction

From the perspective of the development of battery charging technology, there were mainly through the following charge mode [1~5]: (1) constant current charging; (2) constant voltage charging; (3) constant current with limiting charging voltage (charging to limit voltage, by reducing the charging current limiting voltage rise); (4) constant voltage with limited charging current (constant voltage charging, charging current is always less than the limit value); (5) constant current and constant voltage charging (first constant current charging, when charged to a specified voltage transfer constant voltage charging); (6) pulse charging.

---

[1] *Hai-Bo Yu*
[2] *Xiang-Jiang Yang*
[3] *Yu-Pu Wang*
[4] *Zhi-Fu Wang*

The first and second charging mode above were the traditional way and the charging time was long with less battery gassing and small temperature rise. The battery charged with those two method can work well and the charging requirements of the equipment is low. However it's longer charging time cannot meet the current demand for electric vehicles. Constant current with limiting charging voltage charging method merge the constant current charging and constant voltage charging method. Firstly, the battery was charged with constant current charge to a predetermined voltage value, and then to the constant voltage to complete the remaining charging process. Constant current value cannot be too large and the constant voltage was in the vicinity of the gassing. This method was used widely used because of its reliability and the low requirements for charger. The main drawback was that the initial charge current was small, so the charging time was still long.

The key problem to be solved was how to effectively remove battery polarization with the rapid charging method [6~8]. Generally speaking, there are two main ways to present the polarization. One is the nature polarization, such as the intermittent charging method. During the charging process to stop charging to remove the ohmic polarization and partial polarization and electrochemical polarization. Another method is forced polarization, with negative pulse discharge charging method, namely the adding certain negative pulse battery instantaneous discharge way to remove the polarization. It can effectively remove the majority of concentration polarization in the charging process.

The charge mode with negative pulse discharge, which can be removed by adding a certain magnitude and maintaining time of the negative pulse to remove the polarization after the charging process.

Sung Chul Kim researched the intermittent and negative pulse discharging and depolarizing effects on the charging process, and established the mathematical model of charging resistance change in the process of positive and negative plates. And the conclusion was intermittent and discharge pulse can remove a pole, slow charging voltage increasing, and discharge negative pulse in a shorter period of time reduced the internal resistance of the battery, improved the battery charge acceptance ability [9]. Through the above research, we can know that the negative pulse discharge method is better than the intermittent charging method, but it is much more complex than the intermittent charging method. If the negative pulse discharge parameter selection is not reasonable, not only cannot remove the polarization, but also to the battery surface temperature, charging time increases, and fast charge of the target. So it has practical value to study the negative pulse parameters in the charging mode with negative pulse discharge.

It is a difficult problem about the rapid charging method with the negative pulsed discharge, that how to realize the reasonable charge control strategy and how to according to the polarization of fast charging of the selection of appropriate negative pulse parameters in order to achieve good and fast charging. The paper focused on those content.

2. Battery Model and Charging Method

2.1. *Battery model*

The [10] literature analyzed the lead-acid battery polarization from the electrochemical reaction mechanism of internal during the charging process. The polarization voltage was not the result of a single reason, it was mainly caused by ohmic polarization, the concentration polarization and the electrochemical polarization constitute. At the same time, the processing quality of lead-acid battery polarization also has a total to affect that process would lead acid battery has its inherent characteristics, directly determines the ability to charge well or bad. But according to the status of the current lead-acid battery production and processing process, it was very difficult for the production process changes in the inherent characteristics of lead-acid battery. Through the analysis of the inherent characteristics of lead-acid battery, the charging method to improve the performance of lead-acid battery charging is the most effective way to solve the problem.

It is necessary to carry on the theoretical analysis of the battery model. Battery model describes the mathematical relationship between cell factors and job characteristics, the factors considered voltage, current, power, temperature, resistance, cycling and self-discharge. At present, the commonly used battery models include electrochemical model, equivalent circuit model and neural network model [11~13]. It is necessary that the model is accurate enough to reflect the dynamic process of the battery charge and discharge. At the same time, the model is not too complicated so that is can be used in engineering. The three order charge and discharge dynamic model was adopted which can describe the behavior characteristics of lead-acid battery for electric vehicles based on the comparison of various battery models. The three order charge and discharge dynamic model structure is shown in Figure 1.

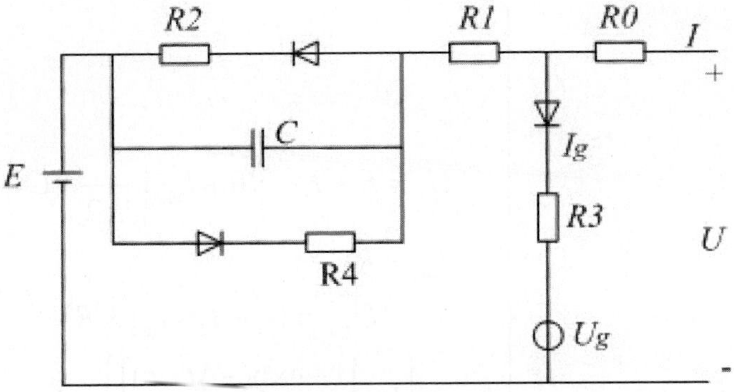

Fig. 1. The model structure about three order.

The charge polarization effect of lead acid battery, polarization effect, concentration polarization effect and gas evolution reaction were considered in the model. The polarization of the battery and the gas evolution reaction under the condition of the I_g branch. The internal resistance of the battery consist of four parts, including ohmic resistance, polarization resistance equivalent, concentration polarization equivalent resistance and condensate gas branch equivalent resistance. Where R_0 is the ohm internal resistance R_1 is concentration polarization equivalent resistance, R_2 for the polarization charging equivalent resistance R_4 for polarization and discharge equivalent resistance R_3 for condensate gas branch equivalent resistance. The C was used to express the polarization equivalent capacitance. In addition, the E was the electric potential of the battery, the U_g is the equivalent of the battery, and the U is the terminal voltage of the battery.

According to the experimental data of the composite pulse of each SOC point, the output equation of the three order equivalent circuit model of the lead acid battery was discretized, and the formula can be obtained (1) and (2).

Charging function:

$$
\begin{cases}
U_i = E + I_i \times R0 + I_{m.i}\left(\dfrac{1}{R2} + C\right) + \\
I_{m.i} \times R1 \\
\qquad U_{g.i} = U_i - I_i \times I_{g.i} \times R3 \\
I_{g.i} = \left\{1 - \dfrac{[1-\exp(-\Delta t/\tau)]}{(\Delta t/\tau)}\right\} \times I_i + \\
\left\{\dfrac{[1-\exp(-\Delta t/\tau)]}{(\Delta t/\tau)} - \exp(-\Delta t/\tau)\right\} \times \\
I_{i-1} + \exp(\Delta t/\tau) \times I_{g.i-1}
\end{cases} \tag{1}
$$

Discharging function:

$$
\begin{cases}
U_i = E - I_i \times R_0 + \\
\qquad I_{m.i}\left(\dfrac{1}{R4} + C\right) + I_{m.i} \times R_1 \\
U_{g.i} = U_i - I_i \times I_{g.i} \times R_3 \\
I_{g.i} = \left\{1 - \dfrac{[1-\exp(-\Delta t/\tau)]}{(\Delta t/\tau)}\right\} \times I_i + \\
\left\{\dfrac{[1-\exp(-\Delta t/\tau)]}{(\Delta t/\tau)} - \exp(-\Delta t/\tau)\right\} \times \\
I_{i-1} + \exp(\Delta t/\tau) \times I_{g.i-1}
\end{cases} \tag{2}
$$

2.2. *Charging method with negative pulse*

The acceptable charging current curves is different for individual lead-acid battery. So the acceptance curve of current for the different lead-acid battery was not the same. Even for the same lead-acid battery, influenced by the environment temperature and a variety of other external factors, the acceptable charging current curve would change. Because of its internal electrochemical reaction, the nonlinear, discrete, and so on were the inherent characteristics of the lead acid battery. During the charging process, almost all of the parameters of the battery will change over time. Thus, the traditional control theory to establish an accurate model of the system cannot be achieved to satisfy the acceptable charging current. The fuzzy control method is suitable for the control of the lead acid battery charging system because of the complex characteristics of the lead acid battery, such as nonlinear and multi parameter uncertainties.

Lead acid batteries with negative pulsed discharge of fast charging fuzzy control system mainly consists of two fuzzy controller, an executive mechanism and corresponding sensor structure composition and its structure as shown in Figure 2.

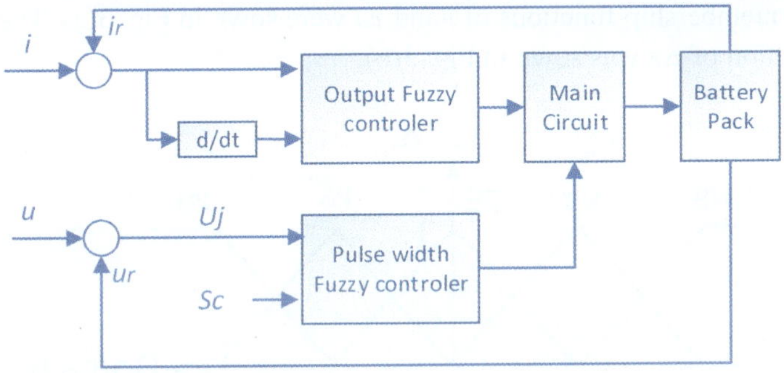

Fig. 2. The fuzzy control structure.

Output fuzzy controller was used to control the output of the charging current. The positive charge pulse was charged by the variable current. Charging current was divided into three stages, respectively using the charging pulse amplitude is 100A, 50A and 30A charging in each stage of the battery terminal voltage reaches 14.4V automatically switches to the charging stage. Because the current sensor was arranged in the control system, so it can be known in real time when the charging current of effective value, fuzzy controller of two inputs can be expressed as a type (3) and (4).

$$i = i_g - i_b \tag{3}$$

$$\Delta i = i_n - i_{n-1} \tag{4}$$

Where i was the current error between the setting value i_g and lead-acid battery charging current feedback current. Δi was the change rate of the current error.

The role of the fuzzy control was to make the charging system in the positive pulse charging process, as far as possible to make the charging current in accordance with the lead acid battery acceptable charging curve. The output can be expressed as a formula (5).

$$\Delta z(n) = z(n) - z(n-1) \tag{5}$$

Among them, $\Delta z(n)$ is the change rate of the control parameter $z(n)$.

The stability of the charge system of lead-acid battery was better, so the fuzzy controller was designed by using the triangle membership function.

The Fuzzy sets of i and Δi as{NB NS ZE PS PB}.

The Fuzzy sets of Δz as {NB NM NS ZE PS PM PB}

The specific meaning of each fuzzy quantity is:

NB-Negative Big, NM-Negative Medium, NS-Negative Small, ZE-Zero, PS-Positive Small, PM-positive Medium, PB-positive Big.

The membership functions of i and Δi were sown in Fig. 3(a). The membership function of Δz was sown in Fig. 3(b).

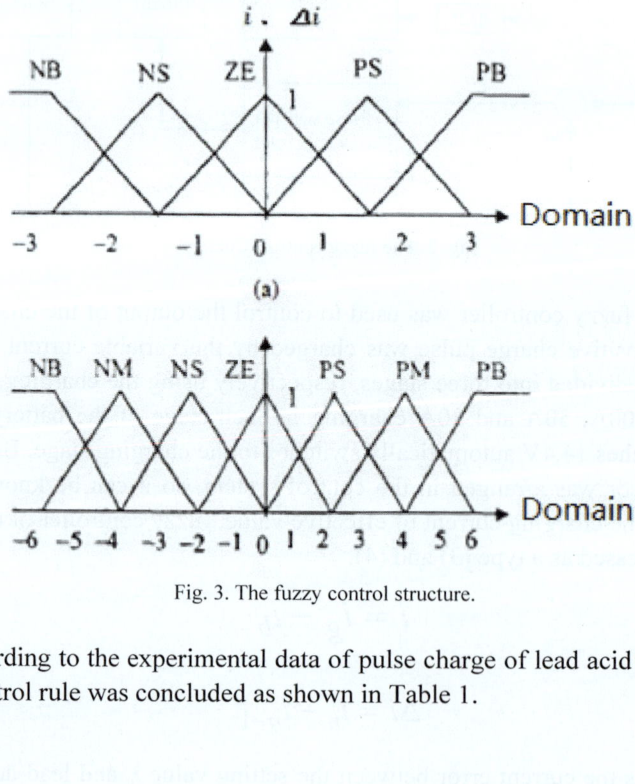

Fig. 3. The fuzzy control structure.

According to the experimental data of pulse charge of lead acid battery, the fuzzy control rule was concluded as shown in Table 1.

Table 1 The fuzzy control rules of *i, Δi and Δz*

i \ Δi	NB	NS	ZE	PS	PB
NB	PB	PB	PB	PM	ZE
NS	PM	PM	PM	ZE	NS
ZE	PM	PS	ZE	NS	NM
PS	PS	ZE	NM	NM	NM
PB	ZE	NM	NB	NB	NB

Based on the above analysis, the fuzzy controller was designed for the negative pulse width of the negative pulse discharge. The fuzzy controller uses the polarization voltage U_j of the lead acid battery and its current charge S_c as the input. The output was represented by t_w. The fuzzy subsets of the charge polarization voltage U_j, the state of charge S_c and the output t of the lead-acid battery were taken for PVS PS PM PB {PVB}. After correction of the corresponding temperature condition, the practical value of the negative pulse width t_w can be obtained.

3. Battery

In order to verify the rapid charging method used in the lead-acid battery negative pulse charging strategy feasibility, there were two experiment were performed. The lead-acid battery charged by varying current 100A, 50A and 30A without negative pulse in the first experiment. At the same time, the lead-acid battery was charge with the fuzzy control negative pulse control method. The experiment field was shown in Figure 4 and the experimental curves were shown in Figure 5 and 6, respectively.

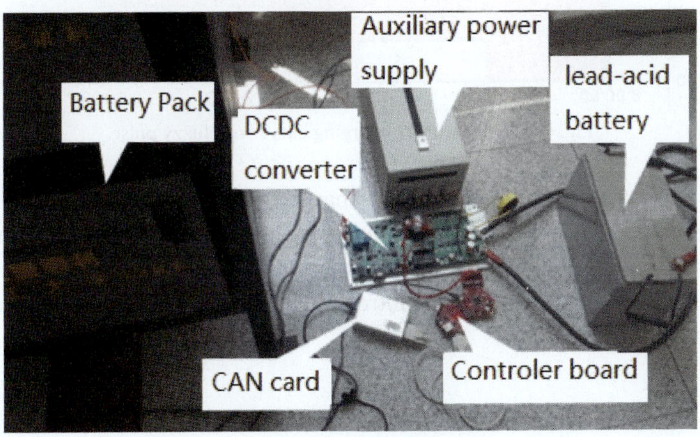

Fig. 4. The experiment field.

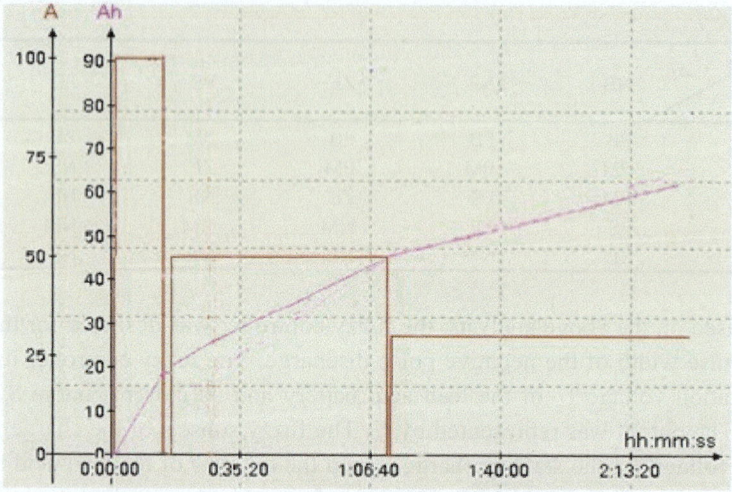

Fig. 5. The experiment charging curve without pulse.

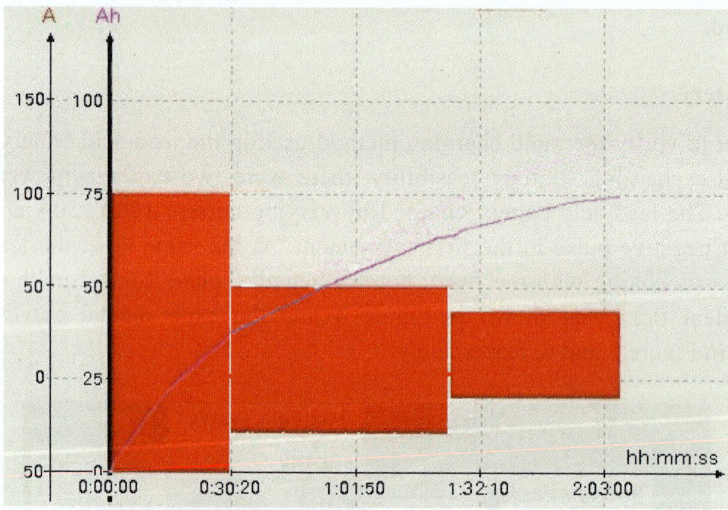

Fig. 6. The experiment charging curve with fuzzy pulse.

It was shown that the battery was charged more 14Ah with the fuzzy pulse control method than with the three phase contest current. The time and volumetric ratio would reach 0.601 and the charging effect was obviously improved. It is proved that the fuzzy pulse control method with the negative pulse charge of lead-acid battery.

4. Conclusion

The rapid charging method of battery used in electric vehicle is importance for the development of the EVS. The charging method with negative current pulse controlled fuzzy controller was an effective mothed to realize the battery rapid charging. The experiment result showed that the charged energy can be more 14Ah in the same charging time. Some work also proved than this method also can improve the battery health state.

Reference

1. Feng Wen, Jiuchun Jiang, Weige Zhang, Huiping Guo. Research on the charge mode of series-connected batteries [C]. 2008 IEEE Vehicle Power and Propulsion Conference, VPPC 2008.
2. Jiapeng Wen, Jiuchun Jiang. Battery management system for the charge mode of quickly exchanging battery package [C]. 2008 IEEE Vehicle Power and Propulsion Conference (VPPC), p 4 pp., 2008.
3. Yong Kon Kwon, Nam S.H., Eui Wan Lee, Soung Soo Park, Yeong Jin Han. Characterization of a pulse-charge-mode multigap pseudospark device as an enhanced electron beam source [J]. Japanese Journal of Applied Physics, Part 1 (Regular Papers, Short Notes & Review Papers), 2003, 42(9A):5753-8.
4. Kim C., Hwang J. The effect of charge mode transition on electrohydrodynamic flow in a multistage negative air corona discharge [J]. Journal of Physics D: Applied Physics, 2012, 45(46):465204 (10 pp.).
5. Si Chuansheng, Xu Xianlei. Research of the charge mode of electric automobile charging station and the key technology of the control system [C]. 2012 2nd International Conference on Consumer Electronics, Communications and Networks (CECNet), p 2788-90, 2012.
6. Chang T.G., Jochim D.M.. Structural changes of active materials and failure mode of a valve-regulated lead-acid battery in rapid-charge and conventional-charge cycling [J]; Journal of Power Sources, 2000, 91(2):177-192.
7. Li Siguang, Zhang Chengning, Xie Shaobo. Research on fast charge method for lead-acid electric vehicle batteries [C]. 2009 International Workshop on Intelligent Systems and Applications, p 5, 2009.
8. Zeng Wen-xian, Ma Jie, Zhai Ying. Rapid charge system for lead-acid battery of solar energy street light based on single-chip microcomputer [C]. 2008 International Conference on Computer Science and Information Technology, p 337-41,2008.
9. Sung Chul Kim, Won Hi Hong. Effect of electrode parameters on charge performance of a lead-acid cell[J]. Journal of Power Sources, 2000, 89(1):15-28.

10. Guo Yifeng. Electric vehicle power battery charging Study of Quasi-resonant switch pulse [D]. Beijing Institute of Technology. 2012.
11. Zhang Cheng, Li Kang, McLoone Sean, Yang Zhile. Battery modelling methods for electric vehicles - A review [C]. 2014 European Control Conference, ECC 2014, p 2673-2678, July 22, 2014.
12. Hammad N.M.. Automotive battery modelling and management [J]. Journal of Engineering Science and Technology Review, 2011,4(2):140-145.
13. Ahmad Hamidi Seyed, Ionel Dan M., Nasiri Adel. Modeling and Management of Batteries and Ultra-capacitors for Renewable Energy Support in Electric Power Systems-An Overview [J]. Electric Power Components and Systems, 2015, 43(12):1434-1452.

Analysis of the surplus rate of power transmission and transformation project cost based on grey correlation analysis

B.J. Li, D.X. Niu, Y. Lu & J.P. Qiu†

School of Business and Management, North China Electric Power University, Beijing, 102206, China.
†Email: jinpeng_q@163.com

L. Liang

State Grid Zhejiang Electric Power Company Economic Research Institute, Hangzhou, 310008, China

Factors affecting the power transmission and transformation project cost savings rate is more and complex relation between indicators, resulting in cost of a larger gap the cost budget and final accounts. In this paper, starting from the individual cost analysis of transmission cost, based on grey correlation analysis to build the surplus rate influence index correlation. Through statistical history engineering cost data of certain province and consulting experts to select the major effect of the surplus rate index as an original analysis set, the correlation for the balance rate, calculate each index, in turn, correlation sort, relational sequence can be analyzed by various factors and the close degree of balance rate to provide basis for managers to control the project cost.

Keywords: Electricity cost; Grey correlation; Surplus rate.

1. Introduction

The project cost is in an indeterminate state from the project was set up to the completion of the final accounts, so that until the completion of the project can the actual cost of the project be determined, resulting in a larger balance estimates. Finally, a shortage of funds, backward technology and other reasons cannot guarantee the successful completion of the project, which not only caused losses to the manager, but also increased the economic losses to the state and social burden, showing the relationship between the study of factors affecting the project cost and project cost savings rate is significant. Based on the theory of gray correlation analysis, the factors affected the cost savings rate of power transmission project were analyzed to provide a theoretical basis for the reasonable cost of power transmission project.

2. Grey Relational Analysis

Grey Correlation Analysis intended to seek numerical relationship between subsystems (or factors) in the system through a certain way. Therefore, gray correlation analysis provides a quantitative measure of the situation for the development of a system change, and is very suitable for dynamic analysis. Grey correlation analysis is actually a quantitative description and compare for the changes of development of the system. The method used gray correlation to describe the strength, size and order of the relationship between the factors, and finally found the biggest factor by sorting relevance.

2.1. *Gray correlation calculation*

Draw the associated graph changing with time of the function $\eta_{ij} = \eta_{ij}(\mathrm{t})$. The horizontal line in the figure shows the correlation coefficient is 1 at any time, and it shows that the associated curve $\eta_{ij} = 1$ between X_i and X_i. The area bounded by the axes by the associated curves $\eta_{ij}(\mathrm{t})$ and $\eta_{ii}(\mathrm{t})$ is η_{ij} and η_{ii}, and the correlation between X_j and X_i is:

$$r_{ij} = \frac{S_{ij}}{S_{ii}} \tag{1}$$

Obviously, $s_{ii} = 1 * M = M$, so:

$$r_{ij} = s_{ij} / M \tag{2}$$

In the actual calculation, the mean of correlation coefficient of comparison sequence and features sequence in each project is:

$$r_{ij} \approx \frac{1}{M} \sum_{i=1}^{M} \eta_{ij}(\mathrm{t}) \tag{3}$$

The correlation shows the relative changes between the factors in the system development process, which is the tendency of proximity. If the two trends in the system development process are basically the same, then the correlation between the two will be great; on the contrary, it is considered a small correlation.

2.2. *Sort of correlation*

Line up the correlation of characteristic sequences which is associated with the comparative sequence X_1, X_2, X_3,...... X_n and is the composition of the associated sequence, which directly reflects the close relationship of comparative sequence for each characteristic sequences, that can clearly be seen various factors affecting the level of the project cost savings rate. The tightness of the factors and engineering cost savings rates can be analyzed by correlating sequence, providing the basis for management control of the project cost.

3. Case study

Take the new 110kv substation engineering project cost from 2013 to 2014 as the study target, and deleted part of the project which does not meet the criteria to collect, finally the effective engineering samples of a total was 77. By consulting experts, list out important indicators of the cost savings rate to the factor analysis from the data table. Selected 19 major indicators of the data collected in the table, and conducted correlation analysis.

3.1. *Index selection*

In the analysis of the main indicators of the balance ratio, excluding the topography and other natural conditions, find out the individual cost which impacts the savings rate greater. According to the actual situation of the construction site and access, collection, analysis, comparison of large amounts of data, and consulting experience engineer, constructed three levels of major influencing factors such as shown in Table 1:

Table 1. Major influencing factors.

First level index	Second level index	Third level index
Balance rate influencing factors	Major equipment	Number and price of main transformers
		Number and price of high voltage circuit breakers
		Number and price of medium voltage circuit breakers
		Number and price of low voltage circuit breakers
		Number and price of low voltage capacitors
	Other equipment	Protection and control equipment costs and other secondary equipments costs
		Intelligent related equipment costs

	Site preparation costs
	Ground handling costs
	Retaining walls and slope protection costs
Other related costs	Pit road costs
	Control building costs
	Construction site land acquisition and compensation costs
	Basic reserve

In the data tables, including a large number of historical data of natural factors, technical conditions, based on past experience and cost guidance of experts, selected a total of 19 significant impact indicators, part of the original data table shown in Table 2.

Table 2. Part of the original data table.

Period number of units	Price of main transformer	Number of high voltage circuit breakers	Price of high voltage circuit breakers
2	253	5	55
2	242.19	3	64.6
2	264.42	3	56.94
1	225.06	2	56.16
2	245.15	5	9.37
1	205.56	1	11.14
……	……	……	……
Number of medium voltage circuit breakers	**Price of medium voltage circuit breakers**	**……**	**Balance rate**
0	0	……	37.05%
0	0	……	32.64%
8	17.25	……	28.15%
4	18.11	……	26.55%
6	6.09	……	22.24%
4	5.96	……	9.27%
……	……	……	……

3.2. *Correlation calculation*

(1) The difference sequence calculation

According to $\Delta = |(X_0(k) - X_i(k))|$, the difference of balance rate and each comparison sequences can be got, and the largest difference is Δmax=6.24658, part of the data is shown as Table 3.

Table 3. The absolute difference between the balance rates with other factors.

X_1	Period number of units	1.4856	1.0537	0.6142	...
X_2	Price of main transformer	1.9942	1.8826	0.7844	...
X_3	Number of high voltage circuit breakers	1.3186	2.0514	1.6120	...
X_4	Price of high voltage circuit breakers	1.3807	0.5574	0.4302	...
X_5	Number of medium voltage circuit breakers	3.1106	2.6787	0.1985	...
X_6	Price of medium voltage circuit breakers	2.9595	2.5275	0.1946	...
X_7	Number of low voltage circuit breakers	0.7690	0.0698	1.1180	...
X_8	Price of low voltage circuit breakers	2.8756	2.4082	2.0278	...
X_9	Number of low voltage capacitors	1.4632	1.0313	0.5919	...
X_{10}	Price of low voltage capacitors	2.4387	1.8000	1.2302	...
...	...	...	...	...	...

(2) Gray correlation calculation

The correlation coefficient is calculated as shown in Table 4, ξ which ranges from (0,1), is calculated as 0.5 here. The correlations between the index and the balance rate are shown in Table 4.

Table 4. The correlations between the index and the balance rate.

No.	factor	Correlation coefficient
X_{18}	Construction site land acquisition and compensation costs	0.8005
X_{17}	Control building costs	0.7859
X_2	Main transformer Price	0.7824
X_{13}	Site preparation costs	0.7817
X_{19}	Basic reserve	0.7812
X_{15}	Retaining walls and slope	0.7793
X_{14}	Ground handling costs	0.7789
X_{11}	Protection and control equipment and other secondary	0.7781
X_3	Number of high voltage circuit breakers	0.7722
X_9	Number of low capacitors	0.7698
X_1	Period number of units	0.7693
X_4	Price of high voltage circuit breakers	0.7672
X_6	Price of medium voltage circuit breakers	0.7668
X_{10}	Price of low voltage capacitors	0.7664
X_{12}	Intelligent related equipment	0.7660
X_{16}	Pit road costs	0.7654
X_7	Number of low voltage circuit breakers	0.7563
X_5	Number of medium voltage circuit breakers	0.7527
X_8	Price of low voltage breakers	0.7402

Observed from the table construction site land acquisition and compensation costs impact the balance rate most, followed by the main control building and main transformer costs, and then various other expenses. The results obtained are basically the same with the actual project cost situation.

4. Conclusions

Due to the indicators impacting power transmission project cost savings rate are too many, this paper selected key indicators from the perspective of the impact on costs. Firstly, conduct a non-dimensional treatment of data of various indicators, and then calculated the gray correlation degree of the balance rate, and used the quantitative method to calculate the tightness of the qualitative indicators and the balance rate. Finally, take the historical data of certain Power Transmission Project Cost as a sample. In the actual grid construction, factors which most associated with the balance rate should be focused on and take effective measures to control, so as to control costs by project managers, and to provide a favorable theory to reduce the balance rate.

5. Preferences

1. Hong Tao. Cost Engineer Practice Manuals [M]. Beijing: China Building Materials Industry Press. 2006:1-2.
2. Wei Daosheng, Li Shuyan. Project Cost Control Based on Gray Relational Analysis [J]. Chongqing Jiaotong University (Natural Science), 2013, 32(2):321-324.
3. Bao Ku, Zhao Jingming, Qu Yu. Development Evaluation of Construction Industry of Liaoning Province Based on Grey Relational Analysis [J]. Construction Economy, 2012(8):19-22.
4. Wang Yin. Application of Data Mining Technology in Power Line Project Cost Management [D]. Chongqing: Chongqing University, 2008.

Chapter 2

Electrical Engineering

Simulation analysis of debonding detection for the honeycomb structure based on infrared thermal waves

H. B. Zhang[1†]
Qingdao Campus of Naval Aeronautical Engineering Institute, Qingdao, China.
†Email: ndt_zhb@126.com

S. T. Wang
Navy Academy of Armament, Beijing, China.

Y. Q. Wang
No. 92514 troops, Yantai, China.

This paper introduced mainly the basic principles of detecting debonding of honeycomb structure that used infrared thermal wave technology, established the debonding defect model of honeycomb structure, analyzed the influence which the choice of thermal excitation method and parameter settings to defects that can be detected, found the optimal choice basis of heated argument under variety of incentives, the analytical results provided some basis for formulating accurate infrared non-destructive testing program of the honeycomb structure debonding.

Keywords: Infrared heat wave; Honeycomb structure; Simulation.

1. Introduction

Honeycomb sandwich structure material is a composite material, which is one of the main materials in aerospace area. Honeycomb sandwich structure is often made up of a layer of honeycomb lattice structures which located between upper and lower skin layers, honeycomb core is firmly bonded with upper and lower skin layers by using a special bonding process, or direct inject or compress to obtain a sandwich structure. The honeycomb sandwich structure composite is expensive and easy to form debonding, water, scratches, cracks, porosity, core short, panel or damage of honeycomb core in the manufacturing and service process, among these the debonding is the most common, the presence of these defects seriously affect the quality and safety of the use of honeycomb sandwich structure of the product.

---

[1] Hai-Bing Zhang

Infrared thermal wave detection technology are widely used in debonding detection of composite because of its advantages of non-contact, one-sided test, test speed and Intuitive test results. Using the finite element method to construct model, studied and analog the detection process using infrared thermal wave, which can provide theoretical basis for true infrared NDT, it can also provide the best detection scheme and improve the detection accuracy, as well as provide an effective means for further detecting quantitatively the defect characteristics.

2. The Principle of Infrared Thermal Wave Detection

The infrared thermal wave detection technology is mainly used for detecting debonding flaw of cellular composite material, which is based on the principle of infrared radiation, using infrared radiation measurement analysis methods, stimulating cellular structure internal defects in the active mode by using different methods to control the heat source. In the process of heat diffusion and transmission in the internal of the honeycomb structure, thermal conduction will produce discontinuity because of the skin debonding, it will produce different temperature distribution on the skin surface. The infrared thermal wave detection technology records infrared radiation of skin surface by is infrared thermal imaging technology, measures and records the temperature distribution of the skin surface, which can obtain the structure information under skin surface of the honeycomb structure according to the measurement information[1].

When the lower surface exists defects or structural inhomogeneities, we simplify the structure into one-dimensional model of two dielectric structure, assumed here that the surface completely absorbs the pulse heat energy and satisfies the boundary conditions and the conservation law of energy , by solving the heat conduction equation we can obtain analytical solution of the surface temperature.

$$\Delta T(0,t)=\frac{Q}{e\sqrt{\pi t}}\left[1+2\sum_{n=1}^{\infty}R^n\exp\left(\frac{-n^2L^2}{\alpha t}\right)\right] \tag{1}$$

Where t is time, e is the heat storage coefficient of material, the heat wave propagation in the honeycomb structure, the heat reflection coefficient R that reflected across the debonding can be expressed as

$$R=\frac{e_1-e_2}{e_1+e_2},\ e=\sqrt{k\rho c} \tag{2}$$

Where k is the thermal conductivity, ρ is density, Q is constant, which means that heat is applied to the unit area, k is the depth of the defect, n is the number of reflections that the heat wave propagation to the cellular interface of the skin [2].

$$\Delta T_C(0,t)=\frac{2Q}{e_1\sqrt{\pi t}}\sum_{n=1}^{\infty}R^n\exp\left(\frac{-n^2L^2}{\alpha_1 t}\right) \tag{3}$$

Subtracting the surface temperature reference value of no debonding area we can get the compare temperature in the actual application, that is:

$$\Delta T_C(0,t)=\frac{2Q}{e_1\sqrt{\pi t}}\sum_{n=1}^{\infty}R^n\exp\left(\frac{-n^2L^2}{\alpha_1 t}\right) \tag{4}$$

Where e_1 and α_1 denote the regenerative coefficient and thermal diffusivity coefficient of the honeycomb structure. When quantitatively identify the defects under the surface, thermal storage coefficient can be obtained by calculating the coefficient of defects in material interfacial debonding heat wave reflection coefficients, in order to achieve the purpose of identification under the skin debonding defects.

3. Create Debonding Model

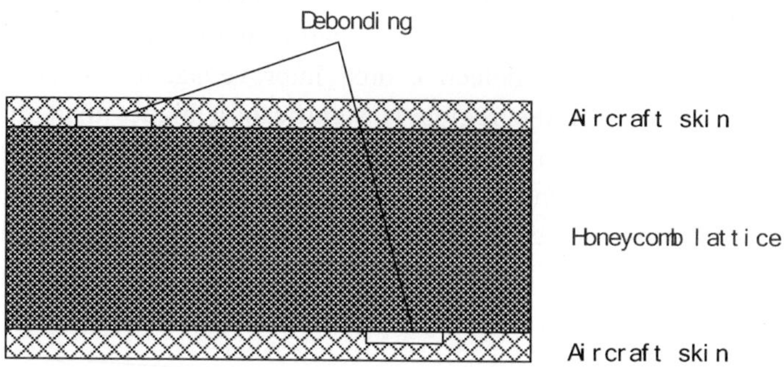

Fig. 1. Simulation model of interfacial debonding.

Use a longitudinal section of the honeycomb structure interfacial debonding specimens as a template, establish two layers interfacial debonding model in the two-dimensional coordinates. As shown in Figure 1. Specimen length is 16cm, skin thickness is 0.3mm, honeycomb sandwich thickness is 12mm, debonding defects as shown in Figure 1, both the thickness are 0.05mm, diameter is 1cm, assumptions internal of defect is air. Thermal excitation is applied to the skin surface at both ends of the specimen cross-section with the outside thermal insulation, another specimen surface is natural convection with air.

4. Simulation Results and Analysis

Based on these assumptions, using finite element software COMSOL, construct finite element model to simulate the use of one-sided method were used continuous radiation excitation pulse impact excitation, ultrasonic lock-in incentives for the detection of skin surface heating process, observed temperature changes in skin surface and record and analyze.

4.1. *Continuous excitation radiation*

The continuous radiation incentive main analyze the effect of incentive time on the detection result, energizing time is set as 1 ~ 60s uniform sampling test. Continuous heat radiated output power is set to 2000W heat setting, when the excitation time is 5s, the simulation results at the monitoring time time = 20s is shown in Figure 2, different colors representing different temperature of each portion of specimen, the right side of the ribbon marked various colors corresponding to the indicated temperature, in the monitoring process can be seen, the surface temperature of the object corresponding to the defective region is significantly higher than the non-defective area, it is particularly evident for defects close to the thermal excitation [3]. Air is defective insulation with respect to the skin, because it is detected by single-sided method, when the radiation heat transfer to the defect, it is hindered and that some radiation is reflected back to the surface monitoring, this part of the radiation make surface temperature corresponding to defective area improve again, so the surface temperature of this part of the region is significantly higher than the non-defective area. Compared with the first layer of debonding defects, temperature in the second layer debonding defects changes little, which is because the second layer debonding defect is far from the heat source and the thermal conductivity of honeycomb sandwich structure is poor, temperature is unresponsive.

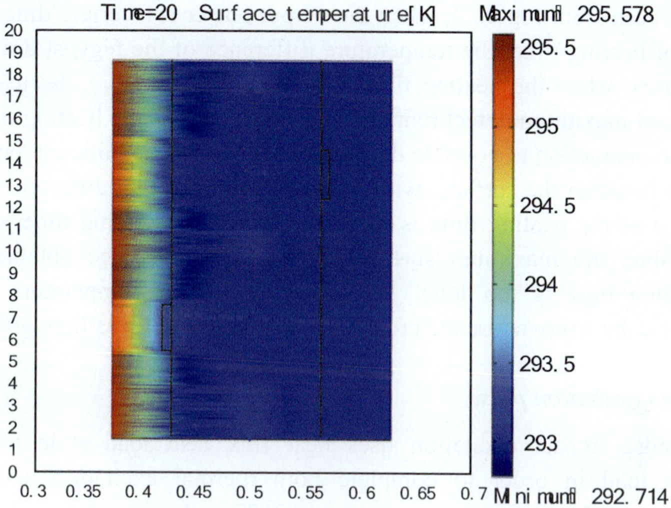

Fig. 2. The simulation renderings of continuous radiation excitation.

Make curve fitting for temperature difference of maximum defect which is detected in different excitation time, we can get the relations of defects can be detected and excitation time, which is shown in Figure 3 (curve 1 represents the first layer of the temperature difference curve of debonding defect, curve 2 represents the second layer of the temperature difference curve of debonding defect).

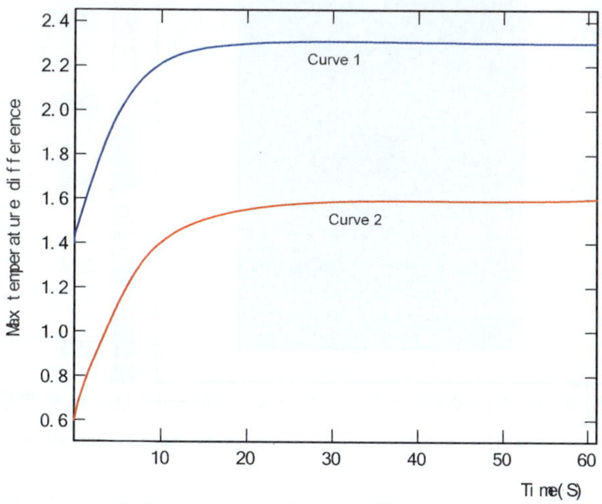

Fig. 3. The relationship of defects that can be detected and excitation time in continuous radiation excitation.

As can be seen from Figure 3, when the output power is a certain time, with the increase of heating time, the temperature difference of the biggest drawback of the increases; when the heating time exceeds a certain value, the temperature difference of maximum defect remains unchanged. When the heating time is too short, heat conduction to possible defects, can not form a significant temperature difference between the surface; with increasing temperature difference between the surface of the heating time is increased, when the heating time to reach a certain value, the maximum surface temperature difference obtained; when heated when time is too long, the specimen surface temperature tends to equilibrium, the temperature difference will become smaller as time goes by.

4.2. *Impact excitation pulse*

Analog pulse impact excitation uses heat flux heat load style to conduct multi-step load in order to complete both thermal excitation and cooling calculations. Ambient temperature is set to 20°C, only considering the thermal conduction process, ignoring thermal convection and thermal radiation influence of load surface. Simulation boundary conditions is set as ideal adiabatic conditions and no energy is lost [4].

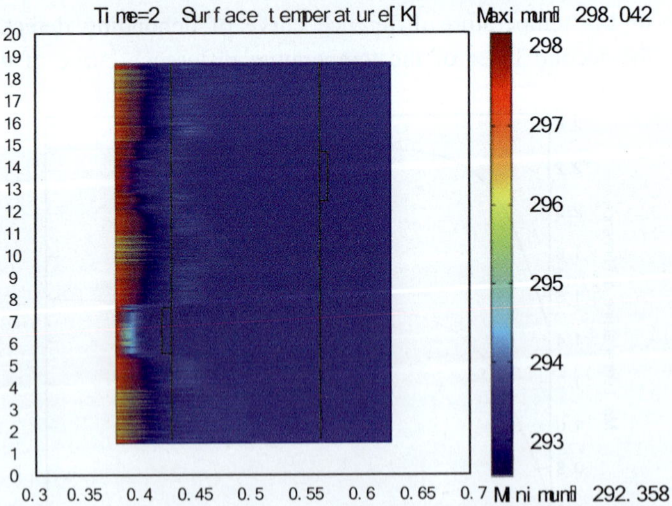

Fig. 4. The simulation renderings of pulse thermal excitation.

The pulse impact excitation is set as pulse width is adjustable, uniform samples from 1 ms to 5 ms for time and from 4000J ~ 8000J for output energy, the output pulse energy and pulse width are not changed simultaneously. General impact excitation pulse detection results appear very fast and very short duration, can be completed in a few seconds the entire testing process. When the output energy of 4000J, a pulse width of 3 ms, the time for monitoring the overall effect Time = 2s when simulation shown in Figure 4, corresponding to the defective region than the defect-free surface area of the corresponding surface also has a higher temperature, continuous thermal excitation, the second layer defects reaction temperature change is not obvious, because the thermal conductivity of honeycomb sandwich material is poor.

Make curve fitting for temperature difference of maximum defect which is detected by pulse width and energy, we can get the relations of defects can be detected and excitation time, by the curve analysis we can draw the conclusion: when the output pulse energy of the impact of certain incentives when the maximum temperature difference increases the pulse width of the defect and then decreased, there is an optimum pulse width so that there is a maximum value; certain time when the pulse width, the greater the energy output, the maximum temperature difference between the maximum defect defect greater the temperature difference . In other words, when a certain energy to stimulate out at different time, the impact may be caused by the specimen is also very different. If the time is too short excitation pulses heat the specimen is difficult to penetrate; increases when the time when the excitation energy is absorbed by the specimen can be reduced, more energy to be lost for distributing various forms. When the pulse width is a certain, the greater the energy output, the more energy is absorbed by the specimen, specimen temperature may rise significantly.

4.3. *Ultrasound lock-in incentives*

Use the halogen which thermal power output is 4000W as excitation value, the pulse width and period are within a range of variation in the periodic rectangular signal, wherein the pulse width is uniformly sampled from 1-4 seconds, period is uniformly sampled 2-8 seconds, one of the two parameters is a constant and another is change in the test. General Ultrasound lock-in incentives before the two detection methods to detect take longer than the time when the cycle is 6s, pulse width is 3s, in the monitoring time Time = 30s, the simulation results is shown in Figure 5, a defective surface area than the corresponding no region corresponding to the surface defects also has higher temperature. The temperature difference of the second layer debonding defect is less than the first layer for the same.

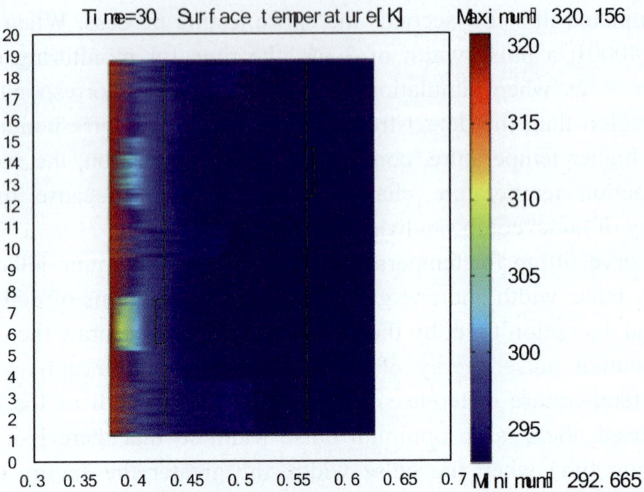

Fig. 5. The simulation renderings of ultrasound lock-in incentives.

Make curve fitting for temperature difference of maximum defect which is detected by different lock-in period and pulse width, we can get the relations of defects can be detected and excitation time curve can be drawn through the pulse width is constant, there is an optimal period of the maximum temperature difference between the defect has a relatively large value; when the cycle is a fixed value, there are specific defects in the pulse width of the maximum temperature difference is relatively large. When the pulse width is increased to the equivalent of a continuous cycle close to the heat flow is not conducive to extract phase-locked signal. Therefore, the parameters of lock-in heat source should be set reasonable in the detection by the circumstances analysis and theoretical analysis.

5. Analysis

In theory, as long as the temperature difference of the specimen surface maximal defects is greater than the temperature resolution of thermal imager, the defect can be detected, and the greater the temperature difference, the detection effect will be more obvious. Thus it can be seen from the simulation results that the article is more reasonable in setting different parameter of thermal excitation and can effectively detect defects [5].

Through the analysis of simulation results, we can calculate the best or more appropriate values of heated argument under various incentives. Contrast detection sensitivity of different excitation methods, we can see that the continuous radiation excitation is smaller than the other two biggest flaw temperature difference and is not easy to detect; pulse impact excitation detection of thin adhesive layer interfacial debonding can be detected best, but it requires the heat energy is relatively large and difficult to achieve; ultrasound lock-in incentives requires relatively smaller heat energy, detection effect is better, but its subsequent processing amounts and data collection amount are large, experiments need to spend more time. In practical tests, we need select the appropriate method according to the specific requirements of thermal excitation and experimental conditions. Further, detecting defects of deeper debonding effect is not obvious because of the heat wave propagation characteristic, we suggested detects both sides of the honeycomb structure in the actual detection process and improve the detection rate of defect. The above simulation and analytical results provide some basis for developing accurate thin adhesive layer for multi-interface debonding nondestructive testing laboratory program.

References

1. Liu Junyan, Dai Jingmin, Wang Yang. Thermal wave detection and defect depth measurement based on lock-in thermography. *Optics and Precision Engineering*, 2010, 18(1):37- 44.
2. Xu Weichao. Ultrasonic infrared thermal wave nondestructive evaluation for crack detection [D]. Beijing: Capital Normal University, 2008.
3. Chen Guicai, Wu Dongliu, Cheng Chayuan. Non-destructive evaluation of composite debonding flaws by infrared thermal wave. *Aerospace Materials & Technology*, 2004, 34(1):55-58.
4. Tian Yupeng. Key technologies research of non-destructive testing of Infrared radiation imaging. Nanjing: *Nanjing University of Aeronautics and Astronautics*. 2009.
5. Li Guohua, Wu Seng. Thickness effect of samples on the non-destructive testing of composite steel plate by infrared thermography [J]. *Infrared and laser engineering*. 2006, 35: 448- 454.

Dynamics of 1550-nm VCSELs with positive optoelectronic feedback

J. Yu[1], L. X. Zhang[2] and S. Y. Chen[3†]

School of Electronic and Information Engineering, Southwest University, Chongqing400715, China

Emails: [1]zlx92@swu.edu.cn, [2]y6638066@swu.edu.cn, [3†]510702776@qq.com

Vertical-cavity surface emitting laser (VCSEL) is the important source for optical networks-on-chip (ONoC). Research on dynamics of VCSEL is very important. We review our theoretical analyses on the nonlinear dynamics of a 1550-nm VCSEL with positive optoelectronic feedback (POEF) based on the spin-flip model extended to account for the optoelectronic feedback. A mapping of dynamical states is presented in the parameter space of feedback delay time and feedback strength, where different states are identified and shown. A bifurcation diagram of the extrema of output peak series versus the feedback delay time is also plotted. By varying the delay time of the feedback loop, the 1550-nm VCSEL subject to POEF follows a quasiperiodic route to chaos, where the various nonlinear dynamical behaviors including regular pulsing, quasiperiodic pulsing and chaotic pulsing are numerically observed.

Keywords: Vertical-cavity surface emitting laser (VCSEL); Optical networks-on-chip (ONoC); Nonlinear dynamics; Positive optoelectronic feedback (POEF).

1. Introduction

With increase the processor data rate requirement, more and more number of integrated processing cores on a single die. Switching information on chip global networks on chip between two processing cores is an attractive solution to solve the communications bottleneck [1]. Several groups propose a large number of solution, such as, electronically interconnection on chip [2] and photonic networks on chip [3]. Several optical on-chip communication architectures have been proposed based on the silicon-on-insulator (SOI) waveguides and microresonators, which are compatible with CMOS technology [4], [5]. Recently, related study focus on the structure of optical network-on-chip (ONoC) [6], the routing rules [7], signal analysis [8] and so on. Optical source, as a very important part in ONoC, has been extensively used. However, theoretically and

---

[1] Jing Yu

[2] Li-Xin Zhang

[3] Si-Yao Chen

experimentally researches on the optical source which especially suitable for ONoC are relatively inadequate. Optical source mainly including edge-emitting semiconductor lasers (EELs) and vertical-cavity surface emitting lasers (VCSELs). VCSELs have exhibited many inherent advantages over conventional EELs such as single longitudinal-mode operation, low threshold current, excellent circular output beam with narrow divergence, low cost, easy large-scale integration into two-dimensional arrays, etc [9]–[11]. As a result, VCSELs are nowadays substituting traditional EELs in many applications such as optical communication, big data center, optical interconnects inter/intra chip and study on the nonlinear dynamics of VCSELs is particularly important. Previous theoretical and experimental researches have demonstrated that VCSELs exhibit a wide range of nonlinear dynamics under external perturbations including optical injection, optical feedback or optoelectronic feedback (OEF) [12]–[38]. Different phenomena of VCSELs subject to optical injection or optical feedback have been studied and reported in recent years. These phenomena include polarization switching (PS) and bistability [12]–[18], injection locking [19], [20], nonlinear dynamics such as regular pulsing (RP), quasiperiodic pulsing (QP), chaos [21]–[30], etc. In comparison to optical injection and optical feedback, OEF is more flexible and reliable because of its convenience to be electronically controlled and its insensitivity to optical phase variations [31]–[33]. Several researches on the polarization dynamics of the short-wavelength VCSELs (~800-1000 nm) with negative optoelectronic feedback (NOEF) [34] and POEF [35] have been investigated, and various dynamical states such as period-1 (P1), two-frequency quasi-periodic (Q2), three-frequency quasi-periodic (Q3) and chaos are also observed. Recently, long-wavelength VCSELs emitting at the important telecom wavelength of 1550-nm have received more and more special attention because of their important applications in long-haul optical networks [17], [18]. However, studies on the nonlinear dynamics of 1550-nm VCSELs subject OEF are relatively inadequate. Thus, we recently focus our attention on it and we have experimentally studied the nonlinear dynamics of 1550-nm with POEF [36] and numerically studied the nonlinear dynamics of 1550-nm with NOEF [37] by varying the feedback strength. The rich variety of the nonlinear dynamics of 1550-nm VCSELs subject to OEF offers exciting prospects for the practical use of these devices in novel applications in optical networks.

In this paper, the nonlinear dynamics of a 1550-nm VCSEL with POEF through varying the feedback delay time is studied numerically. A mapping of the dynamic states is plotted in the parameter space of the feedback delay time and the feedback strength. A bifurcation diagram of the extrema of output peak series versus the feedback delay time is plotted. Various dynamical states, which are determined by combining the output time series, power spectra and phase portraits, have been observed. The POEF system follows a quasiperiodic route to chaotic pulsing.

This paper is organized as follows. Section II describes the rate equation model. In Section III, we present the numerical results. Finally, the conclusions of this work are given in Section IV.

2. Theoretical Model

The schematic diagram of VCSELs with POEF is shown in Fig. 1. The laser is biased at a fixed dc level and no external modulation is applied to the system. The optical output from the VCSEL is detected and converted to an electrical current and then positively fed back to drive the laser together with the dc bias current via a ring cavity composed of a variable attenuator (VA), a photoelectronic detector (PD) and an electronic amplifier (EA). When the bias current density is fixed, the controllable parameters for this POEF system are the feedback strength and the delay time. The VCSEL with POEF can be found to generate chaotic pulses by adjusting the feedback strength and the feedback delay time.

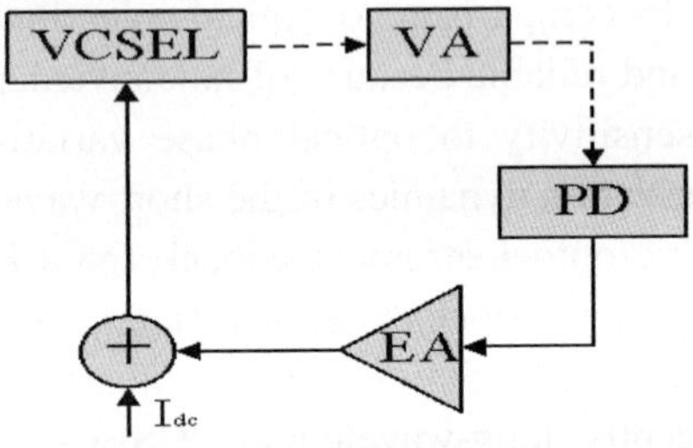

Fig. 1. Schematic diagram of VCSELs with POEF. Dash line: optical path; Solid line: electronic path; VA: variable attenuator; PD: photoelectronic detector; EA: electronic amplifier.

Our rate equation model for VCSEL with POEF is based on the model reported by San Miguel, Feng, and Moloney, also called spin-flip model (SFM) [38], [39]. The SFM assumes a four-level system in which electrons with different spin yield optical transitions with different circular light polarizations. We extended the SFM model including noise to account for the optoelectronic feedback by introducing an additional delayed term in the current [34], [35]. The rate equations of extended SFM for VCSEL with POEF are given by:

$$\frac{dE_x}{dt} = \kappa(1+i\alpha)[(N-1)E_x + inE_y] - (\gamma_a + i\gamma_p)E_x + \sqrt{\beta_{sp}}\xi_x \tag{1}$$

$$\frac{dE_y}{dt} = \kappa(1+i\alpha)[(N-1)E_y - inE_x] + (\gamma_a + i\gamma_p)E_y + \sqrt{\beta_{sp}}\xi_y \tag{2}$$

$$\frac{dN}{dt}=-\gamma_e N(1+P)+\gamma_e\mu[1+\eta\frac{P(t-\tau)}{P_0}]-i\gamma_e n(E_y E_x^*-E_x E_y^*) \tag{3}$$

$$\frac{dn}{dt}=-\gamma_s n-\gamma_e nP-i\gamma_e N(E_y E_x^*-E_x E_y^*) \tag{4}$$

where subscripts x and y represent x and y linear polarized modes, respectively. E is the slowly varied complex amplitude of the field, N is the total carrier inversion between the conduction and valence bands, n accounts for the difference between carrier inversions for the spin-up and spin-down radiation channels. The internal VCSEL parameters are as follows: κ is the photon decay rate, α is the linewidth enhancement factor, γe is the decay rate of total carrier population, γa is the gain anisotropy rate, γp is the linear birefringence rate, and γs is the spin relaxation rate. τ is the feedback delay time, μ is the normalized injection current, η is the feedback index corresponding to the strength of feedback, $P=|E_x|^2+|E_y|^2$ is the normalized output power, and P0 is the free-running output power. The effect of spontaneous emission noise is included by introducing a zero mean Gaussian noise source (as given in [17]). βsp = 10-6 is the spontaneous emission factor [17] and ξx,y are Gaussian white noises of zero mean value.

3. Results

We numerically solved the Equations (1)-(4) of the VCSEL with POEF using the fourth-order Runge-Kutta algorithm. The parameters of the 1550-nm VCSEL used for the simulation are [17]: γe = 1 ns-1, γp = 192.1 ns-1, γa = 1 ns-1, γs = 1000 ns-1, κ = 300 ns-1, and α = 3.

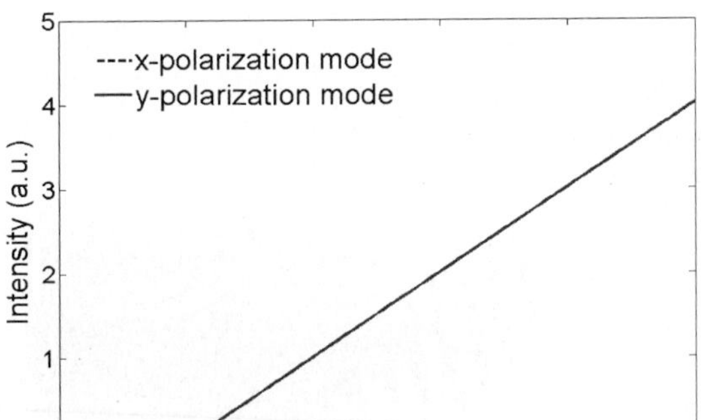

Fig. 2. Numerically calculated P-I curve of a free-running 1550-nm VCSEL. Dash line: x-polarization mode; Solid line: y-polarization mode.

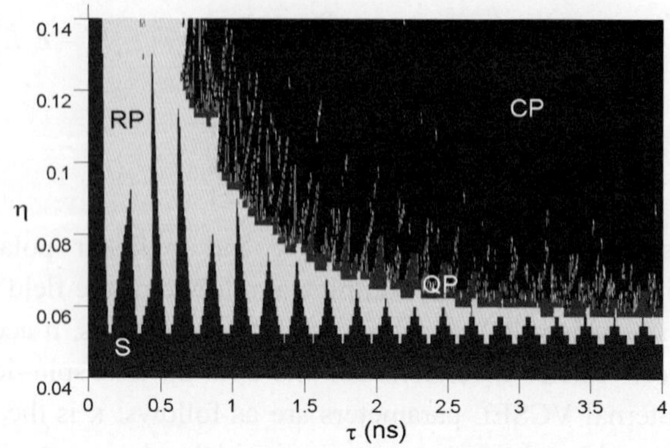

Fig. 3. Mapping of the dynamic states of the 1550-nm VCSEL subject to POEF in the parameter space of (τ, η), where the normalized injection current μ is fixed at 2.5. S: steady states. RP: regular pulsing. QP: quasiperiodic pulsing. CP: chaotic pulsing.

Figure 2 shows the P-I curve of a free-running 1550-nm VCSEL, where the normalized injection current μ is changed from 0 to 5. The free-running 1550-nm VCSEL emits in the y-polarization mode with a threshold current of μth = 1 and does not show polarization switching over the injected current range studied here.

Figure 3 displays the mapping of the dynamic states of the 1550-nm VCSEL subject to POEF in the parameter space of (τ, η), where the normalized injection current μ is fixed at 2.5. As can be seen, the dynamics of this POEF system become more complicated as the feedback delay time τ or the feedback strength η increases, where the chaotic pulsing states are observed in the regions of long delay times and large feedback strengths. Clearly, this POEF system enters chaos through a quasiperiodic route following RP, QP, and finally chaotic pulsing (CP) states.

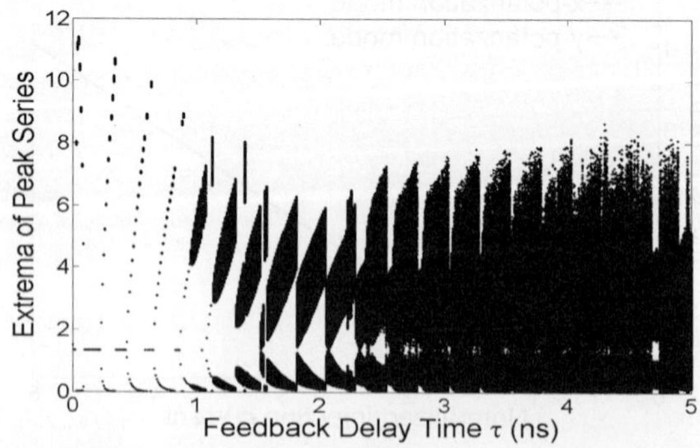

Fig. 4. Bifurcation diagram of the extrema of the peak series versus the feedback delay time for a 1550-nm VCSEL subject to POEF, where the feedback strength and the normalized injection current are fixed at 0.08 and 2.5, respectively.

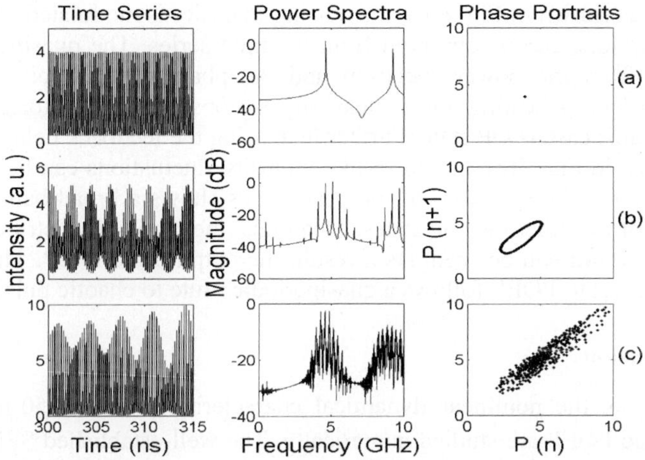

Fig. 5. Numerically calculated time series, power spectra, and phase portraits of different states: (a) RP, (b) QP, and (c) CP states, where the feedback delay times are 0.5 ns, 1.5 ns, and 2 ns, respectively.

Figure 4 shows the bifurcation diagram of the extrema of the peak series of the VCSEL output waveform for a fixed feedback strength $\eta = 0.08$ and a fixed normalized injection current $\mu = 2.5$ while the feedback delay time τ is varied. These extrema of the peak series are obtained by extracting the maxima and minima of the peak series, which make different pulsing states easier to be distinguished. As can be seen in Fig. 4, when the feedback delay time is small, the 1550-nm VCSEL subject to POEF is operated in the RP state. By increasing the feedback delay time, the POEF VCSEL system enters into QP state. Further increasing the delay time, the quasiperiodicity starts to disappear and finally the laser enters into CP state. In general, the complexity of the dynamics increases for this POEF system along with the delay time increasing. In order to see the details of the pulsing states and to identify the route to chaos for the POEF system, the dynamic states, the time series, the power spectra and the phase portraits are needed to be plotted.

Figure 5(a)-(c) display the simulation results for different feedback delay time τ, where the time series, the power spectra and the phase portraits of the 1550-nm VCSEL corresponding to the different nonlinear regions are plotted. The optoelectronic feedback strength and the normalized injection current are fixed at $\eta = 0.08$ and $\mu = 2.5$ while τ is changed. For $\tau = 0.5$ ns, as shown in Fig. 5(a), the output is a RP state, where the time series shows a sequence of regular pulses with a constant pulsing intensity and interval. Meanwhile, the corresponding power spectrum has one fundamental pulsing frequency at $f \approx$ 4.65 GHz, which is close to the relaxation oscillation (RO) frequency of the 1550-nm VCSEL $f_{RO} = [2\kappa\gamma_e(\mu-1)]^{1/2} / 2\pi = 4.8\,GHz$. In the phase portrait, a single dot can be seen for this RP state. For a relative large feedback delay time $\tau = 1.5$ ns, as shown in Fig. 5(b), the pulsing intensity is modulated and a slowly varying envelope can be observed from the time series. The quasiperiodicity is confirmed from the power spectrum and the phase portraits as well, where different orders of beating and a clear ring are observed. Therefore, the output state under this case is QP state. Further increasing the feedback delay time τ to 2 ns, as shown in Fig. 5(c), the noise-like intensity fluctuations can be seen in the time series, and the spectrum of the peak series shows a broad band compared with those of the other pulsing states. Meantime, the randomly distributed dots in the phase portrait can be seen. As a result, the output state is CP. The 1550-nm VCSEL subject to POEF follows a quasiperiodic route to chaotic pulsing.

4. Conclusion

In conclusion, the nonlinear dynamical characteristics of a 1550-nm VCSEL with delayed POEF are studied numerically. The well-established SFM has been extended and generalized to simulate VCSELs subject to OEF by introducing an additional delayed term in the current. Combining the time series, power spectra and phase portraits, it is numerically observed that the 1550-nm VCSEL with POEF follows a quasiperiodic route to chaos. The rich variety of nonlinear dynamics reported in this work for VCSELs operating at 1550-nm offers exciting prospects for novel practical uses of VCSELs in different fields spanning from optical signal processing to encrypted communication systems.

Acknowledgment

Projiect supported by zhe National Undergraduate Training Programs for Innovation and Entrepreneurship of China (Grant No. 201410635006)

References

1. Y. Y. Xie, M. Nikdast, J. Xu, X. W. Wu, W. Zhang, Y. Y. Ye, X. Wang, Z. H. Wang, and W. C. Liu, "Formal worst-case analysis of crosstalk noise in mesh-based optical networkson-chip," *IEEE Trans. Very Large Scale Integr. (VLSI) Syst.*, vol. 21, no. 10, pp. 1823–1836, Oct. 2013.
2. M. J. Kobrinsky, B. A. Block, J.-F. Zheng, B. C. Barnett, E. Mohammed, M. Reshotko, F. Robertson, S. List, I. Young, and K. Cadien, "On-chip optical interconnects," *Intel Technol. J.*, vol. 8, no. 2, pp. 129–141, May 2004.
3. A. Shacham, K. Bergman, and L. P. Carloni, "Photonic networks-on-chip for future generations of chip multiprocessors," *IEEE Trans. Comput.*, vol. 57, no. 9, pp. 1246–1260, Sep. 2008.
4. R. Beausoleil, P. Kuekes, G. Snider, S.-Y. Wang, and R. Williams, "Nanoelectronic and nanophotonic interconnect," *Proc. IEEE*, vol. 96, no. 2, pp. 230–247, Feb. 2008.

5. S. J. Koester, J. D. Schaub, G. Dehlinger, and J. O. Chu, "Germaniumon-SOI infrared detectors for integrated photonic applications," *IEEE J. Sel. Topics Quantum Electron.*, vol. 12, no. 6, pp. 1489–1502, Nov./Dec. 2006.
6. H. Gu, K. H. Mo, J. Xu, and W. Zhang, "A low-power low-cost optical router for optical networks-on-chip in multiprocessor systems-onchip," in Proc. *IEEE Comp. Soc. Annu. Symp. VLSI (ISVLSI)*, 2009, pp. 19–24.
7. N. Sherwood-Droz, H. Wang, L. Chen, B. G. Lee, A. Biberman, K. Bergman, and M. Lipson, "Optical 4×4 hitless silicon router for optical networks-on-chip (NoC)," *Opt. Exp.*, vol. 16, no. 20, pp. 15915–15922, 2007.
8. Y. Vlasov, W. M. J. Green, and F. Xia, "High-throughput silicon nanophotonic wavelength-insensitive switch for on-chip optical networks," *Nature Photon.*, vol. 2, pp. 242–246, 2008.
9. F. Koyama, "Recent advances of VCSEL photonics," *J. Lightw. Technol.*, vol. 24, no. 12, pp. 4502–4513, Dec. 2006.
10. K. Iga, "Surface-emitting laser—its birth and generation of new optoelectronics field," *IEEE J. Sel. Topics Quantum Electron.*, vol. 6, no. 6, pp. 1201–1215, Nov./Dec. 2000.
11. J. Sakaguchi, T. Katayama, and H. Kawaguchi, "All-optical memory operation of 980-nm polarization bistable VCSEL for 20-Gb/s PRBS RZ and 40-Gb/s NRZ data signals," *Opt. Exp.*, vol. 18, no. 12, pp. 12362–12370, Jun. 2010.
12. Z. G. Pan, S. Jiang, M. Dagenais, R. A. Morgan, K. Kojima, M. T. Asom, R. E. Leibenguth, G. D. Guth, and M. W. Focht, "Optical injection induced polarization bistability in vertical-cavity surface-emitting lasers," *Appl. Phys. Lett.*, vol. 63, no. 22, pp. 2999–3001, Nov. 1993.
13. C. Masoller and M. S. Torre, "Influence of optical feedback on the polarization switching of vertical-cavity surface-emitting lasers," *IEEE J. Quantum Electron.*, vol. 41, no. 4, pp. 483–489, Apr. 2005.
14. I. Gatare, M. Sciamanna, J. Buesa, H. Thienpont, and K. Panajotov, "Nonlinear dynamics accompanying polarisation switching in vertical-cavity surface-emitting lasers with orthogonal optical injection," *Appl. Phys. Lett.*, vol. 88, no. 10, pp. 101106-1–101106-3, Mar. 2006.
15. Y. Hong, R. Ju, P. S. Spencer, and K. A. Shore, "Investigation of polarization bistability in vertical-cavity surface-emitting lasers subjected to optical feedback," *IEEE J. Quantum Electron.*, vol. 41, no. 5, pp. 619–624, May 2005.
16. K. H. Jeong, K. H. Kim, S. H. Lee, M. H. Lee, B. S. Yoo, and K. A. Shore, "Optical injection-induced polarisation switching dynamics in 1.5 μm wavelength single-mode vertical-cavity surface-emitting lasers," *IEEE Photon. Technol. Lett.*, vol. 20, no. 10, pp. 779–781, May 2008.

17. M. S. Torre, A. Hurtado, A. Quirce, A. Valle, L. Pesquera, and M. J. Adams, "Polarization switching in long-wavelength VCSELs subject to orthogonal optical injection," *IEEE J. Quantum Electron.*, vol. 47, no. 1, pp. 92–99, Jan. 2011.
18. P. S. Spencer, C. R. Mirasso, and K. A. Shore, "Effect of strong optical feedback on vertical cavity surface emitting lasers," *IEEE Photon. Technol. Lett.*, vol. 10, pp. 191–193, Feb. 1998.
19. C.-H. Hang, L. Chrostowski, and C. J. Chang-Hasnain, "Injection locking of VCSELs," *IEEE J. Select. Topics Quantum Electron.*, vol. 9, no. 5, pp. 1386–1393, Sep. 2003.
20. A. Hurtado, D. Labukhin, I. D. Henning, and M. J. Adams, "Injection locking bandwidth in 1550-nm VCSELs subject to parallel and orthogonal optical injection," *IEEE J. Sel. Topics Quantum Electron.*, vol. 15, no. 3, pp. 585–593, May 2009.
21. M. Sciamanna, K. Panajotov, H. Thienpont, I. Vetennicoff, P. Megret, and M. Blondel, "Optical feedback induces polarization mode hopping in vertical-cavity surface-emitting lasers," *Opt. Lett.*, vol. 28, pp. 1543–1545, Sep. 2003.
22. A. A. Qader, Y. Hong, and K. A. Shore, "Circularly polarized optical feedback effects on the polarization of VCSEL emission," *IEEE Photon. Technol. Lett.*, vol. 24, no. 14, pp. 1200–1202, Jul. 2012.
23. J. Law and G. Agrawal, "Effects of optical feedback on static and dynamic characteristics of vertical cavity surface emitting lasers," *IEEE J. Sel. Topics Quantum Electron.*, vol. 3, no. 2, pp. 353–358, Apr. 1997.
24. S. F. Yu, "Nonlinear dynamics of vertical-cavity surface-emitting lasers," *IEEE J. Quantum Electron.*, vol. 35, no. 3, pp. 332–341, Mar. 1999.
25. J. Buesa, I. Gatare, K. Panajotov, H. Thienpont, and M. Sciamanna, "Mapping of the dynamics induced by orthogonal optical injection in vertical-cavity surface-emitting laser," *IEEE J. Quantum Electron.*, vol. 42, no. 2, pp. 198–207, Feb. 2006.
26. A. Hsu, J.-F. P. Seurin, S. L. Chuang, and K. D. Choquette, "Optical feedback in vertical-cavity surface-emitting lasers," *IEEE J. Quantum Electron.*, vol. 37, no. 12, pp. 1643–1649, Dec. 2001.
27. A. Valle, I. Gatare, K. Panajotov, and M. Sciamanna, "Transverse mode switching and locking in vertical-cavity surface-emitting lasers subject to orthogonal optical injection," *IEEE J. Quantum Electron.*, vol. 43, no. 4, pp. 322–333, Apr. 2007.
28. M. S. Torre, C. Masoller, and K. A. Shore, "Numerical study of optical injection dynamics of vertical-cavity surface-emitting lasers," *IEEE J. Quantum Electron.*, vol. 40, no. 1, pp. 25–30, Jan. 2004.

29. A. Hurtado, A. Quirce, A. Valle, L. Pesquera, and M. J. Adams, "Nonlinear dynamics induced by parallel and orthogonal optical injection in 1550 nm vertical-cavity surface-emitting lasers (VCSELs)," *Opt. Exp.*, vol. 18, no. 9, pp. 9423–9428, Apr. 2010.
30. R. Al-Seyab, K. Schires, N. A. Khan, A. Hurtado, I. D. Henning, and M. J. Adams, "Dynamics of polarization optical injection in 1550 nm VCSELs: theory and experiments," *IEEE J. Sel. Topics Quantum Electron.*, vol. 17, no. 5, pp. 1242–1249, Sep./Oct. 2011.
31. F. Y. Lin, and J. M. Liu, "Nonlinear dynamics of a semiconductor laser with delayed negative optoelectronic feedback," *IEEE J. Quantum Electron.*, vol. 39, no. 4, pp. 562–568, Apr. 2003.
32. F. Y. Lin, and J. M. Liu, "Harmonic frequency locking in a semiconductor laser with delayed negative optoelectronic feedback," *App. Phys. Lett.*, vol. 81, no. 17, pp. 3128–3130, Oct. 2002.
33. S. Tang, and J. M. Liu, "Chaotic pulsing and quasi-periodic route to chaos in a semiconductor laser with delayed opto-electronic feedback," *IEEE J. Quantum Electron.*, vol. 37, no. 3, pp. 329–336, Mar. 2001.
34. W. L. Zhang, W. Pan, B. Luo, X. F. Li, X. H. Zuo, and M. Y. Wang, "Theoretical study on polarization dynamics of VCSELs with negative optoelectronic feedback,". *Appl. Opt.*, vol. 46, no. 29, pp. 7262–7266, Oct. 2007.
35. J. F. Liao, and J. Q. Sun, "Polarization dynamics and chaotic synchronization in unidirectionally coupled VCSELs subjected to optoelectronic feedback," *Opt. Commun.*, vol. 295, pp. 188–196, May 2013.
36. Y. Y. Xie, Z. M. Wu, T. Deng, X. Tang, L. Fan, and G. Q. Xia, "Nonlinear Dynamics of 1550-nm VCSELs Subject to Positive Optoelectronic Feedback," *IEEE Photon. Technol. Lett.*, vol. 25, no. 16, pp. 1605–1608, Apr. 2013.
37. H. J. Che, Y. Y. Xie, and W. L. Zhao, "Numerically Simulate the Polarization Behavior of the 1550 Nm Vertical-cavity Surface-emitting Lasers with Negative Optoelectronic Feedback." *Inform. Technol. J.*, vol. 12, no. 21, pp. 6191–6196, 2013.

38. J. Martin-Regalado, F. Prati, M. San Miguel, and N. B. Abraham, “Polarization properties of vertical-cavity surface-emitting lasers,” *IEEE J. Quantum Electron.*, vol. 33, no. 5, pp. 765–783, May 1997.
39. M. San Miguel, Q. Feng, and J. V. Moloney, “Light-polarization dynamics in surface-emitting semiconductor lasers,” *Phys. Rev. A*, vol. 52, no. 2, pp. 1728–1739, Aug. 1995.

The method of downstream power tracing based on circuit theory

X.K. Ji† and H. Bao

North China Electric Power University, Beijing, China
Emails: †729721750@qq.com, ji_xiaokai@sohu.com

Under the condition of steady-state power flow of electric network being known, this paper traces the power flow of multiple generators on the basis of classical circuit theory. Through the analysis on the transmission relationship of the tracking results, the power transmission path of each generator is obtained, then it turns out that power component of each generator in the same branch is transmitted in parallel. The analysis on the power transmission characteristics, in one hand, connects the power information scattered in each brand in series and formats an organic whole of energy transmission relationship; on the other hand, proves the hypothesis on the form of power injection theoretically and provides the theoretical support for the study of issues related to electric market.

Keywords: Steady-state power flow; Power tracing; Analysis of power transmission characteristics.

1. Introduction

In electric market, for the network with multiple generators, in order to determine the transmission service and wheeling service fee accurately and reasonably, it need to be known which load the power of each generator send to, what the power supply path of each generator is, and how much loss is caused in the process of power transmission. The key to solving these problems is to clear how the power of each generator distributes in the network. The traditional methods are based on the proportional sharing principle, and obtain the relationship between the source and flow by the method of power flow tracing. However, in addition to limited conditions and the use scope in practical applications, this principle is difficult to be proved strictly, and has great errors or even fails in many problems [15]. References [6,7,14] decomposed electrical quantities of nodes and branches according to the basic electrical principle, but this method is complex applied to large power system. Reference [9] put forward a new definition of *power component of the branch whole loss* in determining the contribution of generator or load to the branch loss, deduced the expression of *power component of the whole loss* of source and load in the same branch. and deduced the superposition principle between the power of branch total loss and the power component of the whole loss applying the principle of superposition and the principle of complex power conservation in the network.

References [4,11] analyzed the distribution of power in electric networks from the point of view of power component based on the basic theory of circuit, got the analytical expression of multiple power distribution in any node and branch of power source, this method is strict and reasonable. Firstly, this paper introduces the power tracing method proposed in the references [4,11], secondly, analyzes the transmission characteristics of power in the node and branch, in the end, takes the IEEE 3 machine 9 bus system as an example, traces the power of power source in the network based on the classical circuit theory and the power component theory of the condition of steady-state power being known, the transmission relationships of tracking results are also analysed, it turns out that the power transmission relationship of each source is parallel in the same branch, at the same time, power information scattered in each branch is connected in series through the analysis of power transmission characteristics, forming the organic whole of energy transmission. The analysis method in this paper also gives theoretical support to the solution of related issues in electric power market.

2. Power Tracing

2.1. *The introduction of power tracing method based on circuit theory*

Power tracing, that is, in a specific operating state, to clear which generators the power of a specific load is sent by and which loads the power of a specific generator sends to through the analysis and calculation, and through which transmission elements the two process is completed [10].

According to the search direction, the power flow tracing can be divided to *downstream tracing* whose tracing direction is from power source to load and *upstream tracing* whose tracing direction is from load to power source [2, 10]. This study is about the power distribution of power source in the network, so the category belongs to downstream tracing.

Take the linear network of *N* nodes, *Q* current sources (the node-numbering of current source began to turn from *1* to *q*, the node-numbering later is random) as an example. There are grounding branches in the network, admittance matrix Y whose reference point is ground is non singular matrix, then the nodal voltage equation is:

$$\begin{bmatrix} Y_{11} & Y_{12} & \cdots & Y_{1q} & \cdots & Y_{1n} \\ Y_{21} & Y_{22} & \cdots & Y_{2q} & \cdots & Y_{2n} \\ \vdots & \vdots & \vdots & \vdots & \vdots & \vdots \\ Y_{q1} & Y_{q2} & \cdots & Y_{qq} & \cdots & Y_{qn} \\ Y_{(q+1)1} & Y_{(q+1)2} & \cdots & Y_{(q+1)q} & \cdots & Y_{(q+1)n} \\ \vdots & \vdots & \vdots & \vdots & \vdots & \vdots \\ Y_{n1} & Y_{n2} & \cdots & Y_{nq} & \cdots & Y_{nn} \end{bmatrix} \begin{bmatrix} \dot{U}_1 \\ \dot{U}_2 \\ \vdots \\ \dot{U}_q \\ \dot{U}_{(q+1)} \\ \vdots \\ \dot{U}_n \end{bmatrix} = \begin{bmatrix} \dot{I}_{1s} \\ \dot{I}_{2s} \\ \vdots \\ \dot{I}_{qs} \\ 0 \\ \vdots \\ 0 \end{bmatrix} \tag{1}$$

The expression in matrix form is:

$$Y \cdot \dot{U} = \dot{I}_s \text{ or } \dot{U} = Y^{-1} \cdot \dot{I}_s \tag{2}$$

Any nodal voltage $\dot{U}_i$ can be expressed as:

$$\dot{U}_i = \sum_{m=1}^{q} z_{mi} \dot{I}_{ms}, (i = 1, 2, ...n) \tag{3}$$

According to the basic theory of circuit, power injection can be expressed as:

$$\dot{S}_{ks} = \dot{U}_k \cdot \dot{I}_{ks}^* = \sum_{m=1}^{q} z_{mk} \dot{I}_{ms} \dot{I}_{ks}^*, (k = 1, 2, ..., q) \tag{4}$$

As can be seen from the expression (4), the expresstion for power injection not only includes the power components with the forms of $\dot{I}_{ks} \cdot \dot{I}_{ks}^*$, but also includes the power components with the forms of $\dot{I}_{is} \cdot \dot{I}_{ks}^*$ $(i \neq k)$, these components are clearly shows the ownership of power. The expression (4) shows that the power components with the forms of $\dot{I}_{1s} \dot{I}_{ks}^*, \dot{I}_{2s} \dot{I}_{ks}^*, ..., \dot{I}_{ms} \dot{I}_{ks}^*, ..., \dot{I}_{qs} \dot{I}_{ks}^*$ are sent by power source k, the only difference between them is the ratio coefficient.

2.2. *The formula of power tracing*

By the references [4, 11], the power expression in the grounding branch is:

$$\dot{S}_i = y_i^* \sum_{m=1}^{q} \sum_{l=1}^{q} z_{mi} z_{li}^* \dot{I}_{ms} \dot{I}_{ls}^* \tag{5}$$

The power expression in the tie line branch is:

$$\dot{S}_{ij} = -y_{ij}^* \sum_{m=1}^{q} \sum_{l=1}^{q} z_{mi} (z_{li} - z_{lj})^* \dot{I}_{ms} \dot{I}_{ls}^*, (i, j = 1, 2, ..., n) \tag{6}$$

From the expressions (2-5), (2-6), it can be seen that each power expression includes q^2 power components which are provided by q power sources, each power source provides q power components.

The expressions of the power in the grounding branch and in the tie line branch which are provided only by power source ks are as follows:

$$\dot{S}_i\big|_{\in ks} = y_i^* \sum_{m=1}^{q} z_{mi} z_{ki}^* \dot{I}_{ms} \dot{I}_{ks}^* \tag{7}$$

$$\dot{S}_{ij}\big|_{\in ks} = -y_{ij}^* \sum_{m=1}^{q} z_{mi} (z_{ki} - z_{kj})^* \dot{I}_{ms} \dot{I}_{ks}^*, (i, j = 1, 2, ..., n) \tag{8}$$

The expressions (7) and (8) are the formulas of power tracing. Each expression includes q parts, by the formulas (7) and (8), all power components belonging to the power source ks in the grounding branch and in the tie line branch can be obtained.

3. Analysis of the Power Transmission Characteristics of Branch, Node

3.1. *Analysis of the power transmission characteristics of branch*

Power flow diagram in the branch $i - j$ is as follows, the branch current is $\dot{I}$, the branch input power is $\dot{S}_i$, the branch output power is $\dot{S}_j$.

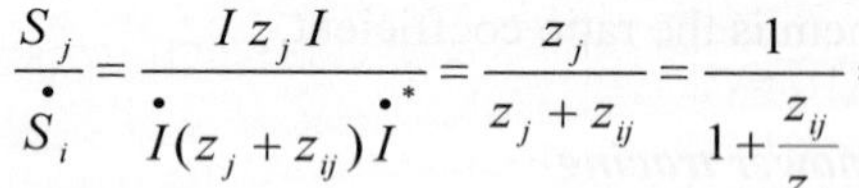

Equivalent circuit can be obtained according to power flow diagram, the branch impedance is z_{ij}, the equivalent impedance of output power is z_j, then

$$\frac{\dot{S}_j}{\dot{S}_i} = \frac{\dot{I} z_j \dot{I}^*}{\dot{I}(z_j + z_{ij})\dot{I}^*} = \frac{z_j}{z_j + z_{ij}} = \frac{1}{1 + \frac{z_{ij}}{z_j}} = \dot{G}_1 \tag{9}$$

3.2. *Analysis of the power transmission characteristics of node*

Power flow diagram in the node j is as follows, the nodal voltage is $\dot{U}$, nodal input power is $\dot{S}_j$, nodal output power is $\dot{S}_k$, nodal grounding power is $\dot{S}_l$.

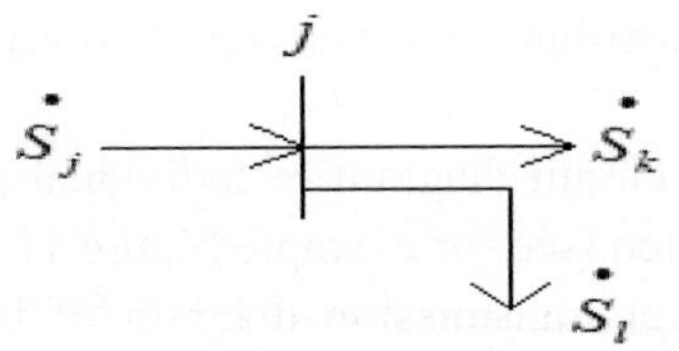

Equivalent circuit can be obtained according to power flow diagram, the equivalent admittance of nodal output power is y_k, the equivalent admittance of nodal grounding power is y_l, then

$$\frac{\dot{S}_l}{\dot{S}_j}=\frac{\dot{U}y_l^*\dot{U}^*}{\dot{U}(y_l^*+y_k^*)\dot{U}^*}=\frac{y_l^*}{y_l^*+y_k^*}=\frac{1}{1+\frac{y_k^*}{y_l^*}}=\dot{G}_2 \tag{10}$$

Connect the power flow diagram of branch with the power flow diagram of node, as shown in the following figure:

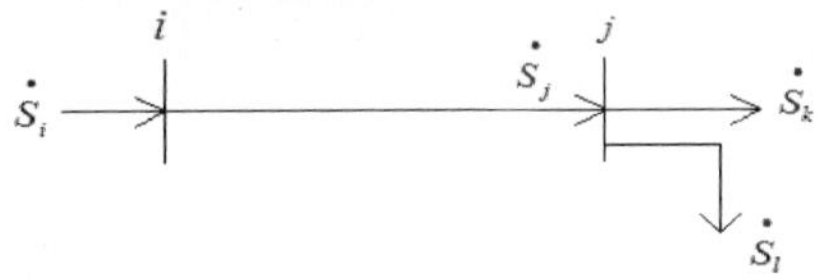

The following can be got:

$$\frac{\dot{S}_l}{\dot{S}_i}=\frac{\dot{S}_j}{\dot{S}_i}\times\frac{\dot{S}_l}{\dot{S}_j}=\dot{G}_1\times\dot{G}_2 \tag{11}$$

Conclusion: for any network, when the power flow in the steady state is known, the power in the network can be connected end to end with the combination form of the impedance (or admittance conjugate) based on equivalent circuit.

4. Describe the Power Tracking Results By the Power Transmission Method

By the formulas (7), (8), power distribution in the network can be obtained for each power source , then equivalent circuit diagram of individual power source can be obtained, finally, it can be verified that individual power source can provide power for electric network through the equivalent network by the analysis method of power transmission characteristics.

The main steps are as follows:

(1) Get the equivalent circuit diagram of power system in the steady state.

(2) Form the node admittance matrix Y of equivalent circuit diagram, then to get the inverse matrix of Y.

(3) Obtain the power distribution of individual power source by the formulas of power tracing.

(4) Get the equivalent circuit diagram of individual power source according to its power distribution (see for example Figure 1).

(5) Work out the power transmission diagram of individual power source based on the combination form of the impedance (or admittance conjugate). See for example Figure 2.

5. The Example Analysis

Take the network of IEEE 3 machine 9 bus as an example, as described in the previous section step (1)~(4), the power supply path of individual power source 1(whose location is in the node 1) can be obtained after tracing the power flow of power source 1(the grounding branch has merged into the node), as shown in Figure 1. The dotted line in the figure represents that there is no power flow there .

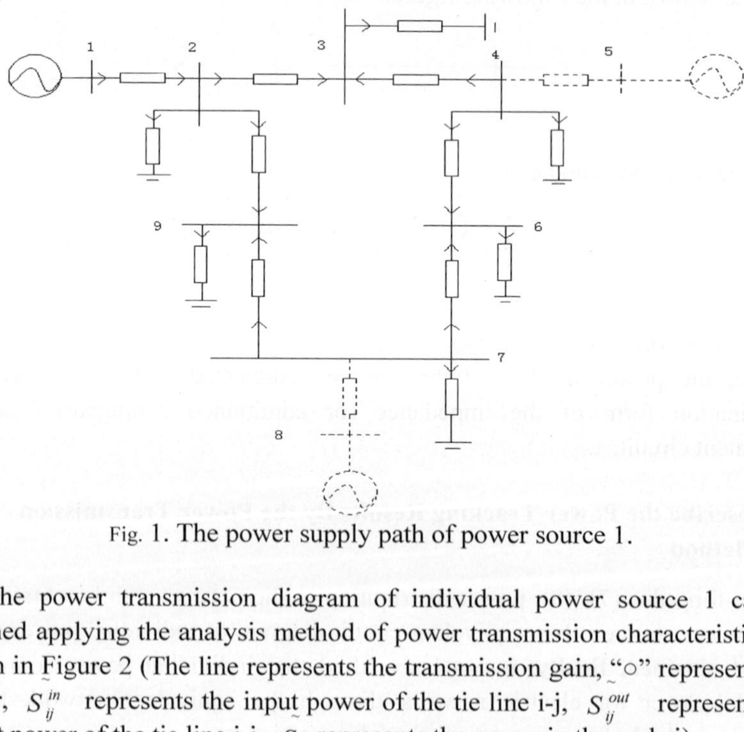

Fig. 1. The power supply path of power source 1.

The power transmission diagram of individual power source 1 can be obtained applying the analysis method of power transmission characteristics, as shown in Figure 2 (The line represents the transmission gain, "o" represents the power, $\tilde{S}_{ij}^{in}$ represents the input power of the tie line i-j, $\tilde{S}_{ij}^{out}$ represents the output power of the tie line i-j, $\tilde{S}_j$ represents the power in the node j).

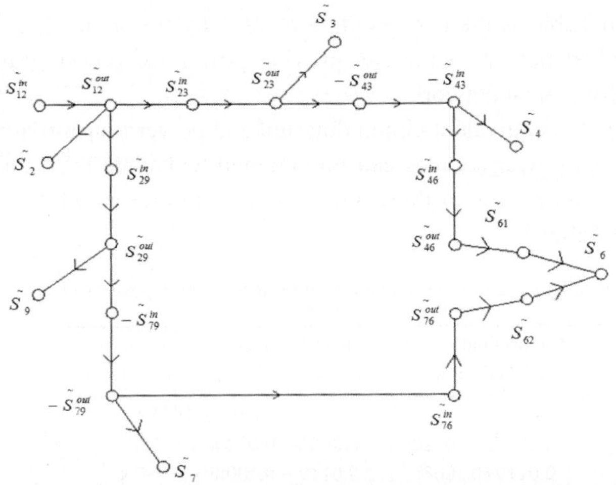

Fig. 2. The power transmission diagram of power source 1.

The ratio of any two power in Figure 2 can be expressed as a combination form of the impedance (or admittance conjugate). The power ratio worked out by actual power values and impedance (or admittance) values are shown in Table 1.

Table 1. The power ratio in Fig. 2.

	Results calculated by actual power values	Results calculated by impedance values
S_{46}^{out}/S_{46}^{in}	0.9761 - 0.0966i	0.9761 - 0.0966i
$S_{46}^{in}/(-S_{43}^{in})$	0.8966 + 0.4528i	0.8966 +0.4528i
$S_4/(-S_{43}^{in})$	0.1034 - 0.4528i	0.1034 - 0.4528i
$(-S_{43}^{in})/(-S_{43}^{out})$	1.0160 + 0.0220i	1.0160 + 0.0220i
$(-S_{43}^{out})/S_{23}^{out}$	0.3815 - 0.0778i	0.3815 - 0.0778i
$S_3/(S_{23}^{out})$	0.6185 + 0.0778i	0.6185 + 0.0778i
S_{23}^{out}/S_{23}^{in}	0.9890 - 0.0517i	0.9890 - 0.0517i
S_{76}^{out}/S_{76}^{in}	0.9869 - 0.0253i	0.9869 - 0.0253i
$S_{76}^{in}/(-S_{79}^{in})$	0.6609 + 0.3801i	0.6609 + 0.3801i
$S_7/(-S_{79}^{in})$	0.3391 - 0.3801i	0.3391 - 0.3801i
$(-S_{79}^{in})/(-S_{79}^{out})$	1.0298 + 0.0319i	1.0298 + 0.0319i
$(-S_{79}^{out})/S_{29}^{out}$	0.1628 - 0.1477i	0.1628 - 0.1477i
S_9/S_{29}^{out}	0.8372 + 0.1477i	0.8372 + 0.1477i
S_{29}^{out}/S_{29}^{in}	0.9618 - 0.1302i	0.9619 - 0.1302i
S_{23}^{in}/S_{12}^{out}	0.5311 + 0.0136i	0.5310 + 0.0136i
S_{29}^{in}/S_{12}^{out}	0.4786 + 0.0643i	0.4786 + 0.0643i
S_2/S_{12}^{out}	-0.0097 - 0.0780i	-0.0097 - 0.0780i
S_{12}^{out}/S_{12}^{in}	0.9960 - 0.0970i	0.9960 - 0.0970i

As seen from Table 1, the two results are basically the same, then it is verified that a individual power source can provide power for power grid end to end through the equivalent network.

Similarly, the equivalent circuit diagram and power transmission diagram of other individual power sources can be obtained, calculate the parallel value of the equivalent impedance of three power sources in the same branch, the results are shown in Table 2.

Table 2. Parallel values of the equivalent impedance of three power sources.

Impedance label	The original impedance value	Parallel value
z_{12}	0.0625i	0.0000 + 0.0625i
z_{23}	0.00855+0.0720i	0.0085 + 0.0720i
z_{34}	0.0119+0.1008i	0.0119 + 0.1008i
z_{45}	0.0586i	-0.0000 + 0.0586i
z_{46}	0.0390+0.16999i	0.0390 + 0.1700i
z_{67}	0.0170+0.0920i	0.0170 + 0.0920i
z_{78}	0.0576i	0.0000 + 0.0576i
z_{79}	0.0100+0.0850i	0.0100 + 0.0850i
z_{92}	0.0320+0.161i	0.0320 + 0.1610i
z_{2}	- 4.3955i	0.0000 - 4.3955i
z_{3}	1.0046+ 0.1660i	1.0046 + 0.1660i
z_{4}	- 3.5273i	0.0000 - 3.5273i
z_{6}	1.1377 + 0.0447i	1.1377 + 0.0447i
z_{7}	- 5.9879i	-0.0000 - 5.9879i
z_{9}	0.7598 + 0.1588i	0.7598 + 0.1588i

As seen from Table 2, parallel results of the equivalent impedance of three power sources in the same branch and the original impedance values of network are consistent.

6. Conclusions

(1) After completing the downstream power tracing of the network with multiple power sources based on the circuit principle, the power flow of individual power source can be described by the combination form of the impedance (or admittance conjugate).

(2) The power components of three power sources are transmitted in parallel in the same branch after power tracing.

References

1. Bialek J. 1996. Tracing the flow of electricity. *IEE Proceedings-C*, 143(2): 405-412.
2. Chaoxia Jing, Yuanyuan Liu & Li Zeng. 2010. Comparison of 4 kinds of analysis methods on the source and flow of power grid. *Proceedings of the CSU-EPSA*, 34(23): 42-51.
3. Dongping Ai & Hai Bao. 2010. Method analysis on working out complex power dynamic distribution factor. *Power System Technology*, 34(10): 169~173.
4. Hai Bao & Qian Ma. 2005. Physical distribution mechanism of power loss. *Proceedings of the CSEE,* 25(21): 82-86.
5. Jianchun Peng, Hui Jiang & Liansheng Cheng. 2001. The branch power component theory of complex power source. *Proceedings of the CSEE*, 21(1): 1-4.
6. Jilai Yu & Zhuo Liu. 2005. Electrical dissecting method of AC branch: (1) Dissection of ordinary branch. *Proceedings of the CSU-EPSA*, 29(1): 22-28.
7. Jilai Yu & Yi Tang. 2007. United electrical dissection of AC branch and node. *Proceedings of the CSEE*, 27(16): 37-42.
8. Kirschen D, Allan R & Strbac G. 1998. Tracing active and reactive power between generators and loads using real and complex currents. *IEEE Trans on Power Systems*, 13(3): 1312-1318.
9. Linqiao Feng & Wenyu Xu, etc. 2002. Conjugate current whole loss component definition and its principle of superposition [J]. *Proceedings of the CSEE*, 22(5): 65~69.
10. Weidong Li & Liming Wei, etc. 2002. Comparative study on power flow tracing methods. *Journal of Dalian University of Technology,* 42(4): 491-196.
11. Wei Li, Hai Bao & Yihan Yang. 2007. A analytical method about determing the relationship between the power supply and the load distribution based on the coefficient matrix. *Relay*, 35(3): 25-28.
12. Wencheng Wu & Yougang Zhang. 2004. Discussion on network loss allocation methods based on complex power tracing. *Relay*, 32(12):82-86.
13. Xiuli Wang & Xifan Wang. 2004. Complex power flow tracing based on the real and imaginary component of current. *Proceedings of the CSU-EPSA*, 28(3): 10-13,23.
14. Yi Tang, Jilai Yu & Xianlin Liu. 2005. Electrical dissecting method of AC branch: (2) Dissection of the branch with phase shifter. *Proceedings of the CSU-EPSA*, 29(2): 25-28.

15. Yi Tang, Jilai Yu & Suquan Zhou. 2006. Electrical dissecting algorithm of the source and flow path of power network. *Proceedings of the CSU-EPSA*, 30(22): 28-33.
16. Yuanhong Liu & Hai Bao. 2010. Analysis of power supply path of wind power generation. *Power System Technology,* 34(11): 188~191.
17. Zhengqiu Wu. 2005. The nonlinear function value contribution factor theory and its application. *Proceedings of the CSU-EPSA,* 25(11): 32-34.

Burr detection system of workpieces based on halcon

D.X. Zuo, X.R. Chen[1] † and Y. Yin
Shanghai Key Laboratory of Modern Optical System &
Engineering Research Center of Optical Instruments and Systems,
Ministry of Education;
University of Shanghai for Science and Technology,
Shanghai, China.
Emails: †*cxrsjtu@163.com*

It's quite important to do deburring work well in the process of mechanical piece production. This paper introduces a burr detection system based on machine vision, which can detect the position of the workpieces' burrs. It can improve the efficiency of deburring work. This system not only uses the burr detection algorithm, but also introduces the sorting algorithm of useless contours. By using these two algorithms, the system can remove the interference of errors, which makes it effective.

Keywords: Burr detection; Halcon; Threshold.

1. Introduction

Burr issue is one of the most difficult problems in modern mechanical production. The burrs usually occur in the edge as the cutting force is not steady enough. The burrs have a bad influence on the precision of the workpiece. The deburring work is becoming more and more important, especially in precision machinery industry [1].

Deburring technology is a very significant step in production process. With the development of production techniques, the deburring technology gradually develops from traditional handwork to automation line [2]. There are several commonly used deburring methods, such as: electrochemical deburring, magnetic finishing deburring, laser deburring and ultrasonics deburring, etc. Considering the different structure features of the workpieces, adopting reasonable deburring methods is the key point in researching this technology.

Machine vision achieves great development in the industrial automation in recent years. The biggest advantage of the machine vision is that it is detached with observed target. Therefore, no damage will cause to observation and observed target. It's safe and reliable. Compared with other ways, it's the best one. It's quite effective and reliable to use image processing technology based on machine vision to detect the burr on workpieces. Halcon, developed by MVTec in German, is the standard machine vision algorithms package. It provides suffi-

[1] Female, associate professor

cient algorithms of machine vision. This paper introduces the burr detection algorithm which is based on the algorithms library of Halcon. It is used for detecting burr and calculating the corresponding coordinates. It can also eliminate interference of useless contours and improve the efficiency of the subsequent deburring process. It makes great sense to do more research on it.

2. Detection System Introduction

The burr detection system includes light source, image acquisition unit and image processing module. It uses parallel ring light to irradiate the workpiece's surface. The image acquisition module is a black-white camera with 5.0 megapixels. The Figure 1 is the image of a workpiece with burr acquired by the camera. We can easily find the burr area. Qualified products should guarantee that there is no burr on the edge of products. Deburring is needed, if the burrs are too excessive.

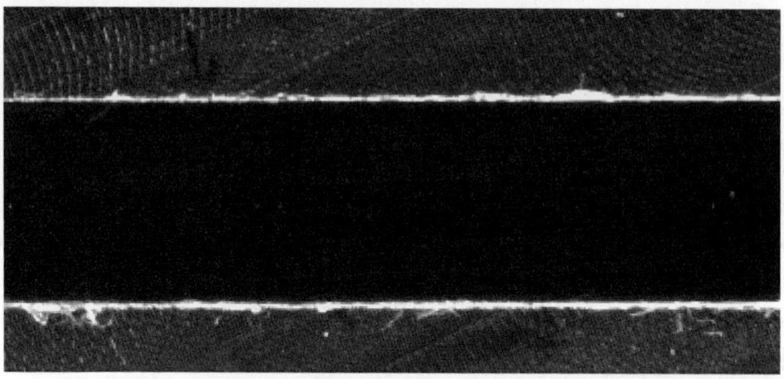

Fig. 1. Image of workpiece.

The image acquired by image acquisition unit is transferred to Halcon, the machine vision software via camera USB, and then processed. The detection system is based on Halcon. It contains image preprocessing algorithm, burr detection algorithm and burr coordinates' calculation.

3. Analysis of the Core Algorithm

3.1. *Image preprocessing algorithm*

The pre-processing work is used to extract the main region of the burr, which can improve efficiency of image processing. The hardware platform of the detection system can keep the inner edge close to the horizontal direction. According to the morphological features of edge orientation, the system extracts the main region of the inner edge by morphological operators provided by Halcon. Also, considering the grey value of the inner edge is relatively high, the system uses threshold value segmentation to get region of interest more accurately. The extracted region is shown in Figure 2.

Fig. 2. Rough region of inner boundary.

3.2. *Edge contour extraction*

Image edge is always the most remarkable part of gray variance of image local area whose gray section may be seen a skip where the gray changes in a small area from low gray level to high gray level [3]. The burr edge is such a region where gray value changes very rapidly. Most burr contour will be included by extracting the inner edge.

There were two commonly used edge detecting methods. The first one is to detect the maximum gradient. The Roberts gradient, Sobel operator and Prewitt operator are the simple and common examples. The second one is to detect the zero-crossing point of second derivative. Canny algorithm and Laplacian algorithm are the common second order algorithms. However, both of the two methods are easily affected by noise.

By observing the image of workpiece, the burrs are concentrated in the outer of inner region edge. However, the edge extraction operator using above may extract excessive region. Although it can extract burr contours, it would also collect too much usefulness information. It causes great inconvenience for the process of eliminating the redundant region. It would be better to use the edge detection method that can be changed parameter by man. The burr detection system use the close_edges operator provided by Halcon. The structure of the close_edges operator is as follows:

close_edges(Edges, EdgeImage : RegionResult : MinAmplitude :).

In this function, parameter Edges is stand for edge extraction area, EdgeImage is stand for the edge image and MinAmplitude is stand for the minimum edge amplitude. The parameter RegionResult is the calculated edge region. The function of the close_edges operator is that it can close edge gaps using the edge amplitude image. This is done by examining the neighbors of each edge point to determine the point with maximum amplitude and adding the point to the edge if its amplitude is larger than the minimum amplitude passed in parameter MinAmplitude.

It can set different minimum gray value amplitudes in different regions when the close_edges operator is used. Because of that, the edge contours are relatively accurate. And the data would not be affected badly by realtime lighting. The method guarantees the robustness of the system effectively .The calculated edge contours are shown in Figure 3.

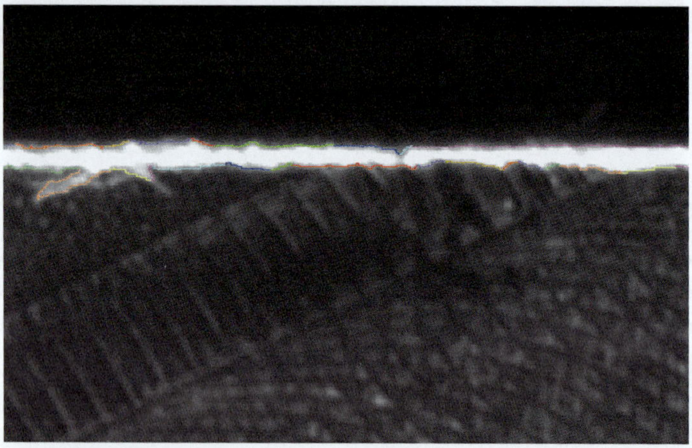

Fig. 3. Extracted edge contours.

3.3. *Elimination of errors*

The extracted edge contours contain large number of useless contours and should be removed. After significant filtering, the left contours are burrs of the workpiece. The methods for eliminating the errors are as follows:

Type 1: contours close to a straight line.

As shown in Figure 4, the contour which is close to a straight line is not the burr obviously. Contrasting fitting degree of the contour and the straight line, the system can extract the type of the useless contours.

Fig. 4. Contour close to a straight line.

Type 2: contours with large distance to the inner edge.

As shown in Figure 5, the contour with large distance to the inner edge belongs to surface textures. The algorithm sets a line located in the edge region and calculates the distance from each contour line to the reference line. Then it sets a threshold of the distance to the line. All the contours whose distance to the edge is more than the threshold should be eliminated.

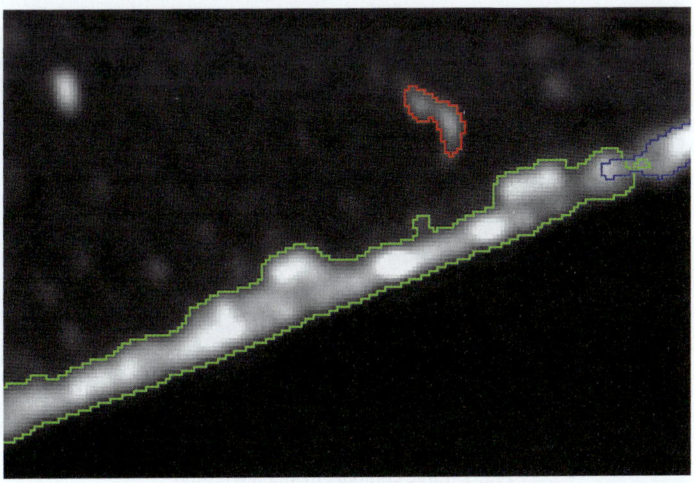

Fig. 5. Contour with large distance to the inner edge.

Type 3: contours in the smooth region.

As shown in Figure 6, the contours of type 1 and type 2 have been eliminated in the picture. The workpiece with smooth surface may be detected with edge contours as the light exposure and it results in gray value gradient in smooth area. However, the contours are not burrs, which should be eliminated.

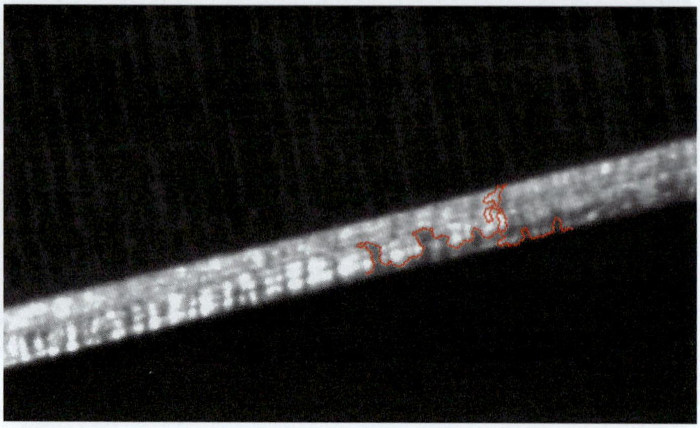

Fig. 6. Contour in the smooth region.

The algorithm obtains the smooth region using threshold segmentation. Figure 7 shows the edge contours of the smooth region. The edge contours are very close to a straight line obviously. Contrasting fitting degree of the contours and straight line, the system can extract the type of the useless contours in the smooth region.

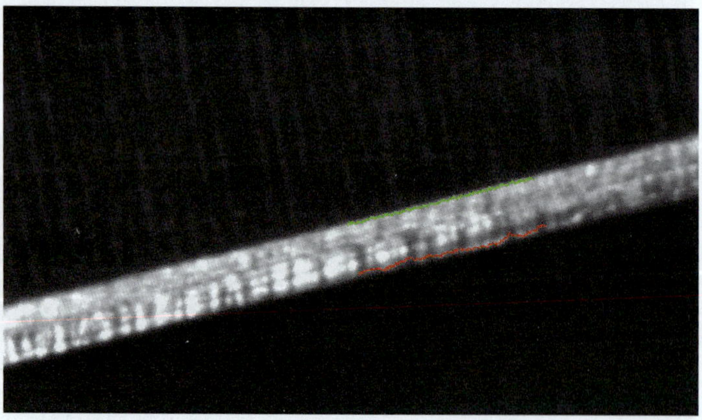

Fig. 7. Edge contours of smooth region.

3.4. *Burr coordinates' calculation*

The above algorithms are designed in Halcon. The burr detection results can be obtained after running the program. As shown in Figure 8, the burr detection has been finished basically. The system calculates the coordinates of the burrs, stored in computer. The data can bring convenience for succedent computer real image treatment. So far, the burr detection work is finished.

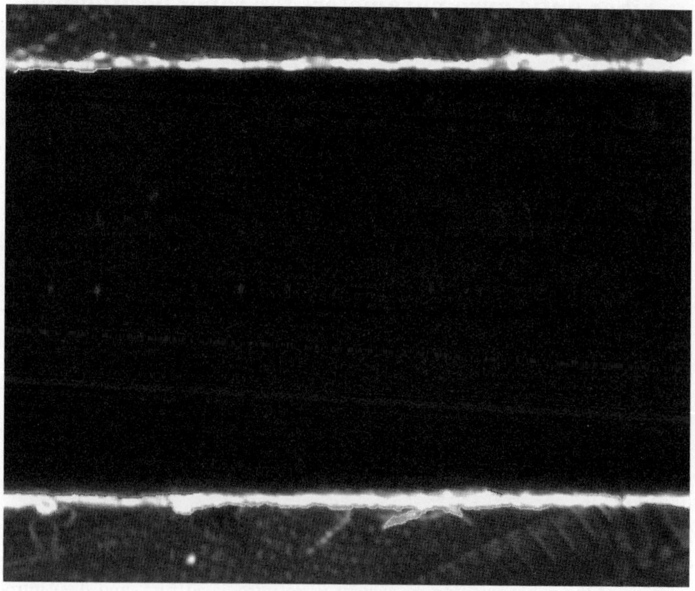

Fig. 8. Burr detection results.

4. Conclusions

The paper presents a burr detection system based on machine vision. To verify the feasibility of the algorithm, the program was written on the platform of Halcon. By using morphological characteristics to preprocess the image, it can reduce the area to be detected. Through edge detection algorithm provided by Halcon, it can also calculate the initial burr contour shape. It introduces a tight algorithm of removing contours that are not burrs, exclude pseudo burrs as more as possible, and effectively improve the feasibility of the algorithm. The system realizes non-contact detection, which is an improved method of detecting burr. The system is in line with the development trends of industrial automation.

References

1. S.P. Chen, J. Luo and J. Shi. 2004. The Burr Problem in Metal Machining. *Machinery Design & Manufacture* 2004(1): 99-101.
2. Z.Y. Chen & G.C. Wang. 2004. Studying status of burr and deburring technology. *Modern Manufacturing Engineering* 2004(2): 126-129.
3. D.P. Xie. 2010. Research in Two Dimensions Digital Image Edge Detection. *Dazhong Keji* 2010(8): 34-35.

Fiber optic gyro parameters for robot positioning

Y. G. Li[1†]
Changchun University, Changchun, China
†Email: liang811@sina.com

W. Liang[2]
Changchun University of Science and Technology, Changchun, China

H. W. Zheng[3] and W. Luo[4]
Institute of Changchun Equipment Technology, Changchun, China

Based on the research on optical fiber gyro realize robot precision localization system, the composition of the robot positioning system were introduced first, and then introduced the basic principle of optical fiber gyro and performance indicators. Finally, the parameter was tested of fiber optic gyroscope. Results show that the fiber optic gyro parameters satisfy the requirement of the robot positioning.

Keywords: Robot; positioning; Gyroscope;.

1. The Introduction of Robot Positioning System

The positioning system of robot is the key to realize the control of the robot, its positioning accuracy will affect the intelligent of the robot, the design and development of positioning system has important practical significance for a robot research. This article mainly introduced the positioning system of robot ode based on optical fiber gyro and light dish ode, positioning system using digital signal processor, to collect signal of fiber optic gyroscope robot relative to the starting point of the angle information, and then the relative position of robot is obtained by odometer information, the robot is obtained by calculating the relative position and angle information relative to the initial coordinate system coordinate information, so as to realize the positioning of the robot. Therefore is the great significance to the study of fiber optic gyroscope.

---

[1] Yue-Guang Li
[2] Wei Liang
[3] Hong-Wei Zheng
[4] Wei Luo

2. The Principle of Fiber Optic Gyro

Fiber optic gyro is based on the Sagnac interferometer [1]. As shown in figure 1, light from a light source is points/beam into two beams, respectively from the optical fiber ring into the optical fiber is sensitive to the ends of the coupling coil, spread across one, counterclockwise. Out of the optical fiber ring at both ends of two beams of light, after stacking beam splitter and points/produce interference. When optical fiber ring in the stationary state, out of the optical fiber ring at both ends of two beams of light, the optical path difference is zero [2].When optical fiber ring with rotation angular rate, due to the Sagnac effect, the cis, counterclockwise to the spread of the optical path difference of this two beams of light, and caused by the phase difference is [3]:

$$\Delta\varphi = \frac{4\pi RLN}{c\lambda}\Omega = \frac{8\pi AN}{c\lambda}\Omega \tag{1}$$

Among them:
R - fiber circle radius
L -fiber ring circumference
A -the area of the optical fiber loop contains,
N - optical fiber coil number of turns
λ- the wavelength of light
C - the light propagation velocity in the medium
By detecting the phase difference (that is, the interference of light intensity) can accelerate the rate of information.For $K = \frac{4\pi RLN}{c\lambda}$ is gyro scale factor.

Sagnac effect can be used as shown in figure 2, the circular ring interferometer to illustrate. From point A to the incident light is separated into two beams of light by beam splitter. Reflected light into a fiber coil propagation along the counterclockwise direction. Transmitted light b is reflected back to later be reflected by splitter, into the fiber coil propagation along the clockwise direction. Two beams of light around a week later, and in the beam splitter rendezvous.

If not consider the influence of the optical fiber core layer refractive index (assuming the core layer of the refractive index n=1). When the interferometer without rotation, propagation along the opposite direction of the two beams of light around a week to equal optical path, is equal to the circumference of the circle loop, namely:

$$L_a = L_b = L = 2\pi R \tag{2}$$

Two beams of light around a week are equal, all divided by the speed of light in a vacuum c, such as:

$$t_a = t_b = \frac{L}{c} = \frac{2\pi R}{c} \tag{3}$$

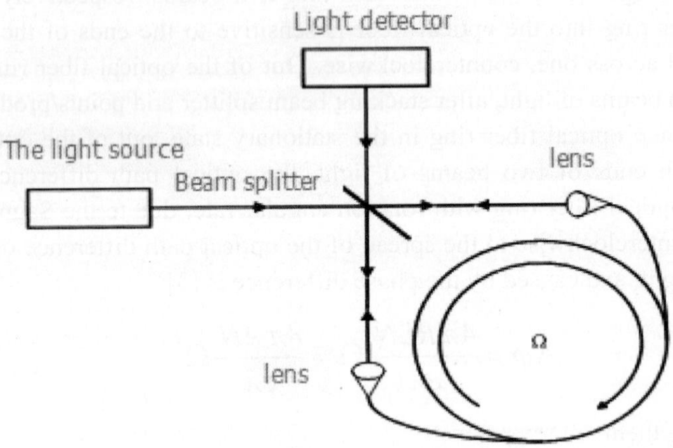

Fig. 1 The principle diagram of fiber optic gyro.

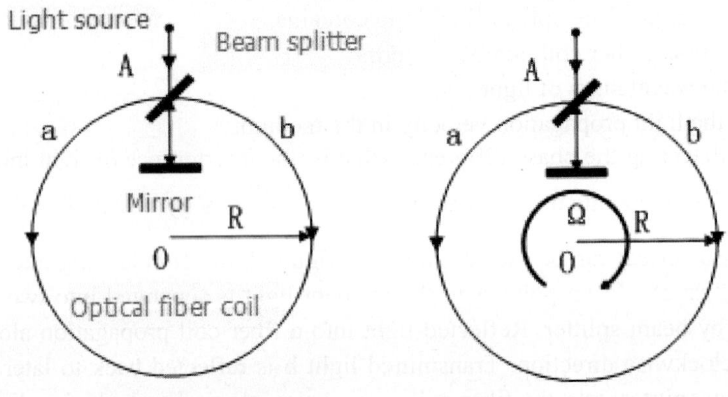

Fig. 2 Circular loop sagnac interferometer.

Reference of previous literature, can know this clear condition in the opposite direction for the spread of two beams of light around a week's time Δt and in the case of optical path ΔL difference and vacuum are exactly the same, that is, nothing to do with the transmission of light medium and refractive index.

Fiber optic gyro inherited Sagnac interferometer, through measuring the phase difference of the phase shift between two beams of light to obtain the angular velocity to be tested. The phase shift $\Delta\varphi$ between the two beams and optical path difference ΔL have the following relations:

$$\Delta\varphi = \frac{2\pi}{\lambda}\Delta L \quad (4)$$

Considering the circumference of the optical fiber ring $L = 2\pi R$, available phase shift when two beams of light around a week.

$$\Delta\varphi = \frac{4\pi RL}{c\lambda}\Omega \quad (5)$$

The above is single turn fiber coil. Fiber optic gyro is number of turns (N) coils of fiber optic fiber coil ring, two beams of light around N weeks when meet again phase shift should be:

$$\Delta\varphi = \frac{4\pi RLN}{c\lambda}\Omega = K\Omega \quad (6)$$

$K = \frac{4\pi RLN}{c\lambda}$ is called the scale factor of fiber optic gyroscope.

On the type, the output phase shift $\Delta\varphi$ is proportional to the output angular velocity Ω .In given conditions, optical fiber coil radius can be achieved by increasing the coil number of turns is the total length of optical fiber to improve the sensitivity of the measurement.

3. The Module of Fiber Optic Gyroscope

3.1. *Analog signal gyro*

This system adopts VG949P, its way is made of the fiber optic gyro, the simplest structure made higher performance and smaller fiber optic gyrscope. Plastic encapsulated VG949P makes this type of gyro is much more economical.

VG949P using open loop design, optical fiber ring optical fiber length of 100 m, two optical coupler made of fiber fusion technology using piezoelectric ceramics was carried out on the phase modulation, the lamp-house is lighting diode of super, testing for the photodiode, main technical parameters are as follows [4]:

Table 1. Analog signal gyro parameter list.

Project	value
The weight of the	45gram
size	60 x 60 x 22mm
Power consumption	0.5W
The power supply	+5V±5%
The startup time	0.01s
Deviation repeatability	0.01deg/s ，1sigma
Deviation change (OTR)	0.1deg/s
Bias stability	0.01deg/s
The scaling factor	6.0mV/deg/s±15%
Frequency range	0 - 1000Hz
The input range	±300°/s
The temperature	-30°C - +71°C
vibration	2g, 20Hz to 500Hz

According to the above parameters, the gyro output signal is the differential signal 3 mV / O / s ,the maximum input is ±300° / s , the output voltage $1.8V$ is biggest, gyro resolution is 0.003deg / s , according to the scale factor, $20.1\mu V$ is the resolution of the gyro output voltage. Due to the deviation for the resolution of the stability factor of the gyro output voltage is $60\mu V$ 。

3.2. *Digital signal gyro*

Analog signal of the gyro need to do your own signal sampling circuit, design of PCB to produce noise and interference is difficult to achieve good results, in order to improve the accuracy of positioning, design the digital signal of the gyro[5].

As the main technical parameters of the analog signal of gyroscope is as follow:

Table 2. Digital signal gyro technology parameters.

Project	value
Power consumption	<1.5W
voltage	+1.5 V
The startup time	<0.5 s
The bias stability (deg/hr)	5
preheating	<0.01°/s
OTR	<0.2°/s
Deviation (20 °C)	<0.4°/s
The scaling factor changes	<0.4(mV/°/s)
The scaling factor	55±20%(mV/°/s)
The stability of the scaling factor	0.10%
Frequency range	0 Hz

4. Fiber Optic Gyroscope Test

4.1. *The measurement of fiber optic gyro parameters*

Main equipment used including, turntable and drive equipment, gyroscope data acquisition module, bluetooth module, PC control platform. Characterization of fiber optic gyro performance indicators mainly including: scale factor, zero bias and zero bias stability, random walk coefficient, fujian value with the resolution, maximum input angular rate, and the preheating time.

Fiber optic gyro (FOG) comprehensive representation need to test the performance of scale factor and zero bias stability. Scale factor shows the output frequency of the relationship with the actual input gyroscope rotation. Ideally, the scale factor has nothing to do with angular rate, time and orientation, and is not sensitive to environmental variation. Zero bias stability test characterization for constant rotation speed (close to zero) input instructions by the rotating velocity of gyroscope drift. Ideally,the zero bias stability also has nothing to do with the time and orientation, and is not sensitive to environmental variation.

4.2. *Scale factor*

y_i as the i input angular rate ω_i gyro output average:

$$y_i = \frac{1}{N}\sum_{j=1}^{N} y_{ij} \tag{7}$$

in the type:

y_{ij} - the j gyro output values

N - the sampling frequency

y_{i0} as the i input angular rate ω_i turntable after resting gyro output average:

$$y_{i0} = \frac{1}{N}\sum_{j=1}^{N} y_{i0j} \tag{8}$$

in the type

y_{i0j} - the j gyro output values

N - the sampling frequency

The theory of the gyro output value is

$$Y_i = y_i - y_{i0} \tag{9}$$

Gyro input and output of the linear regression model is:

$$Y_i = K_0 + K\omega_i + \xi_i \tag{10}$$

In the type:

K_0 - fitting zero

K - scale factor

ξ_i - the fitting error

Using the least squares simulated

($Y_i = K_0 + K\omega_l + \xi_i$) available K_0 and K

$$\begin{cases} \hat{K}_0 = \overline{Y} - \hat{K}\overline{\omega} \\ \hat{K} = \dfrac{\sum_{i=1}^{N_\omega}(Y_i - \overline{Y})(\omega_i - \overline{\omega})}{\sum_{i=1}^{N_\omega}(\omega_i - \overline{\omega})^2} \end{cases} \tag{11}$$

In the type: $\overline{Y}$ - The mean of Y_i observation data

$$\overline{Y} = \frac{1}{N}\sum_{i=1}^{N_\omega} Y_i \tag{12}$$

$\overline{\omega}$ -The mean of ω_i observation data

$$\bar{\omega} = \frac{1}{N} \sum_{i=1}^{N_\omega} \omega_i \tag{13}$$

N_ω - the number of input angular rate

4.3. *Scale factor test*

Here for Russian FIZOPTIKA model for VG949P analog signal of gyro was tested. Use turntable test gyro dynamic characteristic curve, the plus or minus 300 degrees between every 1 c/s/s long time data sampling, the sampling of data processing, after the test data are as follows:

Can see that simulate the gyro angular velocity - 130 ° / s - + 130 ° / s scale factor linearity is better, make its nonlinear is serious.

Table 3. Gyro input and output data.

speed	The AD samples values	speed	The AD samples values	speed	The AD samples values
300	24515.48	270	23099.52	240	21621.30
290	24041.75	260	22609.93	230	21101.71
280	23565.36	250	22120.63	220	20591.20
180	18494.28	150	16827.80	120	13971.52
170	17964.83	140	16187.40	110	12841.33
160	17414.36	130	15126.64	100	11672.50
155	17134.32	125	14555.20	95	11101.52
60	7046.16	30	3516.33	-1	0.50
50	5883.66	20	2353.02	-11	-1302.77
40	4710.87	10	1178.55	-21	-2484.26
35	4107.07	5	569.97	-26	-3063.96
-65	-7646.42	-95	-11096.64	-125	-14521.42
-80	-9412.86	-110	-12833.35	-140	-16170.47
-90	-10517.70	-120	-13986.36	-150	-16840.96
-190	-18986.18	-220	-20512.23	-250	-22099.86
-200	-19496.52	-230	-21072.01	-260	-22597.83
-210	-20024.90	-240	-21603.73	-270	-23072.72

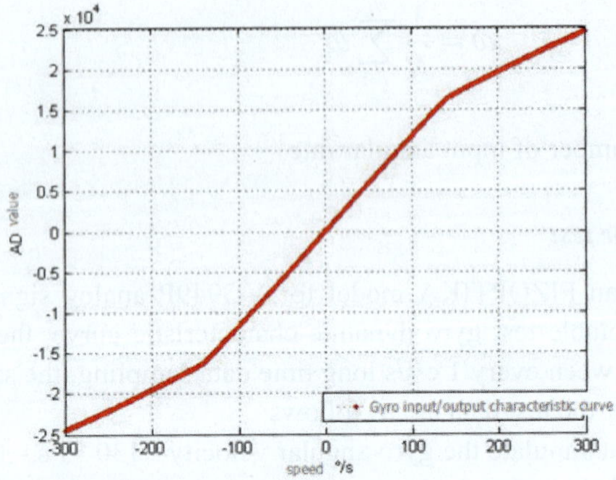

Fig. 3. Gyro input/output characteristic curve.

4.4. *Fiber optic gyroscope input and output characteristics test*

Input and output characteristics of fiber optic gyroscope (static) including zero bias, zero bias stability and zero partial repeatability. fiber optic gyroscope zero bias refers to the fiber optic gyro in the condition of zero input output value, with a long period, the output of the average equivalent to convert the input speed, its unit is $°/h$.

For Russian FIZOPTIKA models for analog signal gyro VG949P repeated tests.Four times after clearing data is as follows:

Table 4. Gyro data input and output characteristics.

	First	Second
Zero bias	0.1012	0.0995
Zero bias stability	0.0209	0.0209
Zero partial repeatability	0.00142	

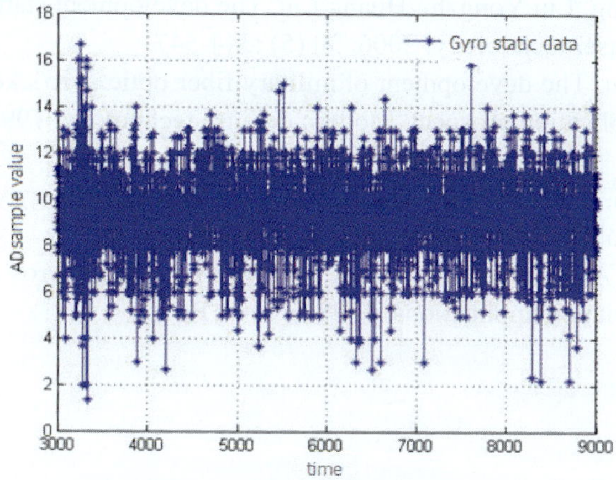

Fig. 4 Gyro static test for the first time.

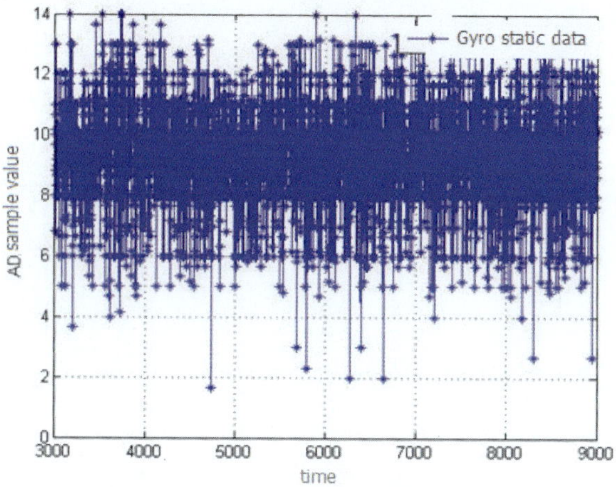

Fig. 5. Gyro static test for the second time.

5. Conclusion

This chapter mainly introduces the robot parameters of fiber-optic gyroscope, and the test of gyro parameters. The test shows the gyro can reach the precision that is enough to satisfy the requirement of the robot positioning.

References

1. Weng Ju, Tian Chijun. The development of high-precision interferometric fiber optic gyroscope technology. Inertial technology in China Journal. 2005, 13 (5) : 92 – 96.

2. Tan Jianrong, Liu Yongzhi, Huang Lin. The development status of fiber optic gyro. Laser technology. 2006, 30 (5) :544-547.
3. Tan Xianyu, The development of military fiber optic gyro's key technology and the application prospect. Modern defense technology. 1998, 26 (4) : 55-61.
4. Chen Saiqi, Yuan Dongli. The reviewed of optical fiber gyro. The optical fiber and cable's application technology. 2005, 6:4-7.
5. Sanders G A, Zanders B S, Liu R Y, et al Fiber optic gyros for space marine and aviation applications [J] 2837:61-71P.

Dynamic channelization based on interpolated FIR filter

Y.X. Wang[†], C.W. Liu, Z.P. Huang and Z. Dong
National University of Defense Technology, Changsha, China
[†]*Email: wangyuxuan263@sina.com*

In the wideband intermediate-frequency (IF) signal, there exist several sub-band signals whose number, bandwidths and locations are unknown and time variant. A method of dynamic channelization is proposed. The minimal bandwidth gap among the signals estimated by spectral analysis determines the channel number of modulated filter bank, and then a prototype filter based on interpolated FIR filter (IFIR) is designed to construct an analysis filter bank. It meets the near perfect reconstruction condition. The location of the sub-band signals is determined by the energy detection, and the corresponding synthesis filter bank is designed to reconstruct signal. When the variation of sub-band signals takes place, the new prototype filter will be acquired by changing the stretch factor and the interpolator filter without iterative optimization. The dynamic channelization when the number of sub-channel varies can be achieved. Theory analysis and computer simulation verify the feasibility of the new method.

Keywords: Dynamic channelization; Modulated filter banks; Multirate signal processing; Interpolated FIR filter (IFIR).

1. Introduction

The channelization technology, as a key element of wideband digital receiver and software-defined radio (SDR), is used to isolate the independent sub-band signal in wideband intermediate-frequency (IF) signal. And how to achieve channelization efficiently is a topic of widespread interest. Nowadays, there are two main channelization methods: one is based on digital down converter (DDC) [6] and the other is based on DFT filter banks with polyphase decomposition [5,7]. Because of the few requirement of signal, the channelization based on DDC is widely used. However, the computation and resource will increase rapidly as the number of sub-channel increase. When the sub-channel signal has the same bandwidth and frequency gap, the DFT filter banks by polyphase decomposition method can realize the multi-channel channelization.

The IF signal in wideband digital receiver and SDR usually contains several sub-band signals, whose bandwidth and frequency gap is different. Abu-Al-Saud & Studen put forward non-uniform channelization based on perfect reconstruction modulated filter banks to achieve the channelization in these signals when the number, bandwidth and location of these signals is known [1]. Li [3] increase the element of energy detection and achieve the dynamic channelization in condition that the minimum gap among sub-band is known and invariable in receive.

In order to realize the dynamic channelization in the case of variable minimum gap, based on Reference [1], this paper increase the element of spectrum estimation to calculate the minimum gap and use interpolated FIR (IFIR) filter to design the prototype filter. When the variation of minimum gap takes place, the new prototype filter will be acquired just by changing the stretch factor and the interpolator filter of IFIR filter without iterative optimization.

2. The Design of Dynamic Channelization

2.1. *The design of modulated filter bank*

The main idea of non-uniform channelization proposed in Reference [1] is analyze and synthesize. Firstly, the IF signal will be divided into a series of narrow-band signals through the analysis filter bank. Then, signal which belongs to the same sub-band will be reconstructed using the synthesis filter banks.If the bandwidth of each sub-band signal x_i in complex IF signal is from ω_i^l to ω_i^u. So, the minimum gap among all sub-band signals is

$$G_{\min} = \min_i \left\{ \omega_{i+1}^l - \omega_i^u \right\} \tag{1}$$

The number of sub-channel in filter bank can be calculated by

$$2M = 2^{\left\lceil \log_2\left(\frac{2\pi}{G_{\min}}\right)\right\rceil} \tag{2}$$

The analysis filter banks are able to divide the spectrum of $X(z)$ into $2M$ narrow-band signals. The upper and lower boundary of the ith narrow-band signal is

$$Q_i^l = \left\lfloor M\omega_i^l / \pi \right\rfloor \quad Q_i^u = \left\lfloor M\omega_i^u / \pi \right\rfloor \tag{3}$$

and let $M_i = Q_i^u - Q_i^l + 1$ be the number of sub-channel in the ith sub-band signal. We can synthesize the reconstructed sub-band signal with the signal in the sub-channel from Q_i^l to Q_i^u.

The reconstructed sub-band signal has the same sampling frequency but narrower bandwidth. In order to reduce the computation of baseband processing, we should decrease the sampling rate depend on the bandwidth of sub-band signal x_i. Define $\hat{M}_i$ to be a power-of-two integer that is related to $\underline{M_i}$ by

$$\hat{M}_i = 2^{\lceil \log_2 M_i \rceil - 1} \tag{4}$$

The sample rate can be safely reduced by a factor of $M / \hat{M}_i$ without causing significant aliasing. So, the entire construction of modulated filter banks is shown in Figure1.

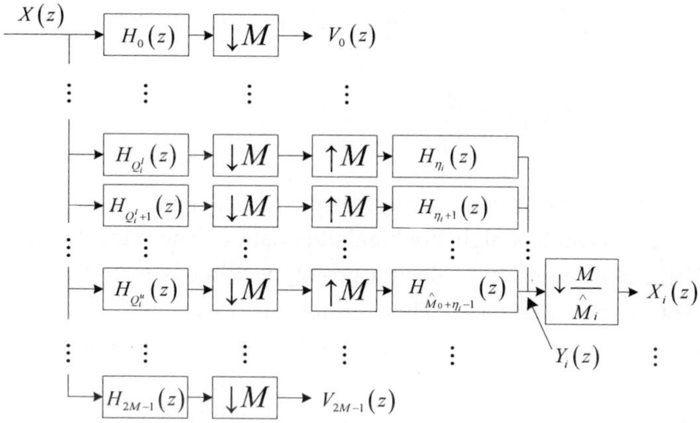

Fig. 1. Diagram of modulated filter bank.

where $\eta_i = Q_i^l \bmod 2$. And the ith channelization output signal $Y_i(z)$ given by

$$Y_i(z) = \frac{1}{M} W_{2M}^{(Q_i^l - \eta_i)\frac{2mM-1}{2}} X\left(zW_M^{-\frac{Q_i^l-\eta_i}{2}}\right) \sum_{k=\eta_i}^{Q_i^u - Q_i^l + \eta_i} H_k^2(z)$$

$$+ \frac{1}{M} W_{2M}^{(Q_i^l - \eta_i)\frac{2mM-1}{2}} \sum_{\substack{p=0 \\ p \neq M-(Q_i^l-\eta_i)/2}}^{M-1} X\left(zW_M^p\right)$$

$$\sum_{k=\eta_i}^{Q_i^u - Q_i^l + \eta_i} H_k\left(zW_M^{p+\frac{Q_i^l-\eta_i}{2}}\right) H_k(z) \tag{5}$$

The first part is the required signal. The second part is aliasing shift due to the downsampling in analysis filter bank. Not all sub-channel signals enter the synthesis filter bank, so that the aliasing cannot be eliminate even if the filter bank has the perfect reconstruction.

2.2. Optimize the design of prototype filter

In order to design modulated filter bank, the prototype filter should meet the following requirements. The signal, passing through the filter, should have no distortion or little distortion. That means the prototype filter H(ω) should meet the following formula [8].

$$|H(\omega)| = 0, \quad \omega > \frac{\pi}{M} \tag{6}$$

$$P(\omega) = |H(\omega)|^2 + \left|H\left(\omega - \frac{\pi}{M}\right)\right|^2 = 1$$

$$0 \le \omega \le \frac{\pi}{M} \tag{7}$$

The wideband IF signal always has large dynamic range, so that the designed filter should has high stopband attenuation. The typical value is about -100dB. The classic method to design prototype filter is to use lattice structure with power complementary. This prototype filter can meet the perfect reconstruction, but if the filter needs -100dB stopband attenuation, the order of the filter is too long. Nowadays, the prototype filter meeting the near perfect reconstruction is designed by using optimized iterative algorithm to solve the non-linear optimization. On this basis, this paper draws on the experience of the idea of interpolation and optimizes the prototype filter by the way of IFIR.

The IFIR filter is cascaded form of component filters, i.e., up-sampled model filter $G(z^L)$ and interpolator $I(z)$. The transfer function of IFIR filter is given as:

$$H(z) = G(z^L)I(z) \tag{8}$$

The spectrum width of up-sampled model filter $G(z^L)$ is relaxed by a factor of L and the L-1 image product is uniformly inserted into spectrum in every period. This L-1 image product can be eliminated by interpolator $I(z)$. Thus, we can use the optimized iterative algorithm to design FIR filter $G(z)$, which has wide pass-band. Then, the prototype filter H(z) can be designed with interpolating and shaping. It can reduce the number of optimize parameters in optimized iterative algorithm and the computation.

The spectrum width of $G(z^L)$ is relaxed by a factor of L , so that the amplitude response of $G(z)$ is L times spread than H(z) and the order of G(z) is reduced by a factor of L. However, if the L is too large, the length of the I(z) will be too long. In order to optimize the computation of design about IFIR filter, the optimal factor L can be calculate as following formula [4].

$$L = \left\lceil \frac{2\pi}{\omega_p + \omega_s + \sqrt{2\pi(\omega_s - \omega_p)}} \right\rceil \quad (9)$$

where ω_p is the passband edge frequency and ω_s is the stopband edge frequency.

The step about the design of prototype filter is:

Step 1. According to the number of channel, the proper initial filter parameter, such as passband frequency and stopband frequency, should be decided. And $G_0(z)$ is designed by window function method.

Step 2. Optimize G0(z) through optimized iterative algorithm and obtain the optimized filter G(z). The objective function during optimizing is shown as [2]:

$$\Phi = \max_{\omega} \left| P(\omega) - 1 \right|, \quad 0 \le \omega \le \frac{\pi}{M} \quad (10)$$

Step 3. $G(z^L)$ is calculated from $G(z)$ by interpolation.

Step 4. Design the interpolated filter $I(z)$ and obtain the prototype filter by cascading $G(z^L)$ and $I(z)$.

2.3. *Process of dynamic channelization*

The sub-band signal in wideband IF signal will vary during receiving, so that the minimum gap among sub-band signals $G_{\min}$ can also be change. When $G_{\min}$ becomes large and the number of channel remain unchanged, the computational efficiency will decrease. If $G_{\min}$ becomes small, some signal cannot be divided because of the gap is too small. So, when the $G_{\min}$ varies, the number of channel should be recalculated and the prototype filter should be redesigned. Because of using IFIR filter as prototype filter, when the number of channel varies, just the factor L and interpolated filter $I(z)$ should be recalculated. Because of avoiding iteration in redesign process, computation and time consumption can be decreased.

According to the aforesaid design flow, let M be the number of channel and the stopband attenuation is -100dB. The factor L equals to $\lceil M/K \rceil$, where K, a positive real number, is less than M. Then L satisfies

$$\frac{M}{K} - 1 < L \le \frac{M}{K} \quad (11)$$

Estimate the order of $G_0(z)$ according to Kaiser equation

$$Ord = \frac{As - 7.95}{14.36 \times \varepsilon / 2\pi} \le \frac{25.2K}{1 - 2K/M} \quad (12)$$

Set $K = 2$ and M is much larger than K. So the order of $G_0(z)$ satisfies

$$ord \le \frac{25.2K}{1-2K/M} \approx 50.4 \tag{13}$$

The order of $G_0(z)$ is nearly invariable and short when M varies. So, in dynamic channelization, just the factor L and interpolated filter I(z) should be adjusted. The whole architecture of dynamic channelization is show in Figure 2.

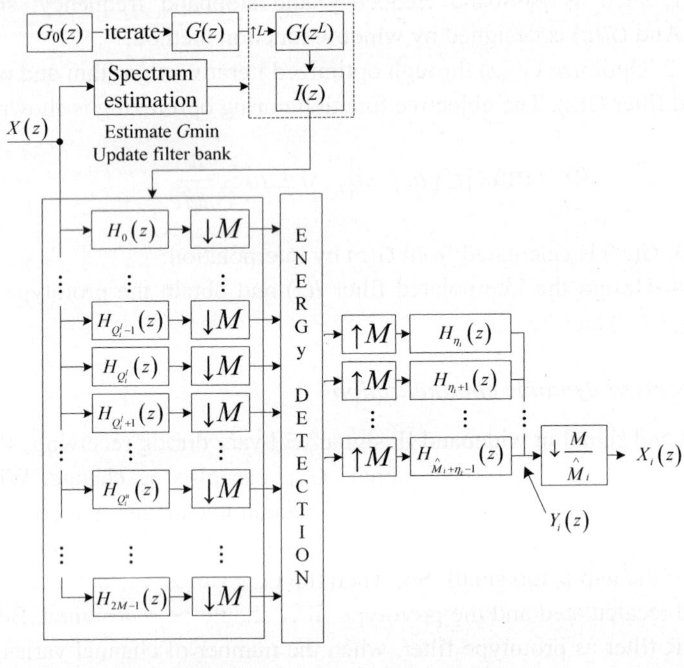

Fig. 2. Diagram of dynamic channelization.

Wideband IF signal will be divided into a series of narrowband signal by analysis filter bank. At that time, we should detect the signal in sub-channel to determine the location of signal. There are several way in signal detection. The simplest way is energy detection. The number of channel is calculated by formula (1), so there must be a sub-channel which contains no signal. And the energy of this sub-channel can be considered as the energy of noise. So, it is convenient to detect the location of sub-band signal and recover the original signal by synthesis filter bank.

3. Simulation examples

3.1. *Performance simulation of prototype filter*

In the condition of same parameters, compare the performance between using IFIR filter and directly using window function to design prototype filter.

Set the number of channel $M = 32$, passband edge frequency $\omega_p=\pi/2M$, stopband edge frequency $\omega_s=\pi/M$ and stopband attenuation $A_s = -100$dB. The amplitude-frequency characteristic of prototype filter, which is designed by iteratively optimize the FIR filter, is shown in Figure 3(a). The order of the prototype is 481. The amplitude-frequency characteristic and aliasing error of the modulated filter bank based on this prototype filter is shown in Figure 3(b) (c).

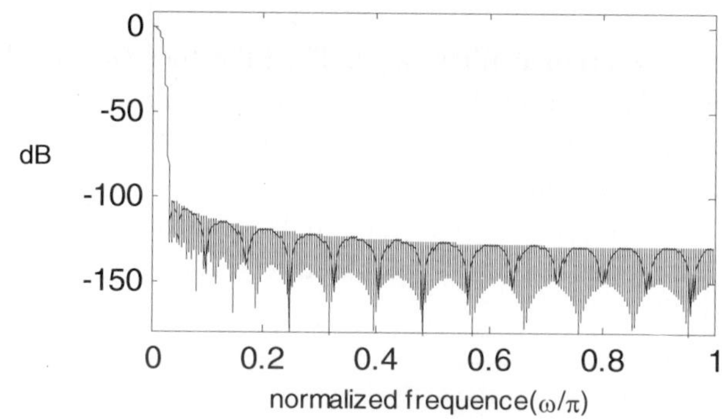

(a) Magnitude-frequency characteristic of pr filter.

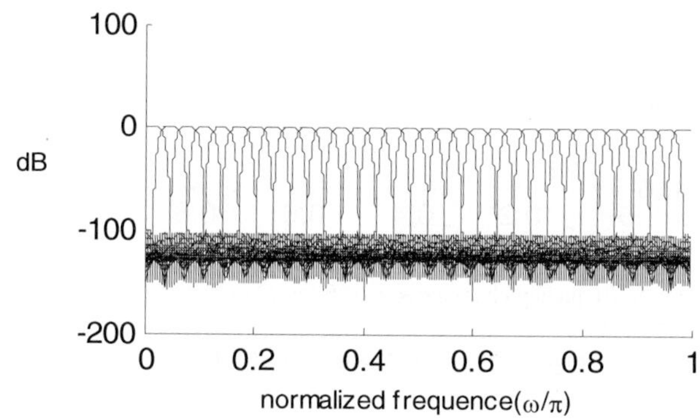

(b) Magnitude-frequency characteristic of analysis filter.

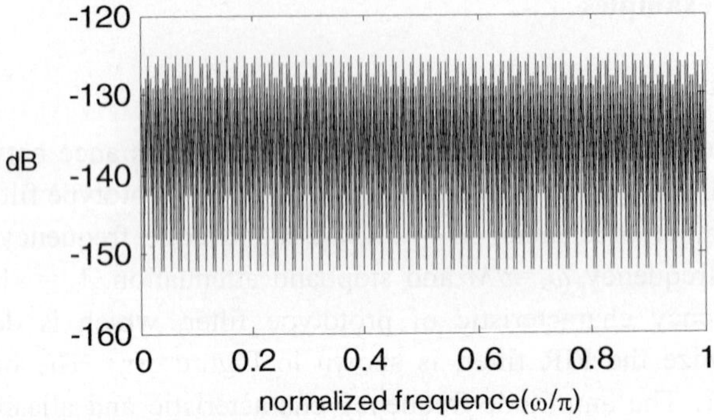

(c) Aliasing error of modulated filter bank.

Fig. 3. Characteristic of filter in direct design method.

Firstly, design the original filter $G_0(z)$. Then the filer $G(z)$ is obtained by iteratively optimize $G_0(z)$ and its frequency response is shown in Fig 4(a). After that, let the $G(z)$ cascade the interpolated $I(z)$ to acquire the prototype filter and its amplitude-frequency characteristic is shown in Fig 4(b). The order of G(z) is 51 and the order of $I(z)$ is 56. The amplitude-frequency characteristic and aliasing error of the modulated filter bank based on this prototype filter is shown in Figure 4(c) (d).

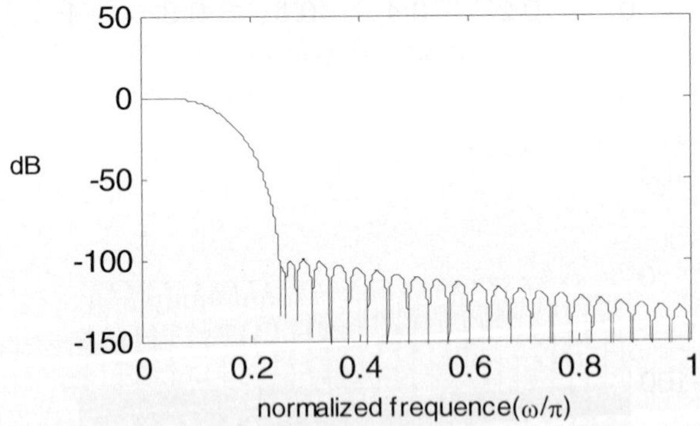

(a) Magnitude-frequency characteristic of filter after iterative optimization.

Fig. 4. The characteristic of IFIR filter.

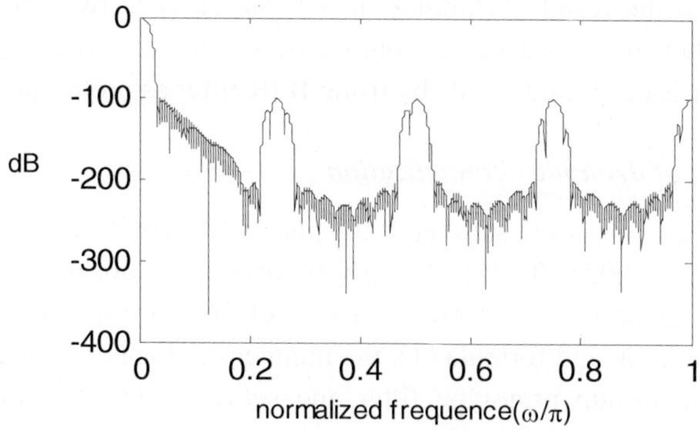

(b) Magnitude-frequency characteristic of pr filter

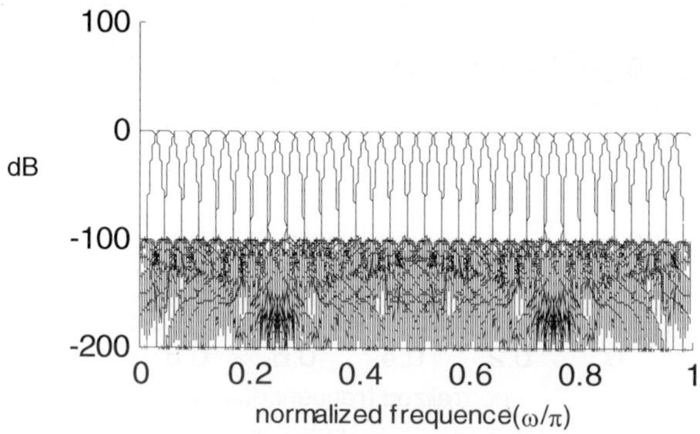

(c) Magnitude-frequency characteristic of analysis filter.

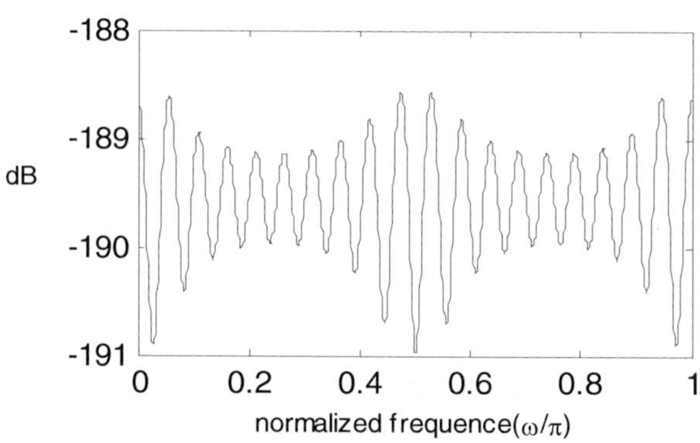

(d) Aliasing error of modualted filter bank.

Fig. 4. (*Cont'd*)

According to the result of simulation, we can cascade two short filters to realize a long filter. It can reduce the computation of iterative optimization. And the aliasing error is decreased 40 dB by using IFIR filter as prototype filter.

3.2. *Simulation of dynamic channelization*

At beginning, there are four sub-band signals in IF signal. Their normalized frequency range is [0.025, 0.175], [0.2550, 0.5050], [0.5900, 0.6900], [0.7450, 0.8950]. The simulation signal is created by MATLAB. According to the result of spectrum estimation and formula (1), the number of channel equals to 64. Use the IFIR filter to design prototype filter and set $L = \lceil M/3 \rceil$, the result of channelization is shown in Figure 5.

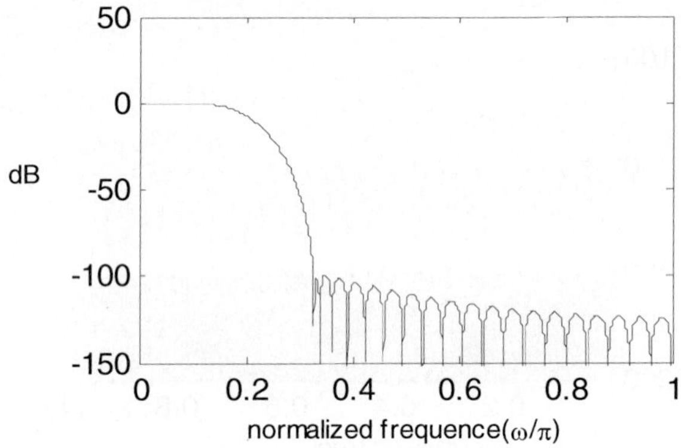

(a) Magnitude-frequency characteristic of filter after iterative optimization.

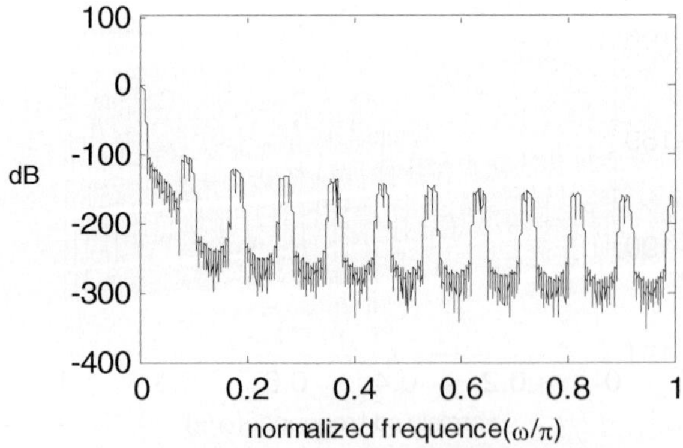

(b) Magnitude-frequency characteristic of pr filter.

Fig. 5. The result of channelization.

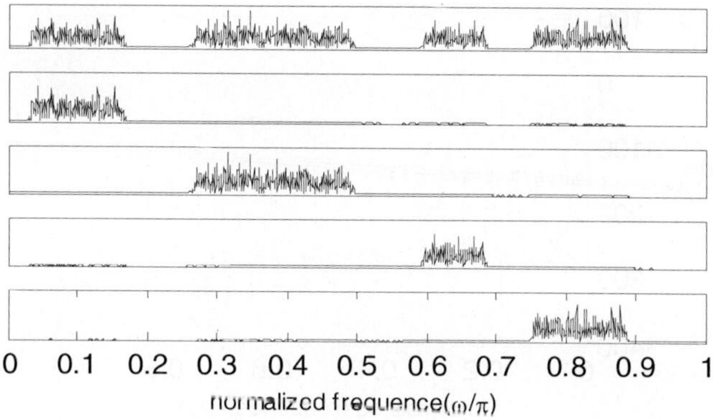

(c) The characteristic of original signal and reconstructed signal.

Fig. 5. (*Cont'd*)

At that time, the sub-band signals vary. The number of signals increased to 5 and the normalized frequency range is [0.025, 0.175], [0.1950, 0.4650], [0.5250, 0.5750], [0.6350, 0.6850], [0.7250, 0.8750]. According to the formula(6), the number of channel become 128. The new prototype filter can be acquired just by calculated L and interpolated filter. The result of channelization is shown in Figure 6.

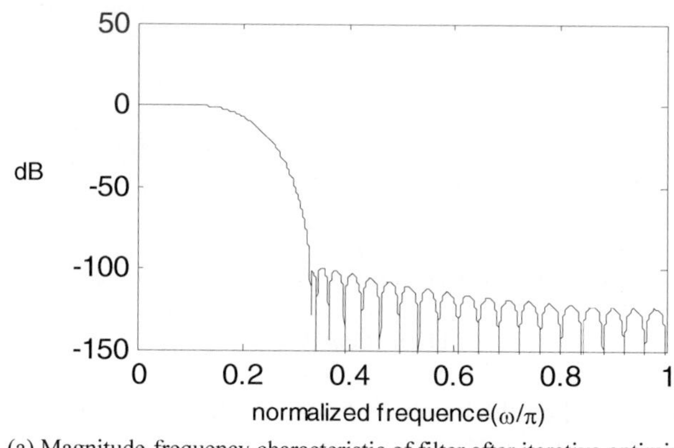

(a) Magnitude-frequency characteristic of filter after iterative optimization.

Fig. 6. The result of channelization.

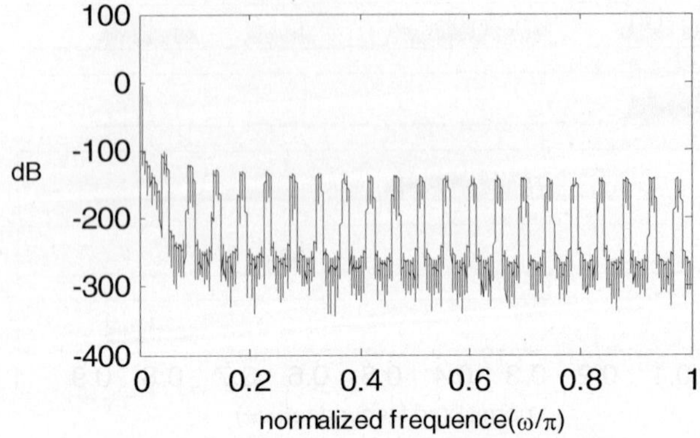

(b) Magnitude-frequency characteristic of pr filter.

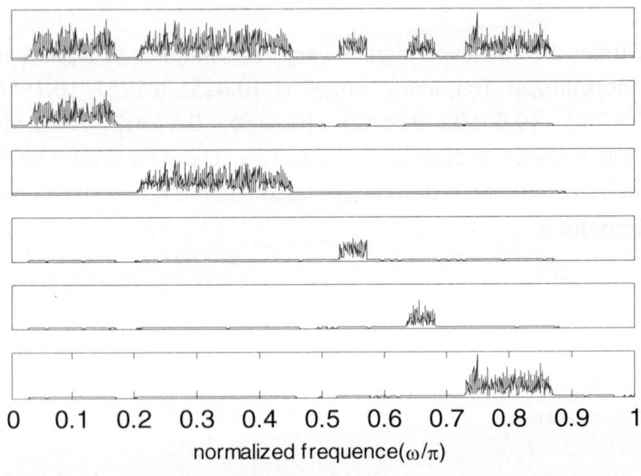

(c) The characteristic of original signal and reconstructed signal.

Fig. 6. (*Cont'd*)

Form the results, when the number of channel M varies, just recalculate the factor L and the interpolated filter $I(z)$ can achieve channelization. So, we can draw a conclusion that the method put forward in this paper is available in dynamic channelization.

4. Conclusion

A new dynamic channelization method that uses IFIR filter as prototype filter is proposed. The proposed method can easily achieve the dynamic channelization when the $G_{\min}$ and the number of channel vary. It just need to recalculate the factor L and interpolated filter $I(z)$ instead of iterative optimizing during updating the prototype, so that the method is shown to have higher computation efficiency in dynamic channelization. Through simulation verification, the dynamic channelization based on IFIR filter can reduce the order of prototype and can be effectively used for dynamic channelization.

References

1. Abu-Al-Saud, W.A. & Studer, G.L. (2004) Efficient wideband channelizer for software radio systems using modulated pr filterbanks. IEEE Transaction Signal Processing, 52(10), 2807-2820.
2. Creusere, C.D & Mitra, S.K. (1995) A simple method for designing high-quality prototype filters for M-band pseudo QMF banks. IEEE Transaction Signal Processing, 43(4), 1005-1007.
3. Li, B. (2007) Research on channelization techniques for software defined radio. Zhengzhou: PLA Information Engineering University. (in Chinese)
4. Mehrnia, A. & Willson, A.N. (2003) On optimal IFIR filter design. In Proceedings of the 2004 International Symposium on Circuits and Systems, Piscataway. IEEE. pp. 133-136.
5. Zahirniak, D.R., Sharpin, D.L. & Fields, T.W. (1998) A hardware efficient, multirate, digital channelized receiver architecture. IEEE Transaction Aerospace and Electronic Systems, 34(1), 137-152.
6. Zangi, K.C. & Koipillai, R.D. (1998) Efficient filterbank channelizers for software radio receivers. In 1998 IEEE International Conference on Acoustics Speech and Signal Processing, Seattle WA, USA. IEEE. pp. 3497-3500.
7. Zangi, K.C. & Koilpillai, R.D. (1999) Software radio issues in celluar base stations. IEEE Journal on Selected Areas in Communications, 17(4), 561-573.
8. Zhang, Z. & Jiao, L. (2001) Design of M-Band Cosine-Modulated Filter Banks. Acta Elecreonica Sinica, 1(29), 84-86.

Research on shortwave dipole curtain phased array

J. P. Sun[1]
Information Department of Navy, Beijing, China

Z. G. Zhang[2†]
Naval University of Engineering, Wuhan, Hubei Province, China
[†]Email: sunxiaona1991@126.com

T. Li[3]
91336 Troop, Qinghuangdao, Hebei Province, China

C. X. Wang[4]
Some Department of Forth Ministry of Navy, Zhanjiang,
Guangdong Province, China

The curtain phased array antenna has the advantages of high gain, wide coverage of communication. But HRS4/4/0.5 is high at the erection in the practical application, and raised the cost. If HRS/6/2/0.5 curtain antenna used in the practical application, nearly half of height can be reduced, the number of array element can be decreased, and cost can be reduced. It analysis whether HRS6/2/0.5 and HRS4/4/0.5 is similar in long distance communication.

Keywords: Shortwave; Curtain phased array; Direction graph.

1. Introduction

The dipole curtain antenna is composed of radiation elements from horizontal symmetry oscillators, and has the advantages of high gain [1], cover small area, and so on. It is widely used in shortwave radio interference antenna. HRS4/4/0.5 is commonly used as dipole curtain antenna, and it means the structure of this antenna is 4 rows and 4 columns, the distance between each layer is 0.5λ, the distance between each oscillator center in horizontal line is 0.5λ, and the distance between lowest layer dipole and ground is 0.5λ too(λ is wavelength corre-

[1] Jian-Ping Sun
[2] Zhi-Gang Zhang
[3] Ting Li
[4] Cong-Xia Wang

sponding to the center frequency) [2]~ [5]. In the practical application, we generally choose two pairs of curtain antenna phased array antenna to cover 9~30MHz shortwave frequency [6]. The frequency of low frequency antenna range 9 to 17MHz, and the center frequency is 12.36MHz. The frequency of high frequency antenna range 17 to 30MHz, and the center frequency is 22.56MHz [7]~ [9]. In this case, the low frequency antenna is 67m wide and 54m high, and cost is high.

It designs a new HRS6/2/0.5 curtain phased array antenna, whose structure is 6 rows and 2 columns [10], and setting parameters basically the same with HRS4/4/0.5 curtain phased array antenna. Though length increased to 97m, height decreased to less than 30m, numbers of array elements reduced from 16 to 12, and cost of phased array antenna reduced. It mainly analyzes whether HRS6/2/0.5 can obtain the similar performance with HRS4/4/0.5.

2. Electrical Characteristics Analysis of Low Frequency Dipole Curtain Antenna

The direction characteristics of dipole curtain antenna in widewave is mainly studied.

2.1. *Function of direction graph*

2.1.1. *The computation formula of function of direction graph in the first half of the space is as follow*

Function of direction graph in horizontal plane:

$$F(\varphi)=\frac{\cos(\alpha l\sin\varphi)-\cos\alpha l}{\cos\varphi}\times\frac{\sin(n_2\frac{ad_1}{2}\sin\varphi)}{\sin(\frac{ad_1}{2}\sin\varphi)}\times\sin(ad_3\cos\varphi) \quad (1)$$

$n_2=\frac{n}{2}$ means the numbers of symmetric oscillator in the same layer; d_1 means the distance between adjacent two symmetric oscillator center; d_3 means the distance between antenna curtain and reflection curtain.

Function of direction graph in vertical plane:

$$F(\Delta) = 2n_2(1-\cos 2\alpha l)\times\frac{\sin(n_1\frac{ad_2}{2}\sin\Delta)}{\sin(\frac{ad_2}{2}\sin\Delta)}\times\sin(ad_3\cos\Delta)$$
$$\times\sqrt{1+|R_\perp|^2+2|R_\perp|\cos(\phi_\perp-2\alpha H_{av}\sin\Delta)} \quad (2)$$

d_1 means the distance between adjacent two layers; H_{av} means the average antenna height . $H_{av}=H_1+(n_1-1)\frac{d_2}{2}$.

2.2.2. *Multiplying function of direction graph in the first half of the space and the following factor is function of direction graph in the after half of the space*

When calculating direction graph in horizontal plane, the factor is $\frac{\sqrt{\delta}}{2\sin(ad_3\cos\varphi)}$,when calculating direction graph in vertical plane, it is $\frac{\sqrt{\delta}}{2\sin(ad_3\cos\Delta)}$. δ is energy permeability coefficient when reflection network calculated by power.

2.2. *Main lobe width of half power point and zero power point in horizontal plane*

$$\text{Width of half power point: } 2\varphi_{0.5}\approx 0.88\sin^{-1}\frac{\lambda}{n_2 d_1} \quad (3)$$

$$\text{Width of zero power point: } 2\varphi_0\approx 2\sin^{-1}\frac{\lambda}{n_2 d_1} \quad (4)$$

2.3. *Direction coefficient, gain coefficient and efficiency*

2.3.1. *Direction coefficient*

The direction coefficient is calculated by the following formula in vertical plane when $\phi=0^\circ$:

$$D=120\frac{f^2(\Delta)}{R_{\Sigma A}} \quad (5)$$

$f(\Delta)$ is direction function in vertical plane; $R_{\Sigma A}$ is the total radiation resistance when antenna come close to antinodes current.

2.3.2. *Gain coefficient*

The gain coefficient in ideal conductive ground is

$$G_d = n_2^{\,2}\frac{1170}{R_{\Sigma\Lambda}}(1-\cos\alpha l)^2 \times \frac{\sin^2(n_1\frac{ad_2}{2}\sin\Delta)}{\sin^2(\frac{ad_2}{2}\sin\Delta)} \times \sin^2(ad_3\cos\Delta)\times\sin^2(H_{av}\sin\Delta) \quad (6)$$

3. Comparison of Direction Graph of Curtain Phased Array Antenna

In order to better compare the advantages and disadvantages of HRS6/2/0.5 and HRS4/4/0.5's performance, FEKO is used for simulation.

3.1. *Comparison of direction graph when no phase shift*

Computing direction graph of HRS6/2/0.5 and HRS4/4/0.5 when no phase shift, we get the horizontal and vertical direction graphs in center frequency (22.6MHz), high frequency (30MHz), low frequency points (17MHz), and list the specific parameters. As shown in figure 1 and table 1.

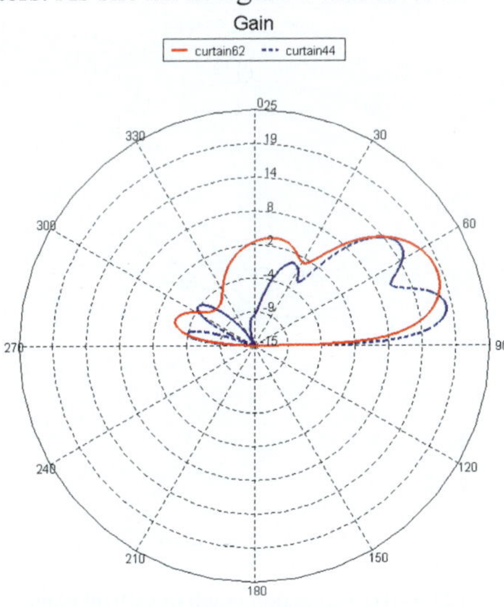

(a)17MHz's direction graph in vertical plane.

Fig. 1. Direction graph of HRS6/2/0.5 and HRS4/4/0.5 curtain array antenna when no phase shift.

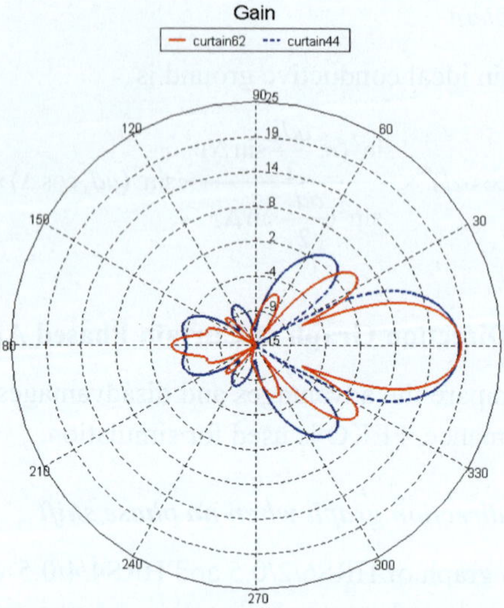

(b)17MHz's direction graph in horizontal plane.

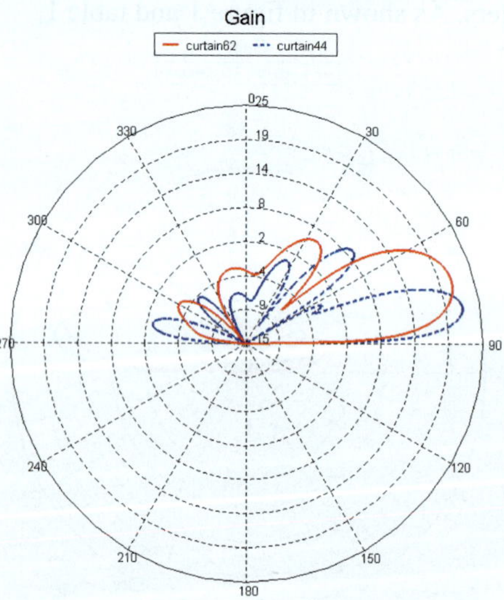

(c) 22.6MHz's direction graph in vertical plane.

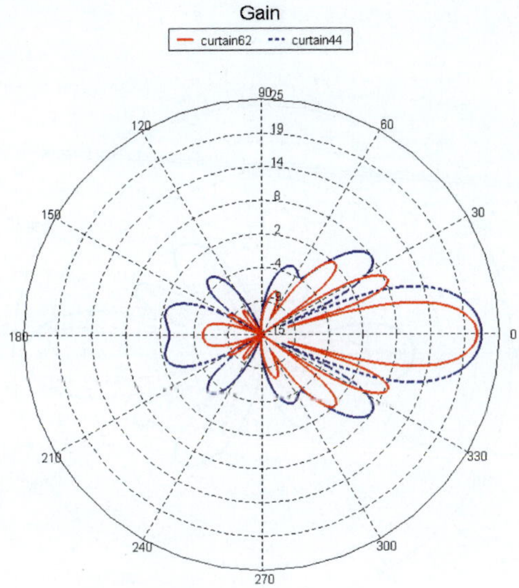

(d)22.6MHz's direction graph in horizontal plane.

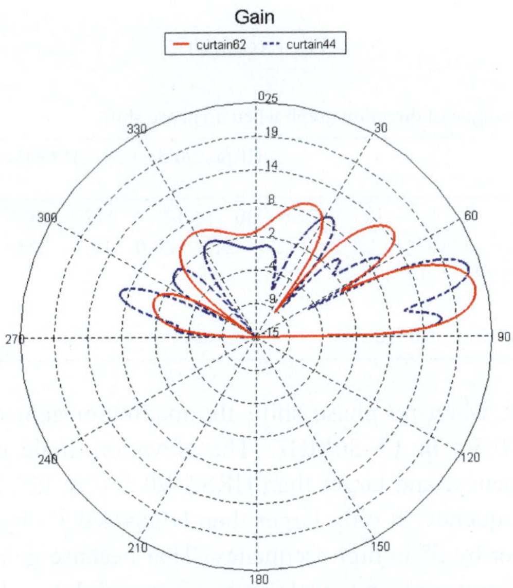

(e)30MHz's direction graph in vertical plane.

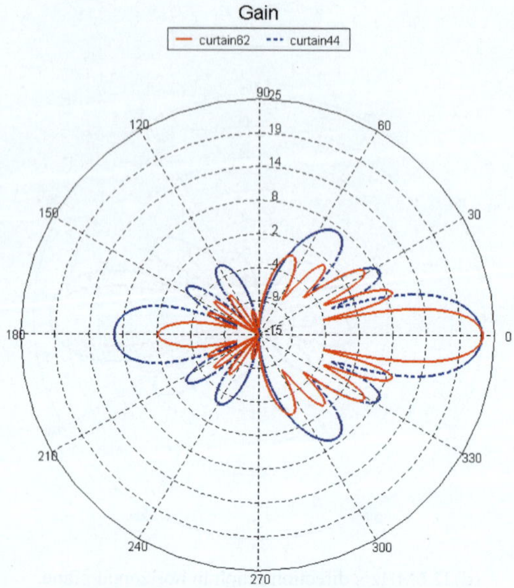

(f)30MHz's direction graph in horizontal plane.

Fig. 1. (*Cont'd*)

Table 1. Eigenvingue of direction graph when no phase shift .

Antenna type	HRS6/2/0.5			HRS4/4/0.5		
Frequency(MHz)	17	22.6	30	17	22.6	30
Max gain(dB)	18.4	21.3	23.0	18.2	22.0	29.2
Gain elevation angle(°)	23	17	13	11	10	18
3dB's mainlobe width(°)	27	20	16	34	28.5	23.5

From table 1, when no phase shift, the maximum gain of HRS6/2/0.5 is close to HRS4/4/0.5's in 17~30MHZ. The elevation angle of HRS6/2/0.5 is large in low frequency, and larger than HRS4/4/0.5's by 12°, but reduced with the increase of frequency. It only larger than HRS4/4/0.5's by 7° in center frequency and smaller by 5° in high frequency. That because gain in elevation angle splits in high frequency, and sidelobe turn into mainlobe, elevation angle becomes larger. 3dB's mainlobe width becomes narrow with the increase of frequency. HRS6/2/0.5's is more narrow than HRS4/4/0.5's by 7.5° in 9~30MHz, but better directivity, and its backlobe is much smaller than HRS4/4/0.5's.

3.2. *Comparison of direction graph when phase shift*

Computing direction graph of HRS6/2/0.5 and HRS4/4/0.5 when phase shifts are 30°,120° and 160°, we get the horizontal direction graphs in center frequency (22.6MHz), high frequency (30MHz), low frequency points (17MHz), and list the specific parameters. As shown in figure 2, table 2 and table 3.

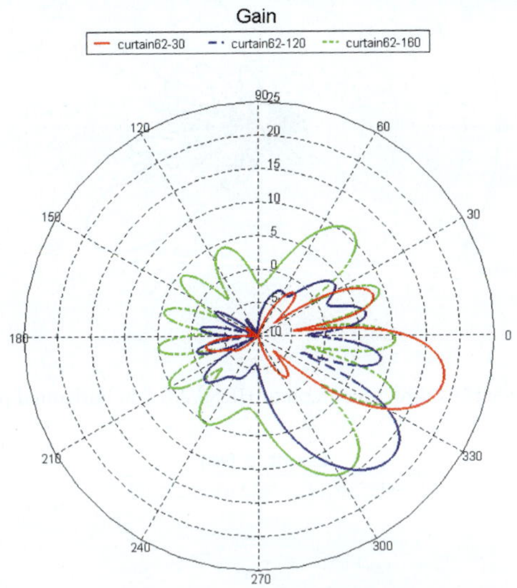

(a)17MHz's direction graph of HRS6/2/0.5 in horizontal plane.

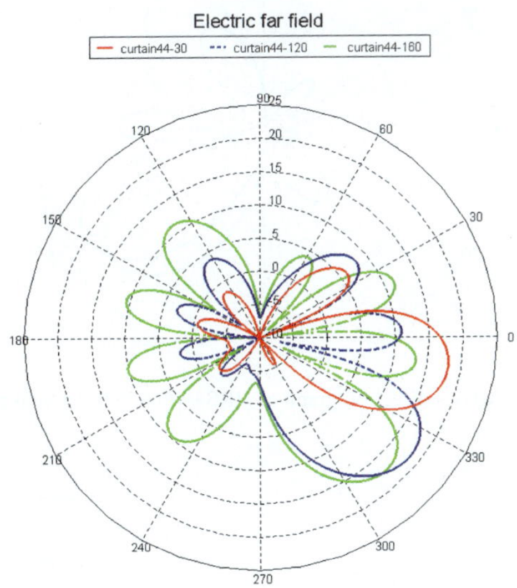

(b)17MHz's direction graph of HRS4/4/0.5 in horizontal plane.

Fig. 2. Direction graph of HRS6/2/0.5 and HRS4/4/0.5 at 30°, 120° and 160° when phase shift.

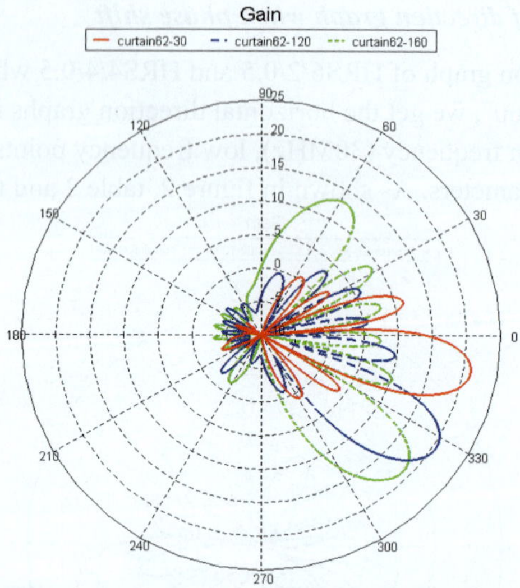

(c)22.6MHz's direction graph of HRS6/2/0.5 in horizontal plane.

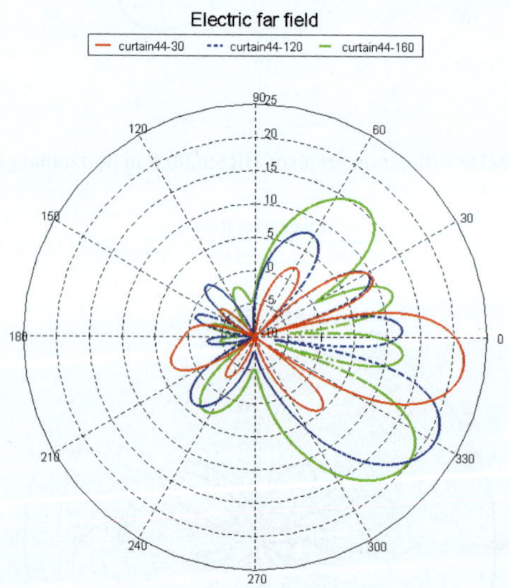

(d) 22.6MHz's direction graph of HRS4/4/0.5 in horizontal plane.

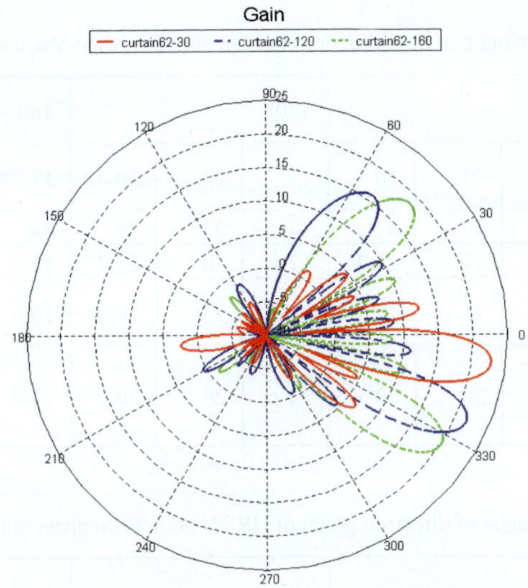

(e) 30MHz's direction graph of HRS6/2/0.5 in horizontal plane.

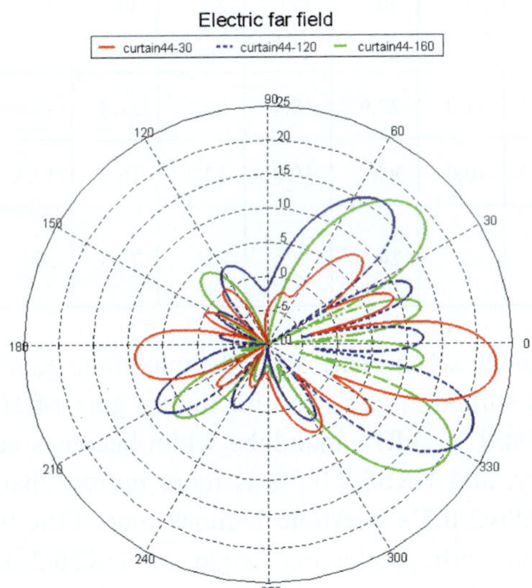

(f) 30MHz's direction graph of HRS4/4/0.5 in horizontal plane.

Fig. 2. (*Cont'd*)

Table 2. Eigenvingue of direction graph of HRS6/2/0.5 when phase shift.

Antenna phase shift	30°			120°			160°		
Frequen-cy(MHz)	17	23	30	17	23	30	17	23	30
Right-skew(°)	14	10	8	43	33	24	56	44	33
Max gain(dB)	18.3	21.1	22.9	17.1	20.5	21.7	14.2	18.9	20.3
Max gain of sidelobe(dB)	8.2	11.1	12.5	8.7	9.9	15.6	10.9	13.6	18.9
3dB's mainlobe width(°)	31	26	20.5	33	28	20.5	29	27	20

Table 3. Eigenvingue of direction graph of HRS4/4/0.5 when phase shift.

Antenna phase shift	30°			120°			160°		
Frequen-cy(MHz)	17	23	30	17	23	30	17	23	30
Right-skew(°)	10	8	6	39	30	24	47	40	33
Max gain(dB)	18.1	21.9	22.9	18.4	21.2	21.8	17.3	19.8	20.5
Max gain of sidelobe(dB)	5.7	10.0	9.2	10.8	12.2	16.5	13.3	15.7	19.1
3dB's mainlobe width(°)	35	30	24.5	37	32	23.5	32	31.5	24.5

From table 2 and 3, when phase shift, the maximum gain increases with decreasing of phased-shifted angles. The maximum gain of HRS6/2/0.5 is little smaller than HRS4/4/0.5's.3dB's mainlobe width becomes narrow with the increase of frequency, and HRS6/2/0.5's is more narrow than HRS4/4/0.5's in 9~30MHz, and HRS6/2/0.5's backhole is much more little than HRS4/4/0.5's. In the same phase shift, deflecting angle of HRS6/2/0.5 is larger than HRS4/4/0.5's at low frequency, and that means its communication range is much farther than HRS4/4/0.5's. But the difference of angle is smaller and smaller with increasing of frequency, and finally can be same at high frequency.

4. Conclusion

HRS6/2/0.5 can obtain approximation maximum gain with HRS4/4/0.5, though its number of arrey is much less than HRS4/4/0.5's. Its 3dB's mainlobe width is more narrow than HRS4/4/0.5's, so it has stronger directivity, and its backlobe much smaller. In the same phase shift, deflecting angle of HRS6/2/0.5 is larger and communication range is farther. It decreases antenna height and reduces cost of manufacture, more convenient in maintenance. Disadvantage is that covering area of HRS6/2/0.5 increases as the result of its increases of length, and bigger flat ground is needed. Any time, HRS6/2/0.5 is suitable for middle and far distance communication though its mainlobe elevation angle is little larger.

References

1. Ma Mangcang. The characteristics of New Short-wave Communication [J]. Dual Use Technologies & Products, 2009(2):44 46.
2. Yin Yalan, Jiang Hua, Huang Jinhui. Shortwave Antenna Engineering [M]. Beijing: Haichao Press, 2004:129-163.
3. NRIET. Manual of Phased Array Antenna [M]. Beijing: Publishing House of Electronics Industry, 2007:1, 180-188.
4. Telecommunications Administration Ministry of Posts. Mentaince Manual of Shortwave [M]. Beijing: Host and Telecom Press, 1991:10, 29-41.
5. Yin Yalan, Deng Jiekun. Design of Shortwave Phased Array Antenna Communication System [J]. Electronics Engineer, 2007, 33(9):31-33.
6. Hansen R C. Intrinsic Features of HF Curtain Arrays [J]. IEEE Transactions on Broadcasting, 1998, 44(3):309-315.
7. Wilensky R. High-power, Broad-bandwidth HF Dipole Curtain Array with Extensive Vertical and Azimuthal Beam Control [J]. IEEE Transactions on Broadcasting, 1988, 34(2):201-209.
8. Jean M M, Frederic G.A. New Design for High Performance HF Curtain Antennas [J]. IEEE Transactions on Broadcasting. 1988, 34(2):180-185.
9. Peter B.HF Curtain Antennas Operable at Very High Wind Speed [R]. Berlin: Antennen Engineering-Consulting, 2006.
10. Yao Yuan, Chen Wenhua. Analysis and Design of Tapered Slot Antenna for Ultra-Wideband Application [J]. Tsinghua Science and Technology, 2009, 14(1):1-6.

A gain enhanced OTA by introducing negative conductance

D. D. Fan, S. X. Song†, P. Q. Jiang and Y. L. Li

College of Electronic Engineering, Guangxi Normal University, Guangxi, China

†E-mail: songshuxiang@mailbox.gxnu.edu.cn

Email: fandongdongmail@163.com

An enhanced recycling folded cascode structure is proposed to enhance the gain of the operational amplifier with an auxiliary amplifier and no additional poles. This enhancement was achieved by introducing negative conductance. Under the supply voltage of 1.2V, the OTA was simulated by Spectre in TSMC 90nm CMOS technology. The results show that, the enhanced amplifier get a DC gain enhancement of 40dB compared with the conventional folded cascode. In addition, the gain-bandwidth also obtains 108.9% improvement, which also leads to a better slew rate performance.

Keywords: CMOS; Transconductance Operational amplifier; High Gain.

1. Introduction

The operational transconductance amplifier (OTA) is one of important modules in the analog circuits and has been widely used in power, ADC, filters and continuous-time oscillators (Dadashi et al. 2012) [4]. Single stage operational amplifier has high speed, moderate gain, and good stability in the old process.

However, with the progressive decrease in the process size and transistor channel length, the MOS transistor intrinsic gain is rapidly scaling down in the current new process. Moreover, the CMOS design has entered the low-voltage and low power consumption regime. The design of high gain amplifier will face greater challenge in such conditions. The gain and bandwidth are the most important two indicators to measure the performance of an amplifier (Roh 2006) [7]. In order to improve these two indicators, some methods have been proposed (Thandri & Silva-Martinez 2006) [9]. High DC gain operational amplifiers can use one or more of the following techniques. These techniques are cascading of gain stages by consuming more current, or the technology of increasing output impedance. Cascading two or more stages will result in a very high DC gain.

However, each cascading brings a high gain and also introduces a low-frequency pole, which produces a negative phase and degrades the phase margin. A lot of phase compensation schemes for operational amplifier is reported (Abdulaziz et al. 2014) [1]. The most common method is Miller compensation. The dominant pole is pushed to the low frequency at the cost of reducing 3-dB bandwidth and gain bandwidth product. The folded cascode (FC) OTA exhibits a single parasitic pole, with good bandwidth characteristics, and is widely used in high frequency circuits. A recycling folded cascode (RFC) configuration is proposed by Assaad & Silva-Martinez (2007)[3]. It modifies from the current source of conventional folded cascode amplifier to improve the DC gain and bandwidth. Recently, extra shunt current sources (Yan et al. 2012)[10] and phase margin network (Zhao et al. 2013)[11] are used for improving the performance of RFC.

In this paper, an enhanced recycling folded cascode (ERFC) amplifier is proposed to enhance the DC gain of OTA without introducing any additional poles. The new proposed technique does not limit the GBW of OTA and, consequently, the frequency performance is identical to the RFC OTA. An overview of conventional recycling OTA structure in section 2. Section 3 describes about the proposed ERFC OTA in this paper. Sections 4 and 5 present the simulation results and the conclusion respectively.

2. Conventional Recycling Structure

The conventional RFC amplifier is shown in Figure 1. The differential input pair and the output bias current source of FC are divided into two halves to produce transistors M_{1a}~M_{4b}. *K* is denoted the size ratio of the current mirrors M_{3a}:M_{3b} (M_{4a}:M_{4b}).

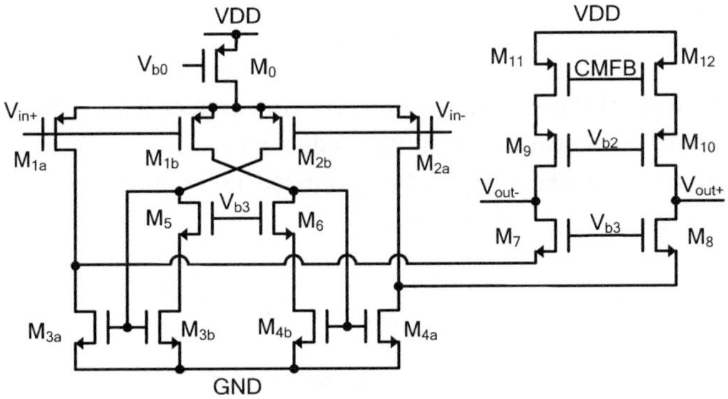

Fig. 1. Recycling folded cascode (RFC).

The transconductance (*Gm*) of RFC OTA is effectively improved by the split transistors and factor *K*, and *Gm* also can be given by,

$$G_{mRFC} = (K+1)g_{m1a} \tag{1}$$

The output impedance of RFC OTA can be described by

$$R_{ORFC} = g_{m7}r_{o7}(r_{o3a}\|r_{o1a})\|g_{m9}r_{o9}r_{o11} \tag{2}$$

which g_{mi} and r_{oi} are the transconductance and output impedance of transistor M_i.

From (1), it can be find that the improved transconductance results in a lager bandwidth compared to the FC in the same power and area. However, the factor K is restricted by the output slew rate in the full differential application, and also the current of input and output should be matched. Those decide that the transconductance of RFC is limited to < 3. In addition, the reduction of the intrinsic gain quickly degenerates the DC gain of the RFC or improved RFC OTA.

3. Architecture of the Proposed ERFC Amplifier

The proposed ERFC is shown in Figure 2c, which is enhanced by an amplifier A_1 between the source and drain of M_7 (M_8) to achieve a negative conductance, as also shown in Figure 2a. The common mode feedback (CMFB) circuit is not exhibited in the schematic, and the detail is described in the literature (Ghosh et al. 2013)[5]. One of schematic diagrams is used to achieve the low gain auxiliary amplifier A_1 as shown in Figure 2b.

As shown in Figure 2a, the operational amplifier A_1 is added between the source and drain of M_7. According to the Miller theorem (Razavi 2001)[6], the output conductance of M_7 becomes $go7'$ (3)

$$g_{o7}' = (1-A_1)g_{o7} \tag{3}$$

If the gain A_1 of the low gain amplifier A_1 is greater than 1, the new output conductance $go7'$ will become negative conductance.

The output impedance r_{oi} can be expressed as reciprocal of conductance, $1/g_{oi}$. The R_{ORFC} (2) also can be given in (4)

$$R_{ORFC} = 1/[\frac{g_{o7}(g_{o3a}+g_{o1a})}{g_{m7}} + \frac{g_{o9}g_{o11}}{g_{m9}}] \tag{4}$$

Using the new conductance g_{o7}' (3) instead of the g_{o7} (4) and assuming $g_{m7}=g_{m9}$, $g_{o7}=g_{o9}$, the output impedance of the ERFC (Figure 2c) can be expressed as ,

$$R_{OERFC} = g_{m7}/[g_{o7}(g_{o3a}+g_{o1a}+g_{o11}) - A_1 g_{o7}(g_{o3a}+g_{o1a})] \tag{5}$$

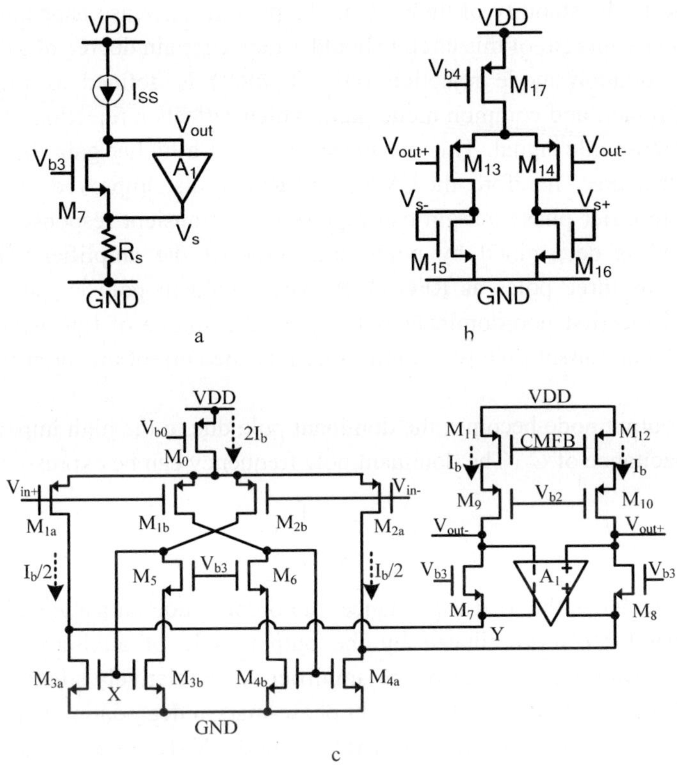

Figure 2. Amplifier A1 to achieve a negative conductance and enhanced recycling folded cascode (ERFC)

a method of obtaining a negative conductance

b auxiliary amplifier A1

c ERFC

The DC gain of ERFC (Figure 2c) also be described by

$$A_{ERFC} = -(1+K)g_{m1a}g_{m7} / [g_{o7}(g_{o3a}+g_{o1a}+g_{o11}) - A_1 g_{o7}(g_{o3a}+g_{o1a})] \tag{6}$$

Note that if

$$A_1 = (g_{o3a}+g_{o1a}+g_{o11})/(g_{o3a}+g_{o1a}) \tag{7}$$

the DC gain of the ERFC can be made to infinity in theory. The A_{ERFC} (6) is a function of A_1, so that a rational designing amplifier A_1 makes the A_1 and g_{oi} match to enhance the gain of ERFC. It is worth to note that the A_1 cannot be made too large, otherwise the denominator of formula (6) becomes negative. It may cause that the output impedance of the circuit is negative, which is not conducive to the stability of the system. To prevent this case happening, the A_1 and the phase margin of this circuit should retain a certain degree of safety.

The common mode rejection ratio (CMRR) is defined as the ratio of differential gain and common mode gain, which reflects a rejection of amplifier for common-mode signal. This enhanced OTA has a higher gain than the RFC and FC structure. Therefore, the CMRR of ERFC is also improved.

An amplifier phase margin can represent the transient response index of a circuit and is determined by poles and zeros of the amplifier. The ERFC structure has three poles as RFC OTA. The dominant pole w_{p1} at the output node, and the first non-dominant pole w_{p2} at the source of transistor M_7. The second non-dominant pole w_{p3} is introduced by the current mirror at the node X in Figure 2c.

The output node becomes the dominant pole due to the high impedance and large capacitance of C_L. The dominant pole frequency can be expressed as (8),

$$\omega_{p1} = \frac{1}{R_{OERFC} C_L} \tag{8}$$

where C_L includes the load capacitance, the output node parasitic capacitor and extra capacitor C_{A1} introduced by the output node of auxiliary operational amplifier. However, the sizes of the amplifier A_1 is relatively smaller , and C_{A1} is far less than C_L, which leads to slight phase-margin degradation in the ERFC.

The second pole appears at cascode node Y (Figure 2c) and can be expressed as (9),

$$\omega_{p2} = \frac{1}{R_{P2} C_Y} \tag{9}$$

Where

$$C_{P2} = C_{gd1a} + C_{db1a} + C_{gd3a} + C_{db3a} + C_{gs7} + C_{sb7} \tag{10}$$

And

$$R_{p2} = \frac{1}{g_{m7}} \| r_{o1a} \| r_{o3a} \approx \frac{1}{g_{m7}} \tag{11}$$

Because of the current mirror M_{3a}-M_{3b}, the parasitic capacitance of the node X not only generates a pole w_{p3} on the non-dominant pole frequency, but also introduces a zero w_{z1}. The zero greatly compensates for the phase shift of poles, and the result is only a smaller change in the phase shift. Consequently, the effect of pole-zero pair can be negligible for the phase margin.

Since the low gain of an amplifier carries a small current at the output, the ERFC OTA has a similar slew rate characteristics as the RFC. The transient characteristics of FC and RFC have been detailed analysis by Akbari et al (2012)[2]. The amplifier A_1 just changes the output impedance, and there is no additional extra pole introduced, which results in excellent bandwidth and slew rate performance. The corresponding slew rate of ERFC can be described by (12)

$$SR_{ERFC} = \frac{2I_b K}{C_L} \tag{12}$$

4. Implementation and Results

In order to compare the performance of three OTAs circuit (FC, RFC and ERFC) in the same power budget, the OTA was simulated by Spectre in TSMC 90nm COMS technology using the supply voltage of 1.2V.

The OTAs frequency characteristics are shown in Figure 3. Less gain is improved in the RFC, however, the gain of the proposed ERFC increases by 40dB. The enhancement of output impedance R_o makes the dominant pole toward to the low frequency. Although the phase margin drops from 76.7° to 70.8°, it still remains a sufficient abundant phase margin to ensure the stability of the circuit. The GBW of ERFC, as shown in Figure 3, is twice as the GBW of the FC. Figure 4 shows the CMRR responses of the three OTAs. Note that the ERFC exhibits about 40/32dB high common mode rejection ratio compared with the FC/RFC. The transient response of three circuit test results are shown in Figure 6. Because the proposed amplifier has a larger GBW and higher gain, it shows more rapid response speed and accuracy under the same test condition. The slew rate of ERFC and RFC is 83.75V/μs and 88.47V/μs, which increases nearly one times compared the slew rate with FC.

The key performance parameters of three OTAs are summarized in Table 1. The ERFC not only has excellent gain and bandwidth, but also obtains better slew rate and reference input noise. The noise characteristic can be achieved by simulating the density of the input referred noise. The noise of FC, RFC and ERFC is 8.2nV/sqrt(Hz), 4.97nV/sqrt(Hz) and 4.98nV/sqrt(Hz), respectively, at the bandwidth of 1 Hz-100MHz. Therefore, the reference input noise is reduced by nearly 40%.

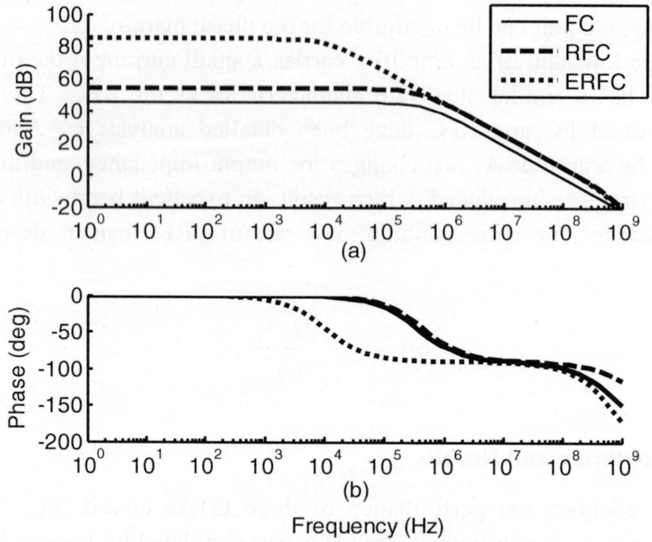

Fig. 3. AC response of RC, RFC and ERFC.

In order to assess the overall performance of the OTA, a figure-of-merits (FoM) comprehensive index is given. The most simple FoM of the amplifier is represented by gain bandwidth product, load capacitance, and power consumption (Sansen 2007) [8]. The FoM of the proposed OTA and the RFC are approximately twice as large as the traditional FC, but the ERFC has also won 57.5% gain boost compared with the RFC.

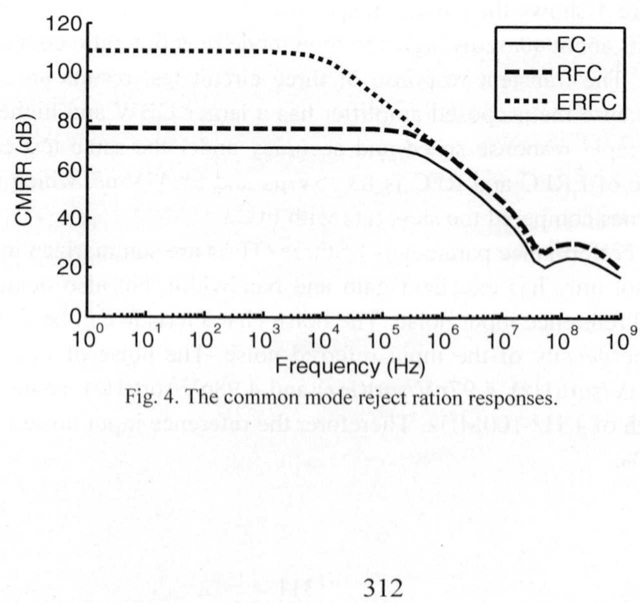

Fig. 4. The common mode reject ration responses.

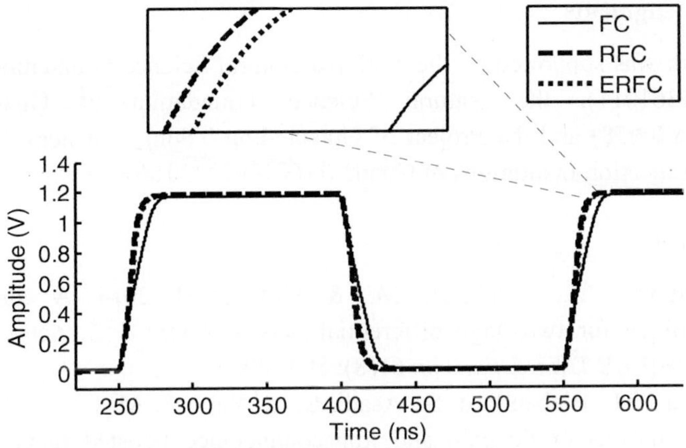

Fig. 5. Transient response of RC, RFC and ERFC.

Table 1: The performance comparison of three OTAs

Parameter	FC	RFC	ERFC
Bias current (μA)	400	400	400
DC Gain (dB)	45.2	53.9	84.9
GBW (MHz)	84.5	180.4	176.5
Phase margin (deg)	87.9	76.7	70.8
Capacitive load (pF)	5	5	5
Slew Rate (average) (V/μs)	44.62	88.47	83.75
Common mode rejection ratio	68.9	77.2	108.4
Input Referred Noise (1Hz-100MHz)(nV/sqrt(Hz))	8.2	4.97	4.98
FoM (MHzpF/mA)	1056	2255	2206

5. Conclusion

An enhanced RFC structure is proposed to enhance the gain of the OTA with an auxiliary amplifier and no additional poles. Three OTAs are designed including FC, RFC and ERFC in the same bias current, using TSMC 90nm CMOS process. The simulation results show that the proposed ERFC is due to the increase of the output impedance, resulting in a 57.5% enhancement in the DC gain compared with the RFC and a 87.8% enhancement compared with the FC. At the same time, the GBW of the ERFC obtains 109.9% enhancement compared with the FC, which leads to a better slew rate performance.

Acknowledgments

This work was supported by the National Natural Science Foundation of China (no. 61361011), the Natural Science Foundation of Guangxi (no. 2014jjAA70058) and the Project of Outstanding Young Teachers' Training in Higher Education Institutions of Guangxi (GXQG022014002).

References

1. Abdulaziz, M., Tormanen, M., & Sjoland, H. 2014. A compensation technique for two-stage differential otas. *Circuits & Systems II Express Briefs IEEE Transactions on* 61(8): 594-598.
2. Akbari, M., Biabanifard, S., Asadi, S., & Yagoub, M. C. E. 2014. Design and analysis of dc gain and transconductance boosted recycling folded cascode ota. *AEU - International Journal of Electronics and Communications* 68: 1047-1052.
3. Assaad, R., & Silva-Martinez, J. 2007. Enhancing general performance of folded cascode amplifier by recycling current. *Electronics Letters* 43(23): 1243-1244.
4. Dadashi, A., Sadrafshari, S., Hadidi, K., & Khoei, A. 2012. Fast-settling CMOS Op-Amp with improved DC-gain. *Analog Integrated Circuits and Signal Processing* 70(3): 283-292.
5. Ghosh, N. N., Todani, R., Chaudhuri, C., & Mal, A. K. 2013. Simplified design method for fully differential gain-boosted folded cascade OTA. *In Information & Communication Technologies (ICT), 2013 IEEE Conference on*: 943-948. IEEE.
6. Razavi, B. 2001. *Design of Analog CMOS Integrated Circuits*. New York: McGraw Hill Higher Education.
7. Roh, J. 2006. High-gain class-AB OTA with low quiescent current. *Analog Integrated Circuits and Signal Processing* 47(2): 225-228.
8. Sansen, W. M. 2006. *Analog design essentials*. Dordrecht: Springer.
9. Thandri, B. K., & Silva-Martinez, J. 2006. An overview of feed-forward design techniques for high-gain wideband operational transconductance amplifiers. *Microelectronics journal* 37(9): 1018-1029.
10. Yan, Z., Mak, P. I., & Martins, R. P. 2012. Double recycling technique for folded-cascode ota. Analog *Integrated Circuits & Signal Processing* 71(1): 137-141(5).
11. Zhao, X., Fang, H., & Xu, J. 2013. Phase-margin enhancement technique for recycling folded cascode amplifier. *Analog Integrated Circuits & Signal Processing* 74(2): 479-483.

Layout research on magnetic navigation system of unmanned electric vehicle

F. Wang†[1] and F. Luo[2]
Clean Energy Automotive Engineering Center, School of Automotive Studies, Tongji University, Shanghai, China
†*E-mail: wangfeitju@163.com*

In this paper, a magnetic navigation system of Unmanned Electric Vehicle (UEV) was described. In the magnetic navigation system, the layout of the magnetic sensor array is important, it is needed to determine the installation height of the magnetic sensor array, the horizontal interval of each magnetic sensor, and the interval of adjacent magnetic markers. To solve this problem, this paper analyses the magnetic field distribution of magnetic marker, and the actual conditions of UEV, then a layout method is presented at last, which is verified by experimental results.

Keywords: Unmanned electric vehicle; Magnetic navigation; Sensor array layout.

1. Introduction

About the navigation of Unmanned Electric Vehicle (UEV), there are many technologies are applied, such as computer vision, magnetic navigation, optical navigation, inertial navigation and global positioning systems [1, 2]. Among the technologies, this paper focuses on magnetic navigation system of the UEV [3, 4].

In the magnetic navigation system, the layout of the magnetic sensor array is very important, which has implications for the magnetic field data collection and vehicle positioning, including determining the installation height of the magnetic sensor array, the horizontal interval of each magnetic sensor, and the interval of adjacent magnetic markers. By analysis to the magnetic field distribution of magnetic marker and the actual conditions of the UEV, a layout method is presented in this paper, which is verified by experimental results.

2. The Unmanned Electric Vehicle Overview

In this paper, the Unmanned Electric Vehicle which is based on magnetic navigation, is modified from a four-seat electric sightseeing car. As shown in

[1] Fei Wang
[2] Feng Luo

Figure 1. The UEV is consist of a power CAN (Control Area Network) which named as CAN_B, and an intelligent CAN which named as CAN_A. All control modules are distributed in the two networks. The Vehicle Management System (VMS) acts as the gateway which connects the power CAN and the intelligent CAN.

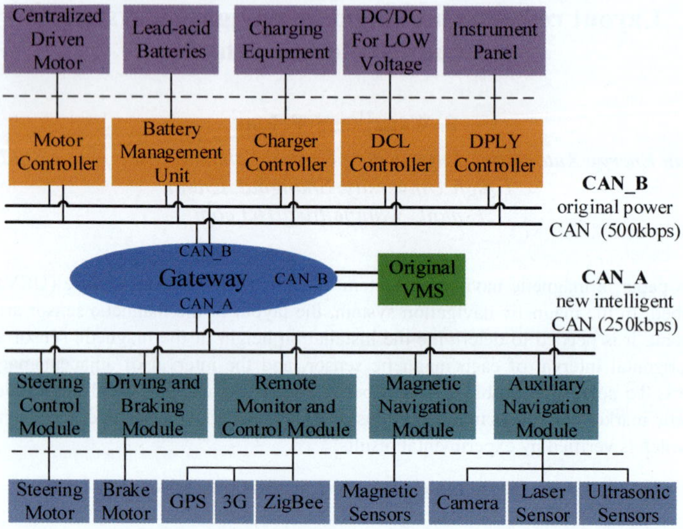

Fig. 1. System structure of the unmanned electric vehicle.

In order to realize automatic travel, the steering control module, the driving and braking module, the remote monitor and control module, the magnetic navigation module, and the auxiliary navigation module are added to the intelligent network. When the vehicle is in autonomous driving, the magnetic navigation module provides current location information to the VMS constantly, the steering control module and the driving and braking module control vehicle's direction and speed according to the location information, the auxiliary navigation module provides more precise location information for the VMS.

3. Magnetic Navigation System

The Magnetic Navigation System (MNS) plays a very important role in the UEV, in this paper, the MNS is mainly consist of magnetic sensor array, ultrasonic sensor, MCU and CAN transceiver, as shown in Figure 2. The magnetic sensor array contains multiple sensors, as shown in Figure 3, can sense the strength of magnetic field, which delivered to the MCU via I^2C, according to the information of magnetic, the MCU can calculate the relative location between the vehicle and the magnetic marker, and log the magnetic code information. The ultrasonic sensor measures the height of magnetic sensor to the ground, which helps the MCU calculation more accurate location information.

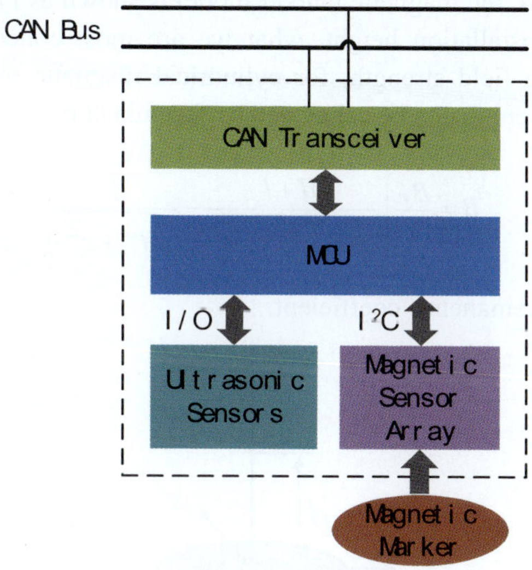

Fig. 2. Structure of magnetic navigation system.

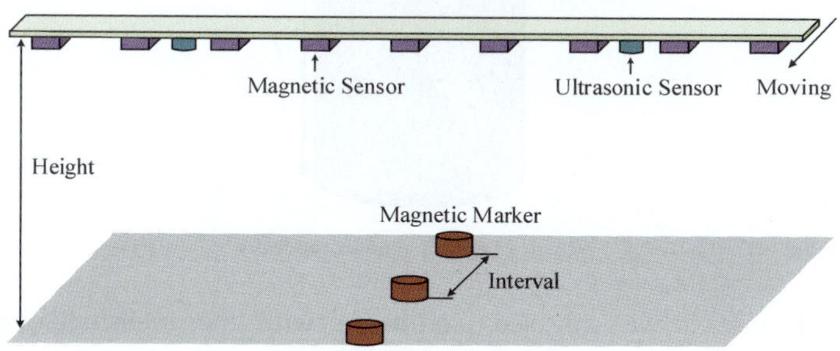

Fig. 3. Magnetic sensor array.

4. Magnetic Sensor Array Layout

The layout of magnetic sensor array mainly includes the installation height from magnetic sensor array to the magnetic markers, the horizontal interval of each magnetic sensor, and the interval of adjacent magnetic markers. Especially the height and the interval of adjacent magnetic markers, which have a strong impact on the measurement and distribution of the magnetic fields.

4.1. *Height of magnetic sensor array*

The installation height of magnetic sensor array affect the measurement of the magnetic field directly. There are several factors which impact the height, such as the measurement range of the magnetic sensor, the magnetic field strength of the magnetic marker itself, the vehicle's running condition -- vehicle body up and down vibration, etc. Firstly, it is necessary to know the distribution rule of the magnetic field, the magnetic marker model is shown as Figure 4. In order to make sure the installation height, what we are most concerned about is the vertical magnetic field strength, for cylindrical magnetic marker, the vertical magnetic field strength can be calculated by formula (1):

$$B = \frac{B_r}{2}\left(\frac{d+l}{\sqrt{(d+l)^2 + r^2}} - \frac{d}{\sqrt{d^2 + r^2}}\right) \tag{1}$$

B_r - Magnetic remanence coefficient.

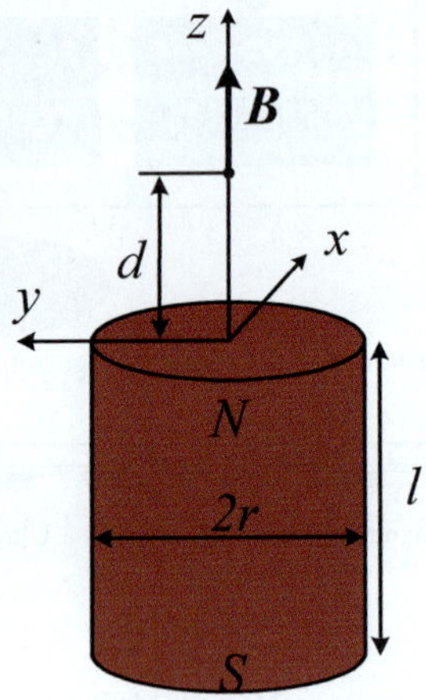

Fig. 4. Magnetic marker model.

From Figure 5 we can see, compared with the measured value and calculated value, the values are so close, and the curve displays at the height of 14cm, the magnetic field is about 5.3 Gauss, with the magnetic sensor we selected in this paper is Honeywell HMC series magnetic sensor, whose measurement range is about ±5.6 Gauss, so the height from magnetic sensor array to the magnetic markers should more than 14cm. On the other hand, with the height increase the strength of magnetic field weakens quickly, and as a result of the geomagnetic field interference, the height should not be too high.

It is determined that the strength of geomagnetic field is about 0.6 Gauss, the Figure 5 shows at the height of 28cm, the magnetic field is about 0.6 Gauss, so the height should no more than 28cm. At the same time, the installation height of magnetic sensor array will impact the effective magnetic field area, the higher the height, the smaller effective magnetic field area, the bigger signal blind area. The practical test in our research shows that when the magnetic field is two times the geomagnetic field (about 1.2 Gauss at the height of 24cm), can ensure that the effective magnetic field area is enough big, which means the height should no more than 24cm.

Based on the above analysis, considering vehicle body up and down vibration (about 3~4cm) in actual vehicle moving, the installation height of magnetic sensor array should within 18~21cm.

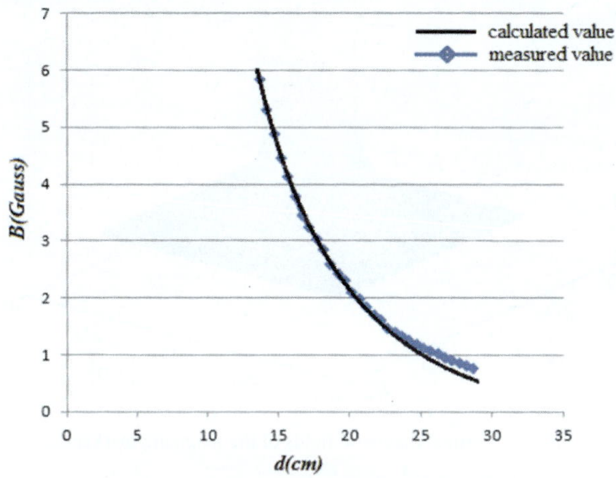

Fig. 5. Axial magnetic field of the magnetic marker.

4.2. *Horizontal interval of each magnetic sensor*

Generally, the smaller the theoretical interval, it can get smaller lateral deviation calculation accuracy, but at the same time, it will increase the number of sensors and the cost, increase burden of data processing, and increase the sampling period.

Formula (1) can only calculate axial magnetic field strength of the magnetic marker, but for determining the horizontal interval of each magnetic sensor and the interval of adjacent magnetic markers, we should know the horizontal magnetic field strength of the magnetic marker, the calculation formula (2) [5] is as follows:

$$\vec{B} = \frac{\mu_0 M}{4\pi r^5}\left[3xz\,\vec{i} + 3yz\,\vec{j} + (2z^2 - x^2 - y^2)\vec{k}\right] \quad (2)$$

μ_0 - Magnetic susceptibility.

According to the formula (2), we can calculate the vertical magnetic field strength at d = 18cm, as shown in the Figure 6, and the horizontal magnetic field strength, as shown in the Figure 7. We can know at the same plane, the vertical magnetic field strength reaches the maximum just above the surface of the magnetic marker, and the horizontal magnetic field reaches the strongest at about 12cm from the center of the magnetic marker, then weakens quickly.

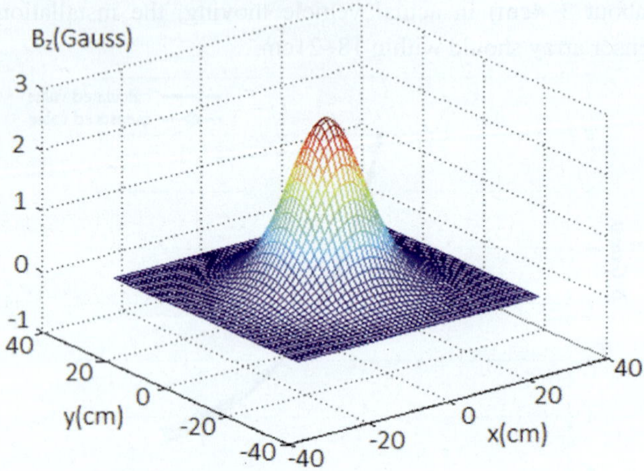

Fig. 6. Vertical magnetic fields of the magnetic marker.

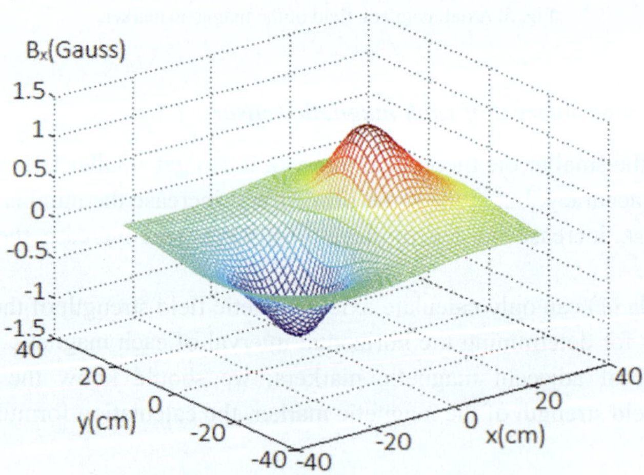

Fig. 7. Horizontal magnetic fields of the magnetic marker.

In order to eliminate the disturbance of the geomagnetic field, ensure the accuracy of the measurement results, we should set a magnetic field effective area, as shown in Figure 8, when out of radius R, we concluded that magnetic field is invalid, according to the distribution rule of horizontal magnetic field above, the R is about 12cm. On the other hand, according to sequence positioning algorithm [6], it should ensure that at least two or more sensors to collect the effective data, so the horizontal interval of each magnetic sensor should be about 12cm.

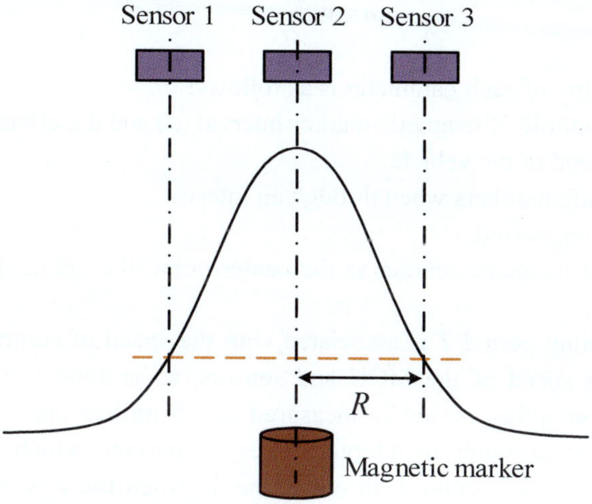

Fig. 8. Magnetic field effective area.

5. Magnetic Markers Layout

In the magnetic navigation system, magnetic markers as the source of magnetic signal, its layout is very important. According to the actual application situation, there are several kinds of schemes about the magnetic marker layout, it's needed to laid two lines for lane departure warning, but for vehicle autonomous navigation, it's needed to track the target trajectory, so we lay one line only, means only one magnetic marker in one line, it can also reduce the cost.

There is a key parameter in the single line layout, it is the interval of each magnetic marker. If the interval is too small, it will increase the cost, at the same time, the magnetic fields of the adjacent markers will overlap, influences the distribution of magnetic field, which is bed for sensors measuring magnetic field.

As shown in the Figure 7, far from the magnetic marker's center, the magnetic field component in horizontal strengthens gradually, and reaches the strongest at about 12cm from the center of the magnetic marker, then weakens quickly. In order to avoid magnetic fields overlap obviously, the interval of adjacent magnetic markers should avoid the point at which the magnetic fields overlap reaches the strongest, so the interval of adjacent magnetic markers should at least greater than 48cm (4×12cm). On the other hand, if the interval of adjacent magnetic markers is too large, there will be a larger signal blind area, it is not conducive to the vehicle control. Then what is the appropriate interval? Here we show a formula where the parameters associated with interval are include:

$$n = v \cdot \frac{kT}{2r} \tag{3}$$

The meaning of each parameter is as follows:

n - The multiple of magnetic marker interval (d) and the effective area ($2r$).

v - The speed of the vehicle.

k - Sampling numbers when through an interval.

T - Sampling period.

r - Regard magnetic marker as the center point, the radius of the effective magnetic field.

The sampling period T is associated with the speed of communication and the processing speed of the MCU and sensors, so is almost constant. In our research, the sampling period is measured as 15 milliseconds. The sampling numbers k when through an interval is very important, which will affect the stability of the vehicle control, in our research, when the k is about 15 times, vehicle automated driving can obtain better control. The radius of the effective magnetic field r, as mentioned before, according to different height, it is from 15cm to 20cm, the magnetic field is effective in this area, and when out of this area, the magnetic field is considered ineffective because it is too weak. Because of the research object in this paper is electric vehicle which is running at low speed, no more than 15 kilometers per hour.

So, with these conditions above, we can calculate the n by formula (3): $n \leq 3$, then the interval of each magnetic marker should no more than 1.2m, within this interval, it can obtain better control. We can also know from formula (3), when the kT stays the same, increasing n, i.e., increasing the interval between each magnetic marker, then the speed will increase, due to the electric vehicle runs at low speed, when the speed is lower, it will increase the signal blind area, which is not conducive to the vehicle control. So considering the vehicle speed, the system sample numbers, and reducing the cost, we take 1m as the interval of each magnetic marker.

6. Test Result

According to the layout method presented above, the magnetic sensor array on experimental unmanned electric vehicle is shown in Figure 9.

Fig. 9. Experimental unmanned electric vehicle.

The installation height of magnetic sensor array is 20cm, the horizontal interval of each magnetic sensor is 10cm, and the interval of adjacent magnetic markers is 1m. The magnetic code is "111101011101111" – the N pole up of the magnetic marker is marked with "1", and the S pole is marked with "0". The experimental results is shown in Figure 10, which shows the magnetic sensor array has better identification on magnetic markers.

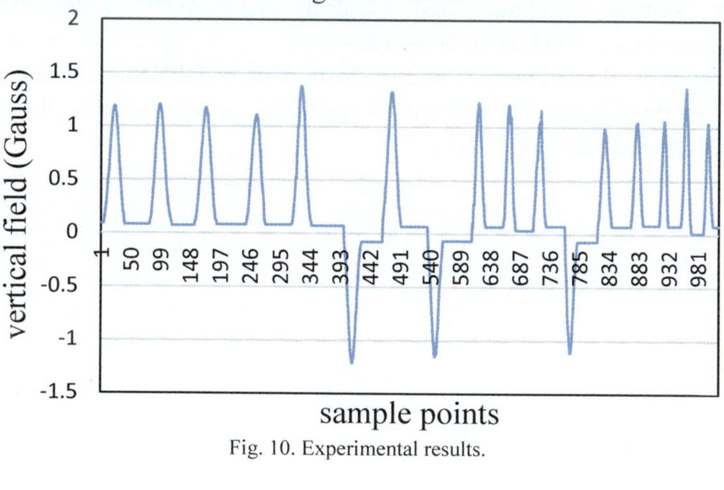

Fig. 10. Experimental results.

7. Conclusion

This paper presented a layout method on magnetic sensor array, by analysis to the magnetic field distribution of magnetic marker and the actual conditions of

UEV, finding out the most appropriate installation height of magnetic sensor array, the horizontal interval of each magnetic sensor, and the interval of adjacent magnetic markers, the experimental results also verified that the layout method played a good effect on magnetic markers identification.

References

1. Dae-Yeong, I. Young-Jae, R. & Sun-Kil, P. 2009. Unmanned Driving of Robotic Vehicle (Robicle) using Magnetic Position Sensor. *ICROS-SICE International Joint Conference*. 2009:5663-5667.
2. Chan, C.Y. 2002. Magnetic Sensing as a Position Reference System for Ground Vehicle Control. *IEEE Transactions on Instrumentation and Measurement*, Vol. 51 (2002): p. 43-52.
3. Sok-Yong, Lee. Hai-Won, Yang. 2012. Navigation of automated guided vehicles using magnet spot guidance method. *Robotics and Computer Integrated Manufacturing*, 2012, 28(3):425-436.
4. William, F. Storms, John, F. Raquet. 2009. Magnetic Field Aided Vehicle Tracking. Proceedings of the 22nd International Technical Meeting of Satellite Division of the Institute of Navigation (ION GNSS 2009). 2009:1767-1774.
5. Stephen, M. Donecker, T.A. Lasky. 2003. A Mechatronic Sensing System for Vehicle Guidance and Control. *IEEE/ASME Transactions on Mechatronics*, 2003, 8(4): 500-510.
6. Soheila, V. Bana. Pravin, Varaiya. 2010. Automated Vehicle Navigation via Magnetic Signals. *IAIN World Congress in association with the US ION 56th Annual Meeting*. 2010: 65-72.

Design and research of corrosion monitor instrument based on a novel coupon-type electrical resistance probe

X.F. Qiao[1†], P.A. Mu[2], S.G. Dai[3] and B. Wang[4]
School of Optical-Electrical and Computer Engineering,
University of Shanghai for Science and Technology, Shanghai 200093
†Email: Qiaoxinfeng1@163.com

In the field of environmental corrosion monitoring, there is a kind of corrosion monitor instrument designed to sensibly and quickly monitor the corrosion of environment. Corrosion monitor instrument based on a novel coupon-type electrical resistance probe, which including 4 coupons, two of them exposed in corrosion environmental and two sealed in epoxy for monitoring the environmental corrosion rates and deviations. Corrosion monitor instrument adopts the technology of Multi-channel switching measurement for the total coupons of resistance to temperature compensation, for the sake of elimination the temperature influence to monitoring environmental corrosion. Eventually corrosion monitor instrument accurately and rapidly realize online monitoring the environmental corrosion.

Keywords: Electrical resistance probe; Corrosion monitoring; Temperature compensation; Multi-channel switch measurement;

1. Introduction

In the field of corrosion monitoring, resistance probe technologies occupies more and more important position in the field of corrosion sensor because its stability. It has been widely used in the petroleum chemical industry, ocean and atmospheric environmental monitoring [1] with its simple, sensitive, and practical features. There are many research center commit to develop and research electrical resistance probe which used in petroleum, chemical industry and the specific corrosive environment at home and abroad [2]. Such as, the serials of CORROCEAN probe of Norwegian, CK-3, ck-4 resistance probe monitoring instrument of TMAG company in USA.

---

[1] Xin-Feng Qiao
[2] Ping-An Mu
[3] Shu-Guang Dai
[4] Bo Wang

A novel electrical resistance probe was researched in this paper, which mainly applied for the quickly monitoring environmental corrosion. Resistance probe which had been one of the most widely used environmental monitoring technologies with the advantages of the reliable methods and the visual results. As we all know, resistance is the energy dissipation element, so there would have an influence on the measurement result in the process of monitoring, because the temperature of resistance will be changed when current passed. The request for the precession of resistance is very high when monitoring the environmental corrosion. So it is necessary for resistance probe to eliminate temperature effect on measurement result when monitoring.

In order to limit the corrosion on the surface of resistance probe, calculation simply, reliable detection. The shape of probe was designed into long thin strips of specimens. The value of coupon is very small, between 0-50 mΩ. To sensitively indicate the change of corrosion, we have to measure the micro level changes on resistance probe. The definition to measuring system is very high, to the level of 1μΩ. So we have to adapt week signal measurement methods. General micro-ohmmeter only measurement one value one time. It was hard to quickly measure 4 coupons resistance meanwhile to the special structure of four specimen resistance. We have to eliminate the effect of temperature and manual switching measurement error.

Current micro-ohmmeter without multi-channel switch measurement function, to realize multi-channel switch measurement [3], we could adapt switch to quickly switch. Achieve the multi-channel measurement meanwhile. So, above the paper we talked about, we propose to design a multi-channel measurement system based on a novel coupon-typed electrical resistance probe. Eventually reduce the temperature drift and manual switch effect on the measurement of the small resistance, and implements rapidly, efficient, sensitive measurement of resistance probe.

2. Principle and Analysis of the Resistance Probe

Coupon-type electrical resistance probe can directly measure metal loss to show the change of corrosion. The change of corrosion and the change of the resistance value, namely the corrosion of resistance probe keep into a certain linear relationship. There are current through probe in the process of measurement, so the temperature of the resistance probe will change, so we have to eliminate the influence on monitoring that caused by temperature.

Figure 1 is the structure of resistance probe which including 4 coupons [4]: there are 4 similar carbon steel coupon on PCB, 2 exposed coupons Rx1, Rx2; 2 contrast coupons which sealed in epoxy Rx1, Rx2. The shape of the resistance test plate is thin lines.

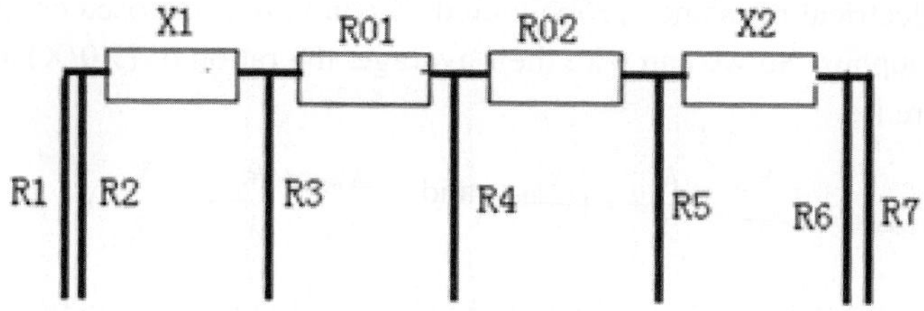

Fig. 1. Probe structure chart.

In the process of corrosion monitoring, to eliminate the influence of temperature, the theory of temperature compensation [5] is:

Relation of resistance (do not contain line) and corrosion (the change of thickness) infer by theory:

According to the formula of resistance that:

$$\frac{R}{R_0} = \frac{d_0}{d} \text{ or } d = d_0 \times \frac{R_0}{R} \tag{1}$$

so,

$$d = d_0 \times \frac{X_0}{X} \quad \text{or} \quad \Delta d = d_0 (1 - \frac{X_0}{X}) \tag{2}$$

Annotation: d_0 is the initial thickness value of exposed coupons; X_0 and X the initial measurement value and the measurement value after corrosion of exposed coupons;

In the actual measurement:

It is hard to measure the initial value and online measure value of exposed coupons at the same time, In the process of the actual measurement, we can obtain the measurement value Xt of exposed coupons and the value R_{0t} contrast coupons at one time t. Considered the temperature influence, $R_{0t} \neq R_0$; $Xt \neq X$. We can get (X_0/X) at the same temperature after multiply the factor (X_0/R_0).

$$\frac{X_0}{X} = \frac{R_{0t}}{X_t} \times (\frac{X_0}{R_0}) \tag{3}$$

Because (R_{0t}/X_t) don't change with the change of temperature, the ratio of the initial value of exposed coupons and the value of corrosion coupon at the same temperature equal to the ration of the contrast coupon and the exposed coupon in the actual measurement multiply the ratio of the initial value of exposed coupon and the contrast coupon. We could call the ratio as the index of corrosion. It represents the ratio of the residual thickness and the original thickness of the exposed coupons.

The electrical resistance probe include 4 coupons, 2 exposed coupons and 2 contrast coupons. So we can take their average, the ration of (X0/X) at the same temperature is:

$$\frac{X_{01}}{X_1} = \frac{R_{01t}}{X_{1t}} \times (\frac{X_{01}}{R_{01}}) \text{ and } \frac{X_{02}}{X_2} = \frac{R_{02t}}{X_{2t}} \times (\frac{X_{02}}{R_{02}}) \quad (4)$$

According to the ratio of the initial value and the measurement value of exposed coupon at the same temperature, this can eliminate the influence of temperature. Take online monitoring to the environmental corrosion.

3. The Key Technology of Corrosion Monitor Instrument

To realize the temperature compensation of Coupon-type Electrical Resistance Probe, it is required to realize the 4 coupon switch measurement at the same time. Electrical resistance probe, including 4 coupons, it can adapt Analog switch devices to achieve multichannel fast switching, then realize 4 coupons measurement at the same time. In the field of measurement, resistance measurements all adopt the method of voltage and current test. The coupon of electrical resistance probe is very small, only dozens of mΩ, definition to the level of 1μΩ. So it is must to eliminate the influence on contact resistance and the measurement lines. Four-wire measurement [6] method was proposed, which can eliminate the contact resistance and the influence of wire resistance on the measurement[7], and can precision measuring small resistance. Figure 2 is system structure.

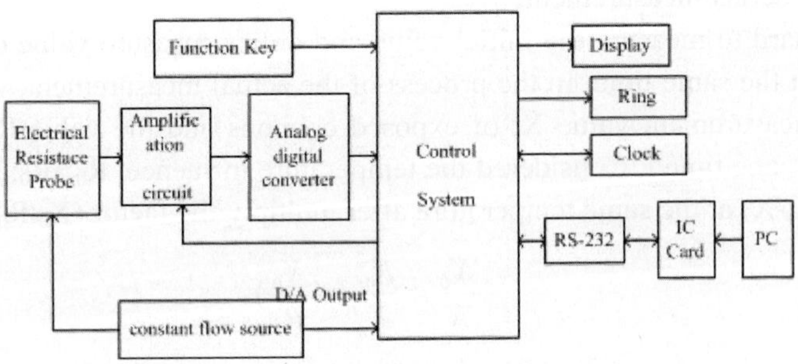

Fig. 2. The structure chart of corrosion monitor instrument.

3.1. *High precision controllable constant current source*

In the field of resistance measurement, resistance measurement realize by the transformation of the voltage – current. So in a certain extent, the design of the constant current source [8] determines the precision of corrosion monitoring.

In this paper, the design of constant current source based on operational negative feedback operational negative feedback constant current, is mainly through the role of negative feedback, made to compare the amplifier two at the input voltage is equal, so as to maintain a constant output current. In addition, operational negative feedback constant current divided into ground load constant current source and float load current source based on the different ways to connect to load, as shown in the figure 3 and figure 4 below. To decrease the input of common mode voltage on the both ends of load, this paper adapt ground load constant current source.

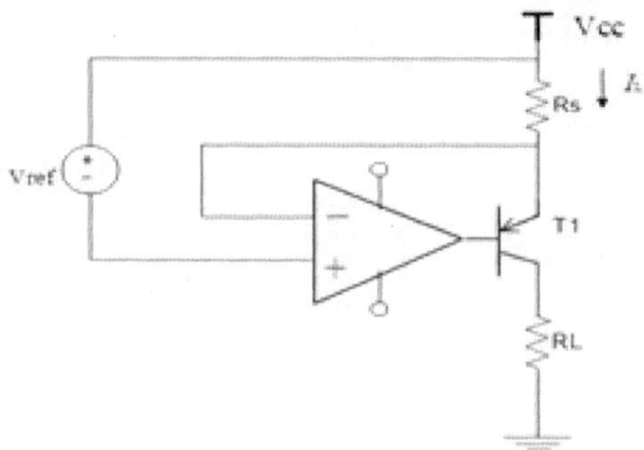

Fig. 3. Ground load constant current source.

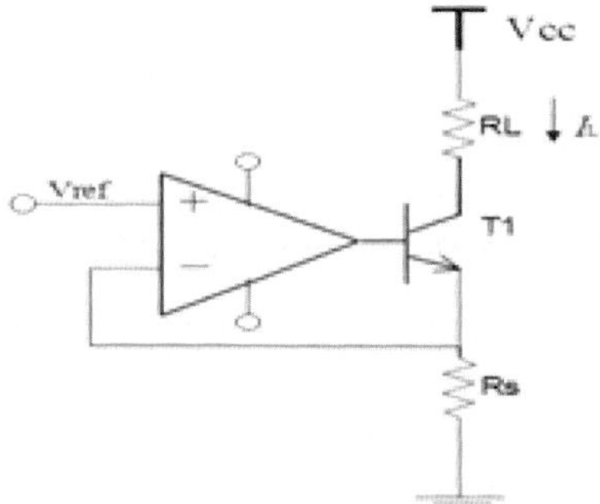

Fig. 4. Float load current source.

In the process of design current source, the basis voltage is also very important, because the precision of the reference voltage determines the precision of constant current source. The design of the constant current source select microprocessor MSP430 DAC output as a reference voltage constant current source, not only because the output is stability, but also is controlled.

3.2. *Automatic measurement of switching*

In the field of measurement, we can use fast on-off control of electronic circuits to complete the Multi-channel measurement of signals. For example, analog switch and small signal relay [9]. In this paper, we choose signal relay to complete the multi-channel measurement. Because the Signal switch resistance is minute, while shut off the impedance is very big, it is very good to small signal. With the feature of small signal relay, small volume, low power consumption, complete isolation between channels, can be well applied in the field of signal measurement. Eventually the multi-channel measurement system of the channel selection switch is a kind of Omron to install the surface of the small signal relay G6K-2P-Y. Through single chip microcomputer to control the relay switching, realize the multi-channel real-time automatic switch, and then realize the corrosion resistance of bolt multi-channel measurement [10]. Figure 5 is the structure of multi-channel real-time automatic switch.

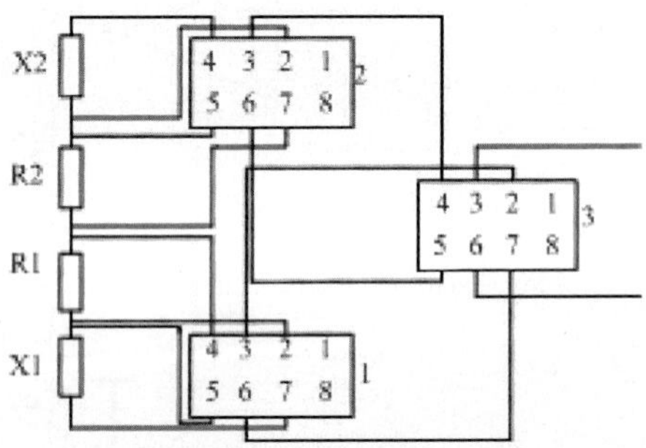

Fig. 5. The structure of multi-channel real-time automatic switch.

3.3. *High precision lead operational amplifier and the AD converter*

In this study, the value of signal mostly in the level of mV, the definition has to get at the level of uV. In order to improve the measuring accuracy and anti-jamming, it is must to choose high input impedance, low output impedance, low offset voltage, temperature drift, high common mode rejection ratio operational amplifier [11]. The instrument amplifier INA141was adopt in this study. The CMRR of INA141 is 117dB, the IB is 5nA, VIO is 50uV/°C. To the amplifier of the multi-channel measurement system it is meet the required. And the gain of INA141 is adjustable, there are 10 and 100 two gain range, can achieve 10 and 100 times magnification effect.

AD circuit is important channel for analog data convert to digital data. ICL7135which is four and a half and dual integral of AD [12]. With the feature of difference input, the input current is 1 pA. And bring their own owe range signal output, signal output polarity etc. It has the very high anti-interference performance, data is stable, and has a lower measurement error.

4. Experimental Analysis of Corrosion Monitor Instrument

The experimental analysis of the design used a medium resistance specimens under corrosion environment (the level of environmental corrosion according to the standard NACE),average corrosion rate is 0.025~0.12mm/a, response time is 1.5 days(the response is the best time that can monitor the change of probe and can obtain the thickness reduction of coupon). The size of the coupon is (width×length×thickness).

According to the standard of environmental corrosion and the formulas of the corrosion rate:

$$V = (d_0 \frac{X_0}{X_t}) / t = [d_0 \times \frac{R_{0t}}{X_t} \times \frac{X_0}{R_0}] / t$$

Table 1 indicate the repetitive of the multi-channel measurement system is very good. Table 2 indicate the influence of temperature on electrical resistance had been largely decreased in the process of the continuous measurement of 10 times.

Table 1. Measurement of the 4 coupons on the electrical resistance probe (unit: mΩ, SN: Serial Number).

SN / coupon	1	2	3	4	5	6	7	8	9	10
X1	4.876	4.876	4.876	4.877	4.876	4.877	4.877	4.878	4.878	4.878
R1	5.35	5.35	5.35	5.35	5.35	5.36	5.37	5.36	5.36	5.36
R2	4.983	4.982	4.983	4.984	4.984	4.985	4.984	4.985	4.986	4.986
X2	5.14	5.14	5.15	5.15	5.15	5.15	5.16	5.16	5.15	5.15

Table 2. The contrast for an resistance coupon before and after temperature compensation (unit: mΩ).

Time (Minute)	0	10	20	30	40	50
X1(Actual measurement)	4.875	4.879	4.885	4.897	5.01	5.12
X1(temperature compensation)	4.875	4.874	4.875	4.876	4.877	4.875

5. Conclusion

In this paper proposed a method which can realization the online monitoring for the environmental corrosion based on the research and design of a novel coupon-type electrical resistance probe. Electrical resistance probe which including 4 coupons realize temperature compensation in the process of measurement. With the technology of multi-channel switch measurement, the design of high precision numerical control constant current source, and the design of high-performance prefix operation amplifier circuit and the application of the conversion technology, implementing the design of micro resistance corrosion monitor instrument who with the feature of strong anti-interference ability and high definition.

Corrosion monitor instrument eliminate the influence of temperature on the measurement with the technology of temperature compensation. Which can quickly response to online monitoring of the environmental corrosion, and react the conditions real-time and efficiently. And it can be widely used in industry, agriculture, environment and other fields, the application value is very high.

References

1. ZHENG L Q. The latest development of petroleum and chemical industrial corrosion monitoring technology [J]. Petrochemical Corrosion and Protection. 2005, 22(1):11-13.
2. ZUO H J, JIN W F. On-line corrosion monitoring applications of MS3500E resistance probe [J]. Corrosion and maintenance. 2000, 21(12):557-558.
3. CUI Z Q, WANG H X, XU Y B, et al. Holding zero switching scheme for multi-channel impedance measurement [J]. Imaging Systems and Techniques, 2009. IST '09. IEEE International Workshop on. Shenzhen. May 2009: 11-12.
4. ZHANG H J, FENG B, WENG Y J, et al. A novel coupon-type electrical resistance probe for environmental corrosion monitoring [J]. Journal of Shanghai University (Natural Science Edition). 2015, 21:88-96.

5. BO R L, DONG Z H, GUO X P, et al. The principle research of corrosion resistance probe based on temperature compensation [J]. Corrosion science and protection technology. 2007, 19 (5):338-341.
6. Noritaka Yusa, Hidetoshi Hashizume. Four-terminal measurement of the distribution of electrical resistance across stress corrosion cracking [J]. NDT & E International, 2011, 44(7):544-546.
7. Dirch H. Petersen, Ole Hansen, Rong Lin, Peter F. Nielsen. Micro-four-point probe Hall effect measurement method [J]. Journal of Applied Physics, 2008, 104(1).
8. XU Y S, GU J H, TAO Z. Design and implementation of a high-precision constant current source [J]. Chinese Journal of Scientific Instrument. 2008, 29(4):725-727.
9. XIAO Z F, HUAN R X, LIU X L, et al. Development of ultra-small intelligent compound switch [J]. Electric power automation equipment. 2010, 30(1):119-121.
10. TAN H, HENNING W, WALBY M.D. Evaluation of Multi-Channel ADCs for Gamma-Ray Spectroscopy [J]. IEEE Transactions on Nuclear Science. 2013, 60(2).
11. TANG X L, LIU K Z, WANG L F, Design of high-performance fully differential operational amplifier [J]. Journal of Hebei University of Science and Technology. 2012, 33(1):50-55.
12. GAO J, YAO S Y, XU J T. Design of A novel low-power structure ADC [J]. Journal of Circuits and Systems. 2011, 16(1):104-107.

A microstrip gysel power divider with uniform characteristic impedance

H. C. Xiao and Y. Quan

Dongguan Power Supply Bureau, Dongguan, Guangdong, China

S.Y. Chen[1†]

College of Communications Engineering, Chongqing University, Chongqing, China
[†]Email: chensy@cqu.edu.cn

A micorstrip Gysel power divider (PD) with arbitrary power dividing ratio based on uniform characteristic impedance of transmission lines is presented in this paper. By using transmission line theory, the design equations are derived to determine the proposed circuit parameters. For verification, two 2 GHz experimental circuits with power dividing ratio of 2:1 and 3:1, respectively, are designed, fabricated and measured. The measured results show return losses are better than 25dB, isolations are better than 20 dB, and insertion losses are about 1.82 dB / 4.97 dB for the first PD and 1.46 dB / 6.47 dB for the second PD at 2 GHz, which are in good agreements with the simulated ones.

Keywords: Gysel power divider; Arbitrary power dividing ratio; Uniform characteristic impedance.

1. Introduction

The power divider (PD) is widely used in microwave communication and radar systems, such as power splitting /combining networks for microwave power amplifier modules or beam-forming networks for the antenna arrays. The Wilkinson and Gysel circuits are the most popular structures. The conventional Wilkinson PD has low insertion loss and high isolation, but its power-handling capability is low because of the isolation resistor between the output ports. The Gysel PD over the Wilkinson PD features advanced heat-sinking capacity, monitoring capability for imbalances at the output ports. Recently, many extensive works concerning the Gysel PD have been reported [1-6]. A modified Gysel PD with arbitrary power dividing ratio for three real terminated loads is discussed and analyzed in [1], which indicates the PD has about 25% fractional bandwidth based on return loss better than 20 dB. In [2], a dual-band Gysel PD with additional shunt stubs is introduced, however it has a narrow bandwidth. As the further work of [2], the design procedure of unequal power divider is discussed in

---

[1] Shi-Yong Chen

[3]. To improve the bandwidth of Gysel PD, several studies have been done. In [4-5], the even and odd-mode method is used to analyze the circuit performance of the multi-section Gysel PD and the least squares method is used to optimize the physical dimensions in a specified wide frequency band to achieve wider operating bandwidth. A compact microstrip broadband Gysel PD with arbitrary power-dividing ratio is presented in [6]. The bandwidth is improved by using microstrip/slotline phase inverter instead of a 180° microstrip transmission line. A Gysel PD employing a rectangular patch structure with etched rectangular pattern in the ground plane to disturb the current distribution and change the patch's characteristics is reported in [7]. It avoids using narrow lines and provides a broad bandwidth. however, it also suffers from a larger physical size.

The power dividing ratio of Gysel PDs mentioned above is achieved by varying the characteristic impedances of the impedance transformers. The major drawback is that high characteristic impedance of transmission line is involved if the power dividing ratio is larger than 4, which leads to the difficulty of physical realization using traditional PCB process. In this paper, a modified microstrip Gysel PD with arbitrary power-dividing ratio by using uniform characteristic impedance transmission lines is proposed. The power dividing ratio of the proposed Gysel PD is controlled by the electrical length of the transmission lines as introduced in [8]. The design parameters of the proposed PD are derived based on transmission line theory. For verification, two prototypes of 2GHz with power dividing ratios of 2:1and 3:1, respectively, have been designed and measured. The measured results show in agreements with the simulated ones.

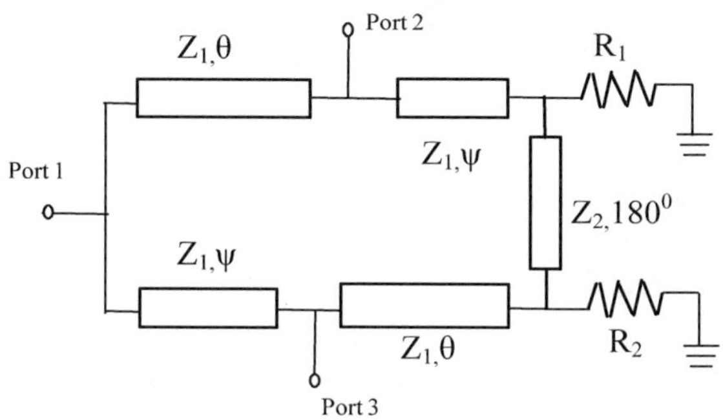

Fig. 1. The proposed PD.

2. Analysis of Proposed PD

The structure of the proposed Gysel PD with arbitrary power-dividing ratio of $K^2=P_2/P_3$ is shown in Figure 1. It basically consists of transmission lines (Z_1, Z_2) with different electrical length and two isolation resistors (R_1, R_2). The electrical length of Z_2 is 180° at operating frequency, which acts as an ideal phase inverter.

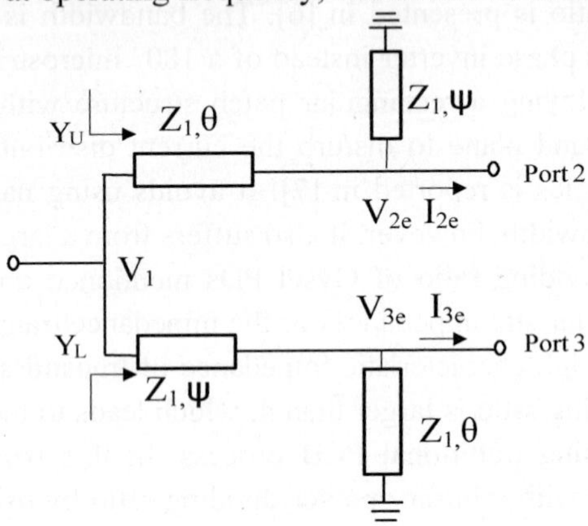

(a)

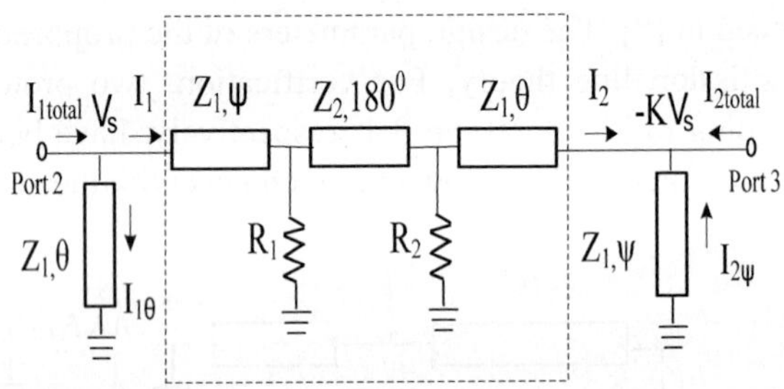

(b)

Fig. 2. The proposed Gysel PD. (a) Even-mode circuit. (b) Simplified odd-mode circuit.

2.1. *Even-mode analysis*

With the assumption that the power division is lossless, the input power is transmitted solely to the terminated loads of port 2 and 3 when input port 1 is excited (all ports of the proposed PD are equal to Z_0). There is no current flowing through the grounded resistors (R_1, R_2) and power dissipation is zero. The circuit structure shown in Figure 1 can be reduced to the equivalent circuit shown in Figure 2(a), where Y_U and Y_L are the input admittances of the upper and lower branches, respectively. By applying the transmission line theory, the two input admittances can then be derived as:

$$Y_U = Y_1 ctg\theta \frac{Y_0 + jY_1(\tan\theta - \cot\psi)}{Y_1(\cot\psi + \cot\theta) + jY_0} \tag{1}$$

$$Y_L = Y_1 ctg\psi \frac{Y_0 + jY_1(\tan\psi - \cot\theta)}{Y_1(\cot\psi + \cot\theta) + jY_0} \tag{2}$$

where Y_0.=$1/Z_0$.

Under the matching condition of the input port, the following equation should be satisfied.

$$Y_0 = Y_U + Y_L \tag{3}$$

By substituting (1), (2) into (3) and equating the imaginary parts of (3) after some algebraic manipulations, the characteristic impedance of Z_1 can be expressed as

$$Z_1 = \sqrt{2} Z_0 \tag{4}$$

The output power-dividing ratio can be calculated from even-mode voltages V_{2e}, V_{3e} as

$$\frac{P_2}{P_3} = \frac{|V_{2e}|^2 / Z_0}{|V_{3e}|^2 / Z_0} = K^2 \tag{5}$$

By employing *ABCD* matrix analysis to both the upper and lower branch networks shown in Figure 2(a), the following relationships are obtained:

$$V_1 = [\cos\theta + \sin\theta\cot\psi]\cdot V_{2e} + jZ_1\sin\theta\cdot I_{2e} \tag{6}$$

$$V_1 = [\cos\psi + \sin\psi\cot\theta]\cdot V_{3e} + jZ_1\sin\psi\cdot I_{3e} \tag{7}$$

where $I_{2e} = V_{2e}/Z_0$, $I_{3e} = V_{3e}/Z_0$

Subsequently, the relationship is obtained by combining (5), (6) and (7):

$$\begin{aligned}&[\cos\theta + \sin\theta\cot\psi]\cdot K\cdot V_{3e} + jZ_1\sin\theta\cdot K\cdot I_{3e}\\ &= [\cos\psi + \sin\psi\cot\theta]\cdot V_{3e} + jZ_1\sin\psi\cdot I_{3e}\end{aligned} \tag{8}$$

By equating the imaginary part of (8), the following result is found:

$$K\sin\theta = \sin\psi \tag{9}$$

From the formula (9), It can be observed that the power dividing ratio can be determined by different electrical length.

2.2. *Odd-mode analysis*

With the assumption that the port 2 and 3 are excited by voltage sources with magnitudes of 2Vs and -2*K*Vs, respectively. By applying the principles of superposition, the total voltage at the input port 1 becomes zero due to the fact that the two contributed voltage components are equal in magnitude and opposite in phase. Moreover, by enforcement of perfect isolation and output port matching condition, the resulted node voltage at port 2 and 3 are equal to Vs and –*K*Vs, respectively. Figure 2(b) shows the simplified odd-mode equivalent circuit. The network currents (I_1 and I_2) shown in Figure 2(b) can be represented as:

$$I_1 = (Y_0 + jY_1 \cot \theta) V_s \tag{10}$$

$$I_2 = (Y_0 + jY_1 \cot\psi) KV_s \tag{11}$$

The relationship between the node voltages and the currents (I_1 and I_2) shown in Figure 2(b) can be expressed in terms of *ABCD* matrix of the network.

$$\begin{bmatrix} V_s \\ I_1 \end{bmatrix} = \begin{bmatrix} A & B \\ C & D \end{bmatrix} \begin{bmatrix} -KV_s \\ I_2 \end{bmatrix} \tag{12}$$

where *ABCD* matrix can be represented as the formula (13).

$$\begin{bmatrix} A & B \\ C & D \end{bmatrix} = \begin{bmatrix} \cos\psi & jZ_1 \sin\psi \\ jY_1 \sin\psi & \cos\psi \end{bmatrix} \begin{bmatrix} 1 & 0 \\ \frac{1}{R_1} & 1 \end{bmatrix} \cdot \begin{bmatrix} -1 & 0 \\ 0 & -1 \end{bmatrix}$$

$$\times \begin{bmatrix} 1 & 0 \\ \frac{1}{R_2} & 1 \end{bmatrix} \begin{bmatrix} \cos\theta & jZ_1 \sin\theta \\ jY_1 \sin\theta & \cos\theta \end{bmatrix} \tag{13}$$

After some simple calculation of (13), the element of *ABCD* matrix can be written as

$$A = D = -\cos\theta\cos\psi + \sin\theta\sin\psi - jZ_1 G\cos\theta\sin\psi$$

$$B = Z_1^2 G\sin\theta\sin\psi - jZ_1(\sin\theta\cos\psi + \sin\psi\cos\theta)$$

$$C = -G\cos\theta\cos\psi - jY_1(\cos\theta\sin\psi - \sin\theta\cos\psi) \tag{14}$$

where $G = (R_1+R_2)/(R_1R_2)$.

By substituting (10) and (11) into (12), the first row of the matrix equation in (13) can be expressed as:

$$1 = -K[-\cos\theta\cos\psi + \sin\theta\sin\psi - jZ_1 G\cos\theta\sin\psi]$$
$$+K[Z_1^2 G\sin\theta\sin\psi - jZ_1(\sin\theta\cos\psi + \sin\psi\cos\theta)] \tag{15}$$
$$\cdot(Y_0 + jY_1\cot\theta)$$

By equating the real and imaginary parts of (15) and employing (9), the following results are found:

$$G = Y_0 \tag{16}$$

$$K\cos\theta\cos\psi = 0 \tag{17}$$

From (17), it can be seen that there exist three cases: $\theta = \pi/2$, $\psi - \pi/2$ or $\theta = \psi = \pi/2$. If $\theta = \pi/2$, it can be got from (9) that

$$\psi = \sin^{-1}(K) \tag{18}$$

which is corresponding to the condition that the power dividing ratio of K^2 is less than 1. If $\psi = \pi/2$, it can be obtained from (9) that

$$\theta = \sin^{-1}(\frac{1}{K}) \tag{19}$$

which is corresponding to the case that the power-dividing ratio of K^2 is larger than 1. If $\psi = \pi/2$ and $\theta = \pi/2$, the proposed Gysel PD is equivalent to the equal Gysel PD as discussed in [3].

As the characteristic impedance of Z_2 is a free parameter, which does not affect the power dividing ratio or return loss of the Gysel PD as discussed in [1], the proposed Gysel PD can be realized by transmission lines with uniform characteristic impedance.

3. Simulated and Measured Results

For verification, two microstrip proposed Gysel PDs with operation frequency of 2 GHz have been designed, fabricated on the substrate with relative dielectric constant of 2.65 and a thickness of 1 mm. The system impedance Z_0 is 50 Ω. The first Gysel PD (Example 1) has the power dividing ratio of $K^2 = 2$. Other parameters are $\theta = 45^0$, $\psi = 90^0$, isolation resistors $R_1 = 200$ Ω, $R_2 = 66.7$ Ω (which is realized by shunting a 100 Ω and a 200 Ω chip resistors), the characteristic impedance of transmission lines $Z_1 = Z_2 = 70.7$ Ω. The second Gysel PD (Example 2) has the power dividing ratio of $K^2 = 3$. Other calculated parameters are $\theta = 35.3°$, $\psi = 90^0$, isolation resistors $R_1 = R_2 = 100$ Ω, the characteristic impedance of transmission lines $Z_1 = Z_2 = 70.7$ Ω. The photographs of fabricated microstrip Gysel PDs are shown in Figure 3.

These simulated results are obtained by using lossless transmission lines and ideal resistors. Scattering parameters are measured by using an HP 8257D network analyzer over the frequency range from 1.8 GHz to 2.2 GHz. Figure 4 shows the insertion loss, return loss and isolation performances of Example 1. The measured results indicate that the input return loss is better than 30 dB and the isolation is greater than 25 dB at 2 GHz, as shown in Figure 4(a). The measured insertion losses at 2 GHz are 1.82 dB (The ideal value is 1.76 dB) and 4.97 dB (The ideal value is equal to 4.77 dB), respectively. The measured output return losses are better than 25dB at 2 GHz as shown in Figure 4(c). Figure 5 shows the insertion loss, return loss and isolation performances of Example 2. The measured input return loss is better than 30 dB and the isolation between two output ports is greater than 22 dB at center frequency. The measured insertion losses at 2 GHz for ports 2 and 3 are 1.46 dB (The ideal value is 1.25 dB) and 6.47 dB (The ideal value is 6.02 dB). The measured output return losses are better than 25 dB at 2GHz as shown in Figure 5(c). In general, the measured results are in good agreements with the simulated ones.

(a)

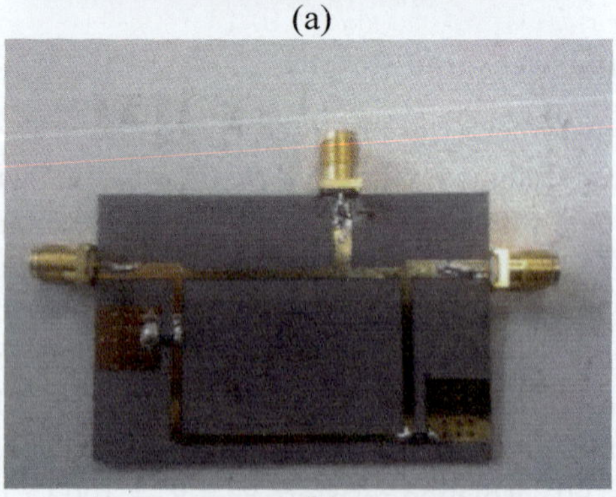

(b)

Fig. 3. Photographs of fabricated Gysel PDs. (a) Example 1. (b) Example 2.

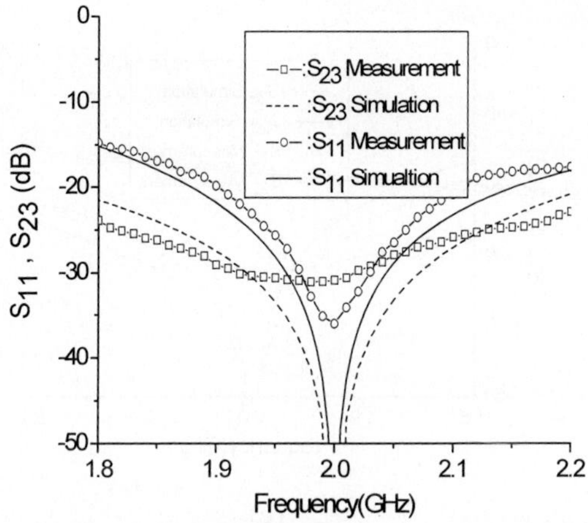

(a)

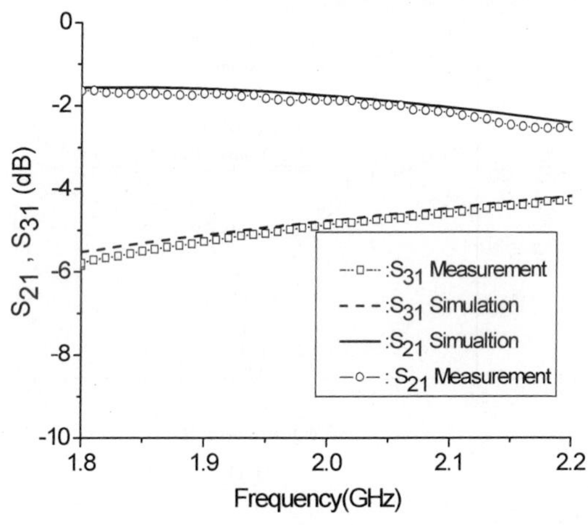

(b)

Fig. 4. Simulated and measured performance of fabricated Example 1. (a) Input return loss and isolation (b) Insertion loss (c) Output return loss.

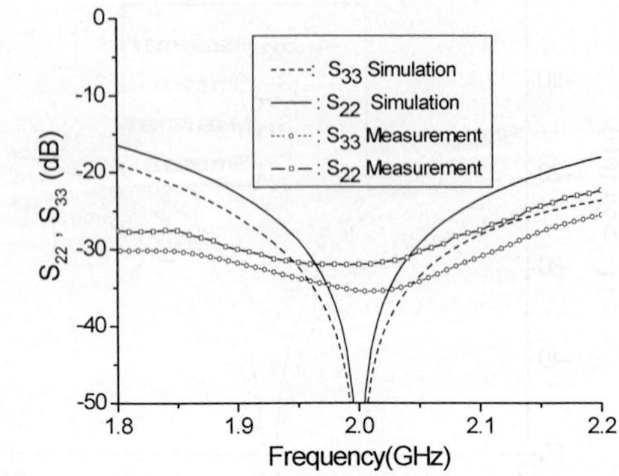

(c)

Fig. 4. (*Con'd*)

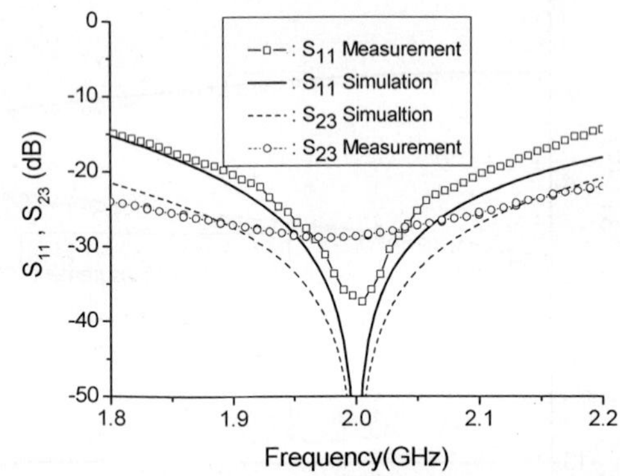

(a)

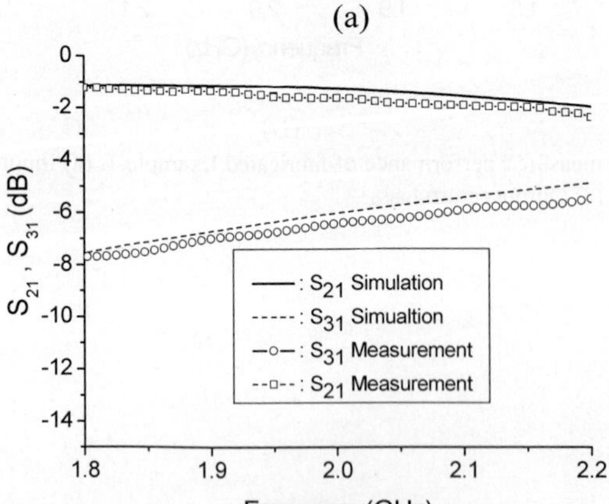

(b)

Figure 5. Simulated and measured performance of fabricated Example 2. (a) Input return loss and isolation (b) Insertion loss (c) Output return loss.

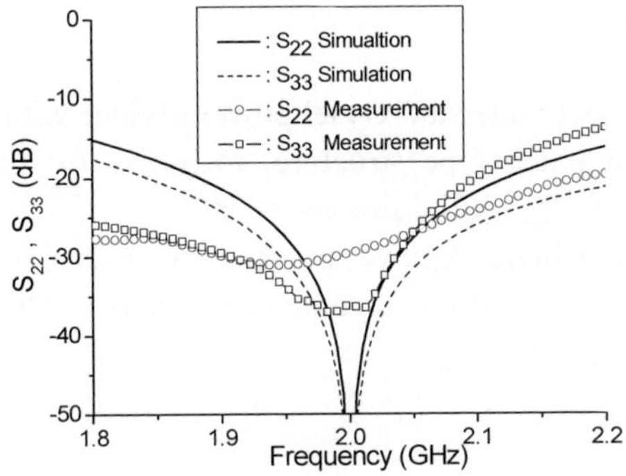

(c)

Fig. 5. (*Con'd*)

4. Conclusion

The circuit topology and design parameters of the proposed Gysel PD has been discussed in this paper. Two experimental circuits have been designed and the measured results have been verified the proposed Gysel PD. As high power dividing ratio of the proposed Gysel PD can be easily realized by using traditional PCB process, the proposed circuit can be widely used in applications that high power dividing ratio is required.

References

1. Yongle Wu, Yuanan Liu, A modified Gysel power divider of arbitrary power ratio and real terminated impedances, *IEEE Microw. Wireless Compon Lett*, 2011(21): 601–603.
2. Z.Y. Sun, L.J. Zhang, Y. Liu, and X. Tong, Modified Gysel power divider for dual-band applications, *IEEE Microw. Wireless Compon. Lett*, 2011(21): 16–18.
3. Z. Y. Sun, L. J. Zhang, and H. Yang, Design of unequal dualband Gysel power divider with arbitrary termination resistance, *IEEE Trans. Microw. Theory Tech*, 2011(59):1955–1962.
4. H. Oraizi, A.-R. Sharifi, Optimum design of a wideband two-way Gysel power divider with source to load impedance matching, *IEEE Trans. Microw. Theory Tech*, 2009 (57): 2238–2248.
5. H. Oraizi, A.-R. Sharifi, Design and optimization of a two-way Gysel power divider with arbitrary power division and impedance matching, *in Proc. Mediterranean Microw Symp (MMS)*, 2009:1–4.
6. Feng Lin, Qing-Xin Chu and Zhi Gong, Compact broadband Gysel power divider with arbitrary Power-Dividing Ratio using microstrip/slotline phase inverter, *IEEE Trans. Microw. Theory Tech*, 2012(60): 1226-1234.

7. H. Zhang, Compact wideband Gysel power divider with arbitrary power division based on patch type structure, *Progress In Electromagnetics Research*, 2011(119):395-406.
8. Kwok-Keung M. Cheng, A novel power-divider Design with unequal power-dividing ratio and simple layout, *IEEE Trans. Microw. Theory Tech*, 2009(57): 1589–1594.

Feature extraction with global-locally preserving projections based on curvelet transform

J. A. Zhao[1]

School of Mechanical Engineering, Southeast University,
Nanjing 211189, PR China
Jiangsu Polytechnic College of Agriculture and Forestry,
Jurong 212400, P.R China
Email: zhaojunai80@163.com

M.P. Jia[2†]

School of Mechanical Engineering, Southeast University,
Nanjing 211189, PR China
†Email: mpjia@seu.edu.cn

A novel feature extraction method with global- locally preserving projections based on multi-scale transform (MGLPP) is proposed. The proposed method consists of the following steps. First, the original images are decomposed using curvelet transform into several elements. Second, the proposed GLPP algorithm is applied to the curvelet coefficients of detailed levels to determine the optimal projection vector W. Finally, the distance between the training samples and testing samples in low-dimensional space is calculated to recognize and classify the target. Moreover, the introduction of degree of belonging functions of the fuzzy c-means (FCM) method to the proposed MGLPP method aims to identify the k neighbors of each sample. The experimental results demonstrate that the MGLPP method performs better in dimensionality reduction and recognition accuracy than the traditional principal component analysis (PCA) and local preserving projection algorithms (LPP).

Keywords: Curvelet transform; Feature extraction; Global-locally feature.

1. Introduction

Efficient and effective feature extraction is an important step in target recognition and classification. The dimensional reduction in feature extraction has always been a difficult research area. Principal component analysis (PCA) [1] is perhaps the most classical tool for linear dimensionality reduction, which can

---

[1] Jun-Ai Zhao
[2] Min-Ping Jia

maintain the global linear structure information of the data set well. However, some drawbacks still exist, such as the inadequacy of PCA for the nonlinear manifold hidden in high-dimensional data space [2]. He et al.(2003) proposed local preserving projection (LPP) to overcome the above limitation of PCA[3,4]. LPP not only provides a novel way to adapt the nonlinear structure but can also better preserve the local dataset structure.

Recent studies [5] show that the recognition accuracy in low-dimensional space is easily affected by noise if the image features were extracted directly from the spatial data by PCA and LPP algorithms. Using a different space by a multi-scale transform (e.g., curvelet transform [6]) can be helpful to decrease the noise sensitivity. Curvelet transform provides a more sparse representation of the image, with the better ability to represent edges and other singularities along curves. Therefore, the curvelet transform is effective in extracting the target features of images. A drawback of curvelet transform is that this algorithm can produce a large number of redundant features. In order to reduce the complexity of the calculation and get higher recognition accuracy, dimension reduction should be carried out to the curvelet transform coefficient.

A novel feature extraction algorithm is proposed in this paper, which is called global–locally preserving projections based on the multi-scale transform (MGLPP). The proposed system adopt curvelet transform, improved PCA and LPP, which not only can preserve the global and local structure information of the dataset but also improve the robustness to noise performance. Furthermore, the introduction of the degree of belonging functions of the fuzzy c-means (FCM) method [7,8] to the proposed MGLPP method aims to identify the *k* neighbors of each sample and improve the recognition accuracy of the target.

2. Global and Local Feature Extraction Based on Multi-Scale Transform

2.1. *Curvelet transform*

The second generation of curvelet transform includes two fast algorithms for its implementation [9], namely, unequally-space fast Fourier transform (USFFT) and wrapping-based transform. USFFT was utilized in feature extraction [10, 11]. A *2D* fast Fourier transform (FFT) of the image was first taken to implement the curvelet transform. The *2D* Fourier frequency plane was then divided into wedges at multiple scales and multiple directions. Finally, an inverse FFT of each wedge was taken to observe the curvelet coefficients at each scale j and angle l. The algorithm is described below.

Using $f[t_1,t_2], 0 \le t_1, t_2 < n$ as input in the Cartesian coordinate, the discrete curvelet transform can be represented as follows:

$$c^D(j,l,k) := \sum_{0 \le t_1, t_2 < n} f[t_1,t_2] \overline{\varphi^D_{j,l,k}[t_1,t_2]} \tag{1}$$

where $\varphi^D_{j,l,k}(\cdot)$ denotes a discrete curvelet function, *j*, *l*, and *k* are the scale, orientation, and position variables, respectively, and *n* is the size of original image.

2.2. *Feature extraction with global-locally algorithm*

The global-locally feature extraction algorithm is combined with PCA algorithm and LPP algorithm. This method attempts to maximize the merits of the PCA and LPP methods, as well as obtain more comprehensive information in the low-dimensional space.

A high-dimensional dataset $X = [x_1, x_2, ..., x_N]$ is defined in Euclidean space R^B, where $x_i (i = 1, 2, ..., N)$ is a *B* dimensional vector. The global objective function of PCA can be written as follows:

$$\begin{aligned} J(W)_g &= \max_W \sum_{i=1}^{n} \| y_i - \bar{y} \|^2 \\ &= \max_W \sum_{i=1}^{n} W^T (x_i - \bar{x})(x_i - \bar{x})^T W \\ &= \max_W W^T C W \end{aligned} \tag{2}$$

where $\bar{y} = (\sum_{i=1}^{N} y_i) / N$, $\bar{x} = (\sum_{i=1}^{N} x_i) / N$, *C* is the covariance matrix of sample *X*, *W* represents the optimal projection vector.

The locally objective function of LPP can be defined as follows:

$$\begin{aligned} J(W)_l &= \min_W \sum_{i,j} (W^T x_i - W^T x_j)^2 S_{i,j} \\ &= \sum_{i,j} W^T x_i D_{ii} x_i^T W - \sum_{i,j} W^T x_i S_{i,j} x_i^T W \\ &= W^T X (D - S) X^T W \\ &= W^T X L X^T W \end{aligned} \tag{3}$$

where *S* is the weight matrix, and which can be defined as follows:

$$S_{i,j}=\begin{cases}\exp(\frac{\left\|x_i-x_j\right\|^2}{t}), & \text{if } x_i (\text{or } x_j) \text{ is among } k \\ & \text{nearest neighbors of } x_j (\text{or } x_i) \\ 0\ , & \text{otherwise}\end{cases} \quad (4)$$

where t is a positive constant, D is a diagonal matrix with $D_{ii}=\sum_j S_{ij}$, and L = D-S is the Laplacian matrix.

Considering the objective function of PCA and LPP algorithms, the objective function of proposed global-locally feature extraction algorithm can be written as follows:

$$\begin{aligned}J(W)_{PCA+LPP} &= \max_W(W^T CW - W^T XLX^T W) \\ &= \max_W(W^T (C - XLX^T)W) \\ &= \max_W W^T \tilde{S}W\end{aligned} \quad (5)$$

where $\tilde{S}=C-XLX^T$. Let eigenvectors $w_1,w_2,...,w_d$ be the solution of $\tilde{S}$, which are ordered according to their eigenvalues $\lambda_1>\lambda_2>\cdots>\lambda_d$. The desired transform matrix $W=[w_1,w_2,...,w_d]$ for preserving both global and local structures of the dataset X can be constructed as follows

$$y_i=W^T x_i,\quad i=1,2,...,N \quad (6)$$

2.3. *Determination of the nearest neighbor number k*

LPP algorithm is sensitive to the selection of the nearest neighbor k. It will produce a large different results of dimensional reduction even if the difference of nearest neighbor is very small, so it needs to be carefully selected. Inspired by the merits of fuzzy c-means (FCM) algorithm, the degree of belonging function is introduced to identify the nearest neighbor of the current sample X_i.

A dataset $X=\left[x_1,x_2,...,x_N\right]$ is defined in Euclidean space R^B, $X_i\in R^B,(i=1,2,...,N)$. The degree of belonging of the sample X_j to X_i $(i=1,2,...,N;\ j=1,2,...,N;i\neq j)$ u_{ij} is defined as follows:

$$u_{ij}=1\Big/\sum_{r=1}^{c}(\frac{d_{ij}}{d_{rj}})^{\frac{2}{b-1}} \quad (7)$$

where u_{ij} is the degree of belonging of the X_j to X_i with the value of between 0 to 1. d_{ij} ($d_{ij} = \|X_j - X_i\|$) denotes the distance between X_j to X_i. d_{rj} is the distance between X_j to X_r $(r = 1, 2, ..., N;\ r \neq j)$. $b(b > 1)$ is the fuzzy weighting exponent, and Ref. [12] shows that the optimal value of b is in domain $[1.5, 2.5]$. To facilitate the calculation, b is set to 2.

For the current sample X_i, if the belonging degree u_{ij} of rest sample $X_j (j = 1, 2, ..., N;\ j \neq i)$ to X_i meet the following requirement, X_j is recognized as the nearest neighbor of X_i.

$$u_{ij} \geq \frac{1}{N-1} \sum_{r=1}^{N} u_{ir} \tag{8}$$

According to (8), the nearest neighbor and nearest neighbor number k can be determined.

The execution steps of the proposed feature extraction method based on curvelet transform are presented as follows:

Step 1: The curvelet transform is applied to decompose the image ($N \times N$) into ($\log_2 N - 3$) levels.

Step 2: The coefficients of the detail levels produced by step (1) are reconstructed, and which can be taken as the high-dimensional space array;

Step 3: Calculating the nearest neighbor and nearest neighbor number k of the current sample X_i according to Equation 7-8;

Step 4: The weight matrix S, diagonal matrix D and Laplacian matrix S in LPP method are calculated using Equation 4;

Step 5: Computing the covariance matrix C of the high-dimensional space array in step (2), and determining the optimal projection vector W using Equation 5;

Step 6: The training and testing samples decomposed by curvelet transform are mapped to the low-dimensional space according to Equation 6, and then the feature vectors of the original images in low-dimensional space are obtained;

Step 7: The distance of the feature vectors received by Step (6) is computed for recognition and classification.

3. Experiments and Analysis

The crack defect images, welding defect images in engineering and the binary images were used in the dimensional reduction test to illustrate the validity of the proposed approach MGLPP based on the curvelet transform. The results of dimensional reduction and recognition accuracy yielded by the proposed MGLPP method were compared with those obtained by four other feature extraction methods: PCA, LPP, PCA+LPP, and PCA+LLE methods.

The experiment images consists of 40 different images, 4 for each of the 10 distinct subjects. Six images from each subject were selected randomly to comprise the training set (24 images), and the remainder made up the testing set (16 images). The size of each image was 256×256. Several sample parts are shown in Figure 1.

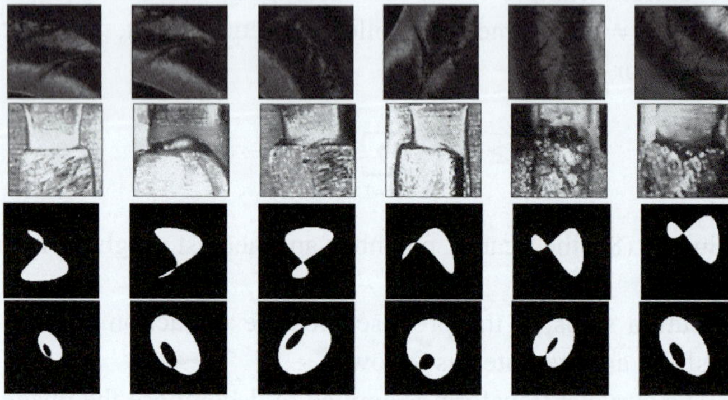

Fig. 1. Examples of experiment images.

The recognition accuracy under different dimensions of the proposed MGLPP based on the curvelet transform approach compared with PCA, LPP, PCA+LPP, and PCA+LLE is shown in Figure 2.

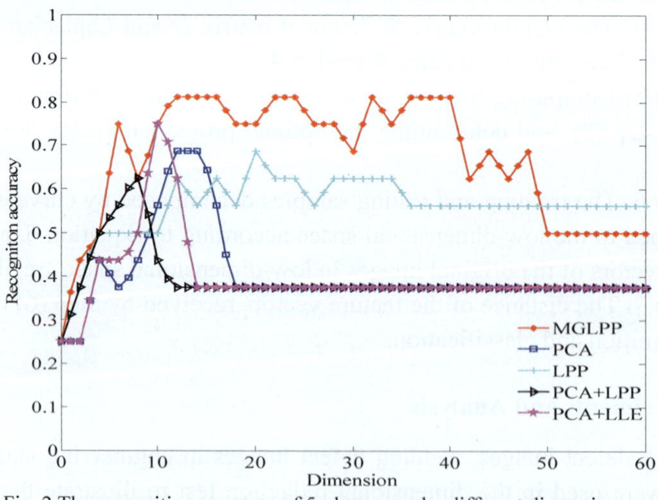

Fig. 2 The recognition accuracy of 5schemes under different dimension.

Figure 2 shows that the proposed MGLPP method performs best compared with the PCA, LPP, PCA+LPP, and PCA+LLE. The reason behind the low recognition of PCA and LPP is that they can only extract global or local information from training samples. Although the PCA+LPP and PCA+LLE algorithms consider the local and global structure of the dataset, they are more sensitive to the nearest *k* neighbors. Therefore, the recognition accuracy of PCA+LPP and PCA+LLE is low. The proposed dimensional reduction approach is the basis for the curvelet decomposition to the original image, which can maintain the anisotropy of the local information and focus on the overall structure. The proposed MGLPP can achieve satisfactory results for the feature extraction on the binary image.

In addition, the number of training samples is also one of the factors that affect recognition accuracy. We selected 3, 5, and 6 images of each subject as the training set and the original 16 images for testing in the experiment. Table 1 shows the maximum recognition rates of different methods and corresponding dimensions.

Table 1. Maximal recognition accuracy (%) and corresponding reduced dimension (shown in parentheses)

Method	PCA	LPP	PCA+LPP	PCA+LLE	GKLPE
12 Training samples	43.75 (14)	50.00 (10)	50.00(12)	43.75(16)	56.25(16)
20 Training samples	50.00 (12)	56.25 (13)	25.00(14)	37.50 (10)	68.75(18)
24 Training samples	68.75 (12)	68.75 (20)	62.50(8)	75.00(12)	81.25(14)

Table 1 shows that recognition accuracy is related to the number of training samples. The proposed MGLPP based on curvelet transform outperforms the other four methods under the same condition, which further demonstrates the effectiveness in feature extraction and recognition.

4. Conclusions

A novel feature extraction method named global-local preserving projections based on curvelet transform is proposed. The curvelet transform is applied to decompose the original image to describe the target comprehensively. A feature reduction method based on the decomposed image (using the curvelet transform) is represented, which attempts to preserve both global and local structures of the dataset in the lower dimensional space. Moreover, the degree of belonging function is adopted to determine the *k* nearest neighbors of each sample. Experimental results show that the proposed MGLPP method based on curevelet transform performed better in feature extraction and recognition accuracy.

5. Conflict of Interests

The authors declare that there is no conflict of interests regarding the publication of this paper.

Acknowledgements

The authors would like to thank the National Natural Science Foundation of China (No. 51075070), and Jiangsu Polytechnic College of Agriculture and Forestry Research Foundation of China (No. 28013011) .

References

1. Kirby, M. & Sirovich, L. 1990. Application of the Karhunen-Loeve procedure for the characterization of human faces [J]. *IEEE Transactions on Pattern Analysis and Machine Intelligence* 12(1):103~108.
2. Richard, O.D., Peer, E.H. & David, G.S. 2001. *Pattern Recognition* [M]. 2nd ed. London: Wiley Interscience.
3. He, X.F. & Niyogi, P. 2003. Locality preserving projections [C]. Proceedings of 16th conference on Neural Information Processing Systems.
4. He, X.F. 2005. *Locality preserving projections* [D]. Illinois, USA: The University of Chicago, Department of Computer Science.
5. Zhang, Y., Han, J. & Yue, J. *et al.* 2014. Weighted KPCA degree of homogeneity amended nonclassical receptive field inhibition model for salient contour extraction in low-light-level image [J]. *IEEE transactions on image processing* 23(6): 2732~2743.
6. Candes, E.J.& Donoho, D.L. 2000. Curvelets, multiresolution representation, and scaling laws [J]. *Wavelet Applications in Signal and Image Processing* VIII , 4119:1~12.
7. Bezdek, J.C. 1981. Pattern recognition with fuzzy objective function algorithms. New York [M]: Plenum Press.
8. Ghosh, A., Mishra, N.S. & Ghosh, V. 2011. Fuzzy clustering algorithms for unsupervised change detection in remote sensing images [J]. *Information Sciences* 181:699~715.
9. Candes, E., Demanet, L. & Donoho, D. *et.al.* 2005. *Fast discrete curvelet transform* [M]. California: California Institute of Technology: 1~44.
10. Xu, K., Ai, Y.H. & Wu, X.Y. 2013. Application of multi-scale feature extraction to surface defect classification of hot-rolled steels [J]. *International Journal of Minerals, Metallurgy and Materials* 20(1):37~41.
11. Mandal, T., Wu, Q.M.J. & Yuan, Y. 2009. Curvelet based face recognition via dimension reduction [J]. *Signal Processing* 89:2345~2353.

12. Gao, X.B. Pei, J.H. & Xie, V. 2000. A study of weighting exponent *m* in a fuzzy c-means algorithm [J]. *Acta Electronica Sinica* 28(4):80~83.

Ship-to-air missile system assignment

X. B. Yang[1]* and Y. H. Lu[2†]

667#, Jiefang Road, Zhongshan District, Dalian City, China
**Emails: doujihua@163.com,yangxingbao@163.com,*
†Emails:luyongh@163.com,Yxb201ds@163.com

Aiming at the issue of constrained optimization problem of assignment for ship-to-air missile system, hybrid genetic algorithm has been proposed in this paper, which can get global optimization results of the assignment, and provide decision support for the assignment.

Keywords: Ship-to-air; Missile system; Algorithm.

1. Introduction

Suppose that ship-to-air missile system uses semi-active radar homing, the-assignment for the system is to assign the illumination radars of the system to the arriving targets. The ship-to-air missile fire effect is affected by the solution quality of theassignment scheme.

The assignment is a typical integer programming problem and belongs to 0-1 Programming problem. Genetic algorithm a searching arithmetic of global optimization (Mitchell Melanie, 1999)[1][2], based on principles inspired by natural genetics "survival of the fittest, excellent win and inferior discard", can be used to solve the assignment. Previous work used standard genetic algorithm to solve the optimization problem of the assignment for the system (Youtang Tong, 2009)[3][4]. However, standard genetic algorithm is of insufficient local optimization, in contrast to hill climbing (Kenekayoro Patrick, 2012)[5], and the two methods can be combined with each other to form hybrid genetic algorithm that can be proposed to get global optimization results.

---

[1] Xing-Bao Yang
[2] Yong-Hong Lu

2. Assignment Model

It is supposed that there are N arriving targets, and the system has M illumination radars, M≥1. Ensuring maximal targets to meet the illumination radar irradiation conditions is considered as optimization objective[6], and then supposing that target number which meet the illumination radar irradiation conditions is defined by f whose maximization is the key of the assignment problem.

$$\max f = \sum_{i=1}^{M}\sum_{j=1}^{N} R_j X_{ij} \tag{1}$$

$$g_j = \sum_{i=1}^{M} X_{ij} - b_j \le 0 \tag{2}$$

$$g_{N+i} = \sum_{j=1}^{N} X_{ij} - 1 \le 0 \tag{3}$$

Where R_j is the j-th target threat value, X_{ij} is a Boolean value used to determine whether the i-th illumination radar is assigned to the j-th target. If the i-th illumination radar is assigned to the j-th target, then $X_{ij}=1$, otherwise $X_{ij}=0$. Eq. (2) represents that the number of illumination radar assigned to the j-th target doesn't exceed b_j. Eq. (3) represents that the i-th illumination radar is assigned to at most one target [7].

The constraints of the assignment model mainly include commander's decision and the illumination radar coverage.

2.1. *Decision constraint*

Because of the complex campaign circumstances, the achieved data is impossible to be complete, determinate and reliable. Commander's decision is necessary, namely, the commander should have definite interference on the assignment scheme. If the j-th target is of very importance, the R_j value of the j-th target given by the commander's decision can have significant impact on the assignment scheme.

2.2. *Illumination radar coverage constraint*

For single illumination radar (let A radar), only when the position of a single target is within the A radar coverage, can the target be illuminated by the A radar.

Let the time for the first target to reach the long-range limit of the A radar coverage be $T_Y(A,1)$, and the time for the first target to stay within the A radar coverage be $T_D(A,1)$, and the time for the first target to be initially illuminated by the A radar be $T_L(A,1)$. The first target can be effectively illuminated by the A radar by satisfying Eq. (4).

$$T_Y(A,1) \le T_L(A,1) \le T_Y(A,1) + T_D(A,1) \quad (4)$$

After the completion of the illumination task of the first target, the second target is illuminated by the A radar which requires a transfer time to initialize work. Let the minimum transfer time for the A radar be T_Z, and then the second target can be effectively illuminated by the A radar by satisfying Eq. (5) and Eq. (6).

$$T_Y(A,2) \le T_L(A,2) \le T_Y(A,2) + T_D(A,2) \quad (5)$$

$$T_Z \le T_L(A,2) - T_L(A,1) \quad (6)$$

Where $T_Y(A,2)$ is the time for the second target to reach the long-range limit of the A radar coverage, $T_D(A,2)$ is the time for the second target to stay within the A radar coverage, $T_L(A,2)$ is the time for the second target to be initially illuminated by the A radar.

3. Genetic Algorithm Based on Adaptive Penalty Function

3.1. *Fitness function design*

Theassignment model is a constrained optimization problem which can't be directly solved by genetic algorithm. By introducing the penalty function, the constrained optimization problem is converted into an unconstrained optimization problem which can be solved by genetic algorithm.

This paper uses adaptive penalty function (Qiongsun Liu etc, 2006), and the penalty factor is a function of the independent variable X_{ij}. Meanwhile, in order to accelerate the convergence speed, this paper draws a "multi-level punishment" thinking which gives bigger constraint violation severer punishment, and smaller constraint violation lesser punishment. The fitness function can be expressed as in Eq. (7).

$$eval = f + C \cdot p \quad (7)$$

$$p = \sum_{k=1}^{M+N} Q_k \cdot \max\{0, g_k\} \quad (8)$$

$$Q_k = \frac{\max\{0, g_k\}}{\sum_{l=1}^{M+N} \max\{0, g_l\}} \tag{9}$$

$$C = 1 + \frac{|f|}{1+p} \tag{10}$$

where p is penalty function, C is the penalty function coefficient.

3.2. *Genetic algorithm design*

Assignment method for ship-to-air missile system based on hybrid genetic algorithm is depicted in Fig.1. The method generates the better individual with the complete global search when the termination condition is met. Then through the use of hill climbing, the method continues to complete the local search process, and best individual is ultimately got.

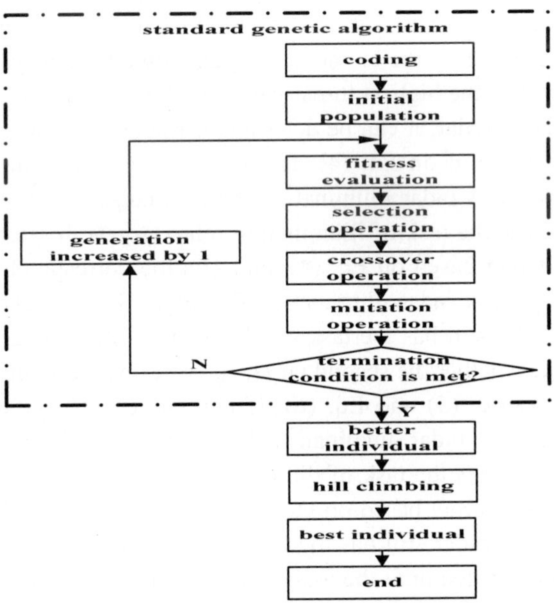

Figure 1. Genetic algorithm design.

3.3. *Standard genetic algorithm design*

Binary encoding. Binary coding handles each chromosome as cord of bits, either 1 or 0. The length of individual chromosome is $M \times N$. The chromosome can be expressed as $A_1^{k_h} A_2^{k_h} \ldots A_M^{k_h}$, $A_i^{k_h} = (X_{i1}^{k_h} X_{i2}^{k_h} \ldots X_{iN}^{k_h})$. $X_{ij}^{k_h}$ is string

k_h of the h-th generation, which represents the i-th illumination radar assigned to the j-th target.

Initial population. Standard genetic algorithm generates initial population by the random method, and each chromosome of initial population might or might not satisfy the constraints defined in Eq. (2) and Eq. (3). The following method can automatically generate initial population which satisfies the constraints. The steps are as follows: first of all, set '0' to all genes in all chromosomes in the population, then each chromosome is transferred to a matrix ranked $M \times N$. In each column, one or two genes are randomly set '1' so that the matrix satisfies the constraint defined in Eq. (3). Next, in each row, several genes are randomly set '1' so that the matrix satisfies the constraint defined in Eq. (2), and the third step is stop.

Fitness evaluation. Each individual represents a assignment scheme, which should meet the illumination radar coverage constraint. This paper implements fitness evaluation based on Eq. (7). The A radar coverage constraint test method is as follows. Based on the sequence of the arriving targets into the A radar coverage, the method determines the A radar illumination sequence for the arriving targets. Which target arrives first, the target is on the first illumination. When two or more targets enter the A radar coverage simultaneously, the priority is given to the target of the highest threat value for the A radar. For the primary illumination of the A radar, it can be determined whether the first target is effectively illuminated based on Eq. (4). If not so, the corresponding gene in the chromosome for the A radar illuminating the first target is set to be '0'. When the A radar still has the task of illuminating other targets, it selects another target as the first target based on Eq. (4), and sets the corresponding gene in the chromosome for the A radar illuminating the target. If so, it should determine whether the A radar still has the task of illuminating the second target. If so, it should determine whether the second target can be effectively illuminated by the A radar based on Eq. (5) and Eq. (6), If not, the corresponding gene in the chromosome for the A radar illuminating the second target is set to be '0', when the the A radar still has the task of illuminating other targets, it selects another target as the second target based on the above procedure. If so, it should determine whether the A radar still has the task of illuminating the third target. The above process is repeated until the assignment scheme satisfies the A radar coverage constraint.

Selection operation. The selection operation is roulette wheel selection in the standard genetic algorithm.This paper adopts the selection strategy of associative proportional selection and distillate model, which selects the individual of maximal fitness value in the H individuals of each generation directly to be the individual of the next generation, while the other individuals of the next generation are engendered by the remaining H-1 individuals of the former generation through the roulette wheel selection method. This selection strategy, on the one hand, can guarantee that the most excellent individual survives to the next generation, on the other hand, it can have the avoidance that the chance to select the next generation is in great disparity because of each individual's different fitness value, which guarantees the diversity of population to a certain extent.

Crossover operation. Selecting two individuals randomly in the population as parent individuals. Suppose that they are chromosome k_{h_1} (which is defined by $A_1^{k_{h_1}} A_2^{k_{h_1}} \dots A_M^{k_{h_1}}$) and chromosome k_{h_2} (which is defined by $A_1^{k_{h_2}} A_2^{k_{h_2}} \dots A_M^{k_{h_2}}$) of the h-th generation. Interchanging $A_1^{k_{h_1}}$ and $A_M^{k_{h_2}}$, interchanging $A_2^{k_{h_1}}$ and $A_{M-1}^{k_{h_2}}$, interchanging $A_3^{k_{h_1}}$ and $A_{M-2}^{k_{h_2}}$, and so on, eventually two new individuals are generated.

Mutation operation. For the chromosome k_h of the h-th generation, $A_{d1}^{k_h}$ and $A_{d2}^{k_h}$ are randomly selected, then interchanged them to form two new chromosomes, $A_{e1}^{k_h}$ and $A_{e2}^{k_h}$. d1, d2, e1, e2=1,2,…,M, the value of d1, d2, e1, and e2 are different. Then the fitness values of the two newly created individuals are compared with the two parent individuals, if the same values, invalid mutation operation is considered to remove the two newly created individuals, and mutation operation is performed according to the above procedure. Otherwise, valid mutation operation is considered, and it substitutes the two newly created individuals for the two parent individuals.

Adaptive probability of crossover and mutation. In order to improve the calculation speed and efficiency of genetic algorithms, the probability of crossover (defined by P_c) and probability of mutation (defined by P_m) can be dynamically adjusted. This paper adjusts the two values by the following equations (M. Srinivas etc, 1994).

$$P_c = \begin{cases} k_1(eval_{\max} - eval_{big}) / (eval_{\max} - eval_{avg}) & eval_{big} \geq eval_{avg} \\ k_3 & eval_{big} < eval_{avg} \end{cases} \tag{11}$$

$$P_m = \begin{cases} k_2(eval_{\max} - eval_g) / (eval_{\max} - eval_{avg}) & eval_g \geq eval_{avg} \\ k_4 & eval_g < eval_{avg} \end{cases} \tag{12}$$

where $eval_g$ is the fitness value of the g-th individual, $eval_{big}$ is the bigger one of two crossover individuals, $eval_{\max}$ and $eval_{avg}$ are the maximum and average fitness value of the current population respectively. g=1,2,…,H. Usually $k_1 = k_3 = 1.0$, $k_2 = k_4 = 0.5$ are taken.

Algorithm termination condition. Let ε be a small positive number, when the absolute difference between the fitness function values of two neighboring generations is smaller than ε, the standard genetic algorithm terminates.

3.4. *Hill climbing design*

Bit hill climbing is adopted because binary coding handles each chromosome, whose bit variation style is a variation of the order from left to right (Xiaozhen Yan etc, 2007). Let the single best individual from the above standard genetic algorithm be k_y with the length $M \times N$. The search process of the bit hill climbing is as follows.

Step one. Suppose the mutation of the d-th bit. The initial value of d is 1.

Step two. The mutation of the d-th bit of k_y inverts the bit value from '0' to '1', or '1' to '0', to obtain new individual (defined by k_x).

Step three. Based on the above fitness evaluation method, the fitness values of k_x (defined by $eval_x$), and k_y (defined by $eval_y$) can be calculated.

Step four. If $eval_x > eval_y$, k_y is replaced by k_x. Otherwise, it is back to step two, and d=d+1. So it is constantly repeated until d= $M \times N$, and the best individual can be obtained.

4. Conclusions

The proposed hybrid genetic algorithm based on adaptive penalty function can solve constrained optimization problem of assignment for ship-to-air missile system. The constrained optimization problem is converted into the unconstrained one through the introduction of adaptive penalty function. Afterwords, hybrid genetic algorithm is proposed by combining standard genetic algorithm with hill climbing to solve the unconstrained optimization problem, and to get global optimization results of the assignment, which helps the assignment decision for the system.

References

1. Kenekayoro Patrick. "Comparison of simulated annealing and hill climbing in the course timetabling problem." African Journal of Mathematics and Computer Science Research, Vol. 5 (11): 176-178, 2012.
2. Mitchell Melanie. An Introduction to Genetic Algorithms. Cambridge: MIT Press, 1999.
3. Youtang Tong. "Research on the Model and Application of Ship-to-air Missile Command Decision Making." PhD diss., Dalian University of Technology, China, 2009.
4. M. Srinivas and L.M. Patnaik. "Adaptive probabilities of crossover and mutation in genetic algorithms." IEEE Trans. Systems, Man Cybernet. Vol. 24 (4): 656–667, 1994.
5. Yang Zhang. "Multi-target Air Defense Fire Distribution Research." Master diss., Dalian University, China, 2012.

6. Qiongsun Liu, Shenghua Zhou. "Hybrid Genetic Algorithm Based on Novel Adaptive Penalty Function." Journal of Chongqing University, Vol. 29 (6): 78-81, 2006.
7. Xiaozhen Yan, Hong Xie. Yao Liu. "Mixed multiuser detection based on genetic algorithm and bit hill climbing algorithm." Applied Science and Technology, Vol. 34 (8): 29-33, 2007.

An improved echo cancellation algorithm for hearing aids

Y.Q. Feng[1a], G.C. Tang[b], R.Y. Liang[c] and Q.Y. Wang[d]
School of Communication Engineering,Nanjing Institute of Technology, Jiangsu Nanjing, China
Email: [a]*fengyueqin@njit.edu.cn*
Emails: [b]*lly1711@163.com,* [c]*tguichen@163.com,* [d]*wangqingyun@vip.163.com*

To improve the performance of the echo cancellation for hearing aids, one double step normalized least mean square - notch filter (DSNLMS-NF) algorithm is proposed. When the echo path is relatively stable, the echo signal is estimated by using the normalized least mean square (NLMS) algorithm and subtracted it from the input signal. When howls occur, the DSNLMS-NF algorithm decreases the step to stabilize the howling frequency and the notch filter with this frequency is dynamically generated based on FFT algorithm to restrain howls. Compared with other echo cancellation algorithms for the hearing aid, the comprehensive performance of DSNLMS-NF is the best. Furthermore, the quality of speech generated by the algorithm is high and the operation time is shorter, which is suitable for the low-power consumption and small volume products such as hearing aids.

Keywords: Hearing aid; Echo cancellation; Notch filter; Normalized least mean square.

1. Introduction

When the output of a hearing aid leaks out through the ear canal, porosity and the edge of the shell, and is captured by the microphone, the acoustic echo will occur. When the gain of the hearing aid is greater than the channel attenuation, howls or rings will be generated. It will damage the hearing aid and even the patient's residual hearing. In the echo cancellation algorithms for hearing aids, the earlier idea is to decrease the gain in the forward path and to increase the attenuation in the feedback path, which does not have a good effect due to the conflict between echoes cancellation and occlusion effect [1]. Currently, there are mainly three echo cancellation algorithms in hearing aids [2], including gain attenuation, notch filter and adaptive filter [3]. The gain attenuation method is to reduce the gain of the howling channel, and the degree of gain reduction is determined by the amplitude of the echo signal. Although this method has the lowest power consumption, it reduces the gain of the desired signal and affect the output gain of hearing aids. Initially, the notch filter method is designed for the echoes caused by static factors. However, there exists the frequency hopping when the numbers of the howling frequencies are more than those of notch filters. The adaptive filter method [4-6] obtains the frequency response of the feedback path by adjusting the parameters of the filter, and then eliminates echoes by subtracting the output of filter from the output of the microphone.

---

[a] Yue-Qin Feng

Although the current method reduces echoes by iterative signal processing, the iterative coefficient is usually not large for improving the speech quality, which results in the slower convergence when the echo path changes rapidly. So, it is very hard to ensure the rapid inhibition of howls.

To solve the previous problems, a double-step normalized least mean square-notch filter (DSNLMS-NF) algorithm is proposed. The algorithm improves the adaptive filter response by monitoring the signal channel to estimate the echo signal. Meanwhile, the howling detector keeps watching the howls, estimating the howling frequencies and generating notch filters to reduce the howling signals. Furthermore, a double-step normalized least mean square (NLMS) algorithm is introduced, that is, different steps are used when the system howls or not. Compared with different echo cancellation algorithms for hearing aids, the proposed algorithm is the best in comprehensive performance.

2. Echo Cancellation Algorithm for Hearing Aids

Fig. 1 shows the echo cancellation model for hearing aids. The gain of the forward path is G, which amplifies the input signal to compensate the hearing of patients. h_k^* is the actual feedback path received by the microphone. h_k is the estimated feedback path whose parameters are adaptively generated from the echo estimation algorithm. s_k is the external speech signal which is super posed with the feedback signal f_k to form the input signal of the microphone e_k is the residual signal and is obtained by subtracting the estimated feedback signal from the input signal of the microphone. It is the real input of hearing aids and has the same statistical characteristic with s_k. d_k is picked by the microphone and $\hat{f}_k$ is the estimated feedback signal generated by the echo cancellation algorithm. When the path suddenly changes, the howling detector and notch filter are used to detect the howls and generate notch filters to reduce them.

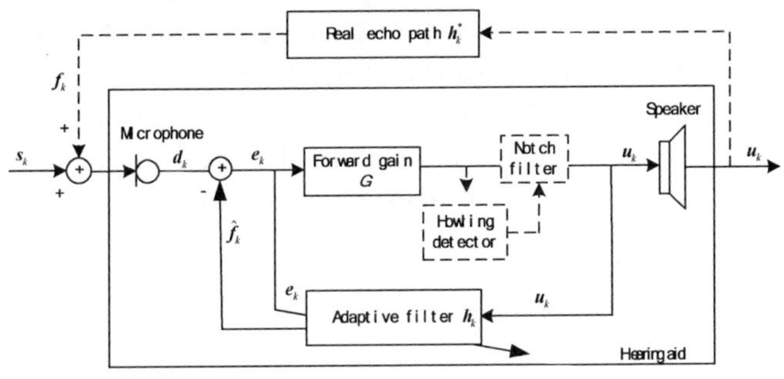

Fig. 1. Echo cancellation model for hearing aids.

The NLMS algorithm is a classic algorithm used in the hearing aids or other echo cancellation systems because of its lower computation complexity and better misalignment performance. Suppose dh_k is the coefficient vector of the adaptive filter H(z) at every iterative. The iteration process of $\hat{h}_k$ is

$$\hat{h}_{k+1} = \hat{h}_k + \frac{\mu}{\|u_k\|^2} u_k e_k \tag{1}$$

where μ is the step factor. u_k is the error signal and e_k is the output of the hearing aid.

When the step factor μ of the NLMS algorithm is small, there will be long convergence time, as well as good speech quality and small critical gain. Otherwise, the convergence will be fast but the speech quality will also be affected. Furthermore, when the system howls, coefficients of the adaptive filter will constantly update to change the system transfer function, which will lead to deviations of the howling frequencies. So the center frequency of the notch filter which is built on the previous frame will not be able to match the howling frequency of the next frame. In order to improve the performance of DSNLMS-NF algorithm, an improved strategy is introduced. When the system howls, a small step factor is used to prevent the deviation of the howling frequency to guarantee the efficiency of the notch filter. When there is no howling, a large step factor is used to reduce the possibility of the howling. The NLMS algorithm can realize echo cancellation and remain good speech quality, but its response speed needs improvement in a rapidly changing environment. So, an improved notch filter is proposed to improve the performance of the NLMS algorithm in order to restrain the howling generated from the sudden changes of the environment.

Here, the process of howling detection[7] is: the signal is firstly transformed using the FFT. Then, the ratio of the maximum value to the mean value in the amplitude-frequency curve is calculated. When the ratio is larger than a threshold value, the howling emerges. Finally, the howling frequency is calculated and the notch filter is designed based on this frequency. The frequency resolution of the normal FFT algorithm is f_s/N. Here, f_s is the sampling frequency, and N is the count of points in FFT. After determining the howling frequency, the notch filter can be used to restrain the howling. The notch filter is a special band-stop filter. In the ideal case, the notch filter suppresses only one frequency. The transfer function of second-order notch filter is:

$$H(z) = \frac{(z-z_1)(z-z_1^*)}{(z-rz_1)(z-rz_1^*)} (0 < r < 1) \tag{2}$$

Assume that the stop band frequency of notch filter is f0, then w0=2*pi*f0/fs, the zero point is $z_1 = e^{\pm j^* w^o}$ and the pole is $r^* e^{\pm j^* w^o}$. The larger the value r is, the deeper and narrower the frequency response curve of the notch filter will be. Substitute z1 into the equation 2 and simplify it.

$$H(z) = \frac{z^2 - 2cosw0^* z + 1}{z^2 - 2r^* cosw0^* z + r^2} \tag{3}$$

If more than one frequency point needs suppressing, multiple notch filters based on different frequency points can be designed respectively according to the equation 10, and then be connected in series. Considering the performance and the efficiency of the algorithm, a double-level notch filter is designed to restrain two howling frequencies at the same time.

3. Experiment Settings and Simulation

To verify the performance of the algorithm, the proposed DSNLMS-NF algorithm is compared with normal NLMS, WAPSM [8] and VLMS [9]. The performance of echo cancellation algorithms for hearing aids are evaluated in subjective [10] and objective test. Besides, the running time of different algorithms is measured to estimate their practicability (based on the MATLAB function etime).

The subjective test is the listening or understanding test for speech quality [10], and the participants could be hearing-impaired patients or people with normal hearing. In order to make the results more objective, only objective measures are used to evaluate the speech quality. The measures of speech quality is segment signal-to-noise ratio (SegSNR) and perceptual evaluation of speech quality (PESQ) [11]. The objective measures of echo cancellation for hearing aids [8] is the misalignment coefficient of adaptive filter(Misalignment), which is defined as:

$$mis = \frac{\|h_k - \hat{h}_k\|^2}{\|h_k\|^2} \tag{4}$$

4. Acoustic Environment

Two acoustic feedback paths are used in this experiment. The path g1 is shown in Fig.2 (a), which is expressed as follows:

$$g1(n) = \begin{cases} 0 & 0 \le n \le 10 \\ b(n)e^{-0.5n} & 11 \le n \le 200 \end{cases} \tag{5}$$

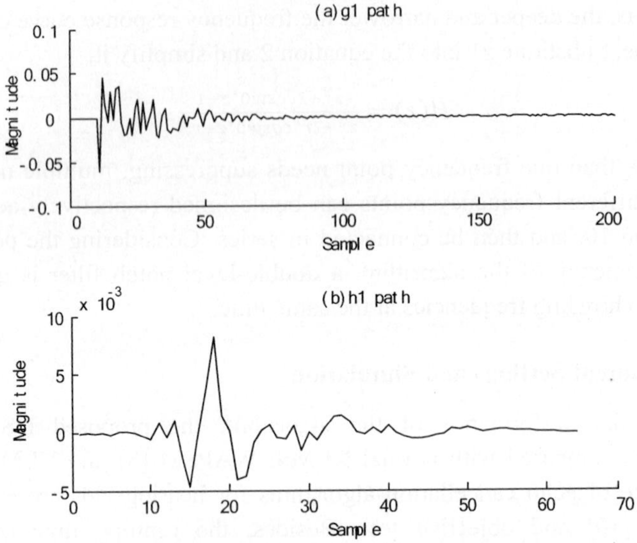

Fig. 2. The feedback path of g1.

where is the Gaussian white noise with mean value of 0 and variance of 0.5. The feedback path is chosen because it is very similar to the real measured feedback path of hearing aids in which the signal leaks from the speaker, goes through the ear canal and ear muff and is captured by the microphone [5]. The path h1, which is shown in Fig. 2(b), is measured in lab and provided by Kim Ngo [3]. The test signals in the experiment are all from the TIMIT speech database. The signals' sampling frequency is transformed into 8 kHz. The types of tested signals include speech signals and non-speech ones (white noise). The selected speech signal is composed of different genders and languages, including mixed language speech, Chinese speech of both genders and English speech of both genders.

5. Comparisons of Echo Performance

The echo cancellation performance of different algorithms is compared in this experiment. The experimental speech is the kdt_001 file in the TIMIT speech database with the content "She had your dark suit in greasy wash water all year". In order to highlight the performance, the gain of the real feedback path is doubled at the sampling point of 5000. In the experiment, the gain of the forward path is $G = 14dB$. The step of NLMS isμ=5×10-3. For WAPSM, γ=0.99 is used. In VLMS, μmax=5×10-3, μmax=5×10-4and λ=0.992 are used. In DSNLMS-NF, μ1=5×10-3 and μ2=5×10-4 are used, and the threshold of howling energy is 10. Results are shown in Fig. 3.

From the misalignment, there is the obvious inflection at the point of 5000 in Fig. 3 (a) and (b) because the echo path changes at that location. From the comparison of the four algorithms, NLMS, VLMS and DSNLMS-NF have better echo cancellation performance and smaller misalignment. And there is less fluctuation in the misalignment curve except the howling point. Comparing the two paths h1 and g1, since no howling occurs in h1, VLMS has better stability and smaller misalignment. NLMS and DSNLMS-NF converges slowly, while they still have the same misalignment as VLMS at the end. In the path g1 where howling exists, the misalignment of VLMS is obviously larger than NLMS and DSNLMS-NF. When howling occurs, the misalignment of DSNLMS-NF drops to a smaller value. For WAPSM, there is huge fluctuation in the misalignment curve, and the overall performance is worse than others three algorithms.

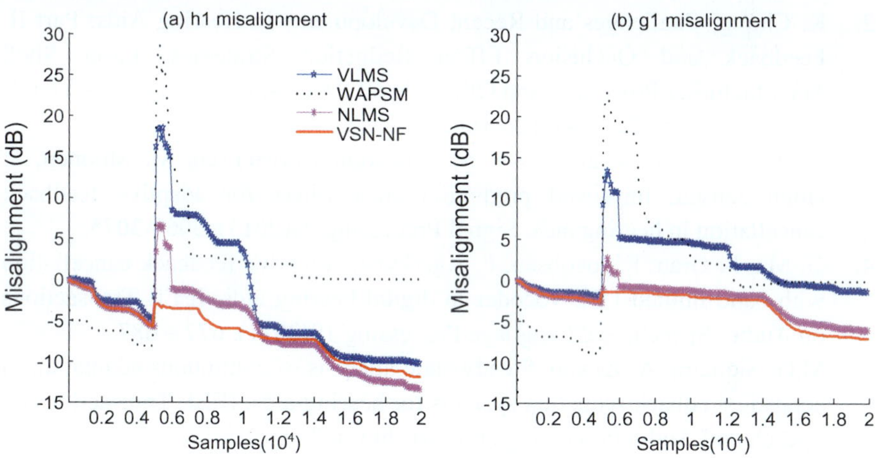

Fig. 3. Comparisons of echo cancellation performance.

6. Conclusion

To improve the performance of echo cancellation for hearing aids, an echo cancellation algorithm with dual-step NLMS filter combined with notch filter is proposed. Compared with NLMS, the algorithm inserts the notch filters to improve the echo cancellation performance when the environment changes suddenly, and to enhance its robustness. And compared with the single notch filter, DSNLMS-NF uses NLMS to estimate the echo path continuously in order to reduce the howling times and the numbers of notch filters, which will benefit the speech quality and increase the signal gain. So, from the comprehensive comparison of performance, DSNLMS-NF is an echo cancellation algorithm which can be used in the low-power devices, such as hearing aids.

Acknowledgment

The work was supported by the National Natural Science Foundation of China under Grant No. 61301219, No. 61375028 and No. 61301295, the Natural Science Foundation of Jiangsu Province under Grant No. BK20130241 and Nanjing Institute of Technology under Grant No. JG201425. The authors would like to thank the reviewers for their valuable suggestions and comments.

References

1. J. Agnew, Acoustic feedback and other audible artifacts in hearing aids, Trends in Amplification 1(1996), 45-82.
2. K. Chung, Challenges and Recent Developments in Hearing Aids: Part II. Feedback and Occlusion Effect Reduction Strategies, Laser Shell Manufacturing Processes, and Other Signal Processing Technologies Trends in Amplification 8(2004), 125-164.
3. K. Ngo, T. Van Waterschoot, M. Græsbøll Christensen, M. Moonen, S. Holdt Jensen, Improved prediction error filters for adaptive feedback cancellation in hearing aids, Signal Processing 93(2013), 3062-3075.
4. G. Ma, F. Gran, F. Jocobsen, F. Agerkvist, Adaptive feedback cancellation with band-limited LPC vocoder in digital hearing aids, IEEE Transactions on Audio, Speech, and Language Processing 19(2011), 677 – 687.
5. M.G. Siqueira, A. Alwan, Steady-state analysis of continuous adaptation in acoustic feedback reduction systems for hearing-aids, IEEE Transactions on Speech and Audio Processing 8(2000), 443-453.
6. C.R.C. Nakagawa, S. Nordholm, W.-Y. Yan, New Insights Into Optimal Acoustic Feedback Cancellation, Signal Processing Letters, IEEE 20(2013), 869-872.
7. T.V. Waterschoot, M. Moonen, Comparative evaluation of howling detection criteria in notch-filter-based howling suppression, Journal of the Audio Engineering Society 58(2010), 923-940.
8. Q. Wang, L. Zhao, J. Qiao, C. Zou, Acoustic feedback cancellation based on weighted adaptive projection subgradient method in hearing aids, Signal Processing 90(2010), 69-79.
9. M. Rotaru, F. Albu, H. Coanda. A variable step size modified decorrelated NLMS algorithm for adaptive feedback cancellation in hearing aids. in 10th International Symposium on Electronics and Telecommunications. 2012. Timisoara, Romania: IEEE Computer Society.

10. M. Guo, S.H. Jensen, J. Jensen, Evaluation of State-of-the-Art Acoustic Feedback Cancellation Systems for Hearing Aids, Journal of the Audio Engineering Society 61(2013), 125-137.
11. K. Wojcicki, M. Milacic, A. Stark, J. Lyons, K. Paliwal, Exploiting Conjugate Symmetry of the Short-Time Fourier Spectrum for Speech Enhancement, IEEE Signal Processing Letters 15(2008), 461-464.

Research on bootkit detection technology of embedded ARM device based on J-Link

D. Li[1†], L.H. Jiang[2], H. Shu[3] and H. Meng[4]
State Key Laboratory of Mathematical Engineering and Advanced Computing, Information Engineering University, Zhengzhou, China
Emails: [†]lgdmgljm@gmail.com

Embedded ARM devices are suffering from the increasing threat caused by Bootkit seriously. Due to the limitation of traditional Bootkit detection techniques, this paper presents an underlying firmware detection method based on J-Link. We establish an attack model of embedded ARM devices considering the key methods of Bootkit attacks and conduct a proof of concept. Given the system booting process and loop identification, our method takes advantage of the information obtained in the process of tracking system booting, and monitors the stages of bootloader and kernel startup. Results show that our method can detect embedded ARM device Bootkit for severe types of embedded devices effectively.

Keywords: Embedded device; Bootkit detection; Bootloader; Staged detection.

1. Introduction

With the development of Internet of Things, intelligent embedded devices are widely used in modern homes, offices and communications infrastructure. Therefore, a growing number of embedded devices combining hardware and software gain access to the Internet. Application systems based on network interfaces provide the available attack surface for attackers. Routers, intelligent printers, mobile phones, switches, smart watches, NSA, industrial control equipment PLC and other equipment can be attack targets [7, 10].

Attack techniques that attackers use are variety [1, 2, 6]. Trapdoor implantation during system startup is the most thorough way to control the device [3, 8]. By implanting trapdoor codes in the device bootloader, described as Bootkit technology, trapdoor programs can execute during or before the system startup and gain system authority. Many techniques have been proposed to detect Boot-

[1] Deng Li
[2] Lie-Hui Jiang
[3] Hui Shu
[4] Huan Meng

kit on computer system: such as selecting some special strings in already known Bootkit as the Bootkit "fingerprint" signature; on equipment board level, adding a credible Platform Module chip TPM (Trusted Platform Module), which performs the integrity tests and measurement of instructions and data and trusted computing [9] during startup and the chip in the system startup code and use the data to perform the detection and measurement of testing based on. However, Bootkit detection technology on the general computer system cannot be fully applied to embedded systems because of limitation of software and hardware resources for embedded device.

This paper focuses on embedded ARM equipment Bootkit detection. We present a Bootkit detection technology based on J-Link emulation debugger. We analyze the underlying peculiarities of embedded system startup, system loading, application execution and trapdoor of firmware to address the issues to apply traditional methods to embedded system to test bootloader security.

The rest of the paper is organized as follows. Section 2 gives a proof of concept for Bootkit attack on embedded devices. Section 3 presents a detailed description of our Implementation of Bootkit detection, which performs staged Bootkit detection on ARM equipment, based on the tracing results. Section 4 shows experimental evaluation of our method, including the effectiveness and performance overhead. Section 5 concludes with a general discussion.

2. Proof of Concept for Embedded Devices Bootkit Attack

Embedded system is actually a dedicated computer system, therefore the design of PC BIOS firmware trapdoor can give guidance to develop embedded device system firmware trapdoor. Bootloader attacks on embedded systems, which called Bootkit, are based on the starting mechanism of bootloader of embedded devices [4]. A successfully malicious code implantation in bootloader for a certain type of embedded Linux system devices requires a series of preparatory work. First we need to extract firmware from the device, or download directly from the official website if firmware is available and collect the information of target device hardware. As the extracted firmware is binary executable, we reverse the firmware to find insertion position malicious code. Then we write trapdoor code, establish proper cross-compiler environment and compile the code to insert into bootloader. Finally, we download the modified firmware to the target device.

By analyzing current Bootkit samples, we find the way is hooking in the startup process mainly to run malicious code on one stage and release those hooked codes gradually to gain control of the system. The normal startup process of ARM equipment is summarized as follows: in the bootloader booting process, Bootkit gets opportunity to execute, releases malicious code carried by

Bootkit into memory or file systems (such as Ramdisk) and hooks the kernel of operating system. After the operating system starts up, malicious programs are invoked on some certain conditions where a hook point is triggered. The process of Bootkit attacks on embedded devices is shown in Figure 1.

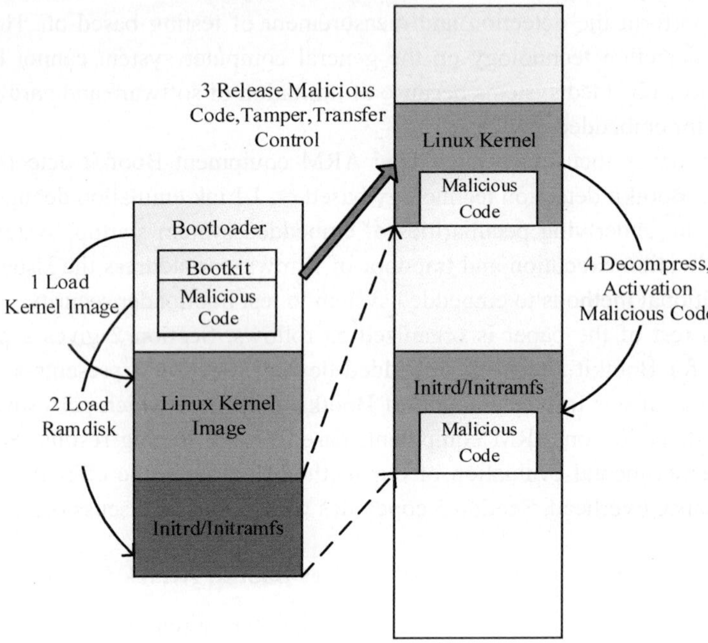

Fig. 1. Bootkit attacks on embedded devices.

3. Bootkit Detection Based on J-Link

Development of embedded systems depends on chip debugging interface to debug and interact with programs in Flash, such as JTAG interfaces. J-Link emulation is the JTAG emulator that the SEGGER Company launched to support ARM core chip. We can step the code and set breakpoints in the device through chip debugging interface. ARM processor chips generally support JTAG emulation debugging functionality. We can track and record the firmware system startup process, using J-Link emulation debugger by leading out JTAG signals through JTAG pins on chips [5]. Then we analyze those signals statically. Dynamic analysis is the primary method to perform staged detection on firmware, and static analysis is the auxiliary. Detection proposed is mainly based on the analysis and partition of embedded ARM Linux booting process analysis and embedded devices division, using a combination of static and dynamic analysis, tracking and recording code information in system startup process and analyzing execution characteristics of the device during the process of bootloader startup and operating system kernel loading to realize Bootkit detection. Embedded ARM Linux devices Bootkit detection model is shown in Figure 2.

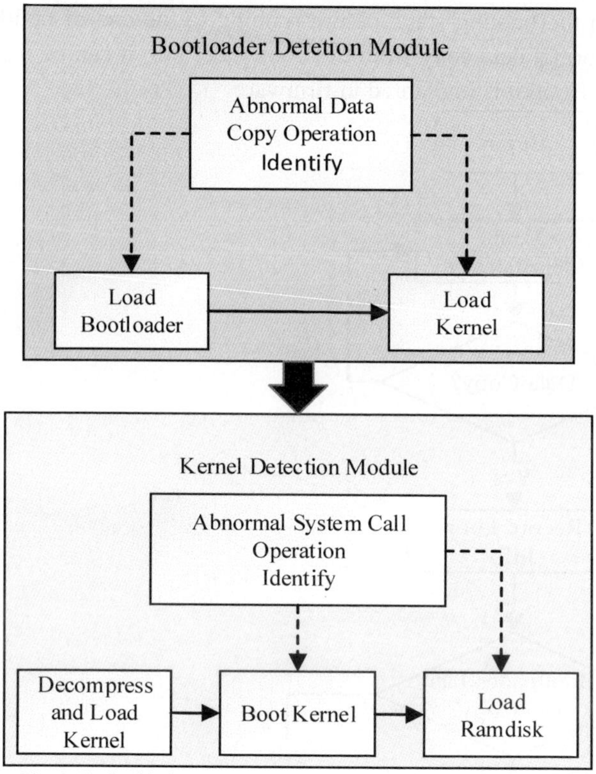

Fig. 2. Embedded ARM Linux devices Bootkit detection model.

This model detects Bootkit on bootloader booting stage and Linux kernel startup stage. On bootloader booting stage, we focus on testing whether there are abnormal mass memory data operations, as Bootkit need to copy itself into memory from bootloader and detecting whether key data are tampered with, such as the system call table.

3.1. *Detection on bootloader booting stage*

Bootkit will release its own malicious code early in system startup process. The time of malicious code release is usually before bootloader transferring control to the operating system kernel. Bootkit copies data to a specific memory space or inserts into file system. Bootloader contains large amounts of data and its data organization is compact, resulting in difficulty for Bootkit designers to find vacant areas inside bootloader to insert Bootkit code. Therefore, Bootkit code is often attached to the tail of bootloader.

According to the above characteristics, the detection monitors data copy operations on the booting stage before running to the end of bootloader. Once there are abnormal data copy from the bootloader tail, it can be regarded as the evidence that Bootkit is implanted in firmware.

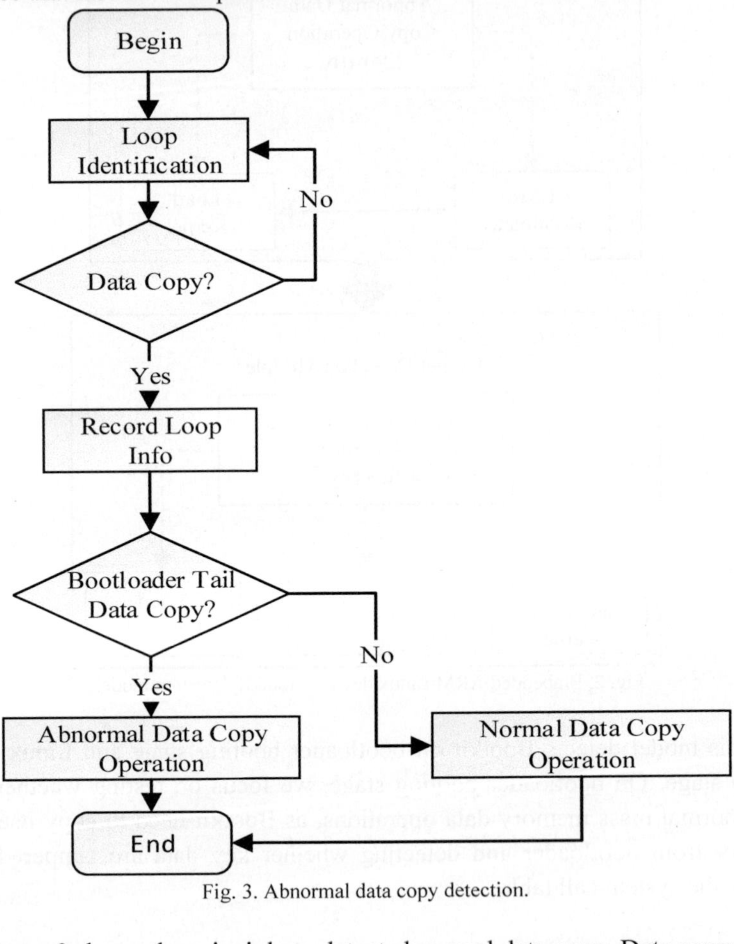

Fig. 3. Abnormal data copy detection.

Figure 3 shows the principle to detect abnormal data copy. Data copy detection mainly tests whether data loading and transferring commands appear in pairs in a loop and the number of loops reaches a certain threshold as well. For the ARM instruction set, common data loading and transferring commands are *str*, *ldr*, *stm*, *ldm* and so on. Determining whether data copy is at the tail of bootloader is to set the address of bootloader boundary in memory as *BEAddr* and set the source address of data copy loop as *CSAddr*. There is abnormal data copy operation if $BEAddr < CSAddr$.

3.2. *Detection on Linux kernel booting stage*

The invocation of malicious code released by Bootkit depends on hooking the operating system kernel. Thus testing whether the key parts of the kernel that are likely hooked have been tampered with is an effective way to detect a firmware Bootkit. Similar to the common Linux Bootkit hooking methods [1], the points that Bootkit on ARM Linux systems hooks the kernel also are system call table, system calls, interrupt descriptor table and so on. Table 1 lists the detection objects.

Table 1. The kernel detection object

Detection Object	Detection Position
Interrupt Descriptor Table	The Value of The Item
System Call Table	The Value of The Item
System Call	Binary Data of Function

We establish the database including the interrupt descriptor table, the system call table of different system versions of Linux (2.4 and above) and the signature library containing system calls. In the interrupt vector, SWI vector is an embedded linux system entry, through it, we can find the entry processing function, and the entry function contains the base address of the system call table. Since we detect at the beginning of Linux kernel startup stage, the system does not completely start up, modify the system call table is bound to hijack SWI interrupt vector.

Detection reads the interrupt descriptor table, the system call table in memory and compares them with the same version in database; detection also reads the interrupt service routines and system calls in memory and extracts signature to match signature library. If inconsistency is detected, we think that a Bootkit is implanted in firmware devices.

4. Experimental Results and Analysis

We ran experiments to evaluate the effectiveness of Bootkit detection technology based on J-Link emulation debugger. This section presents our results. We chose a certain type of development board as the experimental platform with ARM 9 processor, 256MB Nand Flash, 2MB Nor Flash, and 64MB SDRAM, containing JTAG, USB and Ethernet interfaces. We burned the firmware implanted with Bootkit into Nand Flash and then traced and record the startup process of the system. By writing IDA plug-in, we analyzed the record and found abnormal data copy operations after bootloader kernel code copy. The suspected abnormal data copy loop found on the bootloader booting stage is shown in Table 2. The address of bootloader lower boundary in memory was 0x30008128, while R2 was 0x30008380. We can regard this loop as abnormal data copy loop.

Table 2. Abnormal data replication loop.

Address	Instruction
0x30008104	LDMIA R2!, {R9-R12, LR}
0x30008108	STMIA R1!, {R9-R12, LR}
0x3000810C	LDMIA R2!, {R9-R12, LR}
0x30008110	STMIA R1!, {R9-R12, LR}
0x30008114	CMP R2, R3
0x30008118	BCC LOC_8104

R2: the register for storing the source address, R1: the register for storing the target address , R2 = 0x30008380 R1 = 0x306c3dc0

For detection on Linux kernel stage, interrupt descriptor table read from 0xFFFF0000 was same with the table in the database, but the system call table read from memory was not. Figure 4 shows the system call table was tampered. The sys_execve() system call function was replaced with a *jmp* instruction, which can jump to Bootkit functions by assignment to PC.

From the above experimental results and analysis, the proposed firmware Bootkit detection technology can effectively detect the presence of Bootkit.

	0	1	2	3	4	5	6	7	8	9	A	B	C	D	E	F
0000h:	20	39	03	C0	80	C5	02	C0	C8	CD	01	C0	20	97	04	C0
0010h:	84	98	04	C0	34	91	04	C0	CC	92	04	C0	20	39	03	C0
0020h:	20	92	04	C0	38	6B	05	C0	B0	66	05	C0	D8	CD	01	C0
0030h:	5C	86	04	C0	EC	CE	02	C0	08	5D	05	C0	44	8B	04	C0

a) The table in database

	0	1	2	3	4	5	6	7	8	9	A	B	C	D	E	F
0000h:	20	39	03	C0	80	C5	02	C0	C8	CD	01	C0	20	97	04	C0
0010h:	84	98	04	C0	34	91	04	C0	CC	92	04	C0	20	39	03	C0
0020h:	20	92	04	C0	38	6B	05	C0	08	1B	9F	E5	D8	CD	01	C0
0030h:	5C	86	04	C0	EC	CE	02	C0	08	5D	05	C0	44	8B	04	C0

b) The table in memory

Fig. 4. Part of The System Call Table.

5. Discussion and Conclusion

We present the Bootkit detection technology based on J-link emulation debugger to deal with the limitation that applies traditional Bootkit detection to embedded ARM devices. Our detection monitors the equipment stage by stage, records the startup information and detects embedded ARM devices Bootkit.

Although the detection method has some limitations. One is that detection on the booting stage only detects the rear of bootloader; another is that key parts of the Linux kernel cannot contain all possible hook points. Therefore we will find more detection means on bootloader stage and enrich Linux kernel detection objects to further enhance the Bootkit detection capability.

Reference

1. Basnight, Z., Butts, J., Lopez, J. & Dube, T. (2013). Analysis of Programmable Logic Controller Firmware for Threat Assessment and Forensic Investigation. In *Proceedings of the 8th International Conference on Information Warfare and Security*: 9.
2. Costin, A., Zaddach, J., Francillon, A., Balzarotti, D. & Antipolis, S. (2014). A large-scale analysis of the security of embedded firmwares. In *USENIX Security Symposium*.
3. Cui, A., Costello, M. & Stolfo, S. J. (2013). When Firmware Modifications Attack: A Case Study of Embedded Exploitation. In *NDSS*.
4. Gao, H., Li, Q., Zhu, Y., Wang, W. & Zhou, L. (2012). Research on the working mechanism of Bootkit. In *Information Science and Digital Content Technology* 3: 476-479.
5. Hassan, M., Kaushik, A. M., & Patel, H. (2015). Reverse-engineering embedded memory controllers through latency-based analysis. In *Real-Time and Embedded Technology and Applications Symposium (RTAS)*: 297-306.
6. Kornau, T. (2010). Return oriented programming for the ARM architecture. Master's thesis, Ruhr-Universitat Bochum.
7. Michéle, B. & Karpow, A. (2014). Watch and be watched: Compromising all smart TV generations. In *Consumer Communications and Networking Conference (CCNC)*: 351-356.
8. Miller, C. (2011). Battery firmware hacking. *Black Hat USA*: 3-4.
9. Sha, L. T. & Wang, H. X. (2010). R/Bootkit detection based on trusted computing and neural network. In *Information Management and Engineering (ICIME)*: 117-121.
10. Stavrou, A. & Wang, Z. (2011). Exploiting smart-phone USB connectivity for fun and profit. *BlackHat DC*.

Ultrasonic signal recognition for metal creep inspection based on non-creep feature space projection

K. Y. Yuan[a], Z. D. Han[b], Y. F. Chen[c] and Y. X. Zhong[d]
Department of Mechanical Engineering, Tsinghua University, Beijing 100084, China
State Key Laboratory of Tribology, Tsinghua University, Beijing 100084, China
Email: [a]yky@mails.tsinghua.edu.cn, [b]hanzd@tsinghua.edu.cn, [c]chenyifang@mail.tsinghua.edu.cn, [d]zhongyx@tsinghua.edu.cn

Many ultrasonic parameters have been verified to be effective for pure metal or alloy creep inspection. However, these parameters cannot provide adequate sensitivity for weldments creep inspection. It is mainly because of the mono-non-creep standard used in the inspecting procedures. To deal with the complicated microstructure and anisotropy of weldments, a multi-non-creep standard inspecting method named non-creep feature space projection was suggested. This method used all the non-creep inspecting signals to generate a linear space, and calculated distance between any inspecting signal and this space by projection algorithm. Then the distance value could be used to assess creep status, i.e. the larger the distance is, the more serious the creep damage is. This method was verified with pure lead specimen and P91 weldments specimen. Experiments' results showed that the non-creep feature space projection method had a good potential for weldments creep inspection.

Keywords: Creep; Ultrasound; Weld; P91; Space Projection.

1. Introduction

High temperature components used in important industrial field such as refining, petrochemical and fertilizer industries always suffered from creep damage, Which may result in serious disastrous accident. Recent years, to balance the economy and security, techniques that can estimate creep status of high temperature components have been concerned a lot (Touboul M et al. 2013).

Because its advantages in technical and economic, nondestructive testing (NDT) techniques have been researched a lot for metal creep status assessment. Various NDT methods had been tried to inspect creep damage, including ultrasound, electromagnetic, AC potential, radiographic, eddy current, and some new methods such as infrared testing, acoustic emission and digital photograph (Eslami J et al. 2012, Jain P & Godbole M. 2012, Prajapati S et al. 2012, Zhang

Q K & Zhang Z F. 2012, Augustyniak B et al. 2009, Gupta C et al. 2013, Koschack R et al. 2008). In all these NDT techniques, ultrasonic testing had been considered to be the most potential one to completely solve the problem, i.e. assessing metal creep status accurately and industrial practically (Masuyama F. 2011, Sposito G et al. 2010).

The detected objects for creep status estimation are generally creep cavities, precipitates and micro cracks, which are mostly in a size range from 5μm to 0.1mm. Obviously these creep microstructures are too small to reflect ultrasound which is strong enough for traditional ultrasonic testing. In order to achieve adequate inspecting sensitivity, new different ultrasonic inspection parameters had been tried to assess metal creep status. The most popular ultrasonic parameters for creep status estimation include attenuation, ultrasonic velocity, noise analysis, nonlinearity coefficient, acoustical-ultrasonic parameter and acoustic birefringence coefficient, etc (Kim C et al. 2012, Hatanaka H et al. 2007, Szelążek J et al. 2009, Gyekenyesi A L et al. 2002). The validity of these ultrasonic parameters mentioned above had been verified with pure metal or alloy in creep inspecting experiments. However, while these parameters were used for weldments creep status assessment, the inspecting results showed lower accuracy. This result is mainly because of the complicated microstructure of weldments which distorts the propagating of ultrasound and brings down the signal noise ratio (SNR).

Since weldments are indispensable in most high temperature equipments and they are usually the most dangerous positions of creep damage, estimating creep statues in weldments with acceptable accuracy is quite necessary for industrial applied ultrasonic techniques. However, almost all the ultrasonic parameters that have been widely used for creep assessment cannot achieve the desirable accuracy. The limited accuracy of these parameters is resulted from the mono-non-creep-standard used in inspecting procedure. Specifically, all these ultrasonic parameters assess creep status under a relative mode, in which an ultrasonic signal from non-creep specimens is used as a standard and all the inspected ultrasonic signal is compared with this standard signal, the difference between them would be an indication of different creep status. In this inspecting procedure, the standard signal is only a single one. Mono-non-creep standard is adequate for pure metal or alloy because of the homogeneous microstructure in them; but it is insufficient for weldments, because of various grain sizes, inhomogeneous and anisotropic microstructure in them. In some weldments' ultrasonic inspecting experiments, it showed that even ultrasonic incident point or incident angle changed a little, the standard signal should have to change with it, or inspecting accuracy would be affected.

Based on above discussion, mono-non-creep standard is obviously not adequate for weldments creep assessment. In this paper, a new multi-non-creep standards method was suggested, in which all the non-creep signal from a weldment were collected to constitute a non-creep feature space, and projected inspecting signal to the space to calculate distance between the signal and non-creep feature space, then the distance could be used for creep status estimation, i.e. the larger the distance was, the more serious the creep damage was. This non-creep feature space projection method was verified with both pure metal specimens and P91 weldment specimens, and the results showed good coherence to real creep status.

2. Mathmatical Model for Non-Creep Feature Space Projection

Non-creep feature space is a signal set that including all the possible ultrasonic inspecting signal acquired from non-creep specimens. Theoretically it is an infinite set. But it is a reasonable assumption that this infinite set could be expressed by finite signal bases, because the number of modulation sources in ultrasound propagating path is limited. Considering the signal transmission path in an ultrasonic inspecting system, as shown in Figure 1, there are 11 modulation sources in signal transmission path. In an actual inspecting system, all the modulation sources might fluctuate in a small scope for different reasons, but the basic modulation functions should remain static, and could be expressed as transfer functions.

The basic modulation function of each modulation source is indicated in Figure 1, i.e. E_0 for pulser, H_{EP} for emitting probe, H_{SI} for specimen interior, and so on. Then a digital ultrasonic inspecting signal acquired from inspecting system shown in Figure 1 could be expressed as

$$\begin{aligned} x(\mathrm{n}) = {} & E_0(\mathrm{a}_0) \oplus H_{c1}(\mathrm{a}_{c1}) \oplus H_{EP}(\mathrm{a}_{EP}) \oplus H_{CP1}(\mathrm{a}_{CP1}) \\ & \oplus H_{FI}(\mathrm{a}_{FI}) \oplus H_{SI}(\mathrm{a}_{SI}) \oplus H_{BI}(\mathrm{a}_{BI}) \oplus H_{CP2}(\mathrm{a}_{CP2}) \\ & \oplus H_{RP}(\mathrm{a}_{RP}) \oplus H_{C2}(\mathrm{a}_{C2}) \oplus H_{RE}(\mathrm{a}_{RE}) \end{aligned} \qquad (1)$$

where *x(n)* is digital inspecting signal, $E_0(\mathrm{a}_0)$ is modulation function with fluctuation value a_0 which is abstracted from the pulser, other similar expressions are with the same meaning, operator $\oplus$ expresses the transfer and coupling functions between different modulation source, including linear and nonlinear effects.

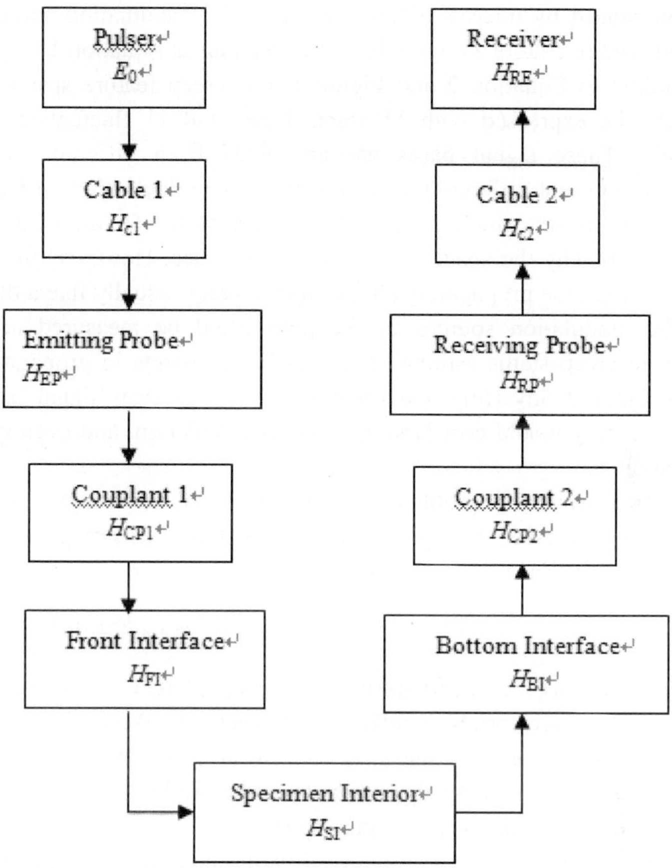

Fig. 1. Modulation sources in signal transmission path of a typical ultrasonic inspecting system.

As discussed before, non-creep feature space is a signal set, then under the inspecting system shown in Figure 1, every element of the signal set could be expressed in the form of Equation 1. The non-creep feature space could be expressed as

$$S_N = \left\{ x(n) \middle| \begin{array}{l} x(n)=E_0(a_0)\oplus H_{c1}(a_{c1})\oplus H_{EP}(a_{EP}) \\ \quad\oplus H_{CP1}(a_{CP1})\oplus H_{FI}(a_{FI}) \\ \quad\oplus H_{SI\_N}(a_{SI\_N})\oplus H_{BI}(a_{BI}) \\ \quad\oplus H_{CP2}(a_{CP2})\oplus H_{RP}(a_{RP}) \\ \quad\oplus H_{C2}(a_{C2})\oplus H_{RE}(a_{RE}) \end{array} \right\} \tag{2}$$

where S_N is non-creep feature space, $H_{SI\_N}(a_{SI\_N})$ is modulation function of the standard specimen with no creep damage, variable a_{SI_N} represents the fluctuation caused by microstructure feature of this modulation source. Other expressions and operators are with the same meaning as Equation 1.

According to Equation 2 and Figure 1, non-creep feature space S_N could theoretically be expressed with 11 signal bases and 11 fluctuation variables respectively. These signal bases are abstracted from different modulation sources and represent different physical features. The fluctuation variables have different value ranges which are also determined by the feature of modulation sources. This is why the space is named feature space. However, for practical ultrasonic testing, the propagating path of ultrasound is usually like a black-box, not all the modulation sources in the path could be measured accurately. Especially in creep status estimation, as nonlinear effects in propagating path cannot be ignored, inverting and obtaining the theoretical signal bases with adequate accuracy would cost large amount of calculation, and even make this work impossible.

In order to solve this problem, here assumed that non-creep feature space could also be expressed by some linear signal bases except physical signal bases, then feature space S_N could be rewrote as

$$S_N = \left\{ x(\mathrm{n}) \middle| x(\mathrm{n}) = \mathrm{a}_1 F_{L1}(\mathrm{n}) + \mathrm{a}_2 F_{L2}(\mathrm{n}) + \cdots + \mathrm{a}_m F_{Lm}(\mathrm{n}) \right\} \quad (3)$$

where, $F_{L1}(\mathrm{n})$, $F_{L2}(\mathrm{n})$, … $F_{Lm}(\mathrm{n})$ are the linear signal bases of non-creep feature space, a_1, a_2, …,a_m are variable coefficients respectively. According to Equation 3, non-creep feature space is translated to a linear space. Based on the mature theory of linear algebra, many algorithms could be used to abstract linear signal bases from the original non-creep feature space.

Considering the ultimate goal of constructing non-creep feature space, that is realizing a multi-non-creep standards method, i.e. estimating creep status by the distance between inspecting signal and non-creep feature space. As non-creep feature space is calculated as a whole in the estimating procedure, replacing physical signal bases by the linear signal bases will not change the distance results.

3. Experiments and Results

Creep status estimating experiments were conducted with pure lead and P91 weldments. The former one is to verify the feasibility of this new multi-non-creep standards method. And another on is for the verification of practicability.

3.1. *Creep status estimating experiments with pure lead*

The melting point and yield strength of pure lead are much lower than high temperature alloy used in practice, but the creep constitutive equations of these metals are just the same except some constant determined by material tissues.

As creep experiment with pure lead can be operated under ambient temperature with low stress and less time, it is usually used to verify whether a new technique is suitable for monitoring creep procedure in metals, e.g. it had been used to verify indentation method when the method was invented. The pure lead specimen used here is shown in Figure 2, which was unloaded after 250 minutes tension.

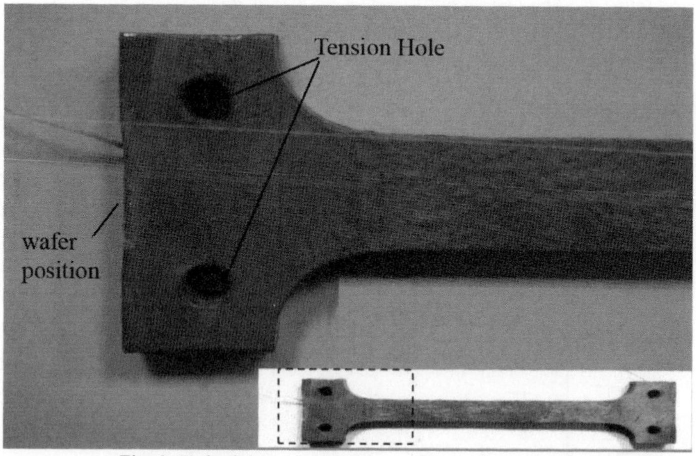

Fig. 2. Uniaxial tension creep specimen of pure lead.

In this continuous creep experiment, tensile stress at the smallest section of the specimen was 7.5 MPa. Total deformation is 17.8%, which was recorded by a digital dial gauge. Two ultrasonic wafers were pasted on each end of the specimen as indicated in Figure 2. Resonant frequency of the wafers was 2.0 MHz. One of them was used as an ultrasonic source which was triggered each minute, while the other one was a receiver, which was connected to an AD card with a sample rate of 100 MHz.

In this experiment, non-creep feature space was generated with 20 inspecting signals that were acquired from the same pure lead specimen under stress-free condition and room temperature. Other inspecting conditions were quite the same as described above. According to non-creep feature space projection method, a creep time VS projection distance curve was obtained after the uniaxial tension creep, as shown in Figure 3. Comparing Figure 3 with the creep deformation curve (Fig. 4), the two curves shows the same tendency and similar profile. Calculating the cross correlation coefficient between these two curves, the coefficient value is 97.97%, which means they had high consistency each other. The result verified that non-creep feature space projection method could be used for metal creep status estimation.

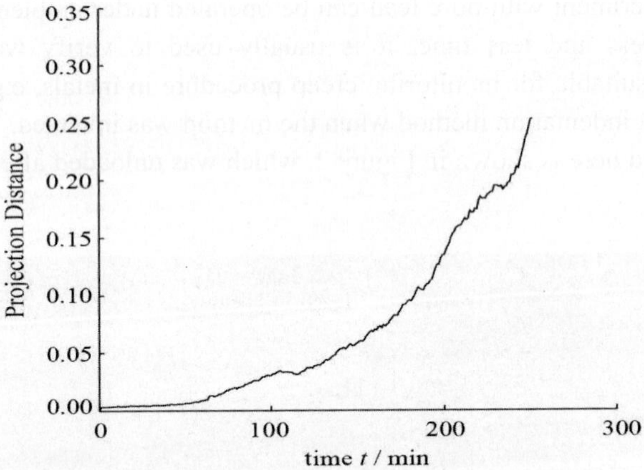

Fig. 3. Projection Distance in creep procedure of pure lead specimen.

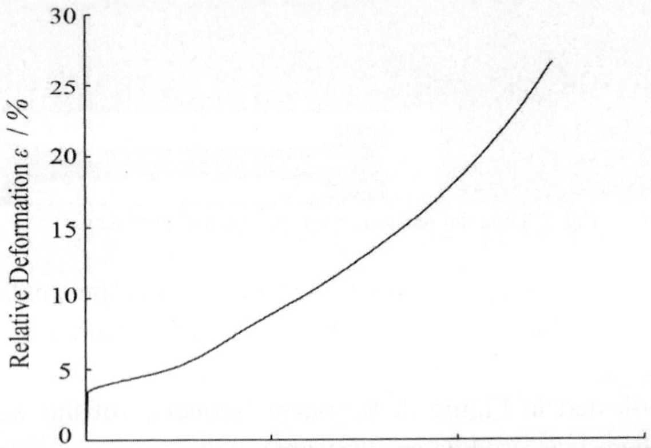

Fig. 4. Relative Deformation in creep procedure of pure lead specimen.

3.2. *Creep status estimating experiments with P91 weldments specimen*

P91 steel is the most popular high temperature material. Many pressure pipings and boilers were made up by P91 steels. NDT methods that can inspect creep damage of P91 steel have been studied a lot. However, because of complex microstructures and stress states, creep damage in P91 weldments is quite difficult to inspect.

Three P91 weldments specimens were used in this experiment. One was original weldment without creep, another was 35% t_r creep status, where t_r is estimated broken time of this type of weldments, and the third one was 70% t_r creep status. One of these specimens was shown in Figure 5. These specimens were inspected with pulse echo method by 2 MHz probes. Ultrasound was incident on cross section surface of the weldments, i.e. the surface shown in Figure 5. Ultrasound would propagate parallel to the weld line while conducting inspection.

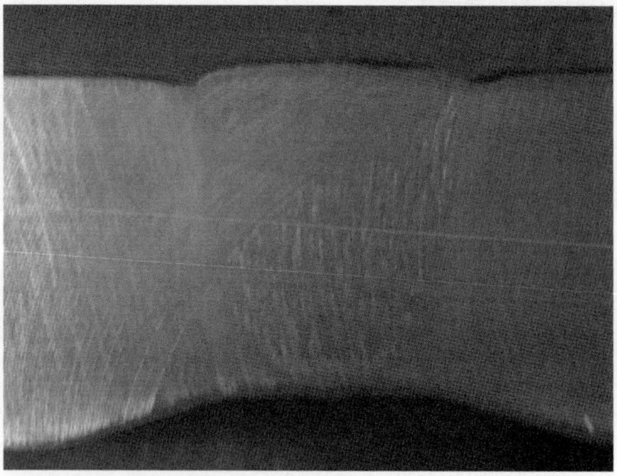

Fig. 5. P91 weldment specimen for creep status estimation.

In this experiment, non-creep feature space was generated with 300 inspecting signals that were acquired from original weldment specimen (creep status 0% tr). The signal acquiring zones included base metal, heat affect zone (HAZ) and welded metal. Then inspecting signal from each point of the specimen was projected to the non-creep feature space, and calculated distance. An ultrasonic scan image could be printed while the grey level of every point was proportional to the projection distance. In this scan image, the brighter the point is, the more serious creep damage it is. The inspected results of three typical creep P91 weldments specimen are shown in Figures 6(a), 6(b), 6(c), which are in the same grey level range. It is obvious that grey level of the image is brighter as creep status increased. As HAZ is the most dangerous positions for creep damage in a weldment, in Figure 6(b) and Figure 6(c), the HAZ positions are also brighter than other positions. And the brightest position is the stress concentration point at HAZ, i.e. the near surface position of HAZ. Theses results showed good consistency with the destructive tests. It is verified that non-creep feature space projection method could be used for P91 weldments creep status estimation.

Fig. 6(a). Projection distance imaging of creep status 0% tr weldments .

Fig. 6(b). Projection distance imaging of creep status 35% tr weldments .

Fig. 6(c). Projection distance imaging of creep status 70% tr weldments.

4. Conclusions

A multi-non-creep standards method for metal creep status estimation in ultrasonic testing was discussed. As this method using non-creep inspecting signals set as a standard projecting space, it was named as non-creep feature space projection method. The mathematical model of this method was described and emphasized on the linearity translation of non-creep feature space. Two ultrasonic inspecting experiments with pure lead specimen and P91 weldments specimens were conducted to verify the feasibility and practicability of this new method, and both the experiments showed expected results. Thus, non-creep feature space projection method should be considered as a good potential method for weldments creep status estimation.

Acknowledgements

The corresponding author of this paper is HAN Zandong (hanzd@tsinghua.du.cn). This paper is supported by National Natural Science Foundation of China (No.: 51375258).

References

1. Augustyniak B, Chmielewski M, Sablik M J, et al. 2009. A new eddy current method for nondestructive testing of creep damage in austenitic boiler tubing. *Nondestructive Testing and Evaluation* 24(1-2): 121-141.
2. Eslami J, Hoxha D, Grgic D. 2012. Estimation of the damage of a porous limestone using continuous wave velocity measurements during uniaxial creep tests. *Mechanics of Materials* 49: 51-65.
3. Gupta C, Toda H, Schlacher C, et al. 2013. Study of creep cavitation behavior in tempered martensitic steel using synchrotron micro-tomography and serial sectioning techniques. *Materials Science and Engineering: A* 564: 525-538.
4. Gyekenyesi A L, Kautz H E, Shannon R E. 2002. Quantifying creep damage in a metallic alloy using acousto-ultrasonics. *Journal of Materials Engineering and Performance* 11(2): 205-210.
5. Hatanaka H, Ido N, Ito T, et al. 2007. Ultrasonic creep damage detection by frequency analysis for boiler piping. *Journal of Pressure Vessel Technology*. 129(4): 713-718.
6. Jain P, Godbole M. 2012. Review of magnetic hysteresis-based NDE of creep damage in power plant steels. *Insight - Non-Destructive Testing and Condition Monitoring* 54(3): 128-133.
7. Kim C, Hyun C, Park I, et al. 2012. Ultrasonic characterization for directional coarsening in a nickel-based superalloy during creep exposure. *Journal of Nuclear Science and Technology* 49(4): 366-372.

8. Koschack R, Passmann N, Imhof R, et al. 2008. Infrared furnace cameras for detection of slag deposits at furnace walls and for life-time monitoring of membrane walls (creep strength). *VGB Powertech* 88(9): 104-110.
9. Masuyama F. 2011. Advanced technology in creep life prediction and damage evaluation for creep strength enhanced ferritic steels. *Materials at High Temperatures* 28(3): 234-244.
10. Prajapati S, Nagy P B, Cawley P. 2012. Potential drop detection of creep damage in the vicinity of welds. *NDT & E International* 47: 56-65.
11. Sposito G, Ward C, Cawley P, et al. 2010. A review of non-destructive techniques for the detection of creep damage in power plant steels. *NDT & E International* 43(7): 555-567.
12. Szelążek J, Mackiewicz S, Kowalewski Z L. 2009. New samples with artificial voids for ultrasonic investigation of material damage due to creep. *NDT & E International* 42(2): 150-156.
13. Touboul M, Crepin J, Rousselier G, et al. 2013. Identification of Local Viscoplastic Properties in P91 Welds from Full Field Measurements at Room Temperature and 625 C. *Experimental Mechanics* 53(3): 455-468.
14. Zhang Q K, Zhang Z F. 2012. In situ tensile creep behaviors of Sn–4Ag/Cu solder joints revealed by electron backscatter diffraction. *Scripta Materialia* 67(3): 289-292.

Application of mine methane gas alarm based on QM-N5 air-sensitive sensor

W.Y. Yu & J. Bai*
Inner Mongolia University of Science and Technology, Baotou, China
**285745601@qq.com*

The thesis introduces mine methane gas alarm by researching QM-N5 air-sensitive sensor, utilizing QM-N5 air-sensitive sensor collects the gaseous concentration, and miner's lamp battery provides it power supply. This alarm applies to the coal mine environment with methane gas, when the gaseous concentration exceeded in mines, it can alarm opportunely. The thesis also introduces the matters needing attention of QM-N5 air-sensitive sensor, in order to use this sensor and device better.

Keywords: QM-N5 Air-sensitive Sensor; Alarm; Detection of gaseous concentration; Miner's lamp;

1. Introduction

Coal, the fundamental energy, is crucial to China's economy development. However, the potential safety hazard in the coal mine imperils the development of mining industry. Gas, the top disaster of the five ones, is easy to cause various safety problems. It brings innumerous economic losses as well as human losses.

Table 1. The number of gas accidents and deaths in 2013~2014.

Years Reference options	The number of gas accidents	The number of deaths
2013	62	367
2014	47	266

Through National Energy Board reports, coal mines happened 47 gas accidents in the country in 2014, 266 people died, compared with 2013, the number of gas accidents and deaths decreased by 24.2% and 27.5%. Although in recent years, coal mine gas accidents decreased year by year, but the total amount of gas accidents is still large, serious gas accidents have not been effectively curbed. Gas prevention and control situation of coal mine is still grim.

Therefore, it is important to have a real-time monitoring of gas density so that gas leakage accident can be avoided. QM-N5 air-sensitive sensor, a semi-conductor air-sensitive sensor taking SnO_2 as the material of the main part, owns several advantages, such as high sensitivity, fast response speed, large output signal and high-reliability and stability when it is used to detect flammable gas, such as CH_4.

2. Introduction to the Air-sensitive Sensor

Air-sensitive sensor is a sensor that could test the category, composition, concentration of gas and the sensor transforms these data into electrical singals. It converts the information about gas into electrical singals. According to the signal strength, we can achieve information of gas in specific environment, thus air-sensitive sensor can test, monitor and alert; it could also consist of analysis system by interface circuit and computer.

Semiconductor air-sensitive sensor is the most frequently used air-sensitive sensor. Many semiconductor materials could be used in semiconductor air-sensitive sensor, such as SnO_2, ZnO_2, Fe_2O_2, $BaTiO_3$. The semiconductor air-sensitive sensor is stable under heating condition, when the gas contacts the surface of the device, and is adsorbed, first the absorption molecules spread on the surface of the device freely, lose their energy of motion, a portion of the molecules evaporate, another portion of the molecules produce by thermal decomposition and adsorb firmly. At this time, because of the gas concentration, the resistance of the device has changed. The gas concentration is obtained from relationship between gas concentration and resistance.

3. QM-N5 Air-sensitive Sensor

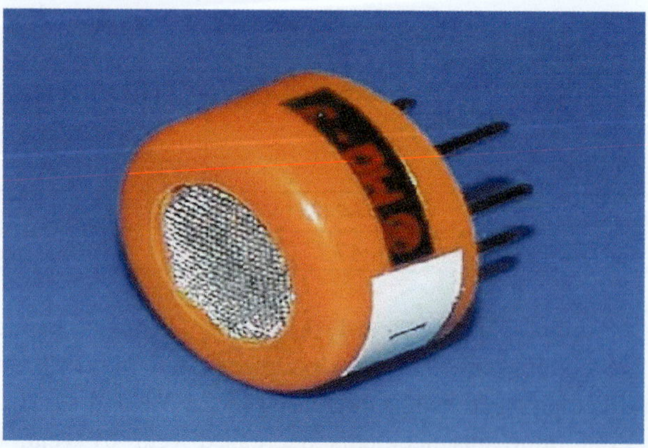

Fig. 1. QM-N5 air-sensitive sensor.

3.1. *The material of QM-N5 air-sensitive sensor*

SnO_2 is a kind of semiconductor materials, is the material of QM-N5 air-sensitive sensor too. Physical properties of SnO_2: SnO_2 is a kind of white powder, its density is 6.16-7.02g/cm^3, its melting point is 1127°C. SnO_2 is not dissoluble in water, is Soluble in strong acid and strong base. The characteristics of SnO_2 semiconductor air-sensitive sensor:

(1) Its resistance changes with the change of gas concentration. This relationship is the change of index. So it applies to test the gas of low concentration.

(2) The stability of SnO_2 is good in physics and chemistry. Compared with other materials sensor, QM-N5 air-sensitive sensor could be used for a long time.

(3) The structure of SnO_2 semiconductor sensor is simple, cheap, high reliability, good mechanical property.

3.2. *The technical indicator of QM-N5 air-sensitive sensor*

The working voltage of QM-N5 air-sensitive sensor is AC or DC (5±0.2V), and the circuit voltage is DC24V at most. The load resistance is 2KΩ, while the resistance is less than or equal to 2000KΩ in clean air condition with the reaction time less than or equal to 10S The detection range is 50-10000ppm.

3.3. *The basic testing circuit of QM-N5 air-sensitive sensor*

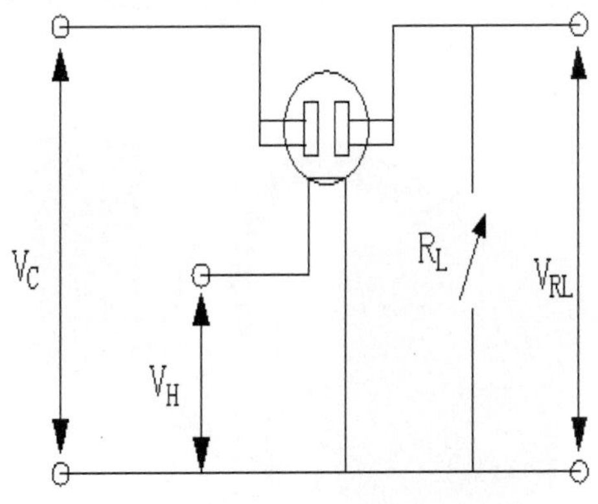

Fig. 2. The basic testing circuit of QM-N5 air-sensitive sensor.

QM-N5 air-sensitive sensor owns two electrodes, one working electrode V_H and the other circuit electrode V_C. R_L is the potentiometer, which is used to adjust the resistence among the sensors of QM-N5. V_{RL} is the output voltage (as Figure 1 shown).

4. Gas Alarm Circuit of the Miner's Lamp

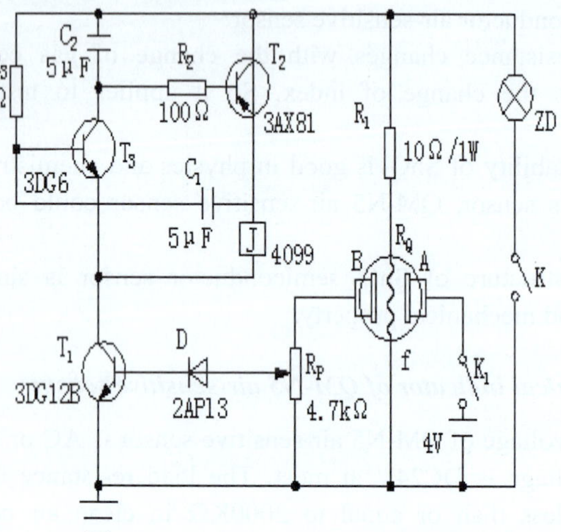

Fig. 3. Gas alarm circuit of the miner's lamp.

As shown in Figure 2, the gas probe is composed of components such as QM-N5 air-sensitive sensor R_Q, R and 4V storage battery of the lamp. The lamp is shown as Figure 3.

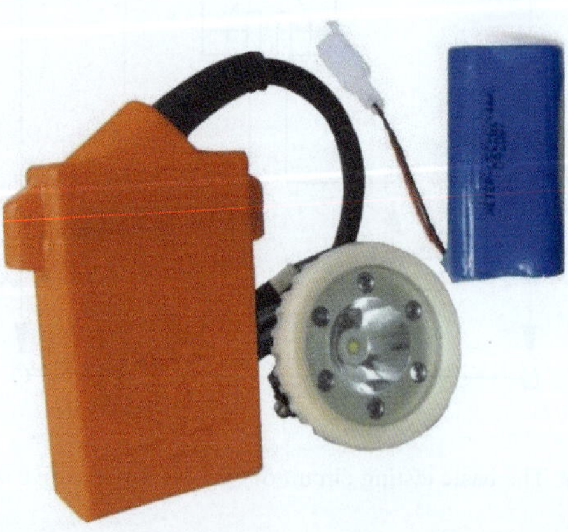

Fig. 4. Commonly used miner's lamp.

As shown in Figure 3, R_1 is a current-limiting resistance. As the air-sensitive sensor will output signal when it is preheated, a false alarm appears. Therefore, the air-sensitive sensor should be preheated for ten minutes or more to avoid the false alarm. Commonly, the alarm is set inside the miner's helmet, taking the storage battery as power. Once gas is transfinite, the lamp will autonomously flash and alarm. As in diagram 2, ZD is the lamp. C_1 and C_2 are the electrolytic condensers of CD10. D is a germanium diodes of type 2APl3 with T_1 as 3DG12B, β=80 while T_2 as 3AX81, β=70 and T_3 as 3DG6, β=20. J is a subminiature mid-power relay of type 4099. All the components are set inside of the helmet. R_P is an alarm setting potentiometer. When the gas is beyond the set point, the output signal R_P will add to the base T_1 through the diode D. At this time, T_1 conducts, thus T_2 and T_3 begin to work. While the gas density is low, the electric potential of the output signal by R_P is low, thus T_1 is cut off, so as T_2 and T_3. T_2 and T_3 are complementary self-exiting multivibrators. When T_1 is on, the power charges C_1 through R_3 and then C_2 begins to be charged by T3 immediately. As a result, T_2 is conducted and the relay J is cut off. When T_2 is conducted, C_1 begins to discharge immediately. The positive pole of C_1 connects to the negative pole through the base of T_3, the emitter junction, collector of T_1, the negative pole of power supply, the positive pole of it and then the collector of T_2 and finally reaches the negative pole of C_1. Therefore, the constant of discharge time is bigger. When voltage of C_1 approaches zero, T3 is cut off. But at the time, T_2 cannot be cut off immediately. The reason is that there is still charge for condenser C_2 and at the moment C_2 charges through R_2 and the emitter junction of T_2 until the voltage of C_2 approaches zero to make T_2 be cut off. As a result, J is discharged. When T_3 is cut off, C_1 begins to charge again and the following course is as the former to make the circuit form a self-exciting vibration and to make J be closed and open continuously. As J and the lamp are set inside the helmet, when J is closed, armature hits the iron core to make the sound "beep, beep", which is transferred to the miners by the helmet. At the same time, the lamp flashes because of the open and close of J to alarm the miners so that they can take ventilation measures in time. For R_Q, wind and dust prevention measures should be taken but at the same time, ventilation is also needed, so it should be set in the front of the helmet. When debugging, adjust R_P in the clean air 15min after an energized state to make the voltage to earth of the positive pole of D lower than 0.5v and to make the T_1 closed; Put the air-sensitive sensor into gas sample and then the alarm works.

When it comes to sensor selection, mental film resistors should be used for R_1, R_2 and R_3. The materials for R_P should be synthetic carbon film potentiometer; For C_1 and C_2, aluminum electrolytic capacitor with high withstand voltage should be chosen while T_1 and T_3 are silicon NPN transistor. J is DC relay with the type of 4099.

5. Matters Needing Attention of QM-N5 Air-sensitive Sensor

When QM-N5 air-sensitive sensor starts to work, it doesn't contact with combustible gas, its conductivity increases sharply and is stable after one minute, it can be use normally at this time. The change of heating voltage will influence the performance of QM-N5, so it is best to use this air-sensitive sensor in specified voltage range. Usage temperature of this air-sensitive sensor is -15~+35°C; its relative humidity is 45%~75%RH; it should be used in atmospheric pressure 80~106KPa; in order to avoid contamination of sensor from the corrosive gas and oil, it should be cleaned after servicing a long time.

6. Conclusions

Coal brings energy to human society, but the gas in mining brings great potential hazards to coal miners. Therefore, it's urgent to escort coal miners' life by utilizing gas-sensitive sensor and the relevant circuit to have a real-time monitoring of the gas density. Through the analysis of gas in mine hazards, the thesis proposes to measure gaseous concentration by utilizing QM-N5 air-sensitive sensor. This scheme is characterized by its simple circuit and low failure rate, which provides reference for the mine gas monitoring.

7. References

1. Rong, X. L. & Xiao, J. (2010). *Electronics and Electronic Technology*. Beijing: Metallurgical Industry Press.
2. Wang, X. M. & Xiao, J. (2008). *Radar and Detection*. Beijing: National Defence Industry Press.
3. Yu, C.B. (2009). *Sensors and Automatic Detection Technology*. Beijing: Higher Education Press.
4. Yu, Y.W. Chang, J & Cheng, J. H. (2003). *Sensors Theory and Engineering Application*. Xi'an: Xian University of Electronic Science and Technology Press.
5. Zhang, H. R., Zhang, Y. F. & Deng H. M.(2008). *Application and Principle of Sensor*.Beijing: Tsinghua University Press.
6. Zheng, H. D.(2003). *Smoke alarm.* Zhejiang: The Electronic Newspaper Press.

Anti-interference capability evaluation of torpedo fuze based on analytic hierarchy process

W. Lin* & L.J. Liu

College of Logistical. Naval University of Engineering,
Tianjin, China
**lw5371@163.com*

J.M. Ma

Base of Navy Test,
Zhanjiang, China

Anti-interference capability evaluation of torpedo fuze is very practical. The index system for evaluation anti-interference capability of torpedo fuze is established based on the torpedo fuze working principle. The weight of index for evaluation is worked out by using analytic hierarchy process. Then the evaluation model of anti-interference capability of torpedo fuze is put forward. The factual example shows that the proposed method can evaluate anti-interference capability of torpedo fuze and have great meaning for reference.

Keywords: Anti-interference; Evaluation; Torpedo fuze; Analytic hierarchy process.

1. Introduction

The non-trigger fuze is an important part of torpedo, the non-trigger fuze is the use of certain physical fields for target information, and through the receiver to the target information processing, processing, the discriminant for the effective signal output signal. The anti-interference ability of the fuze is important aspect of the fuze system working stability and reliability. The pros and cons of fuze anti-interference ability is directly related to the torpedo operational use efficiency. The torpedo fuze anti-interference capability assessment is the important precondition for the torpedo fuze system development, the use, common attention by the researchers and users. [1-2]

The anti-interference ability of radar system evaluation of research work is more, many anti-interference evaluation methods and standards are put forward. The anti-interference capability assessment of main methods are divided into: analytical method, statistical experiment method and computer simulation analysis, these methods are usually mixed applications. Existinganti-interference assessment criteria including power criterion, probability criterion, tactical application code standards (efficiency) and the information code. But for the torpedo fuze anti-interference evaluation is still in its infancy, has not yet formed a mature evaluation system. Torpedo fuze as an electronic equipment, there are many similarities with the radar, sonar and torpedo homing system, but its tactical use and the working characteristics of different makes fuze anti-interference has certain particularity.

In this paper, the application of the analytic hierarchy process (AHP), through the establishment of evaluation index system, put forward the evaluation model of fuze anti-interference ability, for the torpedo fuze anti-interference ability of the comprehensive assessment.

2. The Principle of AHP

Analytic hierarchy process (AHP) is a method of comprehensive evaluation of strong applicability, qualitative analysis and logical judgement can not only absorb evaluation experts, also can pass the test of consistency guarantee the evaluation result uncertainty factors are reduced to the minimum, logic errors of judgment and can overcome the evaluation experts may appear in the process of analysis. Evaluation of specific analytical hierarchy process is as follows [3]:

(1) Constructing judgment matrix, the importance of each element on the same level of a criterion level of two two comparison, and the following relation (i.e. anti symmetric matrix):

$$A = \begin{bmatrix} a_{11} & \cdots & a_{1n} \\ \vdots & \ddots & \vdots \\ a_{n1} & \cdots & a_{nn} \end{bmatrix} \tag{1}$$

In the formula,

$$a_{ij} = 1/a_{ji}; a_{ij} > 0; a_{ii} = 1; i, j = 1, 2, \cdots, n$$

Table 1. The relative importance value.

Definition	Equally important	Slightly important	Obviously important	Strong important	Extremely important	Between adjacent important degree
a_{ij}	1	3	5	7	9	2□ 4□ 6□ 8

Visible, the bigger the a_{ij} value, then u_i u_j the relative importance degree is high.

(2) Calculate the maximum eigenvalue of $\lambda_{\max}$ for judgment matrix A.

$$AW = \lambda_{\max} W \tag{2}$$

In the formula, W is the corresponding eigenvectors of $\lambda_{\max}$,

$$W = [\omega_1, \omega_2, \ldots \omega_n] \tag{3}$$

(3) Determine the index weight, the characteristics of the weight vector normalization can get the weight of u_i corresponding indicators

$$v_i = \frac{\omega_i}{\sum \omega_i}, i = 1, 2 \ldots, n \tag{4}$$

(4) In order to avoid mistakes to make assessment, need to check the consistency of judgement matrix

$$C.R. = \frac{\lambda_{\max} - n}{R.I.(n-1)} \tag{5}$$

In the formula, A is the judgment matrix with random consistency index, when A<0.1, think the judgment matrix satisfy the consistency requirements; B is the average random consistency index, the index based on the order of judgment matrix by lookup Table 2 [4-5].

Table 2. Average random consistency index.

Order	3	4	5	6	7	8	9
$R.I.$	0.52	0.89	1.12	1.26	1.36	1.41	1.46

3. Fuze Anti-Interference Ability Evaluation Index

According to the working principle of the torpedo non-trigger fuze and operational performance, this paper established the torpedo fuze anti-interference evaluation index is: The probability of false alarm, probability of detection and false target jamming probability, signal-to-noise ratio factor, distance decline rate.

3.1. *The probability of false alarm*

False alarm error is that in case no target appears, fuse misoperation. The probability of false alarm is an important index to measure the fuze anti jamming capability. Usually, the false alarm probability is small, fuze anti-interference ability is strong. Its expression is

$$P_f = \int_0^{\infty} P_n(u)du \tag{6}$$

In the formula, $P_n(u)$ is the probability density function of the interference noise.

3.2. *Probability of detection*

Probability of detection refers to the target appears, fuze obtain object information from the received signal, the discriminant probability for the effective target signals of reliable operation. The inspection threshold is high, anti-interference ability is stronger, but the probability of detection will drop. The expression for the probability of detection is.

$$P_d = \int_{u_0}^{\infty} P_{sn}(u)du \tag{7}$$

In the formula, u_0 is the fuze receiver test threshold; $P_{sn}(u)$ is interference noise and deterministic signal probability density function.

3.3. *False target jamming probability*

In order to implement the interference of the fuze, the enemy of man-made some false target signal, the fuze early burst or deviate from the ideal detonation area. Usually the fuze anti-interference capability is strong, the lower the probability of false target jamming. The false target jamming probability is defined as

$$P_j = \frac{Y}{N} \tag{8}$$

In the formula, Y is the number of false target jamming succeeded; N is the total number of the implementation of false target jamming.

3.4. *Signal to noise ratio factor*

The working performance of fuze indexes are usually under certain SNR conditions established, the SNR is small, the larger the interference effect on fuze. But the actual input signal to noise ratio calculated is very difficult, therefore using SNR factor instead of. The greater the signal-to-noise ratio factor, fuze anti-jamming ability is weak. Signal to noise ratio factor is defined as

$$P_s = \frac{SampEn(s)}{SampEn(s_0)} \tag{9}$$

In the formula, $SampEn(s_0)$ is the sample entropy ideal target signal envelope $s_0(t)$; $SampEn(s)$ is the sample entropy receiver output $s(t)$.

3.5. *The average distance between the reduction rate of Fuze*

The effect of distance of Fuze is an important index of operational performance of fuze, is the indicator of Fuze under certain environmental condition value. The role of fuze distance is, the more help to improve the anti interference ability. The average reduction rate of fuze distance smaller, illustrate the fuze anti-interference ability is strong. Fuze distance average reduction rate expressions for

$$P_\gamma = \frac{1}{n}\sum_{i=1}^{n}\left(\frac{r_{i1} - r_{i2}}{r_{i1}}\right) \tag{10}$$

In the formula, r_{i1} is no interference exists fuze distance; r_{i2} is the presence of interference fuze range.

4. Evaluation Model of Anti-Interference Performance of Torpedo Fuze

Comparison of two two $Q = \left[P_f, P_d, P_j, P_s, P_\gamma\right]^T$ was performed to assess the importance index of torpedo fuze anti-interference ability, get the judgment matrix.

$$A=\begin{bmatrix} 1 & 2 & 3 & 5 & 8 \\ \frac{1}{2} & 1 & \frac{3}{2} & \frac{5}{2} & 4 \\ \frac{1}{3} & \frac{2}{3} & 1 & \frac{5}{3} & \frac{8}{3} \\ \frac{1}{5} & \frac{2}{5} & \frac{3}{5} & 1 & \frac{8}{5} \\ \frac{1}{8} & \frac{1}{4} & \frac{3}{8} & \frac{5}{8} & 1 \end{bmatrix} \tag{11}$$

The calculation of available $\lambda_{\max}=5$

The corresponding eigenvectors for

$$W=[0.8401 \quad 0.4201 \quad 0.2800 \quad 0.1680 \quad 0.1050]$$

For normalization can get the weight of evaluation index

$$v_1=0.4633\,, v_2=0.2317$$

$$v_3=0.1544 \; v_4=0.0927\,, v_5=0.0579\,.$$

For consistency check, $C.R.=0<0.1$, therefore the consistency of judgement matrix can be accepted.

Belong to the bigger the better anti-interference capability index, in order to be integrated with other indicators represent the anti-interference ability of the torpedo, contrarian indicator for polarity reversal : $u_i'=1-u_i$. The evaluation index after transformation for Fuze anti-interference is

$$Q'=\left[P_f', P_d, P_j', P_s', P_\gamma'\right]^T.$$

Thus comprehensive evaluation of the anti-interference ability of torpedo fuze model is

$$E=\sum_{i=1}^{5} v_i u_i'=0.4633P_f'+0.2317P_d+0.1544P_j'+0.0927P_s'+0.0579P_\gamma'$$

5. The Example Analysis

According to the evaluation of anti-interference capability of the establishment of the torpedo fuze model, carries on the appraisal to the domestic type two torpedo fuze anti-interference ability. Through calculation and simulation test, evaluation index parameters obtained for the type two torpedo fuze.

Table 3. The anti-interference ability evaluation index of two fuze.

Model	P_f'	P_d	P_j'	P_s'	P_γ'	E
A	0.999	0.995	0.732	0.595	0.371	0.883
B	0.998	0.997	0.809	0.521	0.338	0.886

We can see from Table 3, B type fuse than comprehensive evaluation of A type fuze anti-interference ability, the evaluation results conform to the theoretical analysis and the actual situation. It also proved that the established evaluation index system and evaluation model can accurately in a certain extent different fuze anti-interference ability.

6. Conclusions

Put forward the evaluation index of torpedo fuze anti-interference ability, the weight of each evaluation index was calculated by using the analytic hierarchy process, to establish a comprehensive evaluation model of fuze anti-interference ability, and the anti-interference ability of type two fuze was evaluated. The research results show that the evaluation method is simple and intuitive, practical, reasonable evaluation for the torpedo fuze anti-interference ability to provide the analysis method [6-8].

Acknowledgements

The research work was supported by Education and Research of Logistical College of Navy University of Engineering under Grant No. 2014-5.

References

1. Zhang Tao, Ren Zhi-liang, Sun Chang-cun. 2011. "Anti-interference Evaluation Method of Torpedo Electromagnetic Fuze Based on Synthetic Weight", *Journal of Detection and Control*, Vol. 33(3), pp.14-18.
2. Li Bo-tao, Li Ming, Wu Shun-jun. 2006. "A Disquisition on Radar ECCM Capability", *Modern Radar*, Vol. 28(11), pp.16-19.

3. Guo Jin-yuan, Huang Jun. 2007. "Anti-jamming effectiveness evaluation of airborne radar based on fuzzy AHP", *Electronics Optics and Control,* Vol. 14(4), pp. 151-153.
4. Lu Ying-qi, Wang Ying-long Zhu. Chang-ying.2006. "Evaluation on Radar Anti-jamming Effectiveness based on Neural Network", *Fire Contro1 and Command Control*, Vol. 31(1), pp. 77-79.
5. Li Chao, Zhang Ju-quan. 2003. "The Method, Actuality and Expectation of Radar ECCM Efficiency Evaluation", *Electronic Warfare*, Vol. 5, pp. 41-45.
6. Guo Yu-bing, Sun Lian-bao, Wang Wei. 2009. "A Study on Radar Anti-jamming Effect Evaluation Based on AHP", *Modern Radar*, Vol. 31(9), pp. 20-22.
7. Wang Shao-qing. 1992. *The torpedo design principle of proximity fuze*, Northwestern Polytechnical University Press, Xi'an, China.
8. Hu Fang, Huang Jian-guo, Zhang Qun-fei, 2007. "Study on Torpedo Anti-countermeasure Capability Evaluation Based on Analytic Hierarchy Process" *Journal of Detection and Control*, Vol. 29(2), pp.63-66.

A field investigation system based on android

G. Qin[1†] and H.J. Fu

Computer Network Information Center,
Chinese Academy of Sciences
4,4th South Street, Zhongguancun, Haidian District
Beijing, China
†Email: gqin@cnic.cn, 8634757@qq.com

Field investigation is an important job for scientists and staff of nature reserve. The traditional method for filed investigation is observing animals or plants outside, measuring related data, filling out paper forms in field, and typing into computer for statistics and analysis. This method is error-prone and inefficient. Intelligent mobile terminals running Android system become popular in recent years. They are portable and powerful. In this paper, we implemented a field investigation system based on Android. With this system, people working in filed can record data easily and store data into database automatically. Meanwhile, this system can integrate tracks with photos, videos and audios to illustrate human activity in detail. This system has been put to use and is welcomed by users from Qinghai Normal University. At the end of this paper, we look forward the future work.

Keywords: Field investigation, Database, XML, KML, Web, Android.

1. Introduction

Field investigation has been carried out by scientists studying on zoology, botany, etc. and employees of nature reserves for many years [1]. It is not only an important job for people in these fields but also a major method for data collection. Traditionally, before field investigation, people should discuss about what factors to investigate. They will design investigation forms and print them on paper. They also need to plan the suitable investigation route. During the investigation, they need to record data manually on these paper forms [2] [3]. What they could write on paper forms are some words or numbers. For the data of longitude, altitude, they have to use professional GPS terminals and write down these long complicated numbers on paper forms. For photos, they could only use camera to take photos, and write down the sequence numbers of photos on correspondent place. For the complex species name, it is very difficult to record the scientific name. Figure 1 shows an example of paper investigation form.

[1] Gang Qin

Amphibians Line Transect Method Investigation Form

Number: Y1-1-1 Location: YunNan Date: 2014/8/14

Start time: 21:24 End time: 21:49 Investigator: Wang Wu

Longitude&Altitude_Start: N 48.37210; E 117.52315 Altitude_Start: 570m

Longitude&Altitude_End: N 48.36806; E 117.51158 Altitude_End: 569m

Quadratic Size: 1000m*2m

Weather: Clear Temperature: 17.8℃ : Humidity 73.0%

Water Temperature: Null pH: Null

Habitat Type: Typical Grassland

No.	Species Name	Adult	Sub adult	Young	Egg	Photos Number	Note
1	Bufo raddei	3					

Fig. 1. Example of investigation form.

If we analyze the traditional field investigation method, it is not difficult to find out its drawbacks.

- *Data format is not standard and is error-prone*

Everyone has its own habitat. Data record manually is difficult to be uniform and standard. For example, during data recording, people will use abbreviation for simplicity. While these data is hard to understand for others even for themselves after a few days. Due to the bad work condition, recording data manually is inconvenient and error-prone, especially for those complex data.

- *Paper is flimsy*

As is well known, work environment in filed is hard, paper is liable to go broken or lost, which will cause incomplete or missing data. The lack of original data will be a huge loss to the subsequent data analysis and statistics.

- *Data has to be reprocessed*

Data record on paper is hard to process. In order to use computers to help us to analyze data, we need to type these data into files, which can be identified by computers. On the one hand, input data into files will bring extra work; on the other hand, it may bring new errors during this procedure. Meanwhile, non-standard data has to be standardized before processed by computers.

- *Integration of multisource data is difficult*

In traditional filed investigation, data of different types is separated, such as observation data, GPS data, photos etc. It is necessary to integrate these different types of data to provide a detailed scene of field investigation and to facilitate data analyze. While it is very difficult by means of traditional field investigation.

With the rapid development of mobile Internet, the application of mobile intelligent terminals is becoming more popular. The common mobile intelligent terminals include mobilephones and tablets. Android [4] [5] is currently leading the mobile intelligent terminals segment in terms of market share since its introduction in 2007. Mobile intelligent terminals has become the mainstream network terminals instead of PC (desktops and notebooks).According to [6], at the end of 2014, mobile Intenet users reached 557 million. Some 86% of Chinses users now access the Internet through mobilephones. Compared with PC, mobile intelligent terminals are smaller, portable, and multifunctional. People can use one mobile intelligent terminal with WIFI, Bluetooth, GPS chips, camera, to replace traditional camera and GPS device etc. Morever, with mobile intelligent terminals, it is easy to standardize data format, simplify complex data recording, alleviate burden of personeel working in field, save the troubles to retype data into files. It also can integrate observation data with GPS tracks, photos, videos etc. to make data integration and analysis possible.

2. System Architecture

This system is designed in the model of client/server. Servers include web application server and database server. Intelligent mobile terminals act as clients. They communicate with servers to upload and download data.

According to different functions, this system is composed of three layers, as shown in Figure 2.

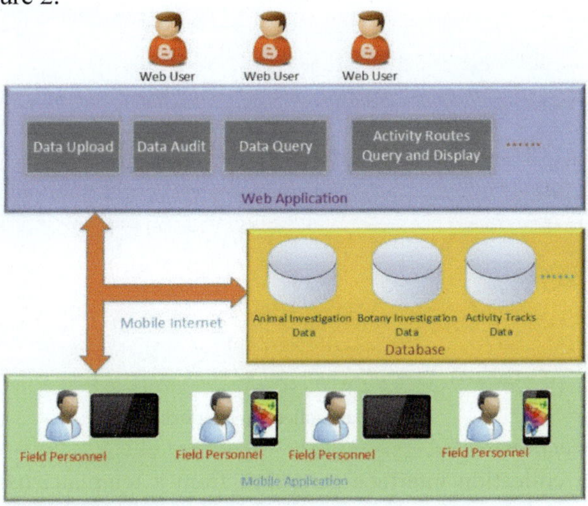

Fig. 2. System architecture.

- Mobile Application: It is running on mobile intelligent terminal. The user interface is self-adaptive to Android mobile phones and Android tablets with different screen sizes. Users fill out observation data, take photos, record videos, and track GPS positions by this application.
- Database: All data associated with field investigation will be stored in database, such as animal investigation data, botany investigation data, and activity tracks data, etc.

• Web Application: It is the interface for web users. They can use web application to implement data upload, data audit, data query, and activity tracks query and display.

3. Mobile Application

The mobile application is running on mobile terminals. The person who works in field will use the mobile application to execute field investigation. Main functions of the mobile applications are as follows.

• *Field investigation template management*

In order to record data on mobile intelligent terminals, we should convert paper forms into electronic format firstly. In this system, we design templates based on paper forms. The templates are based on XML [7]. Requirements of different field investigation correspond to different templates. The mobile application will identify qualified template and generate data collection interface automatically. As shown in Figure 3, a data collection interface is generated from a template. We provide template management function for users to import and delete templates.

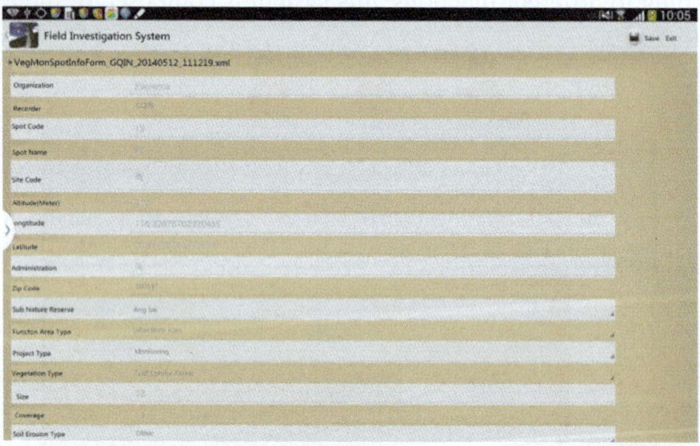

Fig. 3. Data collection interface.

• *Data collection*

With the data collection interface generated from a template, user can record data. For data fields of time, longitude, latitude, altitude, the application will automatically fill them in. For the fields with specific values, especially for the complicated name of species, the application will show the candidates in a list for user to choose. We provide fast index for user to select the target item quickly. For photos data, user can take photos by mobile terminals directly.

• *Activity route tracking*

The application also provides the function to track user's activity route. In most applications, GPS tracks functions are simply record a series of GPS points. In our application, we integrate text description, photos, audios and videos with

GPS tracks. During the field investigation, user can add a key point where he observes some animals or some plants. With the function of adding a key point, he can input some words to descript this spot. In consideration of convenience, he also can record a piece of audio instead of typing words. Moreover, he can attach photos or videos to a key point to reflect the situation of the spot in details. The format of GPS activity route file conforms to KML [8] standard and can be loaded into Google Earth directly. Through GPS tracks, we could know about the details of the investigation carried out. In this system, we correlate observation data with GPS tracks. From investigation data, user can find out a specific record obtained from which point of the correspondent GPS tracks and vice versa. In this application, we use Baidu Map APIs [9] to implement activity routes display and other map related operations. Figure 4 shows an activity route with key points attached with photos.

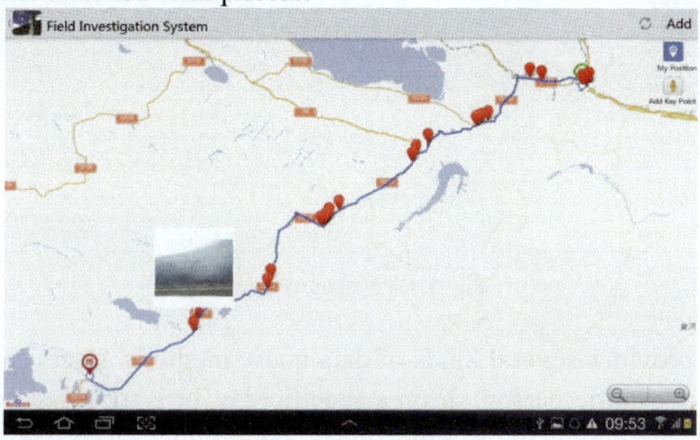

Fig. 4. Activity route.

- *Data upload*

When a field investigation is finished, user can save the investigation data and activity route on the mobile terminal. The application labels these data as un-uploaded. At this phase, users still can modified the data. Then they can upload these data to the remote server through the network of mobile carriers or WIFI. After that, the application labels these data as uploaded. Users can view these data on mobile terminals, but cannot modify them.

4. Web Application

Nowadays, Web has become the most universal portal for user access, data query and visualization display. In this system, we also provide web application for web users. Main functions of the web application are as follows.

- *Data upload*

Although users can upload data from mobile terminals, for big size data or without wireless network connection, it still needs to provide a way to upload data manually from PC or notebooks. Users can login to the web application and upload data from the specific directory of mobile terminals.

• *Data audit*

Data quality is critical for field investigation. Due to the work condition and limited knowledge of the field personnel, it is necessary to perform data audit afterwards. All of the data uploaded by mobile terminals is in the phase of unaudited. Authorized auditor logins to the web application and audit the uploaded data. He can modify the obvious errors of the data and also can replace photos with more clear ones taken by other professional cameras. Once the auditor confirms the data, the data will be imported into the database automatically. Figure 5 shows the data auditing interface.

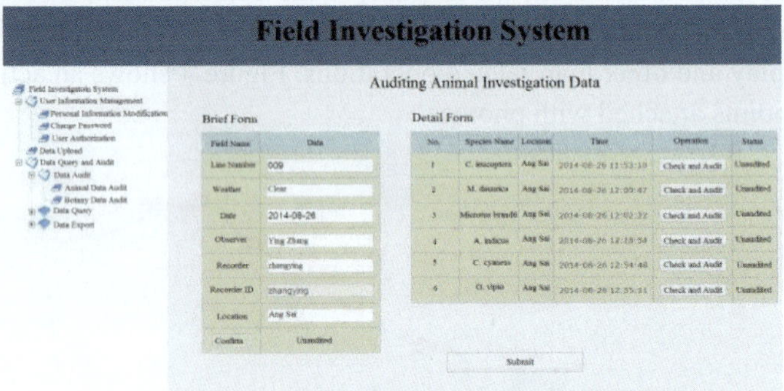

Fig. 5. Data audit webpage.

• *Data query*

This system provides several kinds of data query methods. Users can query data recorded by a specific person, or on a specific day, or a specified period. It also supports a combination of different query conditions. Users can export their query results in the format of excel.

• *Activity routes query and display*

Activity routes give a means to show the history of field investigations. Through analysis on these routes, users can modify or improve the routes of field investigations. Moreover, for managers, they could learn about work load of each field personnel. In this system, web users can query activity routes by username, date, a period of time, and important spot on the track. Users also can select a square range by mouse, and then all tracks pass this range will be retrieved. In web application, we still use Baidu Map APIs to display activity tracks. The multi-type of data including photos, videos, and audios integrated into activity route will show on the webpage. Users can select more than one routes to show on webpage. Figure 6 shows the query and display webpage.

Fig. 6. Activity routes query and display webpage.

5. Summary

In this paper, we introduce a system of field investigation. We design this system in the model of client/server. Clients in this system are intelligent mobile terminals running Android OS. Mobile application on mobile terminals can import field investigation template and generate collection interface automatically. During data collection, it also will record activity route. Both field investigation data and activity route will be stored on mobile terminal. Users can upload these data to server through the network of mobile operator or WIFI. Servers run web application and store data in database. In web application, user can upload investigation data and activity routes manually, audit uploaded investigation data, query data, and query activity tracks and display them on webpage.

Users from Qinghai Normal University have taken this system on trial. They used this system during wild animal investigation in Qinghai Tibet Plateau. After their trial, they gave good praise to this system and believe it facilitate their investigation.

6. Future Work

Tracks recording function in current system integrates original format photos and videos which is in large size and will take a lot of network traffic to upload to server. In future work, we will study how to compress these data efficiently to reduce network traffic and optimize user experience.

Meanwhile, for a group executing field investigation, it is useful to learn about the real-time position of other partners. They could collaborate to finish certain job. It is also meaningful for administrator. He can control overall situation. In case of any emergency occurred, it provides a way to direct rescue operation to the last place reported by the personnel.

Acknowledgement

This research was supported by the project of Knowledge Innovation Program of the Chinese Academy of Sciences under Grant No. CNIC_ZR_2014, Youth Innovation Promotion Association (Chinese Academy of Sciences), Around Five Top Priorities of "One-Three-Five" Strategic Planning, Computer Network Information Center, under Grant No. CNIC_PY-1408 and CNIC_PY-1409, and NSFC Sino-America Software Cooperation Research Project (no. 61361126011).

References

1. Hayne, D. W. An examination of the strip census method for estimating animal populations. J. Wildlife Manage, 1949, 13: 145~157.
2. Southwood TRE, Henderson PA. 2000. Ecological Methods 3rd Edn [M]. London: Blackwell Science Ltd.
3. Sutherland WJ. 1996. Ecological Census Techniques: A Handbook [M]. Cambridge: Cambridge University Press.
4. The Android Open Source Project: Bytecodes for the Dalvik VM. http://source.android.com/tech/dalvik/dalvik-bytecode.html.
5. The Android Open Source Project: Dalvik eXchange (DX). http://android.git.kernel.org/platform/dalvik.git.
6. The 35th Statistical Report on Internet Development in China. http://www.cnnic.net.cn/hlwfzyj/hlwxzbg/hlwtjbg/201502/t20150203_51634.htm.
7. Jon Bosak and Tim Bray, "XML and the Second-Generation Web", Scientific American, May 1999.
8. "Keyhole Markup Language — Google Developers". Developers.google.com. 2012-03-01.
9. Information on http://developer.baidu.com/map/

Research on steel mesh for curtain antenna radiant protection

F. Yuan[1†]
Information Department of East Sea Fleet Headquarter, Ningbo, Zhejiang Province, China
†Email: sunxiaona1991@126.com

Z. G. Zhang[2]
Naval University of Engineering, Wuhan, Hubei Province, China

Y. N. Chen[3]
Ling Yun Science & Technology Group Co.,Ltd, Wuhan, Hubei Province, China

Steel mesh is a common electromagnetic protection material, cheap and durable. Taking Curtain antenna for example, and exploring on the distance from the steel mesh and the antenna, the height of the steel mesh, and steel mesh density. we must ensure that the radiation of steel mesh for antenna is small, and obtain large suppression on field strength, so as to provides practical reference for radiation protection of the high power shortwave antenna.

Keywords: Radiation protection; Distance; Height; Density.

1. Introduction

Shortwave antenna is an important method for long distance wireless communication [1],but the transmission power of shortwave radio sometimes up to several hundred kilowatts, and field strength almost hundred volts per meter, and these affect nearby residents [2]~ [3], so the research on antenna radiant protection is significant. HRS4/4/0.5 curtain antenna is used, it means 4 lines 4 columns, and the distance between lowest layer dipole and ground is 0.5λ, it is horizontal polarization shortwave antenna [4].

The principle of electromagnetic protection is controling the electromagnetic field of internal radiation area is within the designated areas, and preventing

[1] Feng Yuan
[2] Zhi-Gang Zhang
[3] Ya-Nan Chen

foreign radiation comes into [5]. Steel mesh and reinforced concrete are common material for radiant protection, but concrete just reduces oscillation, and steel mesh mainly protects [6]. Based on the simulation software FEKO [7], radiant protection of steel mesh is calculated. We mainly assess how these three factors —the distance from the steel mesh and the antenna, the height of the steel mesh, and steel mesh density— effect antenna radiant protection.

2. Relationship of Electromagnetic Field and Distance

Modeling 4 lines and 4 columns' horizontal dipole curtain antenna by FEKO,and researching the relationship between electromagnetic field produced by high power shortwave antenna and distance. The simulation model is shown in figure 1.

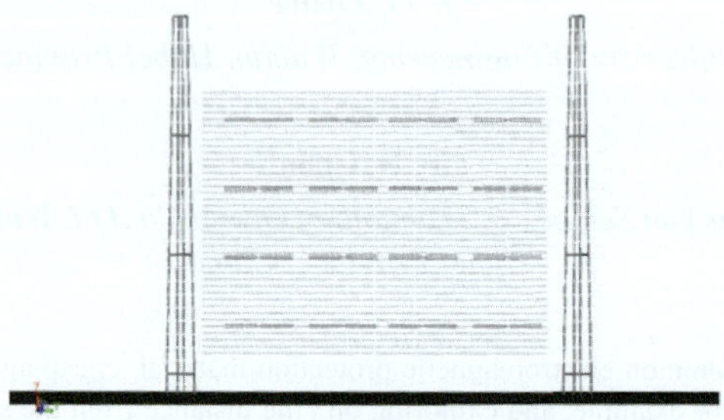

Fig. 1. Model of horizontal dipole curtain antenna.

Figure 2 shows in the ideal conductive ground, the value of field strength on low frequency (17MHz), center frequency (23MHz) and high frequency (30 MHZ) when setting observation plane by four fixed elevation (5 °, 8 °, 15 °, 30°) for 1kW's shortwave curtain antenna, and taking 291 observation points when the distance between steel mesh and antenna is 10m to 300m (the interval is 1m).

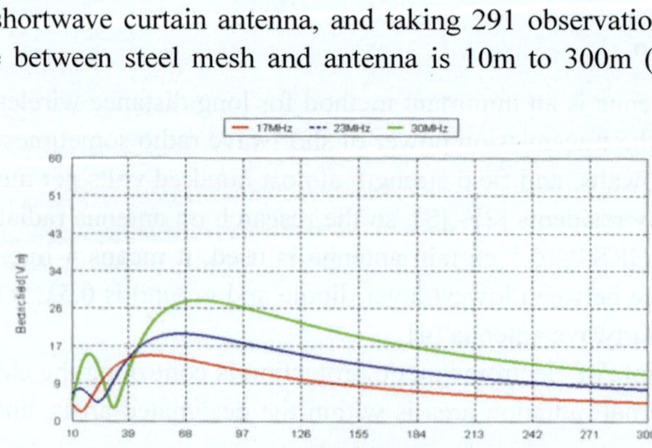

(a) 5° elevation

Fig. 2. Comparison of different elevation.

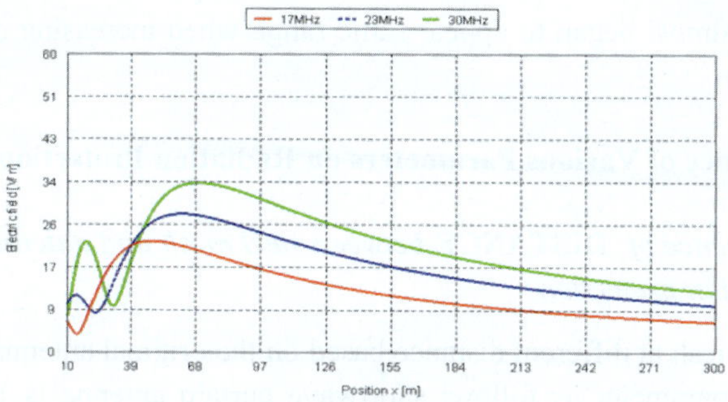

(b) 8° elevation

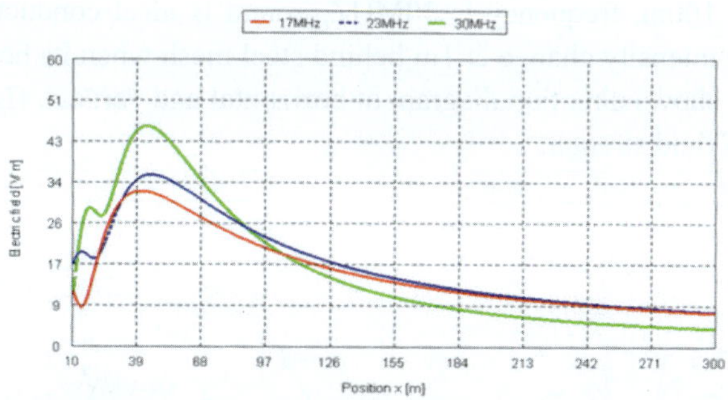

(c) 15° elevation

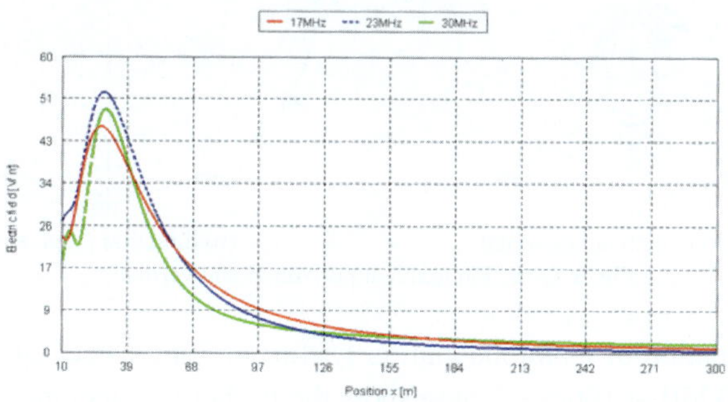

(d) 30° elevation

Fig. 2. (*Cont'd*)

From figure 2, in different elevation and same frequency, with the increase of distance, field strengh rises at a certain extent, decrease at a certain extent after reaching the peak, then maintains stable, as a result of the inverse relationship between field intensity and distance in the far field. In same elevation and different frequency, the lager the frequency, the farther the horizontal distance of turning point of field intensity, and the intervals become narrow with the increase of elevation at different frenquency, then come to close. Three points almost began to appear same range when increasing or decreasing at 30°elevation .

3. The Influence of Various Parameters on Radiation Protection

3.1. *The influence of* DISTANCE *between steel mesh and antenna on radiation protection*

Setting steel mesh at different distance based on the original antenna model. Setting model's parameter as follow: shortwave curtain antenna is 1kw, hight is 11m, mesh density is 0.5m, distance between steel mesh and antenna are 50m,75m and 100m, frequency is 30MHZ,ground is ideal conductive. We observe the field intensity change at 1m behind steel mesh when its height is 2m to 20m. Figure 3 shows direction diagram at horizontal and vertical, figure 5 shows comparison of field strengh.

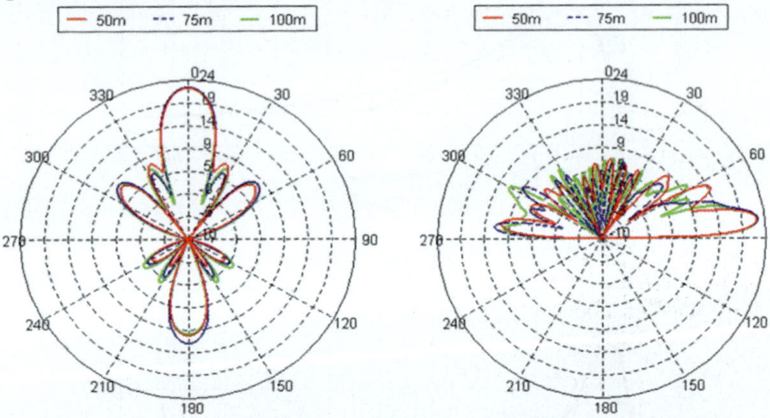

(a) direction graph at horizontal (b) direction graph at vertical

Fig. 3. Direction graph at horizontal and vertical.

The attenuation of mainlobe gain is 1.54dB at 50m's steel mesh, 1.36dB at 75m's, and 1.32dB at 100m's. These show the farther the steel mesh, the smaller the influence on antenna radiation.

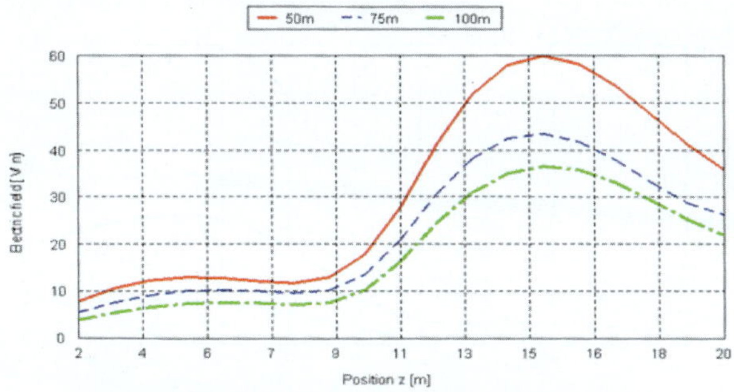

Fig. 4. Comprasion of field strength at different distance.

From figure 4, when the distance between steel mesh and antenna is 50m, 75m and 100m, the field srength decreased gradually. This shows that the farther the steel mesh, the better the effect of suppression at curtain antenna field strength.

Usually, the attenuation of mainlobe gain under 0.5dB is acceptable in engineering, so farther explore on the position of the steel is needed.

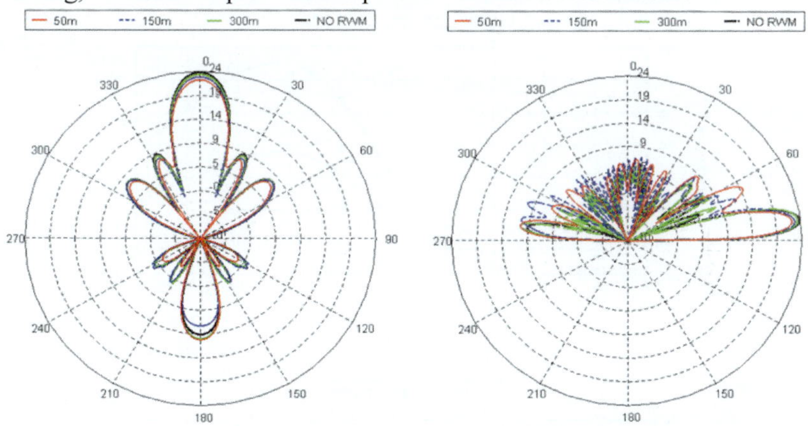

(a) direction graph at horizontal (b) direction graph at vertical

Fig. 5. Direction graph at horizontal and vertical.

Figure 5 shows comprasion of direction graph whose distance of steel mesh is 50m, 150m, 300m and no steel mesh at 30MHz. As we can see, the attenuation of mainlobe gain is 0.92dB at 150m's steel mesh, 0.34dB at 300m's which is qualified.

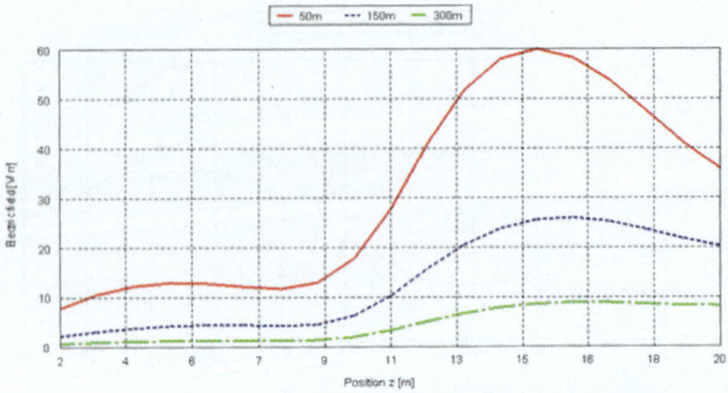

Fig. 6. Comprasion of field strength at different distance.

From figure 6, the influence of antenna radiation is small at 300m's steel mesh, and the suppressed effect is perfect. Therefore, improving the effect of highpower shortwave antenna radiation protection by increasing the steel mesh distance is desirable.

Figure 7 and figure 8 shows in different power and frequency, the relationship between location of steel mesh and radiation protection.

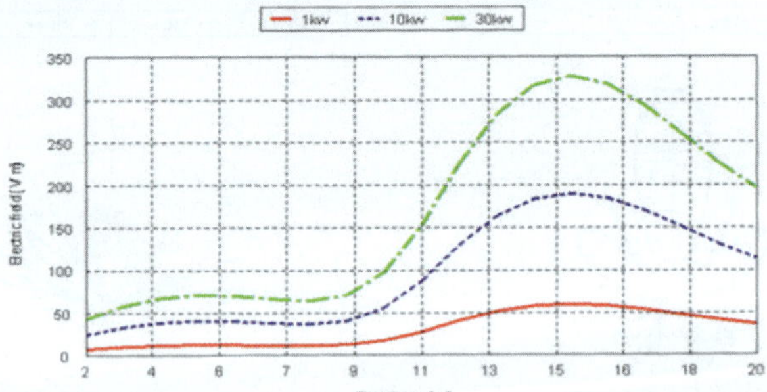

(a) 50m

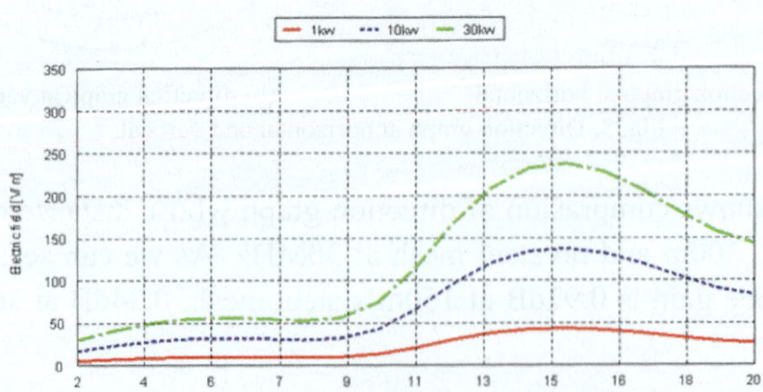

(b) 75m

Fig. 7. Comparison of field strength at different power.

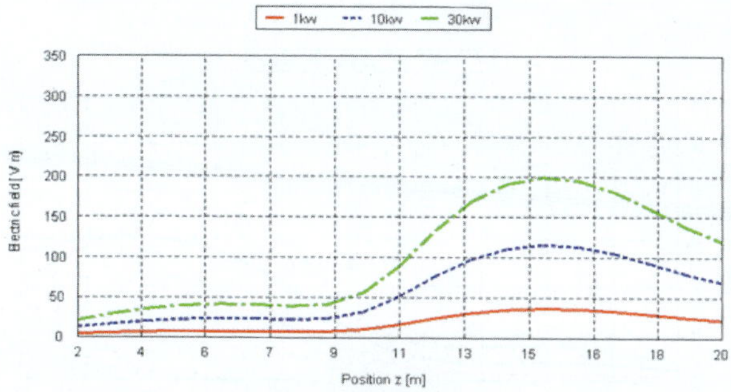

(c) 100m
Fig. 7. (*Cont'd*)

From figure 7, the ratio of field strength at 1kW and 10kw is about $\sqrt{10}$ at any height, and the ratio at 10kW and 30kw is about $\sqrt{3}$. So the proportional relationship between power and field strength can be described as: the ratio of two field strength at different power is the root of ratio of powers. In addition, different powers does not affect the law obtained from figure 7, and that is the farther the steel mesh, the better the suppression for larger power electromagnetic field strength.

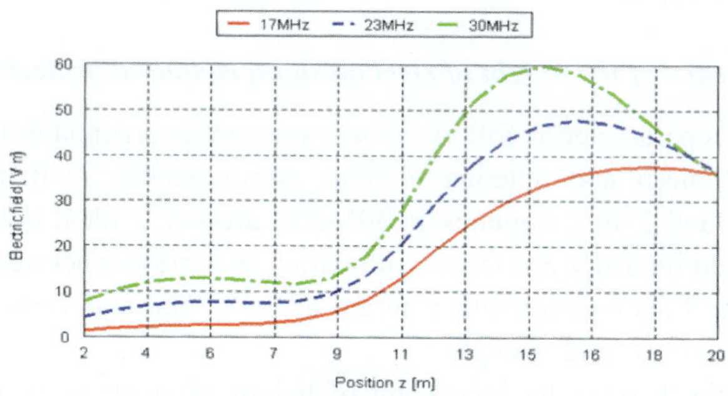

(a) 50m

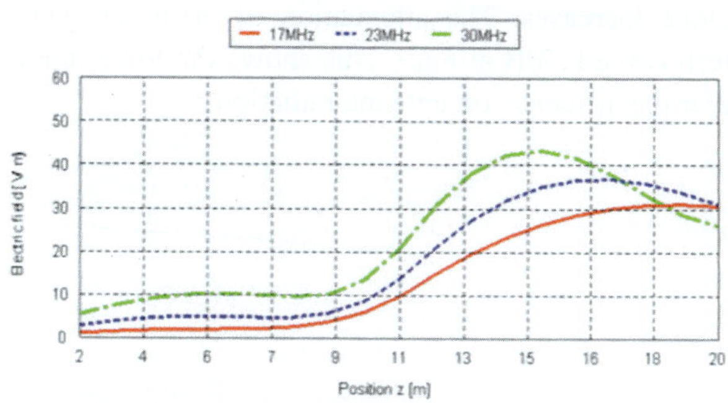

(b) 75m
Fig. 8. Comparison of field strength at different frequency.

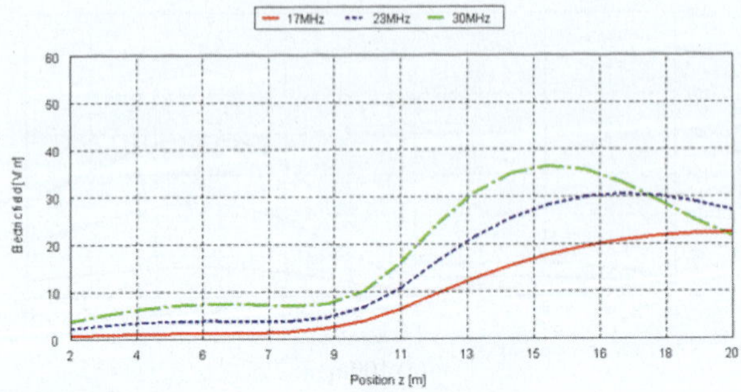

(c) 100m

Fig. 8. (*Cont'd*)

From figure 8, exceeding a certain elevation, 30MHZ's field strength is lower than 17MHZ's and 23MHZ's. Because the observation angles are different at different location, and the numbers of mainlobe and sidelobe contained are different, and value of field strength is different. In addition, the law that the farther the steel mesh, the better the suppression for larger power electromagnetic field strength applies.

3.2. *The influence of the height of steel mesh on radiation protection*

Setting model's parameter as follow: shortwave curtain antenna is 1kw, distance between steel mesh and antenna is 50m, mesh density is 0.5m, height is 6m,11m, 16m and 21m, frequency is 30MHZ, ground is ideal conductive. We observe the field intensity change at 1m behind steel mesh when its height is 2m to 20m. Figure 9 shows direction diagram at horizontal and vertical, figure 10 shows comparison of field strength.

From figure 9, with the increasing of height of steel mesh, the mainlobe gain decreases, the attenuation of mainlobe gain decreases, the elevation is constant, the sidelobe increases. The attenuation of mainlobe gain is 1.54dB at 11m's steel mesh,while 0.5dB at 6m's.This shows the lower the height of steel mesh, the smaller the influence on antenna radiation.

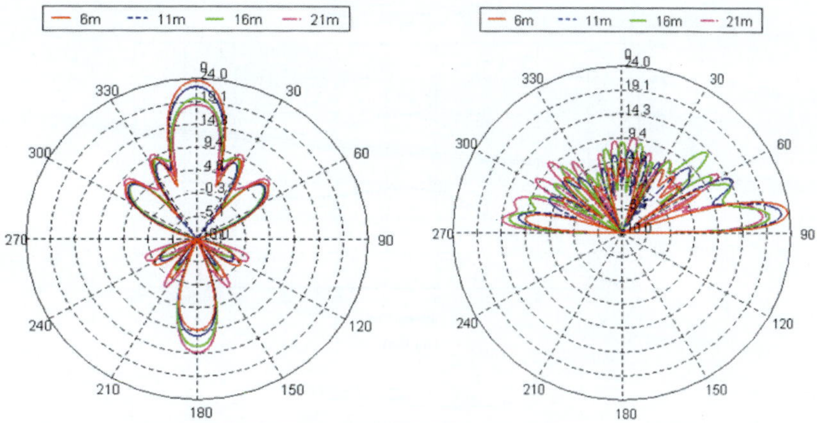

Fig. 9. Direction graph at horizontal and vertical.

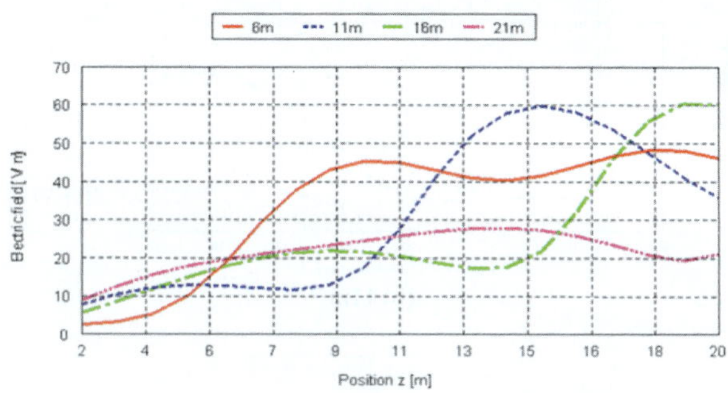

Fig. 10. Comprasion of field strength at different height of steel mesh.

From figure 10, the field strength is small when the distance between 6m's steel mesh and ground is less than 4m ,and the distance of 11m's is 9m, and the distance of 11m's is 14m, and the distance of 21m's is 20m. That is to say, the suppression for field strength is good when height less than the value bounded by height.

Figure 11 and figure 12 shows in different power and frequency, the relationship between height of steel mesh and radiation protection.

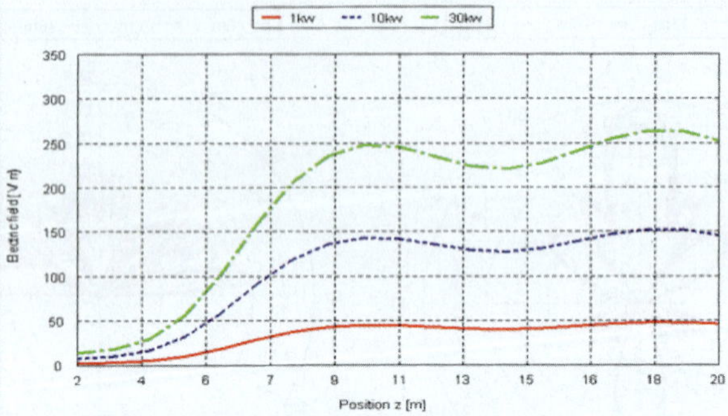

(a) 6m

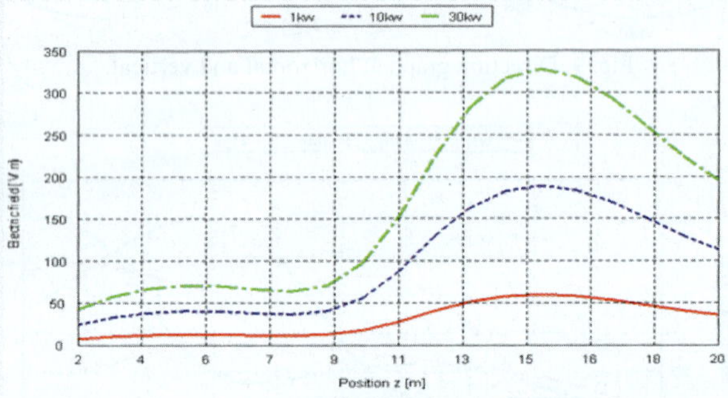

(b) 11m

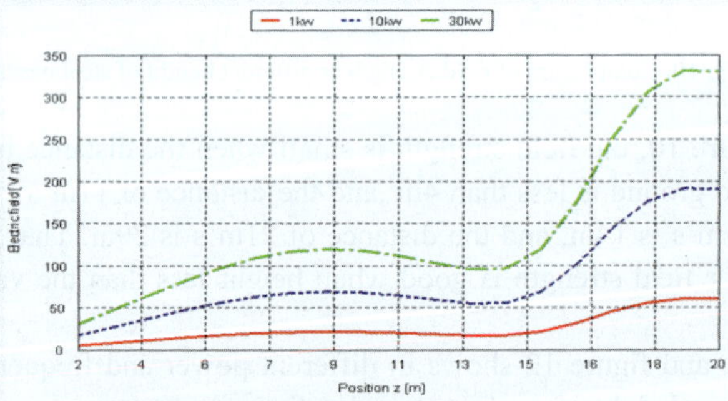

(c) 16m

Fig. 11. Comparison of field strength at different power.

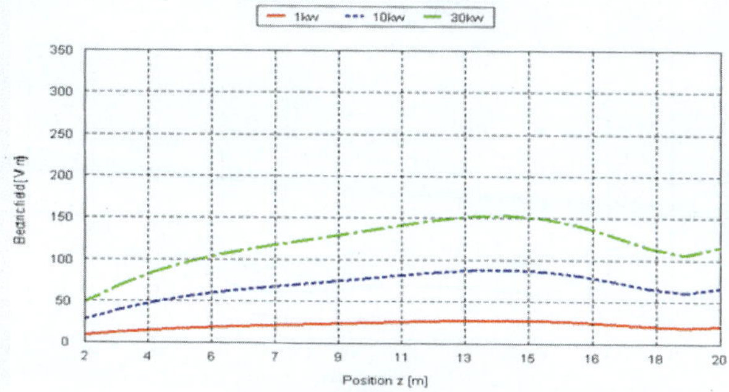

(d) 21m

Fig. 11. (*Cont'd*)

From figure 11, for different height, the ratio of two field strength at different power is still the root of ratio of powers. In addition, the law that the lower the steel mesh, the better the suppression for larger power electromagnetic field strength applies in different power.

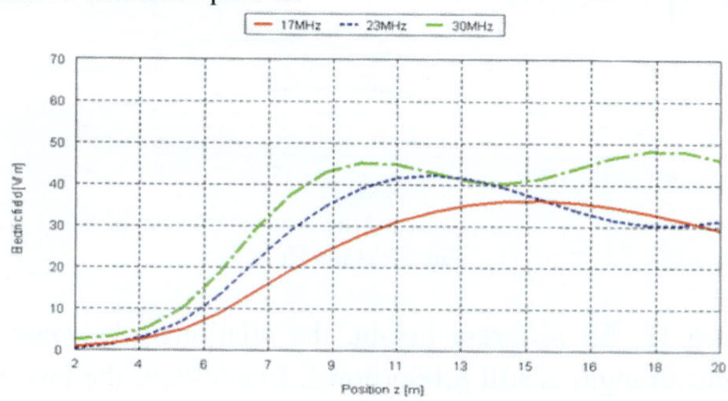

(a) 6m

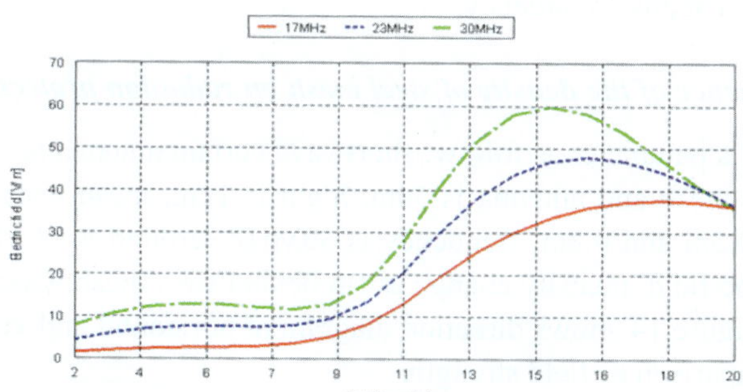

(b) 11m

Fig. 12. Comparison of field strength at different frequency.

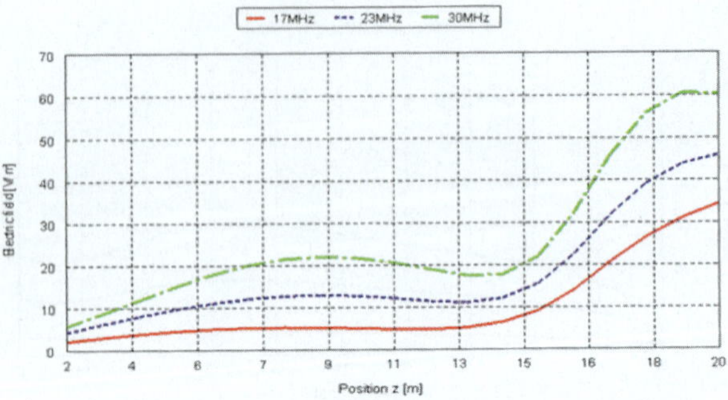

(c) 16m

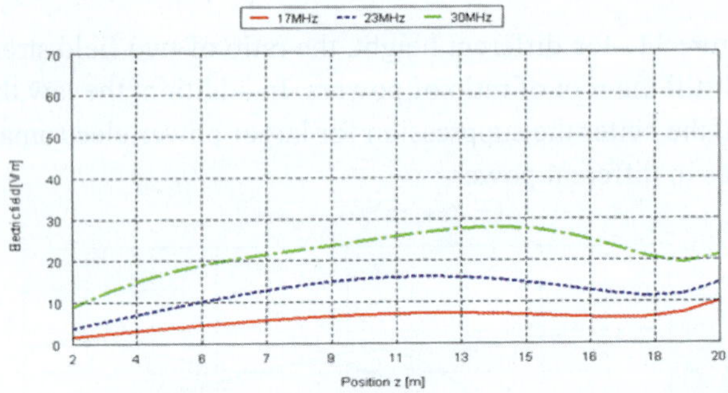

(d) 21m

Fig. 12. (*Cont'd*)

From figure 12, for different height, the relationsip between antenna frequency and field strength is still proportional. In addition, the law that the lower the steel mesh, the better the suppression for larger power electromagnetic field strength applies at any frenquency.

3.3. *The influence of the density of steel mesh on radiation protection*

Setting model's parameter as follow: shortwave curtain antenna is 1kw, distance between steel mesh and antenna is 50m, hight is 11m, mesh density is 0.05m, 0.1m, 0.2m, 0.5m and 0.8m, frequency is 30MHZ, ground is ideal conductive. We observe the field intensity change at 1m behind steel mesh when its height is 2m to 20m. Figure 14 shows direction diagram at horizontal and vertical, figure 15 shows comparison of field strength.

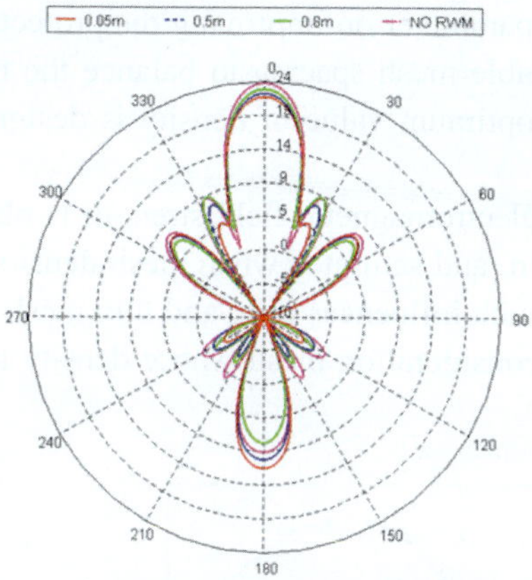

(a) direction graph at horizontal

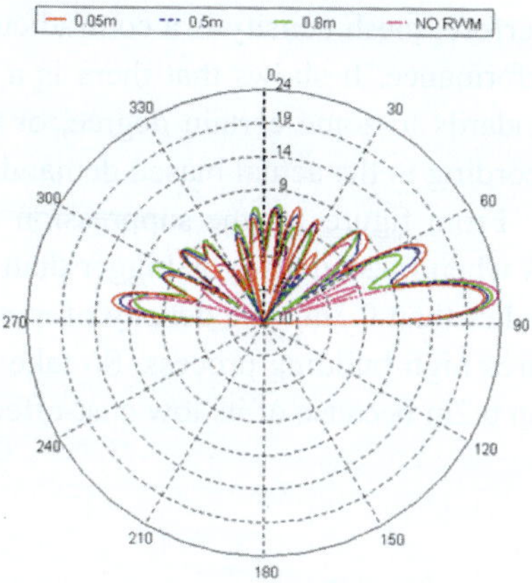

(b) direction graph at vertical

Fig. 13. Direction graph at horizontal and vertical.

From simulation results, when mesh density is 0.05m, 0.1m, 0.2m, 0.5m and 0.8m, their attenuation of mainlobe gain are 2.04dB, 2.02dB,1.95dB,1.54dB and 0.85dB. Figure 14 shows comprasion of direction graph whose mesh density is 0.05m,0.5m and 0.8m and no steel mesh at 30MHz. As we can see, with the increasing of mesh density, the mainlobe gain decreases, the attenuation of mainlobe gain decreases, the elevation is constant. This shows the thiner the steel mesh, the smaller the influence of antenna radiation.

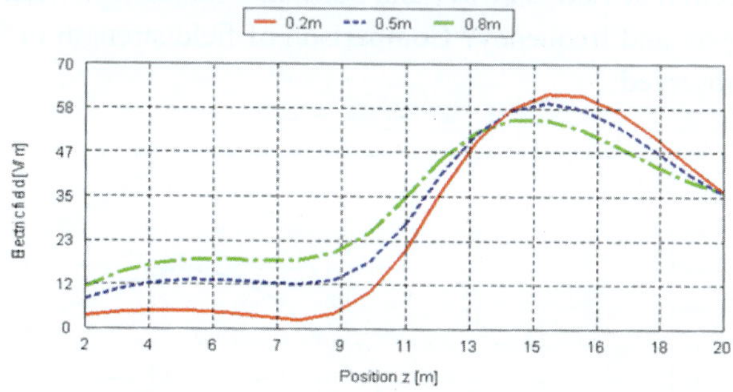

Fig. 14. Comparison of field strength at different density of steel mesh.

Figure 14 shows comparison of field strength at 0.2m's, 0.5m's and 0.8m's density of steel mesh. As we can see,with the increasing of densituy, the value of electromagnetic field strength increases, and the height range that can satisfy with the common exposure standard (30MHz -12.23V/m) decreases. This shows the thiner the mesh,the better the suppression for cuitain antenna electromagnetic field strength.

The thiner the mesh, the better the suppression for cuitain antenna electromagnetic field strength, while the bigger the influence of antenna radiation.

Therfore, mesh density is a contradictory parameter on improving the protective performance. It shows that there is a suitable mesh spacing to balance the two standards to some certain degree, or the optimum value of density is designed according to the actual biased demand.

From figure 15, the suppression for electromagnetic field strength is obvious when mesh density is bigger than 0.2m, and saturated when mesh density is smaller than 0.2m. The density of mesh much influences cost, and thin mesh requires high building process. So take no consideration mesh whose density less than 0.2m because of its low cost-effective.

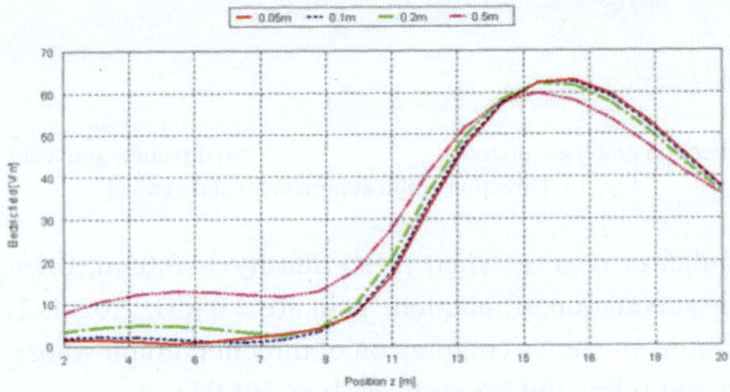

Fig. 15. Comprasion of field strength at different density of steel mesh.

From the above simulation, there are some lows between mesh density and radiation protection at 1kW's power and 30MHz's frequency. What are the lows at different power and frequency? Comparison of field strength in figure 16 and 17 should be observed.

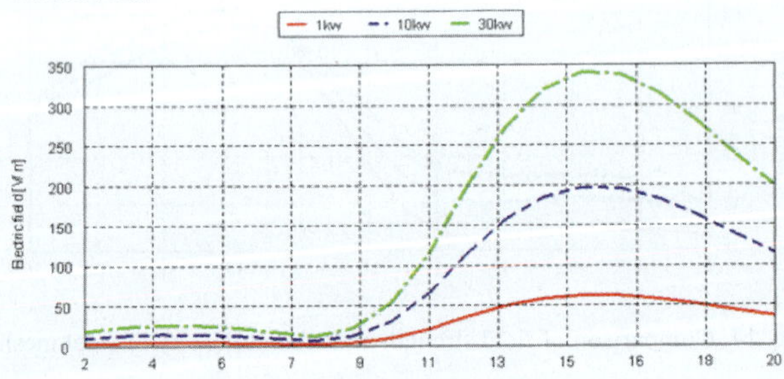

(a) 0.2m

Fig. 16. Comparison of field strength at different power.

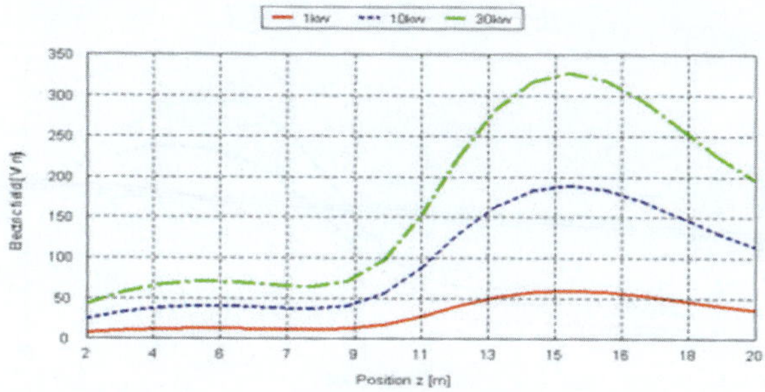

(b) 0.5m

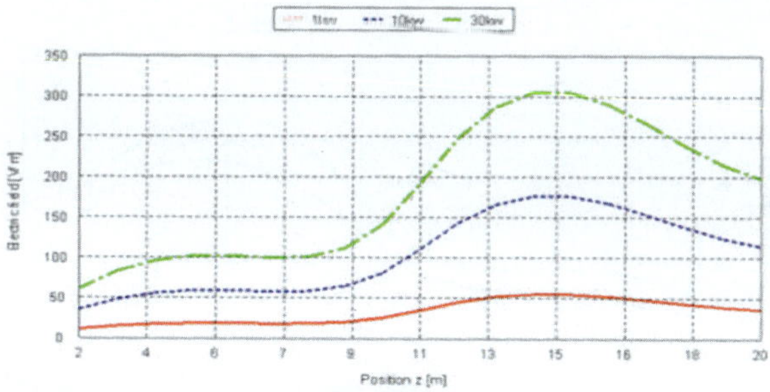

(c) 0.8m

Fig. 16. (*Cont'd*)

From figure 16, for different density, the ratio of two field strength at different power is still the root of ratio of powers. In addition, the law that the thiner the steel mesh, the better the suppression for larger power electromagnetic field strength applies in different power.

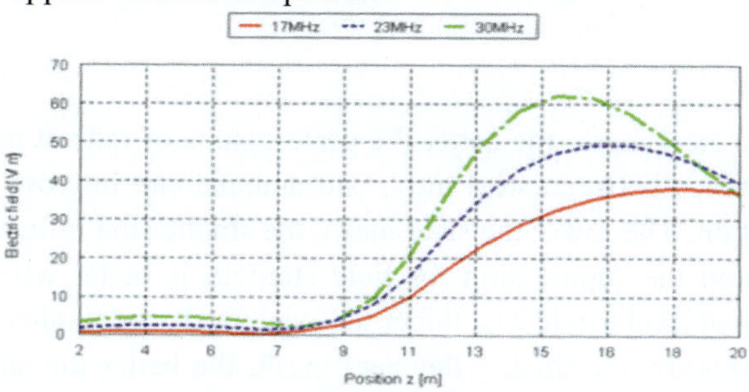

(a) 0.2m

Fig. 17. Comparison of field strength at different frequency.

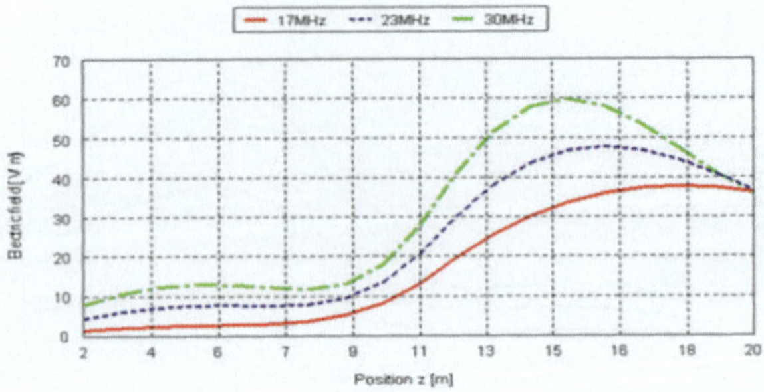

(b) 0.5m

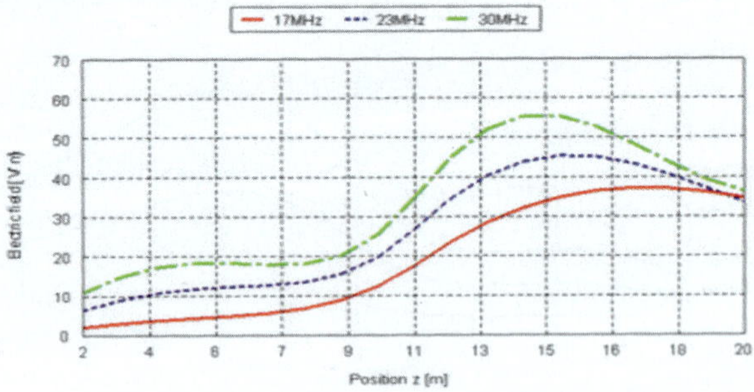

(c) 0.8m

Fig. 17. (*Cont'd*)

From figure 17, for different density, the relationsip between antenna frequency and field strength is still proportional. In addition, the law that the lower the steel mesh, the better the suppression for larger power electromagnetic field strength applies at any frequency.

4. Conclusion

The farther the steel mesh, the better the performance of radiant protection, and increasing distance between steel mesh and antenna can improve the effect of radiant protection. The lower the steel mesh, the smaller the influence on antenna radiation, and the suppression for field strength is better when height less than the height of steel mesh. The thiner the steel mesh, the smaller the influence on antenna radiation, the thicker the steel mesh, the better the suppression for electromagnetic field strength, and exsisting optimum value of density. In addition, relationship between field strength and distance is inverse in the far field. The ratio of two field strength at different power is the root of ratio of powers, and relationship between field strength and frequency is proportional.

References

1. Yee K.S. Numerical Solution of Initial Boundary Value Problems Involving Maxwell's Equations in Isotropic Media. IEEE Trans, On AP, 1966, 14(5): 302~307.
2. Ricktts. L.W, Bridger J.E, Miletta J. EMP Radiation and Protective Techniques [M]. John Wiley & Sons, Inc, 1976.
3. A. Reineix, A. Bound, B. Jecko. Electromagnetic Pulse Penetration into Reinforced-Concrete Buildings [J]. IEEE Trans, EMC1987, 29(1): 72~78.
4. P.E. Wilson, M.T. Ma. Factors Influence Material Shielding Effectivevess Measurement. Electromagnetic Fields Division National Bureau of Standards.
5. M. Kanda. Standards for Electromagnetic Field Measurements. IEEE, 1986, 74(1): 120~128.
6. Taflove A. Computational Eletrodynamics: The Finite-difference Time-domain Method [M]. Boston: Artech House, 2000.
7. J.P. Berenger. A Perfectly Matched Layer For The Absorption of Electromagnetic Waves. J. Computational Physics, 1994, 114: 185~200.

The application of mobile database model based in ad hoc network

T. Zhang[1†]
Information Commission Office, Heilongjiang University, Harbin, China
[†]*Email: Zhangtao2668@126.com*

The traditional mobile database model in Ad hoc network may bring about the problems such as apparent increase in the number of transaction conflicts, increase in communication costs, etc. Therefore, it is necessary to improve the current mobile database model. In the meantime, the paper puts forward data synchronization and algorithms such as transaction redo etc. to enable mobile host to get access to the database efficiently and correctly. The experimental results showed that the new mobile database model has good stability.

Keywords: Ad Hoc; Transaction redo; Mobile database; Database model.

1. Overview

With the development of mobile database and related technologies, mobile database makes people feel the changes it brings to people's life and work – it meets the needs that "people can get the information they need anytime and anywhere." That is to say a new era has opened and it will impact on us in every aspect gradually. Traditional mobile database model mainly includes server, mobile support node and mobile host. The traditional mobile database has the following problems: the location of the mobile node is not fixed, the network communication is not symmetrical, and disconnections of the network occur frequently, the storage capacity and the processing ability are limited. However, the network based on traditional mobile database model has a large number of basic communication facilities such as base stations, mobile switching centers, etc. [1], and most of them have been built around the city. Therefore, it is not necessary to consider too much about the effects of environment. As a multi-hop, center less, self-organizing wireless network, Ad hoc network is adopted

---

[1] Tao Zhang

mostly because the environment or other factors are not conducive to laying infrastructure or the infrastructure has been damaged. In reality, one of the typical applications of Ad hoc network is for the investigation teams to perform field missions, to help disaster relief, and mine surveying, etc. In Ad hoc network, in view of the reasons such as limited capacity of energy and storage of the mobile host, limited bandwidth, and frequent disconnections with the master server, etc., it will bring about a lot of problems to apply the traditional database model to Ad hoc networks, it is necessary to improve the traditional mobile database model to accommodate Ad hoc network. [2]

Many experts both at home and abroad have done in-depth study on mobile databases, including Mallikharjuna Rao, N. Naidu, etc. who have applied fuzzy database to mobile database model, thus it solved the problem of uncertain data storage, improved the performance of mobile database model; Le Gruenwald, Shankar M. Banik, who also provides the methods of managing real-time database transaction in Ad hoc network. Renmin University of China has made a research and development on "Kingbase Lite" which is a successful project based on embedded mobile database system of Pocket PC; Professor Zhou Xingming, etc. of National Defense Science and Technology University has proposed tertiary replication systems structure of mobile database, it divides the data replication into three levels: replication between servers, air replication, replication among move nodes. Besides, they have made a theoretical explanation of three-tiered structure; it is a great contribution to data replication techniques.

Nowadays the related technologies of mobile databases we have studied both at home and abroad mostly based on Ad hoc network. It usually contains a lot of base stations and other infrastructures such as data exchange centers. The communication quality of these facilities is relatively good; the mobile facilities are able to supplement or replace the power timely; network disconnections of mobile host may occur frequently, but the mobile facilities as well as server communication can be restored timely and the cost is not high; the geographical location and environment which human beings often do activities are not very bad. However, with the expansion of the scope of work, the environment is not conducive to building communication infrastructure or the communication facilities in such environment have been damaged. Thus, the existing network can't meet the corresponding requirements, so they need a new network technology to solve these problems. As a center less, no infrastructure, self-organizing wireless network, Ad hoc network is a good choice which can meet such needs. However, the traditional mobile database model didn't design for Ad hoc network, for example, the Ad hoc network nodes may join or exit at any time, changes in network topology occur at any time, the communication cost with the remote server is high, and disconnections with remote server often occur. Besides, the environment may affect the communication quality; the power capacity is limited, etc. Therefore, the traditional mobile database model cannot adapt to Ad hoc networks well. It is necessary to improve the traditional mobile database model, propose some new algorithms, and provide some solutions to the problems. These are the main research content of this paper.

2. Mobile Database Model Based on Ad Hoc Networks

Ad Hoc networks are self-organizing networks without infrastructure. The network communication quality is poor, the bandwidth is limited and the network coverage is small, the energy and storage capacity of the mobile host is limited, and disconnections with the master server often occur, etc. Therefore, to apply the traditional mobile database model to Ad hoc networks will bring many problems such as increase in communication cost and the number of transaction conflicts, longer time in transaction delay [3], and rise in failure or error rate of transactions, increase in false transaction operations, even cause breakdown of the entire network because of long time disconnection with the master server. In order to apply mobile database technology to Ad hoc networks, it is necessary to improve the traditional mobile database model to accommodate the Ad hoc network.

2.1. *Mobile database model*

Mobile database model consists of four major elements, as shown in figure 1:

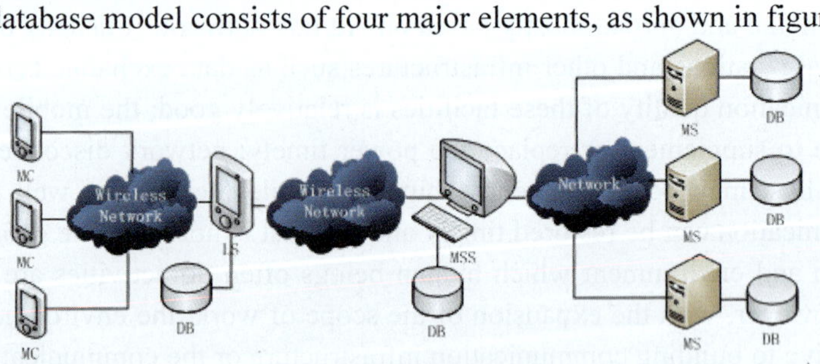

Fig. 1. Newmobile database model.

(1) Master server (MS): - generally it is a fixed node in the network which has its own database that saves all the data for users to use. It only contacts with local server.

(2) Local server (LS): it is selected randomly from a mobile host which belongs to mobile nodes in Ad hoc networks. It saves the copy of partial data of the master server. As an intermediary between the master server and the

mobile host, it is responsible for processing the transaction requests of mobile host. In addition, since it is mobile node itself, the performance will be affected by environment, self-configuration, power and other factors [4].

(3) Mobile host (MC): The processing ability, storage capacity and energy are limited; the cache can store frequently-used data. Since it is in Ad hoc networks, network disconnection often occurs, thus its reliability is relatively low, and the time of network delay is long.

(4) Mobile support node (MSS): it is located in a fixed network. It can be a general host as well as a server which is responsible for data transmission between the mobile host and the server.

The relationship between the mobile host and local server still complies with traditional mobile database model. The difference is that the local server is selected from the mobile host, and it is located in Ad hoc networks. Mobile host still has cache which saves part of frequently-used data in order to reduce the frequency of users' accession to the database server, thereby improving the performance of the system.

2.2. *Operating mode*

Add local server to mobile database model. The local server acts as the role of local master server, connects the mobile host and the main server. The operating mode changes from the original one: mobile host – master server to the existing operating mode: mobile host – local server - master server. This will bring new features which are different from traditional mobile database.

When the mobile host needs to access the data:

Step 1. First, check whether there are data the mobile host need in our own cache. If we have, access to it directly. If we don't have, apply for data requests to the local server, execute Step 2.

Step 2. If there are application data in the local server's database, apply for accessing the data and returning the results to the mobile host [5]; If there aren't any data for the mobile host to access in local server's database, execute Step 3.

Step 3. local server accesses the main server, downloads the data to local server's database, and then offers it for the mobile host to access. If data updates occur on a local server, the updated data will be uploaded to the master server.

The core idea of the improved mobile database model is to take the local server as an intermediary between the master server and mobile host, to act as master server in local Ad hoc network. Its operating principle is shown in Figure 2

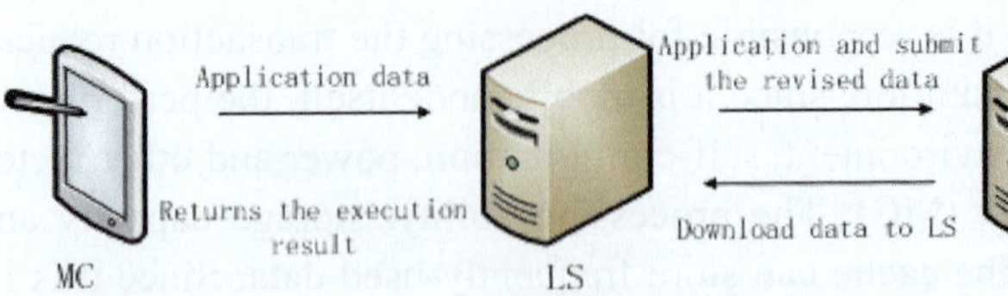

Fig. 2. The operating mode of mobile database.

2.3. *Main features*

(1) Save a copy of partial data of the master server on a local server. It is responsible for obtaining data and submitting data updates to the master server, and when all the mobile hosts want to access the data cache [6], if they don't have, they just send a request to the local server rather than contacting the master server directly.

(2) Considering the characteristics of Ad hoc networks and one of the applications is in the wild, ocean or remote places, the communication will be affected by the environment. So it is not practical for all the mobile hosts to contact the master server. Therefore, in the new mobile database model [7], all the mobile hosts will contact with the local server only. This can avoid that the mobile hosts contact with the master server because they do not have the data needed in the cache and thus bring a large number of data requests, reduce unnecessary bandwidth waste of uplink and the time delay of data requests. Therefore, it saves energy of mobile host, processes transactions faster, and gives feedback of the execution results to the clients as soon as possible.

(3) The master server adopts the combination strategies of the real-time broadcast and periodic broadcast: after updating the data on the master server, the reasonable choice of broadcast strategy will be decided based on data usage frequency; data updates of local server will use immediate update strategy which will upload to the master server immediately. It can reduce the probability of a mobile host reads invalid data and ensure that the mobile host uses the latest version of the data as far as possible. Meanwhile, to guarantee that the data stored in the database of master server is always the correct version and the latest.

(4) The cache of the mobile host can store frequently-used data, but the storage capacity is small. The database of local server stores a copy of all or most of the data of the task, the data in the cache is only a subset in the local server's database. Ad hoc networks can be considered as a small self-running network, the local server is equivalent to the master server, so the updated data in the cache will promptly upload to the local server. For mo-

bile cache and the local server in local Ad hoc networks, always guarantee that the data in the database of the local server is the latest version.

(5) Since the mobile host only contacts with the local server, local data server stores a copy of all or most of the data of the task [8], once disconnection occurs with the remote server, the local Ad hoc networks will not be affected much in a period of time, the mobile host will work properly, all updated data of transactions will store on the local server, after reconnection of the network, these updated data will upload to the master server.

(6) The new mobile database model based on Ad hoc adopts data update strategy of weak consistency. In view of frequent network disconnections, limited communication bandwidth and other conditions, the strategy allows inconsistencies between data copy stored on the local host and partial data of the major version on the master server for a certain time. Then, maintain data consistency between the data of major version and the copy data through corresponding data update.

3. The Study of Data Synchronization Strategy in Mobile Database

3.1. *Outdated data detection*

In Ad hoc networks, due to frequent disconnections of wireless networks as well as network communications etc., it is impossible to make consistency between the data in the local server and the master server, even the immediate updating strategy has been adopted. In addition, the accessed data of mobile host in the new model are data on its own cache or on the local server. It is inevitable for some transactions to use outdated data, but for clients, they expect to use the latest and correct version of the data. [6]

Firstly, give the format of storing data in the database. It is composed of two parts, including the data content and the data timestamp of master server updates respectively. The data timestamp of master server updates refers to recently updated time took place in the master server; data contents also contain several items which have been reflected in the program. [7]

Definition 3 Outdated data refers to data which has not been updated timely and became invalid data.

Clients want to use the latest version of the data, the newer of the version, the more accurate, the greater value it creates for clients who use the data. Therefore, the outdated data should be identified as soon as possible. However, how to judge the data that transaction uses on the local server is outdated [8] This is the problem we're going to discuss below. Assume that the update time of master server of datad on the local server is T_{nu}, set the start time of using data d on the local server T_{ls}, the methods of judging the use of outdated data on the local server are as follows:

1. If $T_{ls}<T_{nu}$, data d used in the transaction is not outdated.
2. If $T_{ls}>=T_{nu}$, data d used in the transaction is outdated.

The algorithm of judging the outdated data used in the transaction is simple. When the start time of using data d in the transaction on the local server happened before the updated time of data d on the master server, the transaction is considered not to use the outdated data; otherwise the transaction is considered to use the outdated data.

3.2. *Redo network transactions*

Algorithm 1: Read transactions redo

Step 1 If $T_{ls}<T_{nu}$, execute Step 3; otherwise, execute Step2.

Step 2 If$T_{ls}>=T_{nu}$, check the level of the transaction:

1. If it is a slight level, you do not need to redo, turn to Step 3.
2. If it is a general level, continue to execute the transaction, meanwhile the local server sends a prompt of outdated data to the mobile host, let (mobile host) users choose whether to redo: If the user decides not to redo, then turn to Step 3; If the user decides to cancel, then cancel the transaction, restart and execute Step l; If feedbacks haven't been received, it defaults not redoing necessarily, execute Step 3.
3. If it is an important level, cancel the transaction, restart and execute Step 1.

Step 3 the end of reading transaction.

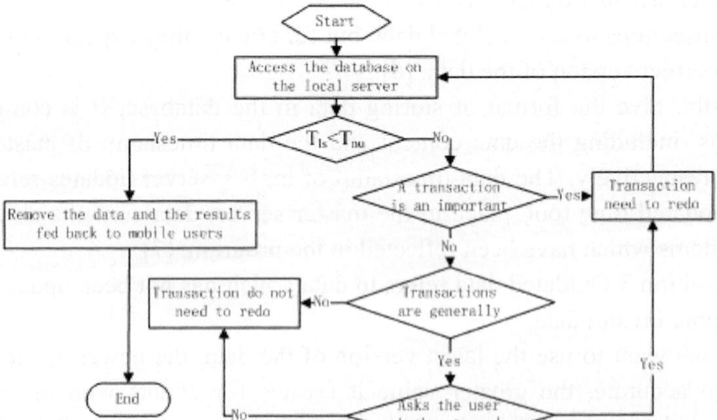

Fig. 3. Read transaction flow chart.

Algorithm 2: write transactions redo

The whole execution process can be divided into two stages, the first stage is to update the data of the local server, and the second stage is to update the data of the master server.[9]

Step 1 In the first stage, when $T_{ls}>=T_{nu}$, cancel the transaction and initiate a new transaction to dothe same work; otherwise, continue to execute Step 2.

Step 2 In the second stage, before updating the data of the master server, to apply for adding a write-lock to the corresponding data of the master server.

1. If the application is failed, it shows that the data is being used, then cancel this transaction, and initiate a new transaction to do the same work, execute Step 1.

2. If the application is successful, submit the local data updates to the master server to conduct the corresponding data update, and then execute Step 3.

Step 3 complete the data updates on the master server, that's the end of write transaction.[10]

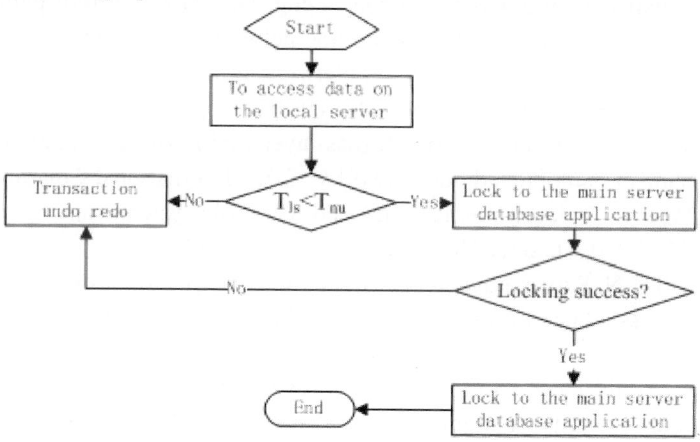

Fig. 4. Write transaction flow chart.

4. Conclusion

Since the traditional mobile database model cannot adapt to Ad hoc network well, the paper has improved the traditional model. In addition, aiming at the problems of data collision and transaction redo because of using outdated data which the new model has brought about, methods have been designed to solve the corresponding problems. Finally, the execution time and data conflicts of the transactions have significantly reduced. It shows that the new model can adapt to Ad hoc network better, and it has achieved an ideal result; besides it explains that the approach to the problem which the new model has brought about is feasible. It guarantees that the new model is able to run steadily in Ad hoc network.

References

1. Li Gang, Sun Hongmei, "Discovering AODV-based multipath routes in wireless ad hoc networks", Security and Communication Networks, Vol. 5, No. 4, (2012), pp 374-383.
2. Kamilov M et al., "A Real-time Routing System for Vehicular Ad Hoc Networks", Proceedings 2009 First International Workshop on Software Technologies for Future.
3. Lei Lei, Wang Dan, Zhou Liang, Chen Xiaoming, Cai Shengsuo, "Link availability estimation based reliable routing for aeronautical ad hoc networks", Ad Hoc Networks, Vol 20, (2014), pp. 53-63. Dependable Distributed Systems, (2009), pp 88-92.
4. Cai Zheng Ce, "Studies on the secure routing protocols of mobile AD hoc networks", Advanced Materials Research, Vol. 734-737, (2013), pp. 3203-3206.
5. Miao Jiansong, Wu Muqing, Zheng Feng, Guo Heng, "Multi-path routing based on link lifetime in mobile ad hoc networks", Proceedings - 2010 2nd IEEE International Conference on Network Infrastructure and Digital Content, IC-NIDC 2010, September 24-26, (2010), pp. 644-648.
6. Letor Nicolas, Blondia Chris, "Cross-layer tuning of the neighbor sensing mechanism in mobile ad hoc networks", 2010 International Congress on Ultra Modern Telecommunications and Control Systems and Workshops, ICUMT 2010, October 18-20, (2010), pp. 784-789.
7. Nagesh K.N, Satyanarayana D, "An analytical expression for k-connectivity of wireless ad hoc networks", Telkomnika, Vol. 12, No. 1, (2014), pp. 179-188.
8. Zarei Mehdi, Faez Karim, "Modified reverse AODV routing algorithm using route stability in mobile Ad hoc networks", IEEE INMIC 2008: 12th IEEE International Multitopic Conference - Conference Proceedings, December 23-24, (2008), pp. 255-259.
9. Song Chao, Liu Ming etc, "Distributed real-time information based routing protocol in vehicular ad-hoc networks", Journal of Software, Vol. 22, No. 3, (2011), pp. 466-480.
10. Qasim Nadia, Said Fatin, Aghvami Hamid, "Performance evaluation of mobile ad hoc networking protocols", Lecture Notes in Electrical Engineering, Vol. 39, (2009), pp. 219-229.

A new MAC protocol for underwater acoustic sensor networks

W. Lin [1†]

Department of Computer Science, Minjiang University, Fuzhou, China,
[†]E-mail: linwen21@163.com

Due to the large propagation delay and low data rates of underwater acoustic channel, the design of media access control (MAC) protocols in underwater acoustic sensor networks (UWASNs) faces many challenges. In these circumstances, MAC protocols designed for terrestrial networks cannot be directly used in UWASNs. Moreover, the large propagation delay causes a serious hidden terminal problem for MAC protocols. In this paper, we proposed a novel MAC protocol based on the MACA-BI protocol, which uses receiver driven handshaking mechanism. It is shown that by using adaptive transmission time and adaptive back-off time, our protocol can solve the hidden terminal problem and improve the channel utilization. The results of simulation are given, which shows that our protocol outperforms in network throughput and end to end delay.

Keywords: Underwater acoustic sensor networks; MACA-BI; MAC protocol.

1. Introduction and Related Work

Underwater acoustic sensor networks (UWASNs) and terrestrial wireless sensor networks have many different characteristics [1-2]. The MAC designed for terrestrial wireless sensor networks cannot be directly used because the long propagation delay and limit bandwidth. The MAC protocol has a great influence on the performance of UWASNs. Therefore, the MAC protocol is a research hotspot in UWASNs.

Recently a lot of attention has been focused on handshaking-based MAC protocols to improve the performance of UWASNs [3-4]. In handshaking-based protocols, a node must exchange the control packet before sending a data packet. Those control packets can notify neighbor nodes about the ongoing transmission, the handshaking mechanism can effectively reduce packet collisions. Hai-Heng et al. propose the multiple access collision avoidance protocol for underwater (MACAU) in [5], which is based on the widely known MACA protocol. MACAU uses 4-way handshake to avoid packet collision. Three aspects of the MACA protocol are improved, namely, the state transition

[1] Wen Lin

rules, the packet forwarding strategy, and the back-off algorithm. However, MACAU cannot solve the hidden terminal problem. Molins et al. introduce S-FAMA protocol in[6]. In the S-FAMA protocol, all control packets and data packets can only be transmitted at the beginning of each time slot. However, S-FAMA requires clock synchronization between all nodes. The long slot length requirement and the handshaking mechanism also affect UWASNs throughput. Chirdchoo et al. propose the MACA-MN protocol in [7], which sends multiple control packets to multiple neighbor nodes, it improves the channel utilization by forming a train of packets destined for multiple neighbor nodes during each round of handshake.

In terrestrial wireless networks, handshaking-based protocols are common. These control packets can notify other neighbors about the ongoing transmission in time. However, the handshaking mechanism was not destined for UWASNs where large propagation delays are prevalent. In the UWASNs, the handshaking mechanism cannot solve the hidden terminal problem completely. Several new MAC protocols have been proposed to solve the hidden terminal problem [8-11]. However, the protocols above which employ multiple way handshaking approach inevitably amplify the effect of large propagation delay, which restricts the throughput and the end-to-end delay.

In this paper, we propose an asynchronous random access MAC protocol based MACA-BI protocol [12]. MACA-BI is a simplified version of the MACA protocol, which is based on the Request-to-send (RTS) and Clear-to-send (CTS) handshake. MACA-BI protocol only uses Ready-to-receive (RTR) control packet which can be viewed as an invitation by the receiver to transmit. As with MACA-BI protocol, our protocol utilizes two-way handshaking mechanism, in order to reduce the time of the control packet using channel. In addition, our protocol can overcome the problems of large propagation delay and hidden terminal by using adaptive transmission time and adaptive back-off time.

The rest of this paper is organized as follow. In section 2, we proposed a novel MAC protocol based on MACA-BI protocol. In section 3, the simulations are carried out, and further discussions are provided. Finally we give our conclusions in section 4.

2. Protocol Design

2.1. *Protocol overview*

In our proposed protocol, we use two-way handshaking mechanism (RTR-DATA) similarly to the MACA-BI protocol. A node who wants to transmit a data packet will waits for an invitation by an intended receiving node, which is a receiver-initiated method. The receiving node uses an RTR control packet to invite intended sending nodes to send DATA packets, the RTR control packet also has to notify neighbor nodes. When neighbor nodes hear the RTR packet, it cannot transmit other control packets to prevent the occurrence of packet conflict. When an intended sending node hears the RTR packet, it will respond with a DATA packet, provided that it is currently not involved in a handshake with another node, and the intended sending node also cannot in the back-off state or silent state. When a no intended sender hears the RTR packet (xRTR), it will enter the back-off state, and sets its back-off timer to $T_{back\text{-}off}$. In the traditional MACA-BI protocol, nodes set their back-off timer to $2T_{max}+T_{data}$, where T_{max} is maximum inter-node propagation delay,T_{data} is the transmission time of each DATA packet. Due to the slow propagation of sound under water, the back-off time of the MACA-BI protocol is too large. Compare to MACA-BI who uses the fixed back-off time, we use a method of adaptive back-off time to minimize the back-off time.

For the MACA-BI protocol, after transmitting receiving an RTR packet, the intended nodesender keeps silent keeps silent and waits for a CTS packet replied from the destinationwill send a DATA packet immediately. A collision between DATA packets and control packets may occurs because of different propagation delays. Figure 1 illustrates failure of the handshaking mechanism of MACA-BI to ensure collision avoidance in UWASNs. In Figure 1, node 1 is close to node 2 and node 3 is hidden from it. We can see that node 3 has sent an RTR packet before receiving node 2's RTR packet, so the node 3's RTR packet collision with node 1's DATA packet. The real reason is that note 2's RTR packets cannot reach potential jamming node (node 3) on time and thus are unable to inform node 3 of the forthcoming DATA transmission. In order to alleviate the hidden terminal problem, we must use a new handshaking mechanism in our proposed protocol. In our protocol, according to the propagation delay between intended sending node and all neighbor nodes, the intended sender decides when to send the DATA packet. We call this method as the adaptive transmission time.

Our protocol is a "receiver driven"mechanism, the timing of sending RTR packets is an important issue. In this paper, we simply set the Poisson distribution for the time between RTR sending.

At last, similar to same conventional handshake protocols, when a source node does not receive returned DATA packet in response to its previous RTR packet, the receiving node shall schedule an RTR re-transmission using binary exponential back off (BEB) algorithm. Our protocol also uses the BEB algorithm.

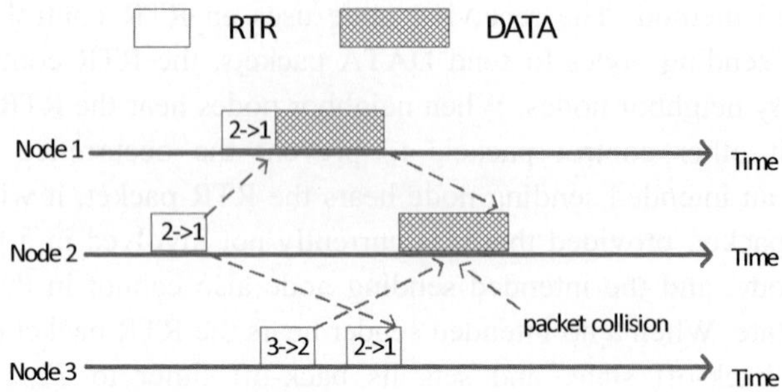

Fig. 1. The hidden terminal problem of MACA-BI.

2.2. *Initialization phase*

First, in our protocol we assume that all nodes know the propagation delay with each one-hop neighbor node. In fact, through a simple exchange of control packets, we can measure the propagation delay between two nodes. However the propagation delay is affected by many factors in the UWASNs. In our protocol, we assume that the propagation delay between two nodes is constant in a short time period. We can periodic update propagation delay through the RTR/DATA packet exchange during the protocol working.

2.3. *Adaptive transmission time*

In contrast to the traditional MACA-BIprotocol, our protocoluses the adaptive transmission time to reduce the probability of packet collisions. The duration of the silent time is determined by the propagation delay between the intended sending node and its neighbor nodes. When an intended sender hears an RTR packet, it sets its silent timer to T_{silent} using the following equations:

$$\begin{cases} \mathrm{T}_{\text{silent}} = 0 & , \quad T_{sr} = \max(T_{rx}) \\ \mathrm{T}_{\text{silent}} = 2\times(\max(T_{rx}) - T_{sr}) & , \quad T_{sr} < \max(T_{rx}) \end{cases} \tag{1}$$

Where, we assume T_{sr} is the propagation delay between a intended sending node and a intended receiving node, T_{rx} is the propagation delay between intended receiving node and its neighbor node x, $max(T_{rx})$ is the maximum value of all of T_{rx}.

As show in Figure 2, when node 1 receives an RTR packet, it must calculate its silent time T_{silent}, and enter the silent state. When the silent time is over, node 1 can send the DATA packet immediately. In our protocol, using the method of adaptive transmission time, the RTR packet sending by node 2 can inform node 3 in time.

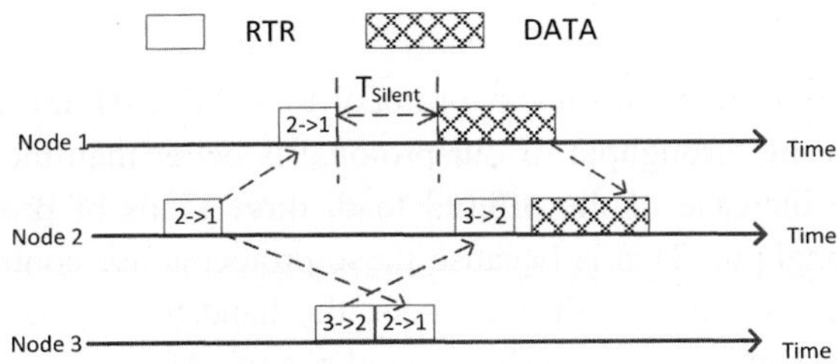

Fig. 2. The method of adaptive transmission time.

2.4. *Adaptive transmission time*

When a no-intended sender hears an xRTR packet, it sets its back-off timer to $T_{back\text{-}off}$ using the following equation:

$$\mathrm{T}_{\text{back-off,x}} = T_{data} + 2\times\left[\max(T_{rx}) - T_{rx}\right] \tag{2}$$

The adaptive back-off time can reduce the back-off time when no intended senders hear an xRTR packet. Then we show the minimum $T_{back\text{-}off}$and the maximum $T_{back\text{-}off}$for our protocol. When a node receives an xRTR packet, if the propagation delay between the node and the receiving node is maximum value of all T_{rx}, it sets the adaptive back-off time to T_{data}. If the propagation delay between the node and the receiving node is minimum value of all T_{rx}, it sets the adaptive back-off time to $T_{data}+2[max(T_{rx})-T_{rx}]$. However, in the traditional MACA-BI protocol, all no intended senders must set the fixed back-off time to $T_{data}+2max(T_{rx})$. Due to the high propagation delay, the back-off time of MACA-BI is too long. We use the adaptive back-off time to minimize the back-off time, it can improve the channel utilization and decrease the end to end delay.

3. Simulations and Results

3.1. *Simulation settings*

The simulation parameters are as follows, we simulate a random network where ten underwater acoustic sensor nodes are uniformly distributed in a 5000m*5000m area. A sending node randomly chooses another node in the network as the destination. We model packets traffic as a Poisson arrival process. The bandwidth of the channel is set to 1kbps and the rate of data transmission is 1kbps. The propagation speed of the acoustic signal is 1500m/s. The transmission range of every node is set to be 1500m. The average data packet length is 1024 bytes and the RTR packets be set 16 bytes. All acoustic modems are used in half-duplex mode.

3.2. *Simulation results*

First we compare the throughput of our protocol to MACA-BI and MACA-U. Figure 3 shows that the throughput of our protocol is better than the other two protocols. With the increase of the offered load, three kinds of protocols can maintain stable throughput. This is because those protocols use control packets handshake before transmitting DATA packets, the handshake mechanism can effectively avoid packets collisions. The throughput of MACA-U and MACA-BI are less than our protocol because of the hidden terminal problem. InMACA-BI and MACA-U protocol, the no intended nodescannotreceive the control packets (RTR, RTS, CTS) in time. Consequently, the no intended nodes may interfere with ongoing packet transmission. Our protocol uses the method of adaptive transmission time to solve the hidden terminal problem, which can reduce the probability of packet collisions and increase the throughput. Specifically, the method of adaptive back-off time also can improve the throughput because all no intended nodes only keep minimum back-off time when they receive an xRTR packet.

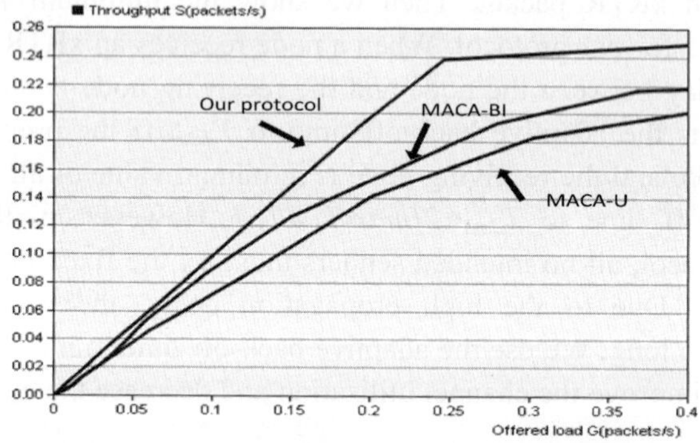

Fig. 3. Throughput versus offered load.

Then we evaluate the end to end delay of the three protocols. Figure 4 shows that MACA-U wastes most time on exchanging control packets than other two protocols. The reason is that the MACA-U protocol uses three-way handshake mechanism and other two protocols only use two-way handshake mechanism. By using the method of adaptive back-off time, our protocol can effectively reduce the impact of large propagation delay.

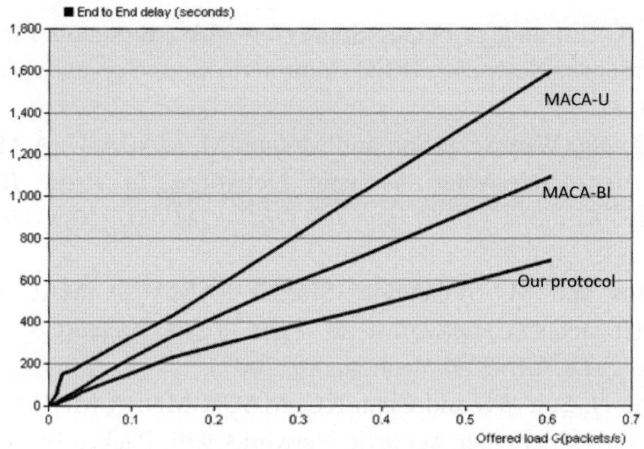

Fig. 4. End-to-end delay versus offered load.

4. Conclusions and Further Work

This paper presents a new MAC protocol for UWASNs. Besides adopting the two-way (RTR-DATA) handshaking mechanism of the MACA-BI protocol, it usesthe method of adaptive transmission time, which significantly alleviates the hidden terminal problem. In addition, our protocol is shown to achieve lowend to end delay, it can be attributed to the method of adaptive back-off time. In order to achieve a high maximum throughput and minimum delay, the sending rate of RTR packets must be carefully chosen. In further work, we will research for how to set an efficient sending rate of RTR packets.

Acknowledgments

This work was supported in part by the Middle and Young aged Educational Research Project of Fujian Province (Grant No.JA13253), and by Scientific Research Project of Minjing University (Grant No.MYK14009), and by Start-up Fund for Talent Introduction of Minjiang University.

References

1. Akyildiz I.F, Pompili. D and Melodia. T, Underwater acoustic sensor networks: research challenges, *Ad Hoc Networks*. **3**, 257(2005).
2. M. Chitre, S. Shahabodeen and M. Stojanovic, Underwater Acoustic Communications and Networking: Recent Advances and Future Challenges, *Marine Technology Society Journal*. **42**, 103 (2008).
3. Y. Noh and S. Shin, Survey on MAC protocols in Underwater Acoustic Sensor Networks, in *Proc. IEEE Int. Symposium on Communications and Information Technologies (ISCIT'14)*, (Incheon, Korea, 2014).

4. C. Keyu, C. En and Y. Fei, A Survey on MAC Protocols for Underwater Wireless Sensor Networks, IEEE Communications Surveys & Tutorials. **16**, 1433 (2014).
5. Hai-Heng. Ng, Wee-Seng Soh and Motani. M, MACA-U: A Media Access Protocol for Underwater Acoustic Networks, in *Proc. IEEE Global Telecommunications Conference (GLOBECOM'08)*, (New Orleans, USA, 2008).
6. M. Molins and M. Stojanovic, Slotted FAMA: A MAC protocol for underwater acoustic networks, in *Proc. IEEE Oceans Asia Pacific Conference (OCEANS'06)*, (Singapore, 2007).
7. Chirdchoo N, Soh WS and Chua KC, MACA-MN: A MACA-Based MAC Protocol for Underwater Acoustic Networks with Packet Train for Multiple Neighbors, in *Proc. Vehicular Technology Conference (VTC'08)*, (Singapore, 2008).
8. WH Liao and CC Huang, SF-MAC: A Spatially Fair MAC Protocol for Underwater Acoustic Sensor Networks, IEEE Sensors Journal. **12**, 1686 (2012).
9. Guo X, Frater MR and Ryan MJ, Design Of A Propagation-Delay-Tolerant Mac Protocol For Underwater Acoustic Sensor Networks, IEEE Journal of Oceanic Engineering. **34**, 170 (2009).
10. Noh Y, Lee U, Han S et al, DOTS: A Propagation Delay-Aware Opportunistic MAC Protocol for Mobile Underwater Networks, Mobile Computing IEEE Transactions on. **13**, 766 (2014).
11. Han S, Noh.Y, Lee.U and et al, M-FAMA: A multi-session MAC protocol for reliable underwater acoustic streams, in *Proc. IEEE INFOCOM, (INFOCOM'13)*, (Turin, Italy, 2013).
12. F. Talucci and M. Gerla, MACA-BI: A wireless MAC protocol for high speed ad hoc networking, in *Proc. IEEE Int. Conference on Universal Personal Communications Record (ICUPC'97)*, (San Diego, USA, 1997).

Design of a zigbee based real-time motor dynamic parameter monitoring system

Z. Y. Zou*, L. J. Xu† and Z. L. Kang
College of Mechanical and Electronic Engineering,
Sichuan Agricultural University, Ya'an,
Sichuan Province, 625014, China
*E-mails: *zouziyong111@163.com, †lijiaxu13@163.com*

The dynamic parameter of motor is an important basis for evaluating the quality of motor as well as a precondition for diagnosing its faults. The real-time motor dynamic parameter monitoring system presented in this paper is aiming at improving the flexibility of motor dynamic parameter acquisition and the convenience and reliability of data storage. The hardware is designed with CC2530 as the core integrating sensor signal acquiring circuits, an embedded controller, Flash storage and a LCD display, etc. In software design, Software realized ZStack protocol stack based wireless communication program, self customized embedded WinCE OS, transplanted embedded SQlite database on WinCE and realized data reading and writing operations. The test result shows that the mean packet drop ratio of the ZigBee network is 0.375%, the maximum deviation of temperature measuring is 0.7°C, the maximum deviation of vibration measuring is 0.9mm/s, the maximum deviation of current measuring is 0.8A, the maximum deviation of rotated speed measuring is 12r/Min.

Keywords: Motor; ZigBee network; Parameter monitoring; Embedded controller; SQlite embedded database.

1. Introduction

The motor may break down or even be damaged as a result of some abnormal operations. People's daily life, more or less, depends on all kinds of motor. And it will bring huge economic loss to industrial production and seriously affect people's normal lives once the motor breaks down and isn't controlled in time, so the motor plays an important role in social economy and people's life. Aiming at monitoring of motor operation parameters, many scholars have proposed a lot of methods [1-6]. In order to improve the flexibility of data acquisition, the reliability of data storage and the convenience of monitoring system installation, a ZigBee technology based monitoring scheme is proposed in this paper.

ZigBee technology is a short-distance wireless network communication technology with low energy consumption, cost, and complexity [7].It is applicable to multipoint data acquired and control application systems.

2. Overall Design of the System

The operation quality of a motor is evaluated mainly with its dynamic parameters, especially the temperature, vibration, current and rotated speed, while how to obtain the precise parameters is the foundation for accurately judging the motor operation quality. As shown in figure 1, the ZigBee based monitoring scheme proposed in this paper consists of two parts that are ZigBee monitoring nodes and ZigBee terminal display. Because the detection module of the temperature sensor, the detection module of the vibration sensor, the detection module of the current sensor and that of the rotated speed sensor shall be installed on different areas of the motor, so the ZigBee monitoring nodes comprise of the ZigBee temperature monitoring node, the ZigBee vibration monitoring node, the ZigBee current monitoring node and ZigBee rotated speed monitoring node. The ZigBee terminal display is mainly used for data displaying and storage.

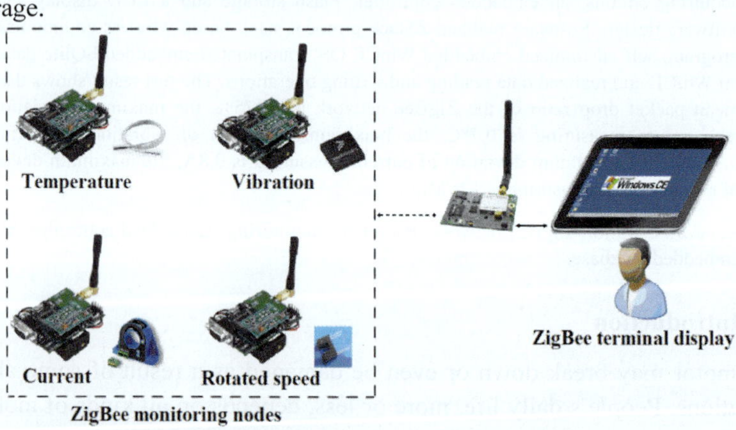

Fig. 1. Structure of the monitoring system.

3. Design of Zigbee Monitoring Nodes

3.1. *Hardware design of ZigBee monitoring nodes*

The ZigBee temperature monitoring node acquires the temperature signals with the PT1000 temperature sensor which forms a bridge detection circuit with 3 resistances. The output induced voltage of the bridge circuit is amplified by the instrumentation amplifier. The output port of the instrumentation amplifier is connected with the port P1.5 of the CC2530 wireless SCM so as to realize the detection of the motor temperature. The port P1.5 is the input port of the internal analog-digital converter of the CC2530 wireless SCM.

The ZigBee vibration monitoring node detects the vibration signals with MMA7660FC vibration detection module. The MMA7660FC module can detect six groups of vibration signals of up and down, left and right and front and back and can provide IIC digital interface output. The ports P1.1 and P1.2 of the CC2530 wireless SCM are respectively connected with the ports of SCL and SDA of the MMA7660FC module; and the port P2.1 of the CC2530 wireless SCM is connected with the port /INT of the MMA7660FC module. When the motor vibrates and the MMA7660FC module detects the vibration signals, the port /INT of the MMA7660FC module outputs falling edge-trigger signals which trigger the external interruption input port P2.1 of the CC2530 wireless SCM, and the CC2530 wireless SCM reads the numeric values of amplitude from the ports SCL and SDA of the MMA7660FC module.

The ZigBee current monitoring node adopts HA2000 Hall sensor detection module. The analog voltage output port of the HA2000 module is connected with the port P1.5 of the CC2530 wireless SCM to realize detection of motor current. Three HA2000 modules are required for three-phase motors for respectively detecting three phase currents of U, V and W.

The ZigBee rotated speed monitoring node acquires motor rotated speed with coding turntable, A11 infrared photocell and LM393 voltage comparator. The A11 infrared photocell outputs a pulse signal when the coding turntable turns a circle, and such pulse signal is output as a standard square-wave signal after shaping by the LM393 voltage comparator. The square-wave signal is linked to the port P2.4 of the CC2530 wireless SCM so as to realize detection of motor rotated speed. The port P2.4 is the input port of the internal counter of the CC2530 wireless SCM. The 4 monitoring nodes are powered by 12 V, 2 AH storage batteries.

3.2. *Software design of ZigBee monitoring nodes*

The secondary application development is performed for ZigBee monitoring nodes program on the basis of ZStackprotocol stack.Its features are: structuring with the thought of operation system, operation in event polling mechanism, and the system enters low power consumption mode after each layer is initialized. When any event happens, the system will be woke up and enters interrupt hander event, and it continues to enter the low power consumption mode upon completion, if there are several events happen simultaneously, it will determine the priority and dispose the events sequentially. The main working flow of the whole ZStack is roughly divided into system starting, drive initialization, OSAL initialization and task polling. The methodof programming is adding the tasks of sending data in the operation system.

4. Design of Zigbee Terminal Display

4.1. *Hardware design of the ZigBee terminal display*

ZigBee terminal display adopts the design concept of the core board with the base board added to improve the electromagnetic compatibility of the circuit design and maintenance convenience.Its structure is shown in Figure 2. The core board consists of a S3C2410 embedded controller, a 256MB SDRAM with two pieces of K4X1G163PC-FGC6 memory, and a 4 GBNAND Flashwith two pieces of K9GAG08U0E-S memory.The base board mainly consists of a power supply module, USB interface, JTAG interface, LCD interface, and URAT interface. The UART interface of the S3C2410 embedded controller is linked to the serial port of the ZigBee coordinator node to achieve communication between the ZigBee monitoring nodes. The IIC interface of the S3C2410 embedded controller is linked to the Wi-Fi communication module to realize communication with otherWi-Fi device. LCD uses seven-inch LCD touch screen AT070TN83V[8].

ZigBee terminal displayis powered by 12 V, 5 AH storage batteries.A 5 V and 3.3 V DC is provided after the power is regulated by AMS2596-5and AMS1117-3.3 regulating chips.

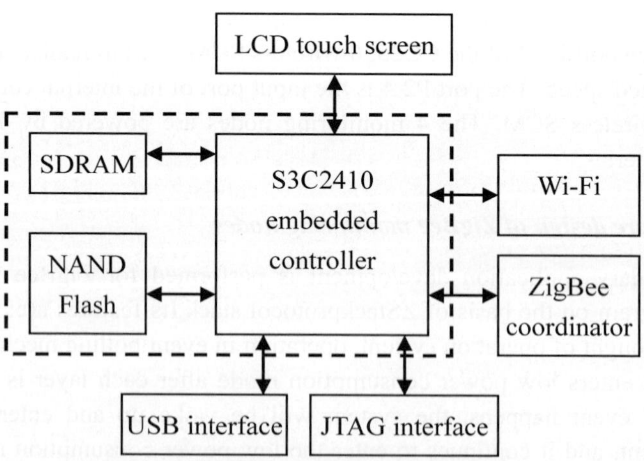

Fig. 2. Terminal display hardware structure.

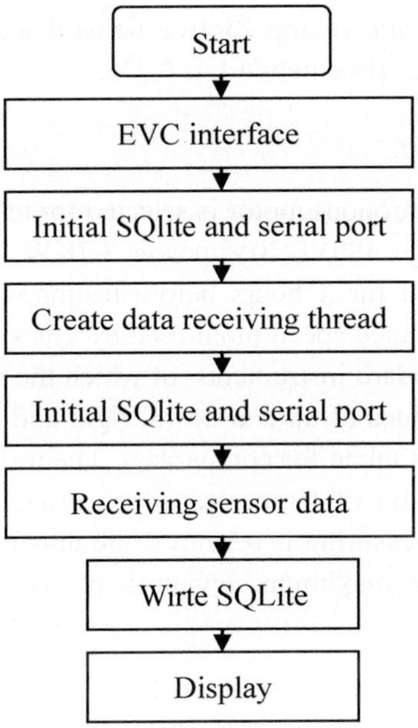

Fig. 3. The program flow chart of the data display.

4.2. *Software design of the ZigBee terminal display*

The embedded OS employs MICROSOFT's Platform6.0. According to the hardware structure of the ZigBee terminal display, the WinCE6.0 graphic OS is self customized[9]. In order to effectively manage the data, the SQlite embedded database is transplanted on WinCE6.0, which has features of small size, open-source, easy operation and requiring only hundreds of Kb operation memory. The SQlite database is realized by adding library files of sqlite.lib and sqlite.dll into application and calling their API function to realize database reading and writing [10].

The application is developed by employing the EVC integrated development environment. The program flow chart of the data display is shown as Figure 3.

5. Testing of System

5.1. *ZigBee network packet drop ratio test*

Given the test of the packet drop ratio of ZigBee network, data packets are sent through ZigBee monitoring nodes. The data packets received by the ZigBee terminal display are then counted to determine the data transfer stability of ZigBee network. The test results are shown in Table 1. The test result shows that the mean packet drop ratio of the ZigBee network is 0.375%, the maximum packet drop ratio of the ZigBee network is 0.5%.

5.2. *Data acquisition test*

A three-phase AC asynchronous motor is selected for testing, and the parameters of the motor are: voltage 380V/220V, power 3.7KW. The testing method: the motor runs without load for 2 hours before testing, the data of temperature, vibration, current and rotated speed monitored by this system are compared with the data detected by standard instruments, of which the vibration is compared by taking the maximum values of up and down, right and left, front and back, and the current of phase U is taken for comparison. The test result is shown as table 1, the maximum deviation of temperature measuring is 0.7°C, the maximum deviation of vibration measuring is 0.9mm/s, the maximum deviation of current measuring is 0.8A, the maximum deviation of rotated speed measuring is 12r/Min.

Table 1. Packet drop ratio of the ZigBee network.

The ZigBee monitoring nodes	Sent data packets	Received data packets	Packet drop ratio %
Temperature	1000	995	0.5
Vibration	1000	996	0.4
Current	1000	998	0.2
Rotated speed	1000	996	0.4

Table 2. Test data of temperature, vibration, current, rotated speed.

Number of times	Temp°C	error	Vibration mm/s	error
1	47.6	-0.3	2.6	0.4
2	47.3	-0.5	2.8	-0.2
3	47.9	0.4	2.2	0.6
4	47.4	0.7	2.4	-0.4
5	47.8	0.5	2.8	0.9
6	48.1	-0.6	2.9	-0.3
7	47.9	0.2	2.5	-0.1
8	48.3	-0.7	2.3	0.7
9	48.2	-0.5	2.6	0.3
10	48.5	0.1	2.7	0.8

Number of times	Current /A	error	Speed r/Min	error
1	5.8	-0.2	1878	11
2	5.3	-0.6	1885	9
3	5.5	0.3	1879	-4
4	5.7	-0.7	1891	15
5	5.4	0.2	1883	-7
6	5.6	-0.8	1887	-8
7	5.8	-0.5	1895	3
8	5.3	0.4	1898	-12
9	5.5	-0.4	1892	4
10	5.9	0.3	1885	9

6. Conclusions

(1) The ZigBee terminal display adopts the design concept of the core board with the base board added to facilitate maintenance. The acquired parameter are managed with the SQlite database and the data processing is reliable and efficient. This system employs an LCD touchscreen with a friendly human–machine interface that is easy to use.

(2) The mean packet drop ratio of the ZigBee network is 0.375%. The maximum deviation of temperature measuring is 0.7°C, the maximum deviation of vibration measuring is 0.9mm/s, the maximum deviation of current measuring is 0.8A, the maximum deviation of rotated speed measuring is 12r/Min, the precision of the motor dynamic parameter acquisition fully meets requirements of practical system operation.

Acknowledgements

The research work was supported Key Projects in Sichuan Province Department of Education No. 12ZA277.

Z. Y. Zou is a Lecturer in College of Mechanical and Electrical Engineering, Sichuan Agriculture University, Ya'an, 625014, China. His research interests include agricultural engineering. (Email: zouziyong111@163.com)

L. J. Xu* (Corresponding author) is a professor in College of Mechanical and Electrical Engineering, Sichuan Agriculture University, Ya'an, 625014, China. Her research interests include agricultural engineering, automatic control technology. (Email: lijiaxu13@163.com)

Z. L. Kang is an assistant professor in College of Mechanical and Electrical Engineering, Sichuan Agriculture University, Ya'an, 625014, China. His research interests include agricultural engineering. (Email: kangzhiliang96@163.com)

References

1. T. P. Banerjee & S. Das. 2012. Multi-sensor data fusion using support vector machine for motor fault detection, J. Information Sciences. (217): 96-107.
2. Z. Sun. 2010. Design of a Data Acquisition and Processing System Based on LABVIEW, J. Electrical & Electronics. (1) :16-18.
3. Y. Liu. 2014. Multi-Parameter Online Identification Algorithm of Induction Motor for Hybrid Electric Vehicle Applications, C. Parallel Architectures, Algorithms and Programming (PAAP), 2014 Sixth International Symposium on. IEEE. 35-39.
4. J. M. Gutierrez-Villalobos, M. A. Martinez-Hernandez & F. Mendoza-Mondragon. 2014. Hybrid artificial neural network for induction motor parameter estimation, C. Education and Research Conference (EDERC), 2014 6th European Embedded Design in. IEEE. 85-89.
5. J. G. Zhu, X. C. Wang & X. D. Hou. 2014. Vector control of permanent magnet synchronous motor based on parameter identification technique, C. Applied Mechanics and Materials. 644: 705-708.
6. K. Sun, Y. Shi & L. Huang L. 2014. An Improved Sensor-less Control Method for IPMSM-Compressor Drives Based on the MRAS with Motor Parameter Variations, J. International Review of Electrical Engineering (IREE). 9(1): 73-82.
7. Y. Jennifer, M. Biswanath & G. Dipak. 2008.Wireless sensor network survey, J. Computer Networks. 52: 2292–2330.
8. Z. Y. Zou, M. Zhou, L. J. Xu & Z. L. Kang. 2014. Mingdan Liu, Design of Methane Fermentation environmental factor detection system based on ZigBee, J. Applied Mechanics and Materials. 539: 567-571.
9. L. Chang, Y. C. Liu & H. Xu. 2013. System design for embedded WinCE6.0 stepper motor, Journal of Shenyang University of Technology. 35(1): 104-108.
10. M. Owens. 2003. Embedding an SQL database with SQLite, Linux Journal. (110): 2.

Design of chaotic generalized synchronization based pseudorandom number generator with application in avalanche encryption

C. Yue[1], L.Q. Min[†2] and D.D. Han[3]
University of Science and Technology Beijing, 100083 Beijing, China
E-mails: [1]392011515@qq.com, [2]minlequan@sina.com, [3]hxu1204@163.com,
[†]E-mails:13501029489@163.com

Based on a generalized synchronization (GS) theorem of discrete chaotic systems, this paper introduces an 8-dimensional discrete chaotic GS system. Simulations show that the trajectories of the GS system have significant characteristics of chaotic attractor. Using the GS system designs a 2^{16} number string chaotic pseudorandom number generator (CPRNG). The FIPS 140-2 test suit/generalized FIPS 140-2 test suit are used to test the randomness of three 1000-key streams consisting of 20000 bits generated by the CPRNG , the RC4 algorithm and the ZUC algorithm, respectively. The test results confirm that there are 100%/98.6%, 99.9%/98.8%, 100%/97.9% key streams passing the tests, respectively. Numerical simulations suggest that there are no significant correlations between the key stream and the perturbed key stream. The key space of the CPRNG is larger than 2^{1195}. Using the CPRNG and a stream encryption scheme with avalanche effect (SESAE) encrypts a text. The results suggest that the CPRNG is able to generate the avalanche effects which are similar to those generated via ideal CPRNGs.

Keywords: Discrete chaotic system; Pseudo-random number generator; Randomness test; Avalanche encryption scheme.

1. Introduction

Chaos is one kind of complex dynamic behaviors and is generated from determined nonlinear discrete or continuous systems. Chaotic behaviors are aperiodic, unpredictable in long term, and extremely sensitive to variation of initial conditions and system parameters ([1], [2]).

Since the last three decades, the research on chaos synchronization has developed rapidly, and can be found in many physical, biological and engineering systems ([3]–[9]). Generalized synchronization (GS) as the generalization of chaos synchronization, the study of GS has also got much attention ([10]–[15]). Chaos synchronization systems may provide new tools in cryptography field ([3], [16]–[21]).

This paper introduces an 8-dimensional discrete chaotic GS system (8DDCGSS), and constructs a 8DDCGSS-based pseudorandom number generator (CPRNG). Using the FIPS 140-2 test suit by INST tests the randomness of the CPRNG. As an application, using the CPRNG and the SESAE encrypts a text.

The rest of this paper is organized as follows. Section 2 introduces the definition and GS theorems. Section 3 presents the 8DDCGSS, and simulates the dynamic behaviors of the GS system. Section 4 designs the CPRNG and uses FIPS 140-2 test suit to test and compare the randomness of the CPRNG, the RC4 algorithm and the ZUC algorithm. Section 5 gives a text encryption example of the CPRNG and the SESAE. Finally, Section 6 states some concluding remarks.

2. Definition and Gs Theorems

First let us remember the following

Definition 1: (GS) (Similar to [22]) Consider two systems

$$X(k+1)=F(X(k)) \tag{1}$$

$$Y(k+1)=G(X(k),Y(k)) \tag{2}$$

Where

$$X(k)=(x_1(k),...,x_n(k))^T \in R^n \tag{3}$$

$$Y(k)=(y_1(k),...,y_m(k))^T \in R^m, m \le n \tag{4}$$

$$F(X(k))=(f_1(X(k)),...,f_n(X(k)))^T \tag{5}$$

$$G(X(k),Y(k))=(g_1(X(k),Y(k)),..., g_m(X(k),Y(k)))^T \tag{6}$$

If there exists a transformation

$$H:R^n \to R^m \tag{7}$$

$$H(X(k))=(h_1(X(k)),...,h_m(X(k)))^T \tag{8}$$

$(X_0,Y_0)\in R^n \times R^m$ and a subset $B=B_X \times B_Y \subset R^n \times R^m$ such that all trajectories of (1) and (2) with initial conditions $(X(0),Y(0))\in B$ satisfy

$$\lim_{k\to+\infty} \|H(X(k))-Y(k)\|=0$$

Then the systems in (1) and (2) are said to be in GS with respect to the transformation $H(X(k))$. System (1) is called the driving system, and system (2) is said to be the driven system.

In order to construct the novel 8DDCGSS, we present the following.

Theorem 1: (Similar to [22]) Let X , Y , $F(X)$ and $G(X,Y)$ be described by Definition 1. Denote

$$X_m = (x_1(k),\dots,x_m(k))^T \in R^m$$

Suppose that

$$H(X_m) = (y_1, y_2, \dots, y_m) \tag{9}$$

is an invertible transformation. If the two systems (1) and (2) are in GS with respect to the transformation $H(X_m)$, if and only if, the function $G(X,Y)$ given in (2) has the following form:

$$G(X,Y) = H(F_m(X)) - q(X_m, Y) \tag{10}$$

Where

$$F_m(X) = (f_1(X), f_2(X), \dots, f_m(X))^T$$

and the function

$$q(X_m, Y) = (q_1(X_m, Y), q_2(X_m, Y), \dots, q_m(X_m, Y))^T$$

guarantees that the zero solution of the following error equation

$$\begin{aligned} e(k+1) &= H(X_m(k+1)) - Y(k+1) \\ &= q(X_m, Y) \end{aligned} \tag{11}$$

is asymptotically stable on the open set B defined by Definition 1.

3. A Novel Discrete Chaotic Map and 8DDCGSS

An 8DDCGSS is introduced. The driving system has the following form:

$$\begin{cases} x_1(k+1) = a\cos(x_1(k)) - \cos(x_1(k))\sin(x_3(k)) \\ x_2(k+1) = b\sin(x_2(k)) - c\cos(x_1(k))\sin(x_3(k)) \\ x_3(k+1) = d\cos(x_4(k)) - \cos(x_1(k))\sin(x_3(k)) \\ x_4(k) = \sin(x_1(k) + x_2(k)) + \sin(x_4(k)) \end{cases} \tag{12}$$

Where $a = 3,\ b = 1.2,\ c = 2,\ d = 2$. The calculated Lyapunov exponents of this chaotic system are {0.46585, 0.18397, -0.069607,-0.54537}. Therefore, systems (12) is a chaotic system.

Now, choose the following vector as initial condition:

$$X(0) = (0.9634,\ -0.8198,\ 0.64548456,\ -4.4405)^{\mathrm{T}} \tag{13}$$

The chaotic orbits of state variables x_1, x_2, x_3, x_4 for the first 5000 iterations are shown in Fig. 1.

The evolution of state variables: (a) $k - x_1(k)$, (b) $k - x_2(k)$, (c) $k - x_3(k)$, (d) $k - x_4(k)$ are shown in Fig. 2. It can be observed that the dynamic behaviors of the chaotic system demonstrate chaotic attractor characteristics.

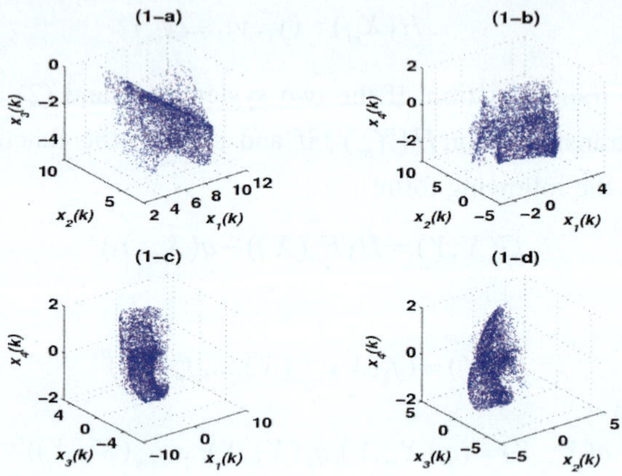

Fig. 1. Chaotic trajectories of variables: (1-a) $x_1(k)-x_2(k)-x_3(k)$, (1-b) $x_1(k)-x_2(k)-x_4(k)$, (1-c) $x_1(k)-x_3(k)-x_4(k)$, (1-d) $x_2(k)-x_3(k)-x_4(k)$.

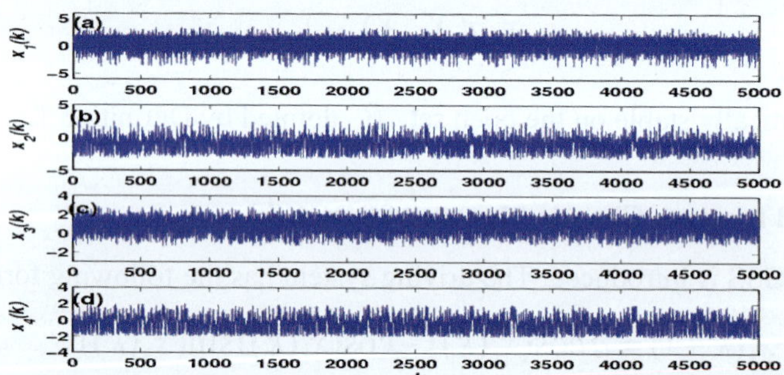

Fig. 2. The evolution of state variables: (a) $k-x_1(k)$, (b) $k-x_2(k)$, (c) $k-x_3(k)$, (d) $k-x_4(k)$.

Second, construct an invertible matrix:

$$A = \begin{pmatrix} 3 & -2 & 4 & 4 \\ 2 & 5 & 0 & -2 \\ -2 & 5 & 3 & 3 \\ -1 & 2 & 5 & -2 \end{pmatrix} \tag{14}$$

with the transformation $H : R^4 \to R^4$ defined as follows:

$$H(X) = AX$$
$$= (h_1(X), h_2(X), h_3(X), h_4(X)) \quad (15)$$

Let

$$q = (X, Y) = 1/8(AX - Y) \quad (16)$$

Where $q(X,Y)$ is used to ensure the error equation (11) be asymptotically stable.

By Theorem 1, the driven system has the following form:

$$Y(k+1) = A[F(X(k))] - q(X(k), Y(k))$$
$$= \begin{pmatrix} y_1(k+1) \\ y_2(k+1) \\ y_3(k+1) \\ y_4(k+1) \end{pmatrix} \quad (17)$$

Then, system (12) and (17) are in GCS with respect to transformation (15).

Moreover, choose (18) as the initial condition:

$$Y(0) = (0.060454664, -2.0499, 3.1908, 0.7360)^T \quad (18)$$

The chaotic trajectories of state variables y_1, y_2, y_3, y_4 for first 5000 iterations are shown in Fig. 3. The evolution of state variables $k - y_1(k)$, $k - y_2(k)$, $k - y_3(k)$, $k - y_4(k)$ are shown in Fig. 4. Fig. 5(a)-(d) show that $X(k)$ and $Y(k)$ are in generalized synchronization with respect to transformation $H = A$, as the theory predicts.

4. Chaotic Pseudorandom Number Generator and Pseudorandomness Tests

4.1. *Pseudorandom number generator*

Denote

$$X_i = \{x_i(k) \mid i = 1, 2, 3, 4\}, \quad (19)$$

$$Y_i = \{y_i(k) \mid i = 1, 2, 3, 4\}. \quad (20)$$

where $x_i's$ $y_i's$ are defined by (12) and (17). First, introduce a transformation $T_1 : R \to \{0, 1, ..., 2^{16} - 1\}$, which transforms the chaotic streams of systems (19) and (20) into key streams. Denote $L = 10^{15}$. Then,

$$P_1 = T_1(X_2) = \mathrm{mod}(round((L(X_2 - \min(X_2)) \\ /(\max(X_2) - \min(X_2))), 2^{16}) \tag{21}$$

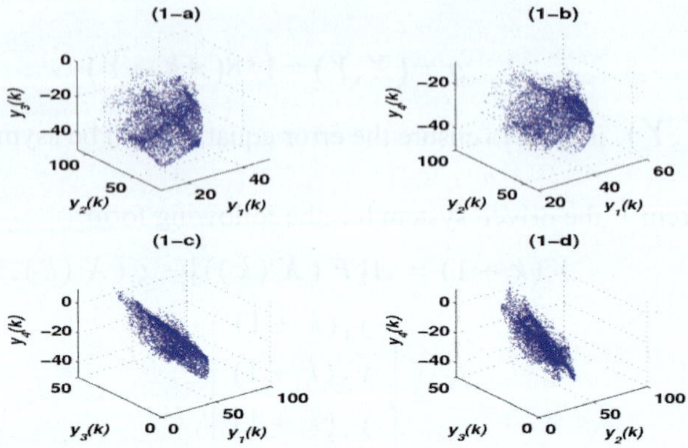

Fig. 3. Chaotic trajectories of variables: (1-a) $y_1(k)-y_2(k)-y_3(k)$,(1-b) $y_1(k)-y_2(k)-y_4(k)$, (1-c) $y_1(k)-y_2(k)-y_4(k)$, (1-d) $y_2(k)-y_3(k)-y_4(k)$ $y_2(k)-y_3(k)-y_4(k)$.

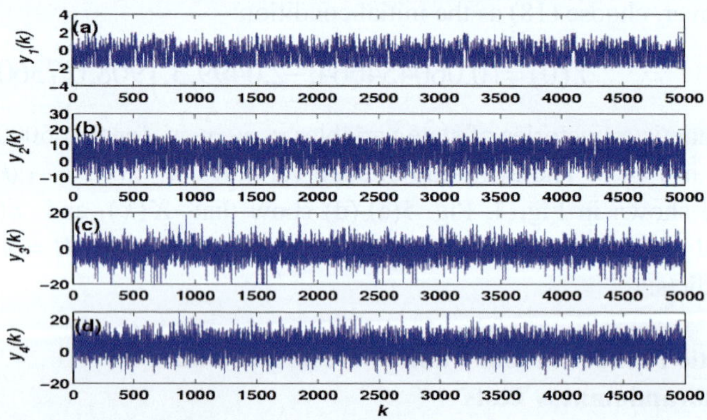

Fig. 4: The evolution of state variables : (a) $k-y_1(k)$, (b) $k-y_2(k)$, (c) $k-y_3(k)$, (d) $k-y_4(k)$.

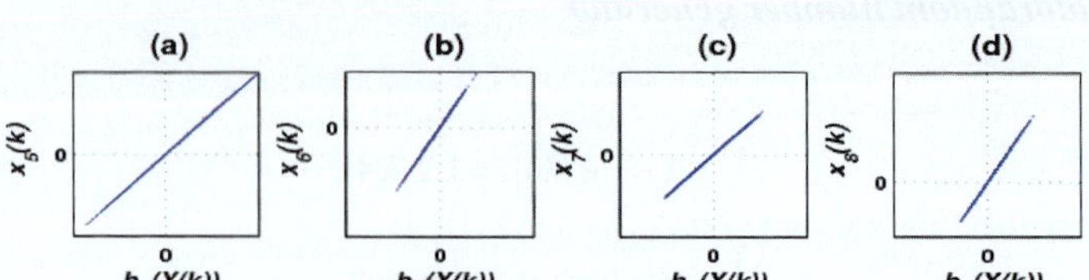

Fig. 5: The state vectors $X(k)$ and $Y(k)$ are in generalized synchronization with respect to the transformation H: (a) $h_1(X(k))-y_1(k)$, (b) $h_2(X(k))-y_2(k)$, (c) $h_3(X(k))-y_3(k)$, (d) $h_4(X(k))-y_4(k)$.

$$P_2 = T_1(Y_3) = \mathrm{mod}(round((L(Y_3 - \min(Y_3)) \\ /(\max(Y_3) - \min(Y_3))), 2^{16}) \quad (22)$$

Therefore, the CPRNG T is defined by

$$T(P_1, P_2) = \mathrm{mod}(P_1 + P_2, 2^{16}). \quad (23)$$

The seeds of the CPRNG are the initial conditions of the 8DDCGSS, which can be chosen via random number generator. Therefore, the output key streams of the CPRNG can be obtained via the transformation (23) on the chaotic streams of the GCS systems (12) and (17).

4.2. *Pseudorandomness test*

The FIPS 140-2 test consists of four sub-tests: Monobit Test, Poker Test, Runs Test and Long Runs Test. Each test needs a single stream of 20,000 one and zero bits from the keystream generator. Any failure in the first three tests means that the corresponding quantity of the sequences falls out the required intervals listed in the second column of Table 1. The Long Runs test is passed if there are no runs of length 26 or more.

It has been pointed out that the required intervals of Monobit test and Porker test correspond significant $\alpha = 10^{-4}$ for the normal cumulative distribution and the χ^2 distribution, respectively, and the required intervals of the Runs tests correspond approximately the significant $\alpha = 1.6 \times 10^{-7}$ for the normal cumulative distribution ([23], [24]). If one selects the significant $\alpha = 10^{-4}$ of all tests, the corresponding accepted intervals are listed in the third column of Table 1. Then, denote the accepted intervals by G FIPS 140-2 test suite.

According to Golomb's three postulates on the randomness, the ideal pseudorandom sequences should satisfy [25], the ideal values of the first three tests should be listed in the forth column of Table 1.

Table 1: The required intervals of FIPS 140-2 Monobit Test, Porker Tests, Runs Test. Here, MT, PT, and LT represent the Monobit Test, the Porker Test and the Long Runs Test. K represents the length of the run of a tested sequence. χ^2 DT represents χ^2 distribution.

Test Item	FIPS140-2 Required Intervals	$\alpha = 10^{-4}$ Required Intervals	Golomb's Postulates
MT	9,725~10,275	9,725~10,275	10000
PT	2.16~46.17	2.16~46.17	χ^2 DT
LT	<26	<26	—
k	Run Test	Run Test	Run Test
1	2,315~2,685	2,362~2,638	2,500

2	1,114~1,386	1,153~1,347	1,250
3	527~723	556~694	625
4	240~384	264~361	313
5	103~209	122~191	156
6+	103~209	122~191	156

Finally, FIPS 140-2 test suite is used to test the randomness performance. One needs to change the keystreams with values $\{0,1,...,2^{16}-1\}$ to binary keystreams via the following transformation:

$$\tilde{T}:\{0,1,...,2^{16}-1\}\to\{0,1\},$$

which is defined by

$$\tilde{T}=T_{22}\circ T_{21}, \tag{24}$$

$$y\in\{0,1,...,2^{16}-1\}^N$$

$$T_{21}(y)=dec2bin(y).$$

Let $z = dec2bin(Y)$. Then

$$T_{22}(z)=z(:),$$

where dec2bin and z(:) are both Matlab commands.

The FIPS 140-2 test is used to check 1,000 keystreams randomly generated, respectively by the CPRNG with perturbed randomly initial conditions (13) and (18) and the matrix (14) in the range $|\varepsilon|\in[10^{-16},10^{-1}]$.

A sequences fail to pass the FIPS 140-2 test and 14 sequences fail to pass the G FIPS 140-2 test. The test results are listed in the third column of Table 2, which the results are described by mean values ± standard deviation (Mean ± SD).

The Rivest Cipher 4 (RC4) has been widely used in popular protocols such as Secure Sockets since it's designed in 1987. The RC4 algorithm as PRNG can be designed via the Matlab commands:

```
N = 20000;
K=randint(1,2^L,[0 2^L-1]);
S=[0:2^L-1];j=0;
for i=1:2^L
        j=mod(j+S(i)+K(i),2^L);
            Sk=S(j+1);
              S(j+1)=S(i);
                 S(i)=Sk;
end
     C=zeros(1,N); j=0;i=0; k=1;
for l=1:N/L
         i=mod(i+1,2^L);
            j=mod(j+S(i+1),2^L);
              Sk=S(j+1);
                S(j+1)=S(i+1);
                  S(i+1)=Sk;
 C(l)=S(mod(S(j+1)+S(i+1),2^L)+1);
end
 C=(dec2bin(C))';
    C=C(:);
      C=bin2dec(C);
```

Here, "randint(1, $2^\wedge L$,[0 $2^\wedge L-1$])" generates a vector of uniformly distributed random integers [0 $2^\wedge L-1$] of dimension 2^L ; "mod" means modulus after division; "zeros(1,N)" is a zero raw vector of dimension N. Consequently, the RC4 algorithm based L-bit segment PRNG is designed.

Next, using FIPS 140-2 test to test the 1,000 keystreams randomly generated by RC4 PRNG. Results show that 1 and 12 sequences fail to pass the FIPS 140-2 test and G FIPS 140-2 test, respectively. The statistic test results are listed in the forth column of Table 2.

Furthermore, ZUC is a stream cipher that forms the heart of the third generation partnership project (3GPP) confidentiality algorithm 128-EEA3 and the 3GPP integrity algorithm 128-EIA3. Then, using FIPS 140-2 test to test the 1,000 keystreams randomly generated by the ZUC algorithm ([26]). It demonstrates that all of the sequences pass the FIPS 140-2 test, and 21 sequences fail to pass the G FIPS 140-2 test. The test results are listed in the fifth column of Table 2.

Finally, compare all test results shown in Table 2. It can be observed, the statistical properties of the pseudorandomness of the sequences generated via the three PRNGs don't have significant differences.

4.3. *Key space*

The key parameters set of the proposed CPRNG includes the initial condition $X(0), Y(0)$ and the matrix $A=(a_{i,j})$ It can be proved that if the perturbation matrix. $\Delta=(\delta_{i,j})$ satisfies

$$\left|\delta_{i,j}\right| < 0.9946$$

the matrix $A+\Delta$ is still invertible. Therefore the CPRNG have 4 + 4 + 16 key parameters denoted by

Table 2: The confident intervals of FIPS 140-2 tested values of 1,000 key streams generated by the CPRNG, the RC4 and ZUC PRNG. Here, SD represents the standard deviation.

Test Item	bits	CPRNG Mean± SD	RC4 Mean± SD	ZUC Mean± SD
MT	0	10005±72.447	9999.7±70.092	9998.4±71.843
	1	9994.7±72.447	10000±0.092	10002±71.843
PT	—	15.231±5.5667	14.87±5.433	15.043±5.5491
LT	0	15.929±2.8855	13.6±1.8214	13.488±1.829
	1	13.611±1.8064	13.642±1.9307	13.595±1.9305
1	0	2498.3±47.55	2500.9±45.568	2501.9±45.735
	1	2499.5±46.152	2501.4±46.398	2502.7±46.121
2	0	1250±32.434	1250.5±31.372	1252.1±32.606
	1	1250.5±31.044	1249 ± 31.048	1249.5±32.221
3	0	625.47±22.786	624.95±22.964	624.09±22.648
	1	624.49±22.655	625.65 ± 22.93	624.64±23.455
4	0	312.32±16.96	311.71±16.548	312.56±16.748
	1	312.05±16.628	312.17±16.822	312.72±16.506
5	0	156.01±11.852	156.41±12.069	155.65±12.097
	1	156.12±11.796	156.6±11.958	156.66±12.369
6+	0	156.37±12.234	156.15±11.792	155.75±11.719
	1	155.86±12.066	155.79±11.979	155.82±11.497

$$K = \{k_1, k_2, \ldots, k_{24}\} \tag{25}$$

The perturbed keys have the forms

$$K(\Delta) = K + [\delta_1, \delta_2, \ldots, \delta_{24}] \tag{26}$$

The Matlab platform uses double precision decimal computations. That means each computed decimal number has 16 bits' accuracy. Therefore, one can select

$$10^{-16} < |\delta_i| \le 10^{-1}, i = 1, \ldots, 24$$

that is,

$$\delta_i = 0.0a_2a_3\ldots a_{16},$$

Where

$$a_i \in [0,1,...,9].$$

Therefore, the 24 keys have a key space which is lager than

$$10^{24\times 15} > 2^{1195}$$

Now, compare the difference between the key stream $S = T(P_1, P_2)$ with 20000 codes length generated by the key set (25) with the key streams S_p generated by the perturbed key set (26), respectively.

The comparison results are shown in the third column of Table 3, where SV denotes the statistic values, DC denotes the different codes, and CC denotes the correlation coefficients.

Observe that the average percent of different codes is 49.99%, which is very closed to the ideal value 50%. And the average of the correlation coefficients is 0.0053, also very closed to the ideal value of 0.

Next, compare the same key stream S with the 1000 streams Sm generated by the Matlab function randi ([0 1], 1, 20000). The comparison results are shown in the forth column of Table 3. Observe that the average percentage of different codes is 50.010% and the the average of the correlation coefficients is -0.0057. The results suggest that the key stream S has no significant correlations with the perturbed key streams S_p and the streams S_m. In summary, the effective key space of the CPRNG is larger than 2^{1195}.

Table 3: The statistic data describes the percentages of the codes of the key stream variations between S and Sp as well as S and Sm.

Item	SV	Sp	Sm
DC	Min	48.8500	48.8850
	Mean	49.9880	50.0109
	Max	51.1100	51.0400
CC	Min	-0.0000	-0.0000
	Mean	-0.0053	-0.0057
	Max	0.0229	0.0223

5. Simulations on SESAE

Consider the avalanche effect of the CPRNG, which is used to encrypt a text "I have a dream" [27] save as a doc format file. The simulation is implemented via the Matlab 7.1 platform.

The SESAE experiments on CPRNG are described as follows:

1) Transfer the text to a binary plaintext steam $M = \{m_1, m_2, \ldots, m_N\}$, where N = 425984.
2) A 16 bit segment key stream of length N + 1000 is generated via CPRNG (23) with initial conditions (13) and (18) by the sender. Then the sender drops the first 1000 iterative values so as to obtain a key stream:

$$P=\{p_1,p_2,...,p_N\}\cdot$$

3) The sender uses the key stream P and the stream encryption scheme with avalanche effect to encrypt the plaintext steam M:

$$c_i=\begin{cases}p_i & if\ m_i=0\\ \sim p_i & if\ m_i=0\end{cases} \tag{27}$$

so as to obtain the ciphertext $C=\{c_1,c_2,...,c_n\}$, denote the ciphertext with:

$$C=E(M,P),$$

where $\sim p_i$ is defined to be the bit string obtained by replacing all '1's in p_i with '0's, and all '0's in p_i with '1's.

4) The receiver uses the key stream P to decrypt the ciphertext:

$$m_i=\begin{cases}0 & if\ c_i=p_i\\ 1 & if\ c_i\neq p_i\end{cases} \tag{28}$$

so as to obtain a decrypted image without errors, which is denoted as $M=E^{-1}(C,P)$.

5) The receiver randomly disturbs the initial conditions (13) and (18), and also the matrix (14), for 1000 times in the range $|\varepsilon|\in[10^{-16},10^{-13}]$, then obtain disturbed key streams:

$$P_i, i=1,2,...,1000. \tag{29}$$

6) The receiver uses $P_i=\{p_1,p_2,...,p_N\}$ to decrypt the ciphertext and obtains a decrypted plaintext:

$$\bar{M}_i=E^{-1}(C,P_i), i=1,2,...,1000.$$

7) The receiver changes all M_i to text.

The corresponding ascii codes of the original text is shown in Fig. 6(a). After changing $\bar{M}_i s$ to text, the ascii codes of all text become almost 255. There are 425984 $\{0,1\}$ codes in each decrypted text. Among the decrypted texts, the minimum of 0′s is 1 and the maximum of 0′s is 16.

The ascii codes of the decrypted text with 1 zero codes and the decriypted text with 16 zero codes are shown in Figs. 6(b)-(c).

Therefore, the percentages of the numbers of "1" codes in the 1000 decrypted texts are within the range [0.999962, 0.999998], which are better than the ideal value $(2^{16}-1)/2^{16}=0.999985$.

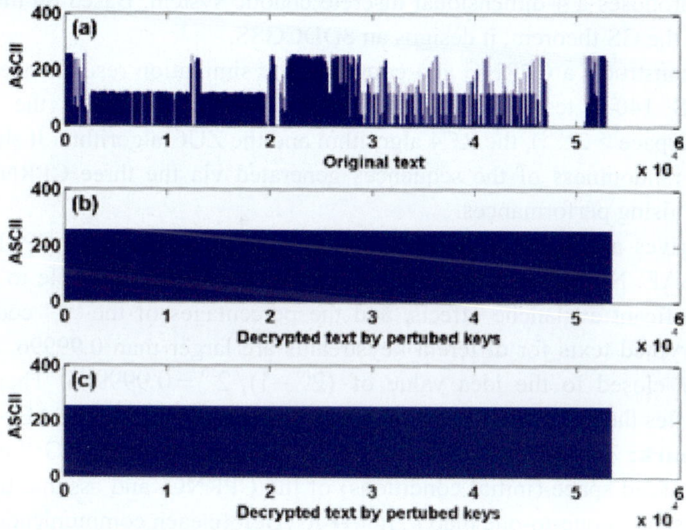

Fig. 6. (a) The ascii codes of original text. (b) The ascii codes of decrypted text with 1 zero codes. (c) The ascii codes of decrypted text with 16 zero codes.

Table 4: Differences between the original keystream S_0 and the keystreams $S_{j,i}$, measured by norm $\|S_0 - S_{j,i}\|$.

	$\|S_0 - S_{j,i}\|(\times 10^{-11})$				
	$S_{0,1}$	$S_{0,1}$	$S_{0,1}$	$S_{0,1}$	$S_{0,1}$
S_0	3.0898	2.5056	2.8068	3.099	2.8498
	$S_{12,1}$	$S_{12,1}$	$S_{12,1}$	$S_{12,1}$	$S_{12,1}$
S_0	2.5552	2.5445	2.2723	2.355	3.1669

Table 4 lists some statistical data of the norms between the original key stream S_0 and the key stream $S_{i,j}$ used in the above ten decrypted texts, respectively. The results suggest that there are no significant correlations between the norms and the corresponding decrypted text.

In summary, the simulation shows using the CPRNG and SESAE to encrypt texts is able to generate encrypted texts with significant avalanche effects.

6. Concluding Remarks

The main results of this paper are summarized as follows:

1) It proposes a 4-dimensional discrete chaotic system. Based on the system and the GS theorem, it designs an 8DDCGSS.

2) It constructs a CPRNG and compares the simulation results tested by the FIPS 140-2 test suite on the keystreams generated via the CPRNG (keyspace > 2^{1195}), the RC4 algorithm and the ZUC algorithm. It shows that the randomness of the sequences generated via the three CPRNGs have promising performances.

3) It gives a text encryption example by using the proposed CPRNG with SESAE. Numerical simulations show that the CPRNG is able to generate significant avalanche effects, and the percentages of the "1" code in the decrypted texts for different keystreams are larger than 0.99996, which is very closed to the idea value of $(2^{16}-1)/2^{16}=0.999985$. Therefore, it verifies the proposed CPRNG is a qualified candidate for SESAE.

Remark: To resisting a "one-time-pad" scheme into CPRNG: Let $\boldsymbol{K}$ be a set in the seed space (initial conditions) of the CPRNG, and assume that Alice and Bob share a one-to-one map $f: R^8 \to R^8$. Before each communication, Alice randomly selects sends (initial condition) w to Bob. Then, they both use $f(w)$ as the seed for one-time encryption.

In summary, the proposed CPRNG is a promising candidate for practical applications. Further comparison with different state-of-the-art PRNG schemes in terms of computational complexity, storage requirement, communication cost, etc., it will be carried out in future research along the same lines.

References

1. T. Y. Li and J. A. Yorke, "Period three implies chaos," Amer Math Monthly, vol. 82, pp. 985–992, 1975.
2. G. Chen and X. Dong, From Chaos to Order: Methodologies, Perspectives, and Applications. Singapore: World Scientific, 1998.
3. M. Pecora and L. Carroll, "Synchronization in chaotic systems," Physical Review Letters, vol. 64, no. 8, pp. 821–825, 1990.
4. D. J and M. G, "Poly-quadratic stability and global chaos synchronization of discrete time hybrid systems," Mathematics and Computers in Simulation, vol. 58, no. 4, pp. 295–307, 2001.
5. Z. Ge and J. Lee, "Chaos synchronization and parameter identification for gyroscope system," Applied Mathematics and Computation, vol. 163, no. 2, pp. 667–682, 2005.
6. H. Wang, Z. Han, Q. Xie, and W. Zhang, "Finitetime chaos synchronization of unified chaotic system with uncertain parameters,"

Communications in Nonlinear Science and Numerical Simulation, vol. 14, no. 5, pp. 2239–2247, 2009.
7. F. Nian, "Module-phase synchronization in complex dynamic system," Applied Mathematics and Computation, vol. 217, no. 6, pp. 2481–2489, 2010.
8. J. Hu, Y. Han, and L. Zhao, "Synchronizing chaotic systems using control based on a special matrix structure and extending to fractional chaotic systems," Communications in Nonlinear Science and Numerical Simulation, vol. 15, no. 1, pp. 115–123, 2010.
9. X. Zhao, Z. Li, and S. Li, "Synchronization of a chaotic finance system," Applied Mathematics and Computation, vol. 217, no. 13, pp. 6031–6039, 2011.
10. K. Murali and M. Laskshmanan, "Secure communication using a compound signal from generalized synchronizable systems," Phys. Lett, vol. 241, pp. 303–310, 1998.
11. Y. L. Zou and J. Zhu, "Control linear and nonlinear generalized synchronization of chaotic systems," in Intelligent Sensing and Information Processing, 2005. Proceedings of 2005 International Conference on, Kokura, Fukuoka Japan, Jan. 2005, pp. 1397–1402.
12. A. Margheri, "Generalized synchronization in linearly coupled time periodic systems," Journal of Differential Equations, vol. 249, no. 12, pp. 3215–3232, 2011.
13. Z. Yuan, Z. Xu, and L. Guo, "Generalized synchronization of two unidirectionally coupled discrete stochastic dynamical systems," Chinese Physics B, vol. 20, no. 7, p. 70503, 2011.
14. X. Wu, H. Hu, and B. Zhang, "Parameter estimation only from the symbolic sequences generated by chaos system," Chaos, Solitons and Fractals, vol. 22, no. 2, pp. 359–366, 2004.
15. B. Nana, P. Woafo, and S. Domngang, "Chaotic synchronization with experimental application to secure communications," Communications in Nonlinear Science and Numerical Simulation, vol. 14, no. 5, pp. 2266–2276, 2009.
16. Y. Wang, K. Wong, X. Liao, and G. Chen, "A new chaosbased fast image encryption algorithm," Applied Soft Computing, no. 11, pp. 514–522, 2011.
17. A. Kanso and M. Ghebleh, "A novel image encryption algorithm based on a 3d chaotic map," Commun Nonlinear Sci Numer Simulat, no. 17, pp. 2943–2959, 2012.
18. C. Guo, C. Chang, and C. Sun, "Chaotic maps-based mutual authentication and key agreement using smart cards for wireless communications," Journal of Information Hiding and Multimedia Signal Processing, vol. 4, no. 2, pp. 99–109, 2013.

19. Y. Liu, X. Tong, and S. Hu, "A family of new complex number chaotic maps based image encryption algorithm," Signal Processing: Image Communication, vol. 28, p. 1548-1559, 2013.
20. B. Du, Q. Ding, and X. Geng, "Analysis and elimination of digital chaotic key sequence's autocorrelation," Journal of Information Hiding and Multimedia Signal Processing, vol. 5, no. 2, pp. 302–309, 2014.
21. E. N. Lorenz, "Deterministic nonperiodic flow," Journal of the atmospheric sciences, 1963, 20(2): 130-141.
22. A. Rukhin, J. Soto, J. Nechvatal, et al., "A statistical test suite for random and pseudorandom number generators for cryptographic applications," NIST Special Publication, 2001: 800-822.
23. L. Min T. Chen, and H. Zang, "Analysis of FIPS 140-2 test and chaos-based pseudorandom number generator. In The Proc. of the 5th Chaotic Modeling and Simulation International Conference (June 2012 Antens Greece)," 345–352, 2012.
24. L. Min, T. Chen, and H. Zang, "Analysis of FIPS 140-2 test and chaos-based pseudorandom number generator," Chaotic Modeling and Simulation no. 2, pp. 273-280, 2013.
25. S. Golomb, Shift Register Sequences. Aegean Park, CA: Laguna Hills, 1982.
26. ETSI/SAGE Specification, Specification of the 3GPP Confidentiality and Integrity Algorithms 128-EEA3 & 128-EIA3. Document 2: ZUC Specification; Version: 1.5, 4 January, 2011.
27. http://baike.baidu.com/subview/376005/5065710.htm?fromtitle=I+have+a +dream&fromid=17501&type=syn

On Max-SINR timing for multicarrier modulation system with hexagonal Time-Frequency lattice over doubly dispersive channel

K. Xu[1†], W. Xie[2], W. F. Ma[3], D. M. Zhang[4], C. Wang[5], N. Li[6] and Y. Y. Xu[7]

PLA University of Science and Technology, Nanjing, China

†Email: lgdxxukui@126.com

In this paper, a novel timing synchronization algorithm for Hexagonal Multicarrier Transmission (HMT) system based on the maximizing Signal-to-Interference-plus-Noise Ratio (Max-SINR) criterion is proposed. The closed form Max-SINR timing offset expressions of the prototype pulse for HMT receiver over DD channel is derived. Simulation results show that the proposed Max-SINR timing offset outperforms traditional first path scheme and obtains an approximation to the theoretical upper bound SINR performance within the full range of channel spread factor. Moreover, the proposed Max-SINR timing synchronization scheme obtains an about 2dB bit error rate (BER) performance gain at SNR=20dB. Meanwhile, the SINR performance of the proposed Max-SINR timing synchronization scheme is robust to the channel RMS delay spread estimation errors.

Keywords: Hexagonal multicarrier transmission system; Maximizing signal-to-interference-plus-noise-ratio (Max-SINR) timing; Doubly dispersive (DD) channel.

1. Introduction

Orthogonal frequency division multiplexing (OFDM) systems with guard-time interval or cyclic prefix can prevent inter-symbol interference (ISI). OFDM has overlapping spectra and rectangular impulse responses. Consequently, each OFDM sub-channel exhibits a sinc-shape frequency response. Therefore, the time variations of the channel during one OFDM symbol duration destroy the orthogonality of different subcarriers, and result in power leakage among subcarriers, known as inter-carrier interference (ICI), which causes degradation in system performance.

---

[1] Kui Xu
[2] Wei Xie
[3] Wen-Feng Ma
[4] Dong-Mei Zhang
[5] Cong Wang
[6] Ning Li
[7] You-Yun Xu

In order to overcome the above drawbacks of OFDM system, several pulse-shaping OFDM systems such as multi-wavelets based OFDM system, OFDM based on offset quadrature amplitude modulation (OFDM/OQAM) system, et al., were proposed ([1] Kumbasar V. & Kucur O. 2007, [2] Das S. & Schniter P. 2007, [3] Xu K. et al. 2014a, [4] Gao X. et al. 2011, [5] Jung P. & Wunder G. 2007, [6] Ma M. et al. 2010, [7] Lin G. et al. 2008, [8] Xu K. et al. 2014b).

It is shown in ([9]Strohmer T. & Beaver S. 2003, [10] Han F. M. & Zhang X.D. 2007, [11] Han F. M. & Zhang X.D. 2009, [12] Han F. M. & Zhang X.D. 2010) that signal transmission through a rectangular lattice is suboptimal for doubly dispersive (DD) channel. By using results from sphere covering theory ([13] Conway J. H. & Sloane N. J. A. 1998), the authors have demonstrated that lattice OFDM (LOFDM) system, which is OFDM system based on hexagonal-type lattice, providing better performance against ISI/ICI ([9] Strohmer T. & Beaver S. 2003). However, LOFDM confines the transmission pulses to a set of orthogonal ones. As pointed out in ([10] Han F. M. & Zhang X.D. 2007), these orthogonalized pulses destroy the time-frequency (TF) concentration of the initial pulses, hence lower the robustness to the time and frequency dispersion caused by the DD propagation channel.

In ([10] Han F. M. & Zhang X.D. 2007), the authors abandoned the orthogonality condition of the modulated pulses and proposed a multicarrier transmission scheme named as hexagonal multicarrier transmission (HMT) by regarding signal transmission as tiling of the TF plane. To optimally combat the impact of propagation channel, the hexagonal lattice parameters and the pulse shape of modulation waveform in HMT system are optimized jointly to adapt to the channel scattering function. It is shown that the HMT system obtains lower energy perturbation by incorporation the best T-F localized Gaussian pulses as the elementary modulation waveform, hence outperforms OFDM and LOFDM system from the robustness against channel dispersion point of view ([17] Xu K. et al. 2009, [18] Xu K. et al. 2011, [19] Xu K. et al. 2013a).

In HMT system, there is no cyclic prefix and data symbols of HMT signal are transmitted at the hexagonal type lattice points in TF plane. The basic mathematical operation on the received signal performed by the demodulator is a projection onto an identically structured function set generated by the prototype pulse function ([5] Jung P. & Wunder G. 2007), i.e. an optimal match filter. It is shown in ([14] Wu J. et al. 2007a, [15] Wu J. et al. 2007b, [16] Wu J. et al. 2005) that the optimum sampling time of wireless communication systems over DD channel is dependent on the power distribution of the channel profiles, and that zero timing offset does not always yield the best system performance.

In our previous work ([19] Xu K. et al. 2013a, [20] Xu K. et al. 2012, [25] Xu K. et al. 2013b), we have presented that traditional HMT receiver proposed in ([10] Han F. M. & Zhang X.D. 2007) using zero timing offset prototype pulse is a suboptimal approach in the view of SINR.

In this paper, theoretical analyses show that there is a timing offset between prototype pulses of the proposed Max-SINR HMT receiver and traditional HMT receiver in the SINR point of view, which is consistent with the conclusion in ([14] Wu J. et al. 2007a, [15] Wu J. et al. 2007b, [16] Wu J. et al. 2005). The closed form timing offset expressions of prototype pulse for Max-SINR HMT receiver over the DD channel is derived. In general, the above mentioned timing offset of the proposed Max-SINR prototype pulses should be matched to the channel scattering function of the DD channel, including channel maximum time delay spread and the root mean square (RMS) delay spread.

Theoretical analyses and simulation results show that the proposed Max-SINR timing synchronization scheme outperforms the traditional first path timing synchronization scheme in ([10] Han F. M. & Zhang X.D. 2007) and obtains an approximation to the theoretical upper bound SINR performance. Specifically, the proposed Max-SINR timing synchronization scheme obtains an about 3dB gain over the corresponding traditional HMT receiver using zero timing offset prototype pulse on the SINR performance over DD channel with exponential power delay profile and U-shape Doppler spectrum. Moreover, the proposed Max-SINR timing synchronization scheme obtains an about 2dB bit error rate (BER) performance gain at SNR=20dB. Meanwhile, the SINR performance of the proposed scheme is robust to the channel RMS delay spread estimation errors.

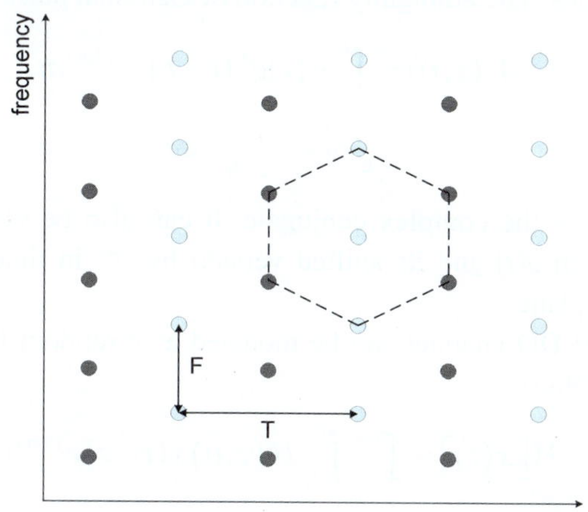

Fig. 1. Partition of the hexagonal lattice into a rectangular sub-lattice V_1 (denoted by ○) and its coset V_2 (denoted by •).

2. System Model

In HMT systems, the original hexagonal lattice can be expressed as the disjoint union of a rectangular sub-lattice V_1 and its coset V_2, as shown in Figure 1. The transmitted baseband signal can be expressed as ([10] Han F. M. & Zhang X.D. 2007)

$$x(t)=\sum_{i}\sum_{m\in\mathcal{M}}\sum_{n\in\mathcal{N}} c_{m,n}^{i} g_{m,n}^{i}(t) \tag{1}$$

where $i=1,2$, $c_{m,n}^{i}$ represent the symbols coming from V_i , respectively. $g_{m,n}^{i}(t)$ denotes the transmitted pulse generated by the prototype pulse *g*(*t*) and $g_{m,n}^{i}(t)=g\left(t-mT-(i-1)T/2\right)e^{j2\pi(nF+(i-1)F/2)t}$. *T* and *F* are the lattice parameters, as shown in Figure 1, which can be viewed as the symbol period and the subcarrier separation, respectively; $c_{m,n}^{i}$ denotes the user data, which is assumed to be taken from a specific signal constellation and independent and identically distributed (i.i.d.) with zero mean and average power σ_c^2 ; $m\in\mathcal{M}$ and $n\in\mathcal{N}$ are the position indices in the TF plane; $\mathcal{M}$ and $\mathcal{N}$ denote the sets from which *m*, *n* can be taken, with cardinalities *M* and *N*, respectively.

The prototype pulse *g*(*t*) is the Gaussian window

$$g(t)=\left(2/\sigma\right)^{1/4} e^{-(\pi/\sigma)t^2} \tag{2}$$

with σ being a parameter controlling the energy distribution in the time and frequency directions. The ambiguity function of Gaussian pulse is defined by

$$\begin{aligned} A_g(\tau,\upsilon) &= \int_{-\infty}^{\infty} g(t)g^*(t-\tau)e^{-j2\pi\upsilon t}dt \\ &= e^{-\frac{\pi}{2}\left(\frac{1}{\sigma}\tau^2+\upsilon^2\right)}e^{-j\pi\tau\upsilon} \end{aligned} \tag{3}$$

where $(\cdot)^*$ denotes the complex conjugate. It can also be viewed as the 2-D correlation between *g*(*t*) and its shifted version by τ in time and υ in frequency in the TF plane.

The baseband DD channel can be modeled as a random linear operator H ([21]Bello P. A. 1963)

$$\mathrm{H}\left[x(t)\right]=\int_0^{\tau_{\max}}\int_{-f_d}^{f_d} H(\tau,\upsilon)x(t-\tau)e^{j2\pi\upsilon t}d\tau d\upsilon \tag{4}$$

where $\tau_{\max}$ and f_d are the maximum multipath delay spread and the maximum Doppler frequency, respectively. The product $\vartheta=\tau_{\max}f_d$ is referred to as the channel spread factor (CSF) ([22]Cohen L. 1995). $H(\tau,\upsilon)$ is called the delay-Doppler spread function, which is the Fourier transform of the time varying impulse response of the channel $h(t,\tau)$ with respect to *t*.

In wide-sense stationary uncorrelated scattering (WSSUS) assumption the DD channel is characterized by the second-order statistics

$$\mathbf{E}\left[H(\tau,\upsilon)H^*(\tau,\upsilon)\right]=S_H(\tau,\upsilon)\delta(\tau-\tau_1)\delta(\upsilon-\upsilon_1) \tag{5}$$

where $\mathbf{E}[\cdot]$ denotes the expectation and $S_H(\tau,\upsilon)$ is called the scattering function, which characterizes the statistics of the WSSUS channel. Without loss R of generality, we use $\int_0^{\tau_{\max}}\int_{-f_d}^{f_d} S_H(\tau,\upsilon)d\tau d\upsilon=1$, which means that the channel has no overall path loss.

The received signal can be expressed as

$$r(t)=\mathbf{H}\left[x(t)\right]+w(t) \tag{6}$$

where $w(t)$ is the AWGN with variance σ_w^2. As shown in equation (4), the propagation channel introduces energy perturbation among the transmitted symbols. It is shown in ([9]Strohmer T. & Beaver S. 2003) that the symbol energy perturbation function is dependent on the channel scattering function and the pulse shape. The properly designed HMT system with Gaussian prototype pulse achieves minimum symbol energy perturbation over DD channel.

3. Max-SINR Timing Synchronization

The basic mathematical operation on the received signal performed by the demodulator is a projection onto an identically structured function set generated by the prototype pulse function, i.e., an optimal match filter ([10] Han F. M. & Zhang X.D. 2007). To obtain the data symbol $\hat{c}^i_{m,n}$, the match filter receiver projects the received signal $r(t)$ on prototype pulse function $\psi^i_{m,n}(t)$, i.e.,

$$\begin{aligned}\hat{c}^i_{m,n}&=\left\langle r(t),\psi^i_{m,n}(t)\right\rangle\\&=\sum_j\sum_{m',n'}c^j_{m',n'}\left\langle \mathrm{H}\left[g^j_{m',n'}(t)\right],\psi^i_{m,n}(t)\right\rangle+\left\langle w(t),\psi^i_{m,n}(t)\right\rangle\end{aligned} \tag{7}$$

where $\psi^i_{m,n}(t)=\psi\left(t-mT-(i-1)T/2\right)e^{j2\pi\left(nF+(i-1)F/2\right)t}$, and $\psi(t)$ is the prototype pulse at the receiver. The energy of the received symbol $c^i_{m,n}$, after projection on the filter function $\psi^i_{m,n}(t)$ over DD channel can be expressed as

$$E_s=\mathbf{E}\left\{\left|\sum_j\sum_{m',n'}c^j_{m',n'}\left\langle\mathbf{H}\left[g^j_{m',n'}(t),\psi^i_{m,n}(t)\right]\right\rangle+\left\langle w(t),\psi^i_{m,n}(t)\right\rangle\right|^2\right\} \tag{8}$$

Under the assumptions of WSSUS channel and source symbols $c_{m,n}^{i}$ are statistically independent, (8) can be rewritten as

$$E_s = \sigma_c^2 \int_\tau \int_\upsilon S_H(\tau,\upsilon)\left[\sum_{m,n}\left|A_{g,\psi}(mT+\tau,nF+\upsilon)\right|^2\right. \\ \left.+\sum_{m,n}\left|A_{g,\psi}(mT+T/2+\tau,nF+F/2+\upsilon)\right|^2 d\tau d\upsilon\right. \\ +\sigma_w^2\left|A_{g,\psi}(0,0)\right| \tag{9}$$

We can see from Eq. (10) that E_s is composed of the expectation symbol energy, ISI/ICI and additive noise. The SINR of the desired symbol $c_{m_0,n_0}^{i_0}$ can be expressed as

$$R_{SIN} = \frac{\sigma_c^2}{E_{IN}} \int_\tau \int_\upsilon S_H(\tau,\upsilon)\left|A_{g,\psi}(\tau,\upsilon)\right|^2 d\tau d\upsilon \tag{10}$$

where the interference-plus-noise energy E_{IN} is the energy perturbation of the desired symbol $c_{m_0,n_0}^{i_0}$ from other symbols over time varying multipath fading channel $H(\tau,\upsilon)$. For presentation simplicity, the desired symbol $c_{m_0,n_0}^{i_0}$ is chosen as $c_{0,0}^{1}$. Hence, E_{IN} can be expressed as

$$E_{IN} = \sigma_c^2 \int_\tau \int_\upsilon S_H(\tau,\upsilon) \sum_{[m,n,i]\neq[0,0,1]} \left|A_{g,\psi}\left(\left(m+\frac{i}{2}\right)T+\tau,\left(n+\frac{i}{2}\right)F+\upsilon\right)\right|^2 d\tau d\upsilon \\ +\sigma_w^2\left|A_{g,\psi}(0,0)\right| \tag{11}$$

Clearly, the interference-plus-noise energy function E_{IN} is dependent on the channel scattering function and the pulse shape (through its ambiguity function).

For the DD channel with exponential power delay profile and U-shape Doppler spectrum, the scattering function can be expressed as ([23]Matthias P. 2002)

$$S_H(\tau,\upsilon) = \frac{e^{-\tau/\tau_{rms}}}{\pi\tau_{rms} f_d\sqrt{1-(\upsilon/f_d)^2}} \tag{12}$$

With $\tau > 0$, $|\upsilon| < f_d$. We assume that $|\upsilon| < f_d$, the theoretical SINR of the received signal over the DD channel with exponential power delay profile and U-shape Doppler spectrum can be expressed as

$$R_{SIN}^{EXP} = \frac{\sigma_c^2}{\pi\tau_{rms} f_d E_{IN}^{EXP}} \int_0^{\infty} e^{-\frac{\tau}{\tau_{rms}}} e^{-\frac{\pi}{\sigma}(\tau-\Delta t)^2} d\tau \int_{-f_d}^{f_d} \frac{e^{-\sigma\pi\upsilon^2}}{\sqrt{1-(\upsilon/f_d)^2}} d\upsilon \tag{13}$$

Substituting (12) into (10), the interference-plus-noise energy function E_{IN}^{EXP} over DD channel with exponential power delay profile and U-shape Doppler spectrum in (13) can be expressed as

$$\begin{aligned} E_{SIN}^{EXP} &= \frac{\sigma_c^2}{2\tau_{\max} f_d} \sum_{[m,n,i]\neq[0,0,1]} \int_0^{\infty} e^{-\frac{\tau}{\tau_{rms}}} e^{-\frac{\pi}{\sigma}\left(\left(m+\frac{i-1}{2}\right)T+\tau-\Delta t\right)^2} d\tau \\ &\times \int_{-f_d}^{f_d} e^{-\sigma\pi\left(nF+\frac{i-1}{2}F+\upsilon\right)^2} \Big/ \sqrt{1-(\upsilon/f_d)^2}\, d\tau + \sigma_w^2 \left|A_{g,\psi}(0,0)\right| \end{aligned} \tag{14}$$

The theoretical SINR upper bound of the received signal over the DD channel with exponential power delay profile and U-shape Doppler spectrum can be expressed as

$$R_{UB}^{EXP} = \arg\max_{\Delta t} R_{SIN}^{EXP} \tag{15}$$

Plugging (13) and (14) in (15), the Max-SINR prototype pulse can be expressed as $\psi(t) = g(t-\Delta t)$ and timing synchronization Δt can be expressed as (see Appendix)

$$\Delta t = \frac{\sigma}{2\pi\tau_{rms}} - \frac{1}{1.76}\sqrt{\frac{\sigma}{2\pi}}\left(\frac{3.28\sqrt{\sigma}}{\tau_{rms}} - \sqrt{\frac{3.28^2\sigma}{\tau_{rms}^2} - 3.52\left(\frac{\sigma}{\tau_{rms}^2} - 4\right)}\right) \tag{16}$$

We can see from (16) that the prototype pulses of the proposed Max-SINR timing synchronization schemes over DD channel are functions of channel maximum delay spread and RMS delay spread, respectively.

4. Simulation and Discussion

In this section, we test the proposed Max-SINR timing synchronization scheme via computer simulations based on the discrete signal model. In the following simulations, the number of subcarriers for HMT system is chosen as N=40, and the length of prototype pulse is set to N_g=600. The center carrier frequency is f_c=5GHz and the sampling interval is set to T_s=10^{-6}s. The system parameters of HMT system are F=25kHz, T=1×10^{-4}s and the signaling efficiency $\rho = 0.8$. σ for prototype pulse $g(t)$ is set to $\sigma = T/\sqrt{3}F$. Traditional projection receiver proposed in ([10] Han F. M. & Zhang X.D. 2007) is named as Traditional Projection Receiver (TPR) in the following simulation results.

4.1. *SINR performance*

The SINR performance of different receivers with the variety of σ_c^2/σ_w^2 for HMT system over DD channels with exponential power delay profile and U-shape Doppler spectrum is depicted in Figure 2. $\tau_{rms}f_d$ in Figure 2 is set to $\tau_{rms}f_d$=0.07 and 0.2, respectively. We can see from Figure 4 that the SINR performance of the proposed Max-SINR timing synchronization scheme outperforms TPR scheme about 1~4dB at $\tau_{rms}f_d$ =0.07 and 1.5~3.5dB at $\tau_{rms}f_d$=0.2, respectively. The SINR gap between the proposed Max-SINR timing synchronization scheme and the theoretical SINR upper bound is smaller than 0.5dB and 0.1dB at _ $\tau_{rms}f_d$=0.07 and 0.2, respectively.

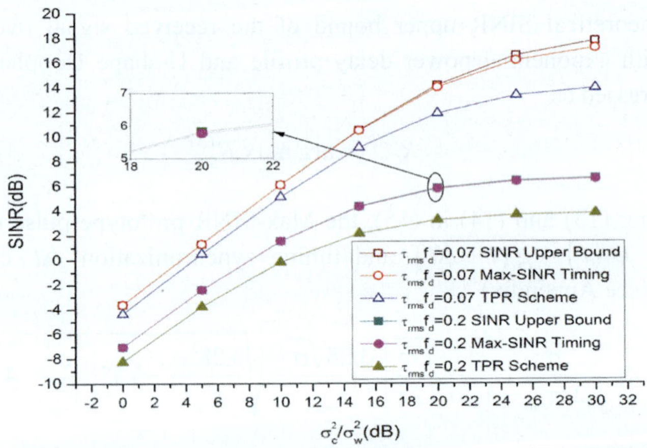

Fig. 2. The SINR performance of different timing synchronization schemes with the variety of σ_c^2/σ_w^2 for HMT system over DD channel with exponential power delay profile and U-shape Doppler spectrum.

The SINR performance with the variety of $\tau_{rms}f_d$ at σ_c^2/σ_w^2 =20dB over DD channels with exponential power delay profile and U-shape Doppler spectrum is depicted in Figure 3. The SINR performance of the TPR scheme is depicted for comparison. It can be seen that there is a degradation of SINR with the increasing of $\tau_{rms}f_d$. The proposed Max-SINR timing synchronization scheme obtains an approximation to the theoretical upper bound SINR performance within the full range of $\tau_{rms}f_d$. The proposed Max-SINR timing synchronization scheme achieves an about 2.5dB maximum SINR gain over TPR scheme at $\tau_{rms}f_d$ =0.35, and the SINR gain increases as $\tau_{rms}f_d$ increases.

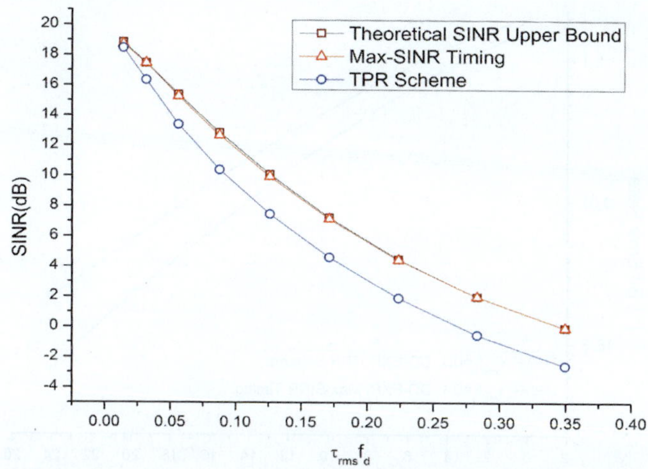

Fig. 3. The SINR performance of different timing synchronization schemes with the variety of $\tau_{rms} f_d$ for HMT system over DD channel with exponential power delay profile and U-shape Doppler spectrum, σ_c^2/σ_w^2 =20dB.

4.2. *BER performance*

The BER performance of the proposed Max-SINR timing synchronization scheme for HMT system over the DD channel with exponential power delay profile and U-shape Doppler spectrum, named as DD-EXP, is given in Figure 4.

$\tau_{rms} f_d$ is set to 0.1. We can see from Figure 6 that the proposed Max-SINR timing synchronization scheme over DD-EXP channel outperforms TPR receiver on the BER performance and the performance gain is about 2.5dB at E_b/N_0=20dB.

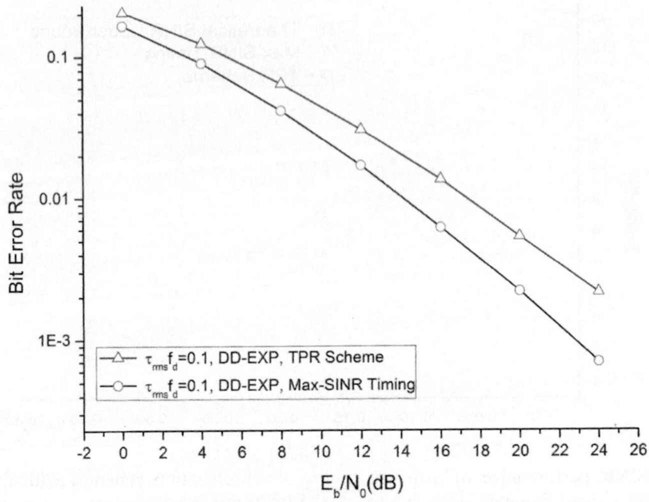

Fig. 4. The BER performance of the proposed Max-SINR timing synchronization scheme for HMT system over the DD channel.

4.3. *Robustness against channel RMS estimation errors*

The robustness of the proposed Max-SINR timing synchronization scheme against channel delay spread estimation errors is depicted in Figure 5. RMS Estimation errors of τ_{rms} for DD-EXP channel is modeled as uniformly distributed random variables in the interval $[-\tau_{rms}/2, \tau_{rms}/2]$.

The SINR performance of the proposed Max-SINR timing synchronization scheme over DD channels with exponential power delay profile and U-shape Doppler spectrum at $\_\tau_{rms}f_d$ =0.1 is given in Figure 5. The SINR gap between the SINR upper bound and the proposed Max-SINR timing synchronization scheme with estimation error of τ_{rms} is within 0.1dB and 0.7dB at σ_c^2/σ_w^2 =0dB and 30dB, respectively. SINR performance of the TPR scheme over DD-EXP channel is also depicted in Figure 5 for comparison. The proposed Max-SINR timing synchronization scheme outperforms TPR scheme when there exists estimation errors of τ_{rms}. In summary, the proposed Max-SINR timing synchronization scheme is robust to the estimation errors of τ_{rms}.

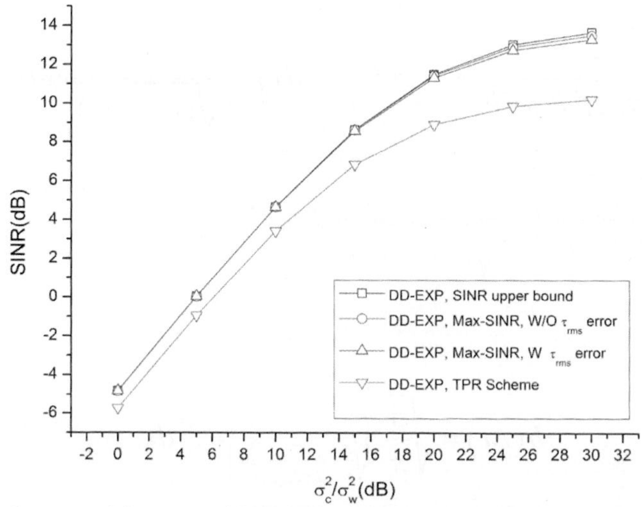

Fig. 5. The robustness of the proposed Max-SINR timing synchronization scheme against channel RMS delay spread estimation errors.

5. Conclusion

A novel timing offset estimator based on Max-SINR criterion for HMT system over DD channel with different channel scattering functions is proposed in this paper. Theoretical analyses show that there is a timing offset between the prototype pulses of the proposed Max-SINR timing synchronization scheme and the traditional projection receiver. The timing offset should be matched to the channel scattering factor of the DD channel. The closed form timing offset expressions of prototype pulse for HMT receiver over DD channel is derived. Simulation results show that the proposed Max-SINR timing synchronization scheme outperforms traditional projection scheme and obtains an approximation to the theoretical upper bound SINR performance within the full range of channel spread factor. Moreover, the proposed Max-SINR timing synchronization scheme obtains an about 2dB BER performance gain at SNR=20dB. Meanwhile, the SINR performance of the proposed Max-SINR timing synchronization scheme is robust to the channel RMS delay spread estimation errors.

6. Appendix

Under the assumption that $\psi(t) = g(t - \Delta t)$, the SINR of the received signal can be expressed as

$$R_{SIN}^{EXP} = \frac{\sigma_c^2}{\pi\tau_{rms} f_d E_{IN}^{EXP}} \int_0^{\infty} e^{-\frac{\tau}{\tau_{rms}}} e^{-\frac{\pi}{\sigma}(\tau-\Delta t)^2} d\tau \int_{-f_d}^{f_d} \frac{e^{-\sigma\pi\upsilon^2}}{\sqrt{1-(\upsilon/f_d)^2}} d\upsilon$$

$$\approx \frac{\sigma_c^2}{\sigma_w^2 \tau_{rms} f_d} \int_{-f_d}^{f_d} \frac{e^{-\sigma\pi\upsilon^2}}{\sqrt{1-(\upsilon/f_d)^2}} d\upsilon \left(\underbrace{e^{\frac{\sigma}{4\pi\tau_{rms}^2}+\frac{\Delta t}{\tau_{rms}}}}_{a(\Delta t)} \underbrace{\int_0^{\infty} e^{-\frac{\pi}{\sigma}\left(\tau-\Delta t+\frac{\sigma}{2\pi\tau_{rms}}\right)^2} d\tau}_{b(\Delta t)} \right) \quad (17)$$

Where

$$b(\Delta t) = \sqrt{\frac{\sigma}{\pi}} \int_{\sqrt{\frac{\pi}{\sigma}}\left(\frac{\sigma}{2\pi\tau_{rms}}-\Delta t\right)}^{\infty} e^{-x^2} dx$$
$$= \frac{\sqrt{\sigma}}{2} erfc\left(\sqrt{\frac{\pi}{\sigma}}\left(\frac{\sigma}{2\pi\tau_{rms}} - \Delta t\right)\right) \quad (18)$$

where $\mathrm{erfc}(\cdot)$ is the complementary error function. If x > 0, we may obtain an approximate solution of the complementary error function $\mathrm{erfc}(\cdot)$ by ([24] Kingsbury N. 2005).

$$erfc\ \frac{x}{\sqrt{2}}\ ;\ \frac{2e^{\frac{x^2}{2}}}{1.64x\ \ \sqrt{0.76x^2\ \ 4}} \quad (19)$$

The optimal timing offset Δt can be obtained by solving the gradient $a(\Delta t)b(\Delta t)$ with respect to Δt to zero,

$$\frac{da(\Delta t)}{d\Delta t} b(\Delta t) + \frac{db(\Delta t)}{d\Delta t} a(\Delta t) = 0 \quad (20)$$

where $da(\Delta t)/d\Delta t = -a(\Delta t)/\tau_{rms}$ and $db(\Delta t)/d\Delta t$ can be expressed as

$$\frac{db(\Delta t)}{d\Delta t} = e^{-\frac{\pi}{\sigma}\left(\frac{\sigma}{2\pi\tau_{rms}}-\Delta t\right)^2} \quad (21)$$

hence, equation (20) can be rewritten as

$$\frac{b(\Delta t)}{\tau_{rms}} = e^{-\frac{\pi}{\sigma}\left(\frac{\sigma}{2\pi\tau_{rms}}-\Delta t\right)^2}$$

$$= \frac{\sqrt{\sigma}}{2\tau_{rms}} erfc\left(\sqrt{\frac{\pi}{\sigma}}\left(\frac{\sigma}{2\pi\tau_{rms}} - \Delta t\right)\right)$$

$$\simeq \frac{\sqrt{\sigma}}{\tau_{rms}} e^{-\frac{\pi}{\sigma}\left(\frac{\sigma}{2\pi\tau_{rms}}-\Delta t\right)^2} \left(1.64\sqrt{\frac{2\pi}{\sigma}}\left(\frac{\sigma}{2\pi\tau_{rms}} - \Delta t\right) + \sqrt{\frac{1.52\pi}{\sigma}\left(\frac{\sigma}{2\pi\tau_{rms}} - \Delta t\right)^2 + 4}\right)^{-1} \tag{22}$$

Equation (22) can be simplified to a quadratic equation. Under the constraint of $\Delta t > 0$, the solution of the quadratic equation can be expressed as

$$\Delta t = \frac{\sigma}{2\pi\tau_{rms}} - \frac{1}{1.76}\sqrt{\frac{\sigma}{2\pi}}\left(\frac{3.28\sqrt{\sigma}}{\tau_{rms}} - \sqrt{\frac{3.28^2\sigma}{\tau_{rms}^2} - 3.52\left(\frac{\sigma}{\tau_{rms}^2} - 4\right)}\right) \tag{23}$$

In summary, for the DD channel with exponential power delay profile and U-shape Doppler spectrum, time offset of the Max-SINR prototype pulse at the receiver can be expressed as equation (23).

Acknowledgement

This work is supported by National Natural Science Foundation of China (No. 61301165), Special Financial Grant of the China Postdoctoral Science foundation (2015T81079), Major Research Plan of National Natural Science Foundation of China (No. 91438115), National Natural Science Foundation of China (No. 61371123), Jiangsu Province Natural Science Foundation (BK2011002, BK2012055), China Postdoctoral Science Foundation (2014M552612) and Jiangsu Postdoctoral Science Foundation (No.1401178C).

References

1. Kumbasar, V. & Kucur, O. 2007. ICI reduction in OFDM systems by using improved sinc power pulse. *Digital Signal Processing* 17(6): 997-1006.
2. Das, S. & Schniter, P. 2007. Max-SINR ISI/ICI-Shaping multicarrier communication over the doubly dispersive channel. *IEEE Transactions on Signal Processing* 55(12): 5782-5795.

3. Xu, K. et al. 2014a. Synchronization approach for dynamic hexagonal time-frequency lattice MCM system. *Digital Signal Processing* 25(2): 72-83.
4. Gao X. et al. 2011. Cyclic prefixed OQAM-OFDM and its application to single-carrier FDMA. *IEEE Transactions on Communications* 59(5): 1467-1480.
5. Jung, P. & Wunder, G. 2007. The WSSUS pulse design problem in multi-carrier transmission, *IEEE Transactions on Communications* 55(10): 1918-1928.
6. Ma, M. et al. 2010. A dual-window technique for enhancing robustness of OFDM against frequency offset, *IEEE Communications Letters* 12(1): 17-19.
7. Lin, G. et al. 2008. Optimal pulses robust to carrier frequency offset for OFDM/QAM systems, *IEEE Communications Letters* 12(3): 161-163.
8. Xu, K. et al. 2014b. Max-SINR receiver for HMCT systems over non-stationary doubly dispersive channel. in *Proc. 31th URSI GASS* 1-4.
9. Strohmer, T. & Beaver, S. 2003. Optimal OFDM design for time-frequency dispersive channels. *IEEE Trans. Commun.* 51(7): 1111-1122.
10. Han, F. M. & Zhang, X.D. 2007. Hexagonal multicarrier modulation: A robust transmission scheme for time-frequency dispersive channels. *IEEE Transactions on Signal Processing* 55(5): 1955-1961.
11. Han, F.M. & Zhang, X.D. 2009. MLSD for hexagonal multicarrier transmission with time-frequency localized pulses, *IEEE Transactions on Vehicular Technology* 58(3): 1598-1604.
12. Han, F.M. & Zhang, X.D. 2010. Asymptotic Equivalence of Two Multicarrier Transmission Schemes in Terms of Robustness Against Time-Frequency Dispersive Channels. *IEEE Transactions on Vehicular Technology* 59(2): 1598-1604.
13. Conway, J. H. & Sloane, N. J. A. 1998. *Sphere Packings, Lattices and Groups, 3rd ed.* New York: Springer-Verlag.
14. Wu, J. et al. 2007a. Performance Analysis of Wireless Systems With Doubly Selective Rayleigh Fading, *IEEE Transactions on Vehicular Technology* 56(2): 721-730.
15. Wu, J. et al. 2007b. On the Error Performance of Wireless Systems with Frequency Selective Fading and Receiver Timing Phase Offset, *IEEE Transactions on Wireless Communications* 6(2): 720-729.
16. Wu, J. et al. 2005. Matched Filter Bound of Wireless Systems over Frequency Selective Channels with Receiver Timing Phase Offset. in *Proc. IEEE Globecom05* 3758-3762.
17. Xu, K. et al. 2009. Effects of carrier frequency offset, timing offset, and channel spread factor on the performance of hexagonal multicarrier modu-

lation systems. *EURASIP Journal on Wireless Communications and Networking* 2009(1-8).

18. Xu, K. et al. 2011. SINR analysis of hexagonal multicarrier transmission systems in the presence of insufficient synchronization for doubly dispersive channel. *Frequenz* 65(5): 149-157.
19. Xu, K. et al. 2013a. Max-SINR Based Timing Synchronization Scheme in Hexagonal Multicarrier Transmission. *Wireless Personal Communications* 68(4): 1789-1805.
20. Xu, K. et al. 2012. On Max-SINR Receiver for Hexagonal Multicarrier Transmission Over Doubly Dispersive Channel. in *Proc. IEEE GLOBECOM12* 3673-3678.
21. Bello P. A. 1963. Characterization of randomly time-variant linear channels, *IEEE Transactions on Communication System* 11(4): 360-393.
22. Cohen, L. 1995. *Time-frequency analysis*. Englewood Cliffs, NJ: Prentice-Hall.
23. Matthias, P. 2002. *Mobile fading channels*. West Sussex, England: John Wiley & Sons, Ltd.
24. Kingsbury, N. 2005. Approximation Formulae for the Gaussian Error Integral, Q(x), Connexions Web site. http://cnx.org/content/m11067/2.4/, Jun 7, 2005.
25. Xu, K. et al. 2013b. Carrier frequency offset estimation approach for multicarrier transmission on hexagonal time-frequency lattice. in *Proc. IEEE WCNC13* 2519-2524.

Telemedicine and health information

F. Wang[1†]

Department of Ophthalmology
China Japan Union Hospital of Jilin University, Changchun, China
Emails: fanwang_hh@yahoo.com, wangf2001@hotmail.com

Z. Miao[2]

Department of Neurology Surgery
China Japan Union Hospital of Jilin University, Changchun, China

L. Wang[3], B. Yang[4] and X. M. Lv[5]

Department of Ophthalmology
China Japan Union Hospital of Jilin University, Changchun, China

Continuing medical education (CME) is an effective way for practicing physicians to acquire up-to-date clinical information. Telemedicine is a new way of delivering health care to people, particularly in remote areas. However, there is little information available in the literature regarding the past and present of telemedicine, the study is to evaluate the state of telemedicine in Germany. Materials and Methods: A literature review of tele-education in Germany was undertaken. The development of the tele-education services at University of Dresden in Germany from 2005 to present is described. The approaches taken are compared with current teaching on eHealth implementation and a retrospective design-reality gap analysis is made. Results: Tele-education has been in use in Germany since the 1970s. Several forms of tele-education are in place at the medical schools and in some Provincial Departments of Health (DOH). Despite initial attempts by the National DOH, there are no national initiatives in tele-education. At University of Dresden in Germany, a tele-education service has been running since 2005 and appears to be sustainable and reaching maturity, with over 1,400 h of videoconferenced education offered per year. The service has expanded to offer videoconferenced education using different ways of delivering tele-education. Conclusions: What it apparent is that it improves access to education and training in resource constrained settings. The development of local and international tele-education has not followed, what is currently considered to be best practice but shows how programs can develop if there is a real need and the solution assists in meeting the need.

---

[1] Fan Wang
[2] Zhuang Miao
[3] Ling Wang
[4] Ben Yang
[5] Xue-Man Lv

Keywords: Telemedicine; E-health; Continuing medical education (CME)

1. Introduction

Telemedicine includes the use of information and communication technology for education in the health sector, tele-education. Tele-education has the potential to provide access to education both formal and continuing medical education. The aims of this study were to review the literature on tele-education in Germany, describe tele-education activities at the University of Dresden in Germany, and to review the development of these programs with respect to current thinking on eHealth project implementation. Telemedicine refers to any application of information and communication technology which removes or mitigates the effect of distance in healthcare [1]. It is a way of delivering healthcare that allows a change from a centralized service to one that is patient centred, resource efficient, and where decisions are made at a local level close to the patient [2]. This is a relatively new concept and its implications on the health system are not fully known. In Germany, various workshops are important methods for continuing medical education (CME) related to training. It is an effective tool for continuing professional development. Typical continuing medical education (CME) workshops in Germany often include lectures delivered by specialists and real-time demonstration of various procedures [3]. Live demonstrations also have a high educational value and contribute significantly to surgery because they are ideal for education and practical teaching of surgery. However, skilled clinicians face fatigue when traveling for real-time demonstrations at away institution [4]. On the other hand, live demonstration via the Internet makes it feasible for specialists from different countries to conduct real-time demonstrations remotely[5], thus eliminating the time and expense of travel. Because of bandwidth limitation, the traditional videoconference system based on the code/decode technique has disadvantages of dropped video frames or distortion of images [6]. This study aims to evaluate the state of telemedicine, compare projects according to the population and area served, address relevant issues, and suggest appropriate changes.

1.1. *Methods*

Relevant information was obtained from the German National Database of Telemedicine Information Exchange. A literature review was undertaken of telemedicine in Germany. An unrestricted Medline literature search was performed and various websites were also consulted. Searches were conducted of Pubmed, German Journals online, and German Index Medicus using search terms such as ("Germany") AND ("tele-education" OR "distance education" OR "videoconference" OR "remote education"). As tele-education falls under the definition of telemedicine the databases were also searched using common keywords relating to telemedicine. For Pubmed, the terms ("telemedicine" OR "ehealth" OR "mhealth" OR "email" OR "electronic mail" OR "telephone" OR "mobile phone" OR "cell phone" OR "videoconference") AND ("Germany") AND ("tele-education" OR "distance education" OR "remote education") were used to search the title, abstract, and keywords of articles indexed. Google Scholar and Google were searched for each of the eight medical and nine Provincial Department of Health (DOH) for VC or tele-education. The abstracts of all papers returned by the searches were read. The full papers of those that appeared relevant were obtained and read. Those reporting the use of ICTs for education in the health sector in South Africa were included in the review. The aim of the review was to document evidence of tele-education in Germany and no attempt was made to assess the quality of the papers.

1.2. *Case study*

A single case study was undertaken of tele-education. The objective of the study was to describe the evolution of the service and compare its development with current thinking about eHealth implementation. The records, reports, and minutes of meetings related of the service were reviewed.

1.3. *Design-reality gap assessment*

A retrospective assessment of the design-reality gap at the time of implementation of video conferenced tele-education was undertaken using Heeks' Design-Reality Gap Assessment model.

2. Results

Eighty-three papers and Web pages relating to tele-education in Germany were found, of which 23 were relevant and summarized. Of these, 16 reported tele-education activities, 5 provided data on tele-education, and 2 referred to the educational benefits of participation in a clinical telemedicine service.

Tele-education in Dresden has followed the evolution of technology and has taken many forms. It can be broadly categorized in terms of continuing medical education, postgraduate education, undergraduate education, and education through participation in telemedicine services.

In 2006, a nationwide radiology education program for trainees was trialed, initially between three medical schools and later with seven participating via a commercial VC bridge service. Problems were noted with the quality of the radiographic images shared over the network as connectivity defaulted to that of the site with the lowest bandwidth, 128 kbps. To overcome this, images of radiographs were saved to PowerPoint and distributed by email before the scheduled sessions and projected locally with VC used for discussion between the sites [7]. A plan to establish shared postgraduate radiology teaching nationally never materialized.

The German government saw the potential of telemed-icine and implemented a National Telemedicine System in 1999. It was to have consisted of three phases. The first, pilot phase, involved establishing telemedicine services at 29 sites in six of the nine provinces. The second phase envisaged connecting the eight medical schools by VC for the provision of both postgraduate and continuing medical education and extending the original project to 76 VC sites. Although there were promising early reports of the system, only a tele-radiology store and forward service in neuro-surgery was sustained, with most services failing within 2 years [8, 9]. As a result, phases two and three were never implemented. In the absence of a Government backed system, various Provincial DOHs, Universities, and the Medical Research Council of German continued to support and develop telemedicine and tele-education.

At the University of Dresden, the Department of Pediatric Surgery has installed video cameras in their operating theaters so that surgical procedures can be broadcast to lecture theaters [10]. They have recently begun sharing weekly seminars with some medical schools in Germany using Internet protocol (IP) based desktop VC. The Department of Gastro-enterology has participated with the Universities of the Berlin in receiving high definition video of endoscopic *surgery* undertaken and streamed from Japan [11]. In 2011, a working group was formed to investigate the use of VC in education at the University and a VC link has been established with a rural hospital.

3. Conclusion

Telemedicine had a relatively late start but has grown steadily over last decade. Tele-education has been used in different forms in the health sector in Germany for a number of years. There is little hard evidence of its educational merit or economic worth. What is apparent is that it improves access to education and training in resource constrained settings. The development of local and international tele-education has not followed what is currently considered to be best practice, but shows how programs can develop if there is a real need and the solution assists the need [12].

Further work is required to analyze the economics of these tele-education endeavors.

The issue of tele-education needs attention as no dedicated project was seen. Technological improvements are essential to expand the current level of usage, particularly in remote areas that are likely to be benefited most [13].

There is also need for well designed multisite trials that would help establish widespread acceptability of telemedicine in future. The government has a key part to play in supporting such moves and providing the necessary infrastructure. Both the Department of Health and the German Executive have already taken useful steps by formulating an elaborate information technology strategy and forming the German Telemedicine Action Forum respectively. The outcomes are awaited.

The health content covers topics such as prevention of mother to child transmission of HIV and infection prevention and control delivered in the form of dramas, discussions, interviews, documentaries, and magazine shows with content aligned to the National DOH policies and guidelines. The programe provides an online certificate course, on the Fundamental Management of HIV and Aids for health-care providers, using a learning management system. The course is integrated into the curriculum of nursing schools and is used for continuing professional development. No data are available evaluating use of Mindset Health [14].

By 2010, this had grown to 35 academic programs broadcasting a total of 123 h of interactive teaching per month with 37 different sites participating in the various programs [15].

In pediatric surgery, weekly seminars have been shared at various times with an academic hospital in Dresden, a regional hospital, neither of which had a pediatric surgeon. In the first 4 years of the service the seminar, attended by an average 13 people, was shared with 63 others at the various sites [16]. Surgeons from several other European countries have requested to be included in the videoconference program. In the absence of VC infrastructure at other medical schools in Germany, the VC seminars have been recorded to digital video disk (DVD) and mailed to four medical schools in Central and East Germany. They have been incorporated into the postgraduate training programs of surgeons and pediatric surgeons and also used in undergraduate medical training. An additional 140 people have access to the seminars through this program[17].

4. Discussion

Use of telemedicine varies widely across the globe. It is therefore relevant to compare the activity of telemedicine in Germany with other countries. However, the lack of a compulsory and comprehensive database of projects makes such comparisons very difficult, if not invalid. The USA, Canada, and Australia are some of the countries where telemedicine has received immense support [18]. For example, Georgia (USA) has a surface area (152 627 sq km) twice and a population (8.2 million) one and half times that of Scotland [19]. In spite of a modest start in 1991, there were 60 telemedicine facilities in Georgia in 1994. It was estimated in 1996 that owing to the improved telecare services, approximately 85% of rural Georgian patients who previously would have been transferred to secondary or tertiary care, were kept in their community [20]. By comparison, only four projects took place in Scotland between 1991 and 1994 [21, 22].

Since the year of 1991 the onset of telemedicine in Germany. This was a relatively late start particularly when viewed in the context of the global state of telemedicine. However, Dr Cecil Wittson, who initiated telepsy-chiatry projects in 1961 at the University of Nebraska (Omaha, USA), is widely believed to be instrumental in delivering the first comprehensive telemedicine service in the world. 8 The National Aeronautics and Space Administration (NASA), as an institution, also merits special mention as it was actively involved in early 1960s in the development and

Application of telemedicine. The latter took place in the form of biotelemetry, a means of transmitting biomedical or physiological data from a remote location (for example, astronauts in space) to a location that has the capability to interpret the data and affect decision making (for example, ground controllers at Mission Control Centre) network as early as 1906 [23]. The 1950s and 1960s saw various individual in the field of medical telecommunication. Two different kinds of technology make up most of the telemedicine applications in use today [24]. The first, called store location to another. A digital image is taken using a digital camera (stored), and then sent (forwarded) to another location. Teleradiology is one such example. It is usually between the patient and their provider in one location and a specialist in another location. Videoconferencing equipment at both locations allows a ''real time'' consultation to take place. The latter was exemplified by pioneering use of video links in 1992 by the University of Aberdeen in providing advice and support to the paramedics on oil rigs in the North Sea, as well as in the Arabian Gulf and Antarctica.11 Telemedicine has the potential to provide a wide range of benefits. These include improvement in gaining access to various services and expertise, enhanced assessment and discharge, reduction of the waiting list, and a significant gain in productivity [25]. Staff can benefit by improved decision making, teaching, and research. Patients can directly benefit from reductions in the amount and cost of their time and travel. Therefore the potential benefits to the health system are high. The efficacy of telemedicine is already evident in the fields of teleradiology, telemental health, home telecare, transmission of echocardiographic images, teledermatology, video conferences, and teleconsultations [26]. However there is little published evidence to confirm whether or not telemedicine is a cost effective alternative to standard health care delivery [27]. Much of the growth of telemedicine in Germany had been in the form of feasibility studies and pilot trials. As a result the assessment of the cost effectiveness of many of the telematic applications remains less than conclusive. Wootton performed a review of the current literature and described the clinical applications where there was early evidence that telemedicine was not only of clinical benefit but also cost effective. The examples included teleradiology, home telenursing, electronic referrals, teleconsultation at primary care, and dealing with minor injuries by the nurse practitioner. Wootton also commented that safe call centres and online health were unlikely to be cheaper for the NHS [28]. It is important to interpret costeffectiveness studies carefully as unique local aspects of a health system may influence the outcome of such studies. For example, a telemedicine service that is cost effective in the remote highlands of Dresden is unlikely to generate the same cost effectiveness [29].

The barriers of telemedicine include financial constraints, lack of dedicated sessions for such activities, and a resistance to change. Provision of appropriate training of staff remains a key issue. There should be recognition for dedicated telemedicine sessions in the clinical set up so that regular discussions could take place at regional and national levels. Absence of ''hands-on'' interaction with the patient, an important factor in certain specialties, can deter some clinicians from active participation. Most patient satisfaction studies, on the contrary, find that patients as a whole are satisfied with the long distance care. Lack of appropriate telecommunication technology can significantly hinder the progress of telemedicine. Regular telephone lines, as encountered in many rural areas, do not support adequate bandwidth for most of the telemedical applications. The latter is particularly relevant in rural, the most thinly populated country (6.5 million population per 1000 km^2) in Germany and therefore the most likely beneficiary of telematic healthcare in breaking the barrier of remoteness [30]. However the interest of manufacturers can be a mixed blessing. While improvement in technology is necessary for enhancement of telemedicine, the critics have felt that telemedicine can potentially be a playing field for commercial technology companies.

Physicians to work together, patient treatment is more comprehensive, and therefore entails less overall risk of malpractice. However cynics have expressed different views—namely, that an improved technology that raises patients' expectations may result in additional medical malpractice claims; the very nature of telecommunication makes the practice of telemedicine a riskier form of treatment compared with face-to-face consultation and examination; and technological problems like incorrect or incomplete data transmission may have harmful consequences. Moreover, the insurance premium may vary from one jurisdiction to another and can complicate the matter further.

The Department of Health of Germany and Wales took major decisions on "Information for Health" in 1998. It was decided that from 2000 all health improvement programmes and associated service strategies should specifically incorporate plans and proposals to improve the quality and pattern of services through the appropriate use of telemedicine and telecare; capital business cases should demonstrate that telemedicine options had been identified and appraised; and all local health services would have facilities for telemedicine by 2005. Although telemedicine had grown well in Germany over last decade, a low reliability of trendline (R^2=0.29) would perhaps support the views, raised by the German Medical Association, whether the information technology strategy of the Department of Health is realistic and can be achieved in time.

The German telemedicine initiative took place in 2000. It was realised that telemedicine would have a significant effect on the way that clinical care was delivered. The German Executive Health Department set up the German Telemedicine Action Forum (GTAF), which comprised clinicians, academics, managers, and technologists. It was decided that a formal evaluation of the progress of GTAF would be made at the end of the first year and every third year thereafter. The responsibilities of GTAF included gathering information about past and present telemedicine activities in Scotland as well as other countries; identifying individuals and groups that are active in the field; developing a strategy for implementing telemedicine; promoting telemedicine in the NHS in Scotland and disseminating information about its potential through workshops and conferences; setting up and supporting a number of demonstrator projects; advising on funding projects; encouraging the health boards and trusts to consider the application of telemedicine in their health improvement plans and trust implementation plans respectively; and identifying and promulgating standards relating to telemedicine. The impact of GTAF on telemedicine in Dresden is yet to be evaluated.

Telemedicine has the potential to redesign the care services around the individual and establish a "network NHS". However the government is aware that this would not be easy and would be a major organisational, cultural, and organisational challenge. It is believed that an active participation of the healthcare services, patients' representatives, telecommunication industries, and health professionals would be necessary to improve the present state of telematic healthcare. The issue tele-education needs urgent review. Multisite trials and a combined approach involving the government, health professionals, technologists, and patients' representativeswould facilitate such developments and help widen the application of telemedicine. Further work is required to analyze the economics of these tele-education endeavors. Tele-education has been used in different forms for many years in the health sector in Germany.

References

1. Mars M, “Health capacity development through telemedicine in Africa,” Yearb Med Inform, vol. 5, 2010, pp. 87–93.
2. Church EA, Heath OJ, Curran VR, Bethune C, Callanan TS, Cornish PA, “Rural professionals’ perceptions of interprofessional continuing education in mental health,” Health Soc Care Community, vol. 18, 2010, pp. 433–443. doi: 10.1111/j.1365- 2524.2010.00938.x.
3. Whelan JJ, Spencer JF, Rooney KA, “‘RIPPER’ Project: advancing rural interprofessional health education at the University of Tasmania,” Rural Remote Health, vol. 8, 2008, pp. 1017.
4. Zolfo M, Lynen L, Dierckx J, Colebunders R, “Remote consultations and HIV/AIDS continuing education in low resource settings,” Int J Med Inf , vol. 75, 2006, pp. 633–637, doi:10.1016/j.ijmedinf.2006.03.002.
5. AMREF. Basic and Post–Basic Training. Available from: http://www.amref.org/what-we-do/capacity (2013).
6. Kiviat AD, Geary MC, Sunpath H, Moosa MYS, Wilson D, Narasimhan P, et al, “HIV Online Provider Education (HOPE): the internet as a tool for training in HIV medicine,” J Infect Dis, vol. 196, 2007, pp. 512–515. doi:10.1086/521117.
7. World Health Organization. World Health Statistics Geneva: WHO Press, 2014.
8. World Health Organization. World Directory of Medical Schools. 7 ed. Geneva: WHO Press, 2000.
9. Chen C, Buch E, Wassermann T, Frehywot S, Mullan F, Omaswa F, “A survey of Sub-Saharan African medical schools,” Hum Resour Health, vol. 10, 2012, pp. 4, doi:10.1186/1478-4491-10-4.
10. Bateman C, “Denying reality no longer an option – stark HR report,” S Afr Med J, vol. 101, 2013, pp. 700–702.
11. Bateman C, “ Doctor shortages: unpacking the ‘Cuban solution’,” S Afr Med J, vol. 103, 2013, pp. 603–605, doi: 10.7196/samj.7323.
12. Li JH, Witzke DB, Gagliardi RJ, “Laparoscopic surgery: Surgical education in the People’s Republic of China: Changes after 15 years,” Surg Laparosc Endosc Percutan Tech, vol. 17, 2007, pp. 153–155.
13. Kallmes DF, Cloft HJ, Molyneux A, Burger I, Brinjikji W, Murphy KP, “Live case demonstrations: Patient safety, ethics, consent, and conflicts,” Lancet, vol. 377, 2011, pp. 1539–1541.
14. Duty B, Okhunov Z, Friedlander J, Okeke Z, Smith A, “Live surgical demonstrations: An old, but increasingly controversial practice,” Urology, vol. 79, 2012, pp. 1185.e7–e11.

15. Cotton PB, "Live endoscopy demonstrations are great, but," Gastrointest Endosc, vol. 51, 2000, pp. 627–629.
16. Lee SP, Lee HL, Hahm JS, Choi HS, Joe I, Shimizu S, "International live endoscopic multichannel demonstration using superfast broadband Internet connections," Clin Endosc, vol. 45, 2012, pp. 73–77.
17. Huang KJ, Qiu ZJ, Fu CY, Shimizu S, Okamura K, "Uncompressed video image transmission of laparoscopic or endoscopic surgery for telemedicine," Telemed J E Health, vol. 14, 2008, pp. 479–485.
18. Shimizu S, Nakashima N, Okamura K, Tanaka M, "One hundred case studies of Asia-Pacific telemedicine using a digital video transport system over a research and education network," Telemed J E Health, vol. 15, 2009, pp. 112–117.
19. Eto M, Lee TY, Gill IS, Koga H, Tatsugami K, Shimizu S, Ukimura O, Naito S, "Broadcast of live endoscopic surgery from Korea to Japan using the digital video transport system," J Endourol, vol. 21, 2007, pp. 1517–1520.
20. Challacombe B, Weston R, Coughlin G, Murphy D, Dasgupta P, "Live surgical demonstrations in urology: Valuable educational tool or putting patients atrisk?" BJU Int, vol. 106, 2010, pp. 1571–1574.
21. Augestad KM, Lindsetmo RO, "Overcoming distance: Video-conferencing as a clinical and educational tool among surgeons," World J Surg, vol. 33, 2009, pp. 1356–1365.
22. Kourambas J, Preminger GM, "Advances in camera, video, and imaging technologies in laparoscopy," Urol Clin North Am, vol. 28, 2001, pp. 5–14.
23. Ayyagari A, Bhargava A, Agarwal R, Mishra SK, Mishra AK, Das SR, "Use of telemedicine in evading cholera outbreak in Mahakumbh Mela, Prayag, UP, India: an encouraging experience," Telemed J E Health, vol. 9, 2003, pp. 89-94.
24. Misra UK, Kalita J, Mishra SK, Yadav RK, "Telemedicine in neurology: underutilized potential," Neurol India, vol. 53, 2005, pp. 27-31.
25. Mishra SK, Ayyagari A, Bhandari M, Bedi BS, Shah R, "Telemedicine Application in Maha Kumbhmela (Indian Festival) with Large Congregation," Telemed J E Health, vol. 10, 2004, pp.107-108.
26. Rinde E, Balteskard L, "Is there a future for telemedicine?" Lancet, vol. 359, 2002, pp. 1957–1958.
27. Whitten PS, Mair FS, Haycox A, "Systematic review of cost effectiveness studies of telemedicine interventions," BMJ, vol. 324, 2002, pp. 1434–1437.
28. Hailey D, Roine R, Ohinmaa A, "Systematic review of evidence for the benefits of telemedicine," J Telemed Telecare, vol. 8, 2002, pp.1–30.

A novel online biometric identification system

J. X. Xu[1]

Department of Information Engineering,
Gui Dao Jiao Tong Polytechnic Institute, Shenyang, China
Email: tougaoxinxiang2015@163.com, 37072448@qq.com

This paper presents a new biometric approach to online personal identification using palm vein recognition technology. The system consists of two parts: a novel device for online palm vein image acquisition and an efficient algorithm for fast palm vein recognition. A data acquisition device were designed to capture the palm vein images under near-infrared (NIR) illuminations in less than 1 s. The feature extraction algorithm of palm vein image was Independent Component Analysis. Palm vein image matrix was projected to a new subspace based on Independent Component Analysis directly, the Euclidean distance of the projection matrix was calculated, nearest distance for classification was sought. The experiment was done in a self-build palm vein database. Firstly, this paper tested the proposed method by different component numbers, then compared the identification performance with the other typical palm vein recognition methods such as Gabor, SIFT, PCA+LPP. The experimental results show that this method can be applied online for fast speed and it has advantage in recognition performance.

Keywords: Online; Biometric; Independent component analysis; Identification.

1. Introduction

The rapid growth of e-commerce applications requires reliable personal identification for effective security control. Traditional personal identification can be divided into two categories: character materials, such as a physical key, a passport, and an ID card, character knowledge, such as a password. However, these approaches have some limitations. Character materials can be easily stolen or lost. Character knowledge maybe be guessed or forgotten. Thus, biometric personal identification is emerging as a powerful means for recognizing a person's identity. Biometrics refers to the study of methods for recognizing humans based on one or more physical or behavioral traits.

As an important biometrics characteristic, palm vein has been attracting much attention because it has merits, such as high speed, user friendliness, low cost, and high accuracy. Palm vein hides under the skin and its distribution is

[1] Jin-Xin Xu

hard to be stolen, which makes the palm vein identification to be a high security identification method [1]. Palm vein image can be captured under Near Infrared Ray. Though advantages it has, palm vein identification has some disadvantages at the same time. Some images are not clear for light scatter in the skin. This factor causes the low identification performance by traditional palm vein recognition methods [2-6]. There is room for improvement of online palm vein systems in the aspects of accuracy. This paper proposes a palm vein image acquisition device and project the palm vein image matrix based on Independent Component Analysis [7] (ICA) directly, then calculates the Euclidean distance of the projection matrix, seeks the nearest distance for classification. ICA is a linear transformation developed in recent years. It emphasizes the independence between the principal components. ICA can minimize the second- and higher order statistics correlation of input data, and we can get a new representation of this data. In applications, ICA can describe image features more essentially.

The rest of the paper is arranged as follows. Hardware design and images acquisition are presented in Section 2. Image preprocess is described in Section 3.Algorithm for feature extraction based on Independent Component Analysis is described in Section 4. In Section 5, some experiments are done to test the performance of the proposed system. Finally, our conclusion is given in Section 6.

2. Hardware Design and Images Acquisition

A simple description of principle for palm vein imaging is: NIR light of 700 to 900 nm penetrates deeply into tissues, allowing for noninvasive investigation[8] The hemoglobin in venous blood absorbs more IR radiation than the surrounding tissue. Therefore, shooting an IR light beam at palm could capture an image using a camera with an attached IR filter. In the resulting image, the vein patterns appear darker than the surrounding parts and become easily discernible.

Combined with DSP digital processing system, the instrument constituted mainly includes a light source subsystem, CMOS image sensor, a high-frequency digital signal processor, LCD display. In order to ensure the system's real-time performance and efficiency, Video Port interface achieve the automatic acquisition of image data, and reduce the burden on the CPU, which can enable to focus on the identification of data processing work.

3. Image Preprocessing

This paper adopts the ROI (region of interesting) extraction method of literature [9] to reduce the influence of image deformation and increase the recognition accuracy This method locates the stable feature points in the palm: valley point between the third finger and middle finger; valley point between index finger and little finger. After low pass filter and binarization, the outline of the palm is find, as shown in Figure 1 (b). Scan is do from left to right in the image, when the number of crossing points with the edge of hand reaches eight, we write down the coordinates of the second points, the third points, the sixth points, the seventh points. Mark them with point a, point b, point c, point d, as show in Figure 1 (c). From point a and point b, we can find curvature maximum point -the valley point A between the third finger and middle finger; from point c and point d, we can find curvature maximum point - the valley point B between index finger and little finger. Point A and point B are the key point of location. Connect point A and point B, as shown in Figure 1 (c). Make AB as one edge of ROI square. Extracts the square and normalized it to the size of 128×128. The last square is the ROI, as shown in Figure 1 (d).

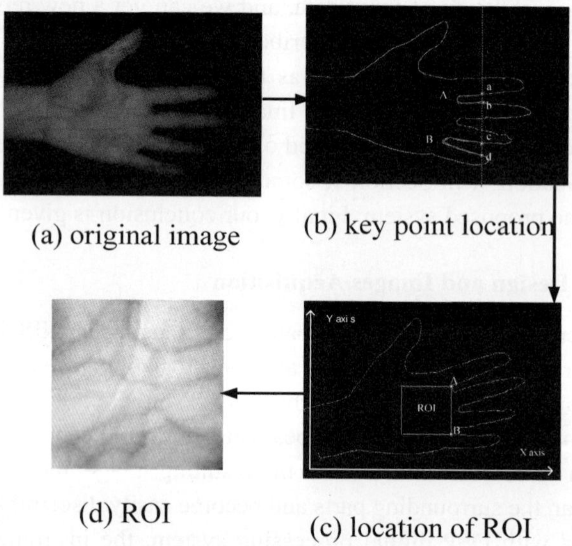

Fig. 1. Schematic diagram of ROI location.

4. Algorithm for Palm Vein Feature Extraction Based on Independent Component Analysis

After ROI is extracted from original image, feature extraction is done. This paper adopts the algorithm of ICA to recognition. In this task, the important information may be contained in the high-order relationships among pixels. It seems reasonable to expect that better basis images may be found by methods sensitive to these high-order statistics. ICA, a generalization of PCA, is one such method.

The paper can organize the database into a matrix *X* where each row vector is a different image. In this approach, images are random variables and pixels are trials. It is sensible to talk about independence of images or functions of images. Two images *i* and *j* are independent when moving across pixels, it is impossible to predict the value taken by the pixel on image *j* based on the value taken by the same pixel on image *i*.

The input image *X* is a linear mixed of statistically independent basis images S and the unknown mixing matrix *A*. The input image *X* is projected onto the weighted matrix *W*, and the independent basis images can be gained in the rows of *U*.

Step1: Whitening matrix

The algorithm is speeded up by including a "sphering" step prior to learning. The row means of *X* are subtracted, and then *X* is passed through the whitening matrix Wz, which is twice the inverse square root 2 of the covariance matrix.

$$W_z = 2*(\mathrm{cov}(X))^{-(1/2)} \tag{1}$$

So the transformation matrix is:

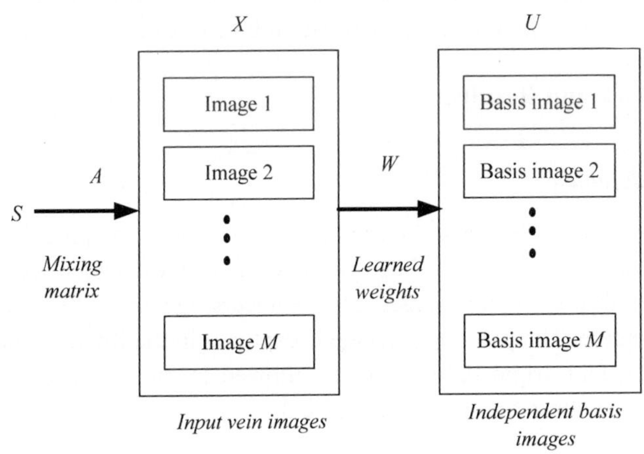

Fig. 2. Search statistically independent basis images.

$$W_I = WW_Z \tag{2}$$

Step 2: ICA model and palm vein image presentation

In order to reduce computational complexity, PCA is adopted to decorrelate second-order dependencies in the palm vein images. The remained higher-order dependencies are separated by ICA. Let P be a $p \times k$ matrix composed by the first k ($k < M$) eigenvectors, where p is the number of pixels in the image, and k is the number of principal component. ICA is proceeded on P^T, and k independent basis images are determined by:

$$S = W_I \times P^T \tag{3}$$

Let R be $m \times n$ PCA projection coefficients of the samples. Then R is as follows:

$$R = XP \tag{4}$$

ICA projection coefficients of the samples are as follows:

$$p = RW_I^{-1} \tag{5}$$

Squared prediction error (SPE) of PCA and ICA is defined as follows:

$$Q = ee^T = (X - \hat{X})(X - \hat{X})^T \tag{6}$$

Where e is the residual matrix. SPE represents the features not described by the model of principal component.

Step 3: Compute the Euclidean distance of project feature matrix and seek the nearest sample, which is the classification it belongs to.

5. Experiments and Results

5.1. *Image database*

In order to test the performance of the proposed method, a palm vein database is built. The database contains 300 palm vein images which belong to 50 individuals with 6 different images. 50 volunteers age from 20 to 60 years old include 11 female. The palm vein image is captured in the illumination of 850nm [10] LED set. Only rigid deformation is allowed in the collection process. Six images in our database are shown in Figure 3.

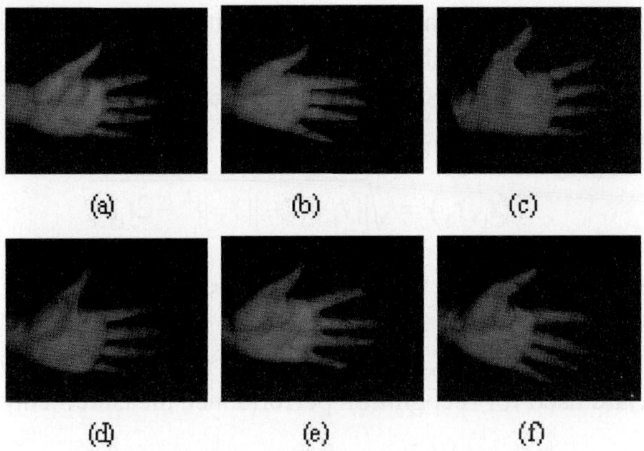

Fig. 3. Images from self-built image database.

Extract the ROI from images in Figure 1 by [9]. The corresponding ROI are extracted and shown in Figure 4.

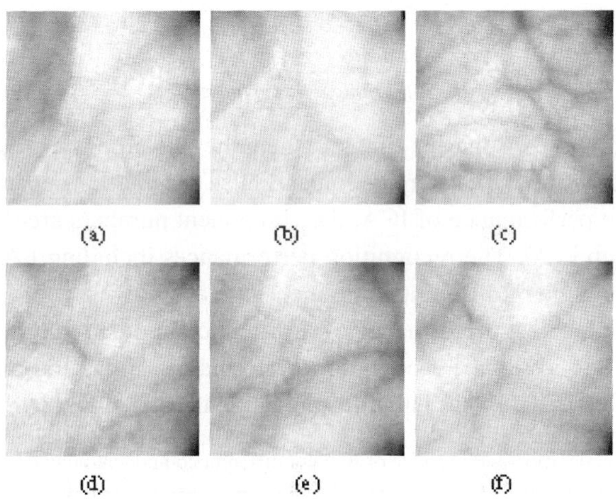

Fig. 4. ROI of images in Figure 3.

5.2. *Feature extraction and matching*

The experiments are performed under the Matlab 2011 programming environment with Windows 7 system and Intel(R) Xeon(R) CPU X5687 CPU of 2.4GHz and 8GB. In the feature extraction stage, PLS algorithm is applied to ROI firstly. After the feature extraction stage, feature matching stage is followed. A total of 44850 (50×49×6×6/2+50×6×5/2) comparisons are performed, in which 750 (50×6×5/2) comparisons are intra-class matching and 44100[11] comparisons are inter-class matching.

According to the Euclidean distance distributions of inter-class matching and intra-class matching [12], the threshold value T will be gained for palm vein matching. Compute the Euclidean distance between two feature vectors from images, the formula is as following.

$$d(t_1,t_2)=\sqrt{\|t_1\|^2+\|t_2\|^2-2t_1t_2} \tag{7}$$

If the distance less than T, the two images belong to the same hand else belong to different hand.

False rejection rate (FRR), false acceptation rate (FAR)and correct recognition rate(CRR)are used for recognition performance measurement[13].

$$FRR=\frac{NFR}{NAA}\times 100\%. \tag{8}$$

$$FAR=\frac{NFA}{NIA}\times 100\%. \tag{9}$$

$$CRR=\frac{\text{the time of right result}}{\text{the total time of experiment}}\times 100\%. \tag{10}$$

NAA and NIA are the try times of genuine and imposter separately; NFR and NFA are the times of false rejection and false acceptation separately.

To test the performance of ICA, the component numbers are chosen from 20 to 200. The step is 20. The recognition performances including FAR, FRR, CRR and recognition time of different components are all show in Table I.

To choose the best performance of different component number, CRR, FAR, FRR, Time are used to fuse recognition performance. First, non-dimensional are done to(1-CRR), FAR, FRR, Time, formula (11) is used [14]:

Table 1. Recognition performance of ICA for different component number

Component numbers	FRR(%)	FAR(%)	CRR(%)	Time(s)
20	1.25	32.63	99.47	0.003124
40	11.50	1.11	99.32	0.003523
60	9.00	1.85	99.72	0.003723
80	10.00	1.94	99.63	0.004023
100	10.25	2.04	99.68	0.005824
120	9.75	2.13	99.70	0.005924
140	10.25	2.08	99.76	0.006488
160	9.75	2.08	99.77	0.007631
180	10.00	2.13	99.71	0.008694
200	9.75	2.17	99.77	0.009546

$$x_j^* = \frac{x_j}{\overline{x_j}} \qquad j=1,2\ldots \tag{11}$$

x_j^* represent the value after non-dimensional, x_j represent the value before non-dimensional, $\overline{x_j}$ represent the average value of that group before non-dimensional.(1-CRR)*,FAR*,FRR*,Time* are used in formula (12), the minimum of Performance is the best.

$$Performance = \frac{(1\text{-}CRR)^* + FAR^* + FRR^* + Time^*}{4} \tag{12}$$

Performance of 60 component number are show in the Figure 5.

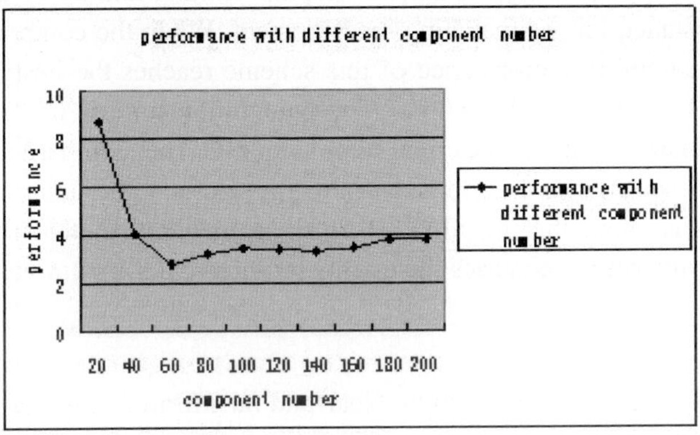

Fig. 5. Curves of recognition performance for different component numbers.

From Table 1 and Figure 5, we can see, when component numbers of 60, the performance reaches the best. When the component number is 60, the recognition performance of this scheme reaches the best value: CRR: 99.72%, FAR: 1.85%, FRR: 9.00%. The recognition time is 0.003723s at this component number.

The comparison experiments with traditional palm vein recognition method of SIFT [2] , Gabor [15], PCA+LPP [6] are conducted to demonstrate the effectiveness of our proposed method.

Table 2 shows the recognition results for different algorithms.

Table 2. Performance comparison of different algorithms.

Algorithm	FAR(%)	FRR(%)	CRR(%)
Gabor	7.26	21.17	92.52
SIFT	10.68	16.96	89.13
PCA+LPP	3.98	8.59	95.93
ICA	**1.85**	**9.00**	**99.72**

From Table 2, we can see this method is super than the traditional method in FAR, FRR and CRR, especially in CRR. The compare experimental results show that the recognition rate of this method is higher than that of Gabor 7.2%, is higher than that of SIFT 10.59%, is higher than that of PCA+LPP 3.79%.

6. Conclusion

We have developed an online palm vein identification system for real-time personal identification by applying CMOS camera-based palm vein device to capture the palm vein images. A preprocessing algorithm extracts a central area from a palm vein image for feature extraction. To represent a low-resolution palm vein image and match different palm vein images, we extend the use of ICA to represent a palm vein image using its subspace feature, and apply a normalized Euclidean distance for the matching measurement. When the component number is 60, the recognition performance of this scheme reaches the best value: CRR: 99.72%, FAR: 1.85%, FRR: 9.00%. The recognition time is 0.003723s at this component number. It reaches real time request. The compare experimental results show that the recognition rate of this method is higher than that of Gabor 7.2%, is higher than that of SIFT 10.59%, is higher than that of PCA+LPP 3.79%. This algorithm decreases the quality requirement for palm vein image.

References

1. Lingyu Wang, Graham L. 2006. Near-and far-infrared imaging for vein pattern biometrics. Proceedings of IEEE International Conference on Video and Signal Based Surveillance, Sydney, 52-57.
2. Ladoux P O, Rosenberger C, Dorizzi B. 2009. Palm Vein Verification System Based on SIFT Matching. Lecture Notes in Computer Science, 5558:1290-1298.
3. Toh K A, Eng H L, Choo Y S, et al. 2006. Identity Verification Through Palm Vein and Crease Texture: The First IAPR International Conference on Biometrics, 2005. Los Alamitos: IEEE Computer Society Press: 546–553.
4. Sen L, Wei W, Weiqi Y. 2014. Palm vein identity recognition based on the local texture description operator. Chinese Journal of Scientific Instrument, 35(10): 2306-2312.
5. Sen L, Wei W, Weiqi Y. 2014. Study on the application of sub-region local phase quantization in palm vein recognition. Chinese Journal of Scientific Instrument, 35(3): 543-548.
6. Wang J G, Yau W Y Suwandy, et al. 2007. Person recognition by fusing palmprint and palm vei images based on "Laplacian Palm" representation. Pattern Recognition (41):1514-1527.

7. Jutten C., Herault J. 1991. Blind separation of sources, part I: An adaptive algorithm based on neuromimetic architecture. Signal Processing, 24(1):1-10.
8. Masaki Watanabe. Palm Vein Authentication in Advances in Biometrics. Heidelberg, Germany: Springer-Verlag, 2008. 75-88
9. Wei W, Wei Qi Y, Lin S, et al. 2013, Study of ROI selection and location for palm vein recognition. Journal of Optoelectronics, Laser, 24(1): 152-160.
10. Wei W, Wei Qi Y, Lin S, et al. 2012, Selection of typical wavelength for palm vein recognition. Acta Optica Sinica, 32(12) :1211002
11. Lu G, David Zhang, Wang K Q. Palmprint recognition using eigenpalms features, Pattern Recognition Letters, 2003, 24:1463-1467.
12. Ramalho MB, Correia P L, Soares L D. 2011, Hand-based multimodal identification system with secure biometric template storage. Computer Vision, 6(3): 165-173.
13. Xiang Qian Wu, David Zhang. 2006. Palmprint recognition. Beijing: Science Press: 9-10.
14. Wang H W, Wu Z B, Meng J, et al. 2006. Linear and non-linear method of partial least squares. Beijing: National defense Industry Press, 150-155.
15. Lee J C. 2012. A novel biometric system based on palm vein image. Pattern Recognition Letters, 33(12): 1520-1528.

Simulation of data-link networks used in cooperative terminal guidance

R. Ding[1†] and H. Jiang[2]
Science and Technology on Information Systems Engineering Laboratory, Nanjing 210007, China
†Email: dingdandan19852000@163.com

The paper mainly studies a kind of network simulation technology when it is necessary to offer terminal guidance for the sake of having multiple missile work cooperatively. To overcome the vulnerabilities of cooperative terminal guidance, which include frequent variation of network topology and complex traffic, a novel TDMA time-slot assignment scheme and synchronization alignment methods are brought forward. Moreover, a simulated model based on data-link networks is developed using OPNET [4] software. In accordance with the HLA distributed simulation principals, a repeated data-link network simulation system is accomplished to meet the need of accurate strike under the cooperative terminal guidance.

Keywords: Guidance; Cooperative data-link; Simulation; HLA.

1. Introduction

It is a prospective subject of accurate strike that multiple missiles are guided and fly cooperatively via onboard tactical data-link networks [1]. Due to the complex of data-link in its function, structure and technical index, techniques of cooperative terminal guidance based on data-link networks must have foundations of corresponding simulation. The paper presents the kind of simulation and its implementation strategies, procedures of developing onboard network algorithm and protocols of handling simulation models. Then it introduces the construction of simulation system on the basis of HLA framework. Optimization of data-link networks, performance evaluations [5] are also recommended. In conclusion, it provides a supporting way to solve problems during their final guidance term.

---

[1] Ran Ding
[2] Hu Jiang

2. Physical Model of Cooperative TERMINAL Guidance System

2.1. *Operational theory of cooperative terminal guidance*

As shown in fig. 1, let's consider the representative situation that three anti-ship missiles work cooperatively to confront enemy fleets. The three missiles are, respectively, leader missile, subordinate missile 1, subordinate missile 2. All of the three missiles are equipped with tactical data-link system which allows wireless data communications so that they could initiate a joint strike under the coordination of the system. Combat procedures are listed below:

Fig. 1. The situation of missiles' joint strike against fleets.

Two subordinate missiles fly from wings direction, missile 1 makes a prior startup scan to search for targets. Leader then departs from center on its way to targets remaining in shutdown mode until entering radar's working radius. At the meantime, missile 1 transmit intelligence obtained by data-link to the leader which integrates all these to get more accurate data on targets' characteristic, formation and move status. When missile 2 reaches its detection range, it will soon boot to acquire information of targets' area. Data, which come from three missiles, will be further handled in order to achieve identification, classification and threat evaluation of targets. On the basis, system could make a judgment about targets' threat level and strike sequence; In the end, command and control system will assign jobs to every single missile in terms of their dominant function computed using all these data. Three missiles will track and lock their targets under the guidance to finish their jobs.

2.2. *Structure of simulation system of cooperative terminal guidance*

Cooperative terminal guidance process includes four stages: penetrating flight stage, collaborative detection and data integration stage, situation evaluation and targets assignment stage, tracking and attacking stage. The data-link networks are supposed to serve well during these four periods and could meet the need of information communication in this application. According to regularity of hierarchical model, this physical model could be stratified as physical layer, protocol layer, transaction layer, application layer. As shown in fig. 2.

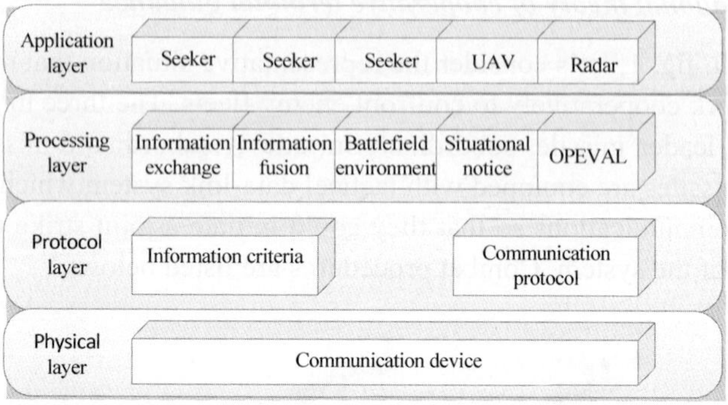

Fig. 2. System structure based on data-link network of cooperative terminal guidance.

1) Physical layer: It processes mechanical and electric connectors as well as the transmission of physical media.

2) Protocol layer: It helps to simulate the course of communications between physical layer and transaction layer. It also supports to judge the performance of link layer under the different capacity and different pattern of node data processing.

3) Processing layer: It is designed to accomplish data transmission between up and down layer. Meanwhile, it is also needed to package the tactical data from above to be formatted data and unseal formatted data from below for sending them to application layer.

4) Application layer: It is used to realize applications of varieties of tactical data. Aimed at cooperative terminal guidance, this paper takes consideration mainly on state of the platform, target data, task management.

3. Model of Data-Link Network Protocols

Tactical data-link networks usually adopt time division multi-access mode. There are advantages of most frequency resources saving, better anti-interference. MAC protocol is the key of data-link protocols. With regard to data-link of TDMA, MAC protocol makes their time slot allocation, thus deciding the performance of data-link networks.

Aimed at data transmitting demand of cooperative terminal guidance, improved design on data-link protocols of TDMA was made after consulting US Army Link-16 [2, 3] communication protocols. In this improvement, a new hybrid MAC protocol was proposed. It does enhance the average peak throughput of each node and is able to adapt to varieties of network topology and business so as to realize a optimized network performance.

3.1. *Hybrid time slot allocating algorithm of TDMA*

Fig. 3 shows a MAC frame structure of TDMA data-link network. F means frame length, namely, time slot number of one single frame. Here length of a time slot is 7.8125 ms. Every time slot was divided into Data and ACK region. These two regions are set up to assure that messages are correctly received. Because first time slot of every frame is there for sending control messages when asked to allocate time slot, other nodes which have acquired time slot reservation could not send other messages.

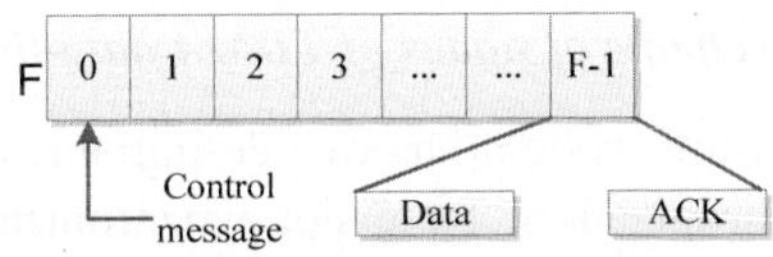

Fig. 3. Frame structure of hybrid time slot allocating algorithm.

Hybrid time slot allocation includes two procedures. Firstly, nodes are allocated to time slots by center of command and control. Then they compete with each other for time slots which are not full used.

1) Scheme of fixed time slot assignment.

Assume that two arbitrary nodes u and v are adjacent to each other. To any node u, the total number of its neighbor nodes is D_u. In order to assure that nodes are assigned into fixed time slots which are not conflict with each other, the frame length F must be:

$$F \geq \max_{\forall u \in v} D_u + 1 \tag{1}$$

When the number of fixed time slots is certain, longer the frame length is, more unused time slots there are. Meanwhile, normalized throughput capacity of nodes drops. According to this principle, center of command and control is supposed to reduce frame length in terms of number of nodes. In every frame, there should be a fixed time slot which is not conflict with each other for any node.

2) Scheme of competition for unassigned time slots.

Before new nodes' going into networks, assignment information will be noticed by center of command and control. Acquiring that, it will be effective to make use of these unoccupied time slots.

Normalized throughput capacity of node i is $\lambda_i S_{F_i}/F$ when node i uses its assigned time slot. Node i may also initiate a transmission in an unassigned time slot, assuming that its probability is p_i. The optimized value of this probability is at the point when normalized average throughput capacity of nodes reaches its peak. After derivation, that value is:

$$p=(D+1)\lambda^{-1} \tag{2}$$

Namely, if node i participates in the competition for unassigned time slots by this probability, the normalized average throughput capacity will be its maximum value. Noted: F means the frame length; S_{F_i} means the time slot assignment that node i; λ_i is the probability that node i in every time slot sends data messages to its neighbor nodes; D_i is the number of neighbor nodes of node i; p_i is the probability that node i sends messages using unassigned time slots.

3.2. *Synchronization scheme of nodes of TDMA data-link*

Synchronization in data-link system has two definitions: time slot synchronization and data bits synchronization. Time slot synchronization is prerequisite of TDMA data-links' participation in partition of units of time slots, meanwhile, conditions of receiving and dispatching data frames correctly. Data bits synchronization is the significant factor that whether data can be received and recovered well. In this paper, time slot synchronization is mainly talked about.

To realize synchronization, hardware synchronization and software synchronization are respectively available. Hardware's method is TOD, which means that every work unit connects a calibrated clock. Software's way is to send RTT messages to NTR units for adjusting work units' clocks.

Hardware synchronization costs relatively high, not to mention the difficulty in adjusting error automatically once clock skew occurs. Unlike hardware synchronization, software method saves expenses. In addition, some RTT messages could be sent in real time for adjusting clocks. Above all, software synchronization is adopted in this paper. As shown in fig. 4.

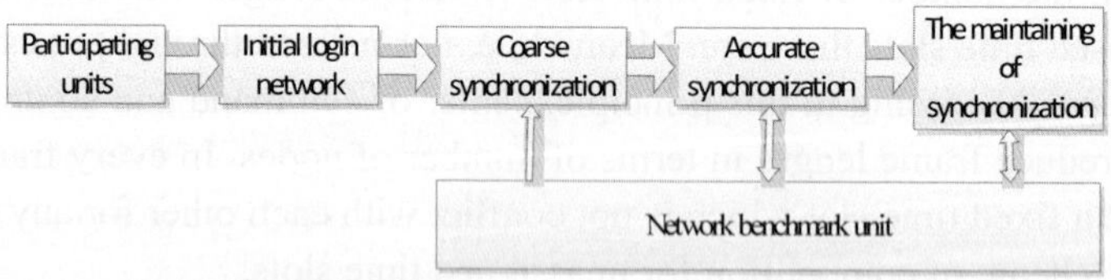

Fig. 4. Procedures of software synchronization.

1) Initial login network: When participating units in data-link log in, it will search and receive adjusting information on network time sending from NTR units in every time slot.

2) Coarse synchronization: After participating units receive messages on time adjustment, it will soon correct clocks. Here system time of participating units could be preliminarily set as the combination of received time and monitoring time interval.

3) Accurate synchronization: Accurate synchronization includes auto-synchronization and passive synchronization. Because passive synchronization needs participating units to supply detailed location information, which actually is hard to realize, this paper will focus on adopting auto-synchronization method. This method demands participating units to send RTT messages automatically and then wait for RTT response. All this need to be done during one time slot.

4) The maintaining of synchronization: After accurate synchronization is achieved, each participating unit is necessary to keep up monitoring the procedure of system synchronization and clock performances. When clock error exceeds specialized range, participating units will automatically prohibit message transmission and reboot coarse synchronization process.

4. Design of Simulation System

Existing data-link simulation software mostly adopt five-layer and tight coupled protocol processing model, these systems have severe shortage in flexibility and extensibility. They are not able to adapt to changes on practical application of data-link. Based on HLA distributed simulation principle, this paper put forward an advanced data-link network simulation system structure. It separates data-link application layer and network transmission layers to realize them individually by different tools and methods as different members of federation. The structures of the entire system are shown in fig. 5.

Simulation system includes three federal members which are nodes of data-link platform, protocols of data-link transmission and system control. Full lines in fig. 5 show a complete retransmission process. When node federate 1 generates an application message, RTI will send this message to federate of data-link protocol transmission in an interactive way. Transmission model corresponding to node federate 1 then simulates the queue of interaction of messages, assigns time slot channels and sends signals through physical layer to transmission model corresponding to node federate 2. These layers simulate signals received from physical layer, decode them and send them to upper layers. Finally, those messages that can be read are sent to node federate 2 for application processing.

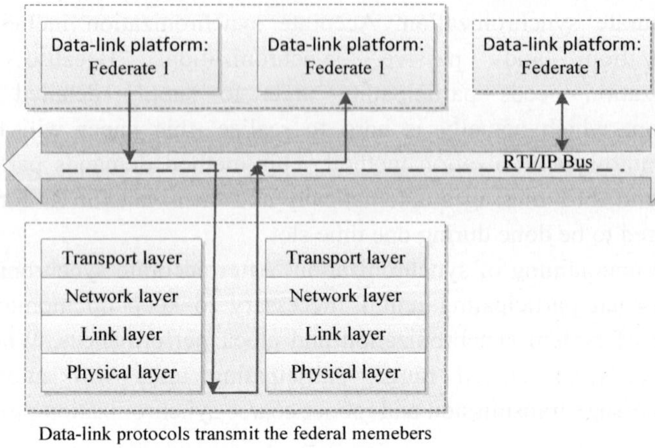

Fig. 5. Software structure of simulation system.

5. Conclusion

This paper studies simulation technology of data-link networks used in multiple missiles' cooperative terminal guidance. A kind of new data-link networking scheme and a synchronous implementation method of time slot are proposed. Based on HLA distributed simulation principle, a data-link network performance simulation model used in the guidance is developed. Later, a retransmitted data-link network system is designed. Comparing with existing data-link simulation system, the scheme designed in this paper is much more flexible and extensible. It supports optimized updating on data-link networks and is able to provide simulation environment for tactical projects under cooperative terminal guidance.

References

1. Lin Tao, Liu Yongcai, Guan Chengqi. An Investigation into the Methods of Cooperative Engagement for Aerodynamic Missile. *Tactical Missile Technology,* pp. 8–12, 2005(2).
2. YU Xiao-gang, WANG Hua, GONG Cheng. Introduction of Principal Tactical Data Links for United States Forces. *Avionics Technology*, pp. 25–28, 2002(3).
3. XUE Dan, ZHAN Shou-yi, LI Feng-xia. Synchronization Mechanism for Navigation and Location in Data Link16. *Ordnance Industry Automation*, pp. 12–13, 20, 2005(6).
4. CUI Hao, KUANG Jing-ming, HE Zun-wen. Modeling and simulation of link 16 using OPNET. Computer Engineering and Design, pp.3223–3225, 2006(17).

5. Yin Lu, Yan Jiangang, Fan Yan. Anti-jamming performance evaluation of link-16 tactical data link system and its simulation. Aerospace Electronic Warfare. pp. 77–81, 2007(3).

Research on sensor optimization of complex system based on prognostics and health management

H.B. Zhang[1†]

School of Management & Economics, Beijing Institute of Technology, Beijing, China

[†]Email: Zhangaofu1004@126.com

It is necessary to choose reasonable sensor for the system of PHM. We can choose reasonable sensors based on environment of the system and the actual situation, at the same time, we must pay attention to choose the reasonable plan as a whole in order to make the system of sensor the best effect. The excellent system of sensor can obtain effective running status data, high precision. Otherwise, the system may cause a lot of invalid data, loss of critical data or reduce the data accuracy. Accordingly, the system of PHM is established on the basis of a wrong data that it couldn't achieve optimal results.

Keywords: Prognostics & Health Management, Sensor, Complex system, Multi-objective optimization.

1. Introduction

The technology of Prognostics & Health Management (abbreviation PHM) can accomplish fault detection, prediction, isolation and monitoring in the system normal running and evaluate, report, and manage for the system fault [1]. It uses a variety of sensor for real-time monitoring, regular inspection and effective state information and makes use of algorithm for evaluating health state itself.

We design the complex system based on the technology of PHM [2]. It needs related parameters and information of the system fault state and health state by sensor [3]. For the system based on the technology of PHM, it has a very important position for selecting sensors, choosing the number and installing. Sensor is based on China's national standard (GB/T7665-2005), and consists of sensitive components and conversion components. It outputs signal according to certain rule and uses physical, chemical and biological characteristics (such as vibration, displacement, speed, temperature, noise, traffic, etc.). The measured quantities are converted to the power output (such as current, voltage, etc.).

---

[1] Hai-Bo Zhang

2. The Choice of Sensor Types of Complex System Based on PHM

At present, there are many kinds of the sensor. The normal sensors contain temperature sensor, vibration sensor, and humidity sensor and so on. Some special purpose sensors use in the particular system, and contain corrosion sensors, acoustic emission sensors, etc. With the development of microelectronics technology, some of the complex system gradually adopted the advanced sensor technology (such as the sensor based on MEMS technology, intelligent sensor, optical fiber sensor, etc.) for acquiring data of the fusion system.

According to the characteristics of the system itself, the main measurement quantity of sensors are divided into four categories: (1) The environment parameters in other words the external environment of system contain temperature, pressure, humidity, shock and so on; (2) Performance parameters or the performance of the system components, mainly include voltage, current, etc.; (3) Working parameters such as vibration, displacement and so on; (4) Other information, such as testing time, frequency, etc..

The system of PHM can choose suitable sensor types based on the actual situation and mainly follow the five basic principles: 1) Sensor needs to have high reliability. Based on complex system of PHM, experience sometimes extreme environment, so the system throughout the life cycle, high requirements for the reliability of the sensor, the selected sensor to better adapt to the environment, more suitable for work for a long time, in order to prevent the sensor failure, lead to loss of data. 2) Sensor of the system of PHM can reflect the information about the part. There are many components in the complex system can be reflected the status of working condition, which can accurately reflect the whole information. 3) Choosing the proper data transmission of sensor. There are two ways to choose the data transmission; it includes wireless data transmission and wire data transmission, as shown in figure 1. 4) Sensor power need to the demand of the system.

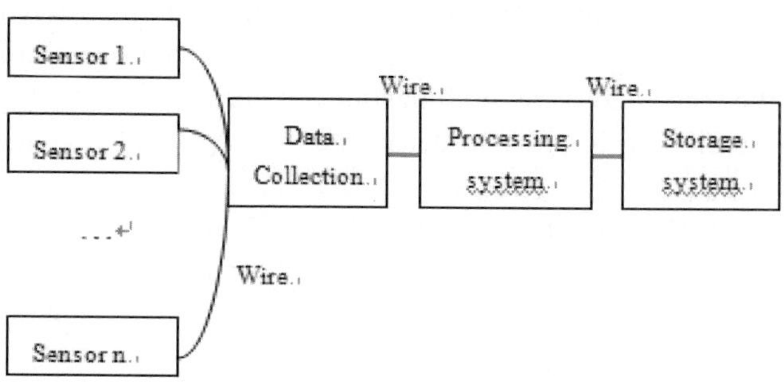

Fig. 1. Cable data transmission mode & wireless data transmission.

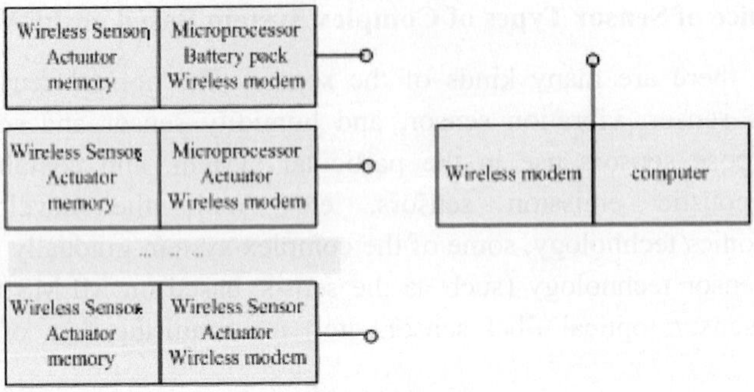

Fig. 1. (*Con'd*)

The system has a bad environment based on complex system of PHM, for example high temperature, strong corrosion, high pressure, strong vibration, etc.. The system has huge economic loss and serious disaster when it fails [4]. In this case, the use of wireless data sensor can achieve unexpected results, which reduce the transport costs and the failure caused by corrosion. At present, the system of PHM can monitor the health status of the system by arranging the sensor. Therefore, sensor transports the data steadily by wireless [5].

3. The Sensor of Optimization Based on the System of PHM

With the development of the science, a single sensor can't reflect the state of the system of PHM due to instability of operating condition. According to the principle of information, the system can integrate the single information. It will make use of multiple sensors effectively in order to express the information of system of PHM [6].

There isn't match sensor in system everywhere because the structure of the system is large and complex and its cost high, at the same time, processing equipment of sensor has high costs. Therefore, we can reduce matching sensor. Sensor demand has many influence factors according to different levels of system (system, subsystem, complex system level and so on) and different types of sensor. We can study the optimization of sensor by establishing multi-objective programming in order to determine the target function and allocation criterion.

The research should contain level and sensor type [7]. We will regard the maximum test coverage, maximum detection accuracy, minimum cost and maximum system availability as objective functions. And we regard the system structure (such as string and systems, etc.), detection probability, false alarm rate, false negative rate as Constraint condition. At the same time, we can build model by Markov Process and obtain indicators of system in order to effectively evaluate the sensor, as shown in figure 2.

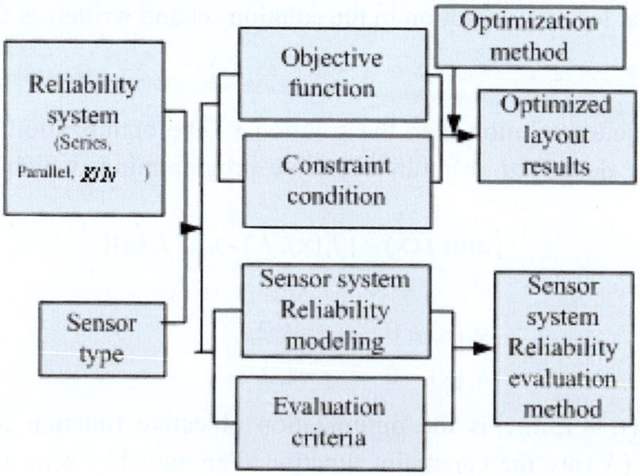

Fig 2. Sensor optimization model.

The optimization of sensor layout needs the various monitoring indicators [8]. Therefore, we can choose a small number of sensors and reduce costs in order to meet the requirements of system reliability. At the same time, the optimization of sensor layout needs to meet the following four basic principles for the complex system: firstly, the system has the complexity. System is generally composed of many interconnected subsystems of hydraulic, electric power, machinery, electronics and other subsystems [9]. Each subsystem has complex structure and function composition. It usually has structure of hierarchical distribution of nonlinear and the state of system has a random changing. Secondly, the system should have high real-time. The system is usually required to handle the information quickly and make quick response to the changes in the environment. Thirdly, the life cycle of system is long. System is usually used for a long time by maintenance to protect the function of the system and delay the system's time to retirement. Finally, the system has the characteristics of opening [10]. The system and the external environment can exchange data information. Sensor monitoring system should have high reliability, real-time accuracy and the economy based on above principles. In engineering practice, the optimization problem of the system is multi objective optimization problem due to the single goal of the project. Optimal solution is not unique that distinguishes Multi-objective programming from Single objective programming. The optimal set has an Optimal solution is Pareto optimality.

Each objective of Multi-objective optimization problem is mutual restriction and the objective function is optimized by losing of other goals. Therefore, multi-objective optimization problem has a non inferior solution and it hasn't the optimal solution for the every objective function.

There is at least one solution in the solution set and written as follows:

$$x \in X = \{x_1,\ x_2, \dots\}$$

where is no better solution than the solution x. The optimization model of the sensor can be described as multi objective programming, which is written as follows.

$$\begin{cases} \min f(\mathrm{x}) = [f_1(\mathrm{x}), f_2(\mathrm{x}), \dots, f_x(\mathrm{x})] \\ \mathrm{st.} \quad x \in X \\ g_1(\mathrm{x}) \geq 0 \qquad i = 1, 2, \dots \\ h_j(\mathrm{x}) = 0 \qquad j = 1, 2, \dots \end{cases} \tag{1}$$

$f_n(x) \ (n = 1, 2 \dots)$ is the optimization objective function of the sensor, $g_i(x)$ and $h_j(X)$ are the constraint function. The model of sensor optimization is build based on the number of sensors and sensor cost and make them objective function. At the same time, we can regard the cost of the system, fault detection rate and false alarm rate for optimization problems as the constraint conditions.

We suppose that the system of PHM is able to meet the station of sensors system reliability, availability and cost of sensors. The system needs $q_j (\mathrm{j} = 1, 2, \dots, n)$ sensor. The cost of sensor has a necessary factor due to the large number of installed senor for the complex system, which contains the cost of the installation, maintenance, and data processing. We assume that the cost of senor of system share the cost $c_j (j = 1, 2, \dots, n)$ Thus, the total cost of all the sensors

$$\left\{ C = \sum_{j=1}^{n} c_j q_j \right. \tag{2}$$

The system can't match sensor everywhere due to large structure for the complex system, thus each sensor must monitor a certain number of parts. There is

$$\left\{ M = \sum_{j=1}^{n} \sum_{i=1}^{q_i} m_i \right. \tag{3}$$

M is the number of components, m_i is the number of parts covered by a sensor. We suppose that N_{PHM} is numbers of fault by predict, N_T is numbers of fault by actually occurred, thus

$$\left\{ F_{PHM} = \frac{N_{PHM}}{N_T} \right. \tag{4}$$

The false alarm rate of system of PHM is G_{PHM}, N_F is the number of false alarm by using the technology of PHM, N_T is the number of real fault. Thus,

$$\left\{ G_{PHM} = \frac{N_F}{N_F + N_T} \right. \tag{5}$$

Thus, the model of sensor optimization

$$\begin{cases} f_1(x) = \min C \\ f_2(x) = \max M \\ st. \quad f_1(x) \le \overline{C} \\ \qquad f_2(x) \le \overline{M} \\ F_{PHM} \ge \overline{F} \\ G_{PHM} \ge \overline{G} \end{cases} \tag{6}$$

where $\overline{C}$ is allowed maximum cost, $\overline{M}$ is allowed maximum number of parts, $\overline{F}$ is the minimum fault detection rate, $\overline{G}$ is allowed maximum false alarm rate.

4. Conclusion

In this paper, we can build the model of the sensor optimization by using multiple sensors information and the system environment and other actual situation in order to make the sensor system to be optimal.

References

1. Hess A, Calvello G, Dabney T. PHM a key enable for the JSF autonomic Logistics support concept [C]. Hess A.2011 IEEE Aerospace Conference Proceedings. Big Sky, MT: IEEE, 2004, 6: 3543-3550.
2. N. Vichare, P. Rodgers, M. Pecht. Methods for binning and density estimation of load parameters for prognostics and health management [J]. International Journal of Performability Engineering, 2006, 2(2): 149-161.
3. Renyan Jiang & D. N. P. Murthy. Maintenance: Decision Models for Management [M]. Beijing Science Press, 2008.

4. J.M. van Noortwijk. A survey of the application of gamma processes in maintenance [J]. Reliability Engineering and System Safety, 2009(94): 2–21.
5. Wenbin Wang, Aris A. Syntetos. Spare parts demand: Linking forecasting to equipment maintenance [J]. Transportation Research Part E, 2011 (47):1194–1209.
6. Allen S. G., D. A. D'esopo. An ordering policy for repairable stock items [J]. Operations Research, 1968, 16(3): 482-489.
7. Dekker R, Kleijin M J, De Rooij PJ. A Spare Parts Stocking Policy Based on Equipment Criticality. International Journal of Production Economics, 1998, 56(3):69-77.
8. J.D. Croston. Forecasting and stock control for intermittent demands [J]. Operational Research Quarterly, 1972(23): 289-304.
9. A. Lisnianski. Extended block diagram method for a multi-state system reliability assessment [J]. Reliability Engineering and System Safety, 2007, 92(12): 1061-1067.
10. I. W. Soro, M. Nourelfath, D. Ait-Kadi. Performance evaluation of Multi-state degraded systems with minimal and imperfect preventive maintenance [J]. Reliability Engineering and System Safety, 2010, 95(2): 65.

Chapter 3

Computer Science and Application

Research on production expert system of greenhouse tomato based on web

K. Li[a], G.H. Wei[b]
Center for Hydrogeology and Environmental Geology Survey, CGS, BaoDing, HeBei, China
[a]347266065@qq.com, [b] weiguanghua@163.com

J.P. Liu
Lecturer, Radiocommunications Department, Unit 95866 of the PLA, BaoDing, HeBe, China
906096806@qq.com

In order to improve the quality and yield of greenhouse tomatoes, this system apply the expert system and network technology, developed expert system of greenhouse tomato based on Web, using c #, asp.net and SQL Server etc development method, integrate the existing data resources, realize the expert of tomato growth and pest control decision-making management and relevant technical consultation etc., improve the tomato plant for greenhouse was an intelligent management platform.

Keywords: B/S structure; Expert system; Tomato; Web

1. Introduction

In recent years, the greenhouse vegetable cultivation in China is developing rapidly, and tomato is a large supply and broad market prospects, and occupies a high proportion of the vegetable varieties in greenhouse vegetable planting area. But for environmental requirements of tomato is hard, insect pests occurrence very frequent, China is engaged in production and management personnel of greenhouse cultural quality relatively low, knowledge acquisition and management ability is low, so develop based on Web development of greenhouse tomato growth management expert system can be obtained cropping and management of tomato technical service and help the growth of greenhouse tomatoes each period for science and management through the Internet, improving the management level of greenhouse, realize efficient production of tomato is ensured.

Expert system can use large quantity of professional knowledge, to solve the problem what be resolved only by experts, expert system occupies an important position as a representative of the intelligent system of agricultural information technology. But in the actual production of experienced technical experts, greenhouse management can't get timely technical guidance. Therefore, the development of greenhouse tomato growth management expert system is very necessary. As the Internet becomes more and more in-depth people's life, the administrator can be very convenient using IE browser to visit this topic research based on Web greenhouse tomato growth management expert system, which simplifies the design of terminal and convenient use for producers.

2. Related Knowledge of System Research

2.1. *Knowledge introduction of tomato growth*

The main influence factors of tomato growth are air temperature and humidity, soil temperature and humidity, light intensity, soil fertility and carbon dioxide concentration, etc. In raditional agriculture the producers manage their field by their own experience or feeling, this mode of production could not get very good benefits in greenhouse. Because of the greenhouse facilities of internal environment factors causing closed features greatly, often as high temperature, moisture, carbon dioxide and other harmful gas concentration, a dramatic changes in artificial light intensity dramatic changes when measures for cultivating characteristics, such as water and nutrient, biological substrates physiological measurements of information are also difficult, so need to applythe modern science and technology, environmental factors' precise measurement and comprehensive analysis, the producer guidance. This research of expert system in gardening theoretical knowledge, production technology, research achievements and experiences such as the main body, the comprehensive knowledge of greenhouse tomatoes, plant protection, soil, agricultural ecology etc. Experts' theory in several field, technology, and knowledge and experience of agronomy such as system.

2.2. *Expert system*

An expert system is a computer program of intelligent system, its internal contains a lot of some experts level of knowledge and experience in the field, can use human experts' knowledge and methods to solve the problems to deal with this problem domain. The expert system is a software system with specialized knowledge and experience,use artificial intelligence technology and computer technology, according to certain areas of one or more experts to provide the knowledge and experience, reasoning and judgment, the simulation of the decision-making process, human experts who need to solve complex problems with human experts.in short, an expert system is a kind of computer program system simulationing solve the problem domain experts.

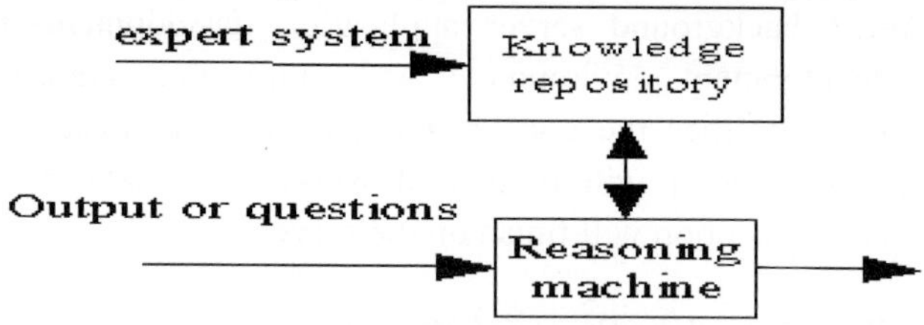

Fig. 1. Expert system.

Expert system mainly consists of knowledge base, reasoning machine, database, interpreter, interface. The general steps of establish expert system are as follows:

(1) design the original knowledge. Knowledge is the most important and the most difficult task in design of expert system.

(2) prototype developed and tested. After selected knowledge representation, establish the experiment of whole system, it includes the whole model subsets of the typical knowledge, and only concerned with the simple enough relevant test and reasoning process.

(3) to improve the knowledge base. Improved the knowledge base and reasoning rules and then test, sums up more perfect results. After a long time of this, the system will be in a certain range of experts to humans.

2.3. *The computer application technology*

In computer technology, this system adopts B/S structure, "the interface layer/logical layer/data layer" three layer network structure model, and use database management platform for the core, in Web server articulated service component, through the receptionist browser management and operation. Using SQL data access interfaces technical, support data batch processing and run speed stable. system management, upgrade and maintenance. At the same time, the system adopts the object-oriented programming technology, visualization technology and ASP technology, realizing the system of layered management. Layered technology ensures the experts in the field, knowledge engineers and computer experts, network administrators and general users are in operation on different levels, ensure the safety and convenience of system maintenance, visualization technology enables the system to a more harmonious human-machine interaction.

Design based on WEB for application system development and operation provides a cross-platform, simple consistent environment. Use a single client browser can realize the database access, and greatly reduced the costs of users' training, management and maintenance, System development focus has transferred WEB background server application development, realizes the separate of development environment and application environment. B/S structure greatly simplifies the client's work, client only install, configure a client software, the server will burden of work, more database access and applications in the execution will finish on the server.

3. The Overall Design of the System Software

This expert system based on the Web adopts B/S structure, "the interface layer/logical layer/data layer" three layer network structure model. Ordinary users can access through the internet browser, to queries, diagnosis, learning, complete greenhouse tomato pests forecasting, diagnosis, and management model of optimization and meteorological disaster etc. This design of system use Microsoft Visual studio.net platform, c #, asp.net technology and SQL Server 2000 database to develop network. System of three layers of B/S structure is shown in figure 2.

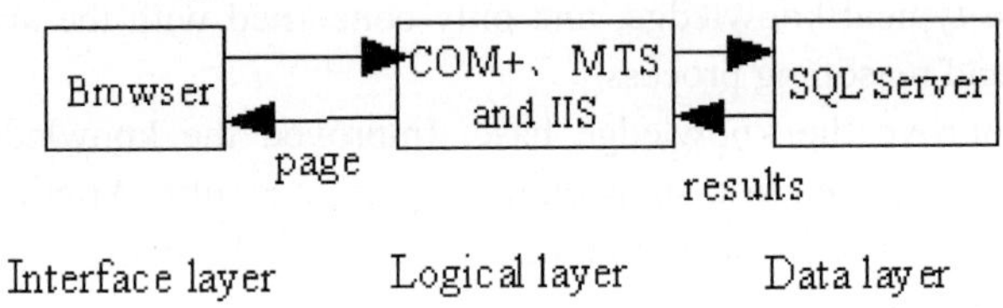

Fig. 2. Three layers of B/S structure.

3.1. *Interface layer*

The interface layer is show the users' interface, use c # and ASP.NET page design, including the greenhouse tomato growth forecast, diagnosis and management of relevant state input and output, etc. System interface is easy to operate, user-friendly, its configuration is the embodiment of the various functions of the system.

(1) the user can input the tomato growth environmental factors by interface, such as temperature, light intensity, soil humidity and comprehensive consideration of the environmental factor influence of tomato growth and judgment of the input for tomato growth environment, the tomato is given by experts and relevant experts reasoning mechanism diagnosis.

(2) the interface layer calls knowledge or database by logical layer, users can call knowledge or database about tomatoes growth by interface, find the

tomatoes growing conditions, such as pest and disease symptoms, and health or not through expert reasoning mechanism gives corresponding expert diagnosis.

(3) by the user interface,user can study tomatoes' management knowledge of multimedia information (text, pictures, videos, etc). Contents include variety selection, seedling technique, technology, field management, common pest control, common data query etc.

3.2. *Logical layer*

Logical layer can also called logic network, including Web server and application server, expert system and explain reasoning function in logic network realization. Figure 2 shows the COM + MTS is component management application server software, IIS (Internet websites server 5.0) is Web server software. This system requires multiple databases transaction between the object and the support, so it need COM + and MTS components. IIS makes asp.net page can distribute, and have some expansion function, IIS can editor environmental interface, fulltext retrieval, multimedia and have other functions. This layer accept user input from the presentation layer interface layer, grab and process database data from the data layer, then return to the user interface .

3.3. *Data layer*

Data layer is mainly the database server, the kownledge of expert system stored in database. in the data layer,can finish the integrity of the data and safety inspection. In this layer mainly including growth management knowledge repository, insect pests, text knowledge base, dynamic database and self-learning module.

4. Knowledge Repository and Reasoning Mechanism

4.1. *Knowledge repository constructing*

Knowledge representation use the object-oriented "rules frame+ rules system" model. The general form is: if Ei (I) then A. in this, Ei (1,2,...n) is "and" relation, in "rules frame + rules system", Ei and A are both called "truth", "concept" or "factor". Rules frame describe tomatoes each phase plane layers of growth management process and the relationship between insect pests, Rules system describe the concrete method. For example, "IF tepal cannot fructifications cavities, small and fruit is decayed leaf-blight, full of insect dung THEN tomato bollworm."

System using SQL Server 2000 building database, using ASP.NET technology access to database, through the COM + and MTS. Component to achieve.

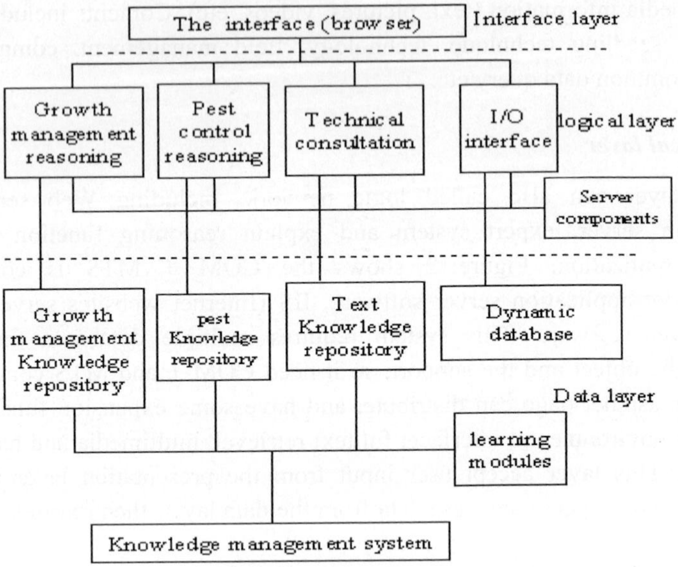

Fig. 3. Overall design structure.

4.2. *Knowledge management system*

(1) Design knowledge operation function: knowledge operation functions including knowledge of add, delete, modify, and statistics, etc.

(2) Design knowledge inspection function: knowledge check including the consistency, integrity, redundancy, etc.

r1: if P then Q;r2: if P then ~Q;

r1: if P then Q; r2: if Q then R ; r3: if R then P

(3) design knowledge repository operation: including operation, delete, knowledge decompose, merger, etc.

4.3. *Reasoning mechanism*

Tomatoes grown and management information has a lot of uncertainty and fuzziness, it is inevitable. Application of expert system in the tomato plant and management, is a collection of logic and instructions and knowledge reasoning for the integration of synthesis reasoning process, so use of reasoning strategy and fuzzy reasoning strategy combining reasoning mechanism can meet the needs of reasoning. In order to reflect the credibility of the conclusion, in fuzzy production rule reasoning add weighted value, through the known condition get the credibility of the right result.

5. Conclusion

The expert system is based on the Web for greenhouse production management guidance of tomatoes. Application three layers of B/S structure can satisfy many users mode access the same server in the same time.applications can centralize management in the server and significantly reduce the platform development costs and maintenance cost. Use Visual C# can easily write interface, more suitable for further expansion of the information age. The hierarchical structure of the system make the database and reasoning machine separation, independent with each other, facilitate system debugging and modification.

In recent years, with the develop of computer technology, network and communication technology, in the rapid development of modern greenhouse vegetable is more mature. Use these techniques to service agricultural research, can reasearch agricultural production mode of reduce input, increase output, add farmers' income, appreciation has great significance.

References

1. Chen Qingyun. Cucumber greenhouse cultivation and management of expert system [J]. Journal of Agricultural Engineering, 2001,17(6):142-146.
2. Zhao Tongmin, Gao Jun. Tomatoes and pest control technology [M]. Beijing: China Agricultural University Press, 2001.06.
3. Zheng Limin, Liu Zhong. The principle and application of agricultural information system [M]. Beijing: Chemical Industry Press, 2006.03.
4. Wang Zichong. Based on the Web of cucumber in greenhouse production management expert system research [J]. North Gardening, 2009, (7): 254-256.
5. Ren Zhenhui, Xie Jingxin. Greenhouse environment management system. Parameters [J]. The Computer Engineering, 2001, 27(8):136-137.
6. Zhu Weixing. Design and realization of tomato expert system based on COM technology [J]. Jiangsu University Journals, 2004, 25(2):93-96.
7. Zhang Baiyi, Cui Shangtai. The auto fault detection based on Web design of expert system [J]. Chang-an University Journals, 2006,26(2):99-102.

Research and application on power of things data DCS algorithm based on geographic relevance

T. Chen† and X. Wang

Xinjiang Electric Power Research Institute; Urumq; Xinjiang; 830011
†*Email: ceylic@foxmail.com*

Compressed sensing is a theory developed in recent years. Firstly, this paper briefly introduces the theory of compressed sensing, which focuses on distributed compressed sensing theory on WSN Based on geographic relevance. It is an effective way to reduce the amount of data transfer that by compressing datain the Power Information Collection System which has the large range of the sensor nodes. Secondly, we compressed the data of the network based on the distributed compressed sensing theory and geographic relevance,which has achieved the desired results.

Keywords: Things; DCS; Compress; Distributed

1. Introduction

Things technology is an important support for the Smart Grid. The depth of integration of smart grids and the Internet of Things will greatly promote the development of intelligent network industry [1-3]. At present, the State Grid Corporation has completed the transmission and distribution of condition monitoring systems, intelligent substation, distribution automation, unified video collection and information collection, etc. Data processing that Largely determine the performance of the network is a key issue to the Power of Things. In this context, redundant data is likely to cause network congestion, it is a problem to face that how to effectively deploy nodesand to ensure comprehensive information based on the actual terrain.

Compressed sensing theory first proposed by the Candes, RomBe who has constructed the theoretical framework of compressed sensing, and formally proposed compressed sensing (CS, compressed sensing) theory in 2006 [5]. Sparse signal, survey design matrix and signal reconstruction algorithm design are a very important part of the theory.

2. Algorithm Model

Standard CS theory use single node perceived relevance of data to reduce the dimension of the node measurement data is mainly based on a single node of the internal data structure design [4]. For the densely distributed nodes, there must be a lot of spatial correlation between the data of adjacent nodes. Using this feature, we combined measurements formed by collaboration of each node. And we make use of the correlation data between the nodes to reduce the overall dimension of the measured data, which would get more compression. Based on this, the Distributed Compressed Sensing (DCS) theory was produced. Theory of DCS suggests that each signal encoding side can use another group that is not associated with the thinning of the base (e.g., a random matrix) for observation and coding, if multiple relevant signals can be expressed on the same sparse group. In this way, encoding that is far less than the original length coding signals would be gotten.

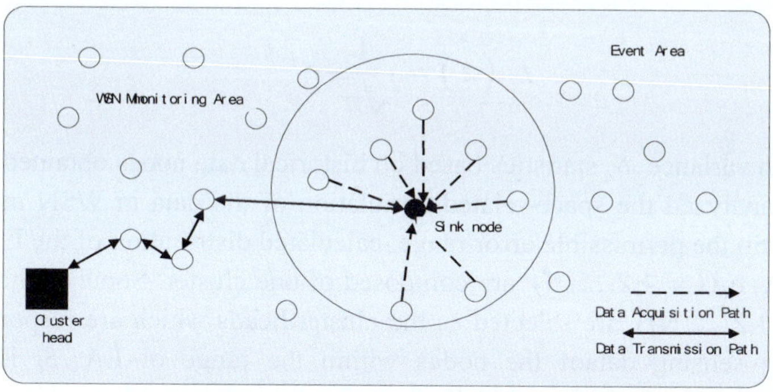

Fig. 1. Geographic correlation-based DCS schematic.

In the things, sensing data of the respective sensor node is gathered up by SINK node. It needs to estimate the event source V in the monitoring area to achieve the purpose that distortion of the estimated results can meet the requirements of power of Things applications. The geographic correlation-based DCS schematic has been given in the fig.1. There are different degrees of Spatially correlation between the nodes that are densely distributed in the things. Assuming the location of the event source S is (0,0), the coordinates of a nodes $n\ (x,y), n\ (x_r, y_r)$ that are distriuted in the event area EA is $(x,y), (x_r, y_r)$, its perception data is $N\ (x,y), N\ (x_r, y_r)$. Variogram is defined as:

$$\gamma(x) = \frac{\left[N\ (x, y) - N\ (x_r, y_r)\right]^2}{2} \tag{1}$$

Here: $(x - x_r)^2 + (y - y_r)^2 = r^2$.

According to the mathematical model proposed in the literature, sensing data of node $n(r,\theta)$ can be defined as $N(r,\theta)$ in polar coordinates. Within the range Event Area (EA), there are the following correlation between the sensing data of node n (0,0) and other nodes around.

$$N(0,0) = I_{(U=T)} + I_{(U=H)}{}^{*}$$

$$\iint_{(\theta,r)} \left(S_{(r,\theta)} + Z_\theta\right)\delta(R = r)\delta(\Theta = \theta \mid R = r) \quad (2)$$

Here: $U = T$ indicates perception data; $N(0,0)$ is obtained by the random variable Y; β represents the probability; Probability density function of random variables Y and Z are $f_Y(y)$ and $f_Z(z)$; Random variables Y, Z and perceptual data $N(r,\theta)$are independent of each other. In the case of spatial correlation exists, random variable Y represents perceived differences between data WSN neighbors. Here variable Z in line with $Z\sim(0,\delta_z^2)$, namely:

$$f_z(z) = \frac{1}{\sqrt{\pi}} e^{\frac{-z^2}{2\delta_z^2}} \quad (3)$$

Mean variance δ_z statistics based on historical data nodes obtained.

We analyzed the space-related correlation of the data in WSN monitoring area, within the permissible error range, calculated distribution of the EA. Nodes in the EA $n_i(i = 1,2,\ldots,\mathrm{N})$ are composed of one cluster. Some of those nodes $n_h(\mathrm{h} \in \{1,2,\ldots,\mathrm{N}\})$ are selected as the cluster heads which are responsible for collecting sensing dataof the nodes within the range of EA. S_i is sensing data n_i.N_i is Observation noise. X_i is defined as sensing data of each sensor node. At the same time. The vector X_N $(X_N = (X_1, X_2, \ldots, X_N)^T \in R^N)$ is defined as sensing data of N nodes. Due to the spatial correlation of WSN, X_N shows K sparsely in wavelets Ψ. Therefore sensing data X_N, wavelets Ψ and transform coefficient vector θ conform to the formula.

$$X_N = \psi\theta \quad (4)$$

Here: $X_i = S_i + N_i, (i = 1,2,\ldots,\mathrm{N})$

$$\theta = \left(\lambda_0, \hat{r}_0 \cdots \hat{r}_{J-1}\right)^T, \|\lambda_0\|_0 + \sum_{i=0}^{J-1}\left\|\hat{r}_i\right\| = k$$

After the cluster head has obtained sensing data vector X_N, it runs DCS encoding algorithm what projected data vector X_N whose sparsity is k onto observation matrix Φ randomly. Thereby we obtain $M(M = ck < N, 2 \le \mathrm{c} \le 4)$observational data vector $Y_M = (Y_1, Y_2, \ldots Y_M)^T \in R^M$, which can draw the following formula:

$$Y_M = \Phi X_N = \Phi\Psi\theta = \Xi\theta \quad (5)$$

Here: $\Phi = (\varphi_1, \varphi_2, \ldots \varphi_N)$, $\varphi_N \in R^M$. And measurement matrix Φ and wavelet moment Ψ irrelevant. Matrix $\Xi = \Phi\Psi$ considered to be the measurement matrix of DCS. After data vector Y_M has been observed by the cluster head, it will be transferred to the sink node in a unlimited multi-hop manner. At this point, Sink node can run DCS decoding algorithm:

$$\min||\theta||_1 subjectto Y_M = \Phi\Psi\theta \tag{6}$$

By finding the sparse optimal solution θ^* of the above formula, sensing data of the each node can be reconstructed.

$$\hat{X}_N = \left(\hat{x}_1, \hat{x}_2, \ldots \hat{x}_n\right)^T = \Phi\,\theta^* \tag{7}$$

Distributed compressed sensing model based on geographic relevance on power of things is shown in Fig. 2.

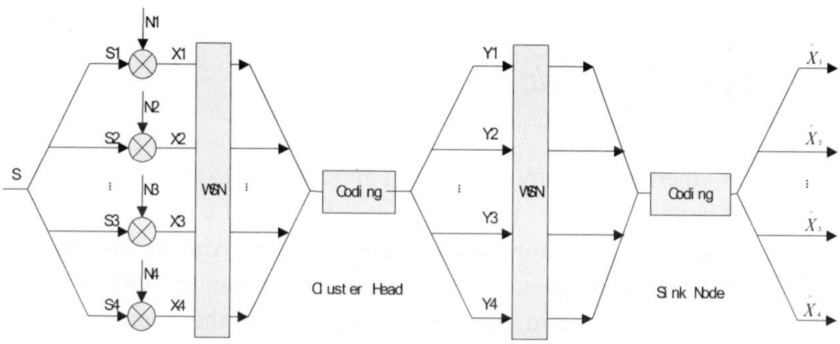

Fig. 2. Distributed compressed sensing model based on geographic relevance.

In practical applications, the cluster head node and sink node generally spaced relatively far, so that after compression algorithm, saving a lot of energy for the entire WSN transmission network.

3. Experiment and Results

Firstly, it is needed to get the distribution range of the event area. Assuming that there is a virtual node $n(0,0)$ in location of Event Source S. Information data of the virtual node is $S(0,0)$. Information data of the boundary node $n(r,\theta)$in the event area is $S(r,\theta)$. They conform to the formula: $|S(r,\theta) - S(0,0) \leq \mu|$, μ is the error threshold. It represents the difference of the nodes sensing information in the different positions with the event source S. R represents the radius of the event area.

According to equation (1) the following formula can be obtained.

$$\gamma(r) = \frac{1}{2} E\left[(S(r,\theta) - S(0,0))^2\right]$$

$$= \frac{1}{2} E\left[Z^2\right] = \frac{1}{2}\int_{-\mu}^{\mu} Z^2 \frac{1}{\sqrt{2\pi}\sigma_z} e^{-\frac{z^2}{2\sigma_z^2}} dz$$

$$= \frac{1}{2}\sigma_z^2 erf\left(\frac{\mu}{\sqrt{2}\sigma_z}\right) - \frac{1}{\sqrt{2\pi}\sigma_z} e^{-\frac{z^2}{2\sigma_z^2}}$$

$$= \psi(\sigma_2, \mu)$$

So: $\gamma(r) = \psi(\sigma_2, \mu)$ (8)

Here $erf(x) = \frac{2}{\sqrt{\pi}} e^{-t^2} dt$

Typically, there are many estimation model about $\gamma(r)$. This estimated model is used in this paper: $\gamma(x) = c(1 - e^{-\lambda t^2})$. Parameter c is an indicator of the degree of strength about the correlation among the data. And parameter λ is an indicator of the degree of speed that data correlation changes with the distance of nodes changing. The two parameters depend on the spatial correlation properties of data field in the monitoring area.

After putting the formula $\gamma(x) = c(1 - e^{-\lambda t^2})$ into formula (1-7), the following can be obtained:

$$c\left(1 - e^{-\lambda t^2}\right) = \psi(\sigma_2, \mu)$$

$$\Rightarrow r = \left[\frac{1}{\lambda}\left(\frac{c}{c - \psi(\sigma_2, \mu)}\right)\right]^{\frac{1}{2}} \tag{9}$$

Secondly, we need to get the wavelet matrix Ψ that based on Lifting wavelet transform. Assuming that perception data vector is X_N. The set of nodes in the event area is $E_J : \{n_1, n_2 \ldots, n_N\}$

Let $\lambda_J = X_N (\lambda_J \in R^N)$

Since the location of each node is different, we split the set of nodes into two sets: Even-collection E_{J-1} and odd collection O_{J-1}. They correspond to the perceptive informatics data set of C and D, respectively.

$$\begin{cases} T(\lambda_J) = (\lambda_{J-1}, \gamma_{J-1}) \\ \gamma_{J-1} = \lambda_{J,0} - P(\lambda_{J,E}) \\ \lambda_{J-1} = \lambda_{J,E} + U(\gamma_{J-1}) \end{cases} \tag{10}$$

Here: γ_{J-1} is wavelet coefficient set; λ_{J-1} is scaling coefficient set; P linear prediction operator. U is linear update operator. We put $\lambda_i(i \in \{1,2,...,J\})$ recursive lifting wavelet transform J times. The following can be obtained:

$$T^J(\lambda_J) = \{\lambda_0, \gamma_0, \gamma_1, ...\gamma_{J-1}\} \tag{11}$$

There are many very small elements in wavelet-coefficient vector set $\gamma_i(i \in \{1,2,...,J-1\})$. After clearing the wavelet technology less than the limit, we can get the set with sparse structure $\hat{\gamma}_i$. We accurately reconstruct the original signal λ_J by the set γ_i. Let $T^{-J}(\bullet)$ to be lifting wavelet inverse transformation. Because the prediction operator and update operator are all the linear operation, $T^J(\bullet)$ and $T^{-J}(\bullet)$ are linear operation. Therefore:

$$T^{-J} = (\lambda_0, \hat{\gamma}_0, \hat{\gamma}_1, ...\hat{\gamma}_{J-1}) = \hat{\gamma} \approx \lambda_J$$

$$\Leftrightarrow \Psi\theta \approx \lambda_J = X_N \tag{12}$$

Here $\theta = (\lambda_0, \hat{\gamma}_0, \hat{\gamma}_1, ...\hat{\gamma}_{J-1})$

$$\|\lambda_0\|_0 + \sum_{i=0}^{J-1} \left\|\hat{\gamma}\right\| = k < N$$

So long as we get the topological structure of each node event in the area, can we calculate the wavelet transform matrix Ψ about data vector perception X_N. And X_N on Ψ is k sparse.

In summary, distributed geographically relevant perceptual compression encoding and decoding algorithm based on WSN can be expressed as follows:

Step 1: According to $T^J(\lambda_J) = \{\lambda_0, \gamma_0, \gamma_1, ...\gamma_{J-1}\}$, sink node calculates the distribution range of event area. Then it activates the nodes $n_i(i = 1,2,...,N)$ in event area by multicast routing mode. After that it selects nodes $n_h(h \in \{1,2,\cdots,N\})$ as the cluster head.

Step 2: The cluster head n_h generates observation matrix $\Phi = R(s_M, \tau_N)$. Among them, $R(\bullet)$ is a pseudo random number generating function. $\tau_N : \{\tau_1, \tau_2, ..., \tau_N\}$, τ_i is number of n_i

Step3: The cluster head communicates with each node in the cluster to obtain the perception data vector the perception data vector $X_N = (X_1, X_i, ..., X_N)^T$ of all nodes in the range of event area in WSN at the same time. Then it get the measure value vector $Y_M = \Phi X_N$ through distributed compressive sensing encoding algorithm that based on the wavelet transform. Turn N dimensional perception data vector $X_N = (X_1, X_i, ..., X_N)^T$ into M dimensional data vector $Y_M = (Y_1, Y_2, ..., Y_M)^T$ by the compression coding. Among them $M = ck$, $2 \le c \le 4$, and $M < N$.

Step 4: The first cluster the coded data Y_M to the sink node by way of wireless multi hop manner

Step 5: Sink node generates the observation matrix $\Phi = R(s_M, \tau_N)$ as same as the cluster head node. And then it get the transformation matrix of wavelet Ψ base on the basis of the network topology structure. After that, it would run the distributed compressed sensing decoding algorithm. According to the formula $Y_M = \Phi\Psi\theta$. We use the sparse wavelet of the the original node sensing data to represent B. The original data is gotten through the OMP reconstruction algorithm. Then according to formula $\Psi\theta^* = X_N$, the sensing data set can be reconstructed accurately.

4. Experiment and Results

4.1. *Example of data fusion*

The preamble introduces clustering limited ad hoc way. That is, all nodes are divided into region-based cluster. The ad hoc network structure is used among the clusters and star, tree or chain structure within them. Assume that we query the meter reading data of the 4 areas. Firstly, we sends a query request. Tree-routing approach would be used and the position of the sensor and the communication path as shown in figure 3.

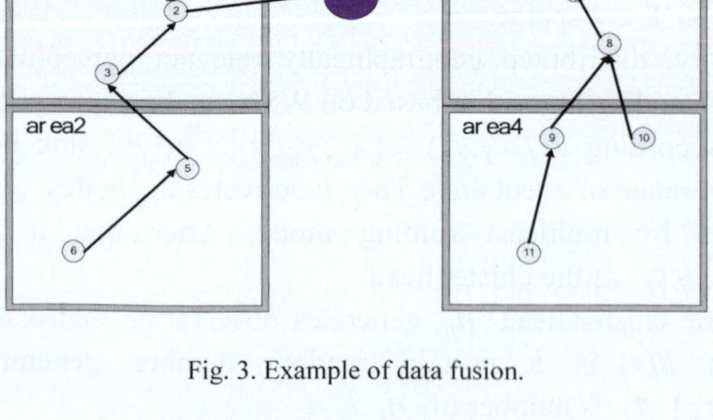

Fig. 3. Example of data fusion.

In Figure 3, each sensor node has prepared a data (Filed, Soil, and moisture) are expressed in the form of. Do the following:

1) Check whether it meets the requirements of each sensor node upload, to decide whether to participate in the transmission;

2) Each node receives other nodes to send data, for local operations, results of operations continue to submit to the upstream node;

3) If intermediate nodes has not received the data that is sent form the neighbor node within a certain period of time, we argue that there is no node below the neighbor request to submit data

According to the above operations, the total net transfer process in 5 copies of data. As shown in Figure 3, the Node 7 send 2 copies of data, Node 8 send 2 copies of data, Node 9 send 1 copies of data. If you do not have any data fusion method, each node independently transmit the data to sink node, and the sink node centralized computing results, the network data transmission quantity is 25, far greater than 5, the corresponding node energy consumption will significantly increase.

4.2. *Experimental and simulation results*

It is important to examine and analyze the relationship between the reconstruction error about which is time-based correlation DCS encoding and decoding in WSN and number of observations. Matlab is used as the policy tool. Two-dimensional Gaussian distribution data were used to simulate the analog spatial correlation data source in WSN. The nodes are distributed in the event area (60m X 60m) randomly.

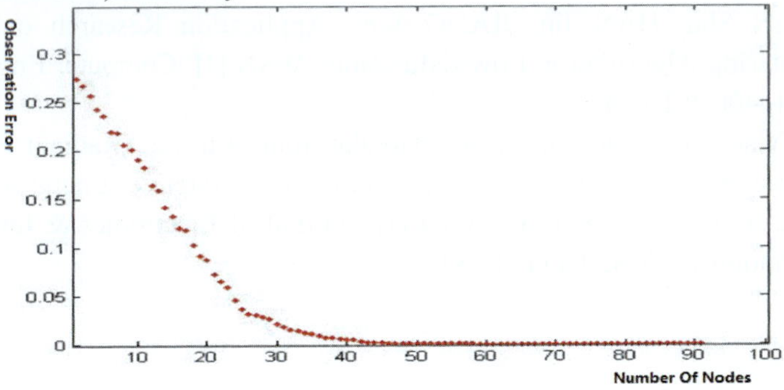

Fig. 4. The relationship between observation error and number of nodes.

In the system, assuming that all sensor nodes generate sensing data at the same time and sends the data to the cluster head. Then Cluster head compresses and coding for the sensing data has been received by DCS coding algorithm. Next, the Cluster head would transmit observation data to the decoder by way of a wireless multi-hop. The sink node reconstruct perceptual data of the nodes in WSN event area accurately also by DCS coding algorithm.

As shown in Figure 4, it is the simulation diagram about the relationship between the reconstruction error and the number of meter data. We can see from the chart when the observed data value M increases,the reconstruction error becomes small.

5. Conclusion

Through the compression algorithm we can reduce the amount of data transmitted to a great extent, which has cleared the obstacle of network congestion and savedthe network energy consumption. In the things, there is a strong spatial correlation in the data of each sensor node. In this paper we consider this factor, do a simulation experiment successfully using the theory of perception of spatial distribution of correlation based compression. The results showed that data algorithm can satisfy the application demand of large power WSN.

References

1. D. Baron, M.B. Wakin, Duarte et al. Distributed compressed sensing [OL]. http://www.dsp.rice.edu/drorb/pdf/DCSl12005/pdf.
2. Z. Sun, C. Tian, J. Chen, et a1. LDPC code design for asynchronous slepian-wolf coding [J]. IEEE Transaction on Communications, 2010,(2), 511-520.
3. GJ Pottie, WJ Kaiser. Wireless integrated network sensors [C]. In: communications of the ACM, 2000, 43(5): 51-58.
4. CHEN Shu, HAN Jin, JIANG Wei. Application Research on Uneven Clustering Algorithm in Low-redundancy WSN [J], Computer Engineering, 2014, 40(8): 10-14.
5. Xu Xiao-rong, Yao Ying-biao, Bao Jian-rong, Lu Yu. Gradient Projection Sparse Reconstruction Approach Based on Adaptive Energy-efficiency Measurement in Cognitive WSN [J]. Journal of Electronics & Information Technology, 2014, 36(1): 27-33.

Research and application on software engineering method based on cloud computing

C.L. Xu[a]
Information Engineering School,
Guangzhou Vocational College of Technology and Business,
Guangzhou 510640, China

W.Y. Zeng[1b], A.M. Yu[c] and Y.X. Zhu[d]
Computer Engineering and Technical College,
Guangdong Institute of Science and Technology, Zhuhai 519090, China
Emails: [a]perfect_xu@qq.com, [b]wyzeng@126.com,
[c]yuaiming@126.com, [d]grasser@126.com

By virtualization and service mode on demand of cloud computing, software engineering based on cloud computing should adapt new mode and method. How to combine traditional software development and deployment with cloud computing platform is still an indefinite problem. To resolve the problem in software development based on cloud computing, software engineering concept and properties are illustrated, and the software engineering methods and strategies are discussed and analyzed, and the implement application development and testing are performed. Through the software engineering based on cloud computing research and practice, some valuable strategies and methods are proposed, and the result shows that software engineering based on cloud computing has much benefits in software development flow and methods.

Keywords: Cloud computing; Software engineering; Method; Strategy; Application; Practice; Intelligent; Cooperative.

1. Introduction

Software engineering is the subject of advice and control for software development procedure. Generally, software development whole thing can be ascribed as 6W1H principal, which means who are the development team, to develop for whom, to develop in what platform, to run at where and when, to get which software product. In the cloud computing age, there exists several software development needs, such as enterprise legal system to be changed to meet with new system requirements, or to be expanded based on the current, or to develop a new service system. There also exist data migration and business continuous assurance measurements. All the above problem need to take advantage of the software engineering based on cloud computing.

---

[1] Wen-Ying Zeng

The rest of the paper is arranged as the follows. In Section 2, related work is illustrated; in the Section 3, the software engineering methods are proposed; in Section 4, the software engineering strategy is discussed; in Section 5, the software development and deployment practice and application based on cloud computing are performed; in Section 6, the evaluation method is discussed. Finally, it is concluded.

2. Related Work

There are some related research on software engineering based on cloud computing.

Zhou Peng etc. in [1] illustrated GIS software architecture, development organization and deployment management to discuss GIS software engineering method based on cloud computing. Software development is becoming more abstract and open to network with the help of network components [2]. Therefore, software engineering mode based on cloud computing includes software architecture, development procedure and deployment management. The paper [3] proposed one kind of application development flow and methods, which are composed of seven steps, i.e. requirements acquired requirement analysis, design, programming, testing, and deployment and maintain. There was much software development or engineering research [4-7] based on cloud computing.

Based on cloud computing software defined network can be constructed [8]. Internet-of-Things/Cyber-Physical Systems, Network Function Virtualization, virtual worlds/multiplayer games, mobile applications, can be constructed by cloud computing platform for deploying and distribution dynamically.

Cloud computing make it possible that the computing and storage resources is obtainable on demand [9]. Cloud computing services include mainly three kinds of types, i.e. IaaS, PaaS and SaaS, which can almost be approachable anytime and anywhere. IaaS examples are Eucalyptus, OpenStack, CloudStack, and OpenNebula. PaaS architectures can alleviate the burdens of resource management for developers who may focus strictly on application development, which examples are Microsoft's Azure, Google App Engine, Heroku, Salesforce, and Engine Yard. Providers of cloud services make their resources available to the general public with certain characteristics which are indicative of cloud computing: scalability, elasticity, and fault tolerance. The amounts of resources which can be attained on-demand are virtually unlimited. There are many private cloud implementations of AWS's IaaS Elastic Compute Cloud (EC2) API including Eucalyptus, OpenStack, and CloudStack. Public offerings includes Heroku, Microsoft Azure, Google App Engine (GAE), Amazon Elastic Beanstalk, and many others, while private cloud PaaS offerings include

AppScale, CloudFoundry, and OpenShift. SaaS shifts the application hosting from the client to remote service, whose physical location may be unknown but addressable through network domain name or IP. Multiple languages such as Ruby-on-Rails (RoR), Java, Python, Go, Node.js, PHP, .NET (Visual Basic and C#). The paper [9] developed a PaaS solution AppScale with web framework, which is an open cloud platform with API-compatible with GAE, and focus on the data management layer for scalable storage, transaction semantics, scalable queries, analysis of Bit Data, and live migration support.

The paper [10] researched on the architecture of participation for cloud computing social biology. Cloud computing can provide interactive social communication and generate massive data from users.

Software architecture deployed in cloud computing is service-oriented and pervasive accessed [11].

The requirements engineering and testing of cloud-based applications and cloud-aided software design is usually used. Software engineering based on cloud computing focus mainly on Platform as a Service and cloud software engineering covering map-reduce, schema-free databases, service-oriented computing, security and compliance and design for resilience [12].

Many of the problems posed by the migration of computation to cloud platforms can be formulated and solved using techniques associated with Search Based Software Engineering (SBSE) [13]. Much of cloud software engineering involves problems of optimization: performance, allocation, assignment and the dynamic balancing of resources.

Much of the innovation in Software Engineering is due to the adoption of cloud computing [14]. Improvement of cloud's lowest layer (infrastructure as a service) is a key process, and a service-oriented architecture is a solution.

This paper [15] presented a new approach on Cloud Computing for Scientific Data Storage and Processing using a hybrid system with both on-site and on the cloud systems.

Related work shows that software development and deployment based on cloud computing have many characteristics, therefore, software engineering based on cloud computing should be discussed and researched deeply. Our creative points lie mainly in proposing some methods and strategies about this area, and combining research with practices.

3. Software Engineering Methods Based on Cloud Computing

In fact, the software development involves the optimization of the resource allocating and scheduling, which should be computing, data, users' experience centralized, or resource restricted at first.

Software service mode has been changed from traditional procedure oriented mode, object oriented mode, component oriented, to service oriented, using on demand. Software development procedure based on cloud computing should be reformed. The reformation is to refer the traditional software engineering, and to be creative in software development ideas, method, flow and management.

Software engineering can be classified as the follows based on the development procedure, which are linear, iteration, evolution, coil, circulation. The software engineering based on cloud computing can also refer this, but may be more open and rapid because cooperation by crowd source or massive intelligence.

The software engineering comparison between traditional and cloud computing is listed in Table 1.

Table 1. The software engineering comparison between traditional and cloud computing

Type	Traditional software engineering	Software engineering based on cloud computing
Methods	Procedure Oriented, Object Oriented, Service Oriented	Object Oriented, Service Oriented
Tools	UML, ER, etc.	UML, ER, etc.
Procedure Model	linear, iteration, evolution, coil, circulation; prototype, waterfall	linear, iteration, evolution, coil, circulation; prototype, cooperation
Testing Model	Unit testing, integration testing	Unit testing, integration testing, cooperative testing, parallel testing

The software engineering mode based on cloud computing may be classified into absolute software engineering, massive software engineering. The software development methods include independent software development, massive cooperative development. The development flows include top to down, and bottom up.

According to the 6W1H illustration, the development team can be composed by the project requirements dynamically and competitively through the network (who), according to the users' characterized needs for massive customers can be customized by self-generated method (to whom), the software can be developed both in local machine and cloud platform (where), the development mode may be massive intelligent and cooperative implemented (how), and at last the software is completed with network complicity system (which).

The software engineering based on cloud computing has reformed in software development team, platform, programming, deployment, maintenance mode, which will be more open, distributed, crowd and massive cooperative, and can be composed by resource virtualization and network components rapidly. The mode also needs integrated management and optimization to reduce cost and promote efficiency.

4. Software Engineering Strategies Based on Cloud Computing

The strategies of software engineering based on cloud computing are related with the opening and scalability of cloud computing technology and environment. Team cooperation and dynamic construction virtual platform may be routings. In cloud computing age, characterization and public conformity will coexists, traditional method and creative application will be combined, local development and network development will be synchronized, parallel computing, data intensive or computing intensive will be usually meet together.

To resolve the complicated requirements above, related characteristics and strategies of software engineering based on cloud computing is discussed and analyzed.

Firstly, software engineering pay attention to the development performed in local machine and debug and deploy in cloud platform.

Secondly, cloud computing platform is to be constructed by virtual machine dynamically based on the lifecycle and mode of application.

Thirdly, the application should meet with multiple and heterogeneous terminal, and compatibility and generality need to be considered.

Fourthly, the integration of the application based on cloud platform is necessary, such as combining with the internet of things and mobile communication devices, so the data stream stored and distributed is very important. The data granularity extraction should be flexible and easy to be managed and utilized. In data defining, the data structure definition should be considered along with the availability.

Fifthly, there are multiple various scales of development by cloud computing platform, and different scales may have related deployment modes. For example, small application for several users can be deployed in one virtual machine, medium size or larger application can be smoothly expanded by dynamic constructed virtual machines and implement to process at parallel and load balance.

Sixthly, collaborative development with huge and complicated application can be easily communicated and shared by cloud computing platform. Cloud computing can provide integrated massive data storage management and insure data consistent.

Furthermore, big data analysis and mining should take use of the unlimited computing and storage resources based on cloud computing. Intelligent push and pop services can be configured automatically in cloud computing according to the customers specified needs.

There are possibly more complicated and extensive team constitution based on cloud computing, and in development methods, the conventional structured, object oriented, former method will be changed to service combination and reconstruction, service distribution will be dynamically virtualized and multiple renters.

The software engineering involves that the project requirement source and analysis, the organization of development team, the selection of development tools, product runtime environment configuration and rent, reform and simplification of maintenance and update mode, and other relevant innovation along with the cloud computing development.

Cloud computing brings the most important optimization to software engineering in the following aspects, i.e., development procedure control and collaboration, flexible outsource, component based construction, service oriented deployment and distribution. Therefore, during the development, there will be meeting with multiple development coordination, multiple users' separation, service vender's migration, various application mode configurations, and so on.

The available resolutions are synchronous updating, virtualization units, standard API, parameters setup, context aware intelligent configuration.

Software engineering based on cloud computing adapted to various scale sized flexibly. To fulfill characteristic configuration, group software engineering and outsource development are usually adapted especially through the Internet. Hierarchical and module software development is often used too.

5. Practice and Application

There are many application types of cloud computing, such as education, finance, insurance, traffic, communication, business, game, and etc.

We take a practice and application of software engineering on the development of education resource share system based on cloud computing. The development procedure is as follows: requirements analysis, scheme design, platform construction, code development, test, deployment and release, maintenance and update.

5.1. *Requirement analysis*

The requirement analysis of education resources share system is performed by considering storage volume capacity, responding performance at parallel, resource management and rights allocation, users' management by register and login. Development and deployment platform selection, access mode is also considered.

The system needs to provide 1000 above customers online to access, limited to 1 milieu second responding delay. The data is stored in database and documents files. The prototype is illustrated with prototypes designing tools such as Axure, Visio, and etc. Then the prototype is shared to the development team and makes everyone comprehend the basic requirement and business processing flow.

5.2. *Architecture design*

The system architecture is composed of three layers, which are data layer, management layer, and application layer. The data layer is setup in cloud storage server, the management layer mainly contains resource, users, rights and security management, the application layer includes user access interface, resource push or pop, etc.

There are three available schemes as the follows.

The first is resource share scheme based on private cloud. The private cloud is built in the enterprise local server clusters, and there is full control by the company administrators. For the limitation of resource in computing and storage resource, the private cloud is often used to developing and testing locally.

The second is resource share scheme based on public cloud. When the software is developed completely, it is necessary to be deployed in the public cloud computing. Through applying for the virtualization resource from the cloud computing platform, the application is easily to be completed by relevant tutor provided by the service platform.

The third is resource share scheme base on hybrid cloud. In fact, the private cloud platform is combined with the public cloud platform is usually to be applied. The hybrid cloud scheme is run practically. Generally, the private cloud is applied in developed and maintained, and the public cloud is used to deployed and service distribution.

5.3. *Programming design*

The software system is developed with Java language and MySQL database to implement the follow functions. The system support users' management and resource management. The user management includes user register, login, modification, system setup and etc. The resource management includes unified access to distributed storage resources, files uploading, files listing, files downloading, files upgrading, and etc.

During the software development procedure, we construct the SVN server based on cloud computing technology to share and collaborate and synchronize the programs modules and data. In fact, there are some open source platform to support the simulate function.

5.4. *Test and deployment*

The system test is performed based on local computer, then in private cloud, and then in public cloud, and hybrid cloud at last.

After being tested successfully, the system is deployed in cloud platform, and the system performance testing is performed by concurrent access from multiple users.

5.5. *Software development and deployment practices based on cloud computing*

We have performed some development practices based on cloud computing. Three times development practices based on cloud computing are listed as the follows for example.

The first practice is performed based on private cloud. The private cloud server is constructed from several servers, which is managed by the middle ware module to load balance. Some video cameras are connected to the software system, and through the software modules we can control to turn on or turn off the cameras. Meanwhile, the data collected from the cameras can be stored, analyzed, processed, and extracted information and knowledge.

The second practice is to development a field communication system based the private cloud and public cloud. At first, we developed the filed communication system in our private cloud servers, and then deployed and tested the software. After testing locally, we deployed the software system to the public cloud such as Ali-cloud and distributed to the customers.

The third practice is to develop a mobile application based on cloud service. The first step is to construct the service side program and database tables, and the second step is to develop the mobile side program which is to communicate with the cloud service side. The mobile application based on the cloud platform has be implemented and the client system has be installed on the mobile phone with android system, and the database and server side is located in the Bmob cloud service.

Through the above practices based on cloud computing, including private cloud, public cloud, or hybrid cloud, we have gained some experiences about software engineering based on cloud computing. Although the development methods and tools are related similarly to the classical or traditional development languages or methods, there are more available, flexible, and scalable benefits than the conventional software development and deployment.

In spite of the different cloud types, from the perspective view of the users, they may be summarized as clients and servers, which take the users' side as clients, the cloud service as the servers. Accordingly, the whole architecture of the software engineering can be described as a hierarchical structure like Figure 1, which means that cloud computing platform can be private, public or hybrid, the basic process of software development and deployment may be transparent to users.

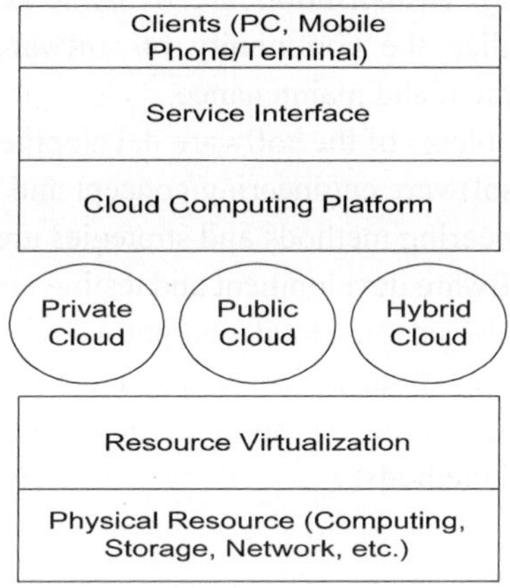

Fig. 1. The software system structure based on cloud computing.

6. Evaluation

Through the software development practice based on cloud computing, the evaluation should consider the follows.

Development cost includes design and development cost, renting fee of platform, communication, deployment and debug cost.

Software testing cost includes the humanity and time of local test, network test, trial test, and etc.

Software life cycle management involves analysis, design, development, operation, upgrading, maintenance and abundance.

Software development and testing evaluation flow is to develop, test, deploy and maintain from local, to private cloud, to public cloud, or to hybrid cloud.

Software development flow evaluation includes predict, development evaluate, testing and deployment evaluate.

Software testing evaluation includes multiple phases testing from local, network, to distributed, self-adaptive environment.

7. Summary

The development of software system is changed with software requirements, goals and deployment mode. Software engineering concepts and methods based on cloud computing have changed from traditional singular layer, centralized, elite, closed mode, to multi-layer, distributed, mass type, collaborative and open. With the help of cloud computing, offline upgrade and maintenance has been converted to online upgrade and maintenance. By the characteristics of collaborative intelligence fully, virtualization and dynamic extensibility, the software development will realize the community of software engineering, collaborative software development and maintenance.

To resolve the problems of the software development and deployment based on cloud computing, software engineering concept and properties are illustrated, and the software engineering methods and strategies are discussed and analyzed, and the application software development and testing are performed. Through the software engineering based on cloud computing research and practice, some valuable strategies and methods are proposed, and the result shows that software engineering based on cloud computing has much benefits in software development flow and methods.

Acknowledgment

This work was supported by the following projects: the Guangdong Province of Integration of Production and Research Project, Project No. 2012B091100499; the Technology Innovation Project of Guangdong Province Science and Technology Small and Medium-sized Enterprise, Project No. 2012CY037; the Higher Vocation Education Teaching Reform Project of Guangdong Province, Project No. 201401091; Guangdong Province Vocational Education Information Technology Project, Project No. XXJS-2013-1008; Guangdong Province Education Research "the Twelfth Five Year Plan" 2013 year research project, Project No. 2013JK256; the Research Institute of Higher Vocation Education of Guangdong Province project, Project No. GDGZ12Y146; the Guangdong Institute of Science and Technology Scientific Research Project, Project No. XJMS201202. The authors are grateful for the above supports; Guangzhou City Education Research “the Twelfth Five Year Plan” 2014 year research project, Project No. 1201420456.

References

1. Zhou Peng, Yin Fei. The cloud computing technology of GIS software engineering model based on [J]. Bulletin of Surveying and Mapping, 2010, (11): 22-24.
2. Shi Jie, Jie Jili & Shi Shaohua. Study on the influence of cloud computing on software engineering. Journal of Kunming University [J], 2011, (6): 67-68, 71.
3. Wu Chengkai. A study on software development and deployment method under a cloud. Huazhong Normal University, Computer Technology, 2013, Master's thesis.
4. Zhang Yong. A preliminary study on software engineering mode under the cloud computing. The third National Software Testing Conference and mobile computing, grid, intelligent Senior Forum papers, 2009-08-15:153-154, 160.
5. Wang Hongchun. The analysis of networked software multi-granularity dynamic characteristic. Wuhan University, Computer Software and Theory, 2010, Doctoral thesis.
6. Yang Yamin. Design and implementation of tool management system based on Web. Dalian University of Technology, Software Engineering, 2012, Master's thesis.
7. Liu Wei. The research of key technology of parallel test based on cloud computing software. Donghua University, Computer Application Technology, 2013, Master's thesis.
8. Subharthi Paul. Software Defined Application Delivery Networking. The Graduate School of Arts and Sciences of Washington University in St. Louis, August 2014.
9. Navraj Chohan. Data and Application Management in an Open Cloud Platform. Computer Engineering, University of California, Santa Barbara, December 2012.
10. Nicholas Williams Brown. An Architecture of Participation for Computational Social Biology. Computer Science, University of California, Los Angeles, 2013.
11. Lijun Mei, Zhenyu Zhang & W.K. Chan. Software Engineering Research Issues from the Cloud Application Perspective. http://wenku.it168.com/d_000062010.shtml.
12. Ian Sommerville. Teaching cloud computing: A software engineering perspective. Volume 86, Issue 9, September 2013, Pages 2330–2332.
13. Mark Harman, Kiran Lakhotia, Jeremy Singer, David R. White, Shin Yoo. Cloud Engineering is Search Based Software Engineering Too.

http://ac.els-cdn.com/S0164121212002853/1-s2.0-S0164121212002853-main.pdf, 2012.

14. J.L. Vazquez-Poletti, R. Moreno-Vozmediano, R.S. Montero, E. Huedo, I.M. Llorente. Solidifying the foundations of the cloud for the next generation Software Engineering. 2013 Journal of Systems and Software, 2013.
15. David Horat, Eduardo Quevedo, & Alexis Quesada-Arencibia. A Hybrid Cloud Computing Approach for Intelligent Processing and Storage of Scientific Data. EUROCAST 2013, Part I, LNCS 8111, pp. 182–188, 2013.

Full custom design of a 12port multi-bank high speed low power register file

Y.X. Jiang

School of Electronic Information and Electrical Engineering, Shanghai Jiao Tong University, Shanghai, China

L.H. Wang*, J. Li†, Z. Bi and P. Liu

Microelectronics R&D Center, Shanghai University, Shanghai, China
Key Laboratory of Advanced Display and System Application, Shanghai University, Shanghai, China
*Emails: *gggyvwxf@163.com, †lijiao@staff.shu.edu.cn*

In this paper, Designed and implemented a 12 port multi-banked register file in Nangate 45nm CMOS dual-well technology. The register file adopted improved memory cell, multi-banked structure and power-gating technology, it can complete 8 read operations and 4 write operations in one clock cycle. Post-layout simulation was completed under the condition of 1v power supply and 1GHz clock frequency, the worst delay is 324ps, total power consumption is 18.9mw. Compared with conventional design method, write 0 latency decreased 52.7%, total power decreased 68.1%, silicon area increased 16.3%.

Keywords: Register file; Multi-banked; Power-gating; Full custom.

1. Introduction

Register file is an significant part of microprocessor [1], it used to store instructions and data for microprocessor or data produced by ALU, register file's performance directly influence the performance of microprocessor. Modern microprocessor usually adopts superscalar structure to attain instruction level parallelism (ILP), thus enhance the processing ability of microprocessor. Superscalar microprocessor requires register file have multi-ports to complete multi read operations and multi write operations simultaneously, whereas it will increase power consumption and access latency. Usually from three aspects to optimize register file's performance: access speed, power consumption and silicon area. In terms of access speed, dynamic logic applied in register file can accelerate the access speed, especially in critical path [2], however the dynamic logic will increase power consumption and reduce robustness. Multi-banked structure can reduce the number of memory cell in bit-line, which can reduce bit-line load significantly, thus accelerate the access speed [3]. In terms of power consumption, multi-threshold transistor and body bias technology can balance power and speed, lower-threshold transistor applied in performance-critical path to accelerate the access speed, higher-threshold transistor applied in non-performance-critical path to reduce power consumption [4][5], but multi-threshold transistor and body bias technology need the support of special technology. In terms of silicon area, multi-banked structure applied to reduce the port number of each bank, which can reduce silicon area and power consumption obviously [3][6], but this structure need more complicated control logic to solve address conflict.

This paper designed a 64*32bit register file which contains 8 read ports and 4 write ports, it can complete 8 read operations and 4 write operations in one clock cycle simultaneously, we improved the conventional memory cell and purposed a new one. The register file was implemented in Nangate 45nm CMOS technology, post-simulation result shows the purposed memory cell can enhance register file's performance.

2. Register File Structure

Register file is mainly composed of memory array, address decoder, and read/write control logic. In nanometer technology, interconnect delay is proportional to the square of wire length, multi-banked structure can accelerate access speed through reduce bit-line load, but the cost of control logic and output logic will be very large with the increase of bank number. In order to make a balance between access speed and cost of control logic or output logic, we divided the memory array into 8 banks, each bank's volume is 8*32bit. Figure 1 shows the register file structure, each bank have their own address decoder and clock generation circuit, read operation and write operation is mutually exclusive in each banks, finally using multiplexers to read data out.

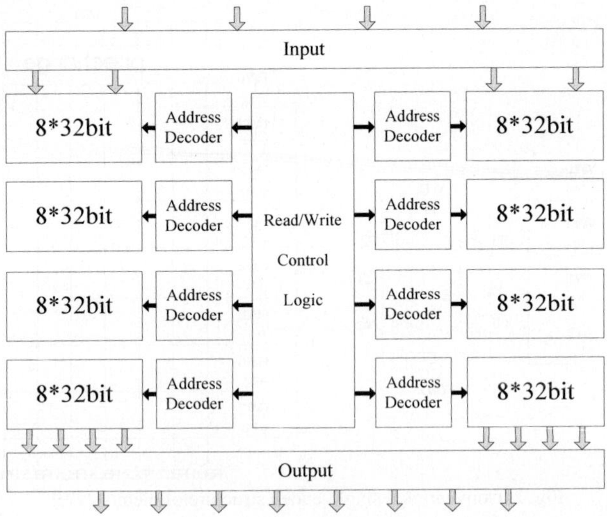

Fig. 1. Register file structure.

3. Memory Cell and Address Decoder

3.1. *Memory cell*

Multi-port register file usually adopts single-ended structure [7], Figure 2 shows the conventional single-ended structure. The core of memory cell is a pair of coupled inverters, when in write operating, 4 write bit-lines write data into memory cell through 4 pass transistors, from which port to write data into memory cell is controlled by 4 write word-lines concretely; When in read operating, data will be read out through 8 pass transistors from node QB, from which port to read data out is controlled by 8 read word-lines concretely. Before read operation, the precharge logic will charge the read bit-line to high level, when in reading logic 0, bit-line will discharge through pass transistor, the voltage of bit-line will decrease, sense amplifier will detect the change of bit-line and read data out; when in reading logic 1, bit-line is already in high level and read data out instantly.

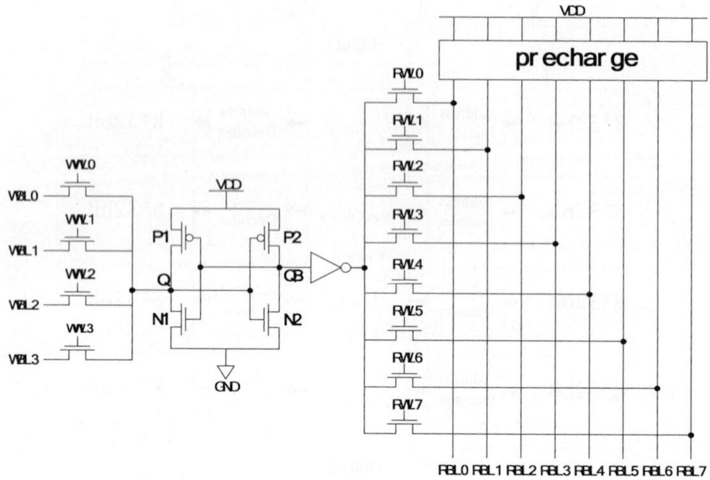

Fig. 2. Conventional single-ended structure of memory cell.

Compared with dual-ended structure, single-ended structure will halve the number of bit-lines, also silicon area, power consumption will be greatly reduced, this structure is suitable for multi-port register file design. However, the conventional single-ended structure also has several disadvantages: firstly, write port use NMOS transistor to write data into memory cell, NMOS transistor exist threshold drop when transmitting high level, the voltage of node Q depend on resistance ratio of N1 to pass transistor, write logic 1 will fail if the voltage of node Q value is less than the threshold voltage. It is often to reduce resistance by increasing the width of pass transistor, but this will increase power consumption and write word-line load; secondly, conventional single-ended structure is rigid to timing in read operation, to avoid process deviation and the effects of crosstalk on timing, it is need to set aside enough margin between different sequential signals, thus enlarge the delay of read operation. In view of the disadvantages of conventional single-ended structure, in this paper, we put forward a new memory cell structure by improving conventional single-ended structure, as is shown in Figure 3.

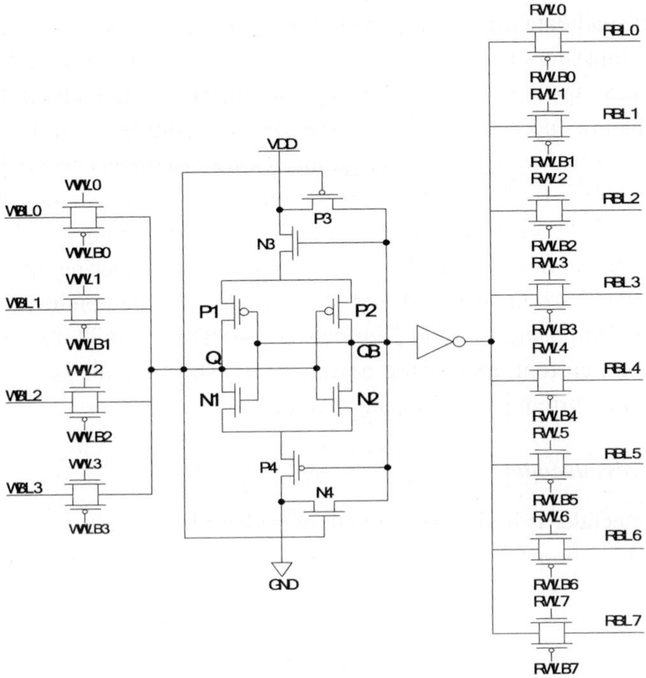

Fig. 3. The improved memory cell structure of register file.

New memory cell structure has made the following improvements:(i) Remove the precharge logic and sense amplifier, node QB drive bit-line through transmission gates to read out the data, this will save the power consumption used in bit-line precharging and sense amplifier detecting, which accounts for a large proportion of total power consumption. Owning to streamline read operation logic makes the read operation timing simpler, less affected by process deviation. (ii) Use power-gating technology in the core of memory cell, add P3, N3 between VDD and P1, P2; add P4, N4 between GND and N1, N2, These 4 transistors construct the power-gating logic. When writing logic 0 into memory cell, node Q convert to 0 and node QB convert to 1, at the same time, P4 and N4 is cut off, coupled inverters disconnect to GND, P3 and N3 is conducted and node QB connect to VDD through P3, which makes the voltage of node QB constantly to be 1; When writing logic 1 into memory cell, node Q convert to 1 and node QB convert to 0, at the same time, P3 and N3 is cut off, coupled inverters disconnect to VDD, P4 and N4 is conducted and node QB connect to GND through N4, which makes the voltage of node QB constantly to be 0. Since single-ended structure just depend on the voltage of node QB to read data out, so it is important to keep node QB stable, node Q will be in a floating status due to power-gating, but it won't affect the accuracy of read operation, because the values of read data only determined by the voltage of node QB. (iii) Replace all the pass transistors by transmission gates, threshold drop of pass transistors will affect access speed and voltage swing, transmission gate without threshold drop can accelerate access speed, and the bit-line voltage is full swing in read operation. (iv) Owning to power-gating, GND disconnect to coupled inverters when node Q is 0, VDD disconnect to coupled inverters when node Q is 1, this helps to reduce power consumption of the core in memory cell, especially in power consumption caused by short circuit current. During the flipping of a inverter, PMOS transistor and NMOS transistor will be conducting at the same time, thus lead a short-circuit path between VDD and GND, resulting to a large short-circuit current, after using power-gating, short-circuit path will be cut off, short-circuit current is significantly reduced.

3.2. *Address decoder*

Address decoder is in the critical path of register file, and its delay accounted for a very large proportion of read/write access latency, there are two types of decoder: static decoder and dynamic decoder, static decoder's advantage is high stability and low power consumption while dynamic decoder performs faster. This paper adopts static decoder and use multi-level structure to accelerate the decoding speed. The register file consist of 8 banks, use a 3-8 decoder to select one of the banks, each bank consist of 8 register, use a 3-8 decoder to select one of the registers, therefore, use two cascade 3-8 decoders can form a 6-64 decoder. We adopted two level structure in each 3-8 decoder, the first level adopted a two input AND gate and a inverter to performs pre-decoding, the second level performs final decoding. Multi-level structure applied in address decoder can avoid using logic gates with high input width, and also reduced transistor number largely. In conclusion, Multi-level decoding structure is benefit for speed, power consumption and silicon area. Figure 4 shows the 6-64 decoder structure.

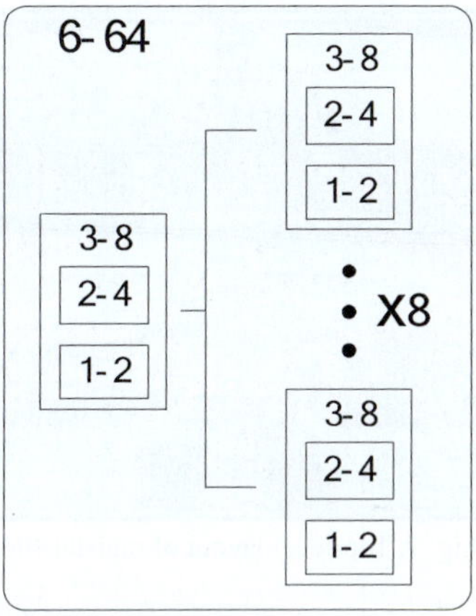

Fig. 4. Address decoder structure.

4. Full Custom Implementation and Post-Layout Simulation

4.1. *Full custom implementation*

Using Nangate 45nm CMOS dual-well technology to design and implemented the register file, Figure 5 shows the improved memory cell layout. Figure 6 shows layout of a bank. Figure 7 shows the top-level layout of the register file.

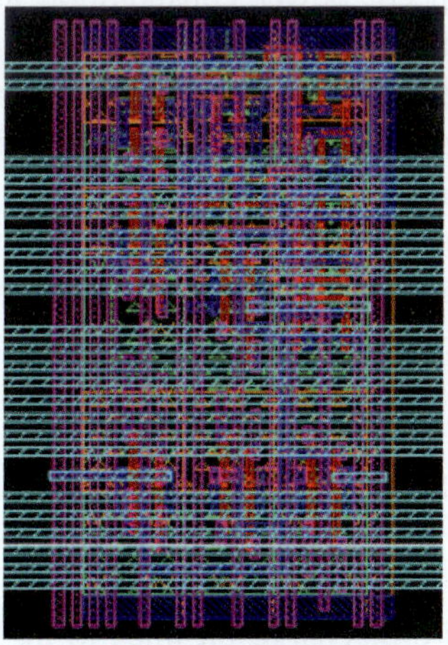

Fig. 5. Layout of memory cell.

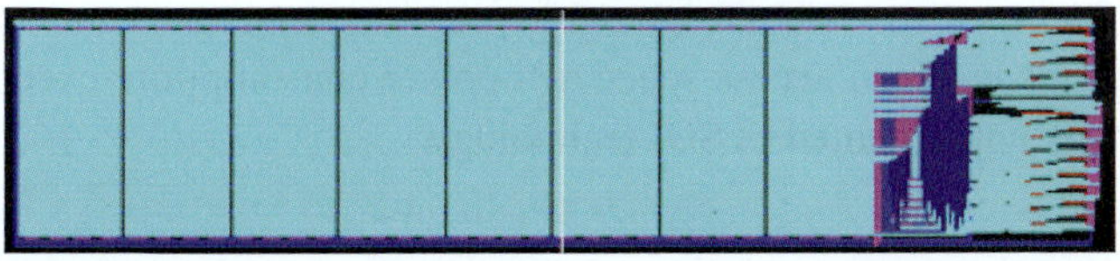

Fig. 6 .Layout of a bank.

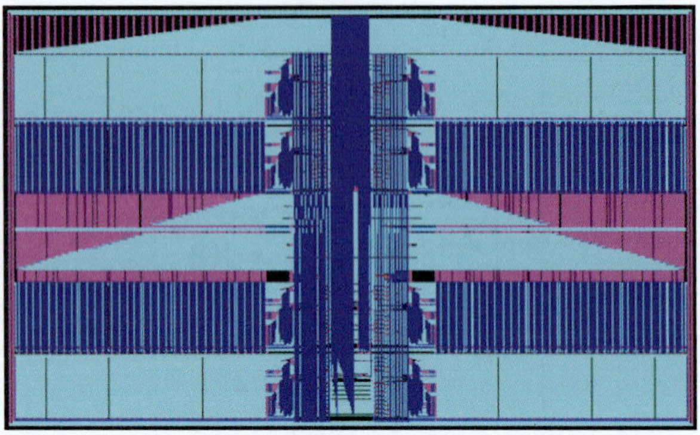

Fig. 7. Top-level layout of register file.

4.2. *Post-layout simulation and performance comparison*

DRC, LVS verification and parasitic parameters extracting was completed in Calibre. Post-layout simulation results shows the register file designed in this paper can complete 8 read operations and 4 write operations in one clock cycle simultaneously. In the condition of 1v power supply and 1GHz clock frequency, write 0 latency is 202ps, write 1 latency is 235ps, read 0 latency is 280ps, read 1 latency is 324ps. Figure 8 shows the waveforms of read/write operation.

Table 1 shows the comparison between conventional design method and ours method in access speed, power consumption and silicon area, it shows performance of write 1 latency and power consumption increased obviously, the decrease of write 1 latency mainly benefit from replace pass transistor with transmission gate, thus avoid threshold drop and speed up write 1 operation, the decrease of power consumption mainly benefit from power-gating technology and removed sense amplifier logic. In terms of silicon area, the transistor number of improved memory cell is larger than conventional one, thus increased the area of memory array, and finally the total silicon area increased 16.3%. It is worth noting that delay of read 1 in table 1 is 0ps, this is mainly determined by its special structure, bit-lines will be charged to high level in conventional single-ended structure, when read operation coming, bit-line voltage is already in high level, thus enable to read data out instantly. The improved single-ended structure's read operation is through node QB to drive bit-lines and read data out, read logic 1 is a procedure of charge bit-lines to high level which takes some time to read data out.

Table 2 compares ours design with previous arts, the results shows our design performs better in access speed and power consumption, except for paper [8], which was implemented in SOI technology.

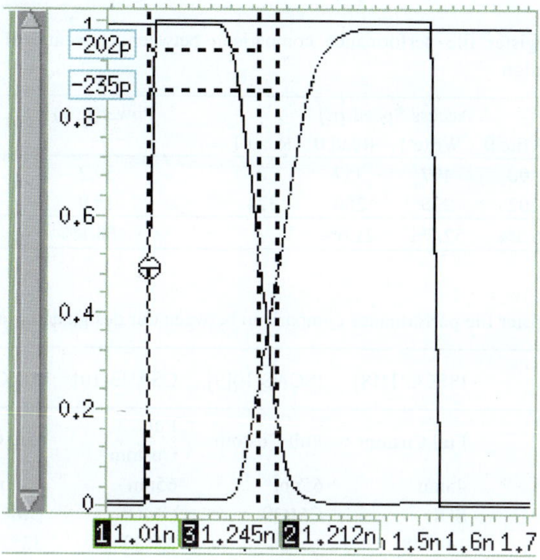

Fig. 8 (a) Post-layout simulation of write operation.

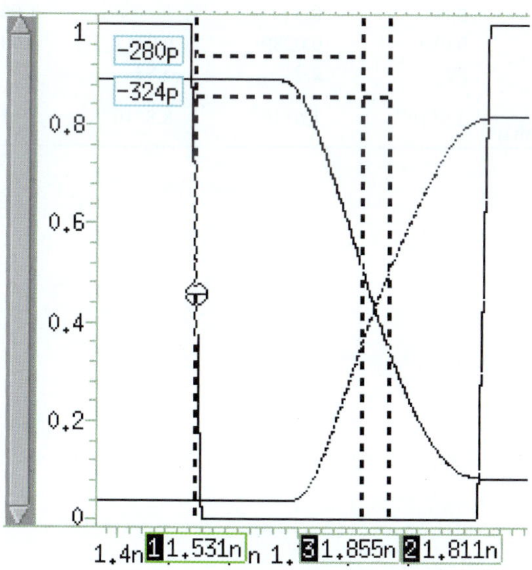

Fig. 8 (b) Post-layout simulation of read operation.

Table 1. Register file performance comparison between conventional design and improved design.

	Access Speed(ps)				Power(mw)	Area(um^2)
	Write 0	Write 1	Read 0	Read 1		
Conventional	190	497	357	0	59.2	100827.40
Improved	202	235	280	324	18.9	117259.64
Gain	-6.3%	52.7%	21.6%	/	68.1%	-16.3%

Table 2. Register file performance comparison between our design and previous art.

	ISSCC'11[8]	ISCAS'10[9]	CSA'13[10]	ASIC'11[11]	Ours Work
Design Method	Full Custom	Full Custom	Full Custom	Full Custom	Full Custom
Process	45nm SOI	65nm CMOS	65nm CMOS	65nm CMOS	45nm CMOS
Volume(bit)	144*78	16*34	32*32	32*32	64*32
Port	4R2W	6R4W	3R3W	4R2W	8R4W
Power Supply(v)	0.9	1.0	1.2	1.2	1.0
Frequency(GHz)	2.76	0.3	0.9	1.0	1.0
Worst Latency(ns)	0.19	0.5	/	1.06	0.324
Area(mm^2)	0.088	0.0289	0.0311	0.0475	0.1173
Power(mw)	59	2.7	5.83	12	18.9
Normalized Power (mw/(GHz*v*port*bit))	$3.5*10^{-4}$	$1.6*10^{-3}$	$8.8*10^{-4}$	$1.6*10^{-3}$	$7.7*10^{-4}$

5. Conclusion

This paper designed and implemented a register file in Nangate 45nm CMOS dual-well technology, it adopted improved memory cell, multi-banked structure and power-gating technology, completed the goal of high speed and low power. In the condition of 1v power supply and 1 GHz clock frequency, it can complete 8 read operations and 4 write operations in one clock cycle simultaneously, the worst latency is 324ps and total power consumption is 18.9 mw.

References

1. Ataur R. Patwary, Hans Greub & Zhong Feng. 2006. Bit-Line Organization in Register Files for Low-Power and High-Performance Application. Electrical and Computer Engineering, Bangladesh: Dhaka, 505-508.
2. Shenglong Li, Zhaolin Li & Fang Wang. 2009. Design of A High-speed Low-Power Multiport Register File. Microelectronics & Electronics, China: Shanghai, 408-411.
3. Jessica H.Tseng & Krste Asanovic. 2003. Banked Multiported Register Files for High-Frequency Superscalar Microprocessors. Computer Architecture, 62-71.
4. Khawar Sarfraz & Mansun Chan. 2014. A Low-Noise Local Bitline Technique for Dual-V_t Register Files. Faible Tension Faible Consommation, Monaco, 1-4.
5. N. Gong, J. Wang & R. Sridhar. 2012. Clock-biased local bit line for high performance register files. Electronics Letters, 48(18):1104-1105.
6. Dao-Ping Wang, Hon-Jarn Lin & Wei Hwang. 2014. Low-Power Multiport SRAM With Cross-Point Write Word-Lines, Shared Write Bit-Lines, and Shared Write Row-Access Transistors. Circuits and Systems II, 61(3):188-192.
7. Georgios K. Konstadinidis, Marc Tremblay & Shailender Chaudhry. 2009. Architecture and Physical Implementation of a Third Generation 65nm, 16Core, 32Thread Chip — Multithreading SPARC Processor. JSSC, 44(1):7-17.
8. Gray S. Ditlow, Robert K. Montoye & Salvatore N. Storino. 2011. A 4R2W Register File for a 2.3GHz Wire-Speed POWER Processor with Double-Pumped Write Operation. ISSCC, San Francisco, 256-258.
9. Johannes Uhlig, Sebastian Hoppner & Georg Ellguth. 2010. A Low-power Cell-Based-Design Multi-port Register File in 65nm CMOS Technology. Circuits and Systems, Paris, 313-316.

10. Youzhong Li, Lijun Zhang & Qixiao Zhang. 2013. A Low Power Area Efficient Full Custom 3-Read 3-Write General Purpose Register in 65nm Technology. Computer Sciences and Applications,Wuhan, 741-744.
11. Baoyu Xiong & XingXing Zhang & Jun Han. 2011. Design of A Single-Ended Cell Based 65nm 32*32b 4R2W Register File. ASIC, Xiamen, 311-314.

Research on the method of backward tracing

B. Ma[1†] and H. Bao

North China Electric Power University, Beijing 102206, China

†Email: eagle_knight@126.com

Method of backward power flow tracing based on separation of power supply path is proposed for getting the relationship of each power quantity in distribution network. The key to realize separation of power supply path of load is to compute the equivalent impedance in each branch of each load. Using the node voltage and load current compute and establish a new form of impedance matrix, the equivalent impedance matrix. The loss of each branch caused by each load and the flow of load power can be calculated based on the calculation the equivalent impedance. Therefore, it can provide the basis for practical problems such as network loss allocation. The case study shows that, the network loss calculated by separation of power supply path method is match with the original power flow, provide a new calculation method for the research of backward tracing.

Keywords: Backward tracing; Equivalent impedance; Separation of power supply path; Line loss.

1. Introduction

With the development of electricity market, people put forward the question of source of power in power system. Therefore, power flow tracing and loss allocation has become a research hot spot. In terms of power flow tracing, various tracking methods coexist.

Power flow tracing was first introduced by Bialek. J [1], in the paper first proposed the Proportional share distribution principles to solve the distribution of node outflow power among each line in power tracing. Later, a variety of tracing methods continue to develop, power tracing method [2-4], current tracing method [5-18], electrical dissecting method [19-20].

The methods described above are in the idea of tracing from the source, tracing the path of energy transferring according to the power flow, called downstream tracing method. Downstream tracing reflected the physical

[1] (1989-) received the B.Sc. degree from North China Electric Power University, Beijing, China, in 2012, and is studying for the Master degree in North China Electric Power University, Beijing, China. His current research interests focus on power flow tracing.

process of the injection of power, power flow and consumption of load. But downstream tracing may change the property of load, if the tracing method changed the property of load, the tracing results will be unacceptable. If the tracing start from the load and to the source, it won't change the load properties and would be more in line with the principle of economy, it's called backward tracing.

In backward tracing we take load as the research subject, under the condition of power flow interface is known, Starting from the load node, According to the power flow path, and determine the relation between load and power flow. The early research about backward tracing is mainly put emphasis on the establishment of Impedance matching relationship between load and line loss [21]. This paper will focus on load current.

In this paper, we take the single power distribution network as an example, starting from the load current, obtained the line loss corresponding to the each load, and then establish the relationship between them. It's the first time to establish a new form of impedance matrix, which is based on the flow path of load current. Then research the loss caused by each load in its passing line and the composition of the line loss. This method takes a new perspective for tracing the power flow in distribution network, can clearly describe the corresponding relation of load and loss, laid the foundation for the operation of the distribution network optimization and parameter identification calculation.

2. The Separation of Power Supply Path of Load

The backward tracing method proposed in this paper is based on the separation of power supply path of load, the process of separation of power supply path are as follows.

Under the condition of power flow interface is known, the power system can be equivalent to a circuit network, power supply can be equivalent to a current source, load can use impedance to represent.

On the premise of maintaining the voltage unchanged of each node, on the basis of the load current, separate the power supply path of each load. The key to realize the separation of power supply path is the equivalent transform of branch impedance.

2.1. *The separation of power supply path of load*

As shown in figure 1, between the node i and node j there is a impedance Zij, in the node j there are two load currents IL1 and IL2. Now we do a equivalent transform to impedance Zij, after equivalent that is shown in figure 2, and can be regarded as in figure 3, thus we achieves separation of power supply path of load.

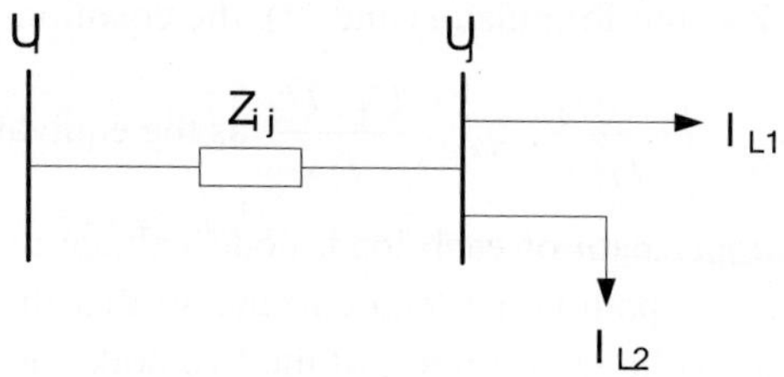

Fig. 1. Branch impedance model.

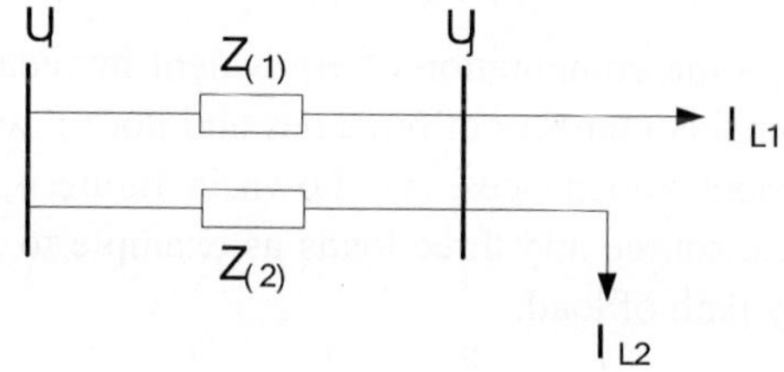

Fig. 2. Branch equivalent model.

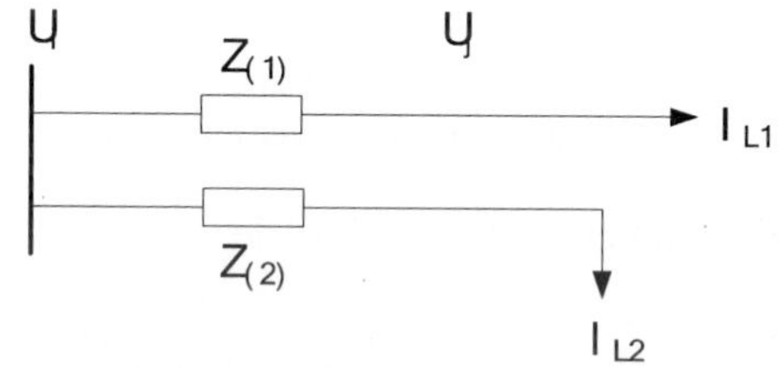

Fig. 3. Load separation model.

To achieve branch equivalent the necessary condition are as follows:

(1) The equivalent impedance $Z_{(1)}$ and $Z_{(2)}$ at both ends of the voltage drop must be equal, that is

$$\dot{I}_{L1} Z_{(1)} = \dot{I}_{L2} Z_{(2)} \tag{1}$$

(2) The parallel impedance of the equivalent impedance $Z_{(1)}$ and $Z_{(2)}$ should be equal the branch impedance Z_{ij}, that is

$$Z_{(1)} \parallel Z_{(2)} = Z_{ij} \tag{2}$$

The two equations are definition formulas of the equivalent impedance.

Mark $Z_{(1)} = \dfrac{\dot{U}_i - \dot{U}_j}{\dot{I}_{L1}}, Z_{(2)} = \dfrac{\dot{U}_i - \dot{U}_j}{\dot{I}_{L2}}$,

Replace $Z_{(1)}$ and $Z_{(2)}$ into formula (1) and (2), the equations are all established, then we can use $Z_{(1)} \quad \frac{\dot{U}_i \quad \dot{U}_j}{\dot{I}_{L1}}, Z_{(2)} \quad \frac{\dot{U}_i \quad \dot{U}_j}{\dot{I}_{L2}}$ as the equivalent impedance.

In the power supply path of each load, node voltage remains constant, the current of power supply path is the load current, so that the equivalent impedance is the ratio of the voltage difference of the two nodes and the load current.

2.2. *The separation of power supply path of radial multi branch*

As shown above, it is the computation of equivalent impedance of one branch, when the system contains numbers of branches and nodes, we can use matrix to calculate the equivalent impedances. As shown in figure 4, we use a network which contains single source and three loads as example to analyze the separation of power supply path of load.

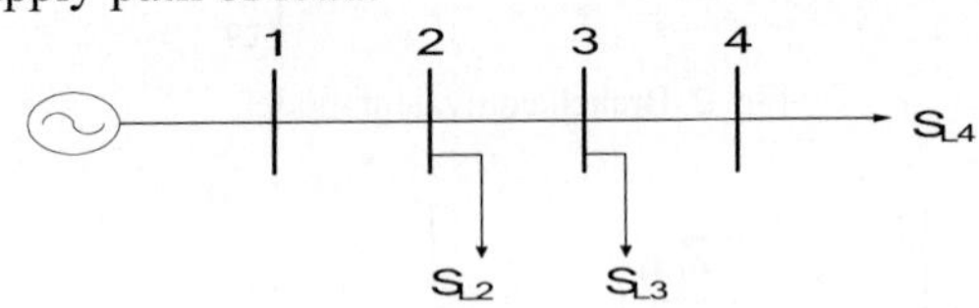

Fig. 4. Single source with three load model.

First, calculate the load current of each node,

$$I_2 = \left(\frac{S_{L2}}{\dot{U}_2}\right)^*, I_3 = \left(\frac{S_{L3}}{\dot{U}_3}\right)^*, I_4 = \left(\frac{S_{L4}}{\dot{U}_4}\right)^* .$$

To load S_{L4}, we can use the product of a matrix and a vector to calculate the equivalent impedance in the each branch, as shown in the following formula.

$$\begin{pmatrix} 1 & -1 & 0 & 0 \\ 0 & 1 & -1 & 0 \\ 0 & 0 & 1 & -1 \\ 0 & 0 & 0 & 1 \end{pmatrix} \begin{pmatrix} \dot{U}_1 \\ \dot{U}_2 \\ \dot{U}_3 \\ \dot{U}_4 \end{pmatrix} * \frac{1}{\dot{I}_4} = \begin{pmatrix} Z_{L4(1)} \\ Z_{L4(2)} \\ Z_{L4(3)} \\ Z_{L4} \end{pmatrix}$$

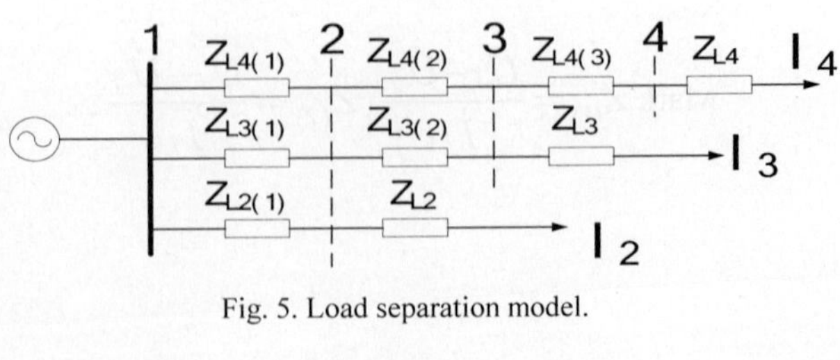

Fig. 5. Load separation model.

Figure 5 is the model of the power supply path after separation, every load's equivalent impedances can be obtained through the above formula, or we can calculate equivalent impedances by a product of matrix and matrix, as shown in the following formula.

$$\begin{pmatrix} 1 & -1 & 0 & 0 \\ 0 & 1 & -1 & 0 \\ 0 & 0 & 1 & -1 \\ 0 & 0 & 0 & 1 \end{pmatrix} \begin{pmatrix} \frac{\dot{U}_1}{\dot{I}_{L4}} & \frac{\dot{U}_1}{\dot{I}_{L3}} & \frac{\dot{U}_1}{\dot{I}_{L2}} \\ \frac{\dot{U}_2}{\dot{I}_{L4}} & \frac{\dot{U}_2}{\dot{I}_{L3}} & \frac{\dot{U}_2}{\dot{I}_{L2}} \\ \frac{\dot{U}_3}{\dot{I}_{L4}} & \frac{\dot{U}_3}{\dot{I}_{L3}} & 0 \\ \frac{\dot{U}_4}{\dot{I}_{L4}} & 0 & 0 \end{pmatrix} = \begin{pmatrix} Z_{L4(1)} & Z_{L3(1)} & Z_{L2(1)} \\ Z_{L4(2)} & Z_{L3(2)} & Z_{L2} \\ Z_{L4(3)} & Z_{L3} & 0 \\ Z_{L4} & 0 & 0 \end{pmatrix}$$

To the right side of the matrix equation, we using column perspective, the elements in each column that are all the equivalent impedance on the power supply path for each load, the last non-zero element in each column is the load impedance. Now we do a little change to the right matrix of the equation, remove the last non-zero element in each column, to form a new matrix.

$$\text{Mark } \Delta Z = \begin{pmatrix} Z_{L4(1)} & Z_{L3(1)} & Z_{L2(1)} \\ Z_{L4(2)} & Z_{L3(2)} & 0 \\ Z_{L4(3)} & 0 & 0 \end{pmatrix},$$

We using row perspective, $Z_{\mathrm{L}4(1)} \Box Z_{\mathrm{L}3(1)} \Box Z_{\mathrm{L}2(1)} = Z_{12}$, the elements in each row that are equivalent impedances of each branch corresponding to the load flow through the branch. In short, each row corresponds to a branch, and each column corresponds to a load. And matrix ΔZ is the equivalent impedances matrix. The increase of load in the same node reflects in matrix ΔZ is the increase of the number of columns.

2.3. *The separation of power supply path of multi ramifications*

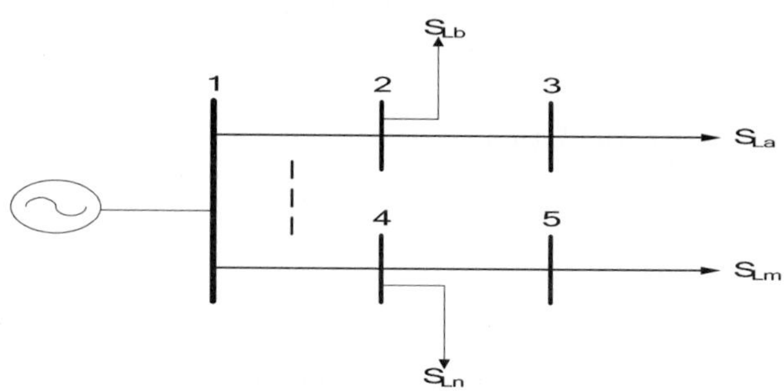

Fig. 6. Multi ramifications model.

As shown in Figure 6, there are X branches in node 1, then the equivalents impedance are calculated as follows.

$$\left[\begin{array}{ccc|c|ccc} 1 & 1 & 0 & & & & \\ & 1 & 1 & & & & \\ & & 1 & & & & \\ \hline & & & \ddots & & & \\ \hline & & & & 1 & 1 & 0 \\ & & & & & 1 & 1 \\ & & & & & & 1 \end{array}\right] \left[\begin{array}{cc|c|cc} \frac{\dot{U}_1}{\dot{I}_{La}} & \frac{\dot{U}_1}{\dot{I}_{Lb}} & & & \\ \frac{\dot{U}_2}{\dot{I}_{La}} & \frac{\dot{U}_2}{\dot{I}_{Lb}} & & & \\ \frac{\dot{U}_3}{\dot{I}_{La}} & 0 & & & \\ \hline & & \ddots & & \\ \hline & & & \frac{\dot{U}_1}{\dot{I}_{Lm}} & \frac{\dot{U}_1}{\dot{I}_{Ln}} \\ & & & \frac{\dot{U}_4}{\dot{I}_{Lm}} & \frac{\dot{U}_4}{\dot{I}_{Ln}} \\ & & & \frac{\dot{U}_5}{\dot{I}_{Lm}} & 0 \end{array}\right] = \left[\begin{array}{cc|c|cc} Z_{La(1)} & Z_{Lb(1)} & & & \\ Z_{La(2)} & Z_{Lb} & & & \\ Z_{La} & 0 & & & \\ \hline & & \ddots & & \\ \hline & & & Z_{Lm(1)} & Z_{Ln(1)} \\ & & & Z_{Lm(2)} & Z_{Ln} \\ & & & Z_{Lm} & 0 \end{array}\right]$$

$$\Delta Z = \left(\begin{array}{cc|c|cc} Z_{La(1)} & Z_{Lb(1)} & & & \\ Z_{La(2)} & 0 & & & \\ \hline & & \ddots & & \\ \hline & & & Z_{Lm(1)} & Z_{Ln(1)} \\ & & & Z_{Lm(2)} & 0 \end{array}\right)$$

When the network contains a number of ramifications, Matrix ΔZ become block matrix, the increase of ramifications, reflects in matrix ΔZ is the increase of the number of blocks.

3. Line Loss Power Calculation

After the separation of power supply path, line loss can be calculated in each branch of each load.

As shown in Figure 4, the model used as an example, mark ΔS as the line loss matrix. Matrix $|I|$ is a diagonal matrix and its elements are the absolute value of the load currents, in the diagonal matrix the load current order corresponding to the column of matrix ΔZ. So the loss matrix ΔS can be calculated as following formula.

$$\Delta S = \Delta Z * |I|^2$$

$$= \begin{pmatrix} Z_{L4(1)} & Z_{L3(1)} & Z_{L2(1)} \\ Z_{L4(2)} & Z_{L3(2)} & 0 \\ Z_{L4(3)} & 0 & 0 \end{pmatrix} \begin{pmatrix} |I_{L4}|^2 & & \\ & |I_{L3}|^2 & \\ & & |I_{L2}|^2 \end{pmatrix}$$

$$= \begin{pmatrix} Z_{L4(1)}|I_{L4}|^2 & Z_{L3(1)}|I_{L3}|^2 & Z_{L2(1)}|I_{L2}|^2 \\ Z_{L4(2)}|I_{L4}|^2 & Z_{L3(2)}|I_{L3}|^2 & \\ Z_{L4(3)}|I_{L4}|^2 & & \end{pmatrix}$$

In matrix ΔS, every element has its meaning. $Z_{L4(1)}|I_{L4}|^2$ represents the line loss caused by load L4 in branch 1-2, the rest were similar.

In column perspective, each column is the line loss caused by corresponding load in its flowing branches. The sum of the column elements is the line loss caused by the load in its flowing path. The first column elements are the line loss caused by load L4 in its flowing branches, the rest were similar.

In row perspective, each row is the line loss caused by the loads which pass through the branch. The sum of the row elements is the line loss caused by all loads passed through the branch, which is the actual line loss in the branch. The elements in first row are the loss caused by load L4, L3 and L2 in branch 1-2, the rest were similar.

The sum of all the elements in the matrix is the total loss in the network. If mark S as power matrix, we can get another equation.

$$S = Z * |I|^2$$

$$= \begin{pmatrix} Z_{L4(1)} & Z_{L3(1)} & Z_{L2(1)} \\ Z_{L4(2)} & Z_{L3(2)} & Z_{L2} \\ Z_{L4(3)} & Z_{L3} & 0 \\ Z_{L4} & 0 & 0 \end{pmatrix} \begin{pmatrix} |I_{L4}|^2 & & \\ & |I_{L3}|^2 & \\ & & |I_{L2}|^2 \end{pmatrix}$$

In matrix S, The sum of the column elements is the initial power injection of corresponding load the initial power injection.

From above we can see the process of power flow, so we can track from the load to the power supply, realize backward tracing.

4. Case Study

The standard distribution system IEEE 33 nodes data is used to verify the proposed backward tracing algorithm in this paper. And the IEEE 33 nodes system is a typical distribution network with just single source.

The topology of IEEE 33 nodes system is illustrated in Fig. 7 with one balance node that is named node1, 32 P-Q nodes, 32 branches that are respectively included in 4 long branches.

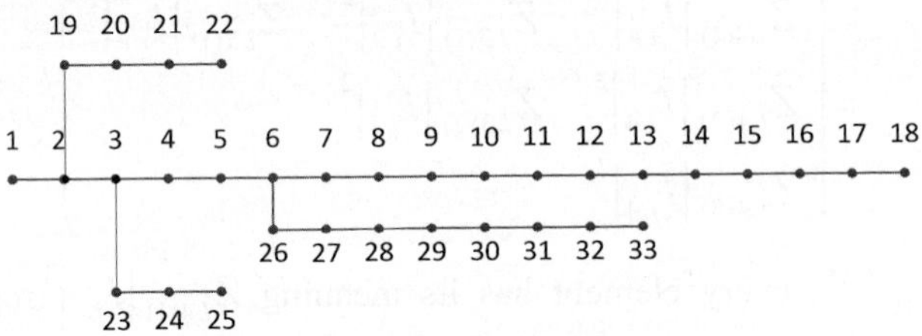

Fig. 7. The topology of IEEE 33 nodes system.

The Newton-Raphson algorithm is selected to calculate the power flow. The reference voltage and power are respectively assumed to be 12.66kV,10MVA. Based on the results of power flow, the line losses of each line are accurately calculated using the proposed backward tracing algorithm.

Because of the limited space, some of the line losses are listed in Table 1, which are the line losses flowing through the branch from node2 to node9. And the composition of losses of line 27-28 is illustrated in Table 2. From Table 2,we can know that the loss of line 27-28 are caused by six nodes, they are node28 to node33. The losses caused by each node are also listed in Table 2. According to compare the results of backward tracing algorithm and power flow, it can verify that the calculation of separation of power supply path is correct.

Table 1. Line losses flowing through the branch of the nodes (MVA).

Line name	1-2	2-3	3-4	4-5	5-6
Node2	0.00032+ 0.00015i				
Node3	0.00029+ 0.00010i	0.00134+ 0.00044i			
Node4	0.00039+ 0.00021i	0.00184+ 0.00097i	0.00108+ 0.00052i		
Node5	0.00019+ 0.00008i	0.00091+ 0.00034i	0.00053+ 0.00018i	0.00047+ 0.00015i	
Node6	0.00020+ 0.00005i	0.00091+ 0.00021i	0.00053+ 0.00010i	0.00047+ 0.00008i	0.00110+ 0.00048i
Node7	0.00066+ 0.00026i	0.00310+ 0.00119i	0.00181+ 0.00061i	0.00160+ 0.00051i	0.00363+ 0.00225i

Node8	0.00067+ 0.00026i	0.00312+ 0.00120i	0.00182+ 0.00062i	0.00160+ 0.00051i	0.00365+ 0.00226i
Node9	0.00020+ 0.00005i	0.00093+ 0.00021i	0.00054+ 0.00010i	0.00047+ 0.00008i	0.00112+ 0.00049i
Line name	6-7	7-8	8-9	Total loss	
Node2				0.00032+ 0.00015i	
Node3				0.00163+ 0.00054i	
Node4				0.00331+ 0.00170i	
Node5				0.00210+ 0.00075i	
Node6				0.00321+ 0.00092i	
Node7	0.00044+ 0.00123i			0.01124+ 0.00605i	
Node8	0.00044+ 0.00123i	0.00113+ 0.00040i		0.01243+ 0.00648i	
Node9	0.00017+ 0.00033i	0.00033+ 0.00007i	0.00036+ 0.00021i	0.00412+ 0.00154i	

Table 2. The composition of losses of line 27-28 (MVA).

	node28	node29	node30	node31	node32	node33	Results of backward tracing	Results of power flow
Line 27-28	0.00074 + 0.00015i	0.00153 + 0.00065i	0.00330 + 0.00685i	0.00190 + 0.00061i	0.00267 + 0.00088i	0.00078 + 0.00039i	0.01092 +0.00953	0.011+ 0.01i

5. Conclusion

(1)A new method is proposed to realize backward tracing by the separation of power supply path of loads, in this paper we take single branch, radial multi branch, multi ramifications as examples to do the separation of power supply path of loads, and put forward the method of calculating the equivalent impedance. Calculation of the equivalent impedance is the key of separation of power supply path of loads.

(2)A new impedance matrix is used to express the equivalent impedance of load in each branch. The number of columns of the matrix is related to the number of loads, the number of rows of the matrix is related to the number of branches.

(3)Calculation of network loss is based on the equivalent impedance matrix, every element has a clear physical meaning in the network loss matrix. From perspective of row and column to observe the equivalent impedance matrix, each row corresponding to a branch, each column corresponds to a load. Without changing the property of the load, realize tracing the power flow from load to the source in single power distribution network, and can also analyze the network loss allocation.

References

1. Bialek J. Tracing the flow of electricity[J]. IEE Proceedings — Generation, Transmission and Distribution, 1996, 143(4): 313-320.
2. Sun H, Yu D C, Zheng Q. AC power flow tracing in transmission networks [C] Power Engineering Society Winter Meeting, 2000. IEEE. IEEE, 2000, 3: 1715-1720.
3. Kirschen D, Allan R, Strbac G. Contributions of individual generators to loads and flows [J]. Power Systems, IEEE Transactions on, 1997, 12(1): 52-60.
4. Ming Z, Liying S, Gengyin L, et al. A novel power flow tracing approach considering power losses [C] Electric Utility Deregulation, Restructuring and Power Technologies, 2004 (DRPT 2004). Proceedings of the 2004 IEEE International Conference on. IEEE, 2004, 1: 355-359.
5. Kirschen D, Strbac G. Tracing active and reactive power between generators and loads using real and imaginary currents [J]. Power Systems, IEEE Transactions on, 1999, 14(4): 1312-1319.
6. Wang Xifan, Wang Xiuli. Research on Current Tracing [J]. Science in China (Series E), 2000, 30(3): 265-270.
7. Bialek J. Tracing the flow of electricity [J]. IEE Proceedings — Generation, Transmission and Distribution, 1996, 143(4): 313-320.
8. Zobian A, Ilic M D. Unbundling of transmission and ancillary services. I. Technical issues [J]. Power Systems, IEEE Transactions on, 1997, 12(2): 539-548.
9. Conejo A J, Galiana F D, Kockar I. Z-bus loss allocation [J]. Power Systems, IEEE Transactions on, 2001, 16(1): 105-110.
10. Feng Linqiao, Xu Wenyu, Dun Changan, et al. Definition of Conjugate Current Whole Power Loss Component and Their Superposition Theorem [J]. Proceedings of the CSEE 2002, 22(5): 65-69.
11. Qiao Zhengyu, Chen Xueyun, Zhang Lizi,et al. A New Concept of The Separation of Electrical Power Flow [J]. Proceedings of the CSEE, 2001, 21(1): 77-79.

12. Peng Jianchun, Jiang Hui, Zhang Jian. Theory of Element Complex Power Components Incurred by Individual Generators and Simulations [J]. Proceedings of the CSEE, 2001, 21(2): 11-15.
13. Li Weidong, Wei Liming, Sun Hui, et al. Research on Comparison of Power Flow Tracing Methods [J]. Journal of Dalian University of Technology, 2002, 42(4): 491-497.
14. Li Zhimin, Pan Minghui, Yang Songyan, et al, Study on Tracing Power Flow [J]. Electric Power, 2000, 33(11): 41-43.
15. Feng Linqiao, Xu Wenyu, Liu Fei. Calculation and Apportionment of Real-time Electric Energy Loss in Power Network [J]. Proceedings of the CSEE, 2004, 24(2): 66-70.
16. Wang Xiuli, Wang Xifan. Complex Power Flow Tracing Via Real and Imaginary Current Components [J]. Automation of Electric Power Systems, 2004, 28(3): 10-13.
17. Daniel J S, Salgado R S, Irving M R. Transmission loss allocation through a modified Ybus [J]. IEE Proceedings — Generation, Transmission and Distribution, 2005, 152(2): 208-214.
18. Wu Xiaorong, Xie Kaigui, Zhou Jiaqi. Transmission Loss Allocation Method Based on Current Decomposition and Bi-directional Trace [J]. Proceedings of the CSEE, 2006, 25(21): 24-29.
19. Yu Jilai, Tang Yi. United Electrical Dissection of AC Branch and Bus [J]. Proceedings of the CSEE, 2007, 27(16): 37-42.
20. Tang Yi, Yu Jilai, Zhou Suquan. Electrical Dissection Algorithm of Electric Power Network [J]. Automation of Electric Power Systems, 2006, 30(22): 28-33.
21. Wu Xuyang, Bao Hai. Method for Backward Power Flow Tracing of Power Quantities of Distribution Network [J]. Power System Protection and Control, 2014,14:87-92.

Construction of MRAM reliability test platform

R.N. Zhou †
School of Electrical and Information Engineering, Beihang University, Beijing, 100191, China
†Email:chinatszrn@163.com

Y.G. Zhang, Q. Shi and K. H. Cao
School of Electrical and Information Engineering, Beihang University, Beijing, 100191, China

Magnetic random-access memory (MRAM) is a kind of desirable emerging memory, which promises non-volatility, high storage density, fast access speed and strong radiation resistivity. However, due to the immature fabrication process, reliability is a significant issue for emerging MRAM technology. In the paper, we propose and implement a MRAM reliability test platform and perform a series of tests of MRAM on this platform to evaluate MRAM reliability in terms of temperature variation, and supply voltage variation. The test results dictate the effectiveness of our proposed testing platform and illustrate relevant reliability metrics of current MRAM technology.

Keywords: Magnetic random access memory; Reliability; Test platform.

1. Introduction

Magnetic Random Access Memory (MRAM) is different from the traditional SRAM, DRAM and Flash. It takes advantage of spin property of electron instead of charge to store information. The core structure of MRAM is the Magnetic Tunnel Junction (MTJ). It consists of one MgO layer sandwiched by two ferromagnetic layers. The resistance value of MTJ which changes with the magnetizations of the top and bottom layers. Except for changing external magnetic field, MTJ can permanently maintain its magnetization unchanged making MRAM non-volatile [1].

The excellent properties of MRAM, especially the strong resistivity against radiation make MRAM be suitable for aerospace applications. However, MRAM has not been in mass production. It requires a long time and a lot of test data, especially in a complex electromagnetic environment, to prove its availability for aerospace applications,. In this work, we designed a novel MRAM reliability test platform testing MRAM reliability in different temperatures, magnetic fields and electric fields. The test methods and test results are given in the paper as well.

2. MRAM Technology

The basic memory cell of MRAM is a magnetic tunnel junction MTJ (Magnetic Tunnel Junction). MTJ structure can be simply regarded as a composing of an insulating layer between the two ferromagnetic layers, which is represented in Fig.1.Generally speaking, one of the ferromagnetic layer is fixed layer or reference layer, the magnetization is fixed; while the other is free layer, its magnetization can be changed by external magnetic field or spin polarized current. The resistance of MTJ is determined by the relative magnetization of the two ferromagnetic layers[3]. When the magnetization of the free layer is opposite to that of the fixed layer, MTJ will be in high resistance state, which is shown in Fig.1 (a). When the magnetization of the free layer is the same as that of the fixed layer, MTJ will change into low resistance state, as shown in Fig.1 (b). Therefore, the resistance state of MTJ can be used to store data information. According to the magnetization changing mechanism, MTJ can also be classified as toggle MRAM and spin torque transfer MRAM (i.e., STT-RAM)[4]. The latter can also be classified into in-plane and perpendicular STT-MRAM. The manufacturing of in-plane Toggle MRAM is relatively easy, and its cost is low thus being a popular MRAM available in commercial market(using another reference). MRAM tested in this work is exactly the Toggle MRAM.

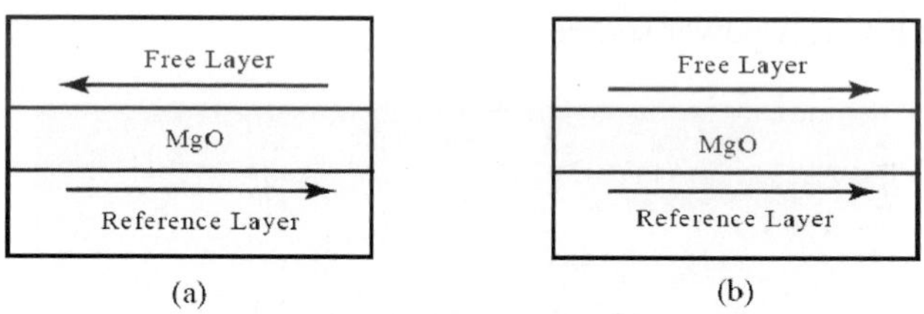

Fig. 1. MTJ sturcture.

The schematic of Toggle MRAM writing operation is presented in Fig.2. Two current lines are used to complete the Toggle MRAM writing operation. According to difference of the direction of the current, we can store 0 or 1 in the MTJ.

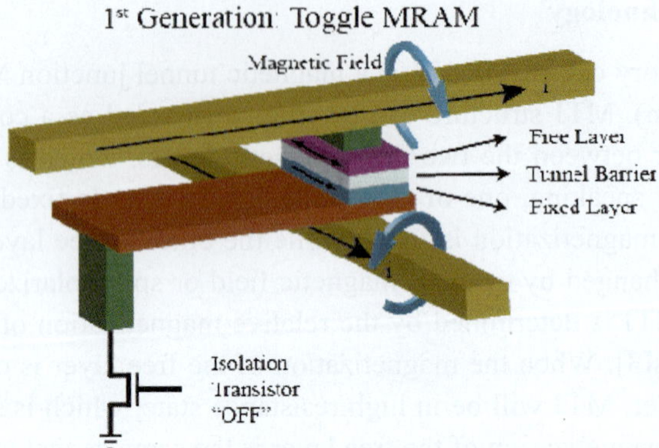

Fig. 2. Toggle MRAM.

3. System Structure

The test chip:

MR4A08BYS35 2M * 8bit MRAM Memory. [6]

Test instrument:

Oriental Morning EM5 electromagnet,

NCV20-H-M high temperature thermostat,

Tektronix2024B oscilloscope,

Agilent E3631A power source.

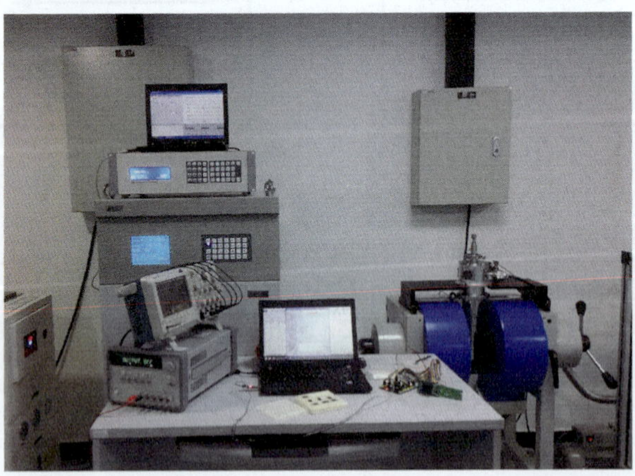

Fig. 3. Test platform.

4. Reliability Test

Like other integrated circuits, MRAM's reliability is determined by many factors. Data integrity can be affected by MTJ memory bit itself, packaging issues,

CMOS reliability as well as the memory organization architecture. we will perform the following tests: magnetic field test, temperature test and electromagnetic compatibility test.

4.1. *Memory testing algorithm introduction*

MRAM read testing algorithm consists of two parts: dynamic testing and static testing. Heidecker et al gave the JPL Endurance Test [7] We propose a brand new dynamic test method based on that work shown as below:

Dynamic testing includes six main steps:

Constant test:

Write 0x55, then read 0x55 and test the error rate

Write 0xaa, then read 0xaa and test the error rate

Write 0x00, then read 0x00 and test the error rate

Write 0xff, then read 0xff and test the error rate

Incremental test:

Write i mod 0xff in address i, then read and test error rate

Random test:

Write random number, then read and test error rate

Static test is performed after per write operation. The chip is powered off for some time. Then the data are read out again to derive the error rate. Watch of the above test procedure was repeated 10 times, the error rate is calculated as the average of the test results.

In the following test process, we use this algorithm.

4.2. *Magnetic field test*

MRAM device can be easily affected by external applied magnetic fields and change its state.

In view of the influence of magnetic field, we conducted two tests, respectively, static test and dynamic test of magnetic field.

The static magnetic field test:

The magnetic field is applied parallel to the surface of the free layer and perpendicular to the surface of the free layer respectively, the data obtained are shown in the following table:

Table 1. Horizontal magnetic field test results.

Horizontal(Oe)	Error rate
0	0%
50	0%
170	0%
180	~10%

Table 2. Vertical magnetic field test results.

Vertical(Oe)	Error rate
0	0%
50	0%
100	0%
450	~10%

In the test, horizontal magnetic field > 180 Oe, and the vertical magnetic field > 450 Oe, when the magnetic field is removed, we can find the error number decreased gradually, returning to normal state after several read and write cycles.

The magnetic field >180 Oe does not exist in real life. Distance be the magnetic field experienced by MRAM in normal use is only 10 Oe, far less than 180 Oe. Therefore MRAM has a strong resistivity against the influence of magnetic field.

Dynamic field test:

The use of magnetic field power control software can provide alternating magnetic field.

MRAM chip is adjusted to the parallel magnetization Then, an alternating magnetic field with amplitude of ± 100 Oe is applied, MRAM read write characteristics are shown in Table 3.

Table 3. Horizontal alternating magnetic field test results.

Horizontal	Magnetic Strength	Frequency	Error
Triangular Wave	±100 Oe	0.05Hz	0%
Rectangle Wave	±100 Oe	0.05Hz	0%
Step Wave	±100 Oe	0.05Hz	0%

Subsequently, MRAM chip is exposed in an external vertical alternating magnetic field of ± 200 Oe magnitude, the test results are shown in the following table.

Table 4. Vertical alternate magnetic test results.

Vertical	Magnetic Strength	Frequency	Error
Triangular Wave	±200 Oe	0.05Hz	0%
Rectangle Wave	±200 Oe	0.05Hz	0%
StepWave	±200 Oe	0.05Hz	0%

From the data listed in above two tables, MRAM has a strong resistance to the alternating magnetic field.

1.1 *Temperature test*

The MRAM chip data sheet shows the normal work-ing temperature range is 0~75°C, so we will test the temperature 0°C, 25°C (normal temperature), 50°C, and 75°C.

The specific process is as follows:

1. Write and read testing in different temperatures. During the temperature testing procedure the chip is attached to a thermal silica, and placed in the thermostat with sapphire substrate on the platform. An insulating ceramics are used to isolate the chip. The vacuum pump is used for temperature adjustment. Firstly, power off the system, and remove the chip from the chip test base. After a while, put MRAM chip on test base and power on the system. If the read operation is correct, power off the system, and remove the MRAM chip.
2. When the temperature control reaches the presetting temperature range, wait for 30min to recover to room temperature.
   We get the results shown in the table below:

Table 5. Temperature test results.

Temperature(T)	0°C	25°C	50°C	75°C
Absolute Temperature(K)	273.5K	-------	323.15K	348.15K
Error Rate	0%	0%	0%	0%

We can see from the table above that the chip works well from 0 to 75°C. Then we add 50 Oe magnetic field, and keep 30min. Then data stored in MRAM are read out to evaluate the error rate. We get the results shown in the table below:

Table 6. Temperature and magnetic field test results.

Temperature(T)	0°C	25°C	50°C	75°C
Absolute Temperature(K)	273.5±5 K		323.15±2 K	348.15±2 K
Magnetic Field	50 Oe	50 Oe	50 Oe	50 Oe
Error Rate	0%	0%	0%	0%

According to the data shown in the two tables, MRAM works well in temperature range 0 ~75°C even applying a <50Oe external magnetic field.

4.4. *Supply voltage variation test*

In the manufacturer's manual, the chip can work normally in the range of 3.0V~3.6V. In order to determine the robustness of the chip under supply voltage variations, we make the chip work at 2.6V, 2.8V, 3.0V, 3.2V, 3.4V, 3.6V, and then test error rate. The results are shown in the following table:

Table 7. Electric test results.

Voltage	2.6V	2.8V	3.0V	3.2V	3.4V	3.6V
Error Rate	100%	0%	0%	0%	0%	0.016%

From the above results, the normal working voltage of the chip is slightly lower than the working voltage. To ensure safety, the working voltage is 2.8V~3.4V.

We separate read and write process under different voltages, it can distinguish specifically where the error occurred. First write chip in 2.6V, and then read in the standard working voltage of 3.3V. We found that the error rate is 100%. If write operation happens in the standard working voltage of 3.3V, and read in 2.6V, there is no mistake. It indicates that MRAM in the voltage for writing can't be less than 2.6V, and the power supply voltage during a read operation may be somewhat smaller.

5. Conclusion

The MRAM is a promising and non-volatile memory. It has a strong resistivity against radiation, low static power consumption and unlimited data retention time. In this work, the chip's sensitivity to the environment is tested, including magnetic field test, temperature test and signal integrity test.

1. Magnetic field test: According to the results of magnetic field test, nominal magnetic field will not affect the data storage of the chip. To avoid the

large magnetic field made by power line and permanent magnet, only a small distance is required to keep the chip working safely. Thus, the chip has a strong resistivity against the influence of magnetic field.

2. Temperature test: In the temperature range of 0°C ~ 75°C, the chip can work very well. Even under the condition of extreme temperatures, with magnetic field less than 50 Oe, there isn't any error observed during testing.
3. Signal integrity test: Seen from the results of oscilloscope, small ripple can be tolerated by MRAM chip.

MRAM technology is developing very fast. The second generation of STT-MRAM products will be available in near future. Its strong resistivity against radiation, low power consumption and other characteristics also make the MRAM an ideal candidate for the next generation memory technology.

References

1. Tehrani, S., et al. "Recent developments in magnetic tunnel junction MRAM." Magnetics, *IEEE Transactions on* 36.5 (2000): 2752-2757.
2. Tehrani, Said, et al. "Progress and outlook for MRAM technology." *Magnetics, IEEE Transactions on* 35.5 (1999): 2814-2819.
3. Engel, B. N., et al. "A 4-Mb toggle MRAM based on a novel bit and switching method." *Magnetics, IEEE Transactions on* 41.1 (2005): 132-136.
4. Mojumder, Niladri Narayan, and Kaushik Roy. "Proposal for switching current reduction using reference layer with tilted magnetic anisotropy in magnetic tunnel junctions for spin-transfer torque (STT) MRAM." *Electron Devices, IEEE Transactions on* 59.11 (2012): 3054-3060.
5. Huai, Y., "Spin-Transfer Torque MRAM: Challenges and Prospects," AAPPS Bulletin, 2008.
6. www.everspin.com/getdatasheet/MR4A08B
7. Heidecker, Jason. "MRAM technology status." (2013).

Body motion tracking and recognition of fracture patients after operation based on Kinect

M. Zhang
School of Electronic Engineering, Dongguan University of Technology, Dongguan, China

L. Lin[1†]
School of Information Engineering, Guangdong Medical University, Dongguan, China
[†]*E-mail: lynwindsent@163.com*

Most of the human body tracking algorithm is not applicable because of false recognition of position of human body. Depth tracking algorithm is designed to solve this problem according to the depth information of human motion based on Kinect sensor. By the analysis of Kinect depth map information, we distinguish the contour of the human body, extract the foreground area and reconstruct the depth histograms of the target area. Based on the depth histogram analysis, the position of joints of human body is obtained The experimental results show that the algorithm to extract the target feature is accurate and robust with avoiding the illumination changes and background under similar circumstances. The research is helpful for doctors to judge the status of fracture patients after operation.

Keywords: Motion tracking; Kinect; Fracture patients.

1. Introduction

Clinical evaluation of the rehabilitation of patients with fracture need long time exercise and correct. The recovery of motor function in patients with a doctor often experience and judgment based on the head, it is difficult to form the objective, the reference standard, also cannot form effective clinical data [1-3]. Therefore, the use of intelligent equipment of motion capture to realize the recognition and tracking of a body, has the important value for continuous tracking and evaluation of fracture recovery of patients after surgery [4,5]. Human motion tracking is based on the image sequences of human motion, object classification and human motion tracking which is a comprehensive technology. It is multi-technology fusion of machine vision, image processing, artificial intelligence and human kinematics knowledge. In recent years, continuously adaptive mean shift algorithm has received extensive attention, because the algorithm is based on the color histogram of the target model, which

[1] Lin Lin, corresponding author

can solve the problem of human body deformation effectively. On the basis of depth map information acquired by Kinect sensor, one can realize real-time tracking of the human body based on Camshift algorithm according to the depth of information. The algorithm is obtained by analysis of Kinect depth map information to distinguish different limbs of the human body and extract the foreground object area to calculate the target area [6-8]. By the depth histogram analysis, the size and location of the human body are determined and distinguished. With the development of three-dimensional human motion tracking technology, one can obtain both RGB information and depth information of the object at low cost [9,10]. In this paper, Kinect is performed to collect the video information of plane images and realize real-time tracking of human motion in image sequences. Compared with the binocular camera, Kinect system needs no marks on the object and is sensitive to human body. The cost of Kinect is only one percent of that of binocular camera such as NDI Polaris.

2. Methods

The structure of Kinect is shown in Figure 1, the RGB camera perform color imaging with a resolution of 300k pixels. The infrared emitter and the sensor camera can capture depth image with a resolution of 100k pixels. Using the depth map of human body, one can overcome the interference of the surrounding noise. Gray pixel depth value in the image depends on the distance of between the human body and the Kinect. The three-dimensional image of the target can be build up with combination of the horizontal and vertical coordinates to express the position of human target in free space. The background and noise information can be removed by setting certain threshold to the depth data. The selection of threshold value and location of the region in space help to complete human target extraction. The human body figure, RGB and depth map by Kinect sensor is shown in Fig. 2.

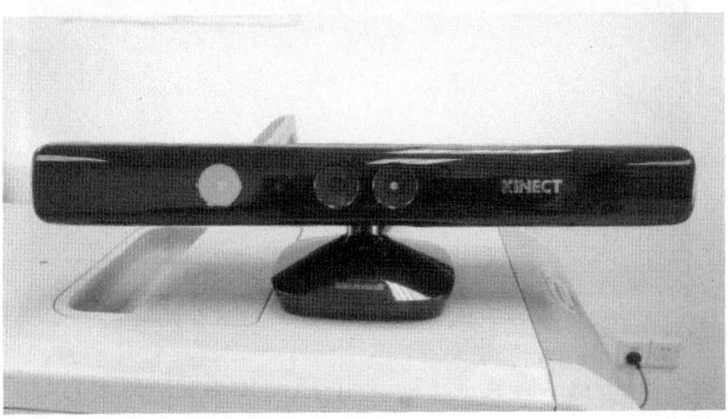

Fig. 1. The Kinect sensor.

Fig. 2. The RGB and depth map by Kinect sensor. (a) The RGB picture, (b) the depth imagine and (c) the skeleton graph.

The reverse process of depth histogram is based on the corresponding pixel in the histogram statistics target depth value instead of pixel values in the original video image. The depth histogram is moving with the target continuously with the color histogram maintaining unchanged. A sub function is used to limit the search scope, determine the target area and eliminate interference.

3. Data Processing

At first, the first frame of video images is extracted to color components. Then the initial target is determined by manual selection, and the search window will be initialized. With the search window center unchanged, the target color histogram is calculated to build up color map. The Camshift algorithm is introduced to calculate the probability plot, the position of the window centroid. The zero order matrix of the search window is defined as:

$$M_{00} = \sum_x \sum_y I(x, y) \tag{1}$$

$$M_{10} = \sum_x \sum_y xI(x, y) \tag{2}$$

$$M_{01} = \sum_x \sum_y yI(x, y) \tag{3}$$

Then the position of the centroid can be determined by $(x_0, y_0) = (\frac{M_{10}}{M_{00}}, \frac{M_{01}}{M_{00}})$. Where I(x,y) represents the pixel value of color coordinates of the probability density distribution. Then, we adjust the center of the search window to the centroid position and judge whether it is convergent.

4. Experimental Results

Tracking algorithms based on the motion of target is programmed in Visual Studio 2010 development environment with combination of Kinect SDK from Microsoft Corp. A single human motion tracking experiment results show that the algorithm is suitable for human motion in the range of Kinect sensor's field of vision. The algorithm uses the depth information tracking with the background being eliminated. The system has strong robustness and will not be affected by illumination and background.

We took two postures, A-Pose and T-Pose, of the volunteer to record the Coordinate information of his 20 joints such as hipcenter, Spine, shouldercenter, Head, shoulderleft, elbowleft, wristleft, handleft, shoulderright, elbowright, wristright, handright, hipleft, kneeleft, ankleleft, footleft, hipright, kneeright, ankleright and footright. The centroid of each joint in x-y plane was also calculated by Camshift algorithm based on color and depth images of the body. The results of Kinect and calculation of algorithm are compared in Fig. 3. The results of Camshift algorithm were verified by the data from Kinect SDK program with root-mean-square error remaining less than 10%. We also took another ten more times measurement to prove the robustness of the method. All the tests showed good agreement with the data of skeleton graph generated by Kinect SDK.

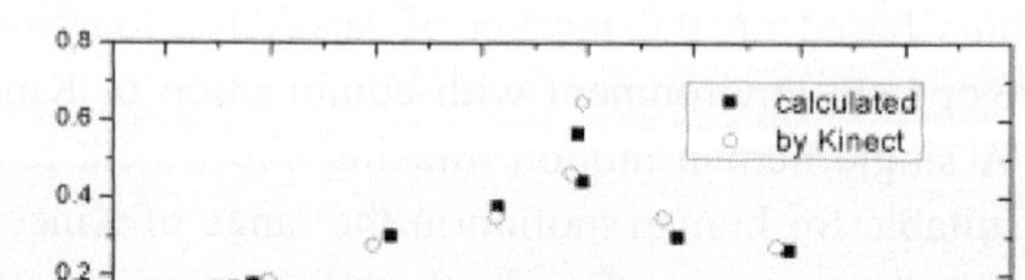

Fig. 3. Results of Kinect and calculation of algorithm.

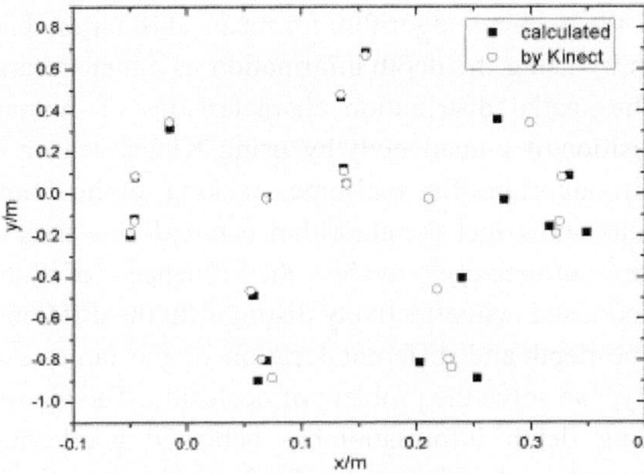
calculated
by Kinect
y/m
x/m
0.8
0.6
0.4
0.2
0.0
-0.2
-0.4
-0.6
-0.8
-1.0
-0.1
0.0
0.1
0.2
0.3
0.4

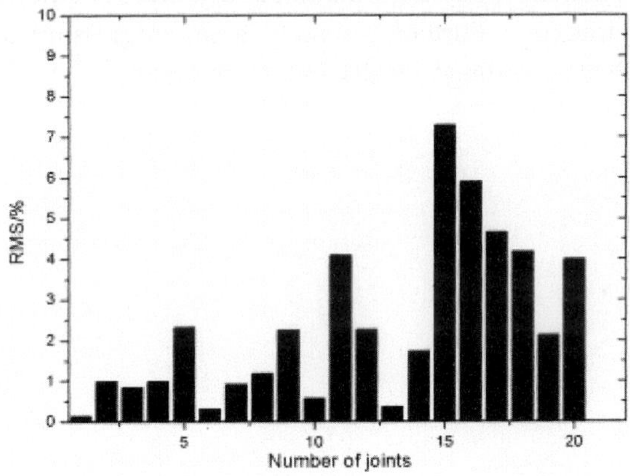
RMS/%
Number of joints
0
1
2
3
4
5
6
7
8
9
10
5
10
15
20

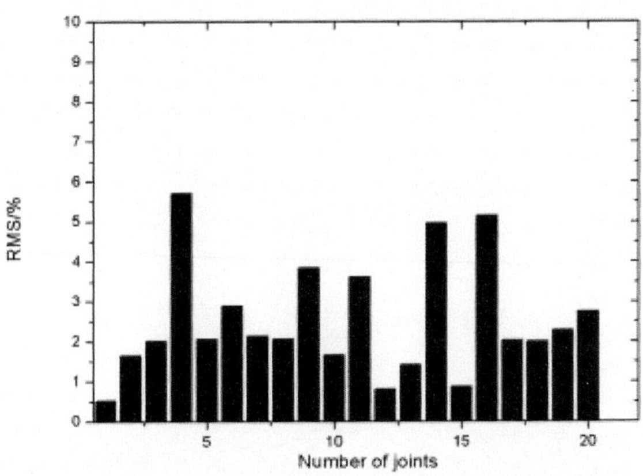
RMS/%
Number of joints
0
1
2
3
4
5
6
7
8
9
10
5
10
15
20

Fig. 3. (*Cont'd*)

5. Conclusion

The algorithm is the tracking algorithm for mean shift target Kinect based on depth information by using the depth information as a human target description which reflects the spatial distribution characteristics of human body target. Depth map acquisition of human body by using Kinect sensor was combined with the Camshift algorithm for real-time tracking of the human body. The experimental results show that the algorithm can real-time tracking of moving human body, can effectively avoid the influence of illumination and complex background, and can effectively distinguish the different depth of the target and the same depth and different location of the target, has strong anti-interference ability, can solve the problem of occlusion. The algorithm of human body tracking using depth information has achieved good effect, the human tracking algorithm and depth information of color and texture features combined with the other is further research. Under the complex background of the target is occluded, the human body deformation and other conditions were studied and experimental tracking. Further research is in progressing to obtain more information of joints according to RGB and depth graphs.

Acknowledgment

The authors acknowledge the support given by the Science and Technological Program for Dongguan's Higher Education, Science and Research, and Health Care Institutions (2012108102001) and Natural Science Foundation of Guangdong Province (2014A030310258).

References

1. R. M. Araujo, G. Grana, A. Virginia, "Towards skeleton biometric identification using the microsoft kinect sensor", Acm Symposium on Applied Computing, NY, pp.21-26, March 2013.
2. G. T. Papadopoulos, A. Axenopoulos, P. Daras, "Real-Time Skeleton-Tracking-Based Human Action Recognition Using Kinect Data", Lecture Notes in Computer Science, Vol. 8325, pp 473-483, January 2014.
3. T. Le, M. Nguyen, T. Nguyen, "Human posture recognition using human skeleton provided by Kinect". International Conference on Computing, Management & Telecommunications. IEEE, January 2013, pp. 340-345.
4. E. D. Aguiar, C. Theobalt, S. Thrun, "Automatic Conversion of Mesh Animations into Skeleton-based Animations", Computer Graphics Forum, vol. 27, pp. 389-397, April 2008.
5. K. Arai, R. A. Asmara, "3D Skeleton model derived from Kinect Depth Sensor Camera and its application to walking style quality evaluations", International Journal of Advanced Research in Artificial Intelligence, vol. 2, pp. 24-28, July 2013.
6. H. T. Pham, J. J. Kim and Y. G. Won, "A Low Cost System for 3D Motion Analysis Using Microsoft Kinect", Innovation for Applied Science & Technology, Vol. 284, pp. 1996-2000, January 2013.
7. H. P. Corporation, "A Review on Technical and Clinical Impact of Microsoft Kinect on Physical Therapy and Rehabilitation", Journal of Medical Engineering, 2014, 2014.

8. B. Lange, C. Chang, E. Suma, et al. "Development and evaluation of low cost game-based balance rehabilitation tool using the Microsoft Kinect sensor", Annual International Conference of the IEEE Engineering in Medicine and Biology Society, Vol. 4, pp.1831 – 1834, 2011.
9. J. Wang, M. She, S. Nahavandi, et al. "A Review of Vision-Based Gait Recognition Methods for Human Identification". Digital Image Computing Techniques & Applications International Conference, pp. 320-327, 2010.
10. K. Arai, R. A. Asmara, "Human Gait Gender Classification in Spatial and Temporal Reasoning", Int J Adv Res Artif Intell, pp. 1-6, 2012.

The design and simulation of arbitrary system counter based on Proteus

L. Zhang[1] and Y.X. Wang[2†]
School of electronics and information engineering,
Lanzhou University of Arts and science, Lanzhou, China.
†Email: zhl_1965@163.com

This paper presents the design scheme of arbitrary system counter by the integrated counter chip, and gives the simulated results based on Proteus.

Keywords: counter; synchronization; simulation;

1. Introduction

Counter is a sequential circuit and most-used in digital system, which can be used not only to count the number of clock pulses, but also be used for timing, frequency division, and digital computing. Therefore, the design of the counter becomes a very important topic.

The counter can be designed based on the basic trigger, and the small and medium size integrated chip.

2. The Introduction of 74ls163 Chip

74ls163 is a kind of synchronous 4 -bit binary counter which possess the functions, such as zero clearing, setting number, counting and keeping, which is shown in Figure 1. [1]

[1] Ling Zhang
[2] Yong-Xiang Wang, corresponding author

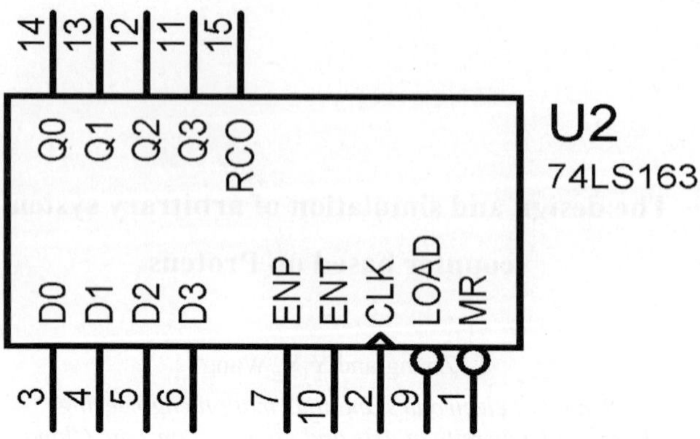

Fig. 1. 74LS163 pin diagram.

(1) The settings of 163 is synchronized, and can realize initialize when the preset control terminal ($\overline{LOAD}$) during the low period and the rising edge of clock (CLK);

(2) The zero clearing of 163 is synchronized, and can realize zero clearing operation only the zero clearing control terminal ($\overline{MR}$) during the low period and the rising edge of clock (CLK);

(3) The counts of 163 is synchronized, and which can realize four binary counting only the count signal (ENT, ENP) zero clearing control terminal ($\overline{MR}$) during the low period and the rising edge of clock (CLK);

(4) 163 possesses the counter -hold function. The counter remains unchanged and the clock (CLK) does not work when the count enable signal (ENT, ENP) exist zero at least one;

(5) 163 possesses the carry-ahead function. The RCO transition is low level when count to Q3Q2Q1Q0=1111, and ENT is equal to 1, and the carry terminal (RCO) generates a high level when the state returns to 0000.

3. The Counter Design of Count Modulo Value is Less Than 16

When the count value is not more than 16, we can achieve the counter design which takes advantage of the setting function of 163 and the clear function of 163. [2]

3.1. *Using the zero clearing method*

The Figure 2 is a 7 band counter with zero clearing method.

Because the 163 is synchronized to zero, so when the Q3Q2Q1Q0=0110 generates a clear signal.

The state is shown in Figure 3 .

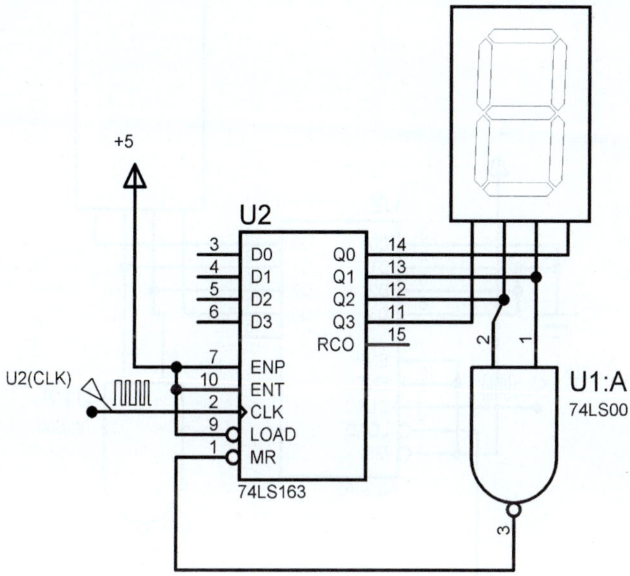

Fig. 2. 7 binary logic diagram of feedback reset method.

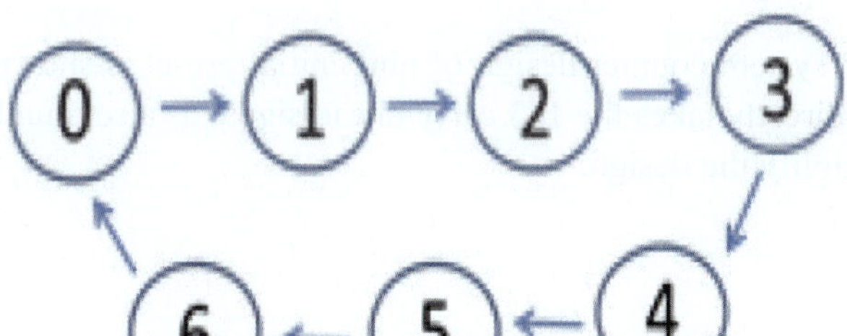

Fig. 3. 7 binary counter main cycle state diagram.

3.2. *Using the method of setting number of realization*

The initial preset can start from zero, and cannot from the zero when designed by Set number method.

(1) The 7 system counter design of zero initial preset is shown in Figure 4, which can generate preset signal $\overline{LOAD}$ when the Q3Q2Q1Q0=0110 since the 163 is a synchronous set number.

In order to match our habits, set the value we set to D3D2D1D0=0000.

Similarly the state is shown in Figure 3.

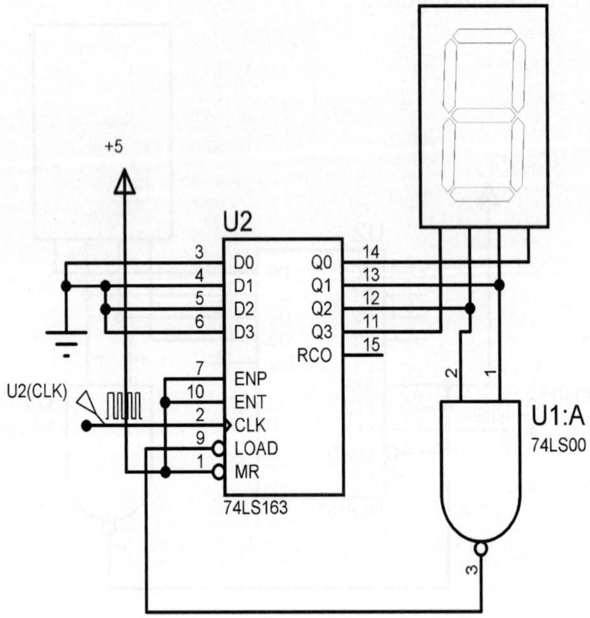

Fig. 4. 7 binary logic diagram of feedback design.

(2) The 7 system counter design of nine initial preset is shown in Figure 5. The program directly takes the 163 carry-ahead signal as a set number of signals, which can simplify the design.

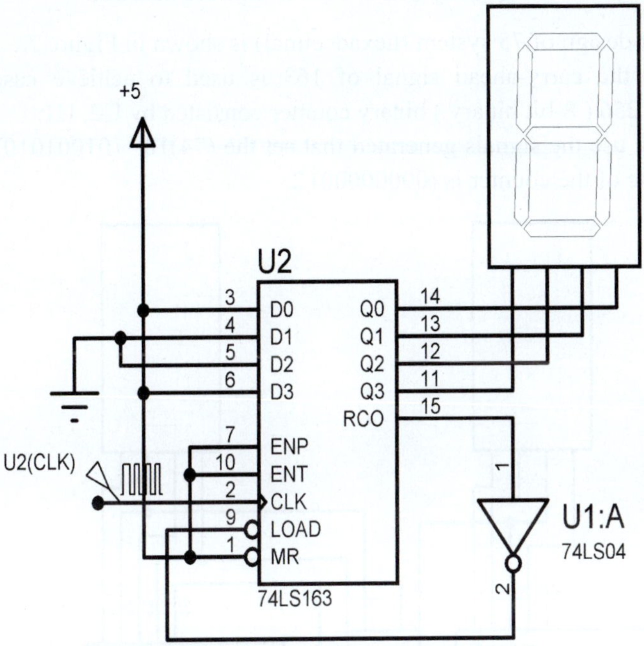

Fig. 5. A logic diagram of 7 binary counters (set 9) designed by feedback.

The state was shown in Figure 6.

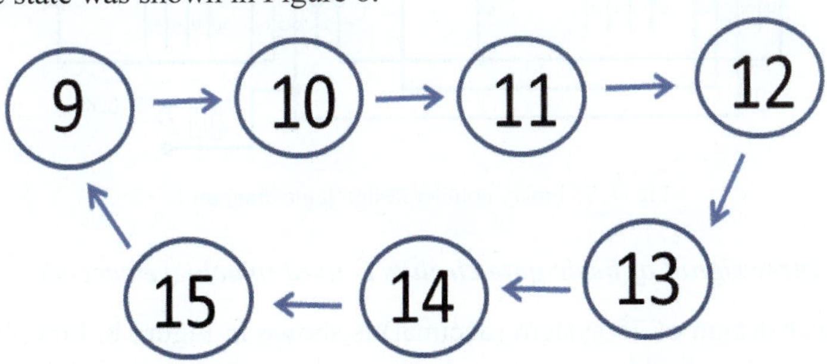

Fig. 6. 7 binary counter main loop state diagram (set 9).

4. The Counter Design of Count Modulo Value is More Than 16

We have to increase the use of cascading modulus value when the count value is greater than 16, thus completing the design. [3]

4.1. *The carry-ahead signal of 163 is used to achieve cascade*

The counter design of 75 system (hexadecimal) is shown in Figure 7 :

Firstly, the carry-ahead signal of 163 is used to achieve cascade, and achieve the 256 (8-bit binary) binary counter consisted by U2, U1.

Second, use the signals generated that set the (74)10= (01001010)2, so that the next state of the counter is (00000000) 2.

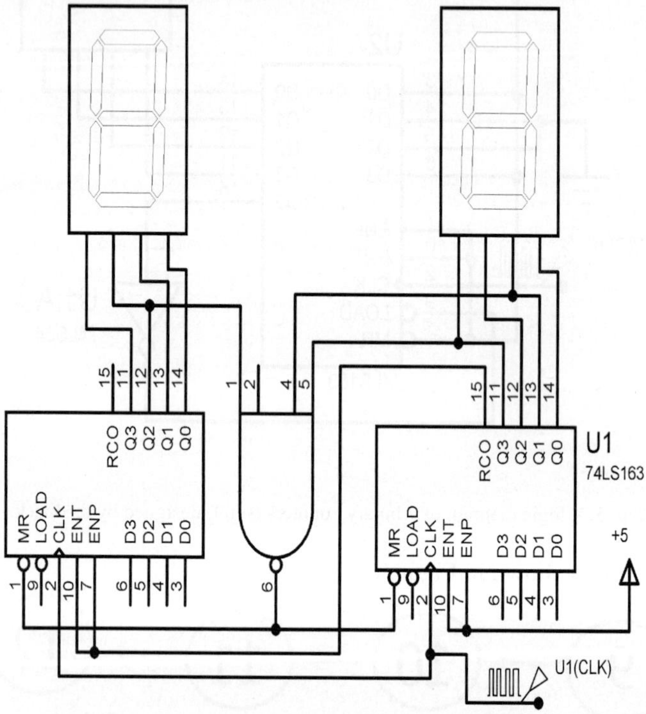

Fig. 7. 75 binary counter design logic diagram 1.

4.2. *The carry signal of basic gate circuits is used to achieve cascade*

The counter design of 75 system (decimal) is shown in Figure 8. First, by clearing and cascading let binary counter in the state of100 counts which composed of U2, U1, where U2 is set as ten, U1 as bits.

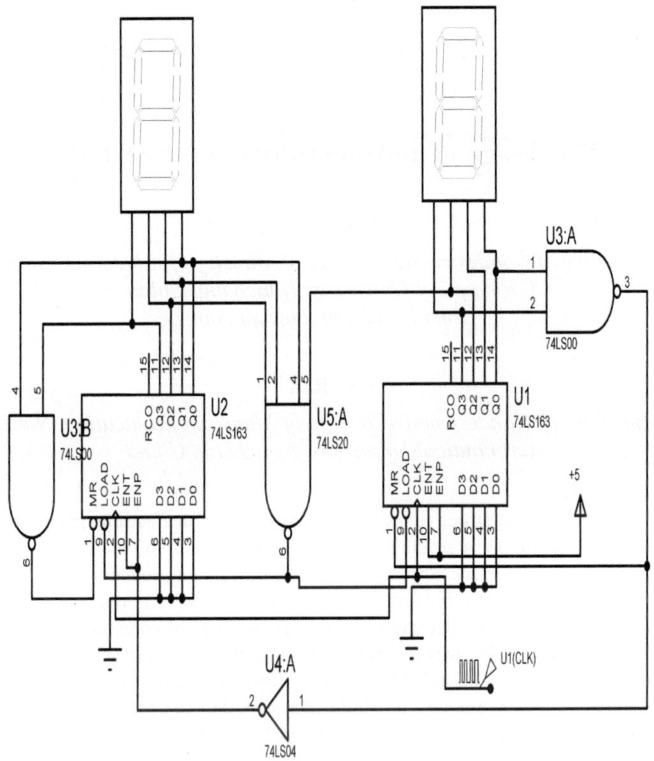

Fig. 8. 75 binary counter design logic diagram 2.

5. Conclusion

According to the simulation, the counter design is often relatively simple based on the 74ls163 and the hexadecimal display, besides, the cost is relatively low, but which is not in line with our daily habits.

References

1. Zhou Liangquan, Fang Xiangqiao. The fundamentals of digital electronic technology (the third edition) [M]. Beijing: Higher Education Press, 2008.
2. Bai Zhongying. Digital logic and digital systems (the fourth edition) [M]. Beijing: Science Press, 2007.
3. John F. Wakerly. The principles and practices of digital design [M]. Beijing: Machinery Industry Press, 2003.

The design of android software for ACFM

S.K. Ye[†]
Key Laboratory of Nondestructive Testing of Ministry of Education, Nanchang Aeronautical University, Nanchang, China
[†]*Email: 365246766@qq.com*

S.K. Ren
Key Laboratory of Nondestructive Testing of Ministry of Education, Nanchang Aeronautical University, Nanchang, China

ACFM is a kind of non-destructive testing technique based on electromagnetic induction principle, whose operation is so simple that can rapidly find surface and near surface defects. However, the interface software of ACFM instrument produced at home needed to run on a computer or laptop for further data analysis and waveform display, which was no convenient for special occasions, such as testing of pressure vessel, large entertainment equipment. In order to develop the advantages of the real-time detection of ACFM, this paper proposes an ACFM interface software based on Android, which can be loaded in portable devices running on Android system and can be operated through the touch screen .Just open Wifi to connect wireless probe can start testing.

Keywords: NDT; ACFM; Android; Wifi.

1. Introduction

With the prompt development of electronic technology, intelligent mobile phones had spread among the general public and the most of smart phones were loaded with Android system [1].On the other hand, ACFM is a new kind of active electromagnetic non-destructive testing technology, which can quickly qualify and quantify surface and near surface defects through observing the change of magnetic field in x, z direction and butterfly diagram [2], with the advantages of real-time detection [3]. However, ACFM instrument developed at home required a computer or laptop as a PC for data analysis and waveform display [4], which is no convenient to field operations, such as large pressure vessel, high entertainment equipment. The design of PC testing software based on Android, makes the smart phone be able to connect wireless probe by Wifi for testing, solved the limits of high detection and reduced greatly cost and power consumption. Furthermore, operating testing instrument through touch screen simplifies the testing operation so that detectors can find surface and near surface defects more quickly by ACFM.

2. Interface Layout Design of ACFM Software

The Android software interface is implemented by XML programming. To design a test interface, it is necessary to make sure what functions the interface need to implement firstly. In order to realize ACFM detection, this software included two interfaces, one is used to set the parameters of the probe, such as connecting wireless sensor, change the excitation frequency of probe. And another interface is used to establish the data transmission with wireless sensor, and display waveform on the interface detection for ACFM testing.

Firstly, design an interface for setting probe parameters, as shown in Fig. 1, each part is added in the XML file, such as background, button, text view, etc. Secondly, set a unique id for each component, which will be convenient for further programming, adding logic functions to components.

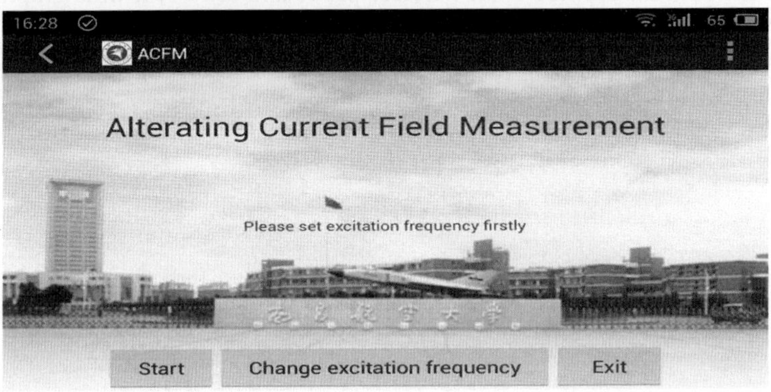

Fig. 1. Interface for probe parameter setting.

The Fig. 1 shows that the "Start" button is used to jump to waveform display interface from this interface. The "Change excitation frequency" button is used to set the excitation frequency of probe for different material such as aluminum, steel, etc. The "Exit" button is used to exit the ACFM software.

Then, due to click on the "Start" button would jump to waveform display interface, so it needs to design a waveform display interface as shown in Fig. 2. Add button, textview, linear layout, etc in the XML file. Especially, it is worth to note that the linear layout is to set size for further real-time graph display, which utilizes the Java library A Chart Engine to realize.

The Fig. 2 shows that "Connect probe" button is used to establish the data transmission with wireless sensor. The button "Stop" is used to stop data transmission with wireless sensor, which also cease to drawing real-time graph. The button "Back" is used to jump back to the interface for probe parameter setting.

3. Programming for ACFM Software

The general steps of ACFM is that: firstly set the excitation frequency of probe according to the measured piece, and then scan to make sure whether defects exist, finally quantify and qualify defects by observing Bx, Bz, Butterfly graph. So, ACFM software should run as the general steps, the specific steps is as follow: Firstly open Wifi to connect the wireless probe named "USR-WIFI232-T", and then open the ACFM software, and click the button "Change excitation frequency" to set the excitation frequency of probe according to the measured piece. Especially, the default frequency is 5kHz, which is used to steel 45. After setting frequency, click the button "Start" to jump into the waveform display interface. And then click the button "Connect probe" to establish data transmission with wireless probe. Finally, the interfaces receive the data transmitted from wireless probe and draw real-time graph.

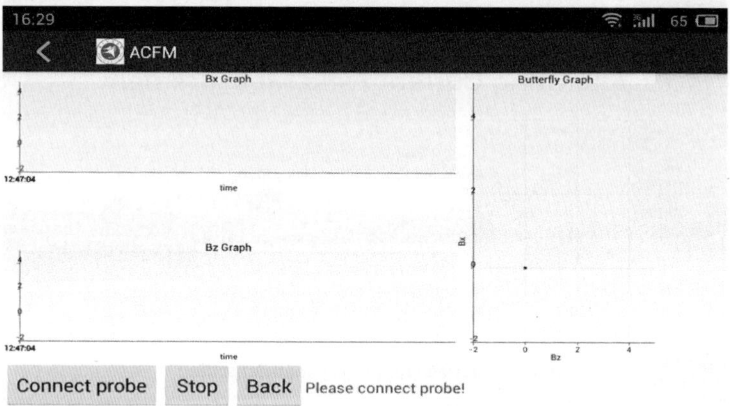

Fig. 2. Interface for waveform display.

According the running steps of ACFM software, add specific functions for the parts of the interfaces by programming, after having designed ACFM interfaces. Since android development is based on Java language, realizing the function of the specific parts need to quote Java library. For example, to achieve the function of button, it need to set a listener for each button which has its unique id. The key program is as follow:

```
    Startbutton=(button)this. Find View by Id
    (R.id. start Button);          start Button. Set on Click Listener (new View.
On Click Listener ()
    {
    @Override
    public void on Click(View v)
    {
    finish();
```

Intent intent = new Intent (); intent. SetClass (MainActivity. this, RtChartsActivity. class);

startActivity (intent);

}

});

This program firstly set up a specific button object for startButton by indexing id in R.java. And the set a listener to this object, which can realize specific functions by overwriting the function onClick(). In the similar way, the other parts of interface also need to find its unique id, and set up the specific object. And then program object to add functions for each part by quoting Java library.

In the next, to introduce how to connect to a wireless probe, and set up data transmission to draw real-time graph. Establish the data transmission with wireless probe need to use the Socket programming. Due to the probe's address of the wireless local area network (LAN) is 10.10.100.254 and the data transmission port is 8899, instantiate a Socket object and receive data from this Socket object through Java class DataStream. The key program is as follow:

```
SocketAddress socketAddress = new InetSocketAddress("10.10.100.254",
8899);
Socket socket = new Socket();
try {
DataInputStream dis = (DataInputStream)socket.getInputStream();
... ...
}
if(len == 9)
{
Bx = temp[3]*256+temp[4];
Bz = temp[6]*256+temp[5];
}
}
} catch (IOException e) {
}
```

This program is used to establish the data transmission with wireless probe, and each frame data is stored in the Java class Datastream after converting into the values of Bx, Bz voltages, which means converting binary numbers into decimal numbers. And then store these numbers into the global variables Bx, Bz, which will used for later graph drawing.

Next, set a timer, which would collect data of global variables Bx, Bz every 5ms. Then, put these data into Time Series for Bx, Bz graph and XY Series for Butterfly graph. Finally, use Java class AChartEngine to draw real-time graph and to refresh it every 2s to realize real-time display. The key program is as follow:

```
handler = new Handler()
{
@Override
public void handleMessage(
Message msg)
{
updateChart(Bx, series_Bx, dataset_Bx, chart_Bx);
updateChart(Bz,        series_Bz,        dataset_Bz,        chart_Bz);
updateChart(Bx,Bz,series_Butterfly,
dataset_Butterfly,chart_Butterfly);        super.handleMessage(msg);
}
};
task = new TimerTask()
{
@Override
public void run()
{
Message message = new Message();
message.what = 200;
handler. SendMessage(message);
}
};
timer.schedule (task, 2*1000,1000);
```

This program is used to set a timer task, which would refresh the graph at regular time, realizing the real-time graph drawing. Inspectors who observe the Bx, Bz and Butterfly graph, can quickly scan defects, then quantify and qualify the defects scanned.

4. Running of ACFM Software

At last, compile this program with Eclipse, which would produce package named ACFM.apk. Then, copy this package and install it into portable devices loaded on Android 4.0. Run this software and open Wifi to connect the wireless probe USR-WIFI232-T. The result of running ACFM software is as Fig. 3.

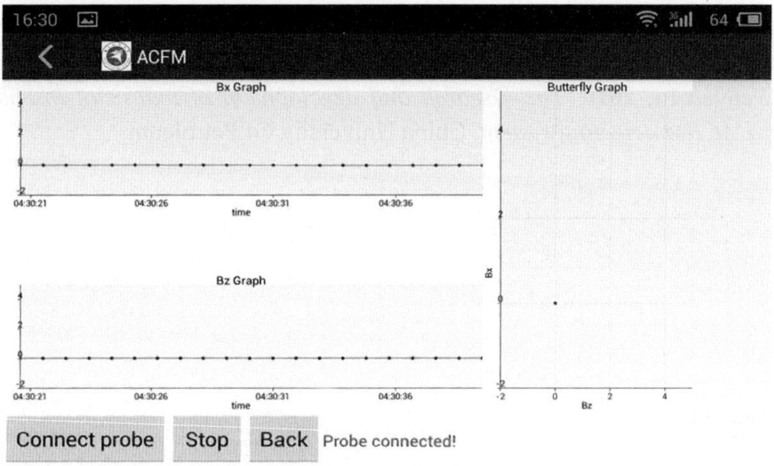

Fig. 3. The running of ACFM software.

The Fig. 3 shows that the ACFM software based on Android can receive data from wireless probe and draw real-time graph in the same time. In addition, this real-time display which is programmed by A Chart Engine can be enlarged or reduced through the touch screen, convenient for defects quantity and quality.

5. Conclusions

This paper design the ACFM software based on the Android system, and successfully installed it on the Android device, realizing the embedded solution that use the wireless intelligent portable device instead of the computer as the upper machine. Experiments shows that: this testing software has a good performance, occupying less capacity, and using small memory. It can be used stably in the ACFM detecting system. With use of supporting wireless probe, wireless transmission will be more conducive to testing on special occasions, such as large pressure vessels and high entertainment equipment. In the meanwhile, make instrument becomes more compact and smart. Use touch screen to operate instrument, simplifying the testing operation, developing better the advantages of ACFM in real-time detection.

References

1. Xiangxiang Wu. 2011. *The research and application of the development for Android software*. Beijing University of Posts and Telecommunications.
2. Yanli Zhao. 2009. *The research of signal processing and application for ACFM.* China University Of Petroleum.
3. Shangkun Ren, Zhibin Zhu, Tianhua Lin, Kai Song, Ren Jilin. 2009 2nd *Design for the ACFM Sensor and the Signal Processing Based on Wavelet*

Denoise. International Congress on Image and Signal Processing (CISP'09), Pressed by IEEE.

4. Wenyan Li. 2010. *The research and development of industrial instrument for ACFM detecting system*. China University Of Petroleum.

One kind of automatic compensation algorithm for Web services transactions

H. S. Deng[1†]
Xi Jing University, Xi'an, China 710123, China
[†]Email: xaut0420@126.com

J. H. Li
School of Computer Science & Engineering,
Xi'an University of Technology,
Xi'an 710048, China

In order to solve the Web Service Transaction Management, the author proposed and realized a kind of compensation-based transaction mechanism, which developed the Transaction engine. It takes advantage of pre-compensation and WS-Transaction specification s' two -phase commitment to parse the database operation statements participant transaction, to do a scene preservation possible changes in the database, and generating the statement set for compensation automatically. When need to cancel the transaction, the management center will rollback automatically according to the client vote and compensate information, which can better solve the problem of Web Service transactions.

Keywords: Web Service; Transaction; Compensation.

1. Introduction

Transaction originated from the DBMS, is key technology to ensure that concurrent access to shared data consistency and correctness failover. X/Open Distributed Transaction Processing Reference Model (Distributed T ransaction P rocessing, DTP), OMG CORBA-based Object Transaction Service (Object Transaction Service, OTS) and so on are widely used in distributed transaction field. [1] Web service has its own characteristics, such as long transaction, autonomy, loose coupling, more potential failure, Web services may span multiple organizations, the previous transaction processing method has been shown to apply to this based on Web applications and services. [2][3] Therefore, Web services have their own unique transaction requires the coordination and implementation mechanism.

[1] Hai-Sheng Deng, male, Master, Associate Professor, the main research directions for mobile computing technology and data mining

Currently has a lot of work in progress, WS-Coordination and WS-transaction is IBM, Microsoft and BEA company's proposed Web services transaction processing model, OASIS WS-CAF model is presented and the Business Transaction Protocol (BTP). [4] [5]

2. Based on the Previous Compensation Transaction Processing Model

The proposed WSTMBC (Web Service Transaction Model Based on Compensation) logical structure shown in Figure 1, not only including Client, Serve and Transaction management center, but also including log management component in order to track the details of the transaction.

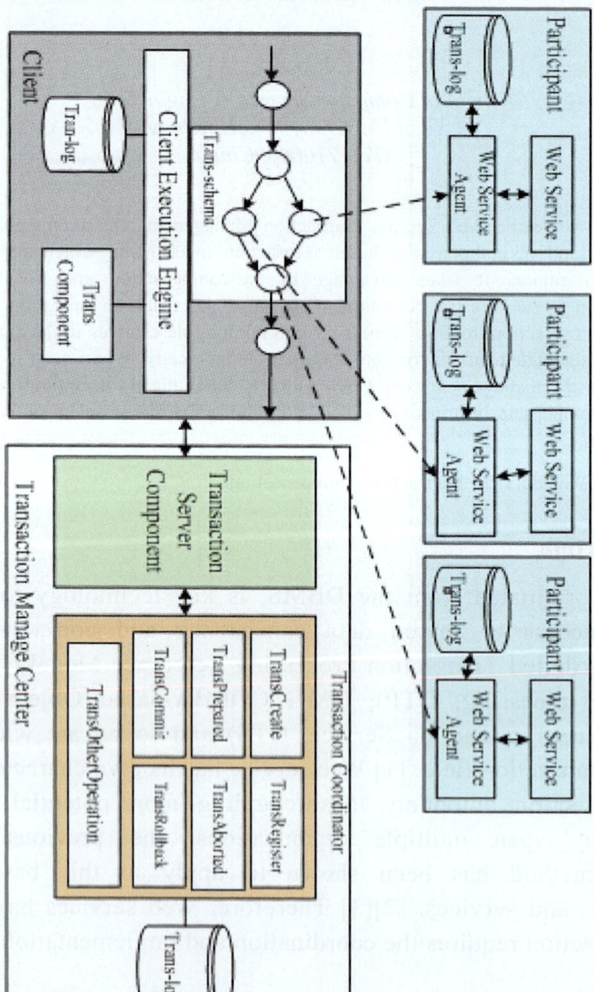

Fig. 1. WEB services transaction processing model.

3. Key Technologies

WEB transaction management services primarily through transaction management client components, server-side component and transaction management center, during a transaction together to ensure data consistency. [6] [7]

3.1. *Achieve client components*

Follows are main functions of Transaction Managerclient component.

(1) Transaction generation and initialization: client component by creating a stack to save the future of Web services transactions require registration information.

(2) Exchange transaction information with Web service client: The client and server-side interactions between the transaction information via a Soap Head Class class to exchange transaction information (Is Transaction, server IP, Port, subTranID), which indicates whether a transaction Is Transaction part of an assignment by the client; The remaining three properties are assigned in the Web service side, ServerIP and Port Management Centre denote the IP address and port number, and subTranID This transaction represents a Web service method sub transaction identifier, in the second transaction processing stage can be used to distinguish the various affairs of Web services transaction information.

(3) Enrollment Web Services Transaction Information: When performing sub-transactional Web service information returns, while returns there Communicate information, to obtain the appropriate information and pressed into (a) in new stack. Then the next sub-transaction until all child transaction completes, all obtained from each sub-transaction transaction information is also pressure to the stack. This will smooth the various Web services in order to save transaction information (Server IP, Port, SubTranID).

(4) Send voting information: Such is the client component of important functions.

3.2. *Implement server components*

Follows are main functions of Transaction Management Server:

3.2.1. *Initialization interactive information services*

Define the same client exchange transaction information SoapHead class, if the client calls Web services include transaction information, generate transaction ID, by calculating a SoapHead class assignment.

3.2.2. *Compensation algorithm*

In a distributed system, because the transaction takes too long to lock concurrent access to database resources is not conducive to the case of compensation compensation algorithm using forward thinking to ensure data integrity. Rationale: if the current database operation statement SQL-A. In SQL-A really executed, based on SQL-A itself provides information, database connection information and the information available to the database analysis to extract, according to some algorithm to construct a sequence of database operations queue to Sequence-A. When a transaction fails need to revoke the SQL-A, press Sequence-A sequence of data manipulation, you can perform on the SQL-A data.

3.2.3. *Web service side local rollback algorithm*

Here is the corresponding transaction rollback algorithms:

```
public void rollBackTran ()
{ if (this._IsTransaction == true)
{while (this._TransactStack.Count> this._preaction)
{this._TransactStack.Pop ();}
while (this._TransactStac k.Count! = 0)
{SqlHelper.ExecuteNonQuery (this._connectionstring, CommandType.Text,
(string) this. Transact Stack. Pop ());
}
}
}
```

3.2.4. *Generating and transmitting transaction information*

Affairs after the successful implementation of Web services, to send information to appropriate compensation specified transaction management center for secondary treatment after the vote. According to the above compensation algorithm, the compensation information is mainly connected with SQL statements, as well as sub-transaction ID stored in the stack set of SQL statements. We define Trans act Message to hold this information(Transact ID, Connection Straing Message Body).

3.3. *Transaction management center main algorithms and implementation*

Transaction Management Center main functions are as follows: Loop to obtain transaction compensation information; listening and receiving transactions signal; right to commit or rollback transaction; view outstanding transaction compensation information. [8,9] The following are its main algorithm described.

3.3.1. *Obtain information about transaction compensation algorithm*

Receive asynchronous message queue data cycle algorithm using activity diagrams as follows:

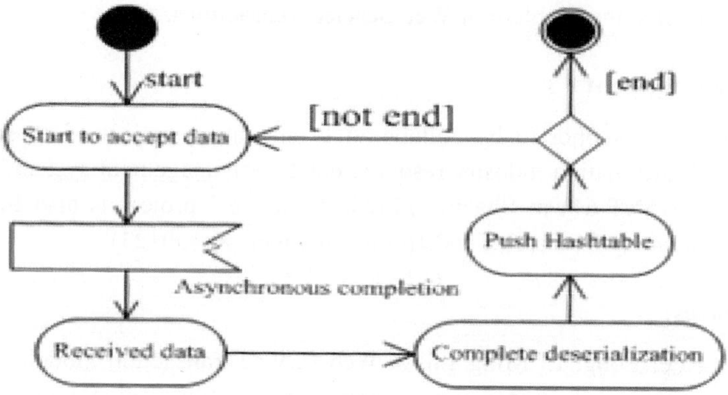

Fig. 2. Activity diagram of receiving transaction compensation information.

3.3.2. *Monitor Transaction Compensation Information*

Meanwhile, in order to monitor transaction management compensation data centers, the program will use the TreeView format transaction data represented compensation information. Specific algorithms are as follows:
Populate Tree Control (System. Xml. Xml Node document, System. Windows Forms, TreeNode Collection nodes)

```
{foreach (System.Xml.XmlNode node in document.ChildNodes)
{
string text = (node.Value != null ? node.Value :
(node.Attributes!=null && node. Attributes. Count > 0) ?
node.Attributes[0].Value : node.Name);
TreeNode new_child = new TreeNode(text);
nodes.Add(new_child);
populateTreeControl(node, new_child.Nodes);
}
}
```

4. Finally

This paper proposed and realized a kind of compensation-based transaction mechanism, which developed the Transaction engine based on pre-compensation and WS-Transaction specifications' second submit. The mechanism includes client, server-side, and transaction management center. Main fuctions are following: Parsing the database operation statements participant transaction, to do a scene preservation possible changes in the database, and generating the statement set for compensation automatically. When need to cancel the transaction, the management center will rollback automatically according to the client vote and compensate information, which can better solve the problem of Web Service transactions.

Acknowledgement

This project is supported by collaborative innovation project of “the Shaanxi provincial information industry resource database management system” (project number: 2015XT-61) in Shaanxi province. And the project is also funded by Xijing University Research Funds(project number: XJ150123).

References

1. Tang Fei Long, Li Ming Lu, A Web service transaction model: structure, algorithm, and transaction compensation, Electronic Journal, 2003, 31(12A):2074-2078.
2. Guo Yu Bing, Wang Xin Zhong, Design and implementation of SOA transaction system, Computer application research, 2007, 24(11).
3. Gong Song Jie, Research on Web service transaction model, Computer Engineering, 2007, 33(10).
4. Vidyasankar K, Vossen Gottfried. Multi-Level Modeling of Web Service Compositions with Transactional Properties [J]. Journal of database management. 2011, 22 (2) :1-31.
5. JE Haddad, et al. TQoS: Transactional and QoS-aware selection algorithm for automatic Web service composition [J]. IEEE Transaction on Service Computing, vol. 2010, 2010.
6. Hahn, K. Adaptive workflow management to ensure transactional service composition [J]. Digital Information Management (ICDIM). 2010:373-378.
7. Nan Cheng, Meina Song, Qian Wang. A Web service Process Transaction Framework Based on Compensation and Proxy [J]. Pervasive Computing (JCPC), 2010:369-372.
8. Xu Feng, Xu Biyun. Web Services Transaction Compensation Mechanism and Implementation [J]. "Computer Science" 2006, 33 (7) :242-244.

A hypercomplex form of polarimetric SAR images for multi-channel joint matching

D. Qu†

The 28th Research Institute of China Electronics Technology Group Corporation, Nanjing 21007, China

†Email: dr_qudi@163.com

W.T. Ma

Science and Technology on Information System Engineering Lab, The 28th Research Institute of China Electronics Technology Group Corporation, Nanjing 21007, China

Most of current multi-channel image matching methods carry out the matching process actually only on single channel images that fused from multi-channel images beforehand, instead of multi-channel joint matching. In this paper, a real multi-channel joint matching method of polarimetric synthetic aperture radar (SAR) images based on hypercomplex number is proposed. Compared with the previous methods, the method makes more use of the polarization information during matching process, and there are two novelties. First, a hypercomplex number form of polarimetric SAR image is proposed. Second, the hypercomplex phase correlation algorithm is induced to realize the multi-channel joint matching of polarimetric SAR images. Experiment using measured polarimeric SAR images at multi-band demonstrated that the proposed joint matching method can achieve the precise matching and the matching probability can reach 97.92%.

Keywords: Hypercomplex number; Multi-channel joint matching; Phase correlation; Polarimentric SAR image.

1. Introduction

Recently decades, along with the great development of polarimetic synthetic aperture radar (SAR), such as the RADARSAT-2, TerraSAR-X, ALOS-2, as well as the deep researches of polarization theory [1] (Wang et al. 2012) [2] (Suri et al. 2010), polarimetric SAR plays a more and more important role in many military and civil applications, such as image matching [3] (Ma et al. 2014), target detection [4] (Ma et al. 2011), image fusion [5] (Alparone et al. 2004), change detection, aircraft location and navigation [6] (Gao et al. 2009) and so on. Therein, polarimetric SAR image matching is a crucial step for many aplications, such as image fusion, change detection, aircraft location and so on.

Each polarimetric SAR image contains four different polarization-channels of SAR images. Compared with single polarization SAR images, they contain distinctive polarization characteristics of targets, reflecting the scattering structure of the targets [7] (Cloude et al. 1996) [8] (Cloude et al. 1997). The researches on SAR image matching have continued for more than twenty years. By contrast, the rsearches on polarmetric SAR image matching are much less. The common idea of polarimetric SAR image matching is firstly fusing the polarimetric SAR images to single-channel images using certain regular, such as SPAN [9] (Chureesampant et al. 2014), PWF, GOPCE and so on, simultaneously improving the signal-to-noise or definition, and then carrying out the matching process on the fused images using the matching methods of single polarization SAR images. However, these methods make use of the distinctive polarization characteristics only during the fused process, instead of the matching process.

In this paper, a multi-channel joint matching method of polarimetric SAR images based on hypercomplex number is proposed. The method carries out the matching at the same time of fusing, making more use of the polarimetric characteristics during the matching process. There are two novelties. First, a hypercomplex number form of polarimetric SAR images is proposed. Second, the hypercomplex phase correlation is induced to realize the muti-channel joint matching of polarimetric SAR images. In addition, the method can be only applied when there are only translations between two polarimetric SAR images.

2. Multi-Channel Jointly Matching of Polarimetric Sar Images

2.1. *Polarimetric scattering matrix*

The single-look polarimetric SAR data is usually represented by a Sinclair matrix as Equation 1:

$$\mathbf{S} = \begin{bmatrix} S_{\mathrm{HH}} & S_{\mathrm{HV}} \\ S_{\mathrm{VH}} & S_{\mathrm{VV}} \end{bmatrix} \tag{1}$$

where H and V represent the horizontal and vertical polarization respectively, and S_{ij} is the backscattering coefficient of target when the polarization states of transmitter and receiver are i and j respectively.

The multi-look polarimetric SAR data is usually represented by a covariance matrix as Equation 2:

$$\mathbf{C} = \frac{1}{L}\sum_{i=1}^{L} \boldsymbol{x}_i \boldsymbol{x}_i^H \tag{2}$$

where L is the number of looks.

$$\boldsymbol{x} = \begin{bmatrix} S_{\mathrm{HH}} \\ \sqrt{2}S_{\mathrm{HV}} \\ S_{\mathrm{VV}} \end{bmatrix} \tag{3}$$

where $S_{\mathrm{HV}} = S_{\mathrm{VH}}$ if reciprocity holds.

2.2. *Multi-channel joint matching of polarimetric SAR images*

After the hypercomplex forms of polarimetric SAR images are obtained, the multi-channel joint matching is realized using the hypercomplex phase correlation algorithm. Assuming that $f(m,n)$ and $g(m,n)$ are the hypercomplex forms of the two polarimetric SAR images to be matched, the phase correlation $ph(m,n)$ can be calculated using the following equation.

$$ph(m,n) = \mathbf{F}^{-R}\left(\frac{R_R(v,u)}{\left|R_R(v,u)\right|}\right) \tag{4}$$

Where m and n are the row and column coordinates, and $\mathbf{F}^{-R}$ represents the hypercomplex inverse Fourier transformation, and $R_R(v,u)$ is the hypercomplex cross-power spectrum calculated using the following equation:

$$R_R(v,u) = \overline{F_L(v,u)}G_{R\parallel}(v,u) + \overline{F_{-L}(v,u)}G_{R\perp}(v,u)$$

where — is conjugate operation. $F_L(v,u)$ and $G_R(v,u)$ are the left and right hypercomplex Fourier transformation of $f(m,n)$ and $g(m,n)$ respectively. $G_{R\parallel}(v,u)$ and $G_{R\perp}(v,u)$ represent the parallel and vertical parts of $G_R(v,u)$ to the hypercomplex Fourier transformation axis.

Finally, the row and column translations between the two images are the row and column coordinates where $ph(m,n)$ is largest. It needs to declare that the method can only deal with the matching that there are only translations between two images.

2.3. *Evaluation of matching results*

If the transformation between two images is known, the matching error is usually used to evaluate the matching result. The matching error is defined as the difference between the real and calculated transformations.

Another intuitionistic evaluation approach is the perspective image [10] (Liang et al. 2014) of the two images using the calculated transformation. In the perspective image, the red and green channels are arranged using the power of the two images respectively, and their positions depend on the calculated translation. The more clear the textures and lines in the perspective image over the overlap areas are, the more precise the matching is. A bad matching behaves as blurred and disordered textures and lines. In addition, under a precise matching, if the powers of the red and green images over the overlap area are nearly same, the areas will present yellow.

3. Experiment

3.1. *Data*

The polarimetric SAR images to be matched in the experiments were all imaged over Taiwan by AirSAR on September 27th. The range and azimuth resolutions are 6.662 m and 12.1 m respectively. In each matching experiment, the imaging conditions of the two images, including the imaging time, incident angle, flight aspect, resolution are same, except that one of them was imaged at L band with 1024×1024 pixel size, and another was imaged at C band with 256×600 pixel size.

3.2. *Match of polarimetric SAR images at multi-band*

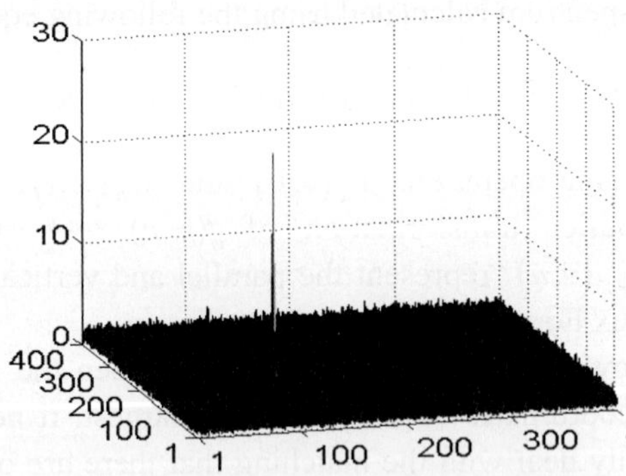

(a) Tridimensional phase correlation surface.

Fig. 1. Phase correlation surface and matching perspective image.

(b) Perspective image.
Fig. 2. (Cont'd)

Table 1. Matching error.

Real translation	Calculated translation	Matching error
(100,100)	(100,100)	(0,0)

Figure 1(a) shows the tridimensional phase correlation surface of the two images, where the axis of the surface are the row and column translations of the two images respectively, and the vertical axis is the value of phase correlation. From the figure, we can see that the points with larger values of phase correlation are concentrated distributed. The correlation peak is very steep, and the sidelobe is very small. Based on the conclusion in the reference [11] (Yanyan et al. 2010) that the matching result will be more precise when the peak of the surface of similarity measure is steeper, it can be inferred that the matching result in figure 1(a) is very precise depending its steep peak.

Figure 1(b) is the perspective image of the two images, and Table 1 shows the matching error. The matching errors of row and column are both zero, presenting very high precise. In Figure 1(b), the overlap areas are very distinct, including the edges of lake and the texture of terrain. And most of the overlap areas present yellow, except some pixels due to their different power caused by different bands. It shows that the two images are precisely matched using the proposed method.

3.3. *Match of a large number of image pairs*

The proposed method is also carried out on 96 pairs of images at multi-band. The imaging conditions of the images are presented in 3.1. The matching errors of row and column are draw using lines in Figure 2. It can be seen that except the 36th and 71th pairs (Figure 3 shows their perspective images behaving as blurred and disordered textures and lines) other 94 image pairs are precisely matched. The matching probability can reach 97.92%.

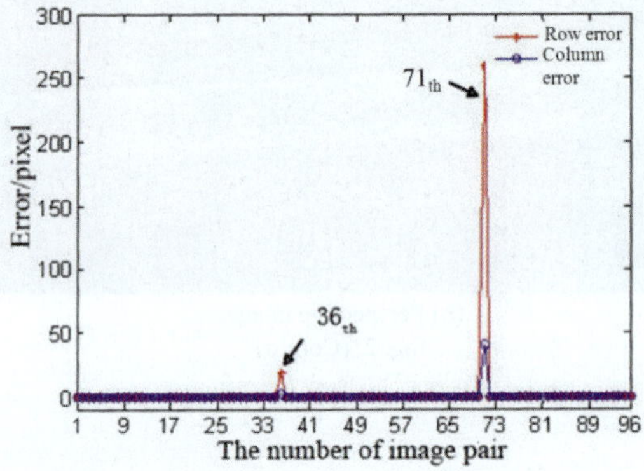

Fig. 2. Matching errors of 96 matching results.

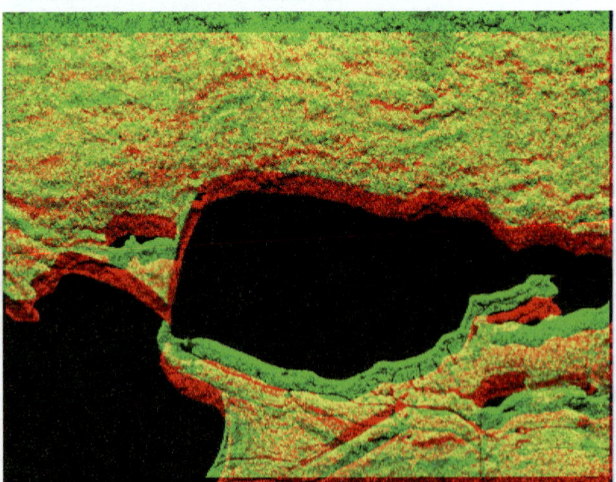
(a) 36th matching perspective image.
Fig. 3. The perspective images of failed matching.

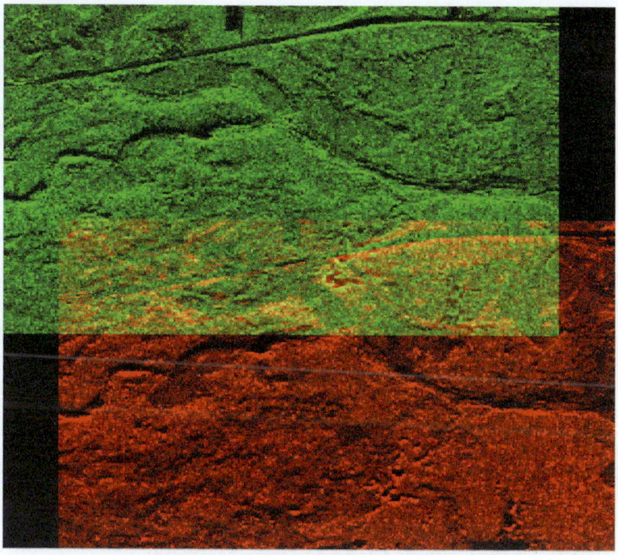

(b) 71th matching perspective image.
Fig. 3. (Cont'd)

4. Conclusion

Polarimetric SAR image matching plays a more and more important role in many military and civil applications. However, most of current multi-channel image matching methods carry out the matching process actually only on single channel images that fused from multi-channel images beforehand, instead of multi-channel joint matching. These methods cannot make full use of the polarization characteristics to improve the matching effectiveness. In this paper, a real multi-channel joint matching method based on hypercomplex number is proposed. In the method, a hypercomplex number form of polarimetric SAR images is proposed, and the hypercomplex phase correlation algorithm is induced to realize the muti-channel joint matching of polarimetric SAR images. Experiment using measured polarimeric SAR images at multi-band demonstrated that the proposed method can achieve the precise matching and the matching probability can reach 97.92%.

References

1. Wang S., You H., Fu K. 2012. BFSIFT: A novel method to find feature matches for SAR image registration. *IEEE Geoscience and Remote Sensing Letter* 9(4): 649–653.

2. Suri S., Schwind P., Uhl J., et al. 2010. Modifications in the SIFT operator for effective SAR image matching. *International Journal of Remote Sensing* 1(3): 243-256.
3. Ma W., Yang J., Gao W., 2014. Automatic GCP extraction in mountainous areas using DEM and PolSAR data. *IEEE Geoscience and Remote Sensing Letter* 11(12): 2075–2079.
4. Ma W., Yang J. 2011. Target detection algorithm for polarimetric SAR images using GOPCE. *Proceedings of 2011 IEEE CIE International Conference on Radar*, Chengdu, China: IEEE Press 1:507-509.
5. Alparone L., Baronti S., Garzelli A., et al. 2004. Landsat ETM+ and SAR image fusion based on generalized intensity modulation. *IEEE Transactions on Geoscience and Remote Sensing* 42(12): 2832–2839.
6. Gao S., Zhong Y., Zhang X., et al. 2009. Multi-sensor optimal data fusion for INS/GPS/SAR integrated navigation system. *Aerospace Science and Technology* 13: 232-237.
7. Cloude S. R., Pottier E., 1996. A review of target decomposition theorems in radar polarimery. *IEEE Transaction on Geoscience and Remote Sensing* 34(2): 498-518.
8. Cloude S. R., Pottier E., 1997. An entropy based classification scheme for land applications of polarimetric SAR. *IEEE Transaction on Geoscience and Remote Sensing* 35(1): 68-78.
9. Chureesampant K., Susaki J., 2014. Automatic GCP extraction of fully polarimetric SAR images. *IEEE Transaction on Geoscience and Remote Sensing* 52(1): 137-148.
10. Liang J., Liu X., Huang K., et al. 2014. Automatic Registration of Multisensor Images Using an Integrated Spatial and Mutual Information (SMI) Metric. *IEEE Transactions on Geoscience and Remote Sensing* 52(1): 603-615.
11. Xuan Y., Huang Y., Yang J., 2010. New approach for improving Mutual Information in POLSAR image registration. *Proceedings of 10th IEEE International Conference on Signal Processing*: 853-856.

Detection of genetically modified components of rice using a membrane-based array method

S. S. Zhang[1*] and L. C. Qiu[2†]

Shengyang Agricultural University, #120, Dongling Rd., Shenyang, 110866, China
**Email: qlccn@126.com, †390890513@qq.com*
Tel/ Fax: +86 24 88487116

G.P. Huang

Sichuan Hua Hansan Bio Technology Co., Ltd. #39, West of fucheng Rd., Chengdu, 610041, China

Y.M. Gao

Beijing Genestone Cooporation, #8, Xueqing Rd., Beijing, 100085, China

Z.Y. Yun

Sub-institute of Food and Agriculture Standardization in China, National Institute of Standardization. #4 Zhichun Rd., Beijing, 100088, China

Y. Liu

Institute of Plant Protection, Sichuan Academy of Agricultural Sciences, #20, JingJuSi Rd., Chengdu, 610066, China

A method for the detection of genetically modified components (GMCs) is reported for detection of GMCs in rice plants and foods using a membrane-based array. A DNA sequence as probes was immobilized on modified paper film-chips to form the sensing array surface. The target DNA from sample was amplified by polymerase chain reaction (PCR) using labeled biotin primers. The PCR products were added to the sensing surface using an automatic membrane-based array instrument (MAI). When the target DNA passed the surface, it was captured by the complementary probes. The signal was visualized with a streptavidin/alkaline phosphatase (AP) and AP color development kit. Compared to manual dot blotting, the MAI could automatically and rapidly finish several GMC types of rice detection, such as BT11, TC1507, Mon88017, and Mon89034. Stable

[1] Sheng-Sheng Zhang
[2] Li-Chun Qiu

and consistent qualitative results were acquired. The sensitivity reached 0.1%. Thus, the automatic MAI might be a valuable GMC detection tool.

Keywords: Genetically modified components; Rice; Dot blot; Detection; Membrane-based array.

1. Introduction

1.1. *Development of GMCs*

According to the International Service for the Acquisition of Agribiotech Applications, GMC crops have been commercially cultivated for over 17 years, and the global planting area has increased 100-fold [1-3]. Along with the rapid development of GMCs and increasing numbers in food, feed, and the environment, GMCs have come under suspicion from governments and citizens because of potential safety risks [4]. Public concern about GMC food safety and potential ecological contamination is becoming widespread [5]. Therefore, strengthening the detection of GMCs and their derivatives is particularly important and urgent.

1.2. *GMC detection methods*

GMC detection methods mainly include nucleic acid-based and protein-based techniques. Due to their high specificity, sensitivity, operability and repeatability, the majority of the previously developed methods for GMC detectio focus on detecting transgenic DNA. The use of PCR, real-time PCR, and loop-mediated isothermal amplification (LAMP) has grown rapidly, and some of these are currently used for the detection of GMCs in food or feed [6-7]. As the quantity of GMCs increased, multiplex GMC detection methods have been developed, such as multiplex PCR, microarrays, suspension array analysis, microdroplet-based PCR, reverse dot blot hybridization (RDBH), and others [8-11]. However, the throughput of these methods is generally low, and only a limited number of target genes can be detected and differentiated in each assay [2].

In comparison, multiplex PCR enables the simultaneous amplification of many target genes in one reaction, thus increasing the assay throughput and allowing more efficient use of each DNA sample. This technology has been successfully applied for the detection of GMCs in food or feed and the identification of pathogenic microorganisms [12]. Therefore, multiplex PCR was used as a reference method in our experiments.

1.3. *Reverse dot blot (RDB) method*

The reverse dot blot (RDB) method was originally developed in the medical field during the 1990s. Capture probes hybridize with homologous amplified target DNA labeled with a fluorescent dye or a chemiluminescence dye, which can be detected by the naked eye or a laser scanner. Although RDB has a lower throughput than a microarray, a 96-well microtiter plate-sized membrane can contain >1000 different hybridization probes, and each array can be washed and reused many times [2,13]. This capacity is much greater than other DNA detection methods, such as PCR and LAMP. Due to its relatively low cost and high throughput, the RDB assay has been applied and adapted for the detection or identification of pathogenic and environmental microorganisms from various sources, including food, animals, and plants [14,15]. However, RDB mainly relies on manual operation. Due to the limited stability and repeatability of RDB methods, RDB is unsuitable for routine analysis of GMC events in imported and exported goods [3]. To solve these issues, we attempted to design and manufacture an automatic membrane-based array dot blot instrument employing the same detection model as DNA arrays for simultaneous detection of GMCs in the current study.

2. Experiment

2.1. *Primers and probes*

Primers were used in PCR to amplify target genes. The primers were designed using Primer 5.0 (Table 1). Probes were used in membrane-based array reverse dot blot (Table 2). Primers and probes can be found in the State Standard of the People's Republic of China (GB/T 31730-2015)[16].

Table 1. Primers used in PCR reaction.

Target gene	Primer name	Sequence	Segment length
BAR	Forward	5'-CACCTGCTGAAGTCCCTGGA-3'	193 bp
	Reverse	5'-biotin-GTACCGGCAGGCTGAAGTCCA-3'	
Cry1Ab	Forward	5'-GACATGAACAGCGCCTTGAC-3'	164 bp
	Reverse	5'-biotin-TTGATGGTTGCAGCATCGAA-3'	
Cry1Ac	Forward	5'-GTTAGACTCAACAGCAGTGGA-3'	161 bp
	Reverse	5'-biotin-GAGAAGATGGATGAATTACCCCA-3'	
CaMV35S-P	Forward	5'-GGCCATCGTTGAAGATGCCTC-3'	144 bp
	Reverse	5'-biotin-TCATCCCTTACGTCAGTGGA-3'	
NOS-3'	Forward	5'-TGCATGACGTTATTTATGAGATGGGTTT-	113 bp

		3'	
	Reverse	5'-biotin-GCGCGCGATAATTTATCCTAGTTT-3'	
CaMV35S-T	Forward	5'-CCAGATAAGGGAATTAGGGTTC-3'	132 bp
	Reverse	5'-biotin-ACTGGATTTTGGTTTTAGGAATTAGA-3'	
GOS9	Forward	5'-GACGATCCGTAGTGGAGC-3'	117 bp
	Reverse	5'-biotin-GTAATCGTAGTTGGATTTCCACC-3'	

Table 2. Probes used in membrane-based array RDB.

Name	Sequence
BAR	5'-CGCGGCATGCTGCGGGCGGCCGGCTTCAAGCACGGGAACTGGCATGACGTGGGTTTCTG-3'
Cry1Ab	5'-GCAGCTAATCTTCACCTCAGCGTGCTTCGAGACGTTAGCGTGTTTGGGCAAAGGTGGGG-3'
Cry1Ac	5'-TCTACCAGATATAGAGTTCGTGTGAGGTATGCTTCTGTGACCCCTATTCACCTCAACGT-3'
CaMV35S-P	5'-AGGAGCATCGTGGAAAAAGAAGACGTTCCAACCACGTCTTCAAAGCAAGTGGATTGATG-3'
NOS-3'	5'-TTATGATTAGAGTCCCGCAATTATACATTTAATACGCGATAGAAAACAAAATATAGCGC-3'
CaMV35S-T	5'-AGCATATAAGAAACCCTTAGTATGTATTTGTATTTGTAAAATACTTCTATCAATAAAAT-3'
GOS9	5'-GCCTTTACATACGTTGGCACGGACGGGAATGAACATCTTGCAGGTCCTTGGGGTGGTGG-3'
PC	5'-GCATCCAGATCAGAAGCAATAATGAGCAGTGCGAGAAGAACGAGTGTCCAAAGTACCAG-3'
NC	5'-GGTTCCTTGAGAAATGTTTTACGGGATTACTTCCATGTTTGTTGGATGATCCTATTTTC-3'

* 5'- Modified group: AminolinkerC6;

2.2. *Manual method*

Multiplex PCR amplification: Multiplex PCR amplification was prepared according to the Table 3 PCR reaction system. Genomic DNA from a transgenic rice standard was used a positive control. Non-genomic DNA from the transgenic rice standard was used as a negative control. A blank lacking nucleic acid extraction was used as a blank control. These samples were used for quality control of the PCR amplification system and membrane chip hybridization reactions. PCR parameters and circulation are shown in Table 4.

Hybrid membrane-based array RDB system: The total volume was 1000 μL, including 979 μL Hybrid buffer solution (2×saline sodium phosphate EDTA (SSPE) buffer and 0.1% SDS), 20 μL PCR amplification product, and 1 μL Positive-Oligo single chain DNA (10 μmol/L). The 1000-μL enzyme incubation system was prepared with 999.5 μL incubation solution (2×SSPE and 0.5% SDS) and 0.5 μL AP-streptavidin. The manual RDB Hybrid process is shown in Table 5.

Table 3. PCR reaction system (50 μL).

Reaction liquid	Test sample	Control		
		Positive	Negative	Blank
Master Mix	25 μL	25 μL	25 μL	25 μL
Primer Mix	5 μL	5 μL	5 μL	5 μL
Sample DNA	200 ng	/	/	/
GMC DNA	/	200 ng	/	/
Non-GMC DNA	/	/	200 ng	/
No DNA	/	/	/	10 μL
Deionized water	up to 50 μL			

Table 4. Multiplex PCR process.

Step	Temperature	Reaction time
1	37 °C	10 min
2	95 °C	10 min
3	95 °C	30 s
4	55 °C	30 s
5	72 °C	30 s
6	Step 3,4,5 circulate	40 times
7	72 °C	10 min
8	4 °C	∞

Table 5. Manual RDB Hybrid process.

Step name	Operation method
Deactivation	Adding deactivation solution 37 °C,8 min.
Rinsing	Removing deactivation solution and adding deactivation rinsing solution 60 °C,5 min.
Hybrid	Removing deactivation rinsing solution and adding hybrid buffer 42 °C,45 min
Rinsing	Removing hybrid buffer solution and adding hybrid rinsing buffer 52 °C,5 min, 2 times.
Enzyme incubation	Removing hybrid rinsing solution and adding enzyme incubation solution 42 °C,30 min.
Rinsing I	Removing enzyme incubation solution and adding enzyme rinsing solution I 42 °C,5 min, 2 times.
Rinsing II	Removing enzyme rinsing solution I and adding enzyme rinsing solution II 42 °C,5 min, 2 times.
Color	Removing enzyme rinsing solution II and adding color liquid 37 °C,15 min.
Rinsing	Removing color liquid and adding deionized water 37 °C,5 min. , 2 times.
Read result	According to the color of the hybridization point, the results were interpreted.

2.3. *Automatic membrane-based array RDB instrument*

The automatic membrane-based array RDB instrument (MAI-RDB) is shown in Figure 1. The automatic hybrid process is listed in Table 6. According to the process, this instrument could perform automatic sampling and add samples onto the micro-reaction membrane with probes. Probes were designed on the surface membrane-based array according to Figure 2.

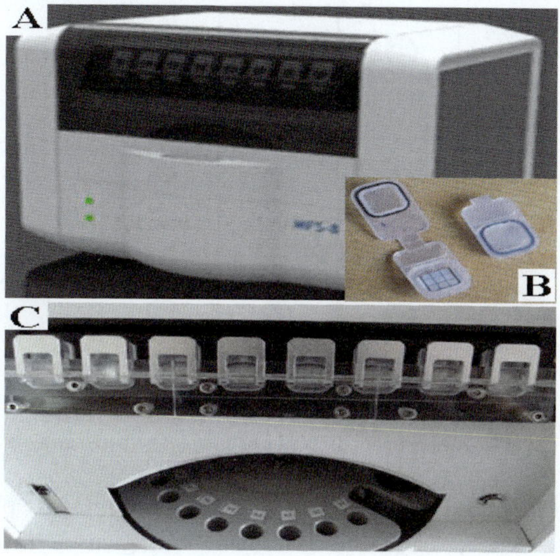

Fig. 1. MAI-RDB. (A) The main body of instrument; (B) Micro-reaction membrane with probes; and (C) Sample plate.

Table 6. Automatic hybrid process.

Process name	Reagent	Temperature °C	Time min
Deactivation	Deactivation solution	37	8
Rinsing	Deactivation rinsing solution	60	5
Hybrid	Hybrid buffer	42	45
Rinsing	Hybrid rinsing buffer	52	5
		52	5
Enzyme incubation	Enzyme incubation solution	42	30
Rinsing	Rinsing I	42	5
		42	5
	Rinsing II	42	5
		42	5
Color	Color liquid	37	15
Rinsing	Deionized water	37	5
	Deionized water	37	5
Read result	According to the color of the hybridization point, the results were interpreted.		

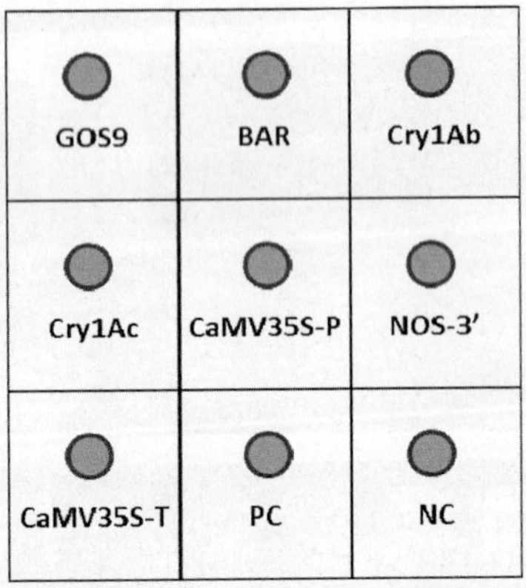

Fig. 2. Arrangement of probes on membrane-based array.

Probe corresponding gene: GOS9: rice gos9 gene; BAR: phosphinothricin acetyltransferase gene; Cry1Ab: Bacillus thuringiensis Insecticidal toxin cry1A(b) gene; Cry1Ac: Bacillus thuringiensis Insecticidal toxin cry1A(c) gene; CaMV35S-P: Promoter from cauliflower mosaic virus; CaMV35S-T: Termination sequence from cauliflower mosaic virus; NOS-3': Termination sequence of the nopaline synthase gene; PC: Positive control; NC: Negative control.

3. Results and Discussion

3.1. *Specificity confirmation and qualitative detection*

The GMC standard LLRice62 (containing GMCs BAR, CaMV35S-P, and CaMV35S-T), GMC standard BT63 (containing GMCs Cry1Ab, Cry1Ac, and NOS-3'), and non-transgenic rice were used. Analysis of each GMC standard was repeated six times, and the results are shown in Figure 3. We found that the point of positive hybridization displayed clear color. Six repeated experimental results were consistent, indicating that the method was specific.

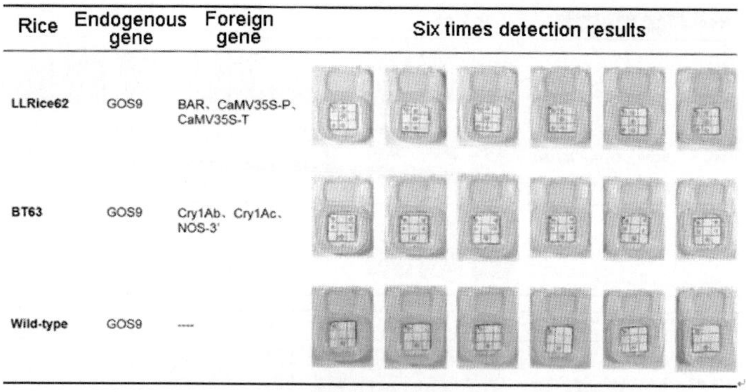

Fig. 3. Specificity confirmation and qualitative detection.

3.2. *Sensitivity*

Three values (2, 0.5, and 0.1% quality percentage) were chosen as the evaluation criteria to determine the sensitivity of the MAI-RDB. Genetically modified standards LLRice62 and BT63, as well as non-transgenic rice, were analyzed. Every rice standard was assayed three times (Figure 4A). The results indicated that the degree of detection was very good. To further confirm the detection sensitivity of the membrane-based array method, a 0.1% standard rice sample was analyzed 10 times (Figure 4B). We found that the sensitivity reached 0.1%.

The above values were checked according to the GMCs identification system of several countries. The European Union regulates that >0.9% GMCs must be identified, the level in Brazil and Australia is 1%, 5% in South Korea, and 3% in Japan.

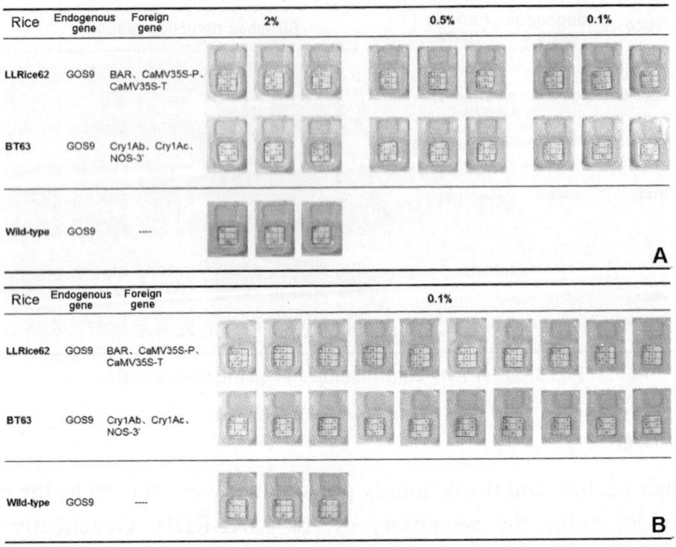

Fig. 4. Sensitivity. (A) Three values (2, 0.5, and 0.1% quality percentage) were assayed; and (B) 0.1% standard rice sample was analyzed.

3.3. *Stability*

Short-term stability assessment: The multiplex PCR- related reagents were placed at 37°C, the AP-streptavidin and ALP color substrate NBT/BCIP were placed at 4°C, and the other reagents were stored at room temperature. MAI-RDB chips stored for 1, 7, and 14 days were tested (Figure 5A). We found that there was no obvious reduction of the signal detected after 14 days of storage, so the MAI-RDB chip displayed short-term stability.

Long-term stability assessment: MAI-RDB chips stored for 1, 3, and 6 months were tested (Figure 5B). The results demonstrated that no obvious reduction of the signal was detected after 6 months, so the MAI-RDB chip displayed long-term stability.

3.4. *Comparison between MAI-RDB and fluore-scence quantitative PCR*

The standard LLRICE62, TT51-1 (BT63) specific cry1Ab/cry1Ac-NOS, and endogenous gene GOS9 were chosen for analysis. Three values (2, 0.5, and 0.1% quality fraction) were used as the standard of comparison. The comparison was repeated three times (Figure 6), and the results were in agreement between the MAI-RDB and fluorescence quantitative PCR methods. Under the detection limit of 0.1%, the MAI-RDB method was equivalent to the detection results of fluorescence quantitative PCR method.

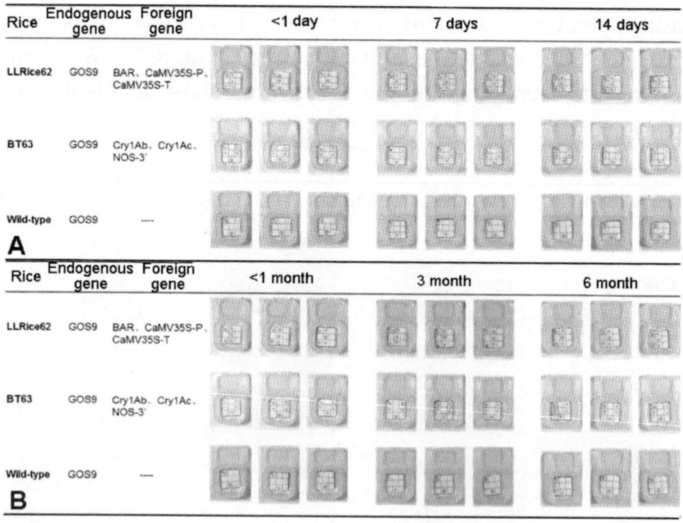

Fig. 5. Stability assessment. (A) Short-term (1-14 days) stability assessment; and (B) Long-term (1-6 months) stability assessment.

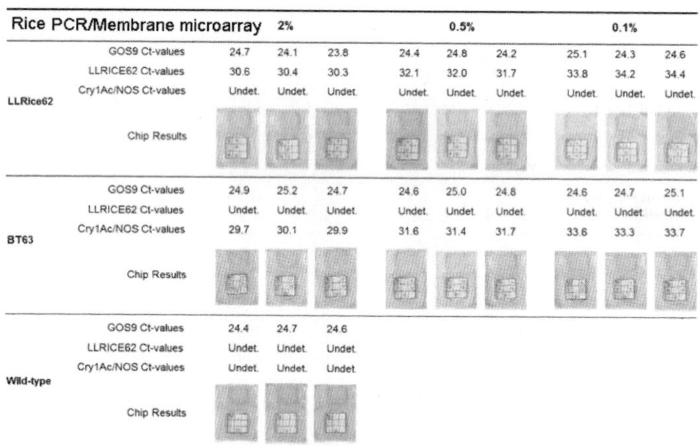

Fig. 6. Comparison between MAI-RDB and fluorescence quantitative PCR.

3.5. *Actual sample detection using automatic membrane-based array RDM instrument*

Ten rice samples or rice products were purchased from a market, including three kinds of bulk rice, three brand bags of rice, two infant rice products, and two rice snow cakes. Each sample was analyzed twice. As positive controls, 0.1% LLRice62 and 0.1% BT63 genomic DNA were added in the extraction of genomic DNA of each sample. The experiments used the genomic DNA of the transgenic rice standards LLRice62 and BT63 as positive controls, non-genomic

DNA of a transgenic rice standard as a negative control, and a sample prepared without nucleic acid extraction as a blank control (Figure 7). We found that the control test results were normal; the 10 commercially available samples did yield GMC signals.

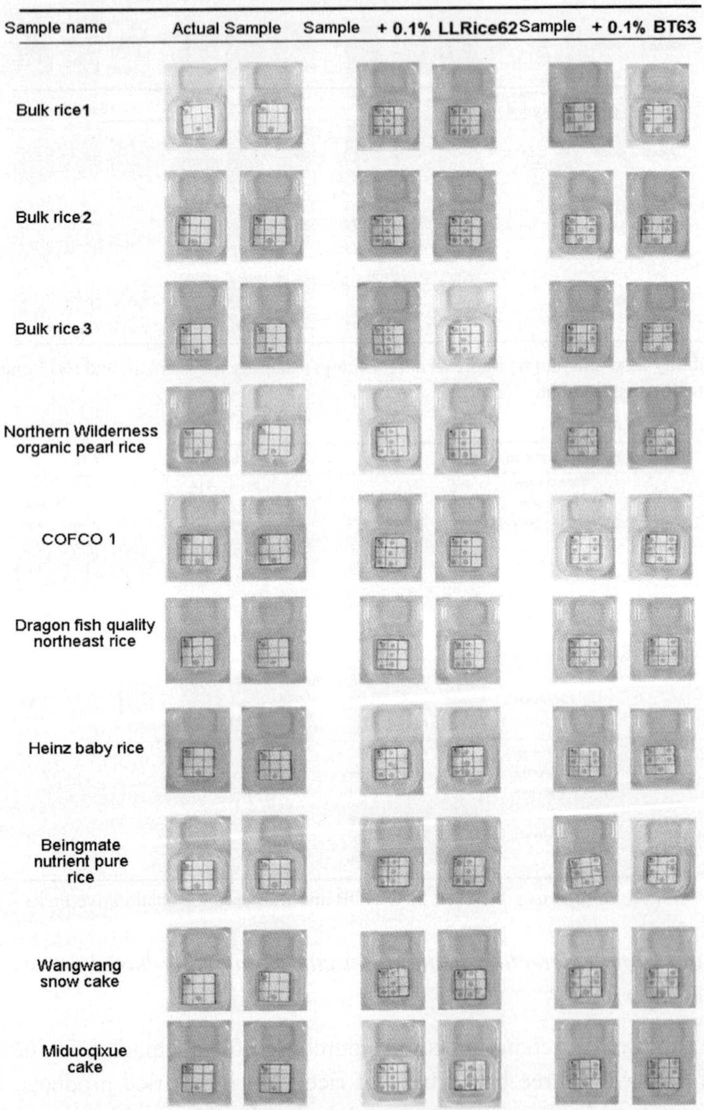

Fig. 7. Actual sample detection.

4. Conclusions

We developed an automatic membrane-based array RDB method. A laboratory prototype instrument was designed and manufactured. This high-throughput detection method is suitable for mass GMC screening of rice products. Comparing traditional manual RDB and PCR methods, we found that the automatic membrane-based array RDB method is rapid, sensitive, convenient, stable, inexpensive, automatable, easily visualized, and high throughput. Thus, the automatic membrane-based array RDB method can be used for qualitative and quantitative detection of multiple GMC types using a simple and rapid procedure.

References

1. James, C. 2013. Global status of commercialized biotech/GM crops: 2013. *ISAAA Brief* 46. Ithaca, NY: ISAAA.
2. Wang. X., Teng, D., Guan, Q., Tian, F. & Wang, J. 2015. Detection of genetically modified crops using multiplex asymmetric polymerase chain reaction and asymmetric hyperbranched rolling circle amplification coupled with reverse dot blot. *Food Chem.* 173:1022-1029.
3. Li, X., Wang, X. X., Yang, J. L., Liu, Y. M., He, Y. P. & Pan, L. W. 2014. A novel quadruplex real-time PCR method for simultaneous detection of Cry2Ae and two genetically modified cotton events (GHB119 and T304-40). *BMC Biotechnol.* 14:43.
4. Bridgers, M. 2004. Genetically modified organisms and the precautionary principle: how the GMO dispute before the world trade organization could decide the fate of international GMO regulation. *Temp. Envtl. L. Tech. J.* 22: 171.
5. Huang, X., Chen, L. L., Xu, J. M., Ji, H. F., Zhu, S. F., & Chen, H. J. 2014. Rapid visual detection of phytase gene in genetically modified maize using loop-mediated isothermal amplification method. *Food Chem.* 156: 184-189.
6. Samson, M. C., Gullì, M., & Marmiroli, N. 2010. Quantitative detection method for Roundup Ready soybean in food using duplex real-time PCR MGB chemistry. *J. Sci. Food Agric.* 90: 1437-1444.
7. Shrestha, H. K., Hwu, K.-K., & Chang, M.C. 2010. Advances in detection of genetically engineered crops by multiplex polymerase chain reaction methods. *Trends Food Sci. Tech.* 21: 442-454.
8. Guo, J., Yang, L., Chen, L., Morisset, D., Li, X., Pan, L. et al. 2011. MPIC: A high-throughput analytical method for multiple DNA targets. *Anal. Chem.* 83:1579-1586.
9. Xu, J., Miao, H., Wu, H., Huang, W., Tang, R., Qiu, M., Wen, J., Zhu, S. & Li, Y. 2006. Screening genetically modified organisms using multiplex-

PCR coupled with oligonucleotide microarray. *Biosens. Bioelectron.* 22: 71-77.

10. Choi, S.H. 2011. Hexaplex PCR assay and liquid bead array for detection of stacked genetically modified cotton event 281-24-236×3006-210-23. *Anal. Bioanal. Chem.* 401: 647-655.
11. Wang, X., Ma, F., Yin, L. L., Huang, Q. N. & Qi, X. T. 2007. A Convenient and Efficient PCR-dot-blot Method for Identification of Transgenic Plants Based on Manual Microarrayer. *Journal of Capital Normal University (Natural Science Edition)* 28: 52-55.
12. Prins, T. W., van Dijk, J. P., Angeline Van Hoef, A. M., Voorhuijzen, M. M., Broeders, S., Trapmann, S., et al. 2010. Towards a multiplex cereal traceability tool using padlock probe ligation on genomic DNA. *Food Chem.* 118; 966-973.
13. Zhang, N., McCarthy, M. L., & Smart, C. D. 2008. A macroarray system for the detection of fungal and oomycete pathogens of solanaceous crops. *Plant Disease* 92: 953-960.
14. Mokrousov, I., Vyazovaya, A., Kolodkina, V., Limeschenko, E., Titov, L. & Narvskaya, O. 2008. Novel macroarray-based method of Corynebacterium diphtheria genotyping: evaluation in a field study in Belarus. *Eur. J. Clin. Microbiol. Infect Dis.* 28: 701-703.
15. Tamaoki, M., Matsuyama, T., Nakajima, N., Aono, M., Kubo, A. & Saji, H. 2004. A method for diagnosis of plant environmental stresses by gene expression profiling using a cDNA macroarray. *Environ Pollut.* 131: 137-145.
16. GB/T 31730-2015. 2015. Detection of Genetically Modified Components in Rice Membrane-based Array Method. *State Standard of the People's Republic of China.*

Research on big data processing technology based on Hadoop

Z. Q. Wang and Y. Y. Lu[†]
School of Control and Computer Engineering,
North China Electric Power University, Baoding, China
[†]*Email: 764484533@qq.com*

In the era of big data, the traditional data storage and computing mode can't meet the growing demand of mass data, and the emergence of Hadoop technology is a good solution to this problem. Hadoop is a software framework for distributed processing of large amounts of data, the core components including HDFS and Map/Reduce, and has the advantages of high reliability, high scalability, high efficiency and high fault tolerance. Now Hadoop has been used by many enterprises to build their own data processing platform. Based on the studies of some basic Hadoop big data processing cases, this paper cards a Hadoop-based big data processing architecture, and specifically optimizes the issues of HDFS for storing small files.

Keywords: Hadoop; Big data processing; storage of small files.

1. Big Data Processing Architecture Based on Hadoop

Since the open source implementation of HDFS and Map Reduce, the two key technologies of Hadoop, due to its ability to achieve high data processing efficient in low cost hardware platform, therefore, Hadoop is rapidly used by enterprises and organizations for big data processing, such as Baidu, Alibaba, Huawei and other companies [1]. In this paper, a hierarchical structure is used to map the current Hadoop mechanism, and a big data processing architecture based on Hadoop is designed, as shown in Figure 1.

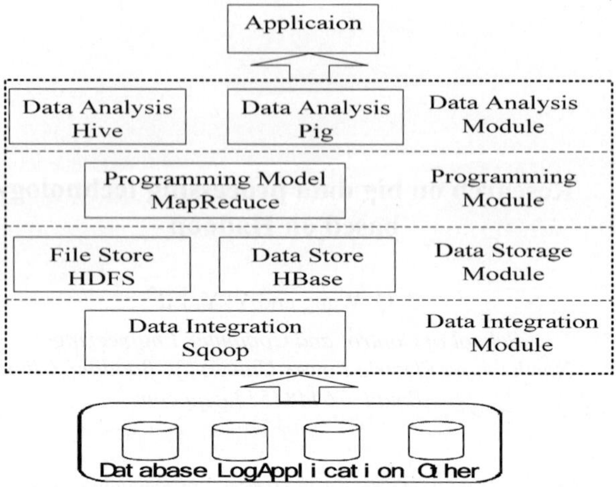

Fig. 1. Big data processing framework based on Hadoop.

(1) Data integration module

Data integration module is in the bottom of the whole framework, it provides the required data to the system, including database data, log data, application data etc. Some data can be directly stored in HDFS[2], such as the formatted log files; some also can be directly processed by MapReduce program. But if the required data is stored in the traditional data warehouse, it can be accessed only through an external API interface. However, this method is not efficient and flexible. So the big data processing framework based on Hadoop introduces the data integration module, using Sqoop[3] tool can import the relational database data into the Hadoop for analysis and processing, it also can export the processed results into the relational database for storage.

(2) Data storage module

In the background of big data, traditional data storage cannot meet the growing demand of massive data, so this architecture introduces a data storage module. Data storage module is based on distributed file system technology, it organizes the numerous storage device together through the network, which is distributed in different geographical locations. It provides data access ability to the upper layers through the interface, and with the increasing amount of data, it can expand storage capacity through a simple hardware expansion. In addition, the data storage module also supports concurrent access of vast amounts of large files, it also has data backup, status monitoring, falut tolerance, security mechanism and other management features.

(3) Programming model module

Programming module provides an abstract, parallel computing programming model for the processing of big data, it is the core part of the framework, and its efficiency determines the efficiency of the whole process. MapReduce[4] is characterized by high performance and simplicity, the combination of MapReduce and computer cluster and network makes it has the function of computing and analyzing. On one hand, programmers can use MapReduce to build their own big data processing program; on the other hand, the data analysis tools of upper layers can use MapReduce to access and analyze data directly, such as Hive [5].

(4) Data analysis module

The main works of data analysis module is to build data model, data mining and provide valuable commercial data to the applications of upper layer. Components for data analysis provided by the Hadoop system are Hive and Pig[6]. Based on MapReduce, Pig provides a higher level function of data processing, including data processing language and its runtime environment; the structured data can be mapped by Hive to a data table for SQL queries, at the same time, the query language is converted into a MapReduce task to execute.

2. Optimization of HDFS Storage Performance

From the above introduction, we know that HDFS is a master/slave structure, and is composed of a NameNode and several DataNodes, showing excellent performance in storage and processing of large files. But if the system has a large number of small files, HDFS will face serious challenges, mainly reflected in: Firstly, HDFS uses a data block(64M) to store the files, when the file size is much smaller than 64M, it will cause a lot of waste of storage space[7]; Secondly, the metadata of all files, existing in memory, are managed by NameNode, and the vast amount of small files will cause the serious shortage of memory space; Finally, the file reading and writing mode of HDFS is designed for large files, when dealing with small files, it needs lots of work to do , such as file data addressing operation and jumping between DataNodes, which will make the speed of file reading and writing reduced greatly[8].

2.1. *Structure design*

In view of the above problems, this paper makes an improvement on the traditional HDFS framework, and proposes a new HDFS structure with File Processing Port, as shown in Figure 2.

In order to optimize the performance of HDFS to store small file, it adds a File Processing Port between the client and HDFS clusters. The File Processing Port includes file type judgment, file merging, file mapping and file prefetching.

In this paper, the Name Node has the same function with the traditional HDFS Name Node, and the Data Node storage area is divided into two parts, the first part is to save the local index information of small files, the other part still serves as the data storage area.

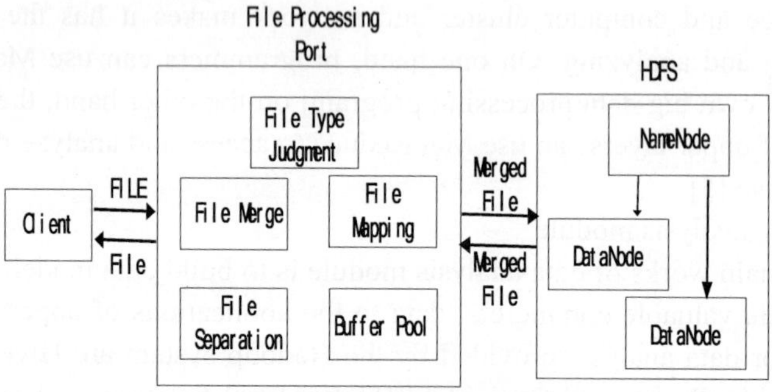

Fig. 2. The improved HDFS framework.

(1) File type judgment module

When a file needs to store, the file type judgment module determines whether the file is a small file (the threshold is set to 1M). If the size is less than 1M, then the file merge module will dispose it; otherwise, the file will be directly uploaded to the HDFS.

(2) File merge module

The file merge module is working at the time, and it will merge all files of a period time to a merged file, taking the time as the name of the merged file. For example, the system starts at ten o'clock in January 1, 2015, and uses eleven o'clock , January 1, 2015 for deadline, all the small files during this period time are written to the same merged file, and named it 2015-1-1:10-11. So far, a file merge operation is completed. In order to merge the small files in time sequence, the small files must be created in time sequence, and written to the merged file in sequence.

When merging the small files, the file merge module will create a local index for each merged file, and each small file has a structure data (smallfielname, offset, length) in it. When merging files, the file merge module will judge the total size of local index and the merged file to ensure that there is no small file data beyond the block storage. For example, the i-th structure data of the index table, specify the starting position and the file length of the small file, thus, the local index information can be used to point to the beginning and ending of the small file. The local index is stored at the starting area in DataNode which contains the merged file, and serves as the index entry for the small file.

(3) File mapping module

File mapping is the process of mapping a small file name to a physical block that contains the merged file which contains the small file. When the client wants to read a small file from the merged file, system can use the file mapping module.

When merging file, the File processing port will create a mapping form a small file name to the merged file name,which is (mergedfilemane, smallfilename), in addition to the local index. When the client needs to read a small file, the system will search the merged file name according to the mapping information firstly, and then the File Processing Port will communicate with NameNode according to the merged file name. Finally, the File processing Port will read data form the corresponding DataNode.

(4) File prefetching module

This paper uses file prefetching mechanism to cache the file data in the buffer pool. When reading a specific small file, system will cache all merged file which was stored at the same DataNode with the target file. When reading the target small file again and other small file which was in the same merged file, system can read it directly form the buffer pool. This method can reduce the number of requests to NameNode, greatly reducing access time of small files.

If the buffer pool saves all file information, it will be a big challenge to the storage capacity of the buffer pool. Therefore, this paper developed an update strategy to drop those pre stored data which is accessed infrequently. In order to facilitate the distinction between which files are infrequently accessed, the file prefetching module set the following structure: (smallfilename, offset, length, value), where value represents the number of the file being accessed, when the value is less than a specific number, the related file information will be removed from the buffer pool to protect the its normal storage ability.

2.2. *File operations*

File writing process:

(1) Firstly, the client upload the file, then the file type judgment module determine its capacity, if it is an ordinary large file, the file is written to HDFS in the traditional way; otherwise, if it is a small file, go to step (2).

(2) The small file is forwarded to file merge module for processing. On one hand, it will be merged according to its arrival time, on the other hand, the file merge module will create a local index (small filename, offset, length), then the local index will be placed at the head of the merged file, after this, this module also creates the mapping information (small filename, merged filename).

(3) The merged file will be written to HDFS in the traditional way.

File reading process:

(1) Firstly, the system will check whether the required file is in the buffer pool or not. If there is, then system will read directly form the buffer pool, if not, go to step(2).

(2) At this step, the system will find the required file in HDFS. There are two types of metadata stored in NameNode, one is ordinary large file metadata, and the other is merged file metadata. When operating, the system will search the ordinary large file metadata firstly, if the search is successful, indicating that the required file is large file, then it will be read out in the traditional way; if not, the system will return to the File port to find the mapping information of small file to the merged file, if this success, then the required file is small file, the system can obtain the merged file name, thus obtain the merged file metadata, go to step(3); if the map search if unsuccessful, the system will return "the required file does not exist" information to the client.

(3) According to the merged file metadata, the File Processing Port will interact with the corresponding DataNode directly. Firstly, system will read the local index information, thus obtain (smallfilename, offset, length), and separate the target file form the data block, then return it to the client. Finally, the system will put the local index and the other files that were stored in the same node with the target file into the buffer pool.

2.3. *Experiment and analysis*

In this paper, the test environment consists of three Hadoop nodes, one of which serves as NameNode configured to: Intel i7 CPU 2.8GHz, 4G memory, 500G hard drive; two serve as DataNodes configured to: Intel Pentium 4 CPU 2.8GHz,2G memory, 320G hard disk. The network environment is fast Ethernet, Hadoop version is 1.0.2 and jdk version is 1.8. In order to validate the efficiency of the combined strategy, this paper designs four data sets, DataBase1 (including 1000 small files), DataBase2 (including 10000 small files), DataBase3 (including 15000 small files), DataBase4 (including20000 small files).

Tradition HDFS design is conductive to large file storage, for small file processing, its performance is very low. For the improved HDFS which contains the File Processing Port, this paper respectively test the reading and writing performance of small files with the above 4 sets, and the test results are shown in Figure 3 and Figure 4.

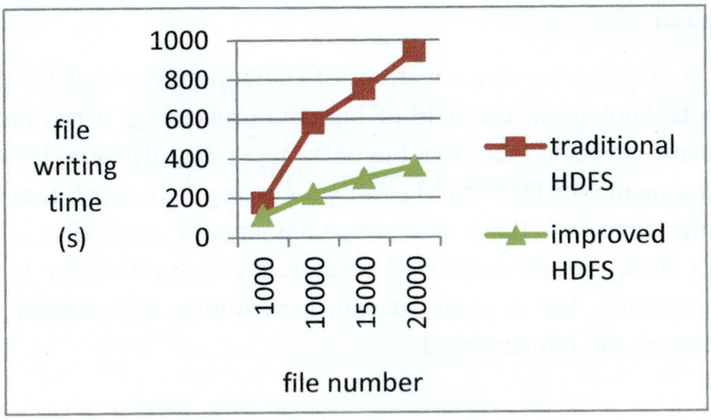

Fig. 3. File writing time comparison chart.

Experimental result analysis:

(1) When storing files, system will check the file size firstly, if it is greater than a predetermined threshold, the file will be stored in the traditional way; if not, the file will be processed by the File Processing Port.

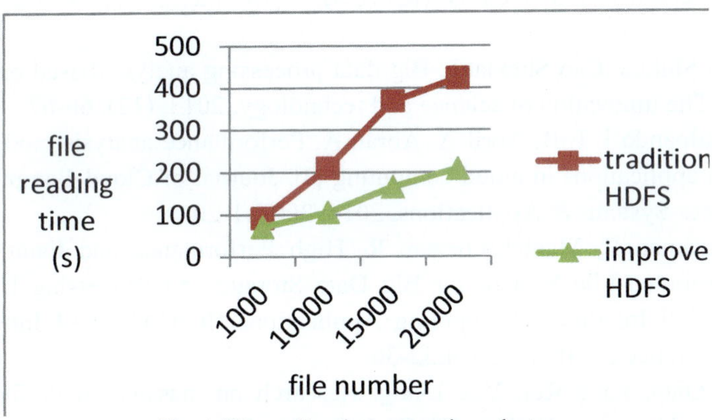

Fig. 4. File reading time comparison chart.

In this way, there may be a number of small files in a file storage process, so the time spent will be shorter than storing each small file separately.

(2) When reading a file, system will cache the whole data block by prefetching strategy. According to the principle of locality, when the file is requested again, this method can effectively reduce the reading time.

But in the file storing and reading process, if the file size is generally greater than the predetermined threshold, then this method does not improve the storage time significantly; and in some cases, the requested files does not exist in the same Data Node, then the read speed will not increase obviously. Therefore, the method introduced in this paper needs to be further studied.

3. Conclusion

The low-cost, high-performance features of Hadoop make it quickly become the mainstream technology in the field of big data processing, it has been used by many companies to build their own big data processing platform. In this paper, a hierarchical structure of Hadoop big data processing framework is consolidated, and the performance of HDFS store large amounts of small files is optimized. Experiments show that the improved scheme has high efficiency for small file reading and writing, but in some special applications, this method need to be further discussed and implemented.

References

1. Pal A, Agrawal S. An experimental approach towards big data for analyzing memory utilization on a Hadoop cluster using HDFS and MapReduce [C] Networks & Soft Computing (ICNSC), 2014 First International Conference on. IEEE, 2014:442-447.
2. Manikandan S G, Ravi S. Big Data Analysis Using Apache Hadoop [C] IT Convergence and Security (ICITCS), 2014 International Conference on. IEEE, 2014:1-4.
3. Dan Shihua, Cao Shexiang. Big data processing analysis based on Hadoop [J]. The innovation of science and technology, 2013, (12): 66-67.
4. Villalpando L E B, April A, Abran A. Performance analysis model for big data applications in cloud computing [J]. Journal of Cloud Computing Advances Systems & Applications, 2014, 3(19):1-20.
5. Sivaraman E, Manickachezian R. High Performance and Fault Tolerant Distributed File System for Big Data Storage and Processing Using Hadoop[C] Intelligent Computing Applications (ICICA), 2014 International Conference on. IEEE, 2014:32-36.
6. Shi Qian, Fang Rui, Yue Liang. Research on massive small file storage method based on Hadoop [J]. Digital technology and application, 2014, (1): 50-50.
7. Huang Suping, Ge Meng. Research on the application of Hadoop platform in big data processing [J]. Modern computer: Professional Edition, 2013, (10): 12-15.
8. Nandimath J, Banerjee E, Patil A. Big data analysis using Apache Hadoop [C] Information Reuse and Integration (IRI), 2013 IEEE 14th International Conference on. IEEE, 2013:700-703.

Research on hierarchical object placement algorithm of object-based storage

Z. Q. Wang and J. B. Cui †

School of Control and Computer Engineering,
North China Electric Power University, Baoding, China

†*Emails: 349262131@qq.com, 654913398@qq.com, cuijb3@sina.com*

Object-based storage is a kind of cloud storage technology, with better data sharing, better security, better intelligence. Mass storage system consists of thousands of devices, and gradually extended to form, therefore, the data placement is one of the key problems in the efficient organization of storage devices. Although consistent hashing has the good adaptability, but it need to increase the number of virtual nodes in heterogeneous environment. This paper introduces the hierarchical data placement algorithm, and improves the consistent hashing and improves the setting weight method and the clustering method.

Keywords: Object-Based storage; Heterogeneous environment; Data placement; Hierarchical; Consistent hashing.

1. Introduction

Object-based storage is one of cloud storage technology. It combined advantages of NAS (Network Attached Storage) and SAN (Storage Area Network)[1], and have better data sharing, security, and intelligence [2] . The core of object-based storage is the object, which is the basic unit that contains the application data and some of the extended storage attributes. Client can access to the object storage device through the standard object interface [3]. Object-based storage may be the new generation of storage standard [4]. At present, the common object-based storage systems include Amazon S3, Swift, Ceph, Gluster FS, etc.

Large-scale storage system is composed of thousands of storage devices, and it is often expanded and scaled down by adding or removing storage devices. Therefore, the data placement is a key issue of efficient organization of storage devices. Data placement standards include: [4] time efficiency, space efficiency, fairness and adaptability. The design of the storage strategy must consider scalability, balance and fairness of the system. Karger et al [5] proposed the classic consistent hashing algorithm, it has good adaptability, but it need introduce a large number of virtual devices. Schindelhauer et al [6] proposed linear method and logarithmic method to distribute the data. Honicky et al [7] designed a family of algorithms, including RUSHp, RUSHt, RUSHr. RUSHp and RUSHr are suitable for small scale systems, and RUSHt is more suitable for large-scale systems. At China, National University of Defense Technology [8,9,10], Huazhong University of Science and Technology [11], Shandong University [12], Wuhan University of Technology [13] and other scientific and technological institutions also have been doing related work on data placement.

Now the data placement algorithm, applied to object-based storage, mostly improves the consistent hashing. It can be divided into two categories: sensitive to device number and sensitive to weight difference. The first kind of algorithm is suitable for the system whose device number is large, and in which weight difference is same. The second kind of algorithm can be very well to adapt to the large weight difference, but too many devices, will result in an increase in complexity of algorithm. This paper introduces hierarchical strategy [8,9], it, On the premise of guarantee time efficiency, fairness and adaptability, improves the space efficiency of consistent hashing and improves the setting weight method and the clustering method in the Ref. [8,9].

2. Model the Problem

Devices and data are abstracted into two sets, problem becomes how to map the both. Given a set of N devices $V=\{v_1,......,v_N\}$ with weights w_v consisted of multiple factors. Device set V, through hash function $h:V\rightarrow[0,1)$, are mapped into some continuous range $[0,1)$. Assuming that device set of the current system is $V_0=\{v_1,......,v_n\}$, $n\leq N$, $V_0\subseteq V$. $X_0=\{x_1,......,x_m\}$ denotes data set of the current system, m the number of data objects and $x_i\in X_0$ one of the data set. Function $f_0:X_0\rightarrow V_0$ maps data set X_0 into device set V_0. The data placement algorithm A can be expressed as a bivariate function $A(x,f)$, $x\in X_0$, a matching function f and $A(x,f)=f(x)$. Data object x is assigned to the device identified by $A(x,f)$. $A_V(v)$ denotes the data set assigned to device v. $M_V(v)$ denotes the number of data objects assigned to device v.

Definition (1). If for a device $v\in V$ with weight w_v, a fair share according to its relative weight, i.e. for m data elements

$$M_V(v)=m\cdot\frac{w_v}{\sum_{v\in V}w_v}. \tag{1}$$

A mapping is called fairness.

Definition (2). A mapping is called consistent, if for $V \subset V'$, all $v \in V$:

$$A_{V'}(v) \subseteq A_V(v) . \quad (2)$$

I.e. a mapping has good adaptability.

3. Hierarchical Data Placement Algorithm

Karger classic consistent hashing algorithm needs to introduce $k \log n$ virtual devices, k is a positive real number. In the heterogeneous environment, storage devices with the weight of w_i need to be allocated $(k \log n)w_i / w_{\min}$ virtual devices. If there is a huge difference between the weight for the devices, the memory space occupied by virtual devices is intolerable.

Therefore, this paper introduces the hierarchical data placement algorithm, which cluster the devices with the their weights to minimize the weight difference between devices. We process the data placement between classes with logarithmic method proposed by Schindelhauer et al[7], while within classes with consistent hashing algorithm[5][8]. This paper focuses on the clustering process and data placement algorithm between classes .

3.1. *Clustering devices*

According to the definition (1), the greater the weight for the device is, the more data it should be allocated to. The weight for the device is usually represented by a real number. In the weight setting process, we consider the storage capacity, computing power and network delay. Assuming that the device i, the weight for the computing speed is w_{i1}, the weight for the storage capacity is w_{i2}, the weight for network delay network is w_{i3}.The faster the computing speed is and the larger the storage capacity is, the greater the amount of data allocated by the device is; The larger the network delay is, the smaller the amount of the data allocated by the device is. Here, we introduce three constants k_1, k_2, k_3, then the weight w_i for the device v_i [14] can be expressed as

$$w_i = k_1 w_{i1} + k_2 w_{i2} + \frac{k_3}{w_{i3}} \quad (3)$$

Set the distance w_{ij} between storage nodes v_i and v_j

$$w_{ij} = \sqrt{\begin{array}{l} k_1^2 (w_{i1} - w_{j1})^2 + k_2^2 (w_{i2} - w_{j2})^2 + \\ k_3^2 (\frac{1}{w_{i3}} - \frac{1}{w_{j3}})^2 \end{array}} \quad (4)$$

Here, we use the maximum and minimum clustering algorithm to cluster the devices [8] according the formula (4). After the clustering, the devices are clustered into g Classes. The device clusters is no longer easy to change to avoid unnecessary data migration.

3.2. *Data placement algorithm between classes*

3.2.1. *Data placement between classes*

Classes and a set of data objects $X_0 = \{x_1, ..., x_m\}$ are mapped to the continuous set M = [0, 1) using appropriate hash function h that behave like uniform independent random variables. Each device v_i is associated with a capacity w_i as defined by the formula (1). Class weight is the sum of the device weight within the class. Assume that *Classi* denotes a cluster i, the hash function h_1 maps each data object x_i into M denoted by $r_i = h_1(x_i)$. Similarly we receive $s_i = h_2(Classi)$ denoting the position of *Classi* in the interval M.

We define a distance function as:

$$D_w(r,s) = \frac{-\ln((1-(r-s) \bmod 1)}{w}, \tag{5}$$

where $a \bmod 1 = a - \lfloor a \rfloor$, we assign each data item r_i to the s_i which minimizes the term $D_{w_v}(r_i, s_i)$.

Redefine a height function:

$$H(r) = \frac{-\ln((1-(r-s_{Class}) \bmod 1)}{W_{Class}}, \tag{6}$$

$H(r) = \min D_W(r, s_{Class}), 1 \le i \le g$ denotes the height of r.

For classes obtained by clustering, they have some positive weights $W_1 ... W_g$. Then according to the height function, assigning a data x to *Classi* with probability $W_i / \sum_{j=1}^{g} W_j$ [6].

The above is in the ideal case, the device and data simultaneously random mapping to the interval M. When data is inserted in the second time, because the device has a fixed position, it will be affected device mapping locations. Here introduing k partitions, i.e. the interval M is divided into a series of continuous partitions $M_1 ... M_k$. M_i = $[(i-1)/k, i/k)$, $1 \le i \le k$. Classes are mapped to each of the continuous partition $M_1 ... M_k$. Similarly, data set $X_0 = \{x_1, ..., x_m\}$ are mapped to each of the continuous partitions. Data object is assigned to a class which is closest in sub-range M_i, and $k = O(\log g)$.

3.2.2. *Data migration between classes*

When expanding storage system, it may add devices in bulk. The performance of these devices are usually very similar, can be classified as a new class, the class algorithm takes into account adding devices in bulk.

The devices are clustered into g classes by clustering algorithm. [0,1) the interval is divided into I sub region. Devices added in bulk are considered as a new class $Class(g+1)$, due to the introduction of new class, The relative weights of all the classes in the system have been changed. Setting $Classi$ the new weight W_i', $1 \le i \le g$, the amount of data on $Classi$ that needs to be migrated is $m(W_i - W_i')/W_i$, $W_i = \sum_{v \in Classi} w_v$.

According to the Formula (1) and the principle of minimum height, each class will take up some certain intervals. When adding a new class, the new class also takes up some certain intervals. at this time, data objects mapped into these intervals will be migrated to the new class. And there is no need for data migration between the other classes.

3.3. *Algorithm performance analysis*

In Hierarchical strategy, devices are clustered into g classes firstly. the interval $[0,1)$ is divided into I partitions. According to the Literature [7] theorem 9, using k partitions has a fragmentation of 2kn–1 by data placement algorithm between classes, and $k = O(\log g)$. The time efficiency of the algorithm between classes is $O(logI)$ [9]. The time efficiency of classic consistent hashing algorithm is $O(logn)$ [5][9], n the number of devices. After clustering, the maximum number of devices in a class is $n_{\max} = \max_{1 \le j \le g} n_j$. The time efficiency of Hierarchical strategy will be $O(\log n_{\max}) + O(\log(glogg))$. Due to the large gap between the number of clusters g and devices total number n, so it approximates $Olog(n)$.

Next, we discuss the space efficiency of hierarchical strategy. The interval $[0,1)$ is divided into I partitions. Each partition needs $\log I$ bits to represent [6]. For any class $Classj$, device v_i^j with wight w_i^j, $w_{\min}^j$ represent the device with minimum wight. The class $Classj$ required $\sum_{i=1}^{n_j} (k \log n) w_i^j / w_{\min}^j$ virtual devices. After clustering, the difference of the device weight in the class is not big. $\sum_{i=1}^{n_j} (k \log n) w_i^j / w_{\min}^j$ approximately equal $n_j k \log n$. The class with the largest number of devices requires $n_{\max} k \log n$ virtual devices. In the device set, the representation of a virtual device needs a maximum of $\log(kglogg) + \log(n_{\max} k \log n)$ bits. In classic consistent hashing, apart from all the virtual devices need $\log(\sum_{i=1}^{n} \log(k \log n) w_i / w_{\min})$ bits. When devices performance have a large difference in the system, for $1 \le i \le n$, most of $w_i / w_{\min}$ has a great value. Since the device number n of large-scale system is very large. Thus $\log(kglogg) + \log(n_{\max} k \log n)$ will be far less than $\log(\sum_{i=1}^{n} \log(k \log n) w_i / w_{\min})$.

4. Experiment and Analysis

In order to test fairness and adaptability of the algorithm, in this paper, the test was carried out by Java program. In the experiment, we assume introducing 200 devices. Weights of storage capacity, computing power and network delay reciprocal of devices are equal to the random positive real number of 1-500. k_1, k_2, k_3 is equal to 10, 2, 1. We firstly cluster devices by formula (4). The results are as shown in figure 1.

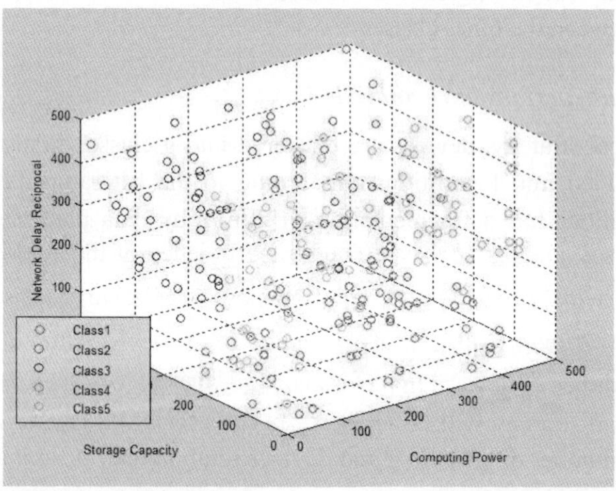

Fig. 1. Clustering according to the formula(4) .

Clustering through the formula (4), The results present zonal placement based on computing power.

Figure 1 shows, after the clustering, the devices are divided into five classes. They are respectively named Class1-5, class weight is the sum of the device weight within the class. The weight ratio of five classes are: 26.33%, 28.43%, 19.69%, 7.95% and 17.59%.

Class 1-5 are randomly mapped to the interval [0,1). According to formula (5) in data placement algorithm between classes, the interval [0,1) is divided into several intervals, as shown in Figure 2.

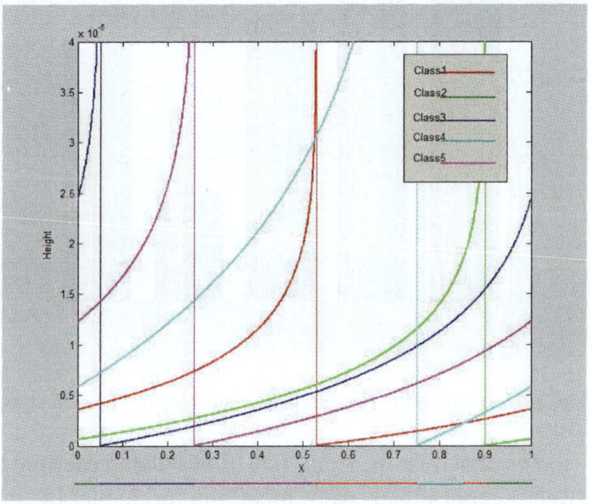

Fig. 2. Data placement algorithm between classes.

If the classes are directly mapped to the interval [0,1), the resulting range fragments cannot meet the principles of fairness. Thus, according to data placement algorithm between classes, it should introduce $O(\log 5)$ partitions. Then, 100000, 200000, 300000, 400000 and 500000 data objects are respectively inserted into the devices collection. Observing the proportion of data objects obtained by each class, testing the algorithm fairness.

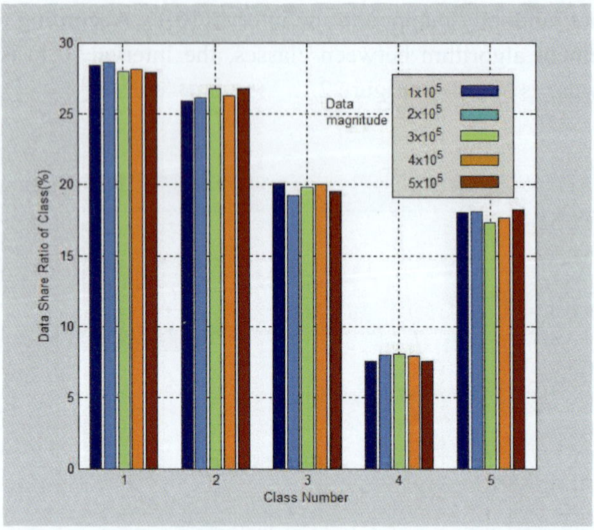

Fig. 3. The proportion of the amount of data obtained by each class.

Figure 3 shows that results from five times data insertion based on the algorithm between classes can well meet the principle of fairness in the definition (1).

To test adaptability of the strategy, we send 300000 data objects to the device set, then adding 40 new devices. In reality, the newly added devices are often more excellent performance. So weights of storage capacity, computing power and network delay reciprocal of these devices are equal to 500. They are clustered into a new class, and named Class6. At this time, weight ratio of Class1-5 becomes: 21.00%, 19.45%, 14.54%, 5.87%, 12.99%, weight ratio of Class6 is: 26.15%. Repeat three times, the actual data migration ratio of Class1-5 is as shown in figure 4.

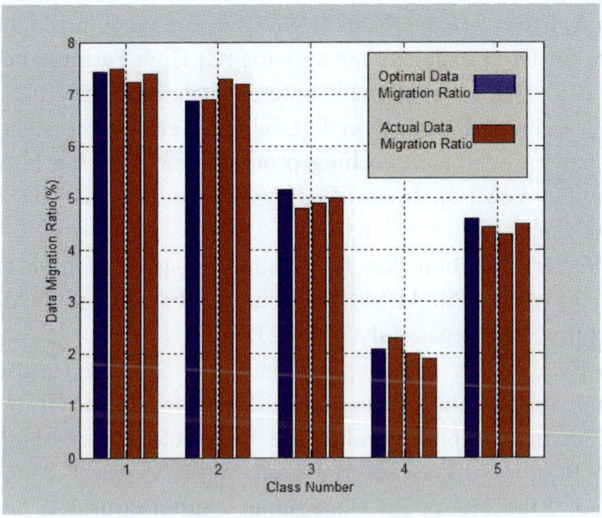

Fig. 4. Data migration ratio of Class1-5.

As it can be seen from Figure 4, the actual data migration ratio and optimal data migration ratio is approximately equal, meet consistent definition in definition(2), the algorithm between classes has good adaptability.

5. Conclusion

This paper introduces hierarchical data placement algorithm, and improves clustering algorithm and data placement algorithm between classes. It has better space efficiency than consistent hashing, and maintain a good time efficiency, fairness and adaptability. It is more suitable for large-scale systems. For small system it is too cumbersome, and yet to be improved.

References

1. Wang YongKui, Zhu LanJuan. Research of Object-based Storage System [J]. Computer Engineering, December 2007: 266-267+270. (in Chinese)
2. Karakoyunlu C, Runde M T, Chandy J A. Using an Object-Based Active Storage Framework to Improve Parallel Storage Systems [C] Parallel Processing Workshops (ICCPW), 2014 43rd International Conference on. IEEE, 2014: 70-78.
3. Nagle D, Factor M E, Iren S, et al. The ANSI T10 object-based storage standard and current implementations [J]. IBM Journal of Research and Development, 2010, 52(4.5): 401-411.

4. Miranda A, Effert S, Kang Y, et al. Reliable and randomized data placement strategies for large scale storage systems [C] High Performance Computing (HiPC), 2011 18th International Conference on. IEEE, 2011: 1-10.
5. Karger D, Lehman E, Leighton T, Levine M, et al. Consistent hashing and random trees: Distributed caching protocols for relieving hot spots on the World Wide Web [C]. In Proceedings of the 29th ACM Symposium on Theory of Computing. 1997: 654-663.
6. Schindelhauer C, Schomaker G. weighted distributed hash tables [C]. In Proceedings of the 17th ACM Symposium on Parallelism in Algorithms and Architectures. Las Vegas, July 2005: 218-227.
7. Honicky R. J., Miller E.L. Replication Under Scalable Hash: A Family of Algorithms for Scalable Decentralized Data Placement [C]. In Proceedings of the 18th International Parallel and Distributed Processing Symposium. 2004: 96-105.
8. Chen Tao, Xiao Nong, Fu Changfang. Clustering-Based and Consistent Hashing Aware Data Placement Algorithm [J]. Journal of Software, 2010, 12:3175-3185 (in Chinese).
9. Chen Tao, Xiao Nong, Liu Fang, et al. Clustering-based and consistent hashing-aware data placement algorithm [J]. Journal of Software. 2010, 21(12): 3175—3185 (in Chinese).
10. Liu Zhong, Zhou Xingming. A data object placement algorithm based on dynamic interval mapping [J]. Journal of Software. 2005. 16(11):1886-1893 (in Chinese).
11. Wang Fang, Zhang Shunda, Feng Dan, et al. Hybrid object allocation policy for object storage systems [J]. Journal of Huazhong University of Science and Technology: Nature Science Edition. 2007, 35(3):46-48 (in Chinese).
12. Zheng Pai, Cui LiZhen, Wang HaiYang, et al. A Data Placement Strategy for Data-Intensive Application in Cloud [J]. Chinese Journal of Computers, 2010, 33(8):1472-1480(in Chinese).
13. Zheng Tong, Li Tong. Data Placement Algorithm for Large-scale Storage System [J]. Computer Science, 2013,S2:270-273 (in Chinese).
14. Song YuChen, Zhang YuYing, Meng HaiDong. Research Based on Euclid Distance with Weight of Clustering Method [J]. Computer Engineering and Application, 2007, 04:179-180 (in Chinese).
15. Cheng Ping, Liu Wei. Method of Determining Attributes Weights Based on Subjective Preference in Multi-attribute Group Decision-making [J]. Control and Decision, 2010, 11:1645-1650 (in Chinese).

A security digital image transmission method based on data segmentation and non-linear encryptions

W. Chen†
Computer Science College Sichuan University, Chengdu, China
†Email: cwcwcwcw2006@163.com

Y. Zhao and E.Y. Zhang
Technology Center Changhong Electric Co., Chengdu, China

To protect the confidentiality and integrity of network digital images, a security image transmission method based on data segmentation STDDS was proposed in the paper. Firstly, the image was divided into segments and encrypted by groups of non-linear encryptions to generate secrets and check sums; the received secrets were verified by the recipient to detect illegal modification. The corresponding two sides dynamically calculated the encryption variables through pre-deployed hash functions, which avoids leaks due to the simultaneously transmission of cipher text and encryption variables. In addition, even if some cipher text was corrupted during the transmission, the plaintext could also be recovered by the redundancy secrets, which ensures the data availability. STDDS has good scalability, such that the data segmentation and polynomial encryption does not depend on the type and structure of the original data, thus STDDS can be combined with other security schemes based on the demands,. Simulation results demonstrated that, compared with traditional Arnold scrambling, chaotic encryption, the data confidentiality can be achieved by STDDS with a smaller time cost, and the integrity and availability of data can be greatly improved in the case of strong noise and data lost.

Keywords: Image security; Encryption; Secret sharing; Data segmentation.

1. Introduction

With the rapid development of network technology and the widespread application of multimedia, digital image has become an important information source. As the most important information carrier Digital, image has been increasingly transmitted over the Internet, but the images are vulnerable to the attackers during the network transmission process [1], therefore, digital image have high security requirements in the military, commercial, medical engineering and other areas [2][3].

In recent years, many image encryption methods have been proposed. Some classical encryption algorithms DES, AES, RSA are utilized to protect the images [4][5][6]. As these algorithms are known to the public, the confidentiality of images depends on the security of the keys and the time cost of the encryption algorithms rapidly increased with the key length and image size [7]. Image scrambling is another common used encryption technology, which randomly changes pixel locations to make the image unrecognizable. There are many scrambling algorithms, such as Arnold [8], Magic Square [9], Gray code [10], Knight-tour [11] and Fibonacci exchange [12]. Unfortunately, scrambling process only changes the position of the pixels, the pixel value and distribution are not modified, therefore, the cipher text is vulnerable to the attacking by histogram Analysis and pixels feature statistics [13]. Other methods, such as Fridich proposed image encryption using symmetric ciphers based on two-dimensional chaotic maps [14]; Pareek designed image encryption using two chaotic Logistic map to make the cipher text scrambling in a wide range [15]. Mazloom proposed image encryption model based on chaotic logistic and algebraic transformations [16]. These methods try to protect the data confidentiality based on various encryption methods. However, an attacker may not only want to acquire plaintext of the images, it may also try to undermine the transmitted ciphertext to prevent the recipient to obtain useful information. Therefore, both the confidentiality and integrity of the images should be taken into consideration when the security model is designed.

In the paper, a security image transmission method based on data segmentation STDDS is proposed. It provides protection of data confidentiality, integrity and availability: the data to be transmitted is divided into some fixed length segments; then each segment is encrypted to generate n sub-secrets and checksums; the receiver first verifies the received sub-secrets based on the corresponding check-sums; then any k verified sub-secrets could be combined to decrypt the ciphertext. As long as there are more than k available sub-secrets arriving at the recipient, then the ciphertext can be decrypted. It should be noted that, STDDS does not relevant to the data structure of the original data, thus it can be easily combined to other security mechanisms to decrypt the ciphertext. As long as there are more than k available sub-secrets arriving at the recipient, then the ciphertext can be decrypted. It should be noted that, STDDS does not relevant to the data structure of the original data, thus it can be easily combined to other security mechanisms to enhance the security.

2. Data Segmentation

The image is divided into small segments, and each segment can be independently encrypted and verified. And then the ciphertext, checksums of segments are coded into blocks and sent to the recipient. Based on segmentation, the data encryption improves the security of images transmission. On the one hand it increases the difficulty of the attackers to crack the ciphertext: they cannot crack the secret without all the secret segments, on the other hand, each segment has different keys and checksum, thus the attackers can hardly make fake communication data to initiate spoofing attacking.

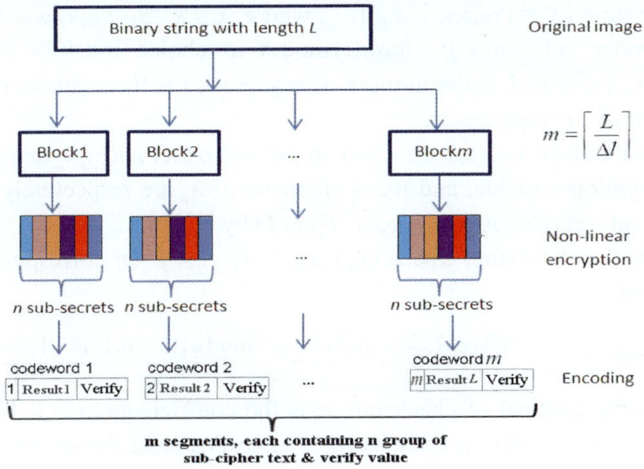

Fig.1. STDDS data partitioning workflow.

As shown in Figure 1, the data partition workflow includes the following 4 steps:

1) Input: read the original binary data of images.

2) Data divide: the binary string is divided into many sub-strings with fixed length Δl, thus the original data with length L will be divided into $m=\lceil L/\Delta l \rceil$ separate segments $\{str_1, str_2 \ldots str_i\}$, $\lceil . \rceil$ represents rounding up.

3) Encryption: each segment str_i is ciphered by non-linear group encryption using verifiable (n, k) threshold secret sharing, which will be described in section 3.1.

4) Coding: the encrypted cipher text of each sub-string str_i (Result), checksum value (verify) as well as the sub-string number i are encoded into codeword and sent to the recipient.

3. The Encryption Scheme

3.1. *The encryption process of sender*

Each segment str_i is encrypted to generate n sub-secrets. To implement non-linear encryption, k coefficients $a_1, a_2 \ldots a_{k-1}$ and an encryption variable x are randomly chosen to construct a k-1 order polynomial $F(x)$ as shown in (1), where $s = a_0 = str_i$ and q is a prime number, $a_i, k \in Z_q$.

$$F_i(x) = s + a_1 x^1 + a_2 x^2 + \ldots + a_{k-1} x^{k-1} \bmod q \tag{1}$$

In STDDS, the encryption variable x is calculated by pre-deployed hash function at both side and does not been transmitted:

1) The receiver and sender hold a same hash function *hash* () and a vector set $encrypt\_array = \{X_j = (x_{j1}, x_{j2}, \ldots x_{jn})\}_{j=1}^{N}$, where $X_j \in R^n, \forall p \neq q, x_{jp} \neq x_{jq}$;

2) The sender calculates $p = hash(i) \bmod N$ to choice index p for the *i-th* segment $str_i, 1 \le i \le L$, L is the number of segments, i is the segment number and N is the size of *encrypt_array*.

3) The *p-th* member $X_p = (x_{p1}, x_{p2}, \ldots, x_{pn})$ in the vector set *encrypt_array* is chosen as the encryption variable, and the n elements of X_p are respectively utilized to generate n sub-secrets ($F_i(x_{p1}), F_i(x_{p2}) \ldots F_i(x_{pn})$) by (1);

4) Calculate the checksum, Check (x_{pj}), $x_{pj} \in X_p, 1 \le j \le n$ corresponding to the n sub-secrets;

$$Check(x_{pj}) = \{c_t \mid c_t = g^{a_t} \bmod q\}, t = 0,1 \ldots k-1 \tag{2}$$

Where g is the cardinal of checksum, a_t is the coefficients in (1). Then $F_i(x_{p1})$, $F_i(x_{p2}) \ldots F_i(x_{pn})$, $Check(x_{p1})$, $Check(x_{p2}) \ldots$ $Check(x_{pn})$ and the segment number i are encoded and sent to the receiver.

3.2. *The decryption of receiver*

For each segment stri, $1 \le i \le L$ the decryption process is as following:

1) An index variable p is calculated using the hash function: p=hash(i), $1 \le p \le N$;

2) Select the *p-th* vector *Xp* from the *encrypt_array*;

3) Verify the integrity of the sub-secrets $F_i(x_{pj})$

$$g^{F_i(x_{pj})} \stackrel{?}{=} \prod_{t=0}^{k-1} c_t \cdot g^{(x_{pj})^t \bmod q}, \quad j=1 \ldots n \tag{3}$$

If the sub-secret $F_i(x_{pj})$ is not modified during the transmission, then the left side of (3) is $g^{F_i(x_{pj})} = g^{(s + a_1 x_{pj}^1 + a_2 x_{pj}^2 + \ldots + a_{k-1} x_{pj}^{k-1}) \bmod q}$, put (2) into the right side of (3) and get $g^{\sum_{t=0}^{k-1} a_t x_{pj}^t \bmod q}$. Therefore, if $F_i(x_{pj})$ is not changed, the equation of (3) holds and $F_i(x_{pj})$ can be reserved for the decryption, otherwise it is abandoned.

4) When there are k verified sub-secrets, then we can construct a group of equations for the decryption. Without loss of generality, suppose the k verified sub-secrets of stri is $F_i(x_{p1}), F_i(x_{p2}) \ldots F_i(x_{pk})$, and get (4).

$$\begin{cases} F_i(x_{p1}) = s + a_1 x_{p1}^1 + a_2 x_{p1}^2 + \ldots + a_{k-1} x_{p1}^{k-1} \bmod q \\ F_i(x_{p2}) = s + a_1 x_{p2}^1 + a_2 x_{p2}^2 + \ldots + a_{k-1} x_{p2}^{k-1} \bmod q \\ \ldots \\ F_i(x_{pk}) = s + a_1 x_{pk}^1 + a_2 x_{pk}^2 + \ldots + a_{k-1} x_{pk}^{k-1} \bmod q \end{cases} \quad (4)$$

Notice that, the encryption variable x_{pi} in (4) is known in step 2), and a_i is the unknown variable to be solved, then (4) can be rewritten as

$$AX = F \quad (5)$$

Where $A(0) = a_0 = str_i$ is the secret to be decrypted and the determinant of X is

$$|X| = \begin{vmatrix} 1 & x_{p1}^1 & x_{p1}^2 & \ldots & x_{p1}^{k-1} \\ 1 & x_{p2}^1 & x_{p2}^2 & \ldots & x_{p2}^{k-1} \\ \ldots & \ldots & \ldots & \ldots & \ldots \\ 1 & x_{pk}^1 & x_{pk}^1 & \ldots & x_{pk}^{k-1} \end{vmatrix} \quad (6)$$

Apparently $|X|$ is a Vandermonde determinant and $|X| = \prod_{1 \le i, j \le k} (x_{pi} - x_{pj})$; as $\forall i \neq j, x_{pi} \neq x_{pj}$, thus $|X| \neq 0$, according to the Clem rule [10], s is calculated by

$$s = a_0 = \frac{\Delta_1}{|X|} \quad (7)$$

Where Δ_1 is shown in (8)

$$\Delta_1 = \begin{vmatrix} F(x_{p1}) & x_{p1}^1 & x_{p1}^2 & \ldots & x_{p1}^{k-1} \\ F(x_{p2}) & x_{p2}^1 & x_{p2}^2 & \ldots & x_{p2}^{k-1} \\ \ldots & \ldots & \ldots & \ldots & \ldots \\ F(x_{pk}) & x_{pk}^1 & x_{pk}^1 & \ldots & x_{pk}^{k-1} \end{vmatrix} \quad (8)$$

At the receiver end, even if some sub-secrets is lost or polluted, as long as the recipient received k different sub-secrets, then the original data a0=s can be solved by(8), which improves the availability of the communication data. For the attackers, on one hand they have to capture k sub-secrets, which makes the intercepting more difficult. On the other hand, as the attackers do not know the encryption vector Xp used by the two corresponding sides, thus they are unable to decrypt the secrets, which ensures the data confidentiality. In addition,

STDDS incorporates verification mechanism using checksum based on secret cardinal, thus the data integrity is ensured.

4. Experiment

In this section, a group of comparison experiments are designed to test the performance of STDDS. The comparison includes three parts, experiment 1 tests the ability to protect the data confidentiality; Experiment 2 and 3 respectively test the ability to maintain data integrity and availability in the case of noise pollution and data loss. The targets include the widely used scrambling algorithm Arnold [17] and the chaotic encryption algorithm [18]. The parameters are set as follows: data segment length 8bits, the sub-secret number of each segment is n=7, polynomial order k=4. The quantitative indicators include pixel histogram, the peak signal to noise ratio and SSIM.

4.1. *The confidentiality test*

In this section, the standard image Lena is encrypted by STDDS and classical image encryption algorithms Arnold and Chaotic, and then the encryption results are evaluated by pixel histogram. Histogram is an important statistical characteristic of images, which describes the distribution of image pixels, and it has been widely used as a measurement of encryption quality. Typically, the ideal histogram of encrypted image should obey uniformly distribution to avoid ciphertext cracking by histogram analysis [19].

Figure 2 is the histogram analysis results, from the results we can see that the original image and encrypted image of Arnold have the same histograms, because Arnold just change the locations of pixels without modifications of pixels value. Therefore the ciphertext can be eavesdropped by histogram comparisons. But the histograms of the encryption images generated by STDDS and Chaotic are totally different from that of the original. Furthermore, the histograms of the images encrypted by STDDS evenly distribute throughout the interval [0, 255], which makes it difficult to eavesdrop data by pixel distribution analysis.

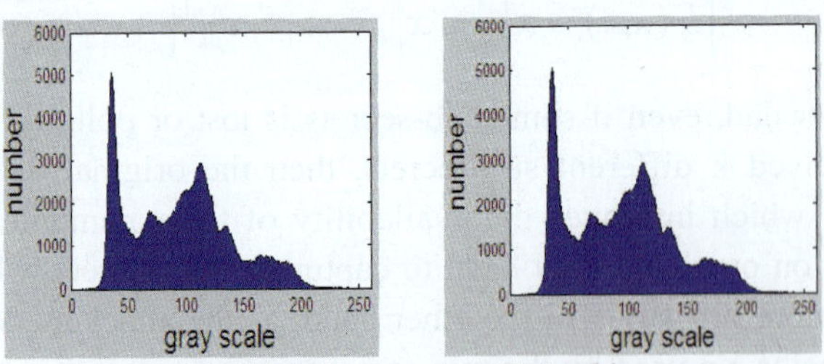

(a) Original (b) Anold

Fig. 2. Histogram analysis.

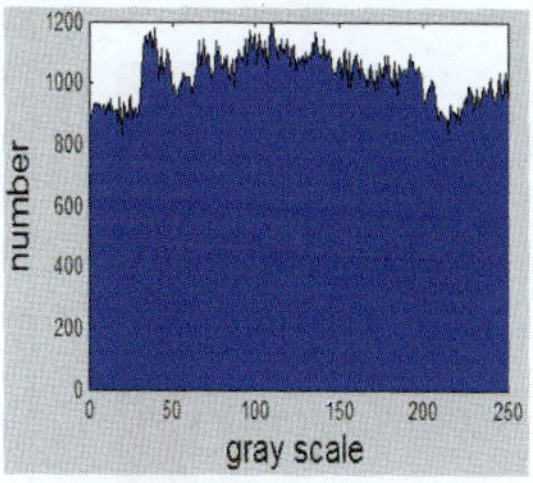

(c) Chaotic

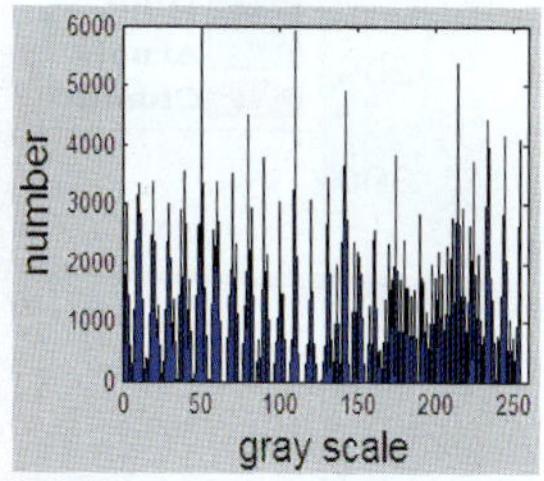

(d) STDDS sub-secret1

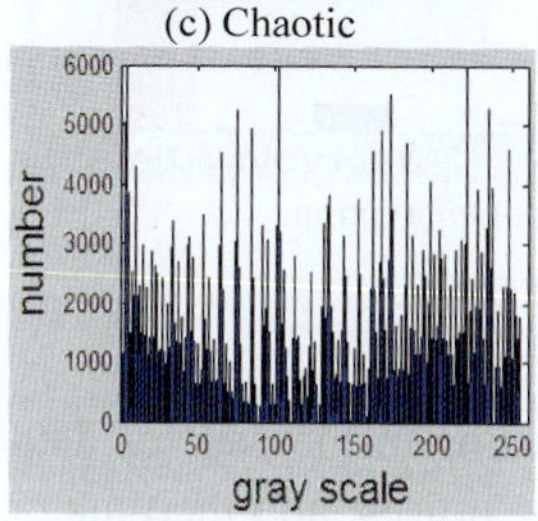

(e) STDDS sub-secret2

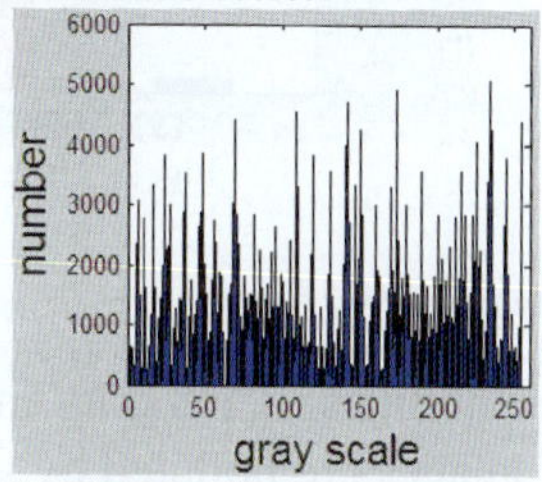

(f) STDDS sub-secret3

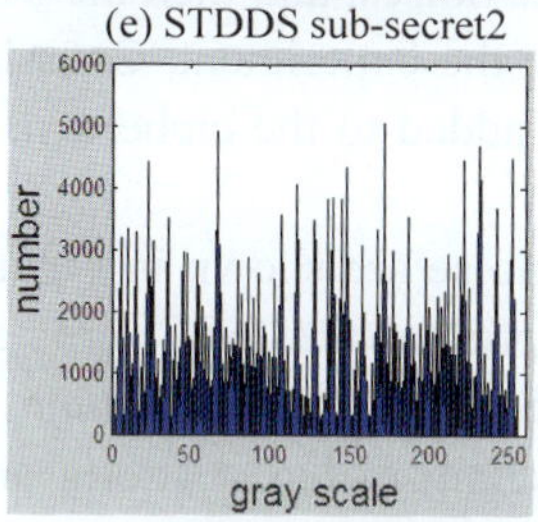

(g) STDDS sub-secret4

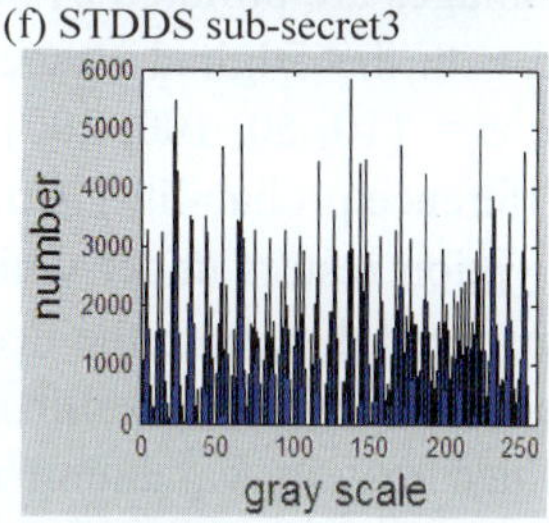

(h) STDDS sub-secret5

Fig. 2. (*Cont'd*).

The time cost of encryption is shown in Figure 3. From the figure we can see that, the time cost of Arnold is the lowest, because it only modifies the location of pixels. Although the polynomial encryption of STDDS increases the time cost, but its encryption and decryption process is mainly performed by polynomial multiplication, thus its cost is similar to Arnold, and far less than that of Chaotic, which based on many XOR operations.

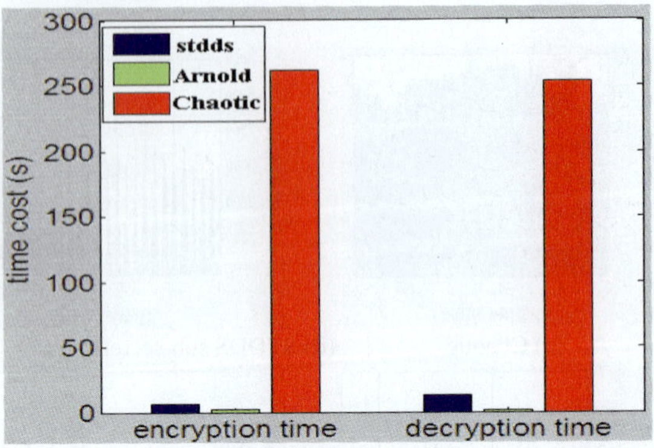

Fig. 3. Time cost of encryption.

4.2. *The integrity test*

To test the capabilities of integrity maintains in the case of data pollution. The ciphertext of images are polluted by white noises, and then the decrypted results are compared with the original image. In the experiment, Gaussian white noise with variance $\sigma = \{10, 50, 100, 200\}$ is added to the ciphertext, and the attack strength (interference probability) is 0.1.

The decryption results under white noise are shown in Figures 4 to 6. The results demonstrated that, both the decryption results of Chaotic and Arnold are distorted. But the decryption result of STDDS maintains high visual quality, which confirmed that its validation mechanism has a strong anti-interference ability to noise.

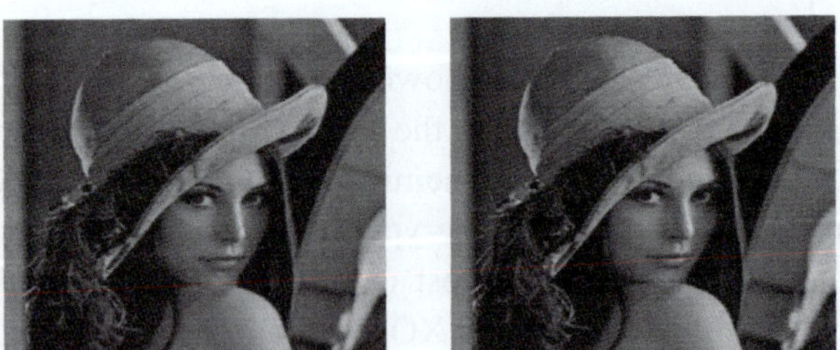
(a) $\sigma = 10$ (b) $\sigma = 50$

Fig. 4. Decryption results under Gauss noise of STDDS.

(c) σ = 100 (d) σ = 200

Fig. 4. (*Cont'd*).

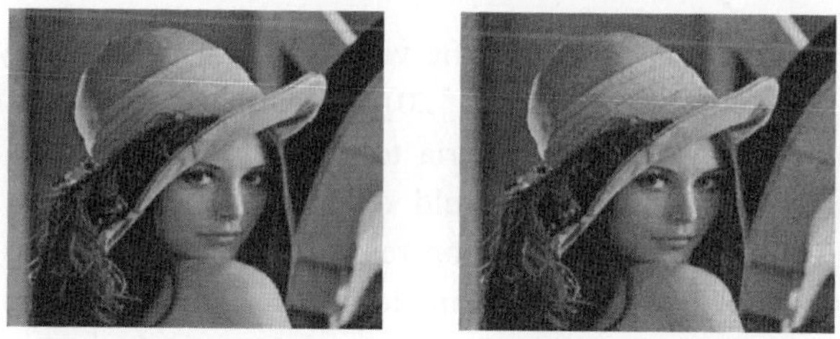

(a) σ = 10 (b) σ = 50

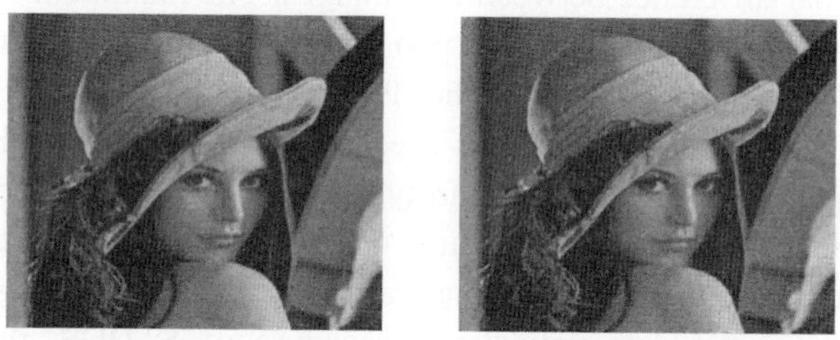

(c) σ = 100 (d) σ = 200

Fig. 5. Decryption results under gauss noise of chaotic.

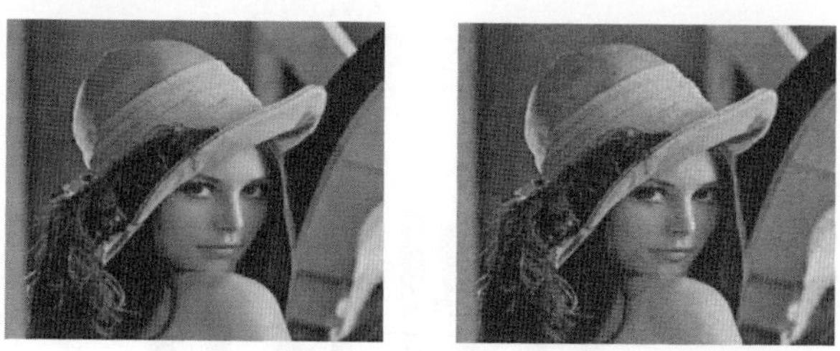

(a) σ = 10 (b) σ = 50

Fig. 6. Decryption results under gauss noise of Arnold.

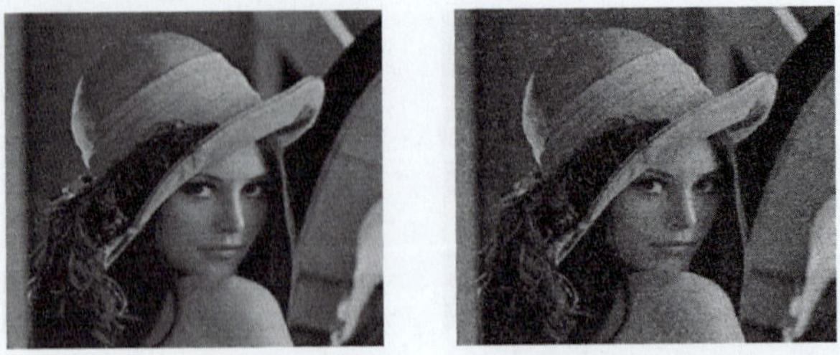

(c) σ = 100　　(d) σ = 200

Fig. 6. (*Cont'd*).

For the quantitative analysis of the visual quality of the decryption results, the peak signal to noise ratio (PSNR) [20] and similarity measure index (SSIM) [21] are employed as objective criteria to compare the integrity performance. PSNR is commonly used in the field of image encryption to evaluate the deterioration of images. The evaluation result of PSNR is expressed in unit dB (decibels). Usually, larger PSNR means less distortion of the decryption result. SSIM is another widely used objective evaluation methods, it can quantitatively reflect the visual difference between two images. The maximum value of SSIM index is 1, which represents the given two images are visually identical [21].

Figure 7 and Figure 8 give the PSNR and SSIM result of different algorithms. From the figures we can see, the visual quality of the decryption images of Arnold and Chaotic sharply degenerates as the attack strength increases. Under the attacking of gauss white noise, the visual quality of the decrypted image from STDDS is maintained even under strong noise. The reason is that, in the data validation procedure of STDDS, the sub-secrets polluted by noise can be found and replaced by other un-polluted sub-secrets to decrypt the image ciphertext, ensuring the restoration of image quality.

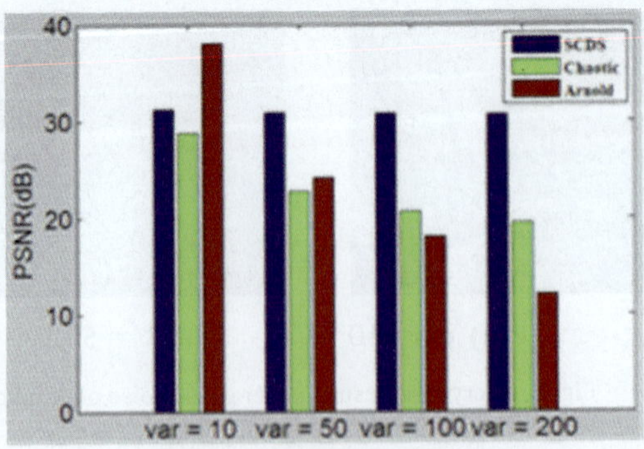

Fig. 7. The histogram of PSNR.

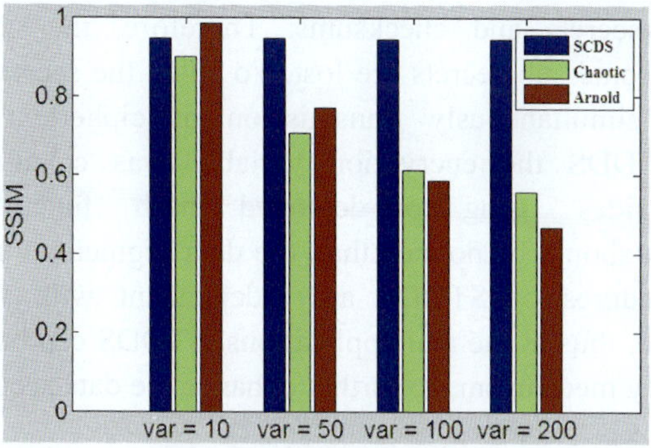

Fig. 8. The histogram of SSIM.

4.3. *The availability test*

We design another couple of comparisons to test the data available ability when part of the ciphertext is lost. In the experiment, the original image is encrypted by STDDS, Arnold and Chaotic respectively, then the ciphertext is randomly removed (the removed pixels are set to 0), and then the ciphertext is decrypted. Again, the peak signal to noise ratio (PSNR) and similarity measure index (SSIM) are utilized to compare the performance.

Table 1. The result of PSNR and SSIM when part of the ciphertext is lost.

Method	PSNR				SSIM			
	1%	5%	10%	5%	1%	5%	10%	15%
STDDS	60.3	43.6	31.8	25.1	0.99	0.99	0.96	0.82
Chaotic	28.7	21.6	18.6	16.9	0.90	0.66	0.51	0.42
Arnold	27.6	20.6	17.7	15.9	0.88	0.66	0.55	0.47

The PSNR and SSIM of different methods are shown in Table 1. The tables show that both the quantitative indicators decrease as the increases of the data lost. The data availability of Arnold and Chaotic is weak, such that even a small ciphertext loss can lead to a poor visual quality. The PSNR and SSIM of STDDS are better than the Chaotic and Arnold. And under the condition that 15% ciphertext data is lost, the SSIM of STDDS is still higher than 0.8 and the PSNR is bigger than 25dB.The results confirmed the STDDS can effectively reduce the defects caused by ciphertext loss, ensuring the data availability.

5. Conclusion

A security digital image transmission method based on data segmentation STDDS was present in this paper. The image was first divided into segments, and then each segment is encrypted by some non-linear equations to generate redundancy sub-secrets and checksums. Therefore, the ciphertext can be decrypted, even some sub-secrets are lost. To solve the secret leaking problem caused by the simultaneously transmission of ciphertext and encryption variables, in STDDS the encryption variable was calculated at the two corresponding sides using pre-deployed hash function to enhance confidentiality. It should be noticed that, the data segmentation and polynomial encryption procedures of STDDS are independent with the original data structure and type, thus in the real applications, STDDS can be easily combined with other security mechanisms to further enhance the data security.

References

1. Jinsoo Shin, Hanseong Son, Rahman Khalil, Gyunyoung Heo. Development of a cyber-security risk model using Bayesian networks [J]. Reliability Engineering & System Safety, 2015, V134, 208-217.
2. Carter, K.M., Idika, N., Streilein, W.W. Probabilistic Threat Propagation for Network Security [J]. IEEE Transactions on Information Forensics and Security, 2013, V9(9):1394-1405.
3. Chasaki, D., Wolf, T. Attacks and Defenses in the Data Plane of Networks. Dependable and Secure Computing [J]. IEEE Transactions on, 2012, V9(6):798-810.
4. El-Zoghdy S F, Nada Y A, Abdo A A. How good is the DES algorithm in image ciphering [J]. Int J Advanced Networking and Applications, 2011, 2(5):796-803.
5. Silva-Garcia V M, Flores-Carapia R. Image encryption based on the modified triple-DES cryptosystem [J]. International Mathematical Forum, 2012, 7(59):2929-2942.
6. A.K Annammal, S. Subha Rani. DICOM Image Authentication and Encryption Based on RSA and AES Algorithm. Trends in Intelligent Robotics, Automation, and Manufacturing, Communication in Computer and Information Science. 2012, v330:349-360.
7. Li C Q, Li S J, Asim M. On the security defects of an image encryption shceme [J]. Image and Vision Computing, 2009, 27:1371-1381.
8. Hou Wen-bin, Wu Cheng-mao. Image encryption and sharing based on Arnold transform [J]. Journal of Computer Application, 2011, 31(10):2682-2686.
9. Liu Ying, Liu Jian-bo. Study and Application of Magic Square Group in Process of Image Scrambling[J]. Computer Technology and Development, 2012, 22(9):0119-0122.

10. Tan Yong-jie, Ma Miao. Image scrambling algorithm based on bit-plane and Gray code [J]. Computer Engineering and Applications, 2010, 46(16):174-177.
11. Bai, S., Cao, C.X. & Han, C.L. 2001. Digital image details hiding technology based on Knight tour transformation. Journal of Image and Graphics 6(11):1096-1100.
12. Zou Jian-cheng, Ward Rabab K, Qi Dong-xu. A new digital image scrambling method based on Fibonacci numbers [C]. In: Proceeding of the IEEE Inter Symposium on Circuits and System, Vancouver, Canada, 2004, 965-968.
13. Lei Zhong-kui, Sun Qiu-yan, Ning Xuan-xi. Image Scrambling Algorithms Based on Knight-tour Transform and its Applications [J]. Journal of Chinese Computers Systems, 2010, 31(5):0984-0989.
14. Fridrich J. Symmetric ciphers based on two-dimensional chaotic maps [J]. Internation Journal of Bifurcat Chaos,1988, 8(6):1259-1284.
15. Pareek N K, Patidar V, Sud K K. Image encryption using chaotic logistic map [J]. Image and Vision Computing, 2006 (24): 926-934.
16. Mazloom S, Eftekhari A M. Color image encryption algorithm based on coupled nonlinear chaotic map [J]. Chaos, Solitons & Fractals, 2009, 42(3):1745 -1754.
17. Khalili, M., Asatryan, D. Colour spaces effects on improved discrete wavelet transform-based digital image watermarking using Arnold transform map [J]. IET Signal Processing, 2013, V7(3): 177-187.
18. A.G. Soriano-Sánchez, C. Posadas-Castillo, M.A. Platas-Garza, D.A. Diaz-Romero. Performance improvement of chaotic encryption via energy and frequency location criteria [J]. Mathematics and Computers in Simulation, 2015, 12,14-27.
19. Y. Zhou, K. Panetta, S. Agaian, C. C. L. Philip, Image encryption using P-Fibonacci transform and decomposition [J]. Optics Communication, 2012, 285(5): 594-08.
20. Soren Forchhammer, Huiying Li, Jakob Dahl Andersen. No-reference analysis of decoded MPEG images for PSNR estimation and post-processing [J]. Journal of Visual Communication and Image Representation, 2011, V22(4):313-324.
21. Z. Wang, A. C. Bovik, H. R. Sheikh, E. P Simoncelli, Image quality assessment: from error visibility to structural similarity [J]. IEEE Transactions On Image Processing, 2004, 13(4):600-612.

SAR image denoising based on partial differential equation

X.T. Chen[1], Y. Cai[2†] and L.Y. Dong[3]
College of Mining Engineering,
North China University of Science and Technology,
Tangshan, China
[†]Email: 550910904@qq.com

In order to filter out the speckle noise in the imaging process of SAR system, this paper puts forward a kind of comprehensive partial differential equation based on P -M andt he high order equation LLT, The algorithm takes advantage of the P - M equation which retains the characteristics of the image edge. Also the high order equation LLT method can recover the smooth area of the SAR image and avoid the block effect. This experiment shows that the combining model can complement each other, which would be a better solution to filter out the speckle noise and maintain the image detail.

Keywords: SAR image; Denoising; P-M equation; LLT equation.

1. Introduction

Inherent speckle noise in SAR systems will seriously hinder extract the features in the image data [1-2]. How to effectively filter speckle noise is key to SAR image preprocessing work, which has very important theoretical value and practical application value.

Image denoising based on partial differential equation are mainly composed of second order partial differential equation. The model has a success in filtering out noise and protecting edge except in the image characteristics. Along with the development of the partial differential equation, the researchers proposed high order partial differential equation model, the model is mainly made up by fourth order partial differential model. The high order partial differential model can protect the block effects caused by image processing. Therefore this paper is based on the theory of SAR image denoising, to set up a combined mathematical

[1] Xing-Tong Chen
[2] Yang Cai
[3] Li-Ya Dong

model of P-M and LLT equation. Then this paper verifies the validity of this model through the simulation results.

2. The Comprehensive Model P-M & LLT Model

2.1. *P-M equation*

Perona and Malik [3] propose the following equation in 1990:

$$\begin{cases} \dfrac{\partial u}{\partial t} = div(g(|\nabla G_0 * u|)\nabla u) \\ u(x,y,0) = u_0(x,y) \end{cases} \tag{1}$$

$$\begin{cases} \dfrac{\partial u}{\partial t} = div(g(|\nabla G_0 * u|)|\nabla u|) \\ u(x,y,0) = u_0(x,y) \end{cases} \tag{2}$$

This is called P-M equation, the g called conduction function in the type generally takes to:

$$g = \frac{1}{1 + \frac{|\nabla u|^p}{k}}, p = 1,2 \tag{3}$$

The k is a critical reference value.

Through the analysis of g, when ∇u is much bigger than k, the variation of this image is large, then at this point the P - M equation stops denoising in the image edge. When ∇u is much less than k, the variation of this image is very small, the P - M equation starts denosing in the image smooth area.

2.2. *LLT equation*

Since the P - M equation can't solve the problem of protecting image characteristics very well, the researchers put forward a new model of partial differential, that is, the high order partial differential [4]. One of the most classic is the LLT equation.

This model can be represented by energy functional minimization problem as following:

$$\min E(u) = \iint_{\Omega} (\sqrt{u_{xx}^2 + u_{xy}^2 + u_{yx}^2 + u_{yy}^2} + \frac{\lambda}{2}|u_0 - u|) \quad d_x d_y \tag{4}$$

The gradient descent flow of variational method is as following:

$$f = \frac{\partial u}{\partial t} = (-(\frac{u_{xx}}{|D^2 u|})_{xx} - (\frac{u_{xy}}{|D^2 u|})_{xy} - (\frac{u_{yx}}{|D^2 u|})_{yx}$$

$$- (\frac{u_{yy}}{|D^2 u|})_{yy} - \lambda(u - u_0)) \tag{5}$$

$$(|D^2 u|)_{i,j} = \sqrt{(u_{xx}^2)_{i,j} + (u_{xy}^2)_{i,j} + (u_{yx}^2)_{i,j} + (u_{yy}^2)_{i,j} + ff} \tag{6}$$

ff is usually taken to 10^{-10} which is a small positive number, its role is to avoid the denominator becoming to zero.

Compared with P-M equation, LLT equation can avoid the block effect and protect the image subtle character. But the efficiency of image denoising and edge protection is not as good as P-M equation.

2.3. *The comprehensive model*

In this paper, getting the P - M and LLT equations together by using θ, and setting P - M equation for f_1, LLT equation for f_2, then the comprehensive model is as follows:

$$u = \theta * f_1 + (1 - \theta) * f_2 \tag{7}$$

From the type (7): When in the edge and smooth region of the image, $\theta \approx 1$, P -M equation plays a main role, namely image is denoising and protecting the edge. When in the image gradient zone, $\theta \approx 0$, then the LLT equation plays a main role, namely image is denoising and on the other side protecting the feature of SAR image. So the θ is equal to the following equation :

$$\theta = \begin{cases} 1, & |\nabla u| \geq c \\ \frac{1}{2} = \cos(\frac{2\pi|\nabla u|}{c}) + \frac{1}{2}, & |\nabla u| < c \end{cases} \tag{8}$$

∇u represents the gradient value of the image. c is always between (0.05,0.2), the θ is shown in Figure 1:

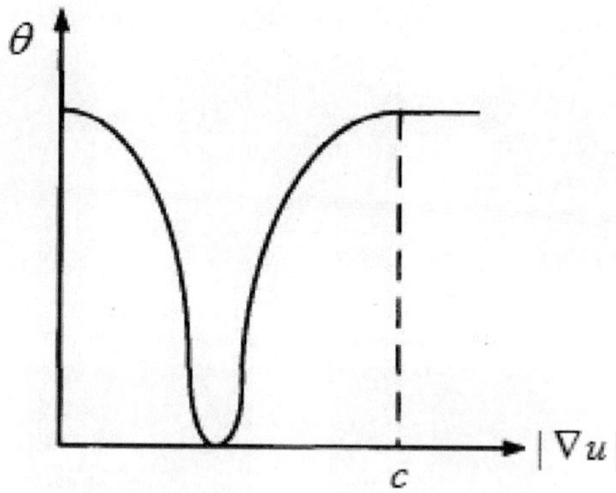

Fig. 1. The weighting function diagram.

According to the outliers σ_{ε} [5], this paper replaces c by σ_{ε} :

$$c = 1.4826 median(|\nabla u| - median(|\nabla u|)) \tag{9}$$

In this paper, calculate the P-M&LLT equation using the numerical calculation. First, calculating $|\nabla u|$ and taking $|\nabla u|$ into θ, then calculating f_1 and f_2 equation, their discrete forms are as follows:

$$\theta_1 = \theta\left[\Sigma \frac{g_p + g_s}{2}(u_p - u_s)\right] + (1-\theta) * \left(\frac{u_{xx}u_y^2 - 2u_x u_y u_{xy} + u_x^2 u_{xy}}{|\nabla u|^2}\right) \tag{10}$$

$$u_t = \sum_{k=1}^{4} \Delta^2 g_k + \lambda(u_0 - u) \tag{11}$$

s and p represent grid point coordinates, the first iteration of image matrix is completed by taking θ into formula (7). Then updates θ and repeats iteration until the upper limit of iterations or algorithm convergence.

$$U = \theta^{n=1} * f_1^{n=1} + (1 - \theta^{n+1}) * f_2^{n+1} \tag{12}$$

3. Application Example

In this paper, setting the SAR image and TM image as experiment object. The SAR-1 image, SAR-2 image, SAR-3 image ,TM-1 image, TM-2 image, and TM-3 image were taken by the ERS satellites of European space agency. As the SAR images are generally above 50 million pixels [6-7], this study only captures part of the image (512 * 512 pixels) to experiment using MATLAB7.0.4 software. The results are as following:

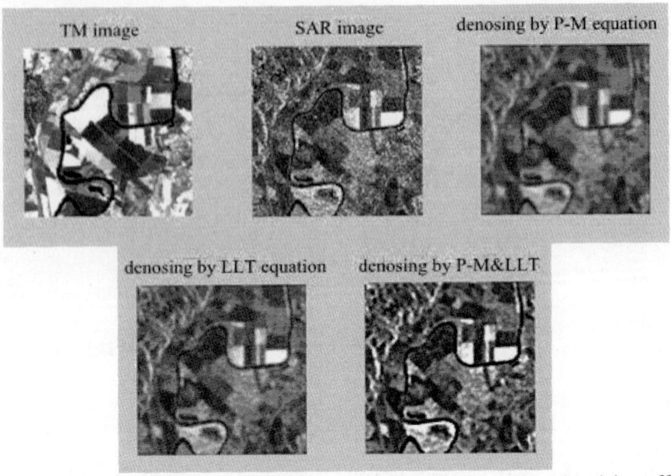

Fig. 2. Three kinds of partial differential equation of SAR-2 image denoising effect.

Table 1. Various partial differential equations denoising of SAR-2 image results comparison.

	Iterations	PSNR	SNR	MSE
SAR image		25.2079	19.4312	330.2414
P-M equation	70	28.2561	22.2098	305.6421
LLT equation	40	29.1285	23.2563	283.4579
P-M& LLT equation	18	29.5278	23.4871	272.4876

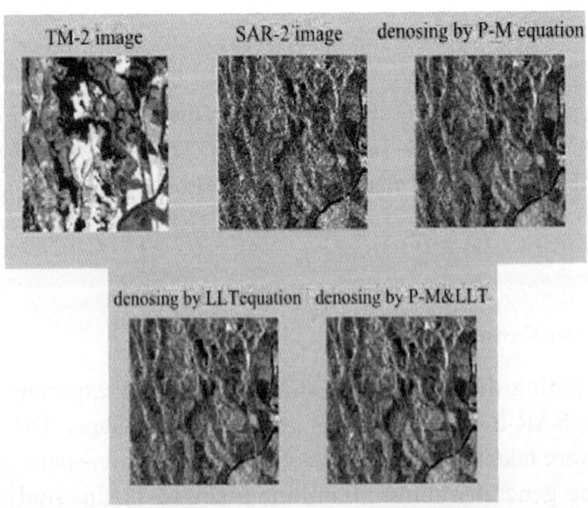

Fig. 3. Three kinds of partial differential equation of SAR-2 image denoising effect.

Table 2. Various partial differential equations denoising of SAR-2 image results comparison.

	Iterations	PSNR	SNR	MSE
SAR image		33.1468	26.4048	229.8637
P-M equation	72	33.1458	26.4489	220.5977
LLT equation	53	33.1474	26.4326	218.4874
P-M& LLT equation	46	33.1493	26.4871	215.7615

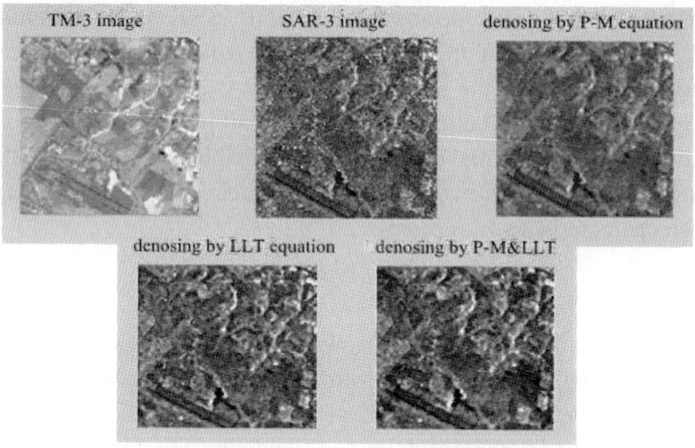

Fig. 4. Three kinds of partial differential equation of SAR-3 image denoising effect.

Table 3. Various partial differential equations denoising of SAR-3 image results comparison.

	Iterations	PSNR	SNR	MSE
SAR image		38.4329	34.6572	180.3217
P-M equation	68	42.8516	36.3238	152.2975
LLT equation	48	44.1387	37.5161	133.1108
P-M& LLT equation	28	44.9623	38.2436	120.5312

4. Conclusion

According to the above three results, denoising effect of the SAR-1 image is the same with SAR-3:

(1) The PSNR of the P – M equation is minimum, that is to say, the denoising effect of the P - M equation is the worst. From the Figs. 2 and 4, after denoising, the SAR image is not clear, so the denoising effect of P - M equation is not obvious, and even appears block effect. and requires iterative many times.

(2) The denoising effect is obviously improved by LLT equation, on the other hand the image details are protected very well.

(3) The P - M&LLT comprehensive model put forward in this paper has two advantages compared with the previous model. First of all, the model improves the denoising effect and the ability of edge protection, at the same time reduces the operation time and the number of iterations. The second, it shows that the P - M&LLT model has better denoising effect for the SAR image having field, rivers and linear features obvious.

For SAR-2 image, after denoising the PSNR, SNR and MSE almost have no change. The reasons for this situation is the alluvial fan characteristics of SAR-2 image. Because of the terrain of the SAR-2 image, most of the speckle noise obeys Rayleigh distribution, and only a small part of them obey Gaussian distribution. Partial differential equation is not suitable for Rayleigh distribution, in conclusion P - M&LLT equation is not suitable for the SAR images with alluvial fan characteristics.

References

1. Bin Zhou, Xiao Lin Yang, Rui Liu, Wei Wei. Image Segmentation with Partial Differential Equations [J]. Information Technology Journal, 2010, 95.
2. Naosheng Qiao, Beiji Zou. Nonlocal orientation diffusion partial differential equation model for optics image denoising [J]. Optik - International Journal for Light and Electron Optics, 2012.
3. Liang Xiao, Lili Huang, Zhihui Wei. Comments on "Staircase effect alleviation by coupling gradient fidelity term" [J]. Image and Vision Computing, 2010,2811.
4. Naosheng Qiao, Beiji Zou. Nonlocal orientation diffusion partial differential equation model for optics image denoising [J]. Optik - International Journal for Light and Electron Optics, 2013, 12414.
5. Yi-Fei Pu, Patrick Siarry, Ji-Liu Zhou, Ni Zhang. A fractional partial differential equation based multiscale denoising model for texture image [J]. Math. Meth. Appl. Sci., 2014, 3712.
6. Yi-Fei Pu, Ji-Liu Zhou, Patrick Siarry, Ni Zhang, Yi-Guang Liu, Juan J. Trujillo. Fractional Partial Differential Equation: Fractional Total Variation and Fractional Steepest Descent Approach-Based Multiscale Denoising Model for Texture Image [J]. Abstract and Applied Analysis, 2013, 2013.
7. Zhenhe Ye, Xin Li, Ying Li. Visual and Artistic Images Denoising Methods Based on Partial Differential Equation [J]. Journal of Multimedia, 2013, 83.

A testability-demonstration software platform based on fault injection

G.H. Wang[1], W. J. Qin[2†] and W. S. Zhang[3]
School of Instrumentation Science and Opto-electronics Engineering,
Beihang University, Beijing, China
Email: wgh@buaa.edu.cn (G.H. Wang)
†Email: mengwuxiy@163.com (W. J. Qin)
Email: 1443908020@qq.com (W. S. Zhang)

Testability is a characteristic of a system or a device to check its status and isolate its inner faults. To check whether the testability meets the requirement before acceptance, testability demonstration is usually conducted by injecting faults into the equipment. However, with the increasing scale and integration of the equipment system, manual testability demonstration takes too much time and man power. To automate the process of the testability demonstration and overcome the data share obstacles, this paper proposed a testability-demonstration software platform based on fault injection. For ten years, our ATS software VITE (Virtual Instrument Test Environment) has been widely used in the aircraft industry. Based on the software, a testability demonstration software platform is proposed to automate the testability demonstration process and realize data sharing. The platform includes the fault sample allocation software (FSAS), the master control software (MCS), the fault injection software (FIS), and the data monitor software (DMS). The practical application has proved that the software platform is able to decline the test time and improve the efficiency significantly.

Keywords: Testability demonstration, Fault sample allocation, Fault injection, Master control, Test case.

1. Introduction

Testability is a design characteristic of a system or a device to check its status accurately and isolate its inner faults, and testability demonstration is to check whether the product meets the standard of the special testability requirement. Nowadays, the primary method of testability demonstration is to inject a number of faults into the unit under test (UUT), detect and isolate the fault in the way provided by the design of the testability, evaluate the level of testability by test result, and finally judge whether the product acceptable [2,9,10] (Bertolino & Strigini, 1997; Shi & Chao 2012; Wang & Qiu, 2014). Lots of experiments has proved that the testability demonstration is able to reflect the real status and the environment, function, action, fault status and test result of the equipment. However, with the increasing scale and integration of the equipment system, the time spent on testability demonstration is much more, and the workload of the testability demonstration becomes even heavier. As a result, improving the automation degree of testability demonstration is more and more important.

Recently, several methods of the testability demonstration have been proposed. Xu proposed a method to complete the testability demonstration of a type

---

[1] Guo-Hua Wang. wgh@buaa.edu.cn
[2] Wei-Jun Qin, corresponding author. mengwuxiy@163.com
[3] Wei-Shuo Zhang. 1443908020@qq.com

of the warship battle system equipment [11] (Xu, 2010), He Yang proposed a method to conduct the testability demonstration of the electromechanical equipment [13] (He, 2011), Xue also propose a method based on virtual instrument to complete testability demonstration [12] (Xue, 2010), and Chang proposed a method to conduct the demonstration experiment of radar equipment [3] (Chang,2011). However, the automation of testability demonstration process appears less applicable the difficulty in realizing interoperability between commercial ATS software.

This paper proposed a testability-demonstration software platform based on fault injection. The proposed software platform is aimed to conduct the testability demonstration test automatically, saving a lot of time.

2. The Testability Demonstration Platform

The proposed software platform includes the following pieces of software: the fault sample allocation software (FSAS), the master control software (MCS), the fault injection software (FIS) and the data monitor software (DMS). The platform works in distributed environment, see Figure 1.

The FSAS is in charge of the allocation of the original fault sample and the generation of test cases. Firstly, it allocates the fault sample using the failure mode information generated by MFECA analysis software. Then, test cases are generated with the allocated fault sample based on the test case models. At last, the TPS and fault injection program are developed based on the test cases.

The MCS is responsible for the synchronization between the pieces of software and the data management, and it is also used to define the map between fault injection programs and the data checking rules. During testability demonstration, the MCS coordinates the FIS and DMS by sending commands, and judge the demonstration result by checking the acquired data from the DMS.

The FIS based on VITE [14] (Yu, Ma & Yang, 2006) I s to conduct fault injection and the DMS is to acquire and check the real-time data.

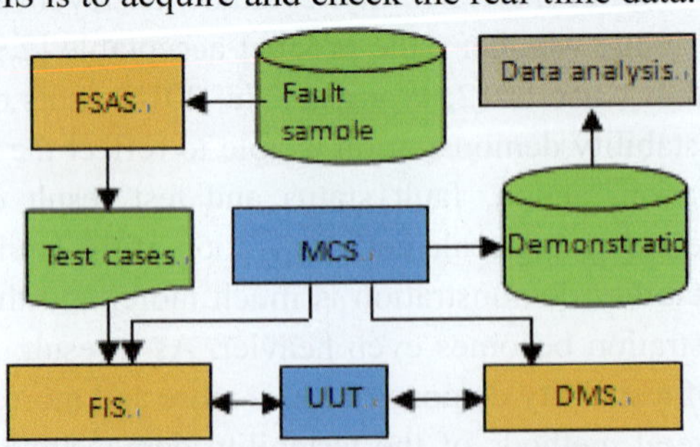

Fig. 1. The structure of the software platform.

3. FSAS

The FSAS is the foundation of the software platform, as it provides the fault sample and test cases for other software [7] (Li & Qiu, 2009).

3.1. *Failure mode information*

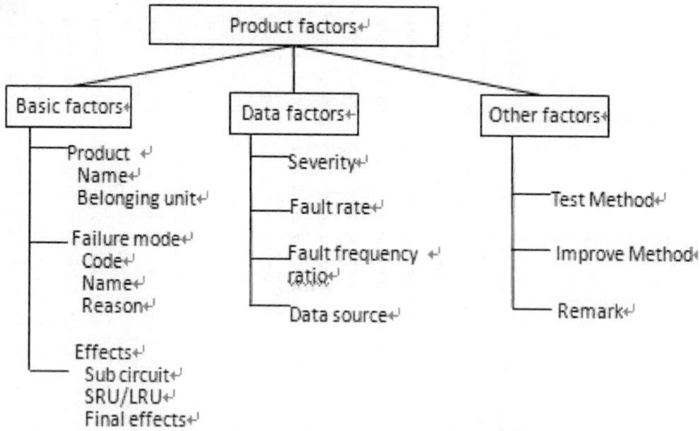

Fig. 2. The failure mode information.

The input of the FSAS is the MFECA result. The information from MFECA result used for fault sample allocation and test cases generation includes: product name, fault-mode code, fault-mode name, fault reason, influence, severity, test method, fault rate, fault frequency ratio, data source, and remark, see Figure 2. This information is useful for fault sample allocation.

3.2. *Fault sample allocation*

To allocate the fault samples, the follow steps must be performed based on the failure mode information.

See Figure 3.

(1)Click the button "Calculate N" and calculate the preliminary sample size N

(2) Input the count x

(3) Click the button "Generate Random Numbers" and generate x groups of random numbers, observe the random numbers and the uncovered number in every group. The number of distributed samples and complement samples is filled automatically

(4) Click the cell Observe and modify some numbers generated in the previous step

(5) Select a group of the random numbers with the minimum uncovered number of the failure modes, save the sample allocation information, generate the alternative sample library

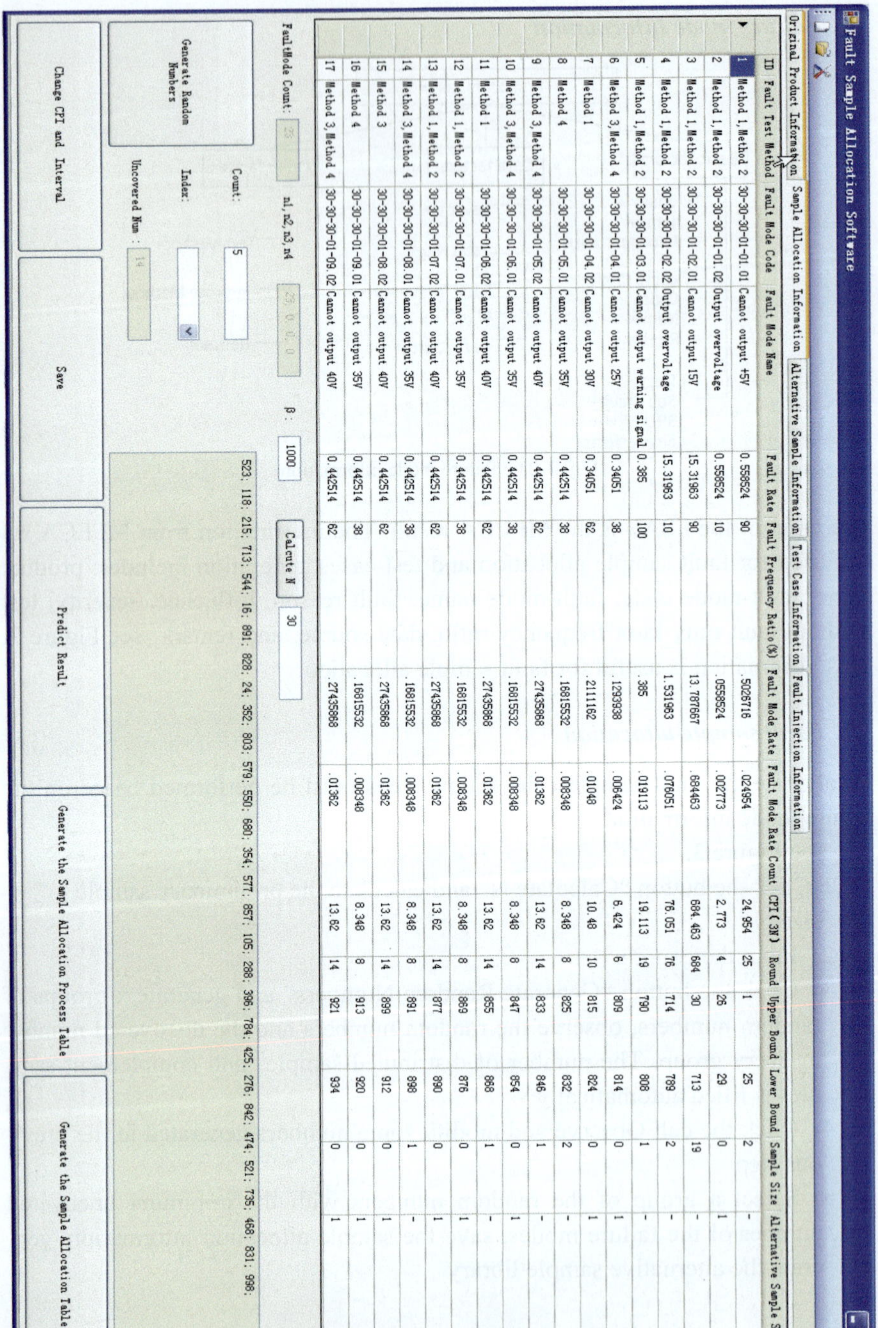

Fig. 3. The process of fault sample allocation.

(6) Click the buttons "Generate the Sample Allocation Process Table" and "Generate the Sample Allocation Table", generate the sample allocation table and the test case table, finishing allocating the fault sample.

3.3. *Test case generation*

Test case can be seen as the principle of the fault injection program. Test case is involved with failure mode, fault injection method, test method and so on. The procedure of the test case generation is as following, see Figure 4.

(1) Fill the necessary information.
(2) Click the button "Add Equipment", add the potential required equipment to the list in the right part.
(3) Select the required equipment.
(4) Click the button "Change Step Model", select the test case template.
(5) Click the buttons "Save" and "Generation the Test Case Table", finishing the test case generation process.

4. MCS

The MCS manages the testability demonstration program which maps the fault injection program to the TPS. The user interface is shown in Figure 5. The left part is the failure modes of the demonstration project, the middle part is the test cases involved with the select failure mode, and the right part displays the map from fault injection program set to TPS and the parameter criterion. During testability demonstration, the MCS coordinates the FIS and the DMS, and save the demonstration result. The scheduling process is shown in Figure 5.

The user interface is as the picture above. The left part presents the failure modes in the opened task, the middle part presents the test cases involved with the selected failure mode, and the right part displays the map from fault injection program set to TPS and the parameter criterion. A failure mode usually corresponds to more than one test case, but the criterion can only be one.

4.1. *Process scheduling*

As the administrator of the software platform, one of the MCS's functions is to dispatch the FIS and the DMS. The scheduling process is as shown above.

The MCS works as a client, and the FIS and the DMS work as servers. When the MCS attempts to execute a demonstration task, it firstly connects to the FIS and the DMS. Only if these connections are created, the MCS is able to perform later steps. The MCS sends an instruction to the FIS to remind the FIS to inject a fault, and the FIS will conduct fault injection and send the result back to the MCS. If the result is successful, the MCS will send an instruction to the DMS to notice the DMS to test and monitor. The DMS will send data acquired back to the MCS.

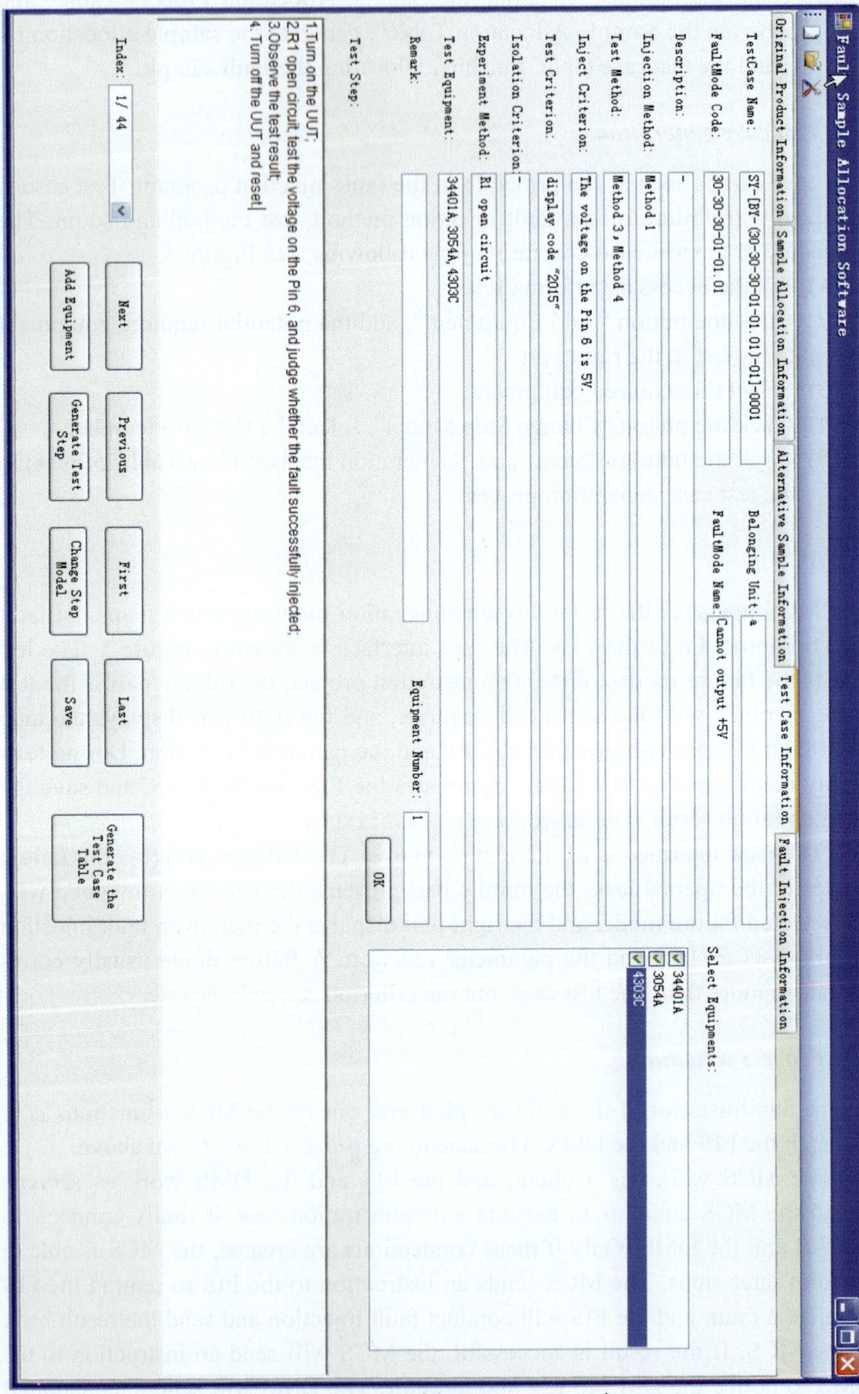

Fig. 4. The process of the test case generation.

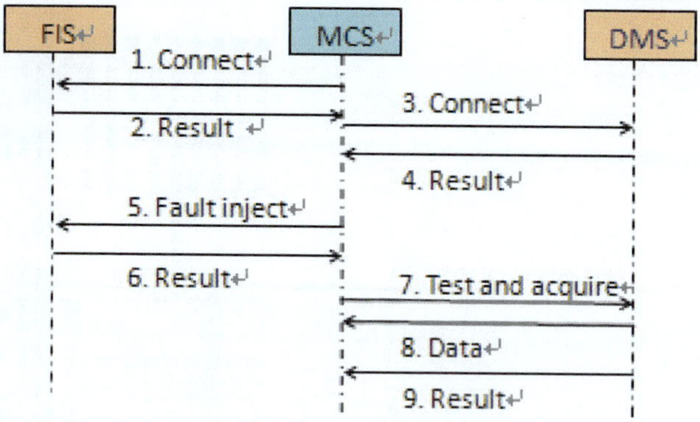

Fig. 5. The scheduling process of the MCS.

When the test finishes, the DMS will send the result back to the MCS, and the MCS deals with the data and judges the demonstration result.

4.1.1. *Fault injection set management*

The MCS is able to manage the fault injection set. A fault injection set includes a set of failure modes, and a failure mode includes a variety of test cases. The fault injection set can be imported. To import the fault injection set, the operator chooses a specialized belonging unit or subsystem to import it, besides, the MCS can also import a single failure mode. If an imported failure mode is removed, the involved fault injection information will be removed at the same time.

4.2. *Project management*

A project is defined as an object to be verified. A project can be seen as a product in a certain degree. It includes several failure modes. Each of the failure modes includes several fault cases, and each case needs defining its fault injecttion information and result parse information. When creating a new project, the operator selects a product and creates an empty project at first, and then selects a failure mode of the product, selects one of the test cases, the involved fault injection program set and the test steps in the selected fault injection program set, later, selects a parameter in the parameter library as the criterion of the fault in jection result, and finally save the project information to the data base. The operator can create, modify or remove a project.

4.3. *Demonstration task management*

A demonstration task is to verify a specialized target. It can be seen as the execution of the project. To create a new task, at first, the operator needs to select an existing project, and then edits the included test cases (add or remove), and the other information will be added automatically.

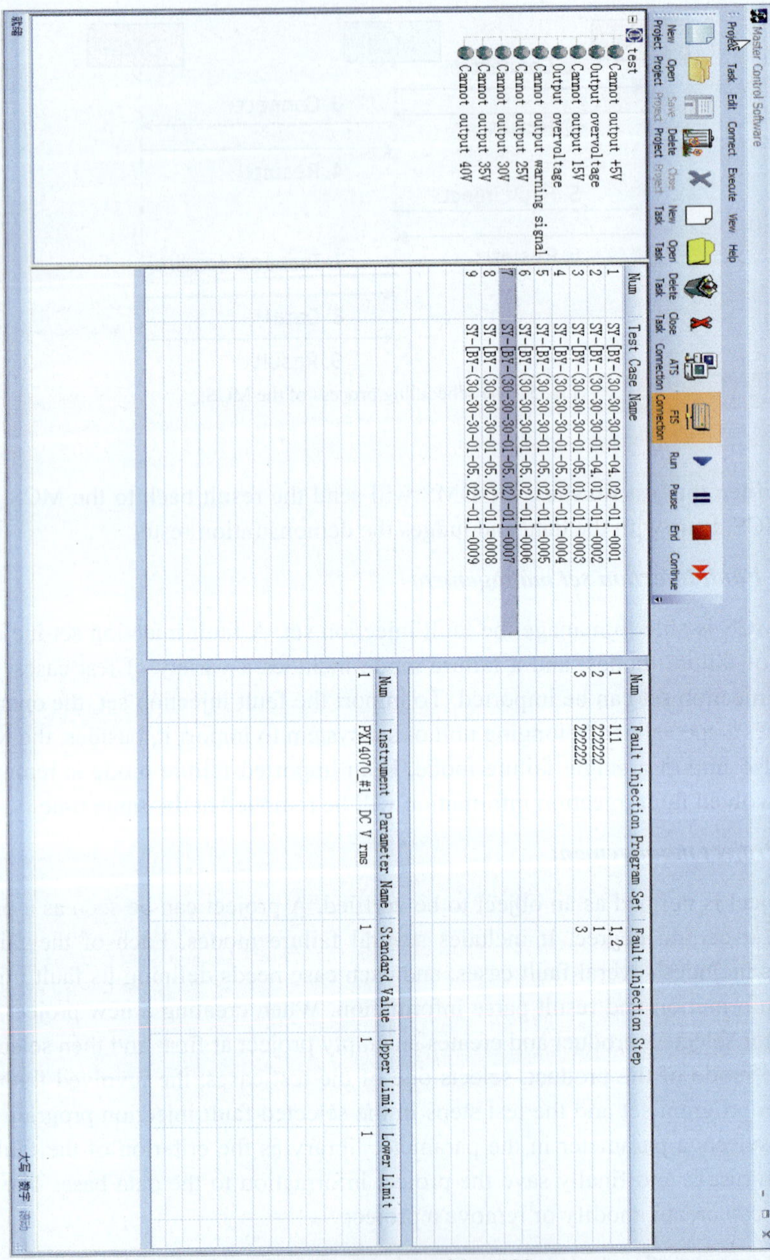

Fig. 5. The interface of the MCS.

Up to now, we can verify the selected product. The demonstration process can be automatic or manual. If in the automatic mode, the fault injection program set is executed in order, additionally, the MSC provides the function of stopping, pausing, resuming, single-step-execute and continuous-execute. The executing step and the real-time data are displayed to the operator when running.

A task can be executed multiple times, and each execution generates a demonstration result. If a task is executed more than one time, only the last validate result is saved. What's more, a task can be executed discontinuously.

5. FIS and DMS

The FIS and the DMS are the real executors in the software platform.

5.1. *FIS*

The fault injection software is to inject a specialized fault into a specialized unit to test the product. The FIS is modified from VITE. As the most important part of the testability demonstration technology, the FIS realizes the fault injection[1,5,6,8,15] (Arlat & Crouzet, 2003; Hsueh & Tsai, 1997; Lei & Qin, 2012; Madeira & Costa, 2000; Zhao & Liu, 2011).

5.1.1. *Fault injection program set management*

A fault injection program is to execute the fault injection and the fault injection program set is defined as a collection of the fault injection programs. A fault injection program is a collection of test points (or fault injection steps).

To create a fault injection program, create an empty fault injection program at first. Later, create an empty step in the fault injection program, select an instrument to inspire the fault signal, and configure the other information in the Figure 7 at the same time. And then, configure the run-time parameter of the selected instrument. Finally, save the current step. Up to now, the fault injection program includes a fault injection step. Repeat this process until the fault injection program is completed. An operator can create, modify or remove a fault injection program.

5.1.2. *Fault injection*

As the name shows, the major work of the FIS is fault injection. FIS can inject a variety of fault signals, for instance, bus signal, source signal and so on. The process of fault injection in this software is to execute a fault injection program to inspire a fault signal, and then observe how the UUT reflects according to the signal. The FIS begins to inject a fault right after it receives the instruction from the MCS.

The FIS is able to work in two different modes. Considering the FIS and the MCS may not exist on a same computer, the FIS needs working in the web mode. When in the web mode, it adjusts itself synchronized with the MCS to keep the data validate via the instructions transported by web. When in the local mode, that is, the FIS and the DMS exist on a same computer, these pieces of software work in line with each other under the management of the MCS.

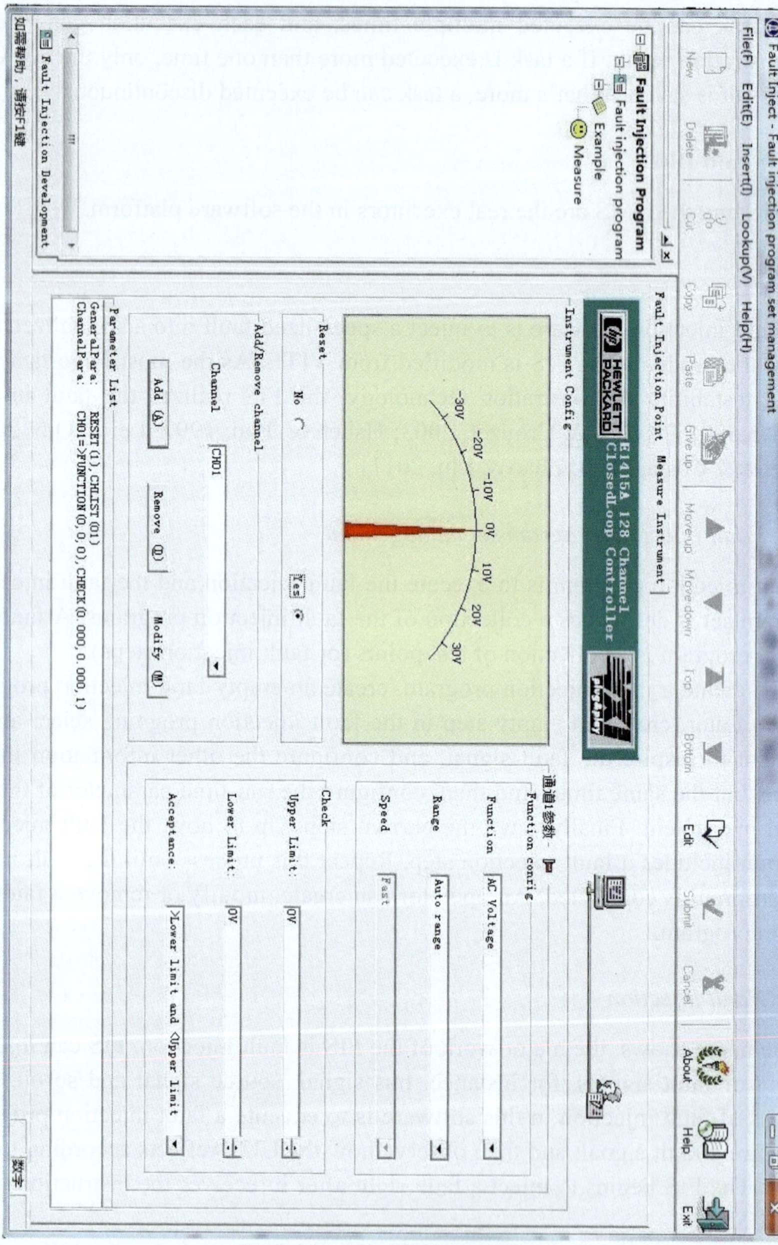

Fig. 7. The fault injection program.

Considering injecting a fault often needs several instruments working at the same time, a common instrument model of all the instruments used is defined. That means, all the instrument drivers have the same structure. In addition, we place the threads into the instrument drivers to make the threads separate. Considering that an instrument may be used in two different pieces of software, the

ActiveX Template Library (ATL) technology is adopted to develop the instrument drivers so that the drivers can be transported through the network [4] (Cui & Chen, 2009). It makes a great contribution to the management of the instruments.

5.2. *DMS*

The DMS is designed to acquire the real-time data to verify the UUT. The DMS is an important part of the software platform.

5.2.1. *Data monitor*

The DMS provides the function of monitoring data from the UUTs. When the operator begins to execute a task, the DMS initializes and tests the required instruments. And then the DMS allots a thread for each of the instruments, avoiding the data to be confusing. When collecting a new data of a parameter required, the DMS will display it to the operator and save it.

For each of the monitor resources on the monitor resource configuration page, the operator selects an instrument and a parameter of the selected instrument to monitor. As the picture below, the operator can add a monitor resource from the right-top corner of the software, and then click on the control to modify the appearance of the work area, where the background color is black. When the thread is started, the operator can see the real-time data.

Additionally, for all these pieces of software, we provide the functions such as user management, instrument management and log service and so on for the maintenance of the software platform. A user must login the software to obtain the right to operate the software. The default user is the so called administrator, which is in charge of the other users. An operator can add, remove or modify an instrument supported after he (or she) logs in the software. And when a piece of software encounters an important event, it saves the involved information to the log file.

6. Conclusion

The platform introduce in this paper is used to conduct the testability demonstration for complex equipment, and the demonstration time is expected to decline from months to weeks. Our applications indicate that the proposed software platform is feasible and the efficiency is improved.

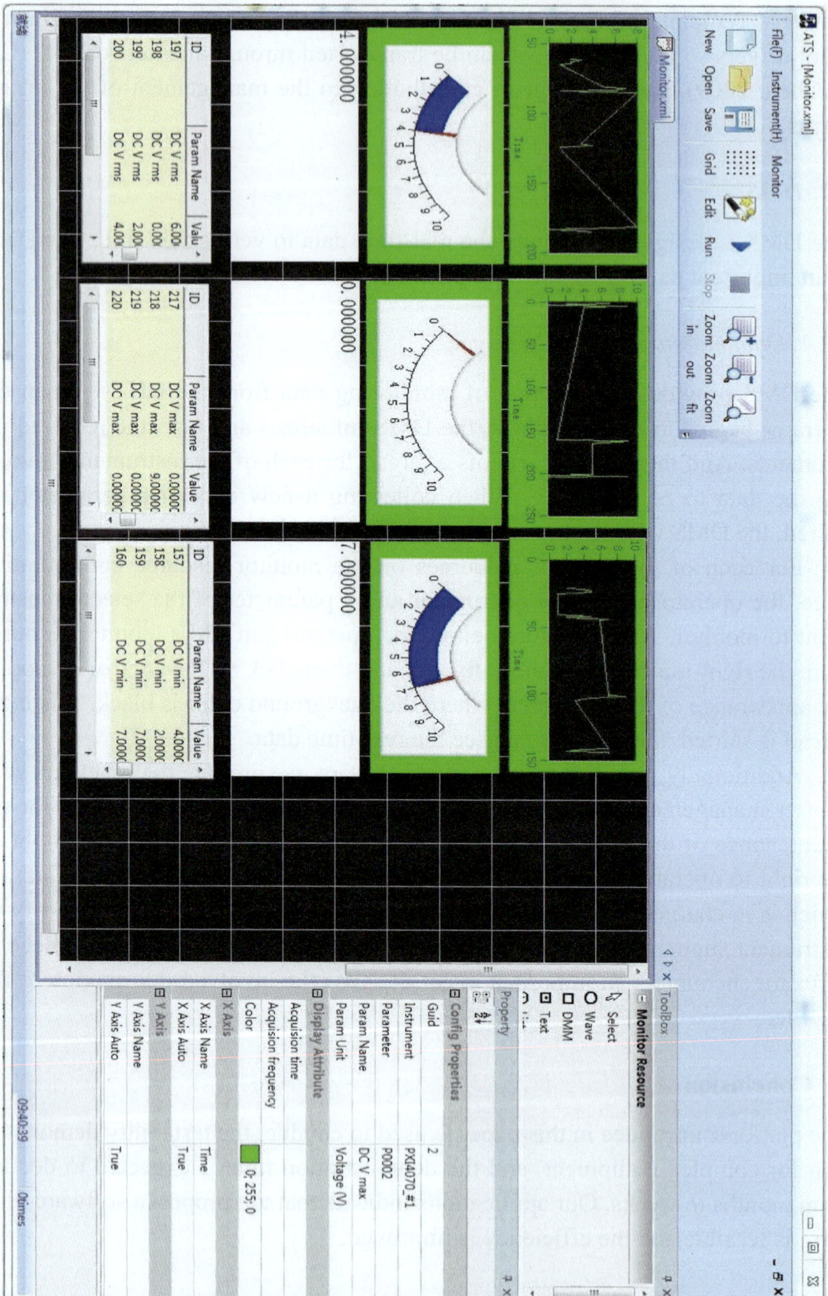

Fig. 8. Data monitor.

References

1. Arlat J, Crouzet Y, Karlsson J, et al. Comparison of Physical and Software-Implemented Fault Injection Techniques [J]. IEEE Transactions on Computers, 2003, 52(9):1115-1133.
2. Bertolino A, Strigini L. Acceptance Criteria for Critical Software Based on Testability Estimates and Test Results [J]. Safe Comp, 1997, 10(4):83-94.
3. Chang C, Yang J, Wang J. Research on Verification and Application of Testability of Radar Equipment [J]. Computer Measurement & Control, 2011, 19(8):1943-1942.
4. Cui K L, Chen Y, Zhang G Y. A study of instrument driver based on COM technology [J]. Electronic Instrumentation Customer, 2009.
5. Hsueh M C, Tsai T K, Iyer R K. Fault Injection Techniques and Tools [J]. Computer, 1997, 30(4):75-82.
6. Lei Hua-jun, Qin Kai-yu. Bayesian method for determination of testability demonstration test scheme [J]. Systems Engineering & Electronics, 2012, 34:2612-2616.
7. Li T, Qiu J, Liu G. Allocation Plan of Failure Samples Based on Failure Rate in Testability [J]. Acta Aeronautica Et Astronautica Sinica, 2009, 30(9):1661-1665.
8. Madeira H, Costa D, Vieira M. On the emulation of software faults by software fault injection [C] Dependable Systems and Networks, 2000. DSN 2000. Proceedings International Conference on. IEEE, 2000:417-426.
9. Shi J Y, Chao J I, Hai-Wei L I. Analysis of Testability Verification Technology and Application Status [J]. Measurement & Control Technology, 2012, 31(5):29-32.
10. Wang C, Qiu J, Lieu G J, et al. Testability demonstration with component level data from virtual and physical tests [J]. Proceedings of the Institution of Mechanical Engineers Part C Journal of Mechanical Engineering Science, 2014, 229(2):265-272.
11. Xu H Q. Testing Verification Technology of Warship Battle System Equipment [J]. Ship Electronic Engineering, 2010, 30(3):136-139.
12. Xue J, Cai J Y, Sa-Sa M A. Testability Verification System Design of Radar ATS Based on Virtual Instrument [J]. Computer measurement & control, 2010, 18(12):2795-2798.
13. Yang H E, Xiaohui Y E, Zhao J. On Testability Verification for Electromechanical Equipment[J]. Electronics Optics & Control, 2011, 11(18):92-95.
14. Yu G, Ma H, Yang G, et al. Architecture of Virtual Instruments Test Environment Soft Bus [J]. Computer Measurement & Control, 2006, 14(2):141-135.
15. Zhao J Y, Li X M, He Y. Present Status of Fault Injection Research for Testability Demonstration and Evaluation [J]. Advanced Materials Research, 2011, 328-330.

Medium and long-term load forecasting based on multi-factors modified PSO-SVM algorithm

Y. K. Cao and H. Shuai†

Shanghai University Of Electric Power, Shanghai, China

†Email: shuaihao23@126.com

Medium and long-term load forecasting is a complex system. The characteristics of the influencing factors with diversity, uncertainty, etc. Such as economic factors, social factors, energy market structure, energy consumption, Electricity elasticity coefficient. This article selection of support vector machine (SVM), which is the core of the medium and long-term load forecasting is presented, using particle swarm optimization (PSO) and cross validation optimization of the modified algorithm to optimize the parameters of the SVM, in a variety of macro factors and historical load as input for load forecasting.

Keywords: Medium and long-term load forecast; Influence factors; PSO and cross validation optimization.

1. Introduction

Along with the rise of power system privatization and deregulation and free competition of the electric power industry, accurately electric load forecasting has become the most important issue in a regional or a national system for market participants to support energy transactions (Bunn, 2000; Douglas et al., 1998). However, the electric load forecasting is complicated and presents nonlinear data patterns by the influencing factors. In the past decades, there are lots of approaches proposed to improve the forecasting performances, the famous approach is: gray prediction (GM), analysis of regression, combination forecast, Neural network algorithm, and the support vector machine (SVM) algorithm. The research object of gray prediction and regression mainly is the historical load data, neural network algorithm and support vector machine (SVM) is the most studied at present.

In the literature [1] although gray algorithm of relying too much on the historical data is improved, Combined with a certain trend of fluctuations in the future, but it can't solve other influencing factors of interference, the stability is not high; The regression prediction in literature [2] also exists the question, can't handle disturbance variable; In the literature [3] combination forecast combines the advantages of various algorithms, but also prone to error. Literature [4] and [5] to improve the convergence speed of the forecast, but high forecasting accuracy is not enough; Literature [6] particle swarm optimization is proposed and a support vector regression method, improve the stability of support vector machine (SVM), but it's easy to have a sample set of fitting; Literature [7] and [8], respectively, puts forward the empirical mode decomposition and principal component analysis and support vector machine, To a certain extent, takes into consideration the weight of influence factors, But not comprehensive, cause corresponding prediction accuracy and stability is limited; Literature [9] season type optimal grey support vector machine (SVM) algorithm, For the quarter load had good prediction, And using the least squares support vector machine (SVM), but the same precision is limited; Literature [10], [11], [12] on the basis of reference [9] have the corresponding improvement, but considering the influential factors of the weight is not enough.

In this paper, the PSO algorithm combined with loop optimization, Through the basic loop optimization to find the best parameters values, While the optimization of parameters of PSO for fine-tuning loop optimization results, avoid falling into local optimum.

2. Diversity of Influencing Factors

Itself has inner medium and long-term load forecasting system, From the outside world and its system of the outside world. Predict the future development of the object is a dynamic development of the system as a whole, With the interaction between each component and influence factors are closely related. The influence factors of long-term power load forecast system has the following kinds.

Literature [13], [14], [15] from regional load complexity, many aspects such as constraint conditions, and data uncertainty had good reasoning, But only do the qualitative analysis.

In this paper, in combination with many influencing factors under the premise of qualitative analysis, To quantify the various influence factors, Economic factors to choose GDP as input; Power structure is mainly selected the first industrial output, the output value of the second industry, the tertiary industry output value of the three main indicators as input. Social factors mainly selected as input population; Energy factors selected as the energy consumption indicators; Electricity consumption elasticity coefficient is selected each year as the input of electrical elasticity coefficient. The above indicators can be obtain in the calendar year statistical yearbook.

Selected to GDP, the output value of primary industry, secondary industry output value, the output value of the third industry, population, energy consumption, electricity elasticity coefficient of a sample of seven indexes such as influencing factors of input, covering almost all effects.

3. Improved PSO-SVM Model

Due to the ordinary double optimization optimization algorithm in the optimization process optimization step must be, Not good traverse the optimal solution.The PSO can find out the optimal solution, but due to the optimal solution is the historical sample, easy to make SVM into fitting. So combining the two methods of optimization, Give priority to with double parameters optimization, PSO to get the optimal solution is used to fine tune, improve the accuracy of double optimization results.

3.1. *Support Vector Machines*

SVM is a new and promising technique for data classification and regression. In this section, we briefly introduce support vector regression (SVR) which can be used for time series prediction. Given training data(x_1, y_1),...,(x_l, y_l), where x_i are input vectors and y_i are are the associated output value of x_i, the support vector regression solves an optimization problem

$$\min_{\omega,b,\xi,\xi^*}\left(\frac{1}{2}\omega^T\omega + C\sum_{i=1}^{l}(\xi_i+\xi_i^*)\right) \tag{1}$$

$$\text{Subject to } y_i - (\omega^T\phi(x_i)+b) \le \varepsilon + \xi_i^*, \tag{2}$$

$$(\omega^T\phi(x_i)+b) - y_i \le \varepsilon + \xi_i \tag{3}$$

$$\xi_i\xi_i^* \ge 0, i = 1,...,l \tag{4}$$

where x_i is mapped to a higher dimensional space by the function ϕ, ξ_i^* is the upper training error (ξ_i is the lower) subject to the ε -insensitive tube $\left|y-(w^T\phi(x)+b)\right| \le \varepsilon$, The parameters which control regression quality are the cost of error C, the width of the tube ε, and the mapping function ϕ.

The constraints of (2) imply that we would like to put most data x_i in the tube $\left|y-(w^T\phi(x)+b)\right| \le \varepsilon$, if x_i is not in the tube, there is an error, ξ_i or ξ_i^* which we would like to minimize in the objective function. SVR avoids underfitting and overfitting the training data by minimizing the training error C $\sum_{i=1}^{l}(\xi_i+\xi_i^*)$ as well as the regularization term (1/2) $\omega^T\omega$. For traditional least-square regression ε is always zero and data are not mapped into higher dimensional spaces. Hence, SVR is a more general and flexible treatment on regression problems.

Since ϕ might map x_i to high or infinite dimensional space, we, instead of solving ω for (2) in high dimension, deal with its dual problem

$$\min_{\alpha,\alpha^*}\left(\frac{1}{2}(\alpha-\alpha^*)^T Q(\alpha-\alpha^*)+\varepsilon\sum_{i=1}^{l}(\alpha_i+\alpha_i^*)+\sum_{i=1}^{l}y_i(\alpha_i-\alpha_i^*)\right) \tag{5}$$

$$\text{Subject to}\quad \begin{aligned}&\sum_{i=1}^{l}(\alpha_i-\alpha_i^*)=0,\\ &0\le\alpha_i,\alpha_i^*\le C, i=1,\ldots,l\end{aligned} \tag{6}$$

where $Q_{ij}=\phi(x_i)^T\phi(x_j)$. However, this inner product may be expensive to compute because ϕ (x) has too many elements. Hence, we apply "kernel trick" to do the mapping implicitly. That is to employ some special forms which are inner products in higher space yet can be calculated in the original space. Some examples are polynomial kernel $\phi(x_i)^T\phi(x_j)=$ $(\gamma x_1^T x_2+c_0)^d$ and RBF kernel $\phi(x_i)^T\phi(x_j)=e^{-\gamma\|x_1-x_2\|^2}$.

3.2. *Particle swarm optimization*

In PSO, a number of particles form a 'swarm' that evolve or fly throughout the feasible hyperspace to search for fruitful regions in which optimal solution may exist. Each particle has two vectors associated with it, the position (X_i)and velocity V_i, vectors. In N-dimensional search space, $X_i=[x_{i1},x_{i2},\ldots,x_{iN}]$ and $V_i=[v_{i1},v_{i2},\ldots,v_{iN}]$ are the two vectors associated with each particle i. During their search, members of the swarm interact with each other in a certain way to optimize their search experience. There are different variants of particle swarm paradigms but the most commonly used one is the gbest model where the whole population is considered as a single neighborhood throughout the flying experience. In each iteration, particle with the best solution shares its position coordinates (gbest) information with the rest of the swarm. Each particle updates its coordinates based on its own best search experience (pbest) and gbest according to the following equations:

$$\begin{aligned}v_1^{k+1}&=wv_1^k+c_1rand_1(pbest_i^k-x_i^k)+c_2rand_2(gbest^k-x_i^k)\\ x_i^{k+1}&=x_i^k+v_i^{k+1}\end{aligned}$$

where c_1 and c_2 are two positive acceleration constants, they keep balance between the particle's individual and social behavior when they are set equal; $rand_1$ and $rand_2$ are two randomly generated numbers with a range of [0,1] added in the model to introduce stochastic nature in particle's movement; and w is the inertia weight and it keeps a balance between exploration and exploitation.

3.3. *Improved PSO-SVM model*

Improved PSO-SVM model mainly by the support vector machine (SVM). Due to the standard support vector machine parameter setting is not the optimal solution, So the prediction precision is not high, Using the parameters obtained from the PSO algorithm optimization is the global optimal solution, is a global optimal training samples, in a overfitting model. Therefore, based on PSO optimization, Using double optimization combined with particle swarm optimization algorithm, Double optimization can find good basic optimal solution, The optimal parameters of PSO algorithm can be used as trimming and double optimization combination, The final result as the optimal parameters of support vector machine (SVM).

Double optimization is set parameters and the optimal step length in advance,Through cross validation,Various parameters combination of traverse parameter range,And optimization precision affected by optimal step length, step length is smaller, the higher the precision, But the longer the running time. In this paper, according to the results of the experiment many times to choose $[2^{-5}, 2^{5}]$ for c, g optimization range,step is $2^{0.5}$.

The main optimization parameters of support vector machine is c and g. The scope of the dual optimization is given c, g, And optimization step at a time. Double precision mainly with optimal parameters c and g optimization scope and step length, Range is too big, easy to make the punishment factor is too large, the model overfitting. While the step length is too long can lead to algorithm running time is too long, At the same time actually improve precision are not obvious. In order to reduce the fitting degree of particle swarm optimization. Set the minimum support vector machine (SVM) of the initial penalty coefficient is 0.1, the maximum is 100.

A double optimization results of c1 respectively with g1, PSO optimization results for c2 and g2. The parameters of SVM for:

$$c=c1+k1*c2 \tag{7}$$

$$g=g1+k2*g2; \tag{8}$$

k1 and k2 fine-tuning weight PSO optimization results respectively.

4. Algorithmic Flow

After determining and quantify influence factors, using the modified PSOSVM for training data set, algorithm design structure as shown in Figure 1:

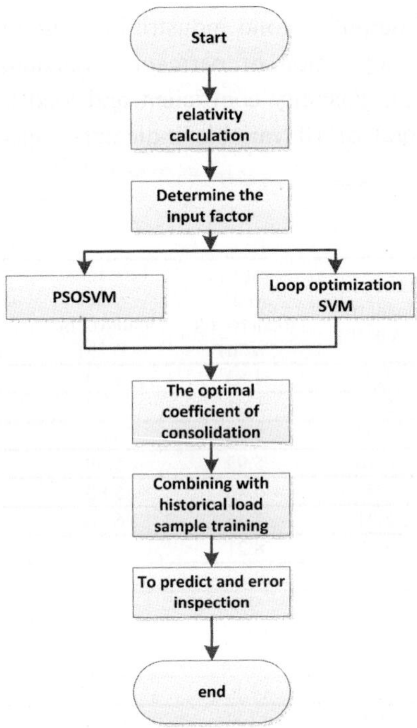

Fig. 1. Algorithm structure.

Step 1:According to the medium and long-term load forecasting system complexity of historical data and the influencing factors of selective extraction, calculate each index of correlation and predictive value, to eliminate low correlation index.

Step 2:The remaining high correlation macro factors index is combined with a forecast years ago, two, three years of historical load as the input samples.

Step 3:Initialize the PSO algorithm and SVM algorithm, respectively using PSO optimization and cycle optimization for SVM parameters selection.

Step 4:Take double loop optimization results of the basic optimal solution, PSO optimization results take corresponding weight value of basic optimal solution for fine-tuning, obtain the optimal solution.

Step 5:The optimal c, g into the SVM algorithm.

Step 6:To carry on the load forecasting, and compared with the actual value, calculate the root mean square error.

5. Simulation Example

In this paper, the data is from 2000 to 2013 in Shanghai. According to the above analysis, This article select samples mainly includes the input variable indicators:

GDP, first industrial output, second industrial output, third industry output, the total number of the population of permanent residents, energy consumption, electricity consumption elasticity coefficient and load forecast years ago, three years of history, a total of 10 variable indicators. Specific data are shown in Table 1:

Table 1. Data set

years	actual load(10^6kW*h)	Last year load(10^6kW*h)	Last two year load(10^6kW*h)	Last three year load(10^6kW*h)	energy consumption(million tons standard coal)	GDP (billion)
2000	5.59	5.01	4.83	4.54	5119.19	4188.73
2001	5.93	5.59	5.01	4.83	5499.48	4771.17
2002	6.46	5.93	5.59	5.01	5894.78	5210.12
2003	7.46	6.46	5.93	5.59	6249.34	5741.03
2004	8.21	7.46	6.46	5.93	6796.34	6694.23
2005	9.22	8.21	7.46	6.46	7405.64	8072.83
2006	9.9	9.22	8.21	7.46	8225.05	9247.66
2007	10.72	9.9	9.22	8.21	8875.7	10572.24
2008	11.38	10.72	9.9	9.22	9670.45	12494.01
2009	11.53	11.38	10.72	9.9	10207.36	14069.87
2010	12.95	11.53	11.38	10.72	10367.38	15046.45
2011	13.39	12.95	11.53	11.38	11201.13	17165.98
2012	13.53	13.39	12.95	11.53	11270.48	19195.69
2013	14.1	13.53	13.39	12.95	11362.15	20181.72

years	first industrial output (billion)	second industrial output (billion)	third industry output (billion)	permanent population (ten thousand)	Electricity consumption elasticity coefficient
2000	74.49	1984.64	2129.6	1608.6	1.06
2001	76.68	2207.63	2486.86	1668.33	0.57
2002	78	2403.18	2728.94	1712.97	0.79
2003	79.68	2622.45	3038.9	1765.84	1.26
2004	81.02	3209.02	3404.19	1834.98	0.71
2005	83.45	3892.12	4097.26	1890.26	1.07
2006	90.26	4381.2	4776.2	1964.11	0.58
2007	93.81	4969.95	5508.48	2063.58	0.55
2008	101.84	5571.06	6821.11	2140.65	0.63
2009	111.8	6085.84	7872.23	2210.28	0.16
2010	113.82	6001.78	8930.85	2302.66	1.24
2011	114.15	7218.32	9833.51	2347.46	0.41
2012	124.94	7927.89	11142.86	2380.43	0.14
2013	127.8	7854.77	12199.15	2415.15	0.55

Because of the influence of various factors index selection is to extract the main indexes for all kinds of macro factors, so in influencing factors was determined, the need for relevance computation, the calculation of correlation experiment is calculated by SPSS software. get the corresponding correlation coefficient values input SPSS software and significant. Show in Table 2.

Table 2. The variable correlation coefficient.

	actual load		
	Pearson Correlation	Significant (both sides)	N
actual load	1		14
Last year load	.993**	0	14
Last two year load	.988**	0	14
Last three year load	.983**	0	14
energy consumption	.996**	0	14
GDP	.985**	0	14
first industrial output	.976**	0	14
second industrial output)	.997**	0	14
third industry output	.970**	0	14
permanent population	.995**	0	14
Electricity consumption elasticity coefficient	-0.447	0.109	14

According to the calculated correlation coefficient, in addition to the electrical elasticity coefficient of correlation, nine other indicators are significantly associated with years of actual load. Therefore, we choose GDP, first industrial output, second industrial output, third industry output. The total number of the population of permanent residents, energy consumption, Electricity consumption, and load forecast years ago, three years of history, from 2000 to 2010 in Shanghai. For SVM training sample input variables, Initialize the particle swarm algorithm, the c optimal range of 0.1 to 100, the g optimal range of 0.01 to 1. Double optimal selection according to the results of the experiment many times $[2^{-5}, 2^{5}]$ for c, g optimization range, step is $2^{0.5}$. Fine-tuning coefficient k1 is 0.5,k2 is 0.001.

At the same time of using modified PSOSVM algorithm to predict, In order to intuitive see the advantages of modified PSOSVM, Selected the gray prediction, neural network and SVM prediction for comparison algorithm, Figure 2 and Table 3 the results for each algorithm training set.

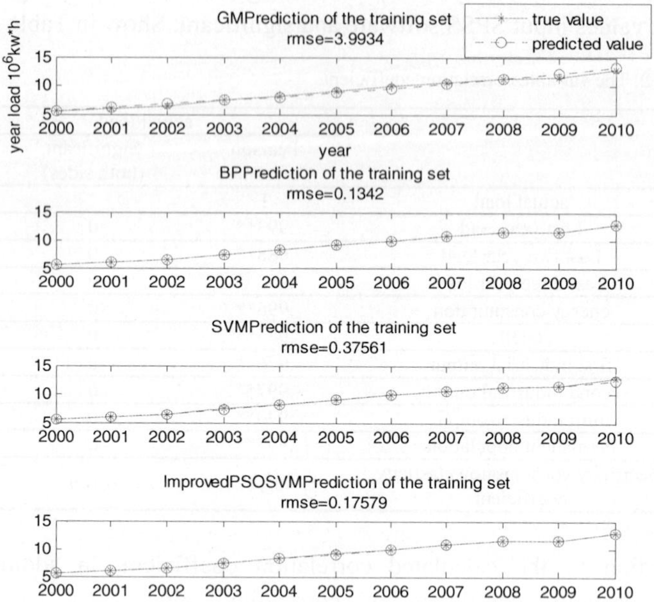

Fig. 2. Each algorithm training set.

Table 3. The algorithm training set results compared.

number	years	actual load	GM	BP	SVM	modifiedPSOSVM
1	2000	5.59	5.59	5.60	5.58	5.58
2	2001	5.93	6.33	5.93	6.03	6.01
3	2002	6.46	6.87	6.48	6.47	6.47
4	2003	7.46	7.46	7.46	7.24	7.27
5	2004	8.21	8.10	8.23	8.20	8.22
6	2005	9.22	8.79	9.21	9.05	9.07
7	2006	9.9	9.55	9.90	9.91	9.90
8	2007	10.72	10.36	10.73	10.78	10.76
9	2008	11.38	11.25	11.39	11.37	11.37
10	2009	11.53	12.21	11.53	11.61	11.55
11	2010	12.95	13.26	12.95	12.52	12.94
RMSE			3.99%	0.15%	0.37%	0.17%

The results can be seen that through the training set. Neural network and support vector machine algorithm relative to gray, To study the degree of training set.

After get the training model, Each algorithm using the model of power consumption of 2011 to 2013 in Shanghai, respectively, to predict, Figure 3 and

Table 4 is the forecasting result of each algorithm, RMSE is the absolute error of root mean square error.

Predicted results from 2011 to 2013, these algorithms are as Figure 3.

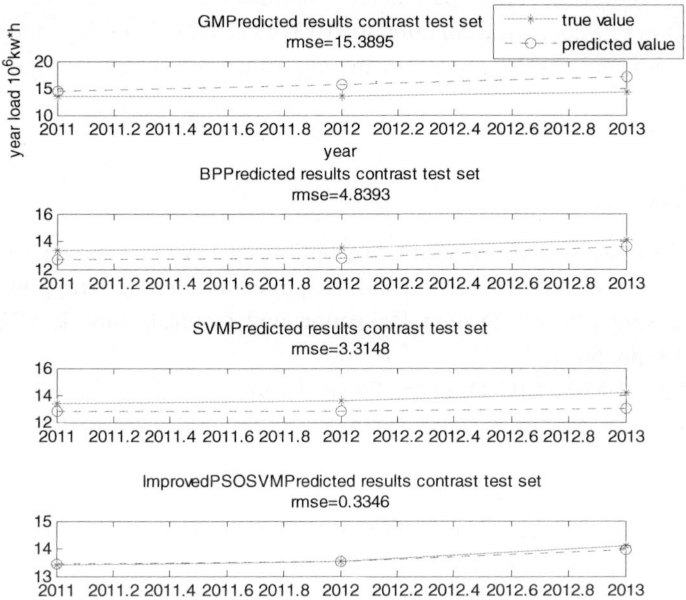

Fig. 3. Each algorithm to predict the results.

Table 4. The algorithm to predict the result comparison.

number	years	actual load	GM	BP	SVM	Modified PSOSVM
1	2011	13.39	14.40	12.81	12.81	13.41
2	2012	13.53	15.63	12.92	12.84	13.50
3	2013	14.1	16.97	13.59	13.03	13.94
RMSE			15.38%	5.29%	3.31%	0.42%

6. Conclusion

Long-term power load forecasting is an organic whole system, in the planning area electric power development blueprint, starting from the actual of region, considering natural, economic, social and cultural elements, Macro to the medium to the microscopic comprehensive and integrated. From the development and utilization of natural resources, industrial development layout, the adjustment of industrial structure optimization to obtain economic benefit, social benefit and ecological benefit is a comprehensive and integrated.

For medium and long-term load forecasting affecting factors, the characteristics of less sample data, this paper presents a modified PSOSVM algorithm based on multi-factor. The algorithm considering a variety of factors. With the modified PSO parameter optimization, to optimize the SVM, combination of Shanghai from 2000 to 2013, electricity conditions are simulated. The experimental results show that the algorithm could improve the prediction precision.

References

1. LI Hong-wei, MAO Wen-jin. An optimized GM(1,1) model based on bi-directional difference method and its application in long-term power demand forecasting. Power System Protection and Control, July 1, 2011, Vol.39 No.13, pp. 56-61.
2. LI Po, LI Min, LIU Di-chen, Power Load Forecasting Based on modified Regression, Power System Technology, Jan. 2006, Vol. 30 No. 1, pp. 99-104.
3. JIANG Yan, WANG Shao-yang, FENG Yun, Medium-long Term Power Load Forecasting Based On Recursive Right Combination Model, Proceedings of the CSU-EPSA, Feb. 2012, Vol. 24 No. 1, pp. 151-155.
4. CHEN Ze-huai, ZHANG Yao, WU Zhi-gang, Application of RBF Neural Network in Medium and Long-Term Load Forecasting Proceedings of the CSU-EPSA, Feb. 2006, Vol. 18 No. 1, pp. 15-19.
5. LU Ning, ZHOU Jian-zhong, HE Yao-yao, Particle swarm optimization-based neural network model for short-term load forecasting, Power System Protection and Control, Jun. 16, 2010, Vol.38 No.12. pp. 65-68.
6. ZHANG Xue-jun, CHEN Gang, ZHOU Jie, MA Ai-jun, ZHANG Zhong-jing, Medium and long-term load forecast based on PSO-RSVR, Power System Protection and Control, Nov. 1, 2009, Vol.37 No.21, pp. 76-80.
7. Luca Ghelardoni, Davide Anguita, Energy Load Forecasting Using Empirical Mode Decomposition and Support Vector Regression, IEEE Transactions on Smart Grid, Vol. 4, No. 1, March 2013, 549-556.
8. WANG Zhi-zheng, YU Yue-feng, YAO Guo-ping, The Application of Principal Component Analysis in Load Forecasting, Power Demand Side Management, Jun 2003, Vol. 5, No. 3, pp. 21-24.
9. WU Yu, Wang Jie, Integrated Optimum Gray Support Vector Machine Model for the Seasonal Power Load Forecasting, East China Electric Power, Jan. 2012, Vol. 40, No. 1, pp. 18-21.

10. LI Hongze, GUO Sen, WANG Bao, Medium and Long-term Power Load Forecasting Based on Grey Prediction Models Combined with Least Squares Method, Water Resources and Power, Aug. 2012, Vol. 30, No. 08, pp. 177-179.

11. Wei-Chiang Hong, Chien-Yuan Lai, Wei-Mou Hung, Yucheng Dong, Electric Load Forecasting by SVR with Chaotic Ant Swarm Optimization, 2010 IEEE Conference on Cybernetics and Intelligent Systems,102-107
12. Qi Wu, Power load forecasts based on hybrid PSO with Gaussian and adaptive mutationand Wv-SVM, Expert Systems with Applications, 37 (2010) 194–201.
13. CUI He-rui, LIU Dong, Complexity Research on Mid-Long Term Forecast of Regional Power Load, East China Electric Power, Aug. 2011, Vol. 39, No. 8, pp. 1233-1237.
14. WANG Ji-Xuan, LIU Xiao-zhen, HAN Zhong-He, Modeling and Researching of Electric Load Forecasting Based on Multi-restrictions, East China Electric Power, Mar. 2013, Vol. 41 No. 3, pp. 0641-0645.
15. ZHENG Zhi-jie, LI Lei, ZHAO Lan-ming, Medium and long term load forecasting considering data uncertainty, Power System Protection and Control, Apr. 1, 2011, Vol. 39 No. 7, pp. 123-127.

Dual prediction based fractional pixel motion estimation for video coding

J. Y. Lu[1†], Y.X. Liu[2], F. Hou[3] and C.H. Huang[4]
Department of Computer Science and Technology,
Guangdong University of Finance, GZ 510521, P. R. China
†Email: Dtc005001@163.com

Fractional pixel motion estimation (FPME) and fractional pixel interpolation (FPI) account for relatively significant amount of computational time in video coding. Fewer search positions FPME checks, the lesser fractional pixels have to be interpolated. Thus, developing fast FPME algorithms is equivalent to save computation on FPME and FPI. Most of fast FPME algorithms are dedicated to motion vector prediction for finding an optimized position to start the successive search. However, these algorithms overlook the minimal cost prediction which can stop the refinement search as early as possible. In this paper, we propose a dual prediction based algorithm for FPME. Spatial and temporal correlations of neighboring blocks are being used to predict motion vectors as well as the optimal costs. According to our experiments, the algorithm not only achieves significant speedup but also provides good R-D performance compared with other algorithms.

Keywords: Video coding; Fractional pixel motion estimation; Motion vector prediction; Minimal cost prediction.

1. Introduction

To further reduce video data rate and improve reconstructed video quality as well, video coding standards e.g. HEVC, H.264, AVS and VC-1 adopt different technologies [1]. Fractional-pixel motion compensation is one of these technologies and it reduces the aliasing of the predicted signal and increases coding efficiency. But this enhancement of R-D performance comes at the very high price of computational complexity. Video encoders enabling this technology have two extra stages, one is fractional-pixel interpolation (FPI) and the other is fractional-pixel motion estimation (FPME). FPME along with FPI account for a significant portion of the CPU usage for video coding [2]. According to our study, fast

---

[1] Ji-Yuan Lu
[2] Yu-Xi Liu
[3] Fang Hou
[4] Chen-Hui Huang

FPME algorithms not only reduce the number of search positions (SPs) but also limit the calculation of fractional-pixels with an on-demand FPI method [2].

Besides Full Fractional Pixel Search algorithm (FFPS) [3], many fast FPME algorithms have been proposed. These fast algorithms focus on the prediction of motion vectors (MVs) only. However, as the minimal costs can be used to decide when to stop the refinement search around the predicted MV, it is also important to predict the minimal costs. We classified the existed algorithms into two categories according to their MV prediction technologies. The first category is model-based FPME algorithm. The idea of these algorithms base on an assumption that mathematical models of fractional cost surfaces can be precisely constructed. The parameters of these models can be computed by the cost values of neighboring integer positions. Wei et al proposed a second order function with five parameters and a two-step modeling strategy to model the error surface in sub-pixel position [4]. This novel algorithm makes full use of 8 neighboring integer-pixels. Dikbas et al used different polynomial models to describe the error surface [5]. Instead of approximating the error model directly, Blasi et al proposed to estimate the curvature of the error model [6]. Because it is unfeasible to model different error surfaces by a single model, Zhang et al proposed to integrate different modeling techniques by classifying error surfaces into well-conditioned and ill-conditioned [7] and Cho et al classified the error surface as plain or textured by the strength of the edge to apply different modeling techniques [8]. However, these models are not exactly matched with the actual error surfaces and these methods overlook the correlation among adjacent blocks which have been proven to be an efficient way to find the best fractional motion vector (MV) [2].

The second category is prediction-based FPME algorithm. CBFPS [9] uses median MVs as start position and a small diamond pattern as refinement search pattern. SFPS [10] uses some fallback and simplified techniques to optimize search process. EPZFPS [11] selects different search patterns of the best matching two positions according to predicted MVs and their relative relationship. Hill et al proposed to use different refinement search techniques to rectify model prediction [12]. Li et al proposed several mode pre-selection rules to provide precise predicted fractional MVs [13]. Some advanced prediction techniques are proposed to even skip the entire integer-pixel search [14]. We also proposed an on-demand algorithm to jointly speed up FPME and fractional pixel interpolation [2]. All of these fast FPME algorithms directly derived the ideas from fast integer pixel motion estimation (IPME) algorithms. And their performance can be further improved if the characteristics specified for FPME are taken into consideration.

In this paper, a dual prediction based algorithm is proposed to accelerate fractional pixel motion estimation. We not only make use of both MVs and costs estimated by IPME to predict the optimal MVs and the minimal costs, but also use a simple pattern for refinement search by considering uni-model characteristics on FPME area. Besides, another advantage of reducing fractional SPs is that a lot of computation requirement can be saved when on-demand FPI algorithms are used.

This paper is organized as follows. A dual prediction fractional pixel algorithm is proposed in Section 2. Simulation results are presented in Section 3. Finally, conclusion is provided in Section 4.

2. Proposed Method

Motion estimation (ME) is to find out a position with the minimal cost in a specific search area. The search area of integer pixel motion estimation (IPME) can be 32x32 or 16x16, but the search area of fractional pixel motion estimation (FPME) is only one pixel around the optimal integer position. The number of candidate search positions (SPs) of FPME relates to fractional resolution, i.e. 1/2 pixel resolution has 8 candidate SPs, 1/4 pixel resolution has 48 candidate SPs, 1/8 pixel resolution has 254 candidate SPs and so on. It is infeasible to check all of these SPs exhaustively. Although FFPS and current fast algorithms have speeded up FPME to a certain extend, there is still room for further improvement in speed with negligible coding lost.

In this paper, an algorithm is proposed with the following structure. Firstly, several predicted MVs are generated as close as possible to the final position by making use of the correlation among spatially neighboring MVs. One of these MVs will be chosen as the initial search position according to their similarities with the integer part of the current MV to conduct the following search. Secondly, we give a minimal cost prediction algorithm to decide when the search should stop. Because fast FPME algorithms are not exhaustive search methods, none of them guarantee to terminate at the global minimal position. The objective of these fast algorithms is to find a position with close enough residual cost to the minimal one.

2.1. *Spatial prediction of fractional MVs*

Most of fast integer pixel motion estimation (IPME) algorithms use MVs from neighboring blocks to predict the final MVs. These MV prediction algorithms can be classified into two categories: spatial MV prediction algorithms and temporal MV prediction algorithms. Many statistical research efforts [15] reveal that spatially predicted MVs are much closer to the final MV than temporally predicted MVs. Thus, fast IPME algorithms prefer to use spatial MV prediction technology. On the other hand, FPME is not exactly the same with IPME. It has a much smaller search area than IPME and only examines a small portion of fractional positions around the best matching integer position. So it will be inefficient to directly apply spatial MV prediction technology of IPME on FPME.

According to our experiments, we find that those spatially neighboring MVs which have the same integer MV are very likely to have the same fractional MV as well. Because FPME is a successive step of IPME, integer MVs must have been known already. The correlation of this integer MV and its spatially neighboring MVs can be revealed before FPME starts. Then, we will select one of them as predicted MV for FPME.

Our prediction method has the following steps. Because the correlation among spatially MVs is stronger than the correlation among temporally neighboring MVs [3], we only take spatially neighboring MVs into consideration. They are up MV, up-right MV, up-left MV and right MV. If multi-block mode is used, then up-layer MV [3] will be added in. All of them are candidate predictors, but only some of them which appear around the optimal integer MV within one pixel will be considered as predictors. We only compute costs of these candidate predictors and the one with the minimal cost will be the start position for further refinement search.

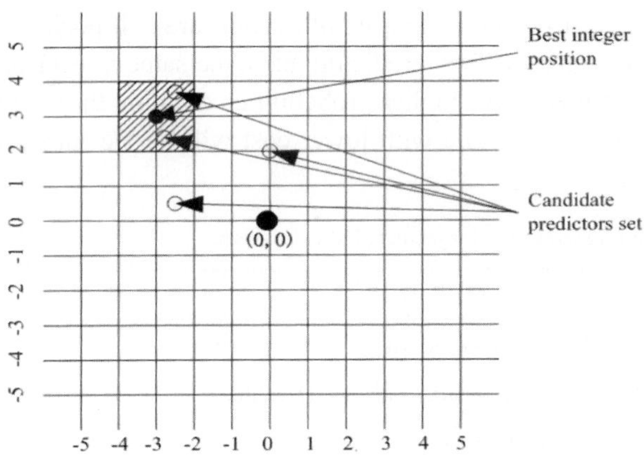

Fig. 1. An example of spatially neighboring MVs and the best matching integer position.

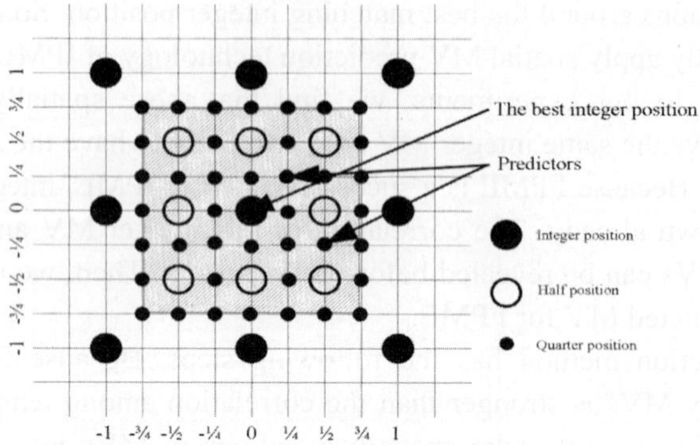

Fig. 2. An example of fractional pixel ME predictors.

Fig. 1 and Fig. 2 illustrate a typical example of our algorithm to decide fractional predictors at 1/4-pixel FPME. Fig. 1 is the initial stage of FPME and the shadow region is the FPME area. The optimal integer position is the center of this region. Four spatially neighboring MVs are candidate predictors for IPME. But only two of them appear inside the shadow region.

Fig. 2 scales up FPME search region i.e. the shadow region in Fig. 1. Large solid dots are integer positions. Large hollow dots are half positions. Small solid dots are quarter positions. Two predictors have the same integer part as the current MV. Both of them have higher probabilities of being the best matching position than others, but the one with lower cost will be the start position for refinement search.

Table 1. Testing sequences for predicted MVs analysis.

Format	Sequences
QCIF	Suzie, Trevor, Akiyo, Bowing
CIF	Bridge_far, Coastguard, Container, Deadline
4CIF	City, Harbour, Crew, Ice
720p	Mocal, Stockholm, Parkrun, Shields
1080p	In_to_tree, Blue_sky, Life, Rush_hour

A comparison of our proposed method with other well-known MV predicted strategies e.g. FFPS and CBFPS [9] is given as follow. Fig. 3 shows the comparison results. The experiments are performed on sequences in Table 1. The accuracy of different prediction methods is evaluated by analyzing the distribution of MV differences related to FFPS. MV difference is a 2-dimensional vector. If this vector is zero vector, then the predicted MV is exactly the MV estimated by FFPS. The higher concentration towards zero vector, the more accurate predictor it gets.

Fig. 3 shows the differences distribution between integer MVs, CBFPS predictors and our proposed predictors with MVs estimated by FFPS. Although using CBFPS predictor is far more accurate than using integer MV directly, our proposed method is the most accurate prediction method. At (0, 0) position, our proposed predictors has much higher superposition rate than other MV predictors.

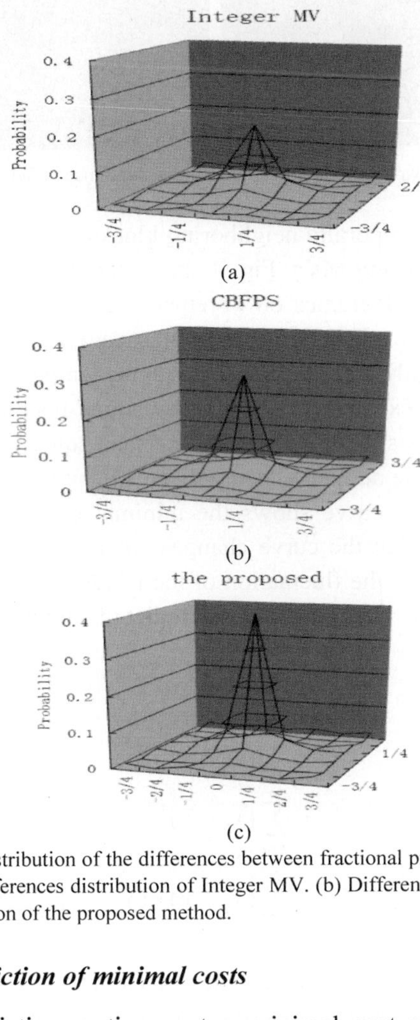

Fig. 3. An example of distribution of the differences between fractional predicted MVs and MVs estimated by FFPS (a) Differences distribution of Integer MV. (b) Differences distribution of CBFPS. (c) Differences distribution of the proposed method.

2.2. *Temporal prediction of minimal costs*

In addition to predicting motion vector, minimal cost prediction is also important. Residual data greatly affects rate-distortion (R-D) performance of video coding. Because residual data is quantized before entropy coding, different residual data might be mapped to a same coefficient. Thus different MVs may produce the same R-D performance. If the difference between the minimal cost and the sub-minimal cost is small enough, their final rate-distortion performance is very likely to be the same. So prediction of minimal cost can be used to assess whether it is worth to search more positions.

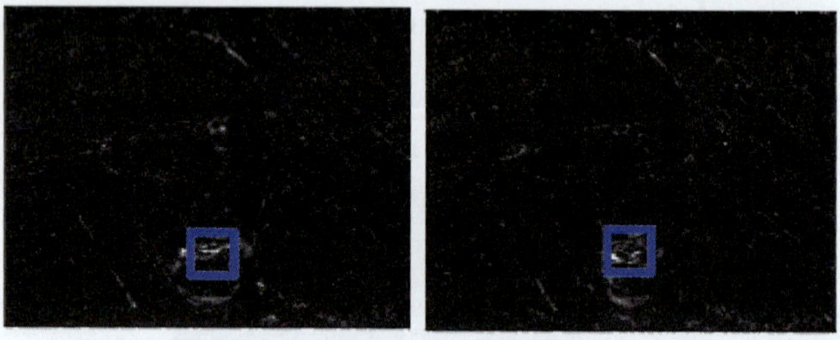

Fig. 4. An example of two temporally neighboring residual image after motion compensation.

Residual data at temporally neighboring blocks has strong correlation similar to spatially neighboring MVs. Fig. 4 shows the displaced frame differences image of two adjacent P frames on Foreman sequence. Large residual coefficients gather around the edges of moving objects e.g. lip, border of the hat. These coefficients are highly correlated along with the temporal direction.

Fig. 5 shows an example of a frame by frame residual cost on a macroblock (MB). The Y-axis is the residual cost, mean square error (MSE). The X-axis describes the frame number. The upper curve shows the minimal cost using IPME only. The lower curve shows the minimal cost using both IPME and FPME. Fig. 5 shows that the curve changes smoother in temporal direction if FPME applied. Because the fluctuation of the residual costs on FPME is lower, we propose a linear regression method to predict the minimal cost:

$$
\begin{aligned}
&Cost_{pred} = \beta_1 \times x + \beta_0, \\
&\beta_1 = l_{xy} / l_{xx} \\
&\beta_0 = \overline{y} - \beta_1 \times \overline{x}, \\
&l_{xx} = \sum_{k=1}^{n} (x_k - \overline{x})^2, \\
&l_{xy} = \sum_{k=1}^{n} (x_k - \overline{x}) \times (y_k - \overline{y}).
\end{aligned}
\tag{1}
$$

x is the frame number and y is the cost value of each frame. k is the number of x and y pairs. $\bar{x}$ and $\bar{y}$ are the average values of x and y. l_{xx} and l_{xy} are k times the variance of x and the covariance of x and y. β_0 and β_1 are parameters of our linear model. This linear model uses the past k cost values to predict the cost value on the next frame.

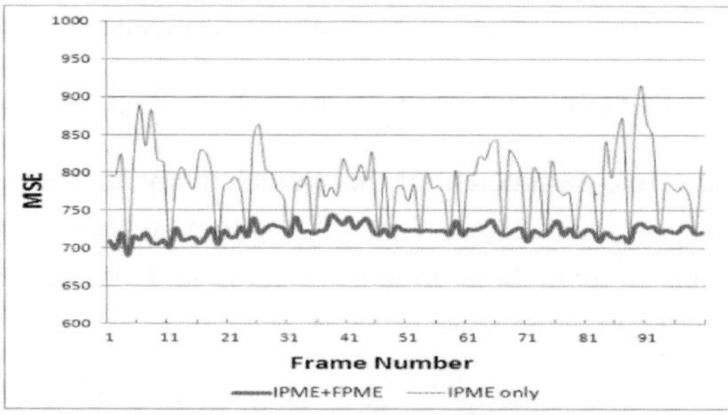

Fig. 5. Frame by frame comparison of residual cost of IPME and FPME.

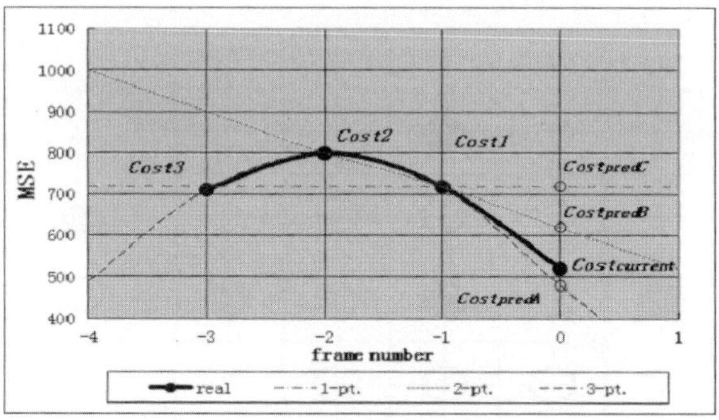

Fig. 6. Minimal cost prediction of temporally neighboring cost value.

In practical application, at most three previous cost values of a same macroblock will be used to predict the current cost. Fig. 6 is an example. Four curves are presented. The solid curve represent the real cost values and three dotted curves are prediction curves. Here, we use *Cost1*, *Cost2* and *Cost3* to represent three previous minimal cost values from the closest one to the farthest one. Because *Cost1*, *Cost2* and *Cost3* are not always available at the same time, we give three formulas to compute the current minimal cost according to different conditions:

$$\begin{aligned} &Cost_{pred}A = \frac{(11\times Cost1 + 2\times Cost2 - 7\times Cost3)}{6} \\ &Cost_{pred}B = 2\times Cost1 - Cost2 \\ &Cost_{pred}C = \mathrm{Cost1} \end{aligned} \quad . \qquad (2)$$

So far, there is only one case left is that the current frame is the first P frame and no previous cost is available. Therefore temporal prediction of minimal cost is inapplicable. At this time, we will make use of neighboring integer cost value to predict the current cost value. Although many fast fractional search algorithms have used parabolic model to predict the best matching position, it will introduce obvious RD penalty [12]. Here, we make use of the parabolic model to predict the minimal cost instead of the best matching position. The parabolic model is:

$$F(x, y) = A \times x^2 + B \times x + C \times y^2 + D \times y + E \quad (3)$$

A, *B*, *C*, *D* and *E* are five parameters of this model. They are the cost values of five neighboring integer positions i.e. top, right, left, bottom and center positions in Fig. 7. If all these parameters of a parabolic model are computed, the following formula can be used to calculate the predicted cost:

$$Cost_{pred} = E - \frac{B^2}{4 \times A} - \frac{D^2}{4 \times C} \quad (4)$$

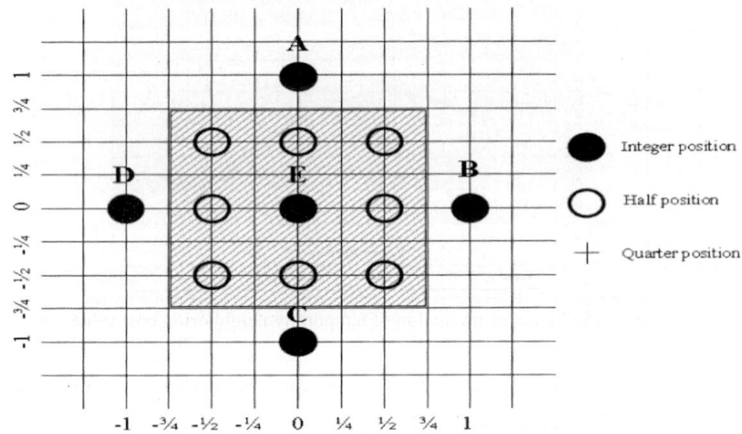

Fig. 7. Minimal cost prediction of spatially neighboring cost value.

Fig. 8 shows a probability distribution of the difference between our predicted cost and the minimal cost attained by FFPS. The X-axis represents their absolute difference value. And the Y-axis represents the percentage of these differences. Zero difference means that our predicted cost is exactly the minimal cost. The probability curve falls off sharply as the difference goes larger and stay close to the Y-axis. It shows that our prediction method is effective. In Fig. 8, different results are shown with Foreman, Silent and Coastguard sequences. Silent contains slow motion with a static background. Foreman and Coastguard capture intensive motion with a moving camera. Fig. 8 shows that our prediction method works better in slow motion sequence than intensive motion sequence. This is because the minimal cost changes more smoothly in slow motion sequences. Even if so, there is still 50% or higher matching rate of our predicted cost with FFPS cost in motion intensive sequences.

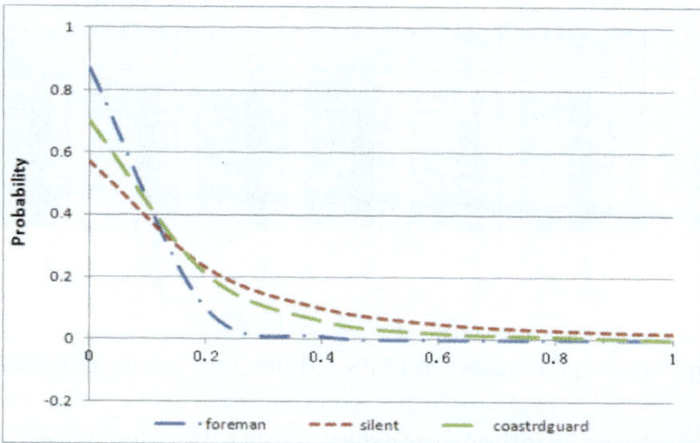

Fig. 8 Difference distribution of proposed prediction cost and FFPS cost.

3. Results and Discussion

Simulated experiments have been completed on HEVC test model (HM 15.0) and x264 i.e. an optimized H.264 encoder. Three quantization parameters such as 16, 20 and 24 are being used. The test has been performed on 22 different sequences in Table 2 with different resolutions and varied motion activities. All of these sequences are encoded in IPPP mode and classified into 6 groups by their resolution. Note that sequences in Table 2 and sequences in Table 1 are not overlapped.

Table 2. 6 groups of testing sequences.

Group	Format	Sequences
A	QCIF	Intros, News, Husky, Bus
B	SIF	Mobile, Stenfan, Garden, Tt, Tennis
C	CIF	Pamphlet, Silent, Sign irene, Students
D	4SIF	Galleon, Vtclnw, Washdc
E	720p	Ducks_take_off, Crowd_run, Old_town_cross
F	1080p	Park_joy, Riverbed, Pedestrian_area

Firstly, experimental results on HM 15.0 will be given. We compared the computational complexity of our proposed algorithm with FFPS, CBFPS [9] and ESA [8] in Fig. 9. The computational complexity of FFPS is the reference mark, 100%. According to the experimental results, our proposed algorithm not only saves about 78% computation from FFPS but also attains the highest speed over other algorithms. Along with the increase of resolution, the computational savings of our proposed algorithm are getting larger. The reason for this is due to the fact that both the MV prediction and the minimal cost prediction are more accurate in high resolution video.

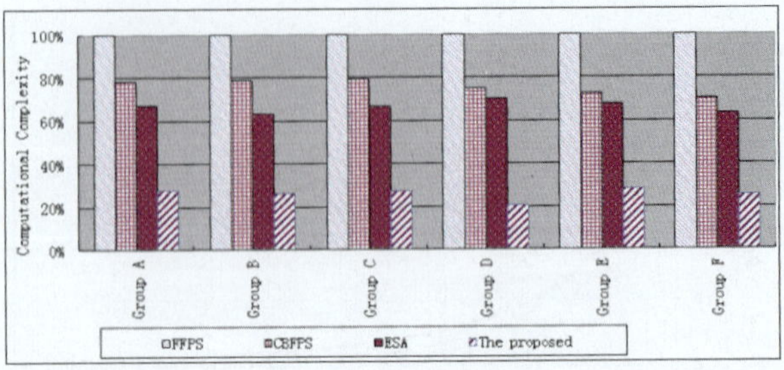

Fig. 9. Computational performance of FFPS, CBFPS, ESA and the proposed algorithm.

Motion activity is another important factor besides video resolution and block mode. In Fig. 9, Groups C and D are inactive and the remaining sequences contain relatively intensive motion. It is important to note that the speedup of our proposed algorithm is obvious even in active video sequences.

Here, we will give the explanation of the source of our computational savings in detail. Fig. 10 compares the number of search positions (SPs) for different FPME algorithms frame by frame. The average number of SPs per macroblock (MB) is used as the Y-axis. Four lines from bottom up in Fig. 10 are the number of SPs of our proposed algorithm, ESA, CBFPS and FFPS respectively. All of fast FPME algorithms save a lot of computation compared with FFPS , but our proposed algorithm always achieves the best performance.

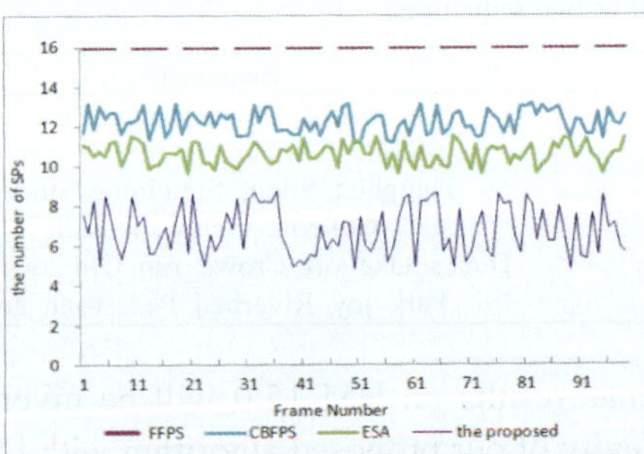

Figure 10. The average number of SPs on Pedestrian area sequence frame by frame.

Table 3 shows the R-D performance for experimental results in Fig 9. We calculate the relative R-D performance of fast FPME algorithms over FFPS:

$$\Delta Bit\ rate = \frac{(Bit\ rate_x - Bit\ rate_{FFPS})}{Bit\ rate_{FFPS}}$$
$$\Delta PSNR = PSNR_x - PSNR_{FFPS} \quad . \tag{5}$$

where $Bit\ rate_{FFPS}$ and $PSNR_{FFPS}$ are the Bit rate and the PSNR of FFPS algorithm. $PSNR_x$ and $Bit\ rate_x$ are the Bit rate and the PSNR of fast FPME algorithms. Table 3 shows that our proposed algorithm has superior coding performance to other fast FPME algorithms. Even compared with FFPS algorithms, only 0.01dB degradation of PSNR and 0.28% increase in Bit rate are collected.

Table 3. R-D performance comparison of CBFPS, ESA and the proposed algorithm over FFPS.

Group	*ΔPSNR(dB)*			*ΔBitrate(%)*		
	CBFPS	ESA	Proposed	CBFPS	ESA	Proposed
A	-0.02	-0.01	-0.01	0.24%	0.32%	0.14%
B	0.02	-0.02	0	0.4%	0.3%	0.03%
C	-0.01	-0.01	-0.01	0.63%	0.23%	-0.28%
D	-0.01	0.01	0.01	0.21%	0.01%	0.64%
E	0	0	0.02	-0.02%	-0.02%	0.35%
F	-0.01	-0.01	-0.03	0.4%	0.46%	0.29%

Secondly, more practical results for our proposed algorithm will be given as follows on x264 platform. And the savings on coding time, PSNR and Bit rate are shown in Table 4. Note that our algorithm drastically reduce fractional SPs and interpolation on these fractional positions should be saved. Thus we used an on-demand FPI method [2] to replace the fixed FPI method in x264. Even compared with an fully optimized H.264 CODEC, our proposed algorithm attain nearly 10% speedup with negligible rate distortion lost.

Table 4. Performance comparison on x264 CODEC.

Group	*Δtime(%)*	*ΔPSNR(dB)*	*ΔBitrate(%)*
A	-11.8%	-0.01	1.24%
B	-9.5%	0	0.4%
C	-8.3%	-0.01	0.43%
D	-10.6%	0.01	1.21%
E	-11.2%	0.02	-2.02%
F	-10.2%	-0.03	0.4%

4. Conclusion

In this paper, we propose a fast FPME algorithm. Firstly, we use a strong correlated condition to identify the optimal spatially neighboring MV. Secondly, we predict the minimal costs by using temporally neighboring minimal cost. And these predicted costs are used to decide when to stop the refinement search. Experimental results show that the proposed algorithm achieves the same R-D performance at the lowest complexity compared to all FPME algorithms.

Acknowledgement

This work was partially supported by the Guangdong Natural Science Foundation of China (2014A030313662, S2012040007847) and the Innovative Project of Guangdong University of Finance (0000-E205010014093).

References

1. Qun Wand, Li Zhuo, Li Zhuo, and Jing Zhang. A rate-distortion model based frame layer rate control algorithm for stereoscopic video coding. Chinese Journal of Electronics, 21(4):678-682, 2012.
2. Jiyuan Lu, Peizhao Zhang, Hongyang Chao, and P. S. Fisher. On combining fractional-pixel interpolation and motion estimation: A cost-effective approach. IEEE Transactions on Circuits and Systems for Video Technology, 21(6):717-728, 2011.
3. Zhibo Chen, Peng Zhou, and Yun He. Fast integer pel and fractional pel motion estimation for jvt, jvt-f017.doc. In Joint Video Team (JVT) of ISO/IEC MPEG & ITU-T VCEG(ISO/IEC JTC1/SC29/WG11 and ITU-T SG16 Q.6) 6th Meeting, 5-13 December 2002.
4. Dai Wei, O. C. Au, Li Sijin, Sun Lin, and Zou Ruobing. Fast sub-pixel motion estimation with simplied modeling in hevc. pages 1560-1563, 2012.
5. S. Dikbas, T. Arici, and Y. Altunbasak. Fast motion estimation with interpolation-free sub-sample accuracy. IEEE Transactions on Circuits and Systems for Video Technology, 20(7):1047-1051, 2010.
6. S. G. Blasi and E. Izquierdo. Residual error curvature estimation and adaptive classication for selective sub-pel precision motion estimation. pages 1193-1196, 2012.
7. Qi Zhang, Yunyang Dai, and C. Kuo. Direct techniques for optimal subpel motion resolution estimation and position prediction. IEEE Transactions on Circuits and Systems for Video Technology, 20(12):1735-1744, 2010.
8. Junsang Cho, J. W. Suh, Jeon Gwanggil, Lee Jooheung, and Jeong Jechang. Error surface-aware modeling algorithm for quarter-pixel motion estimation. IEEE Transactions on Consumer Electronics, 58(3):912-916, 2012.
9. Zhibo Chen, Jianfeng Xu, Yun He, and Junli Zheng. Fast integer-pel and fractional-pel motion estimation for h.264/avc.Journal of Visual Communication and Image Representation, 17(2):264-290, 2006.
10. Xiaoquan Yi, Jun Zhang, Nam Ling, and Weijia Shang. Improved and simplied fast motion estimation for jm, jvt-p021.doc. In Joint Video Team (JVT) of ISO/IEC MPEG & ITU-T VCEG(ISO/IEC JTC1/SC29/WG11 and ITU-T SG16 Q.6)16th Meeting, 24-29 July 2005.

11. Alexis Tourapis and Pankaj Topiwala. Fast subpixel motion estimation support for the enhanced predictive zonal search scheme, jvt-q079, 14-21 October 2005.
12. P. R. Hill and D. R. Bull. Sub-pixel motion estimation using kernel methods. Signal Processing: Image Communication, 25(4):268-275, 2010.
13. G. Li and T. Chang. An efficient mode pre-selection algorithm for fractional motion estimation in h.264/avc scalable video extension. IEEE Transactions on Circuits and Systems for Video Technology, 23(11):1837-1848, 2013.
14. Ling Li, Shaoli Liu, Yunji Chen, Tianshi Chen, and Tao Luo. Motion estimation without integer-pel search. IEEE Transactions on Image Processing, 22(4):1340-1353, 2013.
15. A. M. Tourapis, O. C. Au, and M. L. Liou. Highly efficient predictive zonal algorithms for fast block-matching motion estimation. IEEE Transactions on Circuits and Systems for Video Technology, 12(10):934-947, 2002.

Kinematics analysis of 3-dof (RPS) platform for vehicle simulator using Matlab package

B. F. Z. Ibrahim[1*] and Y. Liu[2†]

Tianjin Key Laboratory of High Speed Cutting & Precision Machining,
Tianjin University of Technology and Education, Tianjin, China
*Emails: *zulfiqarbibifarouk@yahoo.co.uk, † tianjinliuying@163.com*

The paper is aims to provide a solution to the hardware system of vehicle simulator, by providing the simplest approach of solving the kinematics analysis of the 3 degree-of-freedom RPS fully parallel manipulator and authenticating the result by simulation. The combination of joints is used to realize highly precise operation. Linear input is applied to the prismatic joints to drive each limb in the system. Three displacement equations of the active joints are addressed. A study through simulation using MATLAB package is provided for a position and orientation of the limbs and the platform given in the task-space.

Keywords: Vehicle simulator; Singularity; Rigidity/stiffness; Working space.

1. Introduction

Parallel platform is widely used in many industries with the development and application of virtual reality technology which include among others, Vehicle driving Simulator, Parallel Kinematic Machine, Flight Simulator, and Earthquake wave Simulator etc (Z. Bingul and O. Karahan, 2012; Z. Lazarevic). Parallel manipulator is a field of interest for a many researchers due its diversified applications, high accuracy and speed more than serial manipulator (L. W Tsai, 1999). Researchers are battling to come up with new ideas on how to deal with the working space problem of the manipulator without compromising the other design parameters (J. Wu, J. S. Wang, and L. P. Wang, 2008). Its accuracy and repeatability is an attribute to its high stiffness and low inertia and also it has great ability to withstand heavy load due to the fact that the load is merely assumed to be distributed to the platform which in turn the platform is supported by limbs (J.M. Hervé, F. Sparacino, 1991). In some of the manipulators design, the actuators are attached to the limbs to increase stiffness whereas in other de-

[1] Bibi Farouk Zulfiqar Ibrahim
[2] Ying Liu

signs the actuators are attached to the fixed base. The actuator location depends on the designer choice and the manipulator attribute want to achieve (L. W Tsai, 1999). Many researchers have come up with a new idea of reducing the 6-dof (SP) manipulator to a lower degree of freedom manipulator achieving the desired goals of 6-dof (SP) manipulator (P. Vischer, R. Clavel, 2000). This reduction leads to a lot of advantages from an increase in the working space to manufacturing cost and control system complexity. Stewart platform can easily realize 2-6 dof motion in the control algorithm and some of its structure is complex to control compared to serial robots.

Vehicle driving simulator system based on virtual reality technology and parallel platform consists of software system and hardware system. Our concern for this research is hardware system which consists of parallel platform and its attributes only. But other hardware's are driving room and reflection machine. These key technologies play an important part in the study of vehicle driving simulator system. From that point of view, the use of simulation is increasingly relevant and necessary since it allows to display and to test a set of variables in a virtual environment.

These sets of variables include:

The *Driver*: ergonomics, driver-system interaction, behaviour, fatigue, hyper-vigilance, distraction, influence of alcohol and drugs.

Driving Environment: infrastructure, intelligent roads ITS, traffic conditions and type, security.

However, vehicle driving simulator played an important role of solving so many difficulties such as:

➢ Training of small chassis vehicle has formal training venue, but large vehicles, due to the large dimensions of the body and special training courses lack standardized driver training venue.

➢ For Large vehicle driving training, there is high cost, environmental strictness, and long training cycle. These leads to so many shortcomings, such as: the cost of fuel required to put a large number of vehicles on the roads and maintenance, and the safety compromise of other users of the road and the leaner at large. These greatly affect the number of professional big vehicle drivers and there performance during the driving training.

➢ For big vehicles there are special requirements with respect to the ordinary vehicle training, in which they are long, wide and bulks (five-axis vehicle, semitrailer) than ordinary vehicles. The driver must have enough years of ordinary vehicle driving experience, before he can be able to dealt with a large vehicle driving.

➢ Training is limited, under some special circumstances and conditions: rain, snow, weather, night.etc

The study of parallel manipulator together with driving simulator system based on virtual reality technology becomes more essential to *our* daily life. This kind of driving simulator system is inexpensive and has high-efficiency. Due to it effectiveness many institutes pay attention to its research and applications to make thing simple and easy for the stated problems.

The paper is divided into four sections, the first section is an overview of the platform structure, the second section is kinematics analysis, the third section is result and the last section is conclusion.

2. Proposed Mechanism

Several configurations of parallel platforms are used in order to achieve larger workspace, high forces and stiffness, low or higher velocities. In order to achieve higher accuracy, a mechanism of **3-RPS** is proposed. It has linear actuators fixed on the limb to add rigidity to the system. The drawback is a smaller workspace in comparison with a standard **SPS**, for the same travel of the linear actuators (J. Wu, J. S. Wang, and L. P. Wang, 2008).

A sketch of the 3-RPS is given in Figure 1. The platform and base are connected by three limbs which have the prismatic joints as *extensible* joints. It can be seen from Figure 1 that the spherical joint connected the moving platform to the limb and also revolute joints connected the base to the limb, respectively. Points **A** and **B** are geometrical centers of the base and moving platform, Points **b** and **a** are the distance from the geometrical center of the moving platform and base to the point of connection with the limb as shown in Figure 2.

Also coordinates system of *u, v, w* were assigned to the moving platform and *x ,y, z* to the base platform.

Linear motion of the required velocity and acceleration can be *achieved* either by using linear electromotor or rotary electromotor with a converter from rotary to translation motion. Rotary electromotor are relatively cheap and readily available, but their disadvantages are low transmission efficiency and mechanism inaccuracies introduced with each new moving part. Conversion from rotary to translation motion is usually performed by ball-screw transmission or rack-and-pinion transmission. The ball-screw has a useful safety feature that in the case of a power shutdown the mechanism will be held in place. Linear electric motors, on the other hand, are fast and their energy is directly transformed into linear motion (J. Wu, J. S. Wang, and L. P. Wang, 2008). They consist of also ball-screw transmission or rack-and-pinion transmission and it was proposed to be use for this manipulator with ball-screw transmission system.

Fig. 1. Platform model in UG.

3. Kinematics Analysis

3.1. *Inverse Kinematics*

For the inverse kinematics problem, the position vector $\boldsymbol{p}$ and the rotation matrix ${}^{a}\boldsymbol{R}_{b}$ of frame B with respect to A are given and the limb length $\boldsymbol{d_i}$, $\boldsymbol{i}$ = 1, 2, 3 are to be found. Consider the platform in Figure 2.

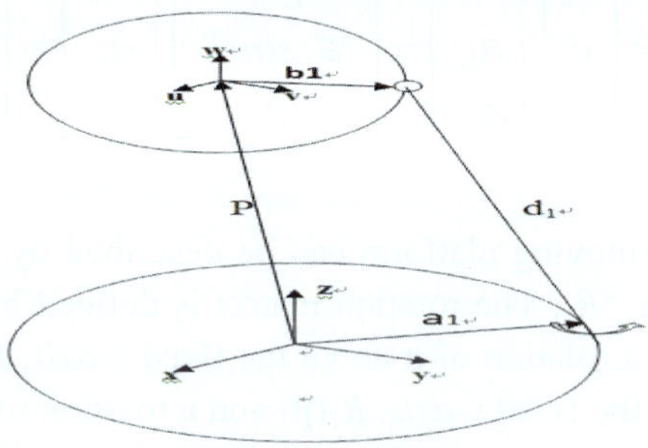

Fig. 2. Vector diagram.

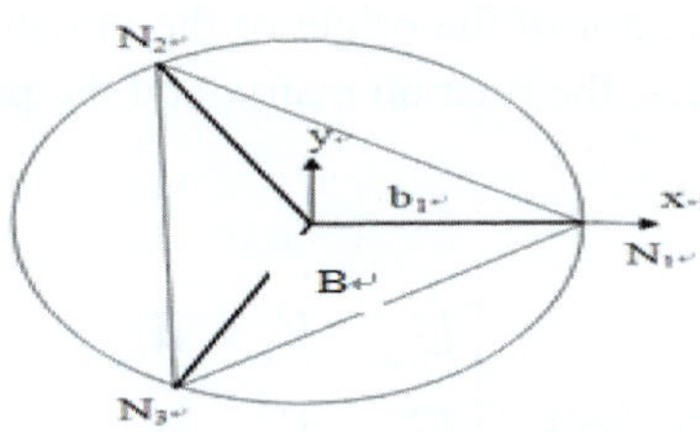

Fig. 3. Moving platform (i=1,2,3).

The length $\boldsymbol{i}^{th}$ of the leg is given by

$$di= p+ aRb.\ bi - ai \tag{1}$$

where $\boldsymbol{a_i}$ is the vector defining the coordinates of the lower fixed point (A), and $\boldsymbol{b_i}$ is the vector defining the coordinates of the moving platform from the fixed point (B), **p** is the vector defining the *coordinates* of the anchor point $\boldsymbol{p_i}$ with respect to the Platform framework, and then $^{a}\boldsymbol{R_b}$ is the rotation matrix.

From Figure 3, the location of the ith attachment point (b_i) on the *moving* platform can be found, $\boldsymbol{b_i}$ and $\boldsymbol{a_i}$ are the radius of the moving platform and fixed base, respectively.

$$\begin{bmatrix} b_{1u} \\ b_{1v} \\ b_{1w} \end{bmatrix} = \begin{bmatrix} b_1 \\ 0 \\ 0 \end{bmatrix}, \begin{bmatrix} b_{2u} \\ b_{2v} \\ b_{2w} \end{bmatrix} = \begin{bmatrix} -b_2 \cos\theta \\ b_2 \sin\theta \\ 0 \end{bmatrix}, \begin{bmatrix} b_{3u} \\ b_{3v} \\ b_{3w} \end{bmatrix} = \begin{bmatrix} -b_3 \cos\theta \\ -b_3 \sin\theta \\ 0 \end{bmatrix} \quad (2)$$

It *can* be seen that $b_{2u} = b_{3u}$ from the limb configuration and also b_{2v} is equal to the b_{3v} with same vector magnitude but opposite in direction.

By using the same approach, the location of the *i*th attachment point (o_i) on the base platform can be also obtained as Equation (3).

$$\begin{bmatrix} a_{1u} \\ a_{1v} \\ a_{1w} \end{bmatrix} = \begin{bmatrix} a_1 \\ 0 \\ 0 \end{bmatrix}, \begin{bmatrix} a_{2u} \\ a_{2v} \\ a_{2w} \end{bmatrix} = \begin{bmatrix} -a_2 \cos\theta \\ a_2 \sin\theta \\ 0 \end{bmatrix}, \begin{bmatrix} a_{3u} \\ a_{3v} \\ a_{3w} \end{bmatrix} = \begin{bmatrix} -a_3 \cos\theta \\ -a_3 \sin\theta \\ 0 \end{bmatrix} \quad (3)$$

The pose of the moving platform can be described by a position vector, *P* and a rotation matrix, $^{a}R_b$. The rotation matrix is defined by the *roll*, pitch and yaw angles, namely, a rotation of $\boldsymbol{\alpha}$ about the fixed *x-axis*, $R_X(\boldsymbol{\alpha})$, followed by a rotation of $\boldsymbol{\beta}$ about the fixed *y-axis*, $R_Y(\boldsymbol{\beta})$ and a rotation of $\boldsymbol{\gamma}$ about the fixed *z-ax1 s*, $R_Z(\boldsymbol{\gamma})$. In this way, the rotation matrix of the moving platform with respect to the base platform coordinate system is obtained. The position vector *P* denotes the translation vector of the origin of the moving platform with respect to the base platform. Thus, the rotation matrix and the position vector are given as the following.

$$^{a}R_b = R_Z(\gamma) R_Y(\beta) R_X(\alpha) = \begin{bmatrix} U_x & V_x & W_x \\ U_y & V_y & W_y \\ U_z & V_z & W_z \end{bmatrix}$$

$$U_x = \cos\beta \cos\gamma \qquad U_y = \cos\beta \sin\gamma \qquad U_z = -\sin\beta$$
$$V_x = \cos\gamma \sin\alpha \sin\beta - \cos\alpha \sin\gamma \qquad V_y = \cos\alpha \cos\gamma + \sin\alpha \sin\beta \sin\gamma$$
$$V_z = \cos\beta \sin\alpha$$

$$W_x = \sin\alpha \sin\gamma + \cos\alpha \cos\gamma \sin\beta \quad W_y = \cos\alpha \sin\beta \sin\gamma - \cos\gamma \sin\alpha$$
$$W_z = \cos\alpha \cos\beta \tag{4}$$

$$P = [p_x \quad p_y \quad p_z]^T \tag{5}$$

The length of the ith limb is obtained by taking the dot product of the vector d_i by its self.

$$d^2_i = [P + {}^aR_b.\ b_i - a_i]^T\ [P + {}^aR_b.\ b_i - a_i]\ (i=1,2,3) \tag{6}$$

$$d^2_i = (p_x - a_{xi} + b_{xi}u_x + b_{yi}v_x)^2 + (p_y - a_{yi} + b_{xi}u_y + b_{yi}v_y)^2 + (p_z + b_{xi}u_z + b_{yi}v_z) \tag{7}$$

Inserting Equation (2) and (3) into (7) and writing the equation for each limb, $i = 1, 2, 3$ yield

$$d^2_1 = p^2_x + p^2_y + p^2_z + a^2_1 + b^2_1 + k_1 \tag{8}$$

where, $k_1 = 2b_1\cos\beta\ (p_x\cos\gamma + p_y\sin\gamma) - 2a_1b_1\cos\beta\cos\gamma - 2bp_z\sin\beta - 2a_1p_x$

$$d^2_2 = p^2_x + p^2_y + p^2_z + a^2_2 + b^2_2 + k_2 \tag{9}$$

where, $k_2 = a(p_x - \sqrt{3}\ p_y) + \sqrt{3}\ b(p_x(\cos\gamma\sin\alpha\sin\beta - \cos\alpha\sin\gamma) + p_y(\cos\alpha\cos\gamma + \sin\alpha\sin\beta\sin\gamma) + p_z\cos\beta\sin\alpha) - b(p_x\cos\beta\cos\gamma + p_y\cos\beta\sin\gamma - p_z\sin\beta) - 0.5ab(u\cos\beta\cos\gamma - \sqrt{3}\cos\beta\sin\gamma) + 0.5ab(\sqrt{3}(\cos\gamma\sin\alpha\sin\beta - \cos\alpha\sin\gamma) - 3(\cos\alpha\cos\gamma + \sin\alpha\sin\beta\sin\gamma))$

$$d^2_3 = p^2_x + p^2_y + p^2_z + a^2_3 + b^2_3 + k_3 \tag{10}$$

where, $k_3 = a(p_x + \sqrt{3}\ p_y) - \sqrt{3}\ b(p_x(\cos\gamma\sin\alpha\sin\beta - \cos\alpha\sin\gamma) + p_y(\cos\alpha\cos\gamma + \sin\alpha\sin\beta\sin\gamma) + p_z\cos\beta\sin\alpha) - b(p_x\cos\beta\ \cos\gamma + p_y\cos\beta\sin\gamma - p_z\sin\beta) - 0.5ab(\cos\beta\cos\gamma + \sqrt{3}\cos\beta\sin\gamma) - 0.5ab(\sqrt{3}(\cos\gamma\sin\alpha\sin\beta - \cos\alpha\sin\gamma) + 3(\cos\alpha\cos\gamma + \sin\alpha\sin\beta\sin\gamma)$

In the same manner the length can be find by using Euclidean norm, *di* can be expressed as

$$d_i = \sqrt{(x_p - x_b)^2 + (y_p - y_b)^2 + (z_p - z_b{}^2)} \tag{11}$$

where, x_p, y_p,z_p are the coordinates of the limb in moving *platform* and x_b, y_b,z_b are the coordinates of the limb in fixed platform

These 3 equations for $i=1, 2, 3$ give the lengths of the 3 limbs to *achieve* the desired position and attitude of the platform.

3.2. *Direct kinematics*

For the direct kinematics problem, the position vector $\boldsymbol{P}$ and the rotation matrix ${}^a\boldsymbol{R}_b$ of frame B with respect to A are to be found, given the limb length $\boldsymbol{d_i}$, $\boldsymbol{i}$ =1,2,3. The position vector contains three scalars unknowns, while the rotation matrix contains nine scalars unknowns.

Consider Figure 2 and Figure 3,

$$k_i = p + {}^{a}R_b . b \tag{12}$$

It's found out that once the limb length is given the resulting, structure of the platform becomes 3RS (J. Wu, J. S. Wang, and L. P. Wang, 2008). We can deduce that;

$$|N_iN_{i+1}| = \sqrt{3}\, b \tag{13}$$

Therefore dot multiplication result in Equation (14),

$$[\ k_i - k_{i+1}]^T\ [k_i - k_{i+1}] = 3b^2 \tag{14}$$

Solving (13) we can be able to obtain vector p which is;

$$P = (k_1 + k_2 + k_3)/3 \tag{15}$$

Equation (14) and (15) can be inserted into Equation (7) to solve for the rotation matrix parameters given the constrain imposed by revolute joints of $q_{1y}=0$, $q_{2y}= -\sqrt{3}\ q_{2x}$, $q_{3y}= \sqrt{3}\ q_{3x}$, for i= 1, 2 ,3.

3.3. *Manipulator synthesis*

The manipulator synthesis is carried out in MATLAB software with the parameters given in Table 1. The manipulator is intended to rotate along x and y axes and translate along z axis.

Table 1. Parameters used to patched the model.

Base radius (mm)	800
Manipulator radius (mm)	400
Limb length from the center line of the R joint to centre of the S joint (mm)	615.7111
Distance between the manipulator centers (mm)	700
Limb angle °	19.4136
Distance from fixed frame to revolute joint (mm)	500
Distance from moving frame to spherical joint (mm)	300
p_z distance (mm)	567.5

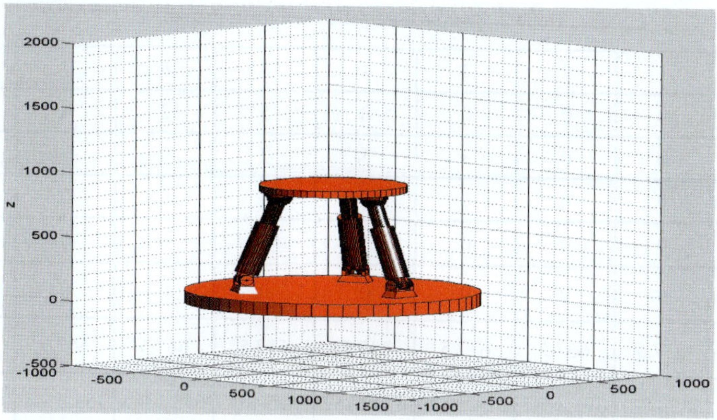

Fig. 4. The graphical picture of the manipulator patched in MATLAB.

4. Result

For the patched Figure 4, inverse kinematics simulation is carried out in MATLAB software. Table 2 below gives the summary of the simulation results and the results obtained by analytical solution, in verifying the inverse kinematics equations (length d_i) of the manipulator. Figure 5 to 12 and the first three graphs in Figures 5-12 was obtained for the global positioning of each limb (d_i=1,2,3) after simulation. The values of the simulation result is directly obtained from the MATLAB graphs and compared with the analytical result. Error between the two is obtained to show the accuracy of the solutions. Equation (8) through (10) is used for analytical result of each limb. The manipulator is designed to rotate in two axes i.e. x-y and translate along z-axis. The maximum angles of rotation are limited to ±16° to achieve high accuracy.

Table 2. Inverse kinematics results in matlab.

S/N	1	2	3	4	5	6	7	8
α°	0	0	0	0	5	2	4	-6
β°	0	0	3	5	7	9	-11	-13
γ°	0	0	0	0	0	0	0	0
P_x/mm	0	0	0	0	0	0	0	0
P_y/ mm	0	0	0	0	0	0	0	0
p_z/ mm	617.5	667.5	617.5	617.5	617.5	667.5	667.5	667.5
Cal d1 /mm	649.0811	696.8187	634.2925	624.6252	615.134	653.1445	753.3174	763.7658
Sim d1 /mm	647.3825	693.3983	633.7243	624.8623	616.3348	653.3181	746.751	756.8522
Error mm	1.6986	3.4204	0.5682	-0.2371	-1.2008	-0.1736	6.5664	6.9136
Cal d2 /mm	649.0777	696.8155	656.582	661.6138	688.7886	728.4251	686.5052	641.3961
Sim d2 /mm	647.3791	693.3951	654.3869	659.1383	684.6494	723.2509	684.4823	643.0931
Error mm	1.6986	3.4204	2.1951	2.4755	4.1392	5.1742	2.0229	-1.697
Cal d3 /mm	649.0777	696.8155	656.582	661.6138	645.1023	710.8259	653.6416	689.8778
Sim d3 /mm	647.3791	693.3951	654.3869	659.1383	644.0721	706.858	654.0528	687.9313
Error mm	1.6986	3.4204	2.1951	2.4755	1.0302	3.9679	-0.4112	1.9465

A. For $p_x=p_y=0$, $p_z=617.5$, $\alpha=\gamma=0$, $\beta=0°$

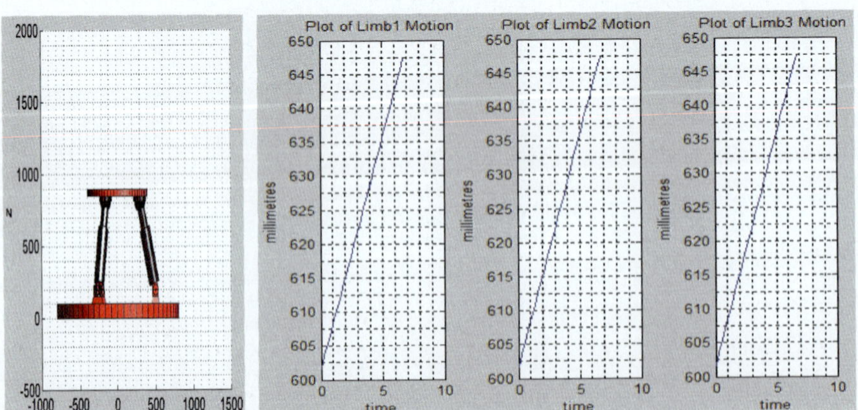

Fig. 5. Platform simulation graph of limb 1,2,3 motion, prismatic joint coordinates and revolute joints angles.

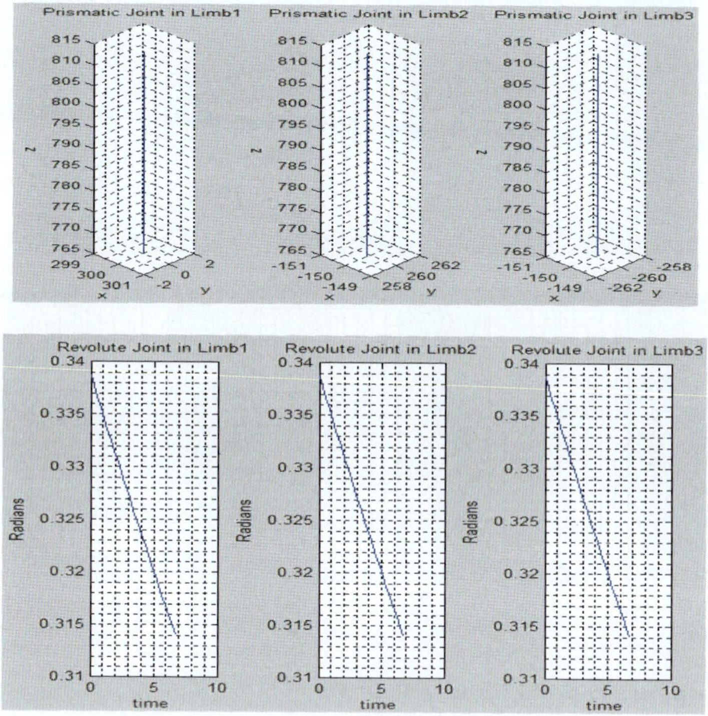

Fig. 5. (*Cont'd*).

B. For $p_x=p_y=0$, $p_z=667.5$, $\alpha=\gamma=0$, $\beta=0^o$

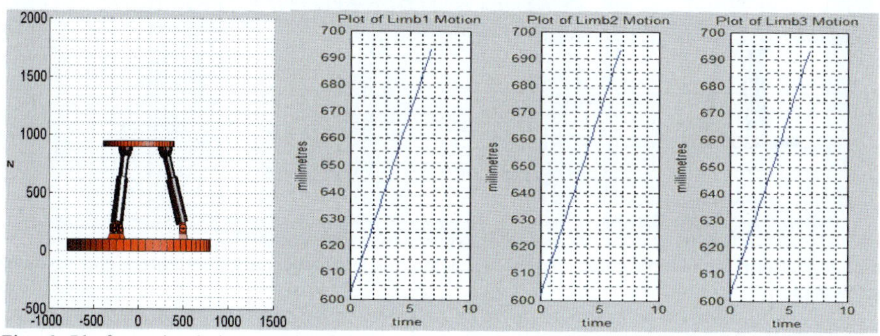

Fig. 6. Platform simulation graph of limb 1,2,3 motion, prismatic joint coordinates and revolute joints angles.

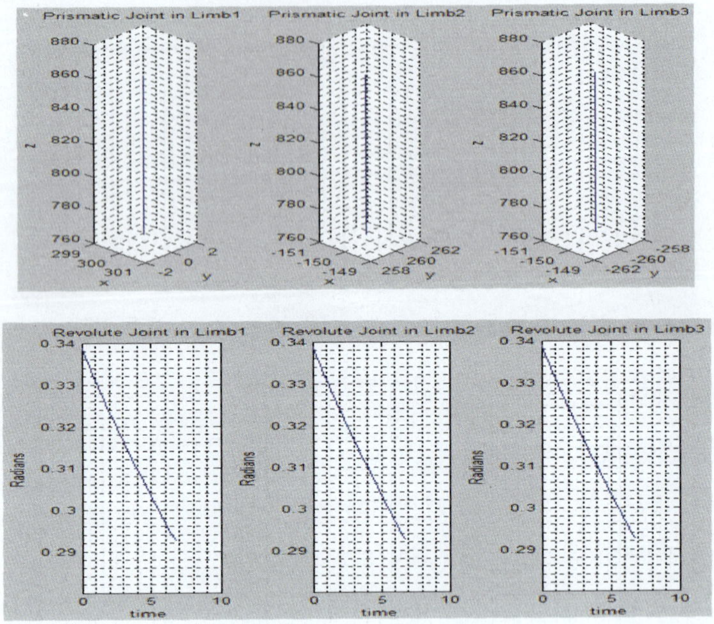

Fig. 6. (*Cont'd*).

C. For $p_x=p_y=0$, $p_z=617.5$, $\alpha=\gamma=0$, $\beta=3^o$

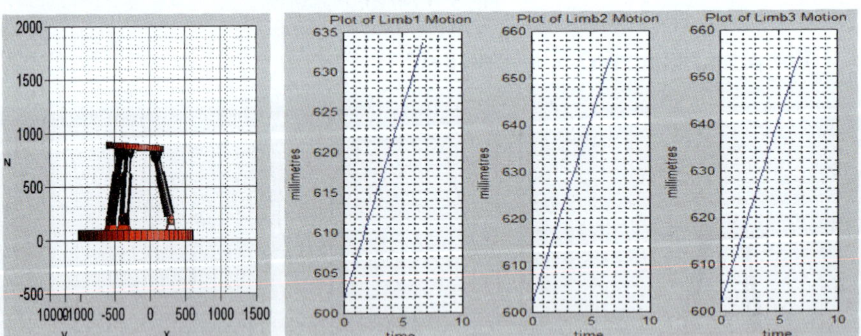

Fig. 7. Platform simulation graph of limb 1,2,3 motion, prismatic joint coordinates and revolute joints angles.

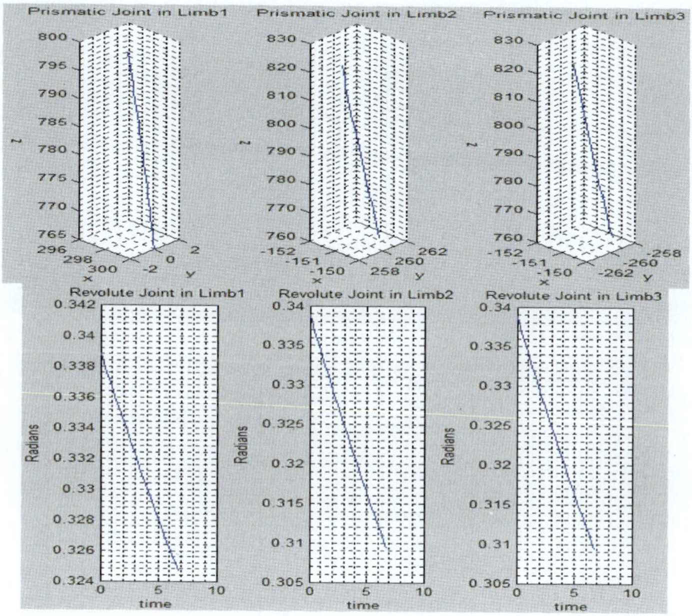

Fig. 7. (*Cont'd*).

D. For $p_x=p_y=0$, $p_z=617.5$, $\alpha=\gamma=0^o$, $\beta=5^o$

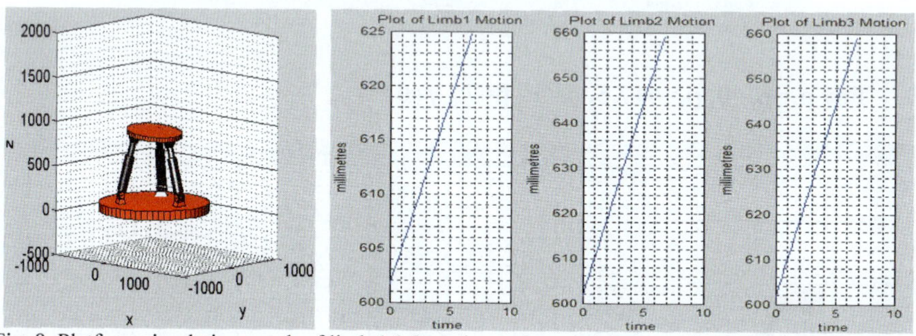

Fig. 8. Platform simulation graph of limb 1,2,3 motion, prismatic joint coordinates and revolute joints angles.

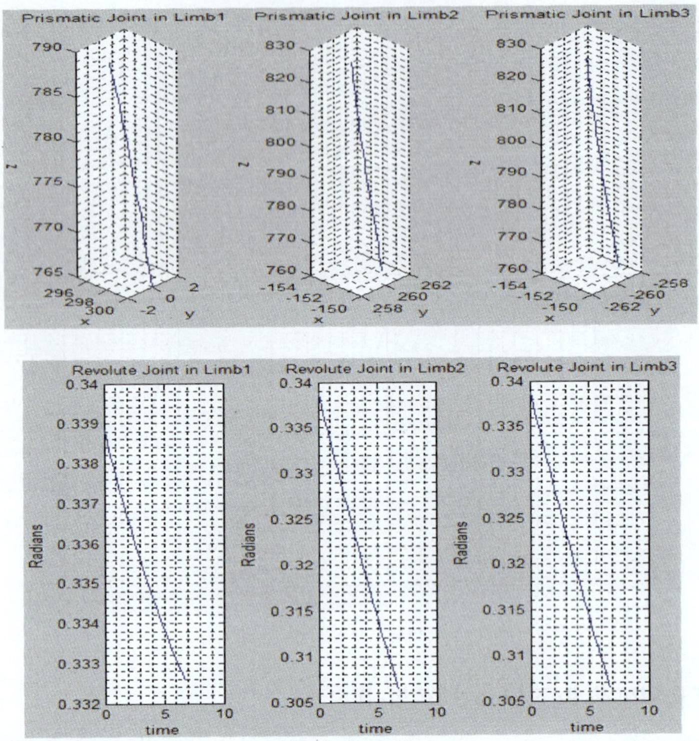

Fig. 8. (*Cont'd*).

E. For $p_x=p_y=0$, $p_z=617.5$, $\alpha=5^o$, $\beta=7^o$, $\gamma=0$ ^{o}

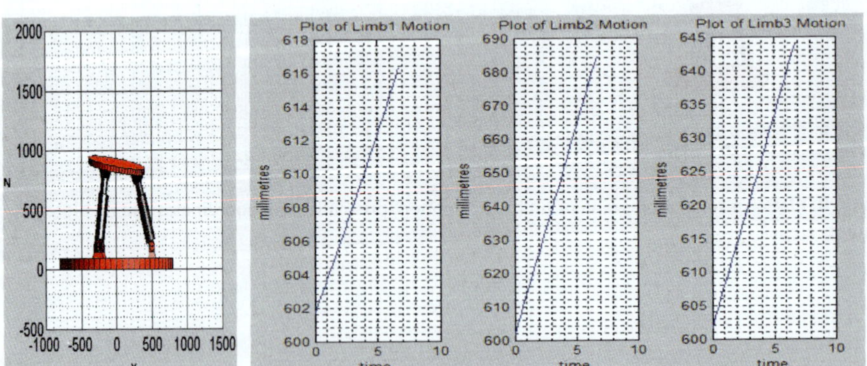

Fig. 9. Platform simulation graph of limb 1,2,3 motion, prismatic joint coordinates and revolute joints angles.

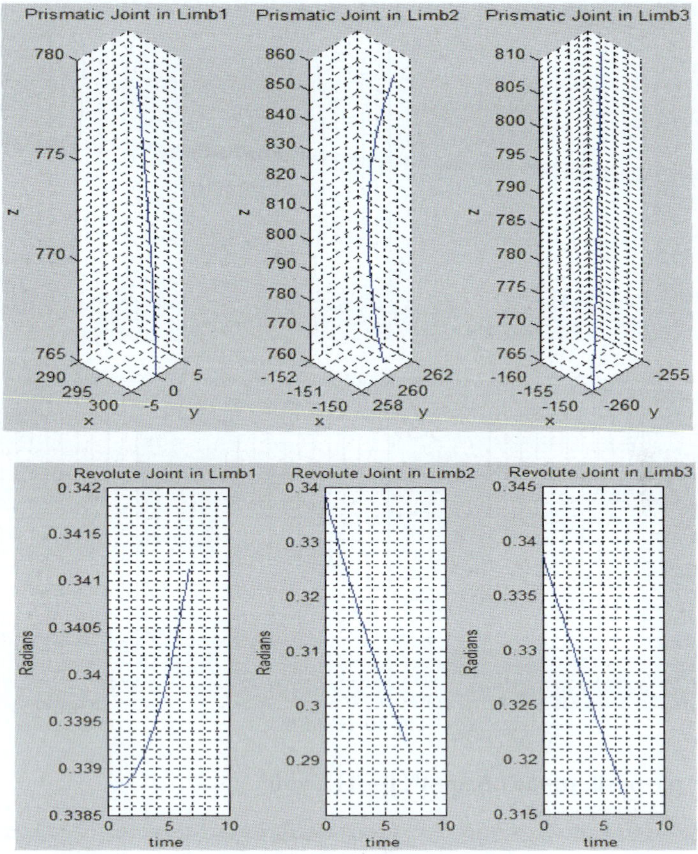

Fig. 9. (*Cont'd*).

F. For $p_x=p_y=0$, $p_z=667.5$, $\alpha=2^o$, $\beta=9^o$, $\gamma=0\ ^o$

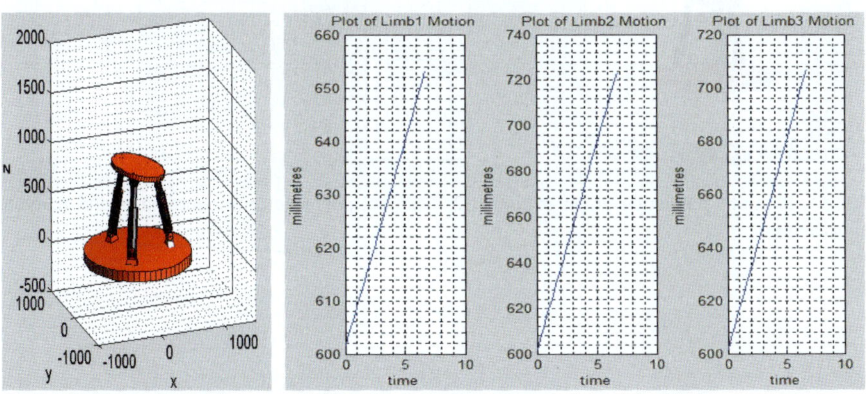

Fig. 10. Platform simulation graph of limb 1,2,3 motion, prismatic joint coordinates and revolute joints angles.

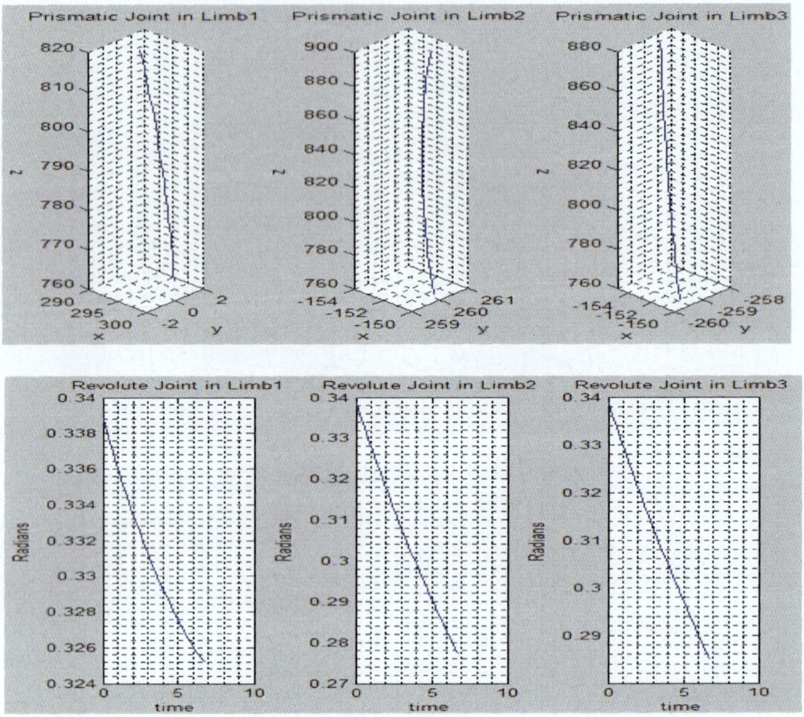

Fig. 10. (*Cont'd*).

G. For $p_x=p_y=0$, $p_z=667.5$, $\alpha=4^o$, $\beta=-11^o$, $\gamma=0^o$

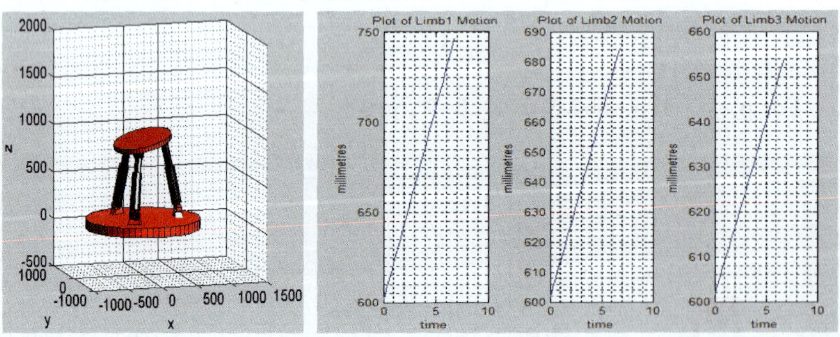

Fig. 11. Platform simulation graph of limb 1,2,3 motion, prismatic joint coordinates and revolute joints angles.

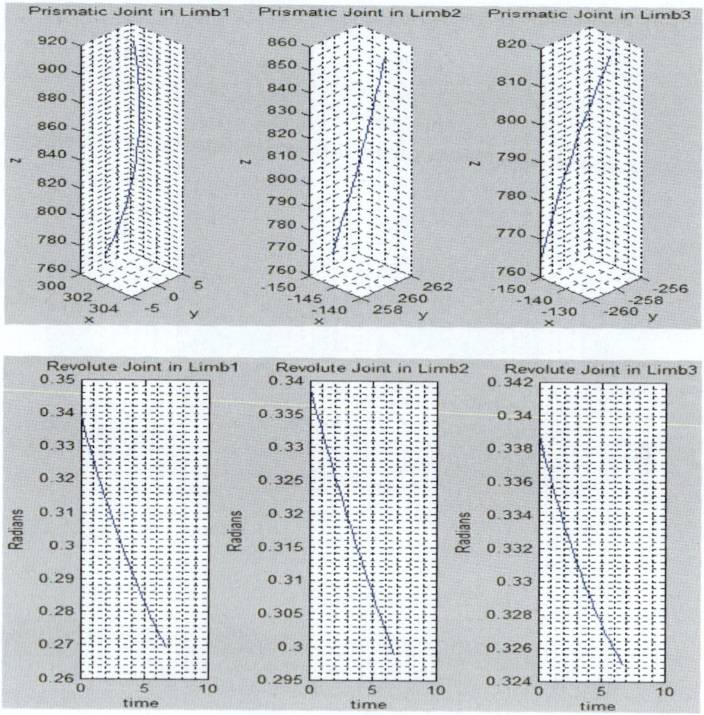

Fig. 11. (*Cont'd*).

H. For $p_x=p_y=0$, $p_z=667.5$, $\alpha=-6^{\circ}$, $\beta=-13^{\circ}$, $\gamma=0$

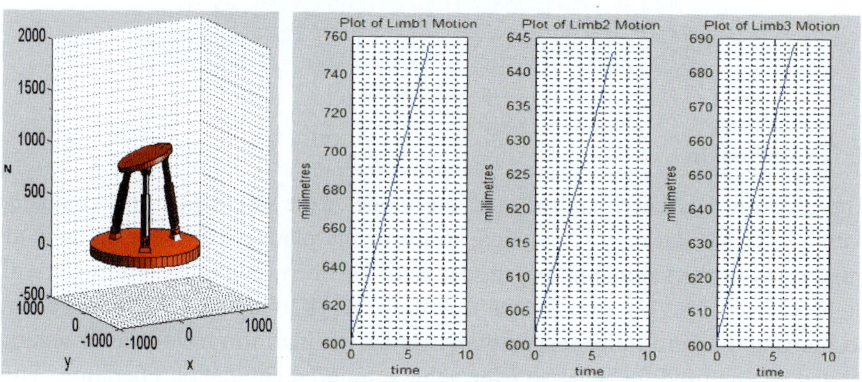

Fig. 12. Platform simulation graph.

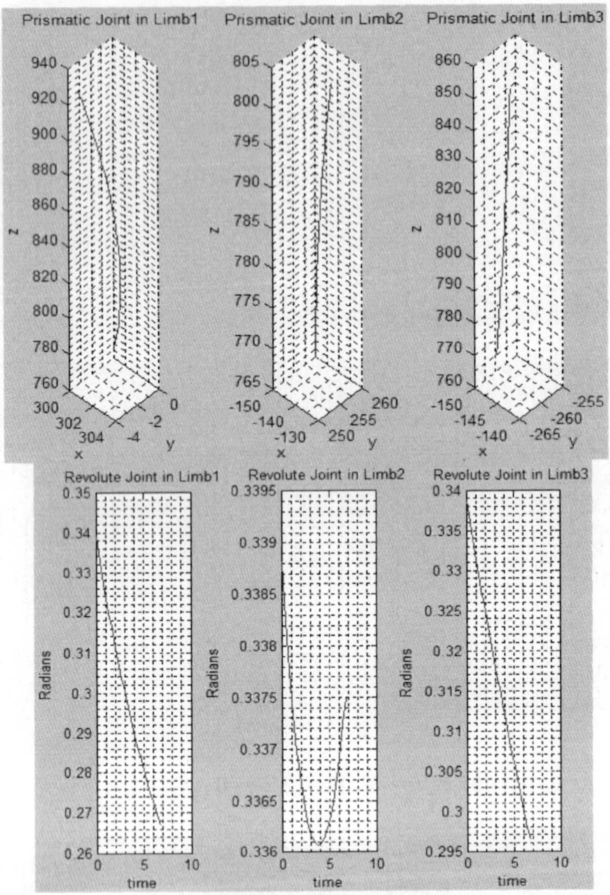

Fig. 12. (*Cont'd*).

5. Conclusion

The platform has been studied, and the model were developed and simulated with MATLAB package. Analytical result of the platform moving limbs is almost equal to the simulated length, the error between the two is not much with highest error of 6.9136 mm for the first limb, 5.1742mm and 3.9679mm for the second and the third limbs respectively. Hence the platform have a good accuracy and can be used for what so ever vehicle training simulation operation. This left to the designer choice for the application intended to perform during the training.

6. Appreciation

The work was supported by Scientific Research Development Fund of Tianjin University of Technology and Education (KJY14-02).

References

1. Zafer Bingul and Oguzhan Karahan (2012). Dynamic Modeling and Simulation of Stewart Platform, Serial and Parallel Robot Manipulators - Kinematics, Dynamics, Control and Optimization, Dr. Serdar Kucuk (Ed.), ISBN: 978-953-51-0437-7.
2. Zoran Lazarevic, "Feasibility of a Stewart Platform with Fixed Actuators as a Platform for CABG Surgery Device" Columbia University Department of Bioengineering.
3. L. W Tsai, Robot Analysis: The Mechanics Of Serial And Parallel Manipulators, Wiley, New York, 1999.
4. J. Wu, J. S. Wang, and L. P. Wang, "Identification of dynamic parameter of 3DOF parallel manipulator with actuation redundancy," J. Manuf. Sci. Eng., vol. 130, no. 4, pp. 041012-1–041012-7, Aug. 2008.
5. J.M. Hervé, F. Sparacino, " Structural Synthesis of Parallel Robots Generating Spatial Translation ", 5th Int. Conf. on Adv. Robotics, IEEE no. 91TH0367-4, Vol.1, pp. 808-813, 1991.
6. P. Vischer, R. Clavel, "Argos : A Novel 3-DOF Parallel Wrist mechanism", The International Journal of Robotics Research, Vol. 19, No. 1, pp. 5-11, January 2000.
7. Ibrahim and W. Khalil "Kinematic and Dynamic Modeling of the 3-RPS Parallel Manipulator" 12th IFToMM World Congress, Besançon (France), June18-21, 2007.
8. Çağdaş BAYRAM "Kinematic and Dynamic Analysis of Spatial Six Degree of Freedom Parallel Structure Manipulator" İzmir Institute of Technology İzmir, Turkey September 2003.
9. Youhong Gong "Design Analysis of a Stewart Platform for Vehicle Emulator Systems" Massachusetts Institute of Technology, January 1992.
10. Nicol´as Rojas and Federico Thomas "A robust forward kinematics analysis of 3-RPR planar platforms" Institut de Rob`otica i Inform`atica Industrial (CSIC-UPC), Barcelona, Spain.
11. Jean-Pierre Merlet, Peter Donelan. "On the Regularity of the Inverse Jacobian of Parallel Robots". ARK, Jun 2006, Ljubljana.
12. Jean-Pierre MERLET "Direct kinematics of planar parallel manipulators" INRIA BP 93 06902 Sophia-Antipolis, France.

13. Ivan J. Baiges-Valentin "Dynamics Modelling of Parallel Manipulators" University of Florida, 1996.
14. Sureyya Sahin and Leila Notash "Force and Stiffness Analyses of WireActuated Parallel Manipulators" 12th IFToMM World Congress, Besançon (France), June18-21, 2007.
15. Félix Majou, Philippe Wenger and Damien Chablat Design of A 3 Axis Parallel Machine Tool For High Speed Machining: The Orthoglide IDMME Clermont-Ferrand, France, May 14-16, 2002.
16. Bibi Farouk Zulfiqar Ibrahim and Liu Ying "Design Of Inward Limbs 3RPS Manipulator For Machining Purpose And Kinematics Simulation Using Adams" Ijrs Volume 4 Issue 2 Febuary 2015.
17. Maurizio Ruggiu Cartesian "Stiffness Matrix Mapping of a Translational Parallel Mechanism with Elastic Joints" International Journal of Advanced Robotic Systems, 06 Aug 2012.
18. G. D. L. Soares Júnior , J. C. M. Carvalho, R. S. Gonçalves "Stiffness Analysis of 6-RSS Parallel Manipulator" 13th World Congress in Mechanism and Machine Science, Guanajuato, México, 19-25 June, 2011.
19. L.W. Tsai, S. Joshi, 2002, "Kinematic Analysis of 3-DOF Position Mechanisms for Use in Hybrid Kinematic Machines", Journal of Mechanical Design, Vol. 124, pp. 245–253.
20. O. Khatib, "A unified approach for motion and force control of robot manipulators: the operation space formulation," IEEE J. Robotics and Automation, vol. 3, no. 1, pp. 43¨C53, 1987.

Identification genes of colorectal cancer with integrated data via Block-sparse NMFL21

D. Wang[1]
School of Information Science and Engineering,
Qufu Normal University, Rizhao 276826, China
Email: dongwshark@126.com,

J.X. Liu[2]
School of Information Science and Engineering,
Qufu Normal University, Rizhao 276826, China
Bio-Computing Research Center, Shenzhen Graduate School,
Harbin Institute of Technology, Shenzhen 518055, China,
Email: sdcavell@126.com

Y.L. Gao[3]
Library of Qufu Normal University, Qufu Normal University,
Rizhao 276826, China
Emails: yinliangao@126.com

J. L. Shang[4]
School of Information Science and Engineering, Qufu Normal University,
Rizhao 276826, China
Emails: shangjunliang110@163.com

Colorectal cancer is the second leading cause of cancer deaths in the word. The assessment of the colorectal cancer is determined by the clinical stage which denotes the number of lymph nodes with metastatic cancer and the identification of metastatic sites in other organs. While the cancer genes associated with colorectal cancer have been characterized, less is known about the identity of special cancer genes that are responsible for advanced clinical stage and metastasis. To characterize genetic aberration in colorectal cancer, we conducted a Block-sparse nonnegative matrix factorization with L_{21}-norm (Block-sparse NMF L_{21}) analysis of multiple genomic data which include mutation, DNA copy number, promoter methylation, messenger RNA and microRNA expression

---

[1] Dong Wang
[2] Jin-Xing Liu
[3] Ying-Lian Gao
[4] Jun-Liang Shang

from The Cancer Genome Atlas (TCGA). With the analysis of the method, we ranked the genes by the scores and identified a set of top gene predictors of advanced clinical stage. The results demonstrate that the analysis of integrated data through block-sparse NMF L_{21} is more comprehensive.

Keywords: Colorectal cancer; Genomics; Clinical stage; Block-sparse; NMF.

1. Introduction

Colorectal cancer (CRC) is one of the most common cancers and the second leading cause of cancer-related death in the world [1, 2]. The distinction between the colon and the rectum is largely based on anatomical [3].

Recently, increasing knowledge of genetic and genomic investigation identified many of the critical genes that are important to CRC development [4-6]. For example, the cancer genes APC, KRAS and TGF have a high frequency of genetic aberrations in CRC and are known to play an important role in CRC development [5]. However, we have not had a fully integrated view of the genetic and genomic changes and their significance for CRC clinical stage [3].

Colorectal cancers are currently treated by surgical resection in locally restricted tumor stages. However, disease recurrence and formation of local and distant metastasis frequently occur even in cases with successful curative resection of the primary tumor with a result of it is a metastatic disease [7].

Metastasis formation is the major cause of death in patients with colorectal cancer, and is depending on tumor stage. Tumor stage divided into four stages: I, II, III and IV which is determined by the TNM criteria. The TNM system rely on the size and extent of the primary tumor, on the metastatic spread to lymph nodes and distant organ sites, and on lymphatic and vascular invasion [7]. Advanced stage (stage III or IV) of CRC has a significantly worse prognosis compared to early stage (stages I or II) that is generally considered curable [8]. Therefore, there is an increasing interest in identification of the specific genes associated with advanced clinical stage [9].

The Cancer Genome Atlas (TCGA) project plans to profile genomic changes in 20 different cancer types and has so far published results on two cancer types [10, 11]. In this paper we used the most comprehensive CRC genomic date to survey the genes [3]. This project also provides clinical information about the metastatic status of individual patients via clinical stage information. The breadth of the TCGA genomic data sets provides a unique opportunity to consider different categories of genetic aberrations at individual gene resolution that other genomic studies have not considered [12-14].

Therefore, the goal of this study is to conducts a method to identify genes are associated with the advanced stage in integrated data. There are two major challenges to the development of the study:

Firstly, in this paper, we conducts an analysis of the integrated TCGA data which are included i) somatic mutations, ii) copy number alterations, iii) gene expression changes and iv) methylation. We analyze this integrated genomic data set about the clinical stage to identify the genes associated with advanced CRC [3].

Secondly, dimension reduction method is also an important part to the identify the genes associated with the clinical stage. Dimension reduction techniques [15, 16] such as principal component analysis [17] and non-negative matrix factorization [18] have been proposed for feature selection. However, these methods didn't perform well because they cannot generate a sparse result. Many sparse methods have been proposed to overcome the disadvantage to generate a sparse result such as sparse principal component analysis (SPCA) [19], nonnegative matrix factorization sparse constraint (NMFSC) [20].

However, the conventional sparse methods impose the same strength sparse constraint on the low-rank matrix which can work well for the same properties data type. But these methods cannot perform well on TCGA dataset which is integrated with four different genomic data (somatic mutations, copy number alterations, gene expression changes and methylation) because different genomic data should impose different sparse constraint [21].

In this paper we propose a novel method Block-sparse nonnegative matrix factorization with L_{21} norm-based (Block-sparse NMF L_{21}) through address the L_{21} norm and block-sparse on the regulation term to generate a sparse and robust result [22-26].

The main contributions of this paper are described as follows:

Firstly, with our integrated TCGA data, we considered a full range of cancer genetic aberrations to identified specific cancer genes associated with advanced clinical stage.

Secondly, facing with the different genomic data, Block-sparse method is proposed to generate a reasonable and feasible result.

2. Methodology

2.1. *Data source*

Clinical stage information genomic data was obtained from the Broad Firehose (http://gdac.broadinstitute.org) which is one of the Genome Data Analysis (GDACs) for TCGA project. The data files from January 2013 analysis/standardization run of colorectal (COADREAD) cancer includes four genomics assays for each sample: DNA copy number variation, mRNA expression level by microarray/RNASeq, somatic mutations by whole exome sequencing, DNA methylation [8]. Clinical information of the samples was obtained from the Broad Firehose and UCSC cancer genome browser [27].

In this paper we conduct an analysis of 197 samples with 5188 genomic features from 1325 genes which integrated exome sequence, DNA copy number, methylation and mRNA expression data. It may have at least one genomic feature and at most four genomic features for each gene.

2.2. *Block-sparse NMFL21 analysis*

The definition of Block-sparse NMF L_{21} is as follows:

$$\min_{\mathbf{Y}} \|\mathbf{X}-\mathbf{AY}\|_{2,1} + \sum_i \lambda_i \|\mathbf{Y}_i\|_{2,1}, \qquad st.\mathbf{Y} > \mathbf{0}, \tag{1}$$

where λ_i is the sparse parameter in different genomic features, we address different sparse parameters on different features because the properties between features are different. It will make overly restrictive about the features if we impose the same sparse constraint on these features. For example the somatic mutations features, copy number alterations features, gene expression changes and methylation with the parameters λ_1, λ_2, λ_3 and λ_4 respectively.

In Eq.(1), the denote L_{21} norm which can defined as:

$$\|\mathbf{M}\|_{2,1} = \sum_{i=1}^{n} \sqrt{\sum_{j=1}^{s} \mathbf{m}_{ij}^2} = \sum_{i=1}^{n} \|\mathbf{m}^i\|_2 \tag{2}$$

The L_{21}-norm emphasis on error function and regulation term can eliminate the noisy data and generate a sparse result respectively [28].

2.3. *The performance of Block-sparse NMF-* L_{21}

In this paper the matrix $\mathbf{X}$ denotes the TCGA integrative data whose size is $\mathrm{m} \times n$, m and n denotes samples and the genomic features, respectively. Then the matrix $\mathbf{X}$ can be rewritten through nonnegative matrix factorization as,

$$\mathbf{X} \approx \mathbf{AY} \tag{3}$$

Where $\mathbf{A}$ is an $m \times d$ matrix, $\mathbf{Y}$ is a $d \times n$ matrix, $d < \min(m,n)$. So, given the suitable parameters of block-sparse NMF-L_{21}, the sparse matrix $\mathbf{Y}$ can be obtained. Then, the special genes can be extracted according to the non-zero entries in $\mathbf{Y}$.

2.3.1. *Ranking scheme*

In the primary results of Block-sparse NMF-L_{21}, a gene can appear at most four times when all four types of genomic data support the gene.

The rank of each genomic feature is determined by its result value in descending order. Then the score of each gene is calculated as follows [8]:

$$\text{score of } g(i)=(1000 \div \sum_{r=1}^{d} r)\times(d-r(i)+1) \tag{4}$$

where r is the rank and d is the total number of ranks in the result from block-sparse NMF-L_{21} analysis, $g(i)$ is the genomic feature, $r(i)$ is the rank of the genomic feature of i.

Finally, the score of a gene will be the sum of scores of genomic features from the gene.

3. Results

The TCGA CRC clinical stage genomic data in this paper include 5188 genomics features from 1325 genes in 197 samples which distribution is as follows: from 1~1117 are Copy Number data, 1118~2030, 2031~4108 and 4109~5188 are MAF, Meth-27 and mRNA-Eigen data, respectively.

3.1. *Application of Block-sparse NMF-L_{21} to integrated genomic features*

The Block-sparse NMF L_{21} algorithm identified genes to delineated clinical information of CRC.

In this section, we run the method with appropriate parameters using individual genomic feature data sets (e.g. mutation alone, mRNA alone, etc.) and the integrated data set, respectively compared to clinical information to validate the performance of the integrated data. The top 16 genes ranked by our score scheme and the results are listed in Table 1.

From Table 1 we can see that, our integrative analysis identified the genes of KDM5C and MAN2B2 as the top ranked clinical stage associated candidates. In addition, the 1th ranked genes DAXX in mutation alone has dropped in 15th ranked in the integrative analysis. The BUB1B gene has extracted in mutation, mRNA data and the integrated data but did not extracted in copy number and Meth data. To this end, the integrative analysis is more comprehensive than the individual analysis.

Table 1. The top 16 genes association with clinical information.

Rank	copy number alone	Mutation alone	Methylation alone	mRNA alone	integrated data
1	sh3bgrl3	daxx	kdm5c	pdcd1lg2	kdm5c
2	aim1l	fhit	42252	man2b2	man2b2
3	cd52	bub1b	pdcd1lg2	sec31a	dctd
4	znf683	naca	hmgcl	kdm5c	muc1
5	dhdds	mrps18a	aff1	mrps18a	npdc1
6	hmgn2	nsd1	mrps18a	mael	sema3b
7	rps6ka1	klhdc3	fam76a	42252	ndufa13
8	pla2g2a	erbb3	nsd1	daxx	taf15
9	rpl11	taf15	ankrd9	bub1b	hoxc11
10	tceb3	man2b2	foxp1	erbb3	hmgcl
11	gale	idh1	c20orf111	bcl3	klhl34
12	hmgcl	htra3	ngef	htra3	mrps18a
13	lypla2	mgmt	fhit	fhit	nsd1
14	fuca1	hmgcl	tceb3	nsd1	runx3
15	cnr2	polr1d	znf683	hmgcl	daxx
16	pnrc2	foxp1	rad51b	taf15	bub1b

3.2. *Genes compared with COSMIC data by block-sparse NMF* L_{21}

Catalogue of Somatic Mutations in Cancer (COSMIC) has been both curated and validated. The COSMIC database contains 484 genes that have been shown to be associated with cancer development and thus are established or candidate cancer genes [29, 30]. In this section, we select 300 genes based on our method with the integrated data and find that there are some genes overlapped with the COSMIC study. The top ten genes are listed in Table 2 and the results demonstrate that our method can select the genes that are important to CRC development.

Table 2. The top ten genes overlapped with COSMIC.

Official Gene Symbol	Gene Name
ALK	anaplastic lymphoma receptor tyrosine kinase
APC	anaplastic lymphoma receptor tyrosine kinase
BCL2	B-cell CLL/lymphoma 2
BRD4	bromodomain containing 4
BRIP1	BRCA1 interacting protein C-terminal helicase 1
BUB1B	budding uninhibited by benzimidazoles 1 homolog beta
C15orf55	chromosome 15 open reading frame 55
CARD11	caspase recruitment domain family, member 11
CBFA2T3	core-binding factor, runt domain, alpha subunit 2; translocated to, 3
CCDC6	coiled-coil domain containing 6

3.3. *Gene terms analysis*

In this section we compare our method with the following two methods on the integrate data: (a) PMD method (proposed by Witten et al. [31]); (b) SPCA method (proposed by Journée et al. [32]); For fair comparison, all the three methods extract 100 genes as the differentially genes associated with the clinical stage on the integrate data. The Gene Ontology (GO) enrichment of functional annotation of the extracted genes by three methods is detected by DAVID Bioinformatics Resources 6.7, NIAID/NIH which is publicly available at [http://david.abcc.ncifcrf.gov/gene2gene.jsp]. Table 3 lists the results the top 5 gene terms with Gene count and P-Value by the three methods. Gene count denotes the gene number selected by the method that associated with the term. From the results we can see that, our method performs well than PMD and SPCA method, for example, from the first term we can see that, our methods can extract 22 associated genes with chromosomal rearrangement, while PMD and SPCA can extract 11 genes, respectively.

4. Conclusion

In this paper, we proposed a new method block-sparse NMF L_{21} to identify the genes associated with clinical stage with the integrated data. Through the integrated data we can considerate a full range of cancer genetic aberrations. Through imposing different sparse constraint on different genomic feature data, our method can generate a more efficient result. The experiments on integrated data and individual data have demonstrated that the analysis of integrated data is more comprehensive. The results compared with other sparse methods can also demonstrate the efficiency of our method.

Acknowledgements

This work was supported in part by the NSFC under grant Nos. 61272339, 61370163 and 61233011; the Shandong Provincial Natural Science Foundation, under grant Nos. ZR2013FL016 and BS2014DX004; Shenzhen Municipal Science and Technology Innovation Council (Nos.JCYJ20140417172417174, JCYJ20140904154645958 and CXZZ20140904154910774).

References

1. P. K. Yalavarthy, B. W. Pogue *et al.*, 2007 "Weight-matrix structured regularization provides optimal generalized least-squares estimate in diffuse optical tomography," *Medical physics,* vol. 34, no. 6, pp. 2085-2098.
2. J. Ferlay, H. R. Shin *et al.*, 2010 "Estimates of worldwide burden of cancer in 2008: GLOBOCAN 2008," *International Journal of Cancer,* vol. 127, no. 12, pp. 2893-2917.
3. C. G. A. Network, 2012 "Comprehensive molecular characterization of human colon and rectal cancer," *Nature,* vol. 487, no. 7407, pp. 330-337.
4. L. D. Wood, D. W. Parsons *et al.*, 2007 "The genomic landscapes of human breast and colorectal cancers," *Science,* vol. 318, no. 5853, pp. 1108-1113.
5. B. Vogelstein, N. Papadopoulos *et al.*, 2013 "Cancer genome landscapes," *Science,* vol. 339, no. 6127, pp. 1546-1558.
6. J. R. Pollack, T. Sørlie *et al.*, 2002 "Microarray analysis reveals a major direct role of DNA copy number alteration in the transcriptional program of human breast tumors," *Proceedings of the National Academy of Sciences,* vol. 99, no. 20, pp. 12963-12968.
7. M. Shibayama, M. Maak *et al.*, 2011 "Prediction of metastasis and recurrence in colorectal cancer based on gene expression analysis: ready for the clinic?," *Cancers,* vol. 3, no. 3, pp. 2858-2869.
8. H. Lee, P. Flaherty *et al.*, 2013 "Systematic genomic identification of colorectal cancer genes delineating advanced from early clinical stage and metastasis," *BMC medical genomics,* vol. 6, no. 1, pp. 54.

9. T. Yoshida, T. Kobayashi *et al.*, 2010 "Clinical omics analysis of colorectal cancer incorporating copy number aberrations and gene expression data," *Cancer Informatics,* vol. 9, pp. 147.
10. R. G. Verhaak, K. A. Hoadley *et al.*, 2010 "Integrated genomic analysis identifies clinically relevant subtypes of glioblastoma characterized by abnormalities in PDGFRA, IDH1, EGFR, and NF1," *Cancer cell,* vol. 17, no. 1, pp. 98-110.
11. C. G. A. R. Network, 2011 "Integrated genomic analyses of ovarian carcinoma," *Nature,* vol. 474, no. 7353, pp. 609-615.
12. R. Shen, A. B. Olshen *et al.*, 2009 "Integrative clustering of multiple genomic data types using a joint latent variable model with application to breast and lung cancer subtype analysis," *Bioinformatics,* vol. 25, no. 22, pp. 2906-2912.
13. E. A. Vucic, K. L. Thu *et al.*, 2012 "Translating cancer 'omics' to improved outcomes," *Genome research,* vol. 22, no. 2, pp. 188-195.
14. L. Chin, W. C. Hahn *et al.*, 2011 "Making sense of cancer genomic data," *Genes & Development,* vol. 25, no. 6, pp. 534-555.
15. D. Chen, X. Cao *et al.*, "Blessing of dimensionality: High-dimensional feature and its efficient compression for face verification." pp. 3025-3032 2013.
16. P. Hall, J. Marron *et al.*, 2005 "Geometric representation of high dimension, low sample size data," *Journal of the Royal Statistical Society: Series B (Statistical Methodology),* vol. 67, no. 3, pp. 427-444.
17. O. Alter, P. O. Brown *et al.*, 2000 "Singular value decomposition for genome-wide expression data processing and modeling," *Proceedings of the National Academy of Sciences,* vol. 97, no. 18, pp. 10101-10106.
18. J.-P. Brunet, P. Tamayo *et al.*, 2004 "Metagenes and molecular pattern discovery using matrix factorization," *Proceedings of the National Academy of Sciences,* vol. 101, no. 12, pp. 4164-4169.
19. H. Zou, T. Hastie *et al.*, 2006"Sparse principal component analysis," *Journal of Computational and Graphical Statistics,* vol. 15, no. 2, pp. 265-286.
20. P. O. Hoyer, 2004 "Non-negative matrix factorization with sparseness constraints," *The Journal of Machine Learning Research,* vol. 5, pp. 1457-1469.
21. J. Huang, F. Nie *et al.*, 2014 "Robust manifold nonnegative matrix factorization," *ACM Transactions on Knowledge Discovery from Data (TKDD),* vol. 8, no. 3, pp. 11.
22. D. Cai, X. He *et al.*, 2011 "Graph regularized nonnegative matrix factorization for data representation," *Pattern Analysis and Machine Intelligence, IEEE Transactions on,* vol. 33, no. 8, pp. 1548-1560.

23. Z. Gao, L.-F. Cheong *et al.*, 2012, "Block-sparse RPCA for consistent foreground detection," *Computer Vision–ECCV* pp. 2012 , 690-703: Springer.
24. F. Zhu, Y. Wang *et al.*, 2014 "Effective spectral unmixing via robust representation and learning-based sparsity," *arXiv preprint arXiv:1409.0685*.
25. H. Zhang, Z. Zha *et al.*, 2014 "Robust (Semi-) Nonnegative Graph Embedding."
26. Y. Jiang, J. Liu *et al.*, 2014 "Semi-supervised Unified Latent Factor learning with multi-view data," *Machine Vision and Applications,* vol. 25, no. 7, pp. 1635-1645.
27. J. Z. Sanborn, S. C. Benz *et al.*, 2010 "The UCSC cancer genomics browser: update 2011," *Nucleic Acids Research*, pp. gkq1113.
28. D. Kong, C. Ding *et al.*, "Robust nonnegative matrix factorization using l21-norm." pp. 673-682.
29. S. A. Forbes, N. Bindal *et al.*, 2010 "COSMIC: mining complete cancer genomes in the Catalogue of Somatic Mutations in Cancer," *Nucleic Acids Research*, pp. gkq929.
30. S. Bamford, E. Dawson *et al.*, 2004 "The COSMIC (Catalogue of Somatic Mutations in Cancer) database and website," *British Journal of Cancer,* vol. 91, no. 2, pp. 355-358.
31. D. M. Witten, R. Tibshirani *et al.*, 2009 "A penalized matrix decomposition, with applications to sparse principal components and canonical correlation analysis," *Biostatistics*, pp. kxp008.
32. G. Nyamundanda, I. C. Gormley *et al.*, 2014 "A dynamic probabilistic principal components model for the analysis of longitudinal metabolomics data," *Journal of the Royal Statistical Society: Series C (Applied Statistics)*.

A chaotic system with one line equilibria and image encryption with avalanche effects

E. Chen[†]
Schools of Mathematics and Physics,
University of Science and Technology Beijing, Beijing China, 100083
[†]*Email: chene5546@163.com,*

L.Q. Min
Schools of Mathematics and Physics,
University of Science and Technology Beijing, Beijing China, 100083
Schools of Automation and Electrical Engineering,
University of Science and Technology Beijing, Beijing China, 100083
Email: minlequan@sina.com

D.D. Han
Schools of Automation and Electrical Engineering,
University of Science and Technology Beijing, Beijing China, 100083
Email: hxu1204@163.com

This paper introduces a 4-dimensional discrete chaotic system with one-line equilibria (DCSLE). Based on the chaotic system and generalized chaos synchronization (GCS) theorem, an 8-dimensional DCSLE GCS system is constructed. The DCSLE GCS system is used to design a chaotic pseudorandom number generator (CPRNG). The FIPS 140-2 test suit / generalized FIPS 140-2 test suit are used to test the randomness of the 1000-key streams consisting of 20,000 bits generated by the CPRNG, the RC4 algorithm and the ZUC algorithm, respectively. The test results confirm there are 100% / 99.2%, 99.9% / 98.8%, 100% / 97.9% key streams passing the tests. Numerical simulations suggest that there are no significant correlations between the key streams and the perturbed key streams generated by the CPRNG. The key space of the CPRNG is larger than 21195. The CPGNG and an avalanche effect encryption scheme are used to encrypt an RGB image. The results demonstrate that the CPRNG is able to generate the avalanche effects which are similar to those generated via ideal CPRNGs.

Keywords: Chaotic system with one-line equilibria; Pseudorandom number generator; Randomness test; Avalanche encryption scheme.

1. Introduction

Chaos is one type of complex dynamic behaviors displaying apparently random and being generated from a determined nonlinear system. Chaos systems are intrinsically sensitive to initial conditions, parameters of chaotic systems, and difficult to predict in a long-term ([1]-[3]). Recently, the research on chaotic systems which have the unusual feature of having a line-equilibrium has attracted much attention. Using a systematic computer search, nine simple chaotic maps with quadratic nonlinearities were found that have one-line equilibria ([4]).

On the other hand, chaos synchronization (CS) is of essential importance for many physical, biological and engineering systems ([5]-[9]). Since the earlier work of Pecora and Carroll ([10]), the research on GS has also developed rapidly ([11]-[14]). Moreover, GS may provide some new tools for cryptography and communications ([15]-[20]).

Considered the cryptography, the avalanche effect means that a small change in the plaintext or key will cause a drastic change in the cipher text which just like an avalanche, which was first used in [21]. Strict key avalanche criterion requires that each binary bit of the cipher text should have a change with the probability of one half when any bit of the key change ([22]). Recently, we presented a d-bit segment stream encryption scheme with avalanche effect (SESAE) ([23]). The main feature of the SESAE is to make each bit of the decrypted plaintext changed to 1 with probability of $(2^d - 1) / 2^d$ instead.

This paper introduces a 4-dimensional discrete chaotic system with one-line equilibria (DCSLE). Then, constructs an 8-dimensional DCSLE GCS system and designs an 8-dimensional DCSLE-based pseudorandom number generator (CPRNG). After simulating and analyzing the complex dynamics of the system, FIPS 140-2 test suite ([24]) is used to test the randomness of the CPRNG, the RC4 and the ZUC algorithm ([25]). Finally, as an application example, the CPRNG and the SESAE ([26]) are used to encrypt an RGB image with numerical analysis.

The rest of this paper is organized as follows. The definition and the theorem on GS are introduced in Section 2. Section 3 presents an 8-dimensional DCSLE GCS, and simulates its complex dynamic behaviors. Section 4 designs the CPRNG and performs statistic tests on the CPRNG, the RC4 and ZUC algorithm, respectively. Section 5 demonstrates an image encryption example by using the CPRNG and the SESAE. Finally, Section 6 concludes this paper.

2. Definition and the GCS Theorem

First, remember the following definition:

Definition 1: (Similar to [8], [27]) Consider two systems,

$$X(k+1) = F(X(k)) \tag{1}$$

$$Y(k+1) = G(Y(k), X(k)), \tag{2}$$

where

$$X(k) = (x_1(k), ..., x_n(k))^T \tag{3}$$

$$Y(k) = (y_1(k), ..., y_m(k))^T, m \leq n \tag{4}$$

$$F(X(k)) = (f_1(X(k)), ..., f_n(X(k)))^T \tag{5}$$

$$G(Y(k), X(k)) = (g_1(Y(k), X(k)), ..., g_m(Y(k), X(k)))^T. \tag{6}$$

If there exists a transformation

$$H = R^n \to R^m \tag{7}$$

$$H(X(k)) = (h_1(X(k)), ..., h_m(X(k)))^T, \tag{8}$$

and $(X_0, Y_0) \in R^n \times R^m$, for any $\varepsilon > 0$, there exists $\delta_1 > 0$ and $\delta_2 > 0$ such that all trajectories of (1) and (2) with initial conditions $(X_0, Y_0) \in B(X_0, \delta_1) \times B(Y_0, \delta_2) \subset R^n \times R^m$ satisfy

$$\lim_{k \to +\infty} \left\| H(X(k) - Y(k)) \right\| = 0,$$

The systems (1) and (2) are said to be in GS with respect to the transformation $H(X(k))$. System (1) is called the driving system, while system (2) is the driven system. In particular, if the two systems are chaotic, then the GS is called generalized chaos synchronization (GCS).

To construct a new discrete chaotic map with the GCS property, the following result is needed.

Theorem 1: ([15]) Let X, Y, X_m, $F(X)$ and $G(Y,X)$ be defined by Definition 1. Denote

$$X_m = (x_1(k), ..., x_m(k))^T.$$

Supposed that

$$H(X_m) = (y_1, y_2, ..., y_m) \tag{9}$$

is an invertible transformation. If the systems (1) and (2) are in GCS via the transformation $Y = H(X_m)$, then the function $G(Y, X)$ given in (2) has the following form:

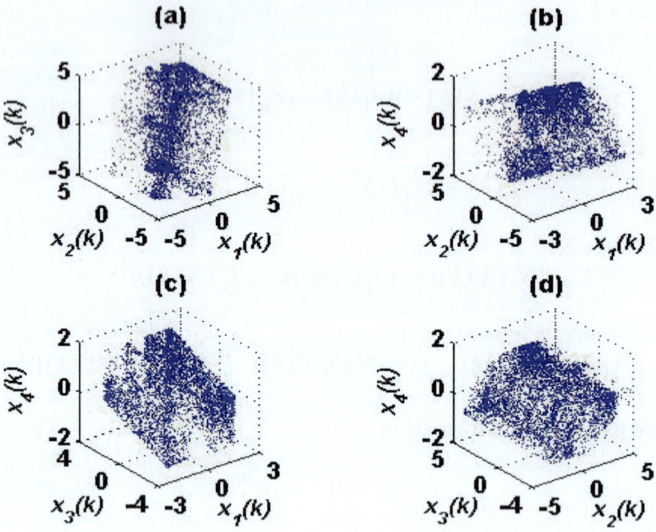

Fig. 1. Chaotic trajectories of variables: (a) $x_1(k) - x_2(k) - x_3(k)$, (b) $x_1(k) - x_2(k) - x_4(k)$, (c) $x_1(k) - x_3(k) - x_4(k)$, (d) $x_2(k) - x_3(k) - x_4(k)$.

$$G(Y, X) = H(F_m(X)) - q(X_m, Y), \tag{10}$$

Where

$$F_m(X) = (f_1(X), f_2(X), ..., f_m(X))^T$$

and the function

$$q(X_m, Y) = (q_1(X_m, Y), q_2(X_m, Y), ..., q_m(X_m, Y))^T$$

guarantees that the zero solution of the following error equation is asymptotically stable:

$$\begin{aligned} e(k+1) &= H(X_m(k+1)) - Y(k+1) \\ &= q(X_m, Y). \end{aligned} \tag{11}$$

3. A DCSLE GCS System

An 8-dimensional DCSLE with the GCS property is introduced. The driving system has the following form:

$$X = \begin{cases} x_1(k+1) = k_1 \sin(x_1(k)) + \sin(x_1(k)x_3(k)) \\ x_2(k+1) = k_2 \sin(x_2(k)) - \sin(x_1(k)) \\ x_3(k+1) = k_3 \sin(x_1(k)x_3(k)) \\ x_4(k+1) = k_4 \sin(x_2(k) + x_4(k)). \end{cases} \tag{12}$$

The equilibria of the system are on x_3 axis. Let $k_1 = 2$, $k_2 = 3$, $k_3 = 5$, $k_4 = 1$. The calculated Lyapunov exponents of the system are {1.1563,0.23062,-0.34383, -0.57884}. Therefore, system (12) is chaotic.

Choose the following vector as initial condition:

$$X(0) = (0.9634, 0.8198, 0.4405, -0.2751)^T, \tag{13}$$

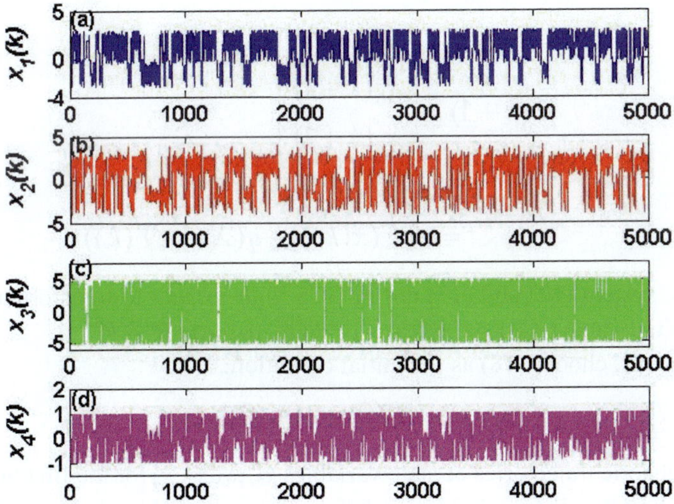

Fig. 2. The evolution of state variables : (a) $k - x_1(k)$, (b) $k - x_2(k)$, (c) $k - x_3(k)$, (d) $k - x_4(k)$.

The chaotic orbits of state variables x_1, x_2, x_3, x_4 for the first 5000 iterations are shown in Figure 1.

The evolution of state variables: $k - x_1(k)$, $k - x_2(k)$, $k - x_3(k)$, $k - x_4(k)$ are shown in Figure 2. Observe that the dynamic behaviors of the chaotic system demonstrate chaotic attractor characteristics.

Second, construct an invertible matrix:

$$A = \begin{pmatrix} -0.99045 & 0.29859 & -0.19287 & 0.98025 \\ 0.99539 & 0.49463 & -0.68629 & 0.39987 \\ 0.00997 & -0.99035 & -0.54915 & 0.22399 \\ 0.49333 & -0.29013 & 0.79999 & 0.69350 \end{pmatrix} \quad (14)$$

with the transformation $H: R^4 \to R^4$ defined as follows:

$$\begin{aligned} H(X) &= AX \\ &\triangleq (h_1(x), h_2(x), h_3(x), h_4(x)) \end{aligned} \quad (15)$$

Let

$$q(X,Y) = \frac{1}{8}(AX - Y), \quad (16)$$

Where $q(X, Y)$ is used to ensure the error equation (11) be asymptotically stable.

By Theorem 1, the driven system has the following form:

$$\begin{aligned} Y(k+1) &= \begin{pmatrix} x_5(k+1) \\ x_6(k+1) \\ x_7(k+1) \\ x_8(k+1) \end{pmatrix} \\ &= A[F(x(k))] - q(X(k), Y(k)). \end{aligned} \quad (17)$$

Then, system (12) and (17) are in GCS with respect to transformation (15) for any initial value $(X_0, Y_0) \in B(X_0, \delta_1) \times B(Y_0, \delta_2) \subset R^4 \times R^4$.

Moreover, choose (18) as the initial condition:

$$Y(0) = AX(0). \quad (18)$$

The chaotic trajectories of state variables x_5, x_6, x_7, x_8 for the first 5000 iterations are shown in Figure 3. The evolution of state variables $k - x_5(k)$, $k - x_6(k)$, $k - x_7(k)$, $k - x_8(k)$ are shown in Figure 4. The simulation results show that the system has chaotic attractor characteristics. Figure 5 show that $X(k)$ and $Y(k)$ are in generalized synchronization with respect to transformation $H = A$, as the theory predicts.

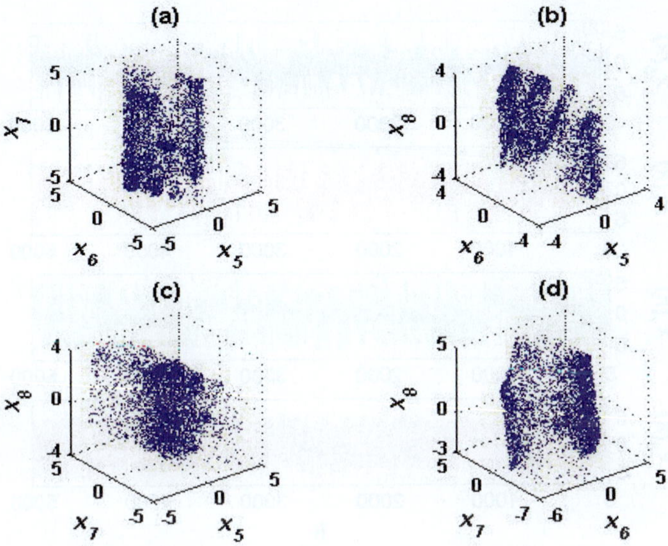

Fig. 3. Chaotic trajectories of variables: (a) $x_5(k) - x_6(k) - x_7(k)$, (b) $x_5(k) - x_6(k) - x_8(k)$, (c) $x_5(k) - x_7(k) - x_8(k)$, (d) $x_6(k) - x_7(k) - x_8(k)$.

4. Chaotic Pseudorandom Number Generator and seudorandomness Tests

4.1. *Pseudorandom number generator*

Denote

$$X_i = \{x_i(k) \mid k = 1, 2, \ldots, N\}, i = 1, 2, \ldots, 8 \tag{19}$$

where x_i is defined by system (12) and (17).

First, introduce a transformation T_1: $R \rightarrow \{0, 1, \cdots, 2^{16} - 1\}$, which transforms the chaotic streams of system (19) into key streams.

Let $L = 10^{15}$. Then, T_1 is defined by

$$\begin{aligned} Y_1 = T_1(X_4) = mod(round(\mathrm{L}(X_4 - \min(X_4)) \\ /(\max(X_4) - \min(X_4))), 2^{16}), \end{aligned} \tag{20}$$

$$\begin{aligned} Y_2 = T_1(X_5) = mod(round(\mathrm{L}(X_5 - \min(X_5)) \\ /(\max(X_5) - \min(X_5))), 2^{16}). \end{aligned} \tag{21}$$

Therefore, the CPRNG T is defined by

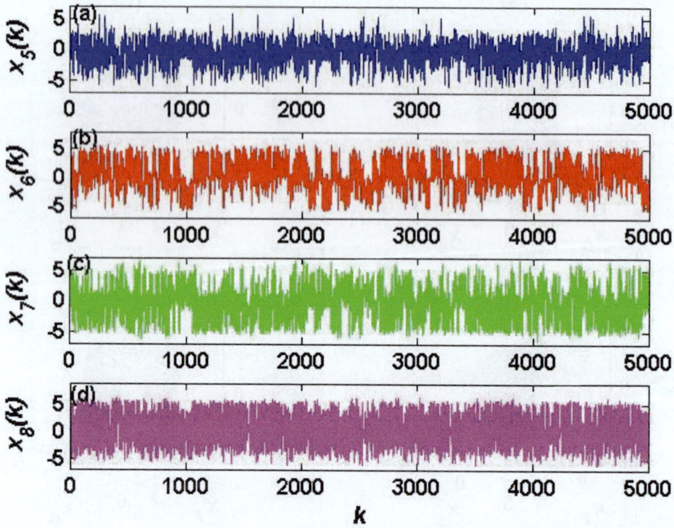

Fig. 4. The evolution of state variables: (a) $k - x_5(k)$, (b) $k - x_6(k)$, (c) $k - x_7(k)$, (d) $k - x_8(k)$.

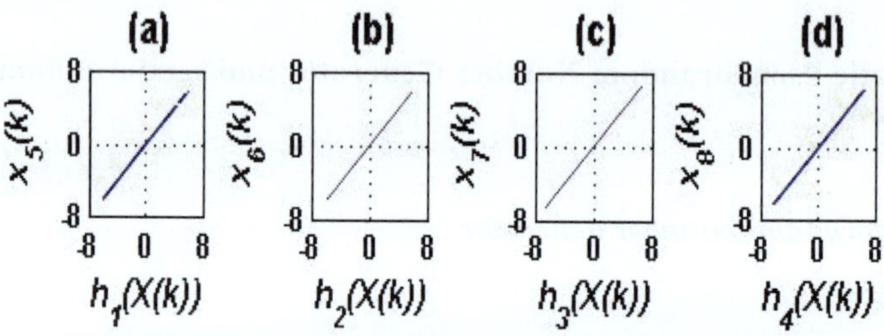

Fig. 5. The state vectors $X(k)$ and $Y(k)$ are in generalized synchronization with respect to the transformation H: (a) $h_1(X(k)) - x_5(k)$, (b) $h_2(X(k)) - x_6(k)$, (c) $h_3(X(k)) - x_7(k)$, (d) $h_4(X(k)) - x_8(k)$.

$$T(X_4, X_5) = mod(Y_1 + Y_2, 2^{16}). \tag{22}$$

The seeds of the CPRNG are the initial conditions of the DCSLE GCS system, which can be chosen via random number generators. Therefore, the output key streams of the CPRNG can be obtained via the transformation (22) acting on the chaotic streams of the GCS systems (12) and (17).

4.2. *Pseudorandomness test*

The FIPS 140-2 test ([24]) consists of four sub-tests: Monobit Test, Poker Test, Runs Test and Long Runs Test. Each test needs a single stream of 20,000 one and zero bits from the key stream generator. Any failure in the first three tests means the corresponding quantity of the sequences falls out the required intervals listed in the second column of Table 1. The Long Runs test is passed if there are no runs of length 26 or more.

It has been pointed out that the required intervals of Monobit test, Poker test and runs test correspond to the confident intervals with significant levels: $\alpha = 10^{-4}$ and $\alpha = 1.6 \times 10^{-7}$ (approximately), respectively ([28], [29]). If one select the significant levels: $\alpha = 10^{-4}$ of all tests, the corresponding required intervals are listed in the third column of Table 1. Then, denote the accepted intervals by G FIPS 140-2 test.

According to Golomb's three postulates on the randomness, the ideal pseudorandom sequences and the ideal values of the first three tests are listed in the fourth column of Table 1 ([30]).

Table 1. The required intervals of FIPS 140-2 Monobit Test, Poker Tests, Runs Test. Here, MT, PT and LT represent the Monobit Test, the Poker Test and the Long Runs Test. *k* represents the length of the run of a tested sequence. χ^2 DT represents χ^2 distribution.

Test Item	FIPS140-2 Required Intervals	$\alpha = 10^{-4}$ Required Intervals	Golomb's Postulates
MT	9725~10,275	9725~10,275	10,000
PT	2.16~46.17	2.16~46.17	χ2 DT
LT	<26	<26	—
k	Runs Test	Runs Test	Runs Test
1	2315~2685	2362~2638	2500
2	1114~1386	1153~1347	1250
3	527~723	556~694	625
4	240~384	264~361	313
5	103~209	122~191	156
6+	103~209	122~191	156

Now, FIPS 140-2 test suite is used to test the randomness performance. One needs to change the key streams with values $\{0, 1, \cdots, 2^{16} - 1\}$ to binary key streams via the following transformation:

$$\tilde{T} : \left\{0,1,\ldots,2^{16}-1\right\} \to \{0,1\},$$

Which is defined by

$$\tilde{T} = T_{22} \circ T_{21}, \tag{23}$$

such that, for any $y \in \{0, 1, \ldots, 2^{16} - 1\}^N$,

$$T_{21}(y) = dec2bin(y).$$

Let $z = dec2bin\ (Y)$. Then,

$$T_{22}(z) = z(:),$$

where *dec2bin* and z(:) are both Matlab commands.

The FIPS 140-2 test is used to check 1,000 key streams randomly generated, respectively by the CPRNG with perturbed randomly initial conditions (13) and (18) and the matrix (14) in the range $|\varepsilon| \in [10^{-16}, 10^{-1}]$. The results show that all 1000 sequences pass the FIPS 140-2 test and 8 sequences fail to pass the G FIPS 140-2 test. The test results are listed in the third column of Table 2, which the results are described by mean values ± standard deviation (Mean ± SD).

The Rivest Cipher 4 (RC4) has been widely used in popular protocols such as Secure Sockets since it's designed in 1987. The RC4 algorithm as PRNG can be designed via the Matlab commands:

```
 N = 20000;
K=randint(1,2^L,[0 2^L-1]);
S=[0:2^L-1];j=0;
for i=1:2^L
      j=mod(j+S(i)+K(i),2^L);
          Sk=S(j+1);
            S(j+1)=S(i);
                S(i)=Sk;
end
     C=zeros(1,N); j=0;i=0; k=1;
for l=1:N/L
      i=mod(i+1,2^L);
          j=mod(j+S(i+1),2^L);
            Sk=S(j+1);
              S(j+1)=S(i+1);
                S(i+1)=Sk;
 C(l)=S(mod(S(j+1)+S(i+1),2^L)+1);
end
 C=(dec2bin(C))';
     C=C(:);
       C=bin2dec(C);
```

Fig. 6. The matlab commands.

Here, "randint (1, 2^8, [0 2^8 − 1])" generates a vector of uniformly distributed random integers $\{0, 1, \cdots, 2^8 - 1\}$ of dimension 2^8; "mod" means modulus after division; "zeros (1, N)" is a zero raw vector of dimension N. Consequently, the RC4 algorithm based 8-bit segment PRNG is designed.

Next, the FIPS 140-2 test is used to test the 1000 key streams randomly generated by RC4 PRNG. Results show that 1 and 12 sequences fail to pass the FIPS 140-2 test and G FIPS 140-2 test, respectively. The statistic test results are listed in the fourth column of Table 2.

Furthermore, ZUC is a stream cipher that forms the heart of the third generation partnership project (3GPP) confidentiality algorithm 128-EEA3 and the 3GPP integrity algorithm 128-EIA3. Then, the FIPS 140-2 test is used to test the 1000 key streams randomly generated by the ZUC algorithm ([25]). It demonstrates all sequences pass the FIPS 140-2 test, and 21 sequences fail to pass the G FIPS 140-2 test. The test results are listed in the fifth column of Table 2.

Finally, compare all test results shown in Table 2. It can be observed, the statistical properties of the pseudorandomness of the sequences generated via the PRNG, the RC4 and ZUC algorithm don't have significant differences.

4.3. ***Key space***

The key parameters set of the proposed CPRNG includes the initial condition $X(0)$, $Y(0)$ and the matrix $A = (\alpha_{i,j})$. It can be proved that if the perturbation matrix $\Delta = (\delta_{i,j})$ satisfies

$$|\delta_{i,j}| < 0.28843,$$

the matrix $A + \Delta$ is still invertible. Therefore the CPRNG have 4 + 4 + 16 key parameters denoted by

$$K_S = \{k_1, k_2, ..., k_{24}\}. \tag{24}$$

The perturbed keys have the forms

$$K_s(\Delta) = K_s + [\delta_1, \delta_2, ..., \delta_{24}]. \tag{25}$$

The Matlab platform uses double precision decimal computations. That means each computed decimal number has 16 bits' accuracy. Therefore, one can select

$$10^{-16} \le |\delta_i| \le 10^{-1}, i = 1, 2, ..., 24,$$

That is, $\delta_i = 0.0a_2 \cdots a_{16}$, where $a_i \in [0,1,\cdots,9]$. Therefore, the 24 keys have a key space which is larger than

$$10^{24\times15} > 2^{1195}.$$

Table 2. The confident intervals of the FIPS 140-2 tested values of 1000 key streams generated by the CPRNG, the RC4 PRNG and ZUC PRNG. Here, SD represents the standard deviation.

Test Item	bits	CPRNG Mean ± SD	RC4 Mean ± SD	ZUC Mean ± SD
MT	0	10000±73.099	9999.7±70.092	9998.4±71.843
	1	9999.5±73.099	10000±70.092	10002±71.843
PT	-	15.125± 5.4236	14.870±5.4333	15.043±5.549
LT	0	13.724±1.8877	13.60±1.8214	13.488±1.829
	1	13.622±1.9475	13.642±1.931	13.595±1.931
1	0	2499.6±47.156	2500.9±45.568	2501.9±45.73
	1	2499.4±46.960	2501.4±46.398	2502.7±46.121
2	0	1252.1±31.452	1250.5±31.372	1252.1±32.606
	1	1251.6±32.380	1249±31.048	1249.5±32.221
3	0	625.39±22.746	624.95±22.964	624.09±22.648
	1	625.61±23.254	625.65±22.93	624.64±23.455
4	0	311.13±16.116	311.71±16.548	312.56±16.748
	1	312.57±16.091	312.17±16.822	312.72±16.506
5	0	156.43±11.762	156.41±12.069	155.65±12.097
	1	156.44±12.648	156.60±11.958	156.66±12.369
6+	0	156.12±12.213	156.15±11.792	155.75±11.719
	1	155.21±11.945	155.79±11.979	155.82±11.497

Now, compare the difference between the key stream $S = T(X_4,X_5)$ with 20,000 codes length generated by the key set (24) with the key streams S_p generated by the perturbed key set (25), respectively.

The comparison results are shown in the third column of Table 3, where SV denotes the statistic values, DC denotes the different codes, and CC denotes the correlation coefficients.

Observe that the average percent of different codes is 50.01%, which is closed to the ideal value 50%. And the average of the correlation coefficients is 0.0055942, also very closed to the ideal value of 0.

Next, compare the same key stream S with the 1000 key streams S_m generated by the Matlab function randi ([0 1], 1, 20000). The comparison results are shown in the fourth column of Table 3. Observe that the average percentage of different codes is 50.011% and the average of the correlation coefficients is 0.0055684. The results suggest that the key stream S has no significant correlations with the perturbed key streams S_p and the streams S_m.

In summary, the effective key space of the CPRNG is larger than 2^{1195}, which will help to increase security.

Table 3. The statistic data describes the percentages of the codes of the key stream variations between S and S_p as well as S and S_m.

Item	SV	S_p	S_m
DC	min	49.000%	48.955%
	mean	50.010%	50.011%
	max	51.080%	51.035%
CC	min	0.0000221	0.0000014
	mean	0.0055942	0.0055684
	max	0.0215920	0.0209230

5. Simulations on SESAE

Consider the avalanche effect of the CPRNG, which is used to encrypt an RGB image with 250 × 140 pixels. The simulation is implemented via the Matlab 7.1 platform.

The SESAE experiments on CPRNG are described as follows:

1) Transfer the image (see Figure 6(a)) to a binary plaintext steam $M = \{m_1, m_2, \cdots, m_N\}$, where $N = 250 \times 140 \times 3 \times 8$.

2) A 16 bit segment key stream of length $N + 1000$ is generated via the CPRNG (22) with initial conditions (13) and (18) by the sender. Then the sender drops the first 1000 iterative values so as to obtain a key stream:

$$P = \{p_1, p_2, \ldots, p_N\}.$$

3) The sender uses the key stream P and the stream encryption scheme with avalanche effect to encrypt the plaintext steam M:

$$c_i = \begin{cases} p_i, & if \quad m_i = 0 \\ \sim p_i, & if \quad m_i = 1, \end{cases} \tag{26}$$

so as to obtain the ciphertext $C = \{c_1, c_2, \cdots, c_N\}$, denote the ciphertext with:

$$C = E(M, P),$$

where $\sim p_i$ is defined to be the bit string obtained by replacing all '1's in p_i with '0's, and all '0's in p_i with '1's.

4) The receiver uses the key stream P to decrypt the ciphertext:

$$m_i = \begin{cases} 0, & if\ p_i = c_i \\ 1, & if\ p_i \neq c_i, \end{cases} \tag{27}$$

so as to obtain a decrypted image without errors , which is denoted as

$$M = E^{-1}(C, P).$$

Table 4. Differences between the original key stream S_0 and the key streams $S_{j,i}$, measured by norm $\| S_0 - S_{j,i}\|$.

	$\| S_0 - S_{j,i}\|$ ($\times 10^{-6}$)				
	$S_{4,1}$	$S_{4,2}$	$S_{4,3}$	$S_{4,4}$	$S_{4,5}$
S_0	2.6078	2.8357	2.8445	3.0177	2.5805
	$S_{23,1}$	$S_{23,2}$	$S_{23,1}$	$S_{24,2}$	$S_{23,1}$
S_0	2.8704	2.8101	2.6508	2.4251	2.1642

5) The receiver randomly disturbs the initial conditions (13) and (18), and also the matrix (14) , for 1000 times in the range $|\varepsilon|\in [10^{-16}, 10^{-13}]$, then obtain disturbed key streams:

$$p_i, i = 1,2,...,1000. \tag{28}$$

6) The receiver uses $P_i = \{p_1, p_2, \cdots, p_N\}$ to decrypt the ciphertext and obtains a decrypted plaintext:

$$\tilde{M}_i = E^{-1}(C, P_i), i = 1,2,...,1000. \tag{29}$$

After changing $\tilde{M}_i$ to RGB images, all images become almost pure white-colored ones. There are 840,000 {0, 1} codes in each decrypted image. Among the decrypted images, the minimum number of 0 is 4 and the maximum number of 0 is 25.

Let $I_{i,j}$ denote the jth image that has i zero codes. The first five images with minimum zero codes and the last five images with maximum zero codes are shown in Figure 6(c)-(l). Here, in order to make the pixels whose brightness is less than 255 be visible in the decrypted images, technically the brightness of the three color planes of these pixels is reduced by 150, respectively.

Therefore, the percentages of the numbers of "1" codes in the 1000 decrypted images are within the range [0.999970, 0.999995], which are very close to the ideal value $(2^{16} - 1) / 2^{16} = 0.999985$.

Table 4 lists some statistical data of the norms between the original key stream S_0 and the key stream $S_{i,j}$ used in the above ten decrypted images, respectively. The results suggest that there are no significant correlations between the norms and the corresponding decrypted image.

In summary, the simulation shows that using the CPRNG and SESAE to encrypt RGB images is able to generate encrypted images with significant avalanche effects.

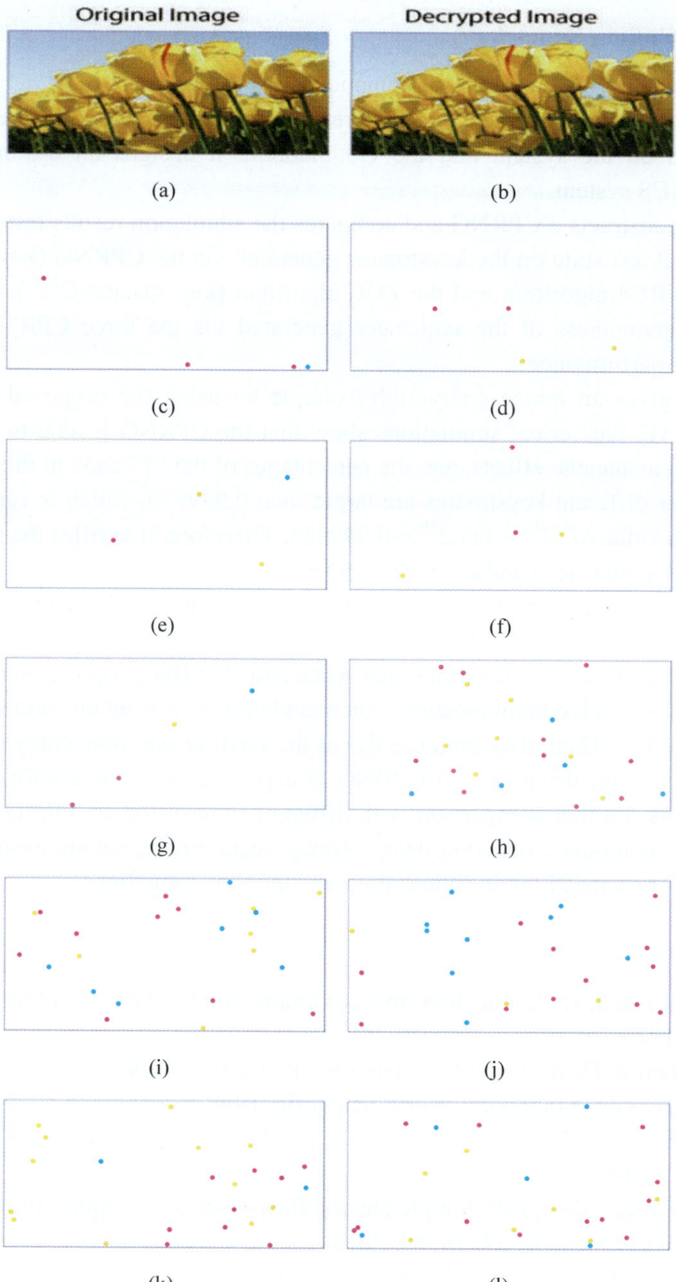

Figure 6. (a) Original image "tulip". (b) Decrypted image via the key streams P. Ten decrypted images via key streams generated with slighted perturbed initial conditions and the matrix within the range $[10^{-16}, 10^{-13}]$: (c) $I_{4,1}$, (d) $I_{4,2}$, (e) $I_{4,3}$, (f) $I_{4,4}$, (g) $I_{4,5}$, (h) $I_{23,1}$, (i) $I_{23,2}$, (j) $I_{24,1}$, (k) $I_{24,2}$, and (l) $I_{25,1}$.

6. Concluding Remarks

The main results of this paper are concluded as follows:

1) It proposes a 4-dimensional discrete chaotic system with one line equilibria. Based on the system and the GS theorem, it designs an 8-dimensional DCSLE GCS system.

2) It constructs a CPRNG and compares the simulation results tested by the FIPS 140-2 test suite on the keystreams generated via the CPRNG (key space > 2^{1195}), the RC4 algorithm and the ZUC algorithm (key space = 2^{128}). It shows that the randomness of the sequences generated via the three CPRNGs have promising performances.

3) It gives an image encryption example by using the proposed CPRNG with SESAE. Numerical simulations show that the CPRNG is able to generate significant avalanche effects, and the percentages of the "1" code in the decrypted texts for different keystreams are larger than 0.999970, which is very closed to the idea value of $(2^{16} - 1) / 2^{16} = 0.999985$. Therefore, it verifies the proposed CPRNG is a qualified candidate for SESAE.

4) Moreover, to resist attacks, one may consider implementing a "one–time–pad" scheme into CPRNG: Let X be a set in the seed space (initial conditions) of the CPRNG, and assume that Alice and Bob share a one-to-one map f: $X \rightarrow X$. Before each communication, Alice randomly selects an element $x \in X$ and sends it to Bob. Then, they both use $f(x)$ as the seed for one-time encryption.

In summary, the proposed CPRNG is a promising candidate for practical applications. Further comparison with different state-of-the-art PRNG schemes in terms of computational complexity, storage requirement, communication cost, etc., it will be carried out in future research along the same lines.

References

1. T.Y. Li & J.A. Period three implies chaos. *Amer Math Monthly* 82: 985–992, 1975.
2. G. Chen & D. X. *From Chaos to Order: Methodologies, Perspectives, and Applications.* Singapore: World Scientific, 1998.
3. J.G. Sprott. *Chaos and Time-Series Analysis.* Oxford: Oxford University Press, 2003.
4. S. Jafari & J.G. Sprott. Simple chaotic flows with a line equilibrium. *Chaos, Solitons & Fractals* 57: 79–84, 2013.
5. H. C. Global chaos synchronization of new chaotic systems via nonlinear control. *Chaos Solition Fract* 23: 1245–1251, 2005.

6. D. J. & M. G. Poly-quadratic stability and global chaos synchronization of discrete time hybrid systems. *Mathematics and Computers in Simulation* 58(4): 295–307, 2001.
7. F. Arecchi. Neuron dynamics and chaotic synchronization. *Fluctuation and Noise Letters* 5(2): 163–173, 2005.
8. A. Breve et al. Chaotic phase synchronization and synchronization in an oscillator network for object selection. *Neural Network* 22: 728–737, 2009.
9. J. Hu et al. Synchronizing chaotic systems using control based on a special matrix structure and extending to fractional chaotic systems. *Communications in Nonlinear Science and Numerical Simulation* 15(1): 115–123, 2010.
10. M. Pecora & L. Carroll. Synchronization in chaotic systems. *Physical Review Letters* 64(8): 821–825, 1990.
11. K. Murali & M. Laskshmanan. Secure communication using a compound signal from generalized synchronizable systems. *Phys. Lett* 41: 303–310, 1998.
12. A. Kadir et al. Generalized synchronization of diverse structure chaotic systems. *Chinese Physics Letters* 28(9): 90(503)–90(506), 2011.
13. Z. Yuan et al. Generalized synchronization of two unidirectionally coupled discrete stochastic dynamical systems. *Chinese Physics B* 20(7): 70503, 2011.
14. A. Koronovskii et al. Generalized synchronization in complex networks. *Technical Physics Letters* 38(10): 924–927, 2012.
15. H. Zang et al. A generalized synchronization theorem for discrete-time chaos system with application in data encryption scheme. In Kokura & Fukuoka Japan, *Proceeding of the 2007 Int. Conf. on Communications, Circuit and Systems*: 1325–1329, 2007.
16. B. Nana et al. Chaotic synchronization with experimental application to secure communications. *Communications in Nonlinear Science and Numerical Simulation* 14(5): 2266–2276, 2009.
17. A. Kanso & M. Ghebleh. A novel image encryption algorithm based on a 3d chaotic map. *Commun Nonlinear Sci Numer Simulat* 17: 2943–2959, 2012.
18. L. Min et al. Study on pseudorandomness of some pseudorandom number generators with application. In Leshan China, *Proceedings of the 9th* International *Conference on Computational Intelligence and Security; IEEE*: 569–574, 2013.
19. Y. Liu et al. A family of new complex number chaotic maps based image encryption algorithm. *Signal Processing: Image Communication* 28: 1548-1559, 2013.

20. B. Du et al. Analysis and elimination of digital chaotic key sequence's autocorrelation. *Journal of Information Hiding and Multimedia Signal Processing* 5(2): 302–309, 2014.
21. H. Feistel. Cryptography and computer privacy. *Sci. Am.* 228(5): 15–23, 1973.
22. R.J. Spillman. *Classical and Contemporary Cryptology.* Upper Saddle River NJ: Pearson Education INC, 2005.
23. L. Min & G. Chen. A novel stream encryption scheme with avalanche effect. European *Physics J. B* 86(11): 459(1) –459 (13), 2013.
24. NIST, FIPS PUB 140-2. *Security Requirements for Cryptographic Modules.* Gaithersburg, MD: NIST, 2001.
25. ETSI/SAGE Specification, Specification of the 3GPP Confidentiality and Integrity Algorithms 128-EEA3 & 128-EIA3. *Document 2: ZUC Specification; Version: 1.5, 4 January,* 2011.
26. L. Min & G. Chen. Generalized synchronization in an array of nonlinear dynamic systems with applications to chaotic cnn. *J. Bifurcat. Chaos* 23(1): 1,350,016(1)–1,350,016(53), 2013.
27. L. Kocarev & U. Parlitz. Generalized synchronization, predictability and equivalence of unidirectionally coupled dynamical systems. *Phys. Rev. Lett* 76: 1816–1819, 1996.
28. L. Min et al. Analysis of fips 140-2 test and chaos-based pseudorandom number generator. In Antens Greece, *The Proc. of the 5th Chaotic Modeling and Simulation International Conference; June* 2012: 345–352, 2012.
29. L. Min et al. Analysis of fips 140-2 test and chaos-based pseudorandom number generator. *Chaotic Modeling and Simulation* 2: 273–280, 2013.
30. S. Golomb. *Shift Register Sequences.* Aegean Park, CA: Laguna Hills, 1982.

Instruction set randomization based on compilation

Y.J. Man[1†], Q. Yin[2] and J. Lin[3]
State Key Laboratory of Mathematical Engineering and Advanced Computing, Information Engineering University, Zhengzhou, China
Emails: † thecodemadmandk@gmail.com

Injection attack is still a serious form of threat to system security. This paper presents an instruction randomization technique based on compilation to effectively prevent code injection attacks. We analyze instruction format and create rules to replace or recover instructions. During compilation, our methods transform instructions according to the random replacement rules and generate the protected target programs. When executing, the protected program is recovered on a runtime code manipulation system in accordance with recovery rules. Results show that our method can correctly capture instructions and effectively thwart injection attacks while introducing an extra runtime overhead of 33% on average compared with execution without protection on the runtime code manipulation system.

Keywords: Instruction randomization; Diversity; Information Security; Obfuscation.

1 Introduction

Standardized interfaces become a double-edged sword with the rapid development of information technologies and attacks. The Internet is a convenience for daily life but at the same time brings serious problems. Homogeneous of software makes development more convenient, it also leads to security threats. Attackers can explore errors to launch attacks with prior knowledge of the system with vulnerabilities. While novel attack technologies are emerging endlessly, code injection attack [8], which is a familiar and ruinous, is a significant threat to security. Attackers inject malicious code into user's process, change control flow by overflow and execute malicious code so as to achieve the goals of the attackers. The most common code injection attacks are buffer overflow attacks, such as stack overflow attacks and heap overflow attacks, which the attackers usually use function stack frame to implement attacks.

Traditional defense against code injection attacks is to use multiple layers of defense to thwart vulnerabilities and malicious code. For example, stack protection prevents most buffer overflow attacks by protecting the return address, DEP forces injected code data area unexecutable by data execution protection, etc. The majority of traditional passive defense patches known vulnerabilities or

---

[1] Yu-Jia Man
[2] Qing Yin
[3] Jian Lin

attempts to eliminate vulnerabilities. However, experience has proved that vulnerability cannot be completely eliminated.

Randomization techniques vary the same environment to exploit vulnerabilities by equivalent replacement of functionally part of programs, obscure control flow information, address information and data information. Existing randomization technologies mainly include address space layout randomization (ASLR) [5], instruction set randomization (ISR) [4, 10], data space randomization (DSR) [3]. ASLR defenses against buffer overflow attacks but is coarse-grained. ISR generates a new set of instructions for the target program to defend against code injection attacks. DSR protects against memory leak attacks by random transformation of critical data in memory. ISR and DSR is relatively fine-grained. These randomization techniques target different types of attacks, therefore different randomization techniques should be selected in order to get effective performance and low overhead.

Existing instruction set randomization techniques mostly analyze the target executable programs and replace instructions randomly according to certain rules. There are several drawbacks nevertheless. First is that disassembly of executable programs cannot guarantee accuracy. If disassembly runs into mistakes, instruction replacement could go wrong, which would cause execution errors; the second is that instruction randomization does not select instructions, which would weaken the efficiency and performance.

We present ISR techniques based on compilation (ISRC), aimed at shortcomings of existing instruction set randomization methods. Instruction analysis is moved to the compilation phase to ensure the correctness of instruction recognition. During the compilation phase, we randomly select instruction replacement rules for the intermediate representation (IR). In addition, we replace instructions in the target programs frequently used by code injection attacks to improve performance and thus reduce the overhead.

2. System Description

Instruction randomization techniques achieve the effect of defense to make malicious code cannot perform properly by changing malicious code instruction semantics during a dynamic execution process. Our system is mainly an extension of a compiler. Based on the procedurals to randomize instructions, Figure 1 shows the framework of our system, which consists of four parts: generation of replacement class; randomized instruction rules' selection; analysis of intermediate representation during compilation and instruction replacement; instruction recovery on dynamic execution tools.

First, we analyze the instruction format, select instructions that have the same length and similar structure, and divide them into a replacement class (RC). The size of instruction RC determines the effectiveness of randomized. So we put instructions that can be replaced with each other as much as possible into one instruction RC. Instruction replacement rules (RPR) can be constructed randomly from instructions that belong to the same RC.

Next, we design instruction randomization rules according to instruction replacement class. The rules determine which instructions should be transformed when replacing instructions, and how to recover randomized instructions properly before the program execute. The complexity of the rules effects randomization performance and cost. Based on instruction replacement class, we use a random key to select instruction replacement rules randomly. In addition, the random key is used to generate the corresponding instruction recovery rules (RCR), in order to ensure the execution of the target. Instruction replacement is in pairs, and we can set a variety of random rules to increase the difficulty of brute force attacks.

Then we compile source code into intermediate representation (IR), randomize IR with RPR, and continue to compile to get an executable program, which cannot execute properly now. We use a dynamic instrumentation platform to load randomized executable program, recover instructions in turn according to the recovery rules to ensure the original operation of the program.

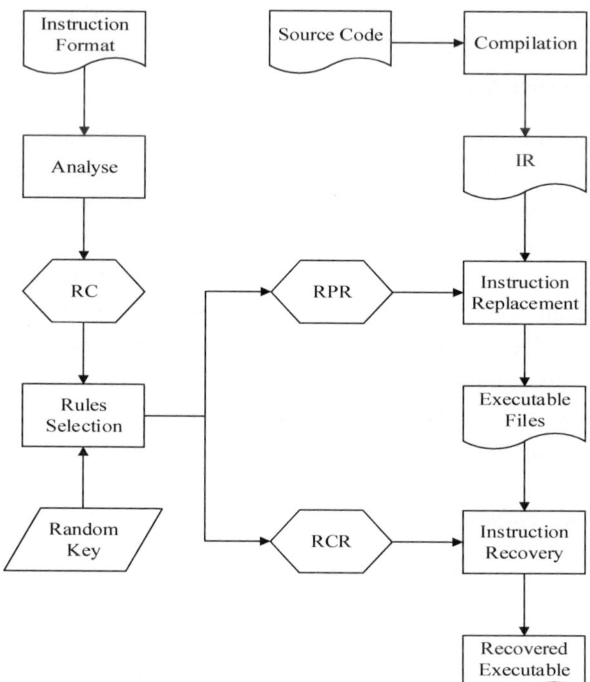

Fig. 1. The framework of ISR based on compilation.

3. Design and Implementation

Random instruction replacement module is implemented on LLVM compiler framework, and instruction recovery and execution are implemented with the help of a binary analysis platform called DynamoRIO [1, 2]. We replaced instructions on the IR level of LLVM [9], and implemented a DynamoRIO plug-in to recover instructions. Next, we discuss on the generation of replacement class of instructions by instruction format analysis, the selection of rules, instruction transformation and recovery.

3.1. *Instruction replacement class*

We put instructions that have the same length and similar structure into one replacement class. Replacement between instructions with different structures can cause two problems: First is instruction truncation or misalign when fetching instructions with different length; the second is the absence or redundancy of operands information if instructions have different number of operands. These make instruction recover incorrectly and lead to execution errors. Structure similarities include the same opcode length, the operands number and the instruction format. Instructions replacement class is the basis for the next stage to select rules.

3.2. *Random instruction rules*

Random instruction rules include instruction replacement rules (RPR) and the corresponding instruction recovery rules (RCR). Using a random key, we select pairs of instructions in replacement class as a rule to replace instructions and generate a corresponding recovery rule using the same key. However, since the instruction recovery is the reverse process of instruction replacement, in order to ensure the program logic unchanged, RCR must strictly correspond to the RPR. But we should notice that RPR is based on IR, while RCR is based on machine code. Thus the two cannot be generated in a manner that easily swaps instruction pairs, and we need to analyze the correspondence between the IR and machine codes, in order to correctly generate rules based on the same random key.

3.3. *Instruction replacement*

Instruction replacement scans IR during compilation, replace instructions according to RPR and generate randomized IR. The basic procedure for the replacement algorithm is to walk through the entire intermediate representation, identify each instruction in turn, and generate a new one. Figure 2 and 3 show the IR fragment before and after replacement. Here we replace the add instruction with sub.

```
%2 = load i32* %a, align 4
%3 = load i32* %b, align 4
%4 = add i32 %2, %3
store i32 %4, i32* %c, align 4
```

Fig. 2. The IR fragment before replacement.

```
%2 = load i32* %a, align 4
%3 = load i32* %b, align 4
%4 = sub i32 %2, %3
store i32 %4, i32* %c, align 4
```

Fig. 3. The IR fragment after replacement.

3.4. *Dynamic instruction recovery*

Randomized executable programs cannot execute directly and require a dynamic binary analysis platform to support. The dynamic platform recognizes and recovers instructions when loading the target program in accordance with RCR. Then these randomized instructions revert back and thus the injected instructions change. Without replacement on the compilation stage, malicious instructions do not need to recover. Therefore, malicious instructions change after instruction recovery and cannot achieve attacker's original goal.

Random replacement is based on IR, and only the instructions appearing in the IR would implement randomization replacement. Nevertheless, compiler would add extra code later, which leads to randomization incompleteness. In order to recover instructions correctly, the additional code cannot be recovered. We mark the source code by adding flags in user function. When loading the target executable on the binary analysis platform, we recover instructions in marked functions to correctly identify the instructions need to be transformed.

4. Experiment

In this section we evaluate the effectiveness of instruction set randomization based on compilation against attacks and the impact on the performance of the target executable programs. Experimental results show that the randomization can successfully prevent code injection attacks. We ran the experiments on Ubuntu 14.04 32-bit operating systems with a 2.53GHz Intel (R) Core (TM) i5 CPU and 4GB of memory. The target program was compiled by LLVM 3.6.0, and the dynamic analysis platform was DynamoRIO 5.1.0.

4.1. *Randomization effectiveness against attacks*

To test the effectiveness of our ISRC, we selected some real attacks. Table 1 shows that our randomization techniques based on compilation can effectively prevent code injection attacks.

Table 1. Results of attacks under ISRC

Application	Vulnerability	Stopped by ISRC
Apache	Buffer overflow	✓
Apache	Format string	✓
BIND	Format string	✓
CVS server	Double free	✓
SAMBA	Buffer overflow	✓
Sendmail	Buffer overflow	✓
SSH	Integer overflow	✓
Rpc.statd	Format string	✓
Wuftpd	Format string	✓

4.2. *Performance overhead*

We measured the overhead added by ISRC to ten programs from the SPEC CPU 2000 benchmark suite [6] and SPEC CPU 2006 [7]. The overhead includes two parts, separately introduced by randomization and DynamoRIO. Figure 4 shows the percentage of execution time overhead using DynamoRIO with our instruction set randomization protection. The extra time overhead added by randomization is 33% on average and the maximum is 72%.

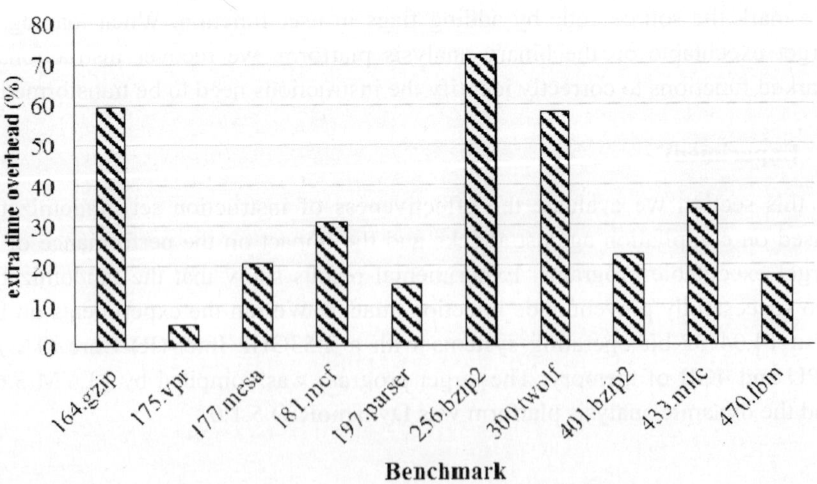

Fig. 4. Extra time overhead added by ISRC.

5. Future Work

While there are many aspects of problems that we have addressed, we realize that there are still some issues for future study. The current randomization rules are relatively simple, and we can continue to explore new ways to replace instructions to gain more possibilities to make it more difficult to attack by brute force; in addition, the performance is far from satisfactory as the recovery need to run on a dynamic platform. We can improve the performance by a better plug-in or a dynamic tool which calls for further research.

6. Conclusion

This paper presents instruction randomization techniques based on compilation, which implemented as an extension of LLVM compilation framework. Our prototype system can recognize instructions correctly during the process of compiling the source code and replace instructions according to random replacement rules; moreover, ISRC use dynamic tools to recover instructions to run properly during program execution. Experimental results show that the system can correctly identify instructions and thwart the vast majority of code injection attacks while introducing an extra runtime overhead of 33% on average.

References

1. Bala, V., Duesterwald, E. & Banerjia, S. 2000. Dynamo: a transparent dynamic optimization system. *ACM SIGPLAN Notices* 35(5): 1-12.
2. Baron, I. 2003. Dynamic optimization of interpreters using DynamoRIO. Doctoral dissertation, Massachusetts Institute of Technology.
3. Bhatkar, S. & Sekar, R. 2008. Data space randomization. In *Detection of Intrusions and Malware, and Vulnerability Assessment*: 1-22.
4. Boyd, S. W., Kc, G. S., Locasto, M. E., Keromytis, A. D. & Prevelakis, V. 2010. On the general applicability of instruction-set randomization. *Dependable and Secure Computing* 7(3): 255-270.
5. Giuffrida, C., Kuijsten, A. & Tanenbaum, A. S. 2012. Enhanced Operating System Security Through Efficient and Fine-grained Address Space Randomization. In *USENIX Security Symposium*: 475-490.
6. Henning, J. L. 2000. SPEC CPU2000: Measuring CPU performance in the new millennium. *Computer* 33(7): 28-35.
7. Henning, J. L. 2006. SPEC CPU2006 benchmark descriptions. *ACM SIGARCH Computer Architecture News* 34(4): 1-17.
8. Johns, M. 2011. Code-injection Vulnerabilities in Web Applications—Exemplified at Cross-site Scripting. *it-Information Technology Methoden und innovative Anwendungen der Informatik und Informationstechnik* 53(5): 256-260.

9. Lattner, C. & Adve, V. 2004. LLVM: A compilation framework for lifelong program analysis & transformation. In *Code Generation and Optimization International Symposium*: 75-86.
10. Portokalidis, G. & Keromytis, A. D. 2010. Fast and practical instruction-set randomization for commodity systems. In *Proceedings of the 26th Annual Computer Security Applications Conference*: 41-48.

The software of astronomical plate digitization and measurement*

D. Yan[1†]
National Time Service Center (NTSC), Chinese Academy of Sciences, Lintong, Shaanxi, China, 710600
University of the Chinese Academy of Sciences, Beijing, China, 100039
†Email: yandanntsc@163.com

R. C. Qiao[2]
National Time Service Center (NTSC), Chinese Academy of Sciences, Lintong, Shaanxi, China, 710600
Email: rcqiao@ntsc.ac.cn

This paper discusses the software for digitization and measurement of astronomical plate. Only by using the advanced commercial scanner, the plate can't be digitized in FITS format but in TIFF format. This software can convert TIFF file to FITS file and add some important keywords in the FITS header. This software can also gives the user a way to automatic measure the digitized image of astronomical plate in batch mode. The measurement of an image is done in three steps: identification of stars, elimination of spurious detections, centering of stars.

Keywords: Astronomical plate; Digitization; Measurement.

1. Introduction

Astronomical plates record more than one hundred years of astronomical data. The digitization of astronomical plates not only helps preserve this scientific heritage permanently, but also benefits the sharing of these data around the world. Moreover, we can process this digital image by some computer software, which maybe improve the history precision of these data. The improvement is especially important for some studies which need long time and accurate observations. Owe to these advantages of the digitization of the astronomical

* The authors are very grateful to Yu Yong of the Shanghai Astronomical Observatory for his help in writing programs. This work was supported by the National Science Foundation of China (NSFC) (Grant No. 11173027 and No.U1431227) and Science and technology foundation work project of Digitization of Astronomical plates (GrantNo.22012FY120500)

[1] Dan Yan

[2] Rong-Chuan Qiao

plate, the software of astronomical plate digitization and measurement is elaborated in this paper. It is programmed by FORTRAN language. The software can be divided into two classes according to the function, includes plate digitizing software and image measurement software.

2. Digitizing Software

It is a simple and low-cost way to digitize astronomical plates by using advanced commercial scanner. Scan each plate by the software provided by the scanner manufacturer. By this way, in general, the plate can't be scanned directly in FITS (Flexible Image Transport System) format but in TIFF format. Then covert TIFF image to FITS image which is the most commonly used data format in astronomical. Each FITS file consists of one or more headers blocks which carry a series of keywords [1]. However, the FITS header obtained by simple converting is incomplete, which only has a few keywords.

This software is designed to convert TIFF image to FITS image and add some keywords in the FITS header. The keywords should contain the complete information of observation and scanning, which are significant for keeping plate information and batch processing of data. The observation information includes target name, observatory, observer name, telescope (diameter, focal length and view field), center direction of plate (RA, DEC) and epoch, observation time, exposure time and filter type. These data are available on the plate jackets and observation logs. The scanning information includes scanner name, author of scan, scan time, scan parameters, the size of plate, and image scale (arcsec/pixel). Table 1 present the comparison of keywords in the FITS image generated by simple converting and generated by our software. From this table ,we can see that there only have 11 keywords in the FITS image generated by simple converting and have 31 keywords in the FITS image generated by our software.

Table 1. The keywords generated by simple converting and the keywords generated by our software.

NO	Keywords generated by simple converting	Keywords generated by our software
1	SIMPLE =T	SIMPLE = T
2	BITPIX = 16	BITPIX = 16
3	NAXIS = 2	NAXIS = 2
4	NAXIS1 = 7559	NAXIS = 7559
5	NAXIS2 = 7559	NAXIS2 = 7559
6	BSCALE=1.000000E+00	BSCALE=1.000000E+00
7	BZERO= 6.553500E+004	BZERO= 6.553500E+004
8	DATAMAX= 6.553500E+004	DATAMAX= 6.553500E+004
9	DATAMINV= 0.000000E+00	DATAMINV= 0.000000E+00
10	HISTORY ImageMagick6.7.0-8 2011-06-13Q16 http://www.imagemagick.org	HISTORY ImageMagick6.7.0-8 2011-06-13Q16 http://www.imagemagick.org
11	END	TELESCOP= yunnan1m
12		TARGET = Uranus
13		PLATE-ID= 199005013
14		EMULSION=Kodak 103ao
15		OBSERVER = Qiao
16		FOCAL= 13000 /mm
17		VIEW = 45 /'
18		RA = 8 41 48.31/h m s
19		DEC=- 23 24 11.88/° ' "
20		EPOCH = J2000
21		DAY = 19900501 /year month day
22		TIME= 20:52:30 /hour minute second
23		FILTER = I
24		EXPOSURE = 1200 /s
25		SCANNER=EPSON 10000XL
26		AUTHOR = Yan
27		SCANTIME= 20140915
28		SCANRESO= 1200/dpi
29		PLATESIZ=16*16 /cm*cm
30		IMASCALE=0.010 / mm/pixel
31		END

3. Measurement Software

Astronomical plate is a specific type of astronomical observation. The images size obtained by digitizing astronomical plate is large than that observed by CCD. For example, scanning a 16 cm by 16 cm plate with 1200dpi and 16 bit Grayscale, the resulting FITS image is about 110MB. And the size of CCD image, which observed by the 1.0m astrometric reflector at the Yunnan Astronomical Observatory, is about 2.0 MB. The plates are usually scanned in negative model. However, sometimes, in order to easy analyze the observed object in plate, such as satellite that near primary star, we scan the plate in positive model which will obtain an anti-color image. In this case, the background is brighter than stars, which is opposite to regular observation (CCD observation). The image of CCD observation is shown in Figure 1(a) and the image obtained by scanning astronomical plate in positive model is shown in Figure 1(b). Due to the differences between CCD image and scanned image, the software that developed to measure CCD image maybe can't be used to measure the digitized image of plate. So this software is designed to automatic measure the digitized image of astronomical plate in batch mode.

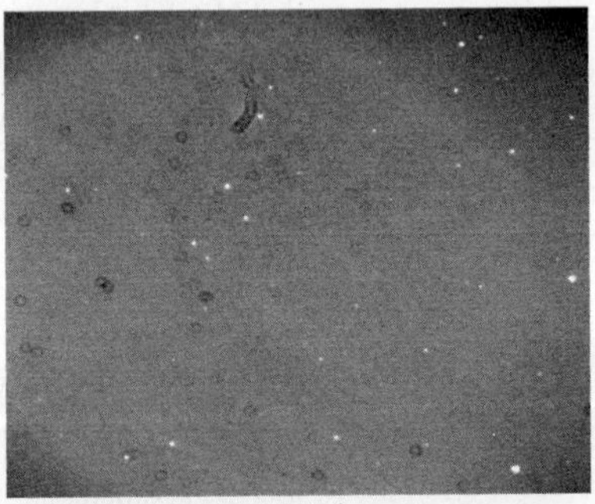

(a)

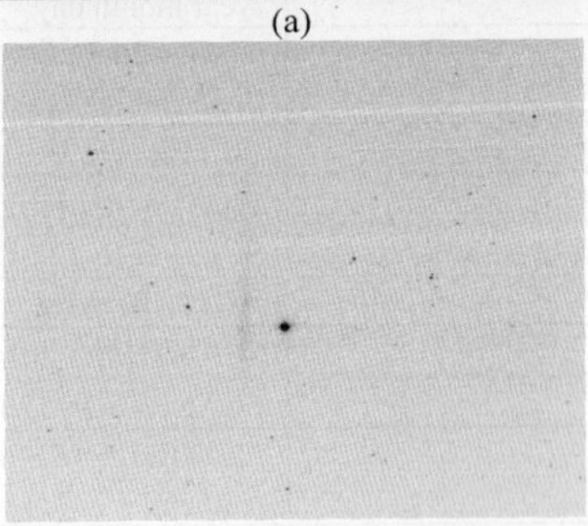

(b)

Fig. 1. The images observed by CCD and obtained by scanning astronomical plate in positive model, which are Figure 1(a) and Figure 1(b) respectively.

The software was programmed by FORTRAN language, which store variables with a contiguous area of memory. In order to avoid the out-of-memory problems in data processing, image data should be divided into several blocks. The number of divided parts is determined by the plate size and scan parameters. Each block was measured sequentially in the same way. The measurement of each block is done in three steps: identification of stars, elimination of spurious detections, centering of stars.

The method of eight connected regions [2-3] is used to identify stars. First, compute the mean and mean variance of image gray value in this block. By the sum of the mean and triple mean variance as threshold level to attain binary image, so that background and object are distinguished. Then scan image pixels according to the order of top-to-bottom and left-to-rights and label pixels by the algorithm of 8-connected regions [2-3]. And eliminate labels inconsistency with the same region. Finally all the connected regions are obtained by this way, which are the identified stars. However, there have some spurious stars in above identification. It maybe come form the fact that the background is higher and the threshold is relative lower. The isolated stain can also lead to the problem. So we eliminate the false detections which only have a single pixel or have low signal noise ratio (SNR) by filtering method.

This software uses the algorithm of two-dimensional modified moment [4] to center stars. The basic equation of two-dimensional moment is given below:

$$x_0 = \frac{\sum_y \sum_x x\left[M(x,y) - B\right]}{\sum_y \sum_x \left[M(x,y) - B\right]}$$

$$y_0 = \frac{\sum_y \sum_x y\left[M(x,y) - B\right]}{\sum_y \sum_x \left[M(x,y) - B\right]} \tag{1}$$

where B is the value of background, M (x, y) is the value of each pixel. This equation has been corrected for the contribution of background. However, this method was sensitive to count levels in the background [5].

So we used another two-dimensional modified moment which is much less sensitive to the background in our software. The basic equations of two-dimensional modified moment [4] is given below, includes equation (2) and equation (3):

$$x_0 = \frac{\sum_y \sum_x x I^{'}(x,y)}{\sum_y \sum_x I^{'}(x,y)}$$

$$y_0 = \frac{\sum_y \sum_x y I^{'}(x,y)}{\sum_y \sum_x I^{'}(x,y)} \qquad (2)$$

$$I^{'}(x,y) = I(x,y) - T \quad \text{if} \quad I(x,y) > T$$

$$I^{'}(x,y) = 0 \quad \text{if} \quad I(x,y) \le T$$

$$T = B + 3\sigma \qquad (3)$$

In this algorithm , T is chosen as the threshold level, which at the 3σ above the background. The σ and background are calculated in the local region of every star.

As stated before, this software can also measure the digitized images which scanned in positive model that the background is brighter than stars. When measure these images, the algorithm of equation (3) should be replaced by the following equation (4).

$$I^{'}\ x,y\quad T\quad I(x,y) \ \text{ if } I\ x,y\quad T$$

$$I^{'}(x,y) = 0 \quad \text{if } I\ x,y > \mathrm{T}$$

$$T = B - 3\sigma \qquad (4)$$

4. Testing of the Software

We take one plate that is a Uranus satellite plate observed in May 1990 at the Yunnan Astronomical Observatory as example to test the software. The FITS header of this plate generated by this software is showed in Table 1. In Figure 2 we present stars that identified by this software, which are marked with green circles. We can see that the identification by our software is reasonable.

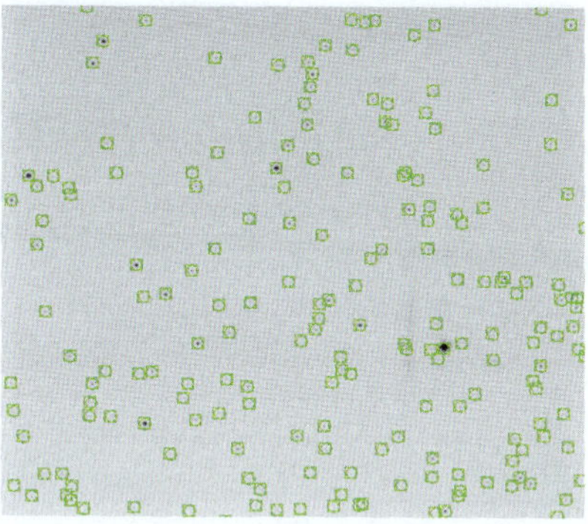

Fig. 2. The stars identified by this software, which are marked with green circles.

In previous, our observation was measured by the software of ASTROMETRICA. ASTROMETRICA is software for astrometric data reduction of CCD image, focusing on measurements of the minor bodies of the solar system (asteroids, comets and dwarf planets). In order to test the measurement of our software, we have made a comparison between ASTROMETRICA and our software. Because the image presented in Figure 2 can't be measured by the software of ASTROMETRICA, so we take a CCD observation to make this comparison. This CCD image was observed by the 1m telescope of Yunnan Astronomical Observatory, which has a field of 7′and a image scale of 0.41″/pixel. Figure 3(a) and Figure 3(b) show the stars identified by ASTROMETRICA and our software, respectively. There have 25 stars identified by ASTROMETRICA and 30 stars identified by our software. A total of 21 stars are both identified by these two software. In order to analyze the measuring accuracy of our software, we have computed the residuals of measurement for the same 21stars, which are the comparison of measurement between our software and ASTROMETRICA. The residuals are plotted in Figure 4. We can see that almost all the absolute residuals are lower than o.5 pixel in both axes. The mean residuals of the same stars are -0.029 pixel in x-axis and -0.083 pixel in y-axis. Moreover, the standard deviations are 0.239 pixel and 0.206 pixel in x-axis and y-axis, respectively. Form the results obtained by above comparison, we can see that the measurement of our software is reliable.

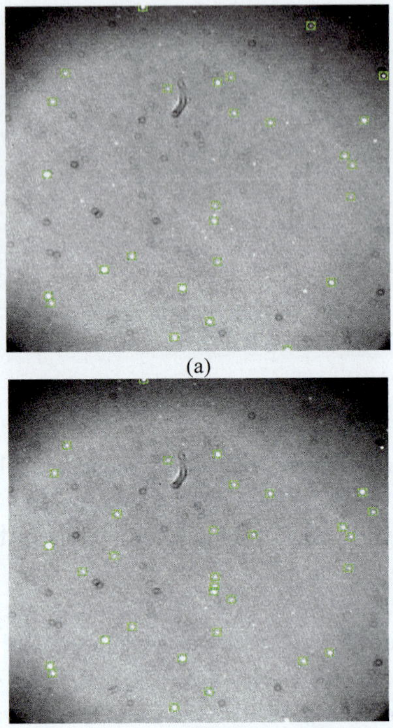

(a)

(b)

Fig. 3. The stars identified by ASTROMETRICA and our software.

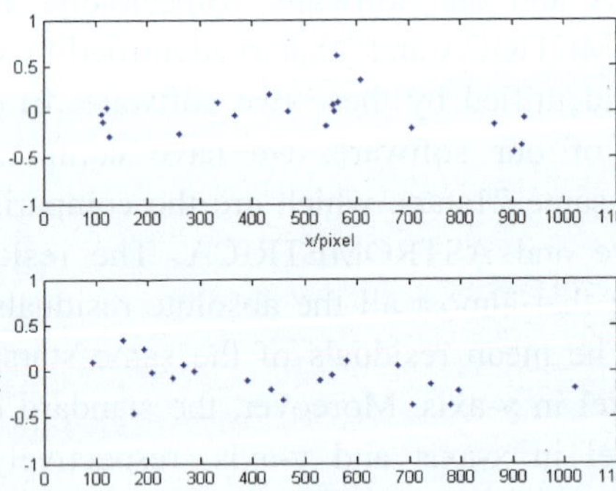

Fig. 4. Plots of residuals which are derived from the comparison of our measurement with the ASTROMETRICA measurement.

5. Conclusion

Astronomical plate is valuable scientific data for the study of astronomy. We described the software for digitization and measurement of astronomical plates. Only by using the advanced commercial scanner, the plate can't be digitized in

FITS format but in TIFF format. This software is a tool can be used not only to convert TIFF file to FIT file, but also to add important keywords to the FITS header, which is significant to keep plate information and automatic process image in batch mode. Due to the scanned image is large than CCD image and the plate maybe scanned in positive model, so the software is also designed to automatic measure the digitized image of astronomical plate in batch mode. The measurement of the digitized image of plate is done in three steps: identification of stars, elimination of spurious detections, centering of stars. The comparison of measurement between our software and ASTROMETRICA shows that the measurement of our software is reliable.

References

1. Pence, W. D. & Chiappetti, L. C. & Page, G. et al. 2010. Definition of the Flexible Image Transport System (FITS), Astronomy & Astrophysics 524(42):1-40.
2. Ishiyama, Y. & Funaka, C. & Kubo, F. et al. 1992. Labeling Board Based on Boundary Tracking. 11th IAPR International Conference on 1992: 34-38.
3. Zuo, W. M. 2006. Study On Connected Regions Extreaction. Computer Applications and sorftware 23(1):75-78.
4. Ji, K. F. Wan, F. 1996. Two Dimensional Modified Moment Centering Algorithm in CCD Images. ACTA ASTRONOMICA SINIA 37(1):85-90.
5. Stone, R. C. 1989. A Comparison of Digital Centering Alogrithms. The Astronomical Journal. 97(4): 1227-1237.

Target segmenting method based on retinal model*

J. Han[1], B. L. Guo[2] and W. Sun[3†]

ICIE Institute, School of Aerospace Science and Technology,
Xidian University, Xi'an, China
[1]E-mail:hanjie289@163.com
[2]E-mail: blguo@xidian.edu.cn
[3†]E-mail: wsun@xidian.edu.cn

For the problem that when doing image segment, the object extraction is susceptible to interference. Based on the saliency map and the model of human retinal receptive field, this paper presents a new method forimage segment. By calculating the map of gazing points gotten using the Kinect, based on human retinal receptive field model, the comprehensive fusion map is obtained, so as to determine the limited area of the segment. Several experiments are done by the method put forward in this paper and the experimental results show that the algorithm is able to solve the problem of susceptible to interference in target segment, while ensuring robustness and meeting the real-time and accuracy needed. And the algorithm in this paper not only improves the speed and robustness on the promise of guaranteeing the segmenting accuracy, but also gives attribution to the field of image segmenting.

Keywords: Kinect; Salience; Retinal field; Fuse; Segment.

1. Introduction

Image segmenting is an important research direction of the field of computer vision, and attracts the interest of many researchers. In many cases, the work of segment is based on manual operation, which brings more difficult for the fulfill of speed [1]. In contrast, Kinect, as a motion sensor, can to some extent complete the interaction. And the principle of human vision gives important scientific ways for intelligent processing of image information [2].

To solve those problems, this paper proposes a new method for target segment based on human retinal receptive field, to achieve better accuracy. First,

* This work was supported by Fundamental Research Funds for the Central Universities under Grant JB141307, National Nature. Science Foundation of China (NSFC) under Grants 61201290 and other Grants 51205301, 61201089, 61305041, 61305040, the China Scholarship Council (CSC) and the National Institutes of Health Grants No.R01CA165225 of the United States.

1 Jie Han
2 Bao-Long Guo
3 Wei Sun, corresponding author

extract the gazing map based on Kinect and the retinal model. Second, extract saliency map based on the phase spectrum [3]. And then fuse the two maps, resulting in a comprehensive map, and ultimately segment target with Grab Cut. Thus, the rapidity, accuracy, and robustness of the system can be optimized.

2. Obtain the Area of Gaze

2.1. *Obtaining the center of gazing area with kinect*

Kinect [4] is a kind of motion sensor of Microsoft, whose RGB and IR camera are respectively responsible for RGB color images (resolution: 640*480) and depth image resolution: 320*240. The location map of face, Kinect, and the screen is as in Fig.1. Combining the two sensors, the data of bone moving and the characteristics of depth information can be obtained. And according to which information about the target and people's eye gaze coordinate values are available.

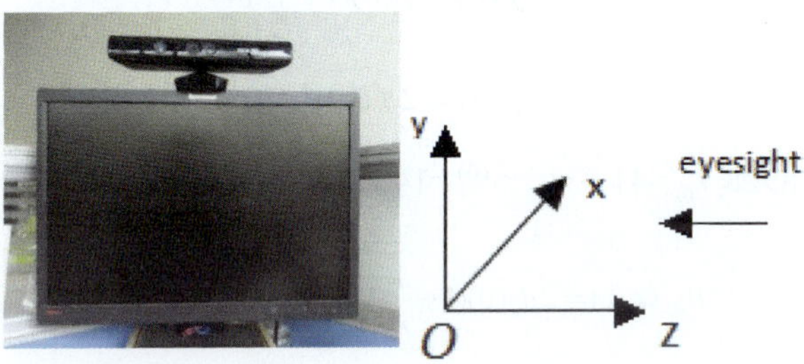

(a) (b)

Fig. 1. Location map

(a) The location of Kinect and the screen

(b) The position relation of eyesight and the system

Table 1. Eyesight points

Number	*x(cm)*	*y(cm)*
1	29.760078	-7.240439
2	27.018982	-5.688334
3	27.436533	-5.977549
4	28.351521	-5.738470
5	30.809893	-5.303936
6	32.642262	-2.544522
7	33.975986	-2.547433
8	35.524555	-2.846047
9	33.256660	-0.341588
10	34.625164	-0.029149

2.2. *Model of human retinal receptive field*

In the human eye, all visual information must traversethe retinal ganglion cells, of which the midget retinal ganglion cells are the most numerous subclasses. Thus the density of their receptivefields [5] plays a key role, which declines exponentially with distance from the fovea. The topology of binocular receptivefields is as following. Density of ganglion cells fieldis:

$$d_{gf}(r,k) = d_{gf}(0) \times [a_k(1+\frac{r}{r_{2,k}})^{-2} + (1-a_k)\exp(-\frac{r}{r_{e,k}})] \tag{1}$$

Where a_k is weight of the first term, and $r_{e,k}$ is thescale factor of the exponential. The meridian isindicated by k. The midget cells [6] constitute a fraction $f(r)$ of allganglion cells at eccentricity r, and its density is given:

$$d_{mf}(r,k) = f(r)d_{gf}(r,k) \tag{2}$$

$$f(r) = f(0)(1+\frac{r}{r_m})^{-1} \tag{3}$$

$$d_{gf}(0) = 2f(0)d_c(0) \tag{4}$$

where f(0)=0.8928, r_m=41.03°, $d_c(0)$=14804.6. Combining 1, 2, and 3, 4 we have:

$$\begin{aligned} d_{mf}(r,k) &= 2d_c(0)(1+\frac{r}{r_m})^{-1} \\ &\times [a_k(1+\frac{r}{r_{2,k}})^{-2} + (1-a_k)\exp(-\frac{r}{r_{e,k}})] \end{aligned} \tag{5}$$

And the spacing of adjacent midget receptive fields is given by:

$$s_{mf}(r,k) = \sqrt{\frac{2}{\sqrt{3}d_{mf(r,k)}}} \tag{6}$$

$$\begin{aligned} s(\{x,y\}) &= \sqrt{(\frac{x}{r_{xy}}s_{mf}(r_{xy},1))^2 + (\frac{y}{r_{xy}}s_{mf}(r_{xy},2))^2} \\ &= \frac{1}{r_{xy}}\sqrt{x^2 s_{mf}(r_{xy},1)^2 + y^2 s_{mf}(r_{xy},2)^2} \end{aligned} \tag{7}$$

Eq. 15 computes mRGC spacing at locations [7] specified in one-eye visual field. Hence, the density model in thebinocular visual can be computed:

$$s_B(\{x,y\}) = \sqrt{2}\sqrt{\frac{s(\{-x,y\})^2 s(\{x,y\})^2}{s(\{-x,y\})^2 + s(\{x,y\})^2}} \tag{8}$$

According to the equations above, the model can be extracted as in Fig. 2 [8]:

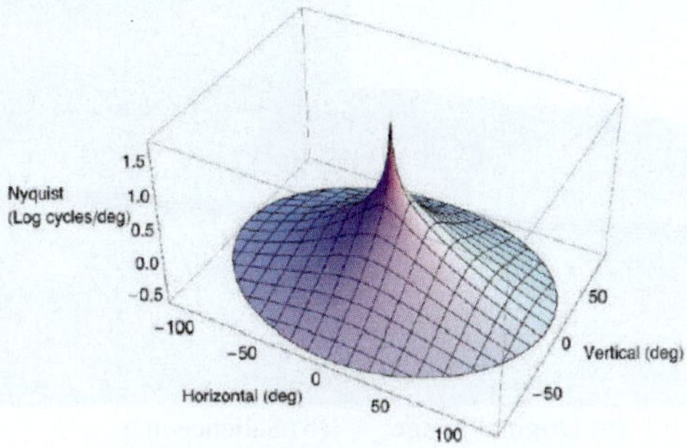

Fig. 2. Model of the receptive fiels.

3. Saliency Detect

More effective and less complex than the Itti model, the visual attention mechanism based on phase spectrum is used to detect the target in the image.

$$E\left(\frac{(\bar{f},\bar{x})}{\|\bar{f}\|\cdot\|\bar{x}\|}\right)\geq 0.5, for|\Omega_b|<\frac{\mathrm{N}}{6} \tag{9}$$

$$\bar{f}=IDCT(sign(DCT(f))) \tag{10}$$

$$\bar{x}=IDCT(sign(DCT(x))) \tag{11}$$

Where $E\left(\frac{(\bar{f},\bar{x})}{\|\bar{f}\|\cdot\|\bar{x}\|}\right)$ represents expectation, Ω_b is the cardinal number of background image in DCT domain[9]. f consists of independent Gaussian distribution points, on the DCT inverse transform.

$$E\left(\frac{\alpha}{\sqrt{\alpha^2+\beta^2}}\right)\geq\sqrt{\frac{2}{\pi}}\approx 0.798 \tag{12}$$

Where $\alpha=\sqrt{\sum_{i\in T_f}\bar{f}_i^2}$, $\beta=\sqrt{\sum_{i\notin T_f}\bar{f}_i^2}$, T_f is non-zero collection of the foreground image.Thus, the image saliency map can be efficiently obtained using the DCT inverse transformation [10]. The Gauss template of an aerial image can be selected as 5×5, δ =2; Its saliency is shown in Fig. 3.

(a) Original image (b) Salience map

Fig. 3. Static salience map.

4. Grab Cut Based on Retinal Model

4.1. *Fusion*

Fuse the maps as shown in Fig. 4:

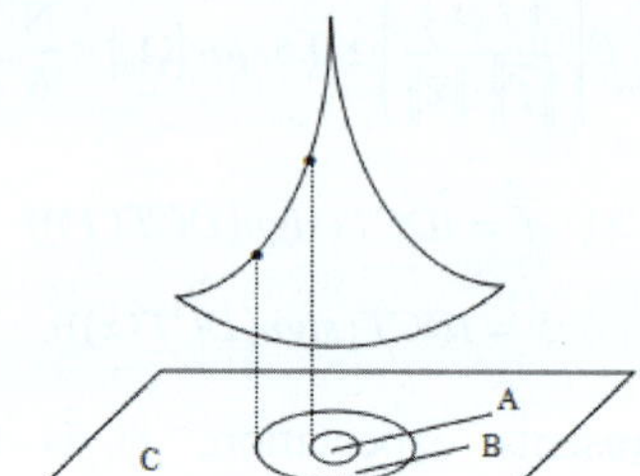

Fig. 4. Fusion method.

Iterative, optimize the Initial radius r_1, r_2 , which can divide the whole image is into three parts A,B and C, and their HSV model are $H_a \; H_b \; H_c$.

$$S_{B-A} = \frac{1}{m}\sum_{i=1}^{m}(H_{bi} - H_{ai})^2 \tag{13}$$

$$S_{B-C} = \frac{1}{m}\sum_{i=1}^{m}(H_{bi} - H_{ci})^2 \tag{14}$$

$$S_k = \frac{S_{(B-A)k}}{S_{(B-C)k}} \tag{15}$$

$$E_{S_k} = \left\| S_k - S_0 \right\|_2^2 \tag{16}$$

$$r_0 = r_1 + \frac{r_2 - r_1}{2} \tag{17}$$

Where, m is the dimension of the vector, S_{B-A}, S_{B-C} are the similarity of different parts, S_k is the ratio of the kth iteration (S_0 is the Final convergence value), in which E_{S_k} is the error. By experiment, $S_0 = 0.6$, and finally r_0 will be calculated. According to the above saliency map, the fusion result is as in Fig. 5:

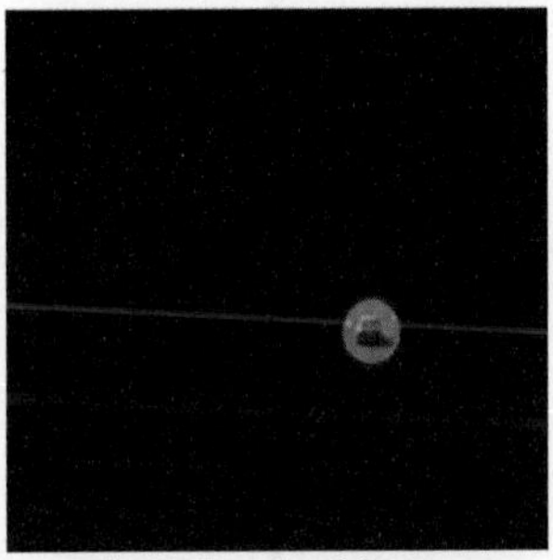

Fig. 5. Result of fusion .

4.2. *Grab cut*

Grab Cut is adopted for image segmentation and using the GMM model it can work without the seed [11] points. Thus, it is more convenient than the Graph Cut. According to above, results can be seen as in Fig. 6:

Fig. 6. Result of grab cut.

5. Algorithm Implementation Process

Based on the above idea, Fig. 7 shows the algorithm implementation process:

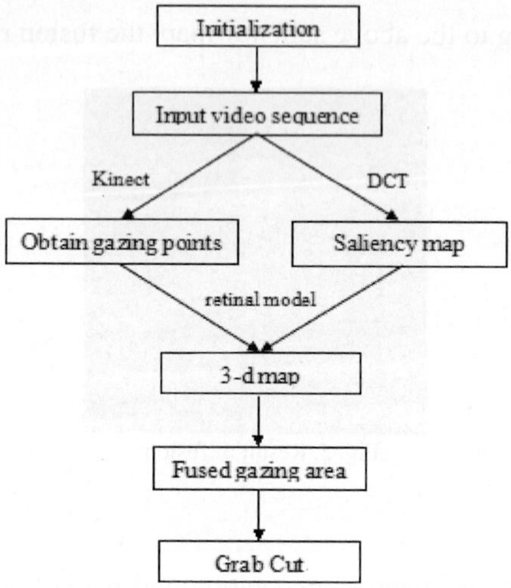

Fig. 7. Algorithm block diagram .

Steps of the algorithm are as follows:

Step 1:Obtain the points of gazing mapwith Kinect;

Step 2: Use retinal model to extract gaze points [12], and obtain the 3-d map;

Step 3: Extractthe saliency map of the video sequence using DCT inverse transform;

Step 4: Fuse the 3-d map and the saliency map;

Step 5:Segment the image with Grab Cut;

Fig. 8 shows the result of two image with the method in this paper, and it can be seen that the target can be accurately segmented [13].

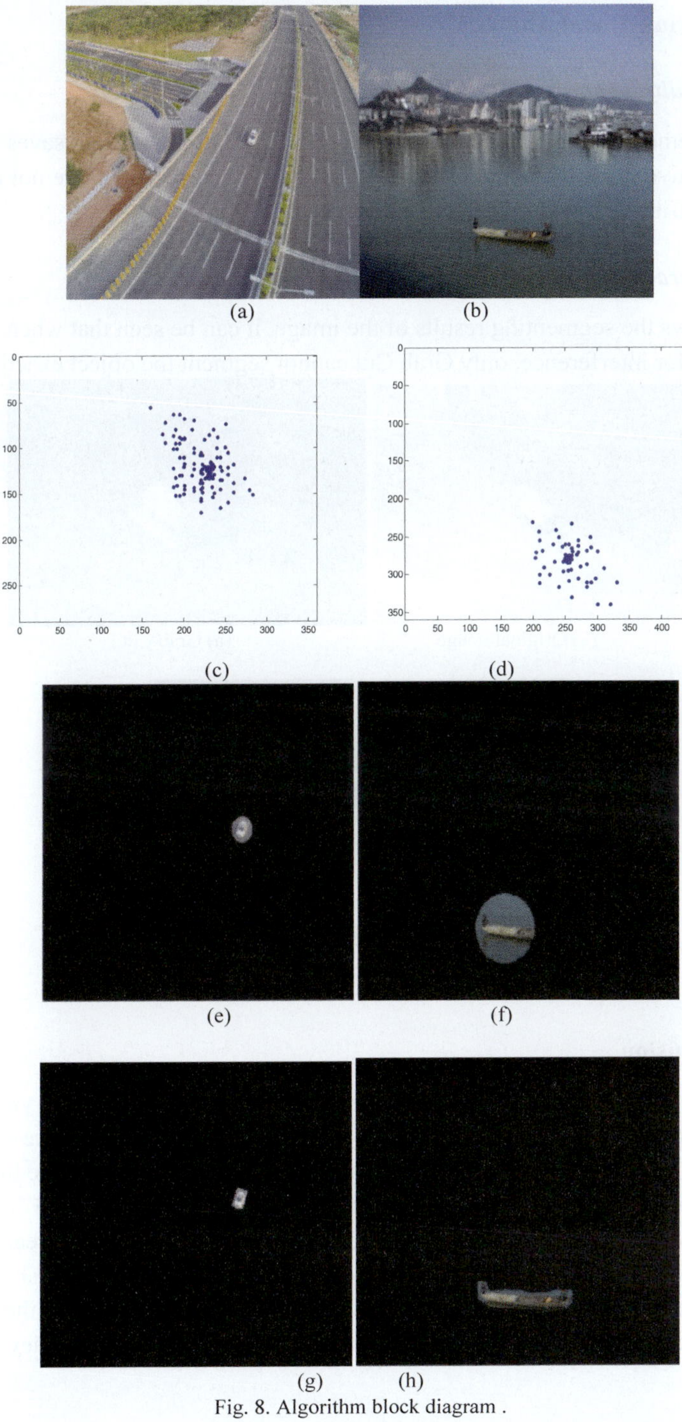

Fig. 8. Algorithm block diagram .

6. Experiment and Analysis

6.1. *Rapidity analysis*

For the method in this paper does not need any manual interfere, it saves a lot of time. Relative to Graph Cut and Lazy Snapping [14], seed points are not needed, which contributes much to the convenience and speed.

6.2. *Accuracy analysis*

Fig. 9 shows the segmenting results of the image, it can be seen that when there is another color interference, only Grab Cut cannot segment the object exactly.

(a) Original image

(a) Grab Cut

(a) Method in paper

Fig. 9. Result of segment.

7. Conclusion

This paper puts forward a new method for target segmenting in aerial video [15] based on human retinal modeland saliency map. First gazing pointsare obtained by Kinect, based on the retinal model, resulting in a 3-d comprehensive map, and then get a saliency map based on DCT inverse transform phase spectrum collection of nonzero. Then fuse them to get the segmenting limit area. Finally segment the image with Grab Cut. The experimental results show that compared to Graph Cut, the algorithm in this paper improves the apeed performance and robustness [16] on the premise of guaranteeing accuracy.

References

1. Wei Sun, Objects detecting and tracking with a new particle filter; CECNet2012, Yichang, pp.3340-3343, 2012.
2. Jie Han, Baolong Guo, Wei Sun. Target Tracking Method in Aerial Video Based on Saliency Fusion. MEIC 2015, Vol. 5, pp. 723-727, 2015.
3. Wei Sun, Chen Long. Binocular Vision-based Position Determination Algorithm and System. CDCIEM 2012, Vol. 1, pp. 170-173, 2012.
4. Baolong Guo, Wei Sun. "Introduction to Digital Image Processing System Engineering". People Post Press, pp. 65-80, 2012.
5. Wei Sun, Xu Zhang, Yunyi Yan. An Objects Detecting and Tracking Method based on MSPF and SVM. International Journal of Soft Computing and Software Engineering [JSCSE], Vol. 2, No. 2, pp. 9-16, 2012.
6. Xumeng Chen, Wei Sun. A Fast Algorithm for Single Image Dehazing Based on Estimating Illumination Veil. MEIC 2015, Vol. 8, pp. 987-991, 2015.
7. P. Viola, M. Jones. "Rapid Object Detection Using a Boosted Cascade of Simple". IEEE Computer Vision and Pattern Recognition, 3(2), pp. 511-518, 2008.
8. Nummiaro K, Koller-Meier E, Van Gool L. "An adaptive color-based particle filter". Image and vision computing, 21(1), pp. 99-110, 2008.
9. Wei Sun, Guo Bao-long. Multiple Objects Tracking with Particle Filter and Mean Shift Clustering. ISDA-2008, Vol. 2, pp. 482- 486, 2008.
10. Gordon NJ, Salmond DJ, Smith AFM. "Novel approach to nonlinear/non-Gaussian Bayesian state estimation" IEE Proceedings F (Radar and Signal Processing). IET Digital Library, 11(2), pp. 37-43. 2003.
11. Evans C. "Notes on the opensurf library". University of Bristol, Tech. Rep., January, pp. 21-33, 2009.
12. WANG Xiao, SUN Zhang, LIN Jing. "Energy-efficient adaptive sensor scheduling for target tracking". Control Theory, 8, pp. 8-92, 2010.
13. XU Jin, LI Jinxing. "State estimation with quantised sensor information in wireless sensor networks". IET Signal Process, 5, pp. 1-26, 2011.
14. SHOU Tiande. "Processing mechanism of visual information". Shanghai Press, pp. 58-80, 1997.
15. L. Itti, P. F. Baldi. "Bayesian Surprise Attracts Human Attention". Vision Research, 49(10), pp. 1295-1306, 2009.
16. Long Chen, Baolong Guo, Wei Sun. A new multiple-objects tracking method with particle filter. IAS 2009, Vol. 1, pp. 281-284, 2009.

Three-image encryption method based on discrete multiple-parameter fractional Fourier transform

H.Y. Mao, F.S. Song*, X.B. Meng

Department of Automation Control Engineering, Shenyang Institute of Engineering, Shenyang, 110136, China

** No. 18 of Puchang Road, Shenyang North New Area, Shenyang, Liaoning, P.R. China.*

26013458@qq.com

A encryption scheme is proposed by using of discrete multiple-parameter fractional Fourier transform. Therefore,the purpose of the paper is to encrypt three images to one image by theta modulation. Three original images were multiplexed into a complex image by utilizing theta modulation after every image being modulated by sinusoidal amplitude grating. The method of encoding three images was multiplying the complex signal by a pixel scrambling operation and random phase mask, and then encrypting by a discrete multiple-parameter fractional Fourier transform. While it could be decrypted using the correct pixel scrambling, random phase mask and the parameters in discrete multiple-parameter fractional Fourier transform. In conclusion, numerical simulations have been done to prove the validity and the security of the proposed method. In this paper, we present a novel three-image encryption algorithm. The computer simulation results show that the algorithm has good performance, can decrypt these three images exactly.

Keywords: Image encryption; Discrete multiple-parameter fractional Fourier transform; Theta modulation.

1. Introduction

With the growth of technology, the security of image is becoming more and more important, more and more image encryption methods have been proposed [1-8]. But, most of them only can encrypt one image [2-4]. In this paper, a encryption method is proposed to encrypt three image into one image. At first, these three images should be multiplexed into one image, and the multiplex method will affect the efficiency of the algorithm and the quality of decrypted image directly. Mosso et al. [7] introduced theta modulation method to image encryption. They encrypted the some images by using double random phase encoding (DRPE) in Fourier domain, and then used theta modulation method to multiplexed these images into a single image. But the encryption key only has two random phase masks, it was not safe enough. To solve this problem, Du Xiangjun and Tao shaohua [8] proposed an optical encryption/decryption method based on fractional Fourier transform. The method keys has not only two random phase masks, but also two orders of fractional Fourier transform. It enhanced the security of the system, but still not safe enough. In our method, the discrete multiple-parameter fractional Fourier transform (DMPFRFT) is presented [9-11] to encrypt image. It has more parameters and can provide and higher security in encryption. This method does not require complex computation. Numerical simulations show that the proposed scheme provides an efficient and secure way for image encryption.

The remaining sections of this paper are organized as follows: Section 2 introduces the proposed three-image encryption method, Section 3 presents the numerical simulation results, and Section 4 states the conclusions.

2. Principle of three-image encryption

The proposed encryption process is shown in Fig. 1. Let $f_i(x, y)$ represent the ith image. At first, we use theta modulation method to multiplex three images [12-15]. A sinusoidal amplitude grating is defined [12-13] by the amplitude transmittance function as follows

$$g(x,y) = [\frac{1}{2} + \frac{m}{2}\cos(2\pi f_0 x)]rect(\frac{x}{L})rect(\frac{y}{L}) \tag{1}$$

where for simplicity we have assumed that the grating structure is bounded by a square aperture of width L. The parameter m represents the peak-to-peak change of amplitude transmittance across the screen, and f0 is the spatial frequency of the grating. $m \leq 1, f_0 >> 2/L$.

Each image can be modulated using grating with a different orientation angle and/or spatial frequency and then multiplexed into one image. The complex signal M(x, y) can be expressed as

$$M(x, y) = \sum_{i=1}^{3} f_i(x, y) g_i(x, y) \tag{2}$$

After that the complex signal M(x, y) is multiplied by pixel scrambling and random phase mask exp $[ip(x, y)]$, where $p(x, y)$ is statistically white sequence uniformly distributed in $[0,2\pi]$. Then by using 2D DMPFRFT with parameters of (M_L, M_R; α_L, α_R; $\mathbf{m}_L$, $\mathbf{n}_L$; $\mathbf{m}_R$, $\mathbf{n}_R$), the final distribution $E(x, y)$ of the output encrypted image can be expressed as

$$E(x,y) = F_{(M_L,M_R)}^{(\alpha_L,\alpha_R)}(\mathbf{n}_L{}',\mathbf{n}_R{}')\left\{\mathrm{J}\left[M(x,y)\right]\exp\left[ip(x,y)\right]\right\}$$
$$= F_{(M_L,M_R)}^{(\alpha_L,\alpha_R)}(x,y)\left\{\mathrm{J}\left[\sum_{i=1}^{3} f_i(x,y)\,g_i(x,y)\right]\exp\left[ip(x,y)\right]\right\} \quad (3)$$

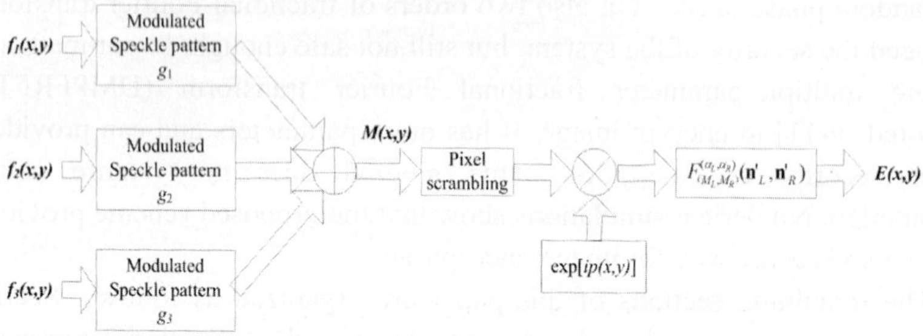

Fig. 1. Schematic of encryption.

The proposed decryption process is shown in Fig. 2. For decryption, the decrypted complex signal *M(x,y)* can be obtained by applying the inverse of 2D DMPFRFT, the conjugate of RPM and inverse pixel scrambling as follows:

$$M(x,y) = \mathrm{J}^{-1}\left\{F_{(M_L,M_R)}^{(-\alpha_L,-\alpha_R)}\left[E(x,y)\right]\exp\left[-ip(x,y)\right]\right\} \quad (4)$$

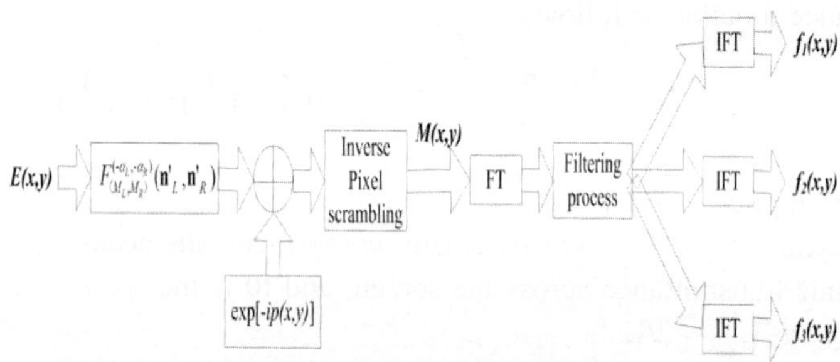

Fig. 2. Schematic of decryption.

Next we analyze the filtering process. By applying Fourier transform (FT) to *M(x,y)*, we can obtain the Fourier plane which has paired spots belonging to each images located a different spatial positions. These spatial positions depend on the spatial frequency f_0, the pitch and orientation of the sinusoidal amplitude grating. Finally, each single image can be obtained by filtering out every spot in order and using the inverse Fourier transform (IFT).

It can be seen that to retrieve three images, the periodicities (M_L, M_R), transform orders (α_L, α_R) and vector parameters ($\mathbf{m}_L$, $\mathbf{n}_{L;}$ $\mathbf{m}_R$, $\mathbf{n}_R$) of DMPFRFT, random phase mask and correct pixel scrambling are essential, they serve as the system keys.

3. Simulation results

Simulations have been carried out to verify the proposed three images encryption method. We take three images with 512×512 pixels and 256 gray levels as the original images, as shown in Fig. 3(a)-(c). In out experiment, the spatial frequency of the sinusoidal amplitude grating are 7.8×104. The angle interval θ is 60°. The system parameters are $(M_L, M_R; \alpha_L, \alpha_R)$=(15,20;0.34,0.73). The vector parameters $(\mathbf{m}_L, \mathbf{n}_L)$ and $(\mathbf{m}_R, \mathbf{n}_R)$ are 1×15 and 1×20 random vectors whose values are independent integer values, respectively. Fig. 3(d) is the complex signal *M(x, y)*. It is the sum of 3 modulated images. Fig 3(e) is the result of pixel scrambling, it can be seen that any information of complex signal can not be obtain. Fig 3(f) is the encrypted image *E(x, y)* with stationary white noise The decrypted complex signal *M(x, y)* with the corrected keys is shown in Fig. 3(g). It is exactly the same as the previous one and without any noise or distortion. Fig. 4(h)-(j) show the three retrieved images. Although they have some noise, we can recognized it without doubt.

(a) (b) (c)

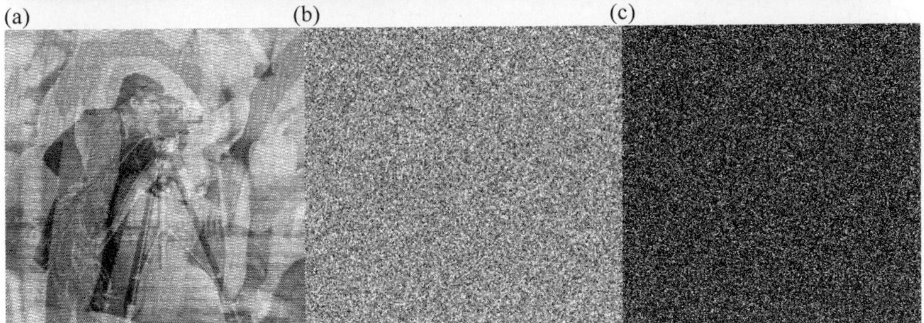

(d) (e) (f)

Fig. 3. (a)-(c) original images; (d) complex signal M(x, y); (e) pixel scrambling result; (f) encrypted image; (g) decrypted complex signal; (h)-(j) decrypted images.

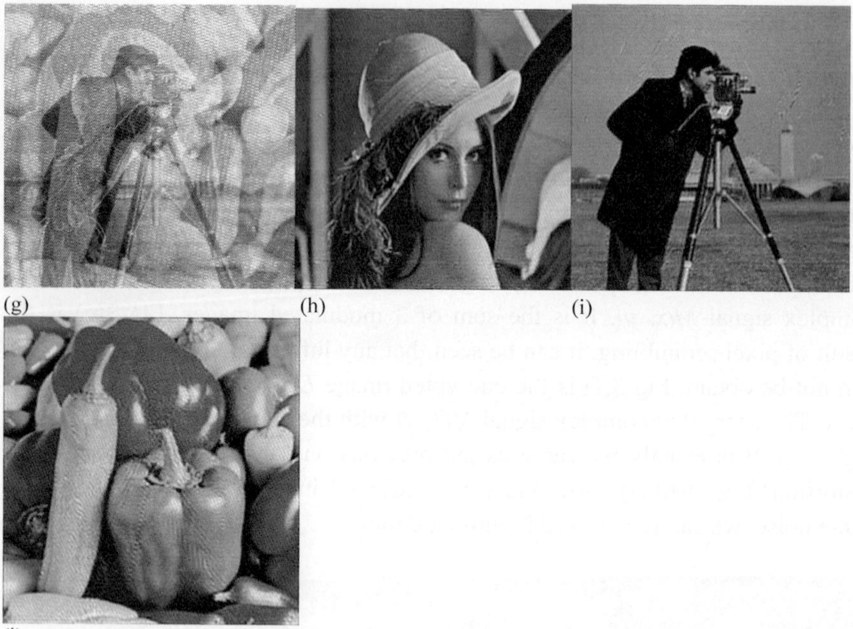

Fig. 3. (*Cont'd*)

In order to further quantitatively evaluate the performance of this method, we also calculated the normalized mean square error (NMSE) between the original images and the decrypted images. Fig. 4 shows the NMSE curves. It can been seen from Fig. 4 that this method has a uniform quality for every frame.

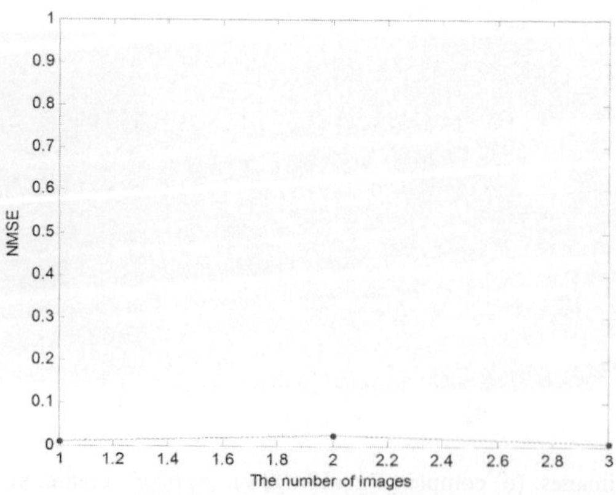

Fig. 4. NMSE for all images.

Image histograms and self-correlation performances are tested to show the statistical properties of the encryption algorithm. Fig. 5(a)-(c) are the image histograms of the three original images, respectively, and Fig. 5 (d) is the image histogram of the encrypted image. There are significant differences between (a)-(c) and (d), so useful information cannot be obtained by attackers according to the statistical properties. Fig. 6(a)-(c) are the self-correlation images of the three original images, respectively, and Fig. 6 (d) is the self-correlation image of the encrypted image. It can be seen that the correlation of the encrypted image is much weaker than that of the original images. It further proves that the algorithm has a strong capability of decorrelation to resist attacks from statistical analysis.

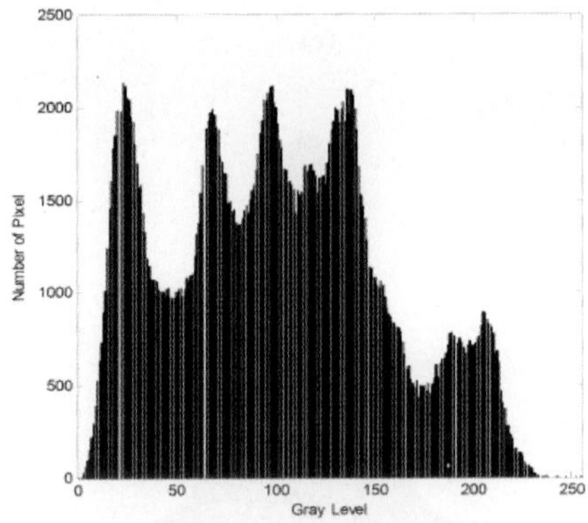

(a)

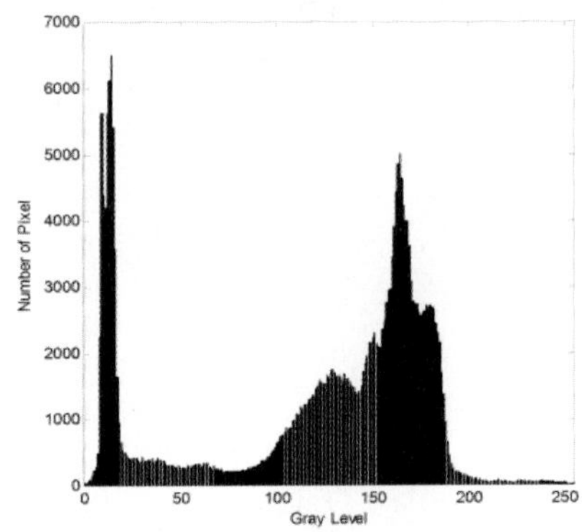

(b)

Fig. 5. (a)-(c) Histogram of the three original images; (d) histogram of the encrypted image.

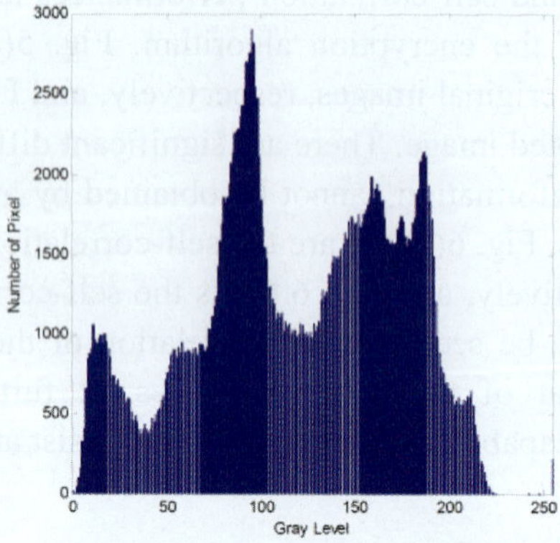

(c)

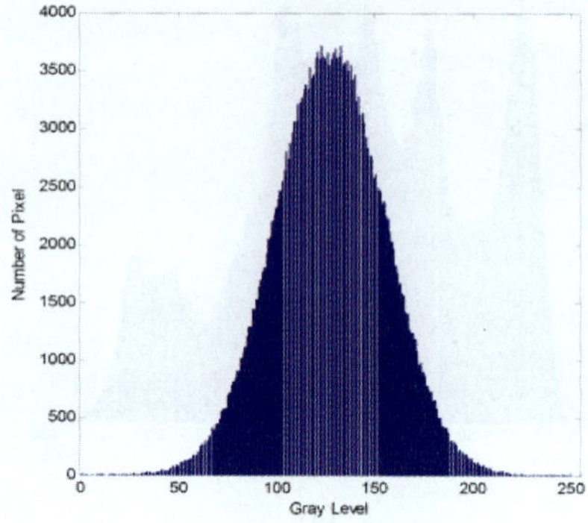

(d)

Fig. 5. (*Cont'd*)

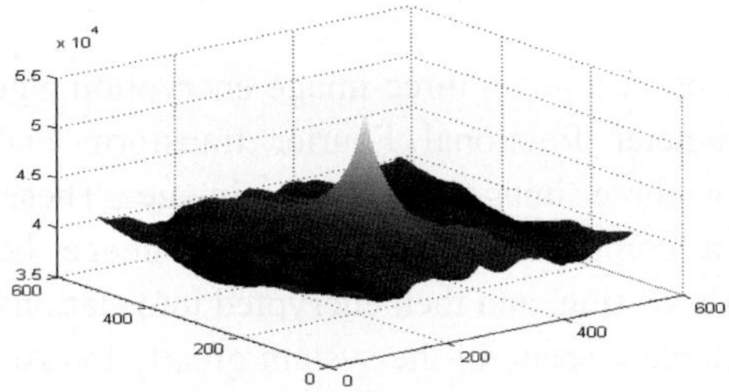

(a)

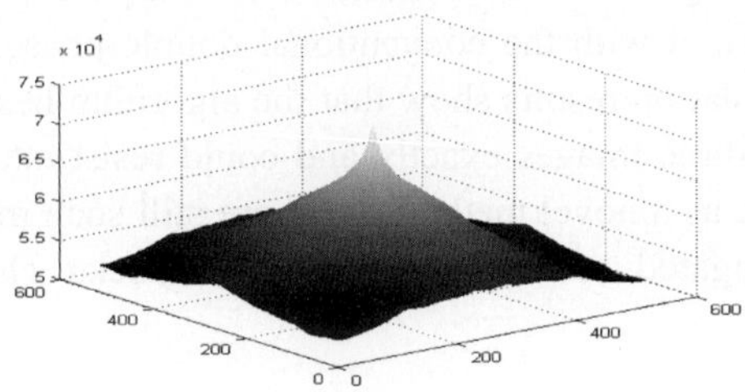

(b)

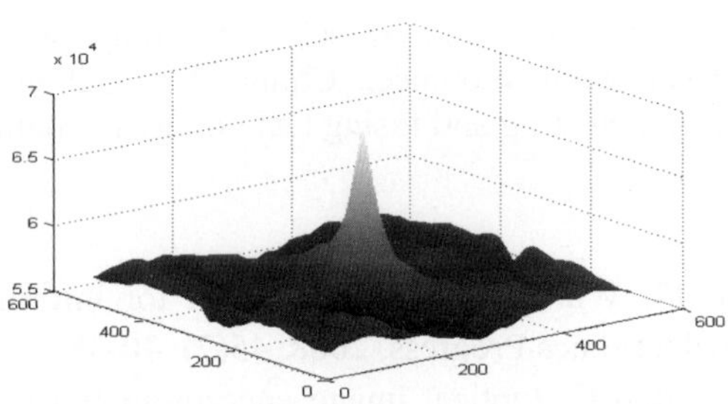

(c)

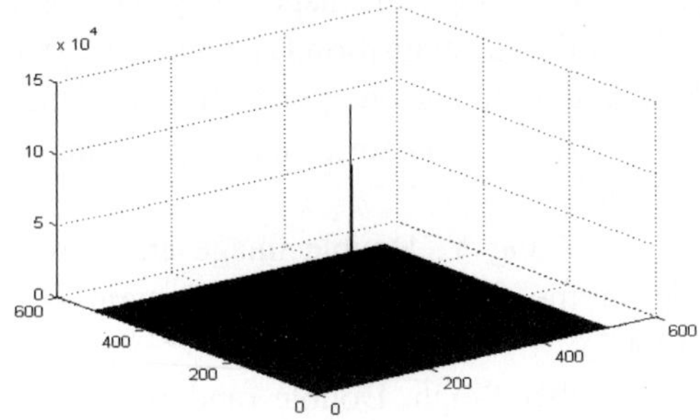

(d)

Fig. 6. (a)-(c) Self-correlation of the three original images; (d) self-correlation of the encrypted image.

4. Conclusion

In this paper, we present a novel three-image encryption algorithm based on the discrete multi-parameter fractional Fourier transform and theta modulation method to encrypt three images into one image. These three images are multiplexed into a complex image after each image being modulated by sinusoidal amplitude grating, and then encrypted into stationary white noise. The method enhances the key space of the system greatly by using theta modulation method and discrete multi-parameter fractional Fourier transform, and further provides a much higher security level, more simplified operation and less computation compared with the conventional double phase encryption system. The computer simulation results show that the algorithm has good performance, can decrypt these three images exactly and could resist attacks from statistical analysis. Naturally, as a novel method, there are still such many other questions to be further investigated, such as its resistance to attacks. This will be our future task.

5. Acknowledgements

The research work was supported by scientific study project of educational commission in Liao Ning province, China, No. L2014523. Research on cooperative spectrum detecting and fusing technology in cognitive radio.

6. References

1. Wu K N, Hu JS, Wu X. Optical encryption for information security [J]. Laser & Optoelectronics Progress, 2008, 45(7): 30-38.
2. Zhu B, Liu S, Ran Q. Optical image encryption based on multifractional Fourier transforms [J]. Opt Lett, 2000, 25: 1159-1161.
3. Tao R, Lang J, Wang Y. Optical image encryption based on the multiple-parameter fractional Fourier transform [J]. Opt Lett, 2008, 33: 581-583.
4. Lang J, Tao R, Wang Y. Image encryption based on the multiple-parameter discrete fractional Fourier transform and chaos function [J]. Opt Comm, 2010, 283: 2092-2096.
5. Tao R, Xin Y and Wang Y. Double image encryption based on random phase encoding in the fractional Fourier domain [J]. Opt. Express 15, 16067-16079 (2007).
6. G. Unnikrishnan, Kehar Singh. Double random fractional Fourier-domain encoding for optical security [J]. Opt Eng. 39(11): 2853-2859 (2000).
7. Mosso F, Barrera J F, Tebaldi M, et al. All-optical encrypted movie [J]. Optics Express, 2011, 19(6):5706-5712.

8. Du X J, Tao S H. All-optical encrypted movie based on fractional Fourier transform [J]. Optical Technique, 2013, 39(1): 68-71.
9. Cariolaro G, Erseghe T, Kraniauskas P, et al. Multiplicity of fractional Fourier transforms and their relationships [J]. IEEE Trans Signal Process, 2000, 48: 227-241.
10. Lang J, Tao R, Wang Y, et al. The multiple-parameter fractional Fourier transform [J]. Sci China Ser F-Inf Sci, 2008, 51: 1010-1024.
11. S.C. Pei, M.H. Yeh, C.C. Tseng. Discrete fractional fourier transform based on orthogonal projections [J]. IEEE Transactions on Signal Processing, 1999, 47(5): 1335-1347.
12. Liang R S, Lv X X. Information optics [M]. Beijing: Publishing House of Electronics Industry, 2008: 71-73, 240-242.
13. J. W. Goodman. Introduction to Fourier Optics [M]. Robert & Company Publishers, 2004: 80-81.
14. Li T F. Amplitude Sine-Grating [J]. Natural Science Edition, 1997, 14(3): 87-91.
15. Sun Y Y, Wang C Y, Zhao M, et al. The research of θ modulate color synthesis system design [J]. Electronic Instrumentation Customer, 2010, 17(1): 12-14.

Research and implementation of information management system based on GIS lake reserve

G. Gou[1†], K. K. Sun[2] and Y. Q. Zhao[3]
College of Computer Science & Technology, Guizhou University, Guiyang 550025, Guizhou, China
†Email:1784788577@qq.com

L. L. Huang[4]
Huaxi people's procuratorate, Guiyang 550025, Guizhou, China

Lake and its drainage basin is an important place for human survival, and the protection of the lake is an important part of environmental protection. Research and establish the information management system of Lake classifying protected areas based on GIS, considerate and plan the lake eutrophication comprehensively by using information technology, on the basis of river basin comprehensive sewage interception, high intensity of pollution, to make artificial intervention and macro control of the restoration of the lake ecosystem.

Keywords: GIS; Division of protected areas; Information system; Lake.

1. Background

Lake is an important land resource, can regulate the runoff, develop irrigation, Provide industrial and drinking water and breed aquatic organisms, shipping communication, improve the regional ecological environment and a variety of functions, plays an import ant role in the development of the national economy [1]. However, with the continuous development of people's life and production, the harm that pollution of life and industry bring to the environment of the lake is becoming increasingly fierce, causing a large number of lakes eutrophication, serious harm the normal lake ecological environment. A survey shows that about 26.9% of the lake belongs to the eutrophication of lakes, 73.1% of the lake belongs to eutrophic lakes, the survey data show a very acute problem of eutrophication [2].

---

[1] Gang Gou
[2] Ke-Ke Sun
[3] Yanqi Zhao
[4] Ling-Li Huang

In practical activities of Eutrophic lakes governance practices and the lake ecosystem restoration, we often tend to headache treatment gap measures, always wanted to make the lakes clear in the short term. In reality we often only consider the governance of the local environment of lakes and neglect of the whole river basin sewage treatment, or only grasping the governance of the point source pollution, and ignore the role of non-point source pollution. The results is that put the large amount of manpower, material resources and financial resources into the lake rich nutrition management, but the lakes eutrophication become more and more serious. We must have a clear understanding of the governance process of lake eutrophication, besides we should consider and plan the governance process of lake eutrophication Systematically comprehensively, on the basis of high intensity of pollution control, to restore the health lakes ecosystems scientifically, through increase the artificial intervention, to improve the efficiency of pollution control and accelerate the repair of the damaged ecological system.

Developing the lake environmental systems based on GIS, dividing the lake protection zone scientifically and rationally, making a macro and overall control about the emissions of various pollutants, so as to effectively maintain the content of phosphorus and nitrogen in lake and curb the lake eutrophication in a timely manner, protect the ecological environment of lakes, maintain sustainable development of our healthy and green society [3].

2. The Current Situation of The Reservoir and the Two Lakes

Guizhou Hongfeng and Baihua Lake (The following two lakes for short) is in Central Guizhou, they are primary and secondary artificial reservoir under the development and construction of cat jump river cascade hydropower that is a tributary of the Wujiang River, designed for the normal high water level storage were 6.01 billion cubic meters and 1.82 billion cubic meters, lake area of 57.2 square kilometers and 14.5 square kilometers, drainage area of 1895 square kilometers [4]. With the rapid development of social economy, the region of two lakes formed the pattern of comprehensive type of industry that contains agriculture, tourism and aquaculture industry, the two lakes became multifunctional reservoirs of power generation, flood control, water supply, irrigation, aquaculture, tourism and water movement; at the same time, with the rapid progress of urbanization, the two lakes is the main drinking water source of Guiyang City, Qingzhen city and Anshun city Pingba county. However, due to the unbalanced development of the society, the two lakes also suffer from the intrusion of eutrophication and have complex pollution emissions [5]. Mainly embodied in the following several aspects.

(1) Sewage pollution from industrial enterprises

Industrial enterprises built by the two lakes in the last century for the historical reasons have poor facilities for environmental protection and extensive pollution treatment, especially the production wastewater of several chemical enterprises has become a major source of industrial pollution of the two lakes.

(2) Urban and rural life sewage pollution

More than 10 cities in the river basin of the two lakes, due to social economic development is relatively backward, only 3 towns start to construct and use the waste treatment facilities in the past few years, the others and major population centers have not yet established a sewage treatment facility, the direct emissions of rural life sewage increased the water pollution load of the two lakes.

(3) Agricultural non-point source pollution

The agricultural population accounts for a large proportion in the river basin of the two lakes, because people with less land, the villagers farming on steep ground, causing great loss of soil, at the same time in the process of planting food crops, a large number of chemical fertilizers and pesticides has been used, increased the emissions of nitrogen, phosphorus and other pollutants, caused a great impact on the water quality of the lake area and destroyed the ecological environment of the lake area.

(4) Tourism pollution

The river basin of the two lakes is the area of rapid economic development in Guizhou province,at the same time is also an important tourist attractions, the annual tourists more than 400000 people. A large number of tourist facilities built along the Lake sewage and travelers' pollution caused serious impacts on water quality of the two lakes.

(5) Irrational exploitation of resources

Since the late 1980s, due to the weak awareness of resources and environmental protection, the river basin of the two lakes has more development and less protection, unreasonable land use, a large number of lakeside zone ecosystem suffer form destruction, forest cover dropped sharply, soil erosion area accounted for 22.5% of the total area of the basin, causes the agricultural non-point source nitrogen, phosphorus and organic matter serious loss, the ecological environment of the two lakes that has been destroyed is difficult to reverse in a long time [1] .

3. Solutions to Problems

The cause of pollution is multi-directional and multi-channel, restoration of ecosystem management and eutrophication of lakes should avoid the headache cure head aches, only consider the governance of the local environment of the lake and neglect the sewage treatment of the whole basin, or simply grasp the point source pollution control and ignore the role of surface pollution. Therefore, improving the efficiency of information management, developing a complete set of lakes GIS management system, can be able to fully open to monitor the lake ecosystem and management of all kinds of pollution sources intuitively [6]. Based on the control of pollution discharge standards, implement the scientific division of reserves, according to the characteristics of the region to perform a targeted and motivated management of the pollution emissions, so as to effectively grasps the ecological environment of the lake on the whole, control phosphorus and nitrogen content of the lake on the total amount, curb the development of lake eutrophication, recovery the clear water to the two lakes.

4. The Design and Implementation of System

The overall task of system development is realize the division of lake grading reserve, on the basis of the pollution source information management function which is developed based on ArcGIS, so as to achieve a targeted and motivated macro control.

4.1. *The overall structure of system*

The system selects the structure which is based on C/S to monitor the management system and use SERVER ArcSDE with SQL database technology work together to access spatial data.

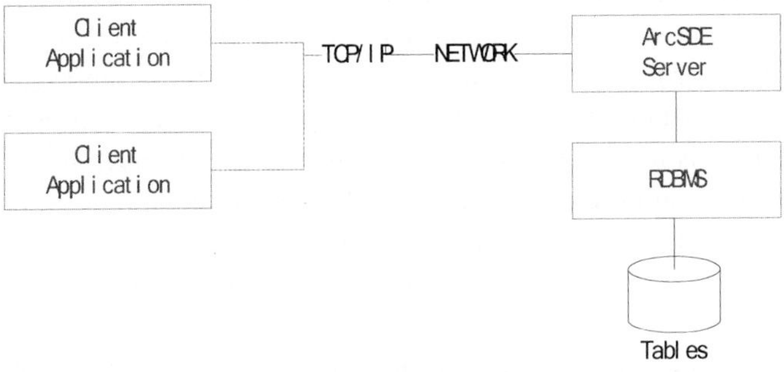

Fig. .1 Client/Server model.

All of the data is stored in RDBMS, Server is responsible for managing and restoring data, Client processing data including Display, Editing, and Analysis.

4.2. *The implementation of function*

System achieves data acquisition, system management, Spatial data query and analysis, layer management, thematic map generation and other functions (such as export images, etc.); alarm, graphics, statistics, information release function.

1) System management module. System management module mainly includes user registration and management, user login, database login, log system log, system help documentation and other functions. Through the" user registration and management "module can add, delete, modify the system and the database users respectively, can also modify the users' password and attribution and set permissions, which makes the reliability and security of system more secure [7].

2) Universal GIS browsing function module. General purpose GIS browse function module is the common function of map operation and data management in GIS. Map operation including: map very scaling (zoom in, zoom out, regional enlarge), map roam, global display, display the map screen width, indicating map coordinates, layer control, eagle eye tracking, thematic mapping. Data management refers to edit and update the attribute data to ensure the accuracy and reliability of the data.

3) Information query service. On the basis of the history of pollution source and real-time information data that have been sorted of the two lakes, construct the information query system of the two lakes' information database and various sources of pollution, to make it convenient for management departments at all levels to perform faster retrieval and query about the various information that is needed. Query includes two parts: historical information and real-time information, the former contains the Ecological and environmental information of the two lakes, the latter contains some information that change over time, such as water, soil, etc.. The information query system which is based on GIS can retrieve and query rainfall information, hydrological information, silt concentration information, water level information, basic information of flood forecast station, various results of statistics and analysis and so on, the user through the client of the above information for a single or conditional combination simulation query, the user will perform a single or conditional combination of the above information to conduct a stimulative query, and the query result is returned in the form of a chart or table [8].

4) Spatial query and visualization analysis module of lake basin. Powerful spatial analysis and query function is on of the basic characteristics of GlS system which is different from other information systems. For spatial information query, the system provides a variety of ways: you can be select (point selection, polygon selection, rectangular box selection, etc.) the object you want to query directly on the map; or conduct the SQL queries based on attribute information (such as "the information of a polluting enterprises in a river basin"); Spatial visualization analysis provides dis-

tance measurement, buffer analysis, overlay analysis and other advanced analysis functions and the foundation for the user's decision.

5) Print output module. The system provides a graphical way of printing, can output the professional layout map that contains scale, compass, mark and other information, to make it convenient for user to perform a printing operation. For spatial information query, the system provides a variety of ways.

6) Classification and decision module of lake grading reserve. On the basic function of the basic GIS system, based on simulation of various kinds of point source pollution, non-point source pollution, pollutants and the situation of the pollutants spread in lake basin, to give the system the different threshold, establish the partition model of lakes grading protection zone, can divide the reserve range of different levels of protected areas of lakes scientifically, thus different sources in protected areas of different levels imposed targeted macro control, which can prevent eutrophication trend in a timely and effective way and protect the lake ecosystem [3].

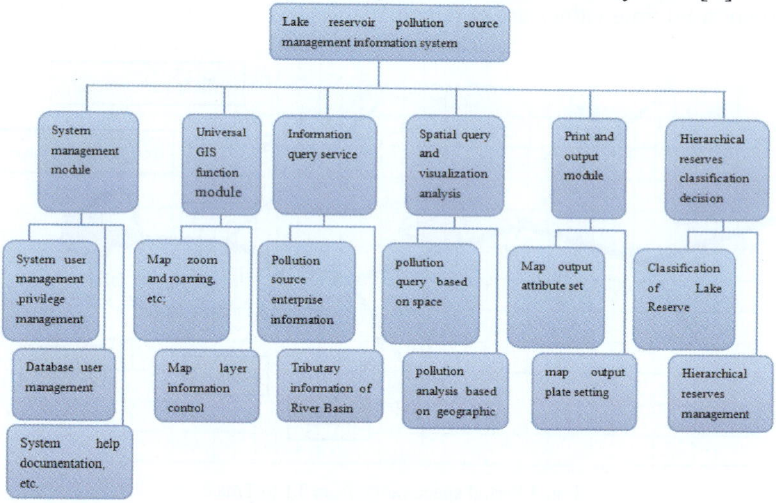

Fig. 2. The HIPO diagram of the system.

5. The Division of Lake Reservoir Classification Reserves

First, we should analyze the historical data of the lake, combined with field measurements, within the application of distributed hydrological calculation for each river basin average monthly volume and flow [9]. Then after the pollutants discharged into the river, according to the average moving flow rate of the river, we can calculate the propagation time that when it arrivals at inlet or outlet section of the reservoir, according to the time limit of propagation, within the space interpolation method to generate the graded water source protection area.

5.1. *Calculate the propagation time of pollutant*

As shown in the following figure, determine a certain storage section, from the section S0 of entering reservoir , according to the flow velocity of the river, to the upstream direction, calculate the cumulative time T. Set the fixed time interval dT, the time value whose cumulative time reach its multiple is defined as Ti[10]. When encounter river branch, first heading toward the tributary to calculate the cumulative time, computing tributaries, after the the calculation of tributary return to the bifurcation section S1, continue the calculation of the cumulative time heading toward the mainstream direction. Repeat the above process, and ultimately meet the maximum response time [11].

5.2. *The generation of contour map*

Covering the Space range of the cumulative time points from T1 to Tmax, portion as shown in the grid, using the cumulative time point data, based on kriging difference method, conducting spatial difference calculation, other grid point total time values are obtained.

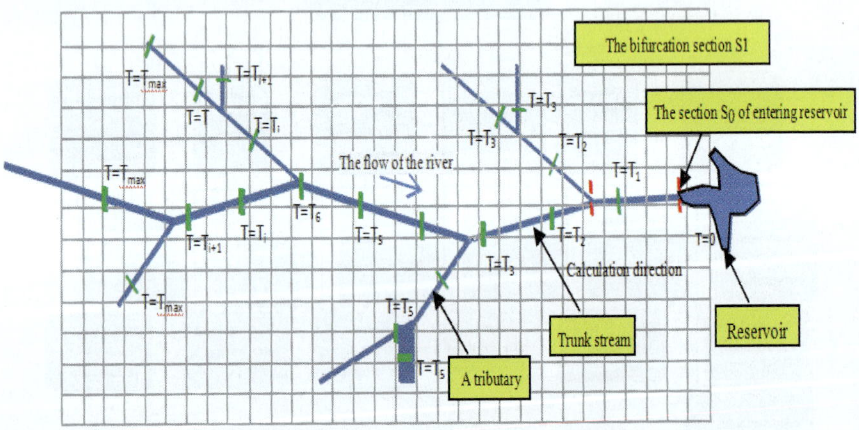

Fig. 3 Partial space range from T1 to Tmax .

5.3. *Division of protected areas*

The difference results are graded, according to the classification threshold (according to the response time to determine), divided into different levels of reserves.

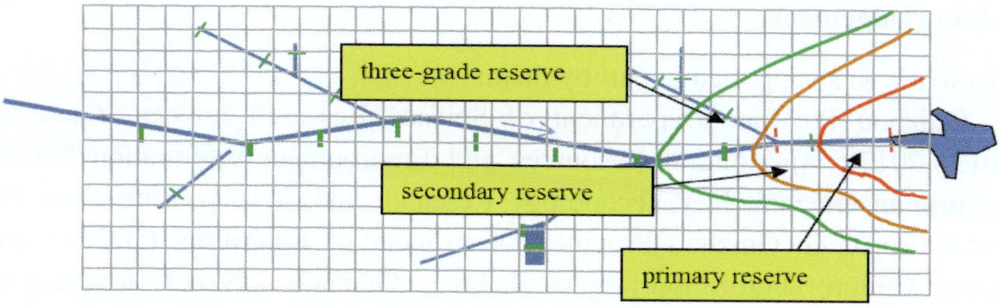

Fig. 4. The division of protected areas.

6. Conclusion

This article puts forward a new method of developing the geographic information system that can manage and control the lake Based on the grading reserve, researches and reconstructs the object model of grading reserve, and using ArcGIS, Sqlserver and other tools to do a detailed analysis and design of the geographic information system of lakes, using Hongfeng Lake and Baihua Lake as the background, constructs the grading reserve geographic information system under the platform of GIS. The use of geographic information systems management of grading reserve will certainly have a positive effect of the control of eutrophication of lakes and the protection of the natural ecology of the lake.

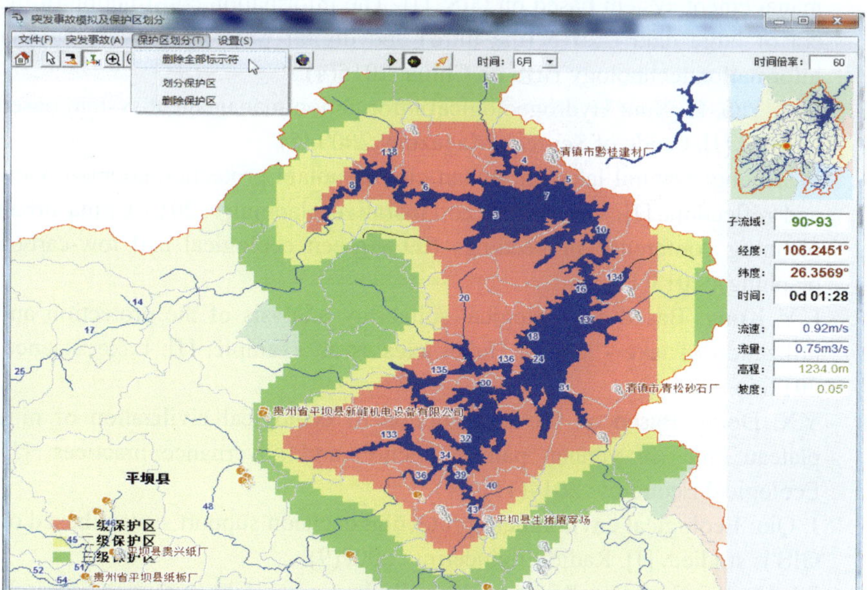

Fig. 5. A demo interface of the system.

Acknowledgements

This work was supported in part by them:

Guizhou Province Department of Science Research Project, (Qian NY [2013] 2013); Agricultural Resources and Environment Information Service Platform in Guizhou Key Technology Research and Development; Base Research Project in Guizhou Province Department of Education, (JD2014226); Tunbao Culture Digitization project planning; Guizhou university teaching reform key projects, (JG2013021); To Pressure and Successful Experience of the Computer Practice Teaching Mode to Stimulate Interest in Guizhou Province Department of Science Research Project, (Qian NY [2013] 3078); Agricultural Resources and Environment Information Service Platform in Guizhou Key Technology Research and Development; Base Research Project in Guizhou Province Department of Education, (JD2014226).

We give our thanks to them!

References

1. B. Li: Wuhan east lake water environment information management system based on GI8S research and development. [J]. Huazhong University of Science and Technology: 2007(6).
2. Y.D. Ba: The Yangtze river basin water resources protection information management system based on GIS. [J]. The information construction of the era of large data - 2015 (the third) and digital water conservancy water information technology BBS on China: 2015(4).
3. W.T. Yao, H. Xun: Hydrogeological information management system based on GIS. [J]. Guide of Sci-tech Magazine: 2009 (5).
4. W. Huang: Natural lakes in wuhan city mountain protection course review and reflection. [J]. Urban age, collaborative planning - 2013 China urban planning conference proceedings (09 - green ecological and low-carbon planning): 2013(11).
5. L.Y. Kong: Based on ecological suitability analysis of the protection and utilization of lake - lake in Hubei axes as an example. [J]. Lake Science: 2012(1).
6. Y.X. Dong: Based on the construction of ecological civilization of nine plateau lakes in yunnan path protection and governance practices. [J]. Ecological Economy: 2014(11).
7. J. Qin: Ecological environment protection decision support system based on GIS is studied. [J]. Radio Engineering: 2007(9).
8. P.J. Huang: Based on RS and GIS in Wuhan urban lake evolution research. [J]. Ecology and Environmental Sciences: 2012(9).

9. Page B, Wohlgemuth V. Advances in Environmental Informatics: Integration of Discrete Event Simulation Methodology With Ecological Material Flow Analysis for Modelling Eco-efficient Systems. Procedia Environmental Sciences. 2010.
10. Editorial: Environmental informatics. Environmental Modelling & Software. 2006.
11. T. Ormsby, E Napoleon, R Burke, C sGroessl, L Feaster: Getting to know ArcGIS desktop: basics of ArcView, ArcEditor, and ArcInfo.. 2008

Chapter 4

Technology for Road Traffic

Multi-depot heterogeneous open incident vehicle routing problem and its solving process

Y.L. Tang[1†]

Dongguan Jinan University Institute, P. R. China
Jinan University Shaoguan Institute, P. R.China
Emails: †tangyalian11@163.com,751224643@qq.com

Aiming at open vehicle routing problem (OVRP) with many extended features is widely used in actual life, multi-depot heterogeneous open incident vehicle routing problem with soft time windows (MDHIVRPSTW) mathematical model is established. An improved ant colony optimization (IACO) is proposed for solving this model. The proposed IACO has been tested on a new 32-customer instance. The experimental results show that IACO is superior to other three algorithms, thus the proposed method is effective and feasible, and the proposed model is better than VRP model.

Keywords: Multi-depot; Incident vehicle routing problem; Improved ant colony optimization; Genetic algorithm.

1. Introduction

The vehicle routing problem (VRP) has attracted high attention since it was proposed for the first time by Dantzig and Ramser in 1959 [1]. In practice, many logistic companies hired vehicles, or when distribution is entrusted to a third party, or vehicles are not required to return to the depot. This situation we called open vehicle routing problem (OVRP), and there is another situation in real life: Different customers need various parts of goods, these goods is part of the product, due to the differences in the properties, characteristics and applications, but there is a link among them, in order to ensure the demand is not affected and will be delivered by a logistic company, and these goods can be compatible, this mode is better for later operation and management. We call this situation incident vehicle routing problem (IVRP). And multi-depot open vehicle routing problem (MDHOIVRP) is an extension of OVRP, it also belongs to NP-h. With the rapid development of logistics, its application has penetrated into kinds of industries, such as transportation, communications, industrial management, electric power and so on. VRP with multiple expansion characteristics are proposed [2~6] in recent years, in this paper, an improved ant colony optimization is proposed for solving MDHOIVRP.

---

[1] *Ya-Lian Tang*

In recent years, many scholars have carried on extensive research to OVRP by heuristic algorithms and evolutionary algorithms. A.D. López-Sánchez et al. [2] proposed a competitive multi-start algorithm for a balanced real-world open vehicle routing problem, and validated the effectiveness of this algorithm. Feiyue Li et al. [3] gave a review for OVRP and the solving algorithm, developed a variant of record-to-record travel algorithm for the standard vehicle routing problem to handle open routes, and report computational results on test problems. P.P. Repoussis et al. [4] presented a hybrid evolution strategy (ES) for solving the open vehicle routing problem, it manipulates a population of μ individuals using a $(\mu+\lambda)$-ES, at each generation, a new intermediate population of λ offspring is produced via mutation, using arcs extracted from parent individuals. The selection and combination of arcs is dictated by a vector of strategy parameters. A multi-parent recombination operator enables the self-adaptation of the mutation rates based on the frequency of appearance of each arc and the diversity of the population. Emmanouil E et al. [5] proposed an innovative local search metaheuristic which examines wide solution neighborhoods, to explore these wide neighborhoods within reasonable computational effort, local search moves are statically encoded into static move descriptor (SMD) entities.

In this paper, we find the solution for MDHOIVRP based on the vehicle cost, maximum route duration constraint, and vehicle capacity constraints etc. The proposed IACO for MDHOIVRP comprised three phases: constructing initial solution, global search and local search. The remainder of the paper was organized as follows. Section 2 stated the problem and mathematical model was established. Section 3 introduced the proposed IACO. The experimental results and analysis were reported in Section 4, followed by a summary and some concluding remarks in Section 5.

2. Problem Formulation

In this section, we formally state the MDHOIVRP, the ultimate goal is to decrease the total distance and the number of vehicles. Customer demands are deterministic, the locations of customers and depots are known, different depots have different kinds of vehicles, different vehicle has different transportation cost, use-cost and load, and the maximum range constraint is known. The notations and mathematical formulation are as follows:

Sets:

I : Set of all customers

H : Set of all depots

V : Set of all types of vehicles

Parameters:

N : Number of all customers

d_{ij} : Distance between point i and j , $i, j \in I \cup H$

g_θ^i : Quality of goods θ for customer i

λ_θ^i : Number of species of goods θ for customer i

$\lambda_1^i g_1^i = \lambda_2^i g_2^i = ... = \lambda_\theta^i g_\theta^i$: Demand relationship for customer i

m_r^h : Number of routes about the depot h , $r = 12,..., m_r^h$

M_r : Total routes to complete all tasks, $r = 12,..., m_r$

k_h : Number of used vehicle for depot h , $h = 1,2,..., H$, $k_h \le K_h$

K_h : Number of vehicles available for depot h , $h \in H$

k_v^h : The vehicle type for depot h , $h \in H$, $v \in V$

n_v^h : The number of k_v^h which used in delivery, $h \in H$, $v \in V$

C_r^h : Number of customers served on the route r , $C_r^h \ge 1$, $h \in H$

n_r^h : The total number of customers on the route r

$Q(k_v^h)$:The capacity of vehicle k_v^h , $h \in H$, $v \in V$

c_{ij} : Unit fuel costs

$c(k_v^h)$: Use-cost

$D_{\max}$: Maximum mileage constraints

Decision variables:

$x_{ijk_v^h}$: 1, if vehicle k_v^h finishes the task at customer i , then travels to j ; 0, otherwise.

$y_{ik_v^h}$: 1, if k_v^h offer service for i ; 0, otherwise.

Mathematical model:

$$\min z = \sum_{i \in I \cup H} \sum_{j \in I \cup H} \sum_{v \in V} \sum_{h \in H} c_{ij} d_{ij} x_{ijk_v^h} + \sum_{v \in V} \sum_{h \in H} c(k_v^h) \cdot n_v^h \tag{1}$$

$$\sum_{v \in V} \sum_{h \in H} \sum_{j \in I \cup H} x_{ijk_v^h} = 1 \quad i \in I \tag{2}$$

$$\sum_{i \in I} g_i y_{ik_v^h} = \sum_{i \in I} \sum_{\theta = 1}^{\theta} g_\theta^i \le Q(k_v^h) \tag{3}$$

$$\sum_{i \in I \cup H} \sum_{j \in I \cup H} d_{ij} x_{ijk_v^h} \le D_{\max} \tag{4}$$

$$\sum_{i \in I} \sum_{j \in H} x_{ijk_v^h} \le 1 \quad h \in H , v \in V \tag{5}$$

$$\sum_{h=1}^{H} \sum_{r=1}^{m_r^h} C_r^h = N \tag{6}$$

$$\sum_{h=1}^{H}\sum_{v=1}^{V} n_v^h \le K_h \tag{7}$$

$$M_r = \sum_{h=1}^{H} K_h = \sum_{r}^{m_r^h}\sum_{h}^{H} k_h \tag{8}$$

$$k_h \le K_h \tag{9}$$

Eq. (1) is the objective function, the goal is to minimize the total cost, including traveling cost and vehicle use-cost. Eq. (2) requires that each customer is assigned to a single route. Eq. (3) is the capacity constraint set for vehicles. Eq. (4) is the maximum route duration constraints. Eq. (5) assures that each customer can be served only once. Eq. (6) assures the number of customers on all routes is equal to the total number of customers. Eq. (7) requires the sum of all types vehicle is equal to the number of vehicles available for a depot. Eq. (8) assures the total number of routes is equal to the total number of used vehicles. Eq. (9) assures the number of used vehicle must less than the number of vehicles available for a depot.

3. Improved Ant Colony Optimization

3.1 *Initialization*

Scanning algorithm (SA) is proposed by Gillett and Miller in 1994. The idea of the algorithm: taking the distribution center as an origin and grouping customers by rotating the ray, until it can't satisfy the constaint conditions, and then starting a new scanning, until all the customers are scanned on the plane.

3.2 *Ant colony optimization*

3.2.1. *State transition rules*

Each ant, denotes as i , is assigned to a depot, the decision-making about combining customers is based on a probabilistic rule which taking into account of both the visibility and the pheromone information on an edge. The ant uses the formula (10) to select the next customer j . Choosing probability between i and j on the route is denoted by P_{ij}^k . Pheromone concentration on the edge (i, j) is denoted by τ_{ij} . The visibility on the edge (i, j) is denoted by η_{ij} . Pheromone trail is denoted by α and visibility value is denoted by β . The feasible customers' set is denoted by $allowed_k$.

$$P_{ij}^{k}=\begin{cases}\dfrac{\tau_{ij}^{\alpha}(t)\eta_{ij}^{\beta}(t)}{\sum_{s\in allowed_k}\tau_{is}^{\alpha}(t)\eta_{is}^{\beta}(t)}, j\in allowed_k \\ 0 \quad otherwise\end{cases} \tag{10}$$

Increasing the probability of random selection appropriately by combining deterministic selection and random selection in order to search the solution space more completely, which can overcome the defects of ACO. Eq. (11) is the rule. And q is a random number among [0, 1], q_0 is a parameter, and $q_0 \in [0,1]$. Ant chooses the next customer as shown in the Eq. (11), or else chooses the edge as Eq. (10).

$$s=\begin{cases}\arg\max_{s\in allowed_k}\left[\tau_{is}^{\alpha}(t)\cdot\eta_{is}^{\beta}(t)\right] if (q\le q_0) \\ S, search \quad probility \quad according \quad to \quad P_{ij}^{k}(t)\end{cases} \tag{11}$$

3.2.2. *Updating pheromone information*

The pheromone increments are assigned to each visited edge. The pheromone updating equations are as Eq. (12), Eq. (13) and Eq. (14). The sum of pheromone increments on the edge (i, j) is denoted as $\Delta\tau_{ij}$. The pheromone increment on edge (i, j) on the k vehicle of the j depot is denoted as $\Delta\tau_{ij}^{kj}$. The parameter that controls the speed of evaporation is denoted as ρ. It can improve the global search ability by decrease ρ but lower the convergence speed, so it's necessary to change ρ adaptively by Eq. (15), and $\rho(t_0)=1$.

$$\tau_{ij}(t+n)=\rho\cdot\tau_{ij}(t)+(1-\rho)\cdot\Delta\tau_{ij} \tag{12}$$

$$\Delta\tau_{ij}=\sum_{j\in J}\sum_{k\in k}\Delta\tau_{ij}^{kj} \tag{13}$$

$$\Delta\tau_{ij}^{kj}(t)=\begin{cases}Q/d_{ijk}, \text{if edge}(i,j) \text{ is serviced by } k \\ 0, otherwise\end{cases} \tag{14}$$

$$\rho(t)=\begin{cases}0.95\rho(t-1) & if\, 0.95\rho(t-1)\ge\rho_{\min} \\ \rho_{\min} & otherwise\end{cases} \tag{15}$$

3.3. *Adaptive crossover probability and mutation probability*

Adjusting the crossover probability and mutation probability adaptively are necessary, as shown in Eq. (16) and Eq. (17), $p_c(t)$ and $p_m(t)$ are measured by average fitness (f_{ave}), maximum fitness ($f_{\max}$) and minimum fitness ($f_{\min}$). The proximity of $f_{\max}$ and $f_{\min}$ reflects the concentration degree of the population, the closer, the more likely to fall into local optimal solution, and $0.5 < a < 1$, p_c and p_m hold the initial value.

$$p_c(t) = \begin{cases} p_c \cdot \dfrac{1}{1 - \dfrac{f_{\min}}{f_{\max}}}, when \dfrac{f_{ave}}{f_{\max}} > a, \dfrac{f_{\min}}{f_{\max}} > b \\ p_c, otherwise \end{cases} \tag{16}$$

$$p_m(t) = \begin{cases} p_m \cdot \dfrac{1}{1 - \dfrac{f_{\min}}{f_{\max}}}, when \dfrac{f_{ave}}{f_{\max}} > a, \dfrac{f_{\min}}{f_{\max}} > b \\ p_m, otherwise \end{cases} \tag{17}$$

Algorithm process of IACO

Step1: Grouping customers.

Step2: Design the GA parameters.

Step3: Generating the optimal solution if satisfy the terminate conditions, otherwise, turn to Step1.

Step4: Generating the initial pheromone distribution according to the optimal solution, and design the ACO parameters, put the ants at all nodes.

Step5: Calculate the probability of the ant moves to the next node, each ant moves to the next node according to choice probability, adjusting α and β, update route pheromone.

Step6: Iterating through all the points and increasing the pheromone for the optimal ant circle.

Step7: If satisfy the termination conditions, reaching the maximum iteration number or obtaining the same result, end of the run, output the best solution; otherwise, turn Step5.

4. Experimental Results and Analysis

In this section, a complex MDHOIVRP mathematical model has been set up. We performed some experiments on Microsoft Windows 7 operating systems, AMD A6-3420M 4 core CPU Processor, 4G memory, 500G Hard disk, using Matlab R2010b, in order evaluate the performance of the proposed algorithm. And then the performance of the algorithms was tested on 32-customer which are generated randomly.

Customers' information and depots' information are in Table 1 and Table 2 respectively. These customers mainly need two categories of goods (I and II) or prefer a certain type, and they are same substantially, but there are difference, goods I is offered by company I, goods II is offered by company II, and $g_1^i = g_2^i = 1$. In this paper, it assumes that the maximum mileage constraint is 200 mileage units.

IACO, GA and ACO are applied. The parameters setting of IACO are as follows: $Pop\_size = 30$, $Gen = 300$, $q_0 = 0.88$, $Q = 100$, $\alpha = 2$, $\beta = 2$, $\rho_{\min} = 0.6$, $p_c = 0.4$, $p_m = 0.004$, $p_{in} = 0.08$. The parameters setting of ACO are as follows: $Pop\_size = 30$, $Gen = 300$, $q_0 = 0.88$, $Q = 100$, $\alpha = 2$, $\beta = 2$, $\rho_{\min} = 0.6$. The parameters setting of GA are as follows: $p_c = 0.4$, $p_m = 0.004$, $p_{in} = 0.08$.

Table 1. Customers' information.

Customer	Location	Demand	Customer	Location	Demand
1	(30,92)	$\lambda_1^1 = 2, \lambda_2^1 = 1$	17	(56,49)	$\lambda_1^{17} = 1, \lambda_2^{17} = 2$
2	(72,53)	$\lambda_1^2 = \lambda_2^2 = 1$	18	(40,60)	$\lambda_2^{18} = 2$
3	(22,56)	$\lambda_1^3 = 3$	19	(60,60)	$\lambda_1^{19} = 2, \lambda_2^{19} = 1$
4	(80,27)	$\lambda_2^4 = 1$	20	(35,24)	$\lambda_1^{20} = 3$
5	(80,37)	$\lambda_1^5 = 1, \lambda_2^5 = 2$	21	(72,59)	$\lambda_1^{21} = 2, \lambda_2^{21} = 1$
6	(14,78)	$\lambda_2^6 = 2$	22	(29,47)	$\lambda_2^{22} = 2$
7	(86,47)	$\lambda_1^7 = 2, \lambda_2^7 = 1$	23	(48,62)	$\lambda_1^{23} = 2, \lambda_2^{23} = 1$
8	(21,95)	$\lambda_2^8 = 2$	24	(58,62)	$\lambda_1^{24} = 2, \lambda_2^{24} = 2$
9	(89,67)	$\lambda_1^9 = 2, \lambda_2^9 = 1$	25	(45,19)	$\lambda_1^{25} = 3$
10	(18,92)	$\lambda_2^{10} = 2$	26	(66,56)	$\lambda_1^{26} = 2, \lambda_2^{26} = 3$
11	(50,3)	$\lambda_1^{11} = 3$	27	(50,29)	$\lambda_1^{27} = 2, \lambda_2^{27} = 1$
12	(44,55)	$\lambda_1^{12} = 2, \lambda_2^{12} = 1$	28	(30,60)	$\lambda_2^{28} = 3$
13	(66,10)	$\lambda_2^{13} = 2$	29	(50,39)	$\lambda_1^{29} = \lambda_2^{29} = 2$
14	(70,70)	$\lambda_1^{14} = 2, \lambda_2^{14} = 1$	30	(70,49)	$\lambda_1^{30} = 2, \lambda_2^{30} = 3$
15	(90,17)	$\lambda_1^{15} = 2, \lambda_2^{15} = 2$	31	(11,11)	$\lambda_1^{31} = 1, \lambda_2^{31} = 2$
16	(34,88)	$\lambda_1^{16} = 3$	32	(1,50)	$\lambda_1^{32} = 3$

Table 2. Depots' information.

Depot	Location	Types and number of vehicles	Load	Unit cost	Use-cost
A	(80,10)	A1:2; A2:2	35;20	0.30;0.25	6;5
B	(50,80)	B1:2; B2:2	35;20	0.30;0.25	6;5
C	(20,40)	C1:2; C2:2	35;20	0.30;0.25	6;5
Depot	Location	Types and number of vehicles	Load	Unit cost	Use-cost
A	(80,10)	A1:2; A2:2	35;20	0.30;0.25	6;5
B	(50,80)	B1:2; B2:2	35;20	0.30;0.25	6;5
C	(20,40)	C1:2; C2:2	35;20	0.30;0.25	6;5
Depot	Location	Types and number of vehicles	Load	Unit cost	Use-cost
A	(80,10)	A1:2; A2:2	35;20	0.30;0.25	6;5
B	(50,80)	B1:2; B2:2	35;20	0.30;0.25	6;5
C	(20,40)	C1:2; C2:2	35;20	0.30;0.25	6;5

The best route networks of 32-customer for MDHOIVRP and MDHOVRP were as shown in Fig. 1 and Fig. 2 respectively, due to the goods were divided into two categories, so there were two distribution networks, and results contrast of two cases as shown in Table 3. We can get the conclusion: the proposed model has superiority, which can save mileage and the number of used-vehicle. Results for these three algorithms tested on MDHOIVRP were as Table 4 shown. From the Table 4, we can get the conclusion that IACO can get satisfied results by comparing with the other two algorithms. IACO is better than ACO; ACO is better than GA. This result attributes to the introduction of genetic operators, and changing crossover probability and mutation probability adaptively can improve the global search ability of the algorithm.

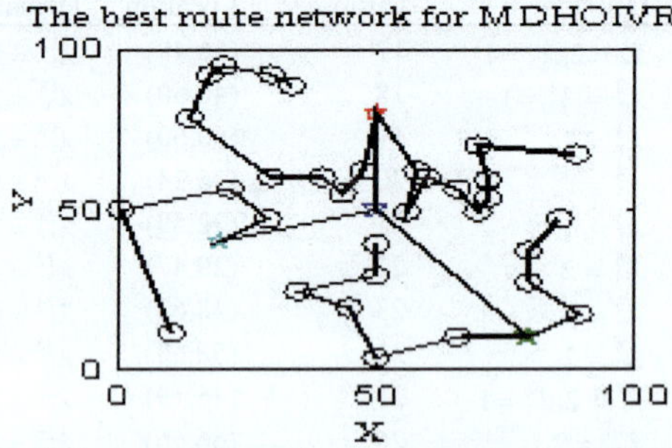

Fig. 1. The best route network of 32-customer (MDHIVRPSTW).

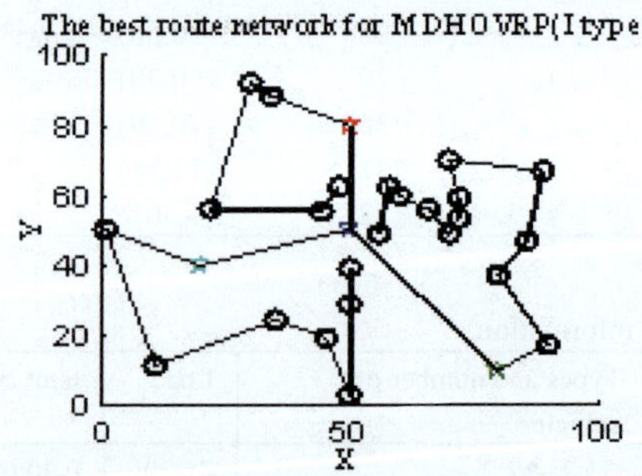

(a)

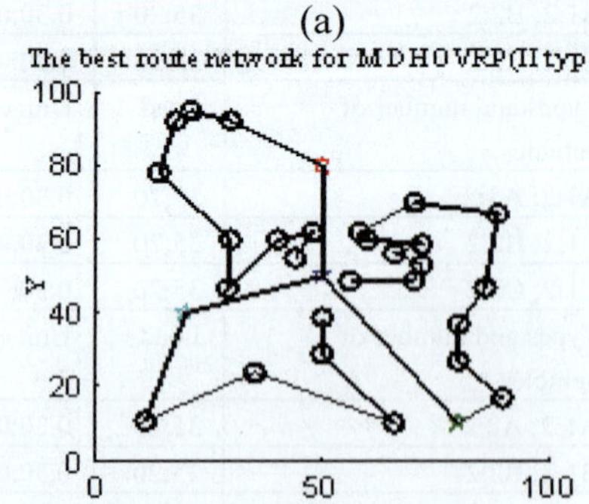

(b)

Fig. 2. The best route network of 32-customer (MDHVRPSTW).

Table 3. Results contrast for two cases.

Cases	Total mileage	used-vehicles
MDHOIVRP	611.09	5
MDHOVRP	1002.56	6

Table 4. Results contrast of algorithms.

Algorithms	IACO	ACO	GA
Convergence time/s	105	123	137
Mileage	611.09	650.64	672.72
Cost	494.21	508.18	519.39

5. Conclusion

In this paper, an improved algorithm was proposed to solve MDHOIVRP. This algorithm provides a feasible method for these kinds of problem. The model is complex but very important in actual life. Considering IVRP with a variety of characteristics and seeking the solution methods will be the future research direction.

References

1. Gulczynski D, Golden B, Wasil E. The multi-depot split delivery vehicle routing problem: An integer programming-based heuristic, new test problems, and computational results. Computers & Industrial Engineering, 2011, 61(3): 794-804.
2. A.D. López-Sánchez, A.G. Hernández-Díaz , D. Vigo, R. Caballero, J. Molina. A multi-start algorithm for a balanced real-world open vehicle routing problem [J]. European Journal of Operational Research. 238(2014), 104-113.
3. Feiyue Li, Bruce Golden, Edward Wasil. The open vehicle routing problem: Algorithms, large-scale test problems, and computational results [J]. Computers & Operations Research, 34 (2007), 2918-2930.
4. P.P. Repoussis, C.D. Tarantilis, O. Braysy, G. Ioannou. A hybrid evolution strategy for the open vehicle routing problem [J]. Computers & Operations Research, 37 (2010), 443-455.
5. Emmanouil E. Zachariadis, Chris T. Kiranoudis. An open vehicle routing problem metaheuristic for examining wide solution neighborhoods [J]. Computers & Operations Research, 37 (2010), 712-723.

Analysis on cascading failure path searching under multiple failures

C. Liang[1†], T. Liu, H.C. Dai and L. Ma
State Grid Energy Research Institute, Beijing, China
†*Email: liangcai00168@163.com*

This paper proposes a transmission line overload assessment model and optimization control method based on transmission line overload risk. It uses probabilistic power flow (PPF) and cumulant method to analyze the uncertainty of system statuses and line flow, applies the severity function to describe the impact of transmission line load fluctuations and its result. Based on that, as an indicator of power system security, overload risk index is defined as the product of the probability and the severity of transmission line overload. A preventive control model was used to optimize the outputs of the system active and reactive resources, which based on optimal power flow (OPF) and primal dual interior point method. It reduces the system overload risks and increases the system security. By test, calculation on the IEEE-16 test system validates the method proposed above.

Keywords: Overload; Risk; Optimization control; Probabilistic power flow; Cumulant.

1. Introduction

As far as traditional assessment to line overload is concerned, transmission line load-based method leads to conservative result. Because that at certain grid structure, line load is dominated by inserting power of nodes, which is fluctuating randomly. As electricity market develops remarkably in China, the uncertainty of power flow increases, leading to an ambiguity of line overload to an extent.

So, on dealing with the condition where power system is at risk of line overload, a new assessment method is needed to overcome the drawback of the known one. As severity-based indicator quantitatively mentions the factor of risk degree, the risk status of bulk power system is also seen. Overload risk is defined as the result of the probability multiplying the severity of transmission line overload and is considered as an indicator of power system security.

---

[1] Cai Liang

The probability is explained by the stochastic probability distributions of that the line load surpassing threshold. As uncertainty is prevailing in system operation, line load is affected and featured as uncertainty. In order to analysis how, a thorough probe to operational condition is basic, which is complex. We turn to probabilistic power-flow. As a power flow analysis method, probabilistic power-flow see both known and unknown quantity of original power flow calculation as random, suitable for analysis of power flow, network loss, probability distribution, expectation, variance, limitation of node voltage, so as to reach a comprehensive evaluation of operational performance, which is useful for both planning and dispatching crew [1,2].

2. The Risk Assessment to Line Overload

2.1. *The cumulant method*

The model of probabilistic power flow is as follow:

$$S - f(X) = 0 \tag{1}$$

Where S denotes active and reactive power inserting into each node; X denotes state-variable (amplitude and phase angle of voltage) of node.

Ignore high order term of Taylor Series of (1):

$$\Delta X = S_{em\,0}\Delta S \tag{2}$$

Where S_{em0} denotes sensitivity matrix.

The model of line flow is as follow:

$$Z - g(X) = 0 \tag{3}$$

Where Z denotes line flow.

Ignore high order term of Taylor Series of (3):

$$\Delta Z = J_{g0}\Delta X \tag{4}$$

Where J_{g0} denotes sensitivity matrix.

Based on (2) and (4), the relationship between line flow and active and reactive power of node is as follows:

$$\Delta Z = S_{fg0}\Delta S \tag{5}$$

Where $S_{fg0} = J_{g0}S_{en0}$.

Stochastic probability distributions of ΔZ can be obtained by that of ΔS. In order to avoid complex convolution, this text analyze stochastic probability distributions of ΔZ by cumulant method [3-5]. Figure 1 is the process of cumulant method, while some points need to be taken into consideration.

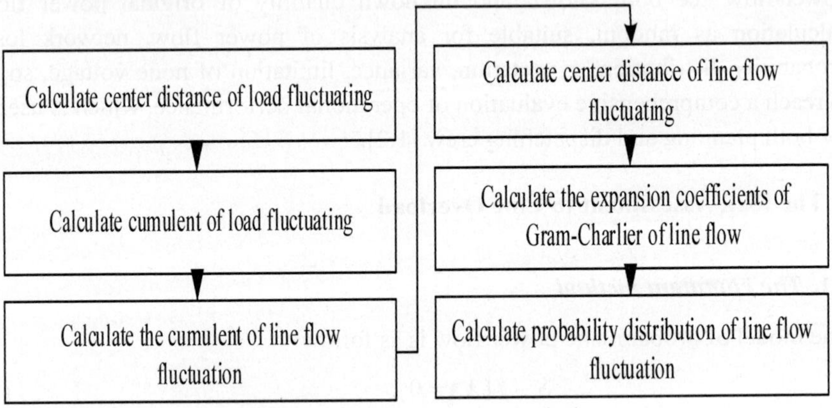

Fig. 1. Flow chart of cumulant method.

(1) Suppose time-varying load as random disturbing variable, which is in the Normal Distribution and independent. Define r-order center distance b_r:

$$b_r = \int_{-\infty}^{+\infty} (x-\mu)^r \, dF(x) \tag{6}$$

Where μ and F denote means and distribution function of the random variable.

(2) The relationship between b_r and its cumulant k_r is as follows:

$$\begin{cases} k_1 = \mu \\ k_2 = b_2 \\ k_3 = b_3 \\ k_4 = b_4 - 3b_2^2 \\ k_5 = b_5 - 10b_2 b_3 \\ k_6 = b_6 - 15b_2 b_4 - 10b_3^2 + 30b_2^3 \\ \ldots \end{cases} \tag{7}$$

$$\begin{cases} b_1 = 0 \\ b_2 = k_2 \\ b_3 = k_3 \\ b_4 = k_4 + 3k_2^2 \\ b_5 = k_5 + 10k_2k_3 \\ b_6 = k_6 + 15b_2b_4 + 10b_3^2 + 15b_2^3 \\ \cdots \end{cases} \tag{8}$$

(3) Calculate the cumulant of line flow fluctuation based on the following two properties of cumulant:

Property 1: The cumulant of sum of independent random variables equals the sum of cumulant of random variables. This holds true for different orders.

Property 2: The r-order cumulant of random variable multiplied by h equals the r-order cumulant of variable multiplied by h^r.

(4) Calculate center distance of line flow fluctuating by recursion formulae below:

(5) Formula of expansion coefficients of Gram-Charlier (c_i) series by b_r :

$$c_1 = 0, c_2 = 0 \ , c_3 = -\frac{b_5}{\sigma^3} \ ,$$

$$c_4 = -\frac{b_5}{\sigma^5} + 10\frac{b^3}{\sigma^3} \ , \quad c_6 = \frac{b_6}{\sigma^6} - 15\frac{b_4}{\sigma^4} + 30 \tag{9}$$

(6) Formula of probability distribution of line flow fluctuation by c_i :

$$p_r(x) = \varphi(x) + \frac{c_1}{1!}\varphi'(x) + \frac{c_2}{2!}\varphi''(x) + \frac{c_3}{3!}\varphi^{(3)}(x) + \cdots \tag{10}$$

Where $\varphi(x)$ denotes probability function of standard normal distribution.

2.2. *The overload severity of transmission line*

The severity depends on the line load. Literature [6] proposes an instant way to define this severity. For each transmission-line, should its active load be 90% of rated, the severity is 0; should its active load be 100% of rated, the severity is 1 and the line-overload severity and its active flow have linear relationship. Should active load of a line be less than 90% of rated, its overload severity is 0.

2.3. *Risk index of overload*

Basing on risk theory and the definition of overload risk above, overload risk is the product of probability of flow-fluctuation (p_r) and severity of fluctuation

(S_e), noted as:

$$r_{isk}(Z_i)=\int_{-\infty}^{+\infty} p_r(Z_i)s_e(Z_i)dZ_i \tag{11}$$

Where Z_i is line flow, $p_r(Z_i)$, $s_e(Z_i)$, $r_{isk}(Z_i)$ are probability and severity of line fluctuation and overload risk of line i.

By rollup of overload risk, overload risk of whole-system can be:

$$r_{all}=\sum r_{isk}(Z_i) \tag{12}$$

Which evaluate security states globally.

3. Preventive Control of Overload

3.1. *Model*

The assessment above provides a way for calculating overload probability of line, which, to some extent, reflect whether the system is secured. At some circumstances, although system works fine, overload risk of whole-system does not match the demand of operator. In order to cleanup hazard, operator can decrease overload risk by preventive measurement. A preventive control mode, based on optimal power flow (OPF) and primal dual interior point method, was used to optimize the outputs of the system active and reactive resources. This reduces the system overload risks and increases the system security.

$$\begin{aligned}
&\min r_{all}(Z)=\sum\int_{-\infty}^{+\infty} p_r(Z_i)s_e(Z_i)dZ_i\\
&s.t. E(P_i)-E(V_i)\sum_{j\in N} E(V_j)(G_{ij}\cos E(\theta_{ij})+B_{ij}\sin E(\theta_{ij}))=0, i,j\in N\\
&E(Q_i)-E(V_i)\sum_{j\in N} E(V_j)(G_{ij}\sin E(\theta_{ij})-B_{ij}\cos E(\theta_{ij}))=0, i,j\in N\\
&\underline{P}_{gi}\le E(P_{gi})\le \overline{P}_{gi}, i\in N_g\\
&\underline{Q}_{gi}\le E(Q_{gi})\le \overline{Q}_{gi}, i\in N_r\\
&\underline{V}_i\le E(V_i)\le \overline{V}_i, i\in N\\
&E(S_i)\le \overline{S}_i, i\in N_l
\end{aligned} \tag{13}$$

Where $r_{all}(Z)$ is the optimal function of overload risk of system; N, N_g, N_r, N_t are number of node, active power, reactive power and line of system; P_{gi}, $\underline{P}_{gi}$, $\overline{P}_{gi}$ are actual value, lower limit and upper limit of active power; Q_{gi}, $\underline{Q}_{gi}$, $\overline{Q}_{gi}$ are actual value, lower limit and upper limit of reactive power; $\underline{V}_i$, $\overline{V}_i$ are lower limit and upper limit of voltage; $\overline{S}_i$ is the rated capacity of line; $E(P_i), E(V_i), E(\theta_{ij}), E(Q_i), E(P_{gi}), E(Q_{gi}), E(S_i)$ are expectation of P_i, V_i, θ_{ij}, Q_i, P_{gi}, Q_{gi} and S_i, respectively.

Optimal power flow is always used for optimization control of system. OPF is an optimal problem subjecting to a set of equation and inequality [7, 8]. The preventive control model of overload risk, as (13) shows, originates from OPF. The innovation lies in that the equation adopts the ones of probabilistic power flow, and that the objective function adopts risk index. The former effectively reflects various operation modes while the latter monitors the risk status, providing smart decision for operational staff.

3.2. *Primal dual interior point method*

Primal dual interior point method is widely used for nonlinearity optimal problem, whose flow chart is shown in Figure 2 Superior to simplified gradient method, quadratic programming method and linear programming method, primal dual interior point method suits for highly nonlinear problem and for problem involving a number of equation and inequality.

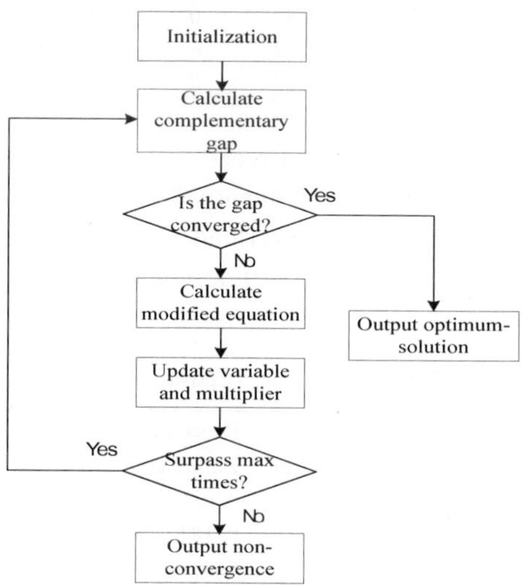

Fig. 2. Chart flow of primal dual interior point method.

4. Test System

Applying line overload risk assessment and risk preventive control on IEEE 14 test system. Suppose different load fluctuate independently, the amplitude is 1%

of initial value. The base power is 100MVA. The upper and lower limit of node voltage is 1.1 and 0.9. The rated capacities of lines are demonstrated on Table 1.

Table 1. Power margin of transmission line.

No.	Margin/p.u.	No.	Margin/p.u.
1	1. 882	11	0. 088
2	0. 906	12	0. 093
3	0. 878	13	0. 213
4	0. 673	14	0. 120
5	0. 498	15	0. 336
6	0. 279	16	0. 065
7	0. 733	17	0. 113
8	0. 336	18	0. 047
9	0. 193	19	0. 019
10	0. 529	20	0. 067

Table 2. Parameter of generators.

Bus	Upper limit/p.u.		Lower limit/p.u.	
	Active	Reactive	Active	reactive
1	3. 4	0. 1	0	- 0. 2
2	1. 4	0. 5	0	- 0. 4
3	1	0. 4	0	0
6	1	0. 24	0	- 0. 06
8	1	0. 24	0	- 0. 06

Table 3 and 4 list the active and reactive power of generators and power flow of lines before and after risk preventive control is applied. It shows that after preventive control is applied, overload risk is decreased to some extent, decreasing from 5.301×10-4 to 4.962×10-4, a decrease by 6.4%, which increase the security of system.

Table 3. Active and reactive power of generators.

Active power/p.u.		Reactive power/p.u.	
Before control	After control	Before control	After control
1. 672	2. 323	0. 088	- 0. 165
0. 473	0. 4	0. 181	0. 435
0. 375	0	0. 4	0. 251
0. 085	0	0. 043	0. 127
0. 069	0	- 0. 009	0. 176

Table 4. Active power of line

Line	Active power		Line	Active power	
	Before control	After control		Before control	After control
1	1. 098	1. 568	11	0. 086	0. 073
2	0. 587	0. 751	12	0. 082	0. 077
3	0. 512	0. 732	13	0. 194	0. 177
4	0. 467	0. 561	14	0. 070	0. 100
5	0. 361	0. 415	15	0. 295	0. 280
6	0. 163	0. 232	16	0. 062	0. 052
7	0. 473	0. 611	17	0. 098	0. 094
8	0. 222	0. 280	18	0. 045	0. 037
9	0. 139	0. 160	19	0. 018	0. 016
10	0. 344	0. 440	20	0. 062	0. 056

5. Summary

Interconnection of power system and development of electricity market changes the way the power system operates. This text focus on line overload problem, proposes a risk assessment and preventive control method. The method adopts probability power flow and cumulant to analyze the probability of line overload. Probability power flow is able to reflect comprehensive operational status of system while cumulant takes place of convolution at solving the probability of line overload. To calculate the severity, the overload severity function Is used. Overload risk is defined as the result of the probability multiplying the severity of transmission line overload and is considered as an indicator of power system security. A preventive control model, based on optimal power flow (OPF) and primal dual interior point method, was used to optimize the outputs of the system active and reactive resources. This reduces the system overload risks and increases the system security. Calculation results of the method in a test system prove the method effective.

References

1. Ding Ming, Li Shenghu, Huang Kai. Probabilistic load flow analysis based on Monte-Carlo simulation [J]. Power System Technology, 2001, 25(11): 10-14 (in Chinese).
2. Wang Chengshan, Zheng Haifeng, Xie Yinghua, et al. Probabilistic power flow containing distributed generation in distribution system [J]. Automation of Electric Power Systems, 2005, 29(24): 39-44 (in Chinese).

3. Wang heaven, Zhong Zhiyong, Xie Zhitang, et al. Ahybrid algorithm using moment and cumulant for power system probabilistic eigenvalue analysis [J]. Proceedings of the CSEE, 2000, 20(5): 37-41 (in Chinese).
4. Wang Chengshan, Yu Xuyang. A method for computing the probability distribution of fault critical clearing time [J]. Proceedings of the CSEE, 2004, 24(1): 6-10 (in Chinese).
5. Feng Yongqing, Zhang Boming, Wu Wenchuan, et al. Fast algorithm of generator maintenance scheduling in electricity market [J]. Automation of Electric Power Systems. 2004. 28(16): 41-44 (in Chinese).
6. Ming Ni, McGalley J D, Vittal V, et al. Online risk-based security assessment [J]. IEEE Transactions on Power Systems, 2003,18 (1): 258-265.
7. Mei Shengwei, Weng Xiaofeng, Xue Ancheng, et al. Blackout model based on OPF and its self-organized criticality [J]. Automation of Electric Power Systems, 2006, 30(13): 1-5 (in Chinese).
8. WU Yan-Juan, LI Lin-chuan. Transient stability constrained optimal power flow based on sensitivity of transient energy margin [J]. Power System Technology, 2005, 29(15): 28-33.

A new nonlinear car-following model for mountain highways based on the traffic safety analysis

H. W. Wang[1], H. Y. Wen[2†], L.Z.Niu[3] and X. H. Guo[4]
School of Civil Engineering and Transportation,
South China University of Technology, Guangzhou 510640, China
[†]*Email: whw2046@126.com*

R. H. Zhang[5] and Y. L. Pei[6]
Xinjiang Academy of Communication Science, Urumqi 830000, China

Mountain highways are restricted by the comprehensive effect of road alignment, longitudinal slope and complex road conditions, which results in a larger accident rate. This article firstly studies the road alignment of mountain highways and attributes of random interference factors, and analyzes the laws of traffic flows in mountain highways. Then by incorporating the impacts of environment factors of mountain highways into the traditional car following model, it puts forward a nonlinear car-following model on the basis of nonlinear dynamics principles. The results conducted on VISSIM proved the correctness and practicability of the newly-established model. This study has provided theoretical support for traffic safety in mountain highways.

Keywords: Mountain highways; Traffic safety; Non-linear dynamics; Car-following model; VISSIM simulation.

1. Introduction

Mountain highways serve as a critical part of transportation system. Compared with the ordinary roads in flat areas, mountain highways are restricted by the comprehensive effect of road alignment, longitudinal slope and complex road conditions, which results in a larger accident rate.

At present, the studies of traffic safety in mountain highway mainly are focused on traffic safety analysis and evaluation, while few refer to the law of traffic flow under complex mountainous environment. Hamdar (2012) [1] summarized a list of human factors and situational environment factors which may affect car-following behavior. In a recent review, Treiber [2] and Kesting (2013) described seven human factors which could affect car-following behavior, and applied them to a car-following model using some hypothetical cases. Antoine [3] and Sylvain, etc., intended to define a continuous car-following

---

[1] Hai-Wei Wang
[2] Hui-Ying Wen
[3] Ling-ZhiNiu
[4] Xiao-HuiGuo
[5] Rong-Hui Zhang
[6] Yu-Long Pei

model exclusively based on the time gap. Nadir [4] presented a piecewise linear car-following model to characterize the stability of the car-dynamics and to determine the stationary regimes. Mohammad [5] and Zheng analyses the benefits and limitations of various model from both the engineering and human behaviors points of view and highlights future research needs in car-following studies. Mohamed [6] and Huang, etc., explored a Bayesian hierarchical model for developing safety performance function for a mountainous freeway. Lena [7] estimated the relationship between accident frequency and the traffic flow empirically treating the hourly traffic flow in two different ways, as consisting of homogeneous vehicles and as consisting of cars and lorries.

Above all, most of the researches focused on a single factor of mountain highway traffic safety, few efforts have been made to apply nonlinear car-following model in the study of it. This paper investigates main random interference factors to mountain highway traffic and hopes to include those factors into nonlinear car-following model, especially road alignment and road environment.

2. Random Interference Factors among Traffic Flow in Mountain Highway

Just as the traffic flow on ordinary roads, the characteristics of traffic flow in mountain highways are also described mainly through the three aspects of speed, volume and density. However, significant distinctions exist between these two kinds of roads. Driving speed on mountain highways varies more drastically because of constraints of complex landform in mountain areas——driving speed here is subject to driver factors, vehicles dynamic performances, road conditions and weather conditions. Besides, traffic density in mountain highways is more vulnerable to surrounding traffic conditions and mountain environments.

Random interference factors refer to unpredictable disturbance with no regular law [8], which means that different results may be obtained from repeated trials under exactly the same conditions, or that what will happen in the future can never been precisely predicted.

In terms of mountain highways, random interference factors to traffic flow can be perceived as the factors which destroy the steady state of traffic flow. They may stem from sudden changes of human behaviors, vehicle conditions or external environment. Such factors comprise of natural conditions in mountain areas, traffic participants and the attributes of the road network in mountain areas. According to the causes of random interference factors, they can be classified into three types: subjective factors, objective factors and other factors [9-10].

3. Nonlinear Car-Following Model for Mountain Highways with the Analysis of Traffic Safety

3.1. *Car-following model*

Based on dynamics, Car-following theory is to explore the driving status of the latter vehicle following the leading vehicle ahead when running on a single lane with no chance of overtaking. It manages to illustrate the various driving statuses through the mathematical model [11].

In 1959, Gazis and others [12] put forward a general formula based on the study of car-following status. The formula is shown as Eq. (1).

$$\ddot{x}_{n+1}\left(t+T\right)=\alpha\frac{\dot{x}^{m}{}_{n+1}\left(t+T\right)}{\left[x_{n}\left(t\right)-x_{n+1}\left(t\right)\right]^{l}}\left[\dot{x}_{n}\left(t\right)-\dot{x}_{n+1}\left(t\right)\right] \tag{1}$$

Wherein $\ddot{x}_{n+1}\left(t+T\right)$ is the acceleration of vehicle (n+1) after the time delay of T; $\dot{x}_{n}\left(t\right)$ is the speed of vehicle (n) in the moment of t; $\dot{x}_{n+1}\left(t\right)$ is the speed of vehicle (n+1) in the moment t; m,l are constants. If m,l both get the value of 0, Eq.(1) represents linear car-following model; if not, Eq.(1) is a non-linear car-following model.

3.2. *Basic hypothesis*

(1) Each vehicle is considered as a rigid body.
(2) Vehicles are running on a single lane with no chance of overtaking
(3) The interaction among vehicles is one-way effect, which means the follower only responds to stimulus coming from the vehicle in front.
(4) Drivers in the study are rational and abide by traffic rules.
(5) There is a common goal for all vehicles—arriving at the destination as soon as possible.

3.3. *The establishment of basic model*

3.3.1. *Nonlinear car-following model based on driver personalities*

Assuming that there are n types of personalities among the drivers, and they respond differently to the stimulus from the leading car because of different personalities, which result in the diverse values of acceleration. So acceleration coefficient [13] is introduced into the model to signify the weight factor for differences of personalities. The basic model ignoring the effect of personality is shown as Eq. (2).

$$a_0 = \ddot{x}_{n+1}(t+T) = \lambda \frac{v_{n+1}(t+T)}{\left[x_n(t) - x_{n+1}(t)\right]^2} \left[\dot{x}_n(t) - \dot{x}_{n+1}(t)\right] \quad n = 1,2,3,\cdots \quad (2)$$

Thus, nonlinear car-following model is defined as Eq.(3).

$$a_1 = a_0 \cdot \sigma(i) \quad (3)$$

3.3.2. *Nonlinear car-following model based on vehicle dynamics performance*

When driving, drivers will keep an expected safety distance as shown Eq. (4)

$$L_e = L_0 + \alpha v_{n+1}(t) \quad (4)$$

Wherein, L_0 is the safety clearance in static status and in this article, it is given by L_0 =2(m); α is reaction coefficient to the change of headway when speed changes, and it is given by α=1 in the paper; $v_{n+1}(t)$ is the speed of the following vehicle at the moment of *t*. Thus, nonlinear car-following model based on vehicle dynamics performance as shown in Eq. (5)

$$a_2 = a_0 \cdot \tanh\left[\Delta x_{n+1}(t) - L_0 - \alpha v_{n+1}(t)\right] \quad (5)$$

3.3.3. *Nonlinear car-following model based on road alignment of mountain highway*

(1) *The influence of curves on car-following movement*

If the preceding vehicle speeds up when driving on a curve, the driver of the following car is prone to making the visual illusion of shortening the change in clearance, which results in a slow response. If the leading vehicle decelerates suddenly, the driver following perceives a drastic decrease in clearance because of visual illusion, which will lead to a significant change in driving behaviors. In this case, let *r* and r_0 denote the real radius of curvature and the minimum safety radius of curvature when driving at a certain speed on curves. The car-following model can be concluded as Eq. (6), as to depict the car-following movement in circumstances where the leading vehicle speeds up or speeds down on curves.

$$a_{3\text{-}1} = a_0 \cdot \left(1 + k_1 \cdot \frac{r_0}{r}\right) \quad (6)$$

Where a_{3-1} represents the acceleration of the latter vehicle when taking the impact of curves into the consideration; k_1 is a constant; $k_1 < 0$ when the leading vehicle speeds up; if not, $k_1 > 0$.

(2) *The impact of gradient on car-following movement*

Compared to running in flat sections, drivers reveal higher reaction intensity to stimulations when running on grade sections, for fear of a sudden deceleration of the preceding vehicle. Besides, the worry of brake failure adds to the intensity of reaction when driving downhill—in such cases, the driver of the following vehicle also reacts more intensively than in the case of flat road. Presume that a series of vehicles are running on a road with a gradient of *i*, then Eq. (7) can be used to indicate the reaction of the driver of the following vehicle.

$$a_{3-2} = a_0 \cdot (1 + k_2 i) \tag{7}$$

Where a_{3-2} represents the acceleration of the latter vehicle when taking the impact of curves into consideration; k_2 is a constant; and i is given by i =Abs (i), because reaction intensify on gradient section is much higher than on flat sections.

(3) *Nonlinear car-following model base on road alignment*

To integrate Eq.(6) and Eq.(7), there obtain the nonlinear car-following model based on road alignments as shown in Eq. (13), where a_3 represents the acceleration of the latter vehicle when taking impacts of road alignments (curves and gradients) into consideration.

$$a_3 = a_0 \cdot \left(1 + k_1 \cdot \frac{r_0}{r}\right) \cdot (1 + k_2 i) \tag{8}$$

3.3.4. *Nonlinear car-following model based on weather conditions in mountain highways*

In this part, a new parameter ϕ is introduced in to signify the complex rate of weather conditions in mountainous areas. It's obvious that ϕ is relevant to bad weather conditions (such as rain, snow, fog and wind) which will impact the driver's visibility and friction coefficient. Because of their critical roles in driving, the article analyzes ϕ mainly from sight distance and friction coefficient. Presume that driver's sight distance and friction coefficient under ideal weather conditions are given respectively by $s = s_0, \varphi = \varphi_0$.

$$\phi = f(s,\varphi) = \frac{s_0}{s} \cdot \frac{\varphi_0}{\varphi} \tag{9}$$

Thus, car-following model under actual weather conditions can be expressed as Eq. (10)

$$a_4 = a_0 \cdot (1 + k_3 \phi) \tag{10}$$

Where a_4 is the acceleration of the latter vehicle when taking the impact of weather into consideration; k_3 is a constant.

3.4. *Nonlinear car-following model suitable for the analysis of traffic safety in mountain highways*

To integrate the aforementioned models together, here we obtain a general nonlinear car-following model suitable for the analysis of traffic safety in mountain highways.

$$a = a_0 \cdot \left[\sigma(i) \cdot \tanh(\Delta x_{n+1}(t) - L_0 - \alpha v_{n+1}(t)) \cdot \left(1 + k_1 \frac{r_0}{r}\right) \cdot (1 + k_2 \cdot i) \cdot \left(1 + k_3 \cdot \frac{s_0}{s} \cdot \frac{\varphi_0}{\varphi}\right)\right] \tag{11}$$

Where, a is the acceleration of the latter vehicle when taking into consideration the impact of driver personalities/ vehicle dynamics performance, road alignment and weather conditions.

Such parameters as the personality of drivers, sight distance and friction coefficient are hard to depict in VISSIM, here we fix $\sigma(i) = 1, k_3 = 0$, as have been shown in the above, L_0 =2(m), α=1. Thus, λ, k_1, k_2 can be given by the outputs of simulation.

$$\lambda = \frac{\ddot{x}_{n+1}(t+T) \cdot [x_n(t) - x_{n+1}(t)]^2}{v_{n+1}(t+T) \cdot \left[x_n(t) - x_{n+1}(t)\right]} = \frac{a(t+T) \cdot (\Delta x(t))^2}{v_{n+1}(t+T) \cdot [v_n(t) - v_{n+1}(t)]} \tag{12}$$

$$k_1 = \frac{r}{r_0}\left(\frac{a_{3-1}}{a_0} - 1\right) \tag{13}$$

$$k_2 = \frac{1}{i}\left(\frac{a_{3-2}}{a_0} - 1\right) \tag{14}$$

4. Simulation Results

A section of Road G105 (Beijing to Zhuhai) in Guangdong province, which is a typical mountain secondary road, is selected for simulation and verification of the model. GPS experiment has been carried out in this paper to collect information of vehicles running on the road such as positional information, speed and acceleration. Besides, the actual geometric data of this road section is used to structure a mountain road in VISSIM. Typical joint segments of flat and steep road sections are selected to simulate running state of traffic flow. According to statistical traffic volume on the selected section, the average traffic volume in off-peak periods is selected here, and the traffic composition includes cars, trucks (substituted by bus in simulation), coaches and motor bikes (substituted by bus in simulation) with a proportion of 55:20:20:5. The designed speed of the road is 60km/h and traffic flow is 500vel/h.

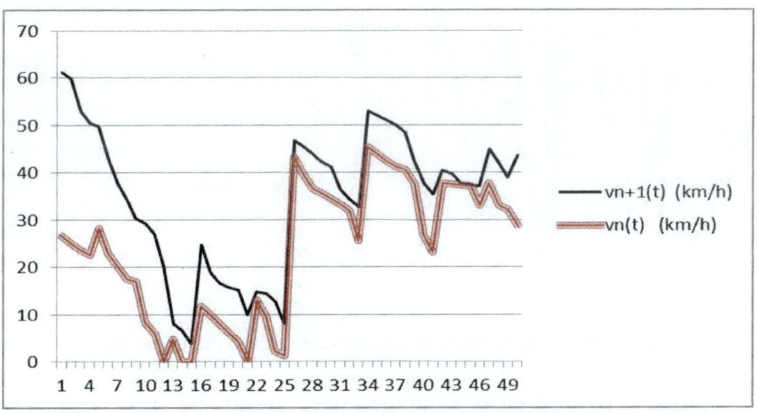

Fig. 1. Comparison of speed between the preceding and following vehicles.

The speed of the preceding vehicles is compared with that of the following vehicles and the calculated acceleration with actual acceleration. Fig. 1 shows a similar change trend in speed of the preceding and following vehicles, which means that the following vehicle changes its speed according to the speed change of the preceding vehicle. And Fig. 2 compares actual accelerations with calculated accelerations, the result of which indicates the acceleration (or deceleration) of the following vehicle decreases with the growth of space headway (which is larger than a certain safety distance). With the same headway, the absolute value of acceleration (the degree of reaction intensity of the following vehicle) increase with the driving speed of the following vehicle.

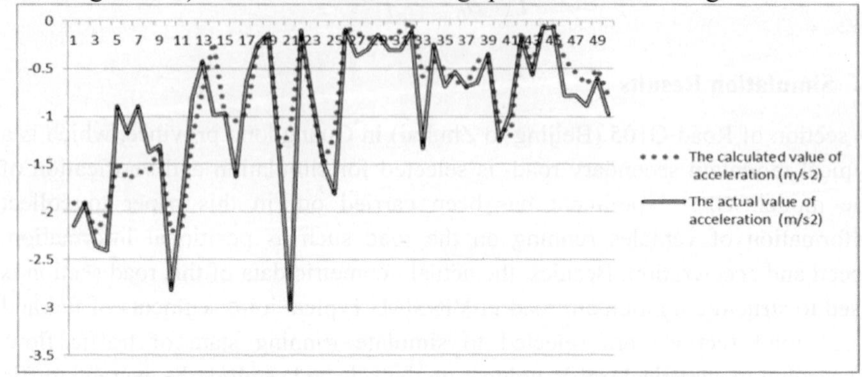

Fig. 2. Comparison between actual accelerations and calculated ones.

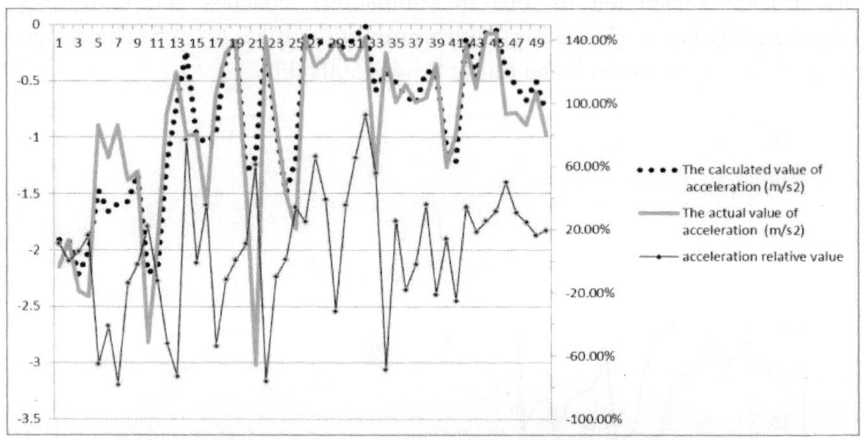

Fig. 3. Difference between actual accelerations and calculated ones.

In normal circumstances, the relative speed of two vehicles with car-following motion is low (it varies within a narrow range of ±10km/h), as a result, the absolute value of acceleration of the following car also shows a small value(varying within±0.5m/s^2). However, if the preceding vehicle brakes or starts suddenly, the relative speed rises and so is the absolute value of acceleration and the acceleration increases with the growth of the relative speed. Fig. 3 indicates that calculated accelerations show no obvious deviation from actual acceleration, except for several groups of data; it demonstrates the consistency of calculated accelerations and actual accelerations and the value of acceleration (or deceleration) varies within the acceptable scope, which validated the correctness and practicability of the established model.

5. Conclusions

Driving in mountain highways will suffer more random interference on account of the special terrain and complex conditions of mountainous areas. As a result, physiological and psychological changes will happen to drivers easily and driving on such sections is becoming more risky. So, the analysis of traffic safety in such areas is very important. This article makes a deep analysis of the complicate conditions of mountain highways and the factors impacting traffic safety. It has established a nonlinear car-following model for mountain highways on the basis of previous theories of car-following model and analyzes the application prospect of the model in traffic safety analysis in the mountain highways. Also, according to the actual road data, several simulations are designed on VISSIM to identify the unknown parameters and attest the validity of the newly established model. The article takes more safety-impacting factors (driver personalities, vehicle dynamic performance, and road geometry and weather conditions) into consideration to perfect the newly-established model for practical application, which provides theoretical and technical reference to the analysis of traffic safety on mountain highways.

Acknowledgments

This work was partially supported by the National Natural Science Foundation of China (Granted No.51378222), Science and technology plan projects of Guangdong Province (Granted No. 2013B010401009) and Chinese Postdoctoral Science Foundation (Granted No. 2013T60904). The first author would like to appreciate Prof. Huiying Wen, for the valuable discussions to improve the quality and the presentation of paper.

References

1. Hamdar, S, 2012. Driver behavior modeling. In: Eskandarian, A. (Ed.), Handbook of Intelligent Vehicles. Springer, pp.537-558.
2. Treiber, M., Kesting, A., 2013. Traffic Flow Dynamics: Data, Models and Simulation. Springer-Verlag, ISBN 978-3-642-32459-8.
3. Antoine Tordeux, Sylvain Lassarre, Michel Roussignol, An adaptive time gap car-following model, Transportation Research Part B: Methodological, 2010 (44): 1115-1131.
4. Nadir Farhi. Piecewise linear car-following modeling. Transportation Research Part C 25(2012)100-112.
5. Mohannad Saifuzzaman, Zuduo Zheng, Incorporating human-factors in car-following models: A review of recent developments and research needs. Transportation Research Part C 48(2014): 379-403.

6. Mohamed Ahmed, Helai Huang, Mohamed Abdel-Aty, Bernardo Guevara, Exploring a Bayesian hierarchical approach for developing safety performance functions for a mountainous freeway, 2011(43):1581-1589.
7. Lena Winslott Hiselius, Estimating the relationship between accident frequency and homogeneous and inhomogeneous traffic flows. Accident Analysis and Prevention 2004(36):985-992.
8. Jianjun Zhang, Existing traffic safety problems of expressways in China. Procedia – Social and Behavioral Sciences, 2013(96):130-135.
9. N. Sugiyama, T. Nagatani. Multiple-vehicle collision in traffic flow by a sudden slowdown. Phisica A, 2013(392):1848-1857.
10. Irene Soria, Lily Elefteriadou, Alexandra Kondyli. Assessment of car-following models by driver type and under different traffic, weather conditions using data from an instrumented vehicle. Simulation Modeling Practice and Theory, 2014(40):208-220.
11. Przybyla J, Taylor J, Jupe J, Zhou X. Simplified, data-driven, errorable car-following model to predict the safety effects of distracted driving. In: Proceedings of the 15th International IEEE Conference on Intelligent Transportation Systems, Alaska, USA. 2012:1149-1154.
12. Kometani & Sasaki. Dynamic Behavior of Traffic with an on Linear Spacing-Speed Relationship. In Proceedings of the Symposium on Theory of Traffic Flow [J]. Research Laboratiories, General Motors, 1959:105-119.
13. WU Chao-zhong, YAN Xin-ping, MA Xiao-feng. Car-following Model Based on Drivers' Characteristics [J]. Journal of Transportation Engineering and Information. 2007, 5(4):18-22.

Urban rail transit comprehensive performance evaluation research based on improved BSC

Q. Li* & Y.B. Lv

School of Traffic and Transportation, Beijing Jiaotong University, Beijing, China

**14120780@bjtu.edu.cn*

As a public product, urban rail transit has brought profound influence to the government, society and related companies in the entire process of construction and operation. To correctly understand the advantages and disadvantages of urban rail transit operation, this article combined the goal of urban rail transit enterprises which is to realize the maximization of the comprehensive benefit of social and economic benefits, and used the improved balanced score card method to build the social benefit, economic benefit and service level, internal business, learning and growth of comprehensive performance evaluation index system and paid more attention to the evaluation of social benefits. It used the analytic hierarchy process (AHP) to set evaluation indicators hierarchically and calculated the index weight, finally built comprehensive performance evaluation system of urban rail transit.

Keywords: Urban rail transit; Balanced scorecard; Analytic hierarchy process.

1. Introduction

With the accelerating process of urbanization and cities expanding, worsening traffic congestion brings a great deal of inconvenience to trip for people. Due to large capacity, urban rail transit can effectively reduce the number of vehicles and save valuable land resources as well as help optimize the layout of the city. Therefore, urban rail transit has become the first choice of most cities to improve the traffic situation. However, most city's rail transit projects suffer seriously financial losses, poor financial benefits have been considered to be characteristic of urban rail transit.

Urban Rail Transit is a form of public benefits that government provides for urban residents, with public welfare and huge externalities. These gains and losses on financial can't fully evaluate the performance of urban rail transit. So how to evaluate the overall performance of urban rail transit is the content of this paper.

2. Balanced Score card and Improvement

2.1. *The basic principles of the Balanced Scorecard*

Balanced Scorecard is a performance evaluation tool proposed by US famous scholars Robert Kaplan and David Norton. It departure from the company's vision and strategy, financial, customer, internal processes and learning and growing levels of decomposition four strategic objectives, the company's vision and strategy into concrete objectives and evaluation indicators to achieve the combination of strategy and performance.

The so-called balance refers to the balance of short-term gains and long-term development of enterprises, outcome indicators and driving indicators, financial indicators and non-financial indicators.

2.2. *BSC index system design considering externality*

A series of externalities generated by urban rail transmit, such as improving the efficiency of urban transportation and urban traffic, reducing environmental pollution, have brought direct or indirect benefit to residents, businesses and public, improving the welfare of entire city. So urban rail transport enterprises not only pursue economic efficiency but also social benefits to achieve economic and social benefits to maximize the overall efficiency, so the urban rail transit enterprise comprehensive performance evaluation have to take social benefits into consideration.

This paper quotes Balanced Scorecard designed by Robert Kaplan and David Norton and increases indicators of social benefits dimension, select urban rail transit business key performance indicators from five levels. According to the characteristics of rail transit in China, from products and services provider perspective, namely business, evaluate comprehensive performance.

In this article, the balanced scorecard includes these five aspects:

(1) Social benefit level. A series of externalities produced by urban rail transit mainly divided into economic benefits bringing for other subjects, environmental benefits and other benefits, etc., so the social benefit level mainly set up evaluation indexes from the three aspects.

(2) Economic benefit level. Economic level, business indicators are commonly used in traditional performance evaluation indicator to show enterprise's strategy and its implementation and enforcement contribute to the improvement of corporate profits.

(3) Service level. In order to improve operational efficiency at the same time and satisfy the "safe, fast, comfortable, economical" passenger demand, enterprise of urban rail transit try best to create a service brand, index data of this level can mainly get by external investigations that subway now provide

service level for passengers to improve the passenger satisfaction to the greatest extent.

(4) Internal operational level. Metro operation management is mainly composed of traffic command, passenger service and equipment maintenance technology system. To achieve the goal of financial and customer and then improve internal management efficiency, subway operators need for a comprehensive internal investigation and planning and improve them.

(5) Learning and development. Learning and development goals provides the infrastructure for the ambitious objectives of the other four areas. It is power driven scorecard above four areas for superior outcomes.

3. The Choice of Evaluation Index

3.1. *Index selection principle*

Accurate analysis of comprehensive benefits of urban rail transit is very important. This paper, from the public interest, select key performance indicators to help companies achieve strategic goals avoid those non-critical indicators to bring interference to the performance evaluation.

3.2. *Social benefit indictors*

For economic and social developed cities, the urban rail transit is a large-scale urban infrastructure projects, with huge externality effect. Urban rail transit externality benefits usually divide into social benefit, economic benefit, environment benefit. Environmental benefit of urban rail transit is an important part of the external benefits, the impact of urban rail transit on urban environment is subtle and difficult to appear in a short time, with the long-term. Environmental Benefits studied in this paper mainly refers to the natural environment. The author selects easing traffic congestion as indicator to suggest social benefit, saving energy, saving urban land resources and reducing buying buses as indicators to suggest economic benefit and reducing air pollution as indicator suggest the environmental benefit.

3.3. *Economic benefit indicators*

Income of metro operating companies is subjected to objective conditions, its operating effectiveness depends on reducing consumption, on the premise of guarantee the quality of transport, saving transport costs, as well as increasing the ticket income for the development and utilization of underground resources. In this paper, from two aspects of enterprise profitability and development ability to evaluate its economic efficiency, specific indicators include operating cost ratio, return on assets, market growth rate and growth rate of total assets.

3.4. *Service level indicators*

When travelers consider whether the use of urban rail transit, they first focus on rail transportation accessibility. Only when accessibility meet the requirements of travelers, travelers will choose rail transit. If the urban rail transit meet accessibility requirements, travelers put forward higher request based on past experience, including convenience, comfort and economy and other factors. Only when the urban rail transit accessibility comfort and convenience meet the needs of travelers, travelers adhere to choose rail transit. So choose accessibility, convenience, comfort and economy as evaluation indicators.

3.5. *Internal operational indicators*

Products and services of urban rail transit can't be stored, producing and consuming the same time. It is impossible to ease the contradiction between supply and demand once the supply shortage. So in the performance evaluation, the product produced in the process of timeliness, accuracy and reliability evaluation are very important, directly related to the quality of performance results. While single urban rail transport large volume, intensive and high operating speed, once the slightest mistake, there are accidents, so security must be fully guaranteed. In this paper, respectively from three aspects, namely the internal efficiency, reliability and safety operating to evaluate performance, these three aspects of index is operation accident rate, running on time rate, equipment intact rate, the overall productivity.

3.6. *Learning and development*

Urban rail transit enterprise is technology-intensive, they need a batch of people of talent who have a good knowledge of both business and management. These are the goal of learning and development level of balanced scorecard. The level indicator is set to the employee satisfaction, employee training, and employee retention, employee ability.

4. Comprehensive Evaluation

This paper choose the analytic hierarchy process (AHP) to evaluate the comprehensive performance of urban rail transit, the analytic hierarchy process is suitable for the scheme selection and weighting, in comprehensive performance evaluation of urban rail transit, it can adjust the weight of every level according to different situation in different cities.

4.1. *Index weight calculation*

According to the proportion of 1 ~ 9 scaling method of analytic hierarchy process (AHP) in the same level of evaluation index system of the importance of the factors relative to the upper compared two quantitative, A judging matrix was established. According to the above method, the weight of five aspects to the total target of A judgment matrix was calculated.

$$A = \begin{bmatrix} 1 & 2 & 1.5 & 3 & 5 \\ 0.5 & 1 & 0.5 & 3.3 & 5 \\ 0.67 & 2 & 1 & 1.5 & 5 \\ 0.33 & 0.3 & 0.67 & 1 & 3 \\ 0.2 & 0.2 & 0.2 & 0.33 & 1 \end{bmatrix}$$

In this paper, it uses standardized method to calculate the biggest characteristic root judgment matrix and eigenvector;

The specification column averaged feature vectors:

$$W = \begin{bmatrix} 0.343 & 0.224 & 0.260 & 0.123 & 0.050 \end{bmatrix} \quad (1)$$

According to $AW = \mathrm{nW}$ largest eigenvalues of judgment matrix is calculated:

$$\lambda_{max} = \frac{1}{5}\sum_{i=1}^{5}\frac{(AW_i)}{W_i} = 5.203 \quad (2)$$

Consistency check

$$\mathrm{CR} = \frac{\mathrm{CI}}{\mathrm{RI}} = 0.045369 \quad (3)$$

Because of CR<1, consistency check meets the requirements.

Similarly, social, economic, service levels, internal operations, learning and growth perspective judgment matrix can be get, calculating the index weight of index layer, the calculation in the table below:

Table 1 Comprehensive evaluation index system of urban rail transit performance.

Target layer	Five levels	Weight	Index	Relative index weight	Index weight
Comprehensive evaluateon index system of urban rail transit performance	Social benefit	0.343	saving energy	0.177	0.0607
			reducing air pollution	0.169	0.0580
			saving urban land resources	0.071	0.0242
			easing traffic congestion	0.462	0.1586
			reducing buying buses	0.121	0.0415
	Economic benefit	0.224	operating cost ratio	0.357	0.0799
			return on assets	0.198	0.0444
			market growth rate	0.221	0.0495
			growth rate of total assets	0.224	0.0502
	Service level	0.260	accessibility	0.201	0.0523
			convenience	0.274	0.0714
			comfort	0.109	0.0283
			economy	0.416	0.1080
	Internal operational	0.123	operation accident rate	0.436	0.0536
			running on time rate	0.179	0.0221
			equipment intact rate	0.091	0.0111
			the overall productivity	0.294	0.0362
	Learning and development	0.050	employee satisfaction	0.156	0.0078
			employee training	0.381	0.0191
			employee retention	0.117	0.0058
			employee ability	0.346	0.0173

4.2. *Index dimensionless processing*

In the comprehensive evaluation, different evaluation index tend to have different dimensions and dimensional units. In order to eliminate the resulting incommensurability, this paper will be dimensionless processing of each evaluation index.

In order to select the appropriate indicators evaluation method, according to the actual situation, the evaluation index is divided into different types. In this article, the indexes can be divided into positive and reverse, reverse index in this paper are: operating cost ratio, accidents, and staff training. After index classification and the dimensionless processing, this article uses the average method to be dimensionless. Average method:

$$y_i = \frac{x_i}{\overline{x_i}} \tag{4}$$

Where: $\overline{x_i}$ is average of x_i, if the index is a positive indicator and y_i less than 1, then the value is lower than the industry standard, and if y_i greater than 1, then the value is higher than the industry standard. If the index is a reverse indicator, the conclusion is opposite.

Because of the positive indicators, reverse index evaluation criterion is different, the indexes must be with the same trend. The reverse indicators is generally translated into positive indicators. Method for reverse index translation in this paper are:

$$y_i^{'} = \frac{1}{y_i} \tag{5}$$

Namely for positive change index, score is the results of the dimensionless processing, for the reverse index, score is inverse of the dimensionless processing. After dimensionless processing, then you can get value of indicators with the same measure, range and direction.

4.3. *Index integrated approach*

According to the analytic hierarchy process (AHP), after dimensionless processing each index of five aspects, then performance evaluation of urban rail transit comprehensive coefficient can be get. The specific calculation steps are as follows:

(1) Calculate the dimensionless Numbers. According to the average method, turn value of positive index into dimensionless value and turn value of negative index into positive dimensionless value.

(2) Calculate the coefficient of single index. According to the weight of each index and the results of the previous step, calculate the product of result of previous step and index weight.

(3) Calculate comprehensive evaluation coefficient. Add all the coefficients of the individual indicators index together, you get a comprehensive evaluation coefficient.

$$W = \sum_{i=1}^{21} w_i y_i \tag{6}$$

Where: W is the coefficient of performance comprehensive evaluation of urban rail transit; w_i is the ith indicators weight; y_i is the ith indicators dimensionless value.

5. Conclusion

This paper, based on analyzing the characteristics of urban rail transit, Balanced score card as the tool of enterprise comprehensive performance evaluation of urban rail transit, according to the characteristics of urban rail transit and the enterprise strategic target, builds performance evaluation index system. And to improve the industry and product characteristics indicator system, add the social indicators in accordance with its attribute of social benefits and economic indexes, so that the performance evaluation results more realistic, fully and accurately reflect the company's operating performance and improvements in indicators, based on the Analytic Hierarchy Process for urban rail transportation enterprises to establish evaluation model to determine the index weight. According to the index weight of comprehensive evaluation system, enterprise can find the future priorities and direction of the work to improve and correct understanding of the pros and cons of enterprises and encourage enterprises to improve the management level to make the performance evaluation system have practical value.

Acknowledgements

The research is supported by Center of Cooperative Innovation for Beijing Metropolitan Transportation; Science & Technology Evaluation and Statistics of Special (2014SE-0205).

Reference

1. Yan Fei, Tang Tao and Gao Chunhai. 2010. Urban rail traffic safety evaluation system research. Urban rapid rail transit 23(3):32-36.
2. Wang Ye. 2007. Introduction to the balanced scorecard in the subway operation of applied in enterprise performance evaluation. Modern urban rail transit (3): 45-46.
3. Zhang Shaoqing. 2007. Urban Rail Transit operating performance indicators and implement ways of building. Shanghai Quality (9): 30-31.

Chapter 5

Mechanical Engineering

Study on a three-dimensional solid platform of the belt drive and the chain drive in mechanical engineering

Y. Sang, F.T. Li[1†], G.H. Zhang, Y.B. Dai and Y.J. Wang
The School of Mechanical Engineering, Dalian University of Technology, Dalian, P. R. China
[†]*Email: lylifengtao@126.com*

Belt and chain are regarded as important parts in mechanical design and they are stereoscopic. Based on this background, this paper presents a belt drive and chain drive interactive simulation platform based on virtual reality technology. Through this platform students can understand the structure, working principles and characteristics of belt drive and chain drive, know various kinds of belts and chains, and learn specific applications of belt drive and chain drive with applied examples. Belt drive and chain drive interactive simulation platform uses browser-based structure, Java3D, virtual reality, Java Applet and other technology to build a Web3D virtual reality learning environment. Students can use mouse to move, rotate, and zoom the solid model, and observe the model from any angle. This interactive simulation platform can satisfy the demands of numerous students study online at the same time, which has great value in mechanical engineering distance education.

Keywords: Belt drive; Chain drive; Virtual reality technology.

1. Introduction

Due to the lack of observable and operable interactive specific laboratory equipment, online education lacks advantages [1] on three- dimensional strong operational knowledge. This issue restricts the further development of modern distance online education, which is the largest challenge online education facing. In order to solve this problem, people introduced 3D virtual reality technology into modern distance education [2].

E-learning is a new teaching model nowadays. Web3D open standards allow the delivery of interactive 3D virtual learning environments through the Internet, reaching potential large numbers of learners worldwide, at any time. The educational use of virtual reality based on Web3D technology was introduced. The main positive and negative results were outlined [3]. Web3D simulation-based virtual worlds have been substantially adopted in educational

---

[1] Address: Room 216-1,The East Building of Mechanical Engineering Institute, The School of Mechanical Engineering, Dalian University of Technology, Dalian, P. R. China,116024

settings worldwide. Since the elaborations on such applications in regard to tourism education are still limited, the application of 3D simulation-based platform has been studied [4]. The prospect for the generation of interactive web-based 3D City Models based on free geo-data available from the Open-Street-Map project was investigated [5]. A cyber-tool that using Java 3D, Web-based tools, and other techniques is developed to support collaborative research in the area of atmospheric nucleation [6]. The paper outlines the Java3D technology in three-dimensional model of interactive design applications and proposes the use of Java3D virtual three-dimensional model for the description of the client to read through the remote server [7]. For the hoisting software's difficulty of version update and the slowness of data interaction in web simulation, a method is presented based on B/S structure. Java3D and the map structure are employed to realize web publishing and cache design [8]. The paper, taking a type of serial robot (MOTOMAN-SV3X) for example, introduces the working principle and systematic architecture of a robot-oriented simulation and monitoring system based on Java3D and closed-loop feedback [9]. The 3D brain imaging has been implemented using texture-mapping function of Java3D, which is reconstruction of 2D cross-section imageries obtained from the scanning devices [10]. This paper studies the integration of VRML and Java3D in the Web3D core technology [11]. The incremental path planner has been implemented in Java and incorporated into a Java3D-based VRML browser [12]. High-end visualizations are usually based on native 3D graphics libraries. The interactive molecular dynamics software for common desktop computers and workstations was presented based on Java3D [13]. A talking head oriented to the creation of a Chatbot is presented. The answer is converted into a facial animation using a 3D face model. The WEB infrastructure has been realized using the Client- Server model delegating the Chatbot [14]. The coupling of Java3D and applet technologies has the potential to revolutionize Web-based simulation visualization.

Belt drive consists of two belt wheels and a tight belt wound around two wheels, which is a flexible friction transmission transmitting motion and power by friction between the belt and wheel contact surfaces. Chain drive is a force transmission which has engaging character and transmits motion and power between two parallel axes by the engagement between the chain and sprockets. A three dimensional solid platform of the belt drive and chain drive has been developed at the Dalian University of Technology, which is a virtual reality learning environment system. This platform applied Java3D, Java Applet, virtual reality technology and other technologies to deploy and integrate network-based 3D virtual reality. This interactive simulation platform can satisfy the requirements of numerous students learning online at the same time, which has great applicable value in mechanical engineering distance education.

2. The Operation Interface of the Three-Dimensional Solid Platform

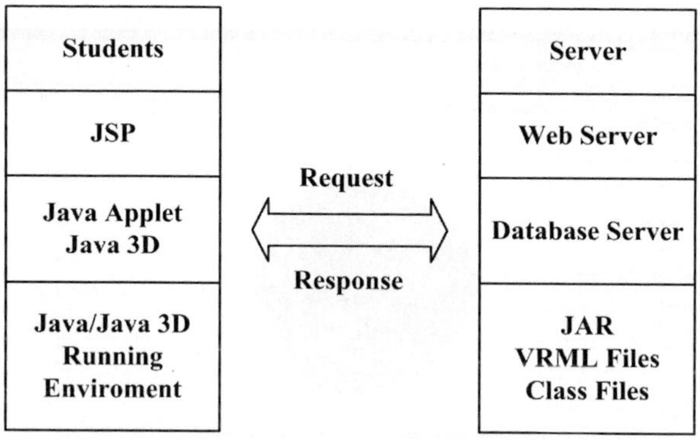

Fig. 1. The working principle of the three-dimensional solid platform.

The working principle of the three-dimensional solid platform is presented in Fig. 1. Students can finish visiting web server and experimental work in any place at any time. Server program provides students network services. It mainly uses Java Server program and JDBC Technology. Before the experiment, students need to download java plug-in software from file server then install the software which includes operating environment supporting Java Applet, VRML and Java 3D, etc. Based on the methods proposed above, a virtual reality learning environment system could be build.

The viewing process starts when the students request to view the three-dimensional solid platform via HTTP. The web server hosting the link will send the web page and the embedded viewer applet. The browser initializes the Java Applet, and then a 3D virtual environment has been founded. The applet allows students to view the interactive belt drive/chain drive system from the desired perspective in the simulation.

The operation interface of the three-dimensional solid platform is presented in Fig. 2. The platform is consisted of the 3D virtual area, control panel area and virtual management area. In the 3D virtual area, the utilize of virtual reality technology provides operative interactivity, three-dimensional quality, highly realistic virtual belt drive/chain drive importation system model, which is the core part including the importing of geometrics in Java3D and models from VRML ;The control panel area lists all the functions of this platform on the right. Control panel applies Java Applet technology to fulfill the control of motion of the 3D virtual belt drive/chain drive system. The belt drive/chain drive system simulation uses virtual reality technology, network audio technology from 3D images, sounds from all angles to impart experimental knowledge to help students to complete virtual experiment, which greatly improve human-machine interactive feature, enhance students' interest on learning and motivate students' enthusiasm to learn.

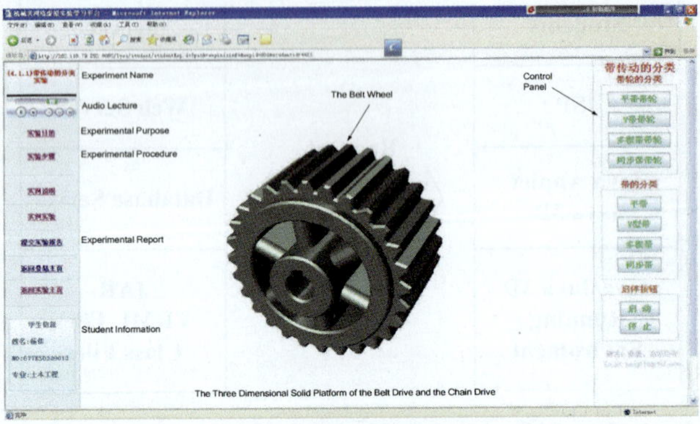

Fig. 2. The operation interface of the Three-Dimensional Solid Platform.

VRML and Java3D are an ideal combination for this three-dimensional solid platform as the Java3D scene graph matches very closely with the VRML structure. The belt drive/chain drive system graphical environments for the web have been created by using VRML in this paper. The Java3D operating environment convert the graph into the data in the memory, then show a three-dimensional effect on the computer. Java 3D can place many objects into a virtual space, and set the values of attributes of the objects such as shape, location, appearance, texture, etc. The three-dimensional solid platform was implemented to interactively explore objects in a virtual universe using models generated by a computer graphics suite and imported using a VRML file loader.

3. The Virtual Belt Drive/Chain Drive System

Students observe the course of action of belt drive/chain drive system dynamically and observe the corresponding operation by translation, rotation and zoom of the virtual scene based on controlling of the mouse, thus it can make you know the working principles and mechanical structure of belt drive/chain drive system more clearly in a virtual reality environment. Fig. 3 shows the shots of different types of virtual pulleys and belts, including flat belt pulley, V belt pulley, ribbed belt pulley, belt pulley, flat belt, V-belt, ribbed belt and timing belt. Click the "flat belt pulley" button, user can observe the shape charact- eristics of flat belt pulley, rotate the view by dragging left mouse button, translate the view by dragging the right mouse button, hold down the middle mouse button and drag to zoom the entire view. Also click on other buttons users can observe the shape characteristics of different pulleys and belts. Click the

"Start" button, users can observe the movement of the pulley. Students learn various kinds of typical belt drive, belt pulleys and the classification of belts through interactive operation.

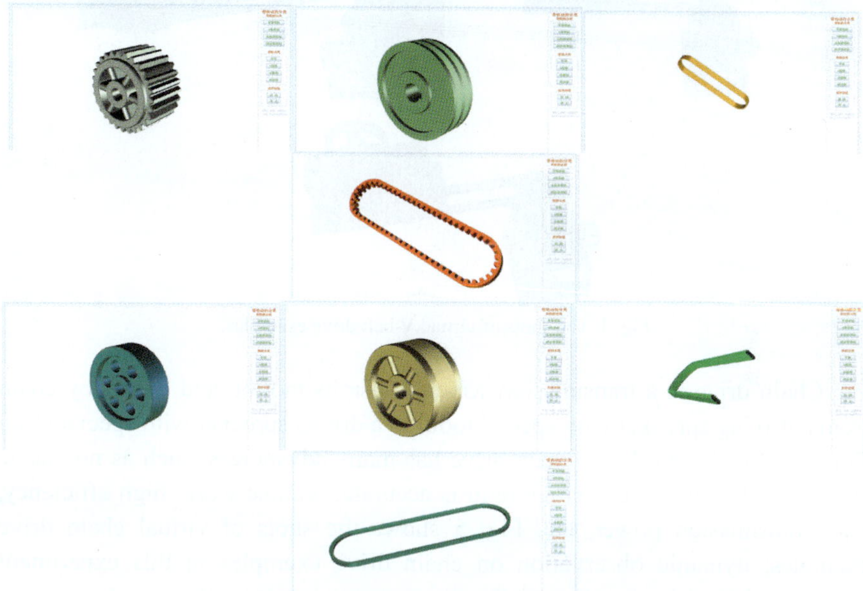

Fig. 3. The shots of different kinds of virtual belt pulleys and virtual belts.

Belt drive is a transmission which uses the flexible belt tensioned on the belt wheel to transmit motion or power. Belt drive has characteristics as simple structure, smooth transmission, can buffer the vibration absorption, can transfer power between the big pitch axis and multi-axis, furthermore its low cost, no need to lubrication, easy maintenance and other advantages make it applied in a wide range of mechanical drive. Fig. 4 shows the shots of virtual belt drive examples. Dynamic observation on V-belt transmission examples by interactive operation makes students learn the transmission characteristics of V-belt and its motion characteristics. Click on the "start" button, users can dynamically observe transmission characteristics of V-belt drive, rotate the view by dragging the left mouse key, translate the view by dragging the right mouse key, hold down the middle mouse button and drag to zoom the entire view. Modify the "speed" value, and then click on the "Settings" button, users can observe the movement of pulleys and gears at different speeds.

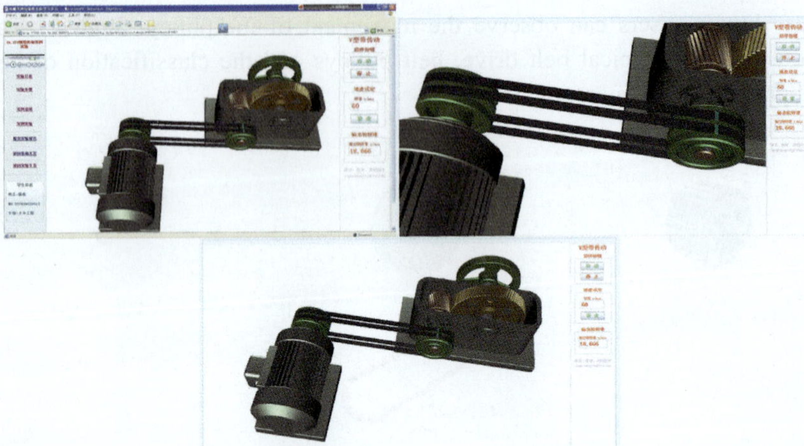

Fig. 4. The shots of virtual V-belt drive examples.

Chain drive is a transmission which transmits motion and power by chain from a driving sprocket with special tooth to a driven sprocket with special tooth. Compared with belt drive, chain drive has many advantages, such as no elastic sliding or skidding, the average ratio is accurate, reliable work, high efficiency, high transmission power, etc. Fig. 5 shows the shots of virtual chain drive examples, dynamic observation on chain drive examples in this experiment makes students clearly observe the engagement between chain and sprocket, better understand how speed fluctuation in transmission and the number of teeth influence the motion stationarity. Users can rotate the view by dragging the left mouse key,translate the view by dragging the right mouse key , hold down the middle mouse button and drag to zoom the entire view. Click on the 'start', users can dynamically observe the transmission characteristics of chain drive. Click on the "stop" button, you can stop the motion.

Fig.5. The shots of virtual chain drive examples.

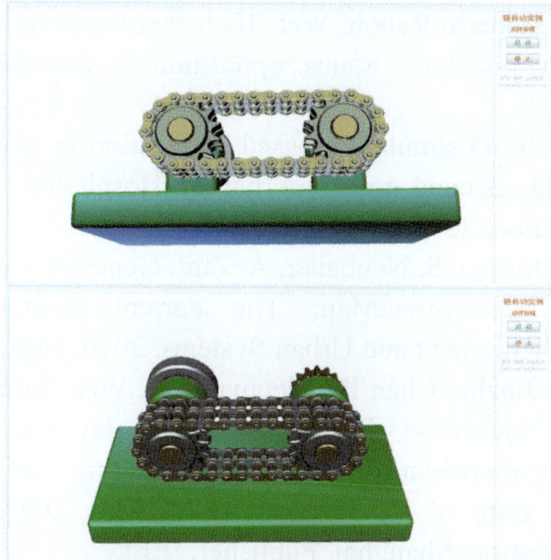

Fig.5. (*Cont'd*)

4. Conclusion

This paper introduces belt drive/chain drive system as an example to describe the developing process of the three-dimensional solid platform, the simulation platform. This interactive simulation platform, belt drive/chain drive system, applies browser-based structure, Java3D, virtual reality and Java Applet, etc. Hands-on interactive operation and observation of the three-dimensional space from any perspectives will greatly enhance students' interest in online learning, stimulate the enthusiasm of students, and encourage students to participate in the learning process actively.

Acknowledgements

We gratefully acknowledge the support by the National Natural Science Foundation of China through the grant number 51275068, Aeronautical Science Foundation of China through the grant number 20130863006 and the support by the Fundamental Research Funds for the Central Universities Grant No. DUT15LK21.

References

1. Mohammad Khosrowjerdi, Gary L. Kinzel, David W. Rosen. Computers in Education: Activities, Contributions, and Future Trends. J. Comput. Inf. Sci. Eng. 2005; 5(3):257-263.
2. Pat Banerjee. Data Interface Software for Windows PC-compatible Virtual Reality Scene Graphs. J. Comput. Inf. Sci. Eng.. 2002, 2(1):72-74.

3. Luca Chittaro, Roberto Ranon. Web 3D technologies in learning, education and training: Motivations, issues, opportunities. Computers & Education. 2007, 49(1):3-18.
4. Liwei Hsu. Web 3D simulation-based application in tourism education: A case study with Second Life. Journal of Hospitality, Leisure, Sport & Tourism Education. 2012, 11(2):113-124.
5. M. Over, A. Schilling, S. Neubauer, A. Zipf. Generating web-based 3D City Models from OpenStreetMap: The current situation in Germany. Computers, Environment and Urban Systems. 2010, 34(6):496-507.
6. Rui Ding, Gao Jinzhu, Chen B., Siepmann J.I. Web-Based Visualization of Atmospheric Nucleation Processes Using Java3D. 9th IEEE/ACM International Symposium on Cluster Computing and the Grid, 2009. CCGRID '09. Date of Conference: 18-21 May 2009, Page(s):597-602 Conference Location: Shanghai, Publisher: IEEE.
7. Qiu Wei. WEB-Based Interactive 3-D Design System Application. Third International Symposium on (Volume: 3) Intelligent Information Technology Application, 2009. IITA 2009. Date of Conference: 21-22 Nov. 2009 Page(s):693-696 Conference Location: Nanchang, Publisher: IEEE.
8. SHU Shilong, WU Min, AN Jianqi, TANG Xiujun. Three-Dimensional Hoisting Simulation Based on B/S Structure and Cache Design. Proceedings of the 31st Chinese Control Conference July 25-27, 2012, Hefei, China. Pp 5524-5529.
9. Du Zhijiang, Yang Donghai, Sun Lining, Fu Lixin. Virtual Robot Simulation and Monitoring System based on Java3D. IEEE International Conference on Robotics and Biomimetics, 2004. ROBIO 2004. Date of Conference: 22-26 Aug. 2004 Page(s): 313–317 Conference Location: Shenyang, Publisher: IEEE.
10. Saladi. S., Pinnamaneni. P., Meyer. J. Texture-based 3D brain imaging. Proceedings of the IEEE 2nd International Symposium on Bioinformatics and Bioengineering Conference, 2001. Date of Conference: 4-6 Nov 2001 Page(s):136 – 143 Meeting Date: 04 Nov 2001-06 Nov 2001 Conference Location: Bethesda, MD Publisher: IEEE.
11. He Li, Peng Yan, Zhang Yingci, Zhang Hong. Website Platform Design and Implementation based on Web3D Virtual Reality Technology. 2011 IEEE 3rd International Conference on Communication Software and Networks (ICCSN), Date of Conference: 27-29 May 2011 Page(s): 618-622 Conference Location : Xi'an, Publisher: IEEE.
12. Tsai-Yen Li, Chih-Ching Chang. Path planning with incremental roadmap update for large environments. IEEE International Conference on Robotics and Automation, 2001. Proceedings 2001 ICRA. (Volume: 2) Date of Conference: 2001Page(s): 1934-1939 vol.2 Publisher: IEEE.

13. Vormoor. Quick and easy interactive molecular dynamics using Java3D. Computing in Science & Engineering. 2001, 3(5):98-104.
14. O. Gambino, A. Augello, A. Caronia, G. Pilato, R. Pirrone, S. Gaglio. A Web-Oriented Java3D Talking Head. Advances in Intelligent and Soft Computing. 2009, 60: 295-311.

Modeling and simulation of electric laminating machine

P.J Mao[†], Y. H. Ding, S. H. Zhang, Y. L. Zhang and Z. M. Qiu
Henan University of Science and Technology, Luoyang, 471003
[†]*Email: 790137007@qq.com*

Z. B. Du
Luoyang, Henan Province Tobacco Company, Luoyang, 471003
Email: zyl217907@126.com

Electric agricultural machinery in the production process, mobile and flexible, easy operation, low noise, less pollution, meeting the requirements of sustainable development and the development of green agriculture. This paper studies the electric laminating machine, through its stress analysis, using MATLAB / Simulink were electric laminating machine system modeling and simulation of visualized mathematical modeling, and characterization. The simulation shows: electric laminating machine with low noise, less pollution, high efficiency, has great advantages and broad application prospects in agricultural production.

Keywords: Electric motor; Agricultural machinery; Electric laminating machine; MATLAB simulation.

1. Introduction

In recent years, with the rapid development of the national economy, living standards and the quality of people is also rising, environmental issues are also increasingly cause for concern. With the further depletion of the world's non-renewable energy sources, in recent years, international oil prices rose significantly higher electricity prices. Currently, both from the point of view energy costs, or from the future energy-oriented point of view, electricity as a new energy replace non-renewable energy is the future trend of development. Compared with the traditional agricultural machinery, electric agricultural machinery low noise, less pollution, high production efficiency, meeting the requirements of sustainable development and the development of green agriculture. Therefore, it is necessary to research the electric agricultural machinery.

Foreign has a long history of study electric farm machinery [1]-[2], 1912

German Siemens company's 36.8kw electric tractor ride, mainly for tillers, follow-up launched 2.9kW electric walk-behind tractors, the series is mainly used for arable operations; 1941 Sweden has developed a interchangeable with multipurpose electric tractor, the tractor is driven by Boveri motor; 1970 US General Electric company launched Elec-Trak series electric tractors, using lead-acid battery powered and permanent magnet brushless motor drived. Domestic electric farm machinery starting point is relatively low, but has been actively developed. For example, 1950 Northeast Agricultural College, Harbin Institute of Technology and Songjiang tractor factory jointly developed the "cow power" series electric tractors. May 2012, China Yituo signed an electric tractor project cooperation agreement with the US General Electric Company. However, researching electric laminating machine is relatively rare at domestic and foreign [3]-[6]. This paper based on the needs of agriculture, by MATLAB / Simulink modeling simulation of electric laminating machine, and characteristic analysis, the results showed that meet the requirements.

2. Electric Laminating Machine Model

2.1. *Characteristics of electric laminating machine*

Electric laminating machine work process: wheels, film work components and vehicle chassis constitute a whole, battery and electric motor mounted in the first half of the vehicle. Motor-driven wheels and film work part by shifting the transmission mechanism. Vehicle chassis for four agencies, two front-wheel for driving , two rear wheels as walking wheels, play stabilizing and balancing role. The laminating machine can complete sraight Ridge jobs. By the design works, combine actual stress analysis [7]-[9], electric laminating machine stress analysis is as Figure 1.

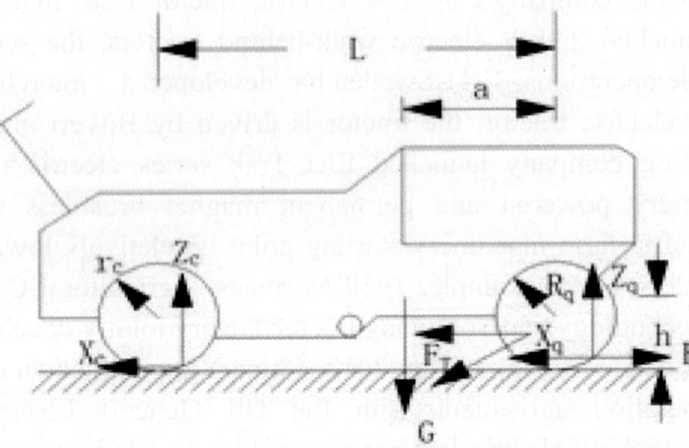

Fig. 1. Stress analysis diagram of the electric laminating machine.

G--quality of the electric laminating machine; F_T --traction resistance; X_q, X_c --soil reaction of the front and rear received parallel floor ; Z_q, Z_c--soil reaction of the front and rear received vertical floor; F_q --driving force; Considered the balance of forces in the horizontal direction, we can get electric laminating machine straight balance equation:

$$F_q = F_T + F_f + G \cdot \frac{\delta_m}{g} \cdot \frac{du}{dt} \quad (1)$$

In the ideal level ground, the wheels and the ground is tangent, on the cut-off point to take moments, get a load of front and rear wheels:

$$Z_c = \frac{Ga - M_q}{L} \quad (2)$$

$$Z_q = \frac{G(L-a) + M_q}{L} \quad (3)$$

As can be seen from the above equation, front and rear vertical ground reaction force by the soil related to the weight of the laminating machine, in addition to, related to the driving torque.

With load distribution coefficient $\lambda = \frac{Z_q}{G}$ indicates that the drive torque to the front and rear vertical load impact suffered, by (3) can be obtained:

$$\lambda = \frac{L-a}{L} + \frac{M_q}{G \cdot L} \quad (4)$$

From the above equation, with electric traction load increases, the vertical load of the laminating machine wheels also followed by increases. According to the above formula, vehicle simulation model of electric laminating machine was built in the Simulink environment.

2.2. *Simulation model of the electric laminating machine*

2.2.1. *Vehicle model*

According to the electric laminating machine stress analysis, establish electric laminating machine driving force balance equation [10]-[11] (as shown in 2.1), thus establishing a simulation model of vehicle, as shown Figure 2. Model, the rolling resistance, grade resistance, traction resistance and acceleration resistance determining a required driving force, while the actual load distribution coefficient decided by the traction resistance.

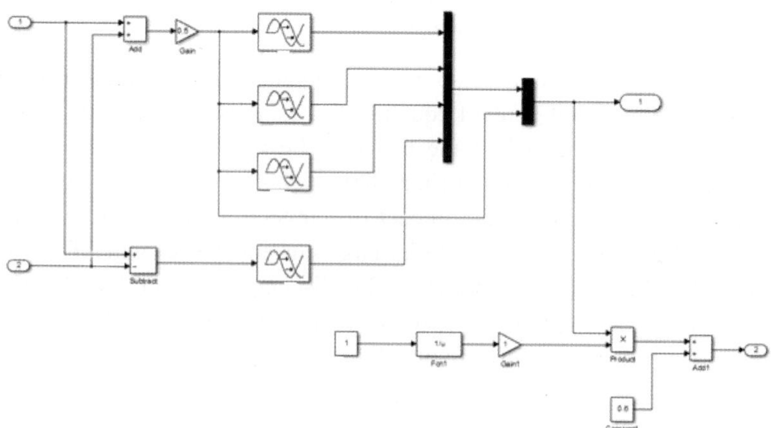

Fig. 2. Vehicle model of the electric laminating machine.

2.2.2. *Simulation model of the shunt DC machine*

To obtain the shunt DC motor operating characteristics [12]-[13], we want to a simulation of the running state. Parameters we have to analyze the armature current, excitation current, speed, electromagnetic torque. The simulation input: voltage of the motor, $U = 220V$, DC power supply; Load torque $T_l = 60$, depending on the jobs of the laminating machine overall stress analysis and a summary of the actual farm, combining the calculated motor parameters. Using oscilloscope to observe the operation of the curve and get the data. The Figure 3 is the established simulation model diagram of the shunt DC machine.

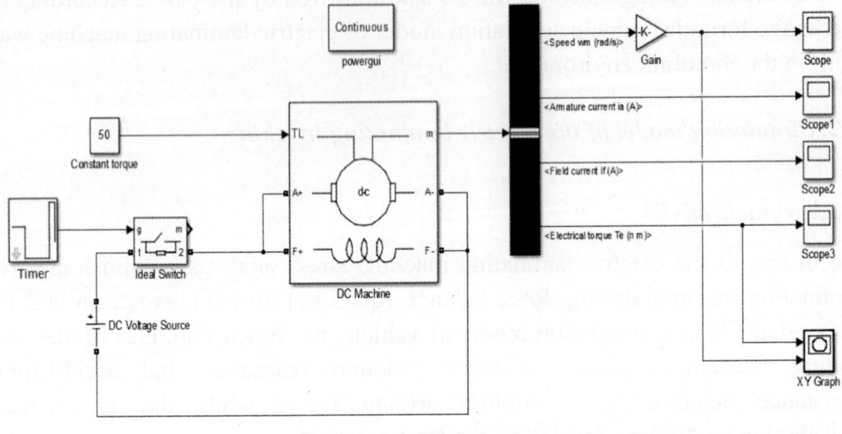

Fig. 3. Simulation model of the DC motor.

2.2.3. *Simulation model of the battery*

According to the circuit equation of the equivalent circuit model of the battery internal resistance, and Peukert equation can establish the state of charge SOC dynamic simulation model of the battery. The internal model is an electrical equivalent circuit of the amount (including the amount of current and charge) to reflect the relationship and the impact of temperature T factors, and when the system load changes, the battery discharge current will change. Therefore, the input of the battery SOC model is the discharge current I and the battery temperature T, the output is the state of charge SOC. SOC calculation simulation model of the Battery as shown Figure 4.

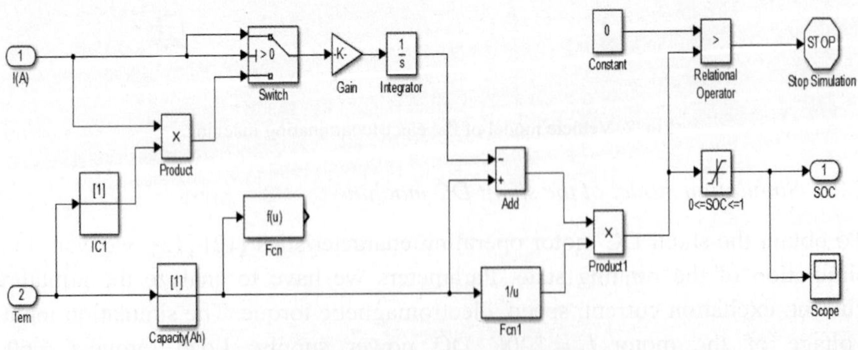

Fig. 4. Simulation model of the battery.

2.2.4. *Simulation model of the driving wheel*

Kinetic analysis of the driving wheels is the design of its slip rate, which determines the actual speed and the drive wheels in the actual process of moving the forces. Due to the existence of the vehicle braking and drive the two conditions in actual driving, and both of the calculated slip ratio is different, so you first need to make use of the tire rolling radius R_c and wheel speed v of the input and the angular velocity ω to do logical judgment, this uses two switch modules (switch) in the system to achieve, to determine whether the wheel is in the drive or braking, subsequently use the corresponding model equations and calculated vertical and horizontal slip rate of the wheel. calculate the slip rate is in order to the actual traveling speed of the wheels and analyze the situation of the force, the Figure 5 is the force driving wheel emulation module, mainly for driving wheel longitudinal force analysis.

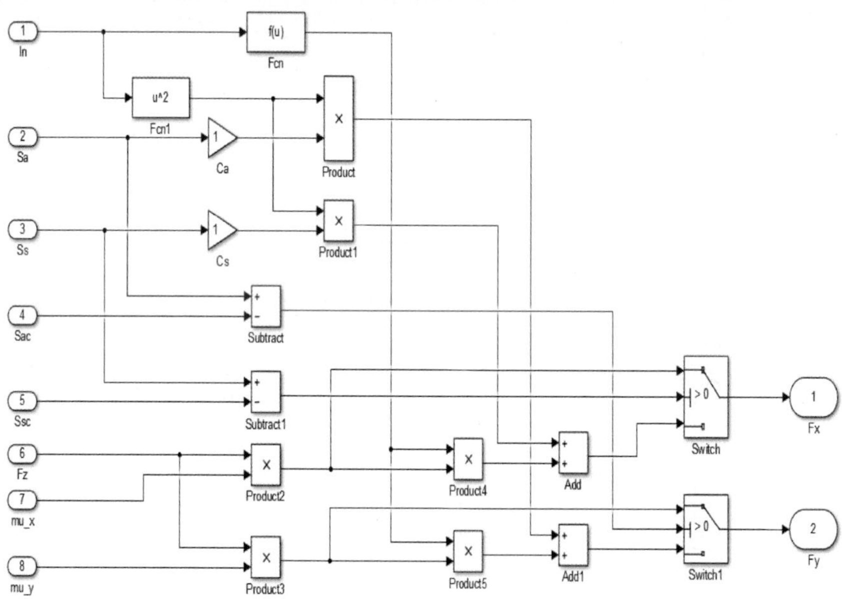

Fig. 5. Simulation model of the driving wheel.

3. Analysis of the Simulation Results

3.1. *Simulation results of the Motor*

Depending on the selected motor voltage $U = 220V$, DC power supply; Load torque $T_l = 60$, depending on the jobs of the laminating machine overall stress analysis and a summary of the actual farm, combining the calculated motor parameters. Using oscilloscope to observe the operation of the curve and get the data, as shown Figure 6-9.

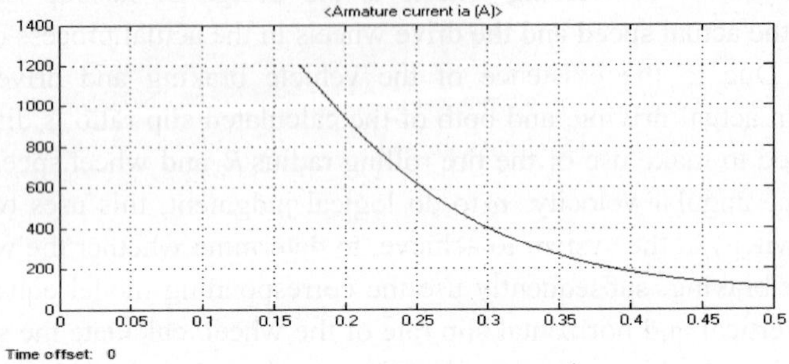

Fig. 6 Armature current.

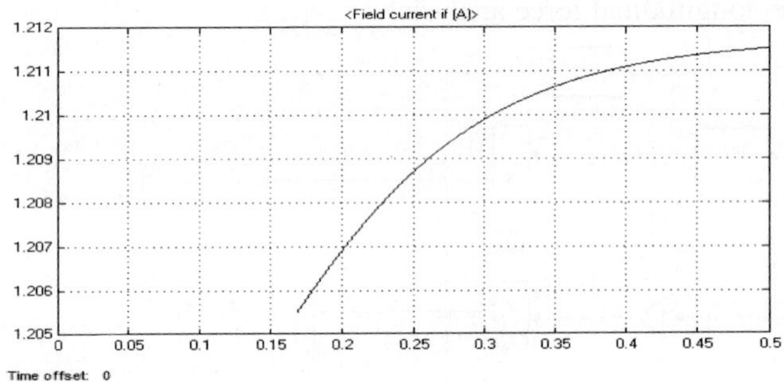

Fig. 7 Excitation current.

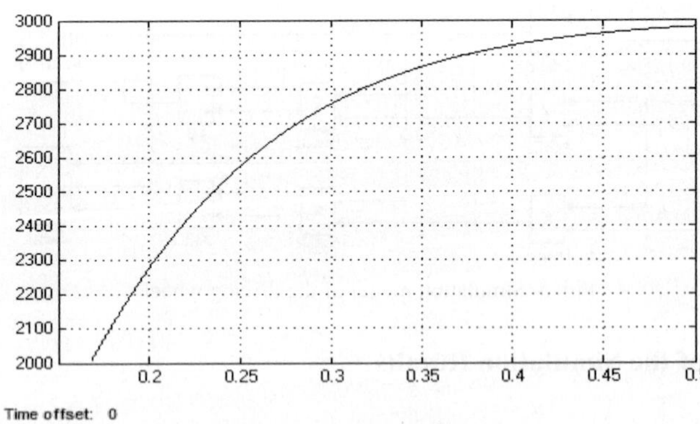

Fig. 8 Motor speed.

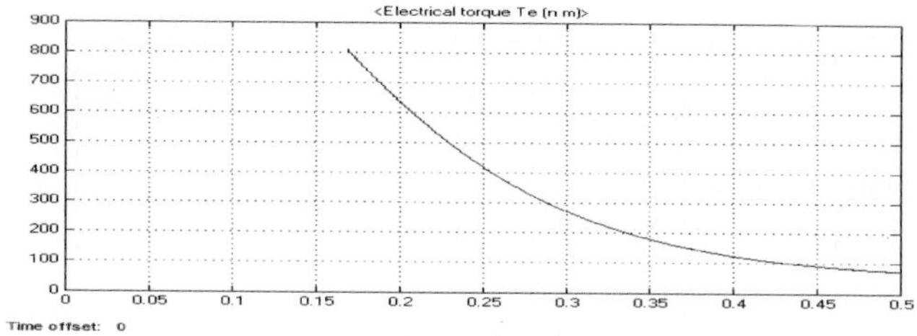

Fig. 9. Electromagnetic torque.

The model simulation parameters of the shunt DC motor: the total resistance of armature circuit $R_a = 0.087\Omega$, armature inductance $L_a = 0.0032H$, the total resistance of excitation circuit $R_f = 181.5\Omega$, mutual inductance between the field and armature windings $L_{af} = 0.56H$, moment of inertia of the motor $J = 0.76kg \cdot m^2$, Viscous friction coefficient of the motor $B = 0.00295$ $N \cdot m \cdot s$, the friction torque of the motor $T_f = 0.516N \cdot m$.

From top to bottom, respectively, as shown in the waveform of the motor armature current, excitation current, motor speed and torque on the map. Seen from the waveform, the electric motor when starting with a load current is large, up to 1200A.In 0.45s, the rotor reaches a maximum of about 2980r / min, while the armature current drops to 160A, also a corresponding change in the electromagnetic torque; from the theoretical analysis and calculation, the excitation current should soon reach a constant value of 1.21A, due to the phenomenon of mutual existence and simulation time setting is not great, so exciting current curve from the chart reflect its reach constant values for some small time.

3.2. *Simulation results of the battery*

Set simulation time is 500s, the simulation algorithm Ode45. Until the end of the simulation, double oscilloscope scope, observe the state of charge SOC battery discharge curve as Figure 10.

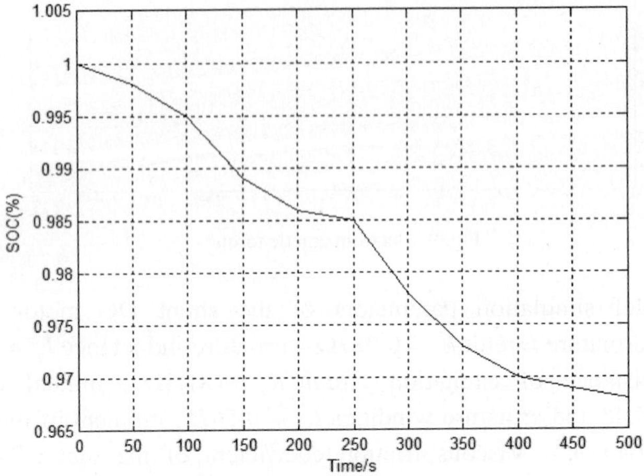

Fig. 10. battery discharge SOC curve.

The internal resistance of the battery state of charge SOC model [14]-[16] reflects the relationship between the amount of remaining battery charge and discharge time. When the battery discharge, the change of residual amount is influenced by the discharge current magnitude and the temperature change. With the battery work, the battery temperature will change; the higher the ambient temperature of battery discharge faster. When the motor load increases, the motor requires more current to the battery, i.e. the battery discharge current to increase. This is the connotation Peukert equation, namely the load changes affect battery state of charge.

3.3. *Simulation results of the wheels force*

Drive wheel force simulation model is a multi-input parameters model, the input parameters for working parameters of the ground and physical parameters of wheels, are known to be set. The simulation model is mainly the relationship between force of the drive wheels and slip rate. The following Figure 11 shows the slip rate changes, wheels suffered longitudinal force changes.

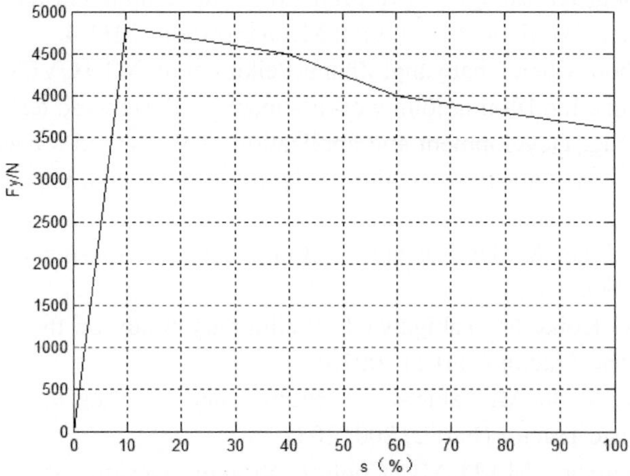

Fig. 11. Wheels suffered longitudinal force change curve.

The Figure 11 reflects the trends over the drive wheels slip ratio, which is a lot of literature findings (including theoretical and experimental results) are consistent. The force on the drive wheels with change of the slip rate that with the actual speed varies with the angle of the wheel. Under the same side of the angle, the braking force of the wheel with the longitudinal slip rate increases rapidly, and soon reaches a maximum, i.e., the longitudinal slip rate at about 10%, preferably from braking performance.

4. Conclusion

Through the analysis of the characteristics of electric laminating machine, modeling and simulation of visualized mathematical modeling, and the test data were analyzed, the study shows that electric laminating machine is reasonable and feasible.

References

1. Zhang-Tiemin, Yan-Guoqi, Wen-Lili, Liao-Lihua. Status and Trends of China's development of electric power agricultural machinery [J]. Agricultural Mechanization Research, 2012(4).
2. Shi-Junlong, Zhu-Sihong, Gao-Huisong, Wang-Yalei. Mini electric tractor drive system design [J]. Tractor and farm transporter, 2012 (2).
3. Yang-Xuliang, Meng-Qinghong, Li Huan et al. Double bar tobacco fertilizer, laminating machine developed and applied. Chinese Tobacco Science, 2003.3

4. Zhang Yong, Xu Liangbin. Multifunction vehicle-mounted electric sprayer development. Xinjiang Agricultural Mechanization, 2011.4.
5. Chen Zhihou, Chen Chaoyang, Zhang Peikun et al. YS-1GVFM-125A type ridge fertilization filming joint working machine design and test. 2013.4.
6. Duan Jixiang. Development and application of flue-cured tobacco fertilizer by Ridge one machine. China Science and Technology of Wealth, 2010 (24).
7. Bodria L, Fiala M. Design and Testing of an Electric-powered Walking Tractor. 1995.
8. Arjham W, Koike M, Takigawa T. Preliminary Study on the Applicability of an Electric Tractor (Part 1) 2001(03).
9. Arjham W, Koike M, Takigawa T. Preliminary Study on the Applicability of an Electric Tractor (Part 2) 2001 (03).
10. Zhang Defeng. MATLAB Control System Design and Simulation. Electronic Industry Press, 2009.
11. Guo Xiuhong. Hybrid Electric Vehicle Engine Modeling and MATLAB Simulation [J]. Chinese Journal of Agricultural Mechanization Chemical, 2013.7.
12. Li Fahai. Electromechanics. Beijing: Science Press, 2007.6.
13. Chen Boshi, Ruan Yi. Electric drive control system. Beijing: Machinery Industry Press, 2012.7.
14. Zhang Shu. Electric vehicle battery management system software design and SOC estimation Strategy [D]. Tianjin University, 2007.1.
15. Gao Huai. Research and design of electric vehicle battery management system [D]. Wuhan University of Technology, 2007.6.
16. Zhong Weiyu. Electric vehicle battery management system [D]. Nanchang University, 2009.10.

Development of electric walking chassis in tobacco field

P. J. Mao[†], Y. L. Zhang, Y. H. Ding and S. H. Zhang
Henan University of Science and Technology
[†]*Email: 790137007@qq.com*

K. Jiang
Luoyang Tobacco Company
Email: zyl217907@126.com

First, the overall structure of the design of the electric harvesting machine are introduced, according to the unit force and establish a mathematical model for the power matching, by working condition requirements, identify the design battery number, transmission system, finally the stability performance analysis of units, meet the performance requirements of the tobacco field.

Keywords: Electric; Machine harvested; Matching.

1. Foreword

Implementation of mechanized agriculture in tobacco, tobacco is the development needs of modern agriculture. By mechanized production, improve labor productivity, the farmers freed from the heavy manual labor in order to achieve sustainable development of tobacco. In this paper, the design and development of the electric tobacco harvesting machine in order to improve the automation of tobacco picked to minimize human involvement, thereby improving labor efficiency and reduce the physical burden of farmers, increase farmers kinds of tobacco initiative.

2. The Overall Configuration and Structure Design

The overall configuration of the electric tobacco harvester is shown in Figure 1, which is composed of the electric drive system, the operating chassis and the other accessories. The whole unit is supplied by the battery pack, and the DC series motor is supplied as the driving motor, and its working speed is adjusted by the gearbox. When the unit walks, the harvesting personnel take the seat, edge picking edge into the inside of the set of the front of the machine, thus finishing auxiliary harvest operation.

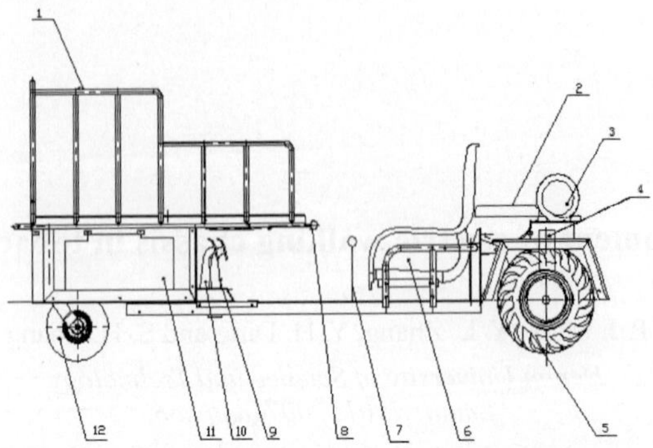

1.tobacco collection basket 2.controller 3.DC motor 4.transmission 5.drive wheel 6.seat 7. The shift operating lever 8.The direction of the operation lever 9 brake pedal 10.The clutch pedal 11.The battery pack 12.steering wheel

Fig. 1. The structure diagram of electric machine harvested.

The main design parameters of the walking sites machine in Table 1.

Table 1. The design parameters of electric harvesting machine.

Parameter	Numerical
Wheelbase (mm)	1270
Tread (mm)	675
Quality of use (kg)	430—460
Operating speed(km/h)	2—5
Steering wheel specification	4.10/3.50-4
Drive wheel specifications	4.00-8

3. Mathematical Model of the Unit by Force

The traction performance of the electric recovery machine is the traction ability under the condition of a certain soil, mainly used the rated traction to evaluate. Fig. 2 the hard working condition of the electric recovery machine is tried.

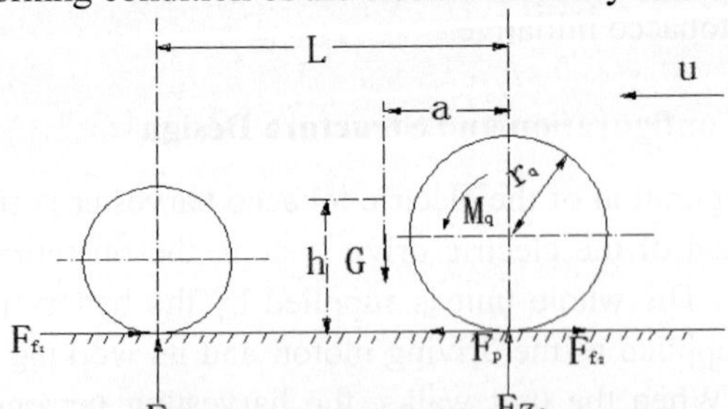

Fig. 2. Electric harvesting machine operation state under the industry by the attempt.

Where G= Use quality walking sites machine; F_{Z1}= Steering wheel vertical reaction force; F_{Z2} =Driving wheel vertical reaction force; F_{f1}=Ground tangential reaction force; F_{f2}=Ground tangential reaction force; F_q=Driving force; M_q=Driving torque; r_q=Driving wheel radius.

The equilibrium equation can be obtained by the equilibrium of the horizontal force.

$$F_q = F_{f1} + F_{f2} \tag{1}$$

From (1) and Table 1 was harvested confidential overcome rolling resistance:

$$F_q = (F_{f1} + F_{f2}) = \mu mg$$

Where μ=Friction coefficient, μ=0.11[1].

After calculation yields: F_q=506N.

4. Parameter Design and Match Motor Selection

In the process of the operation of the machine, the unit has been air drag, acceleration resistance and so on, due to tobacco fields of special terrain and unit low-speed design affect the resistance is relatively small, in which only consider the driving wheel of the resistance, eventually the results can add correction coefficient, harvesting machine power:

$$N = \frac{F_{qv}}{3.6} = 506 \times 2 \times 10^{-3} = 1.012kw$$

The motor power shall not be less than the rated power of the load at the time of operation:

$$P_N = \frac{kN}{\eta_0 \eta_T} \tag{2}$$

where P_N=Motor rated power; k=Motor power reserve coefficient, k=1.2[2]; N=Operation load required rating; η_0=Transfer efficiency of DC motor, η_0=0.85[3]; η_T=Unit traction efficiency, η_T=0.55[4].

Motor rated power can be expressed as:

$$P_N \geq \frac{kN}{\eta_0 \eta_T} = 2.60kW$$

Taking into account the load fluctuation, P_N=3kw, the parameters of the motor are shown in Table 2.

Table 2. The basic parameters of the motor.

Parameter	Numerical
Rated power /kw	3
Maximum power /kw	6
Rated speed /(r/min)	1500
Maximum speed /(r/min)	3300

4.1. *Determination of battery number*

The choice of battery mainly determines the number of batteries and the capacity of the battery. The minimum voltage of the battery should meet the motor the minimum working voltage needs, in order to make sure the battery the minimum number; secondly also according to the continuous working time of the unit to select the number of batteries, and finally select both the maximum value as the final number of cells [5]. The basic parameters of the battery are shown in Table 3.

Table 3. The basic parameters of the battery.

Parameter	Numerical
Battery capacity (Ah)	130
Battery rated voltage (V)	12
The internal resistance of the battery(mΩ)	<12
depth of discharge	0.8

4.2. *Maximum power determination of battery number*

The main parameters of the battery include battery type, capacity and quantity. In the practical application, the maximum power of the battery can be expressed as [6]:

$$P_{b\max} = \frac{2E_0^{\ 2}}{9R_0} \tag{3}$$

Where P_{bmax}=Battery maximum output power; E_0=Battery initial force; R_0=Battery internal resistance.

When the electric chassis is working, the maximum output power of the battery must meet the need of maximum motor power, and the number of batteries can be introduced.

$$n_1 \geq \frac{P_{N\max}}{P_{b\max}\mu_k} \tag{4}$$

Where$P_{N\max}$=Motor maximum power demand;η_0=Controller efficiency,η_0=0.9. After calculation yields:

$$P_{b\max} = 2.67$$

The results of the calculation are brought into the formula (4):

$$n_1 \geq \frac{6}{2.67 \times 0.9} \approx 2.50$$

4.3. ***Determine the number of batteries for a continuous operation time***

Operation time refers to the time of the electric recovery machine in the condition of the saturation of electricity, and use to express the time of continuous operation.. Using the energy of the battery rated power output and the energy consumption of the electric picking machine to calculate the operating time [7]. By formula (1) equal to multiplying both sides by the operation condition of different stalls operating speed, get traction power equation:

$$P_0 = (\frac{F_{f1} + F_{f2}}{3.6})u_j \tag{5}$$

Where U_j=Operating speed of job stalls.

Rated output of energy battery:

$$W = 3600nC_aE_0D_m \tag{6}$$

Where n=Battery Number; c_a= Battery Rated Capacity; D_μ=Depth of discharge. According to the discharge characteristics of the battery-known: the battery discharge current is greater than the nominal discharge current, the total energy output is reduced. The actual amount of discharge[8].

$$W = 3600nC_aE_0D_m(\frac{I_e}{I})^{k-1} \tag{7}$$

The continuous operation time of the electric recovery machine can be obtained by the formula (5) and the type (7).

$$t_n = \frac{3.6nC_a E_0 D_m (\frac{I_e}{I})^{k-1}}{(F_{f1} + F_{f2})u_j} \tag{8}$$

The number of batteries also needs to meet the continuous operating time of the harvesting machine, from the type (8) can export the number of batteries:

$$n_2 \geq \frac{t_n(F_{f1} + F_{f2})u_j}{3.6C_a E_0 D_\mu (\frac{I_e}{I})^{k-1} \eta_T} \tag{9}$$

Final maximum power and continuous operation time are selected to determine the maximum number of batteries. After calculation yields:

$$n_2 \geq 1.12$$

Considering the volatility of the load, the number of batteries is finally taken. n=3.

5. Determination of Transmission Ratio

In field operations, electrical sites are required for different jobs according to the corresponding operating speed. According to the demand of tobacco fields, the basic operating speed and transportation speed are set up [9]. Electric chassis transmission system as shown in Figure 3.

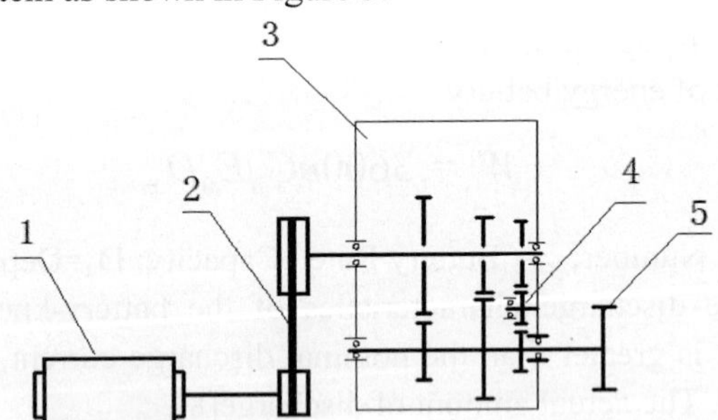

1. DC motor 2 .V type belt 3. Transmission 4. Drive wheel
Fig. 3. The transmission diagram of electric harvesting machine.

It is known that the ratio of transmission has the following relations[10]:

$$u_j = 0.377\frac{n_e r_q}{i} \tag{10}$$

Where n_e=Motor speed rating.

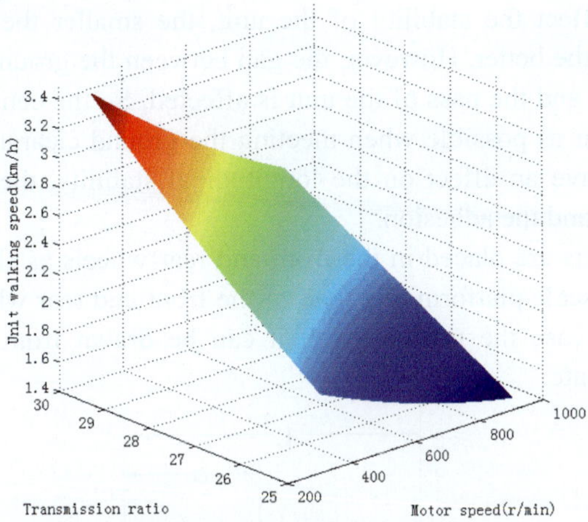

Fig. 4. Relationship transmission ratio, motor speed, speed of the unit.

From the figure, it can be seen that the transmission ratio, the relationship between the motor speed and set speed, gear box bit has two forward gears and one reverse gear, the file transmission ratio calculation results as shown in Table 4 shows satisfying the requirements of the work performance of the unit.

Table 4. Gear ratio of each shift.

Stall	Transmission ratio
I	34.56
II	23.18
Reverse	28.98

6. Stability Analysis

The stability of the oil recovery machine is the performance of the chassis can be stable on a certain slope. The complex operating environment and agronomic requirements of hilly tobacco fields have increased the stability of the units and the difficulties of the stability of the units. According to relevant standards, the stability of the unit is mainly composed of two indicators of longitudinal stability and lateral stability. The lateral stability is mainly composed of slope angle, lateral critical sliding angle, and vertical stability, and the slope angle of slope and the critical angle of slope and slope are evaluated.

6.1. *Determination of gravity center*

The position of the center of gravity of the electric picking machine can be expressed in longitudinal coordinates a and height coordinates h. The height coordinates h affect the stability of the unit, the smaller the vertical and the lateral stability, the better. However, the gap between the ground and the ground will be reduced, and the pass of the unit is affected. So the center of gravity can be reduced as far as possible when meeting the ground clearance. Longitudinal coordinates a have an effect on the longitudinal stability, the maneuverability and the traction and the adhesion.

(1) The units are placed in the front and rear wheels as shown in Figure 4 on the scale platform, the load of the front and rear wheel is Y_c and Y_q, in this case the unit is level, it can be drawn from the longitudinal coordinate.

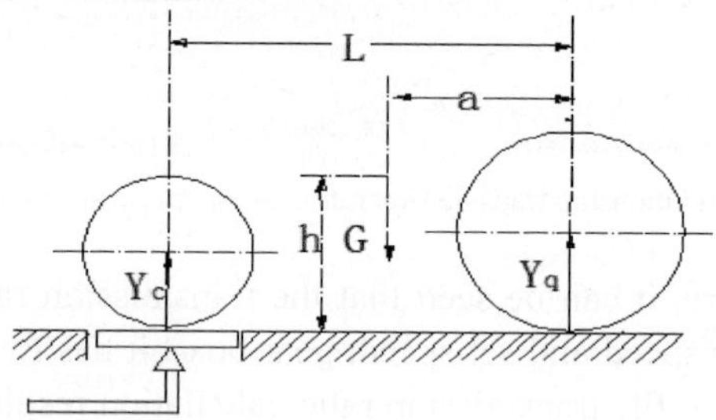

Fig. 5. Determination of unit vertical center of gravity.

$$a=\frac{Y_c}{G}L=\frac{Y_c}{Y_c+Y_q}L \tag{11}$$

The measured data into the formula (11) can be calculated: a=0.45m.

(2) The front unit is placed in a high-profile, shown in Figure 5 at this time is measured as shown in the rear load Y_{q1}, simultaneously measure L_1, H, r_q and r_c get the height of the center of gravity.

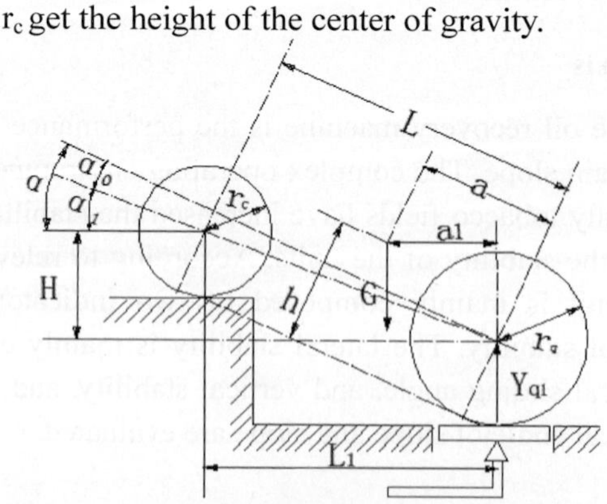

Fig. 6. Determination of the height of the center of gravity of the unit.

$$h = r_q + (a\cot a - \frac{a_1}{\sin a}) \qquad (12)$$

$$a_1 = (1 - \frac{Y_{q1}}{G})L_1$$

$$a = a_0 + a_1$$

$$a_0 = \arctan(\frac{r_q - r_c}{L})$$

$$a_1 = \arcsin(\frac{H\cos a_0}{L})$$

The measured data into the formula (12) can be calculated: h=0.4m.

6.2. *Lateral dip angle*

When the unit parked at an angle on the road, remains stable and will not produce lateral tipping angle is the maximum angle lateral limits of turn. The unit in the transverse direction of the forces ramp shown in Figure 7.

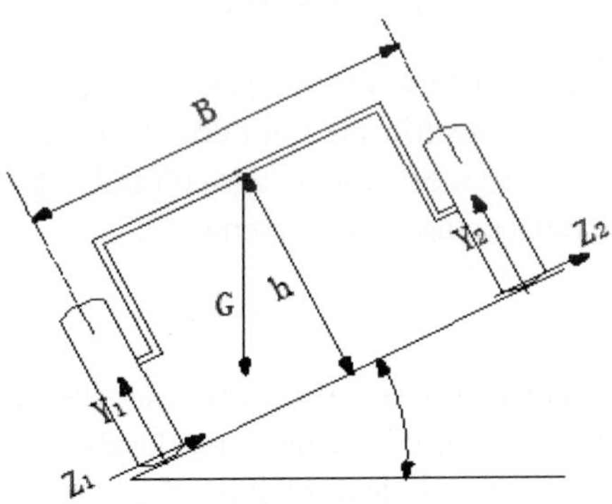

Fig. 7. Harvesting machine lateral ramps on schematic.

$$Z_{1\max} = j_z Y_1$$

$$Z_{2\max} = j_z Y_2$$

Where φ_z=Lateral adhesion coefficient φ_z=0.9φ, φ=0.8.

Electric chassis of the center line of gravity in slope beneath the road wheels connected sites within the unit will remain stable, the center line of gravity if just on the wheel below ground points, the ramp at the top of the two wheels of the vertical reaction force is zero, that is:

$$Y_2 = \left(\frac{B}{2}\cos b_{\lim} - h\sin b_{\lim}\right)\frac{G}{B} \tag{13}$$

where B=Tread wheels.

By the formula (13) lateral limit turning angle is obtained:

$$b_{\lim} = \arctan\left(\frac{B}{2h}\right) \tag{14}$$

The design parameters of electric sites into the formula (14) were: βlim=36.87o.

6.3. *Lateral slip angle*

When the unit is parked in the angle of the road, the maximum angle that is kept stable when the unit is not producing lateral slip is the lateral ultimate slip angle. The condition that the unit does not have the horizontal slip is:

$$Z_{1\max} + Z_{2\max} = j_z(Y_1 + Y_2) \tag{15}$$

This can be calculated from the lateral limit slip angle of the electric site:

$$b_j = \arctan j_z \tag{16}$$

The design parameters of electric sites can be calculated: $\beta\varphi=35.75^{\circ}$.

Through the calculated electric site limit double dip angle and slip angle and agricultural Handbook of mechanical agricultural chassis control, stable performance meet the tobacco use requirements.

7. Conclusion

Currently the actual needs of the specific circumstances of the Chinese tobacco field and in-depth investigation of the farmers, design and development of the walking sites machine, improve leaf picking the degree of automation. The developed electric - picking machine can work in zero emission mode and is safe and reliable. Through the experiment, the electric tobacco harvester design can meet the use requirements.

References

1. Gao Wen-jie. Research on the model of the tractor parametric model for large horsepower tractor sets [D]. Luoyang: Henan University of Science and Technology, 2011.6.
2. Zhou Yi-ming. Automobile tractor theory [M]. Beijing: China Agricultural University press, 2000.

3. Chinese Academy of Agricultural Mechanization Sciences, agricultural mechanical design manual [M]. Beijing: China Agricultural Science and Technology Press, 2007.
4. Zhou Sheng, Zhou Yun-shan. Pure electric vehicle, and Simulation of computer simulation for [J]: 2013,30 (2): 135-139.
5. Zu Yuan-Yang, Qin Da-tong, sun Dong-ye. Electric vehicle power transmission system parameter design and dynamic simulation [J]. China Journal of South China University of Technology: Natural Science Edition, 2006, 34 (4): 33-37.
6. Wang De-zhi. Battery principle and the use of [M]. Beijing: China Railway Press, 1989.
7. Yang Jin-xing, Qiao Wei-gao. Design of power system parameters for electric vehicle [J]. Automotive engineer, 2009(12):31-33.
8. Gao Hui-song electric tractor drive system research [D]. Nanjing: Nanjing Agricultural University, 2008.6.
9. Xue Nian-wen, Gao Xu-Xing. The matching design of power drive system for electric vehicle [J]. Journal of Chongqing Transportation University: Natural Science, 2011, 30 (2): 303-307.
10. Editor of Middle School of Agricultural College of Central China. [M]. Beijing: Agricultural Press, 1983.

Fault diagnosis of motor rotor based on IPSO-LSSVM with hybrid kernel function

L. Wang, W.W. Li† and L. Zhang
Air and Missile Defense College, Air Force Engineering University, Xi an, 710051
†Email: kgdliweiwei@163.com

J. Liu
Unit 95425 of PLA, 655000 Qujing, China

In order to improve the diagnosis accuracy of asynchronous motor rotor fault, based on improved particle swarm optimization (PSO) and least squares support vector machine (LS-SVM) with mixed kernel function, a novel and feasible fault diagnosis method was proposed. First the IPSO with dynamic learning factor and inertia weigh was applied to strengthen the search capability of the standard PSO. Then the mixed kernel function was constructed by combining the polynomial and the radial basis function (RBF). At last, the parameters of the LS-SVM model with mixed kernel function were optimized through IPSO to complete the motor rotor fault diagnosis. The research results show that the proposed method not only overcomes the disadvantage of easily trapping into the local extremum in the later evolution period, but also obviously improves the diagnosis accuracy of asynchronous motor rotor fault.

Keywords: Rotor fault diagnosis; Particle swarm optimization (PSO); Least square support vector machine (LS-SVM); Hybrid kernel function.

1. Introduction

Asynchronous motor is widely used and plays a crucial role in modern industry systems. Its electricity consumption accounts for a considerable proportion of the total power system load. Of all asynchronous motor faults, the ratio of rotor fault is about 10%, so it is one of the important causes resulting in the key asynchronous motor failure in the systems [1]. Therefore, it is of practical significance to detect and diagnose the incipient fault of the asynchronous motor rotor.

At present, the diagnosis process of fault diagnosis system mainly includes fault signal acquisition, fault feature extraction and fault pattern recognition [2]. Many artificial intelligence algorithms are suitable for fault pattern recognition, such as artificial neural network (ANN), expert system (ES), support vector machine (SVM) and fuzzy classification algorithm, etc [3]. SVM is a new learning machine based on statistical learning theory and it can separate different classifications via a small amount of sample data. SVM overcomes the shortcoming that classifiers must be constructed based on a large number of sample data, it has very good prospect in pattern recognition [4]. But the learning ability and classification accuracy of SVM are greatly influenced by parameters. The grid search method is able to achieve optimization in certain conditions, however, if the search range is too large, it will take too much time and the classification accuracy won't be ideal. After the introduction of optimization algorithm, parameter selection for SVM based on genetic algorithm (GA) [5] and SVM based on particle swarm optimization [6] were proposed, which greatly improved the optimization effectiveness.

In this paper, the standard PSO and LS-SVM were both modified. The key parameters of LS-SVM based on hybrid kernel function were optimized by the improved PSO, then the proposed method was applied to the fault diagnosis of asynchronous motor rotor. Compared with other optimization algorithms, it verified that the proposed method avoided the problem of trapping into local extremum and both of the diagnosis efficiency and the diagnosis accuracy were greatly improved.

2. Motor Rotor Fault Theory

Faults of asynchronous motor rotor are commonly classified into broken rotor bar, end ring fault and rotor eccentricity, etc. When the motor just starts, the large impact force and the fast heating up will result in rotor thermal stress. Then broken rotor bar and end ring crack will happen after a long time. Rotor eccentricity fault including dynamic eccentricity and static eccentricity [7] is due to incorrect assembling, bearing bending deformation after a long time running and so on, these reasons will result in an uneven air gap between the stator and the rotor.

Stator current spectrum analysis is an effective method for the rotor fault diagnosis. If the motor is normal, the frequency of the stator current will be fixed which equals to the power supply frequency. But if the rotor is out of order, the sidebands will appear in corresponding position of the spectrum diagram. By finding out the additional frequency components, the cause and the degree of rotor fault can be judged.

If there is something wrong with the rotor bar or the end ring, an additional component with specific frequency will appear in the stator winding current, which can be viewed as the characteristics of the fault. The specific frequency can be formulated as

$$f = (1 \pm 2ks) f_1, k = 1, 2, 3 ..., n \tag{1}$$

Where f_1 is the power supply frequency and s is the slip.

In the condition of rotor eccentricity fault, the air gap permeance along the circumferential direction will appear uneven. Then a relevant component with specific frequency will still appear in the stator winding current and the specific frequency can be formulated as

$$f_{ecc} = f_1[\frac{(R \pm n_d)(1-s)}{p} \pm n_w] \quad (2)$$

where p is the number of motor poles, R is the number of rotor bars, $n_d = 1$ holds when eccentricity is dynamic and $n_d = 0$ holds when eccentricity is static. Also, $n_w = 1,3,5,\cdots$.

3. LS-SVM with Hybrid Kernel Function

3.1. *LS-SVM*

LS-SVM is an improvement of the standard SVM. In order to simplify the calculation, Suykens transformed the problem of quadratic programming in the standard SVM into solving linear equations and established the optimal separating hyperplane. The classification principle of LS-SVM is as follows:

Firstly, given a set of training samples $\{(x_i, y_i)\}$ $(i = 1,2,\cdots N)$, where $x_i \in R^n$ are the input samples and $y_i \in \{-1,1\}$ are the output samples correspondingly, N is the total number of training samples and n denotes the input dimensions of sample space. Based on the principle of structural risk minimization, the objective function and constraint condition of LS-SVM model are

$$\begin{cases} \min R(\omega,\xi) = \frac{1}{2}\omega^T\omega + \frac{c}{2}\sum_{i=1}^{N}\xi_i^2 \\ st.\, y_i[\omega^T \square \varphi(x_i) + b] = 1 - \xi_i \end{cases} \quad (3)$$

where $\varphi(x_i)$ is a non-linear mapping function, ω is the weight vector, b is the offset, ξ_i is the error variable and c is the penalty factor whose value is positive.

Then, the Lagrange function is introduced to solve partial derivatives to ω, b, ξ_i and α_i respectively. Given the requirement that each partial derivative is equal to 0, the equations will be obtained as follows:

$$\begin{bmatrix} 0 & \boldsymbol{Y}^T \\ \boldsymbol{Y} & \boldsymbol{ZZ}^T + c^{-1}\boldsymbol{I} \end{bmatrix} \begin{bmatrix} b \\ \boldsymbol{\alpha} \end{bmatrix} = \begin{bmatrix} 0 \\ \boldsymbol{1} \end{bmatrix} \quad (4)$$

Where

$$\begin{cases} \mathbf{Y} = [y_1, y_2, \cdots y_N]^{\mathrm{T}} \\ \boldsymbol{Z} = \left[\varphi(x_1)^T y_1, \cdots, \varphi(x_N)^T y_N\right]^T \\ \boldsymbol{\alpha} = [\alpha_1, \alpha_2, \cdots, \alpha_N]^{\mathrm{T}} \\ \mathbf{1} = [1, 1, \cdots 1]^{\mathrm{T}} \end{cases} \tag{5}$$

where $\boldsymbol{I}$ is an unit matrix, α_i is the Lagrange multiplier and $\varphi(x_i)^T \square \varphi(x_j)$ is the inner product operation that can be replaced by $K(x_i, x_j)$ which is a kernel function satisfying Mercer condition.

Last, the function estimation of LS-SVM is

$$y(x) = sign[\sum_{i=1}^{N} \alpha_i y_i K(x_i, x_j) + b] \tag{6}$$

3.2. *LS-SVM based on hybrid kernel function*

The fundamental role of LS-SVM kernel function is to transform low dimensional inseparable problem into high dimensional separable problem and to avoid the complex calculation caused by 'large dimension' in the meantime. The kernel function and parameter have a great influence on classification accuracy and training speed of LS-SVM [8]. At present the most commonly used kernel functions are mainly the following:

Polynomial

$$K(x_i, x_j) = [(x_i \cdot y_j) + 1]^q \tag{7}$$

Radial basis function

$$K(x_i, x_j) = \exp(-\frac{\|x_i - x_j\|^2}{2\sigma^2}) \tag{8}$$

Sigmoid

$$K(x_i, x_j) = \tanh[v(x_i, x_j) + c] \tag{9}$$

In Eq.(7), Eq.(8) and Eq.(9), the parameters q, σ, v and c are all real constants. In practical applications, the kernel function and the corresponding parameter values are generally confirmed in accordance with a concrete analysis of specific situations.

Above the common kernel functions, the polynomial is a global kernel function and the RBF is a local kernel function due to the specialties that the global kernel function is good in generalization performance and bad in learning ability, on the contrary, the local kernel function has a strong learning ability and a weak generalization performance [9-10]. We can combine the advantages of two kernel functions respectively and construct a hybrid kernel function with both good generalization performance and learning ability. The hybrid kernel function can be described as

$$K(x_i \cdot y_j) = \lambda K_{poly} + (1-\lambda) K_{RBF} = \\ \lambda[(x_i \cdot y_j) + 1]^q + (1-\lambda)\exp(-\frac{\left\|x_i - x_j\right\|^2}{2\sigma^2}) \quad (10)$$

where λ is the weight coefficient of the hybrid kernel function and $\lambda \in (0,1)$. Its optimal value range is usually (0.55, 0.99).

The proposed kernel function must satisfy Mercer condition. The simple proof process is as follows:

According to the Mercer theorem, the kernel matrix that is defined on any finite set of points based on the kernel function is a positive semi-definite matrix. Given $\boldsymbol{K}_1$ and $\boldsymbol{K}_2$ which are positive semi-definite matrixes on a finite set of points, for any vector $\boldsymbol{a}$, $\boldsymbol{a}^T \boldsymbol{K}_1 \boldsymbol{a} \geq \boldsymbol{0}$ and $\boldsymbol{a}^T \boldsymbol{K}_2 \boldsymbol{a} \geq \boldsymbol{0}$ are true, thus it will be deduced that $\boldsymbol{a}^T (\boldsymbol{K}_1 + \boldsymbol{K}_2)\boldsymbol{a} \geq \boldsymbol{0}$ and $\boldsymbol{a}^T (\lambda \boldsymbol{K}_1)\boldsymbol{a} \geq \boldsymbol{0}$ are true.

Therefore, the hybrid kernel function is a linear combination of two kernel functions and it still satisfies Mercer condition.

4. IPSO-LSSVM with Hybrid Kernel Function

4.1. *PSO*

PSO is proposed as a global search algorithm and its basic concept originates from the research on predatory behavior of bird flock [11-12]. The calculation unit of PSO is particle, which is equivalent to the individual in genetic method. Every particle represents a candidate solution in the solution space, the virtues or the defect degree of the solution is determined by fitness degree function. The velocity of each particle will guide its flight direction and distance. By iteration, the particles will constantly update the optimal solution until the best solution is achieved.

Given a D-dimensional search space, the number of the particles is M and the position of i is $\{\boldsymbol{X}_i = (\mathrm{x}_{i1}, \mathrm{x}_{i2}, \cdots, \mathrm{x}_{id}, \cdots, \mathrm{x}_{iD})\}$ $(i = 1, 2, \cdots M)$. The particle updates itself by the following two extreme values in each iteration. One extreme value named $\boldsymbol{P}_{best}$ is the best one gained by the particle itself and $\boldsymbol{P}_{best} = (p_{i1}, p_{i2}, \cdots, p_{id}, \cdots, p_{iD})$; the other extreme value named $\boldsymbol{G}_{best}$ is the best one obtained by the whole swarm and $\boldsymbol{G}_{best} = (g_{i1}, g_{i2}, \cdots, g_{id}, \cdots g_{iD})$. The updating formulas of velocities and positions are as follows:

$$v_{id}^{k} = wv_{id}^{k-1} + c_1 rand_1 \left(p_{id}^{k-1} - x_{id}^{k-1}\right) + c_2 rand_2 \left(g_d^{k-1} - x_{id}^{k-1}\right) \tag{11}$$

$$x_{id}^{k} = x_{id}^{k-1} + v_{id}^{k} \tag{12}$$

where k denotes the number of iteration, $rand_1$ and $rand_2$ are random constants in the range of [0, 1], w represents the inertia factor, c_1 and c_2 are both learning factors.

4.2. *Improved PSO*

In the standard PSO, the parameters w, c_1 and c_2 are fixed, so it isn't conducive to algorithm optimization. Hence, the improved PSO enhances simultaneously both the search ability and convergence performance of the system by dynamically adjusting the parameter values.

For the model, it requires a great global search capability in the early stage and a good local search ability is essential as it is close to the optimal solution in the later. Because the bigger w is, the algorithm will do better in global search capability and be easier to jump from the local extremum, similarly, the smaller w is, the algorithm will do better in local search ability to accelerate the convergence process. Therefore, to improve the algorithm accuracy and the convergence velocity, the equation that w will change dynamically with the iteration number is provided as follows:

$$w = w_{\max} - (w_{\max} - w_{\min})\frac{T}{T_{\max}} \tag{13}$$

where $w_{\max}$ represents the initial weight and its value is 0.9; $w_{\min}$ indicates the final weight and its value is 0.45; T and $T_{\max}$ respectively represent the current optimization number and the maximum optimization number.

The learning factors c_1 and c_2 affect the self-learning ability and the overall learning ability respectively. In the initial stage, the strong self-learning ability and the weak overall learning ability are necessary, which are just opposite in the later. Therefore, c_1 should decrease gradually and c_2 should increase gradually. The adjustable equations with the iteration number are

$$\begin{cases} c_{1i} = c_{10} - c_{10}\dfrac{T}{T_{\max}} \\ c_{2i} = c_{20} + c_{20}\dfrac{T}{T_{\max}} \end{cases} \tag{14}$$

where c_{1i} and c_{2i} represent respectively the learning factors of the generation i; c_{10} and c_{20} are the initial learning factors.

4.3. *Applying IPSO to optimize LS-SVM with hybrid kernel function*

As the penalty factor, c determines the complexity of the model and the punishment level of the fitting deviation, which can keep the model's generalization ability. The kernel width of RBF is σ which controls the complexity of the sample distribution in the high-dimensional feature space. The weight coefficient of the hybrid kernel function is λ which reflects the proportion of the hybrid kernel function that polynomial accounts for. So c, σ, q and λ directly affect the classification accuracy and generalization ability of LS-SVM. Searching the optimal values of parameters via plenty of experiments will obviously take a lot of time and efforts. Therefore, to improve the classification accuracy of the model fast and efficiently, this paper searched the optimal values of (c, σ, q, λ) by IPSO and the steps were as follows:

Step1 Given the initial parameters of IPSO, such as c_{10}, c_{20}, D, M, $T_{\max}$ and the initial iteration number T=0.

Step2 Given the range of the optimized parameters. The values of (c, σ, q, λ) were given randomly, which was viewed as the initial position.

Step3 Calculate the fitness value of each particle and compare $F(x_i)$ with the optimal solution $F(P_{best})$ which was gained by the particle itself. If $F(x_i) < F(P_{best})$, $\boldsymbol{P}_{best}$ would be replaced by x_i, the fitness function was expressed as

$$F(x_i) = 1 - F(p) \tag{15}$$

Where $F(p)$ is the classification accuracy.

Step4 Search the global optimal solution by comparing $F(P_{best})$ with $F(G_{best})$. If $F(P_{best}) < F(G_{best})$, $\boldsymbol{G}_{best}$ would be replaced by $\boldsymbol{P}_{best}$.

Step5 Update the velocities and locations of the particles according to Eq.(11) and Eq.(12). According to Eq.(13) and Eq.(14), update the inertia factor and the learning factor respectively.

Step6 If the iteration number reached up to $T_{\max}$ or $F(x_i)$ reached a certain precision, output the parameters or return to **Step3**.

Step7 Establish the optimized LS-SVM model. After the parameters search, the model would be applied to fault diagnosis of the motor rotor.

5. Experiment

The essence of fault diagnosis was pattern recognition. In this paper, we diagnose the faults of the motor rotor by IPSO-LSSVM with hybrid kernel function. Four kinds of faults were selected for test, which were listed in Table 1.

Table 1. Typical faults of motor rotor.

Fault ID	Fault description
I	Normal
II	Broken rotor bar
III	Static eccentricity
IV	Dynamic eccentricity

The troubleshooting block diagram was shown in Fig. 1.

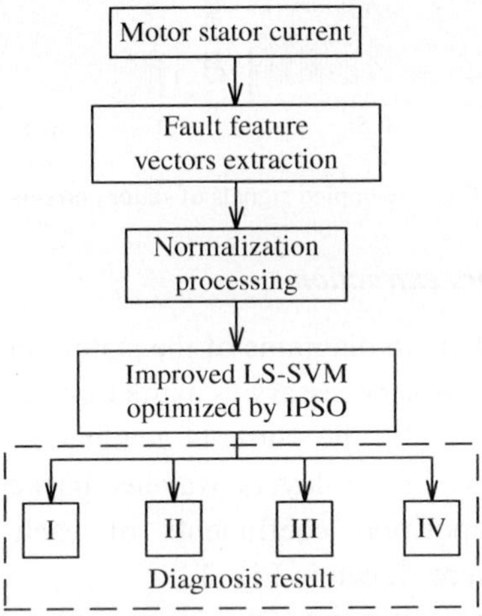

Fig.1. Block diagram of rotor fault diagnosis.

5.1. *Current acquisition*

In the experiment, given a squirrel-cage induction motor whose parameters were 1.1kW, 50Hz, 220V and s=0.02 as the test motor. When it loaded stably, we would measure the stator currents of different states that included normal, broken rotor bar, static eccentricity and dynamic eccentricity. The sampling frequency was set as 1000Hz and the number of the sampling points was 1000. There were 80 groups of samples for each state, so the total sample number of four states was 320.

After the experimental equipment had been ready, the motor stator currents of four states would be acquired. The time-domain diagrams of stator currents were showed in Fig. 2.

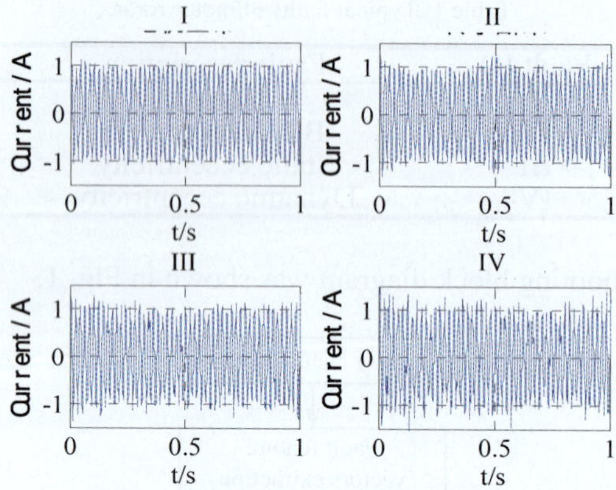

Fig. 2. Sampled signals of stator currents.

5.2. *Fault feature vectors extraction*

According to the time-domain diagrams of the stator currents, the fault type still couldn't be definite, so it was necessary to extract the fault feature vectors which reflected fault feature. Firstly, the currents acquired should be denoised, then decomposed the currents into 3 layers wavelet packet by db3 to obtain the wavelet packet decomposition coefficients of each frequency band. The decomposition results were shown in Fig. 3.

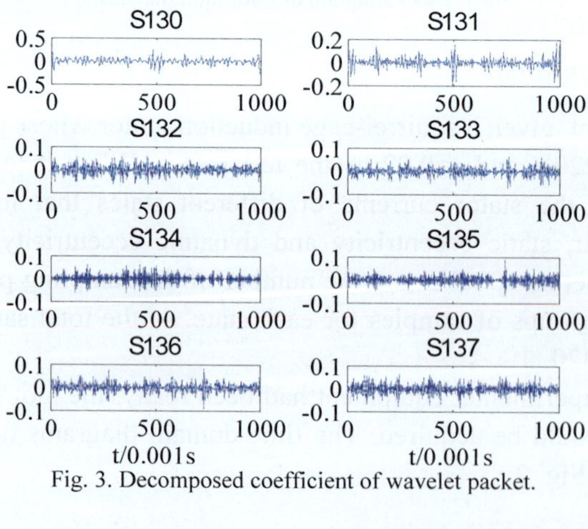

Fig. 3. Decomposed coefficient of wavelet packet.

The energy value of each frequency band in Fig. 3 was calculated and the normalization processing was done to obtain the fault feature vectors as follows

$$T^{'} = \frac{T}{E} = [\frac{E_{30}}{E}, \frac{E_{31}}{E}, \frac{E_{32}}{E}, \frac{E_{33}}{E}, \frac{E_{34}}{E}, \frac{E_{35}}{E}, \frac{E_{36}}{E}, \frac{E_{37}}{E}] \tag{16}$$

Where $E_{3i}(i=0,1,\cdots,7)$ was the energy value of each frequency band, E was the sum of all the energy values, the feature vectors were shown in Table 2 as follows.

Table 2. Feature vectors of different rotor fault.

Fault ID	Feature vector			
I	0.1323	0.1572	0.1194	0.1174
	0.1167	0.1269	0.1147	0.1154
I	0.1343	0.1777	0.1149	0.1109
	0.1316	0.1162	0.1015	0.1129
II	0.2928	0.1864	0.0885	0.0880
	0.0856	0.0841	0.0841	0.0905
II	0.2942	0.1807	0.0908	0.0908
	0.0839	0.0928	0.0864	0.0805
III	0.2123	0.1400	0.1063	0.1029
	0.1069	0.1066	0.1152	0.1100
III	0.2158	0.1349	0.1179	0.1075
	0.0985	0.1069	0.1096	0.1090
IV	0.1718	0.1431	0.1088	0.1184
	0.1086	0.1088	0.1253	0.1151
IV	0.1759	0.1329	0.1164	0.1086
	0.1292	0.1027	0.1188	0.1153

At last, the faults would be diagnosed by optimized LS-SVM after inputting T'.

5.3. *Diagnosis model construction*

LS-SVM was suitable for binary classification, but there were four faults in total need to identify, so it was necessary to combine several models. In this paper, the method to construct multi-classifier by one to one was adopted, whose basic idea was to construct a LS-SVM classifier between any two categories, As a result, the total number of LS-SVMs was 6. After classification, the type of fault was determined by voting. Voting was a commonly used approach in multi-classification [13], which referred to that once the fault type had been determined through the classifier, the number of the corresponding fault votes would be added once. Eventually the fault was which the vote number was the biggest. The rotor diagnosis model was shown in Fig. 4 as follows.

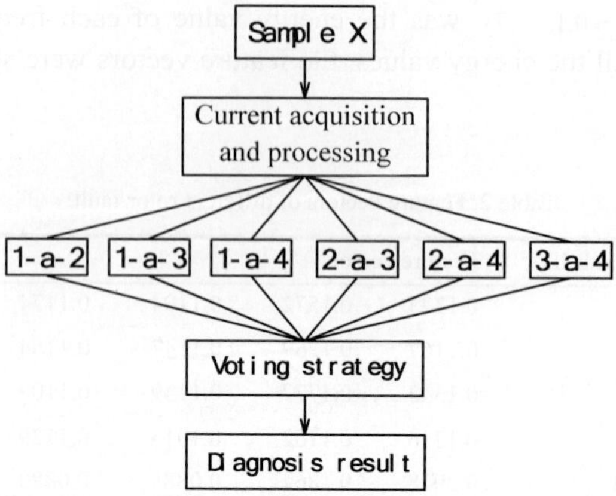

Fig. 4. The model of rotor fault diagnosis based on LS-SVM.

5.4. *Troubleshooting*

The feature vectors of all the motor faults were inputted into the optimized LS-SVM for diagnosis. Of all the vectors, 200 groups were used as training samples and 120 groups were used to test. By adjusting repeatedly, finally the optimal combination of (c, σ, q, λ) was (0.1, 0.01, 1.12, 0.38). Then retrained the training samples to verify the accuracy of the proposed method and checked the generalization ability of the new algorithm through the test samples. The diagnosis results were shown in Table 3, the training and test results both indicate the ratio of the correct identification number to the total number.

Table 3. Diagnosis results of the proposed method.

	I	II	III	IV
Training results	50/50	50/50	49/50	50/50
Training accuracy	100%	100%	98%	100%
Test results	30/30	28/30	28/30	29/30
Test accuracy	100%	93.3%	93.3%	96.7%

We could know from Table 3 that the new method successfully achieved the motor rotor fault diagnosis. To demonstrate the better effectiveness of the proposed method, the feature vectors were inputted into BP neural network, LS-SVM and LS-SVM optimized by the standard PSO respectively for training and testing, the results were shown in Table 4.

Table 4. Compare of diagnosis results.

	c	σ	Iterations	Average accuracy
BP	-	-	308	89.3%
LS-SVM	0.17	0.31	212	92.9%
PSO-LSSVM	0.25	0.16	192	94.7%
IPSO-ILSSVM	0.10	0.01	176	95.8%

We could draw a conclusion from the compare of diagnosis results in Table 4 that other algorithms were easy to trap into the local extremum during parameter optimization and the IPSO-LSSVM with hybrid kernel function had higher diagnosis accuracy. We could also observe that the iteration number of the proposed method was fewer by the convergence curves of improved PSO in Fig. 5 and it improved the diagnosis efficiency effectively.

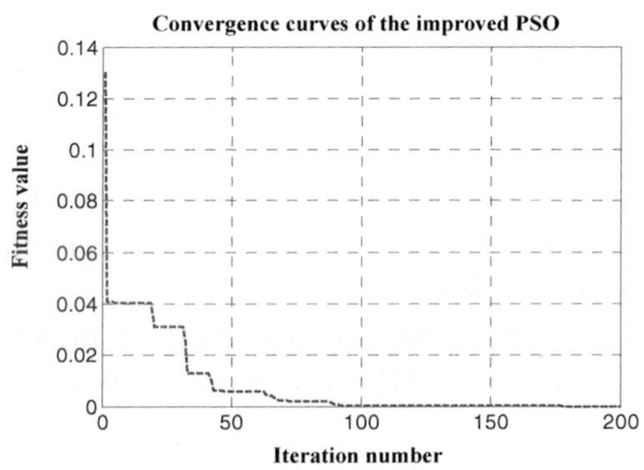

Fig. 5. Convergence curves of the improved PSO.

6. Conclusions

As concerning the problem that LS-SVM parameters were difficult to determine, in this paper, the improved PSO was combined with LS-SVM based on the hybrid kernel function to diagnose the induction motor rotor fault. The diagnosis results show that the new method can find out the optimal parameters of the model with fewer iteration number. It overcomes the shortcomings that the original method is easy to trap into the local extremum and takes less search time, moreover, the diagnosis efficiency and diagnosis accuracy of the proposed method are better.

Funding

Project Category: National Natural Science Foundation of China. Project name: Complex dynamic system fault prediction and optimal maintenance based on semi-quantitative information. Item number: 61370031.

References

1. MA Hong-zhong. Motor performance monitoring and fault diagnosis [M]. Beijing: China machine press, 2008: 786-796.
2. LI Ping, LI Xue-jun, JIANG Ling-li, et al. Fault diagnosis of asynchronous motor based on KPCA and PSOSVM[J]. Journal of vibration, measurement & diagnosis, 2014, 34(4): 616-620.
3. YI Hui, MEI Lei, LI Li-juan, et al. Vibration fault diagnosis for hydroelectric generating units using the multi-class relevance vector machine [J]. Proceedings of the CSEE, 2014, 34(17): 2843-2850.
4. Vapnik V N. The nature of statistical learning theory [M]. New York: Springer-Verlag, 1999.
5. Chen P W, Jungying W, Hahnming L. Model selection of SVMs using GA approach [C]. Proceedings of IEEE International Joint Conference on Neural Networks, Piscataway, 2004: 2035-2040.
6. HUANG Cheng-lung, DUN Jian-fan. A distributed PSO–SVM hybrid system with feature selection and parameter optimization [J]. Applied Soft Computing, 2008, 8(4): 1381-1391.
7. BAO Xiao-hua, Lü Qiang. Review and prospect of air-gap eccentricity faults in induction machines [J]. Proceedings of the CSEE, 2013, 33(6): 93-100.
8. XU Yong-gang, MENG Zhi-peng, LU Ming. Fault diagnosis method of rolling bearing based on dual-tree complex wavelet packet transform and SVM [J]. Journal of aerospace power, 2014, 29(1): 67-73.
9. ZHAO Ji-wen, WANG Ya-hua, CHEN Pan-pan, et al. Research on linear motors models based on the combined kernel function of multiple support vector machine [J]. Electric machines and control, 2014, 18(2): 90-95.
10. ZHU Yong-li, YIN Jin-liang. Study on application of multi-kernel learning relevance vector machines in fault diagnosis of power transformers [J]. Proceedings of the CSEE, 2013, 33(22): 68-74.
11. XU He-ming. Research on multi-objective particle swarm optimization algorithms [D]. Shanghai: Shanghai Jiao Tong University, 2013: 2-15.
12. WANG Xiao-chuan, SHI Feng, YU Lei, et al. The MATLAB neural network analysis of 43 cases [M]. Beijing: Beihang university press, 2013: 127-136.

13. Moreira M, Mayoraz E. Improved pairwise coupling classification with correcting classifiers [C], Processing of the 10th European Conference on Machine Learning. Germany Chemnitz: [S. n.], 1998, 18: 160-171.

Study on passenger risk assessment in coach front impact test

X. Zhang, B. Wang*, X. Yu & L.X. Geng

Key Laboratory of Operation Safety Technology on Transport Vehicles, Ministry of Transport, PRC, Research Institute of Highway Ministry of Transport, Beijing, Courtyard 8#, Xitucheng Road, Haidian District, 100088, China

**b.wang@rioh.cn*

Based on three kinds of coach, the authors study on protection ability of semi-integral and unitary coach body and two point and three point seat belt from driver room deformation, test dummies damage value and the acceleration value of test dummies head by applying test of coach front impact at speed of 30km/h. The result showed that: the driver room deformation of unitary coach body is larger than semi-integral coach body. The acceleration value of test dummies head in passenger seat of unitary coach body is smaller than semi-integral coach body. The maximum acceleration value of passenger head is more than 200g when using two point seat belt and passenger neck is badly hurt because two point seat belt just only fix waist. The maximum acceleration value is less than 35g when using three point seat belt. So three point seat belt can protect passenger better than two point seat belt.

Keywords: Coach; Front impact; Driver; Passenger; Test dummy.

1. Introduction

Coaches are flat head structure in our country and the drivers and passengers are injured easily in coach front impact [1-3]. So it is meaningful to study on coach front impact [4-6].

The paper study on protection ability of coaches from driver room deformation,test dummies damage value and the acceleration value of test dummies head.

The test dummies damage upper threshold is Table1. If the test dummies damage value is less than the threshold, it can be thought that passenger can bear and vice versa.

Table 1. The upper threshold of test dummies [7].

Dummies position	Upper damage threshold of head(HIC)	Chest compression distance(mm)	Left thigh force (kN)	Right thigh force(kN)
Driver rear side dummy	1000	75	10	10
Tour guide rear side dummy	1000	75	10	10
lift side second row dummy	500	/	/	/

The HIC is defined by the following equation:

$$HIC = (t_2 - t_1)\left[\frac{1}{t_2 - t_1}\int_{t_1}^{t_2} a dt\right]^{2.5} \quad (1)$$

Where, a is acceleration of head; t1 and t2 is any time between impact (the unit is s).

2. Test Condition

2.1. *The basic parameter of test coach*

The test used three kinds of coaches as object study. The Table 2 the basic parameter of test coach.

Table 2. Basic parameter of test coach

Coach type	Coach A	Coach B	Coach C
length×width×height(mm)	8995 x 2480 x 3435	12000 x2500 x3720	10490 x2500 x3550
Curb weight	8800kg	12900kg	10675kg
The body form	semi-integral	unitary	semi-integral
Seat belt form	Two point type	Tour guide rear seat three point type, other seats two type point	Three point type

2.2. *Impact speed and method*

The test used 100% front impact, the impact speed is 30km/h.

2.3. *The test dummies position*

The test used Hybrid3 dummies, but the driver seat was installed ECER29 foam dummies because the driver seat dummies are destroyed easily during impact.

The first row passenger seat is installed Hybrid3. The Fig. 1-3 and Table 3 showed the details.

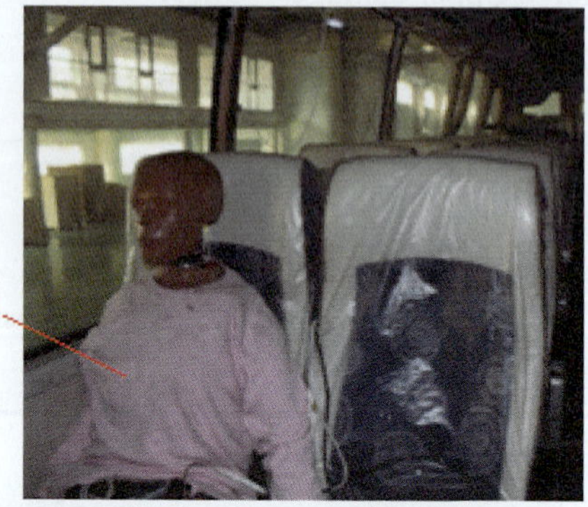

Fig. 1. A dummy position.

Fig. 2. B dummy position.

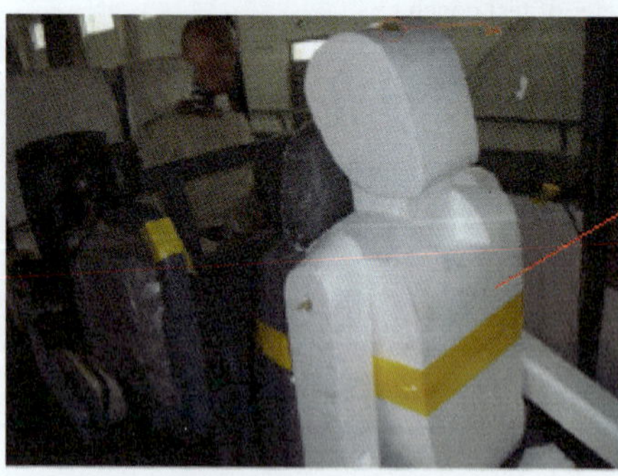

Fig. 3. C dummy position.

Table 3. Dummy type and position.

Dummy No.	A dummy	B dummy	C dummy
Dummy type	Hybrid3 dummy	Hybrid3 dummy	Foam dummy
position	Tour guide rear seat	Driver rear seat	Driver seat

3. Test result and analysis

3.1. *Driver room deformation*

The steering wheel and steering wheel column infringed driver room deformation. The Fig. 4 showed the test result.

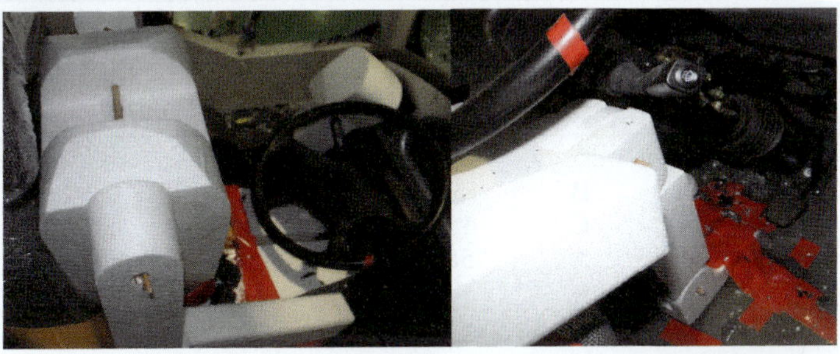

Fig. 4. Driver room deformation.

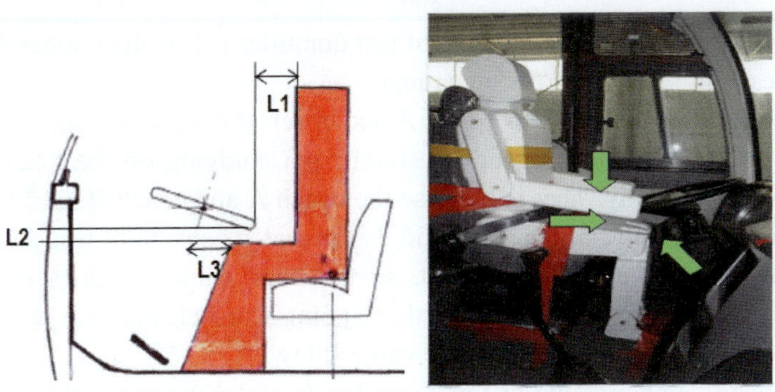

Fig. 5. Driver room deformation measurement diagram.

The driver space is defined by L1, L2 and L3. The Fig. 5 showed the L1, L2 and L3 meaning [8].

The deforming resistance of coach B is worse than coach A and coach C. That is because coach A and coach C are semi-integral coach body and have longitudinal beam. Longitudinal beam can enhance the driver room structural strength. Coach B are unitary coach body without longitudinal beam, so coach B deforming resistance is worse.

The Table 4 showed the driver seat deformation result:

Table 4. Driver room deformation measurement result.

coach	L1	L2	L3
A	33mm	36mm	>160mm
B	contact	contact	contact
C	185mm	78mm	140mm

3.2. *Test dummies damage value and analysis*

The Table 5 showed test dummies damage.

Table 5. Test dummies damage value.

Dummy position	Coach	Head damage	Chest compression distance(mm)	Left thigh force (kN)	Right thigh force (kN)
Driver rear seat dummy	A	1807	0.05	1.2	7.7
	B	1586	37.42	0.7	1.2
	C	156	35.53	0.7	1.5
Tour guide rear seat dummy	A	582	48.67	1.38	0.7
	B	128	0.09	1.2	0.8
	C	253	/	1.8	1.1
Left dummy of the second row	A	/	/	/	/
	B	510	/	/	/
	C	800	/	/	/

The chest and leg damage value of test dummies is less than upper threshold, so the chest and leg are in tolerance range.

The head damage value of coach A and coach B exceed the upper threshold and coach C is far below upper threshold from studying on the head damage value of B test dummy. That is because the coach A and coach B used two point seat belts and there is clapboard behind driver seat. The head of test dummy and clapboard collide during impact test, so the head damage values exceed the upper threshold. The coach C used three point seat belt and the head can be fastened well [9], so the head damage value is low.

The head damage value of three coaches is under the upper threshold from studying on the head damage value of A test dummy. That is because there is no clapboard behind tour guide seat.

The head damage value of coach B is low from studying on the second row passenger, that is because coach B is unitary coach body and the body can absorb energy during testing. The coach C used three point belt, but the belt fasten point is on seat back and the seat back will rock when impact, so the head damage value of coach C is also high [10].

3.3. *The result of acceleration value of test dummies head and analysis*

The acceleration value of test dummies is the important cause of passenger injuring and death. The test dummies get the longitudinal acceleration during test. The following Fig showed the head acceleration value of test dummies.

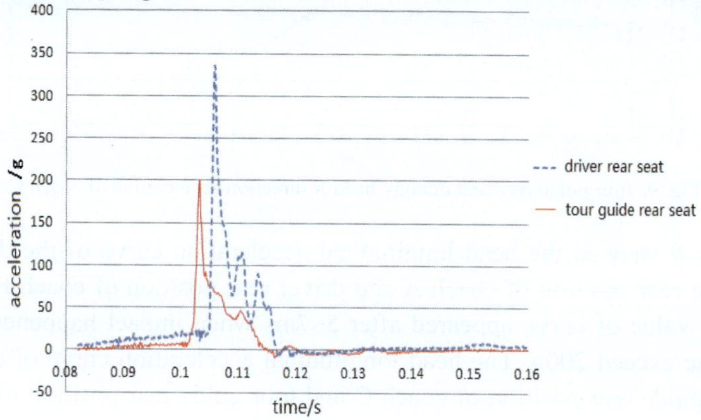

Fig. 6. Dummy head X direction.

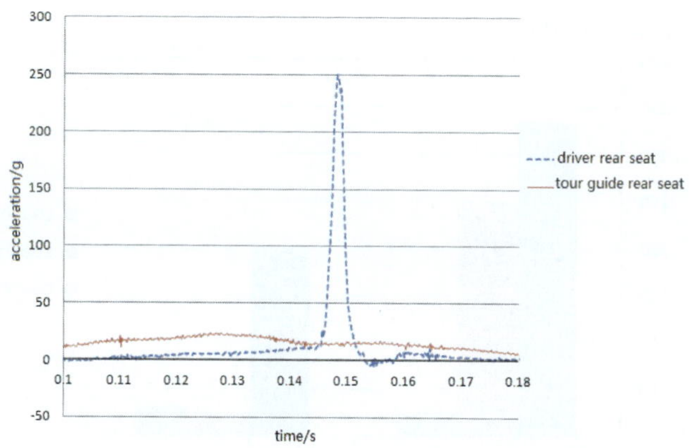

Fig. 7. Dummy head X direction acceleration of coach B.

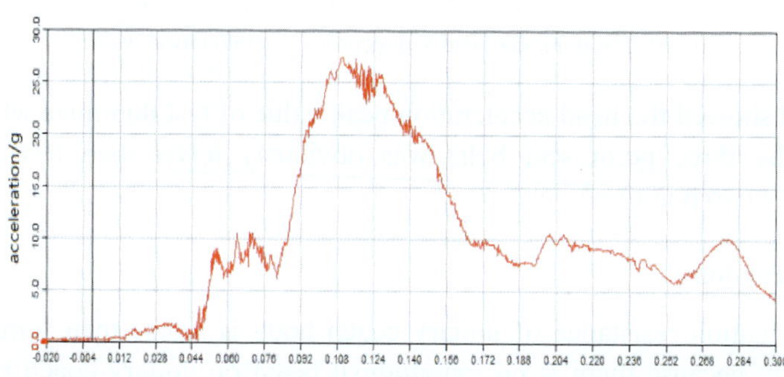

Fig. 8. Driver rear seat dummy head X direction acceleration of coach C.

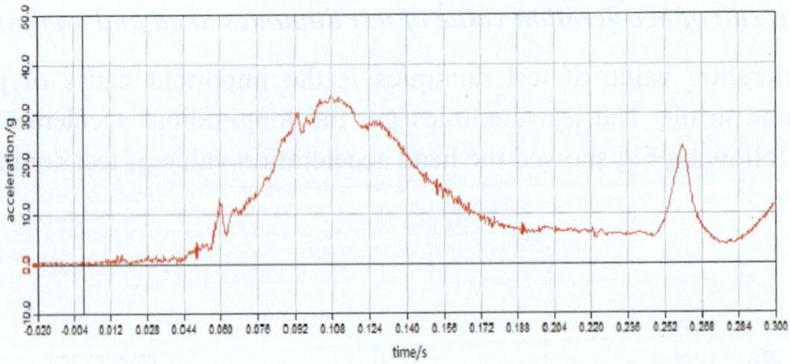

Fig. 9. Tour guide rear seat dummy head X direction acceleration of coach C.

Fig.6~9 showed the head longitudinal acceleration curve of the driver and tour guide rear position of coach A and driver rear position of coach B is taper. The peak value of curve appeared after 5~7ms when impact happened and the peak value exceed 200g. The head longitudinal acceleration curve of the driver and tour guide rear position of coach C and tour guide rear position of coach B changes gently. The peak value of curve appeared after 45~65ms when impact happened and the peak value is less than 35g. That is because three point belt can reduce the head impact.

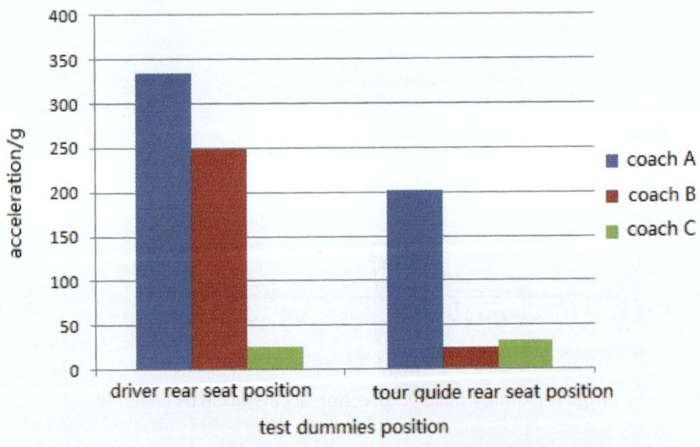

Fig. 10. Test dummies head X direction acceleration peak value.

Fig.10 showed the head acceleration peak value of test dummies which were fastened by three point seat belts was obviously lower than test dummies fastened by two point seat belt.

4. Conclusions

The deformation resistance of unitary coach body is worse than semi-integral coach body because there is no longitudinal beam on unitary coach body. But semi-integral coach body can't absorb energy planarity and the almost impact energy transfer to the passengers. So energy absorbing device can be installed to semi-integral coach body.

The three point seat belt reduced head damage value and head acceleration value of passenger obviously in coach front impact test. The two point seat belt just constraint waist and the head of passengers shake to and fro. It's harmful to neck of passenger. So the three point seat belt is better than two point seat belt.

5. References

1. LIU Zhao-jian, XIONG Xin. Study on Coach Front Impact Simulation Analysis Based on ANSYS/LS-DYNA [J]. Bus & Coach Technology and Research, 2012 (6):1.
2. ZHANG Jian. FAN Ti-qiang. HE Han-qiao. Simulation Analysis Bus Front Safety [J]. Bus & Coach Technology and Research, 2009, 31(3):7-9.
3. SHAO Yi-ming. YU Zhi-gang, Vehicle Impact Safety Present Situation And Trend [J].Highways & Automotive Applications, 2006, (4):1-2.
4. ECE-TRANS-WP29-GRSG-86-11, accident statistics (Frontal collisions of buses).
5. ECE-TRANS-WP29-GRSG-2007-33, Frontal collision of buses.
6. http://www.trcpg.com/whatsnewdetail.asp?id=29
7. GB 11551-2003 The protection of the occupants in the event of a frontal collision for motor vehicle.
8. GB 13057-2003 The strength of the seats and their anchorages of passenger vehicles.
9. NHTSA.Federal mot or vehicle safety standard [EB/OL].[2009.11.20]. http://www.nhtsa.gov/cars/rules /standards /FMVSS .Regs.
10. WANG-Xuan. LI Hong-guang. ZHAO Hang. The Modern Automobile Safety [M]. Beijing: China Communications Press, 1998.

Research of the judgment factors about stepping on the accelerator pedal by mistake

W.Y. Yu & J. Bai*
*Inner Mongolia University of Science and Technology,
Baotou, China*
**Email:285745601@qq.com*

With the increasing number of motor vehicles, the security problems from the motor vehicle has increased,there are many traffic accidents which caused by stepping on the accelerator pedal by mistake. This paper elaborates the judgment method about stepping on the accelerator pedal by mistake. To determine the factors can judge whether to step on the accelerator pedal by mistake.After the research,the data of testing air throttle position sensor and distance sensor determines whether the vehicle is in the state of mistakenly stepped on the accelerator.

Keywords: Motor Vehicle; Stepping on The Accelerator Pedal by Mistake; Air Throttle Position Sensor; Distance Sensor.

1. Analysis of the determining factors about stepping on the accelerator pedal by mistake

In recent years, there were many traffic accidents. The reason of this was that drivers stepped on the accelerator pedal by mistake.A series of problems what are brought by the motor vehicle safety are concerned about by people. According to the causes of this series of accidents, specific situation, I design a system that called “Preventing To Step On The Accelerator Pedal By Mistake And Automatic Brake”. The following describes the function of preventing mistakenly stepped on the accelerator and automatic brake system:The driver encounters emergency situation, when the driver steps on the acceletator pedal as the brake pedal mistakenly, this system can make the auto shut-off circuit and automatic brake. Two action was Involved in the whole process of judging: stepping on the accelerator and automatic brake. To determine whether mistakenly stepped on the accelerator and realize the function of automatic brake, we should be based on the vehicle's engine gasoline supply structure and the principle of braking, to find out the factors of mistakenly stepping on the accelerator and the object of final controlling.

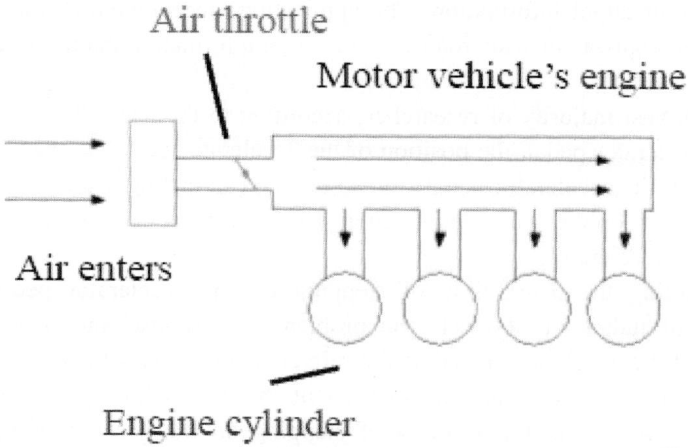

Fig. 1. Engine air intake system diagram.

Figure 1 is engine air intake system diagram, the air enters engine, then air mixes with atomizing gasoline that is supplyed fuel injector, they form the mixed gas. The mixture gas burns to supply the engine's motive. Opening the fuel injector is controlled by the electronic controlling unit of automobile, air throttle position sensor tests air throttle' angle when it opens.

The mixed gas which supplys engine influences engine's rate of work, the air-fuel ratio of mixed gas influences the adequacy of combustion. The research shows that, If you want the full combustion of 1kg gasoline, you need 14.7kg air. There are two ways to make the vehicle slow down. 1. Controls opening angle of air throttle. 2. Controls the amount of gasoline injected. In addition, air throttle is connected with accelerator pedal, The opening angle of air throttle reflects the depth of the accelerator pedal position, there is a linear relationship between the two things. The opening angle of air throttle can reflect velocity or acceleration of the accelerator pedal, stepping on the accelerator mistakenly can be distinguished by velocity or acceleration of stepping on the accelerator pedal. Therefore, to determine whether stepping on the accelerator by mistake can be obtained through the rate of change or twice the rate of change about the opening angle of air throttle. Cutting the fuel injector makes the mixed gas which in fuel injection system not flame fully, this can be used as a method to slow the motor vehicle down.

About the judgement of stepping on the accelerator pedal by mistake. In addition to motor vehicle's throttle opening rate of change related to this element, it also associated with the external condition. When people determine if motor vehicle is dangerous, in addition to considering the error acceleration from the motor vehicle, we also consider that the relative distance of the motor vehicle and the front target information. The application environment of this research topic is the highway or wide roads, so the target information mainly refers to the vehicle ahead.

In the vast majority of researches, according to the using time of stepping on the accelerator pedal, the position of the accelerator pedal, or the velocity of stepping on the accelerator pedal, and so on, we can distinguish two cases what the divers step on the accelerator normally and emergency brake but mistakenly step on the accelerator.

Regarding the using time of stepping on the accelerator pedal as the standard of judgment, it exists the problem of response lag; regarding the position of the accelerator pedal as the reference, this makes the distance of the parking brake lengthen, and the security of this is not high; if we adopt the method of measuring the velocity of stepping on the accelerator pedal, the change of velocity which compared with the change of acceleration is not obvious.

Therefore, there are two main factors that can determine whether mistakenly stepped on the accelerator: the acceleration of accelerator pedal, the distance between the front aim, we need the corresponding sensors, they are air throttle position sensor, distance sensor. In addition, we can also choose a steering sensor to test if the vehicle is in a state which the driver turns on the turn signal and want to overtake the front vehicle, to prevent erroneous judgement of the system.

2. Testing of the acceleration of accelerator pedal when mistakenly stepping on the accelerator

Because of the acceleration sensor is sensitive, It is not suitable to be installed in the vehicle which is oscillatory and noisy, so the acceleration of accelerator pedal can not be directly measured by the acceleration sensor. The acceleration of accelerator pedal is rake ratio of speed variation when stepping on the accelerator, but the velocity of the accelerator pedal can be achieved the variation of the distance from the accelerator pedal longitudinal movement. Therefore, we need to measure the acceleration of the accelerator and the data of air throttle position sensor, then we can find relations between them. Its output characteristics is shown in Figure 2.

The rotary angle A of air throttle and the output voltage of air throttle position sensor is the linear relation (formula (1)), K_1 is a linear constant.

$$V = k_1 A \tag{1}$$

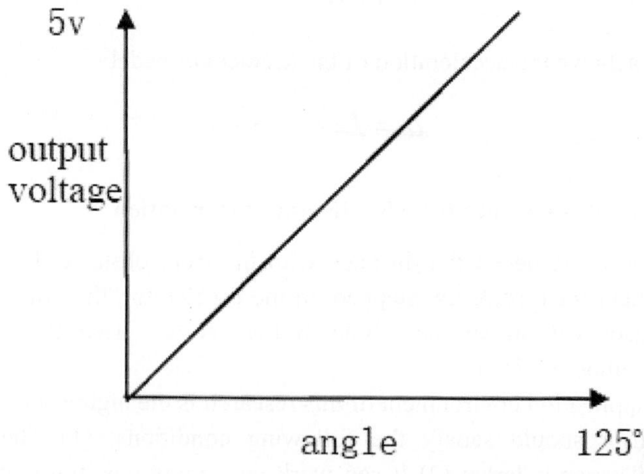

Fig. 2. The output characteristics of air throttle position sensor.

The r is the radius of rotor that is in the air throttle position sensor, the distance of the air throttle's stayguy:

$$L_1 = \frac{A}{360} = 2\pi r \tag{2}$$

Assuming the displacement of the accelerator pedal is L_2, k_2 is a a linear constant, according the relation between L_1 and L_2, a formula is drawn:

$$L_2 = k_2 L_1 \tag{3}$$

According (1), (2) and (3), a conclusion can be known:

$$L_2 = \frac{2\pi k_2 r V}{360 k_1} \tag{4}$$

By the formula known, the displacement of the accelerator pedal and the output voltage of air throttle position sensor is the linear output relation, so L_2 can be reflected by V. Stepping on the accelerator pedal, we can achieve a conclusion by measuring the voltage:

$$V = f(t) \tag{5}$$

If $k = \frac{2\pi k_2 r}{360 k_1}$, so:

$$L_2 = kf(t) \tag{6}$$

We can achieve the acceleration of the accelerator pedal:

$$a = L_2^{''} \tag{7}$$

3. Detection of the front obstacles distance information

Distance sensor can detect the distance with the front obstacle. If we want to design a system for mistakenly stepped on the accelerator, then the function of distance sensor will affect the whole system, so we need to research the measuring method of distance.

As the application environment of this research is the highway or wide road, distance sensor should satisfy the following conditions: (1) The range of measuring distance is large; (2) It can work on a rainy day, fog day and other bad environment; (3) It has high-resolution ratio.

According to the different of mediums of measuring distance, distance sensor can be classified as infrared sensor and microwave sensor, after a lot of consideration, this research should select a microwave sensor, because the microwave sensor has the following advantages: (1) It can detect in harsh environmental condition; (2) It has fast detection speed, high sensitivity, can carry on dynamic detection and real-time processing.

In the working range of distance sensor, its function is to detect the existence of target and to capture the information of target, for example, the relative distance, the relative velocity between the distance sensor and the target, and so on. If the distance sensor receives the reflected signal, this demonstrates that the target is existence; the round-trip time of signal can calculate the distance between distance sensor and target. The distance equation is:

$$R = \frac{\Delta t \cdot c}{2} \tag{8}$$

"Δt" is the round-trip time of signal, "c" is the velocity of mediums.

References

1. Vijay Subramanian, Thomas F. Burks, A. A. Arroyo. Development of machine vision and laser radar based autonomous vehicle guidance systems for citrus grove navigation [J]. Computers and Electronics in Agriculture, 2006, 53(2):130-143.

2. Jurgen Detlefsen, Thomas troll, Michael Rozmann, et al. System aspects and design of an automotive collision warning PN code radar using wave-front reconstruction. IEEE MTT-S Digest, 1992: 625-628.
3. Yung-Chang Cheng. Modeling and nonlinear hunting stability analysis of high-speed railway vehicle moving on curved tracks [J]. Journal of Sound and Vibration, 2009 (324):139-160.
4. Chih-Yung Chen, Hsuan-Ming Feng. Hybrid intelligent vision-based car-like vehicle backing systems design [J]. Expert Systems with Applications, 2009 (36):7500-7509.
5. Marc Heddebaut, Fouzia Elbahhar, Christophe Loye, et al. Millimeter-wave communicating radars for enhanced vehicle-to-vehicle communication [J]. Transportation Research Part C, 2010, 18:440-456.

Research on application in vehicle maintenance teaching of simulated selection of assembly process and methods for planetary steering gear

P. Wang*, X.Y. Wang & X.S. Yan
Changchun Institute of Engineering and Technology,
Changchun, Jilin, China
**Email: 47335774@qq.com*

Two types of simulation models commonly usedfor assembly process for planetary steering gear will be established during vehicle maintenance teaching by using simulation software Flexsim. Comparison and analysis of results from operations of simulation model will find out steering gear assembly process suitable for vehicle maintenance teaching. This has led to good teaching results. Attempts have been successfully made to apply process simulation for selection of assembly process during vehicle maintenance teaching.

Keywords: Assembly process; Simulation; Vehicle maintenance teaching.

Planetary steering gear is heavy, has complex structure, requires long assembly line and needs to fulfill numerous technical specifications. It takes longer time and more energy to complete one cycle of assembly process. Attention has been drawn during vehicle maintenance teaching that most trainees feedback that learning of component assembly process for planetary steering gear is very difficult. In addition, the results are not good. Therefore it is very necessary to research on assembly process for planetary steering gear suitable for vehicle maintenance teaching with good teaching results.

During vehicle maintenance teaching, there are two common methods to be taught on assembly process for planetary steering gear. Main difference is different assembly process for planetary transmission. What assembly process should be used for teaching to enable most trainees to easily gain the knowledge has never been justified. As trainee competence is different from one to another and their technical skill of assembly operation is not stable at beginning, solution to this issue would require plenty of samples available for random test. However, the number of trainees for each turn of component course is far beyond test conditions. In addition, duration of course is limited and cannot be used for experiment. Hence, attempt has been made to use software Flexsim to establish simulation mockup for assembly process. Assembly balance ratio and assembly efficiency are used as justification criteria. With simulation and data analysis, vehicle maintenance teaching specific assembly process suitable for steering gear assembly can be found out.

1. Brief introduction of steering gear

Steering gear is located at end of tank transmission line. It is a rotary component. Such gear consists of many components, heavy, structure complex and requires long assembly line. Its main components are illustrated in figure 1.

Fig. 1. Main component of steering gear.

2. Brief introduction of assembly process for steering gear

Normally, two processes are adopted for teaching on steering gear component assembly:

Process 1: Travelling plate, lockup clutch and planetary transmission are assembled separately before they are put together. Process line is divided into three sections: one short section (travelling plate assembly) and two long sections (separate assembly of lockup clutch and planetary transmission). While putting together lockup clutch and planetary clutch, fit must be guaranteed between the sun gear and four planetary gears, sealing ring of sun gear and inner bore of planetary frame, and inner teeth of inner drum and friction disk of exterior teeth. It is very difficult to put them together. When assembly is smooth, duration of the assembly is very short. If not, rework might be necessary. Duration of assembly cannot be estimated.

Process 2: Assemble travelling plate and lockup clutch separately. Load the components of planetary transmission onto lockup clutch one after another. Process line is divided into two sections: one short section (assembly of travelling plate) and two long sections (assembly of lockup clutch followed by assembly of components of planetary transmission). Assembly of lockup clutch is continued with that of planetary transmission so that patience is required. When process of previous section is smooth, duration of this process is relatively longer. If previous process is not smooth, the duration of this process is relatively shorter.

3. Simulated analysis

3.1. *Simulation mock*

Software Flexsim is used to establish mockup for simulating assembly process of planetary steering gear.

Relationship between simulation mockup and assembly process of planetary steering gear: Generator should correspond to components to be assembled. Processor should correspond to pre-handling. Assembler should correspond to assembly operation. Assembled components should enter into receiver.

Prior to mockup establishment, tighten nuts and lockup nuts. Use processor to simulate tightening of nuts and lockup nuts as pre-assemble. This will simplify mockup while improving speed of simulated operation. In the environment of Flexsim simulation software, drag and drop simulation mockup following the assembly process of planetary steering gear. Specify assembly route between mockups and establish input and output, thereby completing the establishment of simulation mockup, as seen in figure 2. Simulation mockup is coded according to following rule: For combiner and processor in the assembly process, they are prefixed with capital letter "Z". Numerical numbers are used behind the letters to indicate their assembly process (or branch route) and serial number. For example, "Z206" indicates work station 6 in process route 2 (or branch line). That means, code of simulation mockup is the same of work station. The same applies for others.

Fig. 2. Simulation mockup established by Flexsim for assembly line.

3.2. *Set up parameters for mockup*

Common parameters needed to establish mockup for assembly line in the environment of simulation software Flexsim: time of component arrival, capacity of buffer zone, preset time, duration of assembly (or processing time) and duration of simulated operation. For establishment of assembly line mockup for steering gear, parameters are set according to following procedure:

1 Time of component arrival: All components used in reality for assembly process are all qualified. Parameters to be set are normal values lower than assembly speed. This will satisfy the need for simulation.

2 Buffer zone: Sufficient space must be ensured to hold backup parts, that means, assembly process should not stop due to part shortage.

3 Preset time: Preset time will be configured based on duration of actual operations. Arithmetic mean is taken from statistic calculations is used. Therefore, normal value can be used for setting in the simulation process.

4 Duration of assembly (or processing time): In the course of assembly, mockups for processor and comber are used to simulate processing and assembling processes. Detailed data are given in below table:

5 Duration of simulation: Duration of simulation is set as 2160000 seconds. No preheat time.

6 Setup of other parameters: Default value given by the software can be used for the parameters that will not affect simulation results.

Table 1. Process data for lockup clutch assembly.

Work station	**Preset time(S)**	**Processing time (s)**
Z101	0	22
Z102	0	10
Z103	0	36
Z104	0	15
Z105	0	18
Z106	0	90
Z107	0	108
Z108	0	15
Z109	0	20
Z110	0	100
Z111	0	120
Z112	0	20
Z113	0	25
Z114	20	20
Z115	0	10
Z116	0	10

Table 2. Process data for planetary transmission assembly.

Work station	Preset time(S)	Processing time (s)
Z201	0	32
Z202	0	60
Z203	0	60
Z204	0	60
Z205	0	20
Z206	0	50
Z207	0	20
Z208	0	20

Z209	20	20
Z210	0	20
Z211	0	100
Z212	0	120
Z213	0	20

Table 3. Process data for traveling plate assembly.

Work station	Preset time(S)	Processing time (s)
Z301	0	20
Z302	0	25
Z303	10	10

3.3. *Operation of simulation mockups*

Operation of above simulation mockup, as seen in below figure

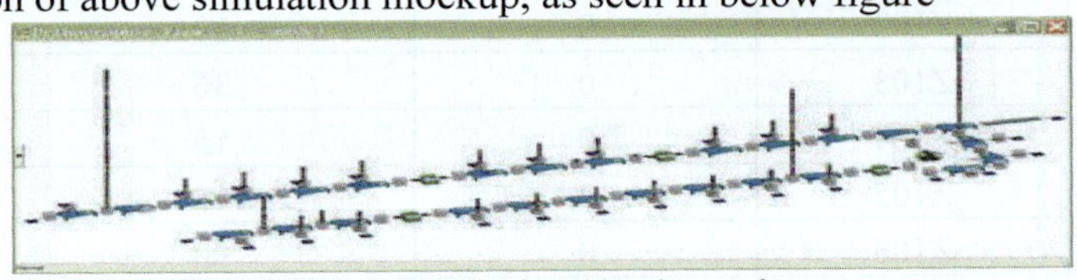

Fig. 3. Diagram of simulated operation.

Table 4. Simulation data report.

Duration of simulation: 216000.008 second; cycle time of assembly: 90 seconds				
Code	Idle	Processing	Collecting	setup
Z101	65.10%	34.90%	0.00%	0.00%
Z102	0.10%	8.60%	91.30%	0.00%
Z103	58.00%	42.00%	0.00%	0.00%
Z104	85.40%	14.60%	0.00%	0.00%
Z105	83.00%	17.00%	0.00%	0.00%
Z106	6.80%	93.20%	0.00%	0.00%
Z107	37.80%	62.20%	0.00%	0.00%
Z108	87.30%	12.70%	0.00%	0.00%
Z109	81.70%	18.30%	0.00%	0.00%
Z110	21.00%	79.00%	0.00%	0.00%
Z111	32.80%	67.20%	0.00%	0.00%
Z112	82.30%	17.70%	0.00%	0.00%
Z113	78.40%	21.60%	0.00%	0.00%
Z114	65.20%	18.00%	0.00%	16.80%
Z115	97.70%	2.30%	0.00%	0.00%

Z116	91.90%	8.10%	0.00%	0.00%
Z201	42.40%	57.60%	0.00%	0.00%
Z202	5.60%	94.40%	0.00%	0.00%
Z203	11.40%	88.60%	0.00%	0.00%
Z204	5.10%	94.90%	0.00%	0.00%
Z205	71.60%	28.40%	0.00%	0.00%
Z206	20.70%	79.30%	0.00%	0.00%
Z207	71.50%	28.50%	0.00%	0.00%
Z208	70.00%	30.00%	0.00%	0.00%
Z209	38.80%	31.30%	0.00%	29.80%
Z210	72.60%	27.40%	0.00%	0.00%
Z211	5.50%	94.50%	0.00%	0.00%
Z212	21.60%	78.40%	0.00%	0.00%
Z213	81.20%	18.80%	0.00%	0.00%
Z301	69.20%	30.80%	0.00%	0.00%
Z302	57.90%	42.10%	0.00%	0.00%
Z303	66.70%	16.70%	0.00%	16.70%

Equilibrium rate of assembly process is calculated using following formula [1]:

$$\eta = \frac{\sum_{i=1}^{n} T_i}{n \cdot T_{\max}} = \frac{T}{n \cdot T_1} \qquad (1)$$

T: Sum of duration of all process operations

n: number of process steps

T_1: duration of bottleneck process step

In data report generated by simulation software Flexsim, idle rate can be obtained for individual process steps. Therefore, formula to calculate equilibrium rate of production line can be converted as below [2]:

$$\frac{\sum_{i=1}^{n} T_i}{n \; T_{\max}} \quad \frac{\sum_{i=1}^{n} T \;\; 1 \;\; T_1}{n \;\; T \;\; 1 \;\; T_2} \qquad (2)$$

T: Total of work time

T_1: idle time of individual process step

n: number of process steps

T_2: idle rate of bottleneck process step

$$\eta = \frac{\sum_{i=1}^{n} T_i}{n \cdot T_{\max}} = \frac{\sum_{i=1}^{n} 1 - T_1}{n \times (1 - T_2)} \quad (3)$$

Assume equilibrium rate for the first type of assembly process is η_1, it can be obtained that η_1 =41.82%. This indicates very low equilibrium rate of assembly process.

Statistic data from simulation is given in below table according to above method while simplifying simulation mockup.

Table 5. Simulation data report.

Simulation time: 216004.680 seconds; cycle time of assemble: 60 seconds				
NO.	idle	processing	collecting	setup
Z101:	62.80%	37.20%	0.00%	0.00%
Z102:	0.50%	35.40%	64.10%	0.00%
Z103:	30.10%	69.90%	0.00%	0.00%
Z104:	22.50%	77.50%	0.00%	0.00%
Z106:	16.50%	83.50%	0.00%	0.00%
Z107:	13.30%	86.70%	0.00%	0.00%
Z108:	18.20%	81.80%	0.00%	0.00%
Z110:	6.50%	93.50%	0.00%	0.00%
Z111:	6.80%	93.20%	0.00%	0.00%
Z112:	7.10%	92.90%	0.00%	0.00%
Z114:	28.50%	40.50%	0.00%	30.90%
Z115:	33.00%	36.20%	0.00%	30.80%
Z116:	8.40%	91.60%	0.00%	0.00%
Z118:	39.10%	30.50%	0.00%	30.50%
Z120:	8.90%	91.10%	0.00%	0.00%
Z121:	9.20%	90.80%	0.00%	0.00%
Z122:	9.50%	90.50%	0.00%	0.00%
Z123:	45.40%	54.60%	0.00%	0.00%
Z201:	47.40%	52.60%	0.00%	0.00%
Z202:	1.20%	98.80%	0.00%	0.00%
Z203:	1.50%	98.50%	0.00%	0.00%
Z204:	1.70%	98.30%	0.00%	0.00%
Z205:	1.40%	94.90%	0.00%	0.00%
Z401:	2.80%	97.20%	0.00%	0.00%
Z403:	69.20%	15.10%	0.00%	15.70%

Equilibrium rate of the second assembly process is $\eta_2 = 75.75\%$ as calculated using formula (3).

After optimization, original assembly line has demonstrated significant achievement: reduced work stations, shortened cycle time, improved equilibrium rate of assembly line. It can be comprehended after comparison on simulation results that the second assembly process is better then the first assembly process for teaching on vehicle maintenance.

4. Actual results of learning

Partial of the learning is divided into two sections: trainer lecture and trainee learning. Acceptance is designed based on this. Acceptance undergoes in the course of vehicle maintenance teaching. Data from totally 5 batches of samples are collected, 2 groups (learning of two different assembly processes) for each batch. 10 trainees are randomly chosen from each group. It is deemed excellent when more than 80% of the assembly process can be understood. Otherwise, the result is not good.

Statistic results are given in table 6 and 7 as below:

Table 6. Data overview of acceptance statistics,

No.	Group 1(the first assembly process)		Group 2 (the second assembly process)	
	Well taught (persons)	Well exercised (persons)	Well taught (persons)	Well exercised (persons)
1	4	6	3	7
2	5	7	4	7
3	3	6	4	9
4	4	6	3	7
5	5	7	4	8

Table 7. Rates of positive feedback.

	Pass rate of group 1	Pass rate of group 2
Effects after lecture	42%	36%
Effects after exercise	64%	76%

Assembly process. However, the second assembly process can yield better results of learning. The reason is the first assembly process is divided into three sections, which are easy to memorize and can show better results after lecture. However, exercise or practice might need combined assembly occasionally.

Rework or incomplete assembly will damage trainee enthusiasm, thereby affecting learning of new knowledge. The second assembly process is divided into two sections, which are hard to memorize. However, practicing is basically smooth. Trainees are proud of their achievement and have good knowledge.

In summary, the second assembly process for planetary steering gear is preferred for teaching in term of better results of learning.

5. Conclusion

In this article, simulation software Flexsim is used in combination of teaching practice and data to establish two types of simulation mockup for assembly process for planetary steering gear. After operations, it is concluded from statistical analysis of simulation results that the second assembly process is more fit for trainee learning. As there is no big difference between two results, there is still long way for justification through relatively longer term of teaching experience. Acceptance data can also be used to simulate original data for cyclic simulation analysis.

For selection of teaching method, advanced simulation process should be selected for simulation analysis. Hence, the conclusion valuable for practical application can be obtained. This is new practice where simulation can be applied for selection of teaching method.

References

1. William Stevenson. Production and operation management [M], translated by Zhangqun, Zhangjie: China Machine Press, 2000.
2. Chen Hongxia, Wang Xiaoyi, Zhou Qinhua. Research on simulation mockup and optimization for continuous extrusion flat aluminum tube production line [J] Forging technology, 2010, issue 3, 153-157.

Research on sewage diffusion from different azimuth and jet angles of QI Dong marine discharge engineering

H. Yu

Tianjin Research Institute for Water Transport Engineering,
Tianjin, 300456,
42717930@qq.com

The sewage diffusion is an environmental material for protecting marine environment. It plays an important role in port engineering. Its effect under deep water was carried out between different diffuser design parameters based on the hydraulic model test, and QI Dong sewage marine disposal project was taken as an example. For this project, jet angle was controlled to be 5 degree, and azimuth angle was controlled to be 0 degree. It not only can ensure that sewage fully diluted mixed, also can avoid the premature convergence and sewage lifting. It is feasible for selecting large angle nozzle under deep water and there are some technical bases have been provided for other sewage marine disposal project.

Keywords: Azimuth angle; Jet angle; Horizontal model.

1. Introduction

Sewage diffusion is a new material which is used to marine environmental protection. The design of jet and horizontal angle plays an important role about diffusion effect of sewage marine discharge. In this article, the QI Dong sewage marine disposal project was taken as an example, diffuser for sewage dilution has been analyzed in deep water based on the physical model test technique. It ensures the implementation of sewage marine disposal project, and put forward to take dip nozzle diffuser structure. It provides a theoretical basis for the future to carry out deep water sewage deep sea discharge study.

2. Research statuses

The diffuser of California Ohno sewage marine disposal project [1] used the single nozzle structure on the rising pipe,the diameter of rising pipe is bigger than the nozzle diameter, it flowed by elbow bend; the project of Australia's North Head [2] used “I” structure diffuser and it is multi nozzle structure, the space of two rising pipe is 21m, every rising pipe had 6 jets; Malabar sewage marine disposal engineering [2] used "L" structure diffuser and every rising pipe had 8 jets; the hydraulic model tests of the United States of America Boston sewage discharged engineering[1] shows that plume discharge nozzle can form a raised ring when the nozzles of rising pipe are more than eight, it will reduce the extent of dilution of sewage; Wilkinson studied the seawater full time of pipe discharge system, it provided reference to predict water wedge formation and development as well as the cleaning time and cleaning process. YAN Zhong-min [3] studied the dilution characteristics of porous diffuser in finite depth waters and the limited width of near-field which combined with the development direction of river discharge in China; according to the physical model test,XU Gao-tian, WEI He-ping [4] improved the design of diffuser which combined with features of Shang hai city sewage treatment project; WANG Chao [5] put forward "T" structure diffuser from shipping, construction and other point of view; ZHANG Guang-yu, ZHAN Shui-fen [6] comprehensively studied the effects of sewage diffusion in near field about horizontal azimuth, jet angle, length, depth of water, jet velocity and other design parameters.

3. Research Example

3.1. *Model design*

Due to the speed difference between the jet and ambient water in the vicinity of the outfall vary greatly, the jet with environmental water blending process has a strong three-dimensional, so it must use the normal physical model. Model designed according to Froude guidelines, the flow pattern in the model must be uniform to the prototype. Reynolds number should be greater than the critical Reynolds number for the flow pattern is similar to the model vents.

$$\mathrm{R}_{ej\min} = \frac{u_{j\min} D_{\min}}{\nu} \geq \mathrm{R}_{ec}$$

$\mathrm{R}_{ej\min}$ —the orifice Reynolds minimum in the model; $u_{j\min}$ —the minimum orifice flow rate in the model(m/s); $D_{\min}$ —The minimum orifice diameter in the model(m); ν —the water sports viscous coefficient; R_{ec} —critical Reynolds number.

3.2. *Analysis of test results*

3.2.1. *Analysis of the results of different azimuth angle*

The vents level azimuth directly related to sewage jet path and diluted diffusion effect on the project, there are three programs which include the azimuth of 0 °, 30 °, 90 ° as a physical model test program. Table 1-Table 2 are the test results of different horizontal angle.

Table 1. 0° azimuth angle test results

Tide	Length of diffuser	Dilute multiple		
		100m	200m	500m
High tide	80m	96.4	145.6	179.3
	100m	102.1	163.3	208.4
Falling tide	80m	121.8	147.9	184.4
	100m	142.9	187.2	214.5
Recreation stream	80m	Diffusion range:0.038km^2		
	100m	Diffusion range:0.051km^2		

Table 2. 30° azimuth angle test results

Tide	Length of diffuser	Dilute multiple		
		100m	200m	500m
High tide	80m	98.4	134.1	163.4
	100m	101.4	158.5	201.8
Falling tide	80m	124.5	121.7	176.9
	100m	135.7	173.1	194.3
Recreation stream	80m	Diffusion range:0.035km^2		
	100m	Diffusion range:0.047km^2		

When the azimuth angle is 0°, the jet perpendicular to the flow direction for the environment, sewage initial dilution is best; the reason is the sewage rapidly mixes with ambient water which is subject to strong perturbations of the environmental flow. Due to the environmental depth of this project is big, it is not easy to be formed the sewage mass on the surface, it is conducive to secondary dilution and diffusion of the sewage, so it has less impact for the environment.

3.2.2.*Analysis of the results of different jet angle*

The jet angle is one of the important factors that affect the dilution of sewage, it is very important for starting dilution, but it is not has a complete computing model currently. On the project, due to the water depth conditions is better, according to the actual situation it can be appropriately increase jet angle, the jet angle of 0°, 5°, 10° three programs, as a physical simulation test program to verify. Table 3-Table 4 are the test results of different jet angel.

Table 3. 0° jet angle test results

Tide	Length of diffuser	Dilute multiple		
		100m	200m	500m
High tide	80m	134.2	168.1	204.1
	100m	201.6	203.4	225.8
Falling tide	80m	157.1	206.3	237.3
	100m	208.5	231.6	251.9
Recreation stream	80m	Diffusion range:0.031km^2		
	100m	Diffusion range:0.035km^2		

Table 4. 5° jet angle test results

Tide	Length of diffuser	Dilute multiple		
		100m	200m	500m
High tide	80m	119.4	146.7	173.2
	100m	147.2	184.1	204.1
Falling tide	80m	115.1	169.7	217.5
	100m	158.3	218.4	238.4
Recreation stream	80m	Diffusion range:0.027km^2		
	100m	Diffusion range:0.032km^2		

The tests showed that longitudinal diffusion related to the shape of jet angle, the greater the angle of jet and perpendicular, after injection of the jet, due to hydraulic around the flow resistance, jet slowly bend, in the meantime, the jet and the cross flow cross mixed slowly, and its width growing. When the jet angle is 5°, due to strong environmental flow disturbance and the sewage has some drift distance, it is difficult to form sewage mass, the project can achieve better dilution and diffusion. Considering the actual water depth in the project, diffuser jet angle is recommended 5°.

3.2.3. *Analysis of different diffuser length*

The length of the diffuser is one of the important factors that affect the dilution and diffusion of sewage. Considering the water quality requirement of sewage discharge field in this project is relatively high, it is selected different programs of diffuser length (80m~100m) for physical model tests, and makes sure the diffuser length program.

Table 5. 80m diffuser; 0° azimuth angle; jet angle 0° test results

Tide	Length of diffuser	Dilute multiple		
		100m	200m	500m
High tide	80m	174.4	209.5	269.5
Falling tide	80m	197.1	214.6	281.9
Recreation stream	80m	Diffusion range:0.036km^2		

Table 6. 100m diffuser; 0° azimuth angle; jet angle 0° test results

Tide	Length of diffuser	Dilute multiple		
		100m	200m	500m
High tide	100m	217.4	269.1	293.1
Falling tide	100m	257.4	278.9	302.4
Recreation stream	100m	Diffusion range:0.043km^2		

4. Conclusions

According to the comprehensive analysis of the physical model tests, it is obtained diffuser length is 100m through research and analysis of the physical model tests. Due to the outfall water depth is about 7m, it is recommended that the horizontal angle is 0° and the jet angle is 5°. In this way it is not also ensure the sewage and the water fully mixed, meet the dilution requirements, and also be able to avoid formed sewage mass in the sea surface. It can be seen that in the deep water conditions, increases the diffuser jet angle and horizontal azimuth angle can ensure better sewage dilution and diffusion through QI Dong sewage marine disposal project physical model test, it can ensure the good operation of the sewage deep discharge project. The sewage diffusion is a key environmental material for port engineering and marine environmental protection.

References

1. XIA Qing. Sewage marine disposal design theory and method [M].1996,5.
2. CHEN Fu, ZHANG Yong-liang, MENG Wei. Experience of urban sewage marine disposal in Australian [J]. Research of environmental Sciences. 1994, 7(5):49-52.
3. WANG Chao, YAN Zhong-min. Experimental study on the dilution characteristics of the near zone blending emission about river diffuser [J]. Hydraulic journal.1993, 6.
4. XU Gao-tian, WEI He-ping. Dilution at the near field of multi-port discharging wastewater at Bailonggang outfall—Shanghai wastewater treatment project II [J]. Marine environmental Science. 2000,19(4):41-45.
5. WANG Chao. Study on flow phenomenon of Tee multiport diffuser discharge [J]. Advances in water sciences. 1995, 6(3):176-181.
6. ZHANG Guang-yu, ZHAN Shui-fen, YUAN Jun-ling. Model Test on Dilution and Diffusion in the Area Near Diffuser [J]. Traffic Environmental Protection. 2005, 26(1):1-6.

The application of orthogonal experimental and integrated balance in injection molding

H.L.Tian[1,a], H.B.Liu[1,b*], X.J.Wang[1,c]

Faculty of Mechanical and Electrical Engineering, Kunming University of Science and Technology ,

Kunming 650500, Yunnan Province, China

[a]594626863@qq.com, [b]lihhong6696@163.com, [c]69509143@qq.com*

The model of mouse base was taken as the object based on the production, Moldflow software and orthogonal experimental. The warpage and volume shrinkage in the injection molding were treated as quality objective, to study the effects of process parameters on part's quality. Two optimum parameter combination for two quality objective were obtained by analysis of range. The process parameter combination giving attention to the two quality objectives was selected by integral balance method and the optimized process parameter was verified by simulation.

Keywords: Orthogonal experimental method; Integrated balance method; Process parameters.

1. Introduction

In this paper, with the mouse base as the research object, using the warping deformation and volume shrinkage as quality indicators, the use of CAE simulation analysis combined with the orthogonal test method, analysis the influence of process parameters on the quality of plastics, ,analysis the degree of influence of process parameters by range method, combined with the integrated balance method, to determine the balance process parameters of two quality indicators, and simulations are carried out to verify.

2. Pretreatment of analysis

In this paper, the research object is a mouse base. Using Pro/E5. 0 to establish mouse base. Mesh grid matching rate is 92.2%. The work piece material is PC+ABS.

Fig. 1. Gating and cooling system.

Due to the structural characteristics of plastic parts, the ejection position in the column legs, supporting rods, wire board and the positions of both sides, the establishment of the cooling system as shown in Figure 1, upper and lower cooling circuits. Considering the smooth surface of parts, ejection and cooling structure position, therefore the use of latent type gate, as shown in Figure 1.

3. Orthogonal test

The orthogonal experiment based on the orthogonal table, with less computing time to obtain many factors influence on the final analysis of target size, each test are highly representative, is the high efficiency the experimental design method to study the effect factors [1,2].

The experimental factors were: A(mold temperature), B(melt temperature), C(injection time), D(pressure holding time), E(holding pressure), according to the plastic parts material selection and injection machine and other objective conditions to determined each factor level value, as shown in Table 1. According to table 1 set of each factor value, can only choose 6 factors and 5 levels orthogonal table L_{25} (5^6), and the last column is null columns [3].

Table 1. Factors and levels of experiment.

levels	factors				
	A(/°C)	B(/°C)	C(/s)	D(/s)	E(/MPa)
1	50	260	1	8	65
2	55	265	1.2	10	70
3	60	270	1.4	12	75
4	65	275	1.6	14	80
5	70	280	1.8	16	85

4. Analysis of warpage

Using the orthogonal test table lists 25 groups test war page results simulated by Moldflow, and calculate the average and range of values, the experimental data and the results are shown in Table 2. The primary and secondary effects of degree: E>C>B>D>A. The optimal sequence is $A_1B_5C_5D_5E_5$. According to the data in Table 2, draw the effect of each factor level trend as shown in Fig. 2 drawing with Origin.

Combined with the analysis of range and Fig. 2, appropriately increasing the mold temperature, warping is decreased; mold temperature is too high will increase the warpage. The high melt temperature made the melt has good mobility, the internal stress of flowing is smaller, The warping deformation gradually decreases. Injection time within 0.8 ~ 1.2s, with the melt filling more and more closely, warpage decreased; in 1.2 ~ 1.4s, the injection time will increase the contraction rate, so that the warping deformation increases; after that, with the increase of injection speed injection time decreases, reduced the shear stress so the warping deformation decreases.

The holding time increased from 8s to 12s, warping deformation gradually increases, the pressure holding time continued, increasing stress difference products in all directions, so that the war page amount increase. Increasing the holding pressure, reduced the warping deformation.

Table 2. Experimental data and results.

	A	B	C	D	E	War. (mm)	Shr. (%)		A	B	C	D	E	War. (mm)	Shr. (%)
1	1	1	1	1	1	0.3488	11.080	14	3	4	1	3	5	0.2350	5.681
2	1	2	2	2	2	0.3035	11.150	15	3	5	2	4	1	0.3078	6.824
3	1	3	3	3	3	0.2647	11.060	16	4	1	4	2	5	0.2397	10.330
4	1	4	4	4	4	0.2522	10.790	17	4	2	5	3	1	0.3085	9.766
5	1	5	5	5	5	0.1834	10.650	18	4	3	1	4	2	0.3125	6.443
6	2	1	2	3	4	0.2651	6.451	19	4	4	2	5	3	0.2683	6.287
7	2	2	3	4	5	0.2258	8.054	20	4	5	3	1	4	0.2255	11.420
8	2	3	4	5	1	0.3155	10.560	21	5	1	5	4	3	0.2713	9.706
9	2	4	5	1	2	0.2655	10.290	22	5	2	1	5	4	0.2724	5.926
10	2	5	1	2	3	0.2752	6.217	23	5	3	2	1	5	0.2301	11.320
11	3	1	3	5	2	0.3164	7.325	24	5	4	3	2	1	0.3059	11.220
12	3	2	4	1	3	0.2737	10.470	25	5	5	4	3	2	0.2697	10.990
13	3	3	5	2	4	0.2316	10.240								
mean W_1	0.271	0.288	0.289	0.277	0.317			mean S_1	10.95	8.978	7.069	10.92	9.890		
mean W_2	0.269	0.277	0.275	0.271	0.294			mean S_2	8.314	9.073	8.406	9.831	9.240		

mean W_3	0.2 73	0.2 71	0.2 68	0.268	0.2 71	mean S_3	8.108	9.9 24	9.8 16	8.7 90	8.74 8
mean W_4	0.2 71	0.2 66	0.2 70	0.274	0.2 49	mean S_4	8.849	8.8 54	10. 63	8.3 63	8.96 5
mean W_5	0.2 70	0.2 52	0.2 52	0.271	0.2 23	mean S_5	9.832	9.2 20	10. 13	8.1 50	9.20 7
Range W	0.0 04	0.0 36	0.0 37	0.009	0.0 95	Range S	2.838	1.0 71	3.5 59	2.7 66	1.14 2

5. Analysis of shrinkage

Using the orthogonal test table lists 25 sets of volume shrinkage test results simulated by Moldflow, and calculate the average and range of data values, as shown in table 2.

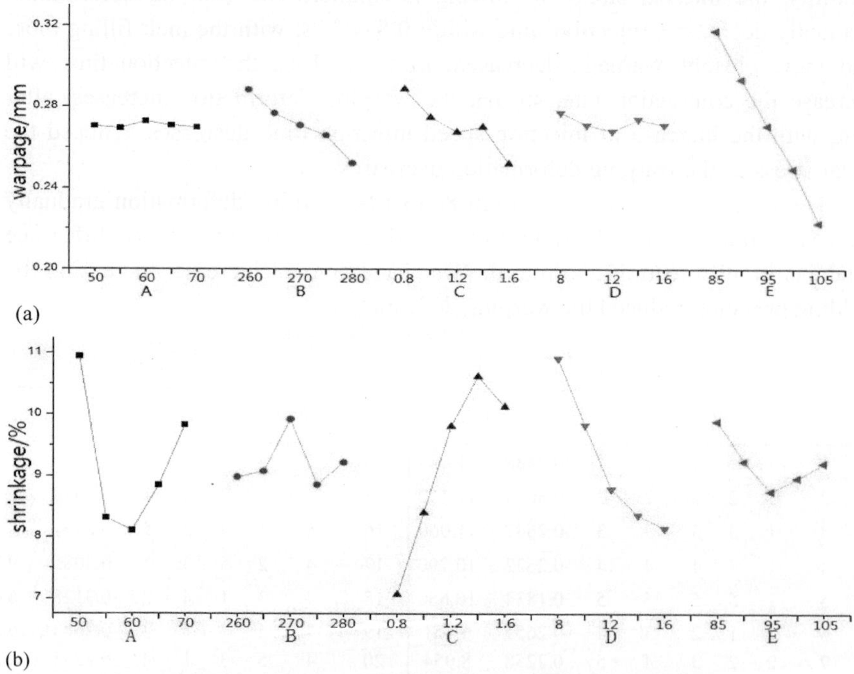

Fig. 2. The tendency of effects.

Combined with table 2 and Fig. 2 indicates that the mold temperature at before 60°C, lower mold temperature accelerates the solidification. Between the melt temperature 260°C ~ 270 °C, shrinkage rate began to decline; the temperature continues to increase with high degree of molecular orientation, volume shrinkage rate began to increase.

The biggest impact to the volume shrinkage is injection time, injection rate decreases with the injection time prolonged, the plastic volume shrinkage increases; injection rate increases with the injection time continued decreases, volumetric shrinkage decreases. The volume shrinkage rate gradually decreased with the holding time prolonged. Holding pressure at 85 ~ 95MPa, shrinkage rate to decline; then shrinkage rate began to rise.

We can see from table 2, the most important factors affecting the volume shrinkage is the injection time and the mold temperature, the minimum influence factors is melt temperature. The volume shrinkage of the optimal sequence is: $A_3B_4C_1D_3E_5$.

6. Comprehensive balance method

Integrated balance method is a common method to handle multi-objective problem. First of all, analysis the various indicators, through the simulation experiment to find out the optimal production conditions of each index; then, comprehensive balance analysis the optimal production conditions on each index, find out the balance of each quality index condition [1].

The primary and secondary effects of the trend and influence show that the influence of mold temperature on the volume shrinkage greater than the amount of warpage, so determine A_3 as the optimal levels. Effect of melt temperature on the warpage amount bigger than the volume shrinkage, so determined that B_5 is the optimal level.

Injection time have the biggest effect on the volume shrinkage rate,so as to determine C_1 as the optimal level. Determined D_3 as the optimal level. Holding pressure has the biggest effect on the warpage, has little effect on the volume shrinkage, selected minimum value 105MPa of warpage is most appropriate, so determine E_5 as the optimal level.

Finally, considered both warping deformation and volume shrinkage ratio selected the optimal combination: $A_3B_5C_1D_3E_5$.

7. Simulation analysis

The simulation analysis of the process combination that obtained by comprehensive balance method carried out with Moldflow, comparison and analysis of the experiments and the optimum results. In the experiment, the minimum value of warping is 0.1834mm, the optimized total warpage amount is 0.2287mm, differs by 0.0453mm from the before optimization result. After optimization, the volume shrinkage rate is 5.575%, the minimum value is 5.681% before optimization experiment, the volume shrinkage rate decreased by 1.87% after optimization. Through comparative analysis, the optimized Parameters received the minimum volume shrinkage, and warpage get control.

8. Conclusions

This paper uses the method of CAE analysis and orthogonal test combination, analyzed warpage and shrinkage of mouse base, optimized process parameters of single target with the method of range analysis, and uses comprehensive balance method giving attention to both technology of the combination of two quality indicators, and validated by Moldflow, the minimum volume shrinkage was obtained, at the same time controlled the warpage effectively.

References

1. K.F. Zhang, Z. Lu. 2008. Analysis of morphology and performance of PP micro-structure by micro injection molding. Micro system Technologies 14(2):209-214.
2. M.L. Wu. 2010. Plastic molding CAE-Moldflow application foundation. BeiJing: Electronics Industry Press.
3. S.T. Wu. 2012. Multi-Objection Optimization of Injection Molding Process of Automobile Bumper on Orthogonal Experimental Method. China Plastic Industry 40(8):60-64.

Design and research of drilling tool for deep hole inner wall blind hole

Z.F. Liu[a], Z. Wang[b]

[1]*School of Mechanical Engineering, Xi'an Shiyou University, Xi'an, Shaanxi (710065), China*

[a]*lzhfg9@163.com*, [b]*ld-wangzhe@163.com*

Based on the perforating gun inner wall blind hole processing in the oil production equipment, designed the drilling tools for deep hole inner wall blind hole. Set up it's theory model, and designed it's structure specifically. It focuses on introducing the feeding structure of cutting tool and guiding support structure. In addition, it also makes dynamics analysis for the tool. Both of them provide a new method and theoretical basis for perforating gun inner wall blind hole processing.

Keywords: Perforating gun; Inner wall blind hole; Deep hole inner wall blind hole; Deep hole drilling.

1. Introduction

In oil fields, the inner wall of perforated barrel (or long tube) needs to drill blind holes. The most simple method is external plugging type [1]. The most disadvantage of this type will produce stress concentration, leading to leakage in the oil well under the position of big formation pressure, affect the normal operation further. There are three processing methods for blind holes both abroad and domestic at present: pressure relief hole type, inner processing type and electrolysis type. The blind holes processing needs plug the hole to use pressure relief hole type, it has possibility of leakage. The other type of drilling inner blind holes is inner processing and (Electron Discharge Machining). Inner processing is used to shallow hole filed and drill in the position of work two terminal. EDM is costly and time-consuming. After analyzing the pros and cons of above-mentioned, this paper presents a new tool for deep hole inner wall blind hole and theory analysis of the tool.

2. The Structure and Working Principle of Drilling System

The drilling system consists of there main components: body systems, drilling tool systems and cooling systems as illustrated in Fig 1. The workpiece is fixed in the Spiral Chuck and Center Rest and the Drill Pipe is installed in the Tailstock. The drilling tool can feed because the Tailstock which can move along the lathe bed driven by electric motors and lead screw give it a leg up. The spindle rotation is drived by servomotor through a pair of spur wheel and bevel gear and the spindle feeds of the cutting tools is drived by feed shaft. The pump supply cooling fluid for the cutting tools through the Coolant Supply Saddle. The cutting chip following the cooling fluid flow into the front and later exhaust chip box to achieve the goal of compulsory cooling and chip removal.

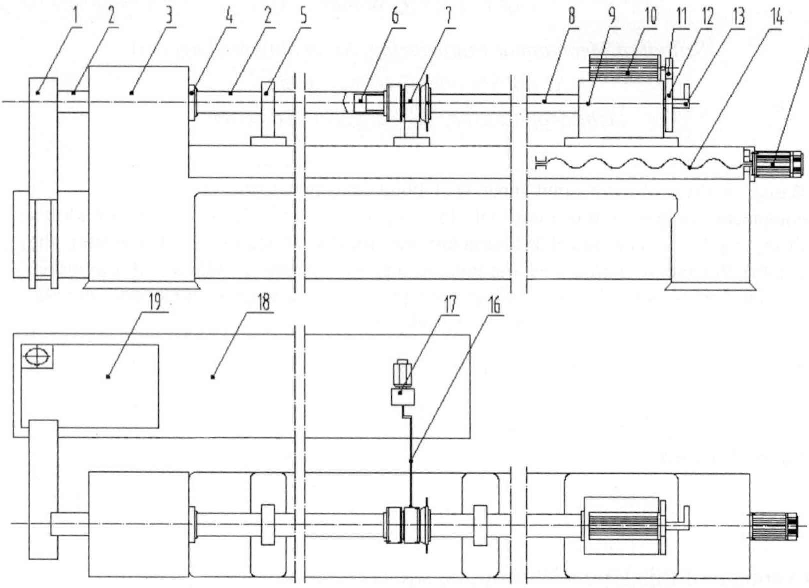

1-exhaust chip box;2-workpiece;3-Lathe Bed;4-spiral chuck;5-center rest;6-cutting tool; 7-coolant supply saddle; 8-drill pipe; 9-tailstock; 10-servomotor; 11-small spur wheel; 12-big spur wheel; 13-feed shaft; 14-leadscrew; 15-servomotor; 16-tube; 17-pump; 18-chip liquid box; 19-exhaust chip box .

Fig. 1. The Drilling System.

3. Design of Drilling Tool

3.1. *Principle analysis of drilling tool*

The axis of blind hole localized in inner wall of slender perforating gun and the axis of workpiece are perpendicular to each other. And the blind hole can't be processed on the normal machine tool. For the above reasons, it is vital to design a drilling tool to machine the blind hole. As illustrated in Fig. 2, the movement of drilling tool is made up of there main parts: linear motion along the OX direction of drilling tool; rotary motion on the plane XOZ and linear motion along the OY direction of the cutting tool. The drilling tool move along the OX direction driven by Tailstock. The cutting tool moves along the OY direction direction driven by feed shaft through the eccenter and rotate on the plane XOZ driven by bevel gear.

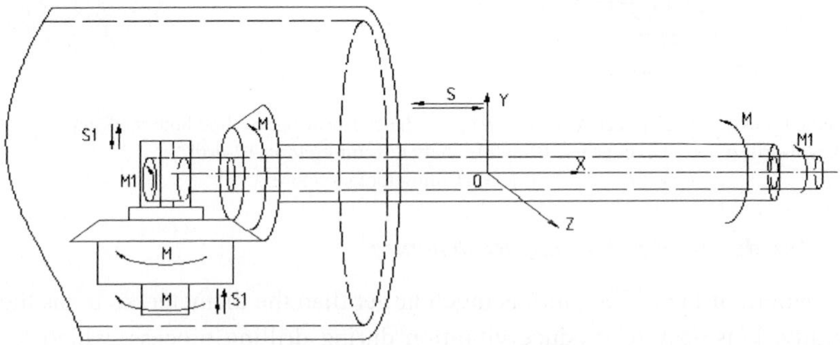

Fig. 2. The Drilling Working Principle.

3.2. *Structure design of drilling tool*

Through analyzing the dilling working principle, this essay intended to devise a reasonable and effective structure of drilling tool, as shown in Fig. 3. Drilling tool consists of drill pipe, drilling and cutting tool feed mechanism. The appearance of drill pipe and drill bits are cylinder. One end of the drill pipe is installed on the tailstock, and the feed motion of the whole tool implements with the aid of the tailstock .The other end of the drill has three guide block along the circumferential direction for supporting on the inner wall of the workpiece. Bevel gear shaft is installed in the drill pipe, and one end is used for external cylindrical gear by equipped with flat key. The inner hole of bevel gear shaft is used for installation tools feeding shaft. Bevel gear is installed inside the dill, and it meshes with bevel gear shaft through the spline. Square groove is used to install the eccentric in the middle of disc, which is axisymmetric body. The bottom hole of disc can fix spline shaft. The end of spline shaft connects to the cutting tool. It installs inside a disc and cooperates with bevel gear by spline. Motor drive the cutter through a pair of cylindrical gear, a pair of conical gear and spline. Tool feed axis drives the cutter to do linear feed movement with the help of eccentric wheel and disc.

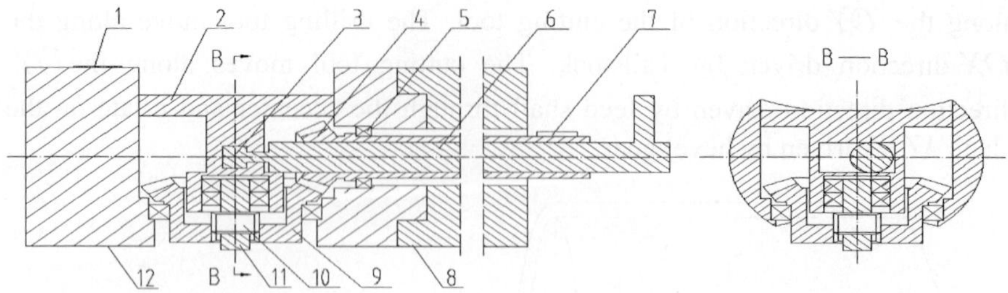

1-guide block; 2- disc; 3-eccentric; 4-bevel gear shaft; 5-bearing; 6-blind hole feed shaft; 7- key;8-drill pipe; 9-bevel gears; 10-spline shaft; 11-cutting tool; 12-drill

Fig. 3. Drilling tool structure.

3.3. *The design of guide support structure*

The length of the drilling tool is much larger than the diameter, so it has the poor rigidity. It is easy to produce vibration during drilling process, which leads to slant drilling, play knife, sticking and other adverse consequences. All of them reduce the processing accuracy and production efficiency. In order to solve the drilling tool rigidity low deviate, drilling, and drilling tool orientation problem, it adds a guiding device andauxiliary support when design the drilling tool.

The role of the guide block is to improve the tool orientation ability. Guide block close to the inner wall of workpiece to make the drilling process stability and reduce vibration in cutting process. Besides, it also can improve the stability of drilling system and increase the precision of blind hole processing. So three focal point round rule is applied, which is that three guide blocks are decorated around the drill. Guide support will pass the cutting force to the workpiece to achieve the goal of reducing vibration and guide positioning. The stress of the drill can be expressed with mechanical model as following Fig. 4.

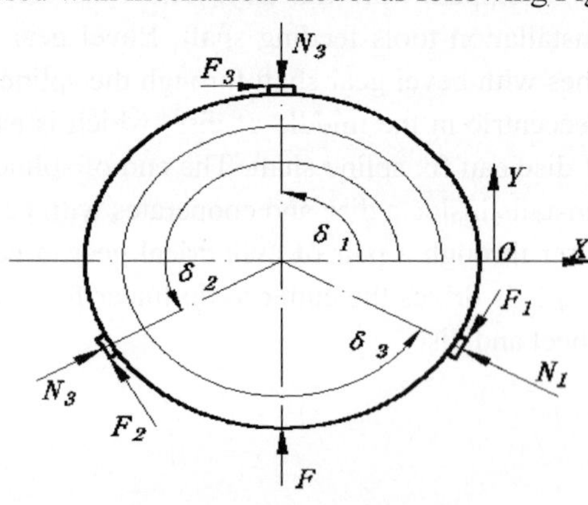

Fig. 4. Drill mechanical model.

The drill mainly has drilling force F , guide block extrusion N_1, N_2, N_3 and frictional force F_1, F_2, F_3. In the diagram, $\delta_1 = 90°$,the drill diameter is d . According to mechanical equilibrium equation can get as following.

Resultant force of X is zero,

$$\sum X=0 \quad F_1 \sin\delta_3 - N_1 \cos\delta_3 + F_2 \cos\delta_2 - N_3 \cos\delta_2 + F_3 = 0$$

Resultant force of Y is zero,

$$\sum Y=0 \quad N_1 \sin\delta_3 + F_1 \cos\delta_3 - F + F_2 \cos\delta_2 + N_3 \sin\delta_2 + N_3 = 0$$

Resultant torque of drill is zero,

$$\sum M=0 \quad F_1 \frac{d}{2} + F_2 \frac{d}{2} + F_3 \frac{d}{2} = 0$$

3.4. *The design of tool feed mechanism*

Reasonable cutting tool feed system is a key part of the deep hole inner blind hole drilling. In the drilling process, the tool of feeding should be smooth and continuous to avoid rigid impact. Eccentric wheel mechanism has advantages of simple structure, easy processing, long service life as well as good compatibility, so tool feed motion is completed by eccentric mechanism. Tool feed principle is shown in Fig. 5. Cutting tool specific motion process is that eccentric rotation drives the cutting tool to do the straight line reciprocating motion on the Y direction to realize the cutting tool into giving back. The eccentric angle and displacement of cutting tool feed can be got by graphic method.

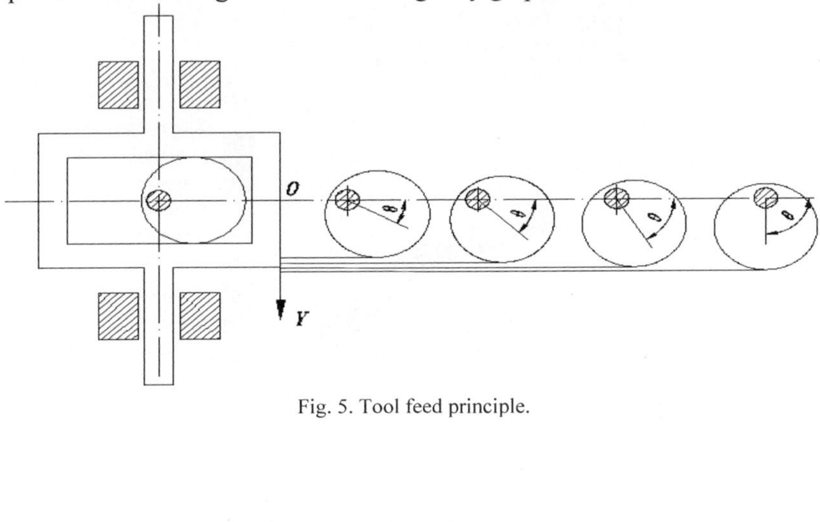

Fig. 5. Tool feed principle.

3.5. *The dynamic characteristic analysis of drilling tool*

Modal analysis is a modern method to study the dynamic characteristics of structure. It also is the basis of the simulation analysis for mechanical systems. The order natural frequency and vibration mode of mechanical systems can be achieved by modal analysis. The result can guide well to avoid the natural frequencies of the mechanical system, so as not to cause resonance phenomenon. What's more, it can provide theoretical basis, fault diagnosis and forecast, structure optimization design for the structure system.

Drill and drill pipe (short for the drill body) are the skeleton parts of the drilling tool. Its stability decides the stability of the drill, so it has a vital role to do modal analysis for drill body. Because of the first few order natural frequency have high impact on the drill body, the first four order natural frequency of drill are extracted.

The material of drill body is 40Cr. Its density is 7.82×10^3 kg/m^3, the elastic modulus is 2.0×10^{11} Pa, poisson's ratio is 0.3. Because of the guide block supports on the inner wall of the workpiece, and one end of drill pipe is installed on the tailstock. So the way we deal with the boundary conditions is that only to the guide block and drill pipe adopts full constraints, regardless of drilling pressure and frictional resistance and other factors in the actual production processing. SolidWorks Simulation is a kind of software, who can provide frequency analysis. Through the modal analysis, it can get drill body's the first four order natural frequency and vibration mode as shown in Fig. 6 and the maximum deformation under different frequency as shown in Table 1.

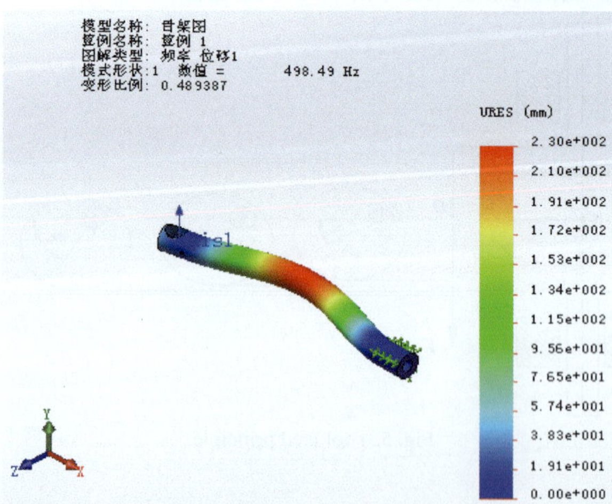

(a) The first order vibration frequency

Fig. 6. The first four order frequency and vibration mode of drill body.

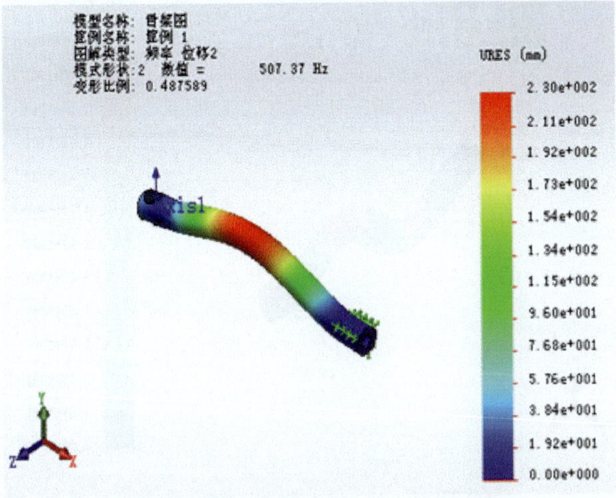

(b) The second order vibration frequency

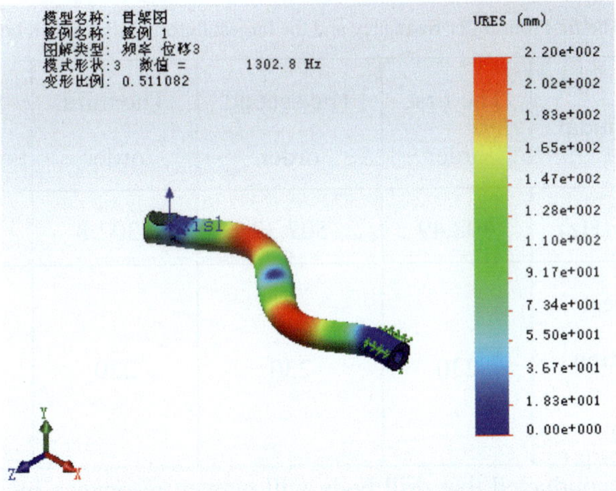

(c) The third order vibration frequency
Fig. 6. (*Cont'd*)

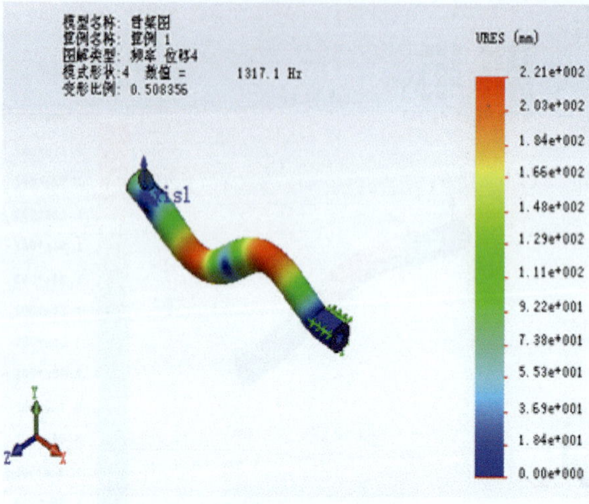

(d) The fourth order vibration frequency

Fig. 6. (*Cont'd*)

Table 1. The first four order frequency and the largest deformation of drill body.

N order modal	The first order	The second order	The third order	The fourth order
frequency(Hz)	498.49	507.37	1302.8	1317.1
the largest deformation (mm)	230	230	220	221

It can be conducted that drill body will happen resonance by observing the first four order vibration mode cloud pictures of the drill body modal analysis. The first order modal mainly show the vibration of *Z* direction. The maximum amplitude locates in the middle of the drill pipe. The second order modal is the vibration of *Y* direction. The largest amplitude is in a quarter of drill body along the *X* axis. The third and fourth order modal mainly present alternating vibration of *X* direction and *Z* direction. The largest amplitude is in a quarter and three-quarters of drill body along the *X* axis. So it should consider increasing the support or the stiffness of drill pipe in the above position, so that to reduce drilling tool system resonance effect on machining accuracy.

4. Conclusions

(1) This paper sets up the theoretical model of drill tool and the mechanical model of cutting tool. It focus on analysing the tool feed mechanism. Both of them are based on the design of drilling tool.

(2) It gets the first four order frequency and the largest deformation by the dynamic characteristic analysis of drilling tool. All above will provide theory basis for design and optimization of drilling tool.

5. Acknowledgement

The cultivation project financial aid of excellent master dissertation of Xi'an Shiyou University.

References

1. Chen Li, Guohua Zhang, Linghai Kong. Inner blind hole perforator application in Jilin oilfield [J]. Well logging technology abroad, 2008, (1).
2. Jiwei Chen, Juan Jin. The current situation and prospect of perforating gun blind hole processing methods. A mechanical engineer, 2013, (11).
3. Shiqing Wang. The deep hole processing technolog [M]. Xi'an: Northwestern Polytechnical University Press, 2003.
4. Yanyan Cao, Dengfeng Zhao. The finite element modal analysis theory and its application [J]. Mechanical engineering and automation, 2007, (1):73-74.
5. Guangxin Huang, Lixin Yu. Reducer gear modal analysis based on SolidWorks and ANSYS [J]. A mechanical engineer, 2014, (2).

Reinforcement of column-damaged frame joint using carbon fiber for seismic analysis

J. Hu

Nanchang Institute of Technology,
Nanchang, China,
496515594@qq.com

The joints of the reinforced concrete frame are important parts in the structure. At present, research on seismic performance of damaged RC frame joints is limited both at home and abroad. This article will study on seismic analysis between undamaged and Column-damaged joints reinforced by CFRP, and it will take load in end of three columns with force-displacement mixed control on the condition of same axial compression ratio by Quasi-static test of cyclic loading. Through reinforcement of Frame Joint with undamaged and Column -damaged using CFRP sheets, it analyzed and compared with two situations in hysteresis curve, ductility, and energy dissipation and etc. The results showed that Column-damaged frame joint reinforced by carbon fiber was mainly to improve seismic behavior such as displacement-ductility coefficient and energy dissipation.

Keywords: Carbon fiber; Frame joint; Column-damaged; Seismic Analysis.

1. Introduction

In the earthquake, seismic behavior of frame joints directly affect the safety of the structures, a large number of buildings are due to failure or damage of frame joints in the earthquake which caused structural damage or collapse. It indicates that reinforcement of joints is an important work of seismic fortification. Experimental study on beam-column joints is common now at home and abroad, there are still some problems unsolved. For example, studying on reinforcement of damaged joints are very limited. [1][2]

This article will study on seismic analysis between undamaged and column-damaged joints reinforced by CFRP.

2. Test

2.1. *Specimens and device*

This test contains a total of two specimens. Specimens are taken from frame structure as model and prototype for 1:2. Specimen is shown in Fig. 1, test devices are shown in Fig. 2, test photos are shown in Fig. 3.

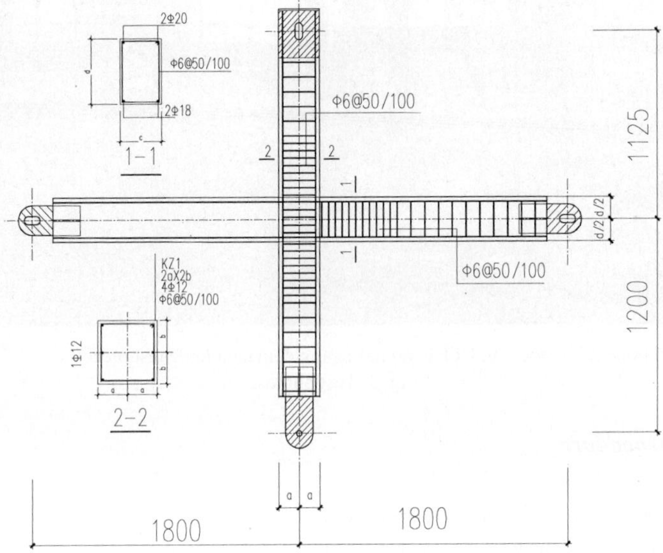

Fig. 1. Sample reinforcement and section detail.

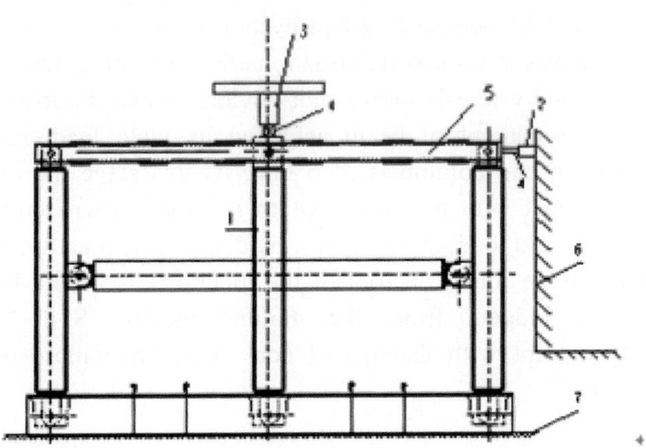

1. Specimen; 2.Horizontal; 3. Vertical load loader; 4. Load sensors; 5. Variable geometry frame; 6.Force walls; 7.Test platform

Fig. 2. Test device.

a. undamaged node test piece b. CFRP strengthening damaged node test piece
Fig. 3. Test photos.

2.2. *Test procedure*

2.2.1. *Joint SJ-1 without reinforcement*

In early loading, hysteresis curve was linear relationship, and no cracks occurred in specimen. When the load reached into 28KN, the first crack occurred around the core area of column. When load increased, Existing cracks were ongoing, and several new cracks occurred. When hysteresis curve was no longer linear relationship, there was a residual deformation after unloading. Due to small load, the formation of the hysteresis loop is not obvious; when the load reached into 60KN, main reinforcement of beam achieved its yield load, and horizontal displacement of the end column is 35.6 mm. At this stage the existing crack increased relative fast with maximum width of 1mm; when reinforcement of beam achieved its yield load, it was loaded by displacement as Δy; When loaded into $2\Delta y$, cracks were found on root of the junction of column-beam, formed plastic hinge; Judged from the failure modes, SJ-1 was mainly Column-damaged joint with damage of core area. The failure of SJ-1 is as shown in Fig. 4 (a).

2.2.2. *Column-damaged joint SJ-1 reinforced by CFRP*

Initially, strain of carbon fiber and beam longitudinal reinforcement increased slowly, at which point end-column horizontal force was mainly assumed by core area of node and stirrups, shear passed from the core area was mainly borne by beam longitudinal reinforcement.

(a) SJ-1 (b) SJ-1(damaged) reinforced by CFRP

Fig. 4. The failures.

As the load increases, tensile strain gradually increased on longitudinal reinforcement and carbon fiber, which were on both sides of core area. Though carbon fiber strain of core area was still not large at this time, core area of shear pass from the column end was still borne by concrete and stirrups together; shear pass from core area was borne by longitudinal reinforcement and carbon fiber together.

When loaded into yield load as Δy, strain of carbon fiber and hoop in core area increased fast; Noise was created by collapse from the specimen surface of carbon fiber, and carbon fiber bear large part force of beam longitudinal reinforcement, meanwhile it reduced shear force of longitudinal reinforcement pass from core area which was borne by carbon fiber, and beam longitudinal reinforcement basically exit hosted.

Loaded into the ultimate stage, the core area of concrete occurred cross shear-cracking, and stirrups yield, at which point core shear was borne mainly by the carbon fiber. Crack increased in the beam without reinforced by CFRP, formed plastic hinge. Specimen was column-damaged failure and test stopped. The failure is as shown in Fig. 4 (b).

2.2.3. *Undamaged joint SJ-2 reinforced by CFRP*

In early stage, hysteresis curve was linear, and the formation of the hysteresis loop was not obvious. When loaded into 120KN as yield load, the horizontal displacement of end column is 27.49mm as yield displacement. After loading by displacement, it continued to load into $2\Delta y$, and the plastic hinge beam appeared.

After the experiment was over, it was found that cross cracks in core area remained by brushing carbon fiber in the surface of joint. Due to the bearing of carbon fiber, it was improved shear capacity of core area of joint which was mainly Column-damaged. The failure of SJ-2 is shown in Fig. 5.

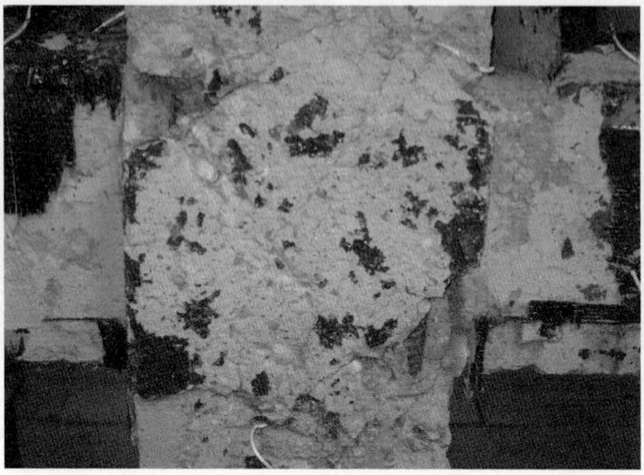

Fig. 5. The failure of SJ-2.

2.3. *Test results and analysis*

It obtains initial cracking load, yield load, ultimate load, destructive load and their corresponding displacement and other data, which are sorted and listed. The data of Specimen before and after reinforcement is as Table 1.

Table 1. Key data of experiment.

NUM STATE	unreinforced	SJ1 reinforced	SJ2 reinforced
Cracking load (kN)	28.00	—	—
Yieldload (kN)	60	50	120
Yield displacement (mm)	35.61	13.3	23.89
Ultimate load (kN)	72	128	136
Ultimate displacement (mm)	70.3	79.3	47.8
Failureload (kN)	67	116	126
Destructive displacement (mm)	106.8	93.1	119.45

2.3.1. *Hysteresis curve*

The hysteresis curve of SJ-1 without reinforcement and the hysteresis curve of SJ-1 with reinforcement are shown in Fig. 6 (a) (b), the hysteresis curve of SJ-2 with reinforcement is shown in Fig. 6 (c). Y shaft represents bearing capacity of end column, x shaft represents displacement of end column.

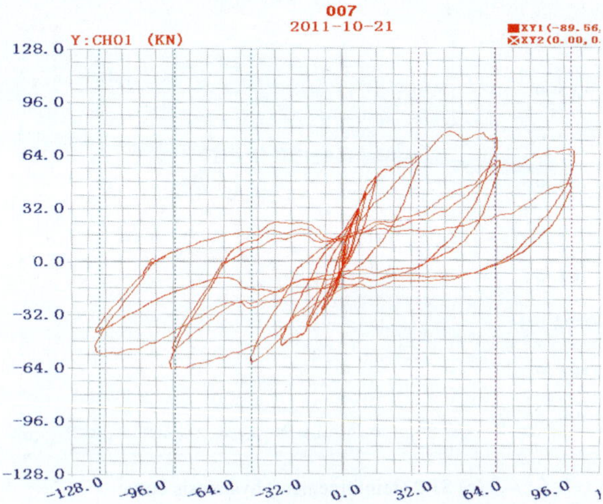

(a) SJ-1 hysteresis curve

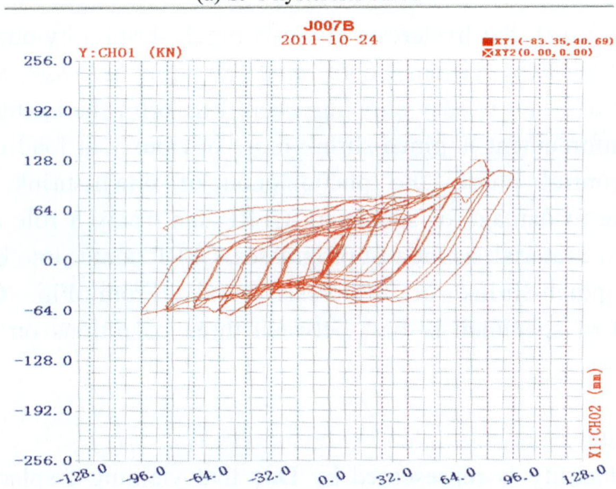

(b) SJ-1(reinforcement) hysteresis curve
Fig. 6. Hysteresis curve.

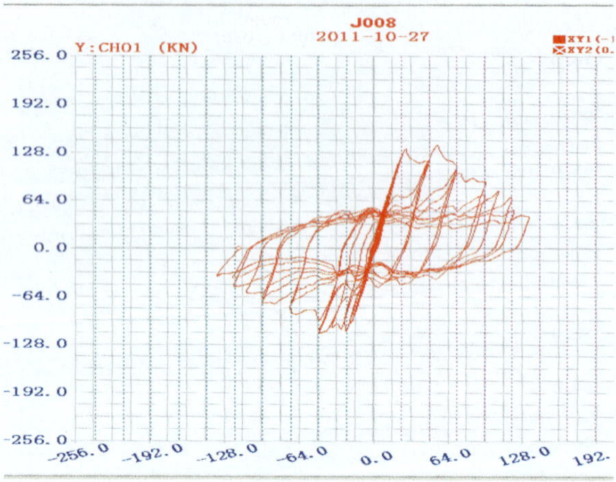

(c) SJ-2 (reinforcement) hysteresis curve
Fig. 6. (*Cont'd*)

From Fig. 6 (a), the hysteresis curve is pinch-shrink obviously during the experiment. For two reasons, core area of joint is easy damaged and reinforcement of beam is easy slipping. From Fig. 6(b), the hysteresis curve of SJ-1 after reinforcement is pinch-shrink more obviously at load control phase. After load control phase, the phenomenon of pinch-shrink is improved gradually. Due to damage of specimen, CFRP can't play a role at early stage which is plastic material; as the test progresses, CFRP comes into bear force that it improves pinch-shrink of hysteresis curve. From Fig. 6 (c), direct reinforcement of specimen is best between three conditions on Saturation of hysteresis curve.

2.3.2. *Ductility*

In this test, ductility is represented by Ddc that yielding displacement of the end-column is divided by limit displacements, and the value is as Equation 1 :

$$\mu_\Delta = \Delta_u / \Delta_y \tag{1}$$

Ddc is larger; ductility of specimen is better, and vice versa. [3][4]Ddc is shown as Table 2.

Table 2. Ductility

NUM	SJ1		SJ2
STATE	unreinforced	reinforced	reinforced
Ddc	1.97	6	5

* Ddc—Displacement-ductility coefficient

Ddc of SJ-1(reinforced) is less than SJ-1 due to damage in core area which result rigidity of specimen smaller than before. Ddc of SJ-2 is less than SJ-1(reinforced) due to yield displacement which is large. For security reasons, the test specimen is not loaded to the maximum load, and displacement-ductility coefficient still has some upside.

2.3.3. *Energy dissipation capacity*

Energy dissipation capacity is measured by energy-dissipation coefficient in this experiment. Energy-dissipation coefficient is defined as follows Equation 2:

$$E = \frac{S_{ABC} + S_{ADC}}{S_{OBE} + S_{ODF}} \tag{2}$$

In the formula: S_{ABC} and S_{ADC} represent area surrounded in the hysteresis curve, S_{OBE} and S_{ODF} represent area of a triangle in Fig. 7. E is larger, energy dissipation capacity of specimen is stronger, [5][6][7] and vice versa. Energy dissipation coefficient is shown as Table 3.

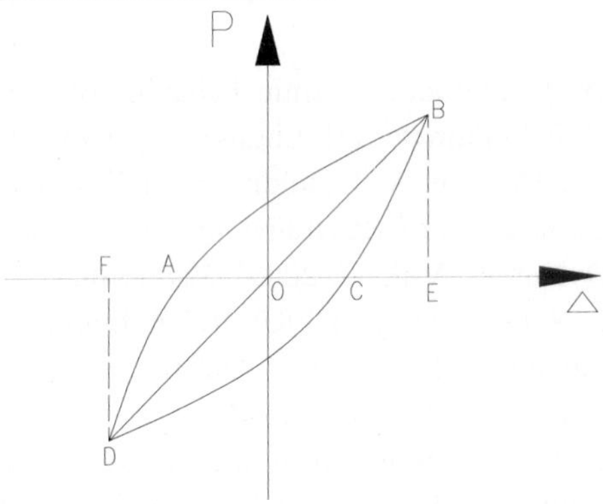

Fig. 7. Calculation of the coefficient.

Table 3. Energy-dissipation coefficient.

NUM	SJ1		SJ2
STATE	unreinforced	reinforced	reinforced
Energy-dissipation coefficient	1.65	1.78	3.2

As the result, energy dissipation coefficient of the test piece has increased after reinforcement, and energy dissipation capacity of damaged specimens after reinforcement is almost the same as specimens without reinforcement.

3. Conclusions

After reinforcement by CFRP, seismic of specimen is improved obviously: Displacements of different stages are larger than specimen without reinforcement, though loads of different stages have a little change; Energy dissipation capacity of reinforced specimen is better than one without reinforcement.

Yield load, maximum load and ultimate load of specimen which is directly reinforced are all larger than specimen which is reinforced after damage, meanwhile yield displacement, ultimate displacement, destructive displacement of the former are smaller than the latter; The performance of hysteretic and energy dissipation capacity of the former is better than the latter, meanwhile ductility of the former is worse than the latter. It can be seen that damage has obvious impact on seismic of specimen reinforced by CFRP: Column-damaged reduces the bearing capacity of specimen reinforced by CFRP at different stages, meanwhile increases the displacement and energy dissipation capacity at each stage.

Reference

1. A. Balsamoa, A. Colombob. Seismic behavior of a full-scale RC frame repaired using CFRP laminates [J]. Engineering Structures. 2005: 769–780.
2. Maria J. Favvata, Bassam A. Izzuddin. Modelling exterior beam-column joints for seismic analysis of RC frame structures. Earthquake Engineering and Structural Dynamics. Vol. 37, 2008:1527-1548.
3. Azadeh Parvin, Selcuk Altay. CFRP Rehabilitation of Concrete Frame Joints with Inadequate Shear and Anchorage Details [J]. Journal of Composites for Construction. 2010: 72-82.
4. Abdelhak Bousselham1. State of Research on Seismic Retrofit of RC Beam-Column Joints with Externally Bonded FRP [J]. Journal of Composites for Construction. 2010: 49-61.
5. Kien Le-Trung, Kihak Lee . Experimental study of RC beam–column joints strengthened using CFRP composites [J]. Composites: Part B. 2010: 76-85.
6. Seyed. S. Mahini. Strength and ductility of FRP web-bonded RC beams for the assessment of retrofitted beam–column joints [J]. Composite Structures. 2010: 1325–1332.
7. Saptarshi Sasmal, Balthasar Novák. Numerical analysis of fiber composite-steel plate upgraded beam–column sub-assemblages under cyclic loading [J]. Composite Structures. 2011: 599–610.

Chapter 6

Material Science and Material Processing Technology

Study on continuous drive friction welding process of aluminum and copper dissimilar materials

J. Wang
School of Mechanical Engineering, Shaanxi University of Technology, Hanzhong, Shaanxi, 723001, China,
E-mail: wangjin63@163.com

Y.C. Yang
School of Materials Science and Engineering, Shaanxi University of Technology, Shaanxi , Hanzhong, 723000, China
E-mail: fenmo163 @163.com

In order to replace copper with aluminum, save the copper material and copper natural resources, aluminum and copper dissimilar materials were joined by continuous drive friction welding process, the hardness, the tensile properties, the bend tests were studied by using hardness tester, WAW-1000 microcomputer control testing machine system, hammer bend tests of aluminum to copper weld. The interface microstructure of aluminum to copper weld is investigated by EPIPHOT-300U inverted microscope model. The intermetallic compound at the weld interface is investigated by scanning electron microscopy. The results show that bending strength of joint is bending at 90 ° by hand and then bends at 90 ° reversely; the joint of aluminum and copper is no craze, copper and aluminum combination is good. The specimens were tensile fracture in aluminum side; this suggests that the weld region had a higher ultimate tensile strength than the parent aluminum. The axial shortening on the aluminum side is more than that on copper side. There is no readily apparent voids flaw and cracking between the aluminum and copper.

Keywords: Continuous drive friction welding; Aluminum and copper dissimilar materials; Microstructure; Mechanical properties.

1. Introduction

With the development of modern science and technology, the electrical appliance industry, the telecommunications industry, the mechanical industry and metallurgical industry needs more and more copper products [1]. Industrial pure copper and copper alloy products have highly conductive, high thermal conductivity, so these are the best materials for manufacturing electric switch, electrical equipment and cable. However, it brings a huge cost pressure to electric enterprise for the shortage of copper and the high price of copper [2]. In order to save copper resources, promote sustainable development of the electrical appliance industry and telecommunication industry. In the application of employing aluminum instead of copper, the joint of the aluminum and the copper is crucial it can not only reduce the production cost, but also reduce the weight of work pieces [3]. There has been a trend of employing aluminum pipe instead of copper pipe in refrigeration industry, because of its similar thermal conductivity at a reduced market price and lower specific density. In industrial applications, some welding methods that may be used on copper and aluminum alloys, including cold welding, friction welding, diffusion welding, explosion welding, resistance pot welding, laser welding, brazing and so on [4]. Regarding potential engineering application of aluminum and copper dissimilar material dissimilar metal components as the background, copper and aluminum was connected by means of continuous drive friction welding process [5]. Continuous drive friction welding process is a competing technology for the welding of aluminum to copper, and welds with high strength can be produced using this process, it is suitable for dissimilar material's joining weld in case of aluminum tube and copper tube, aluminum tube and copper sheet, copper bar and aluminum bar, copper bar and aluminum tube and work pieces made of dissimilar materials [6]. However, it is the primary task to make clear the difficulty of the joining technology and the joining mechanism for the aluminum and the copper, and then to guide the actual production.

2. Experimental material and method

2.1. *Experimental material*

The raw materials are industrial pure copper bar and industrial pure aluminum bar. The chemical compositions of industrial pure copper are shown in Table 1. From the Table 1, industrial pure copper bar consists of a large amount of copper and small number of other impurity elements. The chemical compositions of industrial pure aluminum are shown in Table 2. This chemical composition of copper and aluminum is to meet the requirements of electrical industry, the telecommunications industry and the requirement of the electrical equipment industry.

Table 1. The chemical compositions of industrial pure copper.

element	Cu	Bi	Sb	Fe	Ni	Al
content (wt%)	margin	0.001	0.002	0.005	0.005	0.002

Table 2. The chemical compositions of industrial pure aluminum.

element	Al	Pb	Cd	Fe	Sn	Cu
content (wt%)	margin	0.001	0.002	0.005	0.005	0.002

2.2. *Experimental method*

The continuous drive friction welding process of aluminum and copper dissimilar material was tested on C320 continuous drive friction welding machine. The interface microstructure of aluminum to copper weld was analyzed by EPIPHOT-300U inverted microscope model. The hardness of aluminum to copper weld was analyzed by FEM-7000 type automatic micro hardness tester. The intermetallic compound at the weld interface is investigated by JSM-6390LV canning electron microscopy. The tensile properties of aluminum to copper weld were tested with invariable strain rates on WAW-1000 microcomputer control testing machine system. Hammer bend tests, where one side of the weld (copper half) was held in a vice and the other (aluminum half) subjected to hammer blows until some plastic deformation of the impacted half of the weld, were also conducted on a few samples. The conductivity of aluminum to copper weld was analyzed by 7501 type eddy current conductivity all. C320 continuous drive friction welding machine is shown in Figure 1. The parameters used in the friction welding experiments are friction time is 4s, friction pressure is 60MPa, upset time is 15s, and upset pressure is 120MPa.

Fig. 1. The continuous drive friction welding machine.

3. Experimental result and analyze

3.1. *Mechanical properties of aluminum to copper weld*

The continuous drive friction welding process for dissimilar materials of aluminum and copper is a solid state welding process; the principle of it is to pressure the work piece upward after heating the aluminum and copper material by means of friction heat. The photo of aluminum to copper weld is shown in Figure 2. The bending test specimen of continuous drive friction welding for aluminum to copper is shown in Figure 3, copper is on the left and aluminum is on the right. From the Figure 3, bending strength of joints are bending at 90 ° by hand and then bends at 90 ° reversely, the joint of aluminum and copper is no craze, demonstrating that the joint of aluminum and copper by cold pressure welding has good bending performance. The interface enlarge figure of aluminum and copper is shown in Figure 4. From the Figure 4, there is no cracking between copper and aluminum, copper and aluminum combination is good. In bending test specimen of continuous drive friction weld, one side of the bending test specimen surface has a lot of tensile stress, and other has a lot of compress stress, so bonding strength of copper and aluminum combination is higher than industrial pure aluminum bar. The welding quality of continuous drive friction weld is good.

Fig. 2. The photo of aluminum to copper weld.

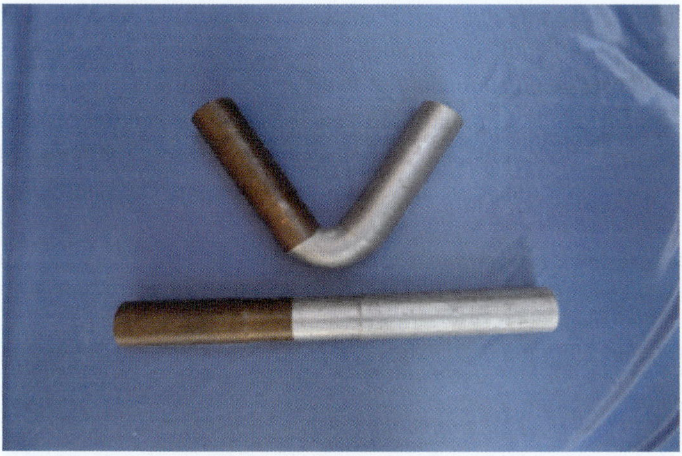

Fig. 3. The bending specimen of aluminum to copper weld.

Fig. 4. The interface enlarge figure of aluminum and copper.

Tensile specimen of aluminum (left) to copper (right) weld is shown in Figure 5, and tensile fracture sample of aluminum (left) to copper (right) weld is shown in Figure 6 From the Figure 6, the specimens were tensile fracture in aluminum side; this suggests that the weld region had a higher ultimate tensile strength than the parent aluminum. In tensile test; cross section of the specimen is subjected to uniform tension stress, aluminum produces the larger plastic deformation, copper produced in very small deformation. The aluminum and aluminum alloy have excellent physical properties, certain mechanical properties, so the fracture area of aluminum is very small, section shrinkage rate is very high, and has the characteristics of good plasticity.

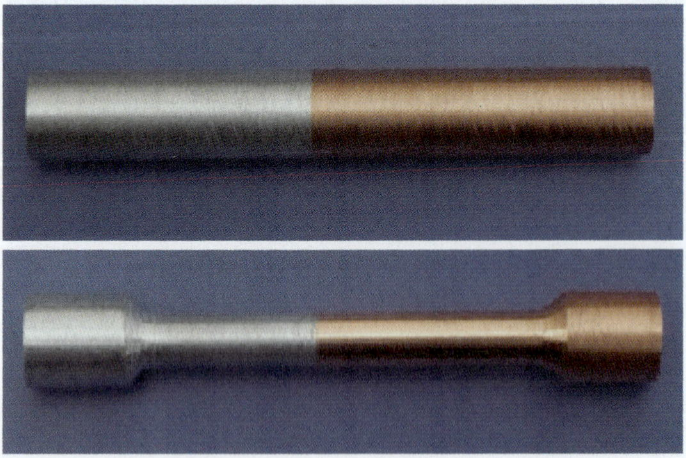

Fig. 5. Tensile specimen of aluminum (left) to copper (right) weld.

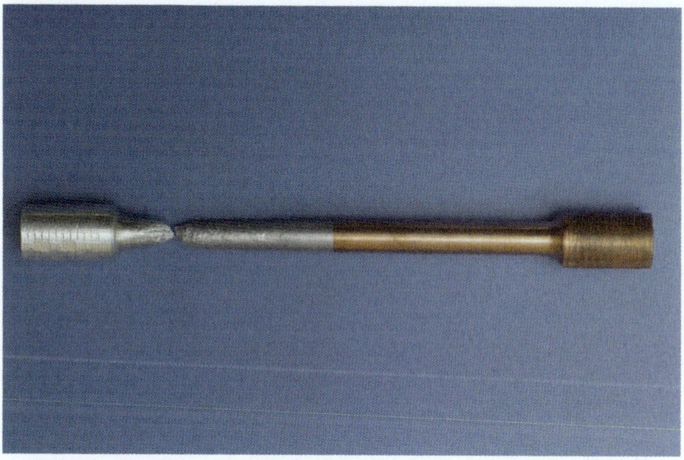

Fig. 6. Tensile fracture sample of aluminum (left) to copper (right) weld.

3.2. *The microstructure of aluminum to copper weld*

The photo of interface region in the friction welded copper and aluminum joint is shown in Figure 7. From the Figure 7, aluminum of interface region produces the larger plastic deformation, copper of interface region produced in very small deformation. Because copper is harder than aluminum. It can be seen that the axial shortening on the aluminum side is more than that on copper side. The microstructure in the interface region of the aluminum to copper weld joints is shown in Figure 8. From the Figure 8, examination of the weld microstructure showed intimate contact between the aluminum and copper with no readily apparent voids flaw and cracking, no large layers of intermetallic compound are present at the weld interface. It is also clear from the JSM-6390LV scanning electron microscopy that no large layers of intermetallic compound are present at the weld interface. A hardness trace across the weld line showed that the hardness increases close to the weld line in both aluminum and copper. With the continuous improvement of the welding technology and welding quality for dissimilar metal of copper and aluminum, the electrical products made from aluminum instead of copper will be more and reliable in quality.

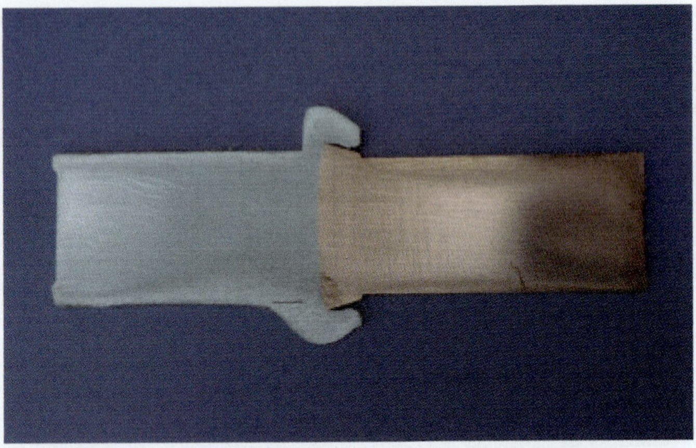

Fig. 7. The photo of interface region of aluminum (left) to copper (right) weld.

Fig. 8. The microstructure of interface region in joint.

4. Summary

(1) When parameters used in the friction welding are friction time is 4s, friction pressure is 60MPa, upset time is 15s, and upset pressure is 120MPa. The welding quality of continuous drive friction weld is good.

(2) There is no readily apparent voids flaw and cracking between the aluminum and copper.

(3) The weld region had a higher ultimate tensile strength than the parent aluminum.

(4) The axial shortening on the aluminum side is more than that on copper side.

References

1. Sahin M, & Akata HE. 2003. Joining with friction welding of plastically deformed steel. *J Mater Process Technol*, 142(1): 239-246.
2. B. Gulenc. 2008. Investigation of interface properties and weldability of aluminum and copper plates by explosive welding method. *Mater Des*, 29 (1): 275–278.
3. Lee WB, & Bang KS, & Jung SB. 2005. Effects of intermetallic compound on the electrical and mechanical properties of friction welded Cu/Al bimetallic joints during annealing [J]. *J Alloy Compd*, 390(2): 212-219.
4. Akata HE, & Sahin M. 2003. An investigation on the effect of dimensional differences in friction welding of AISI 1040 speci-mens. *Ind Lubr Tribol*, 55(5): 223-232.
5. G. M. Workman, & E. D. 1986. Nicholas. Friction welding aluminium and its alloys to various materials. *Met. Mater*, 2 (3): 138–140.
6. Maalekian, M. 2007. Friction welding-critical assessment of literature. *Sci Technol Weld Join*, 12(8): 738–759.

Research on heavy metal pollution with correlation analysis and factor analysis

A. Cao
School of Electrical Engineering, North China Electric Power University
Baoding, Hebei 071003, P.R. China
13373328763@163.com

This paper introduces a method to analyze the reason of heavy metal pollution. At first do correlation analysis between the 8 kind of metal, the results showed that: there are significant positive correlation between *Cu* and *Cr*, *Ni* and *Cr* , *Pb* and *Cd*, *Pb* and *Cu*, which shows they may come from same pollution source. Next do the factor analysis. The results showed that the contribution of the first five factors to total variation accounted for 87.7%. After analysis, we can get that *Cr*, *Ni*, *Cd*, *Pb* is generated by smelting and machining metal in industrial park. *Hg* and *Zn* is mainly generated by automobile exhaust in the trunk road. *As* is generated by fertilization in the mountain farmland.

Keywords: Heavy metal pollution; Correlation analysis; Factor analysis.

1. Introduction

Soil pollution is getting worse these years [1]. Cadmium exceeded rice, blood lead children, arsenic poisoning, mercury poisoning are occurred, which desperately threaten people's health. So it is significant to study the reason of heavy metal pollution.

One kind of activity may generate many kinds of pollutants. In order to find the mainly reason of the pollution, it is needed to classify the pollutants' source and categories. Correlation analysis and factor analysis are usually used to analyze the reason.

2. Correlation Analysis [2]

Correlation analysis is a statistical method which study the correlation between two or more variables. It is usually used to qualitatively analyze the correlation between every environment variable in environmental research. And the Pearson correlation analysis is usually used in environment research. The calculation formula is showed as follow:

$$r = \frac{\sum (X - \overline{X})(Y - \overline{Y})}{\sqrt{\sum (X - \overline{X})^2 \sum (Y - \overline{Y})^2}}$$

Where r is the correlation coefficient, X and Y is the two variables. When $r > 0$, the two variables are positive correlation, and the bigger r is, the stronger the correlation is. When $r < 0$, the two variables are negative correlation, and the smaller r is, the stronger the correlation is. When $r = 0$, there are no correlation between the two variable.

As for analyzing heavy metal pollution source, when the two kinds of heavy metal are positive correlation and the correlation is strong, the two kinds of heavy metal may come from the same pollution source.

This paper use data of problem B of 2013 Higher Education Press Cup National Mathematical Contest in Modeling [3] to analysis. Calculate the correlation coefficient between all heavy metal. When $r > 0.5$, the two kinds of heavy metal is significant. The correlation matrix is showed in Table 1. And the matrix shows that there are significant positive correlation between *Cu* and *Cr* , *Ni* and *Cr*, *Pb* and *Cd*, *Pb* and *Cu*, which shows they may come from same pollution source.

Table 1. The correlation matrix of 8 kinds of heavy metal

Correlation ($\mu g/g$)	As	Cd	Cr	Cu	Hg	Ni	Pb	Zn
As	1	0.255	0.189	0.16	0.064	0.317	0.29	0.247
Cd	0.255	1	0.352	0.397	0.265	0.329	0.66	0.431
Cr	0.189	0.352	1	0.532	0.103	0.716	0.383	0.424
Cu	0.16	0.397	0.532	1	0.417	0.495	0.52	0.387
Hg	0.064	0.265	0.103	0.417	1	0.103	0.298	0.196
Ni	0.317	0.329	0.716	0.495	0.103	1	0.307	0.436
Pb	0.29	0.66	0.383	0.52	0.298	0.307	1	0.494
Zn	0.247	0.431	0.424	0.387	0.196	0.436	0.494	1

3. Factor Analysis [4]

When there are many variables, and there are least or not correlation between them, the information got from analyzing the data may be kind of repetitive. Factor analysis is a method which can get comprehensive index named principal factors by descending dimension.

There are no relationship among principal factors, and they can reflect the main information.

The mathematical model of factor analysis is $X = AF + \varepsilon$, particularly to say:

Table 2. The Contribution Rate of Factor Analysis

Factors	Eigenvalues	Contribution rates	Accumulative contribution rates
The first principal factor	3.560	0.445	0.445
The second principal factor	1.150	0.144	0.589
The third principal factor	0.965	0.121	0.710
The fourth principal factor	0.768	0.095	0.805
The fifth principal factor	0.578	0.072	0.877

The main propose of factor analysis is to make the variables who have similar factor loading at the same factor. The bigger the load between variables and one factor, the closer the relationship between them. According to the factor loading matrix, we can get 5 principal factors, which is showed in Table 3. The first principal factor contained Cr and Ni, or is Zn. The second principal factor contain Cd and Pb.

Where X is principal factors, F is factor variables, $A = (a_{ij})_{p \times m}$ is factor loading matrix, ε is the residual and specific factors which are mutually independent and distributed normally.

For the sake of reducing the subjectivity of explanation, factor rotations is needed to get satisfied principal factors. It can make the square of factor loading coefficient in factor loading matrixes differentiated to 0 and 1. We use the maximum variance cross rotation to rotate the matrixes.

Use $SPSS$ to do factor analysis over the 8 kinds of heavy metal. And the results are showed in Table 2. From Table 2, the first five principal factors indicated 72.0% information, which include the most information, so they may come from the same source. The second principal factor is Cd and Pb. The third principal factor is Hg, the fourth principal factor is As, the fifth principal fact contribution rate of the second, third and fourth principal factor is 43.2%, which is close to the contribution rate of the first one (44.5%). So we can conclude that Cd, Pb, Hg, As and Zn are the main pollution elements.

Table 3. The Elements of Every Principal Factor

Principle factors	The elements
The first principal factor	Cr Ni
The second principal factor	Cd Pb
The third principal factor	Hg
The fourth principal factor	As
The fifth principal factor	Zn

By analyzing the literatures, we can know what producing activity can produce what heavy metal pollution, which is showed in Table 4 particularly.

Table 4. The main source of the soil

Sources	Heavy metal
mining explorations and smelting	Cr, Hg, As, Pb, Ni, Mo
Ashes generated by coal and oil burning	Cr, Hg, As, Pb
Electroplating industrial wastewater	Cr, Cd, Ni, Pb, Cu, Zn
Plastic, batteries, electronic industrial wastewater	Hg, Cd, Pb, Ni, Zn
Hg industrial wastewater	Hg
Dye, chemical leather industrial wastewater	Cr, Cd
Automobile exhaust	Pb
Pesticides and fertilizers	As, Cu, Cd

So we can know that Cr and Ni are generated by mining explorations, smelting and burning coal and oil, Cd and Pb are generated by Plastic, batteries, electronic and electroplating industrial wastewater, Hg is generated by Hg industrial wastewater, ashes after burning coal and oil. As is generated by pesticides and fertilizers, mining explorations and smelting. Zn is generated by Plastic, batteries, electronic and electroplating industrial wastewater too.

4. Conclusion

After the analysis above, we can conclude that Cr, Ni, Cd and Pb are generated by smelting and processing products in industrial area. Hg and Zn are generated by automobile exhaust in the trunk road area, As is generated by fertilization in the mountain farmland.

References

1. Kim K W, Myung J H, Ahn J S, etal. Heavy metal contamination industs and stream sediment in the Taejon area [J]. Korea J Geochemical Exploration, 1998, 64: 409-419.
2. Lin Yan. Pollution Assessment and Prediction of Heavy Metals in Soil based on Geostatistics and GIS [D]. Central South University, 2009.
3. China Society for Industrial and Applied Mathematics. 2011 Higher Education Press Cup National Mathematical Contest in Modeling tournament title [DB/OL]. [2014-10-30]. http://www.mcm.edu.cn/html_cn/node/a1ffc4c5587c8a6f96eacefb8dbcc34e.html
4. Wang Xiongjun, Lai Jianqing, et al. Main source of soil heavy metal pollution based on factor analysis in Taiyuan [J]. Ecology and Environment, 2008, 02:671-676.

Microstructure and properties of Cu-Fe-Co alloy obtained by directional solidification

G. D. Shi, H. Jiang & Z. D. Wang*
University of Science and Technology of Beijing, Beijing 100083, PR China
**wangzidong64@126.com*

X. H. Chen
State Key Laboratory for Advanced Metals and Materials, University of Science and Technology Beijing, Beijing 100083, PR China

Abstract: The microstructure and properties of Cu-Fe-Co alloy obtained by directional solidification were investigated. Results show that the finely dispersed iron-rich nanoparticles can be obtained under the directional solidification condition, with the size of 5-50nm and the coherent interface between the iron nanoparticle and the copper matrix. The Cu-Fe-Co alloy with excellent comprehensive mechanical properties, tensile strength up to 294.35Mpa, the elongation is 33.43%; Iron-rich nanoparticles increase the tensile strength of the alloy reduce the plasticity.

Keywords: Directional solidification; Cu-Fe-Co alloy; Nanoparticles.

1. Introduction

Copper alloy is widely used in social production due to its excellent electrical, thermal conductivity and corrosion resistance. But the strength and hardness of pure copper are low, even by work hardening, the hardness and strength still can't satisfy requirement of use. Since the 70's, in the rapid development of metallurgy, electronics and other industries background, domestic and foreign scholars have developed a number of high strength and conductivity of copper alloy, has been widely used in the manufacture of lining, continuous casting crystallizer of IC lead frame, high speed train overhead wires [1-3]. With the rapid development of science and technology, on the properties of high strength and high conductivity materials are also put forward higher requirements, therefore, the development of new high strength and high conductivity copper alloy materials has become one of the hotspots in the field of copper alloy.

At present, a lot of work to focus on Cu-Nb [4-7], Cu-Ag [8], Cu-Cr [9] alloy. However, because of the price of raw materials is expensive and complicated preparation process, limit the application prospect of copper alloy.

Due to the low price of Fe, low melting point and mother alloy convenient preparation, many scholars have carried out research on deformation of Cu-Fe in situ composite [10-12]. Therefore, intermediate heat treatment and alloying of present research on composite material of Cu-Fe system mainly focus on the deformation of Cu-Fe in situ composite material, and how to improve then as cast Cu-Fe alloy strength and conductivity is rarely reported.

In-situ nanoparticles reinforced copper alloy with excellent mechanical properties was prepared by convection solidification method, the copper alloy strength and plasticity at the same time has been greatly improved, and in situ nanoparticles generated in melt [13]. Single crystal copper alloy with dispersed iron-rich nanoparticles was fabricated by directional solidification, the interface of in-situ iron-rich nanoparticles and matrix was analyzed and the mechanism of iron-rich nanoparticles uniformly embedded in the matrix was described.

2. Experimental procedures

The Cu (1#) and Cu-1wt% Fe-1wt% Co alloy (2#,3# and 4#) were fabricated by directional solidification. The specific experimental parameters set as shown in table 1.

Table 1. Continuous unidirectional solidification process parameters

No.	Melt temperature(°C)	Cooling water(L/h)	Casting speed (mm/min)
1#	1180	900-1000	3
2#	1300	900-1000	3
3#	1250	900-1000	3
4#	1200	900-1000	3

The equipment used to prepare the samples is an induction directional solidification furnace with an ultimate vacuum level 6.0×10^{-2} Pa. The main processing steps are as follows. Cu, Fe and Co metals are first introduced into a graphite crucible located in the vacuum chamber. After being evacuated to 5 Pa, the chamber temperature is increased from room value to 1350°C, and maintained at this value for 30 min. Next, argon is allowed into the chamber to increase the chamber pressure to 0.07MPa.Then constant the melt temperature at value 1180 to 1300°C, Open and adjust the cooling water. Open traction system, adjusting casting speed, complete continuous unidirectional solidification alloy preparation.

The ultimate vacuum level is 6.0×10^{-2}Pa. The main processing steps are as follows. Cu and Fe metals were first introduced into a graphite crucible located in the vacuum chamber. After being evacuated to 5 Pa, the metal temperature was increased from room temperature to 1350°C, and was maintained at this value for 30min. Next, argon was allowed into the chamber to increase the chamber pressure to 0.07 MPa. Then the melt temperature was kept at 1180°C-1300°C, the cooling water flow was opened and adjusted. Finally, traction system was opened and copper alloy bar was prepared with constant casting speed.

The mechanical properties were measured at room temperature in a model CMT4105 TMS, and the microstructures of copper alloys were examined with optical microscope (OM). Furthermore, the morphology of iron-rich nanoparticles, the relationship between the iron-rich nanoparticle and the copper matrix were analyzed with an HRTEM.

3. Results and Discussion

3.1. *Microstructure of the Cu-Fe-Co alloy*

Continuous unidirectional solidification microstructure evolution is divided into three distinct stages: 1) Seeding the initial stage, the solidification microstructure is equiaxed grains; 2) Competitive growth stage, the solidification structure is fan radiating columnar crystal; 3) The stationary growth phase, the solidification microstructure is uniform columnar or single crystal structure. Fig.1 shows the as-cast microstructure of Cu and Cu-Fe alloy with the stationary growth stage. The 1#, 3# and 4# samples of organization are single columnar grain, but the 2# sample is polycrystalline organization. When the casting speed and cooling water are stable value, the increase of melt temperature can make the solid-liquid interface moves down. The crystallizer entrance to solid-liquid interface front inside this paragraph of distance ,due to the high temperature zone of crystallizer and the role of melt temperature, heat conduction is relatively slow; And below the solid-liquid interface, due to the crystallizer temperature and the role of solidified alloy, heat conduction velocity increases, will result nucleation in side wall, formed a polycrystalline microstructure of 2# sample rather than in other specimens of single columnar structure.

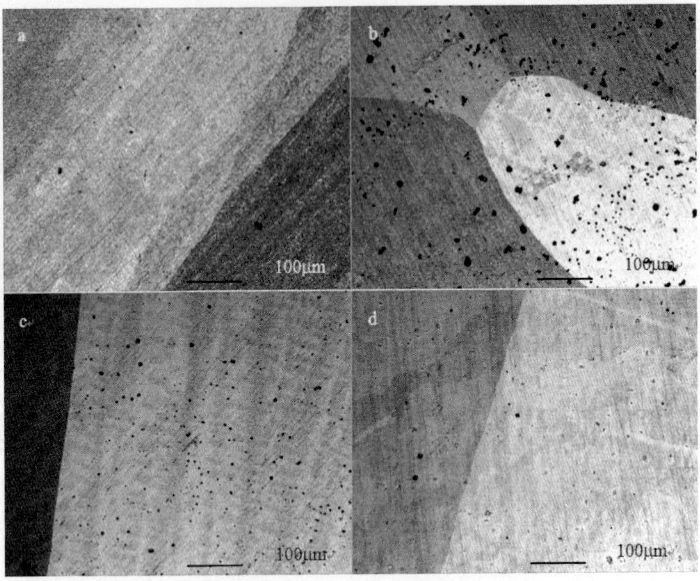

Fig. 1. Microstructures of the Cu alloy: (a)1#;(b)2#;(c)3#;(d)4#.

3.2. *Mechanical properties*

The results from the tensile experiments carried out at room temperature are listed in Table 2. Compared with Cu, the tensile strength of Cu-Fe-co alloy is increased from 131.21 Mpa to 294.35 Mpa, the elongation rate of Cu-Fe-Co alloy is reduced from 43.76% to 33.43%. Fig. 2 shows the Directional solidification of copper and copper alloy stress strain curve. Cu-Fe-Co alloy is a single crystal microstructure from the directional solidification, so the material strength improvement can be thought to be caused by iron-rich nanoparticles which formed in the cooper matrix during the directional solidification.

The enlarged morphologies of fracture surfaces of pure Cu and Cu-Fe-Co alloy were shown in Fig. 3. As shown in Fig. 3, pure copper present fracture characteristics of microvoid growth. Although pure Cu and Cu-Fe-Co alloy are all single crystals, but their fracture morphologies are different. The fractographs of Cu-Fe alloy consisted of micron-sized equal-axis dimples, which is due to the presence of large amounts of the iron-rich nanoparticles in Cu-Fe-Co alloy. During process of plastic deformation, dislocations move at the interface of nanoparticle and matrix, which causes pore forming by decohesion nucleation and leads to fracture.

Table 2. Mechanical properties of directionally solidified Cu and Cu alloy.

No.	Tensile strength(Mpa)	Total elongation(%)
1#	131.21	43.76
2#	179.87	37.2
3#	269.95	31.5
4#	294.35	33.43

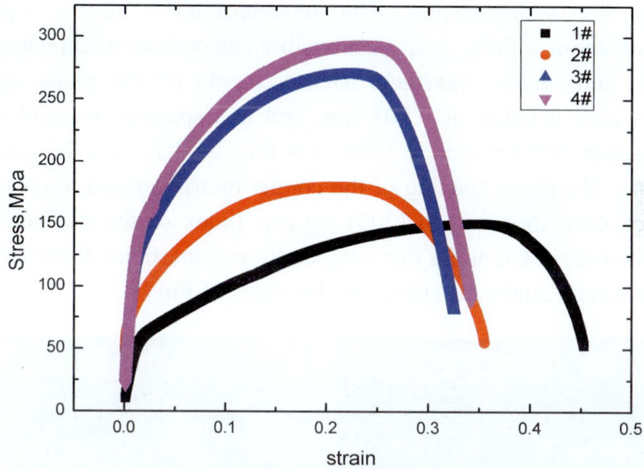

Fig. 2. Stress-strain curves of copper and copper alloy.

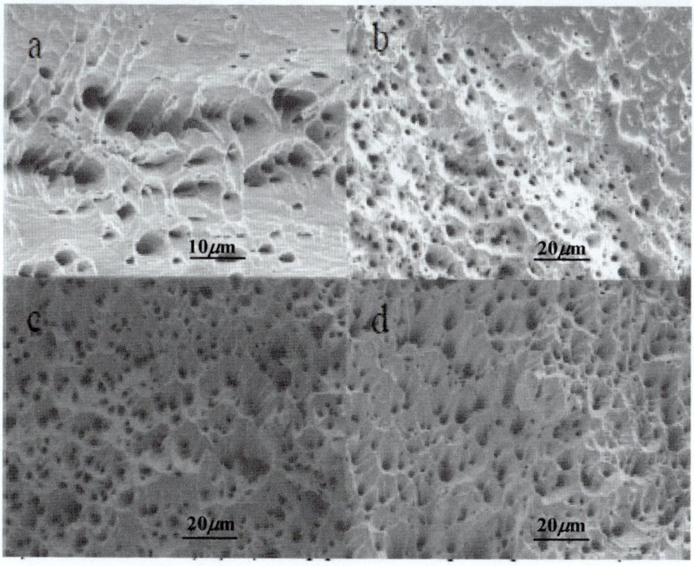

Fig. 3. SEM morphologies of fracture surfaces for Cu and Cu-Fe-Co.

3.3. *Precipitated phase*

Fig. 4 shows the morphology of as-cast Cu-Fe alloy precipitation phase. Precipitated phase is divided into two categories: One is a small nanoparticles with size less than 10nm which is uniformly dispersed in the matrix; Two is a disc-shaped nanoparticles with size less than 100nm.

In order to ascertain the composition and structure of nanoparticles, we examined the nanoparticles at higher magnification, in an HRTEM. Fig. 5 shows small Fe nanoparticles. The nanoparticle is nearly square with dimensions of about 5.5nm, it is clear visible orientation relationship between nanoparticles and matrix interface. Measurements of the plane spacing for nanoparticles give a value of 0.201 nm, and this spacing is equal to spacing between adjacent (110) planes of iron. It is thus evident that the nanoparticles are iron phase. The plane spacing of the copper matrix around the nanoparticles are 0.208 nm, corresponding to (200) copper. Index of the diffraction pattern shows that the matrix Cu and a disc-shaped Fe nanoparticles form a completely coherent relation in Cu||Fe, as shown in the insets of Fig. 5.

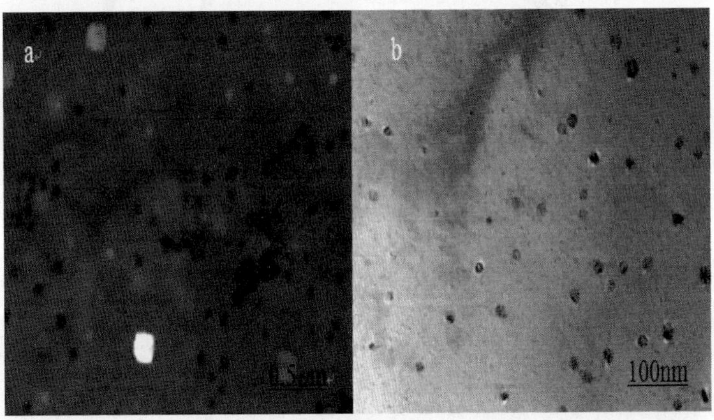

Fig. 4. TEM image of the precipitated phase: (a) 3#; (b) 4#.

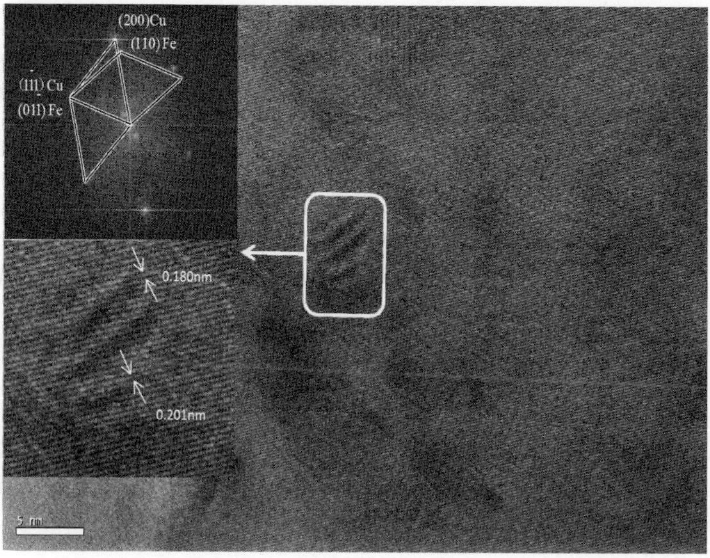

Fig. 5. HRTEM image showing the precipitated phase of 4#.

The Cu-Fe alloy phase diagram shows that: Under the peritectic reaction temperature of 1094 °C, the solubility of iron in copper is about 4%, as the temperature decrease, the solubility of iron in copper dropped significantly, to room temperature, the iron almost is not dissolved in the copper matrix, and intermetallic compound cannot be formed between Fe and Cu. According to the previous analysis, Copper matrix dispersed a large number of precipitates was observed, through the diffraction pattern and calibration analysis, the precipitated phase was determined to be Fe. The solid-liquid interface of directional solidification exists a temperature gradient. Below the solid-liquid interface, thermal conductivity is larger due to the effect of low temperature region of crystallizer and already solidified alloy, so the solute atoms is difficult to diffuse, thus to restrain the growth rate of iron-rich nanoparticles. In directional solidification process, the solidification interface within the tiny frontier region, the liquid melt into the following line, iron-rich nanoparticles nucleation in the solidification front, when the solidification interface to move at a constant speed, the iron-rich nanoparticles will be captured by solidified alloy. Thus to form the uniformly distributed iron-rich nanoparticles in the matrix is formed.

4. Summary

The finely dispersed iron-rich nanoparticles can be obtained under the directional solidification condition, with the size of 5-50nm and the coherent interface between the iron nanoparticle and the copper matrix. Cu-Fe–Co alloy as compared to pure Cu, the tensile strength of Cu-Fe-co alloy is increased from 131.21 Mpa to 294.35 Mpa, the elongation rate of Cu-Fe-Co alloy is reduced from 43.76% to 33.43%

References

1. SUI X.H. & LI H.H. 2007. Analysis on property of oxidiation resistance of Cr-Zr-Cu alloy for continuous crystallizer [J]. Physics Examination and Testing 25(5): 24–26.
2. HUANG G.J. & XIE S.S. & CHENG Z.H. & LI H.Q. 2005. Fish defects of cold rolling copper alloys applied to lead frame. The Chinese Journal of Nonferrous Metals1 5(2): 353–356.
3. JIA, S.G. & LIU, P. & REN, F.Z. & TIAN, B.H. & ZHENG, M.S. & ZHOU, G.S. 2007. Sliding wear behavior of copper alloy contact wire against copper-based strip for high-speed electrified railways. Wear 262(7/8): 772–777.
4. Bevk, J & Harbison, J.P. & Bell, J.L. 1978. Anomalous increase in strength of in situ formed Cu-Nb multifilamentary composites. J Appl Phys 49: 6031 6038.
5. Trybus, C. & Spitzig, W.A. 1989. Characterization of the strength and microstructural evolution of a heavily cold rolled Cu-20%Nb composite. Acta Metall 37: 1971-1981.
6. Verhoeven, J.D. & Spitzig, W.A. & Jones, L.L. 1990. Development of deformation processed copper refractory metal composite alloys. J Mater Eng 12 : 127-139.
7. Snoeck, E. & Lecouturier, F. & Thilly, L. 1998. Microstructural studies of in situ produced filamentary Cu- Nb wires [J]. Scripta Materialia 38(11): 1643-1648.
8. Frommeyer, G. & Wassermann, G. 1975. Microstructure andanomalous mechanical properties of in situ produced silver-copper composite wires. Acta Metall 23: 1353-1360.
9. Kim, S.T. & Berge, P.M. & Verhoeven, J.D. 1995. Deformation processed copper-chromium alloys: optimizing strength and conductivity. Mater Eng Perform 4: 573-580.
10. Hong, S. I. & J.S. Song. 2001. Strength and conductivity of Cu-9Fe-1.2X (X = Ag or Cr) filamentary microcomposite wires. Metall Mater Trans A 32:985-991.
11. Hong, S. I. & J.S. Song. 2001. Adv. Eng. Heavily drawn Cu-2Fe-2Ag and Cu-2Fe-2Cr microcomoposites. Mater 113 :610-616.

12. Biselli, C. & Morris, D.G. 1994. Microstructure and strength of Cu-Fe in situ composite obtained from prealloyed Cu-Fe powders. Acta Metall Mater 42(1) :163-176.
13. Wang, Z.D. & Wang, X.W. & Wang Q.S. & Shih, I. 2009. Fabrication of a nanocomposite from in situ iron nanoparticle reinforced copper alloy. Nanotechnology: 75-83.

Experiment research on dynamic and static stiffness of wj-8c fastener for high-speed railway

J.Y. Yuan*, Z.P. Zeng, D.P. Zeng & X.M. Zhang
School of Civil Engineering, Central South University
Changsha, Hunan, China
**287397263@qq.com*

G.Y. Li & Q.G. Bian
Shandong Jitie Engineering Construction Supervision Co, LTD
Jinan, Shandong, China

G.L. Shi, K.S. Wang, H.B. Zhou & X. Zhao
China Railway 17 bureau Group,
Weihai, Shandong, China

Double-slab ballastless track use the WJ-8C fastening system. Because the fastener assume an important role in the regulation track deformation. Study performance especially dynamic and static stiffness of fasteners for stress analysis of track structure has important theoretical and practical value. Through the dynamic load and static load experiment on track structure. Rational utilization of the experiment of WJ-8C fastener performance appraisal. Determine the WJ-8C fastener reasonable value of dynamic and static stiffness. Designed for the double-slab ballastless track fasteners of dynamic and static stiffness values to provide the reference data and performance testing

Keywords: Dynamic stiffness, Static stiffness, WJ-8C fastener, Experimental analysis

1. Introduction

1.1. *The WJ-8C fastener structure*

In recent years, the rapid development of high speed railway in China. Followed, in our country in order to adapt to the development of high-speed railway is successfully developed several new fasteners. Among them, the WJ-8C fastener is a fastener which is extensively used. The WJ-8C fastener structure as shown in Figure 1.

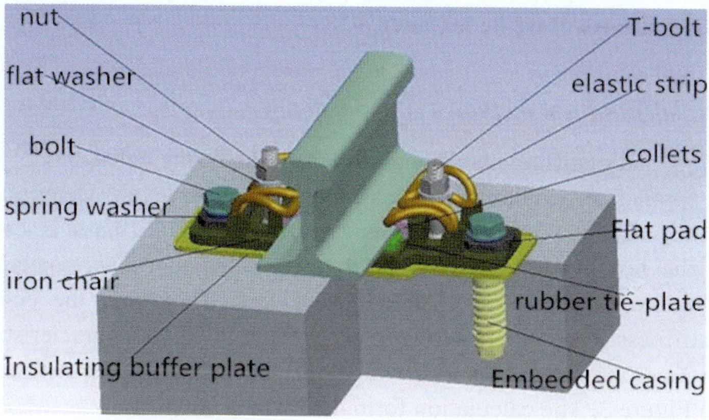

Fig. 1. The WJ-8C fastener structure.

1.2. *Experimental method*

In order to accurately measure dynamic and static stiffness of WJ-8C fastener. The experiment on the roadbed Double-slab ballastless track 1:1 on the full scale model. Which is shown in Figure 2. There have eight fastener, a total of 16 WJ-8C fastener. Arrangement of two actuators for static load and dynamic load test. Rail between the actuator truncation. Each actuator load borne by four fasteners. Using dial indicator test measured the displacement of fasteners, fastener reaction by the actuator load values. The fastener in the dynamic load of the displacement by the IMC collection.

Fig. 2. Double-slab ballastless track 1:1 on the full scale model.

2. Static stiffness of wj-8c fastener

2.1. *Static stiffness test method*

Fasteners of static stiffness is through the machine to the entire model structure of rail is applied perpendicular to the sleeper static load levels. Determination of the rail displacement produced under load. Fastener stiffness is calculated. Because the fastener system is made up of metal and rubber materials. The stiffness curve does not meet the Hooke's law. Therefore, the engineering general stiffness in the characterization by secant stiffness characteristics. That is within a certain range of load average stiffness as a fastener stiffness. As is shown in Figure 3. The calculation formula of static stiffness is

$$K_s = (F_2 - F_1)/(S_2 - S_1) \tag{1}$$

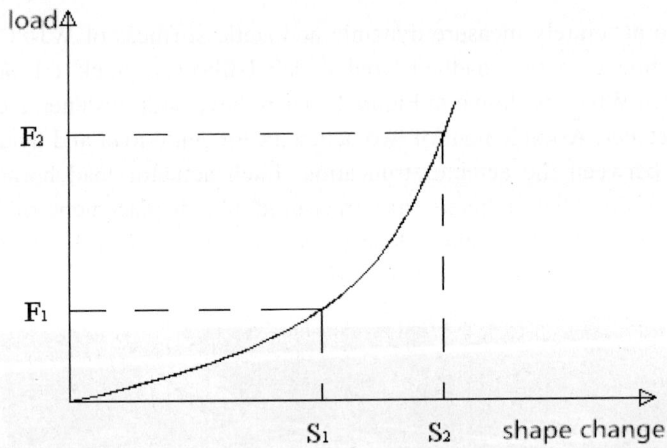

Fig. 3. The static stiffness test method.

Due to an actuator of maximum load of 200 KN. In order to measure displacement of fasteners. Trial by 0 KN, 50 KN, 100 KN, 150 KN and 200 KN load by stage. Actuator arrangement drawing as shown in Figure 4. Here we only use the two actuators.

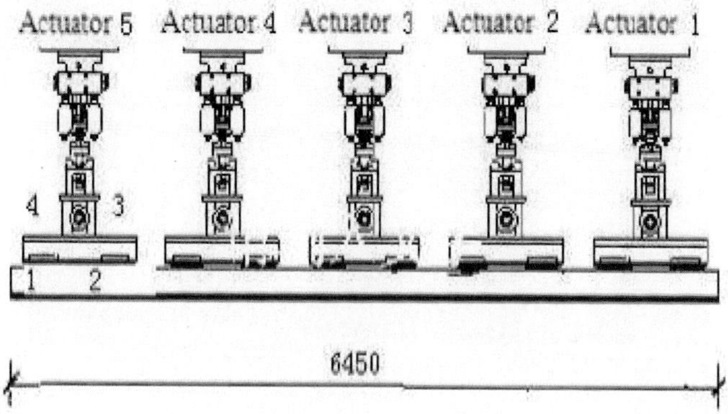

Fig. 4. Actuator arrangement.

2.2. *Procedure of test*

Firstly, preloading of the component 2 times, and then formally tested, and the static displacement meter to zero.

Secondly, according to the hierarchical load solution loaded. Loading speed for 10 KN/min. After each hierarchical load keep constant load records corresponding to the dial indicator displacement after 1 minute.

Thirdly, according to the loading scheme classification to KN 200 after unloading. According to the loading force and the measured displacement in accordance with the scope of reasonable loading fastener static stiffness value.

2.3. *The experimental results and analysis*

By the experimental measured two actuators, a total of eight fasteners of force and displacement. To calculate each fastener static stiffness value and averaged. That as WJ-8C type static stiffness value of the fasteners. The experimental results as shown in Figure 5 and 6.

The Figure 5 and 6 shows that four of each actuator fastener stiffness curve change trends are the same. Along with the rising of the actuator load, displacement of fastener increased gradually. According to theoretical analysis, fastener stiffness curve is nonlinear. But as shown in the figure, within the scope of the test load, test results measured fastener stiffness curve basic linear stiffness change appears.

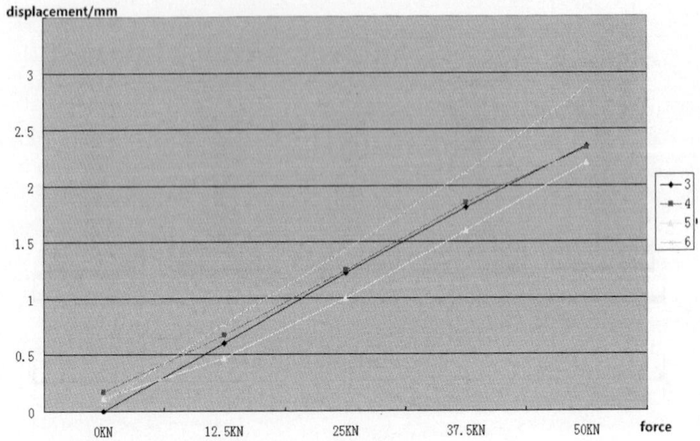

Fig. 5. Actuator 1 test results.

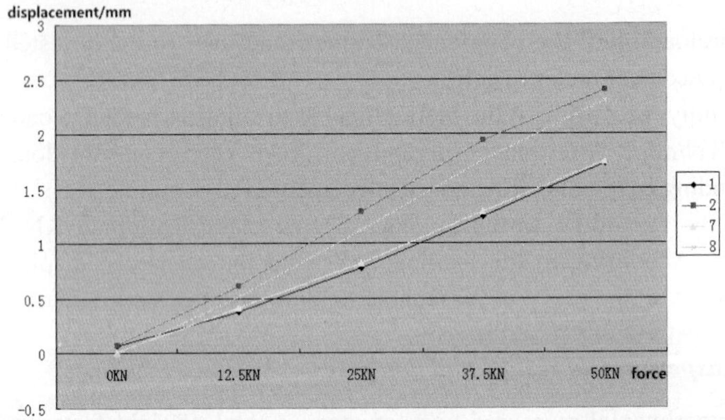

Fig. 6. Actuator 2 test results.

According to the test results, according to the scope of reasonable loading each fastener static stiffness value. With the method of step by step calculation and averaging get average static stiffness. The results are shown in Table 1.

Table 1. Fasteners static stiffness calculation results (KN/mm)

Fasteners number	First level	Second level	average
Fasteners 1	20.97	20.39	20.68
Fasteners 2	25.00	23.15	24.01
Fasteners 3	34.72	28.09	31.41
Fasteners 4	17.86	18.38	18.12
Fasteners 5	40.32	35.21	37.82
Fasteners 6	23.15	21.55	22.35
Fasteners 7	30.49	29.30	29.90
Fasteners 8	23.81	22.26	23.01

As is shown in Table 1, actuator 1 fastener average of static stiffness is 23.55KN/mm, actuator 2 fastener average of static stiffness is 28.04KN/mm. For the average of two actuators is 25.8 KN/mm. Formulated by the ministry of railways "Provisional technical conditions of passenger dedicated line is listed" mentioned: When WJ–8C fasteners use B class elastic plate under the action of an iron plate, Static stiffness for fastener node is 23KN/mm(±m(tatic). So the test results accord with the requirement of the rule. Instructions by the fastener performance test method is feasible.

3. Dynamic stiffness of wj-8c fastener

3.1. *Dynamic stiffness concept*

Dynamic stiffness means structure under the dynamic turbulence of specific resistance to deformation. Under the static load ability to resist deformation is called static stiffness. The ability to resist deformation under dynamic loading, called dynamic stiffness. That caused the unit amplitude dynamic force that is needed. Dynamic stiffness is measured by structural natural frequencies.

If action force change is slow, the action force changes the frequency of the structure is far less than the natural frequency. Can be thought of dynamic stiffness and static stiffness are basically the same. Otherwise, the frequency of the action force greater than the structure of the natural frequency. Structure is not easy to deformation, that is, deformation is small, and the dynamic stiffness of the structure of relative turbulence is bigger.

But the frequency of the action force and structure of the natural frequency is near, possible resonance phenomenon. The minimum dynamic stiffness, deformation is the largest.

Therefore, the dynamic stiffness is measured structure resistance due to the dynamic characteristics of the turbulence ability.

3.2. *Dynamic stiffness computational formula*

Through testing machine at a constant frequency to the assembly of rail fastening system applied perpendicular to the sleeper level cyclic loading. Determination of rail under the load of minimum and maximum displacement. See for Equation 2 below:

$$K_{DYN} = \frac{F_{2a} - F_{1a}}{D_2 - D_1} \tag{2}$$

where F_{1a} = on the rail system being tested actual minimum dynamic load; F_{2a} = on the rail system being tested actual maximum dynamic load; D_1 = the minimum displacement of the rail during dynamic loading; D_2 = the maximum displacement of the rail during dynamic loading.

3.3. *Result and analysis*

We here at speed of 350 km/h and 17t axle load, loading a 10000 cycle dynamic load as the calculating basis. Using IMC collection device. Load curve is shown in Figure 7.

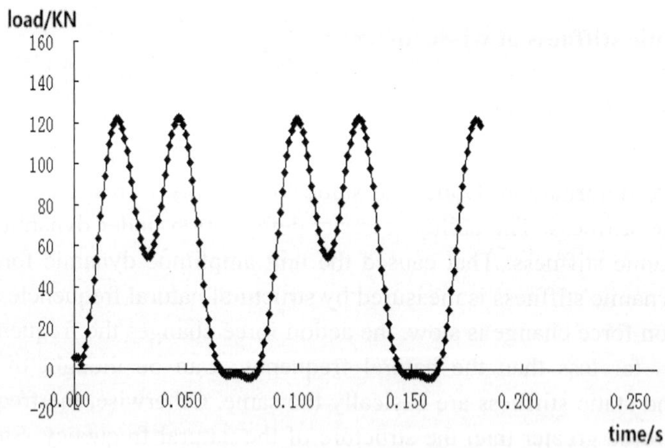

Fig. 7. Load curve in IMC.

According to the load curve, we can find the maximum dynamic load of 121.97 KN, we can find the minimum load of -4.49 KN. Assigned to each fastener node values as follows: 30.49 KN, -1.12 KN. By IMC collection results can be seen that rail during dynamic loading the minimum and maximum displacement is 0.7 mm, respectively is 0.2 mm. So from the formula, it can be concluded. Dynamic stiffness for fastener node is 58.74KN/mm. The ratio of dynamic stiffness and static stiffness is 2.28.

3.4. *Vehicles-line vertical dynamics model*

Rolling stock is by the car body, bogie, wheel peer through the basic components a series of secondary suspension components, the coupling system. In order to simplify the meter Calculate, save calculation time. Usually in the locomotive vehicle dynamics research, these parts approximately regarded as rigid body. There is a play between the basic vehicle parts Constraint or rigid constraint. To limit the structure of the components in the vehicle Relative motion. Therefore, in the vertical vehicle dynamics analysis model is set up, the following assumptions:①Without considering the flexibility of car body, bogie and wheel set Shape, namely, car body, bogie and wheel set are rigid body; ② Wheel and car body along Line direction of uniform motion, don't consider the effect of longitudinal dynamic action; ③ All vehicles suspension system between the damping are calculated according to the viscous damping; ④ Secondary suspension and wheel system and the spring characteristics of positioning is linear; ⑤Train body, bogie and the wheel of the rigid body are small in basic equilibrium position The displacement of the vibration; ⑥Train body on the center of mass before and after the left and right sides is symmetrical, and symmetry.

4. Conclusion

Through the test analysis of experimental data processing. We got the static stiffness of the WJ-8C fasteners is 25.8 KN/mm, dynamic stiffness of the WJ-8C fasteners is 58.74KN/mm. The ratio of dynamic stiffness and static stiffness is 2.28. Through access to relevant data, compared with the results found that we get the result is reasonable. Designed for the double-slab ballastless track fasteners of dynamic and static stiffness values to provide the reference data and performance testing.

Acknowledgements

The research summarized in this paper was supported by the Master Independently Explore Innovative Projects of Central South University (No. 2015zzts246), Science and Technology Foundation of Jinan Railway Administration (No. 2011G19), the Joint Funds of the National Natural Science Foundation of China (Nos. U1334203 and U1134209), the National Natural Science Foundation of China (No. 51378513), the Science and Technology Foundations of China Railway Corporation (Nos. 2014G001-D and 2013G003-

A-3), the Program for Changjiang Scholars and Innovative Research Team in University (No. IRT1296). These supports are gratefully acknowledged.

References

1. Qichang, WANG. 2006. Ballastless track rail fastener, Chengdu: Southwest Jiaotong University Press.
2. The Ministry of Railways of People's Republic of China. 2006. Provisional technical conditions of passenger dedicated line is listed. Beijing: China Railway Press.
3. Zhiping Zeng, Zhiwu Yu, Yangang Zhao, Wentao Xu, Lingkun Chen, and Ping Lou. 2014. Numerical simulation of vertical random vibration of train-slab track-bridge interaction system by PEM. *Shock and Vibration*. vol. 2014, Article ID 304219, 21 pages.
4. Zhiping Zeng, Jiayu Yuan, Xianfeng He, Can Liu, Kunteng, Zhu. 2014. *WJ-8 rail fastener of ballastless track torque test research*. Applied Mechanics and Materials, 2014, 587-589: 1320-1323.
5. Yuyuan Luo, Jinlu Pang, Guangfu Yang, Zhiguo Jiang. 2011. The development of the high-speed rail with elastic plate. Chemical Propellants & Polymeric Materials, 2011, 9(3):14-17.

The study on structural color of porous anodic alumina thin film fabricated in phosphoric electrolyte

S.M. Yang*, J.J. Gu and Y.K. Qi

Department of Physics, Hebei Normal University for Nationalities,
Chengde 06700, Hebei, China
shumin_yang@126.com

The alumina thin films with brilliant structural colors have been fabricated by means of electrochemical oxidation in phosphoric electrolyte under the condition of room temperature. The relationship of the structural colors with the anodization time and the angle of incidence of illuminating light are discussed. The microstructures of the films have been characterized as well. The effective refractive index of film is calculated according to Maxwell-Garnett theory. The maximum reflection wavelength of films is calculated by Bragg's equation, and the generation mechanism structural color has also been discussed. In the end, organic-assist covering method has been applied to obtain the intricate multi-color pattern on the same alumina thin film. This thin film with brilliant colors may have potentials in color displays, decoration and anti-counterfeiting technology.

Keywords: Anodic alumina; Porous material; Structural color.

1. Introduction

Structural color is an optical effect produced by living creatures' periodical microstructures. It is intrinsically different from pigment for its property of being a physical color. The major characteristics of structural color include its high reflectivity, polarization, the property of being never faded in color as well as never disappear in chemicals and hot water.

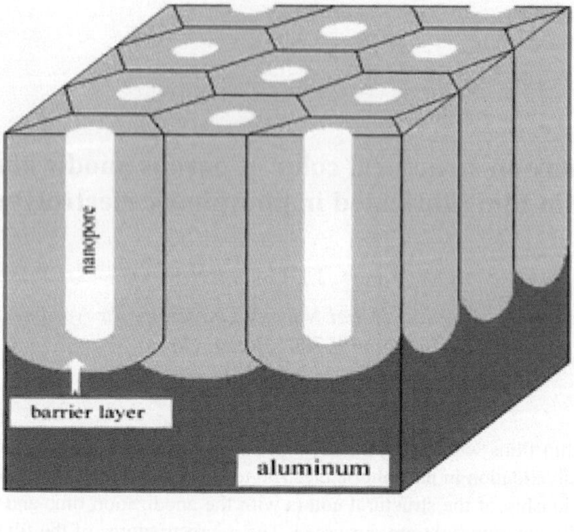

Fig. 1. The structure of AAO template.

Structural colors are divided into three categories: colors produced by diffraction, scattering and interference.

Structural color universally exists in natural world, for instance as that of feather of birds [1] and wings of insects [2]. Recently, the manmade structural color systems have attracted wide attention, for examples, multilayer structure [3], orderly porous surface structure [4] and the super-thin alumina film (AAO) with its pores filled by metals [5].

At the time, porous alumina thin film (AAO) has already been widely used in the fields of decoration material, ultra precise membrane for separation technique, polarizer, solar power selective absorption membrane and beam-shaped microelectrode. Figure 1 shows the structure of AAO thin film. A thin but compact barrier layer is close to the side of aluminum, on which a comparatively thicker and looser porous layer lies. The present researches have already proved that the aperture of AAO thin film is closely related with the decomposition voltage. Cell density is a function of anodization voltage; and thickness of barrier layer is also related with anodization voltage. Owing to the controllable aperture and film thickness as well as its good thermostability, AAO thin film has been used as a good template material and widely applied in the synthesis of orderly nanostructure [6, 7, 8]. The research about the structural color of AAO thin film has already begun with the progress in the study of photonic crystal. In 1969, Diggle et al. [9] reported that within the range of visible light, alumina thin films show bright color when their thickness is less than 1μm with the interference of light. In 2007, Wang et al. [10] reported that they have successfully prepared alumina thin film with comparatively high color saturation by depositing carbon nano-tubes in hole of alumina with CVD technique. Shortly after that, in 2010, Zhao et al [11] realized the subtle control of the colors of alumina thin film by depositing carbon nano-tubes. In 2011, Sun et al [12] prepared alumina thin film with iridescence, using multiple oxidation method.

This paper has been concentrated in the preparation technique of single structural color of AAO thin film by applying anodization method; the theoretical explanation and support have also been given. This research is significant in the bionic study of structural color of birds' feather. Since structural color has the merits as environmental friendliness, being never fades in color and has the property of rainbow effect, it may have wide potentials in color displays, decoration and anti-counterfeiting technology.

2. Experiment

High-purity aluminum foils (99.999%) were annealed at 400°C for 2 hours in an annealing furnace, and then were electropolished in a 1:4 (volume) mixture HClO4 and C2H5OH for 5 min. The foils after electropolishing were cleaned in acetone and deionized water and dried up. After the anodization at the voltage of 100V and constant temperature 5 °C, 6 alumina thin films have been prepare at the different oxidation time of 70s, 80s,100s,105s, 110s, 120s.

The structural and optical features of alumina thin film were characterized by an optical digital camera (Canon-EOS 600D) and field-emission scanning electron microscopy (SEM, Hitachi S-4800).

3. Result and Discussion

Figure 2 shows the structural colors of the six samples (with the angle of incidence of 20°) of the different oxidation time of 70s, 80s, 100s, 105s, 110s, 120s under the voltage of 100V. It shows that the structural color turns from purple to red gradually with the increasing oxidation time.

Firstly, the structural color we have seen is a single color, which is the result of interference of alumina thin film in natural light. The single color indicates that the thickness at everywhere of the thin film is the same. This same thickness indicates that in the forming process of thin film, the current density at everywhere on the surface of aluminum foil as anode is the same, and the current streamlines are well-distributed as shown in Figure 3:

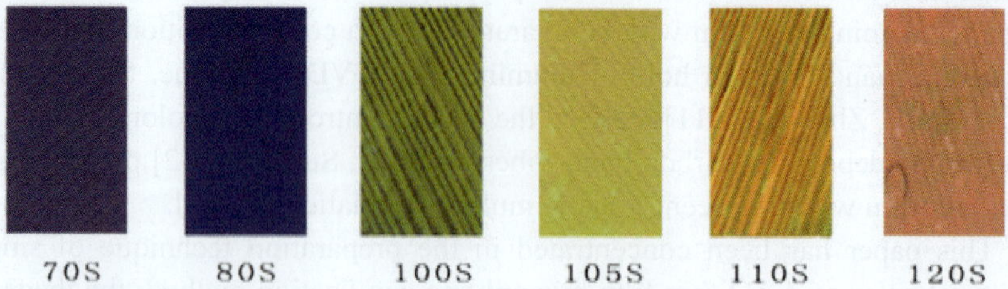

Fig. 2. The structural colors of samples under the oxidation voltage of 100V, oxidation time of 70s, 80s, 100s, 105s, 110s, 120s.

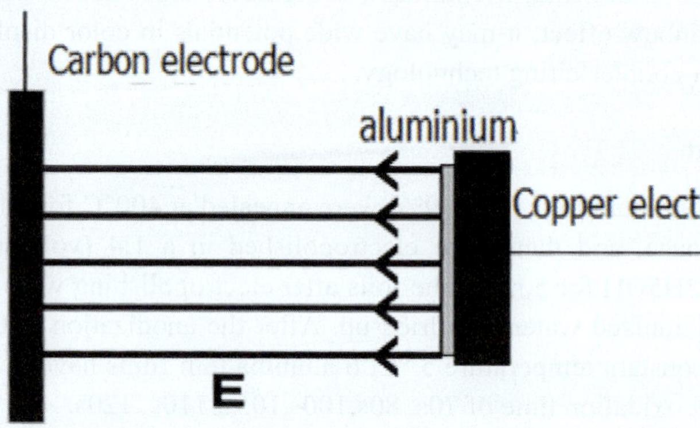

Fig. 3. Current streamlines at the oxidation voltage of 40V.

Figure 4 shows the FE-SEM images (surface image and cross section image) of AAO thin film anodized for 100s at the oxidation voltage of 100V.

It can be seen from the images that the forming of pores of AAO thin film is only at beginning stage, the average radius of holes is about 25nm, and the average pitch of holes is about 161nm. The porosity was thus about 0.068 according to the formula: In this expression, Dint is the pitch of the holes, and r is the diameter of the holes. The dielectric constant of the AAO thin film can be calculated according to Maxwell-Gamett theory: is the effective dielectric constant; and are the dielectric constants of alumina and air respectively; f stands for the porosity of AAO thin film, the relationship between dielectric constant and refractive index can be presented by f =n2 . The average effective refractive index of AAO thin film is about 1.65 according to the calculation. This refractive index is equal with that of pure alumina thin film; that is to say that the influence of pores at their forming stage can be nearly neglected. The thickness of AAO thin film is about 272nm according to the cross section image. The growth rate of AAO thin film in the process of oxidation is about 2.72nm/s. The corresponding thickness of the AAO thin films prepared at the oxidation time from 70s—120s are shown in Table 1. t is oxidation time and d is the thickness of film. Considering that phase difference (half-wave loss) exists between the reflected lights of upper surface and lower surface, the Bragg's equation can be written as. In this equation λ is refraction wavelength; is refraction angle. The maximum reflective wavelengths of this series of sample are illustrated in Table 1. The data of the Table 1 indicates that the thickness of thin film increases with the increasing oxidation time, and the interference wavelength of reflective lights also increases. The calculation results of the wavelength according to Bragg equation are in consistent with the colors of samples observed.

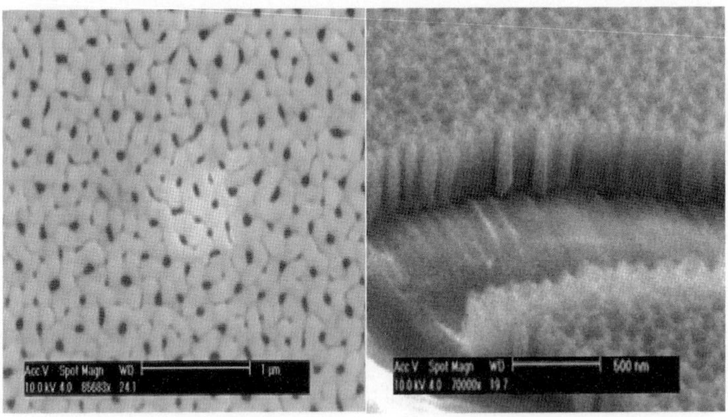

Fig. 4. FE-SEM images of an AAO thin film anodized for 100s (surface image and cross section image).

Table 1. The parameters of samples at the oxidation voltage of 100V, oxidation time of 70s, 80s, 100s, 105s, 110s, 120s.

Oxidation time(s)	70	80	100	105	110	120
Thickness of alumina thin film(nm)	190	218	272	286	299	326
The maximum wavelength of reflective light(nm)	393	450	562	591	618	674
The corresponding colors of reflective wavelength	purple	blue	green	yellow	orange	red

When we observe the color changes of samples in the natural light, we have found that the colors of samples change with the changing viewing angle, as shown in Figure 5. From Figure 5 we see that when viewing angle increases from 20°to 55°, the color of thin film turns from red to purple. To analyze the physical mechanism of this phenomenon, we have calculated the maximum reflective wavelengths (λmax) of corresponding thin films of different viewing angles by Bragg formula. The result is shown in Table 2, which is in consistent with the images of digital photos. This indicates that this change of colors also obeys Bragg formula.

Fig. 5. From left to right are the optical photos of AAO thin films at the viewing angles of 20°, 30°, 45°, 50°, 55°.

Table 2. The parameters samples when the oxidation voltage is 100V, oxidation time is 120s, with the viewing angle from 20°—50°.

Oxidation time(s)			120		
Thickness of AAO thin film(nm)			326		
Viewing angle(▫)	20	30	45	50	55
The maximum reflected wavelength(nm)	674	621	507	461	411
The corresponding color of reflected wavelength	red	Orange yellow	green	Blue purple	purple

From the experimental results above we know that the different colors can be achieved by changing the thickness of AAO thin film. We have also considered if the changing color could be achieved at a same AAO thin film, thus the ideal colorful patterns could be achieved for the practical needs. If we can do this, then the colorful AAO thin film will be very useful in decoration and anti-counterfeiting technology. Following the above idea, we have prepared the different colorful patterns (as shown in figure 6) with the organic-assist covering method. Figure 6 shows the patterns of the same AAO thin film being prepared at different areas of different oxidation time. This indicates that the colorful thin film can be achieved by controlling the oxidation time to the same aluminum foil, and our idea is proved to be correct. The steps are as follow: firstly, we cover parts of surface of AAO film by printing ink with colorful patterns designed earlier, and then we anodize the other parts and repeat the process until the needed pattern is finished. In the end, we cleanse off the printing ink with acetone. The result above indicates that with the organic-assist

Covering method for controlling the thickness of AAO thin film is a simple and practical approach to realize the controlling of colors. And this will be no doubt useful for its application in anti-counterfeiting and decoration material.

This paper has described the process of the one step preparation of alumina thin film with structural colors, which can be controlled by oxidation time. On this basis, the color's change can be realized at a same alumina thin film for achieving the ideal colorful pattern. This paper also discusses the microstructure, optical features and forming mechanism of alumina thin film with structural color. This is significant in the bionic study of structural color of birds' feather. Since structural color has the merits as environmental friendliness, being never fades in color and has the property of rainbow effect, it may have wide potentials in color displays, decoration and anti-counterfeiting technology.

Fig. 6. The colorful patterns of AAO thin film.

Acknowledgements

The project was supported by the Natural Science Foundation of Hebei Province (Grant No. A2012101001), Science and Technology support project of Hebei

Province (Grant No. 132176278), Scientific Research Project of Higher Schools of Hebei Province (Grant No. QN20132013, QN2014340) and Scientific Research Project of Finance Bureau of Chengde City (Grant No. CZ2012007, CZ2013002).

References

1. Kolle M, Pedro M S, Maik R J S, et al. Mimicking the colourful wing scale structure of the papilio blumei butterfly [J]. Nature Nanotechnology, 2010, 5:511-515.
2. Parker A R, Welch V L, Driver D et al. Opal analogue discovered in a weevil nature [J]. Nature, 2003, 426:786-787.
3. Philips R W, Bleikolm A F. Optical coatings for document security [J]. Appl. Opt., 1996, 35:5529-5534.
4. Liu X Y, Zhu S M, Zhang D,et al. Replication of butterfly wing in TiO2 with ordered mesopores assembled inside for light harvesting [J]. Mater. Lett., 2010, 64:2745-2747.
5. Chen H M, Hsin C F, Liu R S, et al. Controlling optical properties of aluminum oxide using electrochemical deposition [J]. J. Electrochem. Soc., 2007, 154:K11-K14.
6. C. R. Martin,1994. Science. 266, 1961.
7. A. Huczko, 2000. Appl. Phys. A., 70, 365.
8. B. Das, S. P. Meginnis, 2000. Appl. Phys. A., 71, 681.
9. Diggle J W, Downie T C, Goulding C W, 1969. Chem. Rev. 69, 365.
10. Wang X H, Akahane T, Orikasa H, 2007. Appl. Phys. Lett., 91, 011908.
11. Zhao X L, Meng G W, Xu Q L, 2010, Adv. Mater., 22, 2637.
12. Qin Xu, Yu-Hua Yang, Li-Hu Liu, 2012. Journal of The Electrochemical Society. 159 (1), C25.

A new spin gapless semiconductor: the quaternary heusler alloy VHfCoGa

X.D. Yang* and X.L. Wu
Institute of Theoretical Physics, Zunyi Normal College,
Zunyi, 563002, China
**Email: yxiude@163.com*

B. Wu
School of Physical Science and Technology, Southwest University,
Chongqing, 400715, China.
Email: fqwubo@163.com

Y. Feng
School of Physical Science and Technology, Southwest University,
Chongqing, 400715, China.

By using generalized gradient approximation (gga) scheme within the density functional theory (dft), the structural, electronic and magnetic properties of limgpdsn-type quaternary heusler alloy vhfcoga were investigated compressively. the results reveal that the vhfcoga alloy at the lowest energy state is (or is nearly) a typical spin gapless semiconductor with a magnetic moment of $3\mu_b$ per primitive cell. as the change of lattice constant is beyond 2.5% the alloy is half-metal ferromagnetic. a 100% spin polarization appears at fermi level (ε_f) in vhfcoga during lattice range of 5.63å-6.89å. The hybridization among magnetic atom v, hf and co plays a crucial role in determining the magnetism.

Keywords: Electronic properties, Magnetism, Spin gapless semiconductor.

1. Introduction

Spin gapless semiconductor (SGS), whose majority-spin band is semiconducting with a energy gap while in the minority-spin band the valence and conduction band touch each other, presents a spin-polarization of 100% at the Fermi level (ε_F). In SGS, the spin-polarized electrons or holes can easily transferee from the valence to conduction band only little energy supplying [1]. Therefore, the spin gapless semiconducting behavior, which bridges the half-metallic and semiconducting properties, has received great attraction from scientific researchers due to potential applications in spintronic devices [2].

As a kind of novel versatile spintronic materials, the Heusler alloy has been early focused because of half-metallic ferromagnetism, such as the classic Cu_2MnAl-type Heusler alloys and the Hg_2CuTi-type Heusler alloy or inverse Heusler alloy. In general, the Cu_2MnAl-type Heusler alloys with composition X_2YZ, where X and Y are transition elements (Ni, Co, Fe, Mn, Cr, Ti, V etc.) and Z is III, IV or V group elements (Al,Ga, Ge, As, Sn, In etc.) [3-5] , are half-metallic ferromagnetism. For Hg_2CuTi-type Heusler alloys, such as Ti_2CoAl, Ti_2NiAl and Ti_2CoGa, in the past third years, are also identified as new half-metallic ferromagnetism from the first-principles calculations [6-7]. In addition, for stabilizing the half-metallicity in Heusler alloy, lots of doped quaternary Heusler alloys have been investigated extensively. It includes $(Co_{1-x}Mn_x)$ MnAl[8] (X-site doping), $Co_2Mn_{1-x}Fe_xSi$[9] (Y-site doping) and $Co_2MnAl_{1-x}Si_x$[10] (Z-site doping) series. Most of them are half-metallic ferromagnetism.

In the past year, the spin gapless behaviors in Heusler alloys attract more researching interest. In 2013, six inverse alloys, involving Ti2MnAl, Ti2CoSi, Ti2VAs, V3Al, Cr2ZnSi, and Mn2CoAl, have been identified as possible spin gapless semiconductor from the first-principle calculation [11]. Especially, S. Ouardi etc. finally confirmed in experiment that the polycrystalline Heusler alloy Mn2CoAl is a new SGS material [12], which in accord with theoretical results. Very recently, spin gapless semiconducting behavior in equiatomic quaternary CoFeMnSi Heusler alloy was also reported [13]. Some new Zr-based quaternary Heusler alloys have been investigated comprehensively for possible spin gapless properties [14].

It is necessary to explore new spin gapless semiconducting materials due to potential applications in spintronics. As a new promising candidate, the Hf-based equiatomic quaternary Heusler alloys may be paid much coming researching attention for spin gapless semiconducting properties and possible high Curie temperature. Hence, in this paper, a comprehensive theoretical investigation on the structure, magnetism and electronic properties of LiMgPdSn-type Heusler alloy VHfCoGa is implemented by GGA calculation scheme.

2. Computational Details

In general, in LiMgPdSn-type equiatomic quaternary Heusler alloy with composition X_1X_2YZ, three crystal structures possibly exist. I: X_1 locates at (0 0 0) , Y at (0.5 0.5 0.5), X_2 at (0.25 0.25 0.25), and Z at (0.75 0.75 0.75) , II: X_1 locates at (0 0 0), X_2 at (0.5 0.5 0.5), Y at (0.25 0.25 0.25), and Z at (0.75 0.75 0.75) and III: X_2 locates at (0 0 0), Y at (0.5 0.5 0.5), X_1 at (0.25 0.25 0.25), and Z at (0.75 0.75 0.75).

All calculations are performed by using CASTEP code within the DFT, and the valence-electron configurations of V($3d^3 4s^2$), Hf ($5d^2 6s^2$), Co ($3d^7 4s^2$) and Ga ($3d^{10} 4s^2 4p^1$) are selected. For dealing with the core-electrons interaction properly, Vanderbilt-type ultrasoft pseudopotentials[15] and the Perdew Bruke Ernzerholf (PBE) correlation potential[16] are used. A 7×7×7 grid of special *k*-points is used for the Brillouin zone sampling in the Monkhorst Pack scheme. For self-consistent calculation, the energy convergence criterion of 1×10^{-5} eV/atom and the energy cutoff of 380eV are satisfied. When the positions of atoms are relaxed the force on each atom and total energy convergence tolerance are less than 0.01eV and 1×10^{-6} eV/atom, respectively.

3. Result and Discussion

In Figure 1, total energy of VHfCoGa as a function of lattice constant is graphed schematically. The ground state structure with the minimal total energy of -5482.080926eV is found when the lattice constant is 6.26084 Å. In inset, density of state (DOS) of VHfCoGa at the minimal energy state is presented. The spin polarization of 100% at ε_F is preserved and in spin-up band the valence and conduction band touch each other, which is or near spin gapless semiconducting behavior. With the decrease of the lattice constant, the Fermi level should be shifted towards high-energy orientation. During 5.63Å-6.89Å of the lattice constant, the alloy is typical half-metallic ferromagnetism (HMF).

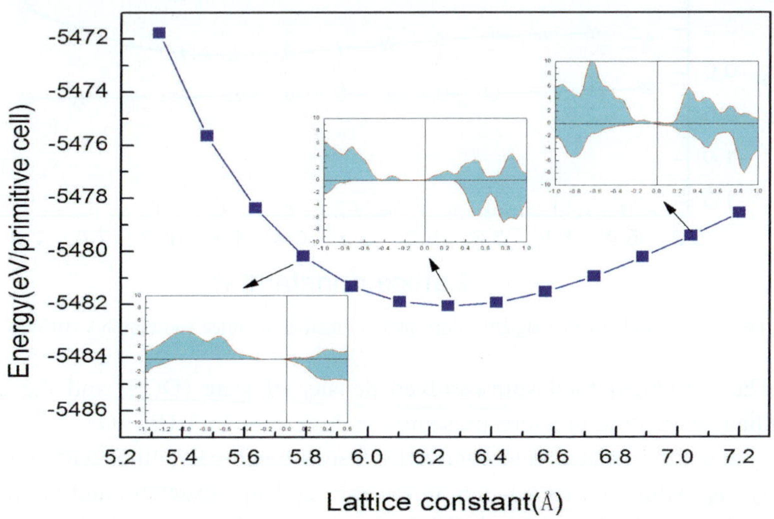

Fig. 1. Total energy and densities of states (inset) as a function of lattice constant in VHfCoGa.

In order to understand the structural dependence of magnetic moment in VHfCoGa, the total magnetic moment (TMM) and atomic magnetic moment (AMM) of V, Hf, Co and Ga as a function of lattice constant are potted in Figure 2. It can be seen that the TMM is about $3\mu_B$ per unit cell, which is in good agreement with Slater Pauling (SP) rule: TMM=N_v-18 (N_v is number of valance electrons per cell. For VHfCoGa alloy N_v is equal to 21). With the increase of lattice constant the V AMM increase while Co AMM decreases, the Ga AMM keeps a small negative value without remarkable change, as well as the Hf AMM encounters a slight decrease firstly and following increase. Considering the contribution of V, Hf, Co and Ga AMM on TMM, the antiferromagnetic coupling relation between V and Co atom is prominent. With the contraction of crystal system, the coupling should be strengthened and V or Co AMM value should be decreased.

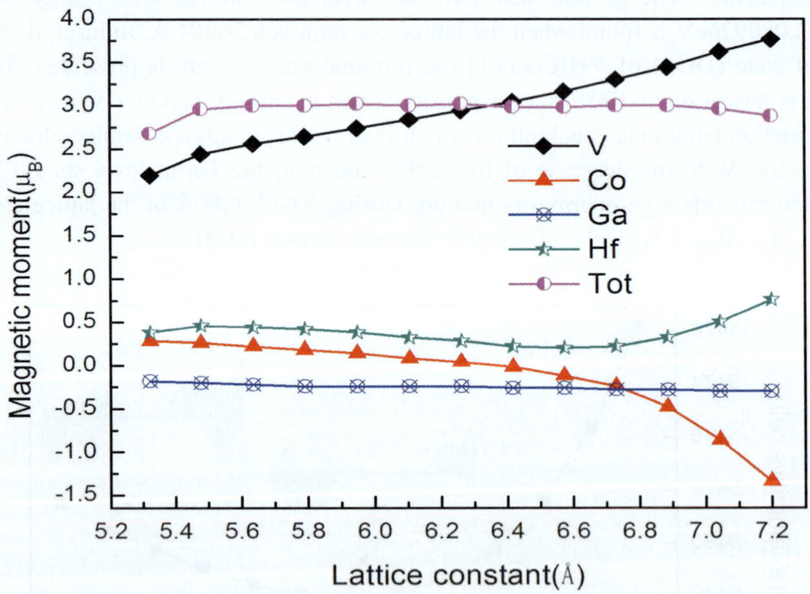

Fig. 2. Total and atomic magnetic moment as a function of lattice constant in VHfCoGa.

The calculated total spin-resolved density of state (DOS) and the corresponding band structures are presented in Figure 3. In VHfCoGa, for spin-up energy state, the valance and conduction bands touch each other within a zero-energy gap, which forms a valley in the spin up band structure, and the joint is extremely near the Fermi level. While for the spin-down energy state, an indirect spin-down gap of 0.9eV is detected. It presents a semiconducting behavior.

Hence, it can be deduced that the VHfCoGa alloy is or near a spin gapless semiconductor. And more interesting is that as excitation energy up to the band gap energy of the spin-up band the electrons can also be 100% spin polarized. Therefore the VHfCoGa alloy can be used as spintronic materials with superior performance to half metals and diluted magnetic semiconductors.

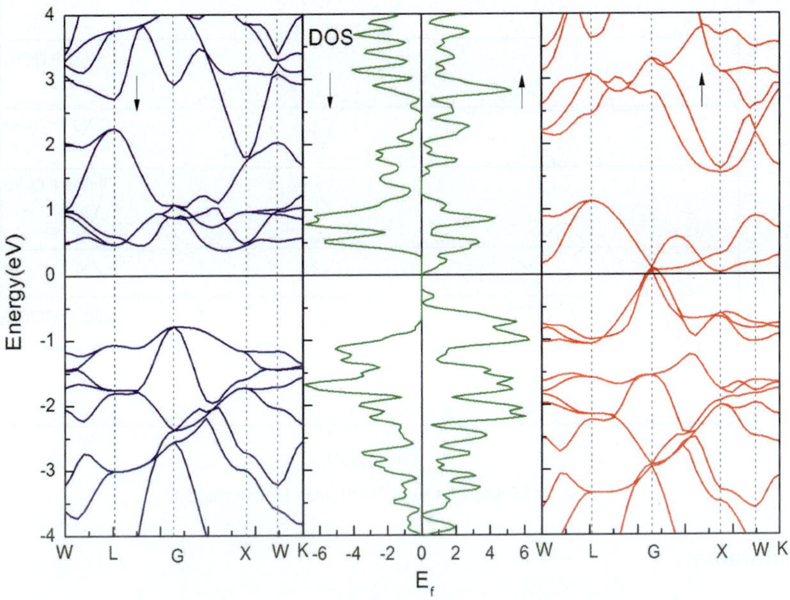

Fig. 3. DOS and band structure of VHfCoGa Heusler alloy.

As far as the electronic properties are concerned, the partial DOS (PDOS) of V, Hf and Co atom are plotted in Figure 4. It can be seen that the spin-down gap origins from d-electrons of V and Hf atom, and in spin-up band the spin gapless properties are contributions of d-electrons of Co atom. From the PDOS, we can also know that there are hybridizations between V-3d, Hf-5d and Co-3d states around the Fermi levels. According to the crystal field theory and associating with DOS characters near the ε_F mentioned above, we therefore deduce that a possible image of hybridization between the d-electronic atoms: The interaction of hybridized between V, Hf and Co orbitals creates 10 *d*-orbitals, among which 5 are below the ε_F and are occupied. In addition, taking 4 occupied sp-bands into account, 9 bands in the system are filled with the valance electrons. As a result, the total magnetic moment of equiatomic quaternary Heusler alloy VHfCoGa is $3\mu_B$ and equal to the number of valence electrons 21 minus 18, which is just the SP rule obeyed by most of inverse Heusler alloys or half-Heusler alloys with Cl_b structure.

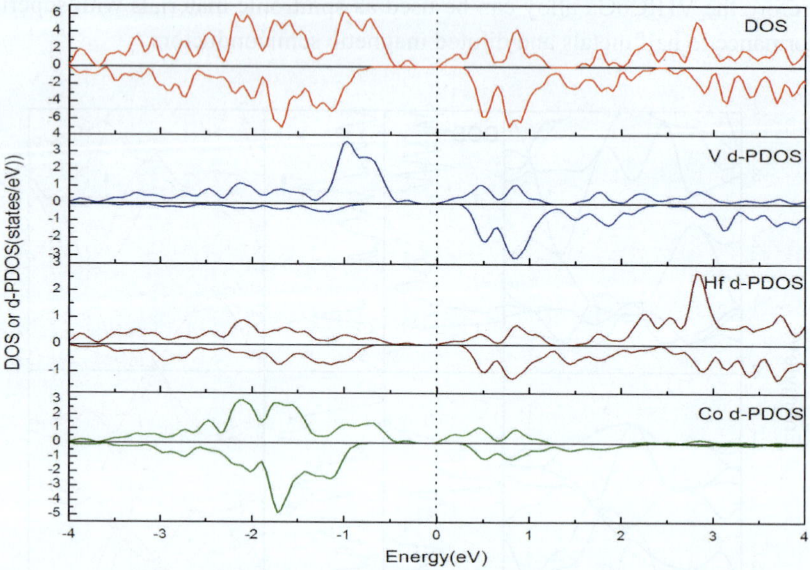

Fig. 4. DOS and PDOS of VHfCoGa Heusler alloy.

4. Summary

We examine the structure, magnetism and energy band properties of equiatomic quaternary Heusler alloy VHfCoGa by using GGA+PBE scheme based on DFT. Our calculations reveal that a 100% spin polarization appears at Fermi level (ε_F) in VHfCoGa during lattice range of 5.63Å-6.89Å. The VHfCoGa alloy with the minimal energy presents an energy zero-gap in spin-up channel and a semiconducting behavior in spin-down band, which is or near spin gapless semiconductor. Moreover, the magnetic moment of $3\mu_B$ per unit cell of VHfCoGa alloy is detected and in accordance with SP rule. As a result, we deduce that the VHfCoGa alloy can be used as spintronic materials with superior performance to half metals and diluted magnetic semiconductors.

Acknowledgment

This work was partly supported by the National Natural Science Foundation of China (11304410), Natural Science Foundation of Technology Department (QJHJZ-LKZS[2012]03, QKHJZ-LKZS[2014]10 and KHJZ [2014] 2170),

Youth Science Foundation of Education Ministry (QJHKZ[2012]084) and the Key Support Discipline ([2011] 275) of Guizhou Province of China.

References

1. X. L. Wang, Phys. Rev. Lett. 100 (2008), 156404.
2. X.L. Wang, S.X. Dou and C. Zhang, NPG Asia Mater. 2 (2010), 31.
3. S. Picozzi and A. Continenza: Phys Rev B 66 (2002), 094421.
4. G. D. Liu, X. F. Dai, H. Y. Liu, J. L. Chen, Y. X. Li, Gang Xiao, and G. H. Wu: Phys Rev B 77 (2008), 014424.
5. I. Galanakis, K. Özdoğan, E. Şaşıoğlu, and B. Aktaş: Phys Rev B 75 (2007), 172405.
6. E. Bayar, N. Kervan and S. Kervan: J. Magn. Magn. Mater. 323 (2011) 2945.
7. L. Feng, C.C. Tang, S.J. Wang, W.C. He: J. Alloys Compd. 509 (2011) 5187.
8. I. Galanakis, K. Özdoğan, E. Şaşıoğlu, and B. Aktas: Phys. Stat. Sol. (a) 205 (2008), 1036.
9. K. Özdoğan, B. Aktaş, I. Galanakis and E. Şaşıoğlu: J. Appl. Phys. 101 (2007), 073910.
10. R.Y. Umetsu, K. Kobayashi, A. Fujita, R. Kainuma, and K. Ishida: Scripta Materialia 58 (2008),723.
11. S. Skaftouros, K. Özdoğan, E. Şaşıoğlu, and I. Galanakis: Appl. Phys. Lett. 102 (2013) , 022402.
12. S. Ouardi, G.H. Fecher, C. Felser, and J. Kübler: Phys. Rev. Lett. 110 (2013), 100401.
13. L. Bainsla, A.I. Mallick, M.M. Raja, A.K. Nigam, B.S.D.Ch.S. Varaprasad, Y.K. Takahashi, A. Alam, K.G. Suresh, and K. Hono: arXiv:1410.0177
14. Q. Gao, H.H. Xie, L. Li, G. Lei, J.B. Deng, X.R. Hu: arXiv:1411.7570.
15. D. Vanderbilt: Phys. Rev. B 41 (1990), 7892.
16. B. Hammer, L. B. Hansen, J. K. Norskov: Phys. Rev. B 59 (1999), 7413.

Progress in revealing latent fingerprints with semi-conductive nano-materials

Y.Y. Xu[†], Z.R. Lin, T. Zhang
Airforce Logistics College, China
[†]*Emails: xuyunyun713@163.com, 517yy@sohu.com*

Since the nanotechnology introduced to the field of forensic science,it has brought many novel approaches for our fingerprint technology workers and received great attentions of forensic scientists both home and abroad for its unique advantages.Nanotechnology based methods have attracted considerable research efforts in recent years.In this paper we introduced the concept of nano-materials in developing fingerprints,and we also reviewed the development of revealing latent fingerprints with semi-conductive nano-materials and its latest progress. It can be foreknown that twenty-first Century relationship between and nano technology and fingerprint technology will become more and more closely, and it will be the focus of forensic scientists attention.

Keywords: Nano-material, Latent fingerprint, Photoluminescence method.

1. Introduction

Nano material is defined in at least one dimension in the nanometer scale (1~100nm) or as a basic element of materials by them [1]. Semiconductor nanocrystals are ultrafine particles whose size in nano level (1nm=10-9m).They are also known as quantum dots, which are atoms or clusters consisted by a very limited number of atoms or molecules. They are usually compounds composed of group IIB-VI or group III-V atoms to silicon atoms. Because of the diameter of semiconductor quantum dot less than the Bohr radius, with the quantum size effect and dielectric confinement effect, which makes the semiconductor nanocrystals exhibit unique fluorescence properties. The unique properties of semiconductor nano materials by the domestic and foreign forensic scientists favor,become a hotspot of current research by fingerprint experts, promoting the development of all countries in the world of criminal investigation technology.

Nano materials show fingerprints refers to the nano technology processing method of fingerprints left samples, making latent fingerprints become visible fingerprint image, and were identified after recorded [2]. Currently nano material specifications used for handprint in the 1~100nm nano powder and nano particle which basic unit constructed by certain rules. The principle of semiconductor nano materials is the combination of fingerprint residue of amino acid, oil or sweatand nano materials, by the use of fluorescence which semiconductor nanocrystals with fingerprint substance, and then get a clear image of the fingerprint. The current research in the world are more concentrated on cadmium, EU and silicon semiconductor nano materials.

2. Nanometer Material Photoluminescence Method of Latent Fingerprints

2.1. *Profiles of nano photoluminescence method to reveal potential finger print*

Photoluminescence of latent fingerprints, which use some of the inherent elements of latent fingerprints in or the formation substances after some chemical reagent treatment in the excitation light source (such as ultraviolet, laser) to produce a fluorescent or phosphorescent under irradiation. According to the application of the excitation light source and the fingerprint fluorescence spectral line range, by selecting a filter which cutoff wavelength between excitation wavelength fluorescence wavelength to filter out the excitation light through fingerprint fluorescence to revealfingerprints [3]. Light induced fluorescence detection technology play an important role in the process of current case. However, these fluorescent substances are mainly organic matter, and some of these organic fluorescent materialweaknesses inherent to make its application limited.

The range of excitation wavelength semiconductor nanocrystal is wide while the range of emission wavelength is narrow, a large Stokes shift, and the stability of semiconductor nanocrystal is better than organic fluorescent dyes, fluorescent spectra are hardly affected by the surrounding environment (such as solvent, pH value, temperature etc..Through controlling the crystal surface of the coating component, semiconductor nanocrystal can be stably dispersed in most solvents. Semiconductor nanocrystals can also specifically for the labeling of biological materials such as cells, protein and nucleic acid, and it has better fluorescence properties. Organic matters in fingerprint residues such as amino acids, protein, fat etc. binding a unique role with semiconductor nanocrystals. By using of the unique physical and chemical properties of semiconductor nanocrystals, get clear fingerprint image under small background interference. Through controlling the crystal surface of the coating component, semiconductor nanocrystal can be stably dispersed in most solvents. Semiconductor nanocrystals can also specifically for the labeling of biological materials such as cells, protein and nucleic acid, and it has better fluorescence properties. Organic matters in fingerprint residues such as amino acids, protein, fat etc. binding a unique role with semiconductor nanocrystals. By using of the unique physical and chemical properties of semiconductor nanocrystals, get clear fingerprint image under small background interference.

2.2. *Application and development of semiconductor material in the fingerprint*

The Menzel research group of Texas University of Technology in the United States in 1999 launched the first cadmium sulfide nano composite photo luminescence studies of latent fingerprints method, pioneered nanotechnology in the application of fingerprint [5]. They appearing details of characteristic fingerprint by nanoparticles [4] covered two octyl sulfide sodium succinate cadmium sulfide processing latent fingerprints left in the top after "502" fuming. But this new material to be dispersed in hexane or heptane, so metal, plastic, glass and other surfaces without the "502" fuming fingerprint cannot use the above method.Otherwise the oil soluble ingredients in the fingerprints will be dissolved, but the fingerprint on the surface of black insulating tape is an exception.Then they control the growth of cadmium sulfide particles through the spatial structure of dendrimer [5]. The grain size was in nanometer, configured solution are used to "fingerprint" after 502 treatment, and get better visualization effect. In addition, research group of Menzel also examined the regulatory effect of two imide ammonia on the process [6]. But this method take too long time, not suitable for field use.

The study of water soluble photoluminescent nanoparticles for handprint is interested by people. In 2008 during the pre experiment of America Dezhou Polytechnic University of Cheng [7], using green (582nm) and red (755nm) two cadmium telluride semiconductor nano crystal with mercaptoacetic acid modified show a latent fingerprint left on aluminum foil and glass, nanoparticle surface carboxyl can and fingerprint substance residue of amino acids and fat effect quantum dots, so that deposition selectivity obtained good effects appear in thefingerprints, the effect of green marker is better than the red, but the red with poroke ratosis to target marker. Two cadmium telluride semiconductor nano crystal with mercaptoacetic acid modified show a latent fingerprint left on aluminum foil and glass, nanoparticle surface carboxyl can and fingerprint substance residue of amino acids and fat effect quantum dots, so that deposition selectivity obtained good effects appear in thefingerprints, the effect of green marker is better than the red, but the red with porokeratosis to targetmarker.

In 2004, Frederick J Rowell research group of the University of Sunderland in the UK[8], synthesized a hydrophobic substance coated silica nanoparticles and used for oil developing latent fingerprints. The diameter of the nanoparticles is about 200~600nm, can be well adsorbed on an extremely small amount of oil, formed strong hydrophobic bonding with fingerprint substance in DNA chain base pairs, thus revealing latent fingerprints, and it is better than the fluorescence powder. In 2007 the research group of Frederick J Rowell synthesized surface coating of various organic fluorescent dye (yellow alkaline, alkaline red, Luo Danming 6G, crystal violet and methylene blue etc.) or color magnetic silica nanoparticles [9], clearly appearing latent fingerprint on non porous surface. In 2009 Frederick J Rowell research group appearing latent sweat fingerprint with hydrophobic nano silica powder [10]. In recent years the domestic fingerprint experts also carried out a series of studies in the fields of court application science of nano technology, and has made gratifying achievements.

The research group of Frederick J Rowell synthesized surface coating of various organic fluorescent dye (yellow alkaline, alkaline red, Luo Danming 6G, crystal violet and methylene blue etc.) or color magnetic silica nanoparticles, clearly appearing latent fingerprint on non porous surface. Frederick J Rowell research group appearing latent sweat fingerprint with hydrophobic nano silica powder. In recent years the domestic fingerprint experts also carried out a series of studies in the fields of court application science of nano technology, and has made gratifying achievements.

In 2006 the Chinese People's Public Security University assumed the project "Research on key technology of latent fingerprints by nano composite materials", the group carried out the research of fingerprint aspects based on the work of Menzel. Research group first designed and synthesized CdS quantum dots fluorescent materials based on PAMAM dendrimer materials [11]. Latent fingerprint on metal surface obtain a better ideal effects appear through the CdS/PAMAM G5.0 nanoparticles solution leaching significantly after long time. The composite have been ideal in application of manifest latent fingerprints on tape [12].

Wang Chunyan etc. synthesis of CdS quantum dots with high quantum yield as the modification reagent with glutathione, and its application to nonosmotic potential fingerprints object glass were revealed and appeared fingerprints clearly.

3. Prospects

Absorption and emission wavelength of nano material can be adjusted based on particle size, long luminous time and combined with fingerprint substance rapid firmly. It can be foreknown that twenty-first Century relationship between and nano technology and fingerprint technology will become more and more closely, and it will be the focus of forensic scientists attention.

For semiconductor nano materials, due to the requirements of semiconductor nanocrystal composite fingerprint, can emit light of different colors in different wavelengths of light, and the light intensity is stronger than the fluorescent dyes in general many times. But at present the biological toxicity problem of most semiconductor nanocrystals is people have been worried about, although the worry remains to be studied to confirm or exclude, therefore non-toxic, high luminous intensity, and fingerprint substance in combination with strong semiconductor nanocrystals will become an important research field.

References

1. Li Qun. Preparation and application of nanomaterials technology [M]. Beijing. Chemical Industry Press. 2008:1.
2. Xiong Xiaobo, Zhang Jian. The progress of nano material manifestation of latent fingerprint [J]. Journal of Hubei University of Police, 2008, 100(1), 110-112.
3. Luo Yaping. Trace examination tutorial [M]. Beijing.Chinese People's Public Security University Press, 2004:43.
4. Menzel E R, Savoy S M, Ulvick S J, et al. Photoluminescent semiconductor nanocrystals for fingerprint detection [J] J. Forensic Sci, 2000, 45(3):545-551.
5. Menzel E R, Takatsu M, Murdock R H, et al. Photoluminescent CdS/ dendrimer nanocomposites for fingerprint detection [J]. J. Forensic Sci, 2000, 45(4): 770-773.
6. Bouldin K K, Menzel E R, Takatsu M, et al. Diimide-enhanced fingerprint detection with photoluminescent CdS/dendrimer nanocomposites [J]. J. Forensic Sci, 2000, 45(6):1239-1242.
7. Cheng K H, Ajimo J, Chen W. Exploration of functionalized CdTe nanoparticles for latent fingerprint detection [J]. J Nanosci Nanotechnol. 2008, 8(3):1170-1173.
8. University of Sunderland, Nanoparticles as Agents for imaging fingerprints, 2004, UK Patent no.GB400234.8.
9. Brenden J, Theaker, Katherine E, Hudson, Frederick J Rowell. Doped hydrophobic silica nano-and microparticlesas novel agents for developing latent fingerprints [J]. Forensic Science International, 2008, 174, 26-34.

10. Frederick J Rowell et al. Detection of drugs and their metabolites in dusted latent fingermarks by mass spectrometry [J]. Analyst, 2009, 134(4):701-707.
11. Yang Ruiqin, Qian Weijie, Zhao Ke. Nano CdS/PAMAM G5.0 shows fingerprints on the metal surface [J]. Journal of Chinese People's Public Security University, 2008, 58(4):2-5.
12. Yang Ruiqin, Zhou Qingying, Wang Yuanfeng et al. Application of nano CdS/PAMAM G5.0 appear latent fingerprint on tape surface [J]. Chinese Journal of Inorganic Chemistry, 2008, 24(11):1874-1879.

A theoretical exploration into durability of carbon nanotubes cement-based materials

Z.R. Lin, Y.Y. Xu† and T. Zhang

Airforce Logistics College, China

†*Emails: xuyunyun713@163.com, 517yy@sohu.com*

Carbon nanotubes have become hotspots in recent several years in the fields of nanomaterials. In construction engineering, the durability of the cement-based materials become salient. The characteristics and performances of carbon nanotubes (CNTs) and the durability of cement-based materials are introduced individually. With these understanding, the possibility of durability improved by multiwalled carbon nanotubes (MWNTs) is also analyzed. It is concluded that the key of the problem are the uniform dispersion of CNTs in the matrix, the compatibility between MWNTs and the matrix, the characteristics and the mechanism of the durability, lay the foundation for the engineering application.

Keywords: Multi-walled carbon nanotubes; Cement-based materials; Durability; Dispersity, Compatibility.

1. Introduction

Carbon nanotubes was found by the Japanese scholar lijima [1] in the process of C60 prepared by the use of arc method. It is a kind of excellent fiber because of excellent mechanical properties, thermal and electrical properties. At present, single wall nanotubes (SWNTs) are generally used in the field of field emission panel display and sensor etc. Multi walled carbon nanotubes (MWNTs) are generally used for reinforced materials research. Research involving base materials are polymer based, metal based, cement and ceramic matrix etc.. Because of multi walled carbon nanotubes in theory is very suitable for the application in strengthening and improving the properties of composite materials, and the production and application of scale has been realized. Academic research on carbon nanotubes cement-based composite materials is just in the initial stage and has made some important achievements in mechanical properties, conductive performance etc. at present. But the study on durability of carbon nanotubes cement-based composite materials is almost a blank area, and the durability of this material is still a lack of research. For the high performance cement-based material, in addition to the conductive properties and mechanical properties, such as durability is particularly important. Therefore, it has important realistic significance to research the durability of carbon nanotubes cement-based composite materials and promote its development and application.

2. The Structure and Properties of Carbon Nanotubes

CNTs is a seamless nanoscale tube by monolayer or multilayer graphite sheet rolled. Each piece of nanotubes is a carbon atom by SP2 hybridization with around three carbon atoms complete bonding, consists of a hexagonal plane composed of cylindrical surface. According to the nanotubes of carbon atoms in different layers, the CNTs can be roughly divided into two classes, single wall nanotubes (SWNTs) and multi wall nanotubes (MWNTs). SWNT is composed of a monolayer of carbon atoms around together, structure is symmetrical and singlewell. Typical diameter and length of SWNTs were 0.75 ~ 3 nm and 1 ~ 50μm.MWTs are composed of layers of carbon atoms in a layer by layer around together, like the shape of acoaxial cable. MWNTs inter layer close to the ABAB stacking, the floors from 2 to 50, the interlayer spacing is about 0.34nm, and graphite (001) crystal plane spacing equal. The typical diameter and length of MWNTs were 2 ~ 30 nm and 0.1 ~ 50μm. Both the MWNTs and SWNTs have high length diameter ratio, generally 100 ~ 1000, highest can amount to 100 ~ 10000, can be considered to be one-dimensional molecular. CNTs has a straight, spiral, bending of different shape.

Except with excellent mechanical properties, electrical properties, thermal properties, Carbon nanotube also has good dielectric and electromagnetic properties [2]. Studies have shown that[3-6] MWNTs with small scale (diameter of micro nano , length of micron grade, length to diameter ratio of 100 ~ 1000), low density, light weight (about 1.33g/cm3, only for 1/6 steel), high hardness, high strength (tensile strength is more than 50GPa, 100 times more than steel, the bending strength is 14GPa), high modulus of elasticity (Young's modulus is more than 1TPa, and the diamond quite), excellent elasticity (bending can be automatically rebound, the corresponding axial elongation rate can reach 20%, the elastic strain up to 6% is about 60 times that of steel) and excellence electrical properties (according to the structure may have the characteristics of metal or semiconductor properties,and the electrical conductivity up to 1000 ~ 2000S/cm, can pass the current density of up to 106A/cm2), is a good fiber, its performance is better than any current fiber materials.

3. Durability of Cement-Based Materials

Durability of cement-based materials, refers to its ability to maintain the long-term strength and the integrity of appearance under actual conditions of use against damaging effect of various factors of internal or external. Most structural designers always pay most attention to the strength properties of concrete, but now for a variety of reasons, designers must have attention to consciousness the durability of concrete. The physical unfavorable effect on durability of concrete including surface wear, cracking caused by pore salt crystallization and deterioration of exposure to freezing or fire extreme temperature environment . Harmful chemical effect on concrete including cement paste by acid liquid filtration, and expansion effect caused by corrosion of reinforced sulfate corrosion, and alkali aggregate reaction in concrete.

Durability and service life of cement based materials can be considered synonymous. This can be reflected from the following two aspects: First, since the cost of structural repair and update of the material increasing, have occupied a large part of the building's total budget, so the social and economic significance of durability is remarkable. In addition, more and more cement based materials used in harsh environments, such as offshore oil platform, concrete tank which store liquefied petroleum gas under the condition of low temperature etc..

4. Possibility of Improving the Durability of the Cement Based Materials by Mwnts

Cement based material is a brittle material, and its toughness, impact resistance, fatigue resistance, antifreeze, and anti fire burst are of poor performance. How to effectively improve the durability is an important issue. In the solid porous, the rate of deterioration is determined by permeable materials. And water is close contact with permeability as the destruction medium and the transmission medium, controlled by the pore size and connectivity. The introduction of MWNTs can partition microporous, improve pore structure. At the same time, it also can inhibit the contraction deformation, effectively prevent the water through the permeable pathway to the material inside, and improve impermeability.

The strength and density are important factors which influence the durability. The early strength and toughness of cement based materials can be improved after doped MWNTs.It can also make the material more compact, thereby improving durability.

5. The Problems of Improving the Durability of the Cement Based Materials by Mwnts

From the above analysis, MWNTs improves the durability of the cement based composite materials is feasible in theory. However, due to the structure characteristics of MWNTs itself, there are still some problems in preparation process, particularly the problems of uniform dispersion of MWNTs.

(1) Uniform dispersion of MWNTs. MWNTs has a high ratio of length to diameter and strong Fan Dehua force between the pipe [7, 8]. Therefore MWNTs is fascicular winding and agglomeration [9-11], and insoluble in water and organic solvents. In the case of composite materials, dispersion uniformity of adding phase is a very important index of material properties. The more uniform dispersion, the more obvious effects, the better the overall performance of the material [18]. Therefore, it need to be uniformly dispersed before adding MWNTs to the matrix, and make it dissolved in the dispersion [12].

(2) MWNTs' compatibility with the matrix. At present, domestic and foreign scholars have few compatibility researches on MWNTs and the other material of matrix. However, it must be considered in preparation process. In addition, the compatibility of the scattered dispersing agent and cement based composite also be considered.

(3) The mechanism and properties of the durability of cement based materials. People's understanding of durability is originated from the actual structure of the case. It is hard to simulate the comprehensive effect of long time in the laboratory that are commonly found in a variety of conditions in the real life.The researchers put the mechanism and characteristic of the durability of each separate consideration and ignored the existence and interaction of many factors at the same time.

6. Conclusions

High performance cement-based composite materials is the dominant direction of twenty-first century materials development, and its research has been a hot spot in the field of materials. The excellent properties of carbon nanotubes can improve the performance of cement based materials in many aspects. And with the multi wall carbon nanotube synthesis technology becomes more mature, low cost mass production of multiwall carbon nanotubes has become possible. Therefore the exploration and research of multi walled carbon nanotubes in cement based materials has become a pressing matter of the moment.

According to the uniform dispersion and compatibility problems of multi walled carbon nanotubes in cement-based composite materials, I think we can cooperate with expert in chemical analysis, research on a specifically targeted at multi wall carbon nanotubes cement-based composite method dispersed in aqueous system, and is committed to improve the compatibility of the two. It can raise the strength of cement based materials, enhanced toughness at the same time, as far as possible, effectively carry out the research on the durability of the cement based composite materials .

References

1. Lijima S. Helical microtubules of graphitic carbon [J]. Nature, 1991, 354(6348): 56-58.
2. G Y Slepyan, S A Maksimenko, A Lakhtakia, O M Yevtushenko. Electomagnet response of carbon nanotubes and nano ropes [J]. Synths Metal 2001, (124):121-123.
3. J P Salvetat, J M Bonard, N H Thomson, et al. Mechanical properties of carbon nanotubes [J]. Appl. Phys, 1999, 69:255.
4. M M J Treacy, T W Ebbesen, J M Gibson. Exceptionally high Young's modulus observed for carbon nanotubes [J]. Nature, 1996, 381(6535): 678-679.
5. Yu M F, et al. Strength and breaking mechanism of multiwalled carbon nanotubes under tensile load [J]. Science, 2000, 287:637.
6. Walters D A, et al. Elastic strain of freely suspended single-wall carbon nanotube ropes, Applied Physics Letters [J]. 1999, 74: 3803-3805.
7. Tatsuhiro Yamamoto, Suguru Noda, Masaru Kato. A simple and fast method to disperse long single-walled carbon nanotubes introducing few defects [J]. Carbon, 2011(49): 3179-3183.
8. D M Delozier, D M Tigelaar, K A Watson, et al. Investigation of ionomers as dispersants for single wall carbon nanotubes [J]. Polymer, 2005, 46:2506-2521.
9. A Cwirzen, K Habermehl-Cwirzen, A G Nasibulin, et al. SEM/AFM studies of cementitious binder modified by MWCNT and nano-sized Fe needles [J]. Materials Characterization, 2009, 60(7): 735-740.
10. T Saito, K Matsushige, K Tanaka. Chemical treatment and modification of multi-walled carbon nanotubes [J]. Physica B-Condensed Matter, 2002, 323:280-283.
11. G Y Li, P M Wang, X H Zhao. Pressure-sensitive properties and microstructure of carbon nanotube reinforced cement composites [J]. Cem. Concr. Compos., 2007, 29:377-382.
12. A Cwirzen, K Habermehl-Cwirzen, V Penttala.Surface decoration of carbon nanotubes and mechanical properties of cement/carbon nanotube composites [J]. Advances in Cement Research,2008,20(2):65-73.

Preparation of nano-Fe69Cr18Ni9Cu4 alloy powder by chemical co-precipitation hydrogen reduction

H.M. Yu* and Z.Y. Li†

College of Materials and Metallurgy, Northeastern University, Shenyang, Liaoning, 110004, P.R. China

*Email: *1329693214@qq.com, †lizy@smm.neu.edu.cn*

In this paper, the composite oxide nano-powder of Fe69Cr18Ni9Cu4 was prepared by co-precipitation method. The Fe69Cr18Ni9Cu4 alloy nano-powder was prepared by the hydrogen co-reduction process. The reaction condition for preparation of alloy nano-powder: reduction temperature was 1000°C, reduction time was 90 minutes.

Keywords: Fe69Cr18Ni9Cu4; Alloy powder; Hydrogen co-reduction; Antibacterial stainless steel.

1. Introduction

The powder metallurgy is a special technology in which combine metal materials manufacturing and parts process in the same processing, it is an effective means to improve the performance of metallic materials and develop new materials, and it also plays an important role in the promotion of scientific and technological progress and national economic development [1].

Fe69Cr18Ni9Cu4 alloy powder, a kind of antibacterial stainless steel materials, is made by the method of the powder metallurgy.When the antibacterial elements of the surface of materials contact with the bacterial cells, cells membrane will be injured. The antibacterial elements will coagulate protein or damage their DNA, and as a result, bacteria will die of the loss of proliferation capacity [2]. This antibacterial stainless steel is widely used in multiple fields, such as kitchen utensils, tablewares, home appliances, medical equipments, food packaging, automotive manufacturing and public facilities [3].

The stainless steel surface is plated with thin antibacterial metal layer. In the process of deep-drawing or surface polishing, the antibacterial surface of the metal is peeled off or fallen with the gradually use, and fail to maintain a long-term. Moreover, the antibacterial properties of antibacterial stainless steel cannot maintain a long time as the antibacterial stainless steel surface is always wiped to clean kitchen supplies, etc [4]. It needs a lot of additional processing for forming metal layer or alloy layer surface, which would increase the cost. The preparation of antibacterial material by the method of powder metallurgy has a series of advantages, such as significant energy-saving, material-saving, excellent performance, high precision and stability, suitable for mass production. Some materials and parts, which could be hard to prepare and process by traditional casting and machining processing, can also be prepared by the technology of powder metallurgy [5].

In this paper, the composite oxide nano-powder of Fe69Cr18Ni9Cu4 was prepared by co-precipitation method, the Fe69Cr18Ni9Cu4 alloy nano-powder was prepared by the hydrogen co-reduction process.

2. Experiments

Experimental apparatus: drying oven, acidometer, electric power blenders, electric thermostat water pot, vacuum pumps, and closed cycle hydrogen-reduction furnace, etc.

Experimental reagents: $NiSO_4 \cdot 6H_2O$, $CuSO_4 \cdot 5H_2O$, $CrCl_3 \cdot 6H_2O$, $FeSO_4 \cdot 7H_2O$, NaOH, anhydrous ethanol, all of which are chemical pure. hydrogen, nitrogen.

Preparation of Fe^{2+},Cr^{3+},Ni^{2+} and Cu^{2+} admixture aqueous solution:Some $NiSO_4 \cdot 6H_2O$, $CrCl_3 \cdot 6H_2O$, $CuSO_4 \cdot 5H_2O$ and $FeSO_4 \cdot 7H_2O$ was solved in 1000mL deionized water and filtered. The Fe^{2+}, Cr^{3+}, Ni^{2+} and Cu^{2+} admixture aqueous solution was obtained.

Preparation of NaOH aqueous solution: By dissolving some NaOH in a certain amount of deionized water, through filtrating the insoluble impurities, NaOH aqueous solution of 4mol·L-1 was obtained.

Preparation of Fe69Cr18Ni9Cu4 composite oxide nano-powder[6]: Taking 1000mL of Fe^{2+},Cr^{3+},Ni^{2+} and Cu^{2+} admixture aqueous solution to a beaker, 4 mol·L-1 of NaOH solution was dripped. The technological conditions was that the reaction temperature 30°C, the pH value of reacting terminal 11, stirring speed 2000 r·min-1 and the reaction time 30 min. Through vacuum filtration, filter cake after washing few times, the Fe69Cr18Ni9Cu4 composite oxide nano-powder was obtained by filtration and dryness. Use EDTA aqueous solution mensurate overall concentration of metal ions in filtrate. The recovery ratios of the Fe69Cr18Ni9Cu4 composite oxide nano-powder were calculated from volume of the filtrated solution and overall contents of the metal ions in the filtrated solution.

Preparation of Fe69Cr18Ni9Cu4 nano-powder [7]: The Fe69Cr18Ni9Cu4 composite oxide nano-powder placed in a porcelain crucible was put into cooling zone (the 3 condensator in Fig. 1) of the closed cycle hydrogen-reduction furnace. The three-way valve was placed in the status of gas replacement, and N_2 gas was using for the elimination of air. Hydrogen was pumped in, and the whole system was set on the closed circulation status, then hydrogen would pass through the system, during which hydrogen and nitrogen was circulated in the system by gas pump. After the reduction furnace was heated to desired constant temperature, the Fe69Cr18Ni9Cu4 composite oxide nano-powder was reduced in constant temperature zone. The change of pressure was also recorded in the system. The reaction will be over when the pressure in the system would not decrease. The reduced powder was quickly pushed to cooling zone (the 6 condensator in figure 1), and then taken out when it was cooled down to room temperature.

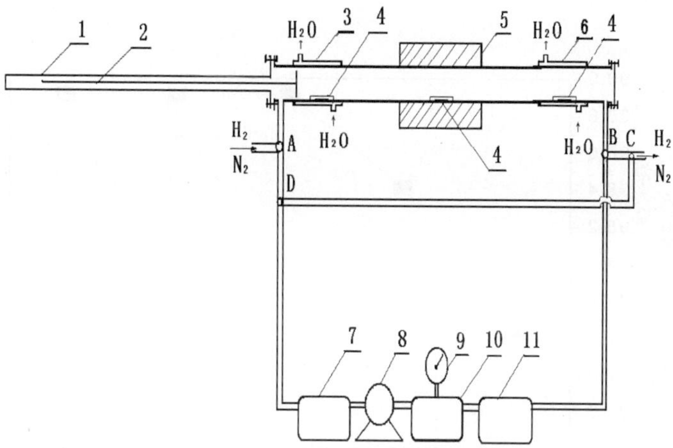

Fig. 1. Schematic diagram of the closed circulation-hydrogen reduction system.

1-Latex sheath, 2- push sample implement, 3and 6-condensator, 4-porcelain crucible and sample, 5-reduction furnace of temperature controller, 7 and 10-buffer bottles, 8-gas pump, 9-manometer, 11-moisture removing apparatus, A, B, C, D- the three-way valve. A→4→B→8→D→C was path of the system gas replacement; A→4→B→8→D→A was path of the system closed circulation.

Characterization: The samples were be writed by XRD (Japan Riko D/max-RB), SEM (Japan SSX-550) and TEM.

3. Results And Discussions

3.1. *Effect of experimental conditions on the recovery of the Fe69Cr18Ni9Cu4 composite oxide powder*

Effect of the pH value of titration end solution, stirring strength, the reaction time and reaction temperature on the recovery of the precursor powder was shown in figure 2.

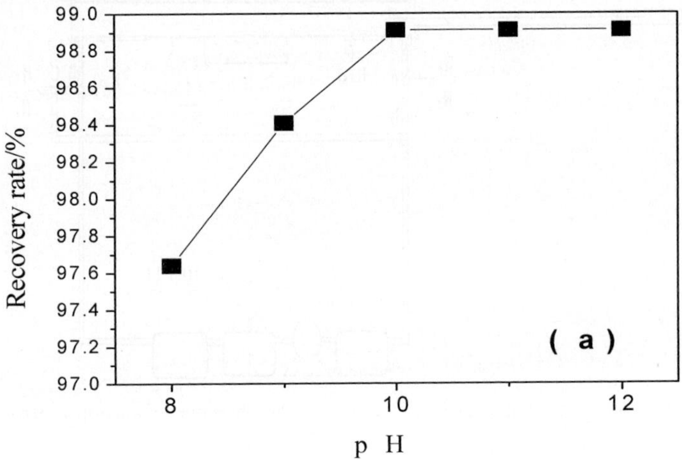

Fig. 2. Effect of experimental conditions on the recovery of the precursor powder.
(a) The curve of the recovery and the pH of the end of the titration (b) The curve of the recovery and stirring speed V (c) The curve of the recovery and reaction time t (d) The curve of the recovery and temperature T.

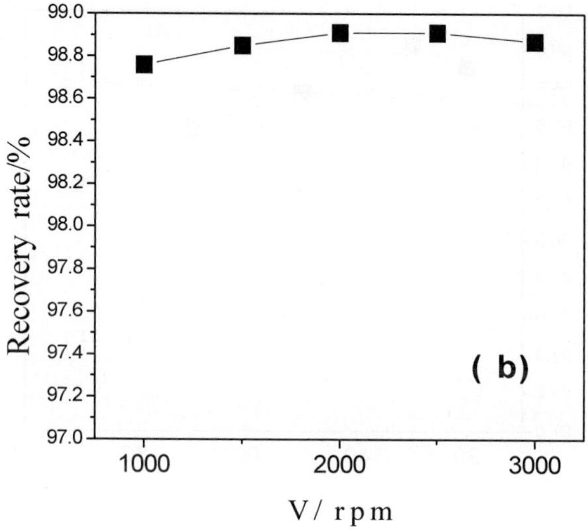
99.0
98.8
98.6
98.4
98.2
98.0
97.8
97.6
97.4
97.2
97.0
Recovery rate/%
(b)
1000
2000
3000
V/ rpm

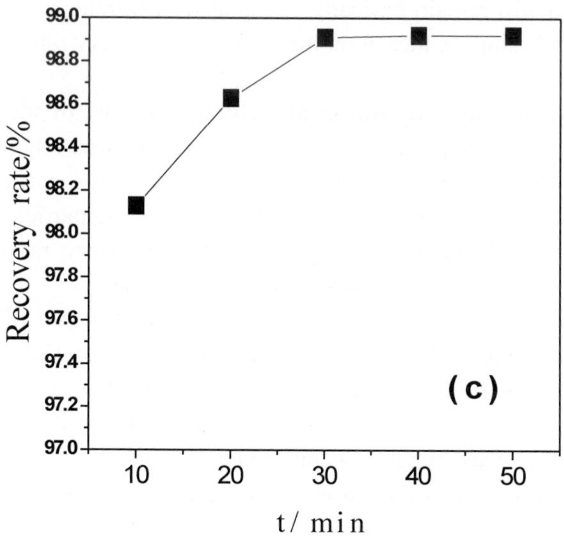
99.0
98.8
98.6
98.4
98.2
98.0
97.8
97.6
97.4
97.2
97.0
Recovery rate/%
(c)
10
20
30
40
50
t/ min

Fig. 2. (*Cont'd*)

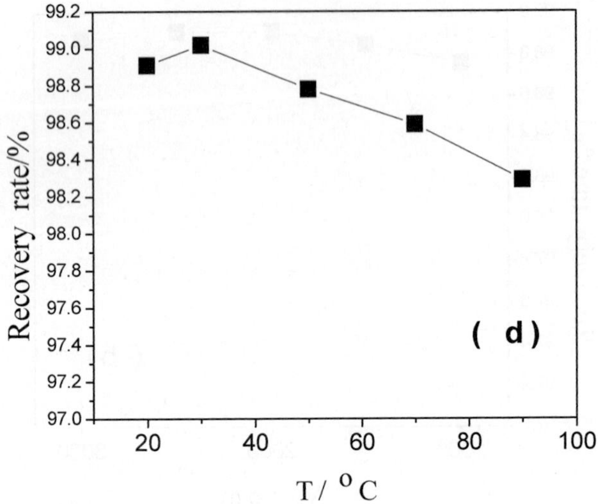

Fig. 2. (*Cont'd*)

Fe, Cr, Ni and Cu metal ions were precipitated completely when pH≥10. The sum of them will remain basically unchanged in filtrate, and the volume of consumed EDTA is similar, therefore, the measured recovery rate did not change significantly.

When the stirring intensity was small, with the increase of stirring intensity, the reaction was more sufficient and the recovery rate was slightly increased. When the stirring intensity to be larger, the reaction was sufficient and the particle size of formation material came to be smaller. The particle was possible to enter the filtrate through the filter paper. The recovery rate was slightly smaller when the stirring intensity was at around 2000rpm.

When the reaction time was less than 30min, the crystal of the reaction products was not formed fully. The crystal grain of some material was small, and it was easy to enter filtrate through the filter paper. The recovery rate was shrinked. When the reaction time was more than 30min, the crystal formed completely, and the recovery rate was improved.

Solubility product constant was a function of temperature. When the reaction temperature was higher than 30°C, with the increase of reaction temperature, solubility product constant KSP change continually, and precipitation reactions had the trend of left. The trend of metal precipitation dissolution was larger, and the ions content in filtrate increased gradually. The recovery rate decreased gradually.

According to above experiment, co-precipitation conditions were selected as follows: the pH value of titration end solution 11, stirring intensity 2000rpm, reaction time 35min, reaction temperature 30°C.

3.2. *Characterization of the Fe69Cr18Ni9Cu4 composite oxide powder*

Under the condition of the pH value of titration end solution 11, stirring speed 2000rpm, reaction time 35min, reaction temperature 30°C, the Fe69Cr18Ni9Cu4 composite oxide powder was prepared. The XRD and TEM analysis results of the Fe69Cr18Ni9Cu4 composite oxide powder were shown in Figure 3 and Figure 4. Figure 3 shows that the Fe69Cr18Ni9Cu4 composite oxide powder was amorphous or the crystal form nanocrystal was incomplete. Figure 4 shows that the morphology of Fe69Cr18Ni9Cu4 composite oxide powder was club-like numerously.

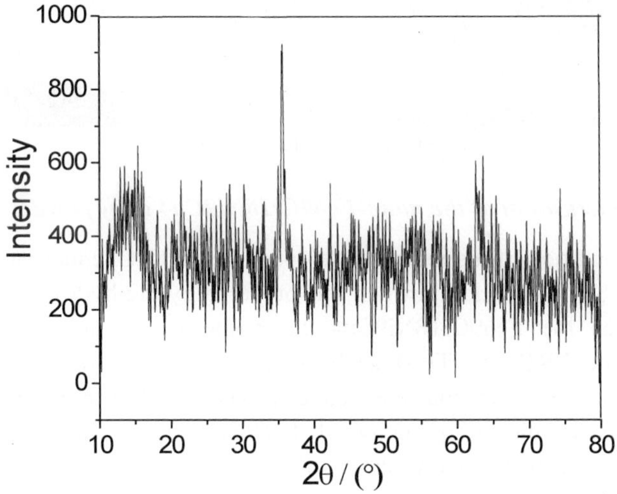

Fig. 3. XRD of the precursor.

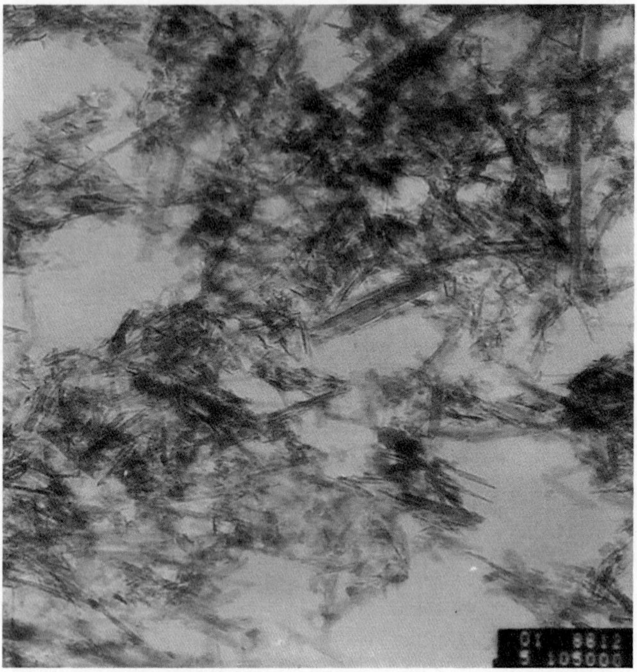

Fig. 4. TEM of the precursor.

3.3. *Characterization of the nano-Fe69Cr18Ni9Cu4 alloy powder*

Under the condition of the reduction temperature 1000°C,reduction time 90min, the Fe69Cr18Ni9Cu4 alloy powder was prepared by the hydrogen co-reduction method taken the Fe69Cr18Ni9Cu4 composite oxide nano-powder as raw materials. The XRD and TEM analysis results were shown in Figure 5 and Figure 6.Figure 5 shows that the Fe69Cr18Ni9Cu4 alloy powder was crystal. Figure 6 shows the morphology of Fe69Cr18Ni9Cu4 alloy powder was spherical, and the diameter was about 100nm.

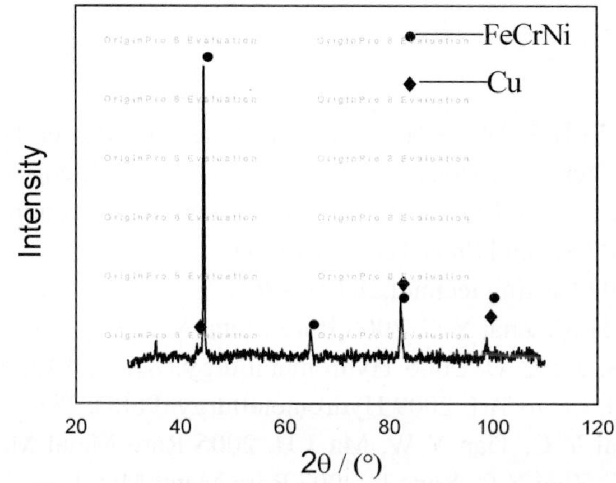

Fig. 5. XRD of the alloy powder.

Fig. 6. SEM of the alloy powder.

4. Conclusions

Under the condition of the pH value of titration end solution 11, stirring speed 2000rpm, reaction time 35min, reaction temperature 30°C, the Fe69Cr18Ni9Cu4 composite oxide powder was prepared. The Fe69Cr18Ni9Cu4 composite oxide powder was amorphous or nanocrystal of incomplete crystal form, the morphology was clublike numerously.

Under the condition of the reduction temperature 1000°C, reduction time 90min, the Fe69Cr18Ni9Cu4 alloy powder was prepared by the hydrogen co-reduction method taken the Fe69Cr18Ni9Cu4 composite oxide nano-powder as raw materials. The Fe69Cr18Ni9Cu4 alloy powder was crystal, the morphology of Fe69Cr18Ni9Cu4 alloy powder was spherical, and the diameter was about 100nm.

References:

1. Gao H Y, Liu B S, Wang H X. 2004 Vehicles and Power Technol. (3),57; Jing H M, Chen S H, Dong J S, Wu X Q, Yu Z M, Zheng Y G, Wei X Y, Yao Z M, Ke W, Lu M Q, Yang K. 2003 Materials protection, 36(10), 9.
2. Liu C F. 2007 Special Steel Technol.13 (4) 4.
3. Liu C F. 2007 Casting technol.28 (9), 1262.
4. Li Z Y, Gu H M, Zhai Y C. 2008 Rare Metal Mat. Eng. 37(3), 966; Gbor P K, Hoque S, Jia C Q. 2006 Hydrometallurgy. 81(2), 130; Mc George B, Gaylard P G, Lewis A E 2009 Hydrometallurgy. 96(3), 235.
5. Li Z Y, Zhai Y C, Tian Y W, Ma J H. 2005 Rare Metal Mat. Eng. 34(6), 965; Li Z.Y., Zhai Y C, Yang F, 2007 Rare Metal Mat. Eng. 36(7), 1245.
6. Tang H, Zhang W, Geng P, Zheng J H, Jin L T, Jin L P, Wu Z R. 2006 Chem. Commun. 10, 772.
7. Wang J L. 2008, Powder Metal. Technol. 26(5), 374; Wang W H, Ma H N, Xiao K, Zheng D. 2007 Powder Metal. Technol. 25(2), 121; Yoshinohu Takeda. 2006 Powder Metallurgy Technol. 24(6), 467; Wei P, Zheng Z M, Liu L. 2005 Powder Metal Technol. 23(6), 445; Zhao Z Y. 2006 Hebei Transportation Technology. 3 (3), 61.

Composite reducing agent synthesized of β-SiAlON

Y.Q. Guo[a] & Y.W. Wang[b]
Huanghe Science and Technology College,
Henan Province, china
[a]*27669932@qq.com,* [b]*407465578@qq.com*

Synthesized β-SiAlON based on bauxite (Al2O3 in 68wt %) by using compound reducing agent silicon /carbon were studied. Calculated mass change ratio of the specimens after firing, Examined Phase composition and microstructure of the specimens by using XRD, SEM and EDS, At the same time contrasted with compound reducing agent nitrided SiO2. The results show that: 1300°C~1400°C, Carbon reduction reaction begins and gradually strengthen, β-SiAlON start generation; 1450°C~1550°C, β-SiAlON formation gradually increased, to 1500°C reached the highest value, the column β–SiAlON′z are about 3.

Keywords: Compound reducing agent; Reduction-nitridation reaction; β-SiAlON.

1. Introduction

Synthetic bauxite based β-SiAlON, not only because of wide source of raw materials, low cost, and reaction to the equipment is not high, but also can greatly reduce the cost of synthesis of SiAlON, therefore research using bauxite raw material for the synthesis of β-SiAlON will expand its application field. China is a big country of bauxite reserves, based on the resource advantages, vigorously develop bauxite based synthetic material is the present trend of development, including the use of high alumina preparation SiAlON [1]. Now, research on Synthesis of nitride bauxite based SiAlON through high temperature reduction achieved certain results[2-6], This work is based on the above work, with Al_2O_3 content of 68% bauxite, using compound reducing agent of aluminum silicon, carbon aluminum, β-SiAlON reduction nitridation effect.

2. Experimental

Experimental materials: bauxite ($Al_2O_3 \cdot 2SiO_2 \cdot 2H_2O$), particle size ≤ 74μm, Loss on ignition contents of Al_2O_3 and SiO_2, respectively 68.53wt% and 24.92wt of%; silica fume (w (Si)≥98%) , particle size ≤ 21μm; aluminum (w (Al) ≥ 99.9%), particle size ≤ 25μm; carbon black (w(C) ≥ 99.9%), particle size ≤ 5μm; Take the z value of 3 β-SiAlON ingredients. Test formulations shown in Table 1.

Table 1. Composition of samples (wt %).

Sample number	Raw bauxite	Silica	Silicon	Carbon
FCS	68		19	13
GCS		78	12	10

Mixing of raw materials, load the crucible into the GN-2003-2$MoSi_2$ high temperature nitriding furnace, respectively, in 1100 °C, 1200 °C, 1300 °C, 1350 °C, 1400 °C, 1450 °C, 1500 °C and 1550 °C insulation 6h reduction and nitridation reaction (N_2 purity of 99.99%,), the sample quality changes before and after the determination of nitride. Ground into fine powder sample after reaction by XRD phase analysis of the nitrided samples were material, and by X-diffraction pattern of the diffraction peak integral area calculation of the relative content of the phase; SEM on the sample microstructure were detected using the EDS for micro-component analysis.

3. Results and discussion

3.1. *Mass change and rate phase change*

Figure 1, Mass change ratios at 1100°C~ 1550°C temperature; Figure 2, Figure 3, XRD patterns of the sample treatment temperature of 1100°C~1550°C.

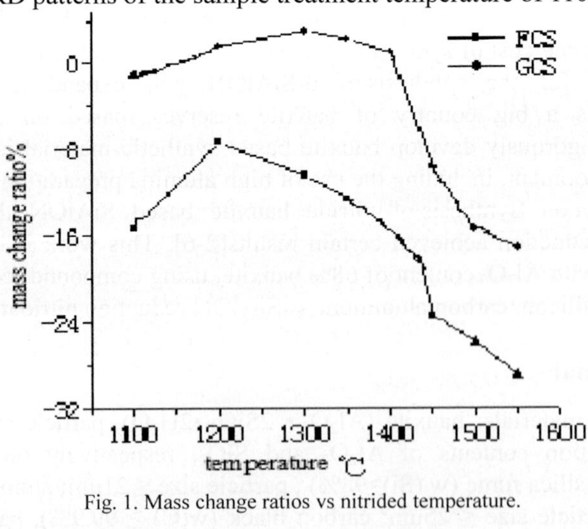

Fig. 1. Mass change ratios vs nitrided temperature.

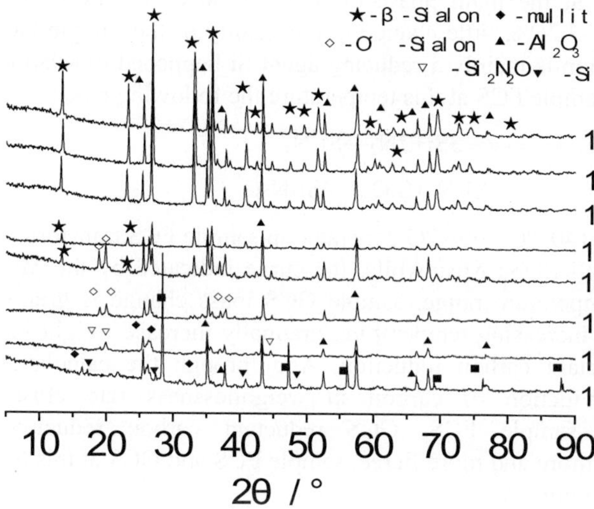

Fig. 2. XRD pattents of sample FCS.

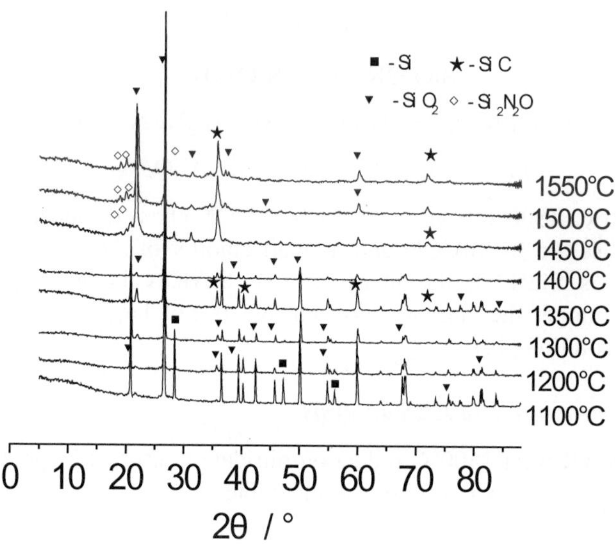

Fig. 3. XRD pattents of sample GCS.

The figure shows:

(1)1100°C, Nitrides are not observed in sample FCS, and the two rate of mass change are: -15.20%, -2.02%, sample FCS mass change is close to the burnin graw materials (13.5%), The raw materials of nitridation reaction is not obvious. 1200°C, nitride Si2N2O generated in the FCS, the main phases of the specimen are Si2N2O, Al2O3, and a small amount of SiO2; No nitride formation in GCS, the main phase of SiO2; 1200°C, the FCS sample mass change rate was -7.26%, little change in the quality of the sample GCS, in this temperature section is mainly a reducing agent Si happened nitridation reaction in ample FCS , sample FCS at this temperature the following reaction are:

$$3Si+2N_2 \rightarrow Si_3N_4 \quad (1)$$

$$Si_3N_4+SiO_2 \rightarrow 2Si_2N_2O \quad (2)$$

(2) 1300°C~1400°C, Si2N2O generated in sample FCS transformed into O′-SiAlON and β- SiAlON; Meanwhile, the curve indicated by the Mass change ratio: in this temperature range, sample GCS mass change is grently, sample FCS, GCC with increasing temperature, gradually increase weight loss, known sample of reductant carbon reduction, And through the calculation of two samples, the reduction of carbon in weightlessness rate close to; The temperature of sample FCS, GCS reductant carbon reducing reaction temperature and more and more fierce, sample FCS and GCS at this temperature the following reaction are:

$$C+SiO_2 \rightarrow SiO+CO \quad (3)$$

$$Si+2SiO_2 \rightarrow 2SiO \quad (4)$$

$$4SiO+2N_2 \rightarrow 2Si_2N_2O+O_2 \quad (5)$$

$$(2\text{-}x)Si_2N_2O(s)+xAl_2O_3(s) \rightarrow Si_{2\text{-}x}Al_xO_{1+x}N_{2\text{-}x}(s) \quad (6)$$

$$Si_{2\text{-}x}Al_xO_{1+x}N_{2\text{-}x}+Al_2O_3 \rightarrow Si_{6\text{-}z}Al_zO_zN_{8\text{-}z} \quad (7)$$

(3) 1450°C~1500°C, sample FCS O′- SiAlON phase disappeared at 1400°C, the final phase are β- SiAlON and a small amount of Al2O3; sample GCS ′s main phase is Si2N2O, the phase of β- SiAlON is not appear, At this temperature range, the sample mass change is not large, indicating that the sample into solution phase.

3.2. *Microstructure and EDS analysis*

Sample FCS, GCS, at 1500°C reduction nitridation after SEM photos and EDS analysis results are shown in Figure 4 and Table 2:

(a) Sample FCS

(b) Sample GCS

Fig. 4 .SEM micrograph at 1500°C.

Sample FCS reduction nitridation generated β-SiAlON are slender cylindrical structure, length of about 4 microns; Sample GCS reduction nitridation generated Si2N2O are granular structure, length of about 2 microns.

Table 2. EDS analysis of the samples reduced 1500°C.

Samples	contents			
	Si	Al	O	N
FAC	24.80	32.69	30.03	12.48
FGS	47.86		13.46	38.74

4. Conclusions

Through comparative test found:

1. The suitable temperature of carbon silicon compound reducing agent reduction nitride preparation of bauxite-based β-SiAlON is 1500°C,the final columnar β–SiAlON'*z* are about 3, the reaction process are as follows:1100°C~1200°C, silicon and silicon dioxide, nitrogen response generated Si_2N_2O; 1300°C~1400°C, Carbon reduction reaction begins and gradually strengthen, β-SiAlON start generation; 1450°C~1550°C, β-SiAlON formation gradually increased, to 1500°C reached the highest value.

2. Reducing agent of carbon in favor of β-SiAlON, Reducing agent silicon in favor of O′-SiAlON generation, Silicon carbon composite is more conducive to the formation of β-SiAlON, It also can reduce the synthesis cost.

References

1. ZHONG Xiangchong. Innovative development of high performance new refractory materials [J]. Refractories, 2005, 39:1~5.
2. GUO Yanqin, LI Suping, ZHONG Xiangchong. Synthesis of bauxite based β-SiAlON by C-Si compound reduction nitridation [J]. Refractories, 2006, 40: 419~421,425.
3. LI Suping, GUO Yanqin, ZHONG Xiangchong. Investigation on synthesis of bauxite β-SiAlON by using aluminum/silicon reducing agent [J]. Non-Metallic Mines, 2008, 31:15-17.
4. HOU Xinmei, ZHONG Xiangchong. Process of synthesizing bauxited based β-SiAlON by aluminum thermal reduction-nitridation [J]. Refractories, 2004, 38: 230~233, 237.

5. GUO Yanqin, Wang Yongwei. Synthesis of bauxite based β-SiAlON by aluminum carbon compound reduction nitridation [J]. Cemented Carbide, 2012, 29 :136~140.
6. GUO Yanqin, Wang Yongwei. Synthesis of bauxite based β-SiAlON by Henan a low grade bauxite [J]. Bulletin of the Chinese Ceramic Society, 2013, 32(8):1510~1514.

Effect of starch on gel properties of silver carp surimi

Z.Z. Shi [a], S.S. Chen [b] & H. Wang [c]

[1]*College of Food Science and Technology, Shanghai Ocean University, Shanghai 201306, China;*

[a] *zhenzhens@163.com,* [b]*sschen@shou.edu.cn,* [c]*wh19001@163.com*

The objective of this study was to determine effects of starch (corn starch and potato starch) on the TPA (Texture Profile Analysis),colour, gel strength, expressible water of silver carp surimi-based product. And microstructure of the surimi gels was used to investigate the corresponding mechanism. The result show that with the two additives, the gel properties of silver carp surimi-based product are obviously improved. As the starch content increased in surimi, the textural parameters such as hardness, Gumminess, chewiness and gel strength of gels was increased($P<0.05$),and the value of whiteness was also increased($P<0.05$). And as the starch content increased in surimi,the expressible water was decrease($P<0.05$). Summary of all the data of texture profile analysis, colour evaluation, gel strength, expressible water, the optimum amount of starch's addition is 60g/kg. Surimi gels with potato starch had higher gel quality comparing to those with corn starch. Surimi gels with potato starch had higher gel strength, Hardness, gumminess, chewiness, whitening and lower expressible water comparing to those with corn starch

Keywords: Silver carp surimi; Starch; Textural properties.

1. Introduction

Surimi, a stabilized fish myofibrillar protein, has unique textural properties and high nutritional value. So surimi product is really a good choice for its characteristic of heath and convenience. The gel properties are the most important functional characteristics of surimi products, which determine the surimi-based products' unique quality such as texture, colour evaluation, Expressible water and sensory characteristics. To better suit the textural preferences of consumers, ingredients must be added to surimi that modify the textural and mobility properties of the surimi (Lee et al.,1992)In China, the starch, such as corn starch, has been added into surimi-based products to enhance the qualities of products.

Silver carp is a popular freshwater fish species in China and usually used for making surimi-based products. Surimi is an elastic food made from low-value fish meat through mincing, washing, rinsing, dehydrating, and adding an amount of cryoprotectants to prevent the freeze denaturation of the protein. Producing surimi is a corking method that low-value fish converte to high-value production. Th Surimi industry has been trying to improve their quality. So this study was designed to assess the effects of different starch on the gelation properties of silver carp surim.

Corn starch and potato starch are major cooking starch in China. Different starch had different effects on the quality and nutritional values of meat products. However, there were few studies in the effects of different starch on the gel properties of surimi-based products. Therefore, the objective of this study was to determine effects of different starch on the texture, colour evaluation, Expressible water and microstructural of silver carp surimi gels.

2. Materials and Methods

2.1. *Materials*

Silver carp was purchased from Guzhong road market (Shanghai, China).Silver carp surimi was produced in surimi processing laboratory. Surimi was cut into approximately 1kg blocks, packaged in plastic bags and stored in a freezer(-20°C). The moisture content was 77.05 g/100 g and the protein was 19.83g/100g for silver carp surimi. Potato starch, corn starch and salt (Hubei Salt Industry Group Co., Ltd, Wuhan, Hubei, China) were purchased from a local supermarket (Shanghai, China).

2.2. *Sample preparation*

The silver carp head and internal organs were removed, and blood was washed away. Put the fish into the meat mining machine for picking. Then surimi was rinsed by 5 time water as much surimi three times. And the water temperature shoud be below 10°C. Cloth was used to pre-dewatering surimi, and surimi was placed in a centrifuge to dewater. Finally, sodium pyrophosphate (3g/kg), sodium tripolyphosphate (3g/kg), sugar (40g/kg), and Sorbitol (40g/kg)were added to surimi and then were chopped. Surimi was stored stored at -20°C for further analyzes.

Following thawing at room temperature (about 25°C) for 2 hour, the silver carp surimi was chopped for 5 min in food processor (CH580, KENWOOD, UK).Salt (30g/kg) was added into surimi and mixed for 5 min.Then starch (corn starch, pitato starch) was added into surimi and mixed for 5 min The concentrations of starch were 30, 60, 90, and 120g/kg, respectively. ice water was added to adjust the final moisture content of the surimi up to 800 g/kg. The surimi were then stuffed into plastic casing (20 mm in diameter),and underwent a two-step treatment: first a 40°C heat treatment for 30 min, then a 90°C heat treatment for 30min. The gels were then cooled in ice water for 30 min and stored at 4°C for 24 hours.

2.3. *Texture profile analysis*

Texture Profile Analysis (TPA) was performed with a TA. XTPlus Texture Analyzer (Stable Micro System corp.,UK). The gels were equilibrated at roomtemperature, and then samples were cut into cylinders (20mm in diameter and 25mm in height). A sample was pressed twice at a speed of 1 mm/s with a Cylinder-ended stainless steel plunger (P36).Hardness (g), gumminess (g) and chewiness (g) were determined.

2.4. *Gel strength*

Multiplying the breaking force and breaking distance ia gel strength. breaking force and breaking distance was performed with a TA. XTPlus Texture Analyzer (Stable Micro System corp., UK).A sample (20mm in diameter and 25mm in height) was penetrated to 15 mm at a speed of 1mm/s with a spherical-ended stainless steel plunger (P/5 S).

2.5. *Expressible water*

Expressible water (EW) was measured by the method of Barreraet al.,(2002). Samples (3±0.2g) were weighed and placed into a filter paper,and then were placed at the bottom of 50 ml centrifuge tubes and centrifuged at 3000g for 15 min. After the centrifugation,the surimi samples were weighed immediately.

2.6. *Colour evaluation scanning electron microscopy*

The color values were determined by a Konica Minolta CR-400 colorimeter (Konica Minolta Holdings, Inc, Tokyo, Japan). lightness (L*),redness (a*) and yellowness (b*) values were recorded by colorimeter. The whiteness (W) can be calculated using the following equation (Park, 1994):

$$W=100-[(100-L^*)^2+a^{*2}+b^{*2}]^{1/2}$$

2.7. *Scanning electron microscopy*

The samples were cut into small cubes(2×2×1mm) immersed in 40g/kg glutaraldehyde for overnight and immersed in 0.1M potassium phosphate buffer for 10 min three times. The specimens were dehydrated by serial concentrations of ethanol solutions. The concentrations of ethanol solutions were 50%, 70%, 80%, 90 %and 100%, Progressively. Then the samples were dried for 12 hours and observed using a scanning electron microscope.

3. Result and Discussion

3.1. *Texture profile analysis*

Table 1 shows hardness, gumminess, and chewiness of silver carp surimi gels with various starch at different concentrations. Starch concentrations significantly affected TPA (hardness, gumminess, and chewiness) of surimi gels ($P < 0.05$). Surimi gels with higher starch below 6% had significantly higher hardness, gumminess, and chewiness values. As the starch increased below 6%, keeping constant moisture level could increase hardness gumminess and chewiness value significantly.

Table 1. Effect of starch on texture of silver carp surimi.

starch	level	TPA		
		Hardness	Gumminess	Chewiness
control	0	793.68±11.5^{d}	654.35±14.5^{f}	616.11±15.8^{e}
corn	3%	827.85±17.1^{c}	686.75±14.5^{ef}	653.00±20.7^{cd}
	6%	890.51±15.9^{b}	771.93±20.2^{ab}	704.5±19.6^{a}
	9%	873.69±10.3^{b}	749.94±16.5^{bc}	694.41±19.7^{ab}
	12%	814.89±18.3^{cd}	709.47±26.5^{de}	638.36±10.7^{de}
potato	3%	841.43±14.3^{c}	700.29±11.1^{de}	672.34±14.2^{bc}
	6%	923.52±10.1^{a}	801.77±22.5^{a}	721.17±16.7^{a}
	9%	890.14±13.5^{b}	766.61±23.2^{bc}	701.07±18.7^{ab}
	12%	834.89±23.3^{c}	732.81±19.5^{cd}	661.70±14.5^{cd}

So: Values in the same column with different letters were significantly different ($P < 0.05$).

The TPA values of silver carp surimi gels with potato starch were higher than corn starch at the same concentration. When starch concentration was 30g/kg, the difference of hardness of gels was not significant. Varieties of starch had no significant effect on the chewiness at the highest value of chewiness. Varieties of starch had no significant effect on the gumminess. The differences among four starch on TPA result might be possibly due to their different amylose concentrtion.

3.2. *Gel strength*

Gel strength is a important judging indicator to silver carp surimi gel. Fig. 1 shows breaking force,breaking distance and chewiness of surimi gels with various starch at different concentrations. Silver carp surimigels with different

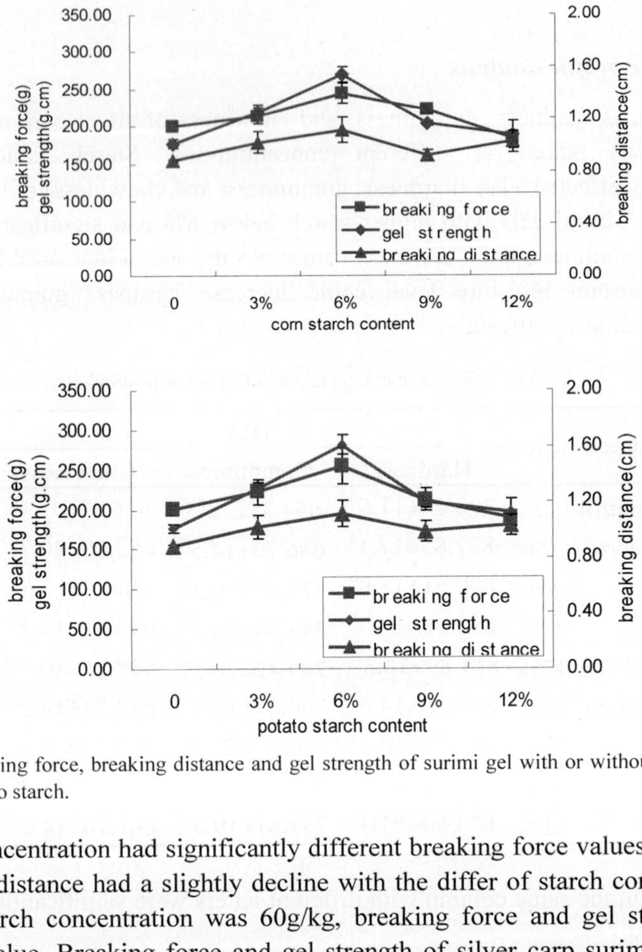

Fig. 1.Breaking force, breaking distance and gel strength of surimi gel with or without starch, corn starch, potato starch.

starch concentration had significantly different breaking force values. However, breaking distance had a slightly decline with the differ of starch concentration. When starch concentration was 60g/kg, breaking force and gel strength had highest value. Breaking force and gel strength of silver carp surimi gels with potato starch were higher than corn starch at the same concentration. However, the difference of breaking force and gel strength of gels was not significant.The highest gel strength is surimi gel with 6% potato starch,and the value of highest gel strength was 282g▪cm. For silver carp surimi gels,gel strength wiith potato starch increased up to 59.42% at same condition compared to surimi gel without starch. But the highest gel strength with corn starch was 271g▪cm.Gel strength of silver carp surimi gels increased up to 53.14% compareing to gel without starch.

3.3. *Colour evaluation*

Table 2 shows L*, a*, b*, and whiteness values of silver carp surimi gels with starch. The addition of starch significantly increased L* values ($P < 0.05$). The whiteness values of gels increased significantly with increasing starch. When the starch concentration is 9%, the higest whiteness value and light value of surimi gels with potato starch is 77.84 and 78.11, respectively. And the higest whiteness value and light value of surimi gels with potato starch is 77.51 and 77.57, respectively. Silver carp surimi gels with different starch had not significantly different whiteness ($P<0.05$).Light value play an important role in whitening value of surimi gels. The whitening effect of starch might be due to the absorbent property effect of amylopectin in the product. The color of the starch which is white may be another reason. Comparing to the color values in Table 2, the color of starch influenced the whitening of surimi gels significantly. Starch with higher lightness had the better whitening effect on surimi gels.

Table 2. Effect of starch on colour of silver carp surimi.

starch	level	colour			
		L*	a*	b*	w
control	0	74.16 ± 0.47^{e}	-2.24 ± 0.065^{a}	1.51 ± 0.56^{c}	74.01 ± 0.46^{e}
corn	3%	75.68±0.25d	-2.17 ± 0.13^{a}	0.91 ± 0.28^{c}	75.56 ± 0.25^{d}
	6%	76.90 ± 0.34^{c}	-2.49 ± 0.11^{abc}	1.81 ± 0.45^{ab}	76.70 ± 0.38^{c}
	9%	77.75 ± 0.52^{a}	-2.62 ± 0.32^{bc}	1.94 ± 0.27^{ab}	77.51 ± 0.51^{a}
	12%	77.39 ± 0.45^{bc}	-2.29 ± 0.06^{ab}	2.08 ± 0.29^{a}	77.17 ± 0.45^{bc}
potato	3%	76.28 ± 0.33^{c}	-2.51 ± 0.44^{a}	1.46 ± 0.30^{c}	76.10 ± 0.35^{c}
	6%	77.50 ± 0.46^{ab}	-2.71 ± 0.28^{c}	1.81 ± 0.45^{ab}	77.26 ± 0.43^{ab}
	9%	78.11 ± 0.17^{a}	-2.69 ± 0.11^{c}	2.15 ± 0.31^{a}	77.84 ± 0.13^{a}
	12%	77.72 ± 0.15^{a}	-2.39 ± 0.18^{abc}	2.07 ± 0.23^{a}	77.17 ± 0.43^{ab}

3.4. *Expressible water*

Expressible water content of silver carp surimi gel added with corn starch and potato starch fish gelatin at different levelsis shownin Fig. 1. Expressible water can indicate the strength of gel network. When the starch was addd, the amount of expressible water in surimi gels was decreased significantly ($P<0.05$). As the starch content increased in the surimi gels, the gel structure was strengthen, resulting in the decrease of expressible water. Comparing with corn starch, surimi gels with potato starch showed the lower expressible water. Surimi gels with potato starch have strengther gel structure. Proportion of amylose and amylopectin might be an important reason. Concentration of amylose is 17% in the corn starch, but concentration of amylose is 11% in the potato starch. Corn starch has high amylose content while potato starch has high amylopectin content. Therefore, surimi gels with corn starch showed the higher expressible water, while surimi gels with potato starch showed the lower expressible water.

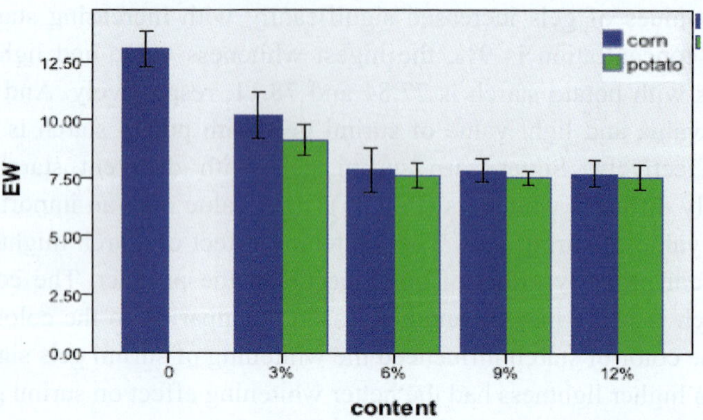

Fig. 2. Effect of starch on expressible water of silver carp surimi

3.5. *Scanning electron microscopy*

Microstructures of silver carp surimi gels added with 6% starch illustrated in Fig. 3. Three figure of microstructure is magnified 1000 times. Surimi gel holes were filled with the addition of 6% starch, as compared with the control gel (without starch). Picture (a) was control figure of surimi gel without starch. Picture (b) and picture (c) were microstructures of surimi gel with 60g/kg corn starch and potato starch, respectively. The cross-linking of myofibrillar protein was never changed by starch. Microstructures of silver carp surimi gels may explain change gel properties such as texture, colour and expressible water. Network with larger voids was not strong and not resistant to the force applied. However, the void can be filled with starch which is absorbent and swelling. This confirmed the positive impact of starch on gelation of surimi.

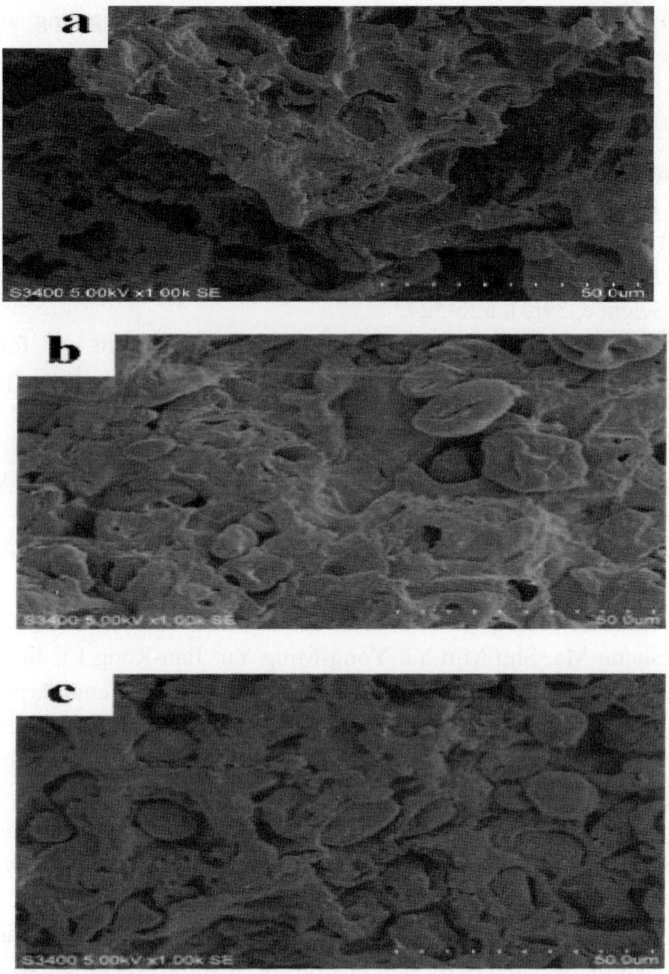

Fig. 3. Electron microscopic images of surimi gel added with different starch (magnification: 10,00×): (a)surimi gel without starch;(b)surimi gel added with 6%corn starch;(c)surimi gel added with 6% potato starch.

4. Conclusion

The results of this study indicate that 60g/kg starch is recommended to be the optimum addition dose to improve the gel properties of silver carp surimi gels. Added with 60g/kg starch, the gel strength values reached the highest. Hardness, chewiness and whitening content of the gel increased significantly than the non-added sample. Adding starch to improve its gel properties could be available for silver carp surimi. Surimi gels with potato starch had higher gel quality comparing to those with corn starch. Surimi gels with potato starch had higher gel strength, hardness, gumminess, chewiness, whitening and lower expressible water.

Reference

1. Barrera, A.M., Ramireza, J.A., Gonzalez Cabrialesa, J.J., Vazquezb, M., 2002. Effect of pectins on the gelling properties of surimi from silvercarp. Food Hydrocoll. 16, 441–447.
2. Park, J.W. (1994). Functional protein additives in surimi gels. Journal of Food Science, 59(3), 525-527.
3. Lee, C.M., Wu, M.-C. & Okada, M. (1992). Ingredient and formulation technology for surimi based products. In Surimi Technology (pp. 2273-302). New York: Marcel Dekker.
4. Hong Yang and Jae W. Park. Effects of Starch Properties and Thermal-processing Conditions on Surimi–Starch Gels [J]. Lebensm.-Wiss. u.-Technol., 1998, 31:344-353.
5. Gipsy Tabilo-Munizaga, Gustavo V, Barbosa-Canovas. Color and texture parameters of pressurized and heat-treated surimi gels as affected by potato starch and egg white [J]. Food Research International. 2004, 37:767-775.
6. Xing-Sheng Ma, Shu-Min Yi, Yong-Ming Yu, Jian-Rong Li, Jin-Ru Chen, Changes in gel properties and water properties of Nemipterus virgatus surimi gel induced by high-pressure processing, LWT - Food Science and Technology, Volume 61, Issue 2, May 2015, Pages 377-384, ISSN 0023-6438.
7. Xianping Lin, Wenge Yang, Dalun Xu, Zhen Jie, Wen Liu, Improving gel properties of hairtail surimi by electron irradiation, Radiation Physics and Chemistry, Volume 110, May 2015, Pages 1-5, ISSN 0969-806X.
8. Fenghui Zhang, Ling Fang, Chenjie Wang, Liu Shi, Tong Chang, Hong Yang, Min Cui, Effects of starches on the textural, rheological, and color properties of surimi–beef gels with microbial tranglutaminase, Meat Science, Volume 93, Issue 3, March 2013, Pages 533-537, ISSN 0309-1740.
9. Torstein Skåra, Svein Kristian Stormo, Dagbjørn Skipnes, Alain Kondjoyan, Agnar Sivertsen, Geert Gins, Eva Van Derlinden, Vasilis P. Valdramidis, Jan F.M. Van Impe, Estimation of surface temperature and thermal load in short-time heat treatment of surimi through reflectance spectroscopy and heat transfer modeling, Journal of Food Engineering, Volume 120, January 2014, Pages 75-80, ISSN 0260-8774.
10. Soottawat Benjakul, Wonnop Visessanguan, Chutima Thongkaew, Munehiko Tanaka, Effect of frozen storage on chemical and gel-forming properties of fish commonly used for surimi production in Thailand, Food

Hydrocolloids, Volume 19, Issue 2, March 2005, Pages 197-207, ISSN 0268-005X.

11. F. Ducept, T. de Broucker, J.M. Souliè, G. Trystram, G. Cuvelier, Influence of the mixing process on surimi seafood paste properties and structure, Journal of Food Engineering, Volume 108, Issue 4, February 2012, Pages 557-562, ISSN 0260-8774.
12. Kong, C. S., Ogawa, H., & Iso, N. (1999). Compression properties of fishm- eat gel as affected by gelatinization of added starch. Journal of Food Science, 64(2), 283–286.
13. Lyons, P. H., Kerry, J. F., Morrissey, P. A., & Buckley, D. J. (1999). The influence of added whey protein/carrageenan gels and tapioca starch on the textural properties of low fat pork sausages. Meat Science, 51(1), 43–52.

Research on the production process of oak barrels for rum aging

Y. Zou
Center for Sugar Technology Research, Guangxi University, Nanning Guangxi, China

W. Gao *
*Forestry College, Guangxi University, Nanning Guangxi, China**
galaxy999@qq.com

W.H. Wu
Guangxi Jinding Furniture Group Ltd., Nanning Guangxi, China

It aims at improving the manufacture of American white Oak barrel in order to better serve the Rum industry in Guaungxi Zhuang Autonomous Region. This paper also makes a comprehensive introduction of the barrel production technique at home and abroad. From the respects of the selecting of the raw material for the barrel, wood splitting and truncation, barrel web fixed length machine and barrel web of arc surface processing, machining Oak barrels type stave, assembly and the hoop frame, bake, steel processing and riveting hoop, pre cooper, processing of double end barrels of the top slot, cover processing and installation, Oak barrels sanding, the final Cooper, sealability, the key points of process packaging and transportation. What's more, there is a comparison of ageing in Oak barrels of Rum and Wine production process.

Keywords: Oak barrels; Microwave treatment; rum; Aging; Mechanism.

1. Introduction

Rum is the world famous liquor which utilize the byproduct of sugar mill such as molasses or sugar cane juice as raw materials. It is full of intense aroma, and its characteristics is the mixture aroma of caramel, sugar cane, subtle sugar sweet and Oak, and it also possesses the tropical and subtropical fruit aromas. Rum is the world's third largest liquor after whiskey and brandy, with the perfect combination style of subtropical romance and sweet sugar cane, crystal clear transparent liquor and cute color (Li & Liu 2012).

Employing the wood of Quercus mongolica from Changbai Mountains as raw materials, radial cut plate in 32 mm thickness was made and to produce the high quality Oak barrel for grape wine brewing by natural air drying, baking, and mechanical processing. The production process of Oak barrel was described in detail. It contrasted and analyzed effective components and properties of grape wine brewed by Oak barrel from self-production and France and discussed the influence of grape variety exert on the quality of the wine tannin concentration(Liu *et al.* 2010).

Barrel-making techniques including material selection, air drying and baking etc. will alter the content of chemical compositions in barrel such as volatile flavoring substances and phenol substances etc., which will influence grape wine quality when it store in Oak barrel. It suggested that American white Oak is the best choice for the production of grape wine with strong Oak aroma, while European Oak is the best choice for the production of grape wine with harmonious Oak aroma, fruit aroma and wine aroma; air drying process of barrel plates would influence the content of total phenol, tannin and volatile substances including vanillin and MOL; and baking intensity will change original content of chemical compositions in Oak, which make Oak barrel have different aroma characteristics and phenol substance content(Lv *et al.* 2006).

He (2009) found that during brewing process the use of an appropriate amount of Oak can bring some vanilla , barbecue and a subtle smoky flavor to the wine, which greatly improves the flavor and quality of the wine. In addition to bringing more complex aroma to the wine, it can also provide the monotonous wine with a good sense of structure, soft tannins and balanced.

The effects of three different factors on quality of dry red wine were studied in aging. Results showed that the necessary of testing different levels to the dry red wine in order to find the best level for quality of wine, thus the best aging method were determined(Shi *et al.* 2007).

List the advantages and disadvantages of wine aging in Oak barrels, application of modern brewing technology and existing problems, discussed the lengthen service life of the Oak barrels, shorten the aging time, improving efficiency, methods to meet the needs of consumers requirements for Oak aged wine sensory such as color, aroma and taste, and verified experiments were carried out(Yang 2013).

Application of micro-oxygenation can make wine color more stable, improve the taste and advance wine sensory quality. So as to find out the optimal treatment conditions, effect of micro-oxygen on red wine aging was studied. Using the technique, we can use much less barrels for wine aging, shorten aging period, reduce the aging cost and increase the productivity and economic benefits for the enterprise(Xia *et al.* 2006).

The wine made by Huaxia winery Co. Ltd. in 2004 was used as raw material to study the changes of flavan-3-ols and proanthocyanidins during aging in American Oak, Europen Oak, or Occident Oak barrel. After the aging in the Oak barrel, the concentration of (-)-epicatechin (EC), (-)-epicatechin gallate (ECG), and (-)-epigallocatechin gallate (EGCG) increased, whereas that of (-)-epigallocatechin (EGC) slightly decreased. The content and mean degree of polymerization (mDP) of proanthocyanidins decreased, and the composition changed accordingly(Wen *et al.* 2006).

Fang and others (2012) consider that China is lack of wood resources, the industry demand for timber are dependent on imports, Oak barrel industry is no exception, the Oak need more than 150 years old for manufacturing Oak barrel , so it is difficult for our country forest to supply such a large number of Oak. A phenomenon needs to pay special attention is that the white Oak commonly used for manufacturing barrel mainly in Europe and the Americas. However, in the last two decades, southern hemisphere countries such as Chile, Argentina, South Africa and Australia where does not produce white Oak emerge the local Oak barrel factory.

Stoichiometric coefficients for reactions of cognac antioxidants with coulometric titrants have been found. Ellagic and gallic acids react with both titrants while aldehydes with electrogenerated bromine only. Furfurals do not show significant reactivity toward both oxidants. Cognac and brandy total antioxidant capacity (TAC) and ferric reducing power (FRP) based on reactions with electrogenerated bromine and hexacyanoferrate (III) ions, respectively, have been evaluated. Both parameters for cognacs are statistically significant higher than for brandies and grow with the age increase. Beverages under investigation has shown relatively high antiradical activity toward2,2-diphenyl-1-picrylhydrazyl (7-92% and 5-93% for cognacs and brandies, respectively). Total phenolics content has been evaluated by Folin-Ciocalteu method(Ziyatdinova *et al.* 2014).

2 Production Process and Technical Points

The basic process of production includes raw material selection, live material cut-off, scraping the bark, aging treatment, lath processing, the combination of softness, barrel shape, mechanical coopers, baking control, chemical testing, barrels machining, assembly pail cover body polish, pressure testing, the barrel molding, packaging, transportation and other sectors. Technical points of each process are as follows:

2.1. *Raw material selection*

The control of country of origin, age and baking of the materials used for Oak barrels is top priority of product's quality and style. The best time for logging is October to December each year, because timber grows slowly in fall and winter during which felling does not affect the future growth of Oak.

Oak at the age of 150 to 250 years, diameter 1 ~ 1.5 m, height generally 30 ~ 36 m was chosen Central timber between 6 ~ 15 m were used to make Oak barrels.

Selection of live tree felling is a very important step, which is related to the wood texture, guarantee of Oak barrels and genuine applications rate of wood.

2.2. *Shortened*

At the beginning of the manufacturing process the raw staves were shortened according to the requested barrel size with the double cross cut saw.

2.3. *Barrel web bench and bucket web internal and external cambered surface processing*

Oak boards were placed laterally on two mutually parallel chains with push hands, started the machine chain to drive the staves forward under the control of two compacting device, in this process, it must be made sure that the staves were firmly suppressed in order to ensure the barrel did not move during board cutting; cutting unit was adjusted according to the required length of the staves.

After the plate length having been fixed, the pusher pushed it to the subsequent equipment to continue processing next surface processing.

2.4. *Rum bucket board processing*

The desire shape was processed by placing the staves into the feed chain, centered them in the chain through four load detection device, then pressed them in the chain with clamping device, the staves went through two inclined cutters which go toward and away from them, the cutters moved under the guidance of template on the detection stick attached to the balance beam.

2.5. *Assembly*

The headband were put on the base of the machine, headband was inserted paralleled to the stave, a rosette formed, paying particular attention to sealing the barrel plates when they were inserted, any little gaps can not be left.

2.6. *Baking oak*

It mainly studies Oak aroma produced by various baking degree which including mild baking, moderate baking, moderate strengthening baking, severe baking. Generally mild baking flavor gives the wine sweet and more aroma. Moderate baking flavor gives wine more licorice, honey and toast aroma. Moderate strengthening characteristics gives wine more dense baked vanilla features, coconut and sugar almond flavor. Severe baking gives wine with smoky spices, baking spices, sometimes even with mineral flavor characteris.

Having been baked in different parameters, weighed a certain quality of raw Oak to detect and analyze the total phenolic content, color, eugenol, T- lactone, C- lactones, and the content of vanillin with GC/MS instrument.

2.7. *Steel hoop processing and hoop riveting*

Riveting was carried out with a hydraulic cylinder, which mixture with the riveting die, the shape depends on the size of the rivet, and pressure was put on the hoop with the pressing block before riveting.

2.8. *Preparatory hoop trussing*

The barrel is placed on the center of the foundation plate of the hooping equipment, through the hydraulic system control rod the plate with driving arm dropping ,the hoop which contacted by the driving arm is tightly pressed to the barrel. Then the barrel was turned and lifted by the barrel gripping device, after it being manual turned over, the other side of barrel hooping is according to the above method.

The barrel joined by the above method was not tight yet, so a loose hoop frame needed to place on the barrel and firmly compressed. With a hoop frame device, the anchor hoop frame was pushed toward the barrel staves until the barrels have been tightly pressed.

2.9. *Groove processing*

Moved the Oak barrels to be processed forward; using the bucket device rose them to the final position and clamped Oak barrels tightly, after the rotating equipment had been turned on, the cutter started and closed to the ends of the barrel, and the barrel milling and top cutting begun, after that, the bucket device pushed barrels to the discharge track, and then removed by manual workers, thus starting the next barrel processing.

2.10. *Barrel head processing and installing*

Each barrel required two heads. Firstly the head boards were processed by a four sided mounding machine to made them into a surface using wooden extrusion equipment pressed them together with the prepared dowel hole drilling and driving machine. For heads with flat joints, reed was used as special sealing material.

2.11. *Sanding barrel body*

Barrels clamped by two chucks, actuated the rotary Rum device and the sanding unit, the operator moved the outer surface of the sanding unit until reached the desired smoothness.

2.12. *Final hoop trussing*

This step is similar to the methods and techniques of preparatory hoop trussing, and need not be repeated here.

2.13. *Sealing performance*

Method of detecting leakage is: injecting 30-40L water, pressurizing to 0.8 bar, held pressure for 10-15 minutes to test whether the barrel is completely sealed without leakage, while adjusting any mechanical defects.
Utilizing the high pressure principle, it researched the expansion sealing performance of barrel under wet conditions.

2.14. *Packaging and transport*

After the barrel had been produced, using the plastic film wrapped it and then put it into a temperature moisture storehouse where should have spray equipment to control temperature and humidity. Finished Rum barrel production should be put into the brewing process as soon as possible.

3. The Difference Between Oak Aging Rum and Wine Production Process

3.1. *Oak selection is not the same*

The alcohol content of Rum is a little bit high, and the mouth-feel is intense and rich with a touch of aromatic fragrance. Strong raw material is used in the aging process, and generally it will choose America white Oak. The alcohol content of wine is a little bit low, and it generally selects France Louvre Oak as its raw materials.

The first difference between American and French Oak barrels. American Oak barrels gives the wine with a more intense flavor of vanilla and cocoa, while French Oak barrels with smoky and toasty flavor. The second difference between American and French Oak barrels: From the taxonomic point of view, the trees used in the manufacture of America Oak barrels and French Oak tree and not the same homologous. American Oak is white Oak, while French Oak belongs to Quercus robur or Sessile Oak. The third difference between American and French Oak barrels: American Oak is typically used artificial drying way, while the French Oak is adopted the natural air dry method. French natural air-dry method makes the Oak remove its bitter phenolic resin naturally, and the appropriate level of dryness is achieved.

3.2. *Sheet cutting process is not the same*

The plates used for Oak barrels for aging Rum required to use radius chop wood method, which is chopping wood along the radial direction, not using multi-saw, double-end cross-sectional anatomy saws for the original timber; while the plates used for Oak barrels for aging wine have not so much requirements.

3.3. *Plate preparation processes is not the same*

The Oak plates for aging Rum need to be placed under the open air, exposed for more than two years to make the contents releases so that they can used for Oak production. The Oak plates for aging wine are not required this complex procedure.

3.4. *Aging time is not the same*

It requires 24 to 36 months for Rum aged in Oak barrels, and the barrel can not be reused. During the aging process, it can use the microwave modified and other equipment to improve the barrel environment, promote the oxidation of wine, so that the tart, jerky wine become soft, mellow and delicate; while it only takes 6 to 12 months for the general wine aged in Oak barrels, and the barrel can be reused.

4. Conclusion

(1) Combined with the research trends and advanced theoretical knowledge on Oak barrels, it systematically and comprehensively verify that the Oak barrel is the best partner brewed Rum which can promote aging and mature, stable wine structure, and make the volatile flavor in the Oak barrel fully integrated into wine, so it has a delicate, sweet taste, fragrant and beautiful alcoholic aroma.

(2) It completely discusses the production process of contemporary Oak barrel from the production process, and other sectors including gathering wood, wood cut off live, bark scraping, aging treatment, lath processing, the combination of softness, barrel shape, mechanical coopers, baking control, chemical testing, barrels machining, assembly barrelhead, sanding barrel body, pressure testing, barrel molding, packaging and transportation.

(3) Finally, it analyzes the differences of Oak barrels production process between Rum aging and wine aging from the material, cutting, sheet processing and aging time. The Oak barrels match with the flavor of Rum product, which ensure the long-term stability of taste and flavor.

Acknowledgements

The authors are grateful for the support of the Special Research Funds of Guangxi Scientific Research and Technological Development Program of China, Grant. No. 1348009-2 of Guangxi Strive to Make Technological Breakthrough.

References

1. Fang, J.S. 2012. Oak barrels local production has mode reference. *Wine World* 7:36-37.
2. He, F. 2009. Oak barrels and alternatives of Oak barrels. *Sino-overseas grapevine and wine* 6:112-114.
3. Li X.Y., Liu J. 2012. Foster the development of Rum sugar industry a new economic growth point. *China S&T Achievements* 13:79-80.
4. Liu Y.L., Liu X.Y., Pang J.Y., Liu Y.J. 2010. Changbai Mountains Quercus mongolica Crape wine brewing production process Oak barrels, *J. Northeast Forestry U.* 38(3): 123-125,130.
5. Lv W., Wang S.S., Zhang C.Y., Xu Z.B. 2006. Effects of barrel-making techniques on chemical compositions content in barrel. *Liquor-making S&T* 12: 43-46.
6. Shi M.L., Xia G.L., Liu C.S., Zhang Q.B., Shao L. 2007. Application of Oak barrel aging technology in red wine aging. *Sino-Overseas grapevine & wine* 2: 16-17.
7. Wen P.F., Chen J.Y., Li J.M., Zhan J.C., Wan S.B., Kong W.F., Huang W.D. 2006. Effect of Oak barrel on the flavan-3-ols and proanthocyanidins in red wine during aging. *Food Sci.* 27(4): 110-114.
8. Xia G.L., Liu C.S., Shi M.L., Zhang Q.B. 2006. Application of micro-oxygenation in red wine aging. *Sino-Overseas grapevine & wine* 3: 16-18.
9. Yang C.Q. 2013. Introduction to Oak barrels aging technology application. *Liquor making* 40(4):49-52.

10. Ziyatdinova, G., Salikhova, I. & Budnikov, H. 2014. Coulometric titration with electrogenerated oxidants as a tool for evaluation of cognac and brandy antioxidant properties. *Food Chemistry* 150: 80-86.

Study on the effects and mechanism on rum aging by microwaving oak barrels

Y. Zou
Center for Sugar Technology Research, Guangxi University, Nanning Guangxi,, China

W. Gao *
Forestry College, Guangxi University, Nanning Guangxi, China
**galaxy999@qq.com*

W.H. Wu
Guangxi Jinding Furniture Group Ltd., Nanning Guangxi, China

In the aging RUM aging process, treated the external appearance of oak barrels using microwave to make the weak cell tissue slightly rupture to rich the microporous of the inner wall of the oak barrels and prevent the blockage of the aging process, this treatment is conducive to the RUM aging. In the aging process, conducted a microwave treatment for oak barrel, measured monthly the changes of phenol, alcohols, aldehydes, acids, esters, conductivity, oxidation-reduction potential, dissolved oxygen and pH value of the RUM, thus analyze the effect of aging process on RUM after microwave treatment to inference oxidation and esterification rate and analyze the influence of the oak barrels pore changes on micro oxygen exchange after microwave treatment. Through exploring the mechanism of microwave affecting RUM aging , it improves the quality of RUM aging and speeds up the aging process.

Keywords: Oak barrels; Microwave treatment; RUM; Aging; Mechanism.

1. Introduction

In RUM aging process, on the effect of the one-way microwave oak exterior treatment and frequently alternated magnetic field, water molecules in the oak in the intensification rotate rapidly and rub each other, moisture in the wood cells rapidly heating and vaporizing, which produces the steam pressure to the inner wall of the cell, and makes the weak cell tissue slightly rupture in order to reduce the inner wall pores blockage of oak barrels and enhance its permeability.

The microwave, different from the traditional microwave aging technology, does not directly treat the RUM on oak barrels. For this reason, after the microwave treatment, the oak barrel aging environment optimized, the release rate of oak aroma improved and the content of phenolic compounds in RUM increased, it ultimately improve the quality of RUM aging, accelerate RUM aging and prolong the service life of the oak barrel.

1.1. *Pore changes and aging influences after microwave treatment of wood*

After the wood had been treated by microwave, different degrees of fractures were found on ray cells (Torgovnikov& Vinden 2004), for example, there was the formation of micro-cracks, pits membrane rupture on larch wood after microwave treatment. (Zhou *et al.* 2007). Its fracture effect can be classified into low, moderate and high degree : for the low-modified material, its wood pit membrane ruptured, resin in the channel exchanged and melt, part of the cell walls of the wood-ray fractured, which can make wood permeability increase 1.1 to 1.5 times; for the moderate modified material, its pit membrane ruptured, resin exchanged and boiled, the cell walls of the wood-ray damaged, which make the wood permeability improve up to a thousand times; for the high modified material, its pit membrane, ray cells, tracheid walls, wood fiber wall and catheters all ruptured, resin exchanged and boiled, which make a cavity formed in the string section (Torgovnikov & Vinden, 2008). Barrel with rich pore will promote RUM extract its active ingredient(Moreira *et al.* 2012). Li *et al* (2012) established a heat transport model when one-way microwave source treated the wood vertically and uniformly.

1.2. *Research progress*

The research status of various kinds of liquor aging technology with microwave treatment Microwave can be applied to Fen flavor liquor quick aging(Qiu *et al.* 2011); the technology of microwave accelerating aging Yellow Wine significantly improved its flavor and quality(Jiang *et al.* 2011); microwave aging wine increased its chroma (Li *et al.* 2010). Microwave treatment had achieved the purpose of aging, and the nutrition of liquor which treated by microwave is safe and reliable, and the operation technology is feasible (Pasha *et al.* 2014).

In a similar study, the oak barrel was treated by electric field, with brandy being its object. The study found that the electric field directly treated on oak barrel can obviously improve the brandy's extraction of polyphenols in oak (Zhang *et al.* 2012a, Zhang *et al.* 2013a), and can reduce the content of fusel oil in brandy, thus upgrade brandy's quality (Zhang *et al.* 2012b, Zhang *et al.* 2013b).

1.3. *Application status of chemical analysis method of oak barrel and liquor in the aging process*

Samples will be taken by methods such as High Performance Liquid Chromatography, HPLC and solid-phase micro extraction, SPME and then analyzed by gas chromatography/mass spectrometry, GC/MS, which known as SPME-GC/MS analysis. These measurement methods are credible and wide used.

Fernandez and others identified 68 kinds of phenolic substances in oak barrels and wine(Fernandez de Simon *et al.* 2014b). The testing of total phenolic content was used Folin - Shaukat reagent method, which can accurately determine the content of total phenolics in oak barrels and in RUM(Ziyatdinova *et al.* 2014).

Not only the phenols in different wine like brandy can be analyzed by HPLC (Canas *et al.* 2013); but also the content of chemical substances in oak barrel such as vanillin, lilac aldehyde and coniferyl aldehyde can be measured using HPLC, which serves as an important basis for baking and aging(Campos *et al.* 2004).

SPME-GC/MS can employed to analyze the aroma in oak barrel, Liu Yanlong and others use this method to measure the substances such as acid, aldehyde and phenol contained in wine in the oak barrel used Changbai Mountain Mongolia wood as materials(Liu *et al.* 2010); it is widely used to analyze the odor in wine, for example, Ma and others applied it to study the flavor components changes of Chinese Fenjiu (Ma *et al.* 2014). And it is often used for chemical analysis in oak barrels and wine. For instance, Fernandez using the method carried out a systematic analysis of oak barrel and Wine (Fernandez de Simon *et al.* 2014a)

2. Research program

2.1. *Microwave treatment on oak barrels program*

260 oak barrels, whose specifications for 225L are adopted, of which 216 for breaking to take samples, 18 for continuously taking sample, and the other 10% for spare samples.

During aging, the oak barrels will be treated by microwave every 10 days, three times a month, and the oak samples and RUM samples will be taken one day before the treatment. Processing frequency is about 10 ~ 24GHz, processing power intensity about 1W / cm2 ~ 20kW / cm2, each lasting time about 1 ~ 600 seconds, and the final parameters will be obtained in the experiment.

Wooden stand is used in order to avoid the metal reflecting on microwave; asbestos cloth is used to cover the steel ring in the surface of oak barrels.

2.2. *Rum sampling plan*

The samples are taken in the liquid level of 50%, 20% and 80%, respectively from the center, the bottom and the top of the sample.

2.3. *The first sampling plans for oak barrels : broken oak barrels to take samples*

The original oak barrel is split and four interval staves are removed along the outer diameter to prepare sample preparation. The original oak barrel and the RUM which being replaced the new container are no longer as the tracking object.

During the 1~24 months aging period(36 months additional track), it will destroy 216 oak barrels for the natural aging and microwave processing each will damage three barrels with three times parallel measurements and the 30 days sampling interval.

2.4. *The second sampling plan for oak barrels : take samples from continuous aging*

The growth cone will be used to continuously take samples for 36 times in the same oak barrel, after sampling, Oak Ding will be taken to repair the barrels without using glue, it will achieve self-sealing employing the theory that the small wood will expand after it contact with water.

Removed the wine samples every 30 days and slanted the oak barrels, samples are taken with the growth cone being perpendicular to staves, and then the barrels are repaired with the same specification oak ding , making sure it will not leak, and then put it upright.

This experimental results serve as a supplement to the broken barrel sampling, and the differences between the two experiments provide further research in this field.

2.5. *Sample preparation plan for oak barrels after sampling*

After removing the bulk sample from the oak barrels, with ring measuring system, samples are hierarchically taken from the thickness of oak barrels. Samples are taken independently every 2 mm from the inner wall in the staves with 0~24 mm thick. Among the Stratified samples, about 5 mm^2 are taken for scanning electron microscopy analysis, about 500 mg for pore structure analysis, and the remaining for detecting cells and pits with solvent replacement method.

2.6. *Experimental method of pore structure on oak samples*

Before testing, Oak samples are placed at a temperature of 105°C degassing with the vacuum for more than 24 h in order to remove the material adsorbed on the surface. The specific surface area calculated with BET equation, the calculation of the micropore volume and pore specific surface area using t-plot method, the total pore volume in the relative pressure of 0.99 converting into the adsorpted liquid volume to calculate. The specific surface area of BET minus the micropore surface area is medium pore specific surface area, and the total pore volume minus the micropore volume obtained mesoporous volume. Pore size distribution is calculated by NLDFT theory.

3. Researches

3.1. *Parameters study of microwave treatment*

Regulating the energy transfer form and decay index of field strength, frequency of microwave and microwave time to analyze microwave vector network, the parameters of microwave treatment are defined through measuring the surface morphology, pore structure, phenolic compounds and aroma content of oak barrels. Achieving moderate modified material requirements, it makes oak pit membrane rupture, resin exchange and boil, wood permeability increase.

3.2. *Pore*

The analysis of the characteristics of micropores, mesopores and macropores in the oak barrels is to study microporous changes caused by the change of pore crack after microwave treatment. Then with the solvent displacement method to characterize the micro structure including cell length, cell diameter, the number of pits, pit occlusion rate, pit membrane thickness and the edge of the pit membrane pores status, the permeability of the inner wall of oak during the aging process is examined.

3.3. *Phenol*

The content change law is rendered through the measurement of the trends of various phenolic contents changing with microwave processing and storage time. The interpretation of increasing or decreasing change rate of condensed tannin content after microwave treatment is to analyze the facilitation that microwave treatment works on the condensed tannin formation. Thus reactions of condensed tannin oxidation, polymerization and condensation were obtained, make a qualitative characterization of how the premise material of condensed tannins such as 3, 4- yellow alkyl diol and 3- of flavanols converting into condensed tannins.

For the other phenolics, it will apply the high performance liquid chromatography (HPLC) to investigate the ordinary phenolics including gallic acid, protocatechuic acid and guaiacol, CIS oak lactones and phenolic substance trans oak lactones, 4- methyl guaiacol, 4- ethyl guaiacol, 4- ethyl phenol, vanillic aldehyde and syringaldehyde.

3.4. *Alcohols*

The influence law of microwave processing is obtained by measuring the phenylethyl alcohol content, analyzing the different aging time and the variation of β-phenylethanol content.

During the microwave processing, pay great attention to the phenomenon that the changes of methanol content as aging time decreasing, and compare the difference between microwave processing and natural ageing, which combined with the pore structure analyze and infer the influence microwave treatment work on methanol evaporation.

Analyzing the effect of microwave treatment on fusel oil (including n-propanol, n-butanol, isobutanol, isoamyl alcohol) whose content change of has been observed, and examining the ability fusel oil converts into the corresponding aldehydes and acids after microwave treatment, thus it can speculate that the transformation of amino acid has influence on the fusel oil content reduction, and the pore structure in oak barrels play influence on the fusel oil volatilization.

3.5. *Aldehyde*

It analyzes the change trend of acetaldehyde which changes with the prolonging of RUM aging time, and studies the effect microwave processing dealing with acetaldehyde volume reduction.

Focus on the analysis of the influence of microwave treatment on the acetals synthesis reactio, by measuring the acetal content in different aging time to analyze its development tendency, thus a better aging line is obtained.

System analysis of the sample (oak sample , RUM sample) with the change of microwave processing and aging time, measuring the variation of furfural, furfure analyzing the chemical reaction law that the processing conditions work on the furan furfural oxidized to formic acid , thus it can summarize that the processing conditions have impact on furfural content changes with microwave treatment .

3.6. *Acid*

Microwave treatment can accelerate the conversion of the acid to the ester, at the same time reinforce the ventilation of the oak barrel inner wall, and accelerate the volatilization of acids, which reduce the impact of acids on the RUM.

The focus is that whether microwave treatment can reduce the content of acetic acid, and has a positive effect on the esterification of acetic acid and alcohols.

The analyzing the change rule of octanoic acid in natural aging and microwave treatment is to summarize the effect that microwave treatment works on esterification of octanoic acid and alcohols, and the volatilization that oak barrels treated with microwave work on octanoic acid.

Measuring capric acid changes with aging time, comparing the two differences and analyzing the rules, it can infer the influence of microwave treatment on esterification reaction of capric acid and alcohols, and obtain a better mechanism of RUM ageing.

3.7. *Ester*

Drawing a diagram that changes with aging time, classifying the change trend of different esters, it summarizes that the change regularity of different esters of RUM which is on the oak barrel treated with microwave, and analyzes the promoting effect of microwave treatment on the ester synthesis.

4. Conclusion

The first is the mechanism that the changes of pore structure in oak barrels with microwave treatment works on the RUM oxidation reaction. Microwave treatment could significantly influence the dissolved oxygen content in RUM, have a far-reaching influence on the oak barrel oxygen replenishment, thus the RUM oxidation mechanism is explored combined with the changes of RUM conductivity, pH and redox potential.

The second is the mechanism that oak barrel permeability works on the content of RUM aromatic substances. Oak barrels treated by microwave, microporous, mesoporous and macroporous form different specifications of the "molecular sieve" to molecular permeability of aromatic substances; and oak hole occlusion, pit membrane edge thickness and pit membrane micropore damage will change greatly. All these will be conducive to the phenol, formaldehyde volatilization, benefit for the extraction of RUM, and can change the RUM flavor. According to a series of measurement results, a correlation analysis of pore structure will be carry out in order to reveal the mechanism that oak barrel permeability effect on RUM aromatic substances.

Finally it summarizes the mechanism of chemical reaction of RUM aging treated by microwave. The aging benefits from vessel environment improvement, the pore structure on the oak barrel with microwave treatment improving the air permeability of the permeability, which provides better conditions for chemical reaction, and increased the alcohol oxidized to the aldehyde, acid to acid; the extraction of aroma substances such as phenol and aldehyde in oak barrel accelerate the dynamic balance that alcohols and esters react into ester,and alcohol and aldehyde integrate into acetal, thus the effect of aging achieves.

Acknowledgements

The authors are grateful for the support of the Special Research Funds of Guangxi Scientific Research and Technological Development Program of China, Grant. No. 1348009-2 of Guangxi Strive to Make Technological Breakthrough.

References

1. Campos, J. O. S., De aquino, F. W. B., Do nascimento, R. F., Da costa, J. G. M., De keukeleire, D. & De casimiro, A. R. S. 2004. Influence and effect of thermal treatment in elaboration of regional wood extracts for cachaca. *Journal of Food Composition and Analysis,* 17: 179-185.
2. Canas, S., Caldeira, I. & Pedro belchior, A. 2013. Extraction/oxidation kinetics of low molecular weight compounds in wine brandy resulting from different ageing technologies. *Food Chemistry,* 138: 2460-2467.
3. Fernandez de simon, B., Martinez, J., Sanz, M., Cadahia, E., Esteruelas, E. & Munoz, A. M. 2014a. Volatile compounds and sensorial characterisation of red wine aged in cherry, chestnut, false acacia, ash and oak wood barrels. *Food Chemistry,* 147: 346-356.
4. Fernandez de simon, B., Sanz, M., Cadahia, E., Martinez, J., Esteruelas, E. & Munoz, A. M. 2014b. Polyphenolic compounds as chemical markers of wine ageing in contact with cherry, chestnut, false acacia, ash and oak wood. *Food Chemistry,* 143: 66-76.
5. Jiang J., Wang W. & Chang P. 2011. Study on Accelerated Ageing of Chinese Rice Wine by Microwave. *Journal of Anhui Agricultural Sciences,* 39: 14374-14376.
6. Li C., Huo X. R., Zheng X. R., Zheng X. Z., Liu C. H., Gao X. C., Deng N. Y., Jin C. J. & Wang H. Y. 2010. Effect of microwave aging conditions on color and pH of dry red wine. *Journal of Northeast Agricultural University* 41: 124-129.

7. Li X. J., Sun W. S., Zhou T. & Lv J. X. 2012. Mathematical Modeling of Temperature Profiles in Wood During Microwave Heating. *Scientia Silvae Sinicae* 48: 117-121.
8. Liu Y. L., Liu X. Y, Pang J. Y. & Liu Y. J. 2010. Production Process and Performances of Oak Barrel Made of Quercus mongolka Wood from Cbangbai Mountains. *Journal of Northeast Forestry University* 38: 123-125,130.
9. Ma, Y., Qiao, H., Wang, W., Chen, T., Du, X., Zhai, X. & Zhang, S. 2014. Variations in physicochemical properties of chinese fenjiu during storage and high-gravity technology of liquor aging. *International Journal of Food Properties,* 17: 923-936.
10. Moreira, R. F. A., Netto, C. C. & De maria, C. A. B. 2012. The volatile fraction of sugar cane spirits produced in brazil. *Quimica Nova,* 35: 1819-1826.
11. Pasha, I., Saeed, F., Sultan, M. T., Khan, M. R. & Rohi, M. 2014. Recent Developments in Minimal Processing: A Tool to Retain Nutritional Quality of Food. *Critical Reviews in Food Science and Nutrition,* 54: 340-351.
12. Qiu Z. Y., Wang W. N., Wang D., Lou G. P., Gong X. L. & Wang P. 2011. Preliminary study on accelerated aging Fen-flavor Liquor. *Liquor Making* 38: 26-29.
13. Torgovnikov, G. & Vinden, P. 2004. New microwave technology and equipment for wood modification. In Austin, TX (ed.). *2004 AIChE Annual Meeting, 7 -12 November 2004.* Rotterdam: American Institute of Chemical Engineers.
14. Torgovnikov, G. & Vinden, P. 2008. Main aspects of microwave wood modification applicable to the timber industry. In Japan (ed.). *Global Congress on Microwave Energy Applications, Proceedings. 4-8 August 2008.* Rotterdam: Otsu.
15. Wang B., Zeng C. A., Yang H. F., Xue Z. G. & Yang M. F. 2012a. Effect of Electric Field on Phenol Compounds of Brandy Aging in Oak Barrels. *Journal of South China University of Technology(Natural Science Edition)* 40: 145-148,154.
16. Zhang B., Zeng X. A., Yu S. J., Yang H. F. & Xue Z. G. 2012b. Effect of electric field on fusel oil of brandy aging in oak barrels. *China Beet & Sugar* 3:9-11.
17. Zhang, B., Zeng, X. A., Sun, D.-W., Yu, S. J., Yang, M. F. & Ma, S. 2013b. Effect of electric field treatments on brandy aging in oak barrels. *Food and Bioprocess Technology,* 6: 1635-1643.
18. Zhang, B., Zeng, X.-A., Lin, W. T., Sun, D.-W. & Cai, J.-L. 2013a. Effects of electric field treatments on phenol compounds of brandy aging in oak barrels. *Innovative Food Science & Emerging Technologies,* 20: 106-114.

19. Zhou Z. F., Jiang T. & Wang Q. W. 2007. Influence of Intensive Microwave Treatment on Mechanical Properties of Larch. *Journal of Northeast Forestry University* :35:7-8, 25.

20. Ziyatdinova, G., Salikhova, I. & Budnikov, H. 2014. Coulometric titration with electrogenerated oxidants as a tool for evaluation of cognac and brandy antioxidant properties. *Food Chemistry,* 150: 80-86.

Research on formulation of puffing brittleness prawn crackers by thermal processing

J. Huang[1,a], S.P. Zhang[2,b], L.Chen[3,c] & S.S. Chen[*4,d]

[1,2,4] *College of Food Science and Technology, Shanghai Ocean University, Shanghai 201306, P.R. China*

[3] *Shanghai Key Laboratory of Agricultural Genetics and Breeding, Agro-Biotechnology Research Institute, Shanghai Academy of Agricultural Sciences, 2901 Beidi Rd, Shanghai 201106, P.R. China*

[a] *jinghuang2CL@163.com,* [b] *626038339@qq.com,* [c] *chenleihefei@163.com,* [d] *sschen@shou.edu.cn*

Penaeus vannamei has high nutrition health care value, but the less processed into new product categories. In recent years,puffed food is a popular snack food among people, It can improve the nutritional value of puffed food to put fresh *Penaeus vannamei* into the puffed food. This paper studied the product formula, and product expansion rate,quality and structure, the sensory score, scanning electron microscopy observations as an index. By single factor experiment, firstiy, find out the suitable range of shrimp, corn starch, cassava modified starch and wheat flour suitable range, and then find out the best formula by the response surface analysis. The formula is 40% of shrimp, 15% of wheat flour, 10% of cassava modified starch, 35% of corn starch. This obtained products is crispy, fragrant and nutrient-rich.

Keywords: *Penaeus vannamei*; Puffing food; Formula; Hot working.

1. Introduction

Penaeus vannamei belonging *Penaeidae*, shrimp genus(*Penaeus*), *Litopenaeus* subgenus [1].There are the world of three major farming shrimp.It includes Penaeus vannamei, penaeus monodon and Chinese shrimp. Penaeus vannamei has a thin shell, thick flesh, grow fast, uniform population growth and high disease resistance, etc. In southern China, It has gradually become the main breeding shrimp species.Some scholars have studied the nutritional value of *P.vannamei* muscle, the results show the essential amino acid index (EAAI) higher than the amount of the new *Penaeus chinensis* and *Metapenaeus ensis*. Phosphorus. Selenium and other minerals content higher than *Penaeus monodon* and *Penaeus chinensis*; It contains 18 kinds of amino acids,essential amino acids accounted for 38.78% of the total; So *P.vannamei* is a nutrient-rich aquatic product[1~2].

Puffed food refers to the grain, potatoes or beans as the main raw material. Adjusted moisture, baked, Fried or extrusion puffing and made, with a range of puffing degree and has a certain looseness of food. Puffed food have wide varieties, crispy, so it is popular among children, adolescents, and some of the elderly, it has become a fashion casual food [3]. However, most studies have focused on the puffed cereal. On the marke, the most of puffed shrimp made by use of powder or other shrimp-flavored, so the majority of the market is less nutrients of puffed food,it can not meet the the basic needs of human nutrition and health, so it is necessary to improve nutrient value of puffed food. Currently, nutrition and preservation of *P.vannamei*, and *P.vannamei* farming technology has been widely reported, but there a few new products processed by use *P.vannamei*. To make full use of nutritional value of it,the raw materials used in this article, including fresh *P.vannamei*, starch and other raw materials. With hot working way, developed a kind of higher levels of shrimp puffed shrimp chips.

2. Materials and Methods

2.1. *Materials and equipment*

Penaeus vannamei bought in market GuZong road, select samples length (9.27±0.54) cm, weight (10.04+1.28) g in -40°C refrigerated storage. Corn starch, wheat flour, Cassava modified starch(acetylated di starch phosphate), salt, sugar, monosodium glutamate, baking soda, corn oil, millet are commercially available.4% glutaraldehyde; 0.1mol/L phosphate buffer,1% osmium tetroxide, different concentrations of ethanol (30%, 50%, 70%, 90%, 100%).

Food physical property tester TA-XT Plus, Chaoji Instruments Co,Ltd;DZF-1B vacuum drying oven, Shang hai Tianheng medical medical Devices Co., Ltd.;HY-81 fried, Shanghai tree new machinery co., Ltd;JTDZ-1000 electrically heated frying pan, Zhucheng good special food Machinery Co.,Ltd.;XMTD-204 digital thermostat water bath, Changzhou Winbond Electric Instrument Co.,Ltd.;S-3400N SEM instrument, Hitachi Co.Ltd.;E-1010 Ion sputtering apparatus, Hitachi Co. Ltd.; AL204 electronic balance, Mettler -Toledo Instruments (Shanghai) Co, Ltd..

2.2. *Technological process [4]*

Chopping: According to the experimental design, weigh a certain amount of shrimp, corn starch, wheat flour, Cassava modified starch and other ingredients.

Firstly, Ingredients with an equal amount of water and dissolved the liquid into ingredients. The shrimp and ingredients were mixed after chopping 3min, then add corn starch and wheat flour slowly, chopping 1min, until the product mixing. Pay attention to the chopping process should be carried out on the ice; Cooking Molding: The mixture was mixed with shrimp, starch and the put it in a mold, 75°C cooking 20min; Aging: Forming of shrimp cake shrimp should age at 4°C for 12h [6]; Sliced: Thickness with a microtome cut into a sheet of about 1mm.Fried: Oil temperature about 180°C, Fried about 1 min until slice floating to the oil level.

2.3. *The product formula design*

2.3.1. *The effect of shrimp content on product quality*

With shrimp and corn starch as the main raw material (Table 1),According to the total weight of the mixture, add 1% sucrose,1% monosodium glutamate,2% salt,0.6% sodium bicarbonate, then add an equal amount of water to the total weight. According to the above method of chopping. Chopped shrimp marinated with starch mixture into the mold cooking forming at 75°C for 20min.Then placed in the refrigerator under 4°C aging for 12h.The aging shrimp cake cut into slices of about 1mm,puting it in a 50 °C thermostat drying oven for 4h and 70°C for 2h,Then the shrimp cake slice in 180 °C fried in oil. Measuring expansion ratio of products and make the right sensory evaluation to products.

Table 1. Design table of factors and levels for shrimp and corn starch

Project	Shrimp (%)	corn starch(%)
1	20	80
2	30	70
3	40	60
4	50	50
5	60	40
6	70	30

2.3.2. *The effect of wheat flour content on product quality*

Fixed 40% shrimp content, and 60% made up of different proportions of corn starch and wheat flour (Table 2), the addition of other ingredients, processing and determination of the content of the finished product are as defined above.

Table 2. Design table of factors and levels for wheat flour and corn starch

project	corn starch(%)	Wheat flour (%)
1	60	0
2	50	10
3	40	20
4	30	30
5	20	40
6	10	50
7	0	60

2.3.3. *The effect of cassava modified starch content on the product quality*

Fixed shrimp content of 40%,wheat flour 10%,by weight of the other half-main ingredient from cassava starch and corn starch configuration (Table 3),the addition of other ingredients, processing and determination of the content of the finished product are as defined above.

Table 3. Design table of factors and levels for cassava modified starch and corn starch

Project	Cassava modified starch(%)	corn starch(%)
1	0	50
2	10	40
3	20	30
4	30	20
5	40	10
6	50	0

2.3.4. *Main ingredient ratio optimization*

Experimental design using Simplex Lattice to A, B, C, D, they refer to shrimp, wheat flour, cassava starch, corn starch respectively. Then puffing brittle shrimp recipe by response surface analysis and optimization of Design Expert 8.0.6 software, determine the optimal ratio of shrimp, corn starch, cassava starch and wheat flour content.

2.4. *Measuring method*

Using millet exclusion; Swelling degree=The volume of puffed after/expanded the volume of the former [5].The volume of an object (cm^3) =millet plus the volume of an object (cm^3)-the volume of millet (cm^3) [6].

2.5. *Observation of scanning electron microscopy*

The sample was cut into a thickness of 1mm,width and length are about 5mm pieces, soaked in 4% glutaraldehyde fixative 24h,then 0.1mol/L phosphate buffer 10min immersion cleaning was repeated three times, then 1% of tetra-tetroxide fixative soaking 2h,then 0.1mol/L phosphate buffer immersion cleaning 10min, repeated three times. Then followed with 30%, 50%, 70%, 90%, 100% ethanol soaked each 10min, 35° Cin a vacuum oven bake 12h;then gilded ion sputtering instrument, giving the sample conductivity, and then observed in the scanning electron microscope[7].

2.6. *Determination of texture*

Determination of brittleness: The probe is HDP-3PB,Test Mode: Compression, Pre-Test Speed 1.00mm/s, Test Speed 3.00mm/s, Post-Test Speed 10.00mm/s, Distance 5.0mm,Target Type Auto, Trigger Type Auto, Trigger Force 25.0g,Break Mode: Level, Break Sensitivity 10.0g,Break Detect: Return.

2.7. *Sensory evaluation*

Using a 10-point scale, selection of 10 organoleptic evaluation member on product appearance, flavor and taste sensory quality evaluation, evaluation criteria as shown in Table 4. 25% of the total score according to the appearance, taste account for 50%,25% of the weighted average calculations.

Table 4. The sensory evaluation criteria.

project	Scoring criteria	score
appearance	Golden yellow color, looks smooth and complete, not broken	8-10
	Pale golden yellow, looks quite smooth, not broken	4-7
	Partial color was black or brown appearance is not smooth, easily broken	0-3
mouthfeel	crispy and delicious, the expansion is very good, no agglomeration ,non-stick teeth	8-10
	A little crispy, some agglomeration phenomenon, and some sticky teeth	4-7
	hard and not crispy, expansion is not good, there is agglomeration, sticky teeth phenomenon obviously	0-3
flavor	Flavor very coordinated, shrimp flavor, taste better	8-10
	Incongruous flavor, shrimp flavor is not obvious, salty	4-7
	Flavor uncoordinated, little shrimp flavor, very salty	0-3

2.8. *Data Processing*

Text data are used repeatedly in parallel, as was the mean of three parallel puffed measured values, the average of six parallel texture measurement values. Significant analysis using SPSS (version19.0) software Duncan method significance level P=0.05 for data analysis and statistics, were compared with each other among all mean[8] using the least significant range method, the results presented as mean ± standard error (Mean values ± CV).These different letters , such as a, b, c, d, different letters represent significant difference. The response surface analysis using the Design -Expert 8.0.6 trials under the software of Simplex Lattice software Design optimization test factors.

3. Results and discussion

3.1. *The product formula optimization*

3.1.1. *Determine the amount of shrimp added*

From Figure 1, it can be seen when the content is less than 50% of the shrimp with the increase of the content of shrimp, prawn crackers puffing rate is increasing, and when the content is 50%,the expansion ratio reaches the maximum; then with the content of shrimp increase, prawn crackers puffing rate decline. The following microstructure chart from scanning electron microscope can be clearly observed in different proportions of shrimp and starch after cooked at 75°C for 20min, as Figure 2 shown in the region, A region indicates that the shrimp protein fibers, B region indicates the starch pasting body; when the shrimp content is 20%,the shrimp is not completely combined with each other with the starch, a very small amount of starch and shrimp combined with each other into a network structure. Whole uniform network structure is not formed in figure a ,the majority of the system all the starch granules, small amounts of shrimp protein fiber; When the shrimp content of 40%,the shrimp and starch essentially combined into a relatively complete network structure, starch pasting body also uniformly embedded in the network structure. When the shrimp content of 70%,and in shrimp cake gel system formed, shrimp large concentration of protein fibers into pieces, wrapped in starch gelatinization body form a tight structure and number of bubbles contained in the chart clearly less than b, a figure, the lack of sufficient power at high temperature expansion, so decreased puffed rate[6].

Experimental operation found that when shrimp content is at least 60%, shrimp cake or has difficulty in forming and is not conducive to slice after aging, and it affect subsequent experiments. Add a certain amount of shrimp in shrimp cakes, after cooking the cake color become red, this phenomenon is due to shrimp this crustacean shells rich in astaxanthin, after heating it turns red [9].In summary, Shrimp content is better between 40%~50%.

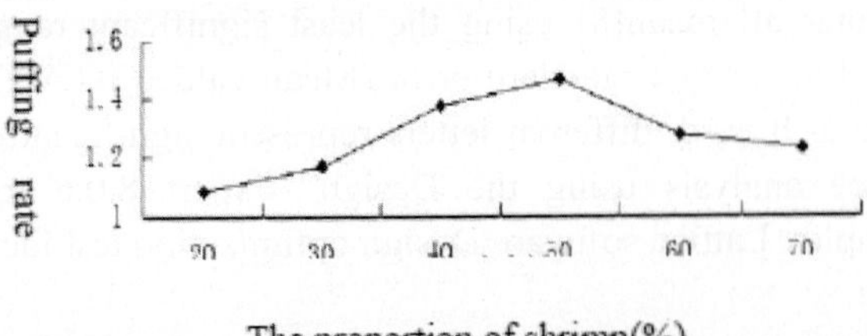

Fig. 1. Effect of different shrimp content on the expansion of the shrimp crackers.

Fig. 2. Electron scanning microscope structure at different content of shrimp cooking shrimp cake after 75°C cooking 20 min (figure a, b, c, refers to 20%, 40%,70%of shrimp respectively).

3.1.2. *Determine the amount of wheat flour added*

It can be seen from Figure 3 that adding wheat flour within a certain range will help improve the expansion ratio. The content of wheat flour is about 10%, this product have higher puffing rate; This may be due to a small amount of wheat flour was added thereto and the shrimp meat protein fibers advantageously combined into a network structure, to avoid irregular shrimp protein fiber is wound so that a network structure can not be formed uniformly with a certain amount of bubbles [6].

Fig. 3. Effect of different wheat flour content on the expansion of the shrimp crackers.

3.1.3. *The determination of cassava modified starch additives*

Add modified starch in puffed food production process can make the moisture in the modified starch can overflow from the starch because of the high temperature, and achieve the goal that make food inflation. And the ordinary starch is short of this effect, so use common starch is difficult to make the puffed food crunchy [10].As can be seen from the Figure 4, modified starch content is 0%~10%, With the increase of cassava modified starch content, puffiing rate continuously improve products, then when the cassava modified starch content increased, the product expansion rate is reduced, the last basic stable as the quantity of added. Cassava modified starch can be concluded from the Figure 4 is around 10%, puffing rate reached the highest. The Figure 5 shows that with the increase of the content of cassava modified starch, the brittleness of products is on the decline. Crispness that is under pressure when the probe distance measured object breaks, so the smaller the value crispness, the greater brittleness. When 10% of modified starch products crispness sharp decline, when content of 20% to 30% starch friability changes slowly; Modified starch is about 50%,the product have low brittleness fragile. It is not conducive to package and storage later. The analysis result shows that adding 10% modified starch, on the swelling rate and brittleness products are greatly improved. Adding suitable amount of modified starch to reduce the cost of production, save raw materials.

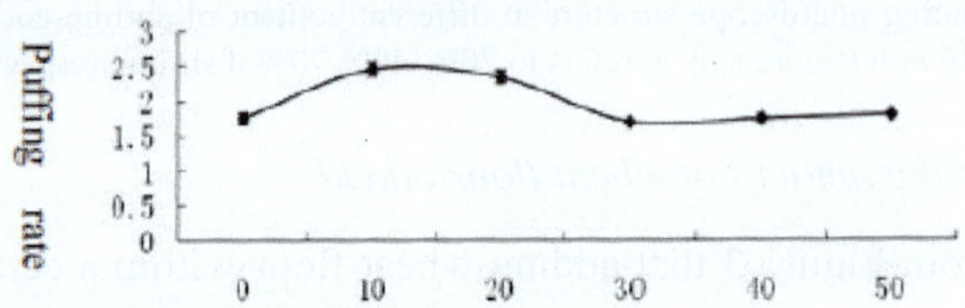

Fig. 4. Effect of different cassava modified starch content on the expansion of the shrimp crackers.

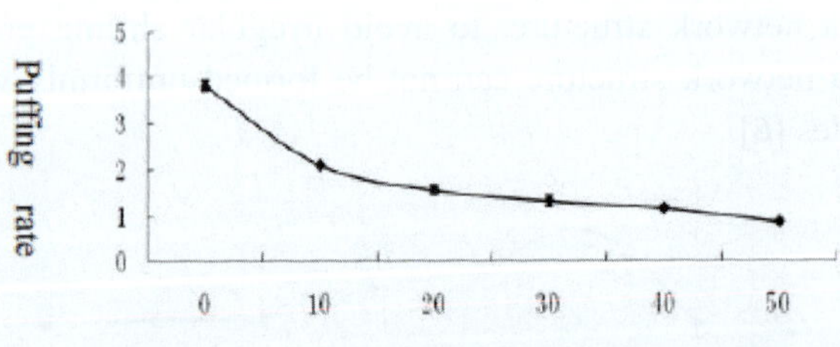

Fig. 5. Effects of different cassava modified starch content on the fragility of the shrimp crackers.

3.1.4. *The main raw material optimization experiments of prawn crackers*

To regression analysis of the data in table 7, get 4 kinds of factors on the product the influence of sensory evaluation of regression equation: $Y=7.82A+7.62B+7.72C+7.69D-0.82AB+2.14AC+0.36AD+1.95BC+3.05BD-4.62CD$, The model F-value of 12.24 implies the model is significant. The regression equation of the $R^2=0.9167$,the better the fit of the regression equation obtained, model P value is

0.0003,it shows the influence of various factors are significant; At the same time of quasi difference is not significant, so analysis of variance and formula optimization to discuss further.

The content of shrimp, corn starch, wheat flour and cassava modified starch on the sensory evaluation value influence the corresponding surface as shown in Figure 6.Can be seen from the figure, at first with the increase of cassava starch content, sensory scores increased, but when the cassava modified starch exceeds a certain proportion, with the decrease of sensory score. In main factors, the content of shrimp meat on the sensory evaluation value influence is larger than cassava modified starch, corn starch is minimal. Mathematical software analysis was carried out on the regression model, according to the results of the shrimp meat content is 40%,wheat flour content is 15%,cassava modified starch is 10%,corn starch content is 35%.In this case, a maximum response can be obtained, namely the organoleptic evaluation value of 8.42.

Table 5. Main ingredient optimization design of experiment and sensory evaluation results.

number	A(%)	B(%)	C(%)	D(%)	sensory evaluation
1	50.00	10.00	10.00	30.00	8.19
2	40.00	20.00	10.00	30.00	8.36
3	40.00	10.00	20.00	30.00	7.51
4	40.00	10.00	10.00	40.00	6.54
5	45.00	15.00	10.00	30.00	7.98
6	45.00	10.00	15.00	30.00	7.48
7	45.00	10.00	10.00	35.00	7.84
8	40.00	15.00	15.00	30.00	8.41
9	40.00	15.00	10.00	35.00	7.67
10	40.00	10.00	15.00	35.00	7.81
11	46.30	11.30	11.30	31.30	7.58
12	41.30	16.30	11.30	31.30	7.83
13	41.30	11.30	16.30	31.30	7.65
14	41.30	11.30	11.30	36.30	7.75
15	42.50	12.50	12.50	32.50	7.86
16	50.00	10.00	10.00	30.00	7.75
17	40.00	20.00	10.00	30.00	7.83
18	40.00	10.00	20.00	30.00	7.42
19	40.00	10.00	10.00	40.00	7.73
20	45.00	15.00	10.00	30.00	7.90

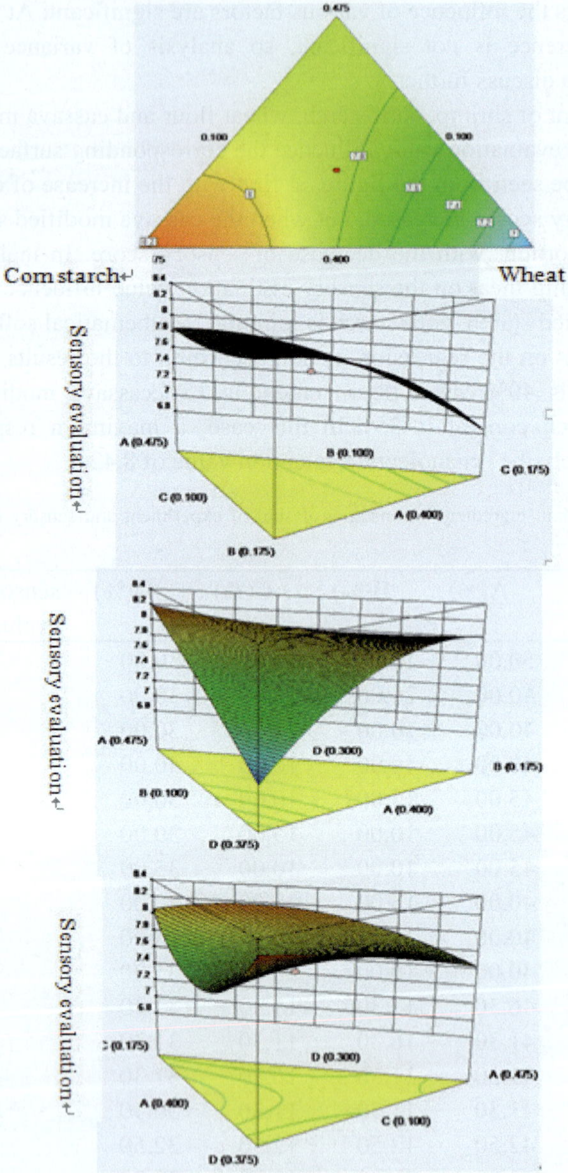

Fig. 6. Response surface plot of the effects of A, B, C and D on the sensory evaluation value formula by hot working high.

4. Conclusion

In this study, determining better range of the shrimp,
corn starch, wheat flour, cassava modified starch preferred by single factor experiment firstly, and then through the response surface analysis to find high puffing rate, high sensory score, good brittleness of prawn crackers main ingredient Finally determined the main ingredient formula is 40% shrimp ,15% wheat flour,10% cassava modified starch,35% corn starch. Complementary makings is the reference of "Study on preparation of silver carp puffed crisps of high the surimi content"[6],namely,1% sugar,1% monosodium glutamate,2% salt,0.6% sodium bicarbonate. Thermal expansion is use oil temperature in a short period of time to make the food of the moisture evaporation and form enough steam to making food puffing [11].Prawn crackers that commercially available, they have added pigment, essence of modulation of shrimp flavor, unfavorable eat frequently; But this experiment product shrimp meat content reached 40%,the high concentration of shrimp, prawn with natural fragrance; Add 10% of cassava modified starch has greatly increased the product brittle, puffing degree; Compared with the prawn crackers that sells on the market, this kind of optimized various factors of prawn crackers, nutrients more reasonable, the cost is not high, for young and old, it is a high content of shrimp, a balanced puffing brittle prawn crackers.

5. References

1. G.J. Zhang, L.P. Han, J.F. Sun, et al. 2003. Nutrition analysis and evaluation of Penaeus vannamei [J]. Journal of Chinese Institute of Food Science and Technology, 13(8): 255~260.
2. X.H. Chen, Qing Chen, D.X Xie, et al. 2001. Meat percentage and muscle nutrition value evaluation of *Penaeus vannamei* [J]. Aquatic Science and Technology Intelligence, 28(4): 165~168.
3. H.X. Xie & H.O Wang. 2006. General situation and development of puffed food processing technologies in China [J]. Journal of Agricultural Products Processing, 9(76): 40~42.
4. Tongdang T, Meenun M, Chainui J. 2008. Effect of Sago Starch Addition and Steaming Time on Making Cassava Cracker (Keropok) [J]. Starch-Stärke, 60(10): 568~576.
5. Y.Z. Li, S.X. Zheng, S.C. Luo. 2003. Characteristics of potato chips by vacuum microwave processing technology [J]. The food and fermentation industry, 29(8): 40~43.
6. Y.P. Shi. 2013. Study on preparation of silver carp puffed crisps of high the surimi content [D]. Master's thesis of Jiangnan University, 1~41.

7. A.M. Zhou, Q.X. Zeng, Xin Liu, et al. 2005. Study on two kinds of protein additives aristichthys nobilis minced can improve fish gel properties [J] Journal of South China University of Technology, 33(4):88~90.
8. Gang Liu, X.S. Hu, Q.H. Li, et al. 2006. Using Colorimeter and Texture Analyzer study of high protein dried noodle color and texture [J]. Science and Technology of Food Industry, 10(27): 99~102.
9. Lu Li, H.L. Shen, Bei Zhang, et al. 2013. Study on extraction and stability of astaxanthin [J]. Journal of Zhejiang Institute of Ocean, 32(3): 245~248.
10. Yan Wang. 2013. The application and prospects of modified starch in the food industry [J]. Value Engineering, 25: 259~260.
11. F.M. Deng, Hu Ying, Q.J. Yang, 2002. Fortified fried puffed food research [J]. Food Science, 23 (1):84~86.

Oxidation behavior of solid lubrication coatings under high temperatures

Z.H. Yu*, X.H. Chen, X.J. Yuan & C.J. Yao
Xi'an High-tech Institute, Xi'an, China,
**yuzhihang0607@163.com*

In this paper, the NiCoCrAlY/BaF2/CaF2/C/Y composite coating prepared by plasma spraying of high temperature oxidation behavior is researched ,which is combined effect of heat and friction environment, as well as simple static study on both the high temperature environment. Results showed that: in high temperature friction environment, coating is oxidized more seriously than simple static oxidation; Degree of coating oxidation under 500°C is lower than 800°C, and oxidized coating will produced peeling in friction process; The state of oxidized film on coating surface is different with friction processing, which with no full of lubrication film, lubrication film alternating appeared, full of lubrication film and variety complex of situation.

Keywords: Friction; High temperature; Static; Oxidation behavior.

1. Introduction

In recent years, a lot of research on solid lubricating materials has made. At present, the more representative research is PS-series coatings (PS100~PS400), adaptive coatings (eg. YSZ–Ag–Mo/TiN multi-layer adaptive coatings by C. Muratore) and nano-composite self-lubricating coatings studied by National Aeronautics and Space(NASA) and Air Force Research Laboratory(AFRL) (Aouadi et al 2010; Stone et al 2010; DellaCorte & Edmonds 2009; DellaCorte et al 2010; Voevodin & Zabinski 2000; Aouadi et al 2009). These coatings have proved that enamel coatings in high temperature friction processes between friction pair surfaces can effectively reduce coating friction and wear (Dellacorte,C. 2003;Ding, C.H.et al. 2008;Muratore, C. & Voevodin, A.A. 2009). But Materials will undergo a complex series of changes at high temperatures, especially in high temperature friction system; surface oxidation will occur to make changes of coating composition and structural and affect the performance of coatings. So study of coating at high temperature oxidation products to implement its own lubrication is of great significance. The composite powders proportion of each coatings are showed in Table 1, and plasma spraying equipment (Sulzer MeCo 9MC plasma spraying system) is showed on Figure 1, which main gas is nitrogen, complemented with hydrogen gas, argon is used as powder feed gas, adjusting parameters and in this process the spraying time and frequency to ensure the coating thickness is between 350~450 μm. Spraying parameters are showed in Table 2.

Fig. 1. Sulzer MeCo 9MC plasma spraying system.

Table 1. Composite powders proportion of each coating.

NO.	NiCoCrAlY	Ni	BaF_2	CaF_2	WSe_2	Ag	C	Y	BN
C1	85.0	--	4.67	2.33	--	--	5.0	3.0	--
C2	65.0	--	13.3	9.8	--	--	3.9	4.0	4.0
C3	--	51.8	10.2	4.0	--	2.0	25.0	3.0	4.0
C4	--	59.5	10.5	5.0	16.0	2.0	--	3.0	4.0
C5	--	69.2	6.2	3.6	--	3.0	15.0	3.0	--
C6	--	80.8	4.31	2.64	8.0	2.0	--	3.0	--

Table 2. Spraying parameters.

Current /A	Voltage /V	N_2flowrate /(scfm)	H_2flowrate /(scfm)	Powder feedrate /(ml/h)	Spraying distance /mm
550	81.2	98	9	190	100

2. Oxidation of High Temperature and Friction

As shown in Figure 2 (b), after high temperature friction, a large number of recrystallization occurred on coating surfaces, while most of the surface is covered with a dense layer of oxidized film, which made a big difference from original coatings, and greatly reduce the amount of coating wear. Experiments found that production of these substances is of important significance to improve coating performance, which realized coatings lubrication under high-temperature and made the coating in a wide temperature range were with ideal mechanical properties and lubricating properties.

As shown in Figure 2, product of oxidation is mainly two forms, granular and needle, which are not the same. After EDS analysis found that elements of the needle-shaped crystals are mainly contain Ni and O in spectrum diagram1, meanwhile granular-shaped crystals are mainly contain Cr and O elements. From the perspective of oxidation kinetics, affinity of O with Cr elements are stronger than Ni elements, so the first oxidation is Cr at the time that NiCoCrAlY alloy oxidation occurred, and diffusion rate of Cr^{3+} ions is greater than that of oxygen ions. That is to say Cr^{3+} ions emit to coated surface continuously. When oxide content is quite high, nickel atoms of alloys will form compounds with Cr_2O_3, which is mainly based on $NiCr_2O_4$ of spinel-shaped structure. Experiment of coating high temperature friction is at 800°C environment, 1h limited. Seen from the coating, oxidation phenomenon is not very full, and granular chromium oxide did not exactly cover coating of oxide film to form on the surface of dense, it may be as a result of low content of Cr. Analysis found that addition of lubrication to increase the antioxidant properties of composite coating might not be very good for.

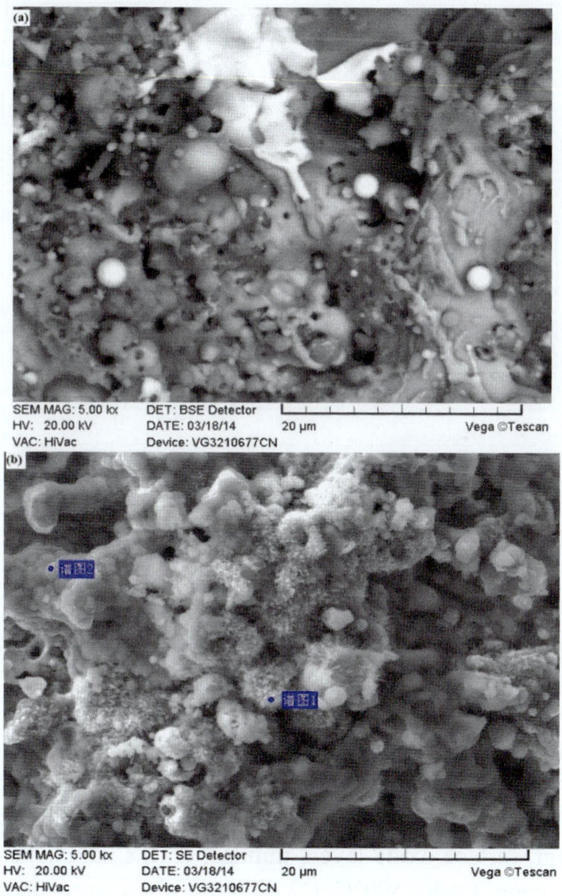

Fig. 2. C2 Surface SEM image and the EDS analysis at 800°C: (a) C2 Original morphology; (b) C2 Oxidized morphology at 800°C.

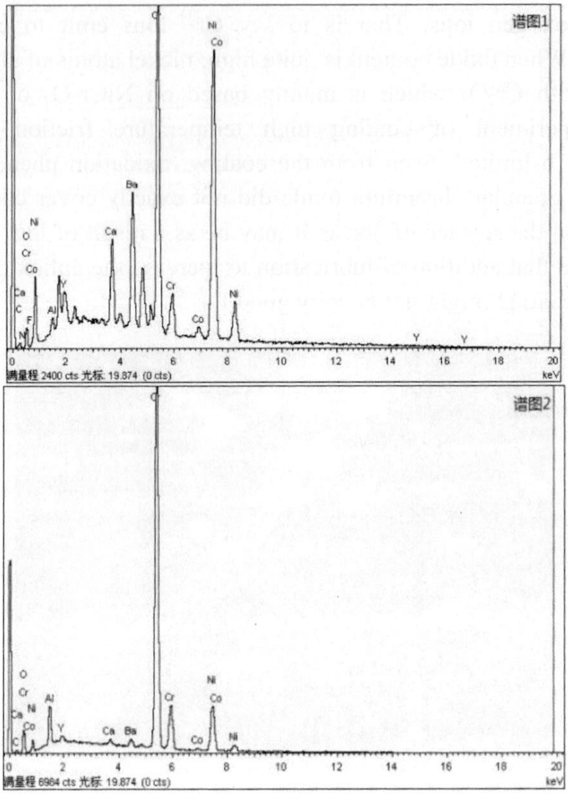

Diagram1	Ni	Cr	B	C	N	O	Al	Ca	Co
Content	37.82	31.42	4.14	14.14	1.44	5.65	1.82	1.96	1.59
Diagram2	Cr	Ni	F	C	N	O	Al	Ca	Co
Content	42.03	33.20	4.08	8.50	0.44	6.68	2.93	0.75	1.39

Fig. 2. (*Cont'd*)

Figure 3 is C1 and C2 SEM image after 500°C and 800°C friction, from which can be found that under 500°C temperatures environment, the oxidation of coating is not as clear as that under 800°C temperatures environment. Through friction, surface oxide layers thinned, cracks, and appeared noticeable flaking phenomenon. Under 500°C, C2 did not form a larger area of lubricating film; main lubrication is from original surface. Under 800°C, coated surfaces appear compact oxide film, which remains better structure after friction. After the above analysis and EDS test, after high temperature oxidation, the main elements of coating surface are O, Cr, Ni and Ba. Compound of several elements at high temperatures can form effective and stable oxide film on coatings surface, and be able to effectively stop the further reaction of oxygen and material below the surface, thus improved lubricating properties and service life of the coatings under high temperature environment.

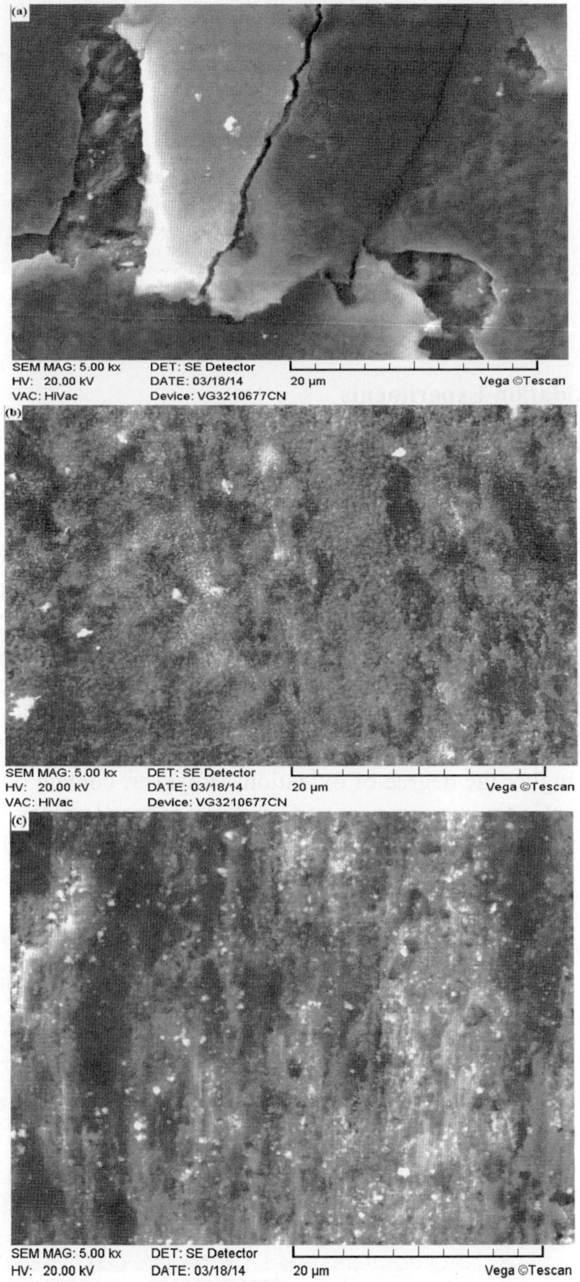

Fig. 3. SEM image of C1 and C2 after 500°C and 800°C friction:
(a) C1 500°C; (b) C1 800°C; (c) C2 500°C; (d) C2 800°C

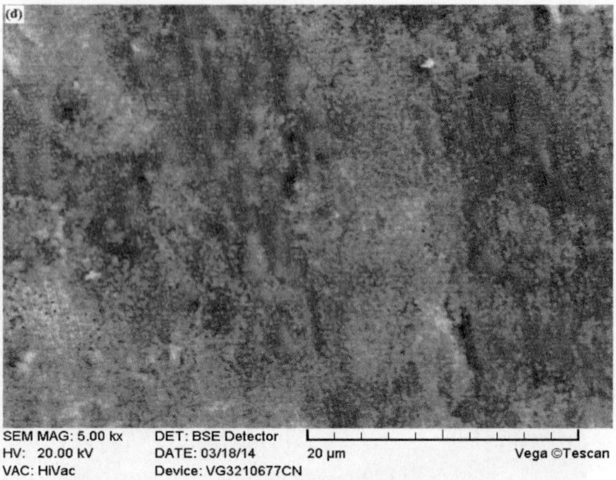

Fig. 3. (*Cont'd*)

3. Static Oxidation Experiments

3.1. *Analysis of experimental results at 500°C temperature*

Figure 4 is C3, C4, C5 and C6 morphology of the specimens, experienced 30min oxidation under 500°C, seen from which the surface oxidation is noticeable, particularly the coatings surface in Figures 4(a) and (b), which is covered with oxide film. By EDS analysis found that the oxygen content in coating surface was significantly higher than the original. Figure 5 is C3, C4, C5 and C6 XRD analysis results of oxidation at 500°C, which showed the oxides on surface is NiO and a little of W substance. That is to say WSe_2 have been decomposed at 500°C, $WSe_2 \rightarrow W + 2Se$, and W element has not been oxidized. Generally speaking more Ni on surface has been oxidized instead, and the more Ni contents the higher the degree of oxidation. However, contain other phase, such as C, BaF_2, eC, the oxygen content will be reduced, other components of oxidation is not obvious. Majority of BaF_2 can be found at such temperatures. They can be found concentrated, mainly coated with Ni centrally.

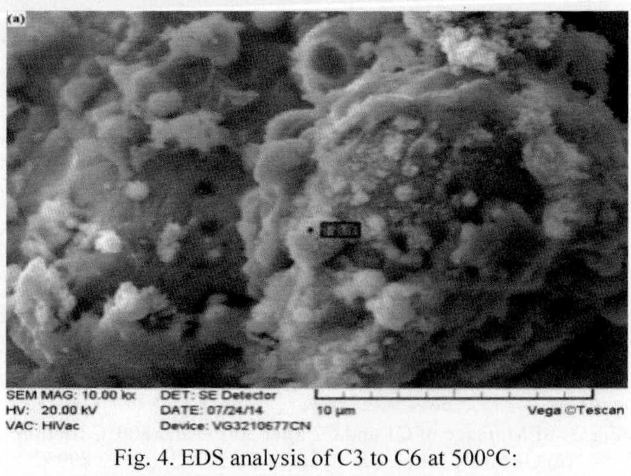

Fig. 4. EDS analysis of C3 to C6 at 500°C:
(a) C3 500°C; (b) C4 500°C; (c) C5 500°C; (d) C6 500°C.

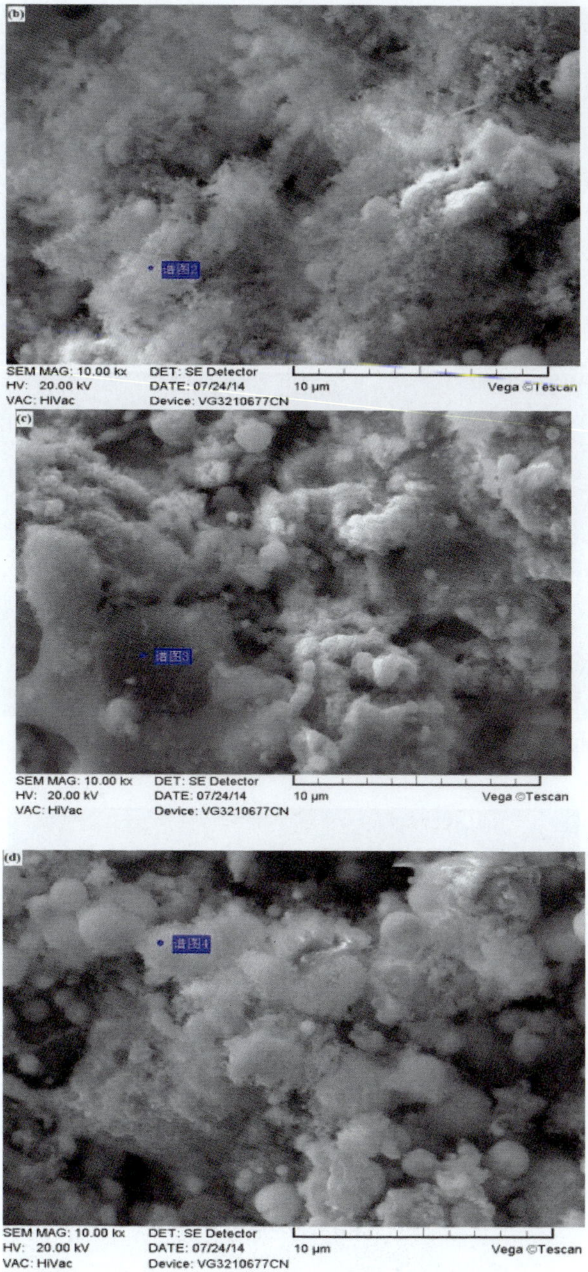
(b)
SEM MAG: 10.00 kx
HV: 20.00 kV
VAC: HiVac
DET: SE Detector
DATE: 07/24/14
Device: VG3210677CN
10 μm
Vega ©Tescan
(c)
SEM MAG: 10.00 kx
HV: 20.00 kV
VAC: HiVac
DET: SE Detector
DATE: 07/24/14
Device: VG3210677CN
10 μm
Vega ©Tescan
(d)
SEM MAG: 10.00 kx
HV: 20.00 kV
VAC: HiVac
DET: SE Detector
DATE: 07/24/14
Device: VG3210677CN
10 μm
Vega ©Tescan

Fig. 4. (*Cont'd*)

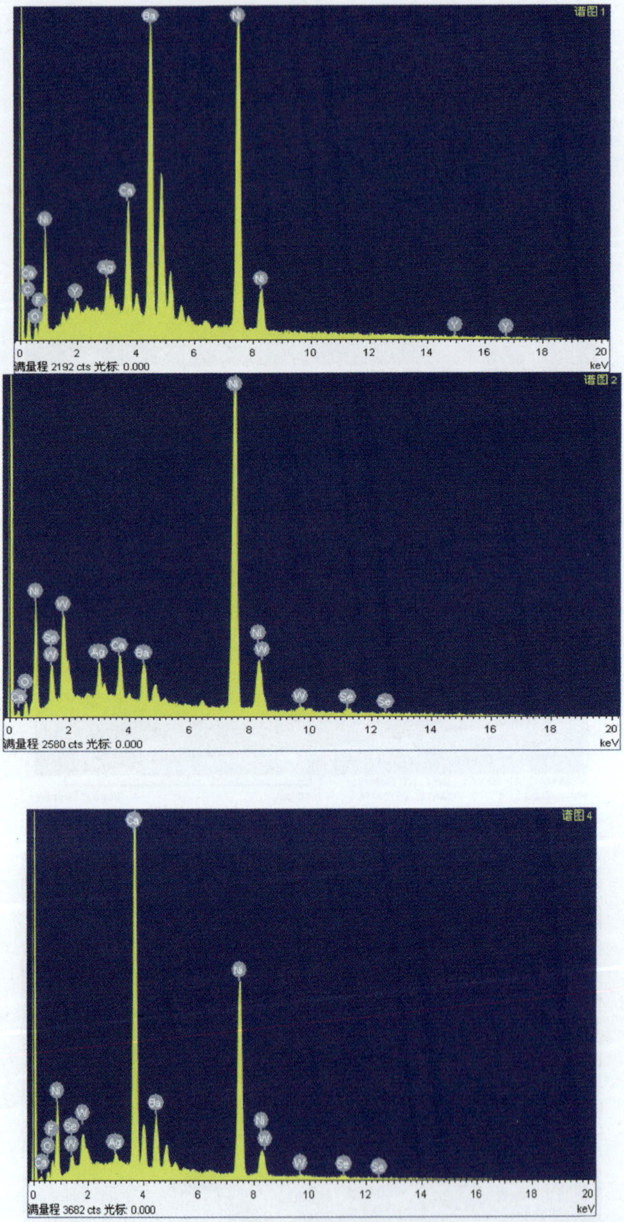
谱图 1
Ba
Ni
Ca
Ni
Ag
Ni
Ca
C
F
Y
O
Y
Y
0 2 4 6 8 10 12 14 16 18 20
满量程 2192 cts 光标: 0.000
keV
谱图 2
Ni
Ni
W
Se
W
Ag
Ca
Ba
Ni
W
O
Ca
W
Se
Se
0 2 4 6 8 10 12 14 16 18 20
满量程 2580 cts 光标: 0.000
keV
谱图 4
Ca
Ni
Ni
W
Se
F
O
W
Ag
Ba
Ni
W
Ca
W
Se
Se
0 2 4 6 8 10 12 14 16 18 20
满量程 3682 cts 光标: 0.000
keV

Fig. 4. (*Cont'd*)

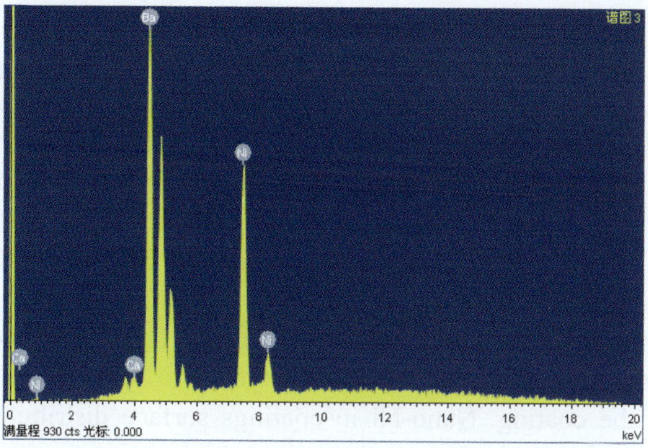

Diagram1	C	O	F	Ca	Ni	Y	Ag	Ba
Content1	18.96	19.89	6.05	4.40	34.87	0.58	1.21	14.03

Diagram2	O	Ca	Ni	Se	Ag	Ba	W
Content2	29.61	3.05	56.97	2.62	2.02	2.96	2.77

Diagram3	C	O	Ca	Cr	Ni	Y	Ag	Ba
Content3	7.63	5.64	1.17	1.08	73.78	7.04	0.49	3018

Diagram4	O	Ca	Ni	Se	Ba	W
Content4	26.83	2.04	66.32	1.01	1.94	1.87

Fig. 4. (*Cont'd*)

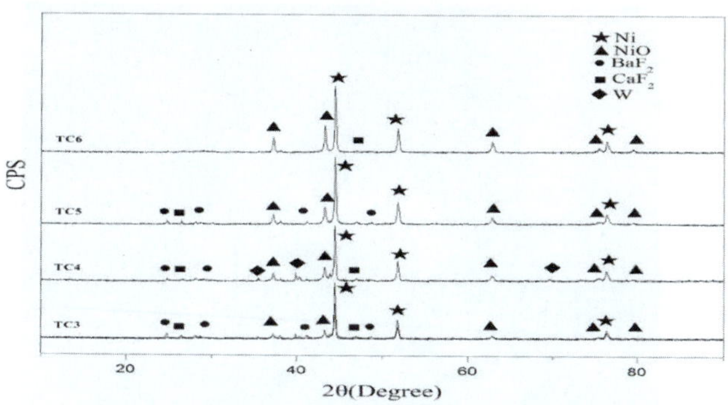

Fig. 5. XRD results of C3, C4, C5, and C6 at 500°C.

3.2. *Analysis of experimental results at 800°C temperature*

Figure 6 is C3, C4, C5 and C6 oxidized at 800°C, serious surface oxidation can be found and the difference between C and the WSe_2 coatings are getting smaller, from the perspective of energy spectrum on the oxygen content, four coatings of oxidation degree are similar. However, different coating composition and structure will lead to oxidation of differences in morphology. Energy spectrum analysis showed that the main elements of coating are Ni and O, and the levels of oxygen content are significantly higher than that of the original sample. Figure 7 is XRD analysis of coating surface after high temperature oxidation, which showed surface oxides are mainly NiO, the oxidation of other components is not obvious. NiO can effectively stop the combination of oxygen and other elements, in particular, can form a dense layer of surface oxide film, also it has some lubrication properties, this would to some extent to ensure the lubrication of the coating. Nano-Ni in coatings surface distribute equality and density, compacted oxide layers formed at high temperatures also greatly improves the oxidation resistance and lubricity of coatings, which improved the overall performance of the coatings.

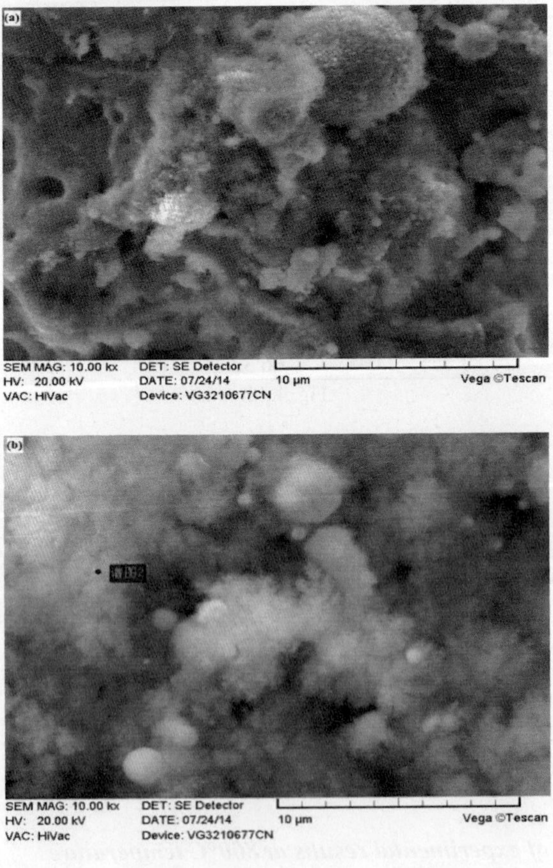

Fig. 6. Surface SEM image and the EDS analysis at 800°C: (a) C3 800°C; (b) C4 800°C; (c) C5 800°C; (d) C6 800°C.

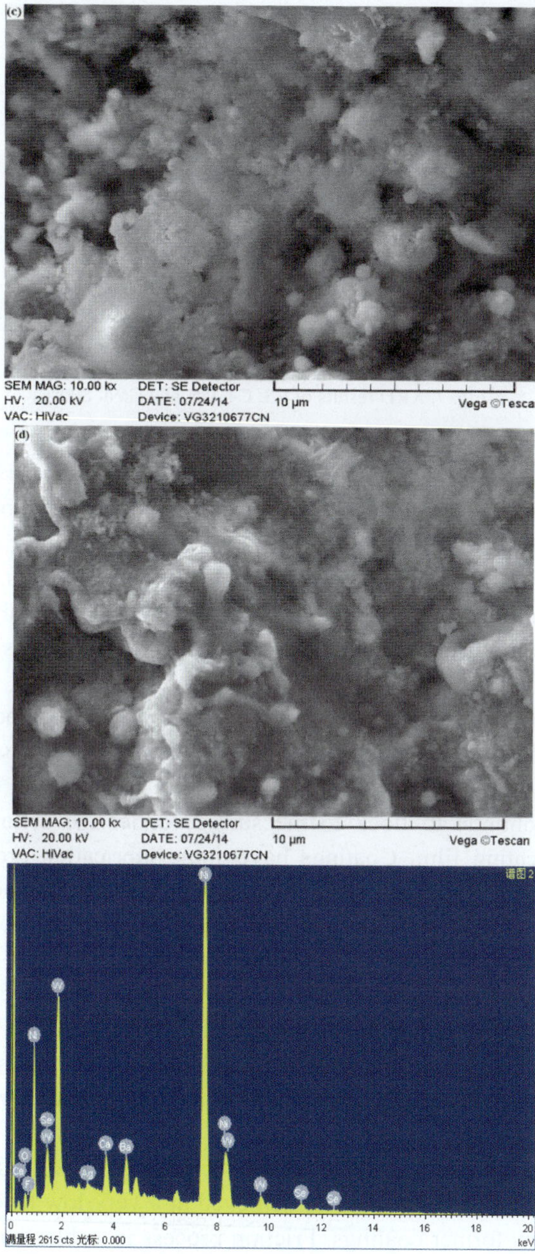

Element	O	F	Ca	Ni	Se	Ag	Ba	W
Content	36.28	4.22	2.16	47.11	1.70	0.24	2.51	5.78

Fig. 6. (*Cont'd*)

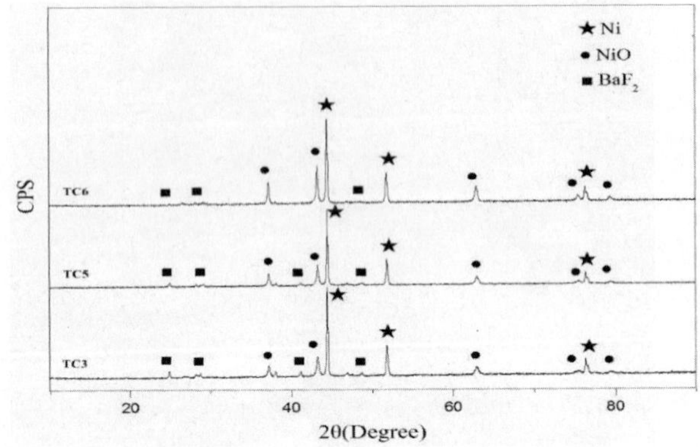

Fig. 7. XRD results of C3, C4, C5, and C6 at 800°C.

4. Research on high temperature lubricating mechanism

High temperature lubricating mechanism has been one focus of discussion and research, fiction system is an open system, a dynamic system. Even if there is no phase change, morphology of coated surface, coating thickness, specific state of contact is changing all the time. This requires concrete analysis of concrete conditions according to the actual situation. Brief discussion is divided into the following:

(1) Do not form a complete oxide films. less lubrication added at Coatings design time, friction without enough of the oxides in the surface to form a complete lubricating film, and lubricant equality distribute in the frame of the base, which continue to flock to the surface under the action of friction, to provides lubricating film. Coatings have some lubrication properties, and the friction coefficient is high. Coating in friction process is often accompanied by a certain degree of spalling adhesion and abrasion phenomena. This is a semi-lubrication coating series.

(2) Form a complete lubricating film in friction in the process goes through a complete failure to complete such a cyclical process, which is stable overall. And average friction coefficient is low, which will be peaks and troughs of alternation. Wear is mainly presented in adhesive and abrasion; this is a lubricating coating series.

(3) Coated surface has a full stable lubricating film. This friction state is best, and friction coefficient can be achieved at a very low level, wear figure is between the tow former coatings. Friction process is stable; surface has a dense oxide coating on it. The main forms of wearing are adhesive and oxidation wear. This is a full-lubricating coating series.

5. Summary

In this article, two environments at high temperature oxidation characteristics were studied in, mainly the following conclusion can be drawn:

(1) On the whole, for Ni-based coatings in high temperature environments, the main elements on surface is O, Ni, if more Cr elements contained, it will form a relatively complete set of Cr2O3 oxide film. And there exist greater affinity of Cr and O, oxide of Ni and Cr can prevent further reaction of O and the surface below, which have good protection.

(2) Degree of coating oxidation under 500°C is lower than 800°C, and hard to form larger oxide film or the thickness of oxide film is thin, and oxidized coating will produced peeling in friction process. The oxide film at 800°C is more dense and stability. NiCoCrAlY with different content of C1 and C2, No friction regions and friction regions oxidation are with similar characteristics. It turns a positive significance of the study on static and oxidation characteristics of coatings

(3) With comparative studies, similar behavior characteristics that oxidation of high temperatures and friction and the pure static are found. However, under the action of friction at high temperatures, coating oxidized more seriously.

(4) Existence state of coating of oxide film on the surface is not the same with the friction processing, which with no full of lubrication film, lubrication film alternating appeared, full of lubrication film and variety complex of situation. It depends on specific coatings and environments to make specific analysis.

Acknowledgement

The research was partially financially supported from the National Natural Science Foundation of China (NO.51405497).

Reference

1. Aouadi, M.S., Luster, B., Kohli, P., Muratore, C., Voevodin, A.A. 2009. Progress in the development of adaptive nitride-based coatings for high temperature tribological applications. *Surfaces & Coatings Technology* 204: 962–968.
2. Aouadi, S.M., Singh, D.P., Stone, D.S., Polychronopoulou, K.., Nahif, F., Rebholz, C., Muratore, C., & Voevodin, A.A. 2010. Adaptive VN/Ag nanocomposite coatings with lubricious behavior from 25 to 1000 °C. *Acta Materialia* 58: 5326–5331.
3. DellaCorte C. & Edmonds B.J. 2009. NASA PS400: A new high temperature solid lubricant coating for high temperature wear applications. NASA-TM-2009-215678.
4. DellaCorte C., Stanford M.K., Thomas F., & Edmonds B.J. 2010. The effect of composition on the surface finish of PS400: a new high temperature solid lubricant coating. NASA-TM-2010-216774.

5. Dellacorte, C. 2003. The effects of substrate material and thermal processing atmosphere on the strength of PS304: a high temperature solid lubricant coating. *Tribology Transactions* 46: 361–368.
6. Ding, C.H., Li, P.L., Ran, G., Zhou, J.N. 2008. PM304 coating on a Ni-based superalloy by PM technique. *Materials Science and Engineering A*, 483/484,755-758.
7. Muratore, C. & Voevodin, A.A. 2009. Chameleon coatings: adaptive surfaces to reduce friction and wear in extreme environments. *Annual Review of Materials Research* 39: 297–324.
8. Stone, D., Liu, J., Singh, D.P., Muratore, C., Voevodin, A.A., Mishra, S., Rebholz, C., Ge, Q., & Aouadi, S.M. 2010. Layered atomic structures of double oxides for low shear strength at high temperatures. *Screening Material*, 62: 735–738.
9. Voevodin, A.A. & Zabinski, J.S. 2000. Supertough wear-resistant coatings with 'chameleon' surface adaptation. *Thin Solid Films*, 370: 223–231.

Fireproof performance study of building exterior wall thermal insulation material

C.L. Huang
Chongqing Vocational College of Architecture Engineering, Chongqing, China
273686535@qq.com

ABSTRACT: As one of the effective means of building energy efficiency, building insulation to achieve "win-win" energy and environmental protection. Exterior insulation is currently promoting a building heat preservation and heat insulation technology. Exterior insulation materials can be divided into organic and inorganic heat preservation material 2 kinds. Different insulation materials have different advantages and defects. For different exterior insulation materials research and application status are expounded. In comparison with organic heat preservation material, inorganic non-combustible insulation materials will be the future environmental protection an important development direction of the fire insulation building materials, but the water resistance and strength remains to be further improved.

Keywords: Exterior wall; Energy saving; Heat preservation material.

1. Introduction

With the development of economy and society, the global demand for energy is also increasing. With increasing of total energy consumption, building energy consumption accounts for about 11% ~ 11% of total energy consumption [1]. Faced with the problem of energy increasingly nervous, building energy conservation problems cause the attention of more and more countries. Building energy efficiency is the reasonable use and efficient use of energy in the buildings, constantly improve the efficiency of energy utilization, reduce building energy consumption. For building energy saving, environmental protection, on the other hand, will also be a direct or indirect impact, use the appropriate heat preservation and heat insulation materials and building energy conservation methods can reduce 50% of CO_2 emissions [2-3]. China's current building energy efficiency level is far lower than developed countries, building energy consumption per unit area in our country is still a climate similar to 3 ~ 5 times that of the developed countries, therefore, building energy conservation is still an important subject of construction industry in our country, strengthen the research and development in the field of construction and application of new building energy conservation and environmental protection materials, is to improve the utilization rate of resources in our country, improve the environment, energy conservation and emissions reduction, the important way to sustainable development, has important practical significance.

2. A Building Energy-Saving Wall Insulation Materials and Technical Performance

2.1. *External wall insulation*

Within the exterior wall insulation construction is convenient, fast construction progress, inferior to the requirement of building external wall vertical degree. In recent years, also often use within the external wall thermal insulation on the project. However, the quality problems of exterior wall insulation is also very obvious. First, the existence of the structure of cold (hot) bridge make the local temperature difference is too big lead to condensation phenomenon. Winter indoor wall temperature and indoor corner (insulated wall and insulation board Angle) around the difference in temperature of 200 °C, and the indoor temperature difference is bigger, if indoor humidity conditions suitable for, in this part of the cold (hot) bridge formation, then the condensation phenomenon is formed here. And dew impregnated or freezing and thawing and easy cause thermal insulation metope mildewy, craze. Second, the walls and roof by the outdoor temperature and the effect of solar heat caused by the temperature change is larger. When indoor temperature higher than the outdoor temperature, the shrinkage of exterior wall thermal insulation system within the range faster than, when the indoor temperature is lower than the outdoor temperature, wall thermal insulation system is lower than the inflation rate, the deformation for internal thermal insulation system has been repeatedly in unstable wall, in the outer walls under the action of stress and deformation is not only easy to destroyed also easily cause the temperature stress in thermal insulation system of the empty drum craze.

2.2. *Both inside and outside heat preservation*

Inside and outside the hybrid heat preservation in the construction ways, hybrid heat preservation can improve the speed of construction, be able to interior wall, external wall insulation board with hot (cold) bridge part of the outer wall junction, so that the building in the middle of a heat preservation. But, suffer severe damage when a hybrid heat preservation building structure. Outside heat preservation practices place let indoor temperature influence the structure of the building wall body, the temperature change is relatively small, so the wall in a relatively stable temperature field, the temperature difference stress and deformation is relatively small; Insulation practice make outdoor environment temperature affected the structure of the building walls, outdoor temperature fluctuation is bigger, so the wall in a relatively stable temperature field, deformation stress produced by temperature difference is relatively large. Local insulation and local mixed exterior insulation heat preservation method, make the structure of the different parts of the outer wall body deformation produce different size and deformation speed, the structure is more unstable environment, for a number of temperature structure deformation cracks, it makes the service life of buildings.

2.3. *Exterior insulation*

Exterior insulation is the thermal insulation system on external wall of the lateral, so that the structure of main body temperature difference effect, a big drop in temperature deformation is reduced, the structure of wall protection and can effectively block the cold (hot) bridge, is conducive to longevity. So from benefits structural stability in terms of outside heat preservation and heat insulation have obvious advantages, in the case of alternative should be preferred outside heat preservation and heat insulation.

(1) The case outside heat preservation

External thermal insulation material of rock (mine) cotton, glass cotton blanket, polystyrene foam board (hereinafter referred to as polystyrene board, EPS, XPS), ceramsite concrete composite polystyrene imitation stone decorative insulation board, wire rack sandwich wallboard, etc. With polystyrene board because it has excellent physical performance and cheap cost, has in the world within the scope of exterior wall thermal insulation technology is widely used in plugin. This plugin technology is to use adhesive mortar or special fixed attach, outer wall heat preservation material, and then wipe anti-crack mortar, pressed into glass fiber grid cloth to form a protective layer, the writing as a decorative surface. Still have a kind of practice is to use special fixed pieces will not easy bibulous of various insulation board fixed outside the wall, then put the aluminum sheet, natural stone, color glass such as plugins in advance to make keel, decorated form directly. Designed by Mr Pei's bank of China head office building of external thermal insulation is used in this design. The case of external thermal insulation installation time consuming, difficult construction, and construction occupies leading period, after waiting for the acceptance of the can to the construction. In the high-rise construction, construction personnel safety guaranteed.

(2) A polystyrene board and wall body casting molding

The technology is in the heart of the concrete frame - shear system of polystyrene in building templates, in the pouring of the wall of the lateral, then pouring concrete, concrete and polystyrene pouring shape for the composite wall at a time. The technology solved the main problems of the external thermal insulation, the advantage is obvious. Because exterior wall body and thermal insulation layer a survival, improve work efficiency, greatly shorten the time limit, and got to ensure the safety of construction personnel. And in the winter construction, polystyrene ACTS as insulation, retaining heat preservation measures can reduce the periphery. But should pay attention to when concreting uniform, continuous casting, or because of the effect of the lateral pressure of concrete will cause of polystyrene in the deformation after dismantle faults and fault, after construction.

(3) Polystyrene particles insulation pulp outer wall heat preservation

Will waste polystyrene plastics (referred to as EPS) processing and broken into 0.5 ~ 4 mm particles, as light aggregate to make thermal insulation mortar. The technology contains the thermal insulation layer and protective layer and crack impermeability protection layer (layer or surface layer impervious crack epoxy resin mortar). The ZL adhesive powder polystyrene particles insulation materials and technology in 1998 by the ministry of construction as a state-level method. This method is the exterior wall thermal insulation technology has been widely recognized.

3. The Energy Saving Wall Exterior Insulation Fire Protection Requirements

3.1. *Outer wall heat preservation material analysis*

Wall of heat preservation and heat insulation layer, there is no ordinary polystyrene flame retardant processing and polyurethane foaming material is a kind of flammable materials, flame retardant polystyrene and pu foam material can achieve combustible or flame retardant grade, and the thermal insulation layer of plaster mortar is commonly incombustible materials [4]. So, thermal insulation layer of plaster mortar.

Quality and the structure of the thermal insulation system is the key of the external thermal insulation system fire resistance and safety. Thermal insulation layer of plaster mortar insulation system of excellent quality and reasonable structure is to ensure that resistance to thermal insulation system fire safety factor. And the development of the global thermal insulation material is in a green pollution-free, waterproof, fire prevention, low fares and friendly in the direction of progress. In the development of new type thermal insulation system at the same time, more emphasis on selection of heat preservation and heat insulation material, improve the efficiency of heat preservation and heat insulation system. In building with good thermal insulation materials and technology, can achieve good effect. In addition to considering the combustion performance of material, water absorption and strength, and integrated standards such as the cost of production. Under the condition of current technology, cannot demand too much polystyrene and flame retardancy of rigid foam, as long as the combustion performance can meet the fire safety requirements.

Over the past few decades, the rapid development of the emergence of a large number of new products, from product structure singlization to steadily improve product quality, has formed with glass wool, rock wool, expanded vitrified beads insulation paste mainly inorganic heat preservation material, give priority to with powder polystyrene particles insulation mortar of organic inorganic composite insulation materials, organic thermal insulation material is given priority to with polystyrene foam three categories of relatively complete variety of industries.

3.2. *Exterior wall thermal insulation system of fire safety requirements*

(1) Exterior wall thermal insulation system fire prevention basic requirements

In the rapid development of China's urbanization process, under the background of social energy is increasingly urgent, all new and existing buildings should conform to the corresponding standards of energy conservation and emissions reduction. Heat preservation system in fire safety hidden trouble using organic heat preservation material, need to take some safety measures. Poor resistance to combustion performance of polystyrene board thin plastering system is the dominant form of external thermal insulation taken in our country, enterprise external thermal insulation system of both domestic and foreign production enterprise external thermal insulation system more influential for fire safety hidden trouble most of the products is evasive attitude, about building the system of external thermal insulation system of fire prevention and standards remains to be perfect. Caused by indoor fire expanding factors to cause the fire, the economic loss and casualties caused by fish is very heavy, exterior what measures the temperature increase and improve outsourcing system of fire safety performance. Must be according to the needs of building energy conservation vigorously develop suspension building external wall of the bearing structure, good fire resistance for the building fire safety bring vital role [5].

To have two aspects of external thermal insulation system has good fire resistance: first, in the part of the insulation system is ignited, system itself can block the spread of flame. Secondly, in the case of a fire exists, heat preservation system whether there is burned.

(2) The particularity of exterior insulation system of fire prevention requirements in our country

With the speeding up of urbanization in our country, the urban land resources become increasingly scarce, most large and medium-sized cities to build more and more high-rise buildings, the higher the longer, it seems to have become a common trend of development [6]. Because our country population density is big, the urban residential most of high-rise buildings, fire danger coefficient is big, fire equipment, fire fighting and rescue is difficult when the fire broke out. The main body in the present situation of external thermal insulation system is the vast majority of heat preservation and heat insulation material is combustible organic material, fire resistance poor polystyrene board thin plastering system is the main form of external thermal insulation system, fire safety situation is very outstanding. High-rise buildings in case of fire, the structure of the staircase various pipelines vertical channel, tens of thousands of workers is not easy to safe evacuation. Add in fire environment people frighten panic psychology, almost not guaranteed safe evacuation. Fire control facilities management system is not perfect, the staff fire safety consciousness, and lack of necessary defense save your common sense and so on, these are the high-rise building fire safety hidden danger, though.

New research and development of external thermal insulation systems, therefore, must conform to our country special national condition and construction features, on the basis of meet the requirements of building thermal insulation performance, satisfies the requirement of building fire safety performance must also be as the first condition. According to the characteristic and the basic national conditions of our country urban construction, external thermal insulation field of fire safety requirements should be more strict, especially to external thermal insulation system of fire safety in the relevant technical standards and specifications reflect higher requirements.

4. Fire Prevention Countermeasures, Energy-Saving Wall Exterior Insulation Materials

4.1. *Strictly outer wall heat preservation material fire safety management*

To strengthen the management, according to the existing and thermal insulation engineering construction fire safety management of fire protection regulations and other relevant rules and regulations, combined with the actual situation, thermal insulation engineering construction safety management measures are put forward, and carefully carried out [7]. To correspond to the structure of external thermal insulation system of the reasonable technical method, such as considering the combustible insulation board before entering the construction site fire-proof interface agent, etc, in order to satisfy the requirement of the construction process of fire safety.

4.2. *Strict construction engineering construction site management*

To strictly implement the provincial construction department issued the construction engineering construction site fire safety standards, strengthening the fire safety management of construction engineering construction site [8]. Especially for the use of open flame must be strict fire safety management. Need, the open flame, hot departments and personnel shall, in accordance with the use of fire management system for examination and approval formalities, to carry out the on-site guardian, after the confirmation no fire and explosion danger when hot construction. When hot construction personnel shall abide by the fire safety regulation, and to carry out the corresponding fire control safety measures, especially closely welding special welding, such as high environment protection requirement. Electric welding, gas welding, electricians and other special type of work staff must hold relevant certificates. Will be easy to fire, severe consequences when a fire breaks out areas identified as key fire prevention area, strict management.

4.3. *Walls set up the common fire protection construction measures*

Set fire belt [9]: level in the construction of the firebreak between layer and layer, generally USES non-combustible or flame retardant insulation material, system is not space between any of the material, has a certain width. Fire damper beam: like fire belt, level set on the edge of the doors and Windows on the mouth of the cave, and the length of the vertical edge of the hole out of doors and Windows must be. Metal fasteners: using thermoplastic insulation system, in order to maintain a fire under the condition of the stability of system, security system is not flame propagation resistance, can consider to use metal firmware.

4.4. *For fire safety training, to enhance staff safety awareness*

Building project construction units and construction units shall establish a sound system of fire control safety education training, before the construction of construction personnel conduct fire control safety education; In the eye-catching place of the construction site, construction personnel focus on accommodation places fire safety publicity column, hanging fire safety wall charts and fire safety warning signs; Regular fire flame operation personnel safety education; Organize fire fighting and emergency evacuation drill, etc.

5. Conclusions

The fire protection problems in recent years, thermal insulation material is widely attention, especially for organic heat preservation material, such as insulation board in the event of fire, the fire will spread rapidly and release harmful gases, bring to people's physical and economic losses. So, had better choose good flame retardant insulation material. In comparison with organic heat preservation material, inorganic non-combustible insulation materials will be the future environmental protection an important development direction of the fire insulation building materials. No matter what kind of exterior wall thermal insulation material has its advantages and disadvantages, only adjust measures to local conditions, reasonable selection of insulation materials, to a greater degree of insulation of exterior wall thermal insulation system to play the performance. Such not only conducive to relieve tension, excessive to reduce greenhouse gas emissions, more conducive to the sustainable development of social economy.

References

1. Yin Xiang, 2010. The building exterior wall thermal insulation fire prevention management, Garden Project Department of Hefei Binhu Hin.
2. Xian Chen Bai, 2011. "Civil Engineering Materials" (second edition), China Building Industry Press.
3. Yang Shuqing editor, 2010. The construction technology, higher education press.
4. Chang-you song. 2011. Exterior insulation system of fire safety analysis [J]. Building Science.
5. Wang Gang, 2011.Combustion performance experiment of organic heat preservation material . Building Science, 22-25 October
6. DB51 / T 5061-2008, 2008. Cement base composite expansion vitrified microsphere, technical code for building insulation system .

7. DB34 / T 1279-2010, 2010. Inorganic insulation mortar wall thermal insulation system .
8. JGJ 144-2004, 2004. The external wall thermal insulation engineering discipline .
9. Wu Bing, 2005. Rich well for a long time. The current situation and future of the ready-mixed mortar: 61-63.

Research on membrane material selection applied for inflatable structure

Z.S. Yang and H.B. Zhang
No.5, East Road, Fengtai District,
Beijing, China
clytze9064@163.com,245601363@qq.com

Inflatable membrane material consists of bearing layer, airtight layer and protective layer. This paper will respectively make the choice of each material layer from the aspects such as strength, impermeability and price. And then, based on software of finite element analysis ABAQUS, it is validated that choice materials (PVDF double-faced coating forming) satisfies the strength requirement of the inflatable membrane structure.

Keywords: Inflatable membrane structure; Membrane material; PVDF; FEM.

1. Introduction

Compared with traditional hard structure, inflatable membrane structure has the advantages of shorter construction cycle, lower construction cost, recyclable and moving quickly, which make it prosperous with every passing day in the world. Kinds of membrane structures has been built in many countries such as elegant and chic large-span stadiums, exhibition halls, airport and so on. Our country is at the top of membrane structure growth.

1.1. *Bearing layer*

Inflatable membrane material has the advantages of high strength, high airtight, lightweight and durable, which is composed of bearing layer, airtight layer and protective layer. The development of inflatable membrane material is also the development of bearing layer, air-tight layer and protective layer and their combination technology. The bearing layer almost support the all strength of inflatable structure. The structure of fabric commonly is plain cloth. The thickness of yarn, twists and arrangement desity in the fabric differs due to airtight layer, and changes with people’s understanding deepening on inflatable structure.

1.2. *Airtight layer*

Before the 1960s, the airtight film was formed by painting natural rubber, annimal glue, wax, nitrate and acetate esters alternately on fabric. With the development of material industry, neoprene, butyl rubber and other synthetic materials had gradually taken the place of previous natural materials to coat the cloth for air barrier, which greatly improves the air tightness. After 1970s, thin films which are coated on the cloth was used as air-tight layer, and these films always are polyester, polyvinylidene chloride, polyethylene, etc. Sometimes, a layer of aluminum was coated on the film, aluminum can further improve the effect of air tightness. Usually high molecular weight materials are stretched by two-way and then can form the film with high crystallinity and high modulus. So it not only is airtight, but also can bear shear stress.

1.3. *Protective layer*

When the protective layer, which also has the effect of airtight was added to membrane materials, the inflatable structure has a significantly longer service life. The protective layer always are PU coating, Hypalon and PVF film coating. The protective layer is put a certain amount of filler such as aluminum, titanium dioxide, etc., which are not only decorative, but also can reflect heat.

When thin film is coated on the cloth, it needs an excellent adhesive to make the cloth and the film as a whole. The adhesive always is polyester or polyurethan.

2. Selection of Inflatable Membrane Material

2.2. *Bearing layer*

Originally natural materials such as hemp and cotton are generally used for inflatable membrane structure as base fabric. Later the base cloth developes to nylon, polyester and other synthetic fabric. These materials can greatly improve the tear strength when ensuring tensile strength. In recent years substrate material gradually always use Kevlar, Vectran such high performance fiber fabric abroad.

Table 1. Contrast of 4 kinds of fabric performance

Name	Nylon (400D)	Polyester (400D)	Kevlar (400D)	Vectran
Thickness/(mm)	0.14~0.15	0.17~0.18	0.18~0.19	0.10~0.12
Areal density/(g/m^2)	90	160	144	52
Tensile Breaking strength /(N/5cm)	499	2895	3128	2365

The base fabric in this artical is applied for inflatable structure, which has the properties of building and tearing down quickly, and removing conveniently. So the membrane material must have a certain strength, and be wear-resisting, easy to fold, easy to process. From Table 1, the strength of nylon (400D) is too low. Vectran has a high strength, but its disadvantages of small thickness, low wear resistance and small elasticity are easy to lead to stress concentration. Kevlar also has a high strength, but a low welding peel strength. On the other hand, Kevlar and Vectran much more expensive than Polyester. After considering all, base material,that is bearing layer, Polyester was chosen.

2.3. *Air-tight layer*

The common airtight layer is PVC, PVF, PVDF, PU, TPU or ETFE. PVC is a clean, colorless, transparent film, which has advantages of good grease resistance, air tightness, light resistance, aging resistance and tear resistance. The main characteristics of PVF film are high wear resistance, moisture penetration resistance and extension ability and excellent anti-aging performance. PVDF film has high mechanical strength and toughness, mildew resistance, high wear resistance, high resistance to permeability of gas and liquid. The most prominent feature of the PU is excellent wear resistance, high hardness, high strength, good elasticity and resistance to low temperature. TPU (thermoplastic polyurethane elastomer) is an emerging plastic variety, which has good process ability, weather ability and environmental protection. ETFE membrane material, which has good durability, easy cleaning property and stability, is a kind of typical fabric type of membrane material, and is currently the most advanced film materials. Considering comprehensively the factors such as the film application occasion (inflatable membrane structure), strength, price, and the processing cycle, we choose PVDF film.

2.4. *Protective layer*

In addition, according to need, we can take special coating treatment for base material, such as adding fireproofing coating, nano antibacterial mouldproof coatings, low temperature resistance and self-cleaning coating on the base material surface. These coating makes the inflatable structure material has high strength, good sealing, fire prevention, mildew proof, weather resistance, self-cleaning and other advantages.

In conclusion, PVDF double-sided coating network material is the most suitable material. Its bearing layer is Polyester, airtight layer is PVDF. Its molecular structure is close and aperture is small, thus gas molecules is not easily get through. So it is very airtight. The internal network structure as the bearing layer can significantly increase inflatable structure bearing capacity. PVDF double-sided coating network material can be processed by high-frequency heat sealing method. Processing technic is simple. It is easy to control quality. So, it is appropriate to be inflatable membrane material.

3. Nonlinear Finite Element Balance Equation

Due to the material of membrane structure is the flexible material, it has no bending and compressive capacity, and shear capacity is also very poor. It keeps the shape completely by applying pretension to the stucture. So when it is under external loads, it only can rely on the change of membrane surface curvature, and cause internal force redistribution to resist the load which is perpendicular to the surface of the membrane. This process has the obvious characteristic of large displacement geometric nonlinearity, and the strain of membrane structure is usually small. Therefore, nonlinear analysis of membrane structure belongs to the large displacement and small strain analysis. Geometric nonlinear problem usually adopt U.L. column virtual work equation, namely, constantly updating the reference configuration. All the statics and kinematics variables always refer to each load or the configuration at the start of time step, to establish the nonlinear finite element equations of the structure.

Using space three-node triangular membrane element, considering the displacement u,v,w and in-plane stress σ_x, σ_y, τ_{xy} of nodes in x,y,z direction, and taking structure shape at time t as a reference configuration, membrane structure shape and displacement increment at t moment can be described by the following equations:

$$\{\Delta x\} = [N]\{X_e\} \tag{1}$$

$$\{\Delta u\} = [N]\{\Delta u_e\} \tag{2}$$

In formula (1) and (2), $[N]$—The shape function matrix of membrane unit; $\{\Delta x\}$, $\{\Delta u\}$—Point coordinates and displacement increment vector in membrane element; $\{X_e\}$, $\{\Delta u_e\}$—Node coordinates and displacement increment vector in membrane element.

Considering it is the geometric nonlinearity problem of large displacement, the quadratic term of strain and displacement relationship should be considered, namely,

$$
\begin{aligned}
{}^{t}\{\Delta\varepsilon\} &= {}^{t}\{\Delta\varepsilon_L\} + {}^{t}\{\Delta\varepsilon_{NL}\} \\
&= \left[{}^{t}_{t}[B_L] + {}^{t}_{t}[B_{NL}]\right]\{\Delta u_e\} = {}^{t}_{t}[B]\{\Delta u_e\}
\end{aligned} \tag{3}
$$

In formula (3),

${}^{t}\{\Delta\varepsilon_L\}$—Linear strain increment vector

${}^{t}\{\Delta\varepsilon_{NL}\}$—Nonlinear strain increment vector

${}^{t}_{t}[B]$—Geometry matrix

${}^{t}_{t}[B_L]$—Linear geometric matrix

${}^{t}_{t}[B_{NL}]$—Nonlinear geometric matrix,is the function of $\{\Delta u_e\}$.

Taking the above linear geometric matrix ${}^{t}_{t}[B_L]$ and nonlinear geometric matrix ${}^{t}_{t}[B_{NL}]$ into U.L. column virtual work equation, we can get the nonlinear finite element equilibrium equation of membrane element, that is

$$
[{}^{t}_{t}[K_L]_e + {}^{t}_{t}[K_{NL}]_e]\{\Delta u_e\} = {}^{t+\Delta t}\{R\}_e - {}^{t}_{t}\{F\}_e \tag{4}
$$

In formula (4), ${}^{t}_{t}[K_L]_e$, ${}^{t}_{t}[K_{NL}]_e$—Linear stiffness matrix and nonlinear stiffness matrix of the element;

${}^{t+\Delta t}\{R\}_e$, ${}^{t}_{t}\{F\}_e$— Column vector of element load and element stress equivalent nodal force vector at time t.

The specific calculation formulas are:

$$
{}^{t}_{t}[K_L]_e = \int_{tv} [B_L]^T [D]^t_t [B_L]^{Tt}\, dV \tag{5}
$$

$$
{}^{t}_{t}[K_{NL}]_e = \int_{tv} [B_{NL}]^T [\sigma]^t_t [B_{NL}]^{Tt}\, dV \tag{6}
$$

$$
{}^{t}_{t}[F]_e = \int_{tv} [B_L]^T \{\sigma\}^t\, dV \tag{7}
$$

In formula (5), (6), (7), $[D]^t_t$—The material constitutive matrix; $[\sigma]^t_t$—Cauchy stress matrix; $\{\sigma\}^t$—The cauchy stress vector.

The above equations are derived in the local coordinate system of the element. Transforming the local coordinate system into the global coordinate system and assembling, we can obtain the finite element balance equation of membrane structure under the global coordinate system:

$$ {}^{t}_{t}[K]\{\Delta u\} = {}^{t+\Delta t}\{R\} - {}^{t}_{t}[F] \tag{8} $$

In formula (8), ${}^{t}_{t}[K], \{\Delta u\}, {}^{t+\Delta t}\{R\}, {}^{t}_{t}[F]$ respectively are total stiffness matrix, the total displacement vector, the general node load vector and total nodal force vector which the initial stress is equivalent to under structural global coordinates.

4. Numerical Example

ABAQUS has excellent ability to solve nonlinear problems. This calculation using ABAQUS/Explicit module solved the problem of nonlinear deformation of inflatable structure by explicitly direct integral.

Element type is M3D4R under ABAQUS/Explicit module, that is 4 nodes decrement integral membrane element, which has both the calculation accuracy and computational efficiency.

For the inflatable membrane structure which span is 38 m, the top height is 13 m, diameter of gas rib is 2.3 m, we choose 0.7 mm PVDF double-sided coating network material, which gram weight is 832 g/m2, breaking strength of 520 N/cm. On the right side of the structure, wind load is 10 level.

The deformation through FEM analysis is shown in figure 1.

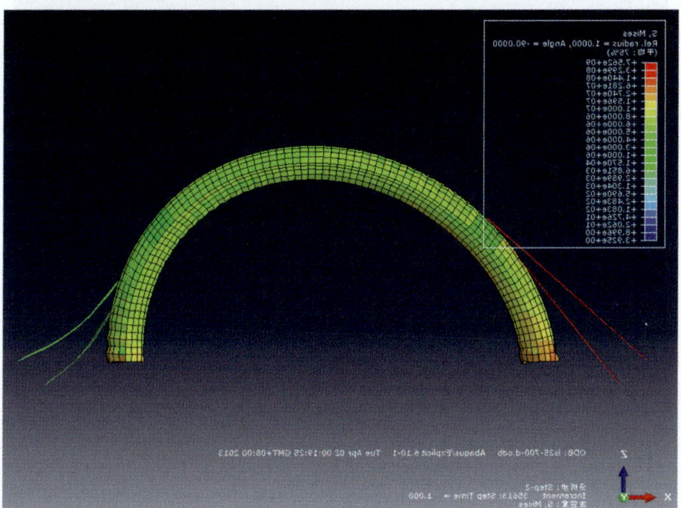

Fig. 1. Deformation analysis results.

Skin stress nephogram as shown in figure 2, maximum stress of the main structure appears in the central position of the lining. The maximum stress is 14Mpa, and membrane materials tensile strength is 65Mpa, so the membrane material under the condition of safety factor of 4.6. Safety coefficient greater than 4 is required for membrane structure.

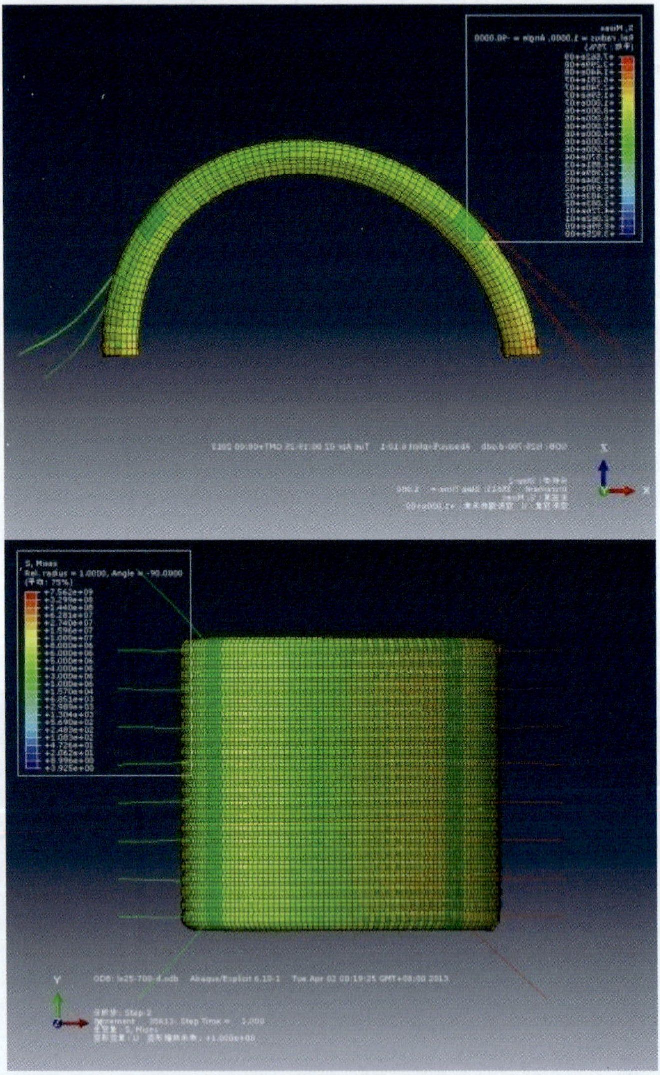

Fig. 2. Stress nephogram.

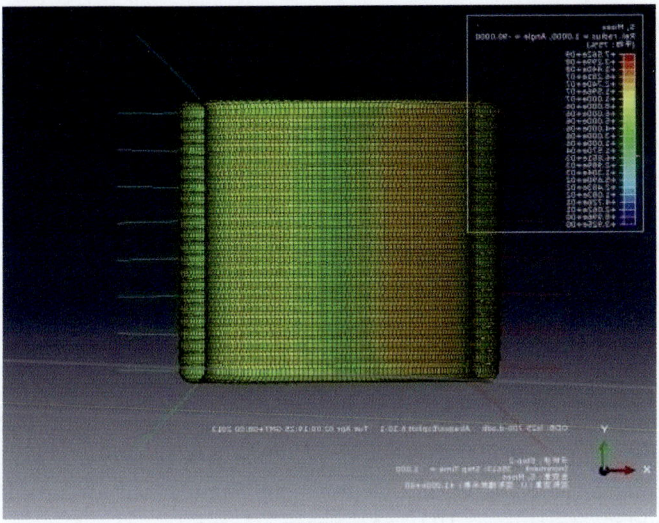

Fig. 2. (*Cont'd*)

From the above finite element analysis, structure displacement and stability satisfies the requirement under the extreme operation condition of inflatable membrane structure, maximum stress is also within the scope of the job requirement, safety coefficient reached the standard of 4 above, therefore the identified material meets the design requirements. To the charging structure of different span, We can choose PVDF double-sided coating network material with different thickness.

5. Conclusion

This article through the analysis and numerical calculation for the inflatable membrane material, and the following conclusions can be obtained:

(1) PVDF double-sided coating network material is very airtight. The internal network structure as the bearing layer can significantly increase inflatable structure bearing capacity. PVDF double-sided coating network material can be processed by high-frequency heat sealing method. Processing technic is simple. It is easy to control quality. So, it is appropriate to be inflatable membrane material.

(2) Through numerical computation, structure displacement and stability satisfies the requirement under the extreme operation condition of inflatable membrane structure, maximum stress is also within the scope of the job requirement, safety coefficient reached the standard of 4 above, therefore the identified material meets the design requirements.

Reference

1. Lin Wang, Yuehua Chen, Wei Li. The development of airship envelop materials [J]. Textile Products, 2008(10):1-4.
2. Zhijun He, Jiemin Ding. Membrane structure initial form design based on nonlinear finite element method [J]. Sichuan scientific research on construction, 2004(30): 15-17.
3. Xiangyang Xiong, Zhenhua Qi. The Influence of Lateral Load Patterns on the Results of Static Pushover Analysis [J]. Building Science, 2001(17): 8-13.
4. Jiaru Qian, Wenbin Luo. Static pushover analysis [J]. Building structure, 2000; (6): 23-26.
5. M Salama, C Kuo, M Lou. Simulation of deployment dynamics of inflatable structures [J]. AIAA Journal, 2000, 38(2):2277-2283.
6. J T Wang, A R Johnson. Deployment Simulation of ultra-lightweight Inflatable structures [R]. American Institute of Aeronautics and Astronautics, AIAA 02-1261, 2002.

Chapter 7

Engineering Management

Discussion on several reliability management problems of inertial confinement fusion facilities in operation

L. Zhang[†], J. L. Zhou , J. Fan and C. Y. Duan
College of Information System and Management,
National University of Defense Technology, Changsha, China
[†] *Email: zldyxdz@163.com*

For the complexity and high reliability demand of inertial confinement fusion facilities, considering the operational situations in main countries, the article discussed several common problems that are important to the reliability management of inertial confinement fusion facilities in operation, which has good reference value to reliability maintenance and reliability growing of inertial confinement fusion facilities.

Keywords: Inertial confinement fusion facilities; Reliability; Reliability management.

1. Introduction

Inertial confinement fusion facility (ICF) is a tool for human beings to explore the mysteries of the universe, aims to make lasers focused to achieve a high level of intensity, and to simulate the temperature, density and pressure of nuclear explosion generated by the state, ultimately create a high level scientific experiment platform with extreme physical conditions for physical experiments. It is a experiment tool for the strategic high technology innovation and basic science, cross frontier science innovation. At present, the NIF device (as shown in Figure 1. [1], builded by USA) and the LMJ device (builded by France) are typical inertial confinement fusion facilities in the world, the SG devices is also a representative of China in this field.

Reliability refers to the ability for a product to complete the required functions in the stipulated conditions and the stipulated time [2]. Reliability management is to take some measures to meet the reliability requirement of the product, whose essence is to do feedback control and run a management system to ensure the reliability target of the product [3]. Inertial confinement fusion facility is a kind of giant system with extremely complicated integrations of optical, mechanical and electrical components, has a very high reliability requirement.

Reliability is not only designed, but also depended on management. Be a huge system engineering of the major countries, the inherent reliability of inertial confinement fusion facilities have been implemented in the periods of design, demonstration, manufacturing and debugging. However, the operational reliability is the the key indicator to measure the effect that inertial confinement fusion facilities meet the expected reliability, the expected goal and task. As the important content of engineering management of equipment system, reliability management of inertial confinement fusion facilities in operation is the essential way to maintain the established reliability and realize the reliability growing. The following contents discussed several typical reliability management problems of inertial confinement fusion facilities in operation.

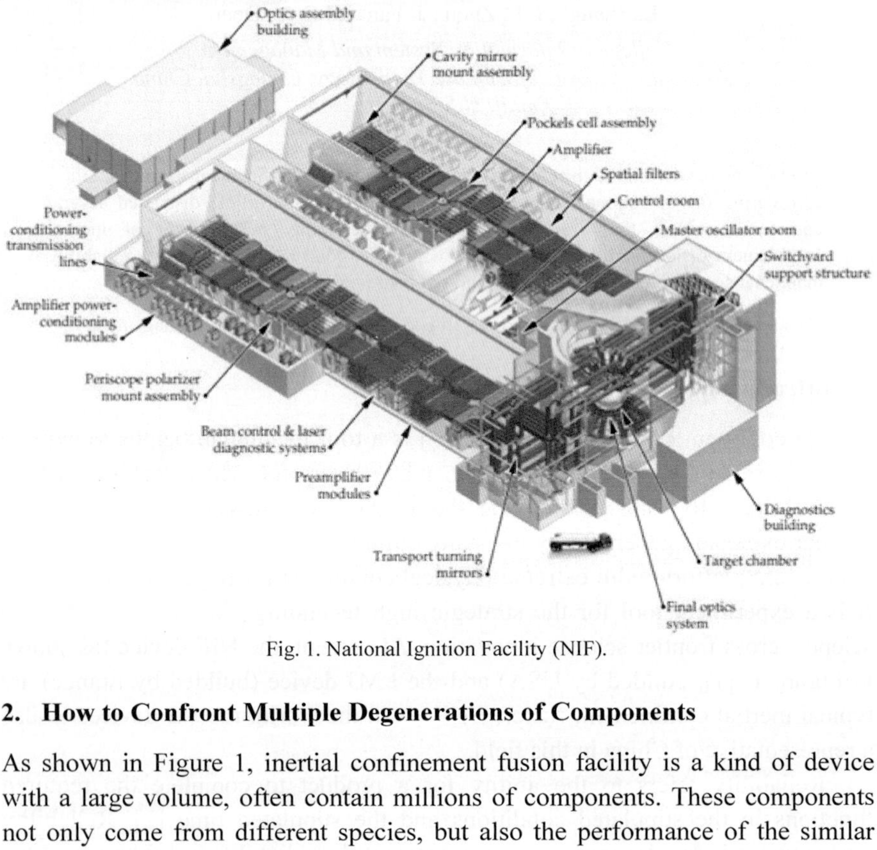

Fig. 1. National Ignition Facility (NIF).

2. How to Confront Multiple Degenerations of Components

As shown in Figure 1, inertial confinement fusion facility is a kind of device with a large volume, often contain millions of components. These components not only come from different species, but also the performance of the similar components are very different. According to the bathtub curve (as shown in Figure 2), with the continuous running of inertial confinement fusion facility, most components will experience three periods of failure, namely: early failure period, incidental failure period and Wear-out failure period. The characteristics of the three periods of failure are respectively as follows:

(1) Early failure period: high failure rate, but drops rapidly with time going by;
(2) Incidental failure period: failure rate is low and stable;
(3) Wear-out failure period: failure rate rises rapidly.

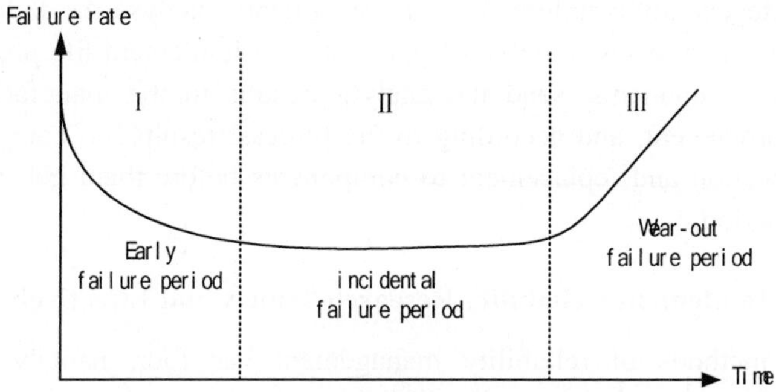

Fig. 2. The bathtub curve.

However, the lengths of the failure periods of different components are not the same, often have very obvious differences in characteristics, which makes it impossible to implement centralized maintenance measures to numerous components at several confirmed times. To do maintenance of components in numerous times will cost a lot of resources, and also greatly limit the operational reliability of inertial confinement fusion facility. How to reduce the differences between components' degenerations according to the task is an important part of the reliability management of inertial confinement fusion facility in operation.

Traditional way of replacement after failures is difficult to solve the problem of the different degenerations of inertial confinement fusion facility' components. Because in the early failure period, a huge amount of components will show unpredictable failure phenomenons, which are difficult to realize in the time domain, so replacement after failures can't achieve the purpose of the preliminary control, also it will be difficult to ensure the operational reliability and the reliable life of inertial confinement fusion facility. The present running situation on inertial confinement fusion facilities indicate that, in the installation and commissioning period, many components showed failures. The failure times of different components are not identical, which suggests that the early failure period of them have different characteristics. With the running of inertial confinement fusion facility, slowly the components will go through the incidental failure period, and then get into the Wear-out failure period, in which the differences on time and scope will be more bigger, so it will be very difficult to control and manage if not well prepared for the work ahead of time.

In the face of such situation, there are two suggestions for the period of early operation: Firstly, implement component procurement procedures and standards according to the outline of the early period of the reliability and reliability plan strictly, do assessment and selection to the low reliable components, shorten the time of early failure period and lengthen the time of Wear-out failure period, so as to reduce the reliability differences between different types of components in the process of operation. Secondly, make full use of the degeneration failure data of components, analyse the failure mode, failure mechanism of low reliable components and implement life prediction of low reliable components, send the analytic results to the manufacturers for process improvement, and according to the forecast results, do comprehensive quality inspection and replacement to components before them get into Wear-out failure period.

3. How to Implement Reliability Research Timely and Effectively

The work methods of reliability management has four, namely planning, organizing, superving and controling [3]. As an important part of the whole life cycle of inertial confinement fusion facilities, the most important means for reliability management in the running stage is to implement controling, and the timely and effective research work is the forerunner for the decision of controling. Generally speaking, the sooner the reliability research work carried out, the more resources can be saved. Figure 3 shows the relationship between the time points for reliability research and the total resource cost that saved.

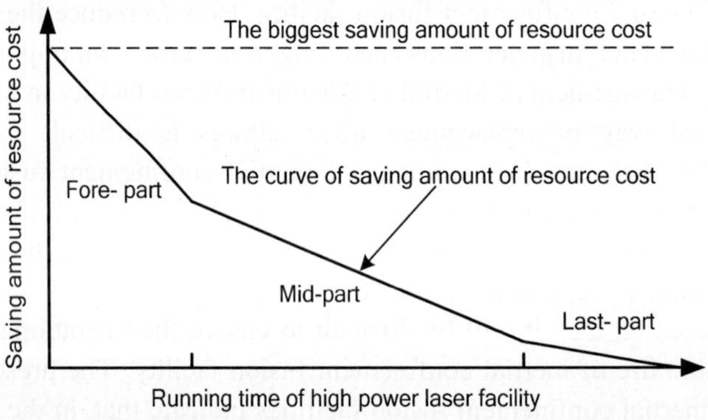

Fig. 3. The relationship between the time points for reliability research and the total resource cost that saved.

To do reliability research work timely and effectively need reliability data which must be real and effective. The data got during the running time are very precious, including the actual external environment of the inertial confinement fusion facility, operational rules and process, failure data, etc. Among them, the early operation failure data is crucial, which contains the most valuable information for reliability research, need to be collected by heart. At present inertial confinement fusion facilities of the main countries have their own computer centralized control decision system, which can implement command and control, fault diagnosis and data collection, also other functions [4]. However, in order to achieve the timely and effective reliability research, it is necessary to put the elements of reliability research into the system design as complete as possible, namely put the automatic collecting and processing of data into system function. Realized the automatic collecting and processing of data, then it will be easy to do real-time monitoring and reliability research of inertial confinement fusion facility, which makes the reliability management become more timely.

In addition, organizational support and adequate funding is the key to implement reliability research work timely and effectively. The United States set up many organizations for reliability management, such as reliability analysis center (RAC),roma laboratory professional bodies, and the research fund for many projects exceed 10% of the total funds [5]. These security measures laid a good foundation for reliability research work, which is worthy for reference, the reliability management of inertial confinement fusion facility in running also need the support of these conditions. On the one hand, there should be a work team for reliability management (the chief engineer should be the team leader) and reliability management tasks should be reasonably designed into the organizational structure of inertial confinement fusion facility; on the other hand, set aside enough money to support reliability management work separately is necessary.

4. How to Ensure the Reliability of Staff

From a macro perspective, inertial confinement fusion facility is a comprehensive system consisting of hardware, software, and staff, its reliability management object is not limited to the content on hardware and software, staff is also included. Staff is the executor of reliability management, also is the object of reliability management. Domestic research on equipment system reliability management focused on hardware and software, few of them considered the reliability of staff. The reliability of staff is an important part of the system reliability management of inertial confinement fusion facility, which can be expressed as follows:

$$R = R_{hw} \times R_{sw} \times R_{sf}$$

Among them: $R_{hw} \rightarrow$ the reliability of hardware;

$R_{sw} \rightarrow$ the reliability of software;

$R_{sf} \rightarrow$ the reliability of staff.

Among the three main ingredients that make up inertial confinement fusion facility, the reliability of staff is the lowest. Statistics have shown that 50%-70% of the system failures in nuclear power plant are caused by human factors [6], 67.86% of the civil aviation flight accidents are caused by the crew [7]. In the operation of inertial confinement fusion facility, command decision-making, fault diagnosis and maintenance tasks must be completed by staff, so the reliability of staff is crucial. Sometimes, an attentive founding by staff can solve a potential failure timely; Conversely, a staff's careless and negligent may cause the failure of inertial confinement fusion facility. How to strengthen the reliability management of staff in the process of running will directly affect the operational reliability of inertial confinement fusion facility. In particular, some basic measures can be done to ensure the reliability of staff, which are shown in Figure 4.

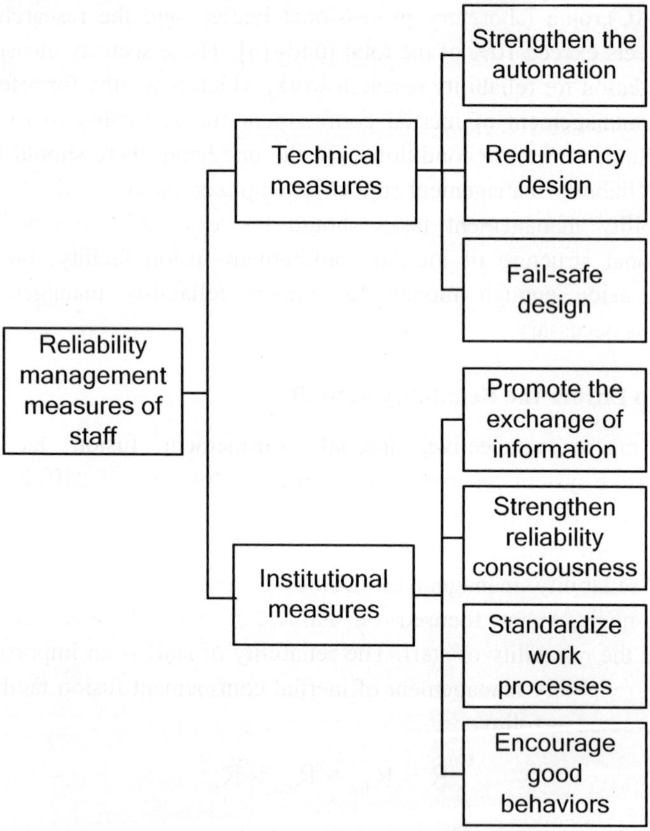

Fig. 4. The measures for reliability management of staff.

Specific contents are as follows:

(1)Strengthen the automation: In the operation and maintenance of inertial confinement fusion facility, many operations need to be done by staff, such as the replacement of damaged pulsed flashlamps, etc., which are likely to produce new unreliabilities. Compared to staff, hardware and software systems have a relatively high reliability for their procedural behavior design. Replace staff' behaviours with hardware and software as far as possible can reduce the influence of human error on the system reliability.

(2)Redundancy design: In the work link of entirely operation by staff, arrange one supervisor that be responsible for the supervision and inspection of work behavior at least; In the work link be operated by both staff and machine, adopt the design of parallel operation, realize the complementary supervision between staff and machine; In the clearance of shooting task, organize review for the results of artificial operation.

(3)Fail-safe design: According to the senses and behavior ability of the staff, do some special designs in the operation place, such as: labeling easily identified labels on control buttons which need to be operated orderly, making special logos on components which are easy to fault.

(4)Promote the exchange of information: In the early running period, rely on reliability project cooperation channels to enhance the exchange of information between staff and reliability professionals; Organize regular discussion about the recent operation status and problems of inertial confinement fusion facility, thus make the mutual information be free.

(5)Strengthen the consciousness of reliability: The leader of the reliability working team should organize reliability education to staff with the experience gained in the operation of inertial confinement fusion facility and other equipments regularly.

(6)Standardize work process: Standard all kinds of staff' behaviors by regulations and documents, make the staff to deal with all kinds of problems in a good way as much as possible. For example, stipulate the steps for field staff to enter the scene and what they can do later in an announcement on the wall which is easy to see.

(7)Encourage good behaviors: Establish a responsibility system for reliability management according to the organization framework of inertial confinement fusion facility, ensure that each module (subsystem) has a real header; carry out rewards and punishment mechanism, implement performance appraisal method, set up a relation between the reliability of inertial confinement fusion facility and staffs' income.

In one word, reliability management of staff is an important and complex work, which can only be achieved by considering various factors and conditions, adopting flexible and effective methods. These measures only give the methods in common use, different inertial confinement fusion facilities have different characteristics and organization structure frameworks, it is reasonable to do emergency treatment when implement reliability management of staff under specific conditions.

5. Conclusions

Reliability management in operation is very important, which has strong influence to the maintenance and promotion of the reliability of inertial confinement fusion facilities. The article discussed three important problems arised in the operation of inertial confinement fusion facilities, which is expected to have a good effect.

References

1. Edward I. Moses. The National Ignition Facility: Exploring ICF Burning Plasmas in the Laboratory [M]. Science in the National Interest, 2005(2): 8430-8488.
2. Guo Bo. System reliability analysis [M]. Changsha: Press of National University of Defense and Technology, 2002.
3. Zhao Huaike. Briefly discussion of reliability management [J]. Electronic Industrial Specialized Equipment, 1995, 24(4):34-38.
4. Lu Junfeng, Yang Cunbang, Chen Ming. Centralized control design of Large ICF experiment' diagnosis system [J]. Laser and Particle Beams, 2006, 17(11):1845-1849.
5. Zeng Jike. America's reliability management mode and its enlightenment [J]. Electronic product reliability and environmental testing, 1997, 1(1): 41-44.
6. Stanton N. Human Factors in Nuclear Safety [M]. London (UK): Taylor & Francis Ltd, 1996.
7. Sun Ruishan, Wang Xin. Application of failure probability of CREAM for judgment and decision-making in the cockpit [J]. Journal of Safety Science and Technology, 2010,6(6): 40-45.

An analytical method for DG placements considering integrated benefit

L. Han[†], J.H. Yu and A.B. Lu
College Of Electric Power Technical, Wuhan, China
[†]*Email: leihan0430@163.com*

Economic considerations are major factors that shape the Distributed Generator (DG)'s type as well as its size and location. In order to obtain the maximal integrated benefits of DG placement, an analytical method for DG placements considering the reliability improvement and environmental benefit is proposed in this paper. The proposed method tries to solve the optimal DG placement from the viewpoint of achieving the best balance between the costs and benefits of DG placement. At the same time, it also sets out a method to make an effective quantification of several indices, which are important to evaluate the reliability of distributed network with DG in different sections, as well as the environmental benefit. Test results have shown that with proper installation type and site selection, DG can be used to improve service reliability, reduce the customer interruption costs and the loss cost, and also to make tremendous contributions to the environment. The most important is that the electric utilities can often obtain the maximal benefit/cost ratio of DG.

Keywords: Customer interruption cost; Distributed generator; Environment benefit; Generating cost; Reliability; Islanding.

1. Introduction

Distributed generation(DG) is normally defined as small generation units (<10MW) installed in distribution systems. It holds more and more shares in electric power in recent years. It utilizes not only traditional power, but also fuel cells and renewable power generation methods such as wind and solar to serve energy to power systems [1]. As the extension of DG, its applications result in positive and negative side effects for both utility and customers [2], [3]. Some will have great impacts especially on distribution system operation and planning because some renewable energy often have strong characteristics of random and intermittent. The degree of impact is closely relative with the site as well as the type and the size of DG. Reasonable planning of DG is very important.

The planners can evaluate the DG performance from the two most important elements—cost and benefit in a general way. But the existing descriptions of two aspects are all insufficient for complete evaluations. Many analytical methods which are available for DG planning have been studied in different references [4-7]. [4] optimized DG position and the capacity in the expansion planning of transmission network based on the method of continuous elimination under the indefinite conditions of its site and size. [5] utilized genetic algorithm to obtain the best location and the size of DG after knowing the total capacity, but not the total amounts and each DG capacity. The result is restricted as the method doesn't care the additions of load node. [6] proposed some processes for DG planning that also permit the incorporation of distribution automation and demand-side management. [7] proposed some guidelines for DG planning.

DG distinguishes itself from conventional generating units, especially for some renewable energy technologies. For that they often cannot supply the loads demand completely when they have the responsibility to energize the load in an island. And their contributions to the environment, which also distinguish DG from substations in integrated evaluation cannot be described. An analytical method is presented in this paper to study the planning of DG. The proposed method not only makes an effective quantification of several important indices, but also expresses the reliability enhancements and the environment cost of DG well. And then considers the benefits and costs of DG placement into account and tries to find the optimal type and location for DG placement.

2. Problem Definition And Solution

2.1. *Problem formulation*

The value-based planning approach attempts to solve the optimal DG placement from the viewpoint of achieving the maximal integrated benefits of DG. It is based on the benefit/cost ratio which balances the cost and the benefit of DG. The benefit of DG includes the power cost, the loss reduction, and reliability enhancement, while the costs of DG placement are showed as the cost of energy considering environmental benefits. The reliability enhancement which is treated as the customer cost can be the customer interruption cost (CIC) [8], [9].The objective function of the valued-based DG placement problem is as follow:

$$Max\ \text{Benefit/cost ratio} = \frac{Benefit^{DG}}{Cost^{DG}}$$

$$= \frac{\frac{\rho(1+\rho)^{N}}{(1+\rho)^{N}-1}\left[K_{DG} * T_{\max} * \left(\Delta P_{loss}^{DG} + W_{DG}\right) + \left(CIC^{0} - CIC^{DG}\right)\right]}{C_{h} * T_{\max} * W_{DG}} \quad (1)$$

$$V^{\min} \leq V_{i} \leq V^{\max}$$

$$Q^{\min} \leq Q_{i} \leq Q^{\max}$$

Where c_a = interface adhesion; δ = friction angle at interface; and k_1 = shear stiffness number. where Benefit^{DG} is the benefit of DG placement; In this part, ρ is the interest rate; N is the expected invest time (y); K_{DG} is the price for power (\$/kWh); T_{max} is the maximum using hours (h); ΔP_{loss}^{DG} is the average power loss reduction per year due to the DG placement; W_{DG} is the average generation of DG per year; CIC^{DG} and CIC^{0} are the customer interrupt cost with DG in the distributed network or not, respectively; Here ρ and N are 0.1 and 30, respectively; In the part of Cost^{DG}, C_h is the cost of energy considering the impacts on environment with different DG type selections.

2.2. *Distribution system reliability assessment and calculation with DG*

DG has some specific characteristics. Their output power depends on the amount of available resource at each moment. For this, a 3-state model of DG is adopted to assess impacts of DG on reliability indices [10]. Therefore the model should consider intrinsic attributes of DG and the way it applied in distribution system.

A sample distribution system with DG is shown in Figure 1 will be analyzed.

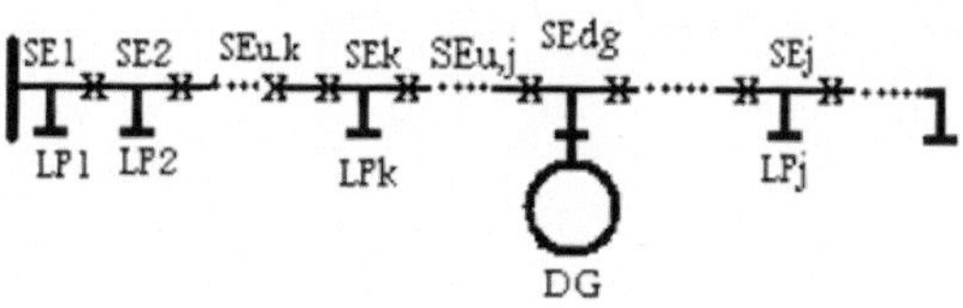

Fig. 1. Sample distribution system with DG.

There are some assumptions in evaluation the proposed method as follows:

a) Both circuit breakers at its ends should be opened immediately in the event of a contingency on any section,
b) DG can continue islanded operation if its output generation is greater than the load demand.
c) Probability density function of DG power is simulated using normal distribution function.

d) Continuity between load point and utility/DG is the sole reliability criterion.

e) The impacts on power flow and protection are not taken into account.

f) In order to permits island formation during the repair time of damaged component, there must be at least one other sectionalizing device between DG and main circuit breaker

It should be noted that DG could not only help load points in the event of faults on the upstream side of it but also load points downstream. In such cases, conception of islanding probability of load point should be put forward. For each load point LP_k, the islanding probability of the load point is the islanding probability of smallest island which contains LP_k. For example, SE_{k-1} which is shown in Figure 1 is the minimal island when LP_k is on the upstream side of the DG or SE_{DG-1} when downstream.

Utilizing short-term analysis, DG output power and load demand as well as DG resource and DG structure remain unchanged in each time interval T_k. Considering the two components: P(Gj Lj) and P(P_{DG} Gj), the islanding probability of the load point LP_k can be calculated using:

$$P(IPLP_k) = P(G_j \geq L_j) * P(P_{DG} \geq G_j) \tag{2}$$

Where P(Gj Lj) is the probability that the DG generation is equal or greater than load demand in the island; P(P_{DG} Gj) is the cumulative probability that DG generation is equal or greater than Gj. Thus the outage time of load point LP_k should be considered in equations (3) and (4) as the islanding is possible.

$$U_{IS,k} = \begin{cases} \sum_{j=1}^{N_D}\sum_{i=1}^{N_U} \lambda_{UK,j}\lambda_{DK,i} r_{UK,j} r_{DK,i} + U_k & \text{upstream} \quad (3) \\ \sum_{j=1}^{N_D}\sum_{i=1}^{N_U} \lambda_{UK,j}\lambda_{DK,i} r_{UK,j} r_{DK,i} + U_k + \sum_{i=1}^{N_D} U_{DK,i} & \text{downstream} \quad (4) \end{cases}$$

Where UK is the set of feeder sections on the upstream side of DG and LP_k; DK is the set of feeder sections between DG and LP_k; λ and γ are the failure rate and repair time of relevant sections, respectively.

With the help of formula of total probability, the associated outage time of load point LP_k in each time interval T_k will be calculated in (5) given below when the load point LP_k can be islanded with the probability of $P(IPLP_k)$. Where $U_{c,k}$ is the outage time of load point LP_k when there is no DG in the distributed network.

$$U_k = P(IPLP_{i,k}) * U_{IS,k} + [1 - P(IPLP_{i,k})] * U_{C,k} \tag{5}$$

As the period has been divided into several shot-term time intervals, and the analysis above is performed for each time interval. Thus, the average outage time for each load point in the time period can be obtained from (6) as follow:

$$\overline{U_k} = \frac{1}{N_T} \sum_{j=1}^{N_T} U_k \tag{6}$$

Where N_T is the number of short-term time intervals contained in test period. Thus the analysis can be performed for all load points on the feeder when DG installed in different locations. Load point indices can be calculated using the modified load point outage times.

2.3. *Customer interruption cost (CIC) calculation*

A proposal method present here is based on amount of investigation to estimate the CIC with the characteristics of the various customers of power.

Through the historic data, customers can be classed with different criteria such as industry, commerce and so on; For each load point, assuming that the time when the interruption cost is the largest is at the peak time, and the degree of power loss is proportional to the load, and then investigating on the peak load and interruption cost (for example 1m.30m.etc) in different outage time of a single customer in all kinds of customers. Calculating the total peak load and the interruption cost in different outage time of each user. And the Sectoral Customer Damage Function (SCDF) can be achieved from the total cost which is caused by different kinds of load in different interrupt time divided by the total peak load at peak time. Combining with the proportion of electricity consumption and load rate, there will be Composite Customer Damage Function (CCDF) and that is the interruption cost function of the load point required.

$$CCDF_{(t)} = \sum_{i=1}^{n} \frac{c_i \times SCDF_{(t)}}{L_i} \tag{7}$$

Where n is the number of load kinds; c_i and L_i are the proportion of electricity consumption and load rate, respectively.

It has been proved that making the use of interpolation would get a perfect simulation result [11]. The outage time of load point obtained in reliability assessment could be utilized to receive the interruption cost function with interpolation. Thus, CIC at each load point can easily be obtained with the total load of each point.

2.4. *Energy cost calculation with environmental benefit*

Without considering the environmental factors caused by different selections of DG types, the cost of energy generated is obtained purely by adding the capital recovery cost and cost per unit of energy [12]. Nevertheless, some renewable energy systems have obviously enormous environmental benefits comparing with the traditional power generation paradigms. It is necessary to add the environment costs into the cost of electric quantities. The calculations are made on an annual basis and thus the cost of energy can be given newly as:

$$C_h = \frac{(1+\rho)^N}{(1+\rho)^N - 1} * \frac{C_{az}}{87.6k} + C_{OM} + C_f + C_e \tag{8}$$

$$k \equiv \frac{\text{Energy generated in KWh per year}}{\text{(system rating in kW)} * \text{(8760 h per year)}} \tag{9}$$

Where C_{az} is installed (capital) cost (\$/kW-h); k is the annual capacity factor in per-unit, (defined in equation (9) above); C_{OM} is the operation and maintenance cost(\$/kW-h); C_{az} and C_{OM} are 1000 and 0.01 in this paper as in reference[12], respectively; C_f is the fuel cost (when the fuel does not need to purchase, this item is zero); C_e is the environment cost whose expenses generally contains two aspects:

a) Environment loss, which is namely the consumption of environment resources, including the drops of environment quality caused by pollutions or the destructions of ecological environment caused by luxus consumption of the natural resource.
b) The penalty received from the emission pollutants.

The expression of C_e can be shown in equation (10) [13].

$$C_e = \sum_{i=1}^{n} (V_{ei} + V_i) * Q_i \tag{10}$$

Where V_{ei} is the environmental value of pollutant emission i (\$/kg); V_i is the penalty for pollutant emission i (\$/kg); Q_i is the total emission quantities of pollutant i; n is the total number of pollutants. Their standard in China will be given in Table 1 [14], [15] as follow:

Table 1. Environmental value standard of pollutant emission in power industry in China.

	$/kg			
	SO_2	NO_x	CO_2	CO
Environmental value standard of pollutant emission	0.75	1.00	0.00288	0.125
Magnitude of penalty for pollutant emission	0.125	0.25	0.00125	0.020

Emission characteristics of several electric generating technologies are shown in Table 2 [16], [17]. And they will be utilized in method applications in next section.

Table 2. Emission characteristics of electric generating technologies.

$/kg

<table>
<tr><td colspan="2">Electric Generating Technologies</td><td>NO_X</td><td>CO_2</td><td>CO</td><td>SO_2</td></tr>
<tr><td rowspan="3">DG with traditional energy</td><td>Coal</td><td>0.1547~3.09383</td><td>86.4725</td><td>—</td><td>0.1083~3.944</td></tr>
<tr><td>Natural gas</td><td>0.0077~1.5469</td><td>49.037 2</td><td>—</td><td>0.464 1</td></tr>
<tr><td>Fuel oil</td><td>0.0077~1.5469</td><td>72.395</td><td>—</td><td>—</td></tr>
<tr><td rowspan="6">DG With Renewable Energy</td><td>Gas</td><td>0.618 8</td><td>184.083</td><td>0.170 2</td><td>0.00093</td></tr>
<tr><td>Fuel gas</td><td>4.795 4</td><td>170.16</td><td>1.222 1</td><td>0.023 2</td></tr>
<tr><td>Diesel oil</td><td>4.331 4</td><td>232.03</td><td>2.320 4</td><td>0.464 1</td></tr>
<tr><td>Fuel cell</td><td>< 0.023</td><td>635.04</td><td>0.054 4</td><td>0</td></tr>
<tr><td>Solar power</td><td>0</td><td>0</td><td>0</td><td>0</td></tr>
<tr><td>Wind power</td><td>0</td><td>0</td><td>0</td><td>0</td></tr>
</table>

3. Method Application

A sample test system which consists of 7 sections, 7 load points and one DG which is prepared to be installed on the candidate section of the feeder is shown in Figure 2.

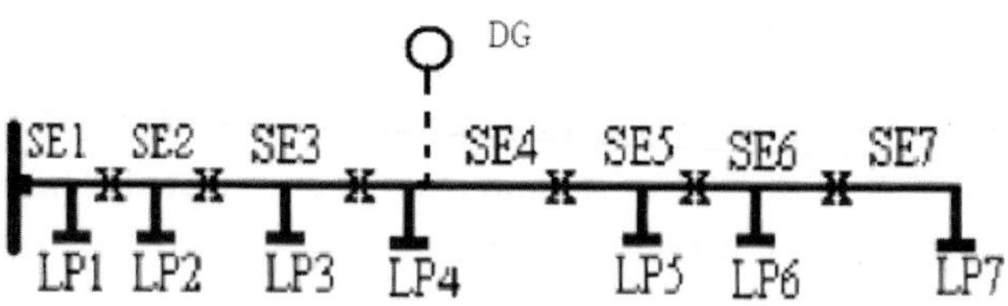

Fig. 2. Distribution test system.

In this section, study period is one day, which is divided into 4 intervals. Analysis is performed for each interval. Reliability parameters for all feeder sections are given in Table 3. The failure rate of each section is assumed to be proportional to the length of the section. Failure rate of the feeder is 0.0650 f/yr per kilometer, and the impedance of the line is $0.17+j0.38\Omega$. S_N=100MVA, U_N=10KV.

Table 3. Reliability parameters of line sections.

Section	Length	Failure rate	Repair time	Load percent
	km	λ(f/yr)	r(hr)	
SE1	1	0.065	5	0.1
SE2	2	0.13	5	0.2
SE3	3	0.195	5	0.3
SE4	1	0.065	5	0.1
SE5	1	0.065	5	0.1
SE6	1	0.065	5	0.1
SE7	1	0.065	5	0.1

The peak load of the feeder is assumed to be 4 MW and the total DG power is 2 MW. The DG has 2 identical generating units. Reliability parameters of the two DG are assumed to be same. Power output of DG is represented by the ratio of actual output to rated capacity. To describe the question well, the proposed method in the paper will take a traditional power generator and a wind energy power plant as examples, respectively.

For wind energy generator, the cumulative probability is the probability that DG generation is equal or greater than actual output. Basic reliability data of the two units are shown in Table 4 below. While for traditional power generation paradigms, this part will be constant 1 no matter how many the output are.

Table 4. Reliability parameters of DG.

DG output (MW)	0.0	0.5	1.0
Cum. probability $P(P_{DG}\geq Gj)$	0.05	0.35	0.65

Power data of the system and two DG are given in Table 5 in which the feeder load and output of DG for a sample day are presented. The power factor of DG and load will be both assumed to be 0.8. T_{max} is 2000.

Table 5. The total load of the system and the output of DG in different intervals.

T_k (hr)	0-6	6-12	12-18	18-24
Lj (MW)	1	2	3	4
Gj (MW)(DG with renewable Energy)	0.5	2	1	1.5
Gj (MW)(DG with traditional Energy)	1	2	2	2

The conditions of the load point in test system are given in Table 6 to Table 7. There will be a composite customer damage function in Table 8. It is assumed that each load point has the same level of load.

Table 6. Outage cost function of all kinds of customer.

Load kind	Cost interrupt (Yuan/KW)					
	1min	30min	1h	2h	4h	8h
Industry	0.97	4.81	10.67	22.00	31.43	59.72
Commercial	3.93	16.50	37.33	75.00	312.97	637.42
Hospital	45.50	75.60	159.98	431.93	826.37	1300.75
Government	0.0	40.50	85.32	141.41	246.29	434.16
Residence	0.0	20.15	42.66	98.12	166.84	275.22

Table 7. The energy consumption ratio of service center.

Load kind	Proportion of electricity consumption (%)	Load rate (%)
Industry	42.55	77.70
Commercial	32.92	68.99
Hospital	7.60	88.99
Government	4.64	59.90
Residence	12.29	78.78
Total	100.00	76.15

Table 8. Composite customer damage function.

Interrupt time	1m	30m	1h	2h	4h	8h
Interrupt cost(Yuan/KW)	6.29	23.24	49.63	110.91	282.5	524.45

According to the data in Table 8, combining with the result of outage time of each load point when DG installed in different sections, and then with the help of interpolation, composite customer damage function in different outage time can be obtained.

The analysis is performed for 2 DG models as the proposed model. Thus the benefit of different locations of DG will be shown in Table 9 and Table 10. Here the price for wind power and traditional power are 0.07 and 0.05, respectively.

Table 9. The benefit of DG in different locations(wind energy generator).

	Power cost and loss reduction	Reliability enhancement	Benefit
	yuan	yuan	yuan
DG in SE1	2268359.6544	8177281.166	1107237.9270
DG in SE2	2317760.3144	9497475.566	1252415.0033
DG in SE3	2368411.2308	9525726.878	1260778.6395
DG in SE4	2375225.0233	9877069.748	1298743.2458
DG in SE5	2373707.6025	11346677.044	1454360.7725
DG in SE6	2373096.8532	11231479.913	1442085.1372
DG in SE7	2367479.0988	10267991.566	1339359.8905

Table 10. The benefit of DG in different locations(traditional power generator).

	Power cost and loss reduction	Reliability enhancement	Benefit
	yuan	yuan	yuan
DG in SE1	1416710.852	20443202.917	2317150.860
DG in SE2	1444202.02	24367030.142	2735990.609
DG in SE3	1469896.34	24435077.862	2745927.265
DG in SE4	1471460.872	25189835.729	2826097.440
DG in SE5	1470795.368	27914099.012	3114798.804
DG in SE6	1468021.052	27016823.946	3019393.577
DG in SE7	1463264.702	25142167.533	2820175.817

It can be seen from the results in Table 9 and Table 10 that with the planning of evaluating the DG performance from benefit without reliability enhancements in test system, section 4 is the optimal location for DG, while section 5 is the optimal location for DG when considering the reliability enhancements no matter what DG type is. It is noticeable that reliability enhancement has played a decisive role in the benefit evaluation of DG.

According to the parameter selections in this paper, we will easily obtain the cost of energy considering the environment cost of DG in Table 11. Thus, there will be a final result about the evaluating indicator, which is proposed in this paper in Table 12.

Table 11. Costs of energy in different DG type.

	Cost of energy	Cost of energy considering environmental cost
	yuan/KWh	yuan/KWh
DG with wind energy	0.53015	0.53015
DG with conventional energy	0.31185	1.16424

Table 12. Results of the evaluating indicator.

	Best Site	Benefit/Cost Ratio
DG With Wind Energy	Section 5	0.6858
DG With Conventional Energy	Section 5	0.6678

The results shown in Table 11 tell that after considering the environmental costs, DG with conventional energy such as coal has to pay for the penalty for its pollutant emission. And this will make the cost of its energy enhance greatly. However, DG with renewable energy such as wind expresses its superiority well from its environmental benefits which help to reduce the cost.

Compared with the same capacity renewable energy DG, conventional generator units have the higher improvement degree in both reliability and total benefit, but the former will have better integrated benefits depending on its particular environmental benefits. This is shown in Table 12 and the conclusion is very important and useful to the planning and design of distributed generations in distributed systems.

4. Conclusion

In this paper, an available DG placement method for distribution systems was proposed. The proposed method has studied the impacts on distribution system reliability indices, and the contributions to the environment with different DG type. And then take the benefits and costs of DG placement into account to evaluate the integrated benefit, according to which the optimal types and sites for DG placement can be found with the maximum benefit/cost ratio. Test results were shown that with proper types, sites and installation sizes selection, DG is one of the best methods to improve the reliability, reduce the customer outage costs and make tremendous contributions to the environment. If the method will be used in engineering practice, results may be over-optimistic. Because the placement of DG is also sensitive to the DG techniques, resource policies, government social policies and utilization rate etc. Any variations on those aspects will actually affect the results obtained in the paper and they will be integrated into the future works.

References

1. Puttgen H.B.; Macgregor P.R.; Lambert F.C.. Distributed Generation Semantic Hype or the Dawn of a New Era IEEE Power and Energy Magazine, 2003, 1(1): 22-29.
2. Daly P.A.; Morrison, J.. Understanding the Potential Benefits of Distributed Generation on Power Delivery Systems. IEEE Power Engineering Society Summer Meeting, pp. A2-1/A2-13.
3. Dugan R.C.; McDermott T. E.. Operating Conflicts for Distributed Generation on Distribution Systems. IEEE Power Engineering Society Summer Meeting, pp. A3-1/A3-6.
4. Brown R.E.; Jiuping Pan; Xiaoming Feng; Koutlev K. Siting Distributed Generation to Defer T&D Expansion. Transmission and Distribution Conference and Exposition, 28 Oct.-2 Nov. 2001, (2):622-627.
5. Celli G; Pilo F. Optimal Distributed Generation Allocation in MV Distribution Networks. 22nd IEEE Power Engineering Society International Conference on Power Industry Computer Applications. 20-24 May 2001:81-86.
6. Dugan, R.C.; McDermott, T.E.; Ball, G.J.. Distribution planning for distributed generation. IEEE Rural Electric Power Conference. pp. C4/1 -C4/7, 2000.
7. Rackliffe, G.. Guidelines for planning distributed generation systems. IEEE PES Society Summer Meeting. Vol. 3, pp. 1666-1667, 2000.
8. Chowdhury, A.A.; Koval, D.O.. Value-based distribution system reliability planning. IEEE Transactions on Industry Applications. Vol. 34, No. 1, Jan.-Feb. 1998, pp.23-29.
9. Teng, J.H.; Lu, C. N,. Feeder Switch Relocation for Customer Interruption Costs Minimization. IEEE Trans. on Power System. Vol. 17, No. 1, pp. 254-259, 2002.
10. In-Su Bae, Jin-O Kim, Jae-Chul Kim, et al. Optimal operating strategy for distributed generation considering hourly reliability worth [I]. IEEE Transactions on Power System. 2004, 19(1); 287-292.

11. R. L. Chen; K. Allen; R.. Billinton. Value-based distribution reliability assessment and planning. IEEE Trans. Power Delivery, vol. 10. pp. 421–429, Jan. 1995.
12. Ramakumar R; Butler N G; Rodriguez A P. Economic aspects of advanced energy technologies [J]. Proceedings of the IEEE, 1993, 81(3):318-332.
13. Sun Ke. Environmental cost analysis and research of different power plants [J]. Energy Engineering. 2004, 23(3):23-26 (in Chinese).
14. Wei Xuehao; Zhou Hao. Evaluating the environmental value schedule of pollutants mitigated in China thermal power industry [J]. Research of Environmental Sciences. 2003, 16(1):53-56 (in Chinese).
15. Zhang Na; Cai Ruixian. The proper order of magnitude of penalty for pollutant emission [J]. Proceedings of the CSEE, 1997, 17(4):286-288 (in Chinese).
16. Wan Y H; Adelman S. Distributed utility technology cost, performance, and environmental characteristic [R]. United States, National Renewable Energy Laboratory, 1995.
17. Greene N, Hammerschlag R. Small and clean is beautiful: exploring the emissions of distributed generation and pollution prevention policies [J]. Electricity Journal. 2000, 13(5):50-60.

Dynamic simulation research about regional protection based on RTDS

Z. Q. Zheng[1] †
State Grid Liaoning Electric Power Research Institute,
Power System Technology Center, Shenyang 110006, China
†Email: zhengzqxx@163.com

H. B. Zhang[2], D.D. Ge[3]
North China Electric Power University,
State Key Laboratory of Alternate Electrical Power System

Based on the existing problems of the traditional main and backup protection, regional protection for intelligent substation is gradually rising as a new security. To make full use of the technology advantage of intelligent substation, a set of protection control device named BDH-811 is brought in and used for 66kV power grid in Chaoyang region. In order to ensure the safe and stable operation of the regional protection system in actual environment, a series of tests have been carried out by building a dynamic simulation testing platform based on RTDS.

Keywords: Regional protection system; Dynamic simulation.

1. Introduction

At present, applications used in domestic intelligent substation are gradually increasing, but its protection configuration is still consistent with the traditional relay protection, which is equipped with main and backup protection. Main protection is used to be a quick trip, while backup protection needs multi-stage cooperation. What's more, protection device and its related circuit with double configuration are fully independent [1]. The current situation shows that the protection configuration of intelligent substation do not make full use of its own technology advantage, there are still the inherent defects that the traditional backup protection have. Due to the truth that its selectivity is determined by coordination of the fixed value and time, its movement speed is slow. Due to various protection configuration that are needed by each device, the structure of

---

[1] Zhi-Qin Zheng
[2] Hai-Bo Zhang
[3] Dan-Dan Ge

protection system is complex and the device investment is high. The traditional protection is realized on the basis of local measurements. The devices are also lack of mutual coordination and cooperation. These may cause successive action between protector and automatic safety device, further the scope of fault effect may be expanded. So in order to make full use of the advantage of intelligent substation that the domain information can be obtained easily, to improve the automation level and reliability of the relay protection, the concept of regional protection is gradually formed.

The regional protection refers to a number of lines closely related as a whole protected area. Protection devices within the protected area communicate with each other through the optical fiber. The regional protection is used to assist the traditional main protection and backup protection. And the final purpose is to form a reliable regional relaying protection system.

A set of domain protection control device named BDH-811 produced by Xuji corporation has been brought in Chaoyang 66kV region of Liaoning province. The device is used to improve the existing main protection function which main criterion is differential protection,complete the backup protection for line, bus, main transformer within the protected area, and the domain backup automatic switch function. In order to ensure the smooth implementation of the project, strengthen the technical supervision and net management of the power grid relay protection and safety automatic device in Chaoyang region, a dynamic simulation system suitable for the test of Chaoyang 66kV regional protection has been built on the basis of RTDS.

2. Introduction of RTDS and the Design of Its Interface with External Devices

RTDS is a digital dynamic simulation system. It is developed by Canada Manitoba high-voltage direct current research center to study the electric power system. Its hardware is based on DSP (digital signal processor). Its simulation step can be up to 50 subtle with the parallel processing mode. Its calculation speed can meet the real-time simulation requirements of power system. In all the simulation results from RTDS are consistent with the actual system [2, 3]. What's more, RTDS supports multiple interface card, such as analog interface card, digital quantity interface card and network interface card, etc. It can be connected to the actual physical device to realize the closed-loop dynamic simulation test. At present, a variety of closed-loop tests about protection device, such as line protection device [4-6], differential bus protection device [7] and transformer protection [8], have been carried out on the basis of RTDS, These test experience determines the direct advantages of RTDS in protection device test.

Combined with the working characteristic of different interface card that RTDS supports, there are three kinds of way to send out analog value: use GTAO card combined with power amplifier; use GTAO card combined with current/voltage collector; use the sampling value card GTNET_SV combined with switches. There are two ways to exchange switch state and instructions with external devices: use GTDI, GTDO combined with PAD; use substation event card GTNET_GSE combined with switches.

According to the above analysis, it is an effective means to build a fixed testing and analysis platform for relaying protection based on RTDS and its flexible software.

3. Build The Regional Protection Dynamic Simulation System

The building of dynamic simulation system mainly reflect in three aspects. One is to determine the study area and build power system simulation model based on RTDS. One is to determine the network mode of regional protection system. The last one is to determine the data interface between RTDS and external devices. The content of the three aspects are analyzed respectively below.

Regional protection depends on the operating conditions of power system, including the sampling values and switch state. So the key to carry out dynamic simulation test is how to provide various operating conditions in the lab that are consistent with the actual power system. Combined with above analysis, RTDS has the incomparable advantage in real-time device-level closed-loop test. So RTDS is adopted as the basic platform in this paper to realize dynamic simulation test.

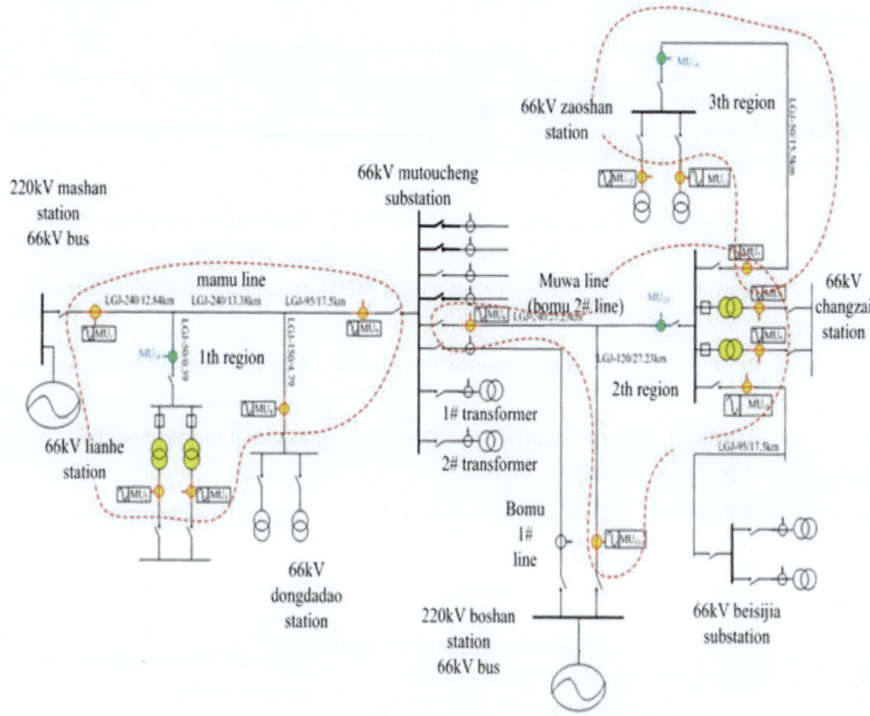

Fig. 1. The network topology of Chaoyang 66kV region.

The regional protection system of Chaoyang 66kV region is centered by 220kV intelligent substation named Mashan. The power system structure that is covered by the regional protection system is shown in Fig. 1. It has two 66kV lines which lead from 220kV substation and six 66kV substations. Build power system simulation model on RTDS software platform according to the structure and the static parameters of devices that are shown in Fig. 1.

Domain protection control device BDH-811 is suitable for the domain level protection of 110kV and 220kV substation. Its oriented object is substation. Its main function is to complete the protection and safety automatic control of the circuit components, such as line, main transformer, bus and so on. Combined with the function modules that are configured in BDH-811, the protection configuration of the area (shown in Fig. 1) is shown in Table 1.

Table 1. The Configuration of Regional Protection.

CPU Number	Protection Module	Protection Scope	Note
1	BDH-811G	Public module	
2	BDH-811L	Over-current protection of MU1	Regional backup protection
3	BDH-811L	Over-current protection of MU5	Regional backup protection
4	BDH-811L	Over-current protection of MU6	Regional backup protection
5	BDH-811L	Over-current protection of MU7	Regional backup protection
6	BDH-811L	Over-current protection of MU10	Regional backup protection
7	BDH-811L	Over-current protection of MU13	Regional backup protection
8	BDH-811B	Regional protection of 1th area (Mamu line and Lianhe transformer)	Regional differential protection (support angle)
9	BDH-811B	Regional protection of 2th area (Muwa line and Changzai transformer)	Regional differential protection (support angle)
10	BDH-811MH	Regional protection of 3th area (Changzao line and Zaoshan bus)	Regional differential protection (do not support angle)
11	BDH-811J	Low cycle load reduction of 66kV side of Mashan transformer	Centralized load reduction
12	BDH-811TL/F	Backup automatic switch of Mashan station, Mutoucheng substation and	Regional backup automatic switch
13	BDH-811D	Malfunction protection of Mutoucheng substation	Accelerate the action time of protection when the switch of MU5 and MU6 is fail.

Combined with the network mode (shown in Fig. 2) of the regional protection system, the dynamic simulation test in this paper was carried out according to the requirements of corresponding rules and technical scheme. As is shown in Fig. 2, the "triple play" technology of 1588 pairs, GOOSE and SV is used for the communication between devices of control and protection layer that are equipped in Mashan station. The switch adopts star connection, and the connection between devices use ST multi-mode fiber. Station level data communication between integrated intelligent units is in the way of point-to-point by using the multiplex channel (SDH communication network). The network connection mode is realized through LC single-mode fiber. In addition, message recording device is connected to the network via a switch through the optical fiber.

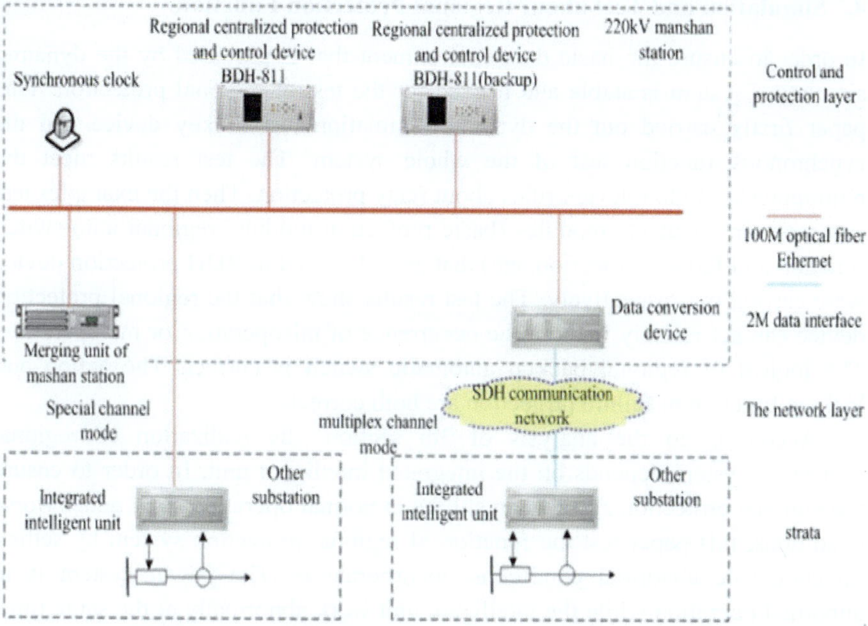

Fig. 2. The architecture diagram of regional protection system.

The regional protection system (shown in Fig. 2) belongs to regional centralized protection mode in view of structure. The collection of power system operating conditions is achieved on the basis of integrated intelligent unit of each substation. So how to design the data interface between RTDS and the external device (merging unit) is the key to correctly send out the RTDS simulation results. There are three ways to send out analog from RTDS. The way with the combination of voltage/current collector is mostly consistent with the actual field data acquisition environment. So this paper adopts a special ratio of linear transformer to interface with merging unit. This way requires that

RTDS simulation model contains electronic transformer model. So the interface connection between RTDS and merging unit is as shown in Fig. 3.

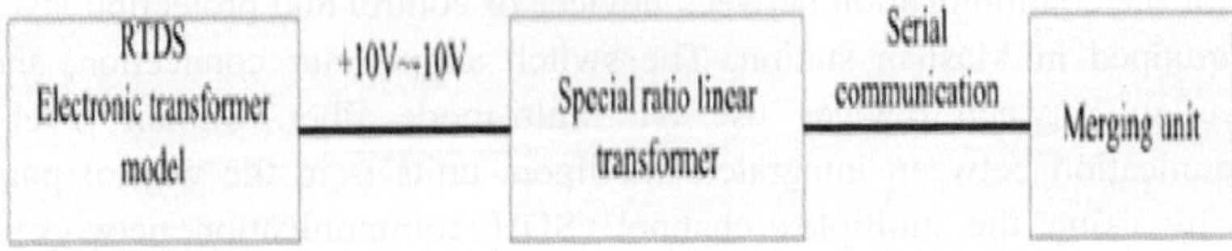

Fig. 3. The data interface between RTDS and merging unit.

4. Simulation and Test about Regional Protection Function

In order to ensure the basic data environment that is provided by the dynamic simulation system is stable and needed for the test of regional protection. This paper firstly carried out the dynamic simulation test of key device and the synchronous function test of the whole system. The test results meet the requirements of the relevant rules about relay protection. Then the examples test and research about the modules (basic protection module, regional auto-switch function and failure protection, etc) that are integrated in BDH protection device were carried out respectively. The test results show that the regional protection device can act reliably without the occurrence of misoperation or maloperation. The logical of regional backup automatic switch is correct. The action and locking function of failure protection are both correct.

According to the analysis of 3th section, the realization of regional protection system depends on the integrated intelligent unit. In order to ensure the regional protection system can still be in normal operation even under worse conditions, this paper test the function of regional protection system by setting up composite abnormal conditions, incorporate is: The power system is in abnormal operation while the intelligent unit work abnormally at the same time. According to the protection configuration given in Table 1, the premise condition of the test is set as MU1, MU5, MU6, MU7, MU11, and MU12 are all closed while MU13 is disconnected. In addition, Mashan station 1# transformer and Boshan station 1# transformer are both in operation. Under the above experimental conditions, A phase CT of MU5 is disconnected, meanwhile A, B two-phase short circuit fault is occurred at Mamu line in 1th region.

Test analysis: The original differential protection of Mamu line will be directly locked. BDH-811 protection control device can obtain all the sampling values and switch quantity from merging units. Therefore, the disconnection of MU5 A phase CT results in the invalid current of the interval that MU5 belongs to, but current from MU1, MU6, MU7 is effective. Current from MU6 plus current from MU7 equals zero. Therefore there is no fault on the Bomu #2line. Current from MU1 plus current from MU6 does not equal zero. Therefore the fault occurred on Mamu line. Because power is next to MU1, The switch that is next to 220kV Mashan station is the only one to be disconnected.

Test results: The differential protection of transformer in 1 th region responses nstantaneously. The switch next to 220kV Mashan station break off. The message recording (achieved from transformer protection module, shown in Fig. 4) is the same as protection.

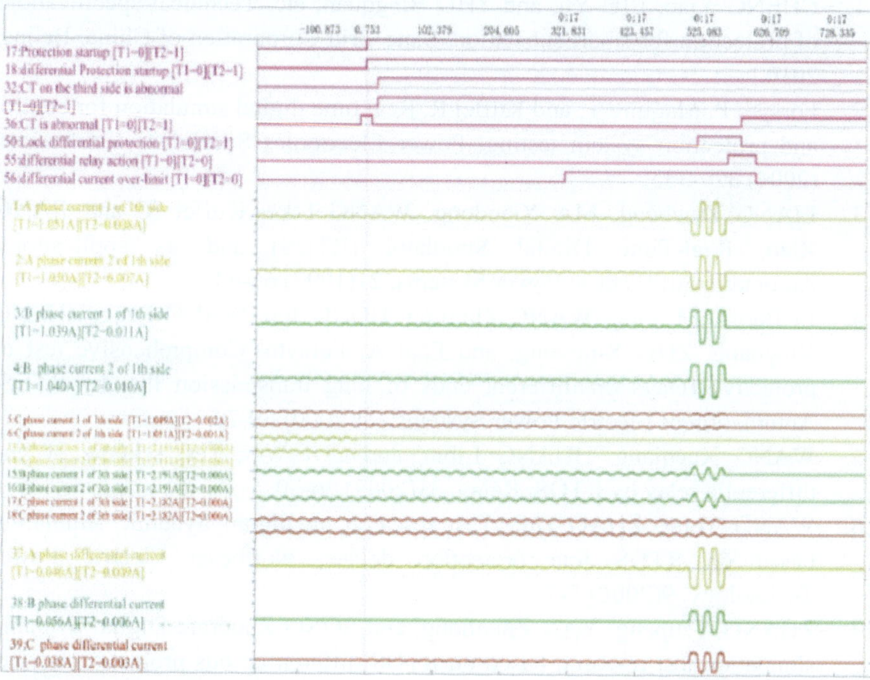

Fig. 4. The action of regional protection when CT is disconnected and power system is in fault condition.

5. Conclusions

A simulation platform suitable for the test of regional protection system has been built on the basis of the study about area protection characteristics in this paper. In order to test the function of regional protection system, various fault conditions that are consistent with the actual system have been reproduced in the lab. The test results show that the regional protection system used for Chaoyang 66kV region meets the requirements of the relay protection procedures and can be implemented for the project.

Acknowledgement

This research was financially supported by State Grid Liaoning Electric Power Research Institute.

References

1. CHENG Xiao, LIU Yu, and ZHU Bingquan, etc, Technical specifications of protection for smart substation, State Grid Corporation of China, Beijng, 2010.
2. Forsyth P, Maguire T, and Kuffel R, Real time digital simulation for control and protection system testing, Power Electronics Specialists Conference. (2004) 329-335.
3. LIANG Zhicheng, MA Xiandong, WANG Like, Kuffel R, and JIANG Xiao, Real-Time Digital Simulator (RTDS) and its applications, Automation of Electric Power Systems. 21(1997) 61-64.
4. LI Baofu, LI Ying, WANG Zhiming, HONG Jun, WANG Peng, ZHENG Xinguang, ZHU Xingyong, and Paul A. Forsyth, Comprehensive test of protective relays on different ends of long transmission line by RTDS, Automation of Electric Power Systems. 24(2000) 69-70.
5. WANG Xueqiang, HUANG Libin, and CUI Wei, Test scheme of line differentia relay by RTDS, Relay. 31(2003) 26-30.
6. WANG Tong, ZHAO Hui, Test research on digital dynamic simulation based on RTDS line protection device, Northeast Electric Power Technology. 9(2008) 5-7.
7. ZHANG Yanpeng, YAN Zhenhong, and SONG Zhuoran, Digital dynamic simulation test research based on RTDS differential bus protection device, Northeast Electric Power Technology. 28(2007) 1-3.
8. GU Bin, TAN Jianchen, A new approach to the discrimination between inrush-current and internal fault current of transformers, Relay. 35(2007) 6-10.

Agricultural pollution assessment in Xiang River basin based on ECM

B. Cao†

Chinese Research Academy of Environmental Sciences, Beijing China, 100012

†E-mail: cbao3000@163.com, Tel:+86-10-84908867

X.B. Wang

Chinese Research Academy of Land & Resources Economics, Beijing China, 101149

Y.H. Zhang

College of Resources Environment & Tourism, Capital Normal University, Beijing China, 100048

H. Luo

Chinese Research Academy of Environmental Sciences, Beijing China, 100012

Export Coefficient Model (ECM) methodology, with improvement on parameters such as chemical fertilizer uptake and loss rates in crop land field, atmospheric deposition rate etc., was applied using county level statistical datasets in large scale watershed like Xiang River Basin (XRB). Geographic Information System was applied to obtain land use information in estimating agricultural NPS pollution loads and its spatial distribution. Research results showed that: i) Total Nitrogen (TN) and Total Phosphorus (TP) loads entered into XRB were 59955, 6375 and 64586, 6162 tons respectively for the year 2000 and 2012; ii) For TN load entering into watercourses, chemical fertilizer losses from cropland contributed 54.7% in year 2000, followed by livestock(26.8%) and rural living waste(18.5%); For the year 2012, TN load contribution rates from fertilizer and livestock increased 4.3% and 11.8% respectively; iii) For TP load entering into watercourses in XRB, chemical fertilizer losses from cropland contributed 70% and 71.5% respectively in year 2000 and 2012.

Keywords: Non point source pollution; Export coefficient methods; Improvement on parameters.

1. Introduction

Non Point Source (NPS) pollution is caused by rainfall or snowmelt moving over and through the ground. As the runoff moves, it picks up and carries natural and anthropogenic pollutants, finally depositing them into lakes, rivers, wetlands, coastal waters and ground water systems [1]. Studies of NPS modeling have always been a core topic in investigations of NPS. NPS studies in western countries were first conducted in the 1960s, and there have since been many NPS models developed [2]. These models can be divided into two categories, empirical or statistical models and physically based or process-based models [3]. Empirical models use monitoring data in typical experimental plots to build empirical relationships between hydrological parameters, such as the export coefficient method, hydrograph separation method, etc. Physically based models, which consider the internal mechanism of the pollution process, are capable of calculating long time series and have clearer spatial characteristics distributions. However, due to the enormous number of parameters, requirements for a large body of input data, and limited available information, it is difficult to calibrate and validate these models, which restricts their use for large-scale regions[4]. Estimating the contribution of different NPS sources in a large watershed is a complex task, yet in watershed without extensive pollutant discharge monitoring data, no adequate soil, land use, atmospheric and hydrodynamic data, makes Export Coefficient Methods (ECM) more practical to simulate a large watershed. In this paper, ECM was applied in Xiang River Basin(XRB) to assess NPS pollution loads and their spatial distribution. Finally proportions of different sources of NPS pollution loads from agriculture were concluded.

2. Methodology

2.1. *The study area*

The Xiang River is an important tributary of the Yangtze River and its basin is located between 110.5ºE-114ºE longitude and 24ºN-29ºN latitude. The length of the Xiang River is 856 km, of which 670 km is in Hunan. The area of Xiang river basin is about 94,600 km^2 and most of its area led in Hunan province with a population of about 40 million. The East-Asian monsoon brings heavy rainfall from the south and more than 70% of the average annual precipitation of 1,512 mm falls between March and August. The average temperature is around 17ºC. The Xiang River basin is effectively the watershed for the Chang-Zhu-Tan (CZT) city cluster lies in the lower section of the basin. The population of CZT is more than 30% of the total basin population. In 2013, the regional GDP in XRB reached CNY 19,129.71 billion, which accounted for about 78% of Hunan province's GDP. Per Capita GDP reached CNY 48,213, 1.3 times Hunan provincial average. The dominance of CZT urban agglomeration of GDP in 2013 reached CNY 10,500 billion, accounted for 43% of Hunan province's gross regional product, or 55% percent of Xiang River basin regional GDP. With rapid development of agriculture, excessive chemical fertilizers was used and huge amounts of livestock and poultries were bred, which made NPS pollution becoming one of considerable contributors influencing water quality in Xiang river basin. The River basin is shown in Figure 1.

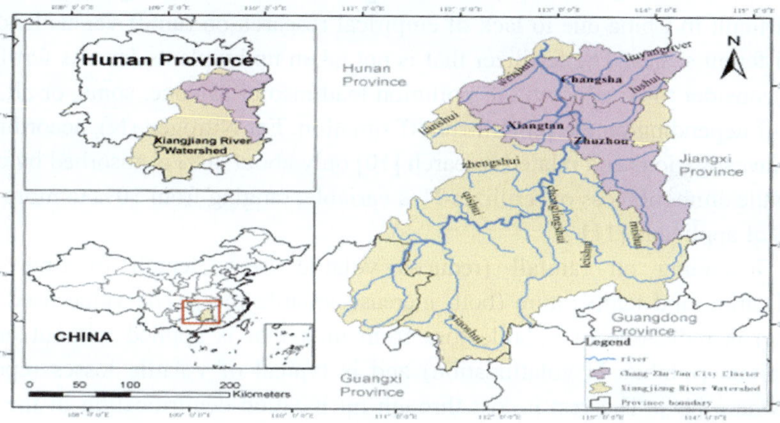

Fig. 1. Location of the study area.

2.2. *Export coefficient model*

The Export Coefficient Model (ECM), which is based on the idea that the nutrient load exported from a watershed is the sum of the losses from individual sources, such as land-use, livestock, rural life, etc., has been in general use because of its simplicity and relative robustness. Moreover, the time step of this method is large (monthly, seasonal or annual), and it allows the use of spatially and temporally based lumped data rather than real-time data, as well as the use of agricultural census data rather than field level data [4]. In recent years, there has been a growing number of ECM based studies of water quality relative to different NPS land uses in China. X. Luo, et al. selected 15 catchments of the Hujiashan Watershed to examine the effect of land use on NPS pollution and found that land use was a controlling factor that determined the amount of nitrogen export [5]. Other studies have used simulated rainfall and field-scale monitoring methods to determine the export coefficients of different land use types as well [6]. The formulation of ECM is:

Where L is pollutants loads discharged from NPS; D_i is discharge coefficient of source i; E_i is pollutant export coefficient for source i; Q_i is quantity of people, livestock, poultry, aquaculture production or chemical fertilizer utilization in the study area.

2.3. *Fertilizer*

Excessive application of fertilizer by Chinese farmers has been widely reported both in Chinese and international literature [7,8]. Nitrogen fertilizer in China is generally in the form of urea or ammonium bicarbonate leading to large losses by ammonia volatilization [9]. Estimating fertilizer runoff using published data is difficult in China due to lack of empirical research on runoff versus land use at different scales. The fertilizer that is not taken up by plants (excess fertilizer) was consider to be a 'potential' pollution load insofar as none, some, or all, will runoff depending on the rainfall-runoff situation. For Nitrogen (N), according to our investigations and related research [10] only about 40% is absorbed by crops. Volatile ammonia loss of fertilizer-N is variable, ranging from 10% to more than 60% of applied N [11].

Depending on rainfall (reduces volatile loss), length of time after application and temperature (both increase volatile loss). The value used here (45%) is indicative of volatile loss from urea that is applied without urease inhibitors (inhibits N volatilization) and is typical of volatile losses reported elsewhere [12]. The rest is lost through nitrification-denitrification in the soil, surface runoff from rain or irrigation, and leaching into groundwater. Therefore, we deduce that the remaining part of fertilizer-N (w15%) is 'potential' load of N. In addition to fertilizer-N, we also address atmospheric deposition of N as noted below.

For P, the utilization rate in the first season is only 10-15% [13] and the remainder will be stored in the soil with gradual loss by soil loss in runoff and leaching (minor). Therefore, we assume that 80 % of applied P is potential load.

Delivered crop pollutant load was calculated as: Delivered pollutant load of cropland = chemical fertilizer application (elemental, in kg/ha) × delivery coefficient. Parameters for calculation of pollution load from fertilizer application to field crops (rice + other field crops) were shown in Table 1.

Table 1. Agricultural and rural NPS Load calculations.

Pollution source	Excreted load	Potential load[a]	Delivered load[b]
Pollution load from fertilizer application to field crops (rice + other field crops)		TN = Total of N application (elemental) × (1 - utilization rate - volatilization rate) Remarks: ●N application(kg/ha) is from investigation and Hunan agricultural statistical year book. ●Utilization rate = 40% ●Volatilization rate = 45% ●TP = Total of P application ●(elemental) × (1-utilization rate) ●Remarks: ●P application(kg/ha) is from investigation and Hunan agricultural statistical year book ●Utilization rate 15%	TN = Total N application (elemental) × delivery coefficient Remarks: Delivery coefficient = 10% TP = Total of P application (elemental) × delivery coefficient Remarks: Delivery coefficient = 1.7%

Table 1. (*Cont'd*)

Pollution load from livestock[c] (excretion in centralized feedlots)	Contaminant (TN, TP) = ∑amount of $livestock_i$ × excretion $coefficient_i$ ×contaminant $coefficient_i$ Remarks: ●i - type of animal Amount of livestock (from investigation and Hunan agricultural statistical year book.) ●Excretion coefficient and contaminant coefficient (see Tables 2 and 3)	Excreted load × untreated rate Remarks: ●Excreted load untreated rate = 50%; ●(Assumes treated manure does not runoff – currently not valid assumption)	Contaminant (TN, TP) = ∑number of $livestock_i$ × excretion $coefficient_i$ × contaminant coefficient × delivery coefficient Remarks: ●i - species of livestock ●Amount of livestock comes from investigation and Hunan agricultural statistical year book. ●Excretion coefficient and contaminant coefficient (see Tables 2 and 3) ●Delivery coefficient = 5%
Pollution load from rural living[c] (waste water)	Contaminant (TN, TP) = Rural population × contaminant coefficient Remarks: ●Rural population comes from Hunan agricultural statistical year book. ●Contamination coefficient (kg $head^{-1}a^{-1}$): TN-0.35, TP-0.04	Excreted load × untreated rate (90%)	Contaminant (TN, TP) = Rural population × contaminant coefficient Remarks: ●Rural population: Hunan agricultural statistical year book. ●Contamination coefficients (kg $head^{-1}a^{-1}$): TN-0.35, TP-0.04) ●Delivery coefficient = 10%

Table 1. (*Cont'd*)

Atmospheric N deposition	N = cropland area × deposition coefficient Remarks: ●Cropland area (from Hunan agricultural statistical year book) ●Deposition coefficient = 45 kg N $ha^{-1}a^{-1}$	TN = cropland area × deposition coefficient × delivery coefficient Remarks: ●Cropland area (Hunan agricultural statistical year book) ●Deposition coefficient = 50 kg N $ha^{-1}a^{-1}$ ●Delivery coefficient = 0.75%

[a] Load at source.
[b] Load delivered to nearest watercourse; for N and P, we use experimentally determined values for the delivery coefficient and apply these to the total fertilizer applied - not to the potential load insofar as the applied fertilizer load is quantified whereas the potential load is our best estimate.
[c] Delivered load is based on excreted load, not potential load, due to large uncertainty in treatment rate.

2.4. *Atmospheric nitrogen*

Values reported for deposition of atmospheric N are quite variable. Zhang et al. report 18.4-38.5 kg/ha/yr with 32.5 kg/ha in the Beijing area that has less air pollution than the Tianjin area[14]. He et al. indicate that[15] total N deposition on maize-wheat cropland near Beijing is 80-90 kg/ ha/yr with about 50 kg/ha/yr being crop-available N. In the absence of our own measurements and the high degree of variability of reported values, 45 kg/ha/yr (shown in Table 1) had been adopted in this study as the atmospheric loading for XRB. The potential load of atmospheric N deposition equals the total area of cropland in XRB times the deposition coefficient.

2.5. *Livestock pollution*

Due to the inadequacy of published methodology for estimation pollution from livestock, we visited many animal raising operations (feedlots) at different scales in XRB to establish the nature of feedlot operations, mainly focusing on methods of waste handling, and relationship with watercourses. We estimated that the rate of 'treatment' is about 50% in XRB however we believe that much of this runs off because of poor manure management. Parameters for excreta and pollutants produce coefficient of livestock and poultry is shown in table 2.

Table 2. Average excreta and produce coefficient of livestock and poultry.

Livestock/poultry	Feeding Period	Excretion kg/year		Pollutants (kg/ton)	
				TP	*TN*
cattle	365	Feces	7300	1.18	4.37
		Urine	3650	0.4	8
pig	199	Feces	398	3.41	5.88
		Urine	656.7	0.52	3.3
goat/sheep	90	Feces	234	1.92	9.74
chicken	210	Feces	25.2	5.37	9.84
duck	210	Feces	27.3	6.2	11

2.6. *Rural living*

According to our interview with villagers, amount of waste water discharged per capita in XRB is about 40 L/d, which is in accordance with findings of Yin et al. Therefore, Yin's contamination coefficients of 0.35 for TN and 0.04 for TP (kg capita-1yr-1) were used in this study [16]. Up to 10% of rural sewage is treated by private biogas facilities in XRB. Pollution loads from rural living might be lower estimated in our study because villages are typically located on the edges of watercourses in south China.

2.7. *Delivered load from fields to watercourse*

It is uncertainty and various from each other very much for pollution loads from cropland entering into watercourse in different seasons and estimated by different researchers, which depends on different circumstances such as terrains, kind of crops, rainfall, composite of chemical fertilizers, time and methods of fertilizer application, runoff etc. To transform the pollutant load at source into delivered load to watercourses, Chen et al. experimented with rice paddies under different rainfall events where the chemical composition of runoff was measured against the total amount of fertilizer applied [17]. We used the delivery coefficient of applied fertilizer for rice and other field crops averages 10% for N and 1.7% for P (Table 1) on condition that inadequate data available in XRB. Long-term abundant of runoff was corroborated by interviews with agricultural officials in a variety of areas within XRB. Therefore the relatively high N and P delivery coefficients are mainly considered for the abundant of runoff-producing rain in monsoon humid environment.

According to the National Survey on Pollution of Livestock and Poultry Industries and its Countermeasures [18] the delivery coefficient of feces at a national scale is 2-8% and liquid waste is 50%. In a Shanghai City study the delivery coefficient was 25-30% [19]; Gu et al. reported that about 25% of all livestock waste goes into watercourses in the Yangtze River Delta area. As we could not check the reliability of any of these published delivery coefficients, nor could we determine the runoff component from 'treated' animal waste, the coefficient of 5% was used for the estimation of TN and TP loads entering into nearby watercourses from untreated livestock breeding pollution loads.

Excreted load and delivered load (TN, TP) calculations are noted in Tables 1 and 2.

3. Estimation Result

Based on above assumptions, ECM models were applied to assess agricultural NPS pollution generation using county level statistical datasets (for the year 2000 and 2012) in XRB, which including rural living wastewater, chemical fertilizer losses, livestock/poultry and aquaculture NPS pollutants generation. And estimated loads from different agricultural NPS entering into watercourse were calculated with county levels and categorized by municipality or regions (CZT / XRB), the estimated results were shown in Table 3.

The research results showed that: The research showed that: i) Total Nitrogen (TN) and Total Phosphorus (TP) loads entered in to XRB were 59955, 6375 and 64586, 6162 tons respectively for the year 2000 and 2012; ii) For TN load entering into watercourses, chemical fertilizer losses from cropland contributed 54.7% in year 2000, which followed by livestock (26.8%) and rural living waste (18.5%); For the year 2012, TN load contribution rates from fertilizer and livestock increased 4.3% and 11.8% respectively in XRB; iii) For TP load entering into watercourses in XRB, chemical fertilizer losses from cropland contributed 70% and 71.5% respectively in year 2000 and 2012; iv) In CZT city cluster, TN contributor is mainly from livestock and chemical fertilizer, however TP contributor is mainly from chemical fertilizer and livestock.

Table 3. Estimated loads from deferent agricultural nps entering into watercourses.

Area	Year 2000					
	Rural living		Chemical Fertilizer		Livestock	
	TN	TP	TN	TP	TN	TP
1 Changsha	1526	133	4513	648	2357	133
2 Zhuzhou	1040	91	2971	341	1339	77
3 Xiangtan	773	68	3058	345	880	47
4 Hengyang	2151	188	6197	775	2295	134
5 Shaoyang	1146	100	2752	379	1836	112
6 Yueyang	453	40	2005	191	669	38
7 Chenzhou	1373	120	3914	629	2297	136
8 Yongzhou	1833	160	5409	854	3430	212
9 Loudi	782	68	1994	301	964	55
CZT	3339	292	10542	1334	4576	257
XRB	11076	968	32812	4463	16067	944
Area	Year 2012					
	Rural living	Chemical Fertilizer	Livestock			
	TN	TP	TN	TP	TN	TP
1	Changsha	70	4405	691	2293	128
2	Zhuzhou	51	3152	356	1543	86
3	Xiangtan	41	3910	388	1026	53
4	Hengyang	120	8220	660	3056	184
5	Shaoyang	67	2866	320	1871	113
6	Yueyang	24	2325	183	954	56
7	Chenzhou	81	4246	760	2422	143
8	Yongzhou	101	6941	845	5040	310
9	Loudi	48	2042	204	1369	79
CZT	1854	162	11466	1436	4863	267
XRB	6907	604	38105	4407	19574	1151

4. Discussion

The export coefficients parameters of ECM are inherently highly variable and reflect particular site conditions for each study, therefore the choice of an export coefficient from the literature is inevitably subject to considerable uncertainty, especially as agricultural land management practices in China differ greatly from those in the US or UK [22]. Nevertheless, ECM are easy to use and can be scaled up to large watersheds and it was still accessible alternatives for the study of large watershed like XRB. Due to the restrict of available data and time, there was not enough calibration work done to validate the accuracy of our study and further research this work should be done to support NPS physical based process modeling in future.

Acknowledgments

This paper was financed by a grant from the national key scientific and technological special project: Research on technologies integrated and benefit assessment of water pollution control and management in watershed (No. 2014ZX07001001-004).

References

1. U.S. EPA, 2003, National Management Measures for the Control of Non-point Pollution from Agriculture [R]. EPA-841-B-03-004, US Environmental Protection Agency, Office of Water, Washington, DC.
2. W. L. Zhang, et al., 2007. A review on non-point source pollution models, Chinese Journal of Applied Ecology, 18, pp. 1886–1890. (in Chinese).
3. F. H. Hao, et al., 2006. The significance, difficulty and key technologies of large scale model applied in estimation of non-point source pollution [J]. Acta Scientiae Circumstantiae, 26, pp. 362–365 (in Chinese).
4. Z. Y. Shen, et al., 2012. An overview of research on agricultural non-point source pollution modeling in China [J]. Separation and Purification Technology, 84, pp. 104-111.
5. X. Luo, et al., 2010. Effects of land use structure on nitrogen export in Hujiashan watershed of Danjiangkou reservoir area [J]. Environmental Science, 31, pp. 58–62 (in Chinese).
6. T. Liang, et al., 2005. Simulation study of non-point source pollution under different land use in Guanting reservoir watershed [J]. Acta Scientiae Circumstantiae, 25, pp. 483–490 (in Chinese).
7. Ju, X.-T., et al., 2009. Reducing environmental risk by improving N management in intensive Chinese agricultural systems. Proceedings of National Academy Sciences of the United States of America 106(9), 3041-3046.
8. Vitousek, P.M., et al., 2009. Nutrient imbalances in agricultural development. Science 324, 1519-1520.
9. Shen, J.L., Tang, A.H., Liu, X.J., Fangmeier, A., Goulding, K.T.W., Zhang, F.S., 2009. High concentrations and dry deposition of reactive nitrogen

species at two sites in the North China Plain. Environmental Pollution 157 (11), 3106-3113.

10. Zhu, Z., 2000. Loss of fertilizer N from plants-soil system and the strategies and techniques for its reduction [J]. Soil and Environmental Sciences 9 (1), 1-6 (in Chinese).
11. Wade, B., 2010. Mitigating Ammonia Volatilization from Urea Fertilizer with Urease Inhibitors: an Overview of Recent Research on Urease Inhibitor Effectiveness and Feasibility. Expert paper submitted to the Expert Panel on Mitigation of Atmospheric Nitrogen Committee in advance of the UNECE Task Force for Reactive TFRN-4 meeting. http://www.clrtap-tfrn.org/webfm_send/236 (accessed 21.06.11.).
12. Rawluk, C.D.L., Grant, C.A., Racz, G.J., 2001. Ammonia volatilization from soils fertilized with urea and varying rates of urease inhibitor NBPT. Canadian Journal of Soil Science 81, 239-246.
13. Baoguo, M.A., Yang, T., Guo, F., 2005. Balance of phosphorus in a rotation system with Winter Wheat and Rice. Journal of Agro-Environment Science 24 (2), 371-374 (in Chinese).
14. Zhang, Y. et al., 2006. Spatial and temporal variation of atmospheric nitrogen deposition in the North China Plain. Acta Ecologica Sinica 26 (6), 1633-1639.
15. He, C.-E., et al., 2010. Determining total N deposition using 15N dilution technique in the North China Plain. Journal of Resources and Ecology 1 (1), 75-82.
16. Yin, J. et al, 2009. Discharge Features of rural domestic wastewater from different types of villages in Water Source Protection Area, Miyun Reservoir, Beijing. Journal of Agro-Environment Science 28 (6), 1200-1207.
17. Chen, Y., et al, 2011. Preliminary study on characteristics of nitrogen and phosphorus runoff losses from rice fields in Hai River Basin. Journal of Agro-Environment Science 30 (2), 328-333 (in Chinese).
18. NEP, 2002. National Survey on Pollution of Livestock and Poultry Industries and Its Countermeasures. Ecology Conservation Department of National Environmental Protection Bureau. China Environmental Science Press, Beijing (in Chinese).
19. Shanghai EPB, 2000. Comprehensive feasibility study report of livestock manure in Shanghai (internal unpublished publication). (in Chinese).

Quaternary reversible logic synthesis based on galois field decision diagram

Z.J. Guan[1†]

College of Computer Science and Technology, Nantong University, Nantong, Jiangsu, China 2260019
[†]Email: guan_g617@163.com

Y.Q. Zhang[2]

College of Science, Nantong University, Nantong, Jiangsu, China 2260019

H.Y. Ma[3]

College of Computer Science and Technology, Nantong University, Nantong, Jiangsu, China 2260019

H. Xu[4]

College of Electronics and Information, Nantong University, Nantong, Jiangsu, China 2260019

P.C. Zhu[5]

College of Computer Science and Technology, Nantong University, Nantong, Jiangsu, China 2260019

In order to realize the synthesis and optimization of quaternary reversible logic functions, we propose quaternary Shannon expansions generation algorithm which is based on quaternary Galois field. Combined Post algebra with Module algebra system and defined single-variable-reduced expression, obtain quaternary Galois field basic Shannon expansion by transform the quaternary function into the combination of four sub-functions which are expressed by each variables, and to get all quaternary Shannon expansions. The algorithm proposes to generate quaternary Galois field decision diagram, and apply all the expansions to express each variable of quaternary functions. The variable selection principles have ensured the SOP is optimal after flattening the decision diagram. The experiment results of some typical Benchmark functions have proved our

[1] Zhi-Jin Guan
[2] Yi-Qing Zhang
[3] Hai-Ying Ma
[4] Hai Xu
[5] Peng-Cheng Zhu

algorithm can get SOP expressions of given functions quickly, and ensure the number of nodes and products is least.

Keywords: Quaternary Logic, Reversible Logic Synthesis, Galios Field, Decision Diagram.

1. Introduction

Multiple-valued reversible logic synthesis is a promising research field. It is an important constitute of logic synthesis research and has important application in some modern scientific research areas such as low-power consumption circuit design, information security and nanotechnology [1-3]. With the increasing requirement in information technology, and its potential application value as well as its scientific theory significance, multiple-valued logic has attracted more and more attention recently years [1-4]. Quaternary logic is a special case in multiple-valued logic, since the inputs/outputs of quaternary function can be realized by traditional Boolean logic without adding any don't care bits, in other words, which means with no redundant encoding [5].

Galois field (also named finite field) contains limited elements which can be operated by addition and multiplication, and these operations satisfy some special rules that make Galois field some unique properties. The algebraic expression of Galois field can describe the multiple-valued circuits which can be realized by basic AND, OR, XOR module very well.

In quaternary reversible logic synthesis, the SOP expression of the function always very complicate if it contains many inputs/outputs, and also not easy to realize or simplify corresponding circuit. In order to solve this problem, some classic expansion methods such as Shannon Expansion, Positive/Negative Davio Expansion and Reed-Muller Expansion are introduced [6]. These methods transformed the complicate function into combination of several single variable functions, and applied BDD method [7, 8] of binary logic synthesis into quaternary logic synthesis at the same time, then we can get simplified SOP expressions by constructing QGFDD (Quaternary Galois Field Decision Diagram).

A lot of research about quaternary logic synthesis have been done around the world. Al-Rabadi use S/D tree to simplify the GFSOP expression [5], however without giving synthesis algorithm and circuit; in [9], Al-Rabadi synthesis ternary logic circuit in using classes of GF(3) reversible fast spectral transforms, this method can be regard as a good reference to quaternary logic synthesis. Green D.H use Reed-Muller expansion to synthesis quaternary logic Galois field function [6]. In this paper, we provide an easy, efficient synthesis method by combining the Galois field Shannon expansion to decision diagram and have great optimization and extend space.

The background and significance of our research is given in Section 1. The basic properties of Galois field is introduced in Section 2. The third Section is the key point of our work, we explain how to create quaternary Galois field Shannon expansion and Quaternary Galois Field Decision Diagram in detail. Synthesis method is presented in Section 4. In Section 5, we analysis the synthesis result of Section 4, and summarize the paper and propose some future research ideas in Section 6.

2. Quaternary Galois Field

A Galois Field GF(p^k), p is prime, k is natural number and k≥1. The binary logic was GF(2), and GF(4) is one of extended fields of GF(2), and also is the smallest non-prime field, the elements of GF(4) can be expressed as 0, 1, 2, 3. According to the special relationship between GF(2) and GF(4), it is possible to represent GF(4) operations and circuit modules by 2-bit binary quantities [5]. The GF(4) circuit and GF(2) circuit can be connected by binary to quaternary transformation circuits or quaternary to binary transformation circuits, which has extensive application space in circuit design in the future.

Generally, GF-based circuit is highly testable due to the facts that the GF operations satisfied the basic properties of group, and the addition and multiplication of GF(4) exhibit in Table 1 constituted 4*4 and 3*3 Latin Square respectively (ignore zeros in multiplication).

Table 1. GF(4) addition and multiplication

+	0	1	2	3
0	0	1	2	3
1	1	0	3	2
2	2	3	0	1
3	3	2	1	0

*	0	1	2	3
0	0	0	0	0
1	0	1	2	3
2	0	2	3	1
3	0	3	1	2

$$(1)\ a+b=b+a$$

$$(2)\ a+(b+c)=(a+b)+c=(a+c)+b$$

$$(4)\ a\times b=b\times a$$

$$(5)\ a*(b*c)=(a*b)*c=(a*c)*b$$

$$(6)\ a*(b+c)=a*b+a*c$$

In this paper, all the "+" and "*" represent Galois field operation if no special instructions.

3. Quaternary Galois Field Reversible Function Expansion Expressions

3.1. *The definition*

The traditional Boolean logic and binary reversible logic synthesis always use Shannon expansion, Reed-Muller expansion i.e to represent function. In this paper, we expand these classic methods to the multiple-valued logic synthesis field, ensure the testability and correctness of the synthesis result.

Definition 1: ${}^{i}x$ is the single-variable reduced expression defined as follows:

$$ {}^{i}x = \begin{cases} 1 & x = i \\ 0 & otherwise \end{cases} \quad (i=0, 1, 2, 3, \ldots, n) $$

In GF(4), every non-zero polynomial can be expressed as follows:

$$ f(x) = a_0 x^3 + a_1 x^2 + a_2 x + a_3 \text{, where } a_n \in \{0,1,2,3\}. $$

If associated these reduced expressions to basic function variable literal, we get following algebraic expressions:

$$ {}^{0}x = x^3 + 1,\ {}^{1}x = x^3 + x^2 + x,\ {}^{2}x = x^3 + 2x^2 + 3x,\ {}^{3}x = x^3 + 3x^2 + 2x $$

3.2. *The reduced sum expression theorem*

Theorem 1: The sum of all the single-variable reduced expressions is 1.

In GF(4), where ${}^{0}x, {}^{1}x, {}^{2}x, {}^{3}x \in \{0,1,2,3\}$ .That is

$$ {}^{0}x + {}^{1}x + {}^{2}x + {}^{3}x = 1 \qquad (1) $$

Proof: According to the definition of single-variable reduced expression of GF(4), we have:

If $x = 0, {}^{0}x = 1, {}^{1}x = 0, {}^{2}x = 0, {}^{3}x = 0$, then ${}^{0}x + {}^{1}x + {}^{2}x + {}^{3}x = 1$;

If $x = 1$ and $f = f_0 + x^3 f_{01} + x(x+1)(x+3) f_{12} + x(x+1)(x+2) f_{13}$, then ${}^{0}x + {}^{1}x + {}^{2}x + {}^{3}x = 1$;

If $x = 2,\ {}^{0}x = 0, {}^{1}x = 0, {}^{2}x = 1, {}^{3}x = 0$, then ${}^{0}x + {}^{1}x + {}^{2}x + {}^{3}x = 1$;

If $x = 3,\ {}^{0}x = 0, {}^{1}x = 0, {}^{2}x = 0, {}^{3}x = 1$, then ${}^{0}x + {}^{1}x + {}^{2}x + {}^{3}x = 1$.

Thus, we have theorem 1. For easy expression, we name the theorem as Reduced Sum Expression Theorem.

3.3. *The Shannon expansion*

Multiple-inputs/outputs function can be represented by its corresponding single-variable functions to express every input variable, and these single-variable functions are formulated by the single-variable reduced expressions in definition 1. For a r-valued n-inputs function, its Shannon expansion is:

$$f(x_1, x_2, \cdots, x_i, \cdots x_n) = {}^0x_i f_0 + {}^1x_i f_1 + \ldots + {}^n x_i f_n$$

In GF(4), the Shannon expansion of any n-inputs function is:

$$f(x_1, x_2, \cdots, x_i, \cdots x_n) = {}^0x_i f_0 + {}^1x_i f_1 + {}^2x_i f_2 + {}^3x_i f_3 \qquad (*)$$

Where f_0, f_1, f_2, f_3 is the output truth vectors when $x_i = 0,1,2,3$ respectively.

$$f_0 = f(x_1, x_2, \cdots, 0, \cdots x_n)$$
$$f_1 = f(x_1, x_2, \cdots, 1, \cdots x_n)$$
$$f_2 = f(x_1, x_2, \cdots, 2, \cdots x_n)$$
$$f_3 = f(x_1, x_2, \cdots, 3, \cdots x_n)$$

3.4. *The quaternary galois field basic expression*

3.4.1. *The quaternary galois field basic expression*

If we substitute the algebraic expressions of ${}^0x$, ${}^1x$, ${}^2x$ and ${}^3x$ to (*), we obtain the QGFBE as follows:

$$f = (x^3 + 1) f_0 + (x^3 + x^2 + x) f_1 + (x^3 + 2x^2 + 3x) f_2 +$$
$$(x^3 + 3x^2 + 2x) f_3 \qquad \text{(QGFBE)}$$

The previous research has preliminary indicated that it's important to find proper expansion for each variable. In this paper, we concentrate on how to create quaternary function Shannon expansion and its application in logic synthesis.

According to the GF(4) operation principles, if we add two same polynomials to the factor of f_0, f_1, f_2, f_3 of QGFBE respectively, the sum of them is 0, which don't change the function output value. In order to obtain different combination form of f_0, f_1, f_2, f_3, we add two 1s which can be replaced by ${}^0x + {}^1x + {}^2x + {}^3x$ according to theorem 1 to arbitrary factor, then extract the common factor and factorization the remain polynomials, we get expansions in demand. The combining algebraic cofactors are defined as follows:

$$f_{01} = f_0 + f_1, f_{02} = f_0 + f_2, f_{03}$$
$$= f_0 + f_3, f_{12} = f_1 + f_2, f_{13} = f_1 + f_3,$$
$$f_{23} = f_2 + f_3, f_{012} = f_0 + f_1 + f_2, f_{013}$$
$$= f_0 + f_1 + f_3, f_{023} = f_0 + f_2 + f_3,$$
$$f_{123} = f_1 + f_2 + f_3, f_{0123} = f_0 + f_1 + f_2 + f_3$$

3.4.2. *The quaternary galois field expansion*

Extract common factors of f_0, f_1, f_2, f_3 in QGFBE without adding any terms, we obtain QGFE1:

$$f = f_0 + x^3 f_{0123} + x^2(f_1 + 2f_2 + 3f_3) + x(f_1 + 3f_2 + 2f_3)$$

Select one factor of f_0, f_1, f_2, f_3 in QGFBE, add two 1s, according to theorem 1, the 1 can be decomposed the sum of all factors, so we can obtain the expansion by extract the common factors. Extract f_0, we have QGFE2-5:

$$f = f_0 + x(x+2)(x+3)f_{01} + x(x+1)(x+3)f_{02} + x(x+1)(x+2)f_{03}$$

$$f = f_0 + x^3 f_{01} + x(x+1)(x+3)f_{12} + x(x+1)(x+2)f_{13}$$

$$f = f_0 + x^3 f_{02} + (x+2)(x+3)f_{12} + x(x+1)(x+2)f_{23}$$

$$f = f_0 + x^3 f_{03} + x(x+2)(x+3)f_{13} + x(x+1)(x+3)f_{23}$$

Extract f_1, we have QGFE6-9:

$$f = f_1 + (x+1)(x+2)(x+3)f_{02} + (x^2+1)(x+1)f_{12} + x(x+1)(x+2)f_{23}$$

$$f = f_1 + (x^2+2)(x+1)f_{01} + x(x+1)(x+3)f_{02} + x(x+1)(x+2)f_{03}$$

$$f = f_1 + (x+1)(x+2)(x+3)f_{03} + (x^2+1)(x+1)f_{13} + x(x+1)(x+3)f_{23}$$

Extract f_2, we have QGFE10-13:

$$f = f_2 + x(x+2)(x+3)f_{01} + (x^2+3)(x+2)f_{02} + x(x+1)(x+2)f_{03}$$

$$f = f_2 + (x+1)(x+2)(x+3)f_{02} + x(x+2)(x+3)f_{12} + x(x+1)(x+2)f_{23}$$

$$f = f_2 + (x+1)(x+2)(x+3)f_{01} + (x^2+3)(x+2)f_{12} + x(x+1)(x+2)f_{13}$$

$$f = f_2 + (x+1)(x+2)(x+3)f_{03} + x(x+2)(x+3)f_{13} + (x^2+3)(x+2)f_{23}$$

Extract f_3, we have QGFE14-17:

$$f = f_1 + (x+1)(x+2)(x+3)f_{01} + x(x+1)(x+3)f_{12} + x(x^2+2)(x+3)f_{13}$$

$$f = f_3 + (x+1)(x+2)(x+3)f_{03} + x(x+2)(x+3)f_{13} + x(x+1)(x+3)f_{23}$$

$$f = f_3 + x(x+2)(x+3)f_{01} + x(x+1)(x+3)f_{02} + (x^2+2)(x+3)f_{03}$$

$$f = f_3 + (x+1)(x+2)(x+3)f_{02} + x(x+2)(x+3)f_{12} + (x^2+2)(x+3)f_{23}$$

Extract f_{012}, we have QGFE18-21:

$$f = f_{012} + (x^2+1)(x+1)f_{01} + (x^2+3)(x+2)f_{02} + x(x+1)(x+2)f_{03}$$

$$f = f_{012} + x^3 f_{03} + (x^2+1)(x+1)f_{13} + (x^2+3)(x+2)f_{23}$$

$$f = f_{012} + x^3 f_{01} + (x^2+3)(x+2)f_{12} + x(x+1)(x+2)f_{13}$$

$$f = f_{012} + x^3 f_{02} + (x^2+1)(x+1)f_{12} + x(x+1)(x+2)f_{23}$$

Extract f_{013}, we have QGFE22-25:

$$f = f_{013} + x^3 f_{01} + x(x+1)(x+3)f_{12} + (x^2+2)(x+3)f_{13}$$

$$f = f_{013} + (x^2+1)(x+1)f_{01} + x(x+1)(x+3)f_{02} + (x^2+2)(x+3)f_{03}$$

$$f = f_{013} + x^3 f_{03} + (x^2+1)(x+1)f_{13} + x(x+1)(x+3)f_{23}$$

Extract f_{023}, we have QGFE26-29:

$$f = f_{023} + x(x+2)(x+3)f_{01} + (x^2+3)(x+2)f_{02} + (x^2+2)(x+3)f_{03}$$

$$f = f_{023} + x^3 f_{01} + (x^2+3)(x+2)f_{12} + (x^2+2)(x+3)f_{13}$$

$$f = f_{023} + x^3 f_{02} + x(x+2)(x+3)f_{12} + (x^2+2)(x+3)f_{23}$$

$$f = f_{023} + x^3 f_{03} + x(x+2)(x+3)f_{13} + (x^2+3)(x+2)f_{23}$$

Extract f_{123}, we have QGFE30-33:

$$f = f_{123} + (x^2+1)(x+1)f_{01} + (x^2+3)(x+2)f_{02} + (x^2+2)(x+3)f_{03}$$

$$f = f_{123} + (x+1)(x+2)(x+3)f_{01} + (x^2+3)(x+2)f_{12} + (x^2+2)(x+3)f_{13}$$

$$f = f_{123} + (x+1)(x+2)(x+3)f_{02} + (x^2+1)(x+1)f_{12} + (x^2+2)(x+3)f_{23}$$

$$f = f_{123} + (x+1)(x+2)(x+3)f_{03} + (x^2+1)(x+1)f_{13} + (x^2+3)(x+2)f_{23}$$

Select all the factors of f_0, f_1, f_2, f_3 in QGFBE, then extract f_{0123}, then extract common factors from the left terms, we have QGFE34:

$$f = f_{0123} + x^3 f_0 + (x^2+1)(x+1)f_1 + (x^2+3)(x+2)f_2 + (x^2+2)(x+3)f_3$$

4. Synthesis Algorithm

The circuit logic function using QGFSOP expression is very complex, and the cost is too high, so it is needed to reduce the circuit size and reduce the number of logic gates.

We can through to each variable to select different Galois field Shannon expansion combined, the number of edges as much as possible for to point to the next level node or terminal node and they are 0s as the basic principle, in accordance with the criteria to solve the problem of extended selection and QGFDD reduction.

The algorithm synthesis:

Step1:Extract output vectors f_0, f_1, f_2, f_3 in the n×n truth table, these vectors corresponding the value of first variable is 0, 1, 2 and 3 respectively in quaternary Shannon expansion;

Step2: calculate combined value f_0, f_1, f_2, f_3 in 34 Shannon expansion;

Step3: To count 0 number of output vector after combined each expansion.

Step4: Select the expansion to express first variable in function. If have more than two expansion for 0 number is simple, then choose one of them.

Step5:the selected variable from step3 as the current variable, and repeat step1, step2, step3 and step4, to selected optimal expansion, and get the second layer output results of QGFDD;

Step6: Repeat step1 to step5, until found right type Shannon expansion for all variables. Here, the last layer node of QGFDD points to the terminal node.

Step6: each non 0 element in output vector final to times corresponding implication term, then to sum for them product, get simple SOP expansion.

5. Experimental Results

5.1. *An example of verification algorithm*

Select a Benchmark function fa, it is a full adder for 3 bit and 4 value. The output vector of full adder is:

$f(x,y,z)$=[0,0,0,0,0,0,0,1,0,0,1,1,0,1,1,1,0,0,0,1,0,0,1,1,0,1,1,1,1,1,1,0,0,1,1,0,1,1,1,1,1,1,1,1,1,2,0,1,1,1,1,1,1,1,1,1,1,2,1,1,2,2]^T

can know, n=3,1≤i≤3 in the function fa.

The corresponding complete QGFDD by the above algorithm, as shown in Figure 1. In order to clearly show that the side of the terminal node 0 is not drawn.

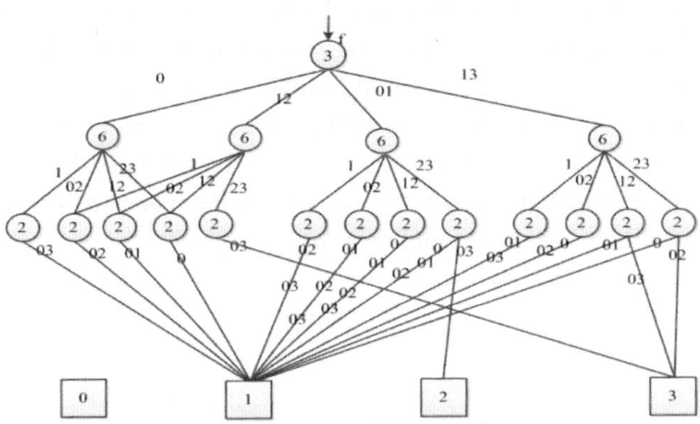

Fig. 1. QGFDD of full adder fa.

5.2. *Experimental results of partial benchmark functions*

The environment of experiment: PC computer, AMD A6-3400M CPU 1.4GHZ, 4G RAM; Windows 7. The above algorithm is realized in Visual C++.

The optimal extended expansion, 0 number of output vector, number of nodes and number of product term for QGFDD of typically Benchmark function is given. As shown in Table 2.

In the Table 2, the function, in, out, Op-Ex, 0-num, Pro- and Opt are input, output, optimal extended expansion, 0 number of output vector, number of product term respectively and optimization efficiency of nodes.

Table 2. Experimental results of partial Benchmark functions.

function	in	out	Op-Ex	0-num	Pro-	Opt
sum2	2	1	1, 1	14	2	44.4%
ha	2	2	2,1	10	12	21.4%
fa	3	2	3,6,2	35	29	12%
rd32	3	2	11,2,2	57	7	44%
4_2cod	4	2	5,2,2,2	253	3	9.09%
4gt4_20	4	1	10,2,5,2	245	11	28.5%
4gt5_21	4	1	10,2,2,2	246	10	15.3%
mini-alu_84	4	2	11,2,2,2	246	10	11.1%

6. Conclusions

In this paper, we propose the method to create quaternary Galois field Shannon expansion and obtain 34 expansion expressions. Then we construct QGFDD and get the GFSOP expressions of the given function by flatten it, and given a quaternary reversible logic synthesis algorithm based on Galois Field Decision Diagram. The experimental results from benchmark functions can be seen that: The reversible quaternary logic synthesis algorithm based on Shannon expansion and QGFDD, it is can be get simple SOP expansion. Thus, we can get optimized quaternary reversible logic circuits.

Acknowledgement

Our research is supported by the National Natural Science Foundation of China under grant No. 61402244; Natural Science Foundation of Jiangsu for colleges and universities under grant No. 14KJB520033.

References

1. Yan, H. Choe, H.S. Nam, S & Hu, Y. 2011. Programmable nanowire circuits for nanoprocessors. *Nature* 470(7333): 240-244.
2. Rowland, B. & Jones, J.A. 2012. Implementing quantum logic gates with gradient ascent pulse engineering: Principles and practicalities. *Physical and Engineering Sciences*, 370(1976): 4636-4650.
3. Michael, A. N & Isaac, C. 2011. Quantum computation and quantum Information: *10th Anniversary Edition*, New York: Cambridge University Press.
4. Curtis E & Perkowski M. 2004. A transformation based algorithm for ternary reversible logic synthesis using universally controlled ternary gates. *In Proc. IWLS*, 2-4 June 2004. California: Tamecula.
5. Green, D. H. 1990. Reed-Muller expansions with fixed and mixed polarities over GF(4). Computers and Digital Techniques. *IEE Proceedings E*. 137(5): 380-388.
6. Große, D. Wille, R. Dueck, G. W. & Drechsler, R. 2009. Exact multiple control Toffoli network synthesis with SAT techniques. *IEEE Trans. on CAD*, 28(5): 703–715.
7. Wille, R & Drechsler, R. 2009. Effect of BDD Optimization on Synthesis of Reversible and Quantum Logic. *Electronic Notes in Theoretical Computer Science*. 253(6): 57-70.
8. Al-Rabadi, A. N. 2004. Quantum Circuit Synthesis Using Classes of GF(3) Reversible Fast Spectral Transforms. *Proc. of the IEEE International Symposium on Multiple-Valued Logic (ISMVL)*, May 19-22, 2004, Canada: Toronto.
9. Muthukrishnan, A & Stroud, C. R. 2000. Multivalued logic gates for quantum computation. *Physical Review A- Atomic, Molecular, and Optical Physics*, 62(5): 052301-052309.

Author Index